Definitions,

16. Bayes' Law:
$$P(A|B) = \frac{P(B|A)P(A)}{P(B|A)P(A) + P(B|A')P(A')}$$

17. Arithmetic Means:

   a. Population mean: $\mu = \dfrac{\sum_{i=1}^{n} x_i}{n}$

   b. Sample mean: $\bar{x} = \dfrac{\sum_{i=1}^{n} x_i}{n}$

18. Variances:

   a. Population variance: $\sigma^2 = \dfrac{\sum_{i=1}^{n} (x_i - \mu)^2}{n}$

   b. Sample variance: $s^2 = \dfrac{\sum_{i=1}^{n} (x_i - \bar{x})^2}{n - 1}$

19. The Normal Conversion Formula: If X is normally distributed with mean μ and standard deviation σ and Z is the standard normal random variable, then
$$Z = \frac{X - \mu}{\sigma}$$

20. The Linear Regression Formulas: The least-squares regression line $Y = mX + b$ has
$$m = \frac{n \sum x_i y_i - (\sum x_i)(\sum y_i)}{n \sum x_i^2 - (\sum x_i)^2},$$
$$b = \frac{\sum y_i - m \sum x_i}{n}$$

21. Arithmetic and Geometric Sums:
   a. If the nth term of an arithmetic sequence is $a_n = a + (n - 1)d$, the sum of its first n terms is
$$S_n = \frac{n[2a + (n - 1)d]}{2}$$

   b. If the nth term of a geometric sequence is $a_n = ar^{n-1}$, $r \neq 1$, the sum of its first n terms is
$$S_n = \frac{a(1 - r^n)}{1 - r}$$

22. Compound Interest Formulas: Future value A, present value P, interest rate $i\%$ per year compounded m times per year, $r = i/100m$, time t years:
   a. $A = P(1 + r)^{mt}$ b. $P = A(1 + r)^{-mt}$

[23.] present ... per year compounded m times per year, $r = i/100m$, time t years:

   a. $S = \dfrac{R[(1 + r)^{mt} - 1]}{r}$

   b. $V = \dfrac{R[1 - (1 + r)^{-mt}]}{r}$

24. The Definition of the Derivative:
$$f'(x) = \lim_{h \to 0} \frac{f(x + h) - f(x)}{h}$$
if this limit exists

25. The Number e:
$$e = \lim_{x \to +\infty} \left(1 + \frac{1}{x}\right)^x$$

26. Formulas for Continuously Compounded Interest: Future value A, present value P, interest rate $i\%$ per year compounded continuously, $r = i/100$, time t years:
   a. $A = Pe^{rt}$ b. $P = Ae^{-rt}$

27. Two Rules of Integration:
   a. For any number k,
$$\int kf(u) \, du = k \int f(u) \, du$$

   b. For any functions f and g,
$$\int [f(u) \pm g(u)] \, du = \int f(u) \, du \pm \int g(u) \, du$$

28. The Definition of the Definite Integral: f continuous on $[a, b]$, $\Delta x = (b - a)/n$, x_i in interval $[a + (i - 1)\Delta x, a + i\Delta x]$:
$$\int_a^b f(x) \, dx = \lim_{\Delta x \to 0} \sum_{i=1}^{n} f(x_i) \, \Delta x$$

29. The Fundamental Theorem of Calculus: If f is continuous on $[a, b]$ and $F(x)$ is an antiderivative for $f(x)$ on $[a, b]$, then
$$\int_a^b f(x) \, dx = F(b) - F(a)$$

30. The Integration by Parts Formula:
$$\int u \, dv = uv - \int v \, du$$

▲ ▼ ▲

MATHEMATICS for Management and the Life and Social Sciences

Index of Referenced Applications

The Wall Street Journal	Corporate Forecasts	10
The Wall Street Journal	Costs	11
The Wall Street Journal	Marriage Age	18
The Wall Street Journal	West German Exports	19
The Wall Street Journal	Chinese Banking	19
The Wall Street Journal, October 12, 1988	Sets	64
Physicians Desk Reference, 1984	Drug Overdose	65
American Scientist, March–April 1986	Memory Modules Kept Busy	89
The Wall Street Journal	Stockbroker Fees	104
Physician's Desk Reference, 1984	Erythromycin Dosage	106
Boston Globe	Ideal Model	117
Kandel and Lee, *Fuzzy Set Switching and Automata: Theory and Applications*, 1979	Employee Relations	118
Higgins, *Analysis for Financial Management*, 1984	Return on Equity	133
Scientific American, March 1986	Visual Reaction Time	134
Scientific American, November 1986	Visual Reaction Time	134
The Wall Street Journal, September 20, 1988	Stockbroker Salaries	136
The Wall Street Journal, September 29, 1988	Worldwide Platinum Usage	137
Higgins, *Analysis For Financial Management*, 1984	Determining Gross Profit	138
Newsweek	Greenhouse Effect	139
Scientific American, January 1983	Growing Mammalian Cells	139
Scientific American, October 1987	Cost in High Speed Computers	160
Stevenson, *Production/Operations Management*, 1986	Plant Costs	161
Scientific American, February 1987	Life Science Ph.D.s by Gender	161
Scientific American, July 1987	Saving the U.S. Steel Industry	162
Rorres and Anton, *Applications of Linear Algebra*, 1977	Sustainable Yields	259
Stevenson, *Introduction to Management Science*, 1989	Optimal Allocation of Budgets	286
Stevenson, *Introduction to Management Science*, 1989	Minimizing Trim Loss	288
1989 World Almanac	Accidental Deaths	353
Aczel, *Complete Business Statistics*, 1989	Lost Luggage on Airlines	354
Harvard Business Review, November–December 1987	Undervalued Stocks	390
Niebel, *Motion and Time Study*, 1988	Machine Down-Time	393
Chase and Aquilano, *Production and Operations Management*, 1985	Quality Control	411
The Wall Street Journal, October 10, 1988	Cola Market Shares	435
Budnick, McCleavey and Mojena, *Principles of Operations Research for Management*, 1988	Credit Rating Transition Matrices	436
Budnick, McCleavey and Mojena, *Principles of Operations Research for Management*, 1988	Hospital Patient Transition Matrices	436
Rorres and Anton, *Applications of Linear Algebra*, 1977	Charity Contributors	437
Journal of Operations Research Society of America, 1954	Military Decision Strategy	446
Rorres and Anton, *Applications of Linear Algebra*, 1977	Influenza Strains	447
Duncan, *Quality Control and Industrial Statistics*, 1986	Quality Control	455
Focus, March–April 1989	Probability of false AIDS test results	456

Index of Referenced Applications

Source		Page
1989 World Almanac	High-school Dropout Rates	487
1989 World Almanac	Per Capita Tax Burdens	487
The Wall Street Journal, October 4, 1988	Company Size vs. Sales per Employee	496
Boston Globe	Money Market Funds	498
The Wall Street Journal, September 26, 1988	GNP Forecast	499
Emory, *Business Research Methods*, 1985	Likert Scales	500
Scientific American, May 1987	Net Worth of U.S. Families	513
Higgins, *Analysis for Financial Management*, 1984	Investment Risk	514
Stevenson, *Production/Operations Management*, 1986	Inventory Control	527
1989 World Almanac	Gold Production	535
Scientific American, June 1988	West German Birth Rate	535
The Wall Street Journal, October 3, 1988	Profits of U.S. Oil Companies	536
Scientific American, March 1986	Nuclear Power Plant Construction Time	537
Scientific American, June 1988	Ozone Levels Over Antarctica	538
The New York Times, July 3, 1988	Voter Participation	539
Higgins, *Analysis for Financial Management*, 1984	Stock Volatility	539
Meigs, Whittington and Meigs, *Principles of Auditing*, 1985	Value of Accounts Receivables	550
Stevenson, *Production/Operations Management*, 1986	Control Charts	550
Stigum, *Money Market Calculations*, 1981	Bond Yields	584
American Scientist, July–August 1988	Parallel Processing Speeds	625
Scientific American, June 1987	Bond Duration	646
Higgins, *Analysis for Financial Management*, 1984	Sustainable Growth Rates	713
Budnick, McCleavey and Mojena, *Principles of Operations Research for Management*, 1988	Useful Life of Production Tools	768
The Wall Street Journal	Yield Curves of Government Securities	781
American Scientist, July–August 1988	Parallel Processing Speeds	783
American Scientist, May–June 1988	Survival of Bald Eagle Colonies	808
Coleman, *Introduction to Mathematical Sociology*, 1964	Income and Consumer Prices	818
Higgins, *Analysis for Financial Management*, 1984	Internal Rate of Return	820
Stigum, *Money Market Calculations*, 1981	Average Daily Yield	821
Scientific American, January 1987	Nuclear Testing	828
Budnick, McCleavey and Mojena, *Principles of Operations Research for Management*, 1988	Length of Advertising Campaigns	835
Budnick, McCleavey and Mojena, *Principles of Operations Research for Management*, 1988	Optimal Holding Period	836
Physician's Desk Reference, 1984	Administering Anesthesia	848
McGann and Russell, *Advertising Media*, 1988	Advertising Learning Curves	849
Scientific American, December 1986	Cancer Drug Toxicity	851
Scientific American, October 1987	Number of Components on Chips	851
Scientific American, September 1988	Nuclear Strategy	851
Chase and Aquilano, *Production and Operations Management*, 1985	Learning Curves	852
Scientific American, January 1987	Nuclear Testing	854
Scientific American, June 1986	Windmill Efficiency	858
Scientific American, April 1988	Energy-Efficient Buildings	866
American Scientist, July–August 1988	Hurricanes	892
American Scientist, July–August 1987	Computer Models of AIDS Epidemiology	942
Scientific American, May 1988	Residential Radon Pollution	970
Coleman, *Introduction to Mathematical Sociology*, 1964	Transition Rates	995
Niebel, *Motion and Time Study*, 1988	Machine Tending Times	1017
Niebel, *Motion and Time Study*, 1988	Object and Background Contrast	1018
McGann and Russell, *Advertising Media*, 1988	Duplication of Advertising Audiences	1036
Scientific American, February 1987	Food Stamp Program Participation	1080
American Scientist, May–June 1988	Survival of Bald Eagle Colonies	1081
American Scientist, July–August 1988	Shorehan, NY Nuclear Power Plant Cost	1081

Applied College Mathematics Series

This text is one of a collection intended for the study of applied finite mathematics and calculus at the introductory level. The only prerequisites assumed are $1\frac{1}{2}$ years of high school algebra or its equivalent. Each text contains a review of algebra which may be used selectively as needed.

Mathematics for Management and the Life and Social Sciences is a comprehensive introduction to applied finite mathematics and calculus appropriate for a two semester or three quarter course.

Calculus for Management and the Life and Social Sciences is intended for a two semester course in applied calculus. The topics in the later portion of the text are independent so that this text would also be appropriate for an ambitious one semester course with some topical omissions.

Brief Calculus for Management and the Life and Social Sciences is appropriate for a one semester course in applied calculus. It covers the first eight chapters of *Calculus*.

MATHEMATICS for Management and the Life and Social Sciences

Donald L. Stancl
Mildred L. Stancl
Both of St. Anselm College

Homewood, IL 60430
Boston, MA 02116

Photos: Journalism Services, Inc.

Cover photo (I.M. Pei Pyramid at the Louvre, Paris), © Imapress/N'Diaye

Chapter 1 (*New York*), © Matthew Rosenzweig, 2
Chapter 2 (*Agra, India*), © Rich Clark, 56
Chapter 3 (*Moscow*), © John M. Nallon, 120
Chapter 4 (*Seattle*), © Dave Brown, 188
Chapter 5 (*San Francisco*), © Snyder Photographic, 266
Chapter 6 (*Rome*), © Fotex/H. Kohls, 344
Chapter 7 (*Sydney, Australia*), © Dirk Gallian, 418
Chapter 8 (*Batavia, IL*), © H. Rick Bamman, 478
Chapter 9 (*Beijing*), © Dave Brown, 552
Chapter 10 (*Chicago*), © Mike Kidulich, 594
Chapter 11 (*Munich*), © Fotex/Taubenberger, 682
Chapter 12 (*Calgary*), © Rinna Borgsteede, 734
Chapter 13 (*St. Louis*), © James Blank, 794
Chapter 14 (*Paris*), © Dave Brown, 870
Chapter 15 (*London*), © Gills J. Copeland, 930
Chapter 16 (*Paris*), © Imapress/N'Diaye, 1006
© RICHARD D. IRWIN, INC., 1990

All rights reserved. No part of this publication may be reproduced, stored in a retrieval system, or transmitted, in any form or by any means, electronic, mechanical, photocopying, recording, or otherwise, without the prior written permission of the publisher.

Sponsoring editor: *Richard T. Hercher, Jr.*
Developmental editor: *Jim Minatel*
Project editor: *Waivah Clement*
Production manager: *Bette K. Ittersagen*
Interior designer: *Lucy Lesiak Design*
Cover designer: *Michael S. Finkelman*
Artist: *Benoit Design*
Compositor: *Arcata Graphics/Kingsport*
Typeface: *10/12 Times Roman*
Printer: *R. R. Donnelley & Sons Company*

Library of Congress Cataloging-in-Publication Data

Stancl, Donald L.
    Mathematics for management and the life and social sciences /
Donald L. Stancl, Mildred L. Stancl.
    p. cm.
    ISBN 0-256-07335-X
    1. Mathematics. I. Stancl, Mildred L. II. Title.
QA37.2.S794   1990   89–35297
510—dc20   CIP

Printed in the United States of America

1 2 3 4 5 6 7 8 9 0 DO 7 6 5 4 3 2 1 0

To George A. Stancl: a loving and lovable father

Preface

Mathematics for Management and the Life and Social Sciences is an applications-oriented text for an introductory college-level mathematics course addressed to students of business, management, economics, the life sciences, and the social sciences. The only prerequisite for understanding the book is a basic course in algebra.

Why We Wrote This Book. We believe that students who take the course for which this book is designed want to see and understand as many practical applications of mathematics as possible. They do not want to learn mathematics for its own sake, nor will they be satisfied with promises that the mathematics they are being taught will be useful at some indeterminate future time. On the other hand, we also believe that a "cookbook" approach to the subject—one that presents applications with little or no attempt to explain the concepts involved or to increase the mathematical sophistication of the reader—will not serve students well in the long run. Students must learn more than just the applications presented in the text: they must learn how to think mathematically so that they will possess the ability and confidence to use mathematics in problem solving. We wrote this book because we felt that other texts in the field did not strike the appropriate balance between the presentation of applications and the development of the student's ability to think mathematically. We have attempted to create such a balance by emphasizing mathematical modeling: each mathematical topic is introduced with and illustrated by as many practical applications as possible, but we pay as much attention to the construction and use of the relevant models as we do to the applications themselves. The result is an approach that not only presents a great variety of applications but does so in a manner that encourages the development of the student's ability to use mathematics in problem solving.

Features of the Book. Some noteworthy features of the book are:

Extensive Development Process. The book has undergone an extensive development process that began with postpublication reviews of the first edition of our text *Calculus for Management and the Life and Social Sciences,* upon which Chapters 10 through 16 of this book are based. This was followed by conversations with knowledgeable mathematicians and surveys of their opinions regarding what should be included in a book of this type. Using these sources as input, we prepared a first draft of the book, which was extensively reviewed. The book was then rewritten to reflect the criticisms and suggestions of the reviewers. The revised version was then reviewed again, and again revised in response to the reviewers, to produce the final manuscript. We believe that because of this careful development process the book reflects the best current thinking as to what a text in this field should be.

Extensive Review Material. We feel that many books in the field do not contain enough review material for today's students with their diverse mathematical backgrounds. Accordingly we have provided in the first three chapters of the book an unusually extensive review, covering algebra, sets, functions, linear functions, and quadratic functions. (For a discussion of the possible uses of this review material, see the section on Content below.)

Examples. We have included more examples than most other books in the field. There are over 650 examples, many of them applied. As soon as a new mathematical topic is introduced we illustrate it with mathematical examples and, when appropriate, with applied examples.

Exercises. Reviewers have commended the variety, choice, and number of exercises, especially the applied ones. There are over 5650 exercises in this book, more than 2000 of them applied. Many of the applied exercises have several parts, and some 220 of them are taken from the literature and are referenced. This is more exercises, and in particular far more applied exercises, than other books in the field. At the end of each section, drill exercises appear first, followed by applied exercises organized by discipline (management, life science, and social science). Drill exercises and those organized by discipline are graded in difficulty and offer problems that will challenge students at every level. The most challenging exercises are marked by asterisks.

Exercises on New Topics. In order to make the exercise sets more interesting and keep the sections to a manageable length, we frequently use the exercises to introduce topics and applications that are not treated in the text proper. Exercises that are concerned with such new topics are also marked by asterisks.

Figures. Whenever we felt that an explanation could be enhanced by a figure, we included the figure. The text contains almost 475 figures used to illustrate concepts, substantially more than other books in the field.

Pedagogical Features. New terms are printed in **boldface.** Important definitions, facts, formulas, theorems, algorithms, and "keys to" are enclosed in boxes for easy reference. The most important definitions, theorems, and formulas are listed in the front endpapers. The rules of differentiation and a table of integrals appear in the back endpapers.

Chapter Summaries. At the conclusion of each chapter we have included a summary that outlines the terms, notations, facts, and techniques that the student should know. These summaries can be used by the student as a study guide and checklist before attempting the review exercises for the chapter.

Review Exercises. Following each summary is a set of review exercises that covers all material discussed in the chapter. These exercises provide a thorough test of the student's mastery of the material covered in the chapter.

Additional Topics. Following the review exercises in each chapter is a list of additional topics. These are topics that are not treated in the text proper but might be of interest to some students. Each additional topic suggests a theorem, a technique, a concept, an application, or a historical subject for the student to investigate. The instructor can use the additional topics as a basis for short written or oral reports instructor can use the additional topics as a basis for short written or oral reports or for longer papers that might be appropriate in a writing across the curriculum program. They are particularly useful as a source of extra-credit assignments.

Chapter Supplements. Each chapter except the first is followed by a supplement containing additional material not usually covered in a textbook of this type. These supplements consider such diverse topics as chaos (Chapter 3), quality control (Chapter 6), and consumption and the multiplier (Chapter 11). Two of them are concerned with the historical development of the calculus. Each supplement is designed to build upon and enrich the material in its chapter. The supplements can be covered in class, but they are also appropriate for assignment as outside reading or as a basis for student presentations to the class. Each of the nonhistorical supplements has its own exercises, graded by difficulty.

Solutions to Exercises. Solutions to most odd-numbered section and supplementary exercises and to all review exercises may be found at the back of the book.

Applications Index. An applications index gives the location of all applications (whether they appear as examples or exercises), grouped by discipline.

Content. The first three chapters of the book are review chapters: Chapter 1 reviews algebra, Chapter 2 reviews sets and functions, and Chapter 3 reviews linear and quadratic functions. These chapters can be used in whatever portion the instructor deems necessary, or they can be omitted entirely. In general, we would not expect Chapter 1 to be covered in class, but rather to be used by students on an individual basis for algebra review as needed. Depending on the backgrounds of the students in the class, the instructor may feel it necessary to cover none, part, or all of Chapters 2 and 3 in class; these chapters are written and organized so that if such coverage is needed it can be accomplished quickly, with much left to the student. Our goal with these chapters has been to provide all the review that students might possibly need while allowing the instructor the maximum amount of flexibility in his or her use of the material.

Following the review chapters, the text falls naturally into four units: matrices and linear programming (Chapters 4 and 5), probability and statistics (Chapters 6, 7, and 8), the mathematics of finance (Chapter 9), and calculus (Chapters 10 through 16). The first three of these units are independent and can be covered in any order; Chapter 13 makes use of some of the material from Chapter 9, and Chapter 15 makes use of some of that from Chapter 6. A dependency diagram may be found on page xiv.

Characteristics of the Book. Many characteristics of this book have been praised by reviewers. Some of these include:

Emphasis on Modeling. As explained above, we emphasize the construction and use of mathematical models as an aid toward development of the student's ability to use mathematics. As part of this process, we strive whenever possible to show the reasoning that goes into the creation of the model so that the student may learn how to analyze practical situations mathematically.

Applications. Wherever possible we illustrate the mathematical topic under discussion with one or more applications. We have made every effort to provide diverse and stimulating new applications as well as the standard ones that must be

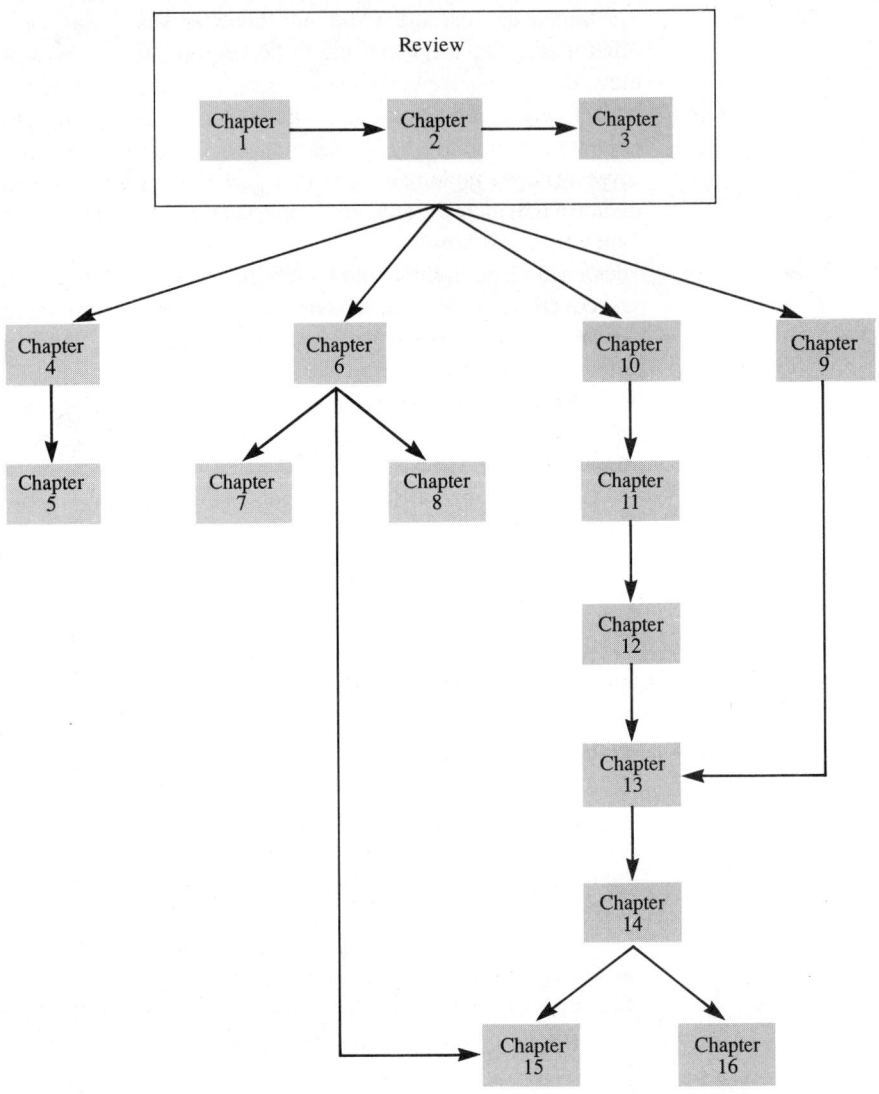

offered in a book of this type. Reviewers have praised both the variety and quantity of our applications.

Flexibility. We feel that this book is more flexible than other texts in the field. As noted above, the large amount of review material and its arrangement allow the instructor to tailor the review to the specific needs of the students. The arrangement of topics allows flexibility in the order in which topics are taken up. The additional topics and chapter supplements contribute to the flexibility of the book by allowing the instructor to individualize the course and vary it somewhat from year to year while still covering all required material.

Coverage. The book covers all the standard topics found in other texts, but offers more than others do in the additional topics and chapter supplements. It has more examples and exercises, and especially more applied exercises, than other texts.

Conversational Style. We have written the book using a straightforward conversational style. We have tried to avoid being overly chatty or cute. Many reviewers have remarked that our writing style is pleasant and appropriate to the material.

Supplements. There are several supplements designed to accompany this text:

Student Solution Manual. The *Student Solution Manual* contains complete, annotated solutions to representative odd-numbered exercises from the text as well as hints for solving other exercises. It is intended to help the student master the problem solving techniques as well as to provide solutions to exercises.

Computer Manual. We have written a computer manual to accompany this book. The manual includes two IBM PC-formatted disks containing BASIC programs that illustrate the material in the text, as well as descriptions of these programs and directions and suggestions for their use. The programs allow students to solve exercises that require a great deal of computation, to check their answers to exercises from the text, and to experiment with many of the mathematical models covered in the text.

Instructor's Solution Manual. The *Instructor's Solution Manual* contains complete solutions to all exercises in the text, lecture notes for each section with examples, notes on material that gives students difficulty, and diagnostic pretests. The diagnostic pretests come in two forms, one covering Chapter 1 of the text and the other covering Chapters 1, 2, and 3. Each form comes in two versions so that the same pretest need not be given in successive years.

Test Bank. A test bank is available in both a printed and computerized version. The printed version includes approximately 70 test questions per chapter. For details on use of the computerized test bank, which is known as COMPUTEST, please contact the publisher.

Complete Instructor's Package. The complete instructor's package consists of the text, the *Instructor's Solution Manual,* and the printed test bank, all packaged together in a handy three-ring binder format. Instructors who order this book will receive the complete instructor's package with a sample of each supplement without charge.

Companion Texts. We have also written two calculus texts for students of management, economics, and the life and social sciences: *Calculus for Management and the Life and Social Sciences* and *Brief Calculus for Management and the Life and Social Sciences,* both published by Richard D. Irwin, Inc., and both of which are now available in their second editions. *Brief Calculus* begins with a review chapter and then consists of Chapters 10 through 16 of this text, with some minor alterations. *Calculus* adds to *Brief Calculus* chapters on differential equations, probability and calculus, trigonometric functions, and sequences and series.

Acknowledgments. Many people contributed to this book. We would particularly like to thank the reviewers: Harold Bennett, Texas Tech University; Larry Bowen, University of Alabama; Chris Burditt, Napa Valley College; Henry Decell, Jr.,

University of Houston; Joe S. Evans, Middle Tennessee State University; Betty L. Fein, Oregon State University; Alice M. Hughey, Brooklyn College/City University of New York; Giles Wilson Maloof, Boise State University; Thomas Miles, Central Michigan University; William Miller, Central Michigan University; Robert Moreland, Texas Tech University; Winston A. Richards, The Pennsylvania State University at Harrisburg; Daniel Scanlon, Orange Coast College; Paula R. Sloan, Vanderbilt University; Robert C. Smith, University of Southern Mississippi; Malcolm Soule, California State University-Northridge; Howard Wilson, Oregon State University; Thomas J. Woods, Central Connecticut State University; Earl J. Zwick, Indiana State University. Their patience in reading the various versions of the manuscript was admirable and their comments, criticisms, and suggestions were of the utmost value to us. Much of what is good about this book is due to the reviewers, and we gratefully acknowledge their contributions. We also thank those who took the time to complete one of the three surveys or provide insights for the development of this text.

We wish also to convey our thanks for their fine work to those who produced the supplements: to James Balch for the *Instructor's Solution Manual* and the *Student Solution Manual;* to Larry Bowen for the diagnostic pretests; to W. H. Howland and Giles Maloof for the Test Bank; and to Babak Ghayour for the lecture notes. Our sincere thanks also to Philip Feinsilver, Anthony Giovannitti, Paul Patten, Winston Richards, Daniel Scanlon, and Mike Shirazi for solving the exercises; to Harold Bennett for verifying the accuracy of the examples; and to Mark Bridger of Bridge Software for the computer-generated artwork in Chapter 16.

Finally, we gratefully acknowledge the contributions of the superb professionals at Richard D. Irwin, Inc., who had so much to do with the making of this book. Sincere thanks to our acquisitions editor, Richard T. Hercher, Jr., for his unfailing support and enthusiasm, not only for this book, but also for our calculus books. Special thanks also to our development editor, James Minatel, whose efforts on behalf of the book have been exceptional and whose patience with his authors has been monumental. We thank the copy editor, Tom Whipple, who did a fine job with a heavily rewritten manuscript. We are indebted also to our project editor, Waivah Clement, who patiently accepted our changes as she put the book together, as well as to the production manager, Bette Ittersagen. Together they have produced an exceptionally attractive book. Our heartfelt appreciation to these people and all the others at Irwin who worked so competently and to such good effect on this book.

Donald L. Stancl
Mildred L. Stancl

Contents

Chapter 1
Review: Numbers and Algebra, 3

1.1 Real Numbers, 4
1.2 Linear Equations, 12
1.3 Linear Inequalities, 19
1.4 Exponents, 25
1.5 Polynomials, 31
1.6 Polynomial Equations, 37
1.7 Rational Expressions, 43
Summary, 49
Review Exercises, 51
Additional Topics, 54

Chapter 2
Sets and Functions, 57

2.1 Sets, 58
2.2 Set Operations, 66
2.3 Functions, 80
2.4 Graphs, 90
Summary, 107
Review Exercises, 108
Additional Topics, 111
Supplement: Fuzzy Sets, 111

Chapter 3
Linear and Quadratic Functions, 121

3.1 Linear Functions, 122
3.2 Linear Models in Business and Economics, 140
3.3 Simultaneous Linear Equations, 152
3.4 Quadratic Functions, 164
Summary, 177
Review Exercises, 178
Additional Topics, 180
Supplement: Chaos, 180

Chapter 4
Matrices, 189

4.1 Gauss-Jordan Elimination, 190
4.2 Matrices and Matrix Operations, 204
4.3 Matrix Multiplication, 217
4.4 Inverse Matrices, 232
4.5 Cryptography and Input-Output Analysis, 245
Summary, 254
Review Exercises, 255
Additional Topics, 258
Supplement: Sustainable Yields, 259

Chapter 5
Linear Programming, 267

5.1 Linear Inequalities in Two Variables, 268
5.2 The Linear Programming Model, 279
5.3 Two-Dimensional Linear Programming, 289
5.4 The Simplex Method: Maximization with $\leq$ Constraints, 302
5.5 The Simplex Method: $\geq$, $=$ Constraints and Minimization, 317
5.6 Duality, 329
Summary, 337
Review Exercises, 338
Additional Topics, 340
Supplement: Shadow Prices and Right-Hand-Side Ranging, 340

Chapter 6
Probability, 345

6.1 Probabilities, 346
6.2 Probabilities of Compound Events, 356
6.3 Conditional Probabilities, 362
6.4 Counting, 376
6.5 The Binomial Formula, 387
6.6 Random Variables, Probability Distributions, and Expected Value, 394
Summary, 405
Review Exercises, 407
Additional Topics, 410
Supplement: Quality Control, 411

Chapter 7
Applications of Probability, 419

7.1 Decision Trees, 420
7.2 Markov Chains, 428
7.3 Two-Person Zero-Sum Games, 438
7.4 Bayes' Law, 449
7.5 Bayesian Decision Trees, 457
Summary, 466
Review Exercises, 466
Additional Topics, 469
Supplement: Game Theory and Linear Programming, 470

Chapter 8
Statistics, 479

8.1 Frequency Distributions, 480
8.2 Measures of Central Tendency, 488
8.3 Measures of Dispersion, 501
8.4 The Normal Distribution, 515
8.5 Linear Regression and Correlation, 529
Summary, 540
Review Exercises, 541
Additional Topics, 544
Supplement: Confidence Intervals, 545

Chapter 9
The Mathematics of Finance, 553

9.1 Arithmetic and Geometric Sequences, 554
9.2 Compound Interest and Present Value, 563
9.3 Annuities, 573
Summary, 586
Review Exercises, 586
Additional Topics, 588
Supplement: Amortization, 588

Chapter 10
The Derivative, 595

10.1 Limits, 596
10.2 One-Sided Limits, Infinite Limits, and Limits at Infinity, 609
10.3 The Derivative, 625
10.4 Rules of Differentiation, 638
10.5 Rates of Change, 646
10.6 The Theory of the Firm, 656
10.7 Differential Notation and Higher Derivatives, 669
Summary, 672
Review Exercises, 674
Additional Topics, 677
Supplement: Newton and Leibniz, 678

Chapter 11
More about Differentiation, 683

11.1 The Product Rule and the Quotient Rule, 684
11.2 The Chain Rule, 691
11.3 Differentials, 704
11.4 Implicit Differentiation, 714
11.5 Related Rates, 719
Summary, 727
Review Exercises, 728
Additional Topics, 730
Supplement: Consumption and the Multiplier, 730

Chapter 12
Applications of Differentiation, 735

12.1 Relative Maxima and Minima: The First Derivative Test, 736
12.2 Relative Maxima and Minima: The Second Derivative Test, 751
12.3 Absolute Maxima and Minima, 757
12.4 Curve Sketching, 769
Summary, 783
Review Exercises, 784
Additional Topics, 787
Supplement: The Qualitative Solution of Differential Equations, 788

Chapter 13
Exponential and Logarithmic Functions, 795

13.1 Exponential Functions, 796
13.2 Logarithmic Functions, 810
13.3 Differentiation of Logarithmic Functions, 821
13.4 Differentiation of Exponential Functions, 829
13.5 Exponential Growth and Decline, 837
Summary, 855
Review Exercises, 856
Additional Topics, 859
Supplement: *S*-curves, 860

Chapter 14
Integration, 871

14.1 The Indefinite Integral, 872
14.2 Finding Indefinite Integrals, 879
14.3 The Definite Integral, 885
14.4 The Definite Integral and Area, 894
14.5 Riemann Sums, 909
Summary, 920
Review Exercises, 921
Additional Topics, 923
Supplement: The Development of the Definite Integral, 925

Chapter 15
More about Integration, 931

15.1 Consumers' Surplus, Producers' Surplus, and Streams of Income, 932
15.2 Integration by Parts and Tables of Integrals, 944
15.3 Improper Integrals, 952
15.4 Continuous Random Variables, 957
15.5 Numerical Integration, 971
15.6 Differential Equations, 984
Summary, 996
Review Exercises, 997
Additional Topics, 1001
Supplement: The Present Value of a Stream of Income, 1002

Chapter 16
Functions of Several Variables, 1007

16.1 Functions of More Than One Variable, 1008
16.2 Partial Derivatives, 1019
16.3 Partial Derivatives as Rates of Change, 1026
16.4 Optimization, 1037
16.5 Lagrange Multipliers, 1047
16.6 Double Integrals, 1054
Summary, 1069
Review Exercises, 1071
Additional Topics, 1073
Supplement: Least-Squares Regression, 1074

Tables

Table 1: Compound Interest, 1084
Table 2: Present Value, 1086
Table 3: Annuities, 1088
Table 4: Present Value of Annuities, 1089

Table 5: Binomial Probabilities, 1090
Table 6: Areas Under the Standard Normal Curve, 1091
Table 7: Powers of e, 1092
Table 8: Natural Logarithms, 1093

Solutions to Exercises, 1095

Applications Index, 1097

Subject Index, 1151

MATHEMATICS
for
Management and the Life and Social Sciences

1

Review: Numbers and Algebra

The major concern of this book is the development of mathematical models for use in business, economics, the life sciences, and the social sciences. For instance, we shall develop models that can be used to analyze

- The cost, revenue, and profit of a firm.
- The interaction between the quantity of a commodity supplied to the market by producers and the quantity of the commodity demanded by consumers.
- The growth or decline of a population.
- The strategies employed by opposing sides in a competitive situation.

All our models will make use of numbers, and most will require some knowledge of algebra. Therefore a familiarity with numbers and algebra will be basic to an understanding of the text. This chapter briefly reviews these important topics.

We begin with an introduction to the number systems we will use, with particular emphasis on the system of real numbers and its properties. This is followed by sections on linear equations, linear inequalities, and exponents and roots. We next consider the arithmetic and factorization of polynomials and the solution of quadratic equations. We conclude with a discussion of the algebra of rational expressions.

1.1 REAL NUMBERS

Throughout this text we will be working with the real number system and its subsystems the natural numbers, the integers, and the rational numbers. In this section we introduce these number systems and discuss the real line, inequalities and intervals on the real line, and the absolute value of a real number.

Number Systems

We will use the following number systems:

the **natural numbers N:**   1, 2, 3, . . . ,
the **integers Z:**   . . . , $-3, -2, -1, 0, 1, 2, 3, \ldots$,
the **rational numbers Q:**   all numbers that can be written in the form m/n, where m and n are integers and $n \neq 0$.

(The three dots . . . mean "and so on.") Note that natural numbers are integers and that integers are rational numbers.

Numbers such as

$$\sqrt{2}, \sqrt{5}, -\sqrt[3]{7}, \text{ and } \pi$$

cannot be written as fractions of integers and hence are not rational; they are called **irrational numbers.** All square roots of integers that are not perfect squares, cube roots of integers that are not perfect cubes, and so on, are irrational.

The system **R** of **real numbers** consists of all rational numbers and all irrational numbers. Thus every real number is either a rational number or an irrational number. Figure 1–1 shows the relationships among the number systems **N, Z, Q,** and **R**.

Every rational number has a decimal representation that either terminates, such as $\frac{5}{8} = 0.625$, or repeats, such as $\frac{1}{3} = 0.333\ldots$. Every irrational number has a decimal representation that neither terminates nor repeats, such as

$$\pi = 3.14159\ldots$$

Therefore every real number is a decimal number. On the other hand, it can be shown that every decimal number is the representation of either a rational number or an irrational number. Hence, we may think of the real number system **R** as consisting of all possible decimal numbers.

The Real Line

The system of real numbers may be represented by a horizontal line called the **real line.** To construct the real line, we first choose a convenient point on a horizontal line to represent the number 0. This point is called the **origin.** We then choose a convenient unit length and, starting at the origin, mark off this length on the line in both directions. The resulting points on the line represent the integers. See Figure 1–2.

REVIEW: NUMBERS AND ALGEBRA 5

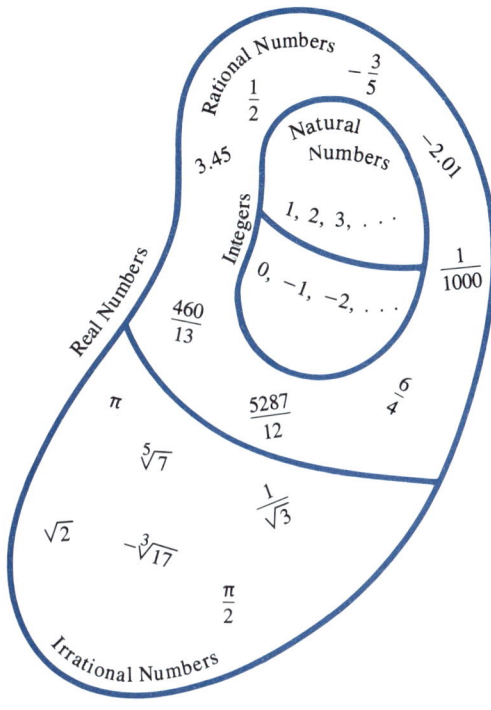

FIGURE 1–1
The Real Number System

Note that each real number corresponds to a point on the real line, and each point on the real line corresponds to a real number.

EXAMPLE 1 In Figure 1–3 we have plotted the numbers $-\sqrt{5}$, $-7/4$, -1, $-1/2$, 0, $2/3$, 1.75, 2.0, 2.8, and $\pi = 3.14 \ldots$ on the real line. ☐

EXAMPLE 2 From left to right, the points plotted on the real line in Figure 1–4 represent the numbers -1.5, -1, $-1/3$, $1/2$, $3/4$, 1.8, $13/6$, and 4. ☐

From now on, when we say "number," we shall mean "real number."

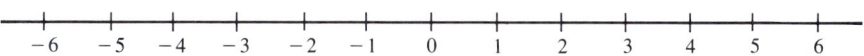

FIGURE 1–2
The Real Line

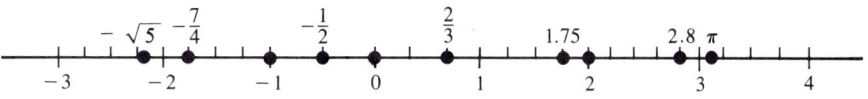

FIGURE 1–3

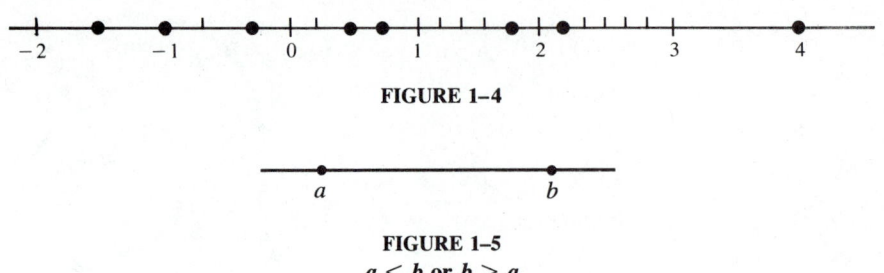

FIGURE 1-4

FIGURE 1-5
$a < b$ or $b > a$

Inequalities

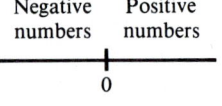

FIGURE 1-6

If a and b are numbers and a is to the left of b on the real line, we say that a is less than b, written $a < b$, or that b is greater than a, written $b > a$. See Figure 1-5. The expressions $a < b$ and $b > a$ are called **inequalities.** If $a < 0$, then a is a negative number; if $a > 0$, then a is a positive number. See Figure 1-6. The number 0 is neither positive nor negative.

An expression of the form $a \leq b$ means that a is less than b or equal to b. Similarly, $b \geq a$ means that b is greater than a or equal to a.

EXAMPLE 3

The following are true:

$$-3 < -2, \quad -1 < -\frac{2}{3}, \quad 1 \leq 1, \quad 1 < 6.5, \quad 2 \geq 0. \quad \frac{5}{4} > \frac{9}{8}.$$

The following are false:

$$-3 < -4, \quad 1 < \frac{2}{3}, \quad 2 \leq -3.25, \quad 2 < 2, \quad 1 \geq 1.01, \quad \frac{5}{4} > \frac{11}{8}. \quad \square$$

Inequalities are often useful in describing practical situations.

EXAMPLE 4

Suppose a hospital uses at least 50 gallons of liquid soap and fewer than 85 gallons of disinfectant per week. If we let S denote the amount of soap and D the amount of disinfectant used per week, S and D in gallons, then we may express these facts by writing

$$S \geq 50 \quad \text{and} \quad D < 85. \quad \square$$

An inequality of the form $a < b < c$ is a **double inequality.** It is true if both $a < b$ and $b < c$ are true, that is, if b is between a and c on the real line.

EXAMPLE 5

The following are true:

$$-3 < -\frac{3}{2} < -1, \quad -\frac{2}{3} < 0.02 \leq 0.021, \quad \frac{1}{2} < \frac{3}{4} < \frac{7}{8}.$$

The following are false:

$$-2 < -3 < -4, \quad 2 < \frac{5}{2} < 2.25, \quad 1 \leq 2 < 2. \quad \square$$

Do not write statements such as 2 < 4 > 0. All inequality symbols in a double inequality should point in the same direction. Therefore the correct way to write the inequality relations among 0, 2, and 4 is 0 < 2 < 4 or 4 > 2 > 0.

Double inequalities are also useful.

EXAMPLE 6

Suppose a hospital uses at least 50, but no more than 60, gallons of liquid soap per week, and each gallon of soap costs $0.80. If S again denotes the amount (in gallons) of soap used per week, and if C denotes the weekly cost of the soap, then

$$50 \leq S \leq 60$$

and

$$\$0.80(50) \leq C \leq \$0.80(60),$$

or

$$\$40 \leq C \leq \$48. \quad \square$$

Intervals

It is often convenient to consider all numbers that lie between two given numbers, or all numbers that are greater than or less than a given number. Such collections of real numbers are called **intervals**. For example, the collection of all real numbers x such that $0 \leq x < 3$ is an interval. Figure 1–7 shows a **line diagram** for this interval; the closed circle at 0 indicates that 0 belongs to the interval; the open circle at 3 indicates that 3 does not belong to the interval. We write [0, 3) to denote this interval; the bracket [at 0 indicates that 0 is in the interval, and the parenthesis) at 3 indicates that 3 is not in the interval. The expression [0, 3) is called **interval notation**.

Figure 1–8 depicts each type of interval with its interval notation and line diagram. The symbols $+\infty$ and $-\infty$ are read as "plus infinity" and "minus infinity," respectively; they are not numbers, but are merely a way of indicating that the interval contains *all* the numbers greater than or less than a given number, as the case may be.

EXAMPLE 7

The interval (1, 5) is the collection of all real numbers x such that $1 < x < 5$. The interval $[1, +\infty)$ is the collection of all real numbers x such that $x \geq 1$. $\quad \square$

EXAMPLE 8

The line diagrams corresponding to the intervals [1, 3], (−1, 2), (−∞, 1), and (3, +∞) are shown in Figure 1–9. $\quad \square$

Note that an interval does *not* consist only of the integers that satisfy its definition, but rather of *all* real numbers, rational and irrational, that satisfy its definition. For example, the interval (1, 5) does not consist only of the integers 2, 3, and 4, but rather of these numbers and all other real numbers that are greater than 1 and less than 5. For instance, the numbers

$$1.001, \quad \sqrt{2}, \quad 7/3, \quad 2.95, \quad \pi, \quad 2\sqrt{3}, \quad 23/5, \quad \text{and} \quad 4.9257$$

The Interval [0,3)

FIGURE 1–7

Interval	Interval notation	Line diagram
$a < x < b$	(a, b)	○————○ a b
$a \leq x < b$	$[a, b)$	●————○ a b
$a < x \leq b$	$(a, b]$	○————● a b
$a \leq x \leq b$	$[a, b]$	●————● a b
$x < a$	$(-\infty, a)$	←————○ a
$x \leq a$	$(-\infty, a]$	←————● a
$x > b$	$(b, +\infty)$	○————→ b
$x \geq b$	$[b, +\infty)$	●————→ b

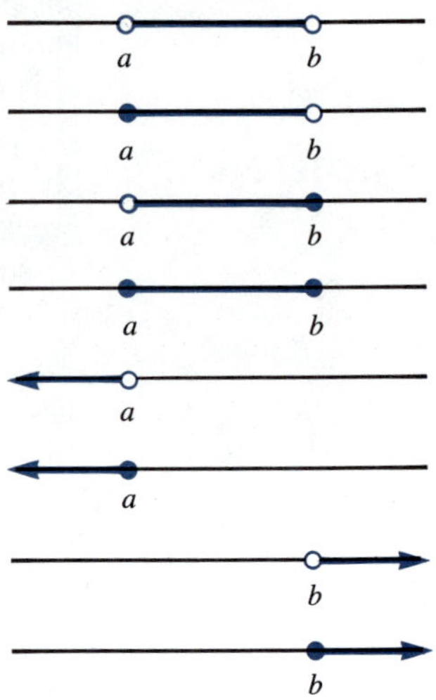

Intervals

FIGURE 1–8

are all in the interval (1, 5) because all of them are greater than 1 and less than 5. (Use a calculator to check this.)

Intervals are often used to describe the ranges over which various alternatives hold.

EXAMPLE 9 Suppose that the recommended dosage of a certain drug is 30 milligrams (mg) if the patient weighs 200 lb or less and 40 mg if the patient weighs more than 200 lb. If we let d denote the recommended dosage in milligrams and w the weight of the patient in pounds, we may express these facts with intervals by writing

$$d = \begin{cases} 30 & \text{if } w \text{ is in } (0, 200], \\ 40 & \text{if } w \text{ is in } (200, +\infty). \end{cases}$$

Of course, we may equally well express them with inequalities by writing

$$d = \begin{cases} 30 & \text{if } 0 < w \leq 200, \\ 40 & \text{if } w > 200. \end{cases} \qquad \square$$

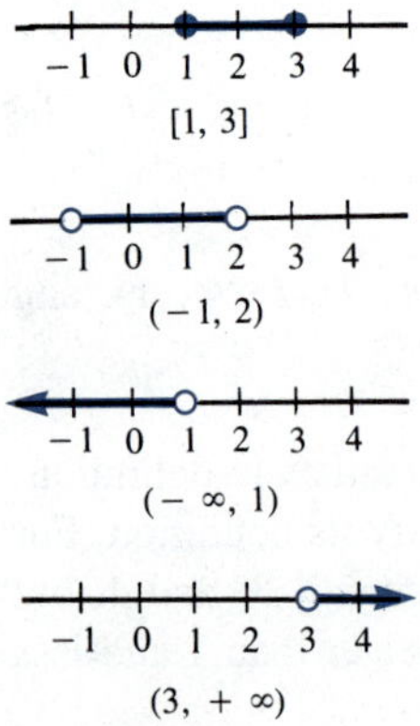

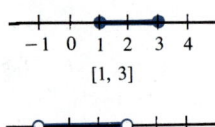

FIGURE 1–9

Absolute Value

The **absolute value** of a real number is its distance from 0 on the real line, without regard to whether the number is positive or negative. For instance, the numbers 3 and -3 are both 3 units from 0 on the real line, so both have an absolute value of 3. A more formal definition of the absolute value of a real number is as follows:

REVIEW: NUMBERS AND ALGEBRA

> **Absolute Value**
>
> The **absolute value** of a real number x, denoted by $|x|$, is defined by
> $$|x| = \begin{cases} x & \text{if } x \geq 0, \\ -x & \text{if } x < 0. \end{cases}$$

EXAMPLE 10 $|2| = 2,\ |-2| = -(-2) = 2,\ |0| = 0,\ |\sqrt{2}| = \sqrt{2},$

$\left|\dfrac{2}{3} - \dfrac{5}{3}\right| = |-1| = -(-1) = 1,\ |-3(8-2)| = |-18| = -(-18) = 18.$ ∎

Absolute value is often used when it is the magnitude of a quantity (its distance from 0), rather than its sign, that is important.

EXAMPLE 11 An automatic pilot will correct a ship's heading if and only if the actual heading differs by at least 0.5 degrees from the programmed heading. Let x denote the actual heading and y the programmed heading, in degrees. Then $|x - y|$ is the difference between the headings, without regard to sign, and hence the automatic pilot will correct the heading if and only if $|x - y| \geq 0.5$. ∎

■ *Computer Manual: Program INTERVAL*

1.1 EXERCISES

Number Systems

In Exercises 1 through 10, state the number systems to which the given number belongs.

1. -5
2. 246
3. $\dfrac{291}{376}$
4. $\dfrac{1}{\sqrt{2}}$
5. 2π
6. $-\dfrac{42}{15}$
7. 82.974
8. $3\sqrt{5}$
9. $42\dfrac{3}{5}$
10. $-2.159737373\ldots$

The Real Line

In Exercises 11 through 13, plot the given numbers on the real line.

11. $-5,\ -4.5,\ -7/4,\ -1,\ -2/5,\ 0,\ 1/3,\ 3/4,\ 1,\ 3/2,\ 17/8,\ 19/6,\ 4.2,\ 5,\ \sqrt{2},\ 2\sqrt{3},\ -\sqrt{5},\ \pi/2,\ -3\pi/4$

12. $0,\ 1/12,\ 1/6,\ 1/4,\ 1/3,\ 5/12,\ 1/2,\ 7/12,\ 2/3,\ 3/4,\ 5/6,\ 11/12,\ 1$

13. $-5.2,\ -11/3,\ -2.1,\ -11/7,\ -9/7,\ -5/8,\ -1/8,\ 1/10,\ 0.3,\ 9/10,\ 1.25,\ 9/5,\ 2.125,\ 24/7$

In Exercises 14 through 16, what numbers are represented by the plotted points?

14.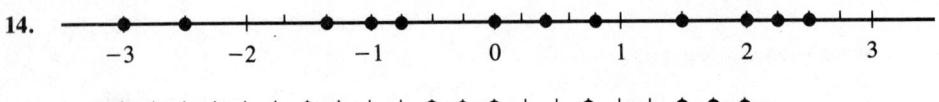

15.

16.

Inequalities

In Exercises 17 through 32, state whether the given inequality is true or false.

17. $-3 < 2$ **18.** $-5 < 0$ **19.** $-6 < -8$ **20.** $2 < 2$

21. $-3 > -4$ **22.** $\frac{2}{3} < \frac{3}{5}$ **23.** $-\frac{2}{3} < -\frac{3}{5}$ **24.** $3 > \pi$

25. $-1 < 0 < 1$ **26.** $-2 \leq -3 < 0$ **27.** $-2 < 3 \leq 3$ **28.** $\frac{2}{3} < \frac{4}{7} < 1$

29. $-\frac{7}{3} \leq -\frac{5}{3} \leq -1$ **30.** $\frac{3}{5} \leq \frac{5}{7} \leq \frac{7}{8}$ **31.** $-\frac{3}{5} \leq -\frac{4}{7} \leq -\frac{7}{8}$ **32.** $-\frac{7}{8} \leq -\frac{4}{7} \leq -\frac{3}{5}$

Management

The statements in Exercises 33 through 41 were suggested by recent articles in *The Wall Street Journal*. Use an inequality to write each statement.

33. A corporation forecasts sales of at least $1,500,000 next quarter.

34. A corporation forecasts sales of no more than $1,500,000 next quarter.

35. A corporation forecasts sales of more than $1,500,000 next quarter.

36. A corporation forecasts sales of less than $1,500,000 next quarter.

37. Construction of manufacturing plants that emit 100 or more tons of air pollutants annually may be prohibited.

38. The price of a corporation's stock ranged from $86.50 to $92.50 during the last six months.

39. A corporation's return on investment during the next quarter is forecast to be more than 10% but less than 12.5%.

40. Monthly automobile production in the United States next month is predicted to be more than 7.7 million units and less than 10.3 million units.

41. A stock analyst expects a firm's earnings to be at least $0.20 per share but not more than $0.22 per share next year.

Life Science

42. A small animal will die unless its body temperature is at least 92°F and no more than 108°F. Express this condition as an inequality.

43. According to a recent article in *The Wall Street Journal*, the current infant mortality rate in the United States is less than one-half of the corresponding rate in the 1960s. If the rate in the 1960s was M, express this fact as an inequality.

Intervals

In Exercises 44 through 51 write the collection of all real numbers x as an interval.

44. $-2 < x \leq 7$ **45.** $4 < x < 8$ **46.** $x > -3$ **47.** $x < \dfrac{2}{3}$

48. $\dfrac{1}{2} \leq x < \dfrac{3}{2}$ **49.** $-3 \leq x \leq \dfrac{3}{4}$ **50.** $x \leq -\dfrac{5}{3}$ **51.** $x \geq 4$

In Exercises 52 through 57 draw the line diagram for the given interval and then write it as the collection of numbers that satisfy an inequality.

52. $(-4, 8)$ **53.** $(-6, 3]$ **54.** $[-5, +\infty)$ **55.** $[-\frac{1}{2}, 3]$ **56.** $(-\infty, -4)$ **57.** $(2, +\infty)$

58. Which of the following numbers are in the interval $[-2, 2)$?
$-2.1, -2, -1.99, -\frac{32}{11}, -\frac{3}{4}, -1, \frac{2}{3}, 2, \pi, \pi/2, 3\sqrt{2}, \frac{9}{5}, \frac{9}{2}$

59. Which of the following numbers are in the interval $[0, 1]$?
$0, \frac{292}{293}, \frac{293}{292}, 0.025, -0.025, 1/\sqrt{2}, 3/\sqrt{5}, \pi/3, \pi/4, 0.9999\ldots, 1.0001, 1$

Management

The statements in Exercises 60 through 63 were suggested by recent articles in *The Wall Street Journal*. Use intervals to write each statement.

60. A corporation's stock paid a dividend of $1.40 per share during the period 1980–1985 and a dividend of $1.60 per share during the period 1986–1989.

61. An industrial chemical costs $0.40 per gallon if less than 1000 gallons are purchased and $0.35 per gallon if 1000 or more gallons are purchased.

62. Sponsors of a TV show will pay $300,000 for each 30-second commercial if the show's rating is at least 21.0; otherwise they will pay $275,000 for each 30-second commercial.

63. A firm pays contract workers $4 per hour if they work fewer than 20 hours per week, $4.50 per hour if they work at least 20 but not more than 40 hours per week, and $5 per hour if they work more than 40 hours per week.

Life Science

In Exercises 64 and 65, use intervals to write the statement about the dosage.

64. The recommended dosage of a drug is 50 mg if the patient is less than 5 years old, 100 mg if the patient is between 5 and 12 years old, and 200 mg if the patient is more than 12 years old.

65. The recommended dosage of a drug is 200 units per day for the first 14 days, 120 units per day for days 15 through 30, and 100 units per day thereafter.

Absolute Value

In Exercises 66 through 80 evaluate the given expression.

66. $|3|$ 67. $|-3|$ 68. $\left|-\dfrac{2}{3}\right|$ 69. $\left|\dfrac{4}{3}\right|$

70. $\left|\dfrac{3}{7}\right|$ 71. $|12 - 7|$ 72. $|7 - 12|$ 73. $-|2(0)|$

74. $-|-8|$ 75. $-|6 + 2|$ 76. $|2(8 - 15)|$ 77. $|-2(8 - 15)|$

78. $2|8 - 15|$ 79. $-2|8 - 15|$ 80. $|-2\pi|$

81. The cruise control mechanism on a car will automatically change the speed of the car if and only if its actual speed is at least 1 mph faster or slower than the desired cruising speed. Express this fact with absolute values.

82. The pressure inside a natural gas tank is in the safe range if it is within ± 2.5 pounds per square inch of the designed pressure of 20 pounds per square inch. Express this fact with absolute values.

Management

83. A production process is in control if the weight of each part it produces is no more than ± 0.25 gram from the designed weight of 220 grams. Express this fact with absolute values.

1.2 LINEAR EQUATIONS

As we construct and analyze mathematical models, we will often encounter situations where it is necessary to solve equations such as

$$2x + 3 = 7 \quad \text{or} \quad 8x + 4 = 2x - 2.$$

Equations such as these are called **linear equations in one variable.** The letter x is the variable. The solution of such an equation is the value of x that makes the equation a true statement. For example, $x = -1$ is the solution of

$$8x + 4 = 2x - 2,$$

since

$$8(-1) + 4 = 2(-1) - 2$$

yields

$$-8 + 4 = -2 - 2,$$

or

$$-4 = -4.$$

In this section we show how to solve linear equations in one variable.

Solving any equation requires that we isolate the variable on one side of the equal sign. In order to do this, we allow the following operations:

> **Operations for Solving Equations**
>
> When solving an equation we may
>
> **1.** Add the same number to or subtract the same number from both sides of the equation.
>
> **2.** Multiply or divide both sides of the equation by the same nonzero number.

We illustrate with an example.

EXAMPLE 1   Let us solve the linear equation

$$8x - 3 = 6x + 7$$

for x.

Add 3 to both sides:
$$\begin{aligned} 8x - 3 &= 6x + 7 \\ + 3 & + 3 \\ \hline 8x &= 6x + 10 \end{aligned}$$

Subtract 6x from both sides:
$$\begin{aligned} 8x &= 6x + 10 \\ -6x & -6x \\ \hline 2x &= 10 \end{aligned}$$

Multiply both sides by $\frac{1}{2}$: $\frac{1}{2}(2x) = \frac{1}{2}(10)$

Simplify: $1x = 5$

Solution: $x = 5.$

Check the solution by substituting 5 for x in the original equation. □

Here is a typical application of linear equations.

EXAMPLE 2   Suppose that Ann's hourly wage rate is $0.50 more than Bob's hourly wage rate. If last week Ann worked 30 hours, Bob worked 40 hours, and Bob made $35 more than Ann, what are their hourly wage rates?

We begin by writing down what we are to find. We must find

$$\text{Ann's hourly wage rate}$$

and

$$\text{Bob's hourly wage rate.}$$

Our next step is to assign a variable to one of the quantities we are to find. Suppose we let

$$x = \text{Ann's hourly wage rate, in dollars per hour.}$$

Now we write mathematical expressions for the rest of the quantities we are to find; in this case, we must write an expression for Bob's hourly wage rate. Since Bob's hourly rate is $0.50 less than Ann's, we have

$$\text{Bob's hourly wage rate} = x - 0.50.$$

The next step is to write an equation that relates the quantities we are to find. For this problem, we know that

$$\text{Ann's earnings last week} = (\text{Ann's hours worked})(\text{Ann's wage rate})$$
$$= 30x$$

and

$$\text{Bob's earnings last week} = (\text{Bob's hours worked})(\text{Bob's wage rate})$$
$$= 40(x - 0.50)$$

Therefore, since Bob earned $35 more than Ann, we have

$$30x + 35 = 40(x - 0.50).$$

We solve this linear equation:

$$\begin{array}{rcr} 30x + 35 = & & 40x - 20 \\ -40x - 35 & & -40x - 35 \\ \hline -10x = & & -55 \end{array}$$

$$-\frac{1}{10}(-10x) = -\frac{1}{10}(-55)$$

$$x = 5.50.$$

Finally, we use the solution of the equation to answer the questions posed by the problem:

$$\text{Ann's hourly wage rate} = x = \$5.50 \text{ per hour}$$

and

$$\text{Bob's hourly wage rate} = x - 0.50 = 5.50 - 0.50 = \$5.00 \text{ per hour.} \quad \square$$

Note the steps we followed in solving the applied problem of Example 2. These steps can be used to solve any applied problem that leads to a linear equation in one variable.

Keys to Solving Applied Problems

1. Write down the quantities that are to be found.
2. Denote one of these quantities by a variable; then use the variable to write mathematical expressions for the remaining quantities.
3. Write an equation relating the expressions of Step 2.
4. Solve the equation.
5. Use the solution of the equation to answer the questions posed by the problem.

EXAMPLE 3 A plane takes off and flies north at 400 miles-per-hour (mph). One hour later a second plane takes off and flies north at 480 mph. How many hours after its takeoff does the second plane overtake the first one?

1. We are to find the time it takes for the first plane to overtake the second.

2. Let

t = Number of hours from takeoff of second plane until it overtakes the first plane.

Then, since the first plane left 1 hour before the second,

$t + 1$ = Number of hours from takeoff of first plane until second plane overtakes it.

3. Since distance traveled equals rate of travel multiplied by time of travel ($d = rt$), we have

Distance flown by second plane = $480t$,
Distance flown by first plane = $400(t + 1)$.

When the second plane overtakes the first, they have both flown the same distance, so

$$480t = 400(t + 1).$$

4. The solution of the equation

$$480t = 400(t + 1)$$

is $t = 5$. (Check this.)

5. Therefore the second plane will overtake the first one 5 hours after the second one takes off. ◻

16   CHAPTER 1

EXAMPLE 4   How many gallons of a solution that is 20% sulfuric acid must be added to 100 gallons of a solution that is 40% sulfuric acid in order to have a solution that is 25% sulfuric acid?

1. We are to find the number of gallons of the 20% solution that must be added.

2. Let x = the number of gallons of the 20% solution that must be added.

3. We must have

    Acid in original solution + Acid added = Acid in final solution.

    The original solution of 100 gallons was 40% acid, so it contains

    $$0.4(100) = 40$$

    gallons of acid. The x gallons of solution added is 20% acid, so it contains

    $$0.2x$$

    gallons of acid. The final solution consists of $100 + x$ gallons and is 25% acid, so it contains

    $$0.25(100 + x)$$

    gallons of acid. Therefore

    $$40 + 0.2x = 0.25(100 + x),$$

    or

    $$40 + 0.2x = 25 + 0.25x.$$

4. The solution of the equation $40 + 0.2x = 25 + 0.25x$ is $x = 300$. (Check this.)

5. Therefore 300 gallons of the 25% sulfuric acid solution must be added. □

■ *Computer Manual: Program LNEAREQN*

1.2 EXERCISES

In Exercises 1 through 22 solve the given equation for x.

1. $2x + 8 = 12$
2. $-2x - 7 = 15$
3. $7 - 2x = 23$
4. $8x + 5 = -20$
5. $-3 - 5x = -8$
6. $10 = -3x + 25$
7. $2 + x = 7 - x$
8. $8 + 3x = 9 + 2x$
9. $-3x + 6 = 5x - 2$
10. $4x + 5 = 7 - 2x$
11. $2x + 9 = 5x$
12. $12x + 6 = -8 - 2x$
13. $2x - \dfrac{1}{2} = \dfrac{3}{2} + x$
14. $\dfrac{1}{3}x - \dfrac{2}{3} = \dfrac{5}{3}x + \dfrac{7}{3}$
15. $\dfrac{2}{5} - \dfrac{3}{4}x = \dfrac{1}{2}x + 5$

Review: Numbers and Algebra

16. $-\dfrac{7}{2}x - \dfrac{1}{2} = 2x + \dfrac{12}{5}$ 17. $x + \dfrac{3}{5}(10 - 2x) = \dfrac{4}{5}x - 2$ 18. $4\left(\dfrac{2}{3}x - \dfrac{4}{3}\right) = 3\left(\dfrac{4}{3}x + \dfrac{2}{3}\right)$

19. $2(x - 3) + 5x = 6x - 7(x + 1)$ 20. $5 + 3(2 + 3x) = 2(4x + 1) - 2x$

21. $10x - 5(4 + 3(x - 1)) = 4[(3x - 2) + 7]$ 22. $12[8 + 4(1 - x)] + 3x = 6x - 3[1 - 2(3 - 2x)]$

If you invest $\$P$ at $i\%$ simple interest per year, then at the end of t years the investment will be worth $\$A$, where

$$A = P(1 + 0.01it).$$

For instance, it you invest $100 at 6% simple interest per year, then at the end of 5 years the investment will be worth

$$\$A = \$100(1 + 0.01(6)(5)) = \$100(1.3) = \$130.$$

Use this simple interest formula in Exercises 23 through 30.

23. Find the worth of an investment of $1000 if it earns 8% simple interest per year for four years.

24. Find the worth of an investment of $5000 if it earns 10.5% simple interest per year for 2.5 years.

25. Find the amount $\$P$ that must be invested at 6% simple interest per year in order to have $1000 at the end of three years.

26. Find the amount that must be invested at 8.25% simple interest per year in order to have $50,000 at the end of 4.75 years.

27. How long will it take for an investment of $10,000 to grow to $12,000 if it earns 8% simple interest per year?

28. How long will it take an investment of $500 to double if it earns 20% simple interest per year?

29. What simple interest rate is required if an investment of $5000 is to double in size in eight years?

30. What simple interest rate is required if an investment of $100,000 is to grow to $110,000 in nine months?

In the sciences temperature is usually measured in degrees Celsius (°C) rather than in degrees Fahrenheit (°F). The two temperature scales are related by the equation

$$C = \dfrac{5}{9}(F - 32).$$

Use this equation in Exercises 31 through 38.

Convert to °C:

31. 212°F   32. 98.6°F   33. 65°F   34. 32°F

Convert to °F:

35. 100°C   36. 50°C   37. 10°C   38. −40°C

Exercises 39 through 60 are applied problems. In each case you are to set up and solve a linear equation.

39. A rectangle is twice as long as it is wide, and its perimeter measures 30 inches. Find its dimensions.

40. The longer side of a rectangle is 4 inches longer than the shorter side, and its perimeter measures 56 inches. Find its dimensions.

41. If you take one-half the length of the longer side of a rectangle and add 1.5 centimeters, you obtain the length of the shorter side of the rectangle. What are the dimensions of the rectangle if its perimeter measures 102 cm?

42. The longest side of a triangle is $5/3$ as long as its shortest side, and the next longest side is $4/3$ as long as the shortest side. Find the lengths of the sides of the triangle if its perimeter measures 120 cm.

43. Suppose Al makes $0.25 more per hour than Bette and works 38 hours each week while Bette works 35 hours per week. If Al makes $27.50 more each week than Bette, what are their hourly wage rates?

44. Arlette makes $6.80 per hour, Boris makes $5.95 per hour, and Boris works 4.5 more hours each week than Arlette does. If their weekly income is the same, how many hours does each of them work per week?

45. Suppose you inherit $50,000 and intend to invest part of the money in gold stocks that pay 14% per year and the rest in certificates of deposit (CDs) that pay 8% per year. How much should you invest in stocks and how much in CDs if your total income from the two investments must be $5410 per year?

46. If a portion of $10,000 is invested at 12.5% per year and the rest at 7.2% per year, and the total income from these investments is $908.31, how much was invested at each rate?

47. If a car starts at noon and travels due east at 40 mph, and a second car starts at 2 P.M. and travels due east at 50 mph, at what time will the second car overtake the first one?

*48. If a plane flies from point A to point B at 450 mph and returns at 400 mph, and the round-trip takes $7\frac{7}{16}$ hours, how far apart are A and B?

49. Suppose cities X and Y are 300 miles apart and are connected by rail. A train starts from X and travels toward Y at 35 mph; 3 hours later another train starts from Y and travels toward X at 45 mph. When will the trains pass one another?

50. A riverboat travels at a constant boat speed. When the boat goes upstream, the current slows it by 5 mph; when the boat goes downstream, the current speeds it up by 5 mph. Suppose that on a round-trip the upstream travel time is 13.2 hours and the downstream return time is 8.8 hours. What is the boat speed?

*51. Two race cars start at the same time, the first one 12 miles ahead of the second. If the first car travels at a speed of 120 mph and the second car catches the first one 50 minutes after they start, what is the speed of the second car?

52. According to an article in *The Wall Street Journal,* eight years ago a prince married a woman who was then five years younger than one-half his age at the time of the marriage. If the woman was 22 years old at the time of the marriage, how old is the prince now?

53. If one solution contains 15% glucose (and therefore 85% water) and another contains 10% glucose, how many gallons of each must be mixed to obtain 20 gallons of a solution that is 12% glucose?

54. How many liters of wine worth $5 per liter should be mixed with 100 liters of wine worth $8.50 per liter to obtain wine worth $7 per liter?

55. Ten liters of a solution contain 35% antifreeze and 65% water. How much pure antifreeze must be added to obtain a solution that is 50% antifreeze and 50% water?

56. Gasohol is made by adding alcohol to gasoline. How many gallons of pure alcohol must be added to 500 gallons of gasoline in order for the resulting mixture to contain 8.5% alcohol?

57. A new cereal consists of a mixture of bran flakes, raisins, and hazelnuts. Bran flakes cost $0.30 per pound, raisins cost $0.75 per pound, and hazelnuts cost $1.12 per pound. How many pounds of hazelnuts must be added to 50 lb of bran flakes and 5 lb of raisins to make a mixture whose total cost is $22?

*58. A recent article in *The Wall Street Journal* reported that a jump in exports propelled West Germany's June trade surplus to a record 14.2 billion marks, thus raising the total trade surplus for the first six months of the year to 59.2 billion marks. What was West Germany's average monthly trade surplus for the first five months of the year?

Management

59. A firm's profits were $1 million; it paid out 20% of its profits in dividends and invested the rest in two new product lines. If one of the new lines earned 6.5% per year and the other earned 9% per year, and if the total earnings of the two lines combined amounted to $63,300 per year, how much was invested in each line?

60. A recent issue of *The Wall Street Journal* reported that a Chinese banking corporation's net income per share was up 19% from the previous year and was $0.02 Hong Kong (HK) more than the previously forecast level of $1.65 HK. What was its net income per share last year?

1.3 LINEAR INEQUALITIES

Inequalities such as

$$2x + 3 < 5 \quad \text{and} \quad 4x + 2 \geq 5x - 1$$

are called **linear inequalities in one variable.** A real number is said to **satisfy** the inequality if substituting the number for the variable in the inequality results in a true statement. For instance, the numbers 0 and $-\frac{1}{2}$ both satisfy the inequality

$$2x + 3 < 5$$

because

$$2(0) + 3 < 5$$

and

$$2(-\tfrac{1}{2}) + 3 < 5$$

are both true statements. On the other hand, the numbers 1 and $\frac{3}{2}$ do not satisfy this inequality, because

$$2(1) + 3 < 5$$

and

$$2(3/2) + 3 < 5$$

are not true statements. The **solution** of an inequality is the collection of all real numbers that satisfy the inequality. In general, the solution will be an interval. In this section, we show how to find the solution interval of a linear inequality in one variable.

Solving an inequality requires that we isolate the variable on one side of the inequality sign. In order to do this, we allow the following operations:

> **Operations for Solving Inequalities**
>
> When solving an inequality, we may
>
> 1. Add the same number to or subtract the same number from both sides of the inequality without changing it.
>
> 2. Multiply or divide both sides of the inequality by the same positive number without changing it.
>
> 3. Multiply or divide both sides of the inequality by the same negative number *provided that* we reverse the direction of the inequality.

We illustrate with examples.

EXAMPLE 1 Let us solve the inequality

$$2x + 3 < 5$$

for x.

Subtract 3 from both sides:
$$\begin{array}{rcr} 2x + 3 & < & 5 \\ -3 & & -3 \\ \hline 2x & < & 2 \end{array}$$

Multiply both sides by $1/2$:
$$\frac{1}{2}(2x) < \frac{1}{2}(2)$$

Solution:
$$x < 1.$$

(Note that the direction of the inequality remained unchanged because we multiplied both sides by the *positive* number $1/2$.) Thus the solution of $2x + 3 < 5$ is the collection of all real numbers less than 1, that is, the interval $(-\infty, 1)$. See Figure 1–10. ☐

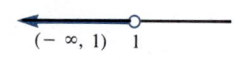

$(-\infty, 1)$ 1

FIGURE 1–10

EXAMPLE 2 Let us solve

$$4x + 2 \leq 5x - 1$$

for x.

Subtract 2 and 5x from both sides:
$$\begin{array}{rl} 4x + 2 \leq & 5x - 1 \\ -5x - 2 & -5x - 2 \\ \hline -x \leq & -3 \end{array}$$

Multiply both sides by −1: $\qquad (-1)(-x) \geq (-1)(-3)$

Solution: $\qquad\qquad\qquad\qquad x \geq 3.$

FIGURE 1–11
(number line showing 3, [3, +∞))

(Note that when we multiplied both sides of the inequality by the *negative* number −1, we reversed the direction of the inequality from ≤ to ≥.) Thus the solution of the inequality is the interval [3, +∞). See Figure 1–11. ☐

We illustrate the use of linear inequalities with some applied problems. The procedure for solving applied problems that lead to inequalities is similar to the one stated in the previous section for applied problems that lead to equations: all we have to do is replace the word "equation" in the procedure of Section 1.2 with the word "inequality."

EXAMPLE 3 If a firm makes x units of its product, its production cost will be $20{,}000 + 45x$ dollars. How many units can the firm make if it can spend no more than $200,000 on production?

Here the problem is partially set up for us: we must find x, given that the production cost must be less than or equal to $200,000; that is, we must solve the inequality

$$20{,}000 + 45x \leq 200{,}000$$

for x. Subtracting 20,000 from both sides of the inequality, we obtain

$$45x \leq 180{,}000,$$

so

$$x \leq \frac{180{,}000}{45} = 4000.$$

Hence, the firm can make at most 4000 units. ☐

EXAMPLE 4 A student has a work-study job that pays $4.00 per hour, and she receives $20 per week from her parents. If she needs at least $75 per week to live on, how many hours must she work each week?

We write down the quantities that are to be found: in this problem, we must find

the number of hours the student must work each week.

We assign a variable to one of the quantities that are to be found: let

x = Number of hours the student must work each week.

We know that

Student's weekly income = Her weekly earnings + $20.

But

$$\text{Weekly earnings} = (\text{Hourly wage rate})(\text{hours worked per week})$$
$$= 4x$$

dollars. Therefore the student's weekly income is

$$4x + 20$$

dollars. Since her weekly income must amount to at least \$75, we have the relationship

$$4x + 20 \geq 75.$$

Solving this inequality, we find that

$$4x \geq 55,$$

so

$$x \geq \frac{55}{4} = 13.75.$$

Thus the student must work at least 13.75 hours per week. □

EXAMPLE 5 One brand of refrigerator sells for \$440 and costs \$26 per month to operate; a second brand sells for \$600 and costs \$18 per month to operate. How long will it take from the time of purchase for the cost of buying and operating the second brand to be less than that of buying and operating the first brand?

Let t be the number of months from the time of purchase until the cost of the second brand is less than that of the first. Then

$$\text{Cost of first brand} = 440 + 26t$$
$$\text{Cost of second brand} = 600 + 18t.$$

We want the cost of the second brand to be less than that of the first, so we must have

$$600 + 18t < 440 + 26t.$$

The solution of this inequality is $t > 20$. (Check this.) Hence, it will be more than 20 months until the cost of the second brand becomes less than that of the first. □

■ *Computer Manual: Program LNERINEQ*

1.3 EXERCISES

In Exercises 1 through 22, find the solution of the given inequality. Write the solution in interval form if possible.

1. $x + 2 < 5$
2. $2x - 3 \leq 7$
3. $5 - 2x \geq 9$
4. $6x + 2 > -4$
5. $x + 3 < 2x - 4$
6. $2x + 1 \leq 5x + 7$

7. $8x + 12 \le 2x$

8. $-4x + 5 \le 8x + 2$

9. $3x + 8 < 15x - 22$

10. $5 - 4x > -8 + 3x$

11. $\frac{1}{2}x + 2 < \frac{3}{2}x - 4$

12. $\frac{1}{2}x + \frac{3}{2} \le \frac{1}{4}x - \frac{3}{4}$

13. $\frac{2}{5}x - \frac{1}{5} \ge \frac{3}{5}x + \frac{4}{5}$

14. $\frac{2}{3}x + \frac{8}{5} > \frac{3}{5}x + \frac{1}{3}$

15. $x + 1 \le x + 2$

16. $x + 1 > x + 2$

17. $\frac{2}{3}(3 - 9x) \ge 4 + 5x$

18. $3\left(\frac{2}{5} - \frac{3}{5}x\right) < \frac{1}{5}x + \frac{4}{5}$

19. $8x + 2(3x - 1) > 4(x - 5) - 2x$

20. $4(5x - 2) \ge 3x + 7(x - 1)$

21. $x + 2[3 - 5(7 - x)] \le 4[3(x + 2) - 1]$

22. $2x - 6\left[\frac{1}{3} - \frac{2}{3}x\right] \ge 8\left[\left(\frac{3}{2}x + \frac{1}{4}\right) - \frac{x}{2}\right]$

Management

23. If a firm produces and sells x units of its product, its profit is $22x - 4400$ dollars. (A negative profit is a loss.) How many units must the firm make and sell in order to attain a profit greater than 0?

24. How many units must the firm of Exercise 23 make and sell in order to attain a profit of at least $99,000?

25. If Omega Corporation makes and sells x units of its product, its profit will be $7.25x - 202,500$ dollars. How many units must Omega make and sell if its profit is to be positive?

26. How many units must Omega Corporation of Exercise 25 make and sell if its profit is to be $1 million or more?

27. Tick-Tock Clock Company has budgeted $2.5 million for producing its line of desk clocks during the coming year. Suppose that the cost of making x desk clocks is $8x + 225,000$ dollars. How many desk clocks can Tick-Tock make during the coming year?

28. Boreas Corporation has budgeted $462,000 for the purchase and installation of new computer terminals. The firm estimates that if it buys and installs x new terminals, it will cost $3200x + 156,800$ dollars. How many terminals can Boreas buy and install?

29. If wheat sells for $$p$ per bushel, farmers will produce $1200p$ million bushels and consumers will buy $5000 - 800p$ million bushels. Find the prices p for which there will be a shortage of wheat. (*Hint:* In order for there to be a shortage, the amount produced must be less than the amount consumers will buy.)

30. Refer to Exercise 29: find the prices p for which there will be a surplus of wheat.

31. If corn sells for $$p$ per bushel production will be $925p + 375$ million bushels and consumers will buy $4848 - 650p$ million bushels. Find the prices p for which there will be a shortage of corn.

32. If compact disks sell for $$p$ each, the recording industry will produce $2.3p - 8.8$ million disks per year and consumers will buy $124.25 - 1.4p$ million disks per year. Find the prices p for which there will be a surplus of compact disks.

Life Science

33. If x birds are hatched at the same time, $0.25x$ of them will still be alive three months later. How many birds must be hatched now in order to have at least 200 survivors three months from now?

34. If a patient takes x pills per day her daily calcium intake will be $24x + 86$ mg. If she needs at least 200 mg of calcium per day, how many pills should she take?

35. A patient undergoing radiation therapy receives $220x + 80$ units of radiation from x treatments. If the upper limit for allowable radiation is 2500 units per year, how many treatments can the patient receive in a year?

In Exercises 36 through 43, set up and solve a linear inequality.

36. If your job pays $6.25 per hour and you need an income of at least $300 per week, how many hours will you have to work each week?

37. Refer to Exercise 36: how many hours will you have to work each week if, in addition to your wages, you receive $55 per week from a trust fund?

38. Adam works in the library for $5.25 per hour and receives $50 per week from his parents. Beth works in the dining hall for $5.85 per hour and receives $30 per week from her parents. Suppose Adam and Beth work the same number of hours per week. How many hours must they work if Adam's weekly income is less than Beth's?

39. What hourly wage rate must your job pay if you work 40 hours each week and you need an income of at least $425 per week?

40. One brand of air conditioner costs $325 plus $1.50 per day to operate; a second brand costs $375 plus $1.25 per day to operate. How many days of operation will it take for the total cost of the second air conditioner to be less than that of the first one?

41. You can rent one apartment for $450 per month, everything included. You can rent another one for $380 per month plus $20 per month for utilities and a nonrefundable security deposit of $760. How many months will it take for the total cost of renting the first apartment to be greater than that of renting the second one?

42. A telephone company offers the following calling plans for instate nonlocal calls:

- plan A: $0.75 per minute plus a $5 monthly service charge;
- plan B: a flat fee of $35 per month.

How many minutes a month must you call in order for plan B to be cheaper than plan A?

*43. Refer to Exercise 42: suppose the plans are as follows:

- plan A: $0.75 per minute for the first 30 minutes or less, $0.60 per minute for all minutes after the first 30, and a $5 monthly service charge;
- plan B: no charge for the first 50 minutes or less, $0.30 per minute for all minutes after the first 50, and a $35 monthly service charge.

Which plan is cheaper when?

1.4 EXPONENTS

If a and r are real numbers, an expression of the form a^r is called an **exponential expression with base a and exponent r.** Examples of such expressions are

$$4^2, \quad 3^{-4}, \quad 10^0, \quad 5^{3/2}, \quad \text{and} \quad (1.005)^{-2.5}.$$

Many of the models that we will encounter later will make use of exponential expressions. In this section, we define such expressions and state some of their properties. We begin with the definition of a^r when r is an integer.

Integral Exponents

Let a be a real number and let n be an integer. Then

1. $a^n = \underbrace{a \cdot a \cdots a}_{n \text{ } a\text{'s}}.$

   In particular, $a^1 = a$.

2. If $a \neq 0$, $a^0 = 1$; 0^0 is not defined.

3. If $a \neq 0$, $a^{-n} = \dfrac{1}{a^n}$.

EXAMPLE 1
$$2^2 = 2 \cdot 2 = 4, \quad (4.3)^1 = 4.3,$$
$$(-3)^5 = (-3)(-3)(-3)(-3)(-3) = -243. \quad \square$$

EXAMPLE 2
$$2^0 = 1, \quad (-5)^0 = 1, \quad -(4/3)^0 = -1. \quad \square$$

EXAMPLE 3
$$2^{-1} = \frac{1}{2^1} = \frac{1}{2}, \quad 3^{-2} = \frac{1}{3^2} = \frac{1}{9},$$
$$\left(\frac{2}{3}\right)^{-3} = \frac{1}{(2/3)^3} = \frac{1}{8/27} = 1 \cdot \frac{27}{8} = \frac{27}{8}. \quad \square$$

We next consider the properties of exponents.

The Properties of Exponents

Let a, b, r, and s be real numbers, and assume that the following exponential expressions are all defined. Then

1. $a^r \cdot a^s = a^{r+s}$.
2. If $a \neq 0$, $\dfrac{a^r}{a^s} = a^{r-s}$.
3. $(a^r)^s = a^{rs}$.
4. $(ab)^r = a^r b^r$.
5. If $b \neq 0$, $\left(\dfrac{a}{b}\right)^r = \dfrac{a^r}{b^r}$.

EXAMPLE 4

$$2^3 \cdot 2^4 = 2^{3+4} = 2^7 = 128,$$

$$\frac{2^3}{2^5} = 2^{3-5} = 2^{-2} = \frac{1}{2^2} = \frac{1}{4},$$

$$(2^3)^2 = 2^{3 \cdot 2} = 2^6 = 64,$$

$$(3 \cdot 5)^2 = 3^2 \cdot 5^2 = 9 \cdot 25 = 225,$$

$$\left(\frac{3}{5}\right)^2 = \frac{3^2}{5^2} = \frac{9}{25}. \quad \square$$

The properties of exponents are often used to simplify algebraic expressions. We illustrate with examples.

EXAMPLE 5

$$(2x^3)(4x^2) = (2 \cdot 4)(x^3 \cdot x^2) = 8x^{3+2} = 8x^5,$$

$$(4x^3y^4z^{-2})(3x^{-5}yz^2) = 12x^{3+(-5)}y^{4+1}z^{-2+2} = 12x^{-2}y^5z^0 = \frac{12y^5}{x^2}. \quad \square$$

EXAMPLE 6

$$\frac{y^2}{y^5} = y^{2-5} = y^{-3} = \frac{1}{y^3},$$

$$\frac{4x^3y^2z^4}{2xy^2z^{-3}} = 2x^{3-1}y^{2-2}z^{4-(-3)} = 2x^2y^0z^7 = 2x^2z^7. \quad \square$$

EXAMPLE 7

$$(x^3)^2 = x^{3 \cdot 2} = x^6,$$

$$(y^{-2})^4 = y^{-2(4)} = y^{-8} = \frac{1}{y^8}. \quad \square$$

EXAMPLE 8

$$(2x^2)^{-2} = 2^{-2}x^{2(-2)} = 2^{-2}x^{-4} = \frac{1}{2^2} \cdot \frac{1}{x^4} = \frac{1}{4x^4},$$

$$(x^2y^{-2}z)^3 = x^6y^{-6}z^3 = \frac{x^6z^3}{y^6}. \quad \square$$

EXAMPLE 9

$$\left(\frac{x^2}{y^3}\right)^2 = \frac{x^4}{y^6}. \quad \square$$

We have defined a^r when r is an integer. Now we would like to define a^r when r is a rational number; in order to do so, we will need the concept of the **nth root** of a number. Recall that

- When n is an odd natural number, its nth root $\sqrt[n]{a}$ is the unique number b such that $b^n = a$.
- When n is an even natural number and $a > 0$, $\sqrt[n]{a}$ is the unique non-negative number b such that $b^n = a$.
- When n is an even natural number and $a < 0$, there is no number b such that $b^n = a$, so $\sqrt[n]{a}$ does not exist.

When $n = 2$, it is usual to write $\sqrt{a}$ instead of $\sqrt[2]{a}$.

EXAMPLE 10

$\sqrt[3]{8} = 2 \quad$ because $2^3 = 8$;

$\sqrt[5]{-243} = -3 \quad$ because $(-3)^5 = -243$;

$\sqrt[2]{16} = \sqrt{16} = 4 \quad$ because $4^2 = 16$ and 4 is positive;

$\sqrt[4]{625} = 5 \quad$ because $5^4 = 625$ and 5 is positive;

$\sqrt[2]{-4} = \sqrt{-4} \quad$ does not exist because there is no real number b such that $b^2 = -4$. $\quad \square$

We use the nth root of a to define a^r when r is rational.

Rational Exponents

Let a be a real number, n a natural number, and m an integer. If the principal nth root of a exists, define

$$a^{1/n} = \sqrt[n]{a}$$

and

$$a^{m/n} = (a^{1/n})^m = (\sqrt[n]{a})^m.$$

Since every rational number may be written in the form m/n, where n is a natural number and m an integer, the preceding statement defines a^r for all rational numbers r.

EXAMPLE 11

$$3^{1/2} = \sqrt[2]{3} = \sqrt{3}, \qquad (-8)^{1/3} = \sqrt[3]{-8} = -2,$$
$$16^{1/4} = \sqrt[4]{16} = 2, \qquad (-16)^{1/4} = \sqrt[4]{-16} \text{ does not exist.} \quad \square$$

EXAMPLE 12

$$4^{3/2} = (4^{1/2})^3 = (\sqrt{4})^3 = 2^3 = 8,$$

$$27^{-4/3} = (27^{1/3})^{-4} = (\sqrt[3]{27})^{-4} = 3^{-4} = \frac{1}{81}. \quad \square$$

Earlier in this section we listed the properties of exponents and illustrated them by using integral exponents. These properties hold for rational exponents also.

EXAMPLE 13

$$2^{1/2} \cdot 2^{3/2} = 2^{(1/2)+(3/2)} = 2^{4/2} = 2^2 = 4,$$

$$\frac{8^{2/3}}{8^{-1/3}} = 8^{(2/3)-(-1/3)} = 8^{3/3} = 8^1 = 8,$$

$$(7^{4/5})^5 = 7^{(4/5)5} = 7^4 = 2401,$$

$$(4 \cdot 81)^{1/2} = 4^{1/2} \cdot 81^{1/2} = \sqrt{4}\sqrt{81} = 2 \cdot 9 = 18,$$

$$\left(\frac{625}{16}\right)^{3/4} = \frac{625^{3/4}}{16^{3/4}} = \frac{(\sqrt[4]{625})^3}{(\sqrt[4]{16})^3} = \frac{5^3}{2^3} = \frac{125}{8},$$

$$\left(\frac{x^{1/2}y^{3/4}}{z^{1/4}}\right)^{4/3} = \frac{x^{1/2 \cdot 4/3}y^{3/4 \cdot 4/3}}{z^{1/4 \cdot 4/3}} = \frac{x^{2/3}y}{z^{1/3}}. \quad \square$$

EXAMPLE 14   Suppose the cost of making x units of a product is $100x^{3/4}$ dollars. Then the cost of making 10,000 units will be

$$100(10,000)^{3/4} = 10^2(10^4)^{3/4}$$
$$= 10^2(10^{4 \cdot (3/4)})$$
$$= 10^2(10^3)$$
$$= 10^5$$
$$= 100,000$$

dollars. The cost of making 15,000 units will be $100(15,000)^{3/4}$ dollars. We can find this cost by using a calculator with an x^y or y^x key. From such a calculator,

$$15,000^{3/4} = 15,000^{0.75} \approx 1355.403.$$

(The symbol $\approx$ means "approximately equal to.") Hence, the cost of making 15,000 units will be approximately

$$\$100(1355.403) = \$135,540.30. \quad \square$$

1.4 EXERCISES

In Exercises 1 through 32, evaluate the given expression.

1. 2^5
2. 2^{-5}
3. 3^{-1}
4. $(-5)^3$
5. 4^0
6. 8^{-3}
7. 1^{100}
8. $(-6)^4$
9. $\left(\dfrac{3}{2}\right)^4$
10. $\left(-\dfrac{3}{4}\right)^{-2}$
11. $(0.9)^3$
12. $(0.02)^{-2}$
13. $2^3 2^5$
14. $4^{-2} 4^5$
15. $7^{-1} 7^{-2}$
16. $\left(\dfrac{2}{3}\right)^3 \cdot \left(\dfrac{2}{3}\right)^4$
17. $\left(\dfrac{3}{4}\right)^{-2} \cdot \left(\dfrac{3}{4}\right)^4$
18. $\dfrac{9^2}{9}$
19. $\dfrac{12^4}{12^2}$
20. $\dfrac{3^9}{3^{12}}$
21. $\dfrac{5^{-2}}{5^{-5}}$
22. $(2^3)^2$
23. $(5^2)^4$
24. $(4^{-1})^3$
25. $(6^2)^{-1}$
26. $(2 \cdot 7)^2$
27. $(3 \cdot 5)^{-2}$
28. $(4 \cdot (-3))^3$
29. $\dfrac{2^3 2^4}{2^5 2^{-2}}$
30. $\dfrac{3^5 3^{-6} 3^3}{3^7 3^{-4}}$
31. $\left(\dfrac{2^3 5^3}{2^4 5^2}\right)^3$
32. $\dfrac{(2^2 3^{-1})^2}{(2^{-2} 3^{-2})^3}$

In Exercises 33 through 42, simplify the given expression.

33. $(5x^2)(7x^3)$
34. $(4x^2 y^{-1})(3xy^3)$
35. $(12x^2 y^3 z^{-4})(3xy^{-4} z^{-2})$
36. $\dfrac{8x^2 y^3}{2xy^4}$
37. $\dfrac{3x^{-5} y^4 z^2}{2x^{-2} y^2 z}$
38. $\dfrac{240 x^4 y^{-3} z^{-5}}{6x^{-4} z^{10}}$
39. $(2x^4)^2$
40. $(3x^2 y^3 z^{-1})^3$
41. $(2x^2 y^3 z)^{-1}$
42. $\left(\dfrac{x^2 yz}{x^3 z^2}\right)^3$

In Exercises 43 through 60, evaluate the given expression.

43. $4^{1/2}$
44. $4^{5/2}$
45. $4^{-3/2}$
46. $27^{-1/3}$
47. $5^{1/2}$
48. $8^{4/3}$
49. $81^{-3/2}$
50. $100^{5/2}$
51. $64^{5/3}$
52. $2^{1/2} 2^{5/2}$
53. $3^{2/5} 3^{-7/5}$
54. $\dfrac{2^{7/5}}{2^{2/5}}$
55. $\dfrac{9^{1/2}}{9^{3/2} 9^{-5/2}}$
56. $(2^{3/5})^5$
57. $(27^{2/3})^6$
58. $(3^{-3/2})^4$
59. $(16 \cdot 25)^{1/2}$
60. $(81 \cdot 16)^{-3/4}$

In Exercises 61 through 66, simplify the given expression. Write your answer using only positive exponents.

61. $(2x^{3/2})(5x^{-5/2})$
62. $(3x^{4/3} y^{2/5})(4x^{-1/3} y^{3/5})$
63. $\dfrac{x^{1/3} y^{2/5}}{2x^{2/3} y^{-3/5}}$
64. $\dfrac{4x^2 y^{2/3} z^{-5/4}}{x^{1/2} y^{1/3} z^{1/4}}$
65. $\left(\dfrac{x^{2/3} y^{5/6}}{z^{1/3}}\right)^3$
66. $\dfrac{(27x^3 y^{-2})^{1/3}}{(3x^{2/3} y^{4/3})^3}$

Management

67. If a firm makes and sells x units of its product, its net income will be $200x^{1/2}$ dollars. Find the firm's net income if it makes and sells 625 units.

68. Find the net income of the firm of Exercise 67 if it makes and sells 1,000,000 units.

69. If a corporation produces x units of its product, its average cost per unit is $40 + 120x^{-2/3}$ dollars. Find the corporation's average cost per unit if it makes 1000 units.

70. Find the average cost per unit for the corporation of Exercise 69 if it produces 125,000 units.

Life Science

Over a certain temperature range, the activity of a cold-blooded animal can be determined from the equation

$$A = kT^{3/2}.$$

Here A is a measure of the animal's activity, T is the temperature in degrees Celsius, and k is a numerical constant that depends on the animal. Exercises 71 through 75 refer to this equation.

71. Suppose that for iguanas the constant k is 1. Find a measure for the activity of iguanas when the temperature is 20°C.

72. Repeat Exercise 71 if the temperature is 30°C.

73. How much more active are iguanas at 30°C than at 20°C? (*Hint:* Divide the result of Exercise 72 by that of Exercise 71.)

*74. Find an expression that describes how much more active a cold-blooded animal is at temperature T_1 than it is at temperature T_0, $T_1 > T_0$.

*75. A 10% rise in temperature increases the activity of a cold-blooded animal by how much? (*Hint:* Use the result of Exercise 74.)

Social Science

Suppose a political pollster asks N voters which candidate they will vote for. Let p_0 be the proportion of these voters who say they will vote for candidate A, and let p be the proportion of all voters who will vote for A. Statistical theory tells us that we can be 95% certain that the value of p lies in the interval

$$p_0 - 1.96 \left[\frac{p_0(1-p_0)}{N} \right]^{1/2} < p < p_0 + 1.96 \left[\frac{p_0(1-p_0)}{N} \right]^{1/2}.$$

Exercises 76 through 80 refer to this equation.

*76. Suppose that 50 of 100 voters polled say they will vote for candidate A. We can be 95% certain that the proportion of all voters who will vote for A lies in what interval? (*Hint:* $p_0 = 50/100$.)

*77. Repeat Exercise 76 if 500 out of 1000 voters polled say they will vote for A.

*78. Repeat Exercise 76 if 5000 out of 10,000 voters polled say they will vote for A.

*79. Suppose that 720 of 1500 voters polled will vote for A and the remaining 780 will vote for B. Can we be 95% certain that A will lose to B?

*80. Repeat Exercise 79 if 705 will vote for A and the remaining 795 will vote for B.

REVIEW: NUMBERS AND ALGEBRA

1.5 POLYNOMIALS

An algebraic expression of the form

$$a_n x^n + a_{n-1} x^{n-1} + \cdots + a_1 x + a_0,$$

where n is a nonnegative integer, $a_n, a_{n-1}, \ldots, a_1, a_0$ are real numbers, and $a_n \neq 0$, is called a **polynomial of degree n in the variable x.** The numbers

$$a_n, a_{n-1}, \ldots, a_1, a_0$$

are the **coefficients** of the polynomial, and

$$a_n x^n, a_{n-1} x^{n-1}, \ldots, a_1 x, a_0$$

are the **terms** of the polynomial.

Thus $7x + 3$ is a polynomial of degree 1 in x and $5q^2 - 3q + 2$ is a polynomial of degree 2 in q. The coefficients of $5q^2 - 3q + 2$ are 5, -3, and 2, and its terms are $5q^2$, $-3q$, and 2. Expressions such as

$$\frac{1}{x}, \quad y^2 - \sqrt{y}, \quad \text{and} \quad \frac{4p - 2}{3p + 1}$$

are not polynomials.

In this section we study polynomials. We begin with a consideration of the arithmetic of polynomials, and then we discuss the problem of factoring polynomials.

Polynomial Arithmetic

Polynomials can be added and subtracted; this is done by adding or subtracting **like terms,** that is, terms in which the variable occurs to the same exponent. We illustrate with examples.

EXAMPLE 1 Let us add the polynomials

$$x^3 + 3x^2 - 7x + 2 \quad \text{and} \quad 2x^3 - 5x^2 + 2x + 1.$$

We have

$$(x^3 + 3x^2 - 7x + 2) + (2x^3 - 5x^2 + 2x + 1)$$
$$= (x^3 + 2x^3) + (3x^2 - 5x^2) + (-7x + 2x) + (2 + 1)$$
$$= 3x^3 - 2x^2 - 5x + 3.$$

Notice how we added like terms: x^3 plus $2x^3$, $3x^2$ plus $-5x^2$, and so on. □

EXAMPLE 2 Let us subtract

$$5k^2 + 3k - 9 \quad \text{from} \quad k^3 - 2k^2 + 6k + 3.$$

We have
$$(k^3 - 2k^2 + 6k + 3) - (5k^2 + 3k - 9)$$
$$= k^3 + (-2k^2 - 5k^2) + (6k - 3k) + (3 - (-9))$$
$$= k^3 - 7k^2 + 3k + 12. \quad \square$$

Polynomials can also be multiplied. This is accomplished by multiplying each term of one polynomial by each term of the other, using the properties of exponents, and then collecting like terms. We illustrate with an example.

EXAMPLE 3 Let us multiply
$$x^3 + 5x^2 + 2 \quad \text{by} \quad x^2 - 3x + 1.$$
We have
$$(x^3 + 5x^2 + 2)(x^2 - 3x + 1) = x^3(x^2) + x^3(-3x) + x^3(1)$$
$$+ 5x^2(x^2) + 5x^2(-3x) + 5x^2(1)$$
$$+ 2(x^2) + 2(-3x) + 2(1)$$
$$= x^5 - 3x^4 + x^3$$
$$+ 5x^4 - 15x^3 + 5x^2$$
$$+ 2x^2 - 6x + 2$$
$$= x^5 + (-3x^4 + 5x^4) + (x^3 - 15x^3)$$
$$+ (5x^2 + 2x^2) - 6x + 2$$
$$= x^5 + 2x^4 - 14x^3 + 7x^2 - 6x + 2. \quad \square$$

Factoring

To **factor** a polynomial is to write it as a product of polynomials of lesser degree. For instance, we may factor the second-degree polynomial
$$2x^2 - 3x - 2$$
by writing it as a product of first-degree polynomials:
$$2x^2 - 3x - 2 = (2x + 1)(x - 2).$$
(Check that this factorization is valid by carrying out the multiplication on the right-hand side.) As we shall see, factoring polynomials is often an aid in problem-solving.

The process of factoring a polynomial begins with a search for the **common factors** of its terms, which can be **factored out** of the polynomial.

EXAMPLE 4 The polynomial $6x^2 + 24x + 24$ has 6 as its only common factor, so
$$6x^2 + 24x + 24 = 6(x^2 + 4x + 4).$$

The polynomial $3x^4 + 12x^2$ has 3 and x^2 as common factors, so
$$3x^4 + 12x^2 = 3x^2(x^2 + 4).$$
The polynomial $2x^2 - 5x - 3$ has no common factors. ☐

Once all common factors (if any) have been factored out of a polynomial, the polynomial that remains may be factorable.

EXAMPLE 5   The polynomial $6x^2 + 24x + 24$ may be factored as follows:
$$6x^2 + 24x + 24 = 6(x^2 + 4x + 4) = 6(x + 2)(x + 2) = 6(x + 2)^2.$$
(Check this.) ☐

In Example 5, how did we know that $x^2 + 4x + 4$ factored as
$$x^2 + 4x + 4 = (x + 2)(x + 2) \,?$$
The answer is that we recognized $x^2 + 4x + 4$ as a special type of polynomial called a **perfect square.** It is often possible to factor a polynomial by recognizing that it has one of the following forms:

Special Types of Polynomials

Perfect Square	$x^2 + 2ax + a^2 = (x + a)^2$
Difference of Two Squares	$x^2 - a^2 = (x - a)(x + a)$
Difference of Two Cubes	$x^3 - a^3 = (x - a)(x^2 + ax + a^2)$
Sum of Two Cubes	$x^3 + a^3 = (x + a)(x^2 - ax + a^2)$

EXAMPLE 6   Since
$$x^2 + 4x + 4 = x^2 + 2 \cdot 2x + 2^2,$$
it is a perfect square with $a = 2$, so it factors as
$$x^2 + 4x + 4 = (x + 2)^2. \ ☐$$

EXAMPLE 7   The polynomial $x^2 - 9 = x^2 - 3^2$ is the difference of two squares, so it factors as
$$x^2 - 9 = (x - 3)(x + 3). \ ☐$$

EXAMPLE 8   The polynomial $x^3 - 8 = x^3 - 2^3$ is the difference of two cubes, so it factors as
$$x^3 - 8 = (x - 2)(x^2 + 2x + 2^2) = (x - 2)(x^2 + 2x + 4). \ ☐$$

EXAMPLE 9 The polynomial $8x^3 + 27 = (2x)^3 + 3^3$ is the sum of two cubes, so it factors as
$$8x^3 + 27 = (2x)^3 + 3^3$$
$$= (2x + 3)((2x)^2 - 3(2x) + 3^2)$$
$$= (2x + 3)(4x^2 - 6x + 9). \quad \square$$

A polynomial that is not one of the special types just considered can sometimes be factored by other methods. To illustrate, let us consider polynomials of degree 2, which are known as **quadratic polynomials.** Every quadratic polynomial has the form
$$ax^2 + bx + c,$$
where a, b, and c are real numbers with $a \neq 0$. In particular, if $a = 1$, the quadratic has the form
$$x^2 + bx + c,$$
and it follows that any factorization of the quadratic must be of the form
$$x^2 + bx + c = (x + r)(x + s),$$
for some numbers r and s. Multiplying out the right-hand side and collecting terms, we have
$$x^2 + bx + c = x^2 + (r + s)x + rs;$$
hence
$$c = rs \quad \text{and} \quad b = r + s.$$
Therefore:

Keys to Factoring $x^2 + bx + c$

To factor
$$x^2 + bx + c,$$
find a pair of integers r and s such that
$$c = rs \quad \text{and} \quad b = r + s.$$
Then
$$x^2 + bx + c = (x + r)(x + s).$$

EXAMPLE 10 Let us attempt to factor $3q^2 - 15q - 72$. We first factor out the common factor of 3 to obtain
$$3q^2 - 15q - 72 = 3(q^2 - 5q - 24).$$
To factor $q^2 - 5q - 24$, we must find a pair of integers whose product is -24 and whose sum is -5. The pairs of integers whose product is -24 are

$$1, -24 \quad 2, -12 \quad 3, -8 \quad 4, -6$$
$$-1, 24 \quad -2, 12 \quad -3, 8 \quad -4, 6.$$

Of these, the only one whose sum is -5 is the pair $3, -8$. Hence,

$$3q^2 - 15q - 72 = 3(q^2 - 5q - 24) = 3(q + 3)(q - 8). \quad \square$$

EXAMPLE 11 Let us attempt to factor $x^2 + 2x + 5$. The only pairs of integers whose product is 5 are the pairs $1, 5$ and $-1, -5$, and neither of these has a sum of 2. Hence this polynomial does not factor. $\square$

Quadratic polynomials $ax^2 + bx + c$ for which $a \neq 1$ are often difficult to factor, because any factorization will be of the form

$$ax^2 + bx + c = (px + r)(qx + s) = pqx^2 + (ps + qr)x + rs,$$

for some numbers p, q, r, and s. It is usually best to look for integers p, q whose product is a and for integers r, s whose product is c, and then proceed by trial and error. We illustrate with an example.

EXAMPLE 12 Let us try to factor $2x^2 + 5x - 3$. Pairs of integers p and q whose product is 2 are $1, 2$ and $-1, -2$. Pairs of integers r and s whose product is -3 are $1, -3$ and $-1, 3$. We try these pairs in possible factorizations until we find a combination that works or until we find that no combination works:

$$(x + 1)(2x - 3) = 2x^2 - x - 3, \quad \text{which is not what we want;}$$
$$(x - 3)(2x + 1) = 2x^2 - 5x - 3, \quad \text{which is not what we want;}$$
$$(-x + 1)(-2x - 3) = 2x^2 + x - 3, \quad \text{which is not what we want;}$$
$$(-x - 3)(-2x + 1) = 2x^2 + 5x - 3, \quad \text{which is what we want.}$$

Hence,

$$2x^2 + 5x - 3 = (-x - 3)(-2x + 1)$$
$$= (-1)(x + 3)(-1)(2x - 1)$$
$$= (x + 3)(2x - 1). \quad \square$$

■ *Computer Manual: Program POLYRITH*

1.5 EXERCISES

Polynomial Arithmetic

In Exercises 1 through 22, perform the indicated arithmetic.

1. $(2x + 5) + (3x - 7)$
2. $(7 - 2x) - (3 - x)$
3. $(x^2 + 6x - 1) + (2x^2 + 3x + 5)$
4. $(5x^2 + 7x - 9) - (2x^2 + 4x - 12)$

5. $(7t^2 + 9t - 3) - (4t + 2)$
6. $(2x^3 + 6x^2 - 3x + 7) + (3x^2 + 8x - 5)$
7. $(5x^3 + 2x + 3) - (7x^3 + 8x^2 - 4)$
8. $\left(\dfrac{2}{3}y^3 - \dfrac{3}{4}y^2 + \dfrac{8}{5}y + \dfrac{2}{7}\right) + \left(\dfrac{1}{3}y^3 - \dfrac{5}{4}y^2 + \dfrac{2}{5}y - \dfrac{9}{7}\right)$
9. $x(4 - 3x)$
10. $(x + 2)(x + 1)$
11. $(t - 7)(t + 3)$
12. $(5y - 2)(y + 1)$
13. $(2x - 1)(3x + 7)$
14. $2x(3x^2 - 2)$
15. $(5p - 2)(p^2 + 1)$
16. $(2 - x^2)(6 + x)$
17. $(3y^2 + y)(4y - 1)$
18. $x(x + 9)(x - 7)$
19. $(2x - 1)(x + 1)(x - 1)$
20. $q(q + 1)(q - 2)(q + 3)$
21. $(x^2 - 2x + 1)(x^2 + 3x - 2)$
22. $(2t^2 + 3t + 5)(1 - 4t + t^2)$

Factoring

In Exercises 23 through 32, factor out all common factors.

23. $2x + 4$
24. $6p^2 - 42$
25. $18y + 54$
26. $35 - 98x$
27. $2x + 1$
28. $6x^2 + x$
29. $y^3 + y^2$
30. $2x^4 + 8x^3 - 12x^2$
31. $210x^5 + 15x^3$
32. $-42x^6 + 70x^5 + 28x^4 + 98x^3$

In Exercises 33 through 75, factor the given polynomial as completely as possible.

33. $x^2 - 5x$
34. $x^2 - 4x + 4$
35. $x^2 - 16$
36. $2x^2 - 6x$
37. $x^2 - 5x - 14$
38. $t^2 - 36$
39. $p^2 + 6p + 9$
40. $4x^2 - 4$
41. $x^2 - 6x - 27$
42. $-3x^2 + 6x + 30$
43. $-x^2 + 11x - 18$
44. $2x^2 + 28x + 98$
45. $x^2 + 1$
46. $50 - 2x^2$
47. $3x^2 + 3x - 60$
48. $x^2 + x + 1$
49. $2x^2 + 4x - 96$
50. $x^2 + 8x - 48$
51. $2x^2 - x - 1$
52. $2x^2 - 11x + 5$
53. $2x^2 - x - 6$
54. $3x^2 - 10x - 8$
55. $4x^2 - 4x - 3$
56. $6x^2 - 7x + 2$
57. $x^3 - 1$
58. $x^3 + 1$
59. $x^3 + 8$
60. $x^3 - 27$
61. $x^3 + 10x^2 + 25x$
62. $2x^3 - 128x$
63. $x^3 - 9x^2 - 22x$
64. $3x^3 + 3x^2 - 90x$
65. $500x - 5x^3$
66. $2x^4 - 24x^3 + 72x^2$
67. $5x^4 - 20x^3 - 60x^2$
68. $-7x^4 + 49x^3 - 84x^2$
69. $x^4 - 8x^3 + 16x^2$
70. $5x^4 + 5x^3 + 5x^2$
71. $10x^4 - 10x^2$
72. $6x^4 + 33x^3 - 18x^2$
73. $27x^4 + 125x$
*74. $-x^4 + 2x^3 - x^2 + 2x$
*75. $12x^5 + 9x^4 + 16x^3 + 12x^2$

POLYNOMIAL EQUATIONS

A **polynomial equation** is one of the form

$$a_n x^n + a_{n-1} x^{n-1} + \cdots + a_1 x + a_0 = 0,$$

where $a_n, a_{n-1}, \ldots, a_1, a_0$ are real numbers and $a_n \neq 0$. A real number is a **solution** of such an equation if it makes the equation true when substituted for x. For instance

$$x^3 - 2x^2 - x + 2 = 0$$

is a polynomial equation with solutions $x = -1$, $x = 1$, and $x = 2$, because

$$(-1)^3 - 2(-1)^2 - (-1) + 2 = 0,$$
$$1^3 - 2 \cdot 1^2 - 1 + 2 = 0,$$
$$2^3 - 2 \cdot 2^2 - 2 + 2 = 0.$$

In this section we consider methods for solving polynomial equations.

Solution by Factoring

Polynomial equations can sometimes be solved by factoring the polynomial and setting each of its factors equal to zero. This method depends on the following property of the real number system:

The Zero Product Law

If r, s are real numbers with $rs = 0$, then either $r = 0$ or $s = 0$.

EXAMPLE 1 Let us attempt to solve the polynomial equation

$$x^3 + 2x^2 - 3x = 0.$$

Since

$$x^3 + 2x^2 - 3x = x(x^2 + 2x - 3) = x(x + 3)(x - 1),$$

the equation becomes

$$x(x + 3)(x - 1) = 0.$$

Hence either $x = 0$, $x + 3 = 0$, or $x - 1 = 0$, and it follows that the solutions are $x = 0$, $x = -3$, and $x = 1$. (Check this by substituting 0, -3, and 1 for x in the original equation.) ☐

Not every polynomial equation has a solution.

EXAMPLE 2   The polynomial equation $x^2 + 1 = 0$ has no solution because it is equivalent to $x^2 = -1$, and there is no real number whose square is -1. ☐

Unfortunately, polynomial equations cannot always be solved by factoring because polynomials cannot always be factored; in fact, there is no method that will solve all polynomial equations. However, there is a method that will solve all equations of the form

$$ax^2 + bx + c = 0,$$

$a \neq 0$; such an equation is called a **quadratic equation.**

Quadratic Equations

A quadratic equation

$$ax^2 + bx + c = 0$$

can always be solved by means of the **quadratic formula.**

The Quadratic Formula

The solutions of the quadratic equation

$$ax^2 + bx + c = 0$$

are given by the **quadratic formula**

$$x = \frac{-b \pm \sqrt{b^2 - 4ac}}{2a}.$$

EXAMPLE 3   Let us use the quadratic formula to solve

$$x^2 - 5x + 6 = 0.$$

We compare $x^2 - 5x + 6 = 0$ with the standard form

$$ax^2 + bx + c = 0$$

of a quadratic equation and find that

$$a = 1, \quad b = -5, \quad \text{and} \quad c = 6.$$

Therefore

$$x = \frac{-b \pm \sqrt{b^2 - 4ac}}{2a} = \frac{-(-5) \pm \sqrt{(-5)^2 - 4(1)(6)}}{2(1)}$$

$$= \frac{5 \pm \sqrt{25 - 24}}{2} = \frac{5 \pm \sqrt{1}}{2} = \frac{5 \pm 1}{2}.$$

REVIEW: NUMBERS AND ALGEBRA   39

Thus the solutions are

$$x = \frac{5+1}{2} = 3 \quad \text{and} \quad x = \frac{5-1}{2} = 2. \quad \square$$

EXAMPLE 4 The quadratic equation

$$x^2 + 8x + 16 = 0$$

has $a = 1$, $b = 8$, and $c = 16$. Thus

$$x = \frac{-8 \pm \sqrt{8^2 - 4(1)(16)}}{2(1)} = \frac{-8 \pm \sqrt{0}}{2} = \frac{-8}{2} = -4.$$

Hence the equation has the single solution $x = -4$. $\square$

Of course, we could have solved the quadratic equations of Examples 3 and 4 by factoring. However, the quadratic formula will always yield the solution of a quadratic equation, even if the quadratic polynomial is not factorable; in addition, if the quantity $b^2 - 4ac$ is negative, the quadratic formula implies that the equation has no solution.

EXAMPLE 5 Consider the quadratic equation

$$2x^2 + 5x - 1 = 0$$

This polynomial is not factorable, but $a = 2$, $b = 5$, and $c = -1$, so

$$x = \frac{-5 \pm \sqrt{5^2 - 4(2)(-1)}}{2(2)} = \frac{-5 \pm \sqrt{33}}{4}.$$

If we wish, we can determine the decimal form of these roots by using a calculator with a square root key to find that

$$\sqrt{33} \approx 5.7445626.$$

Thus, to seven decimal places, the solutions are

$$x \approx \frac{-5 + 5.7445626}{4} = \frac{0.7445626}{4} = 0.1861407$$

and

$$x \approx \frac{-5 - 5.7445626}{4} = \frac{-10.7445626}{4} = -2.6861407. \quad \square$$

EXAMPLE 6 The quadratic equation

$$x^2 + 4x + 5 = 0$$

has $a = 1$, $b = 4$, and $c = 5$. Therefore

$$x = \frac{-4 \pm \sqrt{4^2 - 4(1)(5)}}{2(1)} = \frac{-4 \pm \sqrt{-4}}{2}.$$

Since $\sqrt{-4}$ does not exist, the equation has no solution. $\square$

We sometimes encounter quadratic equations when solving applied problems. We illustrate with examples.

EXAMPLE 7 A ball is thrown into the air, and t seconds later it is $-12t^2 + 60t$ feet above the ground. How long will it stay in the air?

When the ball lands, its altitude is 0 feet, so if we solve

$$-12t^2 + 60t = 0$$

for t, we will find the time t at which it lands. But

$$-12t^2 + 60t = -12t(t - 5) = 0$$

implies that $t = 0$ or $t = 5$. Since the ball is thrown into the air at time $t = 0$, it must land at time $t = 5$. Hence it remains in the air for 5 seconds. □

EXAMPLE 8 Suppose a rectangular field has an area of 15,000 square meters. If the width of the field is 50 meters less than its length, find its dimensions.

We must find the length and width of the field. Let us set

$$m = \text{Length of the field in meters.}$$

Then

$$m - 50 = \text{Width of the field}$$

FIGURE 1–12 (see Figure 1–12) and

$$\text{Area} = (\text{Length})(\text{Width}) = m(m - 50) = m^2 - 50m.$$

Hence, we must have

$$15{,}000 = m^2 - 50m,$$

or

$$m^2 - 50m - 15{,}000 = 0,$$

or

$$(m + 100)(m - 150) = 0.$$

Therefore $m = -100$ and $m = 150$. Since we cannot have a length of -100 meters, our answer is

$$\text{Length} = m = 150 \text{ meters,}$$
$$\text{Width} = m - 50 = 100 \text{ meters.} \quad \square$$

EXAMPLE 9 A firm can sell 700 units of its product at $30 per unit. It intends to raise the unit price of the product, and estimates that for each $1 it raises the price, sales will decrease by 10 units. What unit price must it charge in order to have revenue from sales amount to $24,000?

Suppose we let

$$p = \text{New price of the product}$$

in dollars. Since p will be greater than the present price of $30, we have

$$\text{Rise in price} = p - 30$$

dollars, and therefore

$$\text{Decrease in sales} = 10(p - 30)$$

units. Hence, at the new price,

$$\text{Total sales} = 700 - 10(p - 30) = 1000 - 10p$$

units. Since

$$\text{Revenue} = (\text{Unit price})(\text{total sales})$$
$$= p(1000 - 10p)$$
$$= 1000p - 10p^2$$

dollars, we must have

$$-10p^2 + 1000p = 24{,}000,$$

or

$$-10p^2 + 1000p - 24{,}000 = 0.$$

Solving this equation by factoring or by the quadratic formula, we obtain $p = 40$ and $p = 60$. (Check this.) Therefore the firm can raise the price to $40 or to $60; in either case, its revenue from sales will be $24,000. ∎

■ **Computer Manual: Program QUADFORM**

1.6 EXERCISES

Solution by Factoring

In Exercises 1 through 20, solve the given polynomial equation by factoring, if possible.

1. $x^2 - 4 = 0$
2. $2x^2 - 50 = 0$
3. $2x^2 + 18 = 0$
4. $x^2 - 2x + 1 = 0$
5. $3x^2 + 18x + 27 = 0$
6. $4x^2 + 12x = 0$
7. $x^2 + 3x - 10 = 0$
8. $x^2 + 9x + 14 = 0$
9. $p^2 + 50p = -15p$
10. $3q^2 + 18q = 48$
11. $2x^2 - 3x - 2 = 0$
12. $5x - 3 = 2x^2$
13. $x^3 - x^2 = 0$
14. $x^3 - x^2 - 12x = 0$
15. $x^3 - x^2 + x - 1 = 0$
16. $x^3 + x^2 + x + 1 = 0$
17. $x^3 - 2x^2 + x = 0$
18. $x^4 + 1 = 0$
19. $x^4 + 4x^2 = 0$
20. $x^4 - 2x^2 - 8 = 0$

Quadratic Equations

In Exercises 21 through 42, use the quadratic formula to solve the given quadratic equation.

21. $x^2 - 2x - 24 = 0$
22. $t^2 - 9 = 0$
23. $2x^2 - 32 = 0$
24. $x^2 - 9x + 14 = 0$
25. $2x^2 + 4x - 16 = 0$
26. $x^2 + 6x + 9 = 0$
27. $2q^2 + 13q + 15 = 0$
28. $5x + 1 = x^2$
29. $9x^2 + 3x = 2$
30. $-2x^2 - 34x - 120 = 0$
31. $4x^2 + 1 = 4x$
32. $15x - 10 = 3x^2$
33. $x^2 - 8x + 20 = 0$
34. $4x^2 + 20x + 25 = 0$
35. $5x - 12 = 2x^2$
36. $3x^2 + 7x + 5 = 0$
37. $\frac{1}{2}x^2 - 3x + \frac{1}{2} = 0$
38. $x^2 + \frac{2}{15}x - \frac{8}{15} = 0$
39. $x^3 - 1 = 0$
40. $x^3 + 1 = 0$
41. $x^3 + 4x^2 - 3x = 0$
42. $x^4 + 5x^3 - 2x^2 = 0$

43. If a stone is thrown straight up from ground level, its altitude t seconds later is $-16t^2 + 90t$ feet. Find the time it takes until the stone hits the ground.

44. Refer to Exercise 43: find the time at which the stone is at an altitude of 116 feet.

45. Refer to Exercise 43: find the time at which the stone is at an altitude of 126.5625 feet.

Management

46. If a firm makes and sells x units of its product, its profit will be $-x^2 + 170x - 6600$ dollars. How many units must it make and sell if its profit is to be $0?

47. How many units must the firm of Exercise 46 make and sell if its profit is to be $625?

48. Alef Company intends to depreciate its new headquarters building in the following way: t years after the construction of the building, its book value will be $-t^2 - 24t + 240$ million dollars. When will the book value of the building be zero?

Exercises 49 through 54 are applied problems. In each case, you are to set up and solve a quadratic equation.

49. A rectangular field is twice as long as it is wide, and its area is 3200 square meters. Find its dimensions.

50. A rectangular field has an area of 6500 square feet. If its length is 35 feet more than its width, find its dimensions.

51. An 8-inch by 12-inch block of text is to be printed on a rectangular poster board whose area is 396 square inches. If the top, bottom, and side margins around the block of text are to be of equal width, find this width.

52. A rectangular sheet of tin is twice as long as it is wide. A small square 10 cm by 10 cm is cut from each corner, and the edges are then folded up to make a box. If the volume of the box is 2400 cm³, find the dimensions of the tin sheet.

Management

53. If a firm charges $p for each unit of its product, it can sell $5000 - 25p$ units. What price must it charge in order to have sales revenue of $246,400?

54. Suppose a firm can sell 1000 units of its product at $50 each. If it raises the price, it will sell 5 units fewer for each $1 rise in price. What price must the firm charge in order to have its revenue from sales amount to $58,280?

1.7 RATIONAL EXPRESSIONS

A **rational expression** is a fraction whose numerator and denominator are polynomials, such as

$$\frac{2}{x}, \quad \frac{x+1}{x+3}, \quad \text{or} \quad \frac{p^2 - 5}{2p^3 + 3p - 2}.$$

In this section, we discuss the arithmetic of rational expressions and the solution of equations that involve rational expressions.

The Arithmetic of Rational Expressions

Rational expressions share the fundamental property of all fractions:

The Fundamental Property of Fractions

If P, Q, and S are polynomials, with $Q \neq 0$ and $S \neq 0$, then

$$\frac{PS}{QS} = \frac{P}{Q}.$$

The Fundamental Property says that we may "cancel" common factors from the numerator and denominator of a rational expression. When the numerator and denominator have no common factors remaining, the expression is said to be **reduced to lowest terms**.

EXAMPLE 1 Let us reduce

$$\frac{2x^2 + 4}{2x} \quad \text{and} \quad \frac{n^4 - 4n^2}{n^3 + 4n^2 + 4n}$$

to lowest terms. We have

$$\frac{2x^2 + 4}{2x} = \frac{2(x^2 + 2)}{2x} = \frac{x^2 + 2}{x}$$

and

$$\frac{n^4 - 4n^2}{n^3 + 4n^2 + 4n} = \frac{n^2(n^2 - 4)}{n(n^2 + 4n + 4)} = \frac{n \cdot n(n-2)(n+2)}{n(n+2)(n+2)} = \frac{n^2 - 2n}{n+2}. \quad \square$$

Rational expressions can be added, subtracted, multiplied, and divided. The following rules show how to carry out these arithmetic operations.

The Rules of Rational Arithmetic

Let P, Q, S, and T be polynomials, with $Q \neq 0$ and $T \neq 0$. Then

1. $\dfrac{P}{Q} \pm \dfrac{S}{Q} = \dfrac{P \pm S}{Q}.$

2. $\dfrac{P}{Q} \cdot \dfrac{S}{T} = \dfrac{PS}{QT}.$

3. If $S \neq 0$, $\quad \dfrac{P}{Q} \bigg/ \dfrac{S}{T} = \dfrac{P}{Q} \cdot \dfrac{T}{S}.$

EXAMPLE 2

$$\frac{5}{x} + \frac{3}{x} = \frac{5+3}{x} = \frac{8}{x},$$

$$\frac{x-1}{x+1} - \frac{x+2}{x+1} = \frac{(x-1)-(x+2)}{x+1} = -\frac{3}{x+1},$$

$$\frac{x}{x-1} \cdot \frac{2x+1}{x+1} = \frac{x(2x+1)}{(x-1)(x+1)} = \frac{2x^2 + x}{x^2 - 1},$$

$$\frac{2x}{x+4} \bigg/ \frac{2x-6}{x^2 - 16} = \frac{2x}{x+4} \cdot \frac{x^2 - 16}{2x - 6}$$

$$= \frac{2x(x-4)(x+4)}{2(x+4)(x-3)}$$

$$= \frac{x^2 - 4x}{x-3}. \quad \square$$

Rule 1 of the rules of rational arithmetic requires that the rational expressions have the same denominator. Thus, in order to add or subtract rational expressions that do not have the same denominator, we must put them over a common denominator. A useful common denominator for rational expressions is their **least common denominator** (LCD), which may be found as follows:

REVIEW: NUMBERS AND ALGEBRA

> **The Least Common Denominator**
>
> To find the least common denominator of a collection of rational expressions, factor each of their denominators completely. Their LCD is then the product of all factors that appear in any of the denominators, raised to the highest power to which they appear.

EXAMPLE 3 Consider $\dfrac{5x}{24}$ and $\dfrac{17}{30}$: $24 = 2^3 \cdot 3$ and $30 = 2 \cdot 3 \cdot 5$, so the LCD of these fractions is $2^3 \cdot 3 \cdot 5 = 120$. Now consider

$$\frac{x+1}{x^2-1} = \frac{x+1}{(x-1)(x+1)} \quad \text{and} \quad \frac{x}{4x^2+8x+4} = \frac{x}{4(x+1)^2};$$

The LCD of these rational expressions is $4(x-1)(x+1)^2$. □

Now we can describe how to place two rational expressions over their least common denominator: if L is the LCD of P/Q and S/T, there are expressions V and W such that $L = QV = TW$. Then by the Fundamental Property of Fractions,

$$\frac{P}{Q} = \frac{PV}{QV} = \frac{PV}{L} \quad \text{and} \quad \frac{S}{T} = \frac{SW}{TW} = \frac{SW}{L}.$$

EXAMPLE 4 Suppose we wish to find

$$\frac{5x}{24} + \frac{17}{30}.$$

From Example 3, the LCD of $\dfrac{5x}{24}$ and $\dfrac{17}{30}$ is 120. But $120 = 24 \cdot 5 = 30 \cdot 4$, so

$$\frac{5x}{24} + \frac{17}{30} = \frac{5x \cdot 5}{24 \cdot 5} + \frac{17 \cdot 4}{30 \cdot 4} = \frac{25x}{120} + \frac{68}{120} = \frac{25x + 68}{120}.$$

Similarly,

$$\frac{x+1}{x^2-1} - \frac{x}{4x^2+8x+4} = \frac{x+1}{(x-1)(x+1)} - \frac{x}{4(x+1)^2}$$

$$= \frac{(x+1)4(x+1)}{(x-1)(x+1)4(x+1)} - \frac{x(x-1)}{4(x+1)^2(x-1)}$$

$$= \frac{4(x+1)^2}{4(x-1)(x+1)^2} - \frac{x(x-1)}{4(x-1)(x+1)^2}$$

$$= \frac{3x^2+9x+4}{4(x-1)(x+1)^2}. \quad □$$

Rational Equations

An equation of the form

$$\frac{P}{Q} = 0,$$

where P/Q is a rational expression, is called a **rational equation.** The **solutions** of such an equation are the numbers that make the fraction zero when substituted for the variable. Since a fraction is zero if and only if its numerator is zero while its denominator is not zero, rational equations may be solved by reducing the rational expression to lowest terms and then setting the numerator equal to zero.

EXAMPLE 5 Let us solve the rational equation

$$\frac{x^2 + 8x + 15}{x^2 - 9} = 0$$

for x. Reducing to lowest terms, we have

$$\frac{x^2 + 8x + 15}{x^2 - 9} = \frac{(x + 3)(x + 5)}{(x + 3)(x - 3)} = \frac{x + 5}{x - 3}.$$

Now we set $x + 5 = 0$ to obtain the single solution $x = -5$. □

When solving a rational equation, we must always remember to reduce the rational expression to lowest terms before setting the numerator equal to zero; if this is not done, it is possible to obtain false "solutions." For instance, if in Example 5 we had not reduced the rational expression to lowest terms but had instead set its numerator $x^2 + 8x + 15$ equal to zero and solved, we would have obtained $x = -3$ as a "solution." But $x = -3$ is *not* a solution of the given rational equation, because the denominator $x^2 - 9$ is zero when $x = -3$.

We conclude this section with two examples of applied problems that require the solution of rational equations.

EXAMPLE 6 A firm estimates that t years from now it will be spending

$$\frac{12t + 3}{t + 4}$$

million dollars per year on health insurance for its employees. When will such spending reach $7 million per year?

We must solve

$$\frac{12t + 3}{t + 4} = 7,$$

or

$$\frac{12t + 3}{t + 4} - 7 = 0$$

for t. Performing the subtraction, we obtain

$$\frac{12t + 3}{t + 4} - \frac{7(t + 4)}{t + 4} = 0,$$

or

$$\frac{5t - 25}{t + 4} = 0.$$

Hence, $5t - 25 = 0$, so $t = 5$. Therefore, five years from now the company will be spending $7 million on health insurance.

EXAMPLE 7   It costs a firm $20 to make each unit of its product and $1000 even if it makes no units. How many units must it make in order for its average cost per unit to be $25?

Let

$$x = \text{Number of units the firm makes.}$$

Then

$$\text{Total cost of making } x \text{ units} = 20x + 1000$$

dollars, and

$$\text{Average cost per unit} = \frac{\text{Total cost}}{\text{Number of units made}}$$

$$= \frac{20x + 1000}{x}$$

dollars. Therefore we want

$$\frac{20x + 1000}{x} = 25.$$

We rewrite this equation as

$$\frac{20x + 1000}{x} - 25 = 0,$$

or

$$\frac{20x + 1000 - 25x}{x} = 0,$$

or

$$\frac{1000 - 5x}{x} = 0.$$

Thus $1000 - 5x = 0$, so $x = 200$ units.

■ *Computer Manual: Program RATLRITH*

1.7 EXERCISES

The Arithmetic of Rational Expressions

In Exercises 1 through 12, reduce the given rational expression to lowest terms.

1. $\dfrac{3x}{18}$
2. $\dfrac{15x}{5}$
3. $\dfrac{8x^2}{2x}$
4. $\dfrac{12y^3}{30y^2}$
5. $\dfrac{5x+10}{5x}$
6. $\dfrac{10x-5}{10x+30}$
7. $\dfrac{3q+3}{3q^2-6}$
8. $\dfrac{x^2+x}{x^2-x}$
9. $\dfrac{2x^3-2x}{4x^2}$
10. $\dfrac{x+3}{x^2-9}$
11. $\dfrac{x^2+4x+3}{x^2-1}$
12. $\dfrac{x^2+3x-4}{x^2+7x+12}$

In Exercises 13 through 38, perform the indicated arithmetic. Reduce your answers to lowest terms.

13. $\dfrac{1}{x}-\dfrac{2}{x}$
14. $\dfrac{3}{x^2}+\dfrac{8}{x^2}$
15. $\dfrac{3x}{x-1}+\dfrac{4}{x-1}$
16. $\dfrac{5}{x^2}-\dfrac{x+1}{x^2}$
17. $\dfrac{x}{5}\cdot\dfrac{x}{3}$
18. $\dfrac{x+1}{2}\cdot\dfrac{3}{x-1}$
19. $\dfrac{x-2}{x+3}\cdot\dfrac{1}{x}$
20. $\dfrac{x+2}{x+1}\cdot\dfrac{x+3}{x-2}$
21. $\dfrac{x^2+1}{x^2-4}\cdot\dfrac{x+2}{4}$
22. $\dfrac{2x-2}{x}\cdot\dfrac{x^3+4x^2}{x^2-1}$
23. $\dfrac{x}{3}\Big/\dfrac{x}{5}$
24. $\dfrac{3}{x}\Big/\dfrac{5}{x}$
25. $\dfrac{7x}{8}\Big/\dfrac{2x^2}{14}$
26. $\dfrac{x^2}{4}\Big/\dfrac{1}{x}$
27. $\dfrac{x+2}{x}\Big/\dfrac{x-3}{x}$
28. $\dfrac{x^2-9}{x}\Big/\dfrac{x+3}{x^2+x}$
29. $\dfrac{x}{2}+\dfrac{x}{3}$
30. $\dfrac{x^2}{5}-\dfrac{x^2}{6}$
31. $\dfrac{2}{x}-\dfrac{x}{6}$
32. $\dfrac{t^2}{7}-\dfrac{2}{t}$
33. $\dfrac{2}{x}+\dfrac{7}{x^2}$
34. $\dfrac{3}{x}-\dfrac{5}{x+1}$
35. $\dfrac{x}{x-1}+\dfrac{2x}{x+1}$
36. $\dfrac{x-2}{x^2-1}-\dfrac{2x}{x+1}$
37. $\dfrac{x^2+x}{x^2-x}+\dfrac{x+1}{x^2+2x}$
38. $\dfrac{x}{x^2+x-12}-\dfrac{2}{x^2-9}$

Rational Equations

In Exercises 39 through 50, solve the given rational equation.

39. $\dfrac{x}{2}=0$
40. $\dfrac{x+1}{x}=0$
41. $\dfrac{3-2x}{x^2+1}=0$
42. $\dfrac{2y-6}{3y+1}=10$
43. $\dfrac{x^2-1}{x^2+1}=0$
44. $\dfrac{x^2-4}{x^2-2x}=0$
45. $\dfrac{x^2-4x+3}{x+5}=0$
46. $\dfrac{x^2-6x+8}{x^2+x-6}=0$
47. $\dfrac{x^2+2x-35}{x^2-2x+1}=0$
48. $\dfrac{y^3-1}{y^2-y}=0$
49. $\dfrac{2}{x+1}+\dfrac{1}{x}=0$
50. $\dfrac{x+1}{2}=\dfrac{3}{x}$

Management

51. A firm's sales t years from now will be
$$\frac{20t + 300}{2t + 60}$$
million dollars per year. How long will it be until sales reach an annual level of $8 million?

52. A company's production t years from now will be
$$\frac{3t^2 - 12t + 24}{t^2 + 2}$$
thousand units per year. When will the company's production be 2000 units per year?

53. If it costs a firm $210,000 plus $6 per unit to make its product, how many units must it make in order for its average cost per unit to be $12? in order for it to be $6.01?

54. If it costs a firm $20,000 plus $0.50 per unit to make its product, how many units must it make in order for its average cost per unit to be $3? in order for it to be $0.55?

Life Science

55. The number of ducks using the Atlantic flyway in year t is estimated to be
$$\frac{2t + 20}{t + 4}$$
million. Here t is in years, with $t = 0$ representing 1989. When will the number of ducks using the flyway be 4 million? 3 million?

Social Science

56. A nation has just instituted a campaign to reduce illiteracy among its citizens. The plan is that t years from now the proportion of the citizenry that is literate will be
$$\frac{t + 3}{t + 10}.$$
How long will it be until one-half of the nation's citizens are literate? until nine-tenths are literate?

SUMMARY

This summary consists of terms and symbols whose meaning you should know, facts you should know, and techniques and methods you should be able to employ.

Section 1.1

Natural number, integer, rational number, irrational number, real number. Number systems **N**, **Z**, **Q**, **R**, and the relationships between them. The real line, plotting points. Inequality: $<$, $\leq$, $>$, $\geq$. Positive number, negative number. Double inequality. Interval, line diagram. Interval notation: (a, b), $[a, b]$, $(a, +\infty)$, $(-\infty, a]$, etc. Absolute value of a real number: $|a|$. Describing practical situations in terms of inequalities, intervals, and absolute values.

Section 1.2

Linear equation in one variable. Operations for solving equations. Solving linear equations in one variable. Setting up and solving applied problems that lead to linear equations in one variable.

Section 1.3

Linear inequality in one variable. Operations for solving inequalities. Solving linear inequalities in one variable. Setting up and solving applied problems that lead to linear inequalities in one variable.

Section 1.4

Base, exponent. Integral exponent: a^n, a^{-n}, a^0. The properties of exponents. nth root: $\sqrt[n]{a}$. Square root: $\sqrt{a}$. Rational exponent: $a^{1/n}$, $a^{m/n}$. Simplifying exponential expressions.

Section 1.5

Polynomial in one variable, degree, coefficient, term. Adding, subtracting, multiplying polynomials. Common factor. Special products: perfect square, difference of two squares, perfect cube, difference of two cubes. Factoring polynomials. Quadratic polynomials. Factoring quadratic polynomials.

Section 1.6

Polynomial equation. Solution by factoring. Quadratic equation. Quadratic formula. Setting up and solving applied problems that lead to quadratic equations.

Section 1.7

Rational expression. The Fundamental Property of Fractions. Reducing a rational expression to lowest terms. The rules of rational arithmetic. Adding, subtracting rational expressions having the same denominator. Multiplying, dividing rational expressions. Least common denominator. Finding the LCD. Using the LCD to add and subtract rational expressions having different denominators. Rational equation. Solving rational equations. Setting up and solving applied problems that lead to rational equations.

REVIEW EXERCISES

Answers to all review exercises are at the back of the book.

In Exercises 1 through 16, determine whether the given statement is true or false.

1. $3/5$ is a rational number.
2. -4.265 is an irrational number.
3. $-51/17$ is a negative integer.
4. 2.427 is an integer.
5. $2 < 5$
6. $|-5| = -5$
7. $-3 > -2$
8. $-1 \leq -1$
9. $|-2| = 2$
10. $4 \leq 6 \leq 5$
11. $-|-3| = 3$
12. $-4 < -2 < 1$
13. $\dfrac{13}{15} \leq \dfrac{12}{13} < \dfrac{11}{8}$
14. $|8 - 2| = 6$
15. $-4 \geq -5$
16. $3 < 6 \leq 8$

In Exercises 17 through 20, write the collection of real numbers x as an interval and draw its line diagram.

17. $-2 < x \leq 17$
18. $x \leq \dfrac{4}{9}$
19. $-2 < x < 8$
20. $x > \sqrt{2}$

In Exercises 21 through 24, draw the line diagram for the given interval and write the interval as a collection of real numbers satisfying an inequality.

21. $(5, 6]$
22. $(-2, +\infty)$
23. $(-\infty, 3]$
24. $(-4, 4)$

In Exercises 25 through 28, write the given statement using an inequality.

25. The pressure in a boiler is in the safe range if it is no more than 25 pounds per square inch.
26. A chemical reaction will take place if a solution is at a temperature of at least 20°C.
27. Profits are forecast to be less than $2 per share during the next quarter.
28. Profits are forecast to be at least $2 per share, but not as much as $2.25 per share, during the next quarter.

In Exercises 29 through 34, solve for x.

29. $3x + 5 = 20$
30. $8 - 2x = 9$
31. $3 + 5(2x - 1) = 6$
32. $2x + 6 = 8x - 12$
33. $7x + 3 = 6 + 2(x - 9)$
34. $8 + 5(2x + 4) = 3 - 2(9 - 2x)$

35. If the cost of making x units of a product is $16.25x + 183{,}440$ dollars, find the number of units that must be made if the cost is to be $190,915.

36. Two credit cards charge interest of 1.3% and 1.4% per month, respectively. If you charged a total of $2000 using the two cards and at the end of the month paid $26.75 in interest charges, how much did you charge on each card?

37. A tank contains 200 gallons of a solution that is 95% kerosene and 5% naphtha. How much of a solution that is 90% kerosene and 10% naphtha must be added to the tank to have a solution that is 92% kerosene and 8% naphtha?

38. A drag racer crosses the starting line 2 seconds after his opponent. If the racer's speed is 185 mph and the opponent's speed is 180 mph, how long will it take until

the racer catches his opponent, and how far beyond the starting line will they have gone when he does?

In Exercises 39 through 44, solve for x. Write your answer in interval form.

39. $8x - 3 < 5$
40. $9 - 2x \geq 13$
41. $7x + 3 \leq 12x - 7$
42. $3 - 4x > -2x + 11$
43. $5(6 - 8x) < 9 + 3x$
44. $4x - 3(2x + 1) \geq 0.5(8 - 3x)$

45. If the average annual salary of new Ph.D.'s in mathematics is s dollars per year, graduate schools will produce $0.01s + 20$ of them per year and colleges will hire $800 - 0.02s$ of them per year. Over what range of salaries will there be a shortage of new Ph.D.'s in mathematics in the colleges?

46. If Alpha Company makes and sells x units of its product, its profit will be $12.42x - 225,328$ dollars. How many units must Alpha make and sell in order to have a profit of at least $500,000?

47. You can rent a TV set from one store for $4.24 per day plus a nonrefundable deposit of $55. You can rent the same TV set from a second store for $3.85 per day plus a nonrefundable deposit of $75. If the cost of renting from the second store is to be less than that of renting from the first store, for how many days must you rent the TV set?

48. One runner can run at a speed of 10 meters per second, another at a speed of 9.8 meters per second. If the second runner receives a 2.5-meter head start over the first one, for how long will she lead the race?

In Exercises 49 through 56, evaluate the given expression.

49. 5^{-2}
50. $81^{-1/2}$
51. $9^4 9^{-3}$
52. $\dfrac{4^2 4^3 4^{-1} 3^3}{3^2 3^{-4} 4^5}$
53. $3(2^{-3})^2$
54. $2(5^4)^{-3/4}$
55. $121^{3/2}$
56. $(216 \cdot 1000)^{-2/3}$

In Exercises 57 through 64, simplify the given expression.

57. $(7x^3)(3x^{-5})$
58. $(6x^2 y^{-3} z^4 w)(8x^{-3} y^{-3} z^5 w^{-1})$
59. $\dfrac{4x^3 y^{-2}}{3x^2 y^4}$
60. $\dfrac{6x^2 y^{12} z^{-4}}{5xy^9 z^{-2}}$
61. $(3x^2 y^{-3} z)^2$
62. $(2x^{-2} y^4 zw^{-1})^{-3}$
63. $\left(\dfrac{x^2 y^3 z^2}{x^{-2} y^4 z^4}\right)^{1/2}$
64. $\dfrac{3x^{1/3} y^{-2/3} z^{4/5}}{2x^{-2/3} yz^{1/5}}$

In Exercises 65 through 68, perform the indicated arithmetic.

65. $(5x^2 + 6x + 12) + (8x^2 - 7x - 2)$
66. $(9x^3 + x^2 - 3x + 1) - (6x^2 + 8x + 2)$
67. $(4x - 1)(3x + 9)$
68. $(5x^2 - x + 1)(2x - 3)$

In Exercises 69 through 78, factor the given polynomial.

69. $-3x^2 + 6x$
70. $12x^4 + 2x^2$
71. $x^3 + 27$
72. $27x^3 - 125$
73. $x^2 - 14x + 49$
74. $9x^2 - 25$
75. $x^2 + 2x - 24$
76. $x^2 + 7x - 18$
77. $2x^2 - x - 15$
78. $6x^3 + 11x^2 - 10x$

REVIEW: NUMBERS AND ALGEBRA 53

In Exercises 79 through 88, solve the given quadratic equation.

79. $3x^2 + 27 = 0$ **80.** $2x^2 - 128 = 0$ **81.** $x^2 - 8x = 0$

82. $x^2 - 2x = 99$ **83.** $x^2 + 17x + 52 = 0$ **84.** $4x^2 - 4x - 15 = 0$

85. $x^2 + 18x + 81 = 0$ **86.** $6x + 2 = x^2$ **87.** $2x^2 + 3x + 5 = 0$

88. $-\dfrac{1}{3}x^2 + \dfrac{2}{3}x + \dfrac{5}{3} = 0$

89. A rectangular field is 20 meters longer than it is wide, and its area is 10,000 square meters. Find its dimensions.

90. A square piece of cardboard 10 inches on a side had a small square of side x inches cut out of each of its corners. The area of the piece of cardboard that was left was 64 square inches. What is x?

91. A retailer can sell 400 sweaters at $30 each. For every $1 that the price is lowered, sales will increase by 40 sweaters. How much must the retailer charge in order to sell $15,000 worth of sweaters?

In Exercises 92 through 97, reduce the given expression to its lowest terms.

92. $\dfrac{6x + 2}{4}$ **93.** $\dfrac{3x^2 - 9x}{6x}$ **94.** $\dfrac{7x + 2}{3x - 1}$ **95.** $\dfrac{x^2 + 14x + 24}{x^3 + 12x^2}$

96. $\dfrac{2x^2 + 5x - 3}{2x^2 + 7x - 4}$ **97.** $\dfrac{8x^3 - 27}{4x^2 + 6x + 9}$

In Exercises 98 through 111, perform the indicated arithmetic. Reduce your answers to lowest terms.

98. $\dfrac{x}{5} - \dfrac{2x}{3}$ **99.** $\dfrac{2}{x} + \dfrac{5}{x}$ **100.** $\dfrac{x^2}{3} - \dfrac{2}{x}$

101. $5x^2 + \dfrac{3}{x^2}$ **102.** $\dfrac{2x}{x - 1} + \dfrac{1}{x}$ **103.** $\dfrac{x^2 - 3x}{x^2 + 2} + \dfrac{3x - 5}{x^2}$

104. $\dfrac{x^2}{3} \cdot \dfrac{x^5}{5}$ **105.** $\dfrac{3}{x} \cdot \dfrac{2}{x^2}$ **106.** $\dfrac{x + 3}{x - 1} \cdot \dfrac{2}{x - 1}$

107. $\dfrac{x^2 + 4x}{2x + 1} \cdot \dfrac{3x + 1}{x^2 - x}$ **108.** $\dfrac{x}{2} \bigg/ \dfrac{x}{7}$ **109.** $\dfrac{9x^2}{3} \bigg/ \dfrac{4}{3x}$

110. $\dfrac{x^2 + x}{2} \bigg/ \dfrac{x}{4 + x}$ **111.** $\dfrac{x^2 - 1}{x^2 + 3x} \bigg/ \dfrac{x^2 + 3x - 4}{x^3 + 4x^2 + 3x}$

In Exercises 112 through 117, solve the given rational equation.

112. $\dfrac{3x + 21}{2x} = 0$ **113.** $\dfrac{x^2 - 3x}{x^2 + 4x} = 0$ **114.** $\dfrac{x^2 + 11x + 18}{2x + 4} = 0$

115. $\dfrac{x^2 + 11x + 18}{2x - 4} = 0$ **116.** $\dfrac{x^2 - 2x - 15}{x^2 - x - 20} = 0$ **117.** $\dfrac{x^3 + 5x^2 + 8x}{x^3 + x} = 0$

118. It is estimated that t years from now the cost of reducing the emissions that cause acid rain will be

$$\frac{6t + 20}{0.3t + 4}$$

billion dollars per year. When will the cost be $10 billion per year?

119. If it costs a firm $150,000 plus $40 per unit to produce its product, how many units must it make if its average cost per unit is to be $50?

ADDITIONAL TOPICS

Here are some suggestions for topics not covered in the text that you might want to investigate on your own.

1. A **terminating decimal** is one that stops, such as 2.675. Every terminating decimal is the result of dividing out some rational number p/q. (For instance, $2.675 = 107/40$.) Explain why this is so. For any given terminating decimal, show how to find the rational number that it is equal to.

2. A **repeating decimal** is one in which some block of digits repeats forever, such as 1.2754545454. . . . Every repeating decimal is the result of dividing out some rational number p/q. (For instance, $1.27545454\ldots = 12{,}627/9900$.) Explain why this is so. For any given repeating decimal, show how to find the rational number that it is equal to.

3. Every rational number, when divided out, becomes either a terminating or a repeating decimal. (For instance, $9/16 = 0.5625$ and $13/6 = 2.166666\ldots$) Explain why this is so.

4. In Section 1.5 we discussed addition, subtraction, and multiplication of polynomials, but we did not consider the long division of one polynomial by another. When one polynomial is divided by another, the result is a polynomial quotient plus a remainder that is a rational expression. (The rational expression may be zero.) For instance,

$$(x^3 + 2x + 3) \div (x - 1) = x^2 + x + 3 + \frac{6}{x - 1}.$$

Here $x^2 + x + 3$ is the quotient and $\frac{6}{x - 1}$ is the remainder. Find out how long division of one polynomial by another is carried out.

5. Complex numbers are numbers of the form $a + bi$, where a and b are real numbers and $i^2 = -1$. Find out how complex numbers are added, subtracted, multiplied, and divided. Show that the real number system is a subsystem of the complex number system.

6. Show that if we allow equations to have complex numbers as solutions, then every quadratic equation has as its solutions either two distinct real numbers, one real number, or a pair of complex numbers of the form $a + bi$ and $a - bi$.

7. The quadratic formula gives us the solutions of any quadratic equation $ax^2 + bx + c = 0$. There is an analogous **cubic formula** that gives us the solutions of any cubic equation $ax^3 + bx^2 + cx + d = 0$. The cubic formula was discovered ca. 1535 by Nicolo Fontana and stolen from him and published in 1545 by Girolamo Cardano. Find out about this cubic formula and its colorful history. (A good place

to start would be with the book *Great Moments in Mathematics before 1650* by Howard Eves, published by The Mathematical Association of America in 1980.) You might also find out about the **quartic formula,** which gives us the solutions of any fourth-degree equation $ax^4 + bx^3 + cx^2 + dx + e = 0$. It has been proved that for $n > 4$ there is no formula for finding the solutions of the nth-degree equation $a_n x^n + a_{n-1} x^{n-1} + \cdots + a_1 x + a_0 = 0$.

2

Sets and Functions

A **set** is a collection of objects; a **function** is a rule that associates objects in one set with those in another. Sets and functions are fundamental to mathematics and are extremely important in the creation of mathematical models. For instance, suppose that a businessperson must decide whether or not to market a newly developed product. If the product is not marketed, the business will have to write off as a loss its developmental costs, which amount to $5 million. However, if the product is marketed and is successful, it will cover its developmental costs and generate additional profits of $20 million, whereas if it is marketed and is unsuccessful it will cover its developmental costs and generate additional profits of $1 million. The collection of possibilities faced by the businessperson, namely

- Do not market the product.
- Market the product and it is successful.
- Market the product and it is not successful.

is a set, as is the collection of possible profits

$$-\$5 \text{ million}, \$20 \text{ million}, \$1 \text{ million}.$$

The correspondence between these sets indicated by

$$\text{Do not market the product} \longrightarrow -\$5 \text{ million},$$
$$\text{Market the product and it is successful} \longrightarrow \$20 \text{ million},$$
$$\text{Market the product and it is not successful} \longrightarrow \$1 \text{ million}$$

is an example of a function.

In this chapter we discuss sets and functions. The first section introduces sets, set notation, set equality, and set inclusion; the second section shows how sets may be combined to form new sets by using the operations of complementation, union, and intersection. The last two sections introduce functions and graphing.

2.1 SETS

In this section we define the concept of a set, show how to write sets, demonstrate how to determine when two sets are the same and when one is contained in the other, and illustrate the use of Venn diagrams to depict the relationships among sets.

Sets and Elements

A **set** is a collection of objects. The individual objects that make up the collection are called the **elements** of the set. For example, consider the set S consisting of the objects 1, 2, 3, and 4. The numbers 1, 2, 3, and 4 are the elements of the set S. We write sets by enclosing their elements between braces { }. Thus, for the set S, we write

$$S = \{1, 2, 3, 4\}.$$

This is read "S is the set consisting of the elements 1, 2, 3, and 4."

It is useful to have some notation by which we can indicate whether an object is an element of a given set. For this purpose, we use the symbol $\in$, which is read "is an element of" or "belongs to." For example, if x is an element of the set S, we write

$$x \in S$$

and say "x is an element of S" or "x belongs to S." To indicate that an object is not an element of a set, we use the same symbol with a slash through it. Thus, if x is not an element of the set S, we write

$$x \notin S$$

and say "x is not an element of S" or "x does not belong to S."

EXAMPLE 1   Let $S = \{1, 2, 3, 4\}$. Then $1 \in S$, $2 \in S$, $3 \in S$, and $4 \in S$, but $5 \notin S$, $6 \notin S$, $\pi \notin S$, $\frac{1}{2} \notin S$, $a \notin S$, and so on. ☐

EXAMPLE 2   Let $T = \{$months of the year$\}$. Then January $\in T$, February $\in T$, ..., December $\in T$, but $4 \notin T$, Sunday $\notin T$, and so on. In general, $x \in T$ if and only if x is a month of the year. ☐

The number systems defined in Section 1.1 may be thought of as sets of numbers. Thus, we have

The set **N** of natural numbers.

The set **Z** of integers.

The set **Q** of rational numbers.

The set **R** of real numbers.

Set Notation

Three methods are commonly used to specify sets: the **roster** method, the **descriptive** method, and the **set-builder** method.

> ### The Roster Method
> The **roster method** specifies a set S by listing the elements of S between braces.

EXAMPLE 3 Here are some sets specified by the roster method:

$$S = \{2, 3, 4, 5\},$$
$$T = \{a, b, c, \ldots, z\},$$
$$\mathbf{N} = \{1, 2, 3, 4, 5, \ldots\},$$
$$\mathbf{Z} = \{\ldots, -3, -2, -1, 0, 1, 2, 3, \ldots\}$$

> ### The Descriptive Method
> The **descriptive method** specifies a set by writing a description of its elements between braces.

EXAMPLE 4 We write the sets of Example 3 using the descriptive method:

$$S = \{\text{natural numbers greater than 1 and less than 6}\},$$
$$T = \{\text{letters of the English alphabet}\},$$
$$\mathbf{N} = \{\text{positive integers}\},$$
$$\mathbf{Z} = \{\text{integers}\}.$$

> ### The Set-Builder Method
> The **set-builder method** specifies a set S by writing
> $$S = \{x \mid x \text{ has property P}\}.$$
> This is read "S is the set of all x such that x has property P." The statement of the property P is chosen so that S consists of those objects, and only those objects, that have the stated property.

EXAMPLE 5   We write the sets of Examples 3 and 4 using the set-builder method:

$$S = \{x \mid x \in \mathbf{N} \text{ and } 1 < x < 6\},$$
$$T = \{x \mid x \text{ is a letter of the English alphabet}\},$$
$$\mathbf{N} = \{x \mid x \text{ is a natural number}\},$$
$$\mathbf{Z} = \{x \mid x \text{ is an integer}\}. \quad \square$$

Note that a set can usually be written in many different ways. For instance, the set

$$V = \{1, 2, 3, 4, 5, 6\}$$

can also be written as

$$V = \{x \mid x \in \mathbf{N}, x < 7\},$$
$$V = \{x \mid x \in \mathbf{N}, x \leq 6\},$$
$$V = \{x \mid x \in \mathbf{Z}, 0 < x < 7\},$$
$$V = \{x \mid x \in \mathbf{Z}, x > 0, x^2 < 40\},$$

and so on.

Set Equality and Inclusion

Two sets are considered to be the same if they consist of exactly the same elements, without regard to the order in which the elements occur or possible repetitions of elements. Another way to say this is as follows:

> **Set Equality**
>
> Let S and T be sets. Then $S = T$ if every element of S is an element of T and every element of T is an element of S.

EXAMPLE 6   Let $S = \{1, 2, 3, 4\}$. If $T = \{4, 2, 1, 3\}$, then $S = T$ because every element of S is an element of T and every element of T is an element of S. Similarly, if $V = \{1, 2, 2, 3, 4, 4\}$, then $S = V$ because every element of S is an element of V and every element of V is an element of S. However, if $W = \{1, 3, 4, 6\}$, then $S \neq W$. $\quad \square$

Consider the sets $S = \{a, b, c\}$ and $T = \{a, b, c, d\}$. Then $S \neq T$, but every element of S is an element of T, and thus S is **contained in** T.

SETS AND FUNCTIONS

$S \subset T$

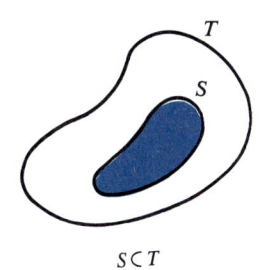

FIGURE 2–1

> ### Set Containment
> Let S and T be sets. If every element of S is an element of T, we say that S is **contained in T,** call S a **subset** of T, and write $S \subset T$.

See Figure 2–1. If S is not contained in T, we write $S \not\subset T$.

EXAMPLE 7 Let $S = \{1, 2, 3, 4\}$, $T = \{1, 2, 3, 4, 5, 6\}$, and $V = \{2, 3, 6\}$. Then $S \subset T$ and $V \subset T$, but $T \not\subset S$, $T \not\subset V$, $S \not\subset V$, and $V \not\subset S$. See Figure 2–2.

EXAMPLE 8 $\mathbf{N} \subset \mathbf{Z} \subset \mathbf{Q} \subset \mathbf{R}.$

EXAMPLE 9 Let S be any set. Then $S \subset S$, since every element of S is an element of S. Thus, every set is a subset of itself.

There is a special set that is a subset of every set. This is the **empty** (or **null**) **set ∅**. The empty set ∅ is the set that has no elements. Using the roster method, we may specify the empty set by writing

$$\emptyset = \{\ \}$$

Using the descriptive or set-builder methods, we may specify ∅ by writing a description that no object satisfies. For example,

$$\emptyset = \{\text{U.S. citizens who are 100 feet tall}\},$$

since there are no such people, or

$$\emptyset = \{x \mid x \in \mathbf{N} \text{ and } x^2 < 0\},$$

since there is no natural number whose square is negative.

The empty set ∅ should not be confused with sets such as $\{0\}$ or $\{\emptyset\}$. Each of these sets has one element and cannot be the empty set because the empty set has no elements.

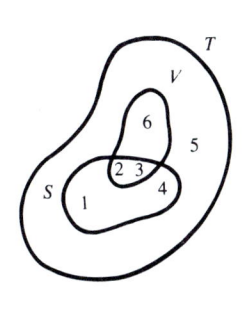

FIGURE 2–2

EXAMPLE 10 The empty set ∅ is a subset of every set. To see this, recall that S is a subset of T if every element of S is also an element of T. Since there is no element of ∅, it must be true that every element of ∅ is also an element of T. Hence, ∅ ⊂ T no matter what set T is. ☐

It is often convenient to have a large set that contains all the elements of the sets under consideration. Such a set is called a **universal set.** Thus, if S, T, and V are sets, a universal set for S, T, and V is any set that contains S, T, and V.

EXAMPLE 11 Let $S = \{1, 2, 3, 4\}$ and $T = \{2, 4, 5, 6\}$. A universal set for S and T is any set that contains both S and T. Thus,

$$\mathcal{U}_1 = \{1, 2, 3, 4, 5, 6\}$$

is a universal set for S and T, as is

$$\mathcal{U}_2 = \{1, 2, 3, 4, 5, 6, 7, 8\}.$$

Other universal sets for S and T are the sets **N, Z, Q,** and **R.** ☐

EXAMPLE 12 Let

$$S = \{x \mid x \text{ is a first-year college student in the United States}\},$$
$$T = \{x \mid x \text{ is a female college student in the United States}\},$$
$$V = \{x \mid x \text{ is a student at a college in Massachusetts}\}.$$

A universal set for S, T, and V is

$$\mathcal{U}_1 = \{x \mid x \text{ is a college student in the United States}\}.$$

Another is

$$\mathcal{U}_2 = \{x \mid x \text{ is a college student anywhere in the world}\}. \quad ☐$$

Venn Diagrams

A **Venn diagram** is a pictorial representation of sets. In a Venn diagram, the universal set is represented by a rectangle, and its subsets under consideration are represented by disks placed inside the rectangle. For example, suppose we have the sets

$$S = \{1, 2, 3\} \quad \text{and} \quad T = \{1, 2, 3, 4, 5\}.$$

Let

$$\mathcal{U} = \{1, 2, 3, 4, 5, 6, 7\}$$

be the universal set for S and T. The Venn diagram representation is shown in Figure 2–3. Note that the disk representing S is contained in the disk representing T because $S \subset T$. Note also the placement of the elements in the diagram. Elements 1, 2, and 3 are within the S disk because $1 \in S$, $2 \in S$, and $3 \in S$. Elements 4 and 5 are inside the T disk and outside the S disk because $4 \in T$ and $5 \in T$ but $4 \notin S$ and

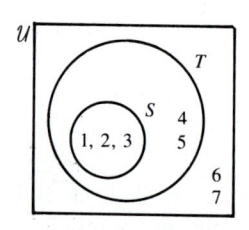

FIGURE 2–3

SETS AND FUNCTIONS 63

$5 \notin S$. Similarly, 6 and 7 are inside $\mathcal{U}$ and outside T because $6 \in \mathcal{U}$ and $7 \in \mathcal{U}$ but $6 \notin T$ and $7 \notin T$.

EXAMPLE 13 The Venn diagram for the sets

$$S = \{1, 2, 3\} \quad \text{and} \quad T = \{3, 4, 5\}$$

with universal set

$$\mathcal{U} = \{1, 2, 3, 4, 5, 6\}$$

is shown in Figure 2–4. ☐

EXAMPLE 14 The Venn diagram for the sets

$$S = \{a, b, c, d\}, \quad T = \{a, c\}, \quad V = \{c, d, e\}, \quad \mathcal{U} = \{a, b, c, d, e, f, g\}$$

is shown in Figure 2–5. ☐

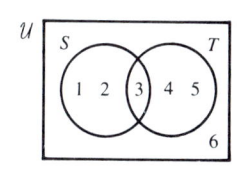

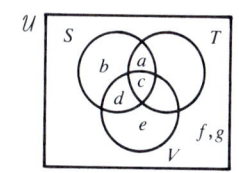

FIGURE 2–4   **FIGURE 2–5**

Venn diagrams are usually impossible to draw when the number of nonuniversal sets involved is greater than 3.

2.1 EXERCISES

Sets and Elements

In Exercises 1 through 8, let $S = \{a, b, c\}$ and $T = \{a, c, e\}$. Indicate whether the given statement is true or false.

1. $a \in S$
2. $a \in T$
3. $b \in S$
4. $b \in T$
5. $c \notin S$
6. $c \notin T$
7. $e \notin S$
8. $e \notin T$

9. Consider the following real numbers:
   $2, \quad \tfrac{2}{3}, \quad \sqrt{2}, \quad -1, \quad 0, \quad \tfrac{3}{2}, \quad 106, \quad \pi, \quad -15\tfrac{7}{8}.$
   Which of these are elements of
   (a) the set **N**?   (b) the set **Z**?

10. Consider the real numbers listed in Exercise 9. Which of them are elements of
    (a) the set **Q**?   (b) the set of irrational numbers?

Set Notation

In Exercises 11 through 16, specify the set using the roster method.

11. The set B consisting of the odd natural numbers
12. The set C consisting of the integers less than 5
13. The set $D = \{$first three months of the year$\}$
14. The set $E = \{$natural numbers greater than 10$\}$
15. $F = \{x \mid x \in \mathbf{N}, x \leq 7\}$
16. $G = \{x \mid x \in \mathbf{Z}, -5 < x < 5\}$

In Exercises 17 through 22, specify the set using the descriptive method.

17. The set A consisting of the first six months of the year
18. The set B consisting of the natural numbers less than 6
19. $C = \{1, 2, 3, 4\}$
20. $D = \{1, 2, 3, \ldots, 99, 100\}$
21. $E = \{x \mid x \in \mathbf{N}, x \leq 8\}$
22. $F = \{x \mid x \in \mathbf{Z}, x^2 < 25\}$

In Exercises 23 through 28, specify the set using the set-builder method.

23. The set A consisting of all female college students in the United States
24. The set B consisting of all natural numbers less than 9
25. $D = \{-2, -1, 0, 1, 2, 3\}$
26. $E = (100, 101, 102, \ldots)$
27. $F = \{$real numbers whose squares are less than 1$\}$
28. $G = \{$natural number multiples of 3$\}$

The sets in Exercises 29 through 33 were mentioned in the October 12, 1988, issue of *The Wall Street Journal*. Write each of them using the set-builder method.

29. The set of Washington lobbyists
30. The set of discount securities brokers
31. The set of colleges that experienced a decline in fund raising in 1986–1987
32. The set of hospitals accredited by the Joint Commission on Accreditation of Healthcare Organizations
33. The set of insolvent insurers

Set Equality and Inclusion

In Exercises 34 through 49, indicate whether the given statement is true or false.

34. $\{1, 2, 3, 4\} = \{1, 3, 4, 2\}$
35. $\{a, b, c\} = \{c, e, b, a\}$
36. $\{2, 3, 5, 0\} = \{2, 3, 5\}$
37. $\{2, 2, 4, 3\} = \{2, 3, 4\}$
38. $\{x \mid x \in \mathbf{N}, x \geq 7\} = \{8, 9, 10, \ldots\}$
39. $\{x \mid x \in \mathbf{N}, x < 5\} = \{x \mid x \in \mathbf{N}, x^2 < 25\}$
40. $\{x \mid x \in \mathbf{N}, x \leq 10\} = \{x \mid x \in \mathbf{Z}, x^2 \leq 100\}$
41. $\{x \mid x \in \mathbf{Z}, |x| > 10\} = \{11, 12, 13, \ldots\}$

SETS AND FUNCTIONS   65

42. $\{x \mid x \in \mathbf{Z}, |x| \leq 2\} = \{-2, -1, 0, 1, 2\}$
43. $\{x \mid x \in \mathbf{R}, -1 < x < 1\} = \{0\}$
44. $\{a, b, c, d\} \subset \{a, b, d\}$
45. $\{a, c, d\} \subset \{a, b, c, d\}$
46. $\{1, 2, 5, 7\} \subset \{x \mid x \in \mathbf{N}, x < 10\}$
47. $\{1, 2, 3, 4\} \subset \{x \mid x \in \mathbf{Z}, |x| < 10\}$
48. $\{1, 2, 3, 0\} \not\subset \{1, 2, 3\}$
49. $\{1, 2, 3\} \not\subset \{1, 2, 3, 0\}$

Let
$$A = \{x \mid x \in \mathbf{N}, x \text{ is an odd integer}\},$$
$$B = \{x \mid x \in \mathbf{N}, x = 2k - 1 \text{ for some } k \in \mathbf{N}\},$$
$$C = \{x \mid x \in \mathbf{Z}, |x| > 10\},$$
$$D = \{x \mid x \in \mathbf{R}, x^2 > 100\},$$
$$E = \{x \mid x \in \mathbf{R}, |x| > 9\}.$$

In Exercises 50 through 62, indicate whether the given statement is true or false.

50. $B \subset A$ 51. $A = B$ 52. $C \subset D$ 53. $D \subset C$ 54. $C = D$ 55. $D \subset E$
56. $E \subset D$ 57. $D = E$ 58. $C \subset E$ 59. $E \subset C$ 60. $C = E$
61. $\mathbf{Z}$ is a universal set for A, B, C, and D.
62. $\mathbf{R}$ is a universal set for A, B, C, D, and E.

In Exercises 63 through 68, indicate whether the given statement is true or false.

63. $\emptyset = \{x \mid x \in \mathbf{N}, x + 2 = 0\}$
64. $\emptyset = \{x \mid x \in \mathbf{N}, x/2 = 1\}$
65. $\emptyset = \{x \mid x \in \mathbf{R}, x^2 + 1 = 0\}$
66. $\emptyset = \{x \mid x \in \mathbf{Z}, 2x = 1\}$
67. $\emptyset = \{x \mid x \in \mathbf{Q}, 2x = 1\}$
68. $\emptyset = \{x \mid x \in \mathbf{N}, x < 0\}$

In Exercises 69 through 73, find all subsets of the given set.

69. $\emptyset$ 70. $\{a\}$ 71. $\{a, b\}$ 72. $\{a, b, c\}$ 73. $\{a, b, c, d\}$

*74. Show that if a set has n distinct elements then it has 2^n distinct subsets. (*Hint:* Let $S = \{x_1, x_2, \ldots, x_n\}$, and consider the following diagram:

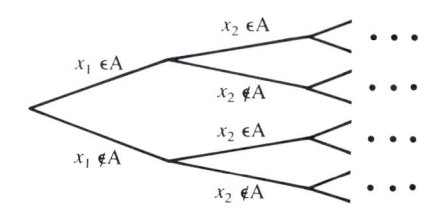

Show that different paths through the diagram yield distinct subsets of S and that distinct subsets of S correspond to different paths through the diagram. Then show that there are 2^n different paths through the diagram.)

Life Science

*75. According to the *Physicians' Desk Reference*, an overdose of a certain drug may cause one or more of the following symptoms:

- Respiratory symptoms: irregular breathing, respiratory depression
- Cardiovascular symptoms: tachycardia, shock, fibrillation
- Gastrointestinal symptoms: nausea, vomiting

(a) How many possible sets of symptoms are there?

(b) Suppose that a victim of an overdose will always exhibit both gastrointestinal symptoms, at least one cardiovascular symptom, and exactly one respiratory symptom. Find the possible sets of symptoms in this case.

Venn Diagrams

In Exercises 76 through 83, draw and label the Venn diagram for the given sets.

76. $S = \{a, b, c\}$, $T = \{a\}$, $\mathcal{U} = \{a, b, c, d\}$
77. $S = \{a, b, c\}$, $T = \{a, c, d\}$, $\mathcal{U} = \{a, b, c, d\}$
78. $S = \{a, b, c\}$, $T = \{a, c, d\}$, $V = \{b, c, e\}$, $\mathcal{U} = \{a, b, c, d, e, f, g\}$
79. $A = \{1, 2, 3, \ldots, 10\}$, $B = \{2, 4, 6, \ldots, 12\}$, $C = \{2, 4, 6, 8\}$, $\mathcal{U} = \{1, 2, 3, \ldots, 15\}$
80. $A = \{1, 2, 3, \ldots, 10\}$, $B = \{3, 5, 7, 12, 15\}$, $C = \{1, 3, 6, 8, 12\}$, $\mathcal{U} = \{1, 2, 3, \ldots, 16\}$
81. $A = \{1, 2, 3, \ldots, 10\}$, $B = \{4, 8, 12, 15, 19\}$, $C = \{2, 4, 8, 15, 20\}$, $\mathcal{U} = \{1, 2, 3, \ldots, 20\}$
82. $G = \{x \mid x \in \mathbf{N}, x < 35, x \text{ is an even integer}\}$,
    $H = \{x \mid x \in \mathbf{N}, x < 35, x \text{ is a multiple of 3}\}$,
    $K = \{x \mid x \in \mathbf{N}, x < 35, x \text{ is a multiple of 5}\}$,
    $\mathcal{U} = \{x \mid x \in \mathbf{N}, x < 35\}$
83. $G = \{x \mid x = 3k \text{ for some } k \in \mathbf{Z}, |x| \leq 25\}$,
    $H = \{x \mid x = 5k \text{ for some } k \in \mathbf{Z}, |x| < 30\}$,
    $K = \{x \mid x \in \mathbf{Z}, 10 < |x| < 30\}$,
    $\mathcal{U} = \{x \mid x \in \mathbf{Z}, |x| \leq 30\}$

2.2 SET OPERATIONS

It is often useful to be able to combine sets to form new ones. For instance, suppose

$$F = \{x \mid x \text{ is a female employee of Alpha Company}\}$$

and

$$M = \{x \mid x \text{ is a male employee of Alpha Company}\}.$$

Then

$$E = \{x \mid x \in F \text{ or } x \in M\}$$
$$= \{x \mid x \text{ is an employee of Alpha Company}\}.$$

SETS AND FUNCTIONS

Note that the set E was constructed from the sets F and M. In this section we show how to construct new sets from old ones by means of the **set operations** of **complementation, union,** and **intersection.** We begin with complementation.

The **complement** of a set S is the set of all elements of the universal set that are *not* elements of S.

Complementation

If S is a set and $\mathcal{U}$ is a universal set for S, the **complement** of S in $\mathcal{U}$ is the set

$$S' = \{x \mid x \in \mathcal{U} \text{ and } x \notin S\}.$$

The notation S' is read "S prime." In translating from set terminology to English, the set operation of complementation corresponds to the word "not." The Venn diagram representing the complement of S in $\mathcal{U}$, with S' shaded, is shown in Figure 2–6.

EXAMPLE 1 Let $S = \{1, 2, 3\}$, $T = \{3, 4, 5\}$, and $V = \{4, 5, 6\}$. Let

$$\mathcal{U} = \{1, 2, 3, 4, 5, 6, 7\}$$

be a universe for these sets. The Venn diagram is shown in Figure 2–7. Since S' consists of all elements of $\mathcal{U}$ that are not elements of S,

$$S' = \{4, 5, 6, 7\}.$$

Similarly,

$$T' = \{1, 2, 6, 7\}$$

and

$$V' = \{1, 2, 3, 7\}. \quad \square$$

EXAMPLE 2 The complement of the empty set $\emptyset$ is the universal set $\mathcal{U}$, since, by definition,

$$\emptyset' = \{x \mid x \in \mathcal{U} \text{ and } x \notin \emptyset\} = \{x \mid x \in \mathcal{U}\} = \mathcal{U}.$$

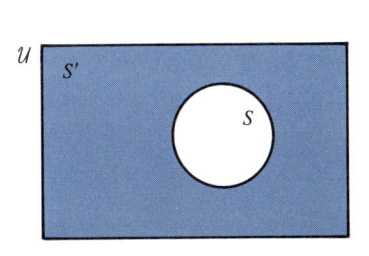

FIGURE 2–6

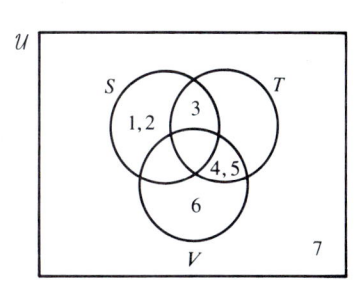

FIGURE 2–7

The complement of the universal set $\mathcal{U}$ is the empty set $\emptyset$, since, by definition,
$$\mathcal{U}' = \{x \mid x \in \mathcal{U} \text{ and } x \notin \mathcal{U}\} = \emptyset.$$

EXAMPLE 3 For any set S and any universal set $\mathcal{U}$ containing S,
$$(S')' = \{x \mid x \in \mathcal{U} \text{ and } x \notin S'\} = \{x \mid x \in \mathcal{U} \text{ and } x \in S\} = S.$$

The next example shows how the operation of complementation corresponds to the word "not."

EXAMPLE 4 Suppose that
$$S = \{x \mid x \text{ is a first-year college student}\},$$
$$T = \{x \mid x \text{ is a female college student}\},$$
$$\mathcal{U} = \{x \mid x \text{ is a college student}\}.$$

Then
$$S' = \{x \mid x \text{ is a college student who is } not \text{ in the first year}\},$$
$$T' = \{x \mid x \text{ is a college student who is } not \text{ a female}\},$$
$$= \{x \mid x \text{ is a male college student}\}.$$

The **union** of two sets is the set of all elements that are in at least one of the sets.

> **Union**
>
> If S and T are sets, the **union** of S and T is the set
> $$S \cup T = \{x \mid x \in S \text{ or } x \in T\}.$$

Thus, $S \cup T$ consists of all elements of S together with all elements of T, including those that are elements of both S and T. The notation $S \cup T$ is read "S union T." In translating from set terminology to English, the set operation union corresponds to the phrase "either . . . or." The Venn diagram for $S \cup T$, with $S \cup T$ shaded, is shown in Figure 2–8.

EXAMPLE 5 Let $S = \{1, 2, 3, 4\}$, $T = \{3, 4, 5, 6\}$, $V = \{2, 5, 7\}$, and
$$\mathcal{U} = \{1, 2, 3, 4, 5, 6, 7, 8\}.$$

Then $S \cup T$ consists of all elements of S together with all elements of T, so
$$S \cup T = \{1, 2, 3, 4, 5, 6\}.$$

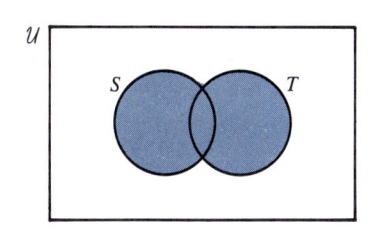

SETS AND FUNCTIONS

FIGURE 2–8 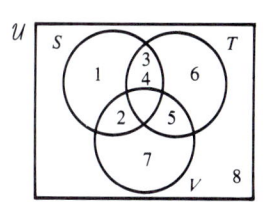 **FIGURE 2–9**

Similarly,
$$S \cup V = \{1, 2, 3, 4, 5, 7\},$$
$$T \cup V = \{2, 3, 4, 5, 6, 7\},$$
and
$$S \cup T \cup V = \{1, 2, 3, 4, 5, 6, 7\}.$$

The Venn diagram is shown in Figure 2–9. Note that
$$(S \cup T)' = \{7, 8\},$$
$$(S \cup V)' = \{6, 8\},$$
$$(T \cup V)' = \{1, 8\},$$
$$S' \cup T = \{3, 4, 5, 6, 7, 8\},$$
$$T \cup V' = \{1, 3, 4, 5, 6, 8\}. \quad \square$$

EXAMPLE 6 For any set S and universal set $\mathcal{U}$ containing S,
$$S \cup \mathcal{U} = \mathcal{U},$$
$$S \cup S' = \mathcal{U},$$
$$S \cup S = S,$$
$$S \cup \emptyset = S.$$

(See Exercise 53.) $\square$

EXAMPLE 7 For any sets S and T, $S \subset (S \cup T)$ and $T \subset (S \cup T)$. If $S \subset T$, then $S \cup T = T$. (See Exercise 54.) $\square$

The next example illustrates how the operation of union corresponds to the phrase "either . . . or."

EXAMPLE 8 Suppose that
$$S = \{x \mid x \text{ is a first-year college student}\}$$
and
$$T = \{x \mid x \text{ is a female college student}\}.$$

CHAPTER 2

Then

$$S \cup T = \{x \mid x \text{ is either a first year college student or a female college student}\}. \quad \square$$

The **intersection** of two sets is the set of all elements that are in both of them.

> **Intersection**
>
> If S and T are sets, the **intersection** of S and T is the set
>
> $$S \cap T = \{x \mid x \in S \text{ and } x \in T\}.$$

Thus, $S \cap T$ consists of all elements that S and T have in common. The notation $S \cap T$ is read "S intersect T." In translating from set terminology to English, the set operation of intersection corresponds to the word "and." The Venn diagram for $S \cap T$, with $S \cap T$ shaded, is shown in Figure 2–10.

EXAMPLE 9   Let $S = \{1, 2, 3\}$, $T = \{2, 4, 5, 6\}$, and $V = \{2, 3, 6, 7\}$, and let

$$\mathcal{U} = \{1, 2, 3, 4, 5, 6, 7, 8\}.$$

Then $S \cap T$ consists of all elements common to S and T, so

$$S \cap T = \{2\}.$$

Similarly,

$$S \cap V = \{2, 3\},$$

$$T \cap V = \{2, 6\},$$

and

$$S \cap T \cap V = \{2\}.$$

The Venn diagram is shown in Figure 2–11. Note that

$$(S \cap T)' = \{1, 3, 4, 5, 6, 7, 8\},$$

$$(S \cap V)' = \{1, 4, 5, 6, 7, 8\},$$

$$S' \cap V = \{6, 7\},$$

$$T \cap V' = \{4, 5\},$$

$$S \cap T \cap V' = \emptyset \quad \square$$

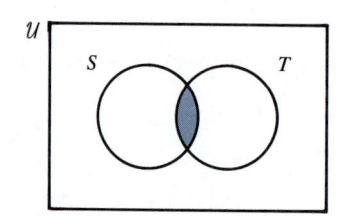

FIGURE 2–10 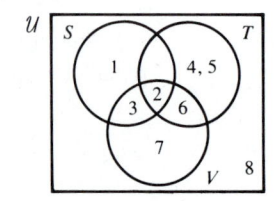   **FIGURE 2–11**

EXAMPLE 10 For any set S and universal set $\mathcal{U}$ containing S,
$$S \cap \mathcal{U} = S,$$
$$S \cap S = S,$$
$$S \cap S' = \emptyset$$
$$S \cap \emptyset = \emptyset.$$
(See Exercise 55.) ☐

EXAMPLE 11 For any sets S and T, $(S \cap T) \subset S$ and $(S \cap T) \subset T$. If $S \subset T$, then $S \cap T = S$. (See Exercise 56.) ☐

The next example illustrates how the operation of intersection corresponds to the word "and."

EXAMPLE 12 Suppose that
$$S = \{x \mid x \text{ is a first-year college student}\}$$
and
$$T = \{x \mid x \text{ is a female college student}\}.$$
Then
$$S \cap T = \{x \mid x \text{ is a first-year college student and a female college student}\}$$
$$= \{x \mid x \text{ is a first-year, female college student}\}. \quad \square$$

EXAMPLE 13 Let
$$\mathcal{U} = \{x \mid x \text{ is a citizen}\},$$
$$F = \{x \mid x \in \mathcal{U}, x \text{ is female}\},$$
$$M = \{x \mid x \in \mathcal{U}, x \text{ is married}\},$$
$$L = \{x \mid x \in \mathcal{U}, x \text{ is a lawyer}\}.$$
Then
$F \cap M \cap L = \{x \mid x \text{ is a female citizen who is a married lawyer}\}$,
$F \cup M \cup L = \{x \mid x \text{ is a citizen who is either female, or married, or a lawyer}\}$,
$F \cup (M \cap L) = \{x \mid x \text{ is a citizen who is either female or a married lawyer}\}$,
$F \cap (M \cup L) = \{x \mid x \text{ is a female citizen who is either married or a lawyer}\}$,
$F \cap (M \cup L)' = \{x \mid x \text{ is a female citizen who is neither married nor a lawyer}\}$. ☐

Sets that have the empty set as their intersection are said to be **disjoint** or **mutually exclusive**.

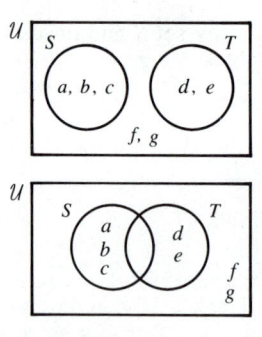

FIGURE 2–12

EXAMPLE 14   Let $S = \{a, b, c\}$ and $T = \{d, e\}$, and let $\mathcal{U} = \{a, b, c, d, e, f, g\}$. The sets S and T are disjoint because $S \cap T = \emptyset$. Figure 2–12 shows two Venn diagrams, both of which depict the relationships among S, T, and $\mathcal{U}$ and the fact that S and T are disjoint. ☐

We conclude this section with an example that shows how Venn diagrams can be used to solve certain counting problems.

EXAMPLE 15   A magazine conducted a market survey of its subscribers in a particular region. Of the 200 surveyed subscribers,

    131 were married.
    108 owned their own homes.
    103 had college degrees.
     68 were married and owned their own homes.
     55 owned their own homes and had college degrees.
     51 were married and had college degrees.
     23 were married, owned their own homes, and had college degrees.

How many of the people surveyed were either married or owned their own home, but did not have a college degree? Let

$$\mathcal{U} = \{\text{surveyed subscribers}\},$$

$$M = \{\text{surveyed subscribers who were married}\},$$

$$H = \{\text{surveyed subscribers who owned their own homes}\},$$

$$D = \{\text{surveyed subscribers who had college degrees}\}.$$

We want to determine the number of people in the set $(M \cup H) \cap D'$. To do this, we restate the given information in terms of the sets $\mathcal{U}$, M, H, and D. There are

    200 people in the set $\mathcal{U}$,
    131 people in the set M,
    108 people in the set H,
    103 people in the set D,

68 people in the set $M \cap H$,
55 people in the set $H \cap D$,
51 people in the set $M \cap D$,
23 people in the set $M \cap H \cap D$,

Since $M \cap H \cap D$ is contained in $M \cap D$, the 23 people in $M \cap H \cap D$ are counted among the 51 people in $M \cap D$. Therefore, there must be
$51 - 23 = 28$ people in $M \cap H' \cap D$. See Figure 2–13. Similarly, there are $55 - 23 = 32$ people in $M' \cap H \cap D$ and $68 - 23 = 45$ in $M \cap H \cap D'$. But then there are 103 people in D, of whom $28 + 23 + 32 = 83$ have already been counted. Therefore, there must be $103 - 83 = 20$ people in $D \cap (M \cup H)'$. See Figure 2–14. Similarly, there are $108 - (45 + 23 + 32) = 8$ people in $H \cap (M \cup D)'$ and $131 - (45 + 23 + 28) = 35$ in $M \cap (H \cup D)'$. Now there are $23 + 28 + 32 + 45 + 20 + 8 + 35 = 191$ people in $M \cup H \cup D$ and, since there were 200 in $\mathcal{U}$ there must be $200 - 191 = 9$ in $(M \cup H \cup D)'$. See Figure 2–15. From this last Venn diagram we see that the number of surveyed people in the set $(M \cup H) \cap D'$ is $35 + 45 + 8 = 88$. ☐

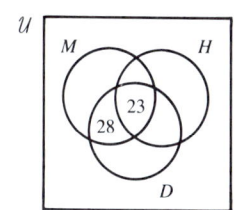

FIGURE 2–13

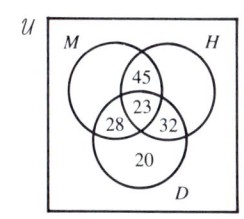

FIGURE 2–14

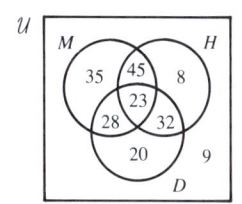

FIGURE 2–15

■ *Computer Manual: Programs SETARITH, SETCOUNT*

2.2 EXERCISES

Let
$$S = \{a, c, d\}, \quad T = \{a, b, d, e, f\}, \quad V = \{a, b, f, g\}, \quad \mathcal{U} = \{a, b, c, d, e, f, g, h, i\}$$
In Exercises 1 through 12, describe the given set using the roster method.

1. S'
2. T'
3. $S \cup T$
4. $T \cup V$
5. $(S \cup V)'$
6. $S \cap T$
7. $T \cap V$
8. $(S \cap V)'$
9. $S \cup T \cup V$
10. $S \cap T \cap V$
11. $S \cup (T \cup V)'$
12. $S' \cap (T \cup V)$

Let
$$X = \{x \mid x \in \mathbf{N}, x \leq 8\}, \quad Y = \{x \mid x = 2k, k \in \mathbf{N}, 1 \leq k \leq 5\},$$
$$W = \{3, 4, 5, 6, 10, 11, 12, 13\}, \quad \mathcal{U} = \{x \mid x \in \mathbf{N}, x \leq 15\}).$$

In Exercises 13 through 24, decribe the given set using the roster method.

13. Y'
14. $X \cup W$
15. $Y \cap W$
16. $X \cap Y$
17. $X \cap Y \cap W$
18. $X' \cap Y \cap W'$
19. $X \cup Y \cup W$
20. $X \cap (Y \cup W)$
21. $Y \cup (X \cap W)$
22. $Y' \cap (X \cup W)$
23. $W \cap (X \cap Y)'$
24. $(W \cap (X \cup Y))'$

Consider the following intervals on the real line:

$I = [8, 10]$, $\quad J = (10, +\infty)$, $\quad K = [10, 20]$, $\quad L = (-\infty, 5)$, $\quad \mathcal{U} = (-\infty, +\infty) = \mathbf{R}$.

In Exercises 25 through 36, describe the given set. Use interval notation where possible.

25. $I \cap K$
26. $I \cup K$
27. $I \cup J$
28. $I \cap J$
29. $I \cap L$
30. $K \cup J$
31. $K \cap J$
32. $K \cap L$
33. J'
34. L'
35. $J' \cap L$
36. $J \cap L'$

Let

$\mathcal{U} = \{x \mid x \text{ is a physician}\}$, $\quad V = \{x \mid x \text{ is an internist}\}$, $\quad W = \{x \mid x \text{ is a retired physician}\}$.

In Exercises 37 through 44, describe the given set using the set-builder method.

37. V'
38. W'
39. $V \cup W$
40. $V \cap W$
41. $V' \cup W$
42. $V \cap W'$
43. $(V \cup W)'$
44. $(V \cap W)'$

Let

$A = \{x \mid x \text{ is a college student taking an art course}\}$,

$B = \{x \mid x \text{ is a college student taking a biology course}\}$,

$C = \{x \mid x \text{ is a college student taking a chemistry course}\}$,

$\mathcal{U} = \{x \mid x \text{ is a college student}\}$.

In Exercises 45 through 52, describe the given set using the set-builder method.

45. A'
46. $A \cup C$
47. $B \cap C$
48. $A \cap B \cap C$
49. $A \cup B \cup C$
50. $A' \cup B$
51. $A \cap C'$
52. $(A \cup B \cup C)'$

*53. If S is any set contained in a universal set $\mathcal{U}$, show that

$$S \cup \mathcal{U} = S \cup S' = \mathcal{U} \quad \text{and} \quad S \cup S = S \cup \emptyset = S.$$

*54. (a) If S and T are any sets, show that

$$S \subset (S \cup T) \quad \text{and} \quad T \subset (S \cup T).$$

(b) If S and T are sets with $S \subset T$, show that $S \cup T = T$.

*55. If S is any set contained in a universal set $\mathcal{U}$, show that

$$S \cap \mathcal{U} = S \cap S = S \quad \text{and} \quad S \cap S' = S \cap \emptyset = \emptyset.$$

*56. (a) If S and T are any sets, show that

$$(S \cap T) \subset S \quad \text{and} \quad (S \cap T) \subset T.$$

(b) If S and T are sets with $S \subset T$, show that $S \cap T = S$.

Management

A company has classified its employees according to the following system:

Classification	Description
A	Employees paid an hourly rate
B	Employees paid a monthly salary
C	Employees who have less than two years' service with the company
D	Employees who have two through five years' service with the company
E	Employees who have more than five years' service with the company

Exercises 57 through 59 refer to this situation. The universal set consists of all employees of the company.

57. Use the descriptive method to write the sets A' and C'.
58. Use the descriptive method to write the sets $C \cup E$ and $C \cap E$.
59. Use the descriptive method to write the sets $(A \cap C)'$ and $(B \cup D)'$.

The company of Exercises 57 through 59 offers the following fringe benefits to eligible employees:

Fringe Benefit	Employees Eligible
Pension plan	Those with two or more years of service
Medical plan	Those who are not paid hourly or who have two or more years of service
Life insurance	Those who are paid a monthly salary or have more than five years' service
Profit sharing	Those who are paid a monthly salary and have two or more years of service
Christmas bonus	Those not eligible for profit sharing

In Exercises 60 through 64, you are to write the given set using *only* the employee classifications *A, B, C, D, E,* of Exercises 57 through 59 and the operations of complementation, union, and intersection.

60. The set of all employees who are eligible for the pension plan.
61. The set of all employees who are eligible for the medical plan.
62. The set of all employees who are eligible for life insurance.
63. The set of all employees who are eligible for profit sharing.
64. The set of all employees who are eligible for the Christmas bonus.

Social Science

A sociologist studying creativity in the arts has classified the people taking part in the study as follows:

Classification	Description
V	Persons creative in the visual arts
M	Persons creative in the musical arts
I	Persons who are immigrants
J	Persons who are native-born but have at least one parent an immigrant
K	Persons who are native-born with both parents native-born

Exercises 65 through 67 refer to this situation. The universal set consists of all those who are taking part in the study.

65. Use the descriptive method to write the sets V' and I'.

66. Use the descriptive method to write the sets $V \cap M$ and $I \cup M$.

67. Use the descriptive method to write the sets $(M \cap K)'$ and $(V \cup J)'$.

The sociologist of Exercises 65 through 67 wishes to single out certain subsets of the universal set for particular study. In Exercises 68 through 73, you are to write the given subset using *only* the classifications $V, M, I, J, K,$ of Exercises 65 through 67 and the set operations of complementation, union, and intersection.

68. The subset consisting of those who are not immigrants and are creative in the musical arts.

69. The subset consisting of those who are creative in the visual arts but not in the musical arts.

70. The subset consisting of those who are either creative in the visual arts or are native-born.

71. The subset consisting of those who are creative in both the visual and the musical arts and are immigrants.

72. The subset consisting of those who are creative in either the visual or the musical arts and are native-born, with at least one parent an immigrant.

73. The subset consisting of those who are creative in neither the visual nor the musical arts and are not immigrants.

Exercises 74 through 83 are counting problems that are to be solved with the aid of Venn diagrams.

74. A college has an enrollment of 1000 students. Of these 1000 students:

250 are taking art courses.

400 are taking biology courses.

300 are taking classics courses.

75 are taking art courses and biology courses.

100 are taking art courses and classics courses.

200 are taking biology courses and classics courses.

50 are taking art courses, biology courses, and classics courses.

How many of the students are taking
- **(a)** art courses and biology courses but not classics courses?
- **(b)** art courses and classics courses but not biology courses?

(c) art courses but neither biology courses nor classics courses?
(d) neither art courses, biology courses, nor classics courses?

75. The town of Dudster is considering inaugurating a bus system that will transport Dudster residents to and from the central business district. In an attempt to determine whether shoppers want or need such a transportation system, the town takes a survey of 1000 shoppers chosen at random. Each shopper is asked if he or she resides in Dudster, owns a car, and favors the inauguration of the bus system. The results of the survey are as follows:

650 reside in Dudster.

490 own a car.

250 favor inaugurating the bus system.

300 reside in Dudster and own a car.

150 reside in Dudster and favor inaugurating the bus system.

140 own a car and favor inaugurating the bus system.

100 reside in Dudster, own a car, and favor the bus system.

How many shoppers
(a) do not favor inaugurating the bus system?
(b) reside in Dudster and do not favor inaugurating the bus system?
(c) reside in Dudster and do not own a car?
(d) reside in Dudster, do not own a car, and favor inaugurating the bus system?
(e) reside in Dudster, do not own a car, and do not favor inaugurating the bus system?
(f) The bus system, if it is inaugurated, will be designed primarily to benefit Dudster residents who do not own a car. On the basis of this survey, should the town provide the bus service or not?

*76. A survey of 230 investors revealed the following:

150 owned automotive stocks.

125 owned oil stocks.

80 owned both automotive and oil stocks but no electronics stocks.

40 owned only electronics stocks.

35 owned only automotive stocks.

15 owned both oil and electronics stocks but no automotive stocks.

10 owned automotive, oil, and electronics stocks.

How many of those surveyed owned
(a) electronics stocks?
(b) only oil stocks?
(c) automotive and electronics stocks but no oil stocks?
(d) either oil or electronics stocks but not both?
(e) either automotive or electronics stocks?
(f) none of these three types of stocks?

Management

77. On a certain weekend, 100 customers made purchases at a Hoopersville garden supply store. Of these 100 customers:

30 purchased tools.

45 purchased fertilizer.

50 purchased seeds.

15 purchased seeds and fertilizer.

20 purchased seeds and tools.

15 purchased tools and fertilizer.

10 purchased tools, seeds, and fertilizer.

How many customers purchased
- **(a)** only tools?
- **(b)** seeds and tools but not fertilizer?
- **(c)** tools and fertilizer but not seeds?
- **(d)** neither seeds, tools, nor fertilizer?

*78. The marketing department of Major Mills is testing three variations of a new cereal to determine which, if any, the company should put on the market. The three variations are Crunchy A, Crunchy B, and Crunchy C. Of a random sample of 70 people who taste-tested the three cereals:

20 liked only A.

2 liked only C.

9 liked A and B.

17 liked A and C.

13 liked B and C.

2 liked all three.

13 liked none of the three.

Major Mills will not market a cereal if less than 40% of the taste testers like it. Based on this survey, which of the three cereals should Major Mills market?

*79. Of the 64 executives at Consolidated Company last year:

30 had more than five years' service with the company.

29 received a performance bonus.

15 were fired.

8 had more than five years' service, received a bonus, and were fired.

3 had more than five years' service, did not receive a bonus, and were fired.

9 received a bonus and were fired

14 had five or fewer years of service, received a bonus, and were not fired.

- **(a)** How many executives with more than five years' service received a bonus and were not fired?
- **(b)** How many executives with more than five years' service were fired?
- **(c)** How many executives who received a bonus were not fired?
- **(d)** How many executives who did not receive a bonus were fired?
- **(e)** How many executives with five or fewer years' of service did not receive a bonus and were not fired?

SETS AND FUNCTIONS 79

Life Sciences

80. A genetics experiment with fruit flies involved classifying the flies as to color (black or albino), eye color (black or red), and wing length (short or long). The results were as follows:

>335 were black.
>
>420 had black eyes.
>
>345 had long wings.
>
>320 had black eyes and were black.
>
>305 had long wings and were black.
>
>300 had black eyes and long wings.
>
>300 had black eyes, long wings, and were black.

These results account for all the flies in the experiment.
- (a) How many fruit flies were involved in the experiment?
- (b) How many were albinos?
- (c) How many were albinos with black eyes?
- (d) How many were albinos with long wings?
- (e) Albinism, red eyes, and short wings are recessive traits. How many flies had at least one recessive trait? all three recessive traits?

*81. A medical experiment was conducted to test a new drug. Some of the patients involved in the experiment received the drug, and others received a placebo. No patient received both the drug and the placebo, but some patients received neither. The following results were recorded.

>400 patients received the drug, and 175 improved.
>
>280 patients received the placebo, and 100 improved.
>
>285 patients received neither the drug nor the placebo, and 85 improved.

These results account for all the patients involved.
- (a) How many patients took part in the experiment?
- (b) How many improved?
- (c) How many did not improve?

Social Science

82. A sociologist conducted a survey to determine the relationships among education, income, and satisfaction. Of the people he surveyed:

>600 had graduated from college.
>
>660 had high incomes.
>
>840 were satisfied.
>
>400 had high incomes and were satisfied.
>
>380 had graduated from college and were satisfied.
>
>120 had graduated from college and had high incomes but were not satisfied.
>
>140 had graduated from college, had high incomes, and were satisfied.

These results account for all the people surveyed.
(a) How many people were surveyed?
(b) How many had not graduated from college, had high incomes, and were satisfied?
(c) How many were dissatisfied?
(d) How many had not graduated from college?
(e) How many had graduated from college and were dissatisfied?

*83. A survey of 830 people in a small town revealed the following:

430 people considered themselves to be conservatives.

380 of those who considered themselves to be conservatives supported capital punishment.

120 of those who considered themselves to be conservatives supported both capital punishment and the food stamp program.

180 of those who supported capital punishment did not consider themselves to be conservatives.

190 supported both capital punishment and the food stamp program.

230 supported the food stamp program but not capital punishment.

220 of those who did not consider themselves to be conservatives supported the food stamp program but not capital punishment.

(a) How many of those surveyed did not consider themselves to be conservatives?
(b) How many supported the food stamp program?
(c) How many supported capital punishment?

2.3 FUNCTIONS

A **function** is a rule that assigns to each element of one set exactly one element of another set, thereby describing a relationship between the sets. For instance, suppose we assign letter grades to examination scores by means of the following scheme:

Exam Score X	Letter Grade
$90 \leq X \leq 100$	A
$80 \leq X < 90$	B
$60 \leq X < 80$	C
$50 \leq X < 60$	D
$0 \leq X < 50$	F

This table defines a function from the set of examination scores to the set of letter grades: each possible examination score is assigned exactly one letter grade.

For another example of a function, suppose that a firm's annual profit is always equal to 10% of its annual sales. Thus, if its annual sales were $10 million, its annual profit would be $1 million, and so on. If we let S denote annual sales and P the annual profit, we may express the relationship between sales and profit by the formula

$$P = 0.1S.$$

SETS AND FUNCTIONS   81

The equation $P = 0.1S$ defines a function from one set of numbers to another: each possible sales number is assigned a single profit number.

Functions are important exactly because they can be used, as we did above, to show how one quantity depends on another. In this section we discuss functions. We begin with the definition of a function from one set to another.

> **Function**
>
> Let S and T be sets. A **function from S to T** is a rule that assigns to *each* element of S a *single* element of T. The set S is called the **domain** of the function. The **range** of the function is the subset of T defined by
>
> range $= \{y \mid y \in T$ and the function assigns y to some $x \in S\}$.

Note that the domain of a function may be thought of as the "input" for the function: it is the set of all objects that the function "uses." The range of a function is the "output" of the function: it is the set of all objects that the function "produces." See Figure 2–16.

EXAMPLE 1 Suppose $S = \{a, b, c\}$ and $T = \{1, 2, 3, 4\}$. We define a rule that assigns elements of T to elements of S as follows:

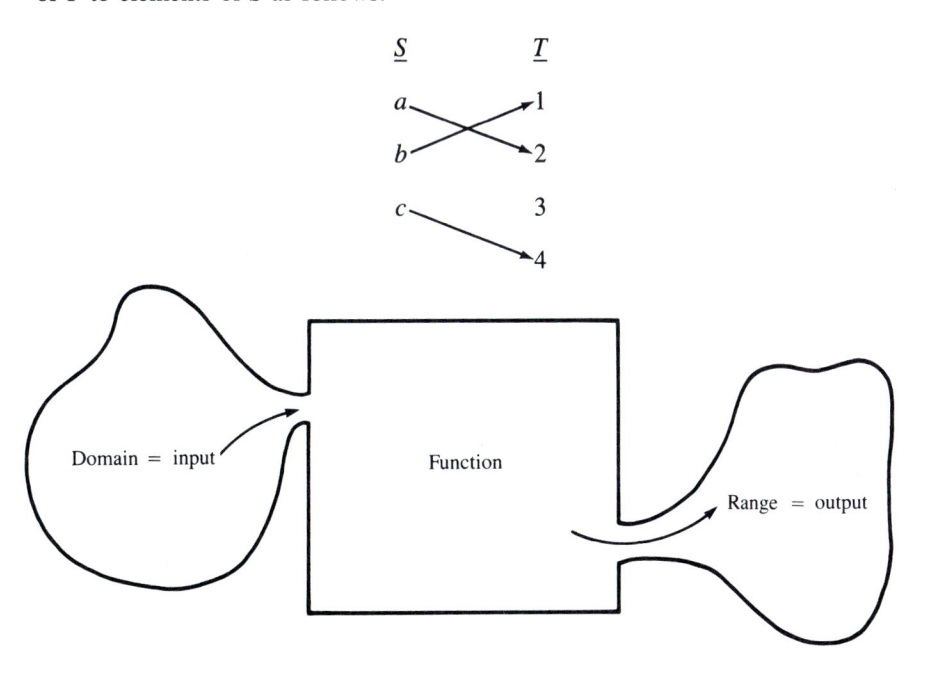

$$\begin{array}{cc} \underline{S} & \underline{T} \\ a & 1 \\ b & 2 \\ c & 3 \\ & 4 \end{array}$$

Domain = input   Function   Range = output

FIGURE 2–16

In other words, the rule assigns 2 to *a*, 1 to *b*, and 4 to *c*. Since *each* element of *S* is assigned a *single* element of *T*, this rule is a function from *S* to *T*; *S* is its domain, and its range is the subset {1, 2, 4} of *T*.

Not every rule of correspondence between sets is a function.

EXAMPLE 2 Let *S* and *T* be the sets of Example 1. We define a rule that assigns elements of *T* to elements of *S* as follows:

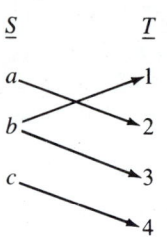

$$\underline{S} \qquad \underline{T}$$

$$a \to 1$$
$$b \to 2$$
$$c \to 3$$
$$ \to 4$$

This rule is *not* a function from *S* to *T*, because the element *b* of *S* is assigned *more than one* element of *T*.

Figure 2–17 is a symbolic depiction of the difference between a rule of correspondence that is a function and one that is not: the first rule in the figure takes *each* element of *S* to just one element of *T*; the second rule takes one element of *S* to two elements of *T*. In general, any rule of correspondence between sets is known as a **relation**. Thus every function is a relation, but not every relation is a function; for instance, the rules of correspondence of Examples 1 and 2 are both relations from the set *S* to the set *T*, but only that of Example 1 is a *function* fron *S* to *T*.

Functions are often denoted by the letters *f*, *g*, or *h*, and written in **functional notation:** in functional notation, the expression $f(x) = y$ is used to indicate that the function named *f* assigns the element *y* of its range to the element *x* of its domain. The expression $f(x) = y$ is read "f of x equals y."

EXAMPLE 3 Let $S = \{a, b, c\}$ and $T = \{1, 2, 3, 4\}$. We define a function *f* from *S* to *T* by setting

$$f(a) = 2, \qquad f(b) = 1, \qquad f(c) = 4.$$

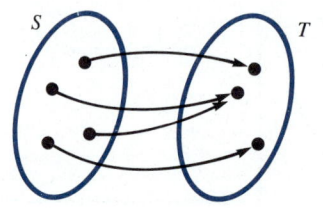

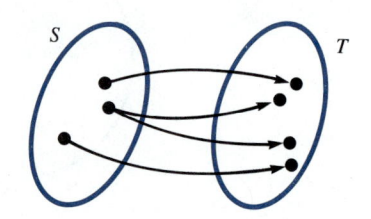

This is a function from *S* to *T*   This is not a function from *S* to *T*

FIGURE 2–17

SETS AND FUNCTIONS 83

Note that f is indeed a function from S to T because it assigns exactly one element of T to each element of S; in fact, f is the function of Example 1. Similarly, if we define g by

$$g(a) = g(b) = g(c) = 2,$$

then g is a function from S to T. Note that the range of g is $\{2\}$. ☐

EXAMPLE 4 Let $S = \{1, 2, 3\}$ and $T = \{2, 4, 6\}$. For each element $x \in S$, define $h(x) = 2x$. Thus

$$h(1) = 2(1) = 2,$$
$$h(2) = 2(2) = 4,$$
$$h(3) = 2(3) = 6.$$

Therefore h is a function from S to T. Note that the range of h is

$$\{2, 4, 6\} = T. \quad \square$$

Most of the functions we will use in this book will be functions from one set of real numbers to another, and they will usually be defined by equations. The function h of Example 4 is one such: it is a function from one set of real numbers to another, and it is defined by the equation $h(x) = 2x$. For another example, consider the statement

$$f(x) = 3x - 2, \qquad x \in \mathbf{R}.$$

This statement defines a function f from $\mathbf{R}$ to $\mathbf{R}$: the equation shows that f assigns to each number x the single number $3x - 2$; the "$x \in \mathbf{R}$" part of the statement tells us that x can be any real number, and hence that the domain of f is the set $\mathbf{R}$ of real numbers. The letter x is called the **independent variable** of the function f. If we substitute a particular real number for x in the defining equation, we can see what value the function assigns to that number. For instance, since

$$f(0) = 3(0) - 2 = -2,$$
$$f(1) = 3(1) - 2 = 1,$$
$$f(-2) = 3(-2) - 2 = -8,$$

we see that f assigns -2 to 0, 1 to 1, and -8 to -2. The numbers $f(0), f(1)$, and $f(-2)$ are called the **functional values** of f at 0, 1, and -2, respectively. Thus $f(x)$ denotes the functional value of f at x. Note that a function f and a functional value $f(x)$ are *not* the same thing: f is a *rule*, but $f(x)$ is an element of the range of f.

EXAMPLE 5 Consider the function g from $\mathbf{R}$ to $\mathbf{R}$ defined by the equation

$$g(x) = 5x + 2, \qquad x \in \mathbf{R}.$$

We have
$$g(0) = 5(0) + 2 = 2,$$
$$g(6) = 5(6) + 2 = 32,$$
$$g(-4) = 5(-4) + 2 = -18,$$
$$g(-3/5) = 5(-3/5) + 2 = -3 + 2 = -1,$$
$$g(2/3) = 5(2/3) + 2 = 10/3 + 2 = 16/3,$$
$$g(2.3) = 5(2.3) + 2 = 11.5 + 2 = 13.5,$$
$$g(\sqrt{3}) = 5\sqrt{3} + 2. \quad \square$$

It is often possible (and useful) to substitute algebraic expressions for the independent variable of a function.

EXAMPLE 6 Let g be the function of Example 5 defined by the equation
$$g(x) = 5x + 2.$$
Then
$$g(t) = 5t + 2,$$
$$g(a^2) = 5a^2 + 2,$$
$$g(b - 1) = 5(b - 1) + 2 = 5b - 5 + 2 = 5b - 3. \quad \square$$

EXAMPLE 7 Suppose it costs a firm $C(x) = x^2 + 5x + 100$ dollars to produce x units of its product. How much will it cost the firm to produce 20 units? The answer is that it will cost
$$C(20) = 20^2 + 5(20) + 100 = 600$$
dollars. In general, if the firm intends to produce b units, it will cost them
$$C(b) = b^2 + 5b + 100$$
dollars. $\square$

Sometimes when we define a function from one set of numbers to another we will state its domain explicitly. For instance, we did this in Example 5, where we defined the function g by the statement
$$g(x) = 5x + 2, \quad x \in \mathbf{R}.$$
For another example, suppose we define a function h by the statement
$$h(x) = x^2 + 1, \quad x > 1.$$
This indicates that the independent variable x is allowed to take on only values greater than 1. Hence, the domain of h is
$$\{x \in \mathbf{R} \mid x > 1\} = (1, +\infty).$$

SETS AND FUNCTIONS

Many times when we define a function from one set of numbers to another we will not explicitly state its domain. In such a case, it is understood that the domain consists of all real numbers that may legitimately be substituted for the independent variable in the defining equation of the function.

EXAMPLE 8 Consider the square root function defined by

$$f(x) = \sqrt{x}.$$

We cannot substitute negative real numbers for x because the square root of a negative number is not defined as a real number. However, we can substitute any nonnegative real number for x. Therefore the domain of the square root function is

$$\{x \in \mathbf{R} \mid x \geq 0\} = [0, +\infty).$$

Similarly, suppose we define a function k by the equation

$$k(x) = \frac{2}{x-1}.$$

Since division by 0 is not possible, we cannot substitute 1 for x in this equation, but we can substitute any other real number for x. Thus the domain of the function k is

$$\{x \in \mathbf{R} \mid x \neq 1\}. \quad \square$$

We will not always use functional notation to describe functions. Sometimes it will be more convenient to define functions by means of equations involving the independent variable and a **dependent variable.** For instance, consider the equation

$$y = 3x - 2;$$

substituting numbers for x in this equation yields numerical values for y:

x	$y = 3x - 2$
0	$3(0) - 2 = -2$
1	$3(1) - 2 = 1$
-2	$3(-2) - 2 = -8$

Thus the equation defines y as a function of x, because substituting any value for x clearly results in exactly one value for y.

Note that the function defined by $y = 3x - 2$ is the same as that defined by $f(x) = 3x - 2$; indeed, $u = 3v - 2$ and $g(t) = 3t - 2$ also define this same function. The letters used are not important; what is important is the relationship between the variables.

EXAMPLE 9 It is estimated that t years from now the size of the bison herd in a national park will be approximately

$$B = 100\sqrt{0.2t + 1}$$

animals. Thus we conclude that at the present time ($t = 0$), the size of the herd is

$$B = 100\sqrt{0.2(0) + 1} = 100\sqrt{1} = 100$$

animals. If we wish to forecast the size of the herd 10 years from now, we need only substitute 10 for t in the equation:

$$B = 100\sqrt{0.2(10) + 1} = 100\sqrt{3} \approx 100(1.73) = 173$$

animals. ☐

Finally, we note that not every equation defines a function. For instance, consider the equation $x^2 + y^2 = 1$: when $x = 0$, this reduces to $y^2 = 1$, which has solutions $y = 1$ and $y = -1$. But an equation cannot define a function unless *each* value of the independent variable yields *exactly one* value for the dependent variable; therefore this equation does not define y as a function of x. However, it does define a relation from the interval $[-1,1]$ to the real numbers, because substituting any number from the interval into the equation for x results in one or more real values for y.

■ *Computer Manual: Program FNCTNVAL*

2.3 EXERCISES

In Exercises 1 through 10, determine whether the given relation from S to T is or is not a function; if it is a function, find its range.

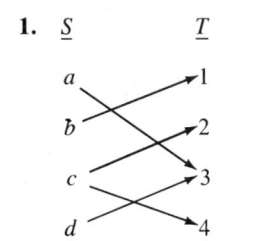

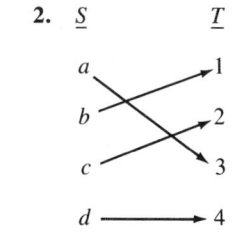

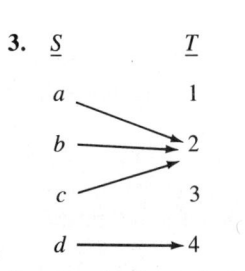

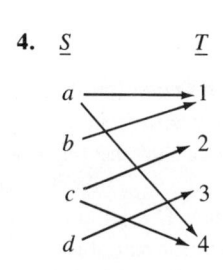

1.

2.

3.

4.

5. $S = \{a, b, c, d\}$, $T = \{2, 4, 6, 8\}$, $f(a) = 4$, $f(b) = f(c) = 2$, $f(d) = 8$
6. $S = \{1, 2, 3, 4, 5\}$, $T = \{p, q, r, t\}$, $g(1) = q$, $g(2) = t$, $g(3) = g(4) = r$, $g(5) = p$
7. $S = \{a, b, m, n\}$, $T = \{-2, -1, 0, 1, 2\}$, $h(a) = 0$, $h(b) = \pm 1$, $h(m) = 2$, $h(n) = -2$
8. $S = \{-3, -1, 1, 3, 6\}$, $T = \{2, 3, 4, 5, 6, 7, 8\}$, $k(-3) = 6$, $k(-1) = 4$, $k(1) = 6$, $k(3) = 2$, $k(6) = 5$

SETS AND FUNCTIONS

9. $S = \{0, 1/2, 3/4, 1\}$, $T = \{0, 1, 2\}$, $f(0) = 1$, $f(1/2) = 1$, $f(3/4) = 1$, $f(1) = 1$
10. $S = \{a, b, c\}$, $T = \{q, r, s\}$, $g(a) = q$, $g(b) = 3$, $g(c) = s$

In Exercises 11 through 19, let f be the function defined by
$$f(x) = 4x - 2$$
and find the indicated functional value of f.

11. $f(0)$ 12. $f(1)$ 13. $f(-3)$ 14. $f(-1/4)$ 15. $f(5/3)$
16. $f(-2.3)$ 17. $f(b)$ 18. $f(b + 2)$ 19. $f(t^2 - 1)$

In Exercises 20 through 23, let g be the function defined by
$$g(x) = \frac{1}{x}.$$

20. What is the domain of g?
21. Find $g(1)$, $g(4)$, $g(2)$, and $g(100)$.
22. Find $g(-3)$, $g(-9)$, $g(1/2)$, and $g(1/20)$.
23. Find $g(c)$, $g(c^3)$, and $g(1/c)$ if $c \neq 0$.

In Exercises 24 through 27, let h be the function defined by
$$h(x) = x^2 - 3x + 2.$$

24. What is the domain of h?
25. Find $h(0)$, $h(2)$, $h(1)$, and $h(-2)$.
26. Find $h(1/2)$, $h(1.4)$, $h(\sqrt{2})$, and $h(\pi)$.
27. Find $h(a)$, $h(a + 1)$, and $h(a^2)$.

In Exercises 28 through 33, consider the function defined by
$$y = x^2 - 4x + 3$$
and find the value of y for the given value of x.

28. $x = 0$ 29. $x = -3$ 30. $x = 1/2$
31. $x = -3/4$ 32. $x = 12$ 33. $x = q + 1$

In Exercises 34 through 37, consider the function defined by
$$y = \frac{x}{x - 3}.$$

34. What is the domain of this function?
35. Find y when $x = -3$, $x = 5$, $x = 4$, and $x = 0$.
36. Find y when $x = 3/2$, $x = -5/3$, $x = 5/2$, and $x = 7/2$.
37. Find y when $x = b$, $b \neq 3$, and when $x = 1/b$, $b \neq 0$.

In Exercises 38 through 41, let k be the function defined by
$$k(x) = 2\sqrt{x + 1}.$$

38. What is the domain of k?
39. Find $k(0)$, $k(3)$, $k(8)$, and $k(-1)$.

40. Find $k(1)$, $k(2)$, $k(-1/2)$, and $k(-3/4)$.
41. Find $k(c-1)$ if $c > 0$; find $k(c^2 - 1)$.

In Exercises 42 through 45, consider the function defined by the equation

$$p = \frac{2}{\sqrt{q}}.$$

42. What is the domain of this function?
43. Find p when $q = 1, 4, 9, 1/4,$ and $1/100$.
44. Find p when $q = 2, 8, 100,$ and $10,000$.
45. Find p when $q = b$ and when $q = b^2$, $b > 0$.
46. Which of the following equations define y as a function of x?

$$y = 2, \quad 2x + y = 6, \quad x + 2y = 6, \quad y = x^2,$$

$$y^2 = x, \quad \frac{x^2}{4} + \frac{y^2}{16} = 1, \quad y = |x|, \quad |y| = x$$

Management

If a firm makes x units of its product, its cost is

$$C(x) = 20x + 30,000$$

dollars. In Exercises 47 through 50, find the firm's cost if it makes the stated number of units.

47. 0 units
48. 10 units
49. 100 units
50. a units

If a firm produces and sells x units of its product, its profit is

$$P(x) = -x^2 + 30x - 200$$

thousand dollars. In Exercises 51 through 54, find the firm's profit if it makes and sells the stated number of units.

51. 0 units
52. 10 units
53. 15 units
54. 25 units

If the price of a product is p dollars per unit, the quantity q demanded by consumers (that is, the number of units consumers will purchase) is given by

$$q = 200,000 - 8000p.$$

In Exercises 55 through 58, find the quantity demanded at the given price.

55. $1
56. $5
57. $15
58. $25

If a firm makes x units of its product, its average cost per unit is

$$A(x) = 32 + \frac{40,000}{x}$$

dollars. In Exercises 59 through 61, find the firm's average cost per unit if it makes the given number of units.

59. 100 units
60. 10,000 units
61. 1,000,000 units

Life Science

The size of a city's population in year t is given by

$$p(t) = \frac{210t + 400}{t + 2}.$$

Here $t = 0$ represents the year 1988 and $p(t)$ is the population of the city in thousands of individuals. In Exercises 62 through 64, find the population of the city in the given year.

62. 1988 ($t = 0$)   **63.** 1990 ($t = 2$)   **64.** 1994

Social Science

The average score of high-school seniors on a standardized mathematics test over a 10-year period is given by

$$M(t) = 480 - 12t + t^2, \quad 0 \le t \le 10.$$

In Exercises 65 through 68, find the average score on the test at the stated time.

65. The beginning of the 10-year period.

66. The middle of the period.

67. Two-thirds of the way through the period.

68. The end of the period.

Consumers are frequently asked to rank products or services using a **Likert scale** such as

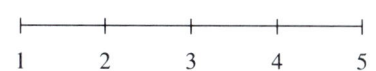

$$\begin{array}{ccccc} \vdash & + & + & + & \dashv \\ 1 & 2 & 3 & 4 & 5 \end{array}$$

Here 1 is the lowest ranking, and 5 is the highest ranking. Suppose, for instance, that students are asked to evaluate a course using this scale. Then the evaluation process can be thought of as a function f from the set of all possible judgments about the course to the interval [1,5]. Thus f will assign to each judgment about the course a single number between 1 and 5, inclusive. Exercises 69 through 71 refer to this evaluation function.

69. Find f(very poor), f(poor), f(average), f(above average), and f(excellent).

70. Find f(halfway between average and above average).

71. Find f(two-thirds of the way from poor toward average).

Computer Science

In an article on computing by Peter Denning in the March-April 1986 issue of *American Scientist*, it is stated that if a computer has n processors and m memory modules and $x = n/m$, then the fraction of modules kept busy per memory cycle is

$$f(x) = 1 + x - \sqrt{x^2 + 1}.$$

Exercises 72 through 77 refer to this situation.

*72. What is the domain of this function? (*Hint: n* and *m* must be natural numbers.)

*73. Find the fraction of memory modules kept busy per memory cycle if a computer has one processor and one memory module.

*74. Repeat Exercise 73 if the computer has one processor and two memory modules.

*75. Suppose a computer has two processors and three memory modules, and then a fourth memory module is added to it. What proportion of memory modules is freed up by this addition?

*76. Suppose a computer has two processors and ten memory modules and an eleventh memory module is added to it. What proportion of memory is freed up by this addition?

*77. Suppose a computer has two processors and five memory modules and another processor is added. How much additional memory is kept busy by this addition?

2.4 GRAPHS

From now on, we will be working almost exclusively with functions from one set of real numbers to another. Therefore let us agree that, unless explicitly stated otherwise, when we say "function" we will mean "function whose domain and range are sets of real numbers."

It is usually possible to depict a function geometrically by **graphing** it. The graph of a function gives us information about the function in a visual way that is easy to understand. For instance, Figure 2–18 is the graph of a function that describes the result of applying an antiseptic to a culture of bacteria. The horizontal scale represents time in hours since the application of the antiseptic, and the vertical scale represents the number of bacteria in the culture, in thousands. We can immediately see from the graph that

- After the application of the antiseptic, the number of bacteria increased for 1 hour and then began to decrease, eventually approaching zero.
- There were 20,000 bacteria in the culture initially.

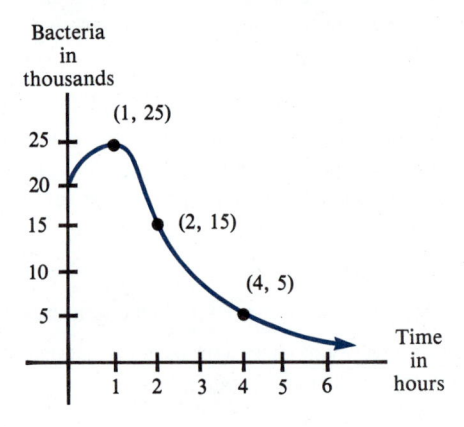

FIGURE 2–18

SETS AND FUNCTIONS 91

- There were 25,000 bacteria in the culture 1 hour after the application of the antiseptic, 15,000 at the 2-hour mark, and 5000 at the 4-hour mark.
- The maximum number of bacteria in the culture was 25,000 at the 1-hour mark.

Of course, before we can analyze the graph of a function in the manner just illustrated, we must be able to draw it. In this section we discuss graphs and graphing.

The process of graphing a function begins with two copies of the real line, one horizontal and one vertical, that intersect at their zero points as in Figure 2–19. The horizontal real line and the vertical real line are called **axes:** the horizontal axis is often called the ***x*-axis,** the vertical one the ***y*-axis.** The two together form the **cartesian coordinate system.** If a and b are any real numbers, the point (a, b) is located on the cartesian coordinate system opposite a on the x-axis and opposite b on the y-axis, as shown in Figure 2–20. The number a is called the ***x*-coordinate** of the point (a, b), and the number b is its ***y*-coordinate.** The point $(0,0)$ where the axes intersect is called the **origin.**

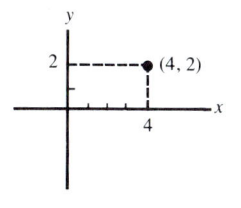

FIGURE 2–21

EXAMPLE 1 The point $(4, 2)$ is opposite 4 on the horizontal axis and opposite 2 on the vertical axis. See Figure 2–21. ◻

It is important to remember that, for a point (a, b), the first coordinate a locates the point relative to the *horizontal* axis and the second coordinate b locates it relative to the *vertical* axis.

The process of locating a point on the cartesian coordinate system, as we did in Example 1, is called **plotting the point.**

EXAMPLE 2 In Figure 2–22, we have plotted the points $(1, 0)$, $(2, 3)$, $(0, 2)$, $(-1, 1)$, $(-3, 0)$, $(-2, -4)$, $(0, -3/2)$, and $(5/2, -5/2)$. ◻

Let f be a function. If a real number x is in the domain of f, we can find its functional value $f(x)$ and form the point $(x, f(x))$.

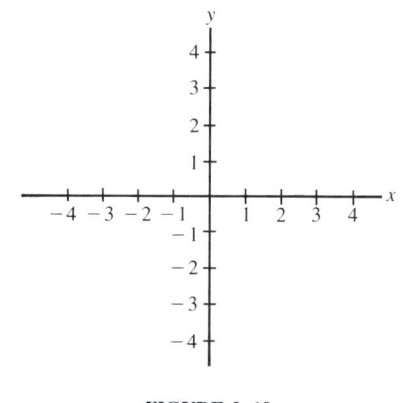

FIGURE 2–19
The Cartesian Coordinate System

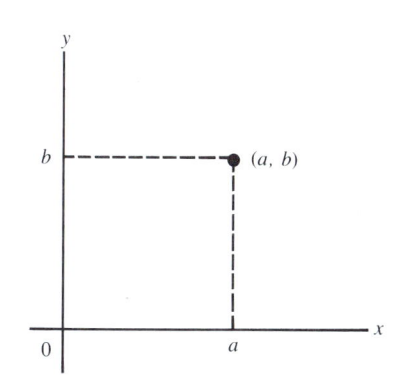

FIGURE 2–20

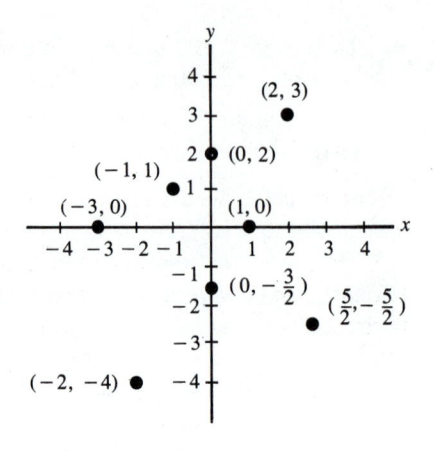

FIGURE 2-22

EXAMPLE 3

Let f be defined by $f(x) = 2x - 1$. If $x = 3$, then $f(3) = 2(3) - 1 = 5$. Thus when $x = 3$,

$$(x, f(x)) = (3, f(3)) = (3, 5). \quad \square$$

The **graph** of a function f is the set of all points $(x, f(x))$ for which x is in the domain of f. To draw the graph of a function, we substitute representative numbers from its domain into its defining equation to obtain points $(x, f(x))$ that lie on the graph; then we plot these points and connect them with a curve. It is also frequently useful to find the **intercepts** of a function. The **y-intercept** of a function f is the point where its graph crosses the y-axis; this is the point $(0, f(0))$. The **x-intercepts** of the function are the points where the graph crosses or touches the x-axis; we find them by solving the equation $f(x) = 0$ for x, if possible.

EXAMPLE 4

Let us graph the function $f(x) = 2x - 1$. Note that since any number can be substituted for x in the equation $f(x) = 2x - 1$, the domain of f is the set **R** of real numbers.

We begin by finding the intercepts of the function f. The y-intercept is the point $(0, f(0)) = (0, -1)$. To find the x-intercepts, we set $f(x) = 0$ and solve for x: we have

$$2x - 1 = 0,$$

or

$$x = 1/2.$$

Therefore the only x-intercept is the point $(1/2, 0)$. Now we select some representative points from the domain of f and find the corresponding points on the graph:

SETS AND FUNCTIONS

x	$f(x) = 2x - 1$	$(x, f(x))$
-2	$f(-2) = -5$	$(-2, -5)$
-1	$f(-1) = -3$	$(-1, -3)$
1	$f(1) = 1$	$(1, 1)$
3	$f(3) = 5$	$(3, 5)$

Plotting these points and the intercepts, we see that the graph of f is the straight line shown in Figure 2–23. ☐

EXAMPLE 5 Suppose that if a firm makes and sells x units of its product, $x \geq 0$, its profit is $P(x) = -x^2 + 60x - 500$ dollars. The domain of this function is the interval $[0, +\infty)$, its y-intercept is $(0, P(0)) = (0, -500)$, and its x-intercepts are the solutions of

$$-x^2 + 60x - 500 = 0,$$

or

$$-(x - 10)(x - 50) = 0.$$

Therefore the x-intercepts are $(10, 0)$ and $(50, 0)$. Now we choose representative numbers x in the interval $[0, +\infty)$ and plot the resultant points $(x, P(x))$ to obtain the graph of Figure 2–24. Since a negative profit is a loss, the graph shows that if the firm makes and sells fewer than 10 or more than 50 units it will suffer a loss, whereas if it makes and sells between 10 and 50 units it will enjoy a gain. Furthermore, if it makes and sells

- 0 units, it will suffer a loss of $500.
- 10 units or 50 units, its profit will be $0 (and thus it will break even).
- 30 units, it will attain its maximum profit of $400. ☐

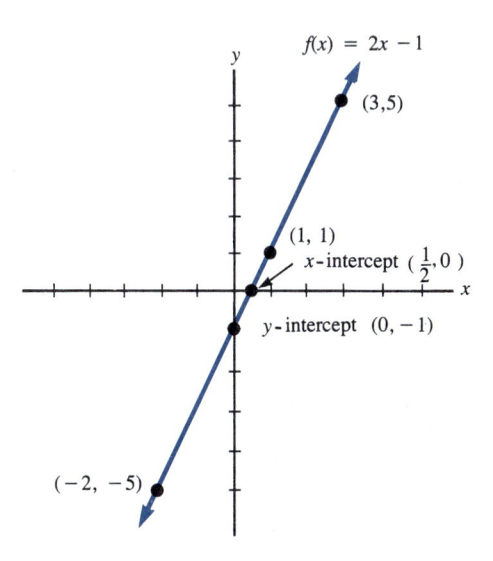

FIGURE 2–23

CHAPTER 2

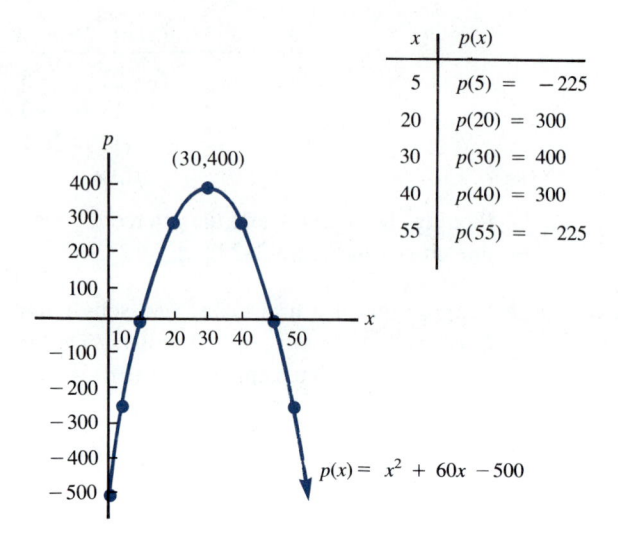

FIGURE 2–24

x	p(x)
5	p(5) = −225
20	p(20) = 300
30	p(30) = 400
40	p(40) = 300
55	p(55) = −225

$p(x) = x^2 + 60x - 500$

A function may have different defining equations on different portions of its domain. When graphing such a function, we must be careful as we select numbers x from its domain to use the defining equation that is appropriate to the choice of x.

EXAMPLE 6   Let us graph the function defined by

$$y = \begin{cases} x/2, & -1 \leq x \leq 2, \\ 3 - x, & 2 < x \leq 4. \end{cases}$$

The domain of this function is the interval $[-1,4]$. If we choose a number x from the domain and $-1 \leq x \leq 2$, then $y = x/2$, while if $2 < x \leq 4$, then $y = 3 - x$. Figure 2–25 shows the graph of this function. ▢

It is often possible to graph relations in the same way as functions. If the relation is described by an equation containing two variables, such as $x^2 + y^2 = 1$, for instance, we choose representative values for x, find the corresponding values for y, and plot the resulting points. Figure 2–26 shows the graph of $x^2 + y^2 = 1$; it is a circle of radius 1 with center at the origin. Note that it is possible for a vertical line to intersect the graph of this relation in more than one point: for example, the y-axis intersects the graph at the points $(0, -1)$ and $(0, 1)$. This means that the single value $x = 0$ is assigned the two numbers 1 and -1 by the equation, and hence the equation does not define y as a function of x. Thus an examination of the graph of an equation can tell us whether the equation defines a function.

SETS AND FUNCTIONS   95

x	$y = x/2$
-1	$-\frac{1}{2}$
0	$\frac{0}{2} = 0$
1	$\frac{1}{2}$
2	$\frac{2}{2} = 1$

x	$y = 3 - x$
$\frac{5}{2}$	$3 - \frac{5}{2} = \frac{1}{2}$
3	$3 - 3 = 0$
$\frac{7}{2}$	$3 - \frac{7}{2} = -\frac{1}{2}$
4	$3 - 4 = -1$

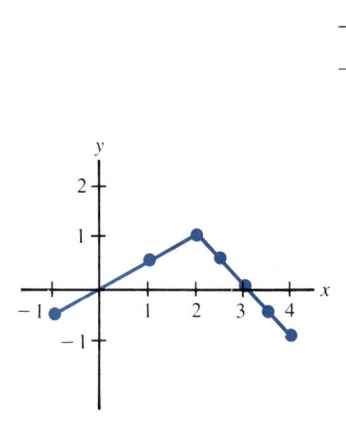

FIGURE 2–25

The Vertical Line Test

A graph is the graph of a function if and only if no vertical line intersects it in more than one point.

EXAMPLE 7 The graph of Figure 2–27 is the graph of a function, but that of Figure 2–28 is not. ☐

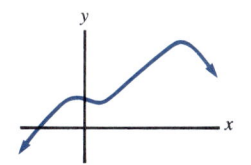

FIGURE 2–27
This Is the Graph of a Function

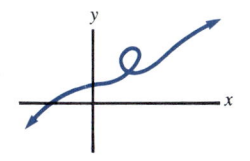

FIGURE 2–28
This Is Not the Graph of a Function

x	$y = \pm\sqrt{1-x^2}$	(x, y)
0	± 1	$(0, \pm 1)$
$\pm\frac{1}{4}$	$\pm\frac{\sqrt{15}}{4}$	$(\pm\frac{1}{4}, \pm\frac{\sqrt{15}}{4})$
$\pm\frac{1}{2}$	$\pm\frac{\sqrt{3}}{2}$	$(\pm\frac{1}{2}, \pm\frac{\sqrt{3}}{2})$
$\pm\frac{3}{4}$	$\pm\frac{\sqrt{7}}{4}$	$(\pm\frac{3}{4}, \pm\frac{\sqrt{7}}{4})$
± 1	0	$(\pm 1, 0)$

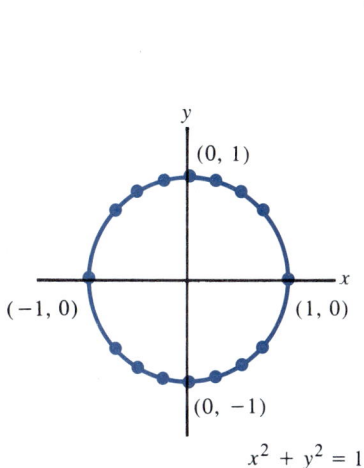

$x^2 + y^2 = 1$

FIGURE 2–26

We must remark that graphing by plotting points is both tedious and potentially unreliable. (Potentially unreliable because we cannot always be sure that we have plotted enough points to give a true picture of the graph.) In the next chapter we will show how to graph some rather simple functions quickly and accurately, but truly efficient methods of graphing require the use of calculus. We will consider the applications of calculus to graphing in Chapter 12.

Graphs often exhibit **symmetries:** Figure 2–29 shows a graph that is **symmetric with respect to the y-axis.** Note that if the portion of the graph to the right of the y-axis is rotated around the y-axis, the two halves of the graph will coincide. Similarly, Figure 2–30 shows a graph that is **symmetric with respect to the x-axis,** and Figure 2–31 one that is **symmetric with respect to the origin.**

As Figure 2–29 indicates, if the graph of an equation is symmetric with respect to the y-axis, then the points (x, y) and $(-x, y)$ must both lie on the graph. But this implies that replacing x by $-x$ in the equation must leave it unchanged. Thus we can test an equation to see if its graph will be symmetric with respect to the y-axis by substituting $-x$ for x in the equation; if the equation remains unchanged, the graph will indeed be symmetric with respect to the y-axis. Similar results hold for symmetry with respect to the x-axis and for symmetry with respect to the origin.

Keys to Symmetries

1. If substituting $-x$ for x in an equation does not change the equation, its graph will be symmetric with respect to the y-axis.

2. If substituting $-y$ for y in an equation does not change the equation, its graph will be symmetric with respect to the x-axis.

3. If substituting $-x$ for x and $-y$ for y in an equation does not change the equation, its graph will be symmetric with respect to the origin.

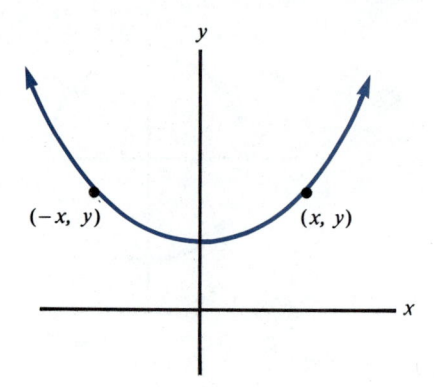

FIGURE 2–29
Symmetry with Respect to the y-Axis

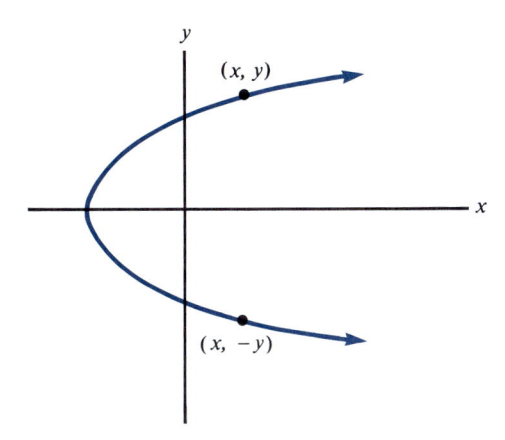

FIGURE 2-30
Symmetry with Respect to the *x*-Axis

EXAMPLE 8 The graph of $y = x^2$ is symmetric with respect to the *y*-axis (see Figure 2–32) because substituting $-x$ for x in the equation results in $y = (-x)^2 = x^2$. The graph of $y^2 = x$ is symmetric with respect to the *x*-axis (see Figure 2–33) because substituting $-y$ for y in the equation results in $(-y)^2 = x$, or $y^2 = x$. The graph of $y = x^3$ (see Figure 2–34) is symmetric with respect to the origin because substituting $-x$ for x and $-y$ for y in the equation results in $-y = (-x)^3$, or $-y = -x^3$, or $y = x^3$. The graph of $x^2 + y^2 = 1$ is symmetric with respect to the *x*-axis, the *y*-axis, and the origin; see Figure 2–26. □

Some functions have the property of always assigning different numbers in their range to different numbers in their domain. Such functions are said to be **one-to-one.**

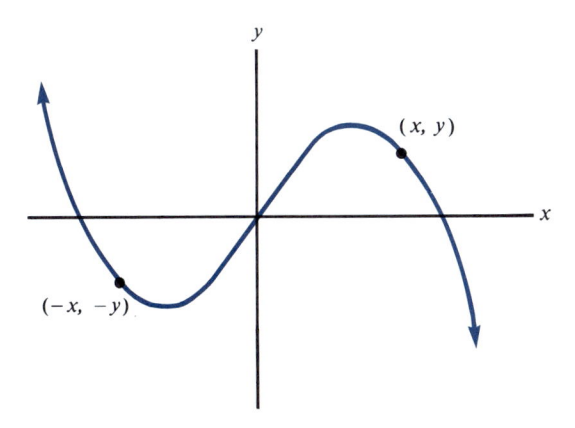

FIGURE 2-31
Symmetry with Respect to the Origin

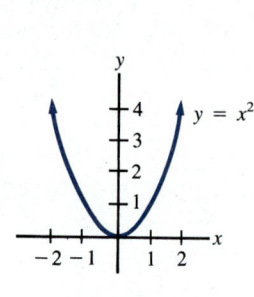

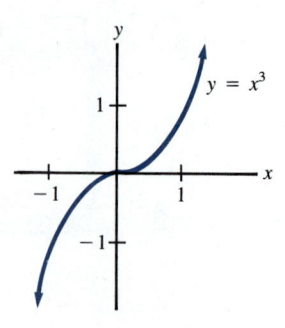

FIGURE 2–32

FIGURE 2–33

FIGURE 2–34

> ### One-to-One Function
> A function f is **one-to-one** if $f(x_1) \neq f(x_2)$ whenever $x_1 \neq x_2$.

Figure 2–35 shows the graph of a function that is one-to-one; Figure 2–36 shows the graph of a function that is not one-to-one.

EXAMPLE 9   The function $f(x) = 2x$ is one-to-one, because if $x_1 \neq x_2$ then

$$f(x_1) = 2x_1 \neq 2x_2 = f(x_2).$$

The function $g(x) = x^2$ is not one-to-one because $-2 \neq 2$, but

$$g(-2) = 4 = g(2). \quad \square$$

If a function f is one-to-one, we can define a function f^{-1} called the **inverse of f**.

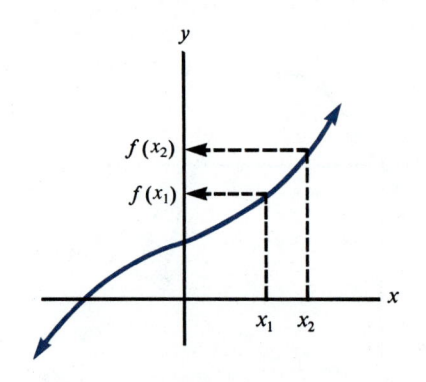

FIGURE 2–35
A One-to-One Function

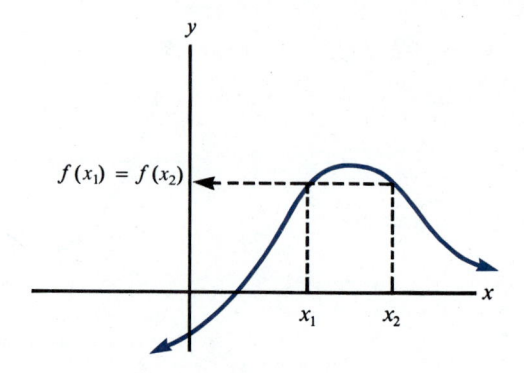

FIGURE 2–36
A Function that Is Not One-to-One

SETS AND FUNCTIONS

> **Inverse Function**
>
> Let f be a one-to-one function. The **inverse function of f** is the function f^{-1} defined as follows: for each y in the range of f,
>
> $$f^{-1}(y) = x \quad \text{if and only if} \quad f(x) = y.$$

See Figure 2–37. Note that the domain of f^{-1} is the range of f, and the range of f^{-1} is the domain of f. Also, a point (x, y) is on the graph of f if and only if $y = f(x)$; but this is so if and only if $x = f^{-1}(y)$, and thus if and only if $(y, f^{-1}(y)) = (y, x)$ is on the graph of f^{-1}. Therefore the graphs of f and its inverse f^{-1} are **symmetric about the line $y = x$**. See Figure 2–38.

If f is a one-to-one function, we can find the defining equation for its inverse function f^{-1} by setting $y = f(x)$, solving this equation for x in terms of y, and interchanging the variables x and y. We illustrate with an example.

EXAMPLE 10   Let $f(x) = 2x$. Since f is a one-to-one function, it has an inverse function f^{-1}. We set $y = 2x$ and solve this equation for x in terms of y, thus obtaining $x = y/2$. If we interchange x and y in this last equation, we have $y = x/2$. Thus the defining equation for f^{-1} is $f^{-1}(x) = x/2$. Figure 2–39 shows the graphs of f and f^{-1}: note the symmetry about the line $y = x$. ☐

We conclude by emphasizing that the inverse function f^{-1} is defined *only* if f is one-to-one. This is so because if f is *not* one-to-one, then there are numbers x_1 and x_2, $x_1 \neq x_2$, such that $f(x_1) = f(x_2)$. Thus if $y = f(x_1) = f(x_2)$, then we would have to define $f^{-1}(y) = x_1$ and $f^{-1}(y) = x_2$, and hence f^{-1} would not be a function.

■ *Computer Manual: Program FNCTNTBL*

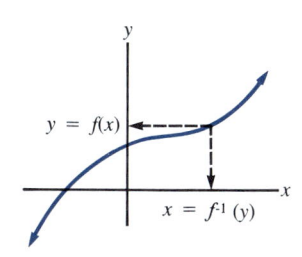

FIGURE 2–37
The Inverse Function

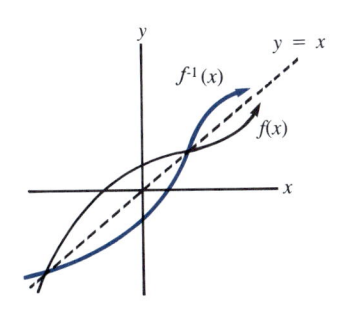

FIGURE 2–38

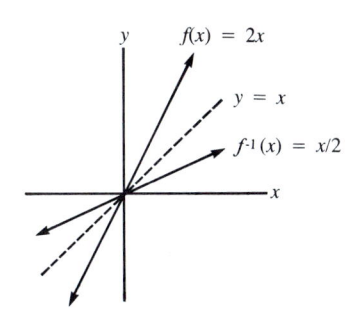

FIGURE 2–39

EXERCISES 2.4

1. Plot the following points on the cartesian coordinate system.
   (a) $(3, 0)$, $(9/2, 0)$, $(-2, 0)$, $(-5.75, 0)$
   (b) $(0, 4)$, $(0, -1)$, $(0, -2.3)$, $(0, 8/3)$
   (c) $(1, 2)$, $(2, 1)$, $(7/3, 3)$, $(3/2, 5/2)$, $(-2, 4)$, $(-3, 3)$, $(-5, 7/2)$, $(-5/4, 12/5)$, $(-2, -2)$, $(-6, -2)$, $(-0.5, -1)$, $(-4.5, -7/3)$, $(3, -7)$, $(5, -2)$, $(1, -2.8)$, $(2/3, -1/3)$
   (d) $(\sqrt{2}, 0)$, $(-\pi, 3)$, $(0, \sqrt{3})$, $(\sqrt{5}, -\sqrt{5})$, $(\pi/2, 3\pi/2)$, $(-2\sqrt{3}, -3\sqrt{2})$

In Exercises 2 through 11, graph the function f defined by the given equation. In each case find the intercepts of the function.

2. $f(x) = 3x$

3. $f(x) = 2x - 3$

4. $f(x) = -x^2$

5. $f(x) = x^2 + 2$

6. $f(x) = x^3 + 1$

7. $f(x) = \sqrt{x}$

8. $f(x) = \begin{cases} 1, & x < 2, \\ 3, & x \geq 2 \end{cases}$

9. $f(x) = \begin{cases} x, & 0 \leq x \leq 2, \\ 5, & x > 2 \end{cases}$

10. $f(x) = \begin{cases} -1, & x < 0, \\ 1, & 0 \leq x \leq 1, \\ x, & x > 1 \end{cases}$

11. $f(x) = \begin{cases} 2x, & 0 \leq x \leq 2, \\ 6 - x, & 2 < x \leq 6 \end{cases}$

In Exercises 12 through 17, state whether the graph shown is the graph of a function.

12.

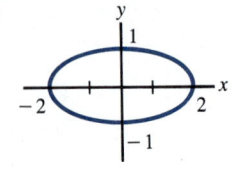

13.

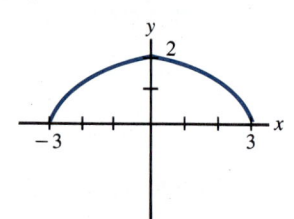

14.

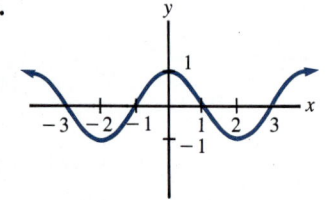

15.

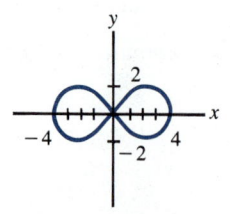

16.

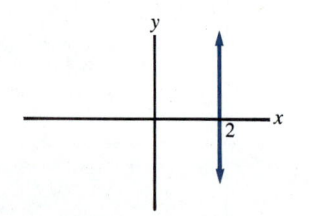

17.

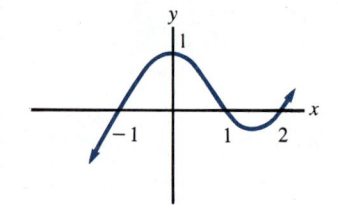

In Exercises 18 through 23, graph the given equation and state whether the equation defines y as a function of x.

SETS AND FUNCTIONS

18. $y = -2x$ **19.** $y = |x|$ **20.** $x^2 + y^2 = 4$ **21.** $4x^2 + y^2 = 4$
22. $y^2 = x + 1$ **23.** $x^2 = y + 1$
*24. For any real number x, define $[x]$ to be the largest integer less than or equal to x.
   (a) Find $[0]$, $[2]$, $[8]$, $[1000]$, $[-3]$, $[-7]$, $[-99]$.
   (b) Find $[1/2]$, $[8/5]$, $[-2/3]$, $[-22/7]$, $[510/509]$, $[509/510]$, $[-510/509]$, $[-509/510]$.
   (c) Find $[0.2]$, $[6.99]$, $[-4.26]$, $[18.01]$, $[-0.45]$, $[-12.98]$, $[-3.01]$.
   (d) Define the **greatest integer function** f by $f(x) = [x]$. Graph the greatest integer function.

In Exercises 25 through 30, state the symmetries of the given graph.

25.

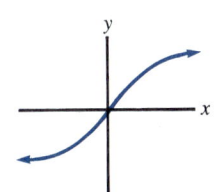

26.

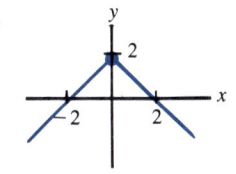

27.

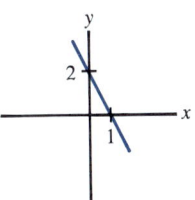

28.

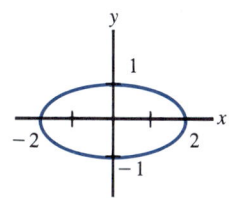

29.

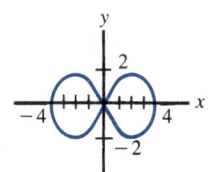

30.

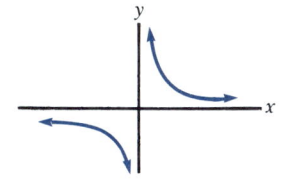

In Exercises 31 through 38, find the symmetries of the graph of the given equation.

31. $y = -x$ **32.** $y = 3x + 1$ **33.** $y = 2 - x^2$ **34.** $y = x^2 + x$
35. $y = x^3 + x$ **36.** $y^2 = x^2$ **37.** $y^2 = 4 - 4x^2$ **38.** $y = |x|$

In Exercises 39 through 44, state whether the graph is that of a one-to-one function; if it is, sketch the graph of its inverse function.

39.

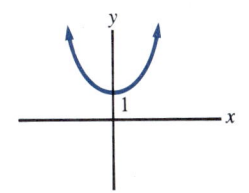

40.

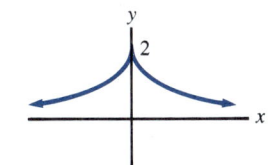

41.

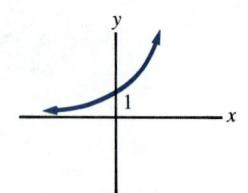

42.

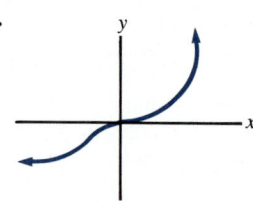

43.

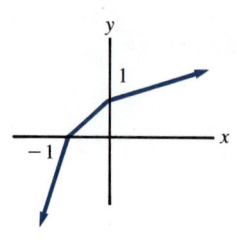

44.

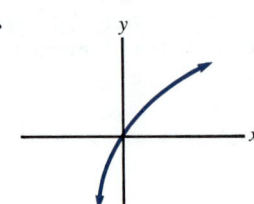

In Exercises 45 through 52, state whether the function f is one-to-one; if it is, find the defining equation for its inverse function f^{-1}.

45. $f(x) = 3x - 1$ **46.** $f(x) = 3 - 4x$ **47.** $f(x) = -x$ **48.** $f(x) = x^2 - 1$
49. $f(x) = x^3$ **50.** $f(x) = 2$ **51.** $f(x) = \sqrt{x}$ **52.** $f(x) = 1/x$

Management

53. The accompanying graph depicts the total cost to Alfa Company if it produces x units of its product.

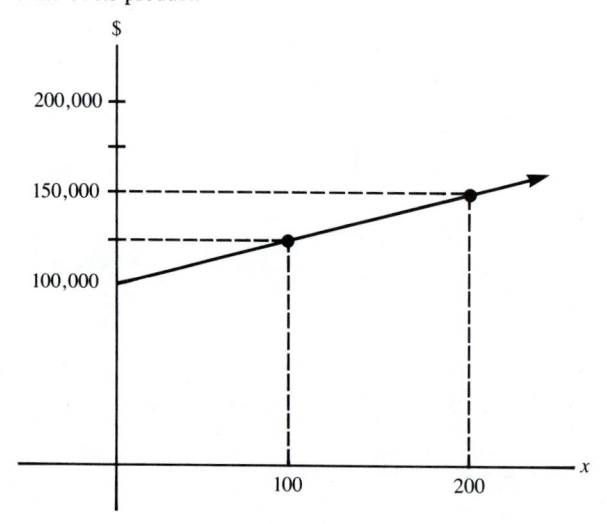

(a) What is the total cost to Alfa if it produces 0 units of its product?
(b) What is the total cost to Alfa if it produces 100 units of its product?
(c) How many units can Alfa produce for a total cost of $150,000?
(d) As the number of units produced increases, what happens to Alfa's total cost?

54. The accompanying graph depicts Bayta Corporation's average cost per unit if it makes x units of its product.

SETS AND FUNCTIONS   103

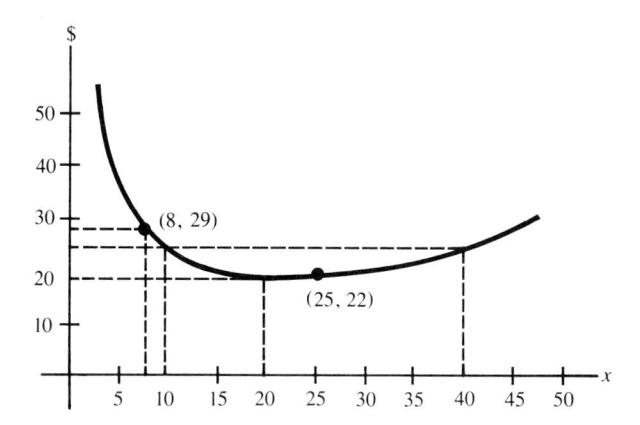

(a) What is Bayta's average cost per unit if it makes 8 units?
(b) What is Bayta's average cost per unit if it makes 25 units?
(c) At what number of units is Bayta's average cost per unit a minimum? What is Bayta's minimum average cost per unit?
(d) How many units must Bayta produce if its average cost per unit is to be $25?
(e) What can you say about Bayta's average cost per unit as the number of units made increases?

55. The accompanying graph depicts the profit of Ghamma Industries, in millions of dollars, if it makes and sells x thousand units of its product.

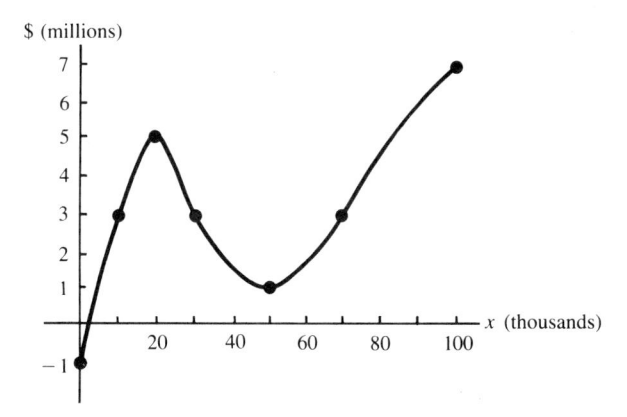

(a) What is Ghamma's profit if it makes and sells 0 units?
(b) What is Ghamma's profit if it makes and sells 20,000 units?
(c) What is Ghamma's profit if it makes and sells 50,000 units?
(d) How many units must Ghamma make and sell if its profit is to be $3 million?
(e) What is Ghamma's maximum profit? How many units must Ghamma make and sell in order to attain its maximum profit?
(f) What can you say about Ghamma's profit as the number of units produced and sold increases from 0 to 100,000?

56. The accompanying graph depicts the average annual sales, in thousands of dollars, of Delda Stores salespeople who have been given t weeks of sales training.

$ (thousands)

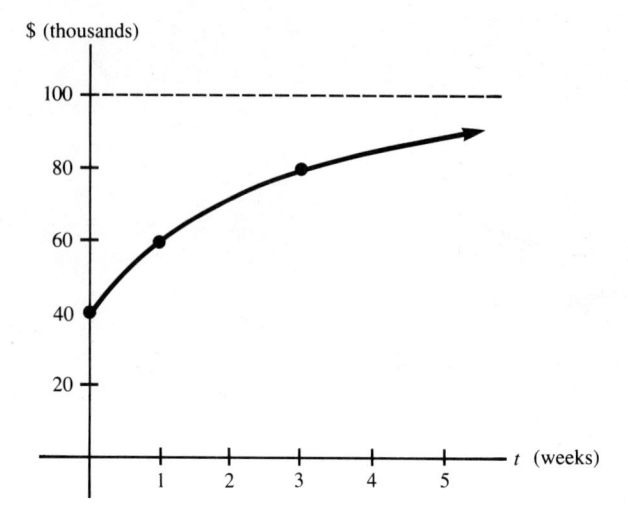

```
100 ┤- - - - - - - - - - - - - - - - -
 80 ┤
 60 ┤
 40 ┤
 20 ┤
    └──┼──┼──┼──┼──┼──── t (weeks)
       1  2  3  4  5
```

(a) What are average annual sales for salespeople who received no training?
(b) What are average annual sales for salespeople who received one week of training?
(c) How many weeks of training must Delda give its salespeople if it wants their average annual sales to be at least $80,000?
(d) What can you say about average annual sales as the number of weeks of training increases?

*57. A firm sells an industrial cleaner at a price of $2 per gallon for the first gallon or part of a gallon, and $1.50 per gallon for each additional gallon or part of a gallon. Let $C(x)$ denote the cost of purchasing x gallons, $x > 0$.
(a) Graph C.
(b) Can you find an equation (or a set of equations) that explicitly defines C as a function of x?

*58. According to information in a recent issue of *The Wall Street Journal*, a discount stockbroker charges a fee of

- $25 plus $0.10 per share traded on trades of fewer than 100 shares.
- $30 plus $0.05 per share traded on trades of at least 100 but fewer than 500 shares.
- $45 plus $0.02 per share traded on trades of 500 or more shares.

Let $B(x)$ be the broker's total fee for executing a trade of x shares.
(a) Graph B.
(b) Can you find an equation (or a set of equations) that explicitly defines B as a function of x?

Life Science

59. The accompanying graph depicts the proportion of bacteria alive in a culture t minutes after it was treated with a bactericide.

SETS AND FUNCTIONS 105

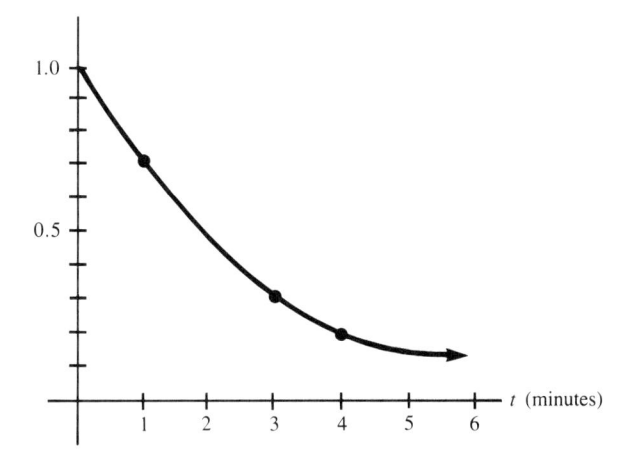

(a) What proportion of the bacteria were alive 1 minute after the application of the bactericide?
(b) What proportion of the bacteria were alive 3 minutes after the application of the bactericide?
(c) How long did it take for the bactericide to kill 80% of the bacteria?
(d) What can you say about the number of bacteria alive in the culture as the time since the bactericide was applied increased?

60. The accompanying graph depicts the planned cleanup of a polluted lake: it shows the concentration of pollutants in the lake, in parts per million (ppm), that are expected to be present t years after the cleanup begins.

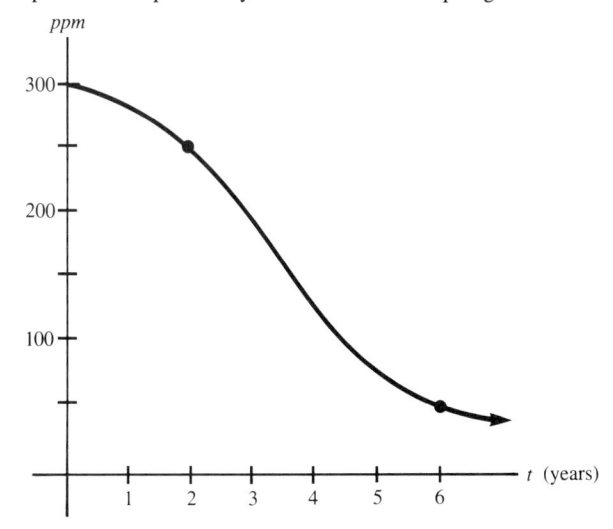

(a) What is the expected concentration of pollutants at the beginning of the cleanup?
(b) What is the expected concentration of pollutants two years after the beginning of the cleanup?
(c) How long will it take until the expected concentration of pollutants is 50 ppm?

(d) What can you say about the concentration of pollutants as the cleanup proceeds?

61. The accompanying graph depicts a patient's temperature t hours after the onset of a disease.

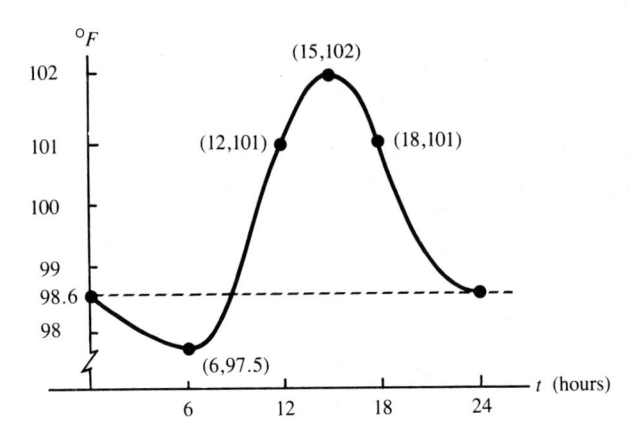

(a) What was the patient's temperature at the onset of the disease?
(b) What was the patient's temperature 6 hours after the onset of the disease?
(c) When was the patient's temperature 101°F?
(d) What was the patient's minimum temperature, and when did it occur?
(e) What was the patient's maximum temperature, and when did it occur?
(f) What can you say about the patient's temperature during the progress of the disease?

*62. The *Physician's Desk Reference* gives the following dosage schedule for an oral suspension of erythromycin:

Body Weight (lb)	Dosage (mg/day)
(0,10)	15 per pound
[10,15)	200
[15,25)	400
[25,50)	800
[50,100)	1200
[100, +∞)	1600

Let $d(x)$ be the dosage for a patient whose weight is x pounds.
(a) Graph the function d.
(b) Can you find an equation (or a set of equations) that explicitly defines d as a function of x?

Social Science

63. The accompanying graph depicts the percentage of children in a group who adopted a new fad within t weeks after its first appearance.

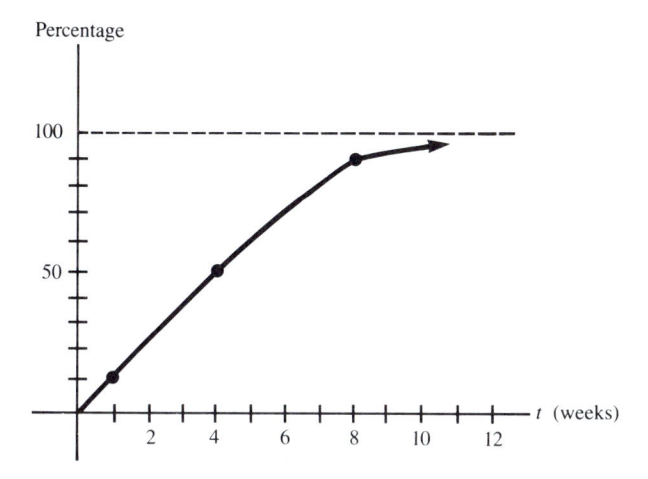

(a) What percentage of the group adopted the fad within one week of its first appearance?
(b) What percentage of the group adopted the fad within eight weeks of its first appearance?
(c) How long did it take until 50% of the group had adopted the fad?
(d) What can you say about the acceptance of the fad by the group?

SUMMARY

This summary consists of terms and symbols whose meaning you should know, facts you should know, and techniques and methods you should be able to employ.

Section 2.1

Set, element, $\in$, $\notin$, { }. Roster method, descriptive method, set-builder method, $\{x \mid x$ satisfies property $P\}$. Writing sets using roster, descriptive, set-builder methods. Set equality. Set containment, subset, $\subset$, $\not\subset$. Empty set, $\emptyset$. Universal set. Venn diagram. Drawing Venn diagrams.

Section 2.2

Complement, $'$. Venn diagram for complementation. Complement as "not." Union, $\cup$. Venn diagram for union. Union as "either . . . or." Intersection, $\cap$. Venn diagram for intersection. Intersection as "and." Translating English phrases into set language using $'$, $\cup$, $\cap$. Disjoint or mutually exclusive sets. Using Venn diagrams to solve counting problems.

Section 2.3

Function, domain, range. Relation, function as relation, not every relation is a function. Functional notation $f(x) = y$. Independent variable. Functional value $f(x)$, distinction between function and its functional values. Finding domains of functions from the reals to the reals. Dependent variable, function defined by an equation involving an independent and a dependent variable. Not every equation defines a function.

Section 2.4

x-axis, y-axis, cartesian coordinate system, origin $(0,0)$. Point (a,b), x-coordinate, y-coordinate, plotting points. Graph of a function. Intercepts of a function: y-intercept, x-intercept. Finding intercepts. Graphing by plotting points. Graphing functions having different defining equations on different intervals. Graphing relations. The Vertical Line Test. Symmetries: with respect to the y-axis, with respect to the x-axis, with respect to the origin. Finding symmetries. One-to-one function. Inverse of a one-to-one function: f^{-1}. Symmetry of the graph of f^{-1} about the line $y = x$. Finding the equation of the inverse function. Analyzing the graph of a function that describes an applied situation.

REVIEW EXERCISES

Let

$A = \{x|x \in \mathbf{N}, x > 6\}$, $\quad B = \{x|x \in \mathbf{Z}, x^2 > 36\}$,
$C = \{x|x \in \mathbf{Z}, x \geq 7\}$, $\quad D = \{x|x \in \mathbf{R}, x \geq 7\}$,
$E = \{x|x \in \mathbf{Z}, 5x = 100\}$, $\quad F = \{x|x \in \mathbf{Z}, 5x = 101\}$,

In Exercises 1 through 15, determine whether the given statement is true or false.

1. $6 \in A$
2. $7 \in B$
3. $15/2 \in C$
4. $\sqrt{89} \in D$
5. $A = B$
6. $A = C$
7. $B \subset C$
8. $B \subset D$
9. $C \subset D$
10. $E = \emptyset$
11. $F = \emptyset$
12. $E \subset F$
13. $F \subset E$
14. D is a universal set for A, B, C, E, and F.
15. $\mathbf{R}$ is a universal set for A, B, C, D, E, and F.

In Exercises 16 and 17, specify the given set using the roster method.

16. The set A consisting of the first five letters of the English alphabet.
17. $C = \{x|x \in \mathbf{N}, x \geq 10\}$

In Exercises 18 and 19, specify the given set using the descriptive method.

18. $B = \{0, 2, 4, 6, \ldots\}$
19. $C = \{x|x \in \mathbf{Z}, x^2 > 100\}$

SETS AND FUNCTIONS

In Exercises 20 and 21, specify the given set using the set-builder method.

20. The set A consisting of the states of the United States.
21. $C = \{-5, -4, -3, -2, -1, 0, 1, 2\}$
22. Let $A = \{a, b, d, e\}$, $B = \{b, c, d, f, z\}$, $C = \{a, c, d, f, h\}$, and $\mathcal{U} = \{a, b, c, d, e, f, g, h, z\}$. Draw and label the Venn diagram.

Let A, B, C, and $\mathcal{U}$ be as in Exercise 22. In Exercises 23 through 31, specify the given set using the roster method.

23. A'
24. $B \cap C$
25. $A \cup B$
26. $A \cap B \cap C$
27. $A \cup B \cup C$
28. $A' \cap C$
29. $A' \cup B$
30. $A \cap (B \cup C)'$
31. $A' \cup (B \cap C')$

Let
$$\mathcal{U} = \{x \mid x \text{ is a registered voter in Ohio}\},$$
$$S = \{x \mid x \in \mathcal{U}, x \text{ is a registered Democrat}\},$$
$$T = \{x \mid x \in \mathcal{U}, x \text{ is less than 30 years old}\},$$
$$V = \{x \mid x \in \mathcal{U}, x \text{ is female}\}.$$

In Exercises 32 through 40, specify each set using the set-builder method.

32. S'
33. $S \cap T$
34. $S \cup V$
35. $S \cup T \cup V$
36. $S \cap T \cap V$
37. $S' \cup T$
38. $S \cap V'$
39. $(S \cap T \cap V)'$
40. $S' \cap (T \cup V')$

Exercises 41 through 44 refer to the following information: Four hundred voters were surveyed and it was found that

- 222 of them were conservatives.
- 120 were Republicans.
- 146 were male.
- 16 were Republican males who were not conservatives.
- 110 were male conservatives.
- 32 were female, Republican, and conservative.
- 40 were male, Republican, and conservative.

41. How many female Republicans were surveyed?
42. How many males surveyed were neither Republican nor conservative?
43. How many females were surveyed?
44. How many males were surveyed who were conservative but not Republican?
45. Is the following rule a function? If so, what are its domain and range?

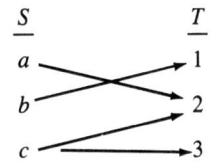

$$
\begin{array}{cc}
S & T \\
a & 1 \\
b & 2 \\
c & 3
\end{array}
$$

46. Same questions as in Exercise 45 for

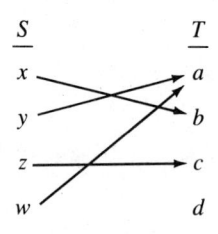

```
     S              T
     x              a
     y              b
     z              c
     w              d
```

Let functions f, g, and h be defined by

$$f(x) = 5 - 3x, \quad g(x) = 2 - x^2, \quad h(x) = \frac{x+1}{x-2}.$$

In Exercises 47 through 60, find the indicated functional value.

47. $f(0)$ **48.** $f(2)$ **49.** $f(4/3)$ **50.** $f(-a)$
51. $g(4)$ **52.** $g(-4)$ **53.** $g(\sqrt{2})$ **54.** $g(\sqrt{b}), b \geq 0$
55. $g(1/\sqrt{b}), b > 0$ **56.** $h(3)$ **57.** $h(0)$ **58.** $h(2)$
59. $h(1/2)$ **60.** $h(t/2)$

61. Let $y = 0.5\sqrt{x+4}$. Find y when $x = 0, 5, 11,$ and 21.
62. Repeat Exercise 61 when $x = -4, -3, -2, 9/16,$ and 4.
63. Let $y = |3x + 2|$. Find y when $x = -3, -1, -2/3, 0, 2/3, 1,$ and 3.
64. Let $y = x^{1/3}$. Find y when $x = 0, 1, -1, 8, -27, 9,$ and -16.
65. Let

$$y = \begin{cases} 0 & \text{if } x \leq -1, \\ x + 1 & \text{if } -1 < x < 2, \\ 3 & \text{if } x \geq 2. \end{cases}$$

Find y when $x = -2, -1, -1/2, 0, 1, 9/5, 2, 13/6,$ and 3.

In Exercises 66 through 70, find the domain of the given function and then graph it.

66. $f(x) = 5 - 3x$ **67.** $g(x) = 2 - x^2$ **68.** $y = 0.5\sqrt{x+4}$ **69.** $y = x^{1/3}$

70. $y = \begin{cases} 0 & \text{if } x \leq -1, \\ x + 1 & \text{if } -1 < x < 2, \\ 3 & \text{if } x \geq 2 \end{cases}$

In Exercises 71 through 74, find the symmetries of the graph of the given equation.

71. $y = 1$ **72.** $y = x^{1/3}$ **73.** $y = x^4$ **74.** $y^2 - x^2 = 1$

In Exercises 75 through 78, state whether the given function is one-to-one; if it is, find a defining equation for its inverse function.

75. $f(x) = 1$ **76.** $f(x) = x^{1/3}$
77. $y = x^4$ **78.** $y = ax + b$, a, b real, $a \neq 0$

ADDITIONAL TOPICS

Here are some suggestions for topics not covered in the text that you might want to investigate on your own.

1. More formal definitions of the concepts of relation and function use the **cartesian product** of two sets. Find out what the cartesian product of two sets is, and how it can be used to define the notions of relation and function. How are these definitions using the cartesian product related to our definitions of relation and function?

2. Find out what an **equivalence relation** on a set is, and show that every equivalence relation on a set partitions the set into disjoint subsets.

3. Points can be plotted by using **polar coordinates** as well as cartesian coordinates; polar coordinates locate points using angles and distances from the origin. Find out how this is done.

4. **Parabolas, ellipses, and hyperbolas** are curves known collectively as **conic sections.** They are the graphs of certain equations. Find out about conic sections and the equations that give rise to them.

5. René Descartes (1596–1650) was a French mathematician and philosopher who discovered the coordinate system named after him (the cartesian coordinate system), invented the mathematical subject of analytic geometry, and founded the philosophical school known as cartesianism. Find out about Descartes' life and his contributions to mathematics and philosophy.

SUPPLEMENT: FUZZY SETS

A set must be **well-defined:** this means that the set's definition must be precise and unambiguous so that, for any object x, it is possible to decide whether x belongs to the set. To illustrate, consider

$$A = \{x \mid x \in \mathbf{N}, x < 5\}$$

and

$$B = \{x \mid x \in \mathbf{N}, x \text{ is small}\}.$$

The set A is well-defined: for each natural number x, we can tell whether or not $x \in A$. However, B is not well-defined, because it is impossible to decide whether some natural numbers belong to B or not. For instance, does $10 \in B$? Does $100 \in B$? Since B is not well-defined, it is not really a set.

The reason that B is not well-defined is, of course, the ambiguity of the word "small" as it applies to natural numbers. In many cases where we would like to apply set theory to practical matters, we run into similar problems of ambiguity. For instance, we might be interested in collections such as

- The days last week that had good weather.
- All new movies.
- The real numbers far from zero.

Since these are not well-defined collections (due to the ambiguity of the words "new" and "good" and the phrase "far from"), they are not sets, and we cannot apply the results of set theory to them. However, a recently developed companion to set theory, called **fuzzy set theory,** does allow us to handle ambiguously defined collections such as these. In this supplement we describe fuzzy sets and fuzzy set operations.

A **fuzzy set** is just a function from a universal set to the interval [0,1] on the real line.

> **Fuzzy Sets**
>
> Let $\mathcal{U}$ be a universal set. Any function f with domain $\mathcal{U}$ and range contained in the interval [0, 1] on the real line is called a **fuzzy set in** $\mathcal{U}$.

We will explain later why such functions are called fuzzy sets.

EXAMPLE 1 Let $\mathcal{U} = \{a, b, c, d\}$. We define a fuzzy set f in $\mathcal{U}$ by setting

$$f(a) = 0, \quad f(b) = 1/4, \quad f(c) = 1/2, \quad f(d) = 1.$$

Similarly,

$$g(a) = g(b) = 0, \qquad g(c) = g(d) = 1$$

defines another fuzzy set in $\mathcal{U}$. ☐

EXAMPLE 2 Let $\mathcal{U} = \mathbf{N}$ and define h by $h(x) = 1/x$ for all $x \in \mathbf{N}$.

Since $1/x \in [0, 1]$ for every $x \in \mathbf{N}$, h is a fuzzy set in $\mathbf{N}$. ☐

Although a fuzzy set in $\mathcal{U}$ is a function and not a subset of $\mathcal{U}$, it does partition $\mathcal{U}$ into three mutually disjoint subsets called the **full members,** the **nonmembers,** and the **partial members** of the fuzzy set.

> **Fuzzy Set Membership**
>
> Let $\mathcal{U}$ be a universal set, and let f be a fuzzy set in $\mathcal{U}$. Then
>
> 1. $\{x \mid x \in \mathcal{U}, f(x) = 1\}$ is the set of **full members of** f.
> 2. $\{x \mid x \in \mathcal{U}, f(x) = 0\}$ is the set of **nonmembers of** f.
> 3. $\{x \mid x \in \mathcal{U}, 0 < f(x) < 1\}$ is the set of **partial members of** f. The closer $f(x)$ is to 1, the closer x is to full membership in f.

EXAMPLE 3 The fuzzy set f of Example 1 has d as a full member, a as a nonmember, and b and c as partial members, with c closer to full membership in f than b. The fuzzy

SETS AND FUNCTIONS

set g of the same example has a and b as nonmembers and c and d as full members; it has no partial members. ☐

EXAMPLE 4 The membership sets for the fuzzy set h of Example 2 are as follows:

$$\{\text{full members}\} = \{x \mid x \in \mathbf{N}, h(x) = 1\} = \{x \mid x \in \mathbf{N}, 1/x = 1\} = \{1\},$$

$$\{\text{nonmembers}\} = \{x \mid x \in \mathbf{N}, h(x) = 0\} = \{x \mid x \in \mathbf{N}, 1/x = 0\} = \emptyset,$$

$$\{\text{partial members}\} = \{x \mid x \in \mathbf{N}, 0 < h(x) < 1\} = \{x \mid x \in \mathbf{N}, 0 < 1/x < 1\}$$
$$= \{x \mid x \in \mathbf{N}, x > 1\}.$$

Furthermore, since $1/x$ gets smaller as x gets larger, the larger a natural number is, the further it is from full membership in h. ☐

Now suppose that we are interested in a poorly defined collection C of elements of $\mathcal{U}$ such as those mentioned at the beginning of this supplement. In such a case, there will usually be some elements of $\mathcal{U}$ that definitely belong to C, some that definitely do not belong to C, and some whose status with regard to C is ambiguous, or "fuzzy." Then we may be able to find a fuzzy set whose full members are the elements that definitely belong to C, whose nonmembers are the elements that definitely do not belong to C, and whose partial members are the "fuzzy" elements. (This is the reason for the name "fuzzy set.") We illustrate with examples.

EXAMPLE 5 Consider the collection of days last week that had good weather. Suppose the weather last week was as follows:

- Sunday, Monday: clear and sunny.
- Tuesday: partly cloudy.
- Wednesday, Thursday: cloudy.
- Friday, Saturday: rainy.

In order to write this collection as a fuzzy set, we must find a function f from

$$\mathcal{U} = \{\text{Sun, Mon, Tue, Wed, Thu, Fri, Sat}\}$$

to $[0, 1]$ such that

- If the weather on day x was definitely good, then $f(x) = 1$.
- If the weather on day x was definitely not good, then $f(x) = 0$.
- If the weather on day x was partially good, then $0 < f(x) < 1$.

One such function might be that defined by

$$f(\text{Sun}) = f(\text{Mon}) = 1,$$
$$f(\text{Fri}) = f(\text{Sat}) = 0,$$

and

$$f(\text{Tue}) = 0.6, \quad f(\text{Wed}) = f(\text{Thu}) = 0.2.$$

This reflects our feeling that Sunday and Monday were definitely good-weather days, that Friday and Saturday were definitely not good-weather days, and that Tuesday, Wednesday, and Thursday were partially good-weather days, with Tuesday the best of the three. The choice of the functional values 0.6 and 0.2 for the partial members of the fuzzy set f is to some extent arbitrary and depends on how close to "good" ($= 1$) we consider each partial member to be.

EXAMPLE 6

Suppose we are interested in the collection of all new movies. We might write this as a fuzzy set in the following way: Let $\mathcal{U} = \{x \mid x \text{ is a movie}\}$, and for all $x \in \mathcal{U}$ let

$$f(x) = \begin{cases} 0 & \text{if } x \text{ was released more than six weeks ago,} \\ 1 - t/6 & \text{if } x \text{ was released } t \text{ weeks ago,} \quad 0 \le t \le 6. \end{cases}$$

See Figure 2–40. By defining the fuzzy set f in this way, we are saying that a movie is definitely not new if it was released six or more weeks ago, that the only movies that can be considered to be entirely new are those that have just been released (that is, those for which $t = 0$), and that movies released during the last six weeks are partially new.

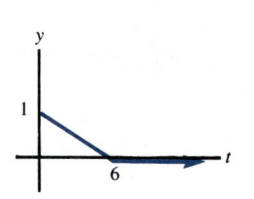

FIGURE 2–40

EXAMPLE 7

Consider the collection of all small natural numbers. We choose to write this as a fuzzy set h in **N** by setting $h(x) = 1/x$ for all $x \in \mathbf{N}$. See Examples 2 and 4. Note that this choice of h implies that the only natural number definitely considered to be small is the number 1, that there are no natural numbers that are definitely not small, and that the larger the natural number the further it is from being small.

EXAMPLE 8

Consider the collection of all real numbers far from zero. We write this as a fuzzy set in **R** by defining

$$f(x) = \frac{x^2}{1 + x^2}$$

for all $x \in \mathbf{R}$. See Figure 2–41. Note that 0 is the only nonmember of this fuzzy set. Note also that the fuzzy set has no full members, but that as x gets larger and larger in absolute value, x comes closer and closer to full membership.

Just as with ordinary sets, we can take complements, unions, and intersections of fuzzy sets.

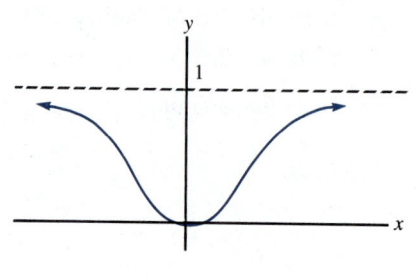

FIGURE 2–41

SETS AND FUNCTIONS **115**

> **Fuzzy Set Operations**
>
> Let $\mathcal{U}$ be a universal set and let f and g be fuzzy sets in $\mathcal{U}$. Then
>
> 1. The **complement of f** is the function f' defined by
> $$f'(x) = 1 - f(x) \quad \text{for all } x \in \mathcal{U}.$$
> 2. The **union of f and g** is the function $f \cup g$ defined by
> $$(f \cup g)(x) = \max\{f(x), g(x)\} \quad \text{for all } x \in \mathcal{U}.$$
> 3. The **intersection of f and g** is the function $f \cap g$ defined by
> $$(f \cap g)(x) = \min\{f(x), g(x)\} \quad \text{for all } x \in \mathcal{U}.$$

Here $\max\{f(x), g(x)\}$ means the larger of the two numbers $f(x)$ and $g(x)$, and $\min\{f(x), g(x)\}$ means the smaller of $f(x)$ and $g(x)$. Note that since $f(x)$ and $g(x)$ are in the interval $[0, 1]$, so are $f'(x)$, $(f \cup g)(x)$, and $(f \cap g)(x)$, for all $x \in \mathcal{U}$. Hence f', $f \cup g$, and $f \cap g$ are fuzzy sets.

EXAMPLE 9 Let f and g be the fuzzy sets of Example 1. Then f, g, f', $f \cup g$, and $f \cap g$ are defined by the following table:

x	$f(x)$	$g(x)$	$f'(x)$	$(f \cup g)(x)$	$(f \cap g)(x)$
a	0	0	1	0	0
b	1/4	0	3/4	1/4	0
c	1/2	1	1/2	1	1/2
d	1	1	0	1	1

EXAMPLE 10 Let h be the fuzzy set of small natural numbers defined in Example 7. Then the complement h' of h is given by
$$h'(x) = 1 - h(x) = 1 - 1/x$$
for all $x \in \mathbf{N}$. See Figure 2–42.

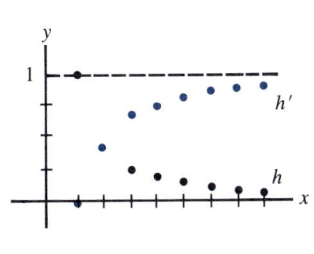

FIGURE 2–42

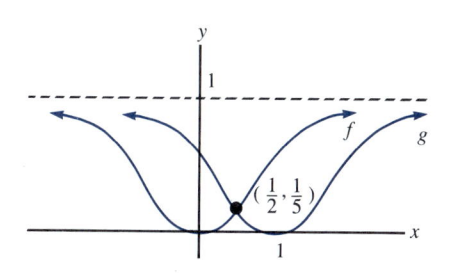

FIGURE 2–43

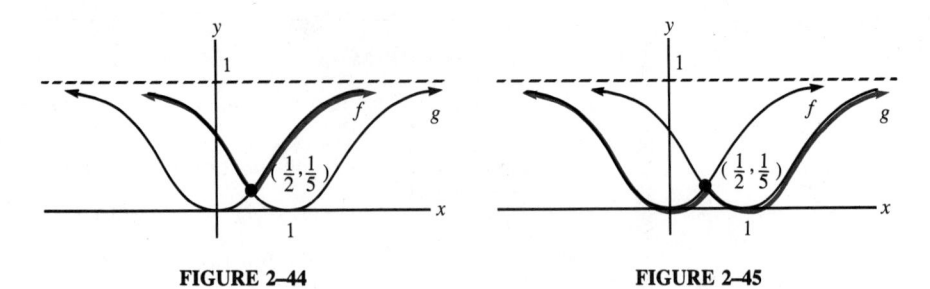

FIGURE 2-44   FIGURE 2-45

EXAMPLE 11   Let f be the fuzzy set of all real numbers far from zero, defined by

$$f(x) = \frac{x^2}{1 + x^2}$$

for all $x \in \mathbf{R}$, and let g be the fuzzy set of all real numbers far from 1, defined by

$$g(x) = \frac{(x - 1)^2}{1 + (x - 1)^2}$$

for all $x \in \mathbf{R}$. See Figure 2–43. (The point of intersection of the curves in Figure 2–43 was found by setting $f(x) = g(x)$ and solving for x to obtain $x = 1/2$.) Thus

$$(f \cup g)(x) = \begin{cases} g(x) & \text{if } x \leq 1/2, \\ f(x) & \text{if } x > 1/2, \end{cases}$$

and

$$(f \cap g)(x) = \begin{cases} f(x) & \text{if } x \leq 1/2, \\ g(x) & \text{if } x > 1/2. \end{cases}$$

See Figures 2–44 and 2–45 for the graphs of $f \cup g$ and $f \cap g$. ☐

Although fuzzy sets are a relatively recent mathematical development, having first appeared in the 1960s, they have already been used in model-building in many different areas, including linguistics, semantics, database design, circuit design, the study of nerve impulses, pattern recognition, and artificial intelligence.

SUPPLEMENTARY EXERCISES

In Exercises 1 through 4, let $\mathcal{U} = \{p, q, r, s, t, u\}$, and state whether the given function f is a fuzzy set in $\mathcal{U}$; if f is a fuzzy set, find its full members, its partial members, and its nonmembers.

1. $f(p) = f(q) = f(u) = 0, f(r) = f(s) = f(t) = 1/2$
2. $f(x) = 3/4$, for all $x \in \mathcal{U}$.
3. $f(p) = 0, f(q) = 1/4, f(r) = 1/2, f(s) = 3/4, f(t) = 1, f(u) = 5/4$
4. $f(p) = 1.0, f(q) = 0.8, f(r) = 0.7, f(s) = 0.6, f(t) = f(u) = 0$

5. Let $\mathcal{U} = \{x | x \in \mathbf{R}, x \geq 1\}$ and $f(x) = 1 - 1/x$, for all $x \in \mathcal{U}$. Graph the fuzzy set f and find its full members, partial members, and nonmembers. (This fuzzy set is the set of all real numbers that are much greater than 1.)

6. Let $\mathcal{U} = \{x | x \text{ is a car}\}$. Which of the following functions could be the fuzzy set of all old cars?

   (a) $f(x) = \begin{cases} 0.2t & \text{if } x \in \mathcal{U}, x \text{ is } t \text{ years old}, t \leq 5, \\ 1 & \text{if } x \in \mathcal{U}, x \text{ is } t \text{ years old}, t > 5, \end{cases}$

   (b) $f(x) = \begin{cases} 1 & \text{if } x \in \mathcal{U}, x \text{ is } t \text{ years old}, t \leq 3 \\ \dfrac{1}{t-2} & \text{if } x \in \mathcal{U}, x \text{ is } t \text{ years old}, t > 3, \end{cases}$

   (c) $f(x) = \begin{cases} 0 & \text{if } x \in \mathcal{U}, x \text{ is } t \text{ years old}, t < 1, \\ 1 - \dfrac{1}{t} & \text{if } x \in \mathcal{U}, x \text{ is } t \text{ years old}, t \geq 1 \end{cases}$

   (d) $f(x) = \begin{cases} 0 & \text{if } x \in \mathcal{U}, x \text{ is } t \text{ years old}, t < 2, \\ 0.25t - 0.5 & \text{if } x \in \mathcal{U}, x \text{ is } t \text{ years old}, 2 \leq t \leq 6, \\ 1 & \text{if } x \in \mathcal{U}, x \text{ is } t \text{ years old}, t > 6 \end{cases}$

7. For the fuzzy sets of old cars in Exercise 6, identify in each case the old cars, the partially old cars, and the new cars.

*8. Write two different fuzzy sets for the collection of integers close to 10.

*9. Write two different fuzzy sets for the collection of real numbers far from 5.

*10. Write two different fuzzy sets for the collection of old magazines.

*11. Write two different fuzzy sets for the collection of students who have good grade point averages.

Let

$$\mathcal{U} = \{p, q, r, s, t, u\},$$
$$f(p) = f(q) = 1.0, f(r) = 0.8, f(s) = 0.5, f(t) = 0.1, f(u) = 0,$$
$$g(p) = 0.2, g(q) = g(u) = 1.0, g(r) = 0.4, g(s) = 0.7, g(t) = 0.$$

In Exercises 12 through 17, write out the given fuzzy set.

12. f' 13. g' 14. $f \cup g$ 15. $f \cap g$ 16. $f' \cup g'$ 17. $f' \cap g'$

Let $\mathcal{U} = \{x | x \in \mathbf{R}, x \geq 1\}$ and

$$f(x) = 1 - \frac{1}{x}, \quad g(x) = \begin{cases} 1 & \text{if } 1 \leq x \leq 2, \\ \dfrac{1}{x-1} & \text{if } x > 2. \end{cases}$$

In Exercises 18 through 23, graph the given fuzzy set.

18. f' 19. g' 20. $f \cup g$ 21. $f \cap g$ 22. $f' \cup g'$ 23. $f' \cap g'$

According to a recent article in *The Boston Globe*, the ideal high-fashion model should have height close to 5 feet 10 inches and weight close to 120 pounds. Exercises 24 through 26 refer to this situation.

*24. Write a fuzzy set f for the collection of models whose heights are close to 5 feet 10 inches.

*25. Write a fuzzy set g for the collection of models whose weights are close to 120 pounds.

*26. Use the fuzzy sets f and g of Exercises 24 and 25 to write
  (a) the fuzzy set of models whose heights are not close to 5 feet 10 inches;
  (b) the fuzzy set of models who are ideal;
  (c) the fuzzy set of models who are not ideal;
  (d) the fuzzy set of models who are either of ideal height or ideal weight.

Exercises 27 through 29 were suggested by material in Kandel and Lee, *Fuzzy Switching and Automata: Theory and Applications*, Crane Russak, 1979. Let

$$\mathcal{U} = \{(x, y) \mid x, y \text{ are employees in a firm}\}.$$

For each pair $(x,y) \in \mathcal{U}$, define

$$f((x, y)) = \begin{cases} 1 & \text{if } x \text{ reports to } y, \\ 0 & \text{if } x \text{ does not report to } y. \end{cases}$$

The fuzzy set f describes the *formal* relationships among employees. However, there are many informal relationships among employees, such as "works with," "socializes with," "advises," and so on.

*27. Write a fuzzy set for the relationship "works with."

*28. Write a fuzzy set for the relationship "socializes with."

*29. Write a fuzzy set for the relationship "advises."

3

Linear and Quadratic Functions

In this chapter we continue our study of functions. Our goal is to demonstrate how **linear** and **quadratic** functions can be used to model common situations in business, economics, and the sciences. For instance, we will show how to use such functions to model the cost, revenue, and profit of a firm, the quantity of a commodity supplied to the market by its producers and the quantity demanded by its consumers, the growth of a population, and many other things. To illustrate:

- Suppose that each unit of product made and sold contributes $40 toward the profit of a firm, and that it costs the firm $100,000 if it makes and sells no units. Then the firm's total profit P from making and selling x units of its product is given by
$$P = 40x - 100,000, \qquad x \geq 0.$$
This is an example of a **linear function.** Its graph is shown in Figure 3–1.

- Suppose that the population of harp seals in a region t years after hunting of them was banned is y, where
$$y = t^2 - 12t + 200, \qquad t \geq 0.$$
This is an example of a **quadratic function.** Its graph is shown in Figure 3–2.

We begin with an introduction to linear functions. We show how to graph linear functions, demonstrate techniques for finding the equation of a line, and apply these ideas to linear models in business and economics. We then consider the solution of simultaneous linear equations and some situations where this is of practical importance. We conclude the chapter with a study of quadratic functions and their applications.

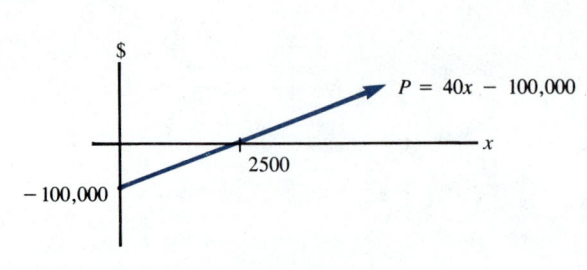

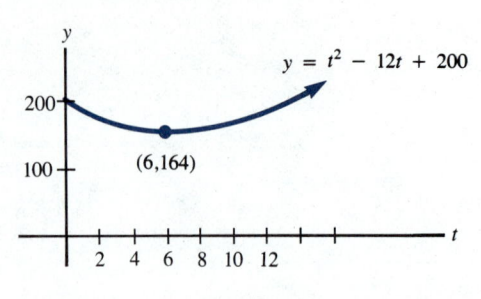

FIGURE 3-1

FIGURE 3-2

3.1 LINEAR FUNCTIONS

A **linear function** is a function whose graph is a straight line. (From now on, when we write "line," we will always mean "straight line.") A linear function can be distinguished by the form of its defining equation.

> **Linear Function**
>
> A function is **linear** if its defining equation can be written in the form
>
> $$y = mx + b$$
>
> for some real numbers m and b.

EXAMPLE 1 The function defined by $y = 2x + 1$ is linear with $m = 2$ and $b = 1$. The equation $12x + 4y = 16$ can be solved for y to yield $y = -3x + 4$; hence it defines a linear function with $m = -3$ and $b = 4$. □

Graphing Linear Functions

Since two points determine a line, in order to draw the graph of a linear function we need only find two points that lie on the line. We can do this by substituting any two distinct values for x in the equation $y = mx + b$. For instance, suppose we wish to graph

$$y = 2x + 1.$$

If

$$x = 0, \quad \text{then} \quad y = 2(0) + 1 = 1;$$

if

$$x = 2, \quad \text{then} \quad y = 2(2) + 1 = 5.$$

Therefore the graph of $y = 2x + 1$ is the line of Figure 3–3.

LINEAR AND QUADRATIC FUNCTIONS

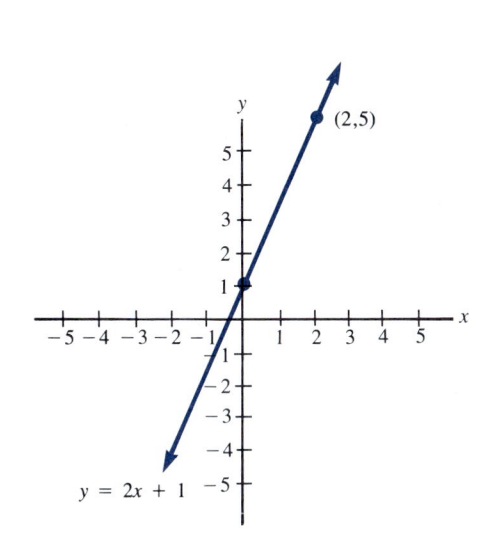

FIGURE 3–3

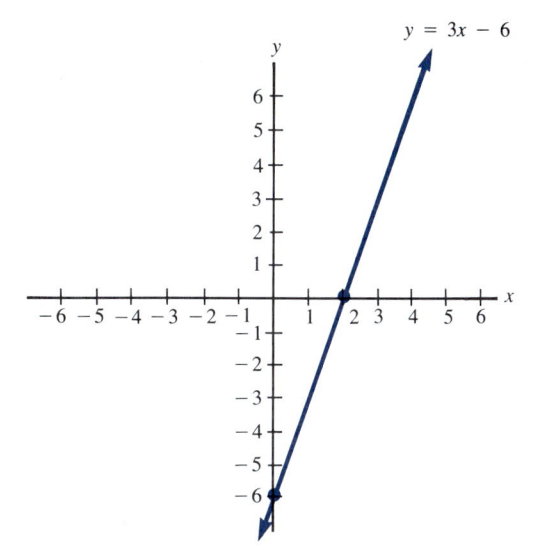

FIGURE 3–4

Recall that the points of intersection of a graph with the coordinate axes are called the **intercepts** of its function. Intercepts of linear functions are easy to find and are useful in graphing the function.

EXAMPLE 2 Find the intercepts of $y = 3x - 6$ and use them to draw the graph.

We find the intercepts by setting x equal to zero and solving for y, then setting y equal to zero and solving for x. Thus, if $x = 0$,

$$y = 3(0) - 6 = -6$$

so $(0, -6)$ is the y-intercept. If $y = 0$, we have

$$0 = 3x - 6$$

so

$$x = 2$$

and the x-intercept is $(2, 0)$. Using the intercepts to draw the graph, we obtain the line pictured in Figure 3–4. ☐

EXAMPLE 3 The only intercept of $y = 2x$ is the point $(0, 0)$; since the point $(1, 2)$ is also on the graph of this function, its graph is the line of Figure 3–5. The linear function $y = 4$ has $(0, 4)$ as its y-intercept, but has no x-intercept; since its graph is the set of points $\{(x, 4) | x \in \mathbf{R}\}$, it is the horizontal line of Figure 3–6. ☐

EXAMPLE 4 Any linear function $y = mx + b$ has y-intercept $(0, b)$. (Why?) ☐

FIGURE 3-5

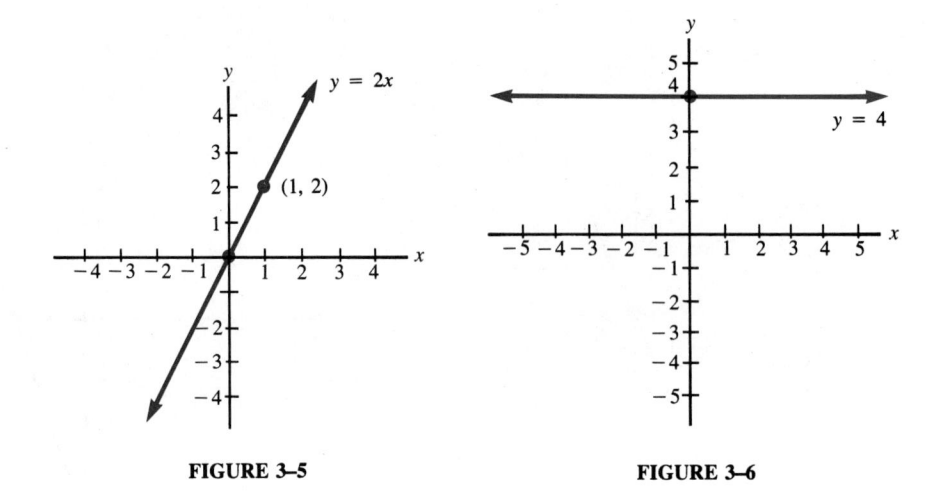

FIGURE 3-6

Note that all the linear functions we have so far considered have had *nonvertical* lines as their graphs. This is because a vertical line is not the graph of a function. (Why not?) An equation of the form $x = a$ has as its graph the set of points $\{(a, y) | y \in \mathbf{R}\}$, which forms a vertical line. See Figure 3-7.

Vertical Line

The graph of the equation $x = a$ is a vertical line with *x*-intercept $(a, 0)$.

EXAMPLE 5   Draw the graph of $x = -3$.
   The given equation is of form $x = a$ with $a = -3$. Hence the graph is the vertical line shown in Figure 3-8. ◻

   Example 4 showed that the number b in the equation $y = mx + b$ gives the *y*-intercept of the linear function. The number m, which is called the **slope of the line,** determines how y changes as x increases.

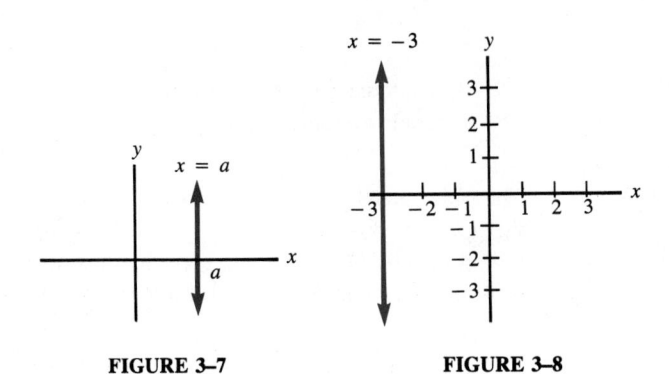

FIGURE 3-7

FIGURE 3-8

LINEAR AND QUADRATIC FUNCTIONS 125

> **The Slope of a Nonvertical Line**
>
> The **slope** of the line with equation $y = mx + b$ is m.

Note that if $y = mx + b$, then an increase of 1 in the value of x changes y by an amount equal to m, because the difference between $mx + b$ and $m(x + 1) + b$ is

$$[m(x + 1) + b] - [mx + b] = mx + m + b - mx - b = m.$$

Therefore the slope of a line measures the slant of the line. If the slope is a *positive* number, then an increase of 1 in the value of x results in an *increase* in the value of y. Hence the line slants upward from left to right, and the larger the slope, the steeper the slant. Figure 3–9 shows two lines, one with slope 5 and the other with slope $\frac{1}{2}$. On the other hand, if the slope of a line is a *negative* number, then an increase of 1 in the value of x results in a *decrease* in the value of y. Hence a line with negative slope slants downward from left to right, and the larger the absolute value of the slope, the steeper the slant. Figure 3–10 shows two lines, one with slope -2 and the other with slope -1. If the slope of a line is 0, then an increase in the value of x does not change the value of y at all, so the line is horizontal. Figure 3–11 shows a line with slope 0.

EXAMPLE 6   The slope of $y = 3x - 6$ is $m = 3$, and hence the line slants upward from left to right. (See Figure 3–4.) To find the slope of $8x + 4y = 24$, we solve this equation for y, obtaining $y = -2x + 6$. Thus the slope of this line is $m = -2$, and the line slants downward from left to right. The equation $y = 4$ may be written as

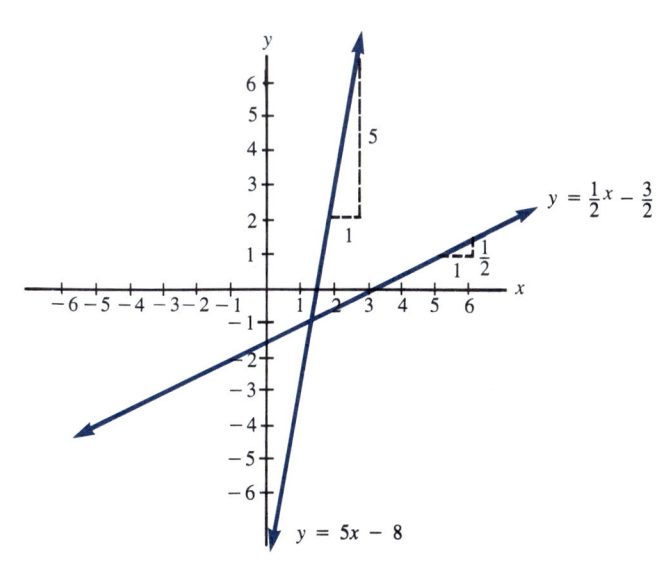

FIGURE 3–9

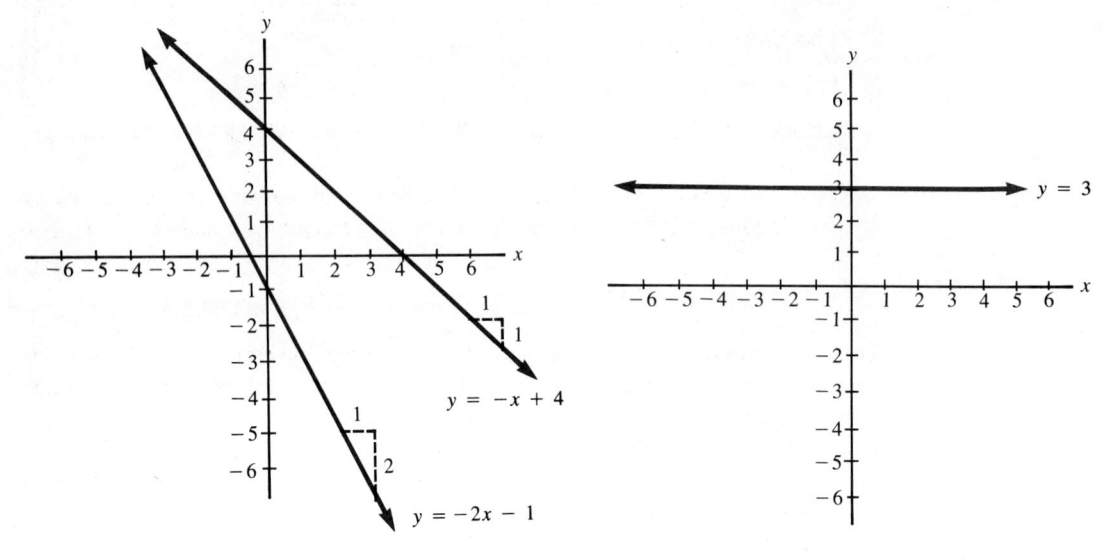

FIGURE 3–10

FIGURE 3–11

$y = 0x + 4$, so its slope is $m = 0$; therefore the line is horizontal. (See Figure 3–6.) □

Before continuing, we must mention an important fact about vertical lines: a vertical line has no slope. This is not the same as saying that a vertical line has slope 0; a horizontal line has slope 0, but a vertical line has no slope at all. This is sometimes expressed by saying that the slope of a vertical line is *infinite*.

The next example shows how a linear model is used to analyze a situation.

EXAMPLE 7   Cappa Corporation's sales are described by the linear model

$$S = 20t + 100,$$

where t is time in years, with $t = 0$ representing 1988, and S is annual sales in thousands of units. See Figure 3–12. From the equation we see that in 1988 ($t = 0$) Cappa's sales were 100,000 units. Since the slope of the equation is $m = 20$,

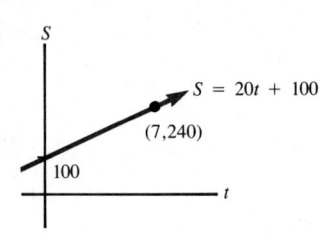

FIGURE 3–12

Cappa's sales are increasing by 20,000 units each year. If the equation continues to describe Cappa's sales, in 1995 ($t = 7$) the corporation's sales will be

$$S = 20(7) + 100 = 140 + 100 = 240$$

thousand units. □

Finding the Equation of a Line

Up until now we have been concerned with drawing the line that is the graph of a given linear equation $y = mx + b$. Now we would like to reverse this procedure: given a line, we would like to find the equation that has the line as its graph. This process is called **finding the equation of the line,** and it is very important in the construction of linear models.

There are several forms that the equation of a line can take. We will consider three of them: the slope-intercept form, the point-slope form, and the general form. We begin with the slope-intercept form.

Slope-Intercept Form

If a (nonvertical) line has slope m and y-intercept $(0, b)$, the **slope-intercept form** of its equation is

$$y = mx + b.$$

We use the slope-intercept form when we know, or can easily find, both the slope and the y-intercept of a line.

EXAMPLE 8 A line has slope 3 and y-intercept $(0, 5)$. Find the equation of the line in slope-intercept form.

Since we know that $m = 3$ and $b = 5$, we write

$$y = 3x + 5. \quad □$$

EXAMPLE 9 Alpha Company's sales were \$100,000 in 1985 and have increased at the rate of \$20,000 per year. Let S denote annual sales in dollars, and let t denote the year, coded so that $t = 0$ represents 1985. Find the equation that relates Alpha's annual sales S to the year t.

Since sales increase by the same amount, namely \$20,000, each year, the relationship between sales and time is linear with $m = \$20,000$. Since annual sales were \$100,000 in 1985 ($t = 0$), we have $b = 100,000$. Therefore, the slope-intercept form of the equation is

$$S = 20,000t + 100,000.$$

If this relationship continues to hold, sales in 1995 ($t = 10$) will be

$$S = 20,000(10) + 100,000 = 300,000,$$

or \$300,000. See Figure 3–13. □

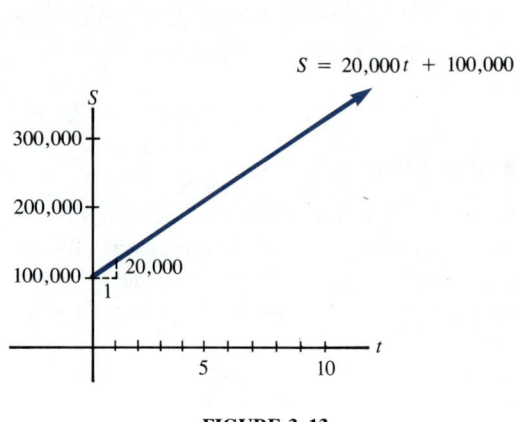

FIGURE 3–13

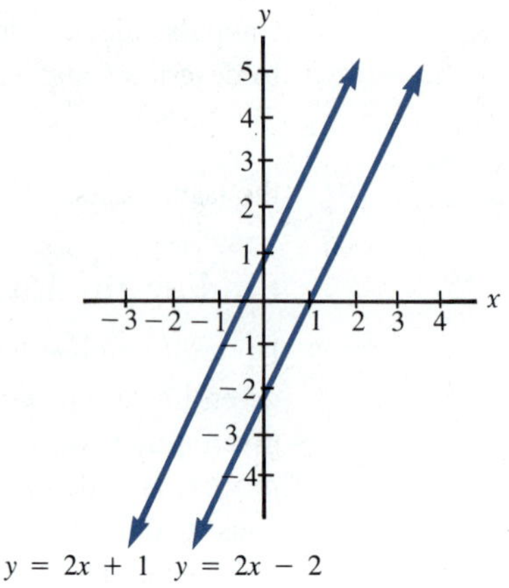

FIGURE 3–14

Two lines are **parallel** if they do not intersect. The lines $y = 2x + 1$ and $y = 2x - 2$ shown in Figure 3–14 are parallel. Note that these lines have the same slope. This is no accident, for it can be shown that

Slopes of Parallel Lines

Two nonvertical lines are parallel if and only if they have the same slope.

Of course, any two vertical lines are parallel.

EXAMPLE 10 Find the equation of the line with y-intercept $(0, 9)$ that is parallel to the line $y = 3x + 5$.

Since the line whose equation we seek is parallel to $y = 3x + 5$, it must have slope $m = 3$. Hence its equation is $y = 3x + 9$. □

If we are attempting to find the equation of a nonvertical line and do not know its slope, we must of course find it. We can find the slope of a nonvertical line if we know two distinct points (x_1, y_1) and (x_2, y_2) on it. See Figure 3–15, where the slope of the line is m. Since the point (x_1, y_1) is on the line, we must have

$$y_1 = mx_1 + b.$$

Similarly, since (x_2, y_2) is on the line,

$$y_2 = mx_2 + b.$$

Subtracting the first of these equations from the second, we have

$$y_2 = mx_2 + b$$
$$- (y_1 = mx_1 + b)$$
$$\overline{y_2 - y_1 = mx_2 - mx_1}$$

LINEAR AND QUADRATIC FUNCTIONS

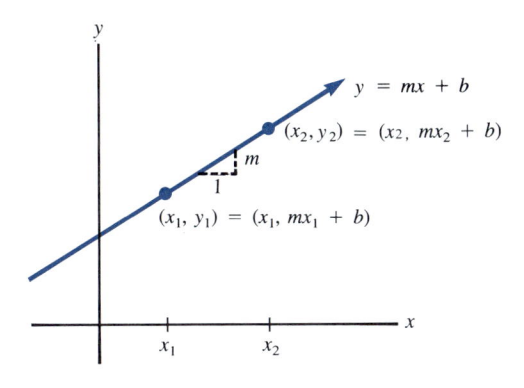

FIGURE 3-15

or

$$y_2 - y_1 = m(x_2 - x_1).$$

Since the line is not vertical, $x_2 \neq x_1$, and thus

$$m = \frac{y_2 - y_1}{x_2 - x_1}.$$

Therefore,

> **The Slope of a Line through Two Points**
>
> A nonvertical line that passes through the points (x_1, y_1) and (x_2, y_2) has slope
>
> $$m = \frac{y_2 - y_1}{x_2 - x_1}.$$

Note that it does not matter which point we label (x_1, y_1) and which point we label (x_2, y_2) since

$$\frac{y_2 - y_1}{x_2 - x_1} = \frac{-(y_2 - y_1)}{-(x_2 - x_1)} = \frac{y_1 - y_2}{x_1 - x_2}.$$

EXAMPLE 11 Find the slope of the line that passes through the points $(2, 5)$ and $(-2, 13)$. Setting $(x_1, y_1) = (2, 5)$ and $(x_2, y_2) = (-2, 13)$, we see that the slope is

$$m = \frac{y_2 - y_1}{x_2 - x_1} = \frac{13 - 5}{-2 - 2} = \frac{8}{-4} = -2. \quad \square$$

The **point-slope form** of the equation of a line is convenient when a point on the line and its slope are known.

Point-Slope Form

If a line has slope m and passes through the point (x_1, y_1) the **point-slope form** of its equation is

$$y - y_1 = m(x - x_1).$$

The point-slope form of the equation may be converted to the slope-intercept form by solving it for y.

EXAMPLE 12 Let us find the point-slope and slope-intercept forms of the equation of the line that passes through the points $(2, 5)$ and $(-2, 13)$. From Example 11, the slope of this line is $m = -2$. Therefore if $(x_1, y_1) = (2, 5)$ we have

$$y - 5 = -2(x - 2)$$

for the point-slope form. Solving this equation for y yields the slope-intercept form

$$y = -2x + 9. \quad \square$$

EXAMPLE 13 Beta Company was founded in 1987, and every year since its founding the cost per unit produced has increased by $0.25. In 1990 the cost per unit was $12.62. Find an equation that describes Beta Company's cost per unit as a function of time.

Since Beta's cost per unit increases at the constant rate of $0.25 per year, we have $m = 0.25$. Letting $C =$ cost per unit in year t and coding t so that $t = 0$ represents 1987, we have $(t_1, C_1) = (3, 12.62)$. Thus, the point-slope form of the equation is

$$C - 12.62 = 0.25(t - 3),$$

which yields

$$C = 0.25t + 11.87. \quad \square$$

EXAMPLE 14 A mathematics professor has recorded the final examination grades of her students and the number of times each student was absent from class during the semester. She finds the following:

$x =$ Number of Absences	$y =$ Average Grade on Final
2	84.6
5	75.0
9	62.2

If we plot the points $(2, 84.6)$, $(5, 75.0)$, and $(9, 62.2)$, we see that the relationship between absences and average grade on the final appears to be linear. See Figure 3-16. To find the equation of the line that passes through the first two of these points, we first find its slope:

$$m = \frac{75.0 - 84.6}{5 - 2} = -3.2.$$

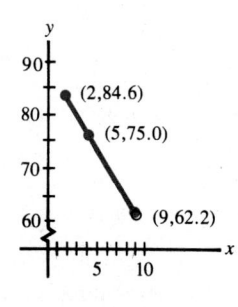

FIGURE 3-16

Thus,
$$y - 84.6 = -3.2(x - 2),$$
or
$$y = -3.2x + 91.$$

Note that substituting $x = 9$ in this equation yields $y = 62.2$, thus proving that the point $(9, 62.2)$ does lie on this line.

The equation $y = -3.2x + 91$ implies that the average grade on the final for those students having no absences is 91 and that each additional absence results in a decrease of 3.2 points in the average grade. ☐

The equation of a vertical line is of the form $x = x_1$, where (x_1, y_1) is any point on the line.

EXAMPLE 15 The equation of the vertical line that passes through the point $(-3, 6)$ is $x = -3$. ☐

The final form of the equation of a line that we will consider is the **general form.**

General Form

The equation of a line is in **general form** if it is written in the form
$$Ax + By = C,$$
where A, B, and C are real numbers.

The equation of every line, vertical or nonvertical, can be written in general form. The slope-intercept form may be obtained from the general form by solving the general form for y. It is sometimes convenient to write the equation of a line in general form.

EXAMPLE 16 Zeta Company produces both small and large snowplow blades. Each small blade requires 10 kilograms (kg) of steel, and each large one requires 20 kg of steel. The company has 2000 kg of steel on hand. Thus if Zeta wishes to use all its steel, it can produce any combination of x small blades and y large blades so long as
$$10x + 20y = 2000.$$

The intercepts of this equation are $(200, 0)$ and $(0, 100)$, which imply that Zeta can produce a maximum of 200 small blades (provided it produces 0 large ones) or a maximum of 100 large blades (provided it produces 0 small ones). See Figure 3–17. ☐

■ *Computer Manual: Program SLPNRCPT*

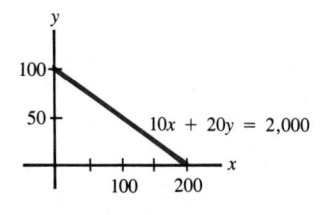

FIGURE 3–17

3.1 EXERCISES

Graphing Linear Functions

In Exercises 1 through 18, find the slope and intercepts of the given linear equation and graph it.

1. $y = 2x - 5$
2. $y = 3x + 7$
3. $y = -x - 2$
4. $y = -2x + 4$
5. $y = 2x$
6. $y = -3x$
7. $y = 5$
8. $y = -3$
9. $x = -1$
10. $x = 10$
11. $x + y = 8$
12. $x - y = 8$
13. $-3x + 8y = 24$
14. $7x - 5y = 35$
15. $\frac{x}{2} + \frac{y}{2} = 1$
16. $\frac{x}{3} + 6y = -1$
17. $2x + 4y = 0$
18. $-5x + 3y = 0$

19. An investor estimates that his risk-reward function is given by the equation
$$y = 2.8x + 2,$$
where x measures the risk of an investment, with $x = 0$ representing a riskless investment, $x = 1$ an investment in U.S. Treasury bills, and larger values of x representing increasingly risky investments. The dependent variable y represents the percentage return on investment required of an investment whose risk is x. Graph this function. What return on investment does the investor require on a riskless investment? on an investment three times as risky as U.S. Treasury bills?

Management

20. Acme Company's annual sales are described by the equation
$$S(t) = 200t + 500,$$
where t is time in years, coded so that $t = 0$ represents 1988, and $S(t)$ is annual sales in year t in thousands of dollars. What were Acme's sales in 1988? Are sales increasing each year? by how much? Graph the function S.

21. Baker Corporation makes both a regular and a deluxe model radio. Twenty worker-hours are required to produce each regular model, and 30 worker-hours are required for each deluxe model. The firm has 60,000 worker-hours of production time available. If x denotes the number of regular radios and y the number of deluxe radios that Baker can produce, then the production possibilities are given by
$$20x + 30y = 60,000.$$
Graph this equation. For every additional regular radio produced, how many deluxe radios must Baker forego producing? Can Baker produce 1200 regular and 1200

deluxe radios? 1800 regular and 800 deluxe? 1600 regular and 1000 deluxe? What is the maximum number of regular radios the firm can produce? the maximum number of deluxe radios?

Exercises 22 through 27 were suggested by material in Higgins, *Analysis for Financial Management*, Irwin, 1984. They concern the effect of leverage on a corporation's return on equity. Return on equity (ROE) is a financial ratio defined by

$$\text{ROE} = \frac{\text{Profit after taxes}}{\text{Owners' equity}}$$

Owners' equity is the amount invested in the company by the shareholders; it is the total ownership interest of the shareholders and is equal to the value of the corporation's assets minus its debts. Return on equity is important because it estimates the return to owners on their investment in the company. Corporations strive to make their ROE as large as possible. Suppose we let

D = outstanding debt,

OE = owners' equity,

L = leverage = D/OE

A = asset value = $D + OE$,

E = earnings before interest and taxes,

i = interest rate paid on debt, in decimal form,

t = tax rate, in decimal form.

Define the return on assets (ROA) by

$$\text{ROA} = \frac{\text{Earnings after interest and taxes}}{\text{Value of assets}} = \frac{E - t(E - iD)}{A}$$

Then ROE can be written in the form

$$\text{ROE} = \text{ROA} + (\text{ROA} - i)L.$$

*22. Find a corporation's ROE if its asset value is $600 million, its debt is $400 million, its earnings before interest and taxes are $200 million, the interest rate it pays on debt is 15%, and its tax rate is 40%.

*23. What is a corporation's ROE if it has no debt?

*24. Graph ROE as a function of L if ROA $>$ i. What happens to ROE when L increases by 1?

*25. Repeat Exercise 24 when ROA $=$ i.

*26. Repeat Exercise 24 when ROA $<$ i.

*27. A corporation will adopt different financial strategies with regard to its debt and owners' equity, depending on whether ROA $>$ i, ROA $=$ i, or ROA $<$ i. To see this, determine how a corporation can increase its ROE if
   (a) its ROA is greater than i;
   (b) its ROA is equal to i;
   (c) its ROA is less than i.

Life Science

28. A certain strain of bacteria is easily destroyed by heat. If T denotes temperature in degrees Celsius and B denotes the percentage of the bacteria alive in a culture at temperature T, then T and B are related by the equation

$$100T + 41B = 4100.$$

Graph this equation. What percentage of the bacteria are alive at 0°C? at 10°C? At what temperature will all the bacteria be destroyed? What percentage of bacteria are destroyed by a 1°C rise in temperature?

29. Pollutants inhibit the hatching of fish eggs. Suppose that if the concentration of pollutants in a lake is x milligrams per liter, then the proportion of fish eggs that will hatch is y, where

$$y = 0.25 - 0.005x.$$

Graph this function. What proportion of the eggs will hatch if there is no pollution in the lake? For every increase of 1 mg of pollutants per liter, what proportion of the eggs will fail to hatch? At what level of pollution will all the eggs fail to hatch?

30. Weber's Law states that if a subject is exposed to a stimulus of level s, the smallest change c in the stimulus that will be noticeable to the subject is given by $c = ms$, for some number m. Suppose that a subject is exposed to a sound level of s decibels (dB) and that $m = 0.02$. Find the smallest noticeable change in the sound level when the sound level is 5000 dB; when it is 50,000 dB. Graph $c = 0.02s$. For every increase of 100 dB in the sound level, what happens to the smallest noticeable change in the sound level?

31. In an article entitled "Mental Imagery and the Visual System" (*Scientific American*, March 1986), Ronald Finke reported on an experiment in which subjects were shown a drawing containing a pattern of dots and an arrow and were asked if the arrow was pointing at any of the dots. It was found that if x was the distance of the arrow from the dots, in centimeters (cm) and y the time it took the subject to make a decision, in milliseconds (ms), then y was given by

$$y = 850 + 26x.$$

Find the reaction time y when the arrow was 2.5 cm from the dots and when it was 25 cm from the dots. Graph the function. What is the significance of its y-intercept? What is the significance of its slope?

In an article entitled "Features and Objects in Visual Processing" (*Scientific American*, November 1986), Anne Treisman reported on an experiment in which a "target" symbol either was or was not placed among a group of "distractor" symbols, and subjects were asked to locate the target symbol. Exercises 32 through 34 refer to this experiment.

32. If there were x distractor symbols and no target symbol present, the average search time until the subject decided that the target was not present was y milliseconds (ms), where

$$y = 40x + 440.$$

Find the average search time when there were 10 distractor symbols present. Graph the function. What is the significance of its slope?

LINEAR AND QUADRATIC FUNCTIONS 135

33. If there were *x* distractor symbols and one target symbol present, the average search time until the subject located the target was *y* ms, where
$$y = 20x + 430.$$
Find the average search time when there were 10 distractor symbols present. Graph the function. What is the significance of its slope?

*34. If subjects examine each symbol individually until they either find the target symbol or decide that it is not present, then on average it should take about half as long to locate a target symbol as it should to decide that one is not present. Do the graphs of Exercises 32 and 33 support the hypothesis that subjects examine each symbol individually?

Social Science

35. In a certain state the proportion of voters who are Independent is given by
$$-\frac{1}{5}t + 20y = 9,$$
where *t* is time in years, coded so that $t = 0$ represents 1988, and *y* is the proportion of Independents in the state in year *t*. Graph this equation. What proportion of the voters in the state were Independents in 1978? in 1988? in 1990? By how much is the proportion of Independents changing each year?

Finding the Equation of a Line

In Exercises 36 through 43, find the slope-intercept form of the equation of the given line.

36. The line with slope -3 and *y*-intercept (0, 2).
37. The line with slope 4 and *y*-intercept (0, 1).
38. The line with slope $-2/3$ that passes through the origin.
39. The horizontal line that passes through the point (0, 6).
40. The line that passes through the point (0, 2) and is parallel to the line $y = 4x - 7$.
41. The line that passes through the point (0, 3) and is parallel to the line $2x + 4y = -6$.
42. The line that passes through the point (4, -6) and is parallel to the line $y = 8$.
43. The line that is parallel to $3x - 2y = 12$ and passes through the point (0, -1).

In Exercises 44 through 53, find the point-slope and slope-intercept forms of the equation of the given line, if possible.

44. The line with slope -2 that passes through the point (-5, -9).
45. The line that passes through the point (1, 3) and is parallel to the line $y = -5x + 2$.
46. The line that passes through the point (2, -1) and is parallel to the line $2x + 2y = 12$.
47. The horizontal line that passes through the point (8, -2).

48. The line that passes through the points (2, 6) and (3, 8).
49. The line that passes through the points (−4, −1) and (−3, −8).
50. The line that passes through the points (5, −2) and (0, −7).
51. The line that passes through the points (7, 3) and (0, 0).
52. The line that passes through the points (6, 2) and (4, 2).
53. The line that passes through the points (2, 6) and (2, 4).

In Exercises 54 through 61, put the given equation into general form and then find its slope-intercept form, if possible.

54. $2x - y = 8$
55. $y = 3x - 7$
56. $5x = 9 - 2y$
57. $9 = 3y - 2x$
58. $x = 2y$
59. $3y = 4x$
60. $y = -4$
61. $x = 7$

Management

62. Epsilon, Inc. has been increasing the size of its sales force by 10 persons per year. If the sales force consisted of 150 salespersons in 1988, find an equation that describes the size of the sales force as a function of time, and forecast
    (a) the size of the sales force in 1995;
    (b) the year in which the sales force will consist of 210 salespeople.

63. Zeta Corporation has been losing market share at the rate of 1.5% per year. If Zeta's market share in 1985 was 45.7%, find an equation that describes Zeta's market share as a function of time, and find
    (a) the corporation's market share in 1990;
    (b) the year in which its market share will be 29.2%.

64. A camera store orders 200 cameras at a time from its supplier and sells them at the rate of 4 per day. Find an equation that expresses the number of cameras the store has on hand as a function of the time since the arrival of the last order.
    (a) How many cameras will the store have on hand 22 days after an order arrives?
    (b) If it takes 10 days from the time an order is sent in until the cameras arrive at the store, what is the latest the store can order cameras if it does not want to run out of them?

65. According to an article in the September 20, 1988 issue of *The Wall Street Journal*, most stockbrokers earn no salary but receive a commission equal to 40% of their sales. Find an equation that describes a stockbroker's commissions as a function of sales.
    (a) What is the commission of a stockbroker who sells $140,000?
    (b) Find the sales that will result in a commission of $100,000.

66. A machine depreciates in value by an equal amount each year of its useful life. If it cost $100,000 new and its useful life is eight years, after which its value is $0, find an equation that describes the value of the machine as a function of the time since it was purchased.
    (a) What will the value of the machine be when it is 5.5 years old?
    (b) When will its value be $34,000?

67. A machine depreciates by an equal amount each year until the end of its useful life, at which time it has a scrap value of $4000. If the machine was purchased for

$28,000 and has a useful life of six years, find an equation that describes its value as a function of the time since it was purchased.
  (a) What is the value of the machine 3.25 years after its purchase?
  (b) When will its value be $16,500?

68. According to an article in the September 29, 1988 issue of *The Wall Street Journal*, worldwide usage of platinum was expected to be 3,320,000 tons in 1988 and 3,860,000 tons in 1993. Find a linear equation that describes worldwide usage of platinum as a function of time.
  (a) Forecast usage in 1991.
  (b) When will usage reach the level of 4,000,000 tons per year?

69. Wizard Wand Company sold 194,000 wands at $36.00 per wand, 193,580 wands at $37.40 per wand, and 193,250 wands at $38.50 per wand. Assuming a linear relationship between sales and price, find the slope-intercept form of the equation that describes this relationship.
  (a) Show that the coordinates of the point you did not use in finding the equation satisfy it.
  (b) What are the significances of the slope and vertical intercept of the equation?
  (c) What will Wizard's sales be if the price is $40 per wand? What must the price be if Wizard wishes to sell 200,000 wands?

70. The advertising director of the Theta Company has the following information concerning TV ads:

Amount Spent on TV Ads	Number of People Who Will See the Ads
$100,000	2,500,000
300,000	3,450,000
550,000	4,637,500

Does the relationship between the amount spent on TV ads and the number of people who will see them appear to be linear? Suppose that the advertising director wants the ads to be seen by at least 5,000,000 people. What should the budget for the ads be? What is the largest audience that can be reached if a maximum of $1,000,000 can be spent on the ads?

71. When Atlantic Hardware, Inc. had 10 stores, its total annual sales were $125 million; when it had 17 stores, they were $212.5 million. Find a linear equation that describes annual sales as a function of the number of stores. What is the significance of the slope of this equation? How many stores would Atlantic need in order to have annual sales of $300 million?

72. Farson Corporation produces two models of clocks, regular and deluxe. The cost of producing a regular model is $25, and that of producing a deluxe model is $35. Farson's production budget is $105,000, and the firm intends to produce enough clocks to use up the entire amount. Find the equation that describes the number of each model that can be produced. What is the maximum number of each model the firm can make? How many regular clocks can be made if the firm intends to produce 1250 deluxe clocks?

73. A haberdasher carries two lines of overcoats: a high-priced line, on which the profit is $80 per overcoat, and a low-priced line, on which the profit is $45 per overcoat. If the haberdasher's profit from sales of overcoats is $18,000, find an equation that describes the number of each type of overcoat sold. Could the haberdasher have sold 150 of the high-priced coats and 200 of the low-priced ones? 135 of the high-priced ones and 160 of the low-priced ones? What is the maximum number of high-priced coats the haberdasher could have sold? the maximum number of low-priced coats?

74. It takes 12 minutes for a worker at a furniture factory to varnish a cocktail table; it takes 30 minutes for the same worker to varnish a dresser. Find an equation that describes the number of cocktail tables and dressers the worker can varnish in one 8-hour day. Can the worker varnish 25 cocktail tables and 6 dressers in one day? 8 cocktail tables and 13 dressers? What is the maximum number of cocktail tables the worker can varnish in one day? the maximum number of dressers?

Exercises 75 through 78 were suggested by material in Higgins, *Analysis for Financial Management*, Irwin, 1984. They concern the **percentage-of-sales method** of determining gross profit.

*75. Suppose a firm's sales are given by a linear equation

$$S = mt + b,$$

where S is sales in dollars and $t = 0$ represents the present. Suppose also that the cost to the firm of the goods it sells (its **cost of goods sold**) is a certain proportion p of its sales. Find an equation that describes the firm's cost of goods sold as a function of time. Assuming that m and b are positive and that $0 < p < 1$, graph the sales equation and the equation for cost of goods sold on the same set of axes.

*76. A firm's **gross profit** is equal to its sales minus its cost of goods sold. For the firm of Exercise 75, find an equation that describes gross profit as a function of time. Then graph that equation on the same set of axes with the equations for sales and cost of goods sold when

  (a) $p > 0.5$;   (b) $p = 0.5$;   (c) $p < 0.5$.

*77. Suppose a firm's sales are given by $S = 8.5t + 22$, where S is in millions of dollars and t is in years, with $t = 0$ representing the present. If the firm's cost of goods sold is 82% of sales, use the result of Exercise 76 to find the firm's gross profit two years from now.

*78. Repeat Exercise 77 if the cost of goods sold is 45% of sales.

Life Science

79. A scientist studying children finds that, on the average, newborn babies are 53 cm tall and grow at the rate of 2.1 cm per month until the age of nine months. Find an equation that describes the relationship between height and age in months, and find
    (a) the average height at age 4.5 months;
    (b) the number of months required for the average height to reach 64 cm.

80. If there were 20 million tons of ozone in the ozone layer in 1987 and the amount of ozone is decreasing at a rate of 1.25 million tons per year, find an equation that describes the amount of ozone present as a function of time.
    (a) How much ozone will be present in 1996?
    (b) How long will it take until 75% of the ozone is gone?

81. According to a recent article in *Newsweek*, the greenhouse effect may increase the average global temperature by approximately 0.6°F per decade. If the average global temperature is now 59.5°F, find an equation that describes average global temperature as a function of time.
    (a) When will the average global temperature reach 61°F?
    (b) How long will it be until the average global temperature is 64°F?

82. It is estimated that the coyote population of New England is increasing by 30 animals per year, and that there were 250 coyotes in New England in 1988. Find an equation that describes the size of the coyote population in New England as a function of time.
    (a) How many coyotes will there be in New England in 1995?
    (b) When will the New England coyote population reach 500?

83. On January 1, 1988, pollutants in Western Lake averaged 0.11 gram per liter of water. At that time a plan was adopted calling for the pollution to be reduced by equal amounts each year, with July 1, 1993, as the target date for zero pollution. At what rate is pollution to be reduced each year? What will be the concentration of pollutants as of January 1, 1991?

*84. In an article entitled "The Large-Scale Cultivation of Mammalian Cells" (*Scientific American*, January 1983), Joseph Feder and William Tolbert reported on an experiment that involved growing cells in a glucose solution: the cells took up the glucose and produced lactic acid. Eighteen hours after the beginning of the experiment, the concentration of glucose in the solution was 3.92 mg/mL and that of lactic acid was 0.90 mg/mL; 32 hours later the corresponding figures were 2.80 mg/mL of glucose and 2.50 mg/mL of lactic acid. Assuming linearity, find equations that describe both the concentration of glucose and the concentration of lactic acid in the solution as a function of time. Graph these equations on the same set of axes and use the graphs to estimate the time when the concentrations were equal.

85. A patient needs to receive exactly 2000 units of a certain drug each day. Brand A pills each contain 80 units of the drug, while brand B pills each contain 125 units. Find an equation that describes the relationship between the number of pills of each brand that the patient can take. How many brand A pills can he take if he takes no brand B pills? if he takes eight brand B pills?

Social Science

86. A town has local elections every two years. In 1986, 62% of the eligible voters actually voted, and since then the percentage of eligible voters who have voted has decreased at the rate of 4.2% per election. Find an equation that relates the percentage of eligible voters who actually vote to time in years. Also find
    (a) the percentage of eligible voters who voted in 1990;
    (b) the first election year in which the percentage who vote will fall below 50%.

87. In a certain city the number of high-school dropouts has been increasing at a rate of 160 individuals per year. If there were 2240 high-school dropouts in the city last year, find an equation that describes the number of dropouts as a function of time.
    (a) How many dropouts will there be six years from now?
    (b) When will the dropout rate reach 3000 per year?

88. If 63.5% of the population of a developing nation was literate in 1980 and 65.8% was literate in 1985, and if literacy has been increasing in a linear fashion, find an equation that describes literacy as a function of time.

(a) What percentage of the population will be literate in 1990?
(b) When will 90% of the population be literate?

89. A sociology professor gives aptitude tests to students who go into sales upon graduation from college and finds the following:

Score on Test	Average Sales First Year on Job
52	$119,000
74	$163,000
87	$189,000

Does the relationship between test score and average sales appear to be linear? Find the equation that describes this relationship. What are the significances of its slope and vertical intercept? Also

(a) What are average first-year sales for those who score 95 on the aptitude test?
(b) If the professor feels that those who can be expected to have first-year sales of less than $150,000 should be advised against going into sales, which students who take the test should be so advised?

3.2 LINEAR MODELS IN BUSINESS AND ECONOMICS

Linear functions are often useful in modeling various situations in business and economics. In this section we discuss linear cost, revenue, and profit models, as well as linear supply and demand models.

Cost, Revenue, and Profit Models

Let C denote the total cost of producing x units of some product. Total cost can be broken down into two parts: fixed costs and variable costs. **Fixed costs** are those costs that accrue even when nothing is being produced. Examples are insurance premiums, payments on outstanding loans, salaries of executives, and so on. The total fixed cost is the sum of all such fixed costs. Total **variable cost,** on the other hand, depends on how many units of the product are produced. If each unit of product costs the same amount to make, say v dollars, then the **variable cost per unit** is v, and the total variable cost is the product vx of the variable cost per unit and the number of units made. Thus,

$$\text{Total cost} = \text{Total variable cost} + \text{total fixed cost,}$$

or

$$C = vx + F,$$

where C = total cost, v = variable cost per unit, x = number of units produced, and F = total fixed cost. The linear function

$$C = vx + F$$

is called a **linear cost function.** See Figure 3–18.

LINEAR AND QUADRATIC FUNCTIONS

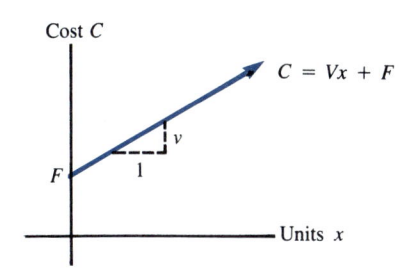

FIGURE 3–18
Linear Cost Function

EXAMPLE 1 Suppose that Acme Company manufactures hammers, and that to produce each hammer requires $2 worth of material and $3 worth of labor. Then the variable cost per hammer is $v = 5$, or $5 per hammer. Suppose that the total fixed cost for Acme Company is $100,000. Then the cost function is defined by

$$C = 5x + 100,000.$$

Note that if no hammers are being produced ($x = 0$), then total cost equals fixed cost ($C = \$100,000$). Total cost C increases by $5 for each increase of 1 in x, the number of hammers produced. The function with defining equation

$$C = 5x + 100,000$$

is the company's cost function. If Acme makes $x = 10{,}000$ hammers, then the total cost will be

$$C = 5(10{,}000) + 100{,}000$$
$$= 150{,}000,$$

or $150,000. See Figure 3–19. ☐

Revenue is income from sales. If R = total revenue, p = sales price per unit, and x = the number of units sold, then the equation

$$R = px$$

defines a **linear revenue function.** See Figure 3–20.

EXAMPLE 2 Suppose that Acme Company sells each hammer at a price of $7, so $p = 7$. Then the company's revenue function is

$$R = 7x,$$

where R is in dollars and x is the number of hammers sold. Note that the revenue is zero if no hammers are sold; that is, $R = 0$ if $x = 0$. Also note that Acme Company's revenue increases by $7 for each increase of 1 in x, the number of hammers sold. If Acme sells 10,000 hammers, then its revenue will be

$$R = 7(10{,}000) = 70{,}000,$$

or $70,000. See Figure 3–21. ☐

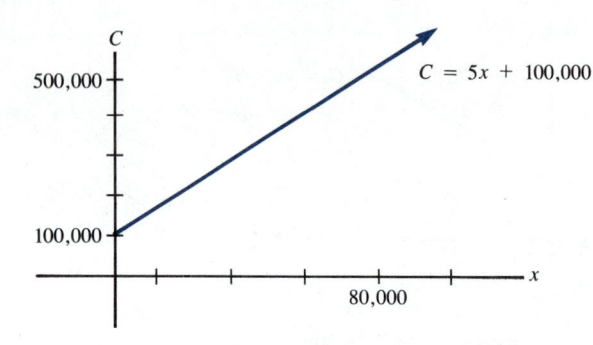

FIGURE 3–19

**FIGURE 3–20
Linear Revenue Function**

(Note that, mathematically speaking, the lines having equations

$$C = 5x + 100{,}000$$

(Figure 3–19) and $R = 7x$ (Figure 3–21) extend indefinitely in both directions. However, since it is not possible to manufacture or sell a negative number of hammers, in reality we cannot allow x to be negative. Therefore, the realities of the situation force us to restrict the domains of these lines to the nonnegative real numbers.)

Profit is revenue minus cost. If $R = px$ and $C = vx + F$ define linear revenue and cost functions, respectively, then the **profit function** P is

$$P = R - C,$$

or

$$P = (p - v)x - F.$$

EXAMPLE 3   If Acme Company's revenue function is given by

$$R = 7x,$$

and its cost function by

$$C = 5x + 100{,}000,$$

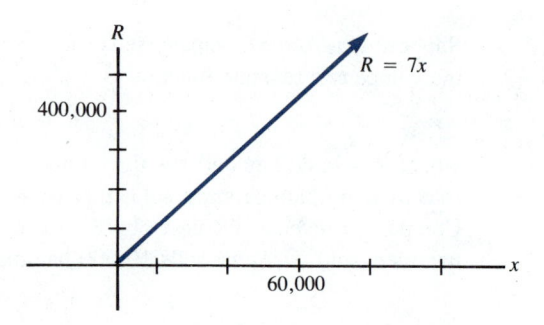

FIGURE 3–21

then its profit function is given by
$$P = 7x - (5x + 100{,}000)$$
or
$$P = 2x - 100{,}000.$$

See Figure 3–22. Thus, if Acme produces and sells 10,000 hammers, its profit will be
$$P = 2(10{,}000) - 100{,}000$$
$$= -80{,}000,$$
or a loss of $80,000. (A negative result for profit denotes a loss.)

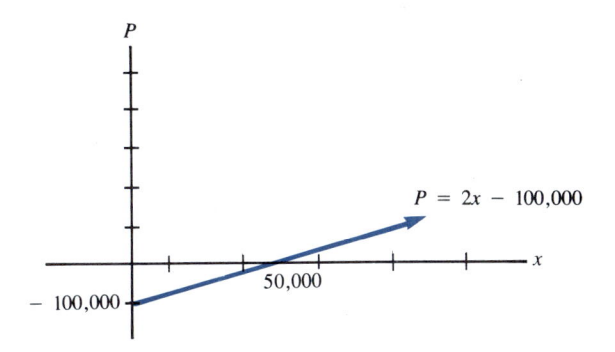

FIGURE 3–22

Let us summarize our knowledge of linear cost, revenue, and profit functions.

Linear Cost, Revenue, and Profit Functions

1. If F is the fixed cost, v the variable cost per unit, and C the total cost of making x units, then the **cost function** is defined by
$$C = vx + F.$$

2. If p is the selling price per unit and R the total revenue from selling x units, then the **revenue function** is defined by
$$R = px.$$

3. If P is the profit from making and selling x units, then the **profit function** is defined by
$$P = R - C,$$
or
$$P = (p - v)x - F.$$

Supply and Demand

Associated with every product or commodity is a **supply curve,** which exhibits the relationship between the price and the quantity of the product supplied to the market by its producers. Typically, the greater the price of an item, the greater the quantity supplied; as the price of an item increases, producers find it more attractive to market the item. Hence the quantity of an item supplied to the market increases as price increases. Supply curves are often linear. Figure 3–23 shows a typical linear supply curve. Note that the quantity q supplied to the market increases as the unit price p increases. Note also that we have drawn the supply curve so that quantity supplied is greater than zero when the price is greater than zero. This will not always be the case.

EXAMPLE 4 When wheat sold for $5.00 per bushel, farmers produced 290 million bushels, and when the price increased to $6.40, farmers produced 371.2 million bushels. Assume that the supply curve is linear. Find its equation and draw its graph.

Letting p = price per bushel and q = quantity supplied in millions of bushels, we have

$$(p_1, q_1) = (5.00, 290.0)$$

and

$$(p_2, q_2) = (6.40, 371.2).$$

Hence

$$q - 290.0 = \left(\frac{371.2 - 290.0}{6.40 - 5.00}\right)(p - 5.00).$$

Solving this equation for q yields

$$q = 58p.$$

See Figure 3–24. Each increase of $1.00 in the price per bushel will result in the additional production of 58 million bushels of wheat. If the price rises to $7.00 per

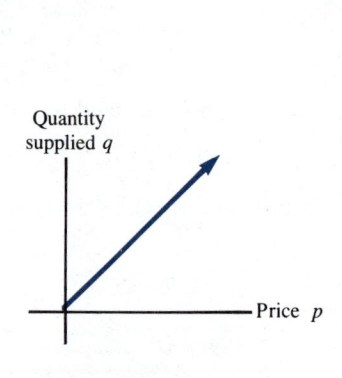

FIGURE 3–23
Linear Supply Curve

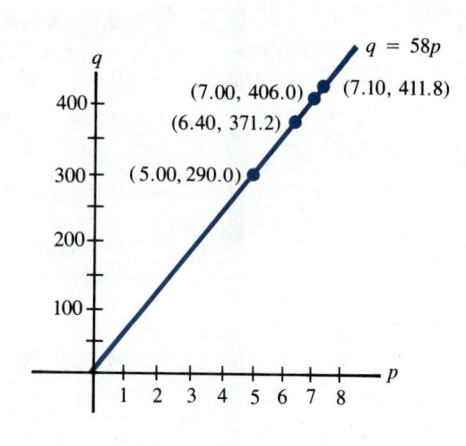

FIGURE 3–24

LINEAR AND QUADRATIC FUNCTIONS

bushel, the quantity supplied will be $q = 58(7) = 406$ million bushels of wheat. If the quantity supplied to the market is known to be 411.8 million bushels, we may solve the equation

$$411.8 = 58p$$

for p to obtain

$$p = 7.10.$$

Thus, a wheat crop of 411.8 million bushels will result in a price of \$7.10 per bushel. ☐

Also associated with every product or commodity is a **demand curve,** which exhibits the relationship between the price of the product and the quantity of the product demanded by consumers. Typically, the greater the price of an item, the less the quantity demanded; as the price of an item increases, consumers are less able to afford the item or less apt to want it even if they can afford it. Hence the quantity of an item demanded by consumers decreases as price increases. Demand curves are often linear. Figure 3–25 shows a typical linear demand curve. Note that the quantity q demanded by consumers decreases as the price p per unit increases.

EXAMPLE 5 When the price of a certain type of electronic calculator was \$100, quantity demanded was 3 million, and when the price dropped to \$30, it was 5.8 million. Assume that the demand curve is linear. Find its equation and draw its graph.

Letting p = price per calculator and q = quantity demanded by consumers in millions of calculators, we have

$$(p_1, q_1) = (100, 3.0)$$

and

$$(p_2, q_2) = (30, 5.8).$$

Hence

$$q - 3.0 = \left(\frac{5.8 - 3.0}{30 - 100}\right)(p - 100).$$

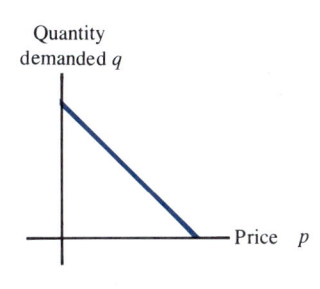

FIGURE 3–25
Linear Demand Curve

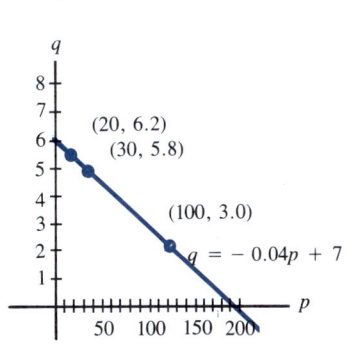

FIGURE 3–26

Solving this equation for q yields

$$q = -0.04p + 7.$$

See Figure 3–26. Each increase of $1 in the price of the calculator will result in a decrease of 0.04 million (40,000) in the quantity of calculators demanded. Note that $q = 7$ when $p = 0$; that is, if calculators were free, the quantity demanded would be 7 million units. If the price should drop to $20, the quantity of calculators demanded would be $q = -0.04(20) + 7 = 6.2$ million units. If we set $q = 0$ and solve the equation

$$0 = -0.04p + 7$$

for p, we obtain

$$p = 175.$$

Thus, if the price per calculator were $175, the quantity demanded would be zero units. ☐

We have adopted the viewpoint that price determines quantity supplied and demanded, but it is equally valid to assume that price is determined by the quantity supplied or demanded. Economists prefer the latter approach. Consequently, instead of writing the supply and demand functions developed in Examples 4 and 5 as

$$q = 58p \quad \text{and} \quad q = -0.04p + 7,$$

respectively, economists would solve these equations for p and write them in the equivalent forms

$$p = \frac{1}{58}q \quad \text{and} \quad p = -25q + 175.$$

The supply and demand curves would then be drawn with price p on the vertical axis and quantity supplied or demanded q on the horizontal axis. See Figures 3–27 and 3–28.

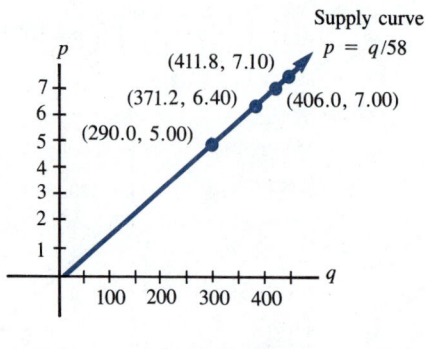

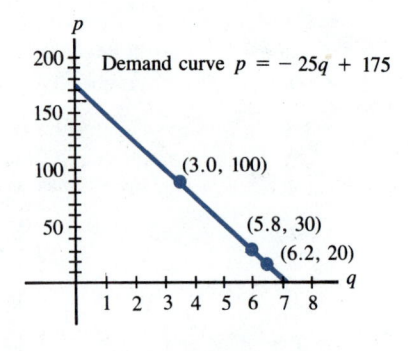

FIGURE 3–27

FIGURE 3–28

■ *Computer Manual: Program SLPNRCPT*

3.2 EXERCISES

Cost, Revenue, and Profit Models

1. A cost function is given by $C = 22x + 4000$. Graph this function. What are the significances of its slope and C-intercept?

2. A cost function is given by $C = 4.75x + 12{,}000$. Graph this function. What are the significances of its slope and C-intercept?

3. A cost function is given by $C = 12.50x$. Graph this function. What are the significances of its slope and C-intercept?

4. A cost function is given by $C = 230x + 247{,}500$. Graph this function. What are the significances of its slope and C-intercept?

5. A revenue function is given by $R = 50x$. Graph this function. What are the significances of its slope and R-intercept?

6. A revenue function is given by $R = 6.34x$. Graph this function. What are the significances of its slope and R-intercept?

7. A revenue function is given by $R = 12.75x$. Graph this function. What are the significances of its slope and R-intercept?

8. A revenue function is given by $R = 225x$. Graph this function. What are the significances of its slope and R-intercept?

9. A profit function is given by $P = 12x - 8400$. Graph this function. What are the significances of its slope and intercepts?

10. A profit function is given by $P = 4.20x - 91{,}980$. Graph this function. What are the significances of its slope and intercepts?

11. A firm's cost function is that of Exercise 1, and its revenue function is that of Exercise 5. Find and graph its profit function. What are the significances of the slope and intercepts of its profit function?

12. A firm's cost function is that of Exercise 2, and its revenue function is that of Exercise 6. Find and graph its profit function. What are the significances of the slope and intercepts of its profit function?

13. A firm's cost function is that of Exercise 3, and its revenue function is that of Exercise 7. Find and graph its profit function. What are the significances of the slope and intercepts of its profit function?

14. A firm's cost function is that of Exercise 4, and its revenue function is that of Exercise 8. Find and graph its profit function. What are the significances of the slope and intercepts of its profit function?

15. A firm's fixed cost is $60,000. It costs the firm $140 to produce each unit of its product, and it sells each unit for $155.
    (a) Find the firm's cost, revenue, and profit functions.
    (b) Find the firm's cost, revenue, and profit if it makes and sells 3000, 4000, and 5000 units.

16. For the firm of Exercise 15:
    (a) Graph its cost, revenue, and profit functions on the same set of axes.
    (b) What happens if the firm makes and sells fewer than 4000 units? more than 4000 units?

17. A firm's fixed cost is $122,500, and its variable cost per unit is $22.50. It sells each unit it makes for $32.99. Find its cost, revenue, and profit functions and graph these on the same set of axes.

18. Repeat Exercise 17 if the fixed cost is $12,820, the variable cost per unit is $6.05, and the selling price is $8.25.

19. A firm's fixed cost is $25,250, its variable cost per unit is $10.30, and it sells each unit it makes for $15.25. Find its cost, revenue, and profit if it makes and sells 5000 units; 5500 units. How much does each unit made and sold contribute toward profit?

20. A firm's fixed cost is $254,000, it costs the firm $77.00 to make each unit of its product, and it sells each unit for $90. Find its cost, revenue, and profit if it makes and sells 15,000 units; 22,000 units. How much does each unit made and sold contribute toward profit?

21. A firm's fixed cost is $40,000 and its variable cost per unit is $120. If its production budget is $520,000, how many units is it planning to make?

22. Repeat Exercise 21 if the firm's fixed cost is $145,000, it costs the firm $14.50 to make each unit, and the production budget is $498,075.

23. A firm sells each unit it produces for $28. How many units must it sell in order to have revenue of $19,152?

24. Gamma Corporation's fixed cost is $250,000, its variable cost per unit is $10, and it sells each unit for $15.
    (a) Find Gamma's cost, revenue, and profit functions.
    (b) Gamma's budget for this year calls for total revenue of $975,000. How many units is Gamma planning to sell this year? What will its profit be if it does sell this many units?

25. Epsilon Company has a fixed cost of $100,000, a variable cost per unit of $20, and sells each unit it produces for $30. If Epsilon has a production budget for the year of $1,000,000, how many units is it planning to make? What will its profit be if it makes and sells this number of units?

26. Zeta Company has a fixed cost of $100,000, a variable cost per unit of $40, and sells each unit it makes for $50. Zeta's profit goal for the year is $200,000. How many units must it make and sell in order to attain this goal?

27. Repeat Exercise 26 if Zeta's fixed cost is $27,500, its variable cost per unit is $17.45, and it sells each unit for $23.75.

28. Theta Associates has budgeted $500,000 for production. Its fixed cost is $80,000, its variable cost per unit is $60, it sells each unit it makes for $75, and its goal is to make at least $50,000 profit. Is this possible?

29. Repeat Exercise 28 if the production budget is $610,000.

30. If Kappa Company's fixed cost is $150,000, its variable cost per unit is $50, and it sells each unit for $60, what will its production budget have to be in order for it to make a profit of at least $100,000?

31. The cost C of making x units of a product is as follows:

x	C
100	$52,000
250	$55,000
325	$56,500

Show that the relationship between cost and units produced is linear, find the cost function, and find the fixed cost and the variable cost per unit. If $100,000 is budgeted for production, how many units can be made?

32. Repeat Exercise 31 if the cost C of making x units is as follows:

x	C
1000	$27,570
2000	$29,840
3000	$32,110

33. Lambda Company's profit at three different production levels was as follows:

Units Produced	Profit
12,000	$35,500
13,500	$60,250
14,200	$71,800

Show that the relationship between units produced and profit is linear, and find the company's profit function and its fixed cost. How many units must it produce in order to make a profit of at least $100,000?

Supply and Demand

34. The supply function for a commodity is given by $q = 220p$. Graph this function. What are the significances of its slope and p-intercept?

35. The supply function for a commodity is given by $q = 12,000p$. Graph this function. What are the significances of its slope and p-intercept?

36. The supply function for a product is given by $q = 1500p - 21,000$. Graph this function. What are the significances of its slope and p-intercept?

37. The supply function for a product is given by $q = 23,000p - 4140$. Graph this function. What are the significances of its slope and p-intercept?

38. The demand function for a commodity is given by $q = -300p + 90,000$. Graph this function. What are the significances of its slope and intercepts?

39. The demand function for a commodity is given by $q = -5000p + 1,000,000$. Graph this function. What are the significances of its slope and intercepts?

40. The demand function for a product is given by $q = -18,250p + 2,500,000$. Graph this function. What are the significances of its slope and intercepts?

41. The demand function for a product is given by $q = -423p + 14,500$. Graph this function. What are the significances of its slope and intercepts?

42. The supply function for a product is given by $q = 300p$, and the demand function for the same product is given by $q = -200p + 150,000$. Graph these functions on the same set of axes. Find the quantity of the product supplied and the quantity demanded when its price is $200 per unit; $300 per unit; $400 per unit.

43. The supply function for a product is given by $q = 2000p - 10,000$, and the demand function for the same product is given by $q = -4000p + 110,000$. Graph these functions on the same set of axes and find the quantity supplied and quantity demanded when the price is $10 per unit; $20 per unit; $30 per unit.

44. Suppose the supply function for a product is given by $q = 1000p$, and the demand function for the same product by $q = -800p + 90,000$. Will there be a surplus or a shortage of the product if its price per unit is
    (a) $20?   (b) $100?   (c) $50?

45. If the supply and demand functions for a commodity are given by $q = 2000p - 4000$ and $q = 100,000 - 1600p$, respectively, will there be a shortage or a surplus of the commodity if it sells for
    (a) $15 per unit?   (b) $25 per unit?   (c) $30 per unit?

46. Suppose the supply function for a commodity is given by $q = 500p - 6000$, and the demand function by $q = -800p + 20,000$.
    (a) Find the price at which the quantity supplied will be 25,000 units. Will there be a shortage or a surplus of the commodity at this price?
    (b) Find the price at which the quantity demanded will be 25,000 units. Will there be a shortage or a surplus of the commodity at this price?

47. For the commodity of Exercise 46:
    (a) Find the price at which the quantity supplied will be 50,000 units. Will there be a shortage or a surplus of the commodity at this price?
    (b) Find the price at which the quantity demanded will be 50,000 units. Will there be a shortage or a surplus of the commodity at this price?

48. For the commodity of Exercise 46, find the price at which
    (a) the quantity supplied is 0 units;
    (b) the quantity demanded is 0 units.

49. Suppose that each increase of $1 in the price of a pound of hamburger results in an increase of 200 million pounds in the quantity supplied to the market and a decrease of 150 million pounds in the quantity demanded by consumers. If quantity demanded when the price was $2 per pound was 350 million pounds, and quantity supplied at this price was also 350 million pounds,
    (a) find the supply and demand functions for hamburger;
    (b) at what price will quantity demanded be zero? at what price will quantity supplied be zero?
    (c) at what price must hamburger sell if the quantity supplied is to be 450 million pounds? if the quantity demanded is to be 450 million pounds?

50. When a product sold for $8.10 per unit, the quantity supplied was 4000 units and the quantity demanded was 41,400 units; when the price was $8.80 per unit, the quantity supplied was 5750 units and the quantity demanded was 37,200 units.
    (a) Find the supply and demand functions for the product, assuming they are linear.
    (b) At what price will quantity supplied be zero? At what price will quantity demanded be zero?
    (c) At what price will quantity supplied be 15,000 units? At what price will quantity demanded be 15,000 units?

(d) Will there be a shortage or a surplus of the product at a unit price of $10? at a price of $12.50? at a price of $14?

51. The following table shows the quantity supplied and quantity demanded of a commodity at certain unit prices:

Unit Price	Quantity Supplied (units)	Quantity Demanded (units)
$2.25	4300	7000
$2.45	4700	6200
$2.80	5400	4800

Show that quantity supplied and demanded are linear functions of price, and find the supply and demand functions. At what prices are quantity supplied and demanded zero? Is there a shortage or a surplus of the commodity at a price of $2.70 per unit?

52. An economist finds that the supply function for a commodity is given by $p = 0.02q$, and the demand function for the commodity by $p = -0.10q + 2500$. (Here p is unit price, and q is the quantity supplied or demanded.)

   (a) Graph these functions with q on the horizontal axis and p on the vertical axis.
   (b) What will the price of the commodity be when the quantity supplied is 10,000 units? when the quantity demanded is 10,000 units?
   (c) What will quantity supplied and quantity demanded be when the price is $1500 per unit?

53. An economist finds that the supply function for a commodity is given by $p = 0.08q$ and the demand function for the commodity by $p = -0.05q + 20$. (Here p is unit price and q is the quantity supplied or demanded in thousands of units.)

   (a) Graph these functions with q on the horizontal axis and p on the vertical axis.
   (b) What will the price of the commodity be when the quantity supplied is 120,000 units? when the quantity demanded is 200,000 units?
   (c) What will quantity supplied and quantity demanded be when the price is $12 per unit?

54. An economist finds that the supply function for a commodity is given by $p = 0.004q + 50$, and the demand function for the commodity by $p = -0.002q + 100$. (Here p is unit price and q is the quantity supplied or demanded.)

   (a) Graph these functions with q on the horizontal axis and p on the vertical axis.
   (b) What will the price of the commodity be when the quantity supplied is 5000 units? when the quantity demanded is 5000 units?
   (c) What will quantity supplied and quantity demanded be when the price is $60 per unit?

55. Solve the equations of Exercise 52 for q in terms of p and graph the resulting supply and demand functions with p on the horizontal axis and q on the vertical axis. Answer questions (b) and (c) of Exercise 52 for these supply and demand functions.

56. Solve the equations of Exercise 53 for q in terms of p and graph the resulting supply and demand functions with p on the horizontal axis and q on the vertical axis. Answer questions (b) and (c) of Exercise 53 for these supply and demand functions.

57. Solve the equations of Exercise 54 for q in terms of p and graph the resulting supply and demand functions with p on the horizontal axis and q on the vertical axis. Answer questions (b) and (c) of Exercise 54 for these supply and demand functions.

3.3 SIMULTANEOUS LINEAR EQUATIONS

If two lines are not parallel, then they intersect in exactly one point. Since that point lies on both lines, its coordinates must satisfy the equations of both lines simultaneously. Finding the point therefore requires the simultaneous solution of two linear equations in two variables. In this section we demonstrate how to solve such a system of equations, and then show how our methods may be used to find a firm's break-even point and to determine the equilibrium price for a commodity. We also briefly discuss the solution of systems of three linear equations in three variables.

Solving Linear Equations Simultaneously

Suppose two nonvertical lines have slope-intercept equations $y = m_1x + b_1$ and $y = m_2x + b_2$. If the lines intersect, the point of intersection is on both of them, so the two expressions for its y-coordinate must be equal; that is, we must have $m_1x + b_1 = m_2x + b_2$. Hence, we can find the point of intersection by solving $m_1x + b_1 = m_2x + b_2$ for x and then back-substituting the result in either of the original equations to find y.

EXAMPLE 1 Let us find the point of intersection of the lines $y = 2x - 7$ and $y = -3x + 3$. We set

$$2x - 7 = -3x + 3$$

and solve for x to obtain

$$x = 2.$$

(Check this.) Substituting 2 for x in either $y = 2x - 7$ or $y = -3x + 3$ yields $y = -3$. Hence the point of intersection of these lines is $(2, -3)$. See Figure 3–29. ◻

EXAMPLE 2 Let us try to find the point of intersection of the lines $y = 2x + 4$ and $y = 2x - 8$. We have

$$2x + 4 = 2x - 8;$$

if we attempt to solve this equation we obtain $4 = -8$, which is never true. Therefore the lines do not intersect. (Notice that both lines have slope equal to 2, so they are parallel.) ◻

EXAMPLE 3 A company makes its product at two plants, A and B. Production at plant A has a fixed cost of $200,000 and a variable cost per unit of $50; that at plant B has a fixed cost of $250,000 and a variable cost per unit of $40. Over what range of production is plant A less costly than plant B?

Suppose we let C_A and C_B denote the cost of producing x units at plant A and plant B, respectively. Then

$$C_A = 50x + 200{,}000 \quad \text{and} \quad C_B = 40x + 250{,}000.$$

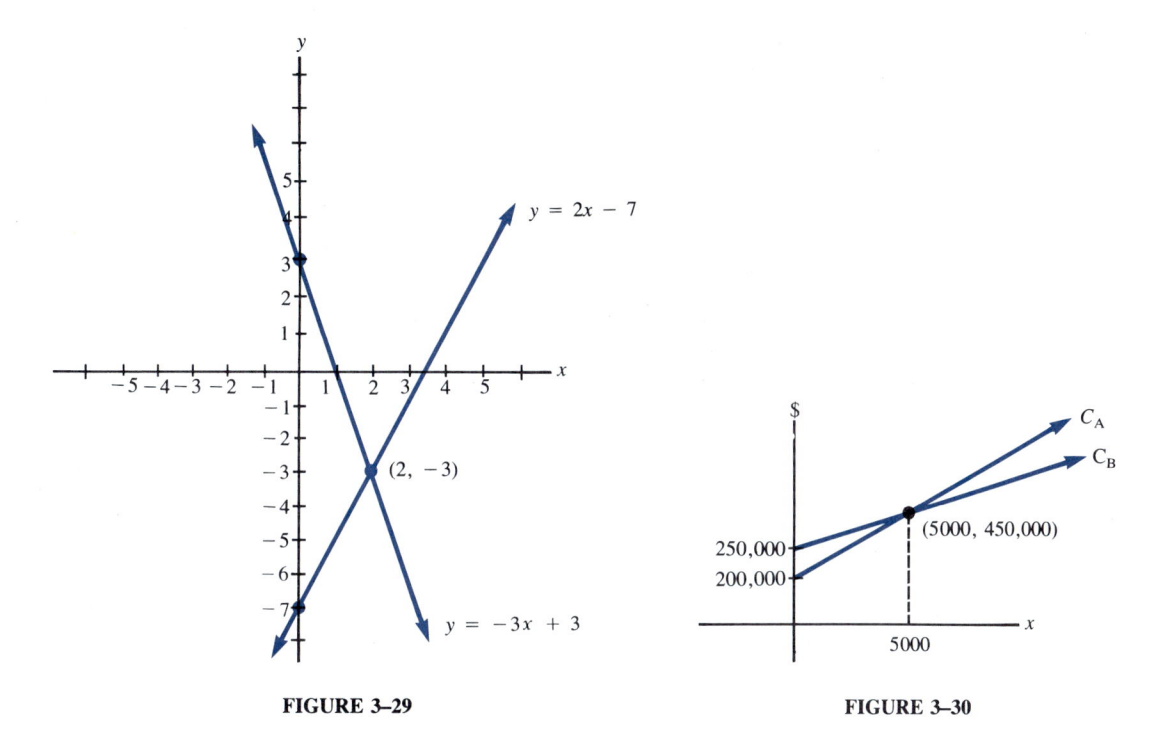

Finding the point of intersection of these lines, we have

$$50x + 200{,}000 = 40x + 250{,}000,$$

which yields $x = 5000$ units and $C_A = C_B = \$450{,}000$. If we now graph the two cost equations on the same set of axes, as in Figure 3–30, we see that

$$C_A < C_B \text{ for } 0 \le x < 5000.$$

Thus plant A is less costly than plant B over the range [0, 5000).

It is always possible to determine the point of intersection of two nonvertical lines by finding the slope-intercept form of their equations and proceeding as in Examples 1, 2, and 3. However, there is another technique for finding the point of intersection of two lines that is useful when the lines are given in general form. This technique involves eliminating one of the variables from the two equations and then solving for the remaining variable. We illustrate with examples.

EXAMPLE 4 Let us find the point of intersection of the lines whose equations in general form are $4x - 2y = 18$ and $-3x + 3y = 9$.

We begin by deciding which variable we intend to eliminate. We choose to eliminate y. Then we must multiply each equation by a nonzero number, choosing the numbers so that adding the resulting equations will eliminate y. In this case we can multiply the first equation by 3 and the second by 2, thus obtaining

$$3(4x - 2y = 18) \rightarrow 12x - 6y = 54$$
$$2(-3x + 3y = 9) \rightarrow \underline{-6x + 6y = 18}$$
$$6x = 72$$

Therefore $x = 72/6 = 12$. Now we find y by substituting 12 for x in either of the original equations: choosing the first of these equations, we have $4(12) - 2y = 18$, which yields $y = 15$. Hence the point of intersection of the lines is $(12, 15)$. □

EXAMPLE 5 Find the point of intersection of the lines having equations $2x - 4y = 8$ and $-6x + 12y = -24$.

Let us eliminate x:

$$3(2x - 4y = 8) \rightarrow 6x - 12y = 24$$
$$1(-6x + 12y = -24) \rightarrow \underline{-6x + 12y = -24}$$
$$0 = 0$$

The result $0 = 0$ indicates that the given equations are equations of the same line. (This is easy to see since both equations are equivalent to the equation $x - 2y = 4$.) Thus there is no single point of intersection; instead, every point on the line $x - 2y = 4$ is a solution of the two simultaneous linear equations. □

EXAMPLE 6 Find the point of intersection of the lines having equations $x + 3y = 5$ and $-3x - 9y = 8$.

We eliminate y:

$$3(x + 3y = 5) \rightarrow 3x + 9y = 15$$
$$1(-3x - 9y = 8) \rightarrow \underline{-3x - 9y = 8}$$
$$0 = 23$$

Since $0 = 23$ is not a true statement, the problem has no solution. There is no point of intersection because the lines having the given equations are parallel. (Both lines have slope $-\frac{1}{3}$.) □

Here is another example of a situation in which it is useful to be able to solve linear equations simultaneously.

EXAMPLE 7 Suppose that each tablet of cold remedy A contains 10 milligrams (mg) of a cough suppressant and 15 mg of antihistamines, and each tablet of remedy B contains 20 mg of cough suppressant and 8 mg of antihistamines. How many tablets of each would you have to take in order to ingest 64 mg of cough suppressant and 41 mg of antihistamines?

Let x denote the number of tablets of remedy A and y the number of remedy B that will be required. Then we have

(cough suppressant) $10x + 20y = 64$,
(antihistamines) $15x + 8y = 41$.

Multiplying through the first equation by 3 and the second by -2 and adding yields

LINEAR AND QUADRATIC FUNCTIONS 155

$$30x + 60y = 192$$
$$-30x - 16y = -82$$
$$44y = 110.$$

Thus

$$y = \frac{110}{44} = 2.5,$$

and

$$10x + 20(2.5) = 64$$

implies that

$$x = \frac{64 - 20(2.5)}{10} = 1.4. \quad \square$$

In the next section we will find it convenient to be able to solve systems of three linear equations in three variables; since the method we will use involves the elimination of variables, we present it here. The idea of the method is to eliminate one of the variables from one pair of the equations, then eliminate the same variable from another pair of the equations. The result will be a system of two equations in two variables, which can be solved as we did above. We illustrate with an example.

EXAMPLE 8 Suppose we wish to solve the system

$$2x + 3y - z = 10$$
$$x - 2y + 2z = -7$$
$$3x + 2y + z = 5$$

for x, y, and z. We decide to begin by eliminating x from the first two equations:

$$\begin{array}{r} 2x + 3y - z = 10 \rightarrow 2x + 3y - z = 10 \\ -2(\ x - 2y + 2z = -7) \rightarrow -2x + 4y - 4z = 14 \\ \hline 7y - 5z = 24 \end{array}$$

Next we eliminate x from another pair of equations; we choose to use the second and third of the original equations:

$$\begin{array}{r} -3(\ x - 2y + 2z = -7) \rightarrow -3x + 6y - 6z = 21 \\ 3x + 2y + z = 5 \rightarrow 3x + 2y + z = 5 \\ \hline 8y - 5z = 26 \end{array}$$

Now we solve the system

$$7y - 5z = 24$$
$$8y - 5z = 26$$

for y and z:

$$\begin{array}{r} -(7y - 5z = 24) \rightarrow -7y + 5z = -24 \\ 8y - 5z = 26 \rightarrow 8y - 5z = 26 \\ \hline y = 2. \end{array}$$

Substituting 2 for y in either $7y - 5z = 24$ or $8y - 5z = 26$ and solving for z now yields $z = -2$. Finally, substituting 2 for y and -2 for z in any one of the three original equations yields $x = 1$. (Check these calculations.) Hence the solution is $x = 1$, $y = 2$, and $z = -2$. □

Break-Even Analysis

Suppose that a firm has a linear cost function with defining equation

$$C = vx + F$$

and a linear revenue function with defining equation

$$R = px,$$

as discussed in Section 3.2. If the graphs of cost C and revenue R intersect as shown in Figure 3–31, the point of intersection is called the **break-even point,** and its x-coordinate is called the **break-even quantity.** If the firm produces and sells its break-even quantity, then cost C is equal to revenue R and the firm will break even. Consequently, since profit is equal to revenue minus cost, the firm's profit will be zero if the firm produces and sells its break-even quantity. Thus we can find the break-even quantity by setting cost equal to revenue or by setting profit equal to zero.

Break-Even Analysis

Let C be the cost function, R the revenue function, and P the profit function for a product. To find the break-even quantity,

set $C = R$ and solve for the independent variable

or

set $P = 0$ and solve for the independent variable.

EXAMPLE 9 Suppose that Acme Company's cost function is given by $C = 5x + 100{,}000$, and its revenue function by $R = 7x$, where both C and R are in dollars and x is the

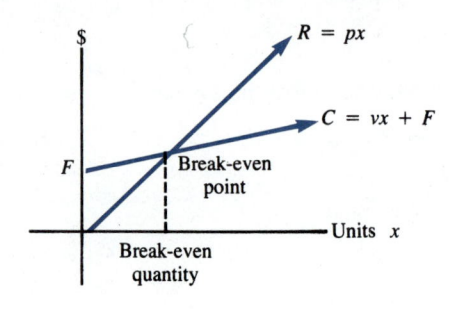

FIGURE 3–31

quantity of Acme's product that is produced and sold. Find Acme Company's break-even point. Graph the cost and revenue functions.

Setting $C = R$ and solving for x, we have

$$5x + 100{,}000 = 7x,$$

so

$$x = 50{,}000.$$

Thus, 50,000 is the break-even quantity and is the first coordinate of the break-even point. To find the second coordinate of the break-even point, we need only substitute 50,000 for x in either the cost equation or the revenue equation. We choose the revenue equation and obtain

$$R = 7(50{,}000) = 350{,}000.$$

Hence the break-even point is (50,000, 350,000). If Acme Company produces and sells 50,000 units of its product, both cost and revenue will be $350,000 and Acme will break even. The graphs are shown in Figure 3–32. Note that Acme Company will lose money if it produces and sells fewer units than its break-even quantity of 50,000 because

$$C > R \quad \text{if } x < 50{,}000.$$

On the other hand, Acme Company will make money if it produces and sells more units than its break-even quantity of 50,000 because

$$C < R \quad \text{if } x > 50{,}000. \quad \square$$

EXAMPLE 10 Suppose that Baker Company's profit function is given by

$$P = 13x - 52{,}000$$

We can find Baker's break-even quantity by setting profit equal to zero and solving for x:

$$13x - 52{,}000 = 0$$

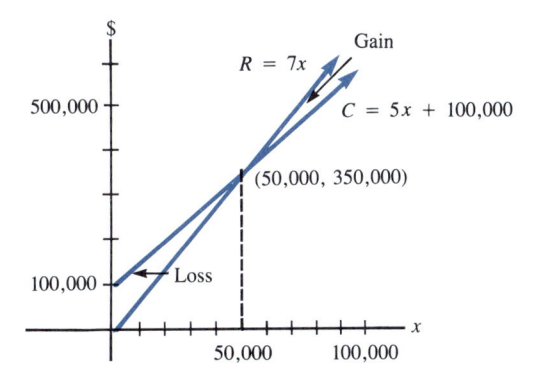

FIGURE 3–32

yields

$$x = 4000.$$

Thus the break-even quantity is 4000 units. The graph of the profit function is shown in Figure 3–33. Note that in this example we cannot find the break-even cost and revenue, since we do not know either the cost or the revenue function. ☐

Market Equilibrium

We discussed linear supply and demand functions in Section 3.2. Suppose the supply and demand curves for a particular product or commodity are linear. Then, since quantity supplied increases as price increases and quantity demanded decreases as price increases, the curves will intersect. The point of intersection is called the **market equilibrium point;** the price coordinate of the point is the **equilibrium price,** and the quantity coordinate is the **equilibrium quantity.** See Figure 3–34. Market equilibrium occurs at the price (equilibrium price) for which quantity supplied equals quantity demanded.

Market Equilibrium

Let p be the price of a commodity, and let $q = f(p)$ be its supply function and $q = g(p)$ its demand function. To find the equilibrium price, set $f(p) = g(p)$ and solve for p.

EXAMPLE 11 Suppose that the supply function for toolboxes is given by $q = 500p$, and the demand function by $q = -1000p + 120,000$, where p is the price in dollars and q is quantity supplied or demanded. Figure 3–35 shows the supply and demand curves graphed on the same set of axes.

To find the equilibrium price, we set quantity supplied equal to quantity demanded and solve for p:

$$500p = -1000p + 120,000,$$

$$p = 80.$$

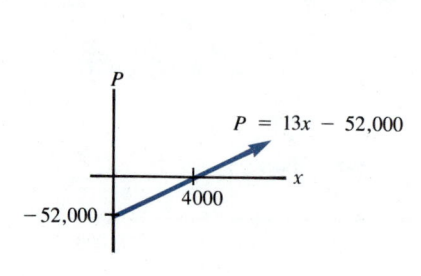

FIGURE 3–33

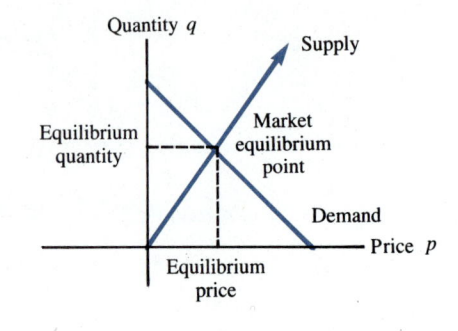

FIGURE 3–34

LINEAR AND QUADRATIC FUNCTIONS

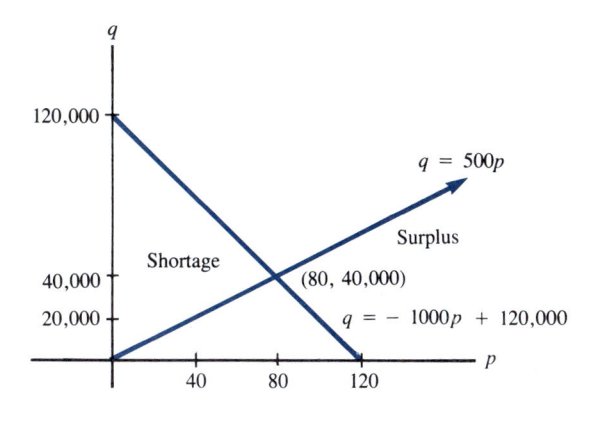

FIGURE 3–35

Thus, the market equilibrium price is $80. To find the market equilibrium point, we find the equilibrium quantity by substituting 80 for p in either the supply equation or the demand equation. Choosing the supply equation, we find

$$q = 500(80) = 40{,}000.$$

Hence, the market equilibrium point is (80, 40,000); that is, if toolboxes sell at a price of $80 each, then producers will supply 40,000 toolboxes to the market and consumers will demand 40,000 toolboxes. If the price is less than $80, there will be a shortage of toolboxes because quantity supplied is less than quantity demanded when $p < \$80$. (See Figure 3–35.) If the price is greater than $80, there will be a surplus of toolboxes because quantity supplied is greater than quantity demanded when $p > \$80$. (See Figure 3–35.) If there is a shortage of toolboxes, consumers will bid up the price, more toolboxes will come onto the market as the price increases, and eventually the equilibrium price will be reached. Similarly, if there is a surplus of toolboxes, suppliers will be forced to lower the price, and this in turn will lead to increased purchases, so eventually the equilibrium price will be reached. Thus, in a free market (a market with many suppliers and consumers and no governmental interference), price will stabilize at the equilibrium price. ☐

■ *Computer Manual: Programs INTRSECT, BREAKEVN*

3.3 EXERCISES

Solving Linear Equations Simultaneously

In Exercises 1 through 8, find the point of intersection of the given lines.

1. $y = 3x$, $y = 2x - 1$
2. $y = 3x - 5$, $y = 4x + 2$
3. $y = 4x + 9$, $y = -2x + 9$
4. $y = -3$, $y = -5x + 1$

5. $y = 2x - 5, y = 5x + 7$
6. $y = -3x + 1, y = -2x + 2$
7. $y = 7x + 3, y = -5x + 1$
8. $y = 5x - 20, y = 5x - 12$

In Exercises 9 through 18, find the point of intersection of the given lines by eliminating one of the variables.

9. $2x + 2y = 6, 5x - 3y = 7$
10. $2x - 2y = 6, 5x + 2y = 9$
11. $-2x + 4y = 8, 2x - 3y = 12$
12. $3x + y = 4, -5x + 4y = 16$
13. $3x - y = 4, x - 6y = 14$
14. $4x - 3y = 11, 8x - 6y = 21$
15. $3x + 2y = 5, -3x - 2y = 5$
16. $x + 3y + 17 = 0, 3x + 2y - 4 = 0$
17. $5x + 2y - 8 = 0, 3x + 2y - 7 = 0$
18. $-2x + 4y = 8, x - 2y + 4 = 0$

In Exercises 19 through 24, solve the given system of linear equations.

19.
$x + 2y - 3z = 14$
$2x - y + z = 0$
$x - 2y + z = -2$

20.
$2x - 3y + 4z = 13$
$-x + 2y - 3z = -8$
$4x - y + 10z = 23$

21.
$x + y - 2z = 0$
$5x + 4y + 3z = 50$
$3x + 2y + 3z = 34$

22.
$5x + 3y = 22$
$2x - y + z = -19$
$2y + 3z = -6$

23.
$2x - 2y + 3z = 1$
$3x + 4y - z = -3$
$5x - 12y + 13z = 7$

24.
$3x + 2y - z = 4$
$-4x - 6y + 7z = 7$
$5x + 4z = 10$

Management

25. Triangle Company manufactures small and large wheelbarrows. Each small wheelbarrow requires 2 worker-hours to produce and each large one requires 3 worker-hours. Each wheelbarrow has one rubber-tired wheel. There are 126 worker-hours and 52 rubber-tired wheels available today. How many of each size wheelbarrow will Triangle produce today?

26. (This exercise draws on material in Peled, "The Next Computer Revolution," *Scientific American*, October 1987.) The speed of computers is measured by how many million instructions per second (MIPS) they can execute. Suppose a business must decide whether to purchase a computer with a single high-speed processor or one with several slower processors. The computer with the single processor will cost $250,000 plus $12 per MIPS, and the one with the multiprocessor will cost $200,000 plus $22 per MIPS.
    (a) Find and graph equations that express the total cost of each computer as a function of the number of MIPS required.
    (b) Over what range of MIPS will the computer with the single processor be the cheaper one?

27. Circle Company intends to purchase a machine for producing a new product. They can purchase any one of three machines, A, B, or C. Machine A costs $20,000 and can make up to 50,000 units of the product per year at a cost of $5.00 per unit. Thus, the cost equation for Machine A is

$$C_A = 5.00x + 20,000,$$

where C_A is in dollars, x is the number of units of the product made, and

$$0 \leq x \leq 50,000.$$

Machine B costs $35,000 and can make up to 40,000 units of the product per year at a cost of $3.50 per unit. Machine C costs $50,000 and can make up to 100,000 units per year at a cost of $3.00 per unit.

(a) Write the cost equations for machine B and machine C.
(b) Graph the cost equations for the three machines on the same set of axes.
(c) For what annual production levels will machine A be the least costly? machine B? machine C?

*28. (This exercise is based on material in Stevenson, *Production/Operations Management*, 2nd edition, Irwin, 1986.) A manufacturer can build a plant at one of four locations. The fixed costs and variable costs per unit for production at each location are as follows:

Location	Fixed Cost	Variable Cost/Unit
A	$100,000	$2.00
B	80,000	2.50
C	86,000	2.20
D	120,000	1.80

For each location, find the range of production for which it will be the best alternative.

Life Science

29. According to an article by Cole and Zuckerman ("Marriage, Motherhood and Research Performance in Science," *Scientific American*, February 1987), during the period 1970–1985 the number of males earning Ph.Ds in the life sciences remained approximately constant at 4000 per year while the number of females earning such Ph.Ds increased from approximately 600 in 1970 to 1700 in 1985. If the increase in female Ph.Ds was linear, when will the annual number of Ph.Ds in the life sciences earned by females equal those earned by males?

30. Each Brand X vitamin pill contains 10 mg of vitamin A and 20 mg of vitamin C. Each Brand Y vitamin pill contains 10 mg of vitamin A and 30 mg of vitamin C. If you require 50 mg of vitamin A and 120 mg of vitamin C each day, how many of each pill must you take?

31. Suppose that biological species A inhabits an area that currently supports 1618 individuals of the species. Suppose also that two individuals of species B have just entered the area, which until now had no individuals of species B. Since species B is more efficient than species A, the population of B will increase by 30 individuals per year, whereas that of A will decrease by 50 individuals per year. How long will it be until the species have equal populations?

32. At the start of a chemical reaction two substances are present in solution, one at a concentration of 3.2 grams per liter (g/L) and the other at a concentration of 0.05 g/L. As the reaction proceeds, the concentration of the first substance decreases linearly and that of the second increases linearly. At the end of the reaction, which requires 4 seconds, the first substance is present in a concentration of 0.01 g/L, and the second is present in a concentration of 3.8 g/L. Approximately how many seconds after the start of the reaction were the concentrations of the substances equal?

Social Science

33. Nationally, Scholastic Aptitude Test (SAT) scores in mathematics have been decreasing linearly at the rate of 4.5 points per year, and the average score in 1987

was 478. A college knows that the average SAT score in mathematics for its entering first-year students has been increasing linearly. The average for the entering first-year students was 385 in 1985 and 397 in 1988. Approximately when will the average SAT score of the college's entering first-year students reach the national average?

*34. In 1988, 48.6% of registered voters were Democrats, 36.4% were Republicans, and the remainder were Independents. The percentage of registered Democrats has been declining at the rate of 2% per year, the percentage of registered Republicans has been declining at the rate of 0.8% per year, and the percentage of registered Independents has been increasing at the rate of 1.9% per year. Approximately how many years will it be until the number of Independents equals the number of

(a) Republicans?
(b) Democrats?
(c) Republicans and Democrats combined?

Break-Even Analysis

35. Alpha Company's fixed cost is $100,000, and its variable cost per unit is $6. It sells each unit for $10. Find the cost and revenue functions and the break-even point.

36. Beta Company's cost function is $C = 15x + 35,000$, and its revenue function is $R = 20x$. Graph these functions on the same set of axes and find the break-even point.

37. Gamma Corporation's profit function is $P = 7x - 280,000$. Graph this function and find Gamma's break-even quantity. What is Gamma's fixed cost?

38. Delta Corporation's cost function is given by $C = 42.50x + 38,500$, and it sells each unit it makes for $55.40. Graph the cost and revenue functions on the same set of axes and find the break-even point.

39. Epsilon, Ltd., has fixed cost of $23,000 and variable cost per unit of $2.40. It sells each unit it makes for $3.20. Find its break-even point.

40. Zeta Company's profit function is given by $P = 2.2x - 319,600$. Graph this function and find the firm's break-even quantity.

41. Repeat Exercise 40 if Zeta's profit function is given by $P = 55.25x - 1,025,000$.

42. A firm's break-even point is (50,000, 145,000), its cost function is linear, its fixed cost is $20,000, and it sells each unit it makes for $2.90. Find its cost, revenue, and profit functions.

43. Repeat Exercise 42 if the break-even point is (80,000, 6,800,000) and the variable cost per unit is $80.

44. If a firm's break-even quantity is 15,000 units, its fixed cost is $12,000, and its profit function is linear, find the profit function.

45. Repeat Exercise 44 if the break-even quantity is 40,000 units and the fixed cost is $150,000.

An article by Julian Szekely ("Can Advanced Technology Save the U.S. Steel Industry?" *Scientific American,* July 1987) contains the following cost information for three types of steel mills:

Linear and Quadratic Functions

Type of Mill	Fixed Cost ($ per ton of capacity)	Variable Cost ($ per ton)
Old integrated	230–280	95–120
Modern integrated	140–190	85–115
Minimill	30–105	110–205

Exercises 46 through 51 refer to this information.

*46. Suppose a mill has 100,000 tons of capacity. Assuming linearity, find its cost function in the best possible case (lowest fixed cost, lowest variable cost per ton) if it is
   (a) an old integrated mill;
   (b) a modern integrated mill;
   (c) a minimill.

*47. Graph the cost functions of Exercise 46 on the same set of axes. If steel sells for $250 per ton, find the break-even quantity for each type of mill and compare it with the mill's capacity.

*48. Suppose steel sells for p per ton. Using the cost functions of Exercise 46, over what range of prices p will each type of mill reach break-even at or before its capacity of 100,000 tons?

*49. Repeat Exercise 46 in the worst possible case (highest fixed cost, highest variable cost per ton).

*50. Graph the cost functions of Exercise 49 on the same set of axes. If steel sells for $250 per ton, find the break-even quantity for each type of mill.

*51. Suppose steel sells for p per ton. Using the cost functions of Exercise 49, over what range of prices p will each type of mill reach break-even at or before its capacity of 100,000 tons?

Market Equilibrium

52. A product has demand function given by $q = -12,000p + 200,000$ and supply function given by $q = 8000p$. Graph these functions on the same set of axes and find the equilibrium price and equilibrium quantity for the product.

53. Repeat Exercise 52 if $q = 8000p - 6000$.

54. A product has demand function given by $q = -15,000p + 500,000$ and supply function given by $q = 25,000p$. Graph these functions on the same set of axes and find the equilibrium price and equilibrium quantity.

55. A commmodity has demand function given by $q = -6250p + 112,000$ and supply function given by $q = 8300p - 4400$. Find the market equilibrium point.

56. Repeat Exercise 55 if $q = -300p + 9348$ and $q = 460p$.

57. A commodity has equilibrium point (9.50, 20,000) and its supply and demand functions are linear. At a price of $4.50 per unit, no units will be supplied and quantity demanded will be 35,000 units. Find the supply and demand functions for the commodity.

58. A product's equilibrium point is (125, 16,250). Its supply and demand functions are linear, and when the unit price is $100, quantity demand is 29,000 units while quantity supplied is 10,750 units. Find the supply and demand functions.

59. Suppose the supply and demand functions for a commodity are given by $p = 0.02q$ and $p = -0.03q + 50$, respectively. Here p is the unit price of the commodity and q is quantity supplied or demanded. Graph these equations on the same set of axes and find the equilibrium point for the commodity.

60. Repeat Exercise 59 if $p = 0.08q$ and $p = -0.06q + 112$.

61. Repeat Exercise 59 if $p = 0.03q + 40$ and $p = -0.025q + 95$.

3.4 QUADRATIC FUNCTIONS

A **quadratic function** is one whose defining equation is of the form
$$y = ax^2 + bx + c,$$
where a, b, and c are real numbers with $a \neq 0$. In this section we study quadratic functions and introduce some useful quadratic models.

Graphing Quadratics

Consider the function with defining equation
$$y = x^2.$$
Writing the equation as
$$y = 1 \cdot x^2 + 0 \cdot x + 0$$
and comparing this with the standard form
$$y = ax^2 + bx + c$$
of a quadratic, we see that $y = x^2$ defines a quadratic function with $a = 1$, $b = 0$, and $c = 0$. Similarly, the equation
$$y = -2x^2 + 8x + 10$$
defines a quadratic function with $a = -2$, $b = 8$, and $c = 10$. The graphs of $y = x^2$ and $y = -2x^2 + 8x + 10$ are shown in Figures 3–36 and 3–37, respectively.

The graph of a quadratic function is called a **parabola.** If we examine the parabolas of Figures 3–36 and 3–37, we note some important facts.

- The parabola of Figure 3–36 is **concave up** (opens upward), and the parabola of Figure 3–37 is **concave down** (opens downward).
- Each parabola has a single **vertex,** that is, a lowest or highest point of the parabola. The vertex of the parabola of Figure 3–36 is the point (0, 0), and the vertex of the parabola of Figure 3–37 is the point (2, 18).
- Each parabola is **symmetric** with respect to a vertical line through its vertex. Thus, if we were to fold the graph along the vertical line through the vertex, the two halves of the parabola would coincide.

The graph of every quadratic function has these properties:

x	$y = x^2$
-5	25
-4	16
-3	9
-2	4
-1	1
0	0
1	1
2	4
3	9
4	16
5	25

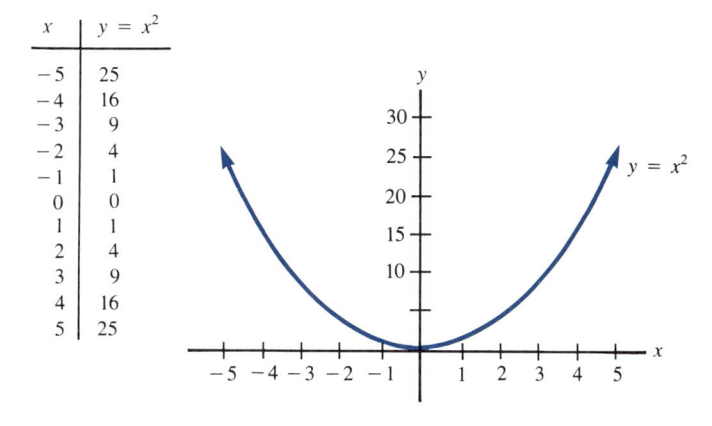

FIGURE 3–36

The Graph of a Quadratic Function

The graph of a quadratic function is a parabola that

1. Is either concave up or concave down.
2. Has a single vertex that is either the highest or lowest point of the parabola.
3. Is symmetric with respect to a vertical line through its vertex.

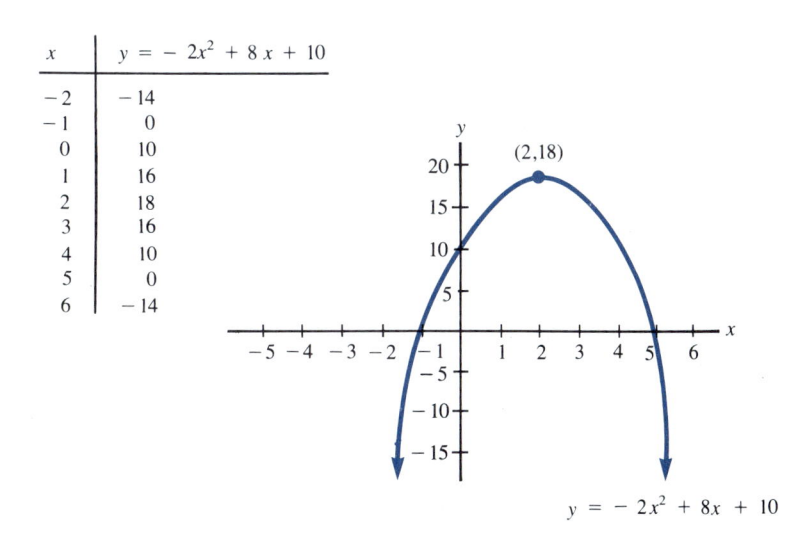

x	$y = -2x^2 + 8x + 10$
-2	-14
-1	0
0	10
1	16
2	18
3	16
4	10
5	0
6	-14

FIGURE 3–37

Consider the standard form

$$y = ax^2 + bx + c$$

of a quadratic. If a is positive, then ax^2 is large and positive when x is large and positive; hence, y is also large and positive, and the parabola that is the graph of $y = ax^2 + bx + c$ is concave up. On the other hand, if a is negative, then ax^2 is negative with a large absolute value if x is large and positive; hence, y is negative with a large absolute value, and the parabola is concave down. Thus, the parabola that is the graph of

$$y = ax^2 + bx + c$$

is concave up if a is positive and concave down if a is negative.

EXAMPLE 1 The graph of $y = -x^2 + 4x + 12$ is a parabola that is concave down, since $a = -1$ is negative. ◻

EXAMPLE 2 The graph of $y = 2x^2 - 7x - 15$ is a parabola that is concave up, since $a = 2$ is positive. ◻

The standard form of the defining equation of a quadratic function can be written as

$$y = a\left(x + \frac{b}{2a}\right)^2 + c - \frac{b^2}{4a}.$$

(You should check the validity of the above identity by simplifying the right-hand side of the equation to obtain $y = ax^2 + bx + c$.) If a is positive, the parabola that is the graph of the equation is concave up and has its vertex at its lowest point. Since

$$\left(x + \frac{b}{2a}\right)^2$$

is nonnegative and since a is positive, the smallest value that y can assume occurs when

$$x + \frac{b}{2a} = 0,$$

that is, when

$$x = -\frac{b}{2a}.$$

Hence, the vertex is at the point with coordinates

$$x = -\frac{b}{2a} \quad \text{and} \quad y = c - \frac{b^2}{4a}.$$

LINEAR AND QUADRATIC FUNCTIONS 167

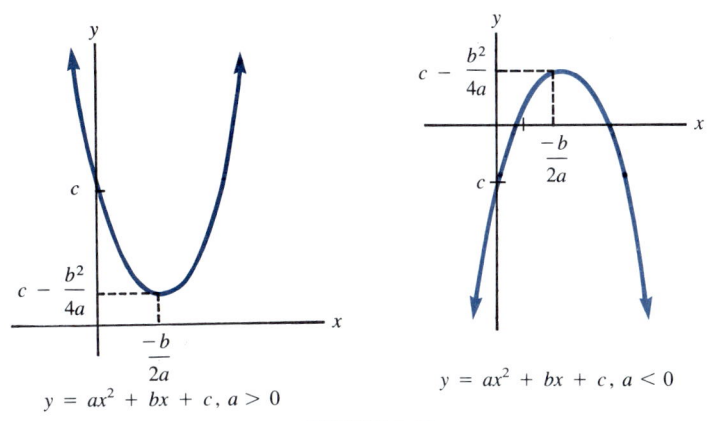

FIGURE 3-38

Similar reasoning shows that the vertex has these same coordinates if a is negative. Consequently, the vertex of the parabola that is the graph of $y = ax^2 + bx + c$ is the point

$$\left(-\frac{b}{2a}, c - \frac{b^2}{4a}\right).$$

See Figure 3-38.

EXAMPLE 3 The quadratic $y = -x^2 + 4x + 12$ has $a = -1$, $b = 4$, and $c = 12$. Therefore

$$-\frac{b}{2a} = -\frac{4}{2(-1)} = 2,$$

and

$$c - \frac{b^2}{4a} = 12 - \frac{16}{4(-1)} = 16.$$

The parabola that is the graph of this quadratic is therefore concave down with vertex at (2, 16). ☐

EXAMPLE 4 The quadratic $y = 2x^2 - 7x - 15$ has $a = 2$, $b = -7$, and $c = -15$. The parabola that is its graph is therefore concave up with vertex at

$$\left(-\frac{b}{2a}, c - \frac{b^2}{4a}\right) = \left(\frac{7}{4}, -\frac{169}{8}\right).$$

(Check this.) ☐

Just as two points determine a line, three points not on a line determine a parabola. Therefore if we wish to graph a quadratic $y = ax^2 + bx + c$, it is necessary to locate at least three points on its parabola. We would like the vertex to be one of these points, and we should also locate the intercepts. The y-intercept is the point $(0, c)$. (Why?) The x-intercepts, if any, may be found by solving the quadratic equation $ax^2 + bx + c = 0$ for x. (Why?)

Let us summarize our results:

Keys to Graphing a Quadratic Function

The graph of the quadratic function defined by

$$y = ax^2 + bx + c$$

is a parabola that

1. Is concave up if a is positive and concave down if a is negative.

2. Has its vertex at $\left(-\dfrac{b}{2a},\ c - \dfrac{b^2}{4a}\right)$.

3. Has y-intercept $(0, c)$.

4. Has x-intercepts, if the parabola intersects the x-axis, at the solutions of the quadratic equation $ax^2 + bx + c = 0$.

EXAMPLE 5 Draw the graph of $y = -x^2 + 4x + 12$.

In Examples 1 and 3 we saw that the parabola is concave down and has its vertex at $(2, 16)$. The y-intercept is $(0, c) = (0, 12)$. To find its x-intercepts we solve the equation

$$-x^2 + 4x + 12 = 0.$$

Factoring this equation, we have

$$-(x - 6)(x + 2) = 0.$$

Therefore,

$$x = 6 \quad \text{and} \quad x = -2,$$

and the x-intercepts are $(6, 0)$ and $(-2, 0)$. The graph is shown in Figure 3–39. ☐

EXAMPLE 6 Draw the graph of $y = 2x^2 - 7x - 15$.

From Examples 2 and 4, the parabola is concave up and has its vertex at $(7/4, -169/8)$. The y-intercept is $(0, -15)$. Using the quadratic formula to solve the equation

$$2x^2 - 7x - 15 = 0,$$

we have

$$x = \frac{7 \pm \sqrt{49 - 4(2)(-15)}}{4} = \frac{7 \pm \sqrt{169}}{4}$$

$$= \frac{7 \pm 13}{4}.$$

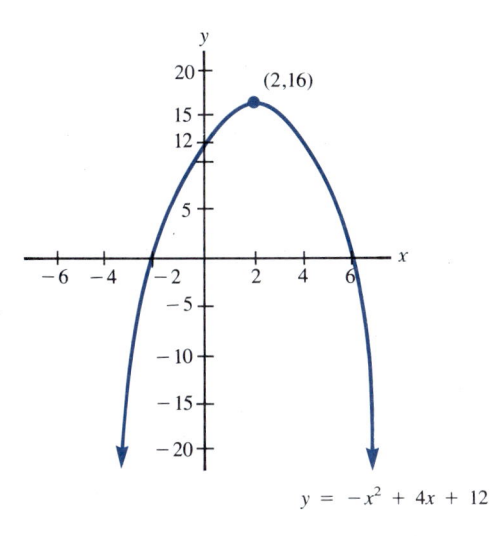

FIGURE 3–39

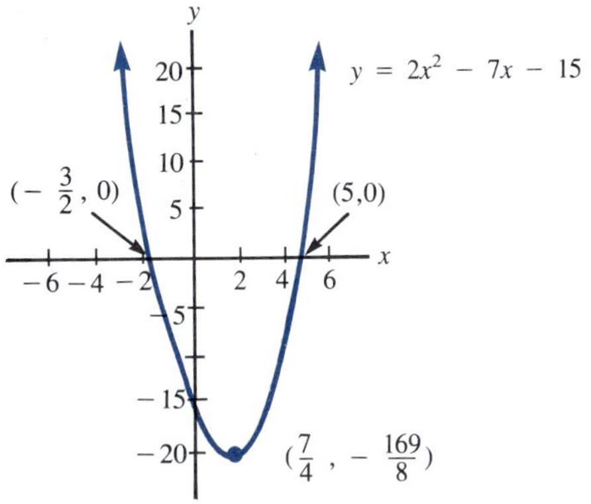

$y = -x^2 + 4x + 12$

FIGURE 3–40

$y = 2x^2 - 7x - 15$

Thus,
$$x = 5 \quad \text{and} \quad x = -\frac{3}{2}.$$

The x-intercepts are (5, 0) and $(-\frac{3}{2}, 0)$. The graph is shown in Figure 3–40.

EXAMPLE 7 Draw the graph of $y = 4x^2 - 20x + 25$.

Here $a = 4$, $b = -20$, and $c = 25$. The parabola is concave up, since $a = 4$ is positive, has its vertex at

$$\left(-\frac{b}{2a}, c - \frac{b^2}{4a}\right) = \left(\frac{5}{2}, 0\right),$$

and has y-intercept (0, 25). Using the quadratic formula to solve the equation

$$4x^2 - 20x + 25 = 0,$$

we have

$$x = \frac{20 \pm \sqrt{400 - 4(4)(25)}}{8} = \frac{5}{2}.$$

Thus the only x-intercept is $(\frac{5}{2}, 0)$. We need another point to draw the graph. Arbitrarily choosing $x = 4$ and substituting this value for x in the equation $y = 4x^2 - 20x + 25$, we obtain $y = 9$. Thus, the point (4, 9) lies on the parabola. The graph is shown in Figure 3–41.

FIGURE 3–41

FIGURE 3–42

EXAMPLE 8   Draw the graph of $y = -3x^2 + 6x - 4$.

The parabola is concave down, has its vertex at $(1, -1)$, and has y-intercept $(0, -4)$. Using the quadratic formula to solve the equation

$$-3x^2 + 6x - 4 = 0,$$

we obtain

$$x = \frac{-6 \pm \sqrt{36 - 4(-3)(-4)}}{-6} = \frac{-6 \pm \sqrt{-12}}{-6}.$$

Thus, the equation $-3x^2 + 6x - 4 = 0$ has no real solutions and the parabola has no x-intercepts. We need another point to draw the graph. Arbitrarily choosing $x = 2$ and substituting this value for x in the equation $y = -3x^2 + 6x - 4$, we find $y = -4$. Hence, the point $(2, -4)$ lies on the parabola. The graph is shown in Figure 3–42. ◻

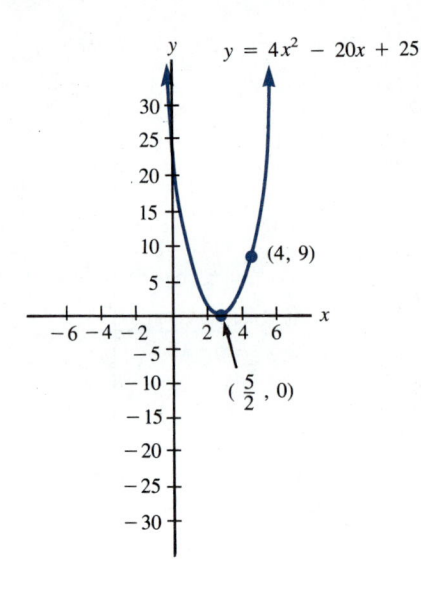

Quadratic Models

Quadratic functions are often useful in creating models for situations that cannot be handled with linear functions. We illustrate with examples.

EXAMPLE 9   Suppose that t months after the inauguration of the President of the United States, y percent of the public approves of the job he is doing, where

$$y = 0.05t^2 - 3t + 77.$$

The graph of this quadratic is the parabola of Figure 3–43. Note that public approval of the president is high (77%) at the time of his inauguration, that it declines

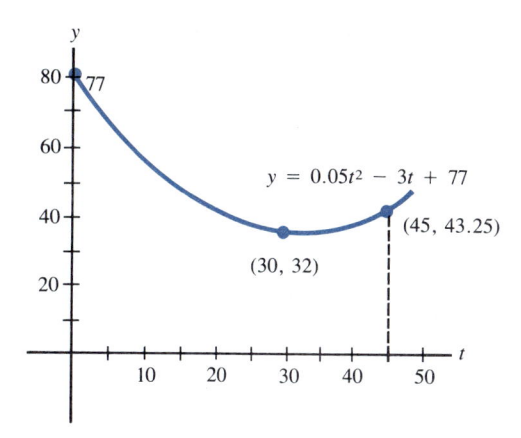

LINEAR AND QUADRATIC FUNCTIONS

FIGURE 3-43

FIGURE 3-44

as his term proceeds, reaching a low point of 32% at 30 months after inauguration, and that it then increases somewhat as the next election approaches. If the next election will take place 45 months after inauguration, we can use the model to predict that at that time the president will have the approval of

$$y = 0.05(45)^2 - 3(45) + 77 = 43.25,$$

or 43.25% of the population. ☐

EXAMPLE 10 Cost, revenue, and profit functions can be quadratics. For instance, suppose that Acme Company's cost and revenue functions are given by

$$C = x^2 + 30x + 500 \quad \text{and} \quad R = 90x.$$

The graphs of these functions are shown in Figure 3-44. We see from the figure that Acme has two break-even points, which may be found by setting cost equal to revenue and solving for x:

$$C = R,$$
$$x^2 + 30x + 500 = 90x,$$
$$x^2 - 60x + 500 = 0,$$
$$(x - 10)(x - 50) = 0.$$

Thus, $x = 10$ and $x = 50$. The break-even points are $(10, 900)$ and $(50, 4500)$, as shown in Figure 3-44. ☐

EXAMPLE 11 Supply and demand functions can also be quadratics. To illustrate, suppose that the supply and demand functions for gadgets are given by

$$q = p^2 + 30p \quad \text{and} \quad q = -p^2 - 100p + 2400,$$

respectively. See Figure 3-45. If we set supply equal to demand and solve for price p, we obtain $p = 15$ and $p = -80$. (Check this.) Discarding the negative solution

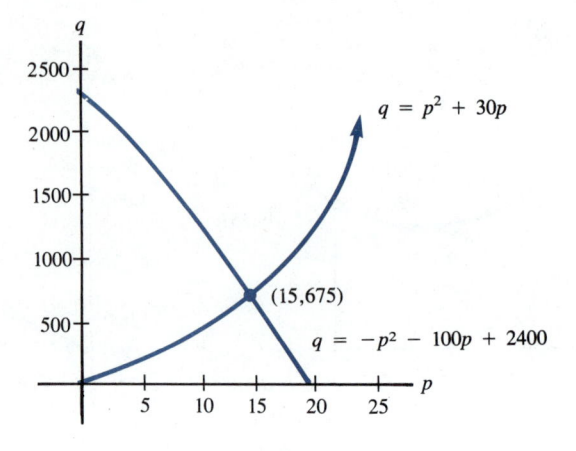

FIGURE 3-45

as meaningless, we see that the market equilibrium point is (15, 675), as indicated in Figure 3–45. □

Any three points not on a line determine a parabola. Thus we can fit a quadratic model to any three observations that do not lie on a line.

EXAMPLE 12 One hour after being admitted to a hospital, a patient's temperature was 2.1° above normal, 2 hours after being admitted it was 3.0° above normal, and 3 hours after being admitted it was 3.7° above normal. Let us fit a quadratic model to these observations.

We let y denote the patient's temperature t hours after being admitted to the hospital, y measured in degrees above normal. Then $y = 2.1$ when $t = 1$, $y = 3.0$ when $t = 2$, and $y = 3.7$ when $t = 3$. If y is a quadratic function, so that $y = at^2 + bt + c$ for some a, b, and c, we have

$$2.1 = a \cdot 1^2 + b \cdot 1 + c,$$
$$3.0 = a \cdot 2^2 + b \cdot 2 + c,$$
$$3.7 = a \cdot 3^2 + b \cdot 3 + c,$$

or

$$a + b + c = 2.1,$$
$$4a + 2b + c = 3.0,$$
$$9a + 3b + c = 3.7.$$

We solve this system of equations for a, b, and c: subtracting the first equation from the second and then the first from the third, we obtain

$$\begin{array}{rl} 4a + 2b + c = & 3.0 \\ -a - b - c = & -2.1, \\ \hline 3a + b = & 0.9 \end{array} \qquad \begin{array}{rl} 9a + 3b + c = & 3.7 \\ -a - b - c = & -2.1. \\ \hline 8a + 2b = & 1.6 \end{array}$$

Linear and Quadratic Functions

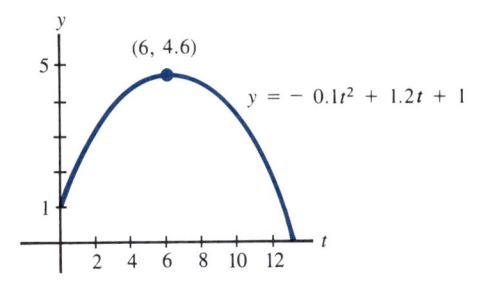

FIGURE 3-46

Then

$$8a + 2b = 1.6$$
$$-2(3a + b = 0.9)$$
$$\overline{2a \qquad = -0.2}$$

Hence,

$$a = -0.1,$$
$$b = 0.9 - 3(-0.1) = 1.2,$$
$$c = 2.1 - (-0.1) - 1.2 = 1.0.$$

Therefore,

$$y = -0.1t^2 + 1.2t + 1.$$

See Figure 3-46. Note that the patient's maximum temperature is 4.6° above normal, and that this occurs 6 hours after admittance. ◻

■ *Computer Manual: Program PARABOLA*

3.4 EXERCISES

Graphing Quadratics

In Exercises 1 through 20, graph the given quadratic, labeling the vertex and intercepts.

1. $y = 2x^2$
2. $y = -x^2$
3. $y = x^2/2$
4. $y = x^2 + 2$
5. $y = -x^2 + 9$
6. $y = -2x^2 - 50$
7. $y = x^2 + 4x$
8. $y = -x^2 + 4x$
9. $y = 3x^2 - 6x$
10. $y = -4x^2 + 8x$
11. $y = x^2 + 2x - 15$
12. $y = -x^2 + 2x - 3$
13. $y = 2x^2 - x - 6$
14. $y = 2x^2 - 16x + 32$
15. $y = x^2 + x + 1$
16. $y = 2x^2 - 6x + 1$
17. $y = -2x^2 + 8x + 9$
18. $y = x^2 + 5x + 10$
19. $y = -3x^2 + 60$
20. $y = -9x^2 + 6x - 1$

Quadratic Models

21. The number of sunspots on the sun varies over an 11-year cycle. Suppose that the average number of sunspots visible t years into the cycle is y, where
$$y = 0.5t^2 - 5.5t + 19,$$
for $0 \le t \le 11$. Graph this quadratic, and find the minimum average number of sunspots visible and the time in the cycle when this minimum occurs.

22. A rocket's altitude, in miles, t minutes after launching is given by the equation
$$y = -\frac{1}{2}t^2 + 18t.$$
Graph this quadratic.
   (a) What is the rocket's maximum altitude? When will it be attained?
   (b) How long will it take after launching for the rocket to return to earth?

23. Repeat Exercise 22 if $y = -0.08t^2 + 8.4t$.

24. A mortar is fired from a hillside into a valley. The height of the mortar shell above the valley floor t seconds after firing is
$$y = -12t^2 + 72t + 100$$
feet. Graph this quadratic.
   (a) How far above the valley floor is the mortar?
   (b) What is the maximum altitude reached by the mortar shell, and when does this maximum occur?
   (c) When does the mortar shell reach the ground?

Management

25. Alpha Company's cost function is given by the equation
$$C(x) = x^2 + 5000,$$
where x is the number of units produced and $C(x)$ is in dollars. Alpha Company sells each unit of its product for $150.
   (a) What is the defining equation of Alpha Company's revenue function?
   (b) Draw the graphs of C and R on the same set of axes.
   (c) Find Alpha Company's break-even quantity or quantities.
   (d) Find Alpha Company's profit function and graph it.
   (e) What is Alpha Company's maximum profit, and how many units must Alpha Company produce and sell in order to realize it?

26. Beta Company's cost function is given by the equation
$$C(x) = x^2 + 40x + 4200,$$
and its revenue function is given by the equation
$$R(x) = x^2 + 100x,$$
where x is the number of units produced and sold and both $C(x)$ and $R(x)$ are in dollars.
   (a) Draw the graphs of C and R on the same set of axes.
   (b) Find Beta Company's break-even quantity or quantities.
   (c) Find Beta Company's profit function and graph it.

Linear and Quadratic Functions

(d) What is Beta Company's maximum profit, and how many units must Beta Company produce and sell in order to realize it?

27. Answer the questions of Exercise 26 if Beta Company's cost function is given by the equation
$$C(x) = 80 + 3x,$$
and its revenue function is given by the equation
$$R(x) = 0.5x^2.$$

28. Gamma Corporation's profit function is given by
$$P = -x^2 + 100x - 1600.$$
Graph this function.
   (a) Over what range of x will Gamma make money?
   (b) What is Gamma's maximum profit and where does it occur?

29. Repeat Exercise 28 if $P = -0.02x^2 + 1.5x - 27$, where x is in thousands of units and P is in millions of dollars.

30. The demand function for a certain product is given by the equation
$$q = -p^2 + 10{,}000,$$
and the supply function for the product is given by the equation
$$q = 150p.$$
Graph the demand and supply functions on the same set of axes. Find the market equilibrium price.

31. Repeat the work of Exercise 30, but assume that
$$q = -p^2 - 20p + 8000 \quad \text{and} \quad q = p^2 + 100p.$$

32. Suppose the supply and demand equations for a commodity are
$$q = p^2 + 20p - 50 \quad \text{and} \quad q = -p^2 - 10p + 150,$$
respectively, where q is in thousands of units. Find the market equilibrium price and quantity for the commodity.

33. Repeat Exercise 32 if
$$q = 8p^2 + 2p - 10 \quad \text{and} \quad q = -4p^2 - 3p + 15,$$
where q is in millions of units.

34. A firm's fixed cost is $50,000, the cost of making 10,000 units of its product is $100,000, and the cost of making 20,000 units is $250,000. Find the firm's cost function if it is known to be a quadratic.

35. When a firm sold 5 units of its product, its revenue was $15; when it sold 10 units, its revenue was $40. Assuming that revenue from sales of 0 units is $0 and that the revenue function is a quadratic, find the revenue function.

36. A firm's fixed cost is $60,000, it lost $32,000 when it made and sold 4000 units, and its profit when it made and sold 16,000 units was $4,000. Find a quadratic model for the firm's profit function, and find the firm's break-even quantities and its maximum profit.

Life Science

37. In certain cases in which patients have rapidly increasing body temperatures, powerful drugs are employed to decrease temperature. Suppose that t minutes after the injection of such a drug a patient's temperature is
$$y = -0.006t^2 + 0.36t + 4.2$$
degrees above normal.
   (a) What will be the patient's maximum temperature above normal, and how long after the injection of the drug will it occur?
   (b) How long will it take for the patient's temperature to return to normal?

38. The concentration of sulfur dioxide in the air of a city from 7 A.M. to 7 P.M. is given by
$$y = -0.3t^2 + 5t + 3,$$
where y is the concentration (measured in parts per million) at time t, $t = 0$ representing 7 A.M. Graph this function. At what time of day is the concentration at its greatest? What is the maximum concentration?

39. A drug is administered to a patient, and t hours later y mg of the drug are present in the patient's liver, where
$$y = -0.025t^2 + 1.8t.$$
What is the maximum amount of the drug present in the liver, and when does this occur? When will all the drug be gone from the liver?

40. The furbish lousewort plant was placed on the endangered species list, and t years later there were
$$y = 2t^2 - 32t + 800$$
living furbish louseworts. How many were there when the plant was placed on the endangered species list? What was the minimum number of plants, and when did this minimum occur?

41. A drug is ingested and t days later $y\%$ of it remains in the body. If y is a quadratic function of t, and it is known that one day after ingestion 87% of the drug remains in the body, two days after ingestion 72% remains, and three days after ingestion 55% remains, find the function. Approximately how long will it take for all of the drug to be eliminated from the body?

42. A census of the bald eagles along an Alaskan river counted 200 eagles; two years later another census counted 276 of them, and two years after the second census a third one counted 384 of them. If the size of the eagle population can be modeled by a quadratic function, find the function and then find the minimum size of the population and when it occurred.

Social Science

43. Altruistic Charities has found that if it sends out x mailings per year soliciting donations, it nets y dollars, where
$$y = -30,000x^2 + 240,000x.$$
How many mailings should Altruistic Charities send out each year in order to maximize net contributions? What will the maximum net contributions be?

44. Voter interest, as expressed by the percentage of eligible voters who state that they intend to vote, usually peaks before election day. Suppose a political scientist finds that t days before an election ($0 \leq t \leq 30$), y percent of the eligible voters intend to vote, where

$$y = -0.07t^2 + 1.26t + 60.$$

Graph this function. How many days before the election is voter interest at its peak? What percentage of eligible voters actually vote?

45. When a governor came into office, her voter approval rating was 60%; six months later it was 78%, and after one year in office it was 74.4%. If voter approval can be modeled by a quadratic function, find the function. Then find the governor's maximum approval rating and when it occurred.

SUMMARY

This summary consists of terms and symbols whose meaning you should know, facts you should know, and techniques and methods you should be able to employ.

Section 3.1

Linear function. Graphing linear functions. Vertical line as graph of relation $x = a$. Slope of a line. Significance of slope. Vertical line has no slope. Finding the equation of a line. Slope-intercept form of the equation of a line. Parallel lines. Nonvertical lines are parallel if and only if their slopes are equal. Finding the slope of a line through two points. Point-slope form of the equation of a line. Equation of a vertical line. General form of the equation of a line. Finding and interpreting linear equations for applied problems that lead to such equations.

Section 3.2

Fixed cost, variable cost, variable cost per unit. Linear cost function. Revenue. Linear revenue function. Profit = Revenue − Cost. Linear profit function. Finding and interpreting linear cost, revenue, profit functions. Linear supply function. Linear demand function. Supply and demand functions with price as a function of quantity. Finding and interpreting linear supply and demand functions.

Section 3.3

Intersecting lines, point of intersection. Finding the point of intersection. Simultaneous linear equations in two and three variables. Solving simultaneous linear equations by elimination. Setting up and solving applied problems that require finding the point of intersection of two lines. Break-even point, break-even quantity. Performing break-even analysis using linear cost, revenue, profit functions. Market equilibrium point, equilibrium price, equilibrium quantity.

Finding the market equilibrium point. Analyzing the relation between quantity supplied and quantity demanded using the market equilibrium point.

Section 3.4

Quadratic function. Parabola. Concave up, concave down. Vertex of a parabola. Properties of the graph of a quadratic function. Graphing quadratic functions: finding concavity, vertex, intercepts. Quadratic models: cost, revenue, profit, supply, demand. Quadratic break-even analysis. Finding market equilibrium for quadratic supply and demand functions. Fitting a quadratic model to three points. Interpreting the graphs of quadratic models, finding maximum or minimum points.

REVIEW EXERCISES

In Exercises 1 through 10, find the slope and intercepts of the given line and draw its graph.

1. $y = 4x - 3$
2. $y = -3x + 9$
3. $y = 6$
4. $x = 5$
5. $4x - 6y = -12$
6. $2x + 5y = 20$
7. $y = -2x$
8. $3x - 11y = -4$
9. $-5x + 3y = 0$
10. $-0.05x - 0.03y = 3$

In Exercises 11 through 22, find the slope-intercept form of the equation of the given line.

11. The line with slope 1/3 that passes through the point (0, 6).
12. The horizontal line that passes through the point (2, 6).
13. The line that passes through the points (2, −7) and (−9, 26).
14. The line with slope −5 that passes through the point (2, −3).
15. The line parallel to $4x + 2y = 30$ that has the same x-intercept as $3x + 5y = 45$.
16. The line with y-intercept (0, −3) and parallel to $y = 5x - 7$.
17. The line parallel to $2x - 6y = 8$ that passes through the point (1, 1).
18. The line with slope 0 that passes through the point (−6, 9).
19. The line that passes through the points (−3/5, 2/5) and (7/4, −11/4).
20. The line that passes through the points (8, 10) and (−10, 10).
21. The line whose equation is $-2x + 10y = 26$.
22. The line whose equation is $8x + 12y = -20$.
23. The number of bacteria in a culture treated with a bactericide declines at a rate of 6000 bacteria per hour. If the culture contained 120,000 bacteria at the time the bactericide was applied, find the equation that describes the number of bacteria present as a function of time t. How long will it take until all the bacteria are dead?
24. A firm's sales were $36,800 in 1985, $74,000 in 1988, and $98,000 in 1990. Find an equation that relates sales to time in years, and find

(a) the firm's sales in 1995;
(b) the year in which sales will reach $148,400.

25. A firm's cost and revenue functions are given by
$$C(x) = 12x + 36,000 \quad \text{and} \quad R(x) = 17x,$$
respectively. Graph these functions on the same set of axes, and find the firm's fixed cost, variable cost per unit, and the price at which it sells each unit.

26. A firm's fixed cost is $84,000. Its variable cost per unit is $120, and it sells each unit for $136.
    (a) Find the firm's cost, revenue, and profit functions, and graph them on the same set of axes.
    (b) What are the significances of the slopes and intercepts of these linear equations?
    (c) Find the firm's cost, revenue, and profit if 6000 units are made and sold.

27. If a commodity sells for $20, the quantity supplied to the market will be 30,000 units and the quantity demanded will be 280,000 units; if it sells for $30, the quantity supplied will be 50,000 and the quantity demanded will be 220,000.
    (a) Assuming linearity, find the supply and demand functions.
    (b) What are the significances of the slopes and intercepts of these linear equations?
    (c) Find the quantity supplied and the quantity demanded when the price is $50 per unit.
    (d) At what price would the quantity supplied be 100,000 units? At what price would the quantity demanded be 200,000 units?

In Exercises 28 through 31, find the point of intersection of the given lines.

28. $y = 3x + 6$, $y = 5x - 12$
29. $4x - 3y = 9$, $2x + 5y = 6$
30. $3x - 2y/5 = 11$, $15x - 2y = 30$
31. $9x - 2y/3 = 7$, $27x - 2y = 21$

32. A farmer wishes to feed cattle a diet supplement containing both calcium and an antibiotic. One brand of diet supplement contains 100 units of calcium and 50 units of antibiotic per pound, and another brand contains 140 units of calcium and 25 units of antibiotic per pound. If the cattle require 446 units of calcium and 160 units of antibiotic per week, how much of each brand should be fed to them?

33. A firm's fixed cost is $92,000, its variable cost per unit is $46, and it sells each unit for $62. Find the break-even point.

34. The supply and demand functions for a commodity are given by
$$q = 6250p - 15,625 \quad \text{and} \quad q = -8500p + 183,500,$$
respectively. Find the equilibrium price and quantity for the commodity.

35. The supply and demand functions for a commodity are
$$p = 0.07q + 70 \quad \text{and} \quad p = -0.05q + 220,$$
respectively. Graph these functions on the same set of axes and find the equilibrium point for the commodity.

In Exercises 36 through 41, graph the given quadratic, labeling the vertex and intercepts.

36. $y = x^2 - 6x - 16$
37. $y = -x^2 + 11x - 28$
38. $y = -2x^2 + 6x$
39. $y = 2x^2 - 10x + 15$
40. $y = -4x^2 + 12x - 9$
41. $y = -x^2/2 + 72x - 22$

42. A firm's cost and revenue functions are given by
$$C(x) = x^2 + 140x + 1050 \quad \text{and} \quad R(x) = x^2/2 + 190x.$$

(a) Find the firm's break-even point or points.
(b) Find and graph the firm's profit function.
(c) What is the firm's maximum profit, and how many units must be made and sold in order for the maximum profit to be attained?

43. A patient was admitted to a hospital with a declining white blood cell count, described by the equation
$$y = t^2 - 30t + 2225.$$
Here t is time in hours since the patient was admitted, and y is the blood count.
   (a) What was the patient's minimum blood count, and when did it occur?
   (b) When the patient's blood count reached 3225, the patient was discharged. When did this occur?

44. The supply curve for a commodity is a parabola. When the unit price of the commodity was $10, 38,000 units were supplied to the market; similarly, when the price was $15, 118,000 units were supplied, and when the price was $20, 238,000 units were supplied. Find the equation of the supply curve and the approximate price at which quantity supplied will be zero.

ADDITIONAL TOPICS

Here are some suggestions for topics not covered in the text that you might want to investigate on your own.

1. Show that the length L of the line segment from a point (x_1, y_1) to a point (x_2, y_2) is given by $L = \sqrt{(x_2 - x_1)^2 + (y_2 - y_1)^2}$.

2. In Section 3.2 we stated that two nonvertical lines are parallel if and only if they have the same slope. Explain why this is so. (You must explain two things: why nonvertical lines that are parallel have the same slope, and why nonvertical lines that have the same slope are parallel.)

3. Lines are **perpendicular** if they meet at a right angle. Thus any horizontal line and any vertical line are perpendicular. Two nonvertical lines with equations
$$y = m_1 x + b_1 \quad \text{and} \quad y = m_2 x + b_2$$
are perpendicular if and only if $m_1 m_2 = -1$. Explain why this is so. (You must explain two things: why nonvertical lines that are perpendicular have slopes whose product is -1, and why nonvertical lines whose slopes have product equal to -1 are perpendicular.)

SUPPLEMENT: CHAOS

In this supplement we introduce the topic of **chaos** by mens of the **logistic equation**
$$y = rx(1 - x), \quad r > 0, \quad 0 \le x \le 1.$$
The logistic equation is often used to model changes in population size; in this context, x and y are interpreted as follows: x is the size of the population, expressed

LINEAR AND QUADRATIC FUNCTIONS

as a proportion of the maximum possible size, in one generation, and y is the size of the population, again as a proportion of the maximum possible size, in the next generation.

EXAMPLE 1 Suppose we take $r = 1$, so that our model is

$$y = x(1 - x).$$

If $x = 0.4$, then

$$y = 0.4(1 - 0.4) = 0.24.$$

Thus if we begin with a population whose size is 40% of the maximum, in the next generation its size will be 24% of the maximum.

We can continue this process to the third generation by letting the computed y-value be our new x-value: if $x = 0.24$, then

$$y = 0.24(1 - 0.24) = 0.1824.$$

Therefore in the third generation the population size will be 18.24% of the maximum. Now if we wish we can continue to the fourth generation, using 0.1824 as our new x-value, and so on. ☐

As suggested in Example 1, we can use the logistic equation to trace the population size over a sequence of generations. We do this by starting with a **seed value** x_1 for the first generation, and then generating x-values for succeeding generations by using the recurrence relation

$$x_n = rx_{n-1}(1 - x_{n-1}), \qquad n \geq 2.$$

The sequence of numbers $x_1, x_2, x_3, \ldots$ is called the **orbit** of x_1 under $y = rx(1 - x)$.

EXAMPLE 2 The orbit of $x_1 = 0.4$ under $y = x(1 - x)$ is

$$0.4, 0.24, 0.1824, 0.1491, 0.1269, \ldots.$$

(The numbers have been rounded to four decimal places.) Notice that the numbers in this sequence are approaching 0; thus we say that the orbit is **attracted to 0.** ☐

EXAMPLE 3 The orbit of $x_1 = 0.2$ under $y = 2x(1 - x)$ is

$$0.2, 0.32, 0.4352, 0.4916, 0.4999, \ldots.$$

The orbit is attracted to 0.5. ☐

What can happen to a population modeled by the logistic equation? On the evidence of Examples 2 and 3, we might guess that the orbit will always be attracted to some number L. Then if $L = 0$, the population will die out, whereas if $L \neq 0$ the population will eventually settle down to a constant size. In fact, as we shall see, these are not the only things that can happen.

Before we continue, we must remark that the calculation of an orbit becomes increasingly burdensome as the generations increase. Therefore when we are finding orbits, it is best to do the calculations on a computer. Here is a BASIC program that calculates orbits:

```
100 'Program ORBIT
110 INPUT "Enter r"; R
120 INPUT "Enter seed x1"; X
130 INPUT "Enter the number of generations n"; N
140 PRINT X;
150 FOR I = 2 TO N
160 Y = R*X*(1 - X)
170 PRINT INT(1000*Y + .5)/1000
180 X = Y
190 NEXT I
200 END
```

This program will find $x_1, x_2, \ldots, x_n$, rounded to three decimal places. (If you want the numbers rounded to m decimal places, $m \neq 3$, replace both occurrences of 1000 in line 170 by 10^m.)

EXAMPLE 4 Here is the result of running ORBIT with $r = 1$, $x_1 = 0.4$, and $n = 12$:

$$\begin{array}{cccccc} .4 & .24 & .182 & .149 & .127 & .111 \\ .099 & .089 & .081 & .074 & .069 & .064 \end{array}$$ □

A more appealing way to analyze orbits uses the graph of $y = rx(1 - x)$ and the line $y = x$. Note that we calculate an orbit by finding $y = rx(1 - x)$ for a given x, then setting the new x equal to y and finding a new y, and so on. This procedure can be depicted graphically by starting with the seed x_1 on the x-axis, moving vertically to the graph of $y = rx(1 - x)$, then horizontally to the line $y = x$, then vertically to the graph of $y = rx(1 - x)$, then horizontally to the line $y = x$, and so on. See Figure 3–47.

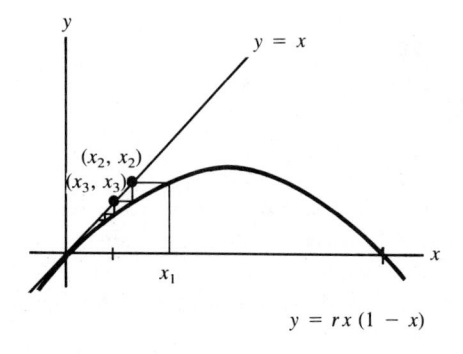

FIGURE 3–47

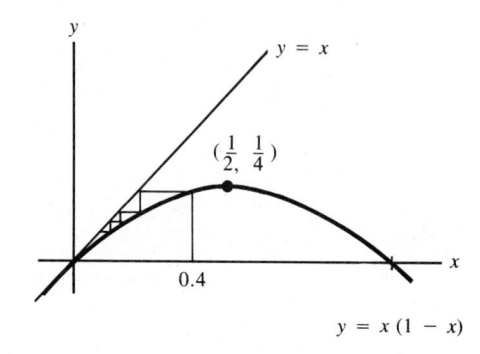

FIGURE 3–48

EXAMPLE 5 Figure 3–48 shows the orbit of $x_1 = 0.4$ under $y = x(1 - x)$. Notice how the orbit is attracted to 0. Notice also that orbits that start out close together stay close together (Figure 3–49). ☐

EXAMPLE 6 Figure 3–50 shows the orbit of $x = 0.2$ under $y = 2x(1 - x)$. Notice how the orbit is attracted to 0.5. Again, it is clear that orbits that begin close to one another remain close to one another. ☐

By representing orbits graphically in this way, we easily see that when $r = 1$, *every* orbit is attracted to 0. Similarly, it is a simple matter to check that when $r = 2$, every orbit except the two with seeds 0 and 1 is attracted to 0.5.

Now let us consider what happens when $r = 3$. As Figure 3–51 shows, here we encounter a new type of orbit, one that eventually settles down to oscillating between two different population levels. (The two levels are represented by the two colored points on the line $y = x$.) Hence, the orbit is attracted, not to a single state, but to an oscillation between two states. Note, however, that, just as before, orbits that start out close together remain close together.

If we increase r further, to 3.5 say, we find even more complicated behavior. See Figure 3–52. Here the orbit eventually oscillates between four states. Again, orbits that begin near one another stay near one another.

As we increase r past 3.5, we can produce orbits that oscillate among 8, 16, 32, . . . different states. The remarkable thing, however, is that if r is made greater than ≈ 3.57, all such patterns can disappear. When $r > 3.57$, the orbit can seem completely random, not being attracted to anything. See Figure 3–53. Furthermore, when $r > 3.57$, orbits that start out close together can get farther and farther apart.

EXAMPLE 7 Let $y = 3.7x(1 - x)$ and consider the first 20 generations for the orbits of 0.3 and 0.31. They are

| .3 | .777 | .641 | .851 | .468 | .921 | .268 | .726 | .735 | .72 |
| .746 | .702 | .775 | .646 | .846 | .482 | .924 | .26 | .713 | .758 |

and

| .31 | .791 | .611 | .88 | .392 | .882 | .386 | .877 | .399 | .888 |
| .369 | .862 | .441 | .912 | .297 | .772 | .651 | .84 | .497 | .925 |

Note how the orbits start out close together but soon diverge. ☐

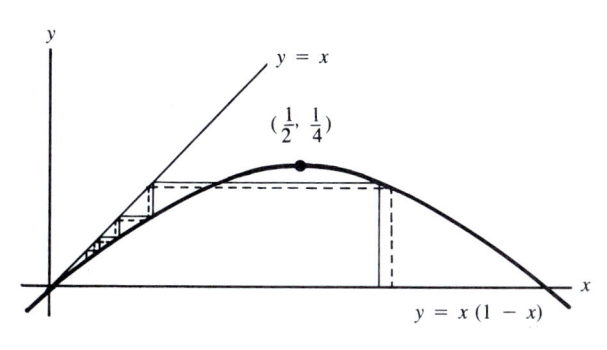

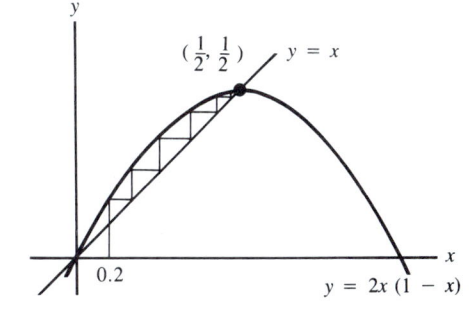

FIGURE 3–49

FIGURE 3–50

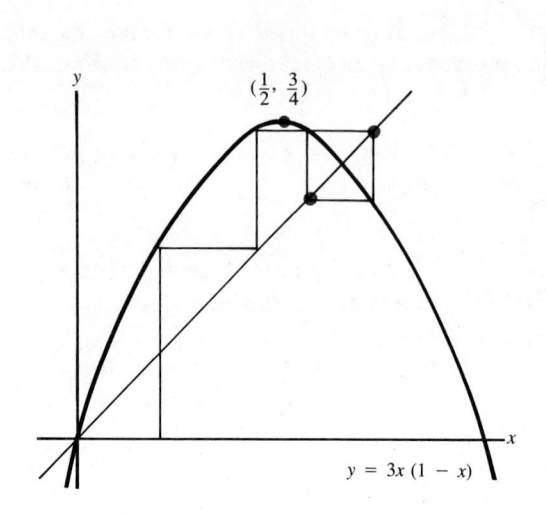

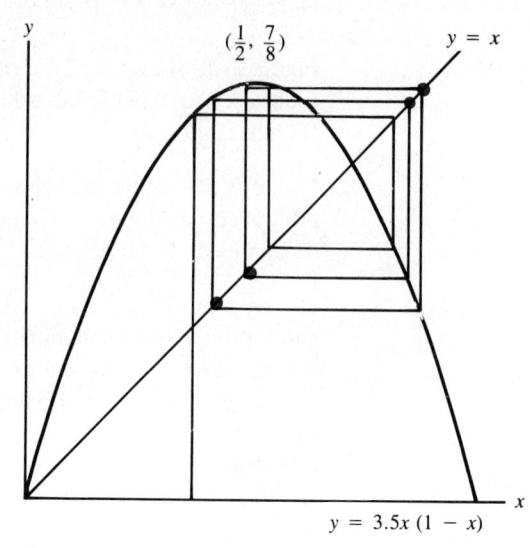

FIGURE 3–51 $y = 3x(1-x)$, point $(\frac{1}{2}, \frac{3}{4})$

FIGURE 3–52 $y = 3.5x(1-x)$, point $(\frac{1}{2}, \frac{7}{8})$

The behavior of the orbital system when $r > 3.57$ is said to be **chaotic.** The essence of chaos is the idea of **sensitivity to initial data.** As we have seen, when $r < 3.57$, the orbital system is not sensitive to initial data, because orbits that start out close together remain close together. However, when $r > 3.57$, orbits that start out close together can get farther and farther apart, and this is what we mean by chaos.

What is the significance of all this for our study of population size? It is this: if the orbits are not chaotic, the population size will either settle down to a single level, zero or nonzero, or it will oscillate back and forth between two or four or eight or . . . different levels. Thus if the orbits are not chaotic, the population size will follow a definite repetitive pattern. Furthermore, since orbits that start out close together stay close together, if we begin with an initial population size x_1 that is

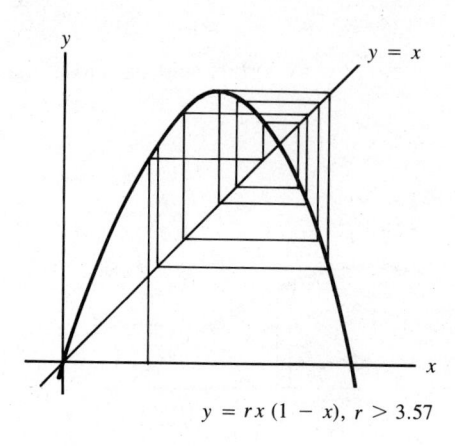

FIGURE 3–53 $y = rx(1-x)$, $r > 3.57$

close to but not quite the same as the actual population size, our predictions for the future will nevertheless be close to what will actually occur.

On the other hand, suppose the orbits are chaotic. Then population size can change wildly from generation to generation, following no definite repetitive pattern. What is more important, if the orbits are chaotic we cannot in the long run reliably predict future population sizes from a knowledge of the current size. This follows from the system's sensitivity to initial data: two seeds that are very close together, so close that in practical terms we cannot even tell the difference between them, can lead to vastly different population sizes in future generations. Therefore, if our seed x_1 differs in even the smallest undetectable amount from actuality, our predictions in the long run can be quite far from what will actually occur. Thus sensitivity to initial data prohibits accurate long-range prediction of the population size when the orbits are chaotic.

Chaos theory is one of the newest developments in mathematics and potentially one of the most important, for many natural systems appear to be chaotic. Examples are changes in population size for gypsy moths, the turbulent flow of water in a brook, fibrillations of the heart, the weather, and perhaps even the movement of stock prices. But as we saw, chaos implies long-range unpredictability, because even the tiniest undetectable differences between the observed initial state and the actual initial state can result in huge differences later on between predicted and actual behavior. (With reference to the weather, this is sometimes referred to as the Butterfly Effect: a butterfly in South America flaps its wings, and the result six months later is a blizzard in New York.) Thus, for instance, if the weather is truly chaotic, accurate long-range weather forecasting must be impossible.

All is not lost, however. It turns out that a three-dimensional chaotic system (such as the weather) can possess a "strange attractor." A **strange attractor** is a surface with the property that all the orbits of the system ultimately lie on it. This does not make the orbits predictable: orbits that are close together on the strange attractor to begin with may be very far apart later on, and they are intertwined on the attractor in a patternless manner. However, the strange attractor does impose an order on the system as a whole rather than on the individual orbits. Thus we may not be able to make reliable long-range weather forecasts, but we may be able to determine the climate.

If you would like to learn more about chaos and its applications, a good place to start is the book *Chaos: Making a New Science* by James Gleick, published by Viking in 1987.

SUPPLEMENTARY EXERCISES

1. Let $y = 1.5x(1 - x)$. Examine the orbits of x_1, $0 \leq x_1 \leq 1$. Can you always predict what an orbit will do?

2. Repeat Exercise 1 using $y = 2.5x(1 - x)$.

3. Repeat Exercise 1 using $y = 3.5x(1 - x)$.
4. Can you find a value of r and a seed x_1 that will yield an eightfold oscillating orbit?
5. Let $y = 3.6x(1 - x)$. Find two seeds that are close together but whose orbits diverge, as we did in Example 7.

Instead of drawing orbits using the graph of $y = rx(1 - x)$ and the line $y = x$, suppose we draw them using the graph of $y = x^2 + c$ and the line $y = x$; in this case, the seed x_1 can be any real number.

6. Examine the orbits of x_1 using $y = x^2 + 1$. Note how they go "off to infinity."
7. Examine the orbits of x_1, using $y = x^2 + \frac{1}{4}$. Is there an attractor? If so, what is it?
8. Examine the orbits of x_1, using $y = x^2$. Is there an attractor?
9. Examine the orbits of x_1, using $y = x^2 - \frac{3}{4}$. Is there an attractor?
10. Examine the orbits of x_1, using $y = x^2 - 1$. Is there an attractor?
11. Examine the orbits of x_1, using $y = x^2 - 2$. Can you see chaos occurring?

4

Matrices

This chapter is devoted to a study of the mathematical objects known as **matrices** (singular: **matrix**). Matrices are important tools for the organization and manipulation of information. They have applications in fields as diverse as inventory control, cryptography, production scheduling, information processing, and economic forecasting. As a relatively frivolous illustration of the use of a matrix, consider the problem of determining the relative strengths of sports teams. For instance, suppose we wish to rank six football teams: the Antelopes (A), Bisons (B), Cougars (C), Dingos (D), Elephants (E), and Ferrets (F). We create an array of numbers that exhibits the strengths of the teams as follows: we label the rows and columns of the array with the team names, and in each position where a row intersects a column we place a 1 if the row team has played and beaten the column team, or a 0 if it has not. Suppose the result of this process is the array

$$\begin{array}{c} \\ A \\ B \\ C \\ D \\ E \\ F \end{array} \begin{array}{cccccc} A & B & C & D & E & F \end{array} \\ \left[\begin{array}{cccccc} 0 & 0 & 1 & 0 & 0 & 0 \\ 0 & 0 & 0 & 1 & 0 & 0 \\ 0 & 1 & 0 & 0 & 0 & 0 \\ 0 & 0 & 0 & 0 & 0 & 0 \\ 0 & 0 & 0 & 1 & 0 & 0 \\ 1 & 0 & 0 & 0 & 1 & 0 \end{array} \right].$$

Thus the Antelopes have played and beaten the Cougars, the Bisons have played and beaten the Dingos, and so on. This array is known as the **dominance matrix** for the teams. From it we see that, as far as head-to-head competition is concerned, the Ferrets are the strongest team: they have two wins, whereas every other team except the Dingos has only one win; the Dingos have no wins.

We begin this chapter by returning to a topic we discussed in Section 3.3: solving systems of linear equations. In the first section we describe how to solve such systems by using matrices and a technique known as Gauss-Jordan elimination. We next consider the arithmetic of matrices, that is, their addition, subtraction, and multiplication. We then introduce the concept of the inverse of a matrix, and conclude with applications of matrices to cryptography and input-output analysis.

4.1 GAUSS-JORDAN ELIMINATION

In Section 3.3 we showed how to solve systems of two or three linear equations in two and three variables by the method of elimination. When carrying out this method, we do not really have to write down the variables at each step; all that is actually required is that we keep track of the changes in the coefficients and constant terms of the system. To illustrate, suppose we wish to solve the system

$$x + 2y = 10$$
$$2x - y = 5$$

for x and y. We store the coefficients and constant terms of this system in the rectangular array

$$\begin{bmatrix} 1 & 2 & | & 10 \\ 2 & -1 & | & 5 \end{bmatrix}.$$

Any rectangular array of numbers is called a **matrix;** this particular array is called the **augmented matrix** of the system of equations. The numbers in the augmented matrix are called its **entries.** Notice that the entries in the first column of the augmented matrix are the coefficients of the variable x, the entries in the second column are the coefficients of the variable y, and the entries in the third column are the constant terms of the system. The vertical bar separates the coefficients from the constant terms. Clearly, each row of the augmented matrix represents one of the equations, and we can reconstruct the equations from the augmented matrix.

Now we intend to solve this system of equations in two ways: on the left by the method of elimination of Section 3.3, and on the right by operating on the rows of the augmented matrix:

$$\begin{array}{cc} x + 2y = 10, & \begin{bmatrix} 1 & 2 & | & 10 \\ 2x - y = 5. & 2 & -1 & | & 5 \end{bmatrix}. \end{array}$$

MATRICES

Add -2 times the first equation to the second equation:

$$-2(\ x + 2y = \ \ 10)$$
$$+\ \ \underline{\ \ \ 2x - \ \ y = \ \ \ \ 5}$$
$$-5y = -15.$$

Add -2 times the first row to the second row:

$$\begin{bmatrix} 1 & 2 & | & 10 \\ 2 - 2\cdot 1 & -1 - 2\cdot 2 & | & 5 - 2\cdot 10 \end{bmatrix}.$$

or

$$\begin{bmatrix} 1 & 2 & | & 10 \\ 0 & -5 & | & -15 \end{bmatrix}.$$

Solve for y:

$$y = 3$$

Solve for y: if we translate the second row of the last augmented matrix back into an equation, it becomes

$$0x - 5y = -15,$$

or

$$-5y = -15,$$

or

$$y = 3.$$

Back-substitute to find x: the equation

$$x + 2y = 10$$

becomes

$$x + 2(3) = 10,$$

so

$$x = 4.$$

Back-substitute to find x: the top row of the last augmented matrix translates into the equation

$$1x + 2y = 10,$$

so

$$x + 2(3) = 10,$$

or

$$x = 4.$$

This shows how a system of linear equations can be solved by working with the rows of its augmented matrix.

In fact, we can avoid the necessity for back-substitution by reducing the augmented matrix even further. Suppose we take the last augmented matrix and multiply its second row by $-\frac{1}{5}$; we obtain

$$\begin{bmatrix} 1 & 2 & | & 10 \\ (-\frac{1}{5})0 & (-\frac{1}{5})(-5) & | & (-\frac{1}{5})(-15) \end{bmatrix},$$

or

$$\begin{bmatrix} 1 & 2 & | & 10 \\ 0 & 1 & | & 3 \end{bmatrix}.$$

Now we take this matrix and add -2 times its second row to its first row to get

$$\begin{bmatrix} 1 - 2 \cdot 0 & 2 - 2 \cdot 1 & | & 10 - 2 \cdot 3 \\ 0 & 1 & | & 3 \end{bmatrix},$$

or

$$\begin{bmatrix} 1 & 0 & | & 4 \\ 0 & 1 & | & 3 \end{bmatrix}.$$

But we can read the solutions of the system directly from this last matrix, for its first row says that $x = 4$ and its second that $y = 3$.

We have just illustrated the method of **Gauss-Jordan elimination** for solving a system of linear equations. In the rest of this section, we will codify the method and show how it is used.

If we are going to solve systems of linear equations by operating on the rows of their augmented matrices, we must make clear what types of operations are allowed. When dealing with a system of equations, we may surely interchange any two of the equations, multiply any one of them by a nonzero number, and add a multiple of any equation to any other one. The analogous operations on the rows of an augmented matrix are known as **row operations.**

Row Operations

There are three types of **row operations** that may be applied to the rows of an augmented matrix:

Type I   Interchanging any two rows of the matrix.

Type II Multiplying every entry in a row of the matrix by the same nonzero number.

Type III Adding a multiple of one row of the matrix to another row.

Let us illustrate the three types of row operations.

EXAMPLE 1   Consider the augmented matrix

$$\begin{bmatrix} 2 & 1 & | & 4 \\ 3 & 0 & | & 6 \\ 4 & -2 & | & 8 \end{bmatrix}.$$

MATRICES

If we interchange row 1 and row 2 of this matrix, we have performed a Type I row operation:

$$\begin{bmatrix} 2 & 1 & 4 \\ 3 & 0 & 6 \\ 4 & -2 & 8 \end{bmatrix} \longrightarrow \begin{bmatrix} 3 & 0 & 6 \\ 2 & 1 & 4 \\ 4 & -2 & 8 \end{bmatrix}.$$

Notice our notation: the arrows show how the Type I row operation is carried out.

If we multiply the second row of the matrix by $\frac{1}{3}$, we have performed a Type II row operation:

$$\begin{bmatrix} 2 & 1 & 4 \\ 3 & 0 & 6 \\ 4 & -2 & 8 \end{bmatrix} \xrightarrow{(\frac{1}{3})R_2} \begin{bmatrix} 2 & 1 & 4 \\ 1 & 0 & 2 \\ 4 & -2 & 8 \end{bmatrix}.$$

Notice the notation: R_2 stands for "row 2," and the arrow and its label show how the Type II operation is carried out.

Finally, if we multiply row 1 of the matrix by -2 and add the result to row 3, we have performed a Type III row operation:

$$\begin{bmatrix} 2 & 1 & 4 \\ 3 & 0 & 6 \\ 4 & -2 & 8 \end{bmatrix} \xrightarrow{-2R_1 + R_3} \begin{bmatrix} 2 & 1 & 4 \\ 3 & 0 & 6 \\ 0 & -4 & 0 \end{bmatrix}.$$

Again, notice the notation: R_1 and R_3 stand for "row 1" and "row 3," respectively, and the arrow and its label show how the Type III row operation is carried out. Note also that in a Type III operation, only the row that is *added to* is changed. Thus, in this case row 3 is changed, but row 1 is not. ☐

Now we are ready to describe Gauss-Jordan elimination for systems that have the same number of equations as they do variables. Thus suppose we have a system of n linear equations in the n variables $x_1, x_2, \ldots, x_n$: the basic idea of Gauss-Jordan elimination is to use row operations to transform the augmented matrix of the system into one of the form

$$\left[\begin{array}{ccccc|c} 1 & 0 & 0 & \cdots & 0 & d_1 \\ 0 & 1 & 0 & \cdots & 0 & d_2 \\ 0 & 0 & 1 & \cdots & 0 & d_3 \\ \vdots & \vdots & \vdots & \vdots & \vdots & \vdots \\ 0 & 0 & 0 & \cdots & 1 & d_n \end{array} \right],$$

if possible. If this can be done, the solution is

$$x_1 = d_1, \quad x_2 = d_2, \quad \ldots, \quad x_n = d_n.$$

> **Gauss-Jordan Elimination**
>
> To solve a system of n linear equations in n variables using Gauss-Jordan elimination, proceed as follows:
>
> 1. Form the augmented matrix of the system.
>
> 2. Move through the augmented matrix from left to right, working on only one column at a time. If you are working on the ith column, do this:
>   a. If the entry in the row i, column i position is not a 1, try to produce a 1 in that position by using Type I and/or Type II row operations. (When performing Type I operations, do *not* interchange row i with any row above it.) If it is not possible to produce a 1 in the row i, column i position, proceed to the next column.
>   b. Use the 1 in row i, column i and Type III row operations to replace all the other entries in column i with zeros. Then proceed to the next column.
>
> 3. When all columns to the left of the vertical bar have been transformed in this way, read the solution from the transformed matrix.

The numbers in the positions of the augmented matrix where we attempt to get 1's are called **pivots.** In the following examples we shall circle the pivot we are working with at any given time.

EXAMPLE 2   We use Gauss-Jordan elimination to solve the system

$$4x - 2y = 14$$
$$x + 2y = 1.$$

The augmented matrix of this system, with its first-column pivot circled, is

$$\begin{bmatrix} ④ & -2 & | & 14 \\ 1 & 2 & | & 1 \end{bmatrix}.$$

The easiest way to get a 1 in the pivot position is to interchange rows 1 and 2:

$$\begin{bmatrix} ④ & -2 & | & 14 \\ 1 & 2 & | & 1 \end{bmatrix} \longrightarrow \begin{bmatrix} ① & 2 & | & 1 \\ 4 & -2 & | & 14 \end{bmatrix}.$$

Now we use the pivot 1 and a Type III operation to "clear" the first column:

$$\begin{bmatrix} ① & 2 & | & 1 \\ 4 & -2 & | & 14 \end{bmatrix} \xrightarrow{-4R_1 + R_2} \begin{bmatrix} 1 & 2 & | & 1 \\ 0 & ⓘ{-10} & | & 10 \end{bmatrix}.$$

MATRICES 195

Now we get a 1 in the second-column pivot position by multiplying the second row by $-\frac{1}{10}$:

$$\begin{bmatrix} 1 & 2 & | & 1 \\ 0 & -10 & | & 10 \end{bmatrix} \xrightarrow{(-\frac{1}{10})R_2} \begin{bmatrix} 1 & 2 & | & 1 \\ 0 & 1 & | & -1 \end{bmatrix}.$$

Finally, we use the pivot 1 to "clear" the second column:

$$\begin{bmatrix} 1 & 2 & | & 1 \\ 0 & 1 & | & -1 \end{bmatrix} \xrightarrow{-2R_2 + R_1} \begin{bmatrix} 1 & 0 & | & 3 \\ 0 & 1 & | & -1 \end{bmatrix}.$$

Therefore $x = 3$ and $y = -1$. □

EXAMPLE 3   We use Gauss-Jordan elimination to solve the following system:

$$x_1 + 2x_2 + x_3 = 2$$
$$2x_1 - x_2 + 2x_3 = 3$$
$$3x_1 + 2x_2 - x_3 = 2.$$

The augmented matrix, with first-column pivot circled, is

$$\begin{bmatrix} ① & 2 & 1 & | & 2 \\ 2 & -1 & 2 & | & 3 \\ 3 & 2 & -1 & | & 2 \end{bmatrix}.$$

(Notice that this matrix already has a 1 as its first-column pivot.) Gauss-Jordan elimination proceeds as follows:

$$\begin{bmatrix} ① & 2 & 1 & | & 2 \\ 2 & -1 & 2 & | & 3 \\ 3 & 2 & -1 & | & 2 \end{bmatrix} \xrightarrow[-3R_1 + R_3]{-2R_1 + R_2} \begin{bmatrix} 1 & 2 & 1 & | & 2 \\ 0 & -5 & 0 & | & -1 \\ 0 & -4 & -4 & | & -4 \end{bmatrix}$$

$$\begin{bmatrix} 1 & 2 & 1 & | & 2 \\ 0 & -4 & -4 & | & -4 \\ 0 & -5 & 0 & | & -1 \end{bmatrix} \xrightarrow{-(\frac{1}{4})R_2} \begin{bmatrix} 1 & 2 & 1 & | & 2 \\ 0 & ① & 1 & | & 1 \\ 0 & -5 & 0 & | & -1 \end{bmatrix} \xrightarrow[5R_2 + R_3]{-2R_2 + R_1}$$

$$\begin{bmatrix} 1 & 0 & -1 & | & 0 \\ 0 & 1 & 1 & | & 1 \\ 0 & 0 & ⑤ & | & 4 \end{bmatrix} \xrightarrow{(\frac{1}{5})R_3} \begin{bmatrix} 1 & 0 & -1 & | & 0 \\ 0 & 1 & 1 & | & 1 \\ 0 & 0 & ① & | & \frac{4}{5} \end{bmatrix} \xrightarrow[-R_3 + R_2]{R_3 + R_1}$$

$$\begin{bmatrix} 1 & 0 & 0 & | & \frac{4}{5} \\ 0 & 1 & 0 & | & \frac{1}{5} \\ 0 & 0 & 1 & | & \frac{4}{5} \end{bmatrix}.$$

Hence the solution is $x_1 = \frac{4}{5}$, $x_2 = \frac{1}{5}$, and $x_3 = \frac{4}{5}$. □

EXAMPLE 4   A warehouse ships parts to three different plants. Suppose that it ships a total of 72 parts to the plants, with twice as many shipped to the second plant as to the first, and three times as many to the third as to the second. How many parts are shipped to each plant?

Let

$$x_1 = \text{Number of parts shipped to plant 1},$$
$$x_2 = \text{Number of parts shipped to plant 2},$$
$$x_3 = \text{Number of parts shipped to plant 3}.$$

Then we have

Total shipped: $\quad x_1 + x_2 + x_3 = 72$

Twice as many to 2nd as to 1st: $\quad x_2 = 2x_1$

3 times as many to 3rd as to 2nd: $\quad x_3 = 3x_2.$

We rewrite these equations as

$$\begin{aligned} x_1 + x_2 + x_3 &= 72 \\ -2x_1 + x_2 &= 0 \\ -3x_2 + x_3 &= 0. \end{aligned}$$

Thus the augmented matrix is

$$\begin{bmatrix} 1 & 1 & 1 & | & 72 \\ -2 & 1 & 0 & | & 0 \\ 0 & -3 & 1 & | & 0 \end{bmatrix},$$

and we have

$$\begin{bmatrix} \textcircled{1} & 1 & 1 & | & 72 \\ -2 & 1 & 0 & | & 0 \\ 0 & -3 & 1 & | & 0 \end{bmatrix} \xrightarrow{2R_1 + R_2} \begin{bmatrix} 1 & 1 & 1 & | & 72 \\ 0 & \textcircled{3} & 2 & | & 144 \\ 0 & -3 & 1 & | & 0 \end{bmatrix} \xrightarrow{(\frac{1}{3}) R_2}$$

$$\begin{bmatrix} 1 & 1 & 1 & | & 72 \\ 0 & \textcircled{1} & \frac{2}{3} & | & 48 \\ 0 & -3 & 1 & | & 0 \end{bmatrix} \xrightarrow[3R_2 + R_3]{-R_2 + R_1} \begin{bmatrix} 1 & 0 & \frac{1}{3} & | & 24 \\ 0 & 1 & \frac{2}{3} & | & 48 \\ 0 & 0 & \textcircled{3} & | & 144 \end{bmatrix} \xrightarrow{(\frac{1}{3})R_3}$$

$$\begin{bmatrix} 1 & 0 & \frac{1}{3} & | & 24 \\ 0 & 1 & \frac{2}{3} & | & 48 \\ 0 & 0 & \textcircled{1} & | & 48 \end{bmatrix} \xrightarrow[(-\frac{2}{3})R_3 + R_2]{(-\frac{1}{3})R_3 + R_1} \begin{bmatrix} 1 & 0 & 0 & | & 8 \\ 0 & 1 & 0 & | & 16 \\ 0 & 0 & 1 & | & 48 \end{bmatrix}.$$

Hence $x_1 = 8$, $x_2 = 16$, and $x_3 = 48$. ∎

Thus far all our systems of linear equations have had unique solutions, but this need not always be the case; when it is not, the Gauss-Jordan method as we have described it will at some stage produce a row of zeros to the left of the vertical bar, that is, a row of the form

$$[0 \ 0 \ 0 \ \ldots \ 0 \mid c].$$

Such a row translates into an equation of the form

$$0x_1 + 0x_2 + \cdots + 0x_n = c,$$

MATRICES   197

or

$$0 = c.$$

Therefore if $c \neq 0$, the system cannot have a solution. On the other hand, if $c = 0$, the system will have infinitely many solutions. We illustrate with examples.

EXAMPLE 5   Let us attempt to solve the system

$$x_1 + 2x_2 + 3x_3 = 4$$
$$3x_1 + 6x_2 + x_3 = 2$$
$$2x_1 + 4x_2 + x_3 = -2$$

by Gauss-Jordan elimination.
We have

$$\begin{bmatrix} ① & 2 & 3 & | & 4 \\ 3 & 6 & 1 & | & 2 \\ 2 & 4 & 1 & | & -2 \end{bmatrix} \xrightarrow[-2R_1 + R_3]{-3R_1 + R_2} \begin{bmatrix} 1 & 2 & 3 & | & 4 \\ 0 & ⓪ & -8 & | & -10 \\ 0 & 0 & -5 & | & -10 \end{bmatrix}.$$

Note that it is not possible to produce a 1 in the pivot position of the second column. We therefore proceed to the pivot position of the next column and continue the process:

$$\begin{bmatrix} 1 & 2 & 3 & | & 4 \\ 0 & 0 & -8 & | & -10 \\ 0 & 0 & ⑤ & | & -10 \end{bmatrix} \xrightarrow{(-⅕)R_3} \begin{bmatrix} 1 & 2 & 3 & | & 4 \\ 0 & 0 & -8 & | & -10 \\ 0 & 0 & ① & | & 2 \end{bmatrix} \xrightarrow[8R_3 + R_2]{-3R_3 + R_1}$$

$$\begin{bmatrix} 1 & 2 & 0 & | & -2 \\ 0 & 0 & 0 & | & 6 \\ 0 & 0 & 1 & | & 2 \end{bmatrix}.$$

The second row of this last augmented matrix tells us that the system has no solution, for it indicates that

$$0x_1 + 0x_2 + 0x_3 = 6,$$

or

$$0 = 6,$$

which is impossible. □

EXAMPLE 6   Let us attempt to solve the following system of equations:

$$x_1 + 2x_2 + x_3 = 4$$
$$2x_1 + 5x_2 - 2x_3 = 3$$
$$3x_1 + 7x_2 - x_3 = 7.$$

We have

$$\begin{bmatrix} \boxed{1} & 2 & 1 & | & 4 \\ 2 & 5 & -2 & | & 3 \\ 3 & 7 & -1 & | & 7 \end{bmatrix} \xrightarrow[-3R_1+R_3]{-2R_1+R_2} \begin{bmatrix} 1 & 2 & 1 & | & 4 \\ 0 & \boxed{1} & -4 & | & -5 \\ 0 & 1 & -4 & | & -5 \end{bmatrix} \xrightarrow[-R_2+R_3]{2R_2+R_1}$$

$$\begin{bmatrix} 1 & 0 & 9 & | & 14 \\ 0 & 1 & -4 & | & -5 \\ 0 & 0 & \boxed{0} & | & 0 \end{bmatrix}.$$

It is not possible to produce a 1 in the pivot position of the third column, so we are done. The last augmented matrix implies that

$$x_1 + 9x_3 = 14$$

and

$$x_2 - 4x_3 = -5.$$

If we solve these equations for x_1 and x_2 in terms of x_3, we obtain

$$x_1 = 14 - 9x_3$$

and

$$x_2 = -5 + 4x_3.$$

Choosing any real number and substituting it for x_3 will now result in a solution for the system. For instance,

- If $x_3 = 1$, then $x_1 = 5$, $x_2 = -1$, and $x_3 = 1$ is a solution;
- If $x_3 = 2$, then $x_1 = -4$, $x_2 = 3$, and $x_3 = 2$ is a solution;
- If $x_3 = 0$, then $x_1 = 14$, $x_2 = -5$, and $x_3 = 0$ is a solution;

and so on. Hence, the system has infinitely many solutions. ☐

Gauss-Jordan elimination can also be used to solve systems of linear equations in which the number of variables and the number of equations are not the same. We illustrate with examples.

EXAMPLE 7   Let us solve the following system of three equations in four variables:

$$x_1 - x_2 + 2x_3 + x_4 = 3$$
$$2x_1 - 2x_2 + 4x_3 + 2x_4 = 6$$
$$-3x_1 + 3x_2 - 4x_3 - 5x_4 = 1.$$

We have

$$\begin{bmatrix} \boxed{1} & -1 & 2 & 1 & | & 3 \\ 2 & -2 & 4 & 2 & | & 6 \\ -3 & 3 & -4 & -5 & | & 1 \end{bmatrix} \xrightarrow[-3R_1+R_3]{-2R_1+R_2} \begin{bmatrix} 1 & -1 & 2 & 1 & | & 3 \\ 0 & \boxed{0} & 0 & 0 & | & 0 \\ 0 & 0 & 2 & -2 & | & 10 \end{bmatrix}.$$

MATRICES 199

Since we cannot produce a 1 in the pivot position of the second column, we proceed to the third column:

$$\begin{bmatrix} 1 & -1 & 2 & 1 & | & 3 \\ 0 & 0 & 0 & 0 & | & 0 \\ 0 & 0 & ② & -2 & | & 10 \end{bmatrix} \xrightarrow{(½)R_3} \begin{bmatrix} 1 & -1 & 2 & 1 & | & 3 \\ 0 & 0 & 0 & 0 & | & 0 \\ 0 & 0 & ① & -1 & | & 5 \end{bmatrix} \xrightarrow{-2R_3 + R_1}$$

$$\begin{bmatrix} 1 & -1 & 0 & 3 & | & -7 \\ 0 & 0 & 0 & 0 & | & 0 \\ 0 & 0 & 1 & -1 & | & 5 \end{bmatrix}.$$

Since there is no fourth row, there is no pivot position in the fourth column, and we are done. Thus

$$x_1 - x_2 + 3x_4 = -7$$

and

$$x_3 - x_4 = 5,$$

or

$$x_1 = x_2 - 3x_4 - 7$$

and

$$x_3 = x_4 + 5.$$

Any choice of x_2 and x_4 yields a solution for the system; hence, it has infinitely many solutions. ☐

EXAMPLE 8 Let us solve the following system of equations:

$$x_1 + 3x_2 = 7$$
$$-2x_1 - x_2 = -4$$
$$3x_1 + x_2 = 5.$$

We have

$$\begin{bmatrix} ① & 3 & | & 7 \\ -2 & -1 & | & -4 \\ 3 & 1 & | & 5 \end{bmatrix} \xrightarrow[-3R_1 + R_3]{2R_1 + R_2} \begin{bmatrix} 1 & 3 & | & 7 \\ 0 & ⑤ & | & 10 \\ 0 & -8 & | & -16 \end{bmatrix} \xrightarrow{(⅕)R_2}$$

$$\begin{bmatrix} 1 & 3 & | & 7 \\ 0 & ① & | & 2 \\ 0 & -8 & | & -16 \end{bmatrix} \xrightarrow[8R_2 + R_3]{-3R_2 + R_1} \begin{bmatrix} 1 & 0 & | & 1 \\ 0 & 1 & | & 2 \\ 0 & 0 & | & 0 \end{bmatrix}.$$

Hence, the solution is $x_1 = 1$ and $x_2 = 2$. ☐

EXAMPLE 9 A firm makes two products. Each unit of the first uses 2 lb of plastic and 4 lb of tin. Each unit of the second uses 4 lb of plastic and 5 lb of tin. The firm has 600 lb of plastic and 500 lb of tin on hand, and it must make a total of exactly 110 units of product. Find the number of units of each product that it will make if it is to use up all the plastic and all the tin.

Let

$$x_1 = \text{Number of units of first product made,}$$
$$x_2 = \text{Number of units of second product made.}$$

Then we have

Plastic: $2x_1 + 4x_2 = 600$

Tin: $4x_1 + 5x_2 = 500$

Total: $x_1 + x_2 = 110.$

Thus

$$\begin{bmatrix} ② & 4 & | & 600 \\ 4 & 5 & | & 500 \\ 1 & 1 & | & 110 \end{bmatrix} \to \begin{bmatrix} ① & 1 & | & 110 \\ 4 & 5 & | & 500 \\ 2 & 4 & | & 600 \end{bmatrix} \xrightarrow[-2R_1 + R_3]{-4R_1 + R_2} \begin{bmatrix} 1 & 1 & | & 110 \\ 0 & ① & | & 60 \\ 0 & 2 & | & 380 \end{bmatrix} \xrightarrow[-2R_2 + R_3]{-R_2 + R_1}$$

$$\begin{bmatrix} 1 & 1 & | & 50 \\ 0 & 1 & | & 60 \\ 0 & 0 & | & 260 \end{bmatrix}.$$

Hence, the problem has no solution: the firm cannot produce the required number of units and still use up all the raw materials. ∎

■ *Computer Manual: Program SMLTEQNS*

4.1 EXERCISES

All the exercises in this set are to be solved using Gauss-Jordan elimination:

In Exercises 1 through 26, solve the given system of linear equations.

1. $x_1 + 4x_2 = 3$
   $3x_1 + 11x_2 = 5$

2. $-2x_1 + x_2 = 0$
   $x_1 - 3x_2 = 5$

3. $2x_1 - x_2 = 6$
   $8x_1 - 4x_2 = 24$

4. $-x_1 + 7x_2 = 1$
   $3x_1 - 21x_2 = 0$

5. $4x_1 + x_2 = 2$
   $-2x_1 - 0.5x_2 = -1$

6. $2x_1 + 3x_2 = -2$
   $-x_1 - 1.5x_2 = 4$

7. $x_1 + x_2 + x_3 = 0$
   $2x_1 - x_2 + 2x_3 = 3$
   $3x_1 + x_2 + x_3 = 2$

8. $2x_1 + 3x_2 + 4x_3 = -1$
   $x_1 - x_2 + 5x_3 = 10$
   $-x_1 + 4x_2 - x_3 = -15$

9. $2x_1 + 4x_2 - x_3 = -3$
$6x_1 - x_2 - 3x_3 = -5/2$
$5x_1 - 3x_2 + 4x_3 = 12$

10. $x_1 + 2x_2 - x_3 = 4$
$-x_1 + 3x_2 - 2x_3 = 5$
$2x_1 - x_2 + x_3 = 0$

11. $2x_1 + x_2 = 2$
$x_1 - x_2 + x_3 = 4$
$3x_1 + x_3 = 6$

12. $4x_1 - 3x_2 + x_3 = -5$
$-x_1 + 2x_2 - 3x_3 = -3$
$5x_1 - 5x_2 + 4x_3 = -2$

13. $x_1 - x_2 + x_3 - x_4 = -1$
$x_1 + x_2 - x_3 - x_4 = 3$
$2x_1 + x_2 + x_4 = 5$
$-x_1 + x_3 = -4$

14. $2x_1 - x_2 + x_3 = 1$
$x_2 + x_3 + x_4 = 1$
$x_1 - x_3 + 2x_4 = 2$
$x_1 - 3x_2 + x_4 = -11$

15. $5x_1 + 6x_2 - 3x_3 + x_4 = 7$
$2x_1 + 2x_4 = 0$
$x_2 + x_3 - 2x_4 = 4$
$3x_1 + x_2 + 5x_2 = 9$

16. $x_1 + x_2 + x_3 + x_4 = 9$
$2x_1 - x_2 - x_3 - x_4 = -7$
$2x_2 + x_3 - 2x_4 = -1$
$-x_1 + x_2 - 3x_4 = 13$

17. $x_1 + x_2 + x_3 = 5$
$2x_1 - x_3 = 4$

18. $2x_1 + x_2 + 3x_3 = -6$
$4x_1 + x_2 - 2x_3 = 0$

19. $x_1 + x_2 = -3$
$x_1 - x_2 = 7$
$2x_1 + 3x_2 = -11$

20. $2x_1 + x_2 = 16$
$-x_1 + 3x_2 = -22$
$4x_1 + 8x_2 = 8$

21. $x_1 + x_2 = 0$
$3x_1 - x_2 = 8$
$2x_1 + 4x_2 = 3$

22. $2x_1 - x_2 + x_3 + x_4 = 2$
$5x_1 - 3x_2 + x_4 = 5$
$x_1 + 5x_2 - 2x_3 - x_4 = 2$

23. $-x_1 + 2x_2 + x_3 + 2x_5 = 0$
$2x_1 - x_3 + 4x_5 = 2$
$x_2 + x_4 + x_5 = -3$

24. $3x_1 - 2x_2 + x_3 + x_4 = 6$
$2x_1 + 3x_2 - 5x_3 + x_4 = 0$
$4x_1 - 33x_2 + 41x_3 - x_4 = 10$

25. $2x_1 - x_2 + 2x_3 = 15$
$x_1 + 2x_2 - 5x_3 = 12$
$6x_1 + x_2 + 3x_3 = 39$
$-x_1 - x_2 + 4x_3 = -13$

26. $2x_1 + x_2 + x_3 = 2$
$-x_2 + 4x_3 = 11$
$-4x_1 - 3x_2 + 2x_3 = 7$
$-2x_1 - 3x_2 + 7x_3 = 20$

27. You invested a total of $100,000 in three mutual funds. At the end of the year, one fund had returned 12% on your investment, the second 6%, and the third −2%. If the total return on your $100,000 investment was $6760 and you had $6000 more invested in the fund that returned 12% than you did in the fund that returned 6%, how much did you invest in each fund?

Management

28. Ajax Company makes three sizes of wheelbarrows: small, medium, and large. Each wheelbarrow has one rubber wheel. Each small wheelbarrow and each medium wheelbarrow have one strut for bracing, and each large one has two struts. Each small wheelbarrow takes 3 worker-hours, each medium one 4 worker-hours, and each large one 6 worker-hours to assemble. On a given day, Ajax has 60 rubber wheels and 90 struts on hand and has 280 worker-hours available for assembly. How many of each size wheelbarrow can the company make that day?

29. Repeat Exercise 28 if the time it takes to assemble wheelbarrows is 4 worker-hours each for the small and medium ones, and 5 worker-hours for the large ones.

30. Gamma Corporation knows from experience that it costs $12,200 to make 100 units of their product, $43,200 to make 200 units, and $94,200 to make 300 units. Assuming that Gamma's cost function is a quadratic $y = ax^2 + bx + c$, find it.

31. When widgets sold for $20 each, quantity demanded was 9200 units and quantity supplied was 40,200 units. When they sold for $5 each, quantity demanded was 9950 and quantity supplied was 2250. If they were free, quantity demanded would be 10,000 and quantity supplied would be 0. Assuming that the demand and supply functions are both quadratics, find their equations and then find the equilibrium price of widgets.

32. A carpet company ships carpets from its warehouse to its retail stores. Suppose the company has three stores and that during one week a total of 100 carpets are to be shipped from the warehouse to the stores. Suppose also that the first store requires 11 more carpets than the second, and that the second requires 5 fewer than the third. How many carpets will be shipped to each store?

33. Repeat Exercise 32 if the company has four stores, a total of 245 carpets are to be shipped to them, and the second store requires twice as many carpets as the first, the third one and one-half times as many as the first, and the fourth two-thirds as many as the second.

*34. Suppose four airlines share the Atlanta, Boston, Chicago, and Dallas markets, and that

The first airline has 20% of the Atlanta market, 20% of the Boston market, 20% of the Chicago market, and 30% of the Dallas market;
The second airline has 30% of the Atlanta market, 40% of the Boston market, 30% of the Chicago market, and 40% of the Dallas market;
The third airline has 30% of the Atlanta market, 10% of the Boston market, 50% of the Chicago market, and 20% of the Dallas market;
The fourth airline has 20% of the Atlanta market, 30% of the Boston market, 0% of the Chicago market, and 10% of the Dallas market.

Find the number of passengers who flew out of each city if 40,450 flew on the first airline, 61,650 flew on the second, 60,850 flew on the third, and 18,050 flew on the fourth.

35. Repeat Exercise 28, assuming that the wheelbarrows do not have any struts.

36. Repeat Exercise 28, assuming that, in addition to the requirements stated there, each small wheelbarrow sells for $60, each medium one for $100, and each large one for $150, and the firm wishes to sell exactly $6700 worth of wheelbarrows.

*37. A firm ships its product to three different customers. Suppose the three customers require a total of 400 units of the product, and that the second customer needs exactly as many units as the first and the third twice as many as the first. It costs the firm $10 to ship each unit to the first customer, $15 to ship each unit to the second, and $20 to ship each unit to the third. How much must the firm budget for shipping costs if it is to meet the demands of its customers? (*Hint:* Assume the firm spends $c on shipping.)

38. When a commodity sold for a price of $10 per unit, the quantity supplied to the market was 60,000 units; when it sold for $20 per unit, quantity supplied was

160,000 units; when it sold for $30 per unit, quantity supplied was 300,000 units; and when it sold for $40, quantity supplied was 500,000 units. Is it possible to describe quantity supplied as a quadratic function of price? If so, find the function.

39. Repeat Exercise 38 if the quantity supplied when the price was $40 per unit was 480,000 units.

Life Science

40. Each Brand X vitamin pill contains 10 mg of vitamin A, 10 mg of vitamin C, and 20 mg of vitamin D. Each Brand Y pill contains 15 mg of A, 12 mg of C, and 18 mg of D. Each Brand Z pill contains 25 mg of A, 10 mg of C, and 16 mg of D. Suppose you need exactly 120 mg of A, 78 mg of C, and 137 mg of D. How many pills of each brand should you take?

41. Repeat Exercise 40, assuming you need exactly 120 mg of A, 84 mg of C, and 148 mg of D.

42. A recent article in the *Boston Globe* stated that in 1961 there were between 400 and 500 mountain gorillas in central Africa, in 1970 there were 231, and in 1988 there were 400. Find a quadratic model that describes the size of the mountain gorilla population as a function of time if in 1961 there were
  (a) 400 gorillas;
  (b) 450 gorillas;
  (c) 500 gorillas.

43. Repeat Exercise 40, assuming that you do not need any vitamin A.

44. Repeat Exercise 40, assuming that, in addition to the information given there, each Brand X pill contains 10 mg of vitmin E, each Brand Y pill contains 16 mg of E, each Brand Z pill contains 0 mg of E, and you need exactly 100 mg of E.

Social Science

45. A town has 2700 registered voters, each of whom is either a Republican, a Democrat, or an Independent. In the last primary election, 45% of the Republicans, 25% of the Democrats, and 30% of the Independents voted, for a total of 895 voters. If the number of Independent voters in the town is 80% of the number of Republicans and Democrats combined, how many Republicans, Democrats, and Independents are there?

46. A professor discussed a certain concept in class on day 0 and announced a test on the material two weeks thereafter, that is, on day 14. An unannounced quiz on day 1 showed that 81% of the students had satisfactory understanding of the concept. Another unannounced quiz on day 7 showed that at that time 33% of the students had satisfactory understanding of the concept. On the announced test, 66% of the students demonstrated satisfactory understanding of the concept. Find a quadratic model that describes the relationship between the time elapsed since the concept was first discussed and the percentage of students who show satisfactory understanding of it.

47. Repeat Exercise 46 if, in addition to the information given there, an unannounced quiz on day 10 showed that 45% of the students had a satisfactory understanding of the concept.

4.2 MATRICES AND MATRIX OPERATIONS

As we saw in the previous section, a **matrix** is a rectangular array of numbers arranged in rows and columns. Thus,

$$\begin{bmatrix} 1 & 2 & 3 \\ 0 & -1 & 4 \end{bmatrix}, \quad \begin{bmatrix} -2 & 5 \\ 6 & 12 \\ 9 & 2 \end{bmatrix}, \quad \text{and} \quad \begin{bmatrix} 0 & -2/3 & 1 \\ 4 & -3/4 & 4 \\ 0 & 2 & 5 \end{bmatrix}$$

are matrices. In this section we demonstrate how matrices are used to store information and show how they can be added, subtracted, and multiplied by numbers.

Matrices

Matrices are classified by the number of their rows and columns: if a matrix has m rows and n columns, it is said to be an $m \times n$ **matrix** or to be of **order $m \times n$**. Notice that the number of rows is given first. The notation $m \times n$ is read "m by n."

EXAMPLE 1   Let

$$\mathbf{A} = \begin{bmatrix} 4 & 0 & 2 \\ 3 & 1 & -1 \end{bmatrix}, \quad \mathbf{B} = [1, 3, 5],$$

$$\mathbf{C} = \begin{bmatrix} 2 \\ 1 \\ 0 \\ 3/2 \end{bmatrix}, \quad \text{and} \quad \mathbf{D} = \begin{bmatrix} 4 & 2 & 16 \\ 3 & 1 & 7 \\ 5 & 2 & 0 \end{bmatrix}.$$

Then $\mathbf{A}$ is a 2×3 matrix, $\mathbf{B}$ is a 1×3 matrix, $\mathbf{C}$ is a 4×1 matrix, and $\mathbf{D}$ is a 3×3 matrix. Said another way, $\mathbf{A}$ has order 2×3, $\mathbf{B}$ has order 1×3, $\mathbf{C}$ has order 4×1, and $\mathbf{D}$ has order 3×3. Note the use of commas to separate the numbers in $\mathbf{B}$. Commas should not be used to separate the numbers in a matrix that has more than one row. ◻

EXAMPLE 2   If every number in an $m \times n$ matrix is zero, the matrix is then called the $m \times n$ **zero matrix**. Thus,

$$\begin{bmatrix} 0 & 0 & 0 \\ 0 & 0 & 0 \end{bmatrix}$$

is the 2×3 zero matrix,

$$\begin{bmatrix} 0 & 0 & 0 & 0 \\ 0 & 0 & 0 & 0 \\ 0 & 0 & 0 & 0 \\ 0 & 0 & 0 & 0 \end{bmatrix}$$

is the 4×4 zero matrix, and so on. ◻

An $m \times 1$ matrix is called a **column vector;** a $1 \times n$ matrix is called a **row vector.** Thus a column vector is a matrix that has only one column, and a row vector is a matrix that has only one row.

EXAMPLE 3 The matrices

$$\begin{bmatrix} 1 \\ -2 \end{bmatrix}, \quad \begin{bmatrix} 2 \\ 1 \\ 3 \end{bmatrix}, \quad \text{and} \quad \begin{bmatrix} 5 \\ 2 \\ 4 \\ 0 \\ 1 \end{bmatrix}$$

are, respectively, a 2×1, a 3×1, and a 5×1 column vector. ◻

EXAMPLE 4 The matrices $[1, -2]$, $[2, 1, 3]$, and $[5, 2, 4, 0, 1]$ are, respectively, a 1×2, a 1×3, and a 1×5 row vector. ◻

A **square matrix** is an $m \times n$ matrix for which $n = m$. Thus, a square matrix has the same number of rows as it has columns and is therefore an $n \times n$ matrix for some positive integer n.

EXAMPLE 5 The matrix

$$\begin{bmatrix} 4 & 3 & 0 \\ 6 & 0 & 0 \\ 0 & 1 & -2 \end{bmatrix}$$

is a 3×3 square matrix. The matrix

$$\begin{bmatrix} 2 & 1 & 3 \\ 4 & 2 & 1 \end{bmatrix}$$

is not a square matrix. ◻

Matrices are often used to store information. For example, they are widely used in connection with inventories of products. To illustrate, suppose that an automobile dealer sells three models: the subcompact (S), the compact (C), and the regular-sized (R). Suppose further that each model comes in four styles: four-door sedan (4D), two-door sedan (2D), hatchback (HB), and station wagon (SW). The dealer's **inventory matrix** might be as follows:

$$\begin{array}{c} \\ \\ \text{model} \begin{array}{c} \text{S} \\ \text{C} \\ \text{R} \end{array} \end{array} \begin{array}{c} \text{style} \\ \begin{array}{cccc} \text{4D} & \text{2D} & \text{HB} & \text{SW} \end{array} \\ \begin{bmatrix} 2 & 6 & 4 & 3 \\ 5 & 7 & 2 & 4 \\ 4 & 3 & 0 & 2 \end{bmatrix} \end{array}.$$

The numbers in the matrix tell us how many cars of each model and style the dealer has in his inventory. Thus, he has six subcompact two-door sedans in stock, four compact station wagons in stock, and so on. If he sells one subcompact station wagon, two compact four-door sedans, one regular-sized four-door sedan, and one regular-sized station wagon, then his inventory matrix will be as follows:

$$\text{model} \begin{array}{c} \\ S \\ C \\ R \end{array} \overset{\overset{\text{style}}{\begin{array}{cccc} 4D & 2D & HB & SW \end{array}}}{\begin{bmatrix} 2 & 6 & 4 & 2 \\ 3 & 7 & 2 & 4 \\ 3 & 3 & 0 & 1 \end{bmatrix}}.$$

Many businesses computerize their inventory records by storing an inventory matrix in a computer. Each time a unit is sold, the computer automatically changes the inventory matrix to update it, just as we did above. At any given time the computer can print out the current inventory matrix. By comparing the current inventory matrix with a previous inventory matrix, the manager or executive can tell how many units of each type of product have been sold in a given time period.

Matrix Equality, Addition, Subtraction, and Scalar Multiplication

The numbers in a matrix are called its **entries.** We define equality of matrices in terms of their entries.

> **Matrix Equality**
>
> Two matrices are equal if and only if they have the same order and their corresponding entries are equal.

EXAMPLE 6   The matrices

$$\begin{bmatrix} 4 & 5 \\ 3 & 1 \\ 2 & 0 \end{bmatrix} \text{ and } \begin{bmatrix} 4 & 5 \\ 3 & 4 \\ 2 & 0 \end{bmatrix},$$

have the same order, but they are not equal because they have corresponding entries (namely, those in row 2, column 2) that are not equal. The matrices

$$\begin{bmatrix} 3 & 6 \\ -1 & 0 \end{bmatrix} \text{ and } \begin{bmatrix} 3 & 6 \\ -1 & 0 \\ 0 & 0 \end{bmatrix}$$

cannot be equal because they do not have the same order. □

MATRICES 207

EXAMPLE 7 The matrices

$$\begin{bmatrix} 2 & 1 & 3 \\ x & 0 & y \end{bmatrix} \quad \text{and} \quad \begin{bmatrix} u & v & 3 \\ 5 & 0 & 4 \end{bmatrix}$$

will be equal if and only if $x = 5$, $y = 4$, $u = 2$, and $v = 1$. ☐

We may add or subtract matrices by adding or subtracting their corresponding entries; however, this can be done only if the matrices are of the same order.

> **Matrix Addition and Subtraction**
>
> If **A** and **B** are matrices of the same order, their sum $\mathbf{A} + \mathbf{B}$ is obtained by adding the corresponding entries of **A** and **B**; their difference $\mathbf{A} - \mathbf{B}$ is obtained by subtracting each entry of **B** from the corresponding entry of **A**. If **A** and **B** are not of the same order, then $\mathbf{A} + \mathbf{B}$ and $\mathbf{A} - \mathbf{B}$ are not defined.

EXAMPLE 8 Let

$$\mathbf{A} = \begin{bmatrix} 3 & -2 & 5 \\ 1 & 6 & -3 \end{bmatrix} \quad \text{and} \quad \mathbf{B} = \begin{bmatrix} 4 & 2 & 6 \\ -3 & 2 & -5 \end{bmatrix}.$$

Then

$$\mathbf{A} + \mathbf{B} = \begin{bmatrix} 3 & -2 & 5 \\ 1 & 6 & -3 \end{bmatrix} + \begin{bmatrix} 4 & 2 & 6 \\ -3 & 2 & -5 \end{bmatrix}$$

$$= \begin{bmatrix} 3+4 & -2+2 & 5+6 \\ 1+(-3) & 6+2 & -3+(-5) \end{bmatrix}$$

$$= \begin{bmatrix} 7 & 0 & 11 \\ -2 & 8 & -8 \end{bmatrix}.$$

Also,

$$\mathbf{A} - \mathbf{B} = \begin{bmatrix} 3-4 & -2-2 & 5-6 \\ 1-(-3) & 6-2 & -3-(-5) \end{bmatrix}$$

$$= \begin{bmatrix} -1 & -4 & -1 \\ 4 & 4 & 2 \end{bmatrix}. \quad ☐$$

EXAMPLE 9 Let

$$\mathbf{C} = \begin{bmatrix} 3 & -2 & 1 \\ 6 & 2 & 2 \end{bmatrix} \quad \text{and} \quad \mathbf{D} = \begin{bmatrix} 4 & 6 & 1 \\ 3 & 2 & 8 \\ 5 & -1 & 7 \end{bmatrix}.$$

We cannot form $C + D$ or $C - D$ because these matrices are not of the same order. ◻

There are several rules for addition and subtraction of matrices. They follow from the corresponding properties for addition and subtraction of real numbers.

Rules for Addition and Subtraction of Matrices

Let A, B, and C be $m \times n$ matrices, and let 0 be the $m \times n$ zero matrix.

1. *Commutative Law:* $A + B = B + A$.
2. *Associative Law:* $A + (B + C) = (A + B) + C$.
3. $A + 0 = A$.
4. $A - 0 = A$.
5. $A - A = 0$.

(See Exercises 41 and 42.)

Matrices can also be multiplied by numbers. This process is known as **scalar multiplication.**

Scalar Multiplication

Let A be a matrix, and let c be a real number. The **scalar multiple of A by c** is the matrix cA obtained by multiplying every entry of A by c.

EXAMPLE 10 Let

$$A = \begin{bmatrix} 2 & 3 & 2 \\ 4 & -1 & 0 \end{bmatrix}$$

and $c = 2$. Then

$$2A = 2 \begin{bmatrix} 2 & 3 & 2 \\ 4 & -1 & 0 \end{bmatrix} = \begin{bmatrix} 2 \cdot 2 & 2 \cdot 3 & 2 \cdot 2 \\ 2 \cdot 4 & 2(-1) & 2 \cdot 0 \end{bmatrix} = \begin{bmatrix} 4 & 6 & 4 \\ 8 & -2 & 0 \end{bmatrix}.$$

Similarly,

$$-3A = \begin{bmatrix} -6 & -9 & -6 \\ -12 & 3 & 0 \end{bmatrix} \quad \text{and} \quad \tfrac{1}{2} A = \begin{bmatrix} 1 & \tfrac{3}{2} & 1 \\ 2 & -\tfrac{1}{2} & 0 \end{bmatrix}. \quad ◻$$

EXAMPLE 11 The operations of matrix addition, subtraction, and scalar multiplication may be combined. For instance, suppose

$$\mathbf{A} = \begin{bmatrix} 2 & 3 & 2 \\ 4 & -1 & 0 \end{bmatrix} \quad \text{and} \quad \mathbf{B} = \begin{bmatrix} 1 & -1 & 0 \\ 2 & 3 & 5 \end{bmatrix}.$$

Then

$$2\mathbf{A} + 3\mathbf{B} = 2\begin{bmatrix} 2 & 3 & 2 \\ 4 & -1 & 0 \end{bmatrix} + 3\begin{bmatrix} 1 & -1 & 0 \\ 2 & 3 & 5 \end{bmatrix}$$

$$= \begin{bmatrix} 4 & 6 & 4 \\ 8 & -2 & 0 \end{bmatrix} + \begin{bmatrix} 3 & -3 & 0 \\ 6 & 9 & 15 \end{bmatrix}$$

$$= \begin{bmatrix} 7 & 3 & 4 \\ 14 & 7 & 15 \end{bmatrix}.$$

Similarly,

$$5\mathbf{A} - 4\mathbf{B} = 5\begin{bmatrix} 2 & 3 & 2 \\ 4 & -1 & 0 \end{bmatrix} - 4\begin{bmatrix} 1 & -1 & 0 \\ 2 & 3 & 5 \end{bmatrix}$$

$$= \begin{bmatrix} 10 & 15 & 10 \\ 20 & -5 & 0 \end{bmatrix} - \begin{bmatrix} 4 & -4 & 0 \\ 8 & 12 & 20 \end{bmatrix}$$

$$= \begin{bmatrix} 6 & 19 & 10 \\ 12 & -17 & -20 \end{bmatrix}. \quad \square$$

Here are the rules that govern scalar multiplication:

Rules for Scalar Multiplication

Let $\mathbf{A}$ and $\mathbf{B}$ be $m \times n$ matrices and let $\mathbf{0}$ be the $m \times n$ zero matrix. Let c and d be numbers.

1. *Distributive Laws:* $(c + d)\mathbf{A} = c\mathbf{A} + d\mathbf{A}.$
   $c(\mathbf{A} + \mathbf{B}) = c\mathbf{A} + c\mathbf{B}.$
2. $1\mathbf{A} = \mathbf{A}.$
3. $\mathbf{A} + (-1)\mathbf{B} = \mathbf{A} - \mathbf{B}.$
4. $0\mathbf{A} = \mathbf{0}.$

(See Exercises 59 and 60.)

■ *Computer Manual: Program MATRICES*

4.2 EXERCISES

Matrices

Let

$$A = [1, 2, 4, 7], \quad B = \begin{bmatrix} 2 & 3 \\ -2 & 6 \end{bmatrix}, \quad C = \begin{bmatrix} 6 & 1 \\ 5 & 2 \\ 4 & 0 \end{bmatrix}, \quad D = \begin{bmatrix} 3 \\ 5 \end{bmatrix}, \quad E = [2],$$

$$F = [0, 0], \quad G = \begin{bmatrix} 0 \\ 0 \\ 0 \end{bmatrix}, \quad H = \begin{bmatrix} -3 & -2 & \frac{1}{2} \\ 8 & 0 & 0 \\ 0 & 4 & 0 \end{bmatrix}, \quad J = \begin{bmatrix} 0 & 0 & 0 \\ 0 & 0 & 0 \end{bmatrix},$$

$$K = \begin{bmatrix} 1 & 2 & 3 & 4 \\ 5 & 6 & 7 & 8 \\ -1 & 2 & -3 & 5 \\ 6 & -8 & 9 & 0 \end{bmatrix}, \quad L = [0, 0, 1], \quad M = \begin{bmatrix} 1 \\ 1 \\ 1 \\ 1 \end{bmatrix},$$

$$N = \begin{bmatrix} 1 & 0 & 0 & 0 \\ 0 & 1 & 0 & 0 \\ 0 & 0 & 1 & 0 \\ 0 & 0 & 0 & 1 \\ 0 & 0 & 0 & 0 \end{bmatrix}, \quad P = \begin{bmatrix} 2 & 3 & 5 & 7 & 9 \\ 1 & 4 & 6 & 8 & 9 \\ -1 & 0 & 0 & 0 & 1 \end{bmatrix}, \quad Q = \begin{bmatrix} 0 & 1 \\ -1 & 0 \end{bmatrix}.$$

Exercises 1 through 5 refer to these matrices.

1. What are the orders of these matrices?
2. Which of them are square matrices?
3. Which of them are zero matrices?
4. Which of them are vectors?
5. Which of them are row vectors? Which of them are column vectors?

Management

6. A hardware store carries three brands of garden hose, A, B, and C. Each brand comes in 10-, 20-, and 40-foot lengths. Of Brand A hoses, the store has in stock six of the 10-foot ones, seven of the 20-foot ones, and three of the 40-foot ones. The corresponding figures for Brand B are 2, 0, and 5, and for Brand C they are 12, 25, and 12. Write two inventory matrices for the store's supply of garden hoses.

Mom's Market stocks three brands, X, Y, and Z, of canned vegetables, and each brand has peas (P), beans (B), carrots (C), and tomatoes (T). At the start of business on a particular day, the inventory matrix was as follows:

MATRICES   211

$$\text{Vegetable} \begin{array}{c} \\ P \\ B \\ C \\ T \end{array} \overset{\text{Brand}}{\begin{array}{c} X \quad Y \quad Z \end{array}} \left[\begin{array}{ccc} 25 & 13 & 16 \\ 18 & 10 & 12 \\ 5 & 15 & 3 \\ 7 & 21 & 14 \end{array} \right].$$

Exercises 7 through 9 refer to this matrix.

7. How many cans of Brand X carrots were in stock at the beginning of the day? how many cans of Brand Y tomatoes? how many cans of Brand Y peas? how many cans of beans? how many cans of Brand Z vegetables?

8. During the course of the day, Mom's Market sold 6 cans of Brand X peas, 3 cans of Brand X tomatoes, 4 cans of Brand Y beans, 10 cans of Brand Y tomatoes, 3 cans of Brand Z beans, 1 can of Brand Z carrots, and 4 cans of Brand Z tomatoes. What was the inventory matrix for canned vegetables at the end of the day?

9. At the end of the day, Mom herself counted the number of cans of each vegetable in stock. She found that there were 47 cans of peas, 30 cans of beans, 22 cans of carrots, and 22 cans of tomatoes. Does this agree with the inventory matrix of Exercise 8? What is going on here?

Matrices can be used to depict competitive situations. To illustrate, suppose Gamma Corporation plans to introduce a new product, whose price will be set either high (H) or low (L). Gamma's competitor either will respond with a similar product (R) or it will not (NR). Gamma estimates the dollar amounts it stands to gain under each possible set of circumstances and embodies this information in a **payoff matrix**:

$$\text{Price of new product} \begin{array}{c} \\ H \\ L \end{array} \overset{\text{Competitor's response}}{\begin{array}{c} R \quad\quad NR \end{array}} \left[\begin{array}{cc} 5000 & 50,000 \\ 25,000 & 40,000 \end{array} \right].$$

Thus, for example, if Gamma sets the price of the product low (L) and its competitor does not respond with a similar product (NR), Gamma stands to gain $40,000. Exercises 10 through 12 refer to this matrix.

10. Suppose Gamma knows that its competitor will not respond. What should Gamma do?

11. Suppose Gamma knows that its competitor will respond with a similar product. What should Gamma do?

12. Suppose Gamma is not sure what its competitor will do. What do you think would be the best course of action for Gamma in this case?

Life Science

13. A pediatrician studying premature babies finds that males born prematurely average 41.2 cm in height if they are two weeks premature, 35.4 cm if they are one month premature, and 32.0 cm if they are six weeks premature. Females average 38.2 cm, 34.3 cm, and 32.1 cm if they are two weeks, one month, and six weeks premature, respectively. The pediatrician also finds that when the babies are dis-

charged from the hospital, those who were two weeks premature average 43.4 cm if male and 42.2 cm if female; those who were one month premature average 42.6 cm if male and 42.8 cm if female; and those who were six weeks premature average 41.8 cm if male and 42.2 cm if female. Write two matrices that embody these facts.

Social Science

A sociologist is studying status among a group of executives in a firm. Formal status is determined by the reporting relationships among the executives: if A reports to B, then B has greater formal status than A. Informal status depends on the perceptions of the members of the group as to which individuals are of higher or lower status. Exercises 14 and 15 refer to this situation.

14. Suppose that executives B, F, and E all report directly to C, that D reports directly to F, and that A and G report directly to E. Write a matrix that exhibits the formal status among these executives, using the following scheme: if A reports directly to B, put 1 in row A, column B; if A does not report directly to B, put 0 in row A, column B. Assume that no one reports to himself or herself.

15. The sociologist has recorded informal status among the executives of Exercise 14 in another matrix, in which 1 in row A, column B means that B has higher informal status than A:

$$\begin{array}{c} \\ A \\ B \\ C \\ D \\ E \\ F \\ G \end{array} \begin{array}{c} \begin{array}{ccccccc} A & B & C & D & E & F & G \end{array} \\ \left[\begin{array}{ccccccc} 0 & 0 & 1 & 0 & 0 & 1 & 0 \\ 0 & 0 & 1 & 0 & 0 & 1 & 0 \\ 0 & 0 & 0 & 0 & 0 & 0 & 0 \\ 1 & 1 & 1 & 0 & 0 & 1 & 1 \\ 1 & 1 & 1 & 0 & 0 & 1 & 1 \\ 0 & 0 & 0 & 0 & 0 & 0 & 0 \\ 0 & 0 & 1 & 0 & 0 & 1 & 0 \end{array} \right] \end{array}.$$

Who has the highest informal status among the group? Who has the lowest informal status? Compare this matrix with the one of Exercise 14. Is there a difference between the formal and informal status of any of the executives?

Matrix Equality, Addition, Subtraction and Scalar Multiplication

In Exercises 16 through 23, let

$$A = \begin{bmatrix} 2 & 6 & 1 \\ 3 & 5 & 2 \end{bmatrix}, \quad B = \begin{bmatrix} 2 & x & y \\ 3 & -5 & z \end{bmatrix}, \quad C = \begin{bmatrix} 2 & 6 & x \\ y & z & 2 \end{bmatrix}, \quad D = \begin{bmatrix} u & v & 3 \\ w & -1 & 2 \end{bmatrix},$$

$$E = \begin{bmatrix} -3 & -2 & 5 \\ 6 & x & y \\ 0 & z & 0 \end{bmatrix}, \quad \text{and} \quad F = \begin{bmatrix} u & -2 & w \\ v & 1 & 1 \\ 0 & 0 & 1 \end{bmatrix}.$$

Can values be assigned to x, y, z, u, v, and w so that the given equality is true?

16. A = B
17. A = C
18. B = C
19. C = D
20. B = D
21. A = E
22. F = C
23. E = F

MATRICES

In Exercises 24 through 40, perform the indicated matrix arithmetic, if possible.

24. $\begin{bmatrix} 2 & 1 \\ -3 & 0 \end{bmatrix} + \begin{bmatrix} 4 & 2 \\ 1 & -1 \end{bmatrix}$

25. $\begin{bmatrix} 4 & 4 \\ 2 & 6 \end{bmatrix} - \begin{bmatrix} 2 & 1 \\ -3 & 1 \end{bmatrix}$

26. $[1, 3, 2] + [0, 4, 5]$

27. $\begin{bmatrix} 1 \\ 4 \\ -3 \end{bmatrix} + \begin{bmatrix} 2 \\ 0 \\ 5 \end{bmatrix}$

28. $\begin{bmatrix} 8 & 3 & 1 \\ 3 & 0 & 0 \\ 4 & -1 & 5 \end{bmatrix} - \begin{bmatrix} 6 & 3 & 2 \\ 8 & 4 & 1 \\ 5 & -1 & 9 \end{bmatrix}$

29. $\begin{bmatrix} 8 & 3 \\ 2 & 1 \\ 2 & 4 \end{bmatrix} + \begin{bmatrix} 8 & 4 \\ -1 & -2 \\ 0 & 6 \end{bmatrix}$

30. $\begin{bmatrix} 2 & 2 \\ 1 & 3 \end{bmatrix} + \begin{bmatrix} 0 & 7 \\ 1 & -1 \\ 3 & -2 \end{bmatrix}$

31. $\begin{bmatrix} 6 & 2 & 4 & 3 \\ -1 & 0 & 0 & 1 \\ 1 & 3 & 2 & 0 \end{bmatrix} - \begin{bmatrix} 5 & 6 & 1 \\ 1 & 9 & 0 \\ 2 & 3 & 1 \end{bmatrix}$

32. $\begin{bmatrix} 1 & 2 & -4 \\ 3 & 1 & -6 \\ 2 & 2 & 0 \\ 9 & 3 & 1 \end{bmatrix} - \begin{bmatrix} 1 & 2 & -8 \\ 3 & 1 & 7 \\ 2 & 2 & 1 \\ 6 & 3 & -1 \end{bmatrix}$

33. $\begin{bmatrix} 4 & 2 & 0 & 1 & 3 \\ 6 & 1 & 1 & -1 & 2 \end{bmatrix} - \begin{bmatrix} 4 & 3 & 1 & 2 & 1 & 6 \\ 4 & 0 & 1 & -1 & 2 & 2 \end{bmatrix}$

34. $\begin{bmatrix} 6 & 0 & 2 \\ 3 & 1 & -1 \\ 2 & 4 & 5 \end{bmatrix} + \begin{bmatrix} 0 & 0 & 0 \\ 0 & 0 & 0 \\ 0 & 0 & 0 \end{bmatrix}$

35. $\begin{bmatrix} 5 & 2 & 1 \\ 3 & 9 & 2 \end{bmatrix} + \begin{bmatrix} 0 & 0 \\ 0 & 0 \end{bmatrix}$

36. $\begin{bmatrix} 2x \\ y \\ z \\ -w \end{bmatrix} - \begin{bmatrix} 2x \\ 3y \\ 4z \\ 5w \end{bmatrix}$

37. $\begin{bmatrix} x & y \\ z & w \end{bmatrix} + \begin{bmatrix} r & s \\ t & u \end{bmatrix}$

38. $\begin{bmatrix} x & y \\ z & w \end{bmatrix} + \begin{bmatrix} 0 & 0 \\ 0 & 0 \end{bmatrix}$

39. $\begin{bmatrix} 2x+y & 3x-z \\ 4y & 8x \end{bmatrix} + \begin{bmatrix} 2y-x & z-3x \\ -5z & 5x+2y \end{bmatrix}$

40. $\begin{bmatrix} 3x & 4y & -2z \\ 0 & x+y & -y \\ 5z+x & 2y & 1 \end{bmatrix} - \begin{bmatrix} 2x+2y & 3x+y & -2z \\ x+y & x-y & -3 \\ 2 & 2y & 1 \end{bmatrix}$

*41. Verify that rules 1 and 2 for addition and subtraction of matrices hold for the following matrices:

$$\mathbf{A} = \begin{bmatrix} a & b \\ c & d \end{bmatrix}, \quad \mathbf{B} = \begin{bmatrix} e & f \\ g & h \end{bmatrix}, \quad \mathbf{C} = \begin{bmatrix} i & j \\ k & l \end{bmatrix}.$$

*42. Verify that rules 3, 4, and 5 for addition and subtraction of matrices hold for the following matrices:

$$\mathbf{A} = \begin{bmatrix} a & b \\ c & d \end{bmatrix}, \quad \mathbf{0} = \begin{bmatrix} 0 & 0 \\ 0 & 0 \end{bmatrix}.$$

In Exercises 43 through 50, let

$$\mathbf{A} = \begin{bmatrix} 4 & -2 \\ 2 & 6 \end{bmatrix} \quad \text{and} \quad \mathbf{B} = \begin{bmatrix} 1 & -1 \\ 2 & 0 \\ 3 & 5 \end{bmatrix},$$

and find the given scalar multiple.

43. $3\mathbf{A}$.

44. $1\mathbf{A}$

45. $-1\mathbf{A}$

46. $-4\mathbf{A}$

47. $\dfrac{1}{2}\mathbf{A}$

48. $\dfrac{1}{3}\mathbf{B}$

49. $-\dfrac{2}{3}\mathbf{B}$

50. $0\mathbf{B}$

In Exercises 51 through 58, let

$$C = \begin{bmatrix} 4 & 3 & -1 & 2 \\ 5 & 0 & 3 & 1 \end{bmatrix} \quad \text{and} \quad D = \begin{bmatrix} 2 & -3 & 4 & 7 \\ 0 & 2 & 4 & 0 \end{bmatrix},$$

and find the given matrix.

51. $2C + 3D$

52. $-4C + 5D$

53. $6C - \dfrac{1}{2}D$

54. $3D + 5C$

55. $2(C + D)$

56. $\dfrac{1}{2}(2C - 4D)$

57. $(5 - 5)C$

58. $(6 + 8)(D - D)$

*59. Verify rules 1 and 3 of the rules for scalar multiplication for the matrices

$$A = \begin{bmatrix} x & y \\ z & w \end{bmatrix} \quad \text{and} \quad B = \begin{bmatrix} r & s \\ t & u \end{bmatrix}.$$

*60. Verify rules 2 and 4 of the rules for scalar multiplication for the matrices **A** and **B** of Exercise 59.

Management

61. Inventory at the end of a period is equal to inventory at the beginning of the period plus purchases during the period, less sales during the period; that is,

   Ending inventory = Beginning inventory + purchases − sales.

   An automobile dealer sells three models; subcompact (S), compact (C), and regular-sized (R). Each model comes in four styles: four-door (4D), two-door (2D), hatchback (HB), and station wagon (SW). Suppose the

$$\text{Beginning inventory matrix} = \begin{array}{c} \\ S \\ C \\ R \end{array} \begin{bmatrix} 4D & 2D & HB & SW \\ 10 & 3 & 8 & 6 \\ 2 & 5 & 3 & 2 \\ 6 & 8 & 5 & 3 \end{bmatrix},$$

$$\text{Purchase matrix} = \begin{array}{c} \\ S \\ C \\ R \end{array} \begin{bmatrix} 4D & 2D & HB & SW \\ 6 & 2 & 8 & 1 \\ 3 & 4 & 0 & 2 \\ 0 & 3 & 1 & 5 \end{bmatrix},$$

and

$$\text{Sales matrix} = \begin{array}{c} \\ S \\ C \\ R \end{array} \begin{bmatrix} 4D & 2D & HB & SW \\ 2 & 4 & 10 & 2 \\ 4 & 1 & 2 & 0 \\ 3 & 7 & 0 & 8 \end{bmatrix}.$$

   Find the ending inventory matrix.

62. Suppose rows and columns are labeled as in Exercise 61, and

$$\text{Beginning inventory matrix} = \begin{bmatrix} 8 & 2 & 9 & 3 \\ 0 & 4 & 7 & 1 \\ 6 & 2 & 8 & 4 \end{bmatrix},$$

$$\text{Purchase matrix} = \begin{bmatrix} 3 & 0 & 2 & 1 \\ 2 & 4 & 1 & 1 \\ 3 & 6 & 5 & 4 \end{bmatrix},$$

and

$$\text{Ending inventory matrix} = \begin{bmatrix} 6 & 2 & 1 & 4 \\ 1 & 3 & 3 & 0 \\ 7 & 4 & 10 & 6 \end{bmatrix}.$$

Find the sales matrix.

63. Again, suppose rows and columns are labeled as in Exercise 61 and

$$\text{Beginning inventory matrix} = \begin{bmatrix} 5 & 4 & 5 & 7 \\ 2 & 3 & 0 & 0 \\ 8 & 4 & 9 & 12 \end{bmatrix},$$

$$\text{Sales matrix} = \begin{bmatrix} 2 & 3 & 5 & 1 \\ 4 & 2 & 2 & 3 \\ 1 & 1 & 0 & 2 \end{bmatrix},$$

and

$$\text{Ending inventory matrix} = \begin{bmatrix} 6 & 3 & 1 & 7 \\ 2 & 1 & 0 & 5 \\ 13 & 6 & 14 & 13 \end{bmatrix}.$$

Find the purchase matrix.

64. The following matrix gives the value of the milk in a store's dairy case, in dollars:

$$\text{Size} \begin{array}{c} \text{Quart} \\ \text{Half gallon} \\ \text{Gallon} \end{array} \begin{bmatrix} \overset{\text{Regular}}{200} & \overset{\text{Lowfat}}{150} & \overset{\text{Skim}}{200} \\ 400 & 300 & 300 \\ 700 & 200 & 100 \end{bmatrix}.$$

Suppose all milk prices increase by 10%. Use scalar multiplication to write a new matrix that gives the value of the store's milk after the price increase.

65. A video rental store's inventory of tapes depreciates at a rate of 2% per month. Suppose the value of the store's current inventory is given by the following matrix (figures in thousands of dollars):

$$\text{Videos} \begin{array}{c} \text{Movie} \\ \text{Exercise} \\ \text{Sports} \\ \text{Instructional} \end{array} \begin{bmatrix} \overset{\text{VHS}}{50} & \overset{\text{Beta}}{25} & \overset{\text{Super VHS}}{5} \\ 10 & 4 & 2 \\ 14 & 1 & 0 \\ 8 & 3 & 2 \end{bmatrix}.$$

Assuming no new purchases, find the value of the store's inventory one month from now; two months from now.

66. Suppose the video store of Exercise 65 will receive a shipment of tapes one month from now, and that the shipment will have the following value (figures in thousands of dollars):

Format

$$\text{Videos} \begin{array}{c} \text{Movie} \\ \text{Exercise} \\ \text{Sports} \\ \text{Instructional} \end{array} \begin{array}{ccc} \text{VHS} & \text{Beta} & \text{Super VHS} \\ \begin{bmatrix} 6 & 3 & 1 \\ 3 & 1 & 0 \\ 2 & 0 & 2 \\ 4 & 0 & 1 \end{bmatrix} \end{array}.$$

Assuming no other purchases by the store, what will the value of its inventory be right after this shipment is received? What will the value be two months from now?

Life Science

The Environmental Protection Agency monitored the emissions of three plants in the same region. At the start of the monitoring period, emissions information about the plants was stored in a matrix:

$$\textbf{S:} \quad \text{Plant} \begin{array}{c} \text{A} \\ \text{B} \\ \text{C} \end{array} \begin{array}{ccc} \text{CO}_2 & \text{SO}_2 & \text{NO} \\ \begin{bmatrix} 25 & 400 & 80 \\ 15 & 500 & 60 \\ 30 & 100 & 90 \end{bmatrix} \end{array}.$$

At the end of the monitoring period emissions information was stored in another matrix:

$$\textbf{E:} \quad \text{Plant} \begin{array}{c} \text{A} \\ \text{B} \\ \text{C} \end{array} \begin{array}{ccc} \text{CO}_2 & \text{SO}_2 & \text{NO} \\ \begin{bmatrix} 18 & 300 & 100 \\ 12 & 250 & 50 \\ 60 & 120 & 110 \end{bmatrix} \end{array}.$$

Exercises 67 through 72 refer to this situation.

67. Find **S** − **E**. What do its entries represent?
68. Use the matrix **S** − **E** of Exercise 67 to find the plant that reduced CO_2 emissions the most.
69. Use the matrix **S** − **E** to find the plant that reduced SO_2 emissions the most.
70. Use the matrix **S** − **E** to find the plant that reduced NO emissions the most.
71. Which plant appears to have made an effort to reduce emissions of all three pollutants?
72. Which plant appears to have made no effort to reduce emissions of any of the pollutants?

Decision Matrices

A **decision matrix** is a matrix that gives the payoffs of alternative strategies under different states of nature. For instance, suppose you can invest in either a money market fund, a stock fund, or a bond fund. Your payoff may then depend on whether interest rates fall, stay the same, or rise. The possibilities can be embodied in a decision matrix such as the following:

$$\text{Strategy: invest in} \begin{array}{c} \text{Money Market} \\ \text{Stocks} \\ \text{Bonds} \end{array} \overset{\begin{array}{c} \text{States of Nature:} \\ \text{Interest Rates} \end{array}}{\begin{array}{ccc} \text{Fall} & \text{Stay Same} & \text{Rise} \end{array}} \begin{bmatrix} -100 & 20 & 150 \\ -200 & -100 & 200 \\ 100 & 0 & -75 \end{bmatrix}.$$

The entries in the matrix are the dollar amounts you will gain or lose under each possible investment strategy and state of nature. Let us denote this decision matrix by **D**. Exercises 73 through 76 refer to the matrix **D**.

*73. There are several different criteria for using a decision matrix to select a strategy. One of them is the maximax criterion: select the strategy that has the best payoff. What is the maximax strategy for the matrix **D**?

*74. Another criterion for selecting a strategy is the maximin criterion: find the worst payoff for each strategy and then select the strategy having the best of these worsts. What is the maximin strategy for the matrix **D**?

*75. A third criterion for selecting a strategy is the average value criterion: total the payoffs for each strategy and divide the total by the number of states of nature, thus obtaining an average payoff for each strategy, and then select the strategy having the best average payoff. What is the average value strategy for the decision matrix **D**?

*76. A manufacturer must decide whether to build a small new plant, build a large new plant, expand the old plant, or do nothing. The payoffs will depend on whether demand for the manufacturer's product is high, medium, or low. The decision matrix, with payoffs in millions of dollars, is as follows:

$$\text{Strategy:} \begin{array}{c} \text{Small new plant} \\ \text{Large new plant} \\ \text{Expand old plant} \\ \text{Do nothing} \end{array} \overset{\begin{array}{c} \text{States of Nature:} \\ \text{Demand is} \end{array}}{\begin{array}{ccc} \text{High} & \text{Medium} & \text{Low} \end{array}} \begin{bmatrix} 20 & 15 & -10 \\ 40 & -5 & -30 \\ 20 & 5 & -15 \\ 5 & 0 & -5 \end{bmatrix}.$$

Find the manufacturer's maximax, maximin, and average value strategies.

4.3 MATRIX MULTIPLICATION

In the previous section we discussed scalar multiplication, which is the multiplication of a matrix by a number. In this section we discuss **matrix multiplication,** which is the multiplication of one matrix by another to produce a third matrix. Matrix multiplication is a more complex operation than scalar multiplication, and it has properties that are unlike those of ordinary multiplication of numbers, but it is extremely useful.

As noted above, matrix multiplication is the process of multiplying two matrices **A** and **B** together to obtain a product matrix **AB.** Before we define matrix multiplication in general, we illustrate the technique involved by examining the special case where **AB** is the product of a row vector **A** and a column vector **B**.

Let $\mathbf{A} = [a_1, a_2, \ldots, a_n]$ be a $1 \times n$ row vector, and let

$$\mathbf{B} = \begin{bmatrix} b_1 \\ b_2 \\ \vdots \\ b_r \end{bmatrix}$$

be an $r \times 1$ column vector. We wish to form the product **AB**, in that order. The product **AB** is defined if and only if **A** has the same number of columns as **B** has rows, that is, if and only if $n = r$. If $n = r$, the matrix product **AB** is the 1×1 matrix defined by

$$[a_1, a_2, \ldots, a_n] \begin{bmatrix} b_1 \\ b_2 \\ \vdots \\ b_n \end{bmatrix} = [a_1 b_1 + a_2 b_2 + \cdots + a_n b_n].$$

EXAMPLE 1   Let

$$\mathbf{A} = [1, 2] \quad \text{and} \quad \mathbf{B} = \begin{bmatrix} 3 \\ 4 \end{bmatrix}.$$

Then **A** is a 1×2 row vector, and **B** is a 2×1 column vector. Therefore, the matrix product **AB** is defined and

$$\mathbf{AB} = [1, 2] \begin{bmatrix} 3 \\ 4 \end{bmatrix} = [1(3) + 2(4)] = [11]. \quad \square$$

EXAMPLE 2   Let

$$\mathbf{A} = [1, 3, 0, -2, 5] \quad \text{and} \quad \mathbf{B} = \begin{bmatrix} 2 \\ -1 \\ 3 \\ 5 \\ 2 \end{bmatrix}.$$

Then

$$\mathbf{AB} = [1, 3, 0, -2, 5] \begin{bmatrix} 2 \\ -1 \\ 3 \\ 5 \\ 2 \end{bmatrix}$$

$$= [1(2) + 3(-1) + 0(3) + (-2)(5) + 5(2)] = [-1]. \quad \square$$

MATRICES

EXAMPLE 3 Let

$$A = [1, 2, 3] \quad \text{and} \quad B = \begin{bmatrix} 1 \\ 2 \end{bmatrix}.$$

Then A is a 1×3 row vector and B is a 2×1 column vector. Thus, since $n = 3$ and $r = 2$, $n \neq r$ and the product AB is not defined. □

Now that we know how to multiply a row vector by a column vector, we are ready to define matrix multiplication in general:

Matrix Multiplication

Let A be an $m \times n$ matrix and B be an $r \times s$ matrix. The **matrix product** AB is defined if and only if $n = r$. If $n = r$, the matrix product is an $m \times s$ matrix C. The entry in row i, column j of C is obtained by performing the vector multiplication

$$[\text{row } i \text{ of } A] \begin{bmatrix} \text{column} \\ j \\ \text{of} \\ B \end{bmatrix}.$$

We illustrate matrix multiplication with an example.

EXAMPLE 4 Let

$$A = \begin{bmatrix} 2 & 1 \\ 3 & -1 \end{bmatrix} \quad \text{and} \quad B = \begin{bmatrix} 3 & 2 & 0 \\ 5 & -1 & 1 \end{bmatrix}.$$

We will find the matrix product AB. Note that A is 2×2 and B is 2×3, so $n = 2 = r$. Therefore the product matrix AB is defined. Also, since $m = 2$ and $s = 3$, AB will be a 2×3 matrix.

We find the entry in row i, column j of AB by performing the vector multiplication

$$[\text{row } i \text{ of } A] \begin{bmatrix} \text{column} \\ j \\ \text{of} \\ B \end{bmatrix}.$$

The entries of AB are thus obtained as follows:

Location of Entry in **AB**	Entry
row 1, column 1	$[2, 1]\begin{bmatrix} 3 \\ 5 \end{bmatrix} = 2(3) + 1(5) = 11$
row 1, column 2	$[2, 1]\begin{bmatrix} 2 \\ -1 \end{bmatrix} = 2(2) + 1(-1) = 3$
row 1, column 3	$[2, 1]\begin{bmatrix} 0 \\ 1 \end{bmatrix} = 2(0) + 1(1) = 1$
row 2, column 1	$[3, -1]\begin{bmatrix} 3 \\ 5 \end{bmatrix} = 3(3) + (-1)(5) = 4$
row 2, column 2	$[3, -1]\begin{bmatrix} 2 \\ -1 \end{bmatrix} = 3(2) + (-1)(-1) = 7$
row 2, column 3	$[3, -1]\begin{bmatrix} 0 \\ 1 \end{bmatrix} = 3(0) + (-1)(1) = -1$

Therefore

$$\mathbf{AB} = \begin{bmatrix} 11 & 3 & 1 \\ 4 & 7 & -1 \end{bmatrix}. \quad \square$$

Before presenting more examples of matrix multiplication, we reiterate three points:

1. The requirement that n be equal to r means that for the matrix product **AB** to be defined the number of columns in the matrix **A** must be the same as the number of rows in the matrix **B**. If $n \neq r$, the product **AB** is not defined.
2. If the matrix product $\mathbf{AB} = \mathbf{C}$ is defined, then **C** has the same number of rows as **A** does and the same number of columns as **B** does.
3. To carry out the matrix multiplication **AB** (assuming it to be defined), we multiply the ith row of **A** by the jth column of **B** to obtain the entry in row i, column j of the product matrix **C**:

$$\text{row } i \begin{bmatrix} & & & \\ a_1 & a_2 & \cdots & a_n \\ & & & \end{bmatrix} \begin{bmatrix} b_1 \\ b_2 \\ \vdots \\ b_n \end{bmatrix} = \text{row } i \begin{bmatrix} \vdots \\ \cdots c \cdots \\ \vdots \end{bmatrix},$$

where $c = a_1 b_1 + a_2 b_2 + \cdots + a_n b_n$.

A diagram such as

$$\underset{m \times n}{\mathbf{A}} \quad \underset{r \times s}{\mathbf{B}}$$

is useful in keeping the first two points listed above in mind. If $n \neq r$ in the diagram, then the product **AB** is not defined:

MATRICES

$$\begin{array}{cc} \mathbf{A} & \mathbf{B} \\ m \times n & r \times s \end{array}$$
$$\boxed{\neq}$$

If $n = r$, however, the product is defined and is an $m \times s$ matrix:

$$\begin{array}{cc} \mathbf{A} & \mathbf{B} \\ m \times n & r \times s \end{array}$$
$$\boxed{=}$$
$$\mathbf{AB} \text{ is } m \times s$$

In Example 4 we wrote out all entries of the matrix product explicitly. It is not necessary to do this every time we wish to find a matrix product. If the product **AB** is defined, we merely multiply the first row of the left-hand matrix **A** by each of the columns of the right-hand matrix **B** in succession, thus obtaining the first row of the product matrix. We then repeat the process using the second row of **A** in order to obtain the second row of the product, and so on. When all rows of **A** have been used, we are done.

EXAMPLE 5 The matrix product

$$\begin{bmatrix} 2 & 3 \\ 6 & 1 \\ 1 & 7 \\ 0 & 9 \end{bmatrix} \begin{bmatrix} 2 & 4 & -1 \\ 3 & 6 & 2 \end{bmatrix}$$

$$\begin{array}{cc} 4 \times 2 & 2 \times 3 \end{array}$$
$$\boxed{=}$$
$$4 \times 3$$

is defined and is the following 4×3 matrix:

$$\begin{bmatrix} [2,3]\begin{bmatrix}2\\3\end{bmatrix} & [2,3]\begin{bmatrix}4\\6\end{bmatrix} & [2,3]\begin{bmatrix}-1\\2\end{bmatrix} \\ [6,1]\begin{bmatrix}2\\3\end{bmatrix} & [6,1]\begin{bmatrix}4\\6\end{bmatrix} & [6,1]\begin{bmatrix}-1\\2\end{bmatrix} \\ [1,7]\begin{bmatrix}2\\3\end{bmatrix} & [1,7]\begin{bmatrix}4\\6\end{bmatrix} & [1,7]\begin{bmatrix}-1\\2\end{bmatrix} \\ [0,9]\begin{bmatrix}2\\3\end{bmatrix} & [0,9]\begin{bmatrix}4\\6\end{bmatrix} & [0,9]\begin{bmatrix}-1\\2\end{bmatrix} \end{bmatrix}$$

$$= \begin{bmatrix} 2(2)+3(3) & 2(4)+3(6) & 2(-1)+3(2) \\ 6(2)+1(3) & 6(4)+1(6) & 6(-1)+1(2) \\ 1(2)+7(3) & 1(4)+7(6) & 1(-1)+7(2) \\ 0(2)+9(3) & 0(4)+9(6) & 0(-1)+9(2) \end{bmatrix} = \begin{bmatrix} 13 & 26 & 4 \\ 15 & 30 & -4 \\ 23 & 46 & 13 \\ 27 & 54 & 18 \end{bmatrix}.$$

EXAMPLE 6 If possible, find the following matrix product:

$$\begin{bmatrix} 2 & 6 & 8 & -1 \\ 3 & -5 & 2 & 4 \end{bmatrix} \begin{bmatrix} 1 \\ 6 \\ 2 \\ 3 \end{bmatrix}.$$

We apply the diagram

$$\underbrace{2 \times 4 \quad \underset{=}{4 \times 1}}_{2 \times 1}$$

to find that the matrix product is defined and is a 2×1 matrix:

$$\begin{bmatrix} 2 & 6 & 8 & -1 \\ 3 & -5 & 2 & 4 \end{bmatrix} \begin{bmatrix} 1 \\ 6 \\ 2 \\ 3 \end{bmatrix} = \begin{bmatrix} 2(1) + 6(6) + 8(2) - 1(3) \\ 3(1) - 5(6) + 2(2) + 4(3) \end{bmatrix} = \begin{bmatrix} 51 \\ -11 \end{bmatrix}. \quad \square$$

EXAMPLE 7 Let

$$\mathbf{A} = \begin{bmatrix} 2 & -1 & 3 \\ 0 & 1 & 2 \end{bmatrix} \quad \text{and} \quad \mathbf{B} = \begin{bmatrix} 2 & 1 \\ 4 & 3 \end{bmatrix}.$$

We cannot find the product **AB**, because

$$\begin{array}{cc} \mathbf{A} & \mathbf{B} \\ 2 \times 3 & 2 \times 2 \\ \underset{\neq}{} \end{array}$$

However, we can find the product **BA**, and it will be 2×3, because

$$\underbrace{\underset{\mathbf{B}}{2 \times 2} \quad \underset{=}{\underset{\mathbf{A}}{2 \times 3}}}_{2 \times 3}$$

We have

$$\mathbf{BA} = \begin{bmatrix} 2 & 1 \\ 4 & 3 \end{bmatrix} \begin{bmatrix} 2 & -1 & 3 \\ 0 & 1 & 2 \end{bmatrix}$$

$$= \begin{bmatrix} 2(2) + 1(0) & 2(-1) + 1(1) & 2(3) + 1(2) \\ 4(2) + 3(0) & 4(-1) + 3(1) & 4(3) + 3(2) \end{bmatrix}$$

$$= \begin{bmatrix} 4 & -1 & 8 \\ 8 & -1 & 18 \end{bmatrix}. \quad \square$$

EXAMPLE 8   Let

$$A = \begin{bmatrix} 2 & -1 & 0 \\ 3 & 2 & 5 \\ 1 & 4 & -2 \end{bmatrix}$$

and

$$B = \begin{bmatrix} 3 & 1 & 2 \\ -1 & 2 & 4 \\ 5 & -3 & 1 \end{bmatrix}.$$

Find **AB** and **BA** if possible.

Both matrix products are possible:

$$\begin{array}{cc} \mathbf{A} & \mathbf{B} \\ 3 \times 3 & 3 \times 3 \end{array} \qquad \begin{array}{cc} \mathbf{B} & \mathbf{A} \\ 3 \times 3 & 3 \times 3 \end{array}$$

$$\underbrace{}_{3 \times 3} \qquad \underbrace{}_{3 \times 3}$$

In each case the matrix product is a 3×3 matrix:

$$\mathbf{AB} = \begin{bmatrix} 2 & -1 & 0 \\ 3 & 2 & 5 \\ 1 & 4 & -2 \end{bmatrix} \begin{bmatrix} 3 & 1 & 2 \\ -1 & 2 & 4 \\ 5 & -3 & 1 \end{bmatrix}$$

$$= \begin{bmatrix} 2(3) - 1(-1) + 0(5) & 2(1) - 1(2) + 0(-3) & 2(2) - 1(4) + 0(1) \\ 3(3) + 2(-1) + 5(5) & 3(1) + 2(2) + 5(-3) & 3(2) + 2(4) + 5(1) \\ 1(3) + 4(-1) - 2(5) & 1(1) + 4(2) - 2(-3) & 1(2) + 4(4) - 2(1) \end{bmatrix}$$

$$= \begin{bmatrix} 7 & 0 & 0 \\ 32 & -8 & 19 \\ -11 & 15 & 16 \end{bmatrix},$$

$$\mathbf{BA} = \begin{bmatrix} 3 & 1 & 2 \\ -1 & 2 & 4 \\ 5 & -3 & 1 \end{bmatrix} \begin{bmatrix} 2 & -1 & 0 \\ 3 & 2 & 5 \\ 1 & 4 & -2 \end{bmatrix}$$

$$= \begin{bmatrix} 3(2) + 1(3) + 2(1) & 3(-1) + 1(2) + 2(4) & 3(0) + 1(5) + 2(-2) \\ -1(2) + 2(3) + 4(1) & -1(-1) + 2(2) + 4(4) & -1(0) + 2(5) + 4(-2) \\ 5(2) - 3(3) + 1(1) & 5(-1) - 3(2) + 1(4) & 5(0) - 3(5) + 1(-2) \end{bmatrix}$$

$$= \begin{bmatrix} 11 & 7 & 1 \\ 8 & 21 & 2 \\ 2 & -7 & -17 \end{bmatrix}.$$

Note that $\mathbf{AB} \neq \mathbf{BA}$. ◻

Let **A** be an $n \times n$ square matrix. Then the matrix product **AA** is defined and is another $n \times n$ square matrix, which we denote by $\mathbf{A}^2$. The matrix product $\mathbf{A}^2\mathbf{A}$

is also defined and is still another $n \times n$ square matrix, which we denote by $\mathbf{A}^3$. Hence, for any square matrix $\mathbf{A}$, we define

$$\mathbf{A}^1 = \mathbf{A},$$
$$\mathbf{A}^2 = \mathbf{AA},$$
$$\mathbf{A}^3 = \mathbf{A}^2\mathbf{A},$$
$$\mathbf{A}^4 = \mathbf{A}^3\mathbf{A},$$

and so on. In general,

$$\mathbf{A}^{k+1} = \mathbf{A}^k\mathbf{A}$$

for any $k \in N$. For example, let

$$\mathbf{A} = \begin{bmatrix} 3 & -1 \\ 2 & 5 \end{bmatrix}.$$

Then

$$\mathbf{A}^1 = \mathbf{A} = \begin{bmatrix} 3 & -1 \\ 2 & 5 \end{bmatrix},$$

$$\mathbf{A}^2 = \mathbf{AA} = \begin{bmatrix} 3 & -1 \\ 2 & 5 \end{bmatrix}\begin{bmatrix} 3 & -1 \\ 2 & 5 \end{bmatrix} = \begin{bmatrix} 7 & -8 \\ 16 & 23 \end{bmatrix},$$

$$\mathbf{A}^3 = \mathbf{A}^2\mathbf{A} = \begin{bmatrix} 7 & -8 \\ 16 & 23 \end{bmatrix}\begin{bmatrix} 3 & -1 \\ 2 & 5 \end{bmatrix} = \begin{bmatrix} 5 & -47 \\ 94 & 99 \end{bmatrix},$$

$$\mathbf{A}^4 = \mathbf{A}^3\mathbf{A} = \begin{bmatrix} 5 & -47 \\ 94 & 99 \end{bmatrix}\begin{bmatrix} 3 & -1 \\ 2 & 5 \end{bmatrix} = \begin{bmatrix} -79 & -240 \\ 480 & 401 \end{bmatrix},$$

and so on.

You have noticed by now that matrix multiplication differs from ordinary multiplication of real numbers. In the first place, it is always possible to multiply any two real numbers, but, as we have shown in some of our examples, it is not always possible to multiply two matrices. Second, when real numbers are multiplied, their order does not affect the result: 3(5) is the same as 5(3). However, this is not true for matrices. Even if the matrix product **AB** is defined, it is possible that the product **BA** is not defined, and even if both **AB** and **BA** are defined, they need not be equal, as in Example 8.

Matrix multiplication does have some similarities to ordinary multiplication. One is that every $n \times n$ matrix has a multiplicative identity, called the **$n \times n$ identity matrix $\mathbf{I}_n$**. The $n \times n$ identity matrices are defined as follows:

MATRICES

$$\mathbf{I}_1 = [1], \qquad \mathbf{I}_2 = \begin{bmatrix} 1 & 0 \\ 0 & 1 \end{bmatrix},$$

$$\mathbf{I}_3 = \begin{bmatrix} 1 & 0 & 0 \\ 0 & 1 & 0 \\ 0 & 0 & 1 \end{bmatrix}, \qquad \mathbf{I}_4 = \begin{bmatrix} 1 & 0 & 0 & 0 \\ 0 & 1 & 0 & 0 \\ 0 & 0 & 1 & 0 \\ 0 & 0 & 0 & 1 \end{bmatrix},$$

and so on.

The matrices $\mathbf{I}_n$ act for matrix multiplication in the same manner that the number 1 acts for multiplication of real numbers. If $\mathbf{A}$ is any $m \times n$ matrix, then

$$\mathbf{I}_m \mathbf{A} = \mathbf{A} \quad \text{and} \quad \mathbf{A}\mathbf{I}_n = \mathbf{A}.$$

If $\mathbf{A}$ is any $n \times n$ matrix, then

$$\mathbf{I}_n \mathbf{A} = \mathbf{A}\mathbf{I}_n = \mathbf{A}.$$

EXAMPLE 9 Let

$$\mathbf{A} = \begin{bmatrix} 2 & 1 \\ 3 & 2 \\ 4 & -1 \end{bmatrix} \quad \text{and} \quad \mathbf{B} = \begin{bmatrix} 2 & 3 \\ 5 & 4 \end{bmatrix}.$$

Then

$$\mathbf{I}_3 \mathbf{A} = \begin{bmatrix} 1 & 0 & 0 \\ 0 & 1 & 0 \\ 0 & 0 & 1 \end{bmatrix} \begin{bmatrix} 2 & 1 \\ 3 & 2 \\ 4 & -1 \end{bmatrix}$$

$$= \begin{bmatrix} 1(2) + 0(3) + 0(4) & 1(1) + 0(2) + 0(-1) \\ 0(2) + 1(3) + 0(4) & 0(1) + 1(2) + 0(-1) \\ 0(2) + 0(3) + 1(4) & 0(1) + 0(2) + 0(-1) \end{bmatrix}$$

$$= \begin{bmatrix} 2 & 1 \\ 3 & 2 \\ 4 & -1 \end{bmatrix} = \mathbf{A}$$

and

$$\mathbf{A}\mathbf{I}_2 = \begin{bmatrix} 2 & 1 \\ 3 & 2 \\ 4 & -1 \end{bmatrix} \begin{bmatrix} 1 & 0 \\ 0 & 1 \end{bmatrix} = \begin{bmatrix} 2(1) + 1(0) & 2(0) + 1(1) \\ 3(1) + 2(0) & 3(0) + 2(1) \\ 4(1) + (-1)(0) & 4(0) + (-1)(1) \end{bmatrix}$$

$$= \begin{bmatrix} 2 & 1 \\ 3 & 2 \\ 4 & -1 \end{bmatrix} = \mathbf{A}.$$

Similarly,

$$\mathbf{I}_2 \mathbf{B} = \mathbf{B}\mathbf{I}_2 = \mathbf{B}.$$

(Check this.) ☐

Now we are ready to state the rules of matrix multiplication:

The Rules of Matrix Multiplication

Let **A**, **B**, and **C** be matrices. Let **0** be the zero matrix and **I** the identity matrix. We assume that all matrices are of the appropriate sizes for the products and sums listed to be defined.

1. *Associative Law:* **A(BC) = (AB)C.**
2. *Distributive Laws:* **A(B + C) = AB + AC.**
   (B + C)A = BA + CA.
3. **IA = A and AI = A.**
4. **0A = 0 and A0 = 0.**

(See Exercises 47 and 48.)

We conclude this section with an application of matrix multiplication. Suppose that a store sells two brands of soap, Brand A and Brand B, and that each brand comes in two sizes, regular (R) and large (L). Suppose further that the following is the inventory matrix for the soap, where the entries are the number of boxes of soap the store has in stock:

$$\text{Brand} \begin{array}{c} \\ A \\ B \end{array} \begin{array}{c} \text{Size} \\ \begin{bmatrix} R & L \\ 100 & 50 \\ 150 & 80 \end{bmatrix} \end{array}.$$

Now we assume that each regular-sized box of soap of either brand sells for $1.00 and each large-sized box sells for $2.00. Thus, we have the following **price vector:**

$$\begin{array}{c} \\ R \\ L \end{array} \begin{array}{c} \text{Price} \\ \begin{bmatrix} 1.00 \\ 2.00 \end{bmatrix} \end{array}.$$

We multiply the inventory matrix by the price vector:

$$\begin{bmatrix} 100 & 50 \\ 150 & 80 \end{bmatrix} \begin{bmatrix} 1.00 \\ 2.00 \end{bmatrix} = \begin{bmatrix} 200 \\ 310 \end{bmatrix}.$$

Thus, the store has on hand $200 worth of Brand A soap and $310 worth of Brand B soap, for a total of $510 worth of soap on hand.

Inventory matrices are often multiplied by price vectors to determine the value of the inventory. In doing this, it is important that we set up the multiplication so that it makes sense; that is, we must multiply quantities by related quantities. Thus, in our illustration, we set up the multiplication so that we multiplied quantities of the sizes (regular and large) in stock by the prices for those sizes. Suppose that initially we had set up the inventory matrix in the following form.

$$\text{Size} \begin{array}{c} \\ R \\ L \end{array} \overset{\text{Brand}}{\begin{bmatrix} A & B \\ 100 & 150 \\ 50 & 80 \end{bmatrix}}.$$

Then we would *not* perform the multiplication

$$\begin{bmatrix} 100 & 150 \\ 50 & 80 \end{bmatrix} \begin{bmatrix} 1.00 \\ 2.00 \end{bmatrix}$$

because this would involve multiplying quantities of brands in stock by prices of the sizes, which does not make sense. If we had set up the inventory matrix in this form, we would have made the price vector a row vector and performed the following multiplication:

$$[1.00, 2.00] \begin{bmatrix} 100 & 150 \\ 50 & 80 \end{bmatrix} = [200, 310].$$

This makes sense, because we are multiplying prices of the sizes by quantities of those sizes.

■ *Computer Manual: Program MATRICES*

4.3 EXERCISES

Let

$$A = [2, 4], \quad B = \begin{bmatrix} 6 \\ 2 \end{bmatrix}, \quad C = [3, 0, -1], \quad D = \begin{bmatrix} 5 \\ 2 \\ 1 \end{bmatrix},$$

$$E = \begin{bmatrix} 2 & 1 \\ -1 & 0 \end{bmatrix}, \quad F = \begin{bmatrix} 1 & 3 \\ 0 & 4 \end{bmatrix}, \quad G = \begin{bmatrix} 1 & 2 \\ 3 & -5 \\ 0 & 2 \end{bmatrix},$$

$$H = \begin{bmatrix} 2 & 1 & 1 \\ 0 & 2 & 3 \end{bmatrix}.$$

In Exercises 1 through 22, find the given product if possible.

1. AB
2. CD
3. EF
4. FE
5. AC
6. AD
7. EB
8. AE
9. GB
10. AH
11. CG
12. HD
13. FG
14. HF
15. GH
16. HG
17. BA
18. DC
19. E^2
20. F^2
21. G^2
22. H^2

In Exercises 23 through 36, find the given matrix product.

23. $\begin{bmatrix} 2 & 3 & 1 \\ 2 & 5 & -3 \end{bmatrix} \begin{bmatrix} 3 & 4 \\ 2 & 6 \\ -1 & 0 \end{bmatrix}$ 24. $\begin{bmatrix} 3 & 4 \\ 2 & 6 \\ -1 & 0 \end{bmatrix} \begin{bmatrix} 2 & 3 & 1 \\ 2 & 5 & -3 \end{bmatrix}$ 25. $\begin{bmatrix} 2 & 1 \\ -1 & 0 \end{bmatrix} \begin{bmatrix} 4 & -1 \\ 1 & 3 \end{bmatrix}$

26. $\begin{bmatrix} 4 & -1 \\ 1 & 3 \end{bmatrix} \begin{bmatrix} 2 & 1 \\ -1 & 0 \end{bmatrix}$ 27. $[2, 6, 1, 3] \begin{bmatrix} 4 \\ 1 \\ 0 \\ -1 \end{bmatrix}$ 28. $\begin{bmatrix} 4 \\ 1 \\ 0 \\ -1 \end{bmatrix} [2, 6, 1, 3]$

29. $\begin{bmatrix} 1 & 0 \\ 0 & 1 \end{bmatrix} \begin{bmatrix} 2 & 3 \\ -5 & 17 \end{bmatrix}$ 30. $\begin{bmatrix} 8 & 1 & -3 \\ 9 & -1 & 7 \\ 11 & 6 & 3 \end{bmatrix} \begin{bmatrix} 1 & 0 & 0 \\ 0 & 1 & 0 \\ 0 & 0 & 1 \end{bmatrix}$ 31. $\begin{bmatrix} 1 & 0 & 0 \\ 0 & 1 & 0 \\ 0 & 0 & 1 \end{bmatrix} \begin{bmatrix} a & b & c \\ d & e & f \\ g & h & i \end{bmatrix}$

32. $\begin{bmatrix} 0 & 0 & 0 \\ 0 & 0 & 0 \\ 0 & 0 & 0 \end{bmatrix} \begin{bmatrix} a & b & c \\ d & e & f \\ g & h & i \end{bmatrix}$ 33. $\begin{bmatrix} r & 0 & 0 \\ 0 & r & 0 \\ 0 & 0 & r \end{bmatrix} \begin{bmatrix} a & b & c \\ d & e & f \\ g & h & i \end{bmatrix}$ 34. $\begin{bmatrix} a & b \\ c & d \end{bmatrix} \begin{bmatrix} e & f \\ 0 & 0 \end{bmatrix}$

35. $\begin{bmatrix} e & f \\ 0 & 0 \end{bmatrix} \begin{bmatrix} a & b \\ c & d \end{bmatrix}$ 36. $\begin{bmatrix} a & b \\ c & d \end{bmatrix} \begin{bmatrix} e & 0 \\ f & 0 \end{bmatrix}$

In Exercises 37 through 46, let

$$A = \begin{bmatrix} 2 & -1 \\ 3 & 1 \end{bmatrix}, \quad B = \begin{bmatrix} 0 & 1 & 0 \\ 0 & 0 & 1 \\ 1 & 0 & 0 \end{bmatrix}, \quad \text{and} \quad C = \begin{bmatrix} 4 & 3 \\ 2 & 0 \end{bmatrix},$$

and find the indicated matrix if possible.

37. A^2 38. A^3 39. A^4 40. B^2 41. B^3
42. B^4 43. A^2C^2 44. $(AC)^2$ 45. C^2A^2 46. $(CA)^2$

*47. Verify rules 1 and 2 of matrix multiplication for the matrices

$$A = \begin{bmatrix} a & b \\ c & d \end{bmatrix}, \quad B = \begin{bmatrix} e & f \\ g & h \end{bmatrix}, \quad \text{and} \quad C = \begin{bmatrix} i & j \\ k & l \end{bmatrix}.$$

*48. Verify rules 3 and 4 of matrix multiplication for the matrices

$$A = \begin{bmatrix} a & b \\ c & d \end{bmatrix}, \quad 0 = \begin{bmatrix} 0 & 0 \\ 0 & 0 \end{bmatrix}, \quad \text{and} \quad I = \begin{bmatrix} 1 & 0 \\ 0 & 1 \end{bmatrix}.$$

Management

49. A hardware store sells three brands of paint: Brand A, Brand B, and Brand C. Each brand comes in three colors: white (W), red (R), and yellow (Y). The following is the store's inventory matrix for paint, where the entries represent the number of gallons of paint in stock.

$$\text{Brand} \begin{array}{c} A \\ B \\ C \end{array} \begin{bmatrix} 30 & 6 & 15 \\ 10 & 8 & 9 \\ 12 & 0 & 2 \end{bmatrix} \begin{array}{c} \text{Color} \\ \text{W R Y} \end{array}$$

Each gallon of Brand A costs the store $8; each gallon of Brand B, $10; and each gallon of Brand C, $12. Form the price vector, and use matrix multiplication to find the value of the store's paint inventory. What is the value of each color of paint in the inventory?

MATRICES

50. The following is the inventory matrix for canned vegetables at Mom's Market, where each entry represents the number of cans in stock:

$$\text{Brand} \begin{array}{c} \\ A \\ B \\ C \\ D \end{array} \begin{array}{c} \text{Vegetable} \\ \begin{array}{cccc} W & X & Y & Z \end{array} \\ \begin{bmatrix} 20 & 15 & 0 & 25 \\ 40 & 32 & 20 & 8 \\ 10 & 5 & 13 & 19 \\ 6 & 10 & 45 & 22 \end{bmatrix} \end{array}.$$

Each can of vegetable W costs $0.75; each can of X costs $0.60; each can of Y costs $0.45; and each can of Z costs $0.40. Find the value of the inventory using matrix methods.

51. Let the inventory matrix for canned vegetables at Mom's Market be as in Exercise 50. Suppose that each can of Brand A vegetables costs $0.50, each can of Brand B costs $0.55, each can of Brand C costs $0.48, and each can of Brand D costs $0.53. Find the value of the inventory using matrix methods.

52. Suppose a video store sells all movie videotapes for $60, all exercise videos for $45, all sports videos for $30, and all instructional videos for $40. If the store has the following inventory of videotapes, find the retail value of its inventory. (Figures are the number of tapes.)

$$\text{Videos} \begin{array}{c} \\ \text{Movie} \\ \text{Exercise} \\ \text{Sports} \\ \text{Instructional} \end{array} \begin{array}{c} \text{Format} \\ \begin{array}{ccc} & & \text{Super} \\ \text{VHS} & \text{Beta} & \text{VHS} \end{array} \\ \begin{bmatrix} 200 & 80 & 20 \\ 75 & 10 & 20 \\ 120 & 40 & 0 \\ 50 & 10 & 5 \end{bmatrix} \end{array}.$$

53. Suppose the video store of Exercise 52 purchases each movie videotape for $45, each exercise tape for $30, each sports tape for $25, and each instructional tape for $25. Using the result of Exercise 52, find the gross profit the store will realize on the sale of its inventory.

54. Suppose the video store of Exercise 52 sells all VHS tapes for $40, all Beta tapes for $60, and all Super VHS tapes for $75. Find the retail value of its inventory.

55. Suppose the video store of Exercise 52 purchases all VHS tapes for $40, all Beta tapes for $55, and all Super VHS tapes for $70. Using the result of Exercise 54, find the gross profit the store will realize on the sale of its inventory.

An appliance store keeps price information on its TV sets in a matrix **X**:

$$\mathbf{X} = \begin{array}{c} \\ \text{Brand A} \\ \text{Brand B} \\ \text{Brand C} \end{array} \begin{array}{c} \text{Screen size} \\ \begin{array}{ccc} 13 \text{ in} & 17 \text{ in} & 21 \text{ in} \end{array} \\ \begin{bmatrix} 210 & 240 & 280 \\ 230 & 270 & 300 \\ 200 & 240 & 280 \end{bmatrix} \end{array}.$$

The entries of **X** are the prices of the sets, in dollars. Exercises 56 through 60 refer to this situation.

56. Let

$$\mathbf{P} = \begin{bmatrix} 1.1 & 0 & 0 \\ 0 & 1 & 0 \\ 0 & 0 & 1.05 \end{bmatrix}.$$

Find **PX**. What does the matrix **PX** represent?

57. Suppose the store has a sale on Brand B sets only: 20% off the price of all Brand B sets. Find a matrix **Q** such that **QX** gives the prices of the store's TV sets during the sale.

58. Suppose the store has a sale on all 13-inch and 17-inch sets: 10% off the prices of all 13-inch sets and 15% off the prices of all 17-inch sets. Find a matrix **V** such that **XV** gives the prices of the store's TV sets during the sale.

*59. Let **W** = [6, 4, 7] and find **WX**. What might **W** and **WX** represent?

*60. Let

$$\mathbf{Z} = \begin{bmatrix} 6 \\ 4 \\ 7 \end{bmatrix}$$

and find **XZ**. What might **Z** and **XZ** represent?

Directed Graphs

A diagram such as the following is called a **directed graph.**

Every directed graph has associated with it a matrix **G**, obtained by placing a 1 in row i, column j if there is an arrow directly from v_i to v_j and a 0 in row i, column j if there is no arrow from v_i to v_j. Thus the matrix for the directed graph in the figure is

$$\mathbf{G} = \begin{bmatrix} 0 & 1 & 1 & 0 & 0 \\ 0 & 0 & 0 & 1 & 1 \\ 0 & 0 & 0 & 0 & 1 \\ 0 & 0 & 0 & 0 & 1 \\ 0 & 0 & 0 & 0 & 0 \end{bmatrix}.$$

The matrix **G** may be thought of as exhibiting all one-step (i.e., one-arrow) paths in the graph; similarly, the matrix $\mathbf{G}^m$ exhibits all m-step paths in the directed graph. For instance, since

$$\mathbf{G}^2 = \begin{bmatrix} 0 & 0 & 0 & 1 & 2 \\ 0 & 0 & 0 & 0 & 1 \\ 0 & 0 & 0 & 0 & 0 \\ 0 & 0 & 0 & 0 & 0 \\ 0 & 0 & 0 & 0 & 0 \end{bmatrix},$$

there is one two-step path from v_1 to v_4, two two-step paths from v_1 to v_5, and one two-step path from v_2 to v_5. Similarly, $\mathbf{G}^3$ will exhibit all three-step paths in the directed graph. If $m \geq 4$, $\mathbf{G}^m = \mathbf{0}$, because there are no m-step paths when $m \geq 4$. Thus, the total number of paths in the directed graph can be found from the matrix

MATRICES

$$G + G^2 + G^3 = \begin{bmatrix} 0 & 1 & 1 & 1 & 3 \\ 0 & 0 & 0 & 1 & 2 \\ 0 & 0 & 0 & 0 & 1 \\ 0 & 0 & 0 & 0 & 1 \\ 0 & 0 & 0 & 0 & 0 \end{bmatrix}.$$

For instance, there are a total of three paths from v_1 to v_5, a total of two from v_2 to v_5, and so on. Exercises 61 through 63 refer to directed graphs and their associated matrices.

*61. The following directed graph is a **parts assembly diagram**; it shows the order in which parts v_1, v_2, v_3, and v_4 are assembled to form a final product v_5.

(a) Find and interpret the matrix **G** for this parts assembly diagram.
(b) Find the matrices G^2, G^3, G^4, and G^5. Use these matrices to find all two-step, three-step, and four-step paths through the parts assembly diagram.
(c) Find the matrix $G + G^2 + G^3 + G^4$. Find the total number of paths from the first part v_1 to the final product v_5 and from the third part v_3 to v_5.

*62. Here is the matrix **G** of a parts assembly diagram:

$$G = \begin{bmatrix} 0 & 1 & 0 & 1 & 1 & 0 \\ 0 & 0 & 0 & 1 & 1 & 1 \\ 0 & 0 & 0 & 1 & 0 & 1 \\ 0 & 0 & 0 & 0 & 1 & 0 \\ 0 & 0 & 0 & 0 & 0 & 1 \\ 0 & 0 & 0 & 0 & 0 & 0 \end{bmatrix}.$$

(a) Use **G** and its powers to find the total number of paths from the first part to the finished product.
(b) Draw the parts assembly diagram and use it to check your answer to part (a).

*63. Every organization chart defines a directed graph: if person v_i reports directly to person v_j, there is an arrow from v_i to v_j. Suppose that

$$G = \begin{bmatrix} 0 & 0 & 0 & 0 & 1 & 0 \\ 0 & 0 & 0 & 1 & 1 & 0 \\ 0 & 0 & 0 & 1 & 0 & 0 \\ 0 & 0 & 0 & 0 & 0 & 1 \\ 0 & 0 & 0 & 0 & 0 & 1 \\ 0 & 0 & 0 & 0 & 0 & 0 \end{bmatrix}$$

is the matrix of an organization chart.
(a) Use **G** and its powers to find all chains of command in the organization. (A chain of command is a path from any person at the bottom of the organization chart to the person at its top.)
(b) Draw the organization chart and use it to check your answers to part (a).

Dominance Matrices

*64. At the beginning of this chapter we considered a dominance matrix for football teams. The matrix was

$$\mathbf{D} = \begin{bmatrix} 0 & 0 & 1 & 0 & 0 & 0 \\ 0 & 0 & 0 & 1 & 0 & 0 \\ 0 & 1 & 0 & 0 & 0 & 0 \\ 0 & 0 & 0 & 0 & 0 & 0 \\ 0 & 0 & 0 & 1 & 0 & 0 \\ 1 & 0 & 0 & 0 & 1 & 0 \end{bmatrix},$$

where a 1 in row *i*, column *j* signifies that team *i* played and beat team *j*. Thus **D** gives the results of head-to-head competition among the teams. The matrix $\mathbf{D}^2$ gives the result of competition "through" common opponents: a 1 in row *i*, column *j* of $\mathbf{D}^2$ signifies that team *i* beat a team that beat team *j*. Similarly, a 1 in row *i*, column *j* of $\mathbf{D}^3$ signifies that team *i* beat a team that beat another team that beat team *j*. Thus **D** and its powers can be used to rank the teams. Begin by summing each row of **D**: if row *i* has a greater sum than row *j*, then team *i* is stronger than team *j*. If this ranking results in ties, attempt to refine it by taking into account common opponents and summing the rows of $\mathbf{D} + \mathbf{D}^2$; again, the greater the sum, the stronger the team. If ties still remain, proceed to $\mathbf{D} + \mathbf{D}^2 + \mathbf{D}^3$, and so on. Use this system to rank the teams whose dominance matrix is **D**.

*65. Use $\mathbf{D}$, $\mathbf{D}^2$, and $\mathbf{D}^3$ to rank the teams whose dominance matrix is

$$\mathbf{D} = \begin{bmatrix} 0 & 0 & 1 & 0 & 1 & 0 \\ 1 & 0 & 0 & 0 & 0 & 0 \\ 0 & 0 & 0 & 0 & 0 & 1 \\ 0 & 0 & 1 & 0 & 1 & 1 \\ 0 & 1 & 0 & 0 & 0 & 0 \\ 0 & 1 & 0 & 0 & 0 & 0 \end{bmatrix}.$$

4.4 INVERSE MATRICES

Given any real number $r \neq 0$, we can solve the equation

$$rx = 1$$

for the real number *x*. The solution, of course, is $x = 1/r$. The real number $1/r$ is called the **multiplicative inverse** of *r* and is frequently denoted by r^{-1}. The number 1 is the identity for multiplication of real numbers, and for any real number $r \neq 0$ and its multiplicative inverse $r^{-1} = 1/r$, we have

$$r \cdot r^{-1} = r^{-1} \cdot r = 1.$$

Some square matrices also have multiplicative inverses; that is, if **A** is an $n \times n$ square matrix, it is sometimes possible to solve the matrix equation

$$\mathbf{AX} = \mathbf{I}_n$$

for the matrix **X**. In this case, the matrix **X** is called the **multiplicative inverse of A** and is denoted by $\mathbf{A}^{-1}$. Furthermore, the product

$$\mathbf{AA}^{-1} = \mathbf{A}^{-1}\mathbf{A} = \mathbf{I}_n$$

MATRICES

is the $n \times n$ identity matrix $\mathbf{I}_n$. In this section we discuss inverse matrices, show how to find them when they exist, and demonstrate how to use them to solve systems of n linear equations in n variables.

The Inverse of a Matrix

Let $\mathbf{A}$ be an $n \times n$ matrix. If there is an $n \times n$ matrix $\mathbf{X}$ such that

$$\mathbf{AX} = \mathbf{XA} = \mathbf{I}_n,$$

then $\mathbf{X}$ is the **inverse** of matrix $\mathbf{A}$, and we denote $\mathbf{X}$ by $\mathbf{A}^{-1}$. If $\mathbf{A}^{-1}$ exists, it is unique; for if $\mathbf{B}$ is also a matrix such that

$$\mathbf{AB} = \mathbf{BA} = \mathbf{I}_n,$$

then $\mathbf{B}$ must be an $n \times n$ matrix and

$$\mathbf{B} = \mathbf{BI}_n = \mathbf{B}(\mathbf{AA}^{-1}) = (\mathbf{BA})\mathbf{A}^{-1} = \mathbf{I}_n\mathbf{A}^{-1} = \mathbf{A}^{-1}.$$

Hence it is permissible to speak of $\mathbf{A}^{-1}$ as *the* inverse of $\mathbf{A}$.

EXAMPLE 1 Let

$$\mathbf{A} = \begin{bmatrix} 5 & 2 \\ 2 & 1 \end{bmatrix}.$$

Then

$$\mathbf{A}^{-1} = \begin{bmatrix} 1 & -2 \\ -2 & 5 \end{bmatrix},$$

because

$$\mathbf{AA}^{-1} = \begin{bmatrix} 5 & 2 \\ 2 & 1 \end{bmatrix} \begin{bmatrix} 1 & -2 \\ -2 & 5 \end{bmatrix} = \begin{bmatrix} 5-4 & -10+10 \\ 2-2 & -4+5 \end{bmatrix} = \begin{bmatrix} 1 & 0 \\ 0 & 1 \end{bmatrix} = \mathbf{I}_2,$$

and similarly, $\mathbf{A}^{-1}\mathbf{A} = \mathbf{I}_2$. □

EXAMPLE 2 Let

$$\mathbf{A} = \begin{bmatrix} 1 & 2 & 5 \\ 1 & 3 & 0 \\ 0 & 0 & 1 \end{bmatrix}.$$

Then

$$\mathbf{A}^{-1} = \begin{bmatrix} 3 & -2 & -15 \\ -1 & 1 & 5 \\ 0 & 0 & 1 \end{bmatrix}$$

because $\mathbf{AA}^{-1} = \mathbf{A}^{-1}\mathbf{A} = \mathbf{I}_3$. (Check this.) □

Not every square matrix has an inverse. For example, the matrix

$$\begin{bmatrix} 0 & 0 \\ 0 & 0 \end{bmatrix}$$

cannot have an inverse since, for any 2×2 matrix $\mathbf{X}$,

$$\begin{bmatrix} 0 & 0 \\ 0 & 0 \end{bmatrix} \mathbf{X} = \mathbf{X} \begin{bmatrix} 0 & 0 \\ 0 & 0 \end{bmatrix} = \begin{bmatrix} 0 & 0 \\ 0 & 0 \end{bmatrix}.$$

Less trivially, the matrix

$$\mathbf{A} = \begin{bmatrix} 1 & 2 \\ 2 & 4 \end{bmatrix}$$

does not have an inverse. To see this, suppose $\mathbf{A}^{-1}$ does exist. Then $\mathbf{A}^{-1}$ is a 2×2 matrix, say

$$\mathbf{A}^{-1} = \begin{bmatrix} w & x \\ y & z \end{bmatrix},$$

and

$$\mathbf{A}\mathbf{A}^{-1} = \begin{bmatrix} 1 & 2 \\ 2 & 4 \end{bmatrix} \begin{bmatrix} w & x \\ y & z \end{bmatrix}$$

$$= \begin{bmatrix} w + 2y & x + 4z \\ 2w + 4y & 2x + 4z \end{bmatrix} = \begin{bmatrix} 1 & 0 \\ 0 & 1 \end{bmatrix}.$$

Thus

$$w + 2y = 1$$

and

$$2w + 4y = 0.$$

However, $w + 2y = 1$ implies that $2(w + 2y) = 2$, or that

$$2w + 4y = 2.$$

Hence we have simultaneously that $2w + 4y = 0$ and $2w + 4y = 2$. Obviously this situation is not possible because $0 \neq 2$. It follows that the matrix $\mathbf{A}$ cannot have an inverse.

A square matrix that does not have an inverse is called a **singular matrix,** and a square matrix that does have an inverse is called a **nonsingular matrix.**

Finding Inverses

In order to find the inverse of an $n \times n$ nonsingular matrix $\mathbf{A}$, we must solve the matrix equation

$$\mathbf{AX} = \mathbf{I}$$

for the $n \times n$ matrix **X**. For instance, if

$$\mathbf{A} = \begin{bmatrix} 5 & 2 \\ 2 & 1 \end{bmatrix},$$

then $\mathbf{A}^{-1}$, if it exists, will be a 2×2 matrix

$$\mathbf{X} = \begin{bmatrix} x_1 & x_2 \\ y_1 & y_2 \end{bmatrix}$$

such that

$$\mathbf{AX} = \begin{bmatrix} 5 & 2 \\ 2 & 1 \end{bmatrix} \begin{bmatrix} x_1 & x_2 \\ y_2 & y_2 \end{bmatrix} = \begin{bmatrix} 5x_1 + 2y_1 & 5x_2 + 2y_2 \\ 2x_1 + y_1 & 2x_2 + y_2 \end{bmatrix} = \begin{bmatrix} 1 & 0 \\ 0 & 1 \end{bmatrix}.$$

Thus we can find $\mathbf{X} = \mathbf{A}^{-1}$ if we can solve the systems of equations

$$\begin{array}{cc} 5x_1 + 2y_1 = 1 & 5x_2 + 2y_2 = 0 \\ 2x_1 + y_1 = 0 & \text{and} \quad 2x_2 + y_2 = 1 \end{array}$$

for x_1, y_1, x_2, and y_2. But we can solve these systems of equations by Gauss-Jordan elimination. In fact, since the systems have the same coefficients, we can solve them both at the same time. We do this by setting up the augmented matrix in the form

$$\begin{bmatrix} 5 & 2 & | & 1 & 0 \\ 2 & 1 & | & 0 & 1 \end{bmatrix}$$

and performing Gauss-Jordan elimination; at the conclusion of the process, we can read the values of x_1 and y_1 from the first column of the augmented matrix and those for x_2 and y_2 from the second column of the augmented matrix. Thus

$$\begin{bmatrix} \circled{5} & 2 & | & 1 & 0 \\ 2 & 1 & | & 0 & 1 \end{bmatrix} \xrightarrow{(1/5)R_1} \begin{bmatrix} \circled{1} & 2/5 & | & 1/5 & 0 \\ 2 & 1 & | & 0 & 1 \end{bmatrix} \xrightarrow{-2R_1 + R_2}$$

$$\begin{bmatrix} 1 & 2/5 & | & 1/5 & 0 \\ 0 & \circled{1/5} & | & -2/5 & 1 \end{bmatrix} \xrightarrow{5R_2} \begin{bmatrix} 1 & 2/5 & | & 1/5 & 0 \\ 0 & \circled{1} & | & -2 & 5 \end{bmatrix} \xrightarrow{(-2/5)R_2 + R_1}$$

$$\begin{bmatrix} 1 & 0 & | & 1 & -2 \\ 0 & 1 & | & -2 & 5 \end{bmatrix}.$$

Therefore (as we saw in Example 1),

$$\mathbf{A}^{-1} = \begin{bmatrix} 1 & -2 \\ -2 & 5 \end{bmatrix}.$$

Notice how we found $\mathbf{A}^{-1}$: we performed Gauss-Jordan elimination on the augmented matrix $[\mathbf{A} \mid \mathbf{I}_2]$ and obtained the matrix $[\mathbf{I}_2 \mid \mathbf{A}^{-1}]$. This technique works in general.

Finding the Inverse Matrix

If $\mathbf{A}$ is an $n \times n$ matrix, we can find $\mathbf{A}^{-1}$ (if it exists) by performing Gauss-Jordan elimination on the augmented matrix $[\mathbf{A} \mid \mathbf{I}_n]$ to obtain $[\mathbf{I}_n \mid \mathbf{A}^{-1}]$. If Gauss-Jordan elimination cannot transform the left-hand side of the augmented matrix from $\mathbf{A}$ into $\mathbf{I}_n$, then $\mathbf{A}^{-1}$ does not exist.

EXAMPLE 3   Let

$$\mathbf{A} = \begin{bmatrix} 2 & 0 & -1 \\ -1 & 1 & 2 \\ 1 & 2 & 1 \end{bmatrix}.$$

Suppose we wish to find $\mathbf{A}^{-1}$.

The augmented matrix is

$$[\mathbf{A} \mid \mathbf{I}_3] = \begin{bmatrix} 2 & 0 & -1 & | & 1 & 0 & 0 \\ -1 & 1 & 2 & | & 0 & 1 & 0 \\ 1 & 2 & 1 & | & 0 & 0 & 1 \end{bmatrix},$$

and Gauss-Jordan elimination proceeds as follows:

$$\begin{bmatrix} ② & 0 & -1 & | & 1 & 0 & 0 \\ -1 & 1 & 2 & | & 0 & 1 & 0 \\ 1 & 2 & 1 & | & 0 & 0 & 1 \end{bmatrix} \longrightarrow \begin{bmatrix} ① & 2 & 1 & | & 0 & 0 & 1 \\ -1 & 1 & 2 & | & 0 & 1 & 0 \\ 2 & 0 & -1 & | & 1 & 0 & 0 \end{bmatrix} \xrightarrow[-2R_1 + R_3]{R_1 + R_2}$$

$$\begin{bmatrix} 1 & 2 & 1 & | & 0 & 0 & 1 \\ 0 & ③ & 3 & | & 0 & 1 & 1 \\ 0 & -4 & -3 & | & 1 & 0 & -2 \end{bmatrix} \xrightarrow{(\frac{1}{3})R_2} \begin{bmatrix} 1 & 2 & 1 & | & 0 & 0 & 1 \\ 0 & ① & 1 & | & 0 & \frac{1}{3} & \frac{1}{3} \\ 0 & -4 & -3 & | & 1 & 0 & -2 \end{bmatrix} \xrightarrow[4R_2 + R_3]{-2R_2 + R_1}$$

$$\begin{bmatrix} 1 & 0 & -1 & | & 0 & -\frac{2}{3} & \frac{1}{3} \\ 0 & 1 & 1 & | & 0 & \frac{1}{3} & \frac{1}{3} \\ 0 & 0 & ① & | & 1 & \frac{4}{3} & -\frac{2}{3} \end{bmatrix} \xrightarrow[-R_3 + R_2]{R_3 + R_1} \begin{bmatrix} 1 & 0 & 0 & | & 1 & \frac{2}{3} & -\frac{1}{3} \\ 0 & 1 & 0 & | & -1 & -1 & 1 \\ 0 & 0 & 1 & | & 1 & \frac{4}{3} & -\frac{2}{3} \end{bmatrix}.$$

Therefore

$$\mathbf{A}^{-1} = \begin{bmatrix} 1 & \frac{2}{3} & -\frac{1}{3} \\ -1 & -1 & 1 \\ 1 & \frac{4}{3} & -\frac{2}{3} \end{bmatrix}.$$

(Check this by showing that $\mathbf{A}\mathbf{A}^{-1} = \mathbf{A}^{-1}\mathbf{A} = \mathbf{I}_3$.) ☐

EXAMPLE 4   Suppose we attempt to find the inverse of the matrix

$$\begin{bmatrix} 1 & -2 & -2 \\ -1 & 1 & 0 \\ 0 & -1 & -2 \end{bmatrix}.$$

MATRICES   237

We have

$$\begin{bmatrix} ① & -2 & -2 & | & 1 & 0 & 0 \\ -1 & 1 & 0 & | & 0 & 1 & 0 \\ 0 & -1 & -2 & | & 0 & 0 & 1 \end{bmatrix} \xrightarrow{R_1 + R_2} \begin{bmatrix} 1 & -2 & -2 & | & 1 & 0 & 0 \\ 0 & ⊖① & -2 & | & 1 & 1 & 0 \\ 0 & -1 & -2 & | & 0 & 0 & 1 \end{bmatrix} \xrightarrow{-1R_2}$$

$$\begin{bmatrix} 1 & -2 & -2 & | & 1 & 0 & 0 \\ 0 & ① & 2 & | & -1 & -1 & 0 \\ 0 & -1 & -2 & | & 0 & 0 & 1 \end{bmatrix} \xrightarrow[R_2 + R_3]{2R_2 + R_1} \begin{bmatrix} 1 & 0 & 2 & | & -1 & -2 & 0 \\ 0 & 1 & 2 & | & -1 & -1 & 0 \\ 0 & 0 & ⓪ & | & -1 & -1 & 1 \end{bmatrix}.$$

Since we have a row of zeros to the left of the vertical bar, it is impossible to produce the identity matrix I_3 to the left of the bar, so A^{-1} does not exist. ☐

The technique of using Gauss-Jordan elimination to find the inverse of a matrix may be applied to a general 2×2 matrix

$$A = \begin{bmatrix} a & b \\ c & d \end{bmatrix}$$

with the following result:

The Inverse of a 2 × 2 Matrix

Let

$$A = \begin{bmatrix} a & b \\ c & d \end{bmatrix}.$$

1. If $ad - bc \neq 0$, then A^{-1} exists and

$$A^{-1} = \frac{1}{ad - bc} \begin{bmatrix} d & -b \\ -c & a \end{bmatrix}.$$

2. If $ad - bc = 0$, then A^{-1} does not exist.

(See Exercises 27 and 28.)

EXAMPLE 5   Let

$$A = \begin{bmatrix} 4 & 3 \\ 2 & -1 \end{bmatrix} \quad \text{and} \quad B = \begin{bmatrix} 5 & 4 \\ 10 & 8 \end{bmatrix}.$$

For the matrix A, $ad - bc = 4(-1) - 3 \cdot 2 = -10 \neq 0$. Hence, A^{-1} exists and

$$A^{-1} = -\frac{1}{10} \begin{bmatrix} -1 & -3 \\ -2 & 4 \end{bmatrix} = \begin{bmatrix} 1/10 & 3/10 \\ 1/5 & -2/5 \end{bmatrix}.$$

For the matrix B, $ad - bc = 5 \cdot 8 - 4 \cdot 10 = 0$, so B^{-1} does not exist. ☐

Solving Matrix Equations Using the Inverse Matrix

Any system of linear equations can be written as a **matrix equation** $AX = B$, where A is the matrix consisting of the coefficients of the system, X is a column vector consisting of the variables of the system, and B is a column vector consisting of the constant terms of the system.

EXAMPLE 6 Consider the system of equations

$$\begin{aligned} 2x_1 \quad\quad\quad - x_3 &= 2 \\ -x_1 + x_2 + 2x_3 &= 5 \\ x_1 + 2x_2 + x_3 &= -3. \end{aligned}$$

We may write this system as the matrix equation

$$\begin{bmatrix} 2 & 0 & -1 \\ -1 & 1 & 2 \\ 1 & 2 & 1 \end{bmatrix} \begin{bmatrix} x_1 \\ x_2 \\ x_3 \end{bmatrix} = \begin{bmatrix} 2 \\ 5 \\ -3 \end{bmatrix},$$

since performing the matrix multiplication yields

$$\begin{bmatrix} 2x_1 \quad\quad - x_3 \\ -x_1 + x_2 + 2x_3 \\ x_1 + 2x_2 + x_3 \end{bmatrix} = \begin{bmatrix} 2 \\ 5 \\ -3 \end{bmatrix},$$

which is clearly equivalent to the original system of equations. $\square$

Now suppose that a system of n linear equations in n variables is written in matrix form as $AX = B$. If A^{-1} exists, we may solve this matrix equation for X by multiplying on the left by A^{-1}:

$$AX = B$$

implies that

$$A^{-1}(AX) = A^{-1}B,$$

or

$$(A^{-1}A)X = A^{-1}B,$$

or

$$I_n X = A^{-1}B,$$

or

$$X = A^{-1}B.$$

Solving Matrix Equations Using the Inverse Matrix

If A is an $n \times n$ matrix and A^{-1} exists, the solution of the matrix equation $AX = B$ is $X = A^{-1}B$.

EXAMPLE 7

Let us solve the system of linear equations

$$\begin{aligned} 2x_1 \phantom{{}+x_2} - x_3 &= 2 \\ -x_1 + x_2 + 2x_3 &= 5 \\ x_1 + 2x_2 + x_3 &= -3. \end{aligned}$$

From Example 6, this system may be written in matrix form as

$$\begin{bmatrix} 2 & 0 & -1 \\ -1 & 1 & 2 \\ 1 & 2 & 1 \end{bmatrix} \begin{bmatrix} x_1 \\ x_2 \\ x_3 \end{bmatrix} = \begin{bmatrix} 2 \\ 5 \\ -3 \end{bmatrix}.$$

Since

$$\begin{bmatrix} 2 & 0 & -1 \\ -1 & 1 & 2 \\ 1 & 2 & 1 \end{bmatrix}^{-1} = \begin{bmatrix} 1 & 2/3 & -1/3 \\ -1 & -1 & 1 \\ 1 & 4/3 & -2/3 \end{bmatrix}$$

(see Example 3), we have

$$\begin{bmatrix} x_1 \\ x_2 \\ x_3 \end{bmatrix} = \begin{bmatrix} 1 & 2/3 & -1/3 \\ -1 & -1 & 1 \\ 1 & 4/3 & -2/3 \end{bmatrix} \begin{bmatrix} 2 \\ 5 \\ -3 \end{bmatrix} = \begin{bmatrix} 19/3 \\ -10 \\ 32/3 \end{bmatrix}.$$

Hence the solution is $x_1 = 19/3$, $x_2 = -10$, and $x_3 = 32/3$. □

Thus we have shown that we can attempt to solve a system of n linear equations in n variables by writing it in matrix form $\mathbf{AX} = \mathbf{B}$, finding $\mathbf{A}^{-1}$, and setting $\mathbf{X} = \mathbf{A}^{-1}\mathbf{B}$. In general, this is not an efficient way to proceed, for it is usually a greater chore to find $\mathbf{A}^{-1}$ than it is to solve the system directly by Gauss-Jordan elimination. However, if $\mathbf{A}^{-1}$ is already known, this method reduces to a simple matrix multiplication, and hence is quite easy to perform. Even if $\mathbf{A}^{-1}$ is not known, if the equation $\mathbf{AX} = \mathbf{B}$ must be solved repeatedly using the same $\mathbf{A}$ but different choices of $\mathbf{B}$, it may be worthwhile to find $\mathbf{A}^{-1}$, set up the multiplication $\mathbf{X} = \mathbf{A}^{-1}\mathbf{B}$, and substitute the different constant vectors $\mathbf{B}$ as required.

EXAMPLE 8

The scientific lens division of an optical company has a special order for lenses. The lenses are of two types: polarized and unpolarized. Each polarized lens requires 5 hours of grinding and 2 hours of polishing. Each unpolarized lens requires 3 hours of grinding and 1 hour of polishing. Because of previously scheduled work, this week the grinding and polishing departments can allot only the amounts of time shown in Table 4–1 to make the lenses.

TABLE 4–1

	Hours Alloted Each Day		
	Monday	Tuesday	Wednesday
Grinding	30	60	85
Polishing	10	22	33

If we let

$$x_1 = \text{Number of polarized lenses made each day},$$
$$x_2 = \text{Number of unpolarized lenses made each day},$$

then we have the system of equations

$$5x_1 + 3x_2 = b_1$$
$$2x_2 + 1x_2 = b_2,$$

where

$$b_1 = \text{Grinding hours available each day},$$
$$b_2 = \text{Polishing hours available each day}.$$

Since

$$\begin{bmatrix} 5 & 3 \\ 2 & 1 \end{bmatrix}^{-1} = \begin{bmatrix} -1 & 3 \\ 2 & -5 \end{bmatrix},$$

we can find the daily production of lenses by finding

$$\begin{bmatrix} x_1 \\ x_2 \end{bmatrix} = \begin{bmatrix} -1 & 3 \\ 2 & -5 \end{bmatrix} \begin{bmatrix} b_1 \\ b_2 \end{bmatrix}$$

for the three possible constant vectors

$$\begin{bmatrix} b_1 \\ b_2 \end{bmatrix}.$$

Thus,

Monday: $\begin{bmatrix} x_1 \\ x_2 \end{bmatrix} = \begin{bmatrix} -1 & 3 \\ 2 & -5 \end{bmatrix} \begin{bmatrix} 30 \\ 10 \end{bmatrix} = \begin{bmatrix} 0 \\ 10 \end{bmatrix}$

Tuesday: $\begin{bmatrix} x_1 \\ x_2 \end{bmatrix} = \begin{bmatrix} -1 & 3 \\ 2 & -5 \end{bmatrix} \begin{bmatrix} 60 \\ 22 \end{bmatrix} = \begin{bmatrix} 6 \\ 10 \end{bmatrix}$

and

Wednesday: $\begin{bmatrix} x_1 \\ x_2 \end{bmatrix} = \begin{bmatrix} -1 & 3 \\ 2 & -5 \end{bmatrix} \begin{bmatrix} 85 \\ 33 \end{bmatrix} = \begin{bmatrix} 14 \\ 5 \end{bmatrix}.$

Therefore the firm can produce no polarized lenses on Monday, 6 on Tuesday, and 14 on Wednesday, and 10 unpolarized lenses on Monday, 10 on Tuesday, and 5 on Wednesday. ☐

■ *Computer Manual: Program MATRICES*

4.4 EXERCISES

The Inverse of a Matrix

In Exercises 1 through 8, verify that $AA^{-1} = A^{-1}A = I$.

1. $A = \begin{bmatrix} 3 & 1 \\ 11 & 4 \end{bmatrix}$ and $A^{-1} = \begin{bmatrix} 4 & -1 \\ -11 & 3 \end{bmatrix}$

2. $A = \begin{bmatrix} 2 & 0 \\ 0 & 2 \end{bmatrix}$ and $A^{-1} = \begin{bmatrix} 1/2 & 0 \\ 0 & 1/2 \end{bmatrix}$

3. $A = \begin{bmatrix} 2 & 3 \\ 6 & 10 \end{bmatrix}$ and $A^{-1} = \begin{bmatrix} 5 & -3/2 \\ -3 & 1 \end{bmatrix}$

4. $A = \begin{bmatrix} 6 & 3 \\ -1 & 5 \end{bmatrix}$ and $A^{-1} = \frac{1}{33}\begin{bmatrix} 5 & -3 \\ 1 & 6 \end{bmatrix}$

5. $A = \begin{bmatrix} 1 & 0 & -1 \\ 0 & 1 & 1 \\ 1 & 1 & 1 \end{bmatrix}$ and $A^{-1} = \begin{bmatrix} 0 & -1 & 1 \\ 1 & 2 & -1 \\ -1 & -1 & 1 \end{bmatrix}$

6. $A = \begin{bmatrix} 1 & 2 & 1 \\ 2 & -1 & 0 \\ 0 & 1 & -1 \end{bmatrix}$ and $A^{-1} = \frac{1}{7}\begin{bmatrix} 1 & 3 & 1 \\ 2 & -1 & 2 \\ 2 & -1 & -5 \end{bmatrix}$

7. $A = \begin{bmatrix} 1 & 0 & 0 & 1 \\ 0 & 1 & 1 & 0 \\ 0 & 1 & 0 & 0 \\ 0 & 0 & 0 & 1 \end{bmatrix}$ and $A^{-1} = \begin{bmatrix} 1 & 0 & 0 & -1 \\ 0 & 0 & 1 & 0 \\ 0 & 1 & -1 & 0 \\ 0 & 0 & 0 & 1 \end{bmatrix}$

8. $A = \begin{bmatrix} 2 & 1 & 0 & -1 \\ 3 & 2 & 0 & 0 \\ -1 & 0 & 5 & 2 \\ 0 & 1 & 2 & 1 \end{bmatrix}$ and $A^{-1} = \frac{1}{10}\begin{bmatrix} -2 & 6 & 4 & -10 \\ 3 & -4 & -6 & 15 \\ 4 & -2 & 2 & 0 \\ -11 & 8 & 2 & -5 \end{bmatrix}$

Finding Inverses

In Exercises 9 through 26, find the inverse of the given matrix if it exists.

9. $\begin{bmatrix} 1 & 4 \\ 2 & 7 \end{bmatrix}$

10. $\begin{bmatrix} 1 & 2 \\ -1 & 0 \end{bmatrix}$

11. $\begin{bmatrix} 3 & -1 \\ 6 & -2 \end{bmatrix}$

12. $\begin{bmatrix} -3 & 15 \\ -1 & 5 \end{bmatrix}$

13. $\begin{bmatrix} 3 & 4 \\ 2 & 5 \end{bmatrix}$

14. $\begin{bmatrix} -2 & -1 \\ 3 & 8 \end{bmatrix}$

15. $\begin{bmatrix} 1 & 1 & 2 \\ 1 & -2 & 1 \\ 0 & 1 & -1 \end{bmatrix}$

16. $\begin{bmatrix} 1 & 2 & -2 \\ 3 & 0 & 0 \\ 1 & -4 & 4 \end{bmatrix}$

17. $\begin{bmatrix} 2 & 4 & 6 \\ 3 & -1 & 5 \\ -1 & -9 & -7 \end{bmatrix}$

18. $\begin{bmatrix} 2 & 2 & 2 \\ -2 & 1 & 1 \\ 2 & -2 & 8 \end{bmatrix}$

19. $\begin{bmatrix} 2 & 1 & 0 \\ 2 & 1 & 3 \\ 1 & -1 & 1 \end{bmatrix}$

20. $\begin{bmatrix} 4 & -5 & 1 \\ 3 & 6 & 7 \\ 5 & -16 & -5 \end{bmatrix}$

21. $\begin{bmatrix} 1 & 2 & 0 & 0 \\ 0 & 1 & 3 & 1 \\ 0 & -1 & 1 & 2 \\ 3 & 4 & 4 & 0 \end{bmatrix}$

22. $\begin{bmatrix} 1 & 2 & -1 & 2 \\ 1 & 0 & 1 & 3 \\ 2 & 5 & -3 & -3 \\ 2 & 1 & 1 & -1 \end{bmatrix}$

23. $\begin{bmatrix} 1 & 0 & 2 & 5 \\ -1 & 1 & 1 & 0 \\ 2 & -1 & 0 & 4 \\ 2 & 0 & 1 & 9 \end{bmatrix}$

24. $\begin{bmatrix} 2 & -1 & 1 & -2 \\ 1 & 2 & -3 & 0 \\ -1 & -1 & 1 & 4 \\ 3 & 1 & -1 & -6 \end{bmatrix}$
25. $\begin{bmatrix} 1 & 0 & 2 & 0 \\ -1 & 3 & 5 & 1 \\ -4 & 7 & 12 & 0 \\ 3 & 2 & 7 & 3 \end{bmatrix}$
26. $\begin{bmatrix} 4 & 5 & 1 & 1 \\ -2 & 0 & 3 & -7 \\ 1 & 0 & 0 & 2 \\ 3 & 1 & 4 & 1 \end{bmatrix}$

*27. Let
$$A = \begin{bmatrix} a & b \\ c & d \end{bmatrix}.$$
Use Gauss-Jordan elimination to show that if $ad - bc \neq 0$ then
$$A^{-1} = \frac{1}{ad - bc} \begin{bmatrix} d & -b \\ -c & a \end{bmatrix}.$$

*28. Let
$$A = \begin{bmatrix} a & b \\ c & d \end{bmatrix}.$$
Use Gauss-Jordan elimination to show that if $ad - bc = 0$ then A^{-1} does not exist.

*29. Let A be a matrix such that $A^{n-1} \neq 0$ but $A^n = 0$ for some n. Show that
$$(I - A)^{-1} = I + A + A^2 + \cdots + A^{n-1}.$$

In Exercises 30 through 35, use the result of Exercise 29 to find the inverse of the given matrix. (*Hint:* Write the given matrix as $I - A$ for some A.)

*30. $\begin{bmatrix} 1 & -2 \\ 0 & 1 \end{bmatrix}$
*31. $\begin{bmatrix} 1 & 4 \\ 0 & 1 \end{bmatrix}$
*32. $\begin{bmatrix} 1 & -1 & -1 \\ 0 & 1 & -1 \\ 0 & 0 & 1 \end{bmatrix}$

*33. $\begin{bmatrix} 1 & -2 & 0 \\ 0 & 1 & -5 \\ 0 & 0 & 1 \end{bmatrix}$
*34. $\begin{bmatrix} 1 & 0 & 0 \\ -6 & 1 & 0 \\ 8 & -1 & 1 \end{bmatrix}$
*35. $\begin{bmatrix} 1 & 1 & 2 & 3 \\ 0 & 1 & -1 & 4 \\ 0 & 0 & 1 & -1 \\ 0 & 0 & 0 & 1 \end{bmatrix}$

Solving Matrix Equations Using the Inverse Matrix

In Exercises 36 through 47, solve the given system of equations by using the inverse of the coefficient matrix.

36. $x_1 + 4x_2 = 3$
    $2x_1 + 7x_2 = 5$

37. $2x_1 + x_2 = 0$
    $x_1 + x_2 = 6$

38. $4x_1 - 2x_2 = -3$
    $5x_1 - x_2 = -4$

39. $2x_1 + 3x_2 = 6$
    $-5x_1 + 2x_2 = 10$

40. $x_1 + x_2 + 2x_3 = -2$ (see Exercise 15)
    $x_1 - 2x_2 + x_3 = 5$
    $x_2 - x_3 = 3$

41. $2x_1 \quad\quad + x_3 = 1$
    $-x_1 + x_2 + 2x_3 = -1$
    $x_1 \quad\quad + x_3 = 5$

42. $2x_1 + x_2 \quad\quad = 5$ (see Exercise 19)
    $2x_1 + x_2 + 3x_3 = 7$
    $x_1 - x_2 + x_3 = 4$

43. $2x_1 + 2x_2 + 2x_3 = 6$ (see Exercise 18)
    $-2x_1 + x_2 + x_3 = -8$
    $2x_1 - 2x_2 + 8x_3 = 0$

MATRICES

44. $x_1 \quad\quad + 2x_3 + 5x_4 = -3$ (see Exercise 23)
$-x_1 + x_2 + x_3 \quad\quad = 1$
$2x_1 - x_2 \quad\quad + 4x_4 = 2$
$2x_1 \quad\quad + x_3 + 9x_4 = 5$

45. $2x_1 - x_2 + x_3 - 2x_4 = 7$ (see Exercise 24)
$x_1 + 2x_2 - 3x_3 \quad\quad = -4$
$-x_1 - x_2 + x_3 + 4x_4 = 4$
$3x_1 + x_2 - x_3 - 6x_4 = 0$

46. $2x_1 + x_2 \quad\quad - x_4 = 3$
$3x_1 + 2x_2 \quad\quad = 1$
$-x_1 \quad\quad + 5x_3 + 2x_4 = 5$
$x_2 + 2x_3 + x_4 = -3$

47. $x_1 + 2x_2 - x_3 \quad\quad = 7$
$2x_1 - x_2 \quad\quad + x_4 = 3$
$3x_1 + 2x_2 \quad\quad + 4x_4 = \dfrac{55}{2}$
$x_2 - x_3 + x_4 = \dfrac{17}{2}$

48. You have a total of $54,000 invested in stocks and bonds. The stocks pay 8% per year, the bonds 12% per year. How much do you have invested in each if the total annual payment is
  (a) $5000?
  (b) $5400?
  (c) $5680?

Management

49. A company makes two products, A and B. It takes 4 worker-hours to assemble each unit of A and 3 worker-hours to paint it. It takes 6 worker-hours to assemble each unit of B and 2 worker-hours to paint it. Find the number of units of each product the company can produce in a day if it has
  (a) 10 assemblers and 6 painters, all working a 10-hour day;
  (b) 30 assemblers and 15 painters, all working an 8-hour day;
  (c) 78 assemblers and 55 painters, all working an 8-hour day.

50. A manufacturer ships regular, deluxe, and super units by air express. Information concerning the three types of units is given in Table 4–2. How many units of each type are there in a shipment that
  (a) weighs 710 lb, occupies 265 ft^3 of space, and is insured for its full value of $2900?
  (b) weighs 1670 lb, occupies 690 ft^3 of space, and is insured for its full value of $9100?
  (c) weighs w lb, occupies s ft^3 of space, and is insured for its full value of $$v$?

TABLE 4–2

Type	Size (ft^3)	Weight (lb)	Value per unit ($)
Regular	2	3	20
Deluxe	2	5	30
Super	1	4	10

51. Ace Job Shop intends to purchase new milling machines. Information on three brands of machines is given in Table 4–3. How many machines of each brand should Ace buy if
  (a) it must mill 11,500 parts each day, has 2250 ft^2 to allot to the new machines, and can spend $260,000 to purchase them?

244 CHAPTER 4

TABLE 4–3

Brand	Capacity (units milled per day)	Size (ft²)	Cost ($)
X	1000	200	20,000
Y	1500	300	35,000
Z	2500	450	60,000

(b) it must mill 12,000 parts per day, has 2200 ft² to allocate to the new machines, and can spend $280,000 to purchase them?
(c) it must mill a parts each day, has s ft² to allot to the new machines, and can spend $C to purchase them?

You may use the fact that

$$\begin{bmatrix} 1 & 1.5 & 2.5 \\ 0.2 & 0.3 & 0.45 \\ 20 & 35 & 60 \end{bmatrix}^{-1} = \begin{bmatrix} 9 & -10 & -0.3 \\ -12 & 40 & 0.2 \\ 4 & -20 & 0 \end{bmatrix}.$$

Life Science

52. Two species of birds share the same ecological niche. Each individual of species A requires 200 square meters (m²) of territory and 40 kilograms (kg) of food per year, and each individual of species B requires 150 m² and 48 kg. An ecologist is studying these species in three different habitats: one has 12,000 m² of territory and approximately 3120 kg of food available for the birds; the second has 14,500 m² and 3440 kg; and the third has 40,000 m² and 9872 kg. How many of each species of bird can the ecologist expect to find in each habitat?

53. A doctor prescribes vitamins for a patient. The patient has a choice of three brands of vitamin pills, X, Y, and Z, whose composition is given in Table 4–4. How many pills of each brand should the patient take if the doctor prescribed
(a) 18 mg of vitamin A, 18 mg of vitamin B, and 4 mg of vitamin E?
(b) 10 mg of A, 15 mg of B, and 5 mg of E?
(c) a mg of A, b mg of B, and c mg of E?

You may use the fact that

$$\begin{bmatrix} 5 & 2 & 3 \\ 3 & 6 & 3 \\ 0 & 2 & 1 \end{bmatrix}^{-1} = \begin{bmatrix} 0 & 1/3 & -1 \\ -1/4 & 5/12 & -1/2 \\ 1/2 & -5/6 & 2 \end{bmatrix}.$$

TABLE 4–4

	Milligrams of Vitamin per Pill		
Brand	A	B	E
X	5	3	0
Y	2	6	2
Z	3	3	1

4.5 CRYPTOGRAPHY AND INPUT-OUTPUT ANALYSIS

In this section we present two important applications of inverse matrices: cryptography and input-output analysis.

Cryptography

Cryptography is the science of making and breaking codes. Matrices are widely used in cryptography. To illustrate, we show how a nonsingular matrix and its inverse may be used to encode and decode a message. The process requires that each letter of the message be assigned a numerical value. The message is then broken up into groups of numbers, and a nonsingular matrix $\mathbf{A}$ is used to encode it. After the message is transmitted as a string of numbers, the matrix $\mathbf{A}^{-1}$ is used to decode it. The sender and receiver of the message must agree beforehand on the numerical value of each letter, and the sender must possess the matrix $\mathbf{A}$ and the receiver the matrix $\mathbf{A}^{-1}$.

EXAMPLE 1 Suppose that a sender and a receiver assign numerical values to letters and to punctuation as follows:

$$A = 1, B = 2, \ldots, Z = 26, ! = 27, \text{space} = 28.$$

Suppose also that the encoding matrix is

$$\mathbf{A} = \begin{bmatrix} 2 & 1 & 3 \\ 0 & 1 & -1 \\ 1 & 2 & -2 \end{bmatrix},$$

and that the decoding matrix is therefore

$$\mathbf{A}^{-1} = \frac{1}{4} \begin{bmatrix} 0 & -8 & 4 \\ 1 & 7 & -2 \\ 1 & 3 & -2 \end{bmatrix}.$$

The sender wishes to transmit the message

FLEE AT ONCE!

Since the encoding matrix $\mathbf{A}$ is 3×3, she groups the letters and spaces of the message into groups of three, assigning them their numerical values:

F	L	E	E		A	T		O	N	C	E	!		
6	12	5	5	28	1	20	28	15	14	3	5	27	28	28

Note that there must be two blanks at the end of the message to fill out the last group of 3. She then constructs the code matrix

$$\mathbf{C} = \begin{bmatrix} F & L & E \\ E & & A \\ T & & O \\ N & C & E \\ ! & & \end{bmatrix} = \begin{bmatrix} 6 & 12 & 5 \\ 5 & 28 & 1 \\ 20 & 28 & 15 \\ 14 & 3 & 5 \\ 27 & 28 & 28 \end{bmatrix},$$

and multiplies **C** by **A** to obtain

$$\mathbf{CA} = \begin{bmatrix} 17 & 28 & -4 \\ 11 & 35 & -15 \\ 55 & 78 & 2 \\ 33 & 27 & 29 \\ 82 & 111 & -3 \end{bmatrix}.$$

The message is transmitted as the string of numbers

$$17, \ 28, \ -4, \ 11, \ldots, 82, \ 111, \ -3.$$

Since the receiver's decoding matrix $\mathbf{A}^{-1}$ is 3 × 3, he divides this string of numbers into groups of 3 and thus reconstructs the matrix **CA**. He then multiplies **CA** by $\mathbf{A}^{-1}$ to obtain the message:

$$(\mathbf{CA})\mathbf{A}^{-1} = \mathbf{CI}_3 = \mathbf{C} = \begin{bmatrix} 6 & 12 & 5 \\ 5 & 28 & 1 \\ 20 & 28 & 15 \\ 14 & 3 & 5 \\ 27 & 28 & 28 \end{bmatrix} = \begin{bmatrix} F & L & E \\ E & & A \\ T & & O \\ N & C & E \\ ! & & \end{bmatrix},$$

or

FLEE AT ONCE! □

If a large matrix is used, the method of encoding and decoding a message illustrated in Example 1 is usually quite difficult for an outsider to break, because it is an extremely difficult task to find the inverse of a large matrix.

Input-Output Analysis

An important application of matrix theory is found in a technique known as **input-output analysis.** Input-output analysis is concerned with the interrelationships among the sectors that make up an economy. To illustrate, let us consider the nation of Fredonia. Fredonia has a very simple economy with only two sectors: manufacturing (M) and agriculture (A).

The current **input-output table** for the Fredonian economy is Table 4–5 (the figures are in millions of dollars).

TABLE 4–5

				Output		
				Internal Demand	External Demand	Total Input
Input	M A	$\begin{bmatrix} M & A \\ 2 & 6 \\ 10 & 3 \end{bmatrix}$		8 13	12 17	20 30
Value added		8	21			
Total output		20	30			50 GNP

The matrix

$$\begin{bmatrix} 2 & 6 \\ 10 & 3 \end{bmatrix}$$

is called the **input-output matrix** for the economy.

The rows of Table 4–5 tell us where an industry's input goes. Thus, M uses $2 million worth of its own products and sells $6 million worth to the agricultural sector A. The total internal demand for the products of M is thus

$$\$2 \text{ million} + \$6 \text{ million} = \$8 \text{ million}.$$

Also, M sells $12 million worth of its products to final consumers, and this is the external demand for the products of M. Note that total input equals internal demand plus external demand. The sum of all total inputs is the country's gross national product (GNP). Thus, in our example, Fredonia has a current GNP of $50 million.

Since the total output of a sector must be equal to its total input, we see that the total output for M is $20 million and for A it is $30 million. The columns of Table 4–5 tell us where the value of an industry's output comes from. For example, of the $20 million total output of M, $2 million was due to its use of its own products and $10 million was due to its use of products of A. The difference

$$\$20 \text{ million} - (\$2 \text{ million} + \$10 \text{ million}) = \$8 \text{ million}$$

is M's **value added.** That is, M takes $2 million worth of its own products and $10 million worth of the products of A and does something to them to create products worth $20 million. The $8 million difference is the value M has added to the raw materials by doing whatever it does to them.

We can now form the **technological matrix T** (also called the **direct requirements matrix**) for the Fredonian economy. We do this by starting with the input-output matrix and dividing each entry in a column by the total output for the column:

$$\mathbf{T} = \begin{matrix} \\ M \\ A \end{matrix} \begin{matrix} M & A \\ \begin{bmatrix} 2/20 & 6/30 \\ 10/20 & 3/30 \end{bmatrix} \end{matrix} = \begin{matrix} \\ M \\ A \end{matrix} \begin{matrix} M & A \\ \begin{bmatrix} 0.10 & 0.20 \\ 0.50 & 0.10 \end{bmatrix} \end{matrix}.$$

The entries of the matrix **T** may be interpreted as follows: for each $1.00 worth of output produced by M, M needs $0.10 worth of its own output and $0.50 worth of the output of A; similarly, for every $1.00 worth of output produced by A, A needs $0.20 worth of the output of M and $0.10 worth of its own output.

Note that

$$\begin{bmatrix} 8 \\ 13 \end{bmatrix} = \begin{bmatrix} 0.1 & 0.2 \\ 0.5 & 0.1 \end{bmatrix} \begin{bmatrix} 20 \\ 30 \end{bmatrix};$$

that is,

$$\begin{bmatrix} \text{Internal} \\ \text{demand} \\ \text{vector} \end{bmatrix} = \mathbf{T} \begin{bmatrix} \text{Total} \\ \text{output} \\ \text{vector} \end{bmatrix}.$$

This will be true for any total output: if

$$x_M = \text{Total output of sector M}$$

and

$$x_A = \text{Total output of sector A,}$$

then

$$\mathbf{X} = \begin{bmatrix} x_M \\ x_A \end{bmatrix} = \begin{bmatrix} \text{Total} \\ \text{output} \\ \text{vector} \end{bmatrix},$$

and the internal demand vector will be equal to **TX**. (This is so because multiplying **T** by total output gives the proportion of output allocated to filling the internal demand for each sector's production.) But it is always the case for any economy that

$$\text{Total output} = \text{Total input} = \text{Internal demand} + \text{external demand.}$$

Suppose we let

$$d_M = \text{External demand for the products of sector M,}$$

and

$$d_A = \text{External demand for the products of sector A,}$$

so that

$$\mathbf{D} = \begin{bmatrix} d_M \\ d_A \end{bmatrix} = \begin{bmatrix} \text{External} \\ \text{demand} \\ \text{vector} \end{bmatrix}.$$

Then

$$\text{Total output} = \text{Internal demand} + \text{external demand}$$

becomes

$$\mathbf{X} = \mathbf{TX} + \mathbf{D}.$$

We solve this matrix equation as follows:

$$\mathbf{X} = \mathbf{TX} + \mathbf{D},$$
$$\mathbf{X} - \mathbf{TX} = \mathbf{D},$$
$$[\mathbf{I}_2 - \mathbf{T}]\mathbf{X} = \mathbf{D},$$
$$[\mathbf{I}_2 - \mathbf{T}]^{-1}[\mathbf{I}_2 - \mathbf{T}]\mathbf{X} = [\mathbf{I}_2 - \mathbf{T}]^{-1}\mathbf{D},$$
$$\mathbf{I}_2\mathbf{X} = [\mathbf{I}_2 - \mathbf{T}]^{-1}\mathbf{D},$$

or

$$\mathbf{X} = [\mathbf{I}_2 - \mathbf{T}]^{-1}\mathbf{D}.$$

Check that this is so by showing that

$$\begin{bmatrix} 20 \\ 30 \end{bmatrix} = [\mathbf{I}_2 - \mathbf{T}]^{-1} \begin{bmatrix} 12 \\ 17 \end{bmatrix}.$$

The analysis we have just carried out for the Fredonian economy holds in general.

The Input-Output Model

Given the technological matrix **T** and the external demand vector **D** for an economy, let **X** be the total output vector for the economy. Then

$$\mathbf{X} = [\mathbf{I} - \mathbf{T}]^{-1}\mathbf{D},$$

where **I** is an identity matrix of the appropriate size, and the vector **TX** is the internal demand vector for the economy. The matrix $[\mathbf{I} - \mathbf{T}]^{-1}$ is known as the **Leontief matrix** for the economy.

Thus, we can quickly determine the total output of an economy if we know its technological matrix **T** and its external demand vector **D**. Furthermore, the matrix equation

$$\mathbf{X} = [\mathbf{I} - \mathbf{T}]^{-1}\mathbf{D}$$

may be used for purposes of prediction, if we can assume that the matrix **T** will not change in the future. For Fredonia, for example, we must be able to assume that it will continue to be the case that M will require $0.10 of its own output and $0.50 worth of the output of A in order to produce $1.00 worth of its output, and similarly for A. Assuming that this is the case, suppose economic forecasts for Fredonia indicate that next year external demand for the products of M will increase by 10%, whereas external demand for the products of A will decrease by 10%. Then we expect the external demand vector for next year to be

$$\mathbf{D} = \begin{bmatrix} 12 + 0.10(12) \\ 17 - 0.10(17) \end{bmatrix} = \begin{bmatrix} 13.2 \\ 15.3 \end{bmatrix},$$

and we can use the matrix equation

$$\begin{bmatrix} x_M \\ x_A \end{bmatrix} = [\mathbf{I} - \mathbf{T}]^{-1} \begin{bmatrix} 13.2 \\ 15.3 \end{bmatrix}$$

to forecast the total output for the two industries next year. Since

$$[\mathbf{I} - \mathbf{T}]^{-1} = \begin{bmatrix} 0.90 & -0.20 \\ -0.50 & 0.90 \end{bmatrix}^{-1} = \begin{bmatrix} 1.27 & 0.28 \\ 0.70 & 1.27 \end{bmatrix}$$

CHAPTER 4

(rounded to two decimal places), we see that

$$\begin{bmatrix} x_M \\ x_A \end{bmatrix} = \begin{bmatrix} 1.27 & 0.28 \\ 0.70 & 1.27 \end{bmatrix} \begin{bmatrix} 13.2 \\ 15.3 \end{bmatrix} = \begin{bmatrix} 21.0 \\ 28.7 \end{bmatrix}.$$

Therefore, the total output for next year is forecast to be $21.0 million for the products of M and $28.7 million for the products of A.

If we wish, we may complete the forecasting process by producing a forecasted input-output table for Fredonia. We do this by first filling in the forecasted external demand, total input, internal demand (equal to total input minus external demand), and total output in Table 4–6. Then the direct requirements matrix **T** can be used to fill in the rest of the table. Recall that we obtained **T** from the input-output matrix by dividing each column of the input-output matrix by its total output. Therefore if we multiply each column of **T** by the new total output, we will obtain the new input-output matrix:

$$\begin{bmatrix} 0.1(21.0) & 0.2(28.7) \\ 0.5(21.0) & 0.1(28.7) \end{bmatrix} = \begin{bmatrix} 2.1 & 5.7 \\ 10.5 & 2.9 \end{bmatrix}$$

(rounded to one decimal place). Therefore, the forecasted input-output table for Fredonia is Table 4–7.

TABLE 4–6

				Output		
		M	A	Internal Demand	External Demand	Total Input
Input	M			7.8	13.2	21.0
	A			13.4	15.3	28.7
Value added						
Total output		21.0	28.7			49.7 GNP

TABLE 4–7

				Output		
		M	A	Internal Demand	External Demand	Total Input
Input	M	2.1	5.7	7.8	13.2	21.0
	A	10.5	2.9	13.4	15.3	28.7
Value added		8.4	20.1			
Total output		21.0	28.7			49.7 GNP

■ *Computer Manual: Program INOUTANL*

4.5 EXERCISES

Cryptography

In Exercises 1 through 6, assign numerical values to letters as follows:

$$A = 1, \quad B = 2, \quad C = 3, \quad \ldots, \quad Z = 26, \quad \text{space} = 27.$$

Use the matrix

$$\mathbf{A} = \begin{bmatrix} 2 & 0 & 1 \\ -1 & 1 & 2 \\ 1 & 0 & 1 \end{bmatrix}$$

to encode the messages in Exercises 1 through 3.

1. LOOK OUT BEHIND YOU
2. HELP IS ON THE WAY
3. MATRIX THEORY CAN BE USEFUL

In Exercises 4 through 6, use the inverse of the matrix $\mathbf{A}$ to decode the given message.

4. 47, 5, 43, −11, 27, 66, 14, 12, 45
5. 29, 21, 74, 63, 6, 54, 34, 27, 97, 27, 21, 75, 51, 12, 60, 34, 5, 43
6. 4, 15, 46, −9, 27, 67, 22, 13, 46, 65, 1, 41, 6, 27, 75, 17, 27, 79, 19, 18, 58, 1, 22, 58, 28, 27, 95

Input-Output Analysis

7. The technological matrix $\mathbf{T}$ for a two-sector economy with sectors A and B is

$$\begin{array}{c} \\ A \\ B \end{array} \begin{bmatrix} A & B \\ 0.30 & 0.20 \\ 0.40 & 0.10 \end{bmatrix},$$

and the external demand vector $\mathbf{D}$ is

$$\mathbf{D} = \begin{bmatrix} 100 \\ 200 \end{bmatrix}.$$

Find the total output and GNP.

8. Repeat Exercise 7 for a three-sector economy having

$$\mathbf{T} = \begin{array}{c} A \\ B \\ C \end{array} \begin{bmatrix} A & B & C \\ 0.20 & 0.30 & 0.10 \\ 0.10 & 0.00 & 0.50 \\ 0.10 & 0.20 & 0.20 \end{bmatrix}$$

and

$$\mathbf{D} = \begin{bmatrix} 50 \\ 80 \\ 120 \end{bmatrix}.$$

Hint: The Leontief matrix for this exercise is

$$\frac{1}{509} \begin{bmatrix} 700 & 260 & 250 \\ 130 & 630 & 410 \\ 120 & 190 & 770 \end{bmatrix}.$$

A country has the input-output table shown in Table 4–8 (figures are in billions of dollars). Exercises 9 through 12 refer to this input-output table.

9. Fill in the table.
10. Find and interpret the technological matrix **T**.
11. Find the Leontief matrix.
12. It is forecast that external demand for the products of both agriculture and manufacturing will increase next year by 10%. Prepare a forecasted input-output table for the country's economy next year.

TABLE 4–8

	Agr.	Mfrg.	Internal Demand	External Demand	Total Input
Agriculture	10	5		35	
Manufacturing	2	3		25	
Total output					GNP

Exercises 13 through 16 refer to the input-output table of Table 4–9.

13. Fill in the table.
14. Find the technological matrix **T**.
15. Find the Leontief matrix.
16. Prepare a forecasted input-output table, assuming an increase of 10% in external demand for the products of manufacturing and a decrease of 20% in the external demand for the products of service.

TABLE 4–9

	Mfrg.	Serv.	Internal Demand	External Demand	Total Input
Manufacturing	0	8		42	
Service	10	4		46	
Total output					GNP

Exercises 17 through 19 refer to the input-output table of Table 4–10.

TABLE 4–10

	Mfrg.	Agr.	Serv.	Internal Demand	External Demand	Total Input
Manufacturing	20	5	50		100	
Agriculture	10	10	40		200	
Service	30	15	60		300	
Total output						GNP

17. Fill in the table.
18. Find the technological matrix **T**.
19. Prepare a forecasted input-output table, assuming an increase of 5% in external demand for the products of manufacturing, an increase of 10% in external demand

MATRICES

for the products of agriculture, and a decrease of 10% in external demand for services. The Leontief matrix is

$$\begin{bmatrix} 1.1646 & 0.0335 & 0.1721 \\ 0.0936 & 1.0498 & 0.1339 \\ 0.2398 & 0.0771 & 1.2136 \end{bmatrix}.$$

Parts Matrices

The following directed graph is a **parts diagram**:

Here p_5 is the finished product, and p_1, p_2, p_3, and p_4 are parts used to make p_5, and in some cases each other as well. The number attached to each arrow in the diagram tells us how many units of each part are required to make one unit of the next one; for instance, each unit of p_2 requires 2 units of p_1 and 1 unit of p_3. The information contained in the parts diagram can be embodied in a **parts matrix P**:

$$\mathbf{P} = \begin{bmatrix} 0 & 2 & 3 & 0 & 0 \\ 0 & 0 & 0 & 1 & 0 \\ 0 & 1 & 0 & 0 & 2 \\ 0 & 0 & 0 & 0 & 2 \\ 0 & 0 & 0 & 0 & 0 \end{bmatrix},$$

where the entry in row i, column j of **P** is the number of units of p_i needed to make 1 unit of p_j. The total output of the production system must be equal to the sum of the parts used internally to make other parts and the parts shipped outside the system to satisfy external demand. By the same reasoning as that used in the input-output analysis of this section, if **X** is the total output vector for the system and **D** is its external demand vector, then $\mathbf{X} = [\mathbf{I} - \mathbf{P}]^{-1}\mathbf{D}$. Exercises 20 through 22 refer to the preceding parts diagram.

*20. Find $[\mathbf{I} - \mathbf{P}]^{-1}$.

*21. Find the total output of the production system if there is no external demand for parts p_1, p_2, p_3, and p_4, and the demand for the finished product p_5 is 1000 units.

*22. Find the total output of the production system if external demand is 100 units of p_1, 200 units of p_2, 300 units of p_3, 400 units of p_4, and 1000 units of p_5. Of the total output of each part, how many units are used internally?

Exercises 23 and 24 refer to the following parts diagram:

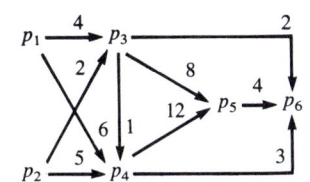

*23. Write the parts matrix **P** for the parts diagram.

*24. Find the total output of the production system if external demand is 450 units of p_1, 0 units of p_2 and p_3, 180 units of p_4, 440 units of p_5, and 2500 units of the finished product p_6. Find the internal demand vector for the system.

SUMMARY

This summary consists of terms and symbols whose meaning you should know, facts you should know, and techniques and methods you should be able to employ.

Section 4.1

Matrix, entry of a matrix. Augmented matrix of a system of linear equations. Row operations: Type I, Type II, Type III. Gauss-Jordan elimination. Solving systems of n linear equations in n variables using Gauss-Jordan elimination. Gauss-Jordan elimination and systems having no solution. Gauss-Jordan elimination and systems having infinitely many solutions. Solving systems of m equations in n variables, $m \neq n$, using Gauss-Jordan elimination. Setting up systems of linear equations for applied problems and solving them using Gauss-Jordan elimination.

Section 4.2

$m \times n$ matrix, order of a matrix. Zero matrix. Column vector, row vector. Square matrix. Inventory matrix. Matrix equality. Matrix addition and subtraction. Rules for matrix addition and subtraction. Scalar multiplication. Combining scalar multiplication, addition, and subtraction. Rules for scalar multiplication.

Section 4.3

Multiplication of a row vector by a column vector. Matrix multiplication: when and when not defined, order of the product matrix, entry in row i, column j of the product matrix. Finding the product of two matrices. Matrix multiplication not commutative (**AB** not necessarily equal to **BA**). Powers of a square matrix, $\mathbf{A}^n$. Identity matrix, $\mathbf{I}_n$. Properties of identity matrices. Rules of matrix multiplication. Using matrix multiplication to value inventories.

Section 4.4

Inverse of a square matrix, $\mathbf{A}^{-1}$. Uniqueness of the inverse. Inverse need not exist. Singular, nonsingular matrix. Finding inverses using Gauss-Jordan elimination. How Gauss-Jordan elimination indicates that the inverse does not exist. How to tell whether the inverse of a 2×2 matrix exists. Formula for finding the inverse of a nonsingular 2×2 matrix. Matrix equation $\mathbf{AX} = \mathbf{B}$. Solving

MATRICES

matrix equations using inverse matrices. Setting up matrix equations for applied problems and solving them using inverse matrices.

Section 4.5

Cryptography. Using a matrix and its inverse to encode and decode messages. Input-output analysis. Input-output table. Input-output matrix. Internal demand, external demand. Technological matrix. Finding the technological matrix. Total output vector. External demand vector. The input-output model. The Leontief matrix. Forecasting total output. Preparing a forecasted input-output table.

REVIEW EXERCISES

In Exercises 1 through 5, solve the system of equations using Gauss-Jordan elimination.

1. $x_1 + 2x_2 - x_3 = 23$
   $2x_1 + 5x_2 + x_3 = 43$
   $-x_1 + 4x_2 + 2x_3 = 10$

2. $2x_1 + x_2 - x_3 = 2$
   $3x_1 + 4x_2 + 6x_3 = 4$
   $-x_1 + 2x_2 + 8x_3 = 1$

3. $x_1 + 2x_2 - x_3 + x_4 = 3$
   $-x_1 + 2x_2 + 3x_3 + 2x_4 = 0$
   $2x_2 - x_4 = 2$
   $x_1 + 2x_2 + x_3 + 6x_4 = 2$

4. $x_1 + x_2 - x_3 = -5$
   $2x_1 - 3x_2 + x_3 = 17$
   $3x_1 + 4x_2 - 6x_3 = -30$
   $x_1 + x_3 = 6$

5. $2x_1 + x_3 + x_4 = 7$
   $x_1 - x_2 + 3x_4 = -5$
   $-x_1 + 2x_2 + x_3 = 0$

In Exercises 6 through 8, use Gauss-Jordan elimination.

6. A firm makes three products. Each unit of the first requires 5 kg of steel and 3 kg of tin, each unit of the second requires 2 kg of steel and 1 kg of tin, and each unit of the third requires 3 kg of steel and 4 kg of tin. If the firm has 240 kg of steel and 205 kg of tin on hand, can it schedule production so as to use up all the steel and tin? If so, how?

7. Suppose that in addition to the information given in Exercise 6, each unit of the first product requires 5 ounces of chrome, each unit of the second requires 1 ounce of chrome, and each unit of the third requires 2 ounces of chrome. If the firm has 185 ounces of chrome on hand, can it schedule production so as to use up all the steel, tin, and chrome? If so, how?

8. Suppose that in addition to the information given in Exercises 6 and 7, each unit of the first product requires 1 gallon of paint, each unit of the second requires 2 gallons of paint, and each unit of the third requires 1 gallon of paint. If the firm has 120 gallons of paint on hand, can it schedule production so as to use up all the steel, tin, chrome, and paint? If so, how?

In Exercises 9 through 13, let

$$A = \begin{bmatrix} 2 & 1 & x \\ y & z & -1 \end{bmatrix}, \quad B = \begin{bmatrix} u & 1 & 0 \\ 2 & 2 & v \end{bmatrix}, \quad C = \begin{bmatrix} 3 & 6 \\ -2 & 1 \end{bmatrix}, \quad D = \begin{bmatrix} 4 & 2 \\ 1 & 0 \end{bmatrix},$$

$$E = \begin{bmatrix} 2 & 0 \\ 2 & 3 \\ 1 & -1 \end{bmatrix}, \quad F = \begin{bmatrix} 1 \\ 0 \\ 0 \\ 1 \end{bmatrix}, \quad G = [2, -3, 5], \quad \text{and} \quad H = \begin{bmatrix} 3 & 1 \\ -1 & 0 \\ 2 & 6 \end{bmatrix}.$$

9. What are the orders of these matrices? Which of them are vectors? Which are square matrices?

10. (a) Can values be assigned to x, y, z, u, and v in such a way as to make **A** = **B**?
    (b) Can values be assigned to x, y, and z in such a way as to make **A** = **C**?

Find, if possible,

11. **C** + **D**   12. **C** − **D**   13. **E** + **D**

In Exercises 14 through 17, let

$$A = \begin{bmatrix} 2 & 6 & -1 \\ 3 & 0 & 4 \end{bmatrix} \quad \text{and} \quad B = \begin{bmatrix} 1 & 5 & 1 \\ 0 & -2 & 2 \end{bmatrix}.$$

Find

14. 2**A**   15. −3**B**   16. 4**A** + 2**B**   17. 5**B** − 3**A**

18. An appliance store sells two brands of television sets, A and B, in both black-and-white (BW) and color (C) models. The store's inventory and purchase matrices for the month are as follows:

$$\text{Beginning inventory} = \begin{array}{c} \\ A \\ B \end{array} \begin{matrix} BW & C \\ \begin{bmatrix} 10 & 8 \\ 2 & 12 \end{bmatrix} \end{matrix},$$

$$\text{Ending inventory} = \begin{array}{c} \\ A \\ B \end{array} \begin{matrix} BW & C \\ \begin{bmatrix} 13 & 6 \\ 4 & 8 \end{bmatrix} \end{matrix},$$

$$\text{Purchases} = \begin{array}{c} \\ A \\ B \end{array} \begin{matrix} BW & C \\ \begin{bmatrix} 5 & 0 \\ 2 & 4 \end{bmatrix} \end{matrix},$$

Find the store's sales for the month by brand and model.

In Exercises 19 through 27, let

$$C = \begin{bmatrix} 2 & 1 \\ -1 & 5 \end{bmatrix}, \quad D = \begin{bmatrix} 4 & 1 & -1 \\ 2 & 0 & 6 \end{bmatrix}, \quad E = [2, 1, 3], \quad \text{and} \quad F = \begin{bmatrix} -2 \\ 1 \\ 7 \end{bmatrix}.$$

Find, if possible,

19. **CD**   20. **DC**   21. **CE**   22. **DF**   23. **EF**
24. **FE**   25. C^2 26. C^3 27. C^4

28. Suppose that the appliance store of Exercise 18 sold each black-and-white set for $200 and each color set for $500. Use matrix methods to find the value of the store's sales for the month.

In Exercises 29 through 34, find A^{-1}, if possible.

29. $A = \begin{bmatrix} 4 & 7 \\ 5 & 9 \end{bmatrix}$ 30. $A = \begin{bmatrix} 8 & 4 \\ 4 & 2 \end{bmatrix}$ 31. $A = \begin{bmatrix} 5 & -2 \\ 3 & 7 \end{bmatrix}$

32. $\mathbf{A} = \begin{bmatrix} 2 & -1 & 5 \\ 3 & 2 & -2 \\ 7 & 0 & 8 \end{bmatrix}$ **33.** $\mathbf{A} = \begin{bmatrix} 1 & 6 & -4 \\ 3 & 2 & 0 \\ 2 & 3 & 1 \end{bmatrix}$ **34.** $\mathbf{A} = \begin{bmatrix} 1 & 0 & 2 & 2 \\ 4 & 3 & -1 & 0 \\ 0 & 1 & 1 & 0 \\ 5 & 4 & 2 & 4 \end{bmatrix}$

In Exercises 35 through 37, solve the given system of equations using the inverse of the coefficient matrix.

35. $4x_1 + 7x_2 = 2$
$5x_1 + 9x_2 = 6$

36. $8x_1 + 4x_2 = 3$
$4x_1 + 8x_2 = -1$

37. $x_1 + 6x_2 - 4x_3 = -3$
$3x_1 + 2x_2 = 4$
$2x_1 + 3x_2 + x_3 = 0$

38. A firm wishes to buy a mix of small, medium, and large lathes. Table 4–11 shows the number of operators and the floor space needed for each size lathe, as well as the cost.

TABLE 4–11

Size	Number of Operators	Floor Space (ft^2)	Cost ($)
Small	1	10	20,000
Medium	2	18	40,000
Large	3	24	72,000

How many of each size lathe can the firm purchase if it has
(a) nine operators and 80 ft^2 of floor space available, and $192,000 budgeted for purchasing the lathes?
(b) twenty-three operators and 202 ft^2 of floor space available, and $496,000 budgeted for purchasing the lathes?
(c) twenty-six operators and 224 ft^2 of floor space available, and $568,000 budgeted for purchasing the lathes?

39. Use the assignment
$$a = 27, \quad b = 26, \ldots, \quad z = 2, \quad \text{blank} = 1$$
and the matrix
$$\mathbf{A} = \begin{bmatrix} 2 & 1 & 0 & -1 \\ 3 & 2 & 0 & 0 \\ -1 & 0 & 5 & 2 \\ 0 & 1 & 2 & 1 \end{bmatrix}$$
to encode the message
THIS IS HOW MATRICES ARE USED IN CRYPTOGRAPHY

40. Use the assignment of Exercise 39 and the inverse of the matrix **A** of Exercise 39 to decode the message

66	55	35	3
33	34	117	34
77	57	39	-7
34	44	125	44

70	54	106	16
80	64	76	6
77	73	145	35
72	90	189	72

The technological matrix for a two-sector economy with sectors A and B is

$$T = \begin{array}{c} \\ A \\ B \end{array} \begin{array}{cc} A & B \\ \begin{bmatrix} 0.20 & 0.50 \\ 0.60 & 0.10 \end{bmatrix} \end{array}.$$

Exercises 41 and 42 refer to this economy.

41. If the external demand vector for the economy in 1988 was

$$D = \begin{bmatrix} 1000 \\ 2000 \end{bmatrix},$$

find total output and GNP for 1988.

42. Suppose that in 1989 external demand for A increased to 1200, and for B it decreased to 1800. Prepare an input-output table for the 1989 economy.

ADDITIONAL TOPICS

Here are some suggestions for topics not covered in the text that you might want to investigate on your own.

1. The **determinant** of a square matrix is a number: for instance, the determinant of the matrix

$$\begin{bmatrix} 1 & 2 \\ 5 & 3 \end{bmatrix}$$

is the number -7. Find out how to calculate determinants of 2×2 and 3×3 matrices.

2. Find out how to calculate determinants using Type I, II, and III row operations.

3. **Cramer's Rule** is a method for solving systems of n linear equations in n variables using determinants. Find out what Cramer's Rule says and how to use it.

4. The Gauss-Jordan method for solving systems of linear equations is named in part for Carl Friedrich Gauss (1777–1855), possibly the greatest mathematician of all time. Gauss, who has been called "the prince of mathematicians," was a child prodigy whose mathematical genius did not fade when he became an adult. Find out about Gauss's life and some of his contributions to mathematics and astronomy.

5. Use of the Gauss-Jordan method to solve large systems of linear equations can often lead to serious computational error. Therefore when large systems must be solved, the Gauss-Jordan method is usually modified in such a way as to minimize the buildup of computational errors. One such modification uses a method known as **partial pivoting** to find the **echelon matrix** of the system and then back-substitutes to determine the solution. Find out about this method and why it is less prone to error buildup than the Gauss-Jordan method.

6. The Leontief matrix is named for the American economist Wassily Leontief, who pioneered input-output analysis in economics (and won a Nobel prize for doing so). Find out about Leontief's work in input-output analysis and why it is important.

SUPPLEMENT: SUSTAINABLE YIELDS[1]

In this supplement we address the problem of harvesting a crop in such a way as to sustain its yield over time. To illustrate, consider a forest: suppose we wish to harvest trees from the forest at regular intervals, each time cutting down the same number of trees and having the forest be completely replenished by the time of the next harvest. If we can do this, we can sustain the yield of the forest indefinitely.

Let us be more specific. Suppose the forest consists of 10,000 trees, partitioned into classes. The classes might be based on the heights of the trees, their ages, their species, or their value as timber. We will assume that the forest is partitioned into four classes by height:

	Class of Trees	Height Range (feet)
	1	$[0, 1)$
	2	$[1, 4)$
	3	$[4, 8)$
	4	$[8, +\infty)$

The trees grow during the spring and summer and some of them are harvested each winter. Each tree that is harvested is immediately replaced by a seedling. (Seedlings are class 1 trees.) Thus the total number of trees in the forest remains constant at 10,000.

Although the number of trees remains constant, their distribution among the height classes will change as they grow, are harvested, and are replaced by seedlings. At any particular time, let

$$p_i = \text{Number of trees in class } i, \quad i = 1, 2, 3, 4.$$

The vector

$$\mathbf{P} = \begin{bmatrix} p_1 \\ p_2 \\ p_3 \\ p_4 \end{bmatrix}$$

is called the **profile** of the forest at this time.

---

[1] Suggested by material in Rorres and Anton, *Applications of Linear Algebra*, Wiley, 1977.

EXAMPLE 1 Suppose that just after a harvest the profile of the forest is given by

$$\mathbf{P} = \begin{bmatrix} 5000 \\ 2000 \\ 1250 \\ 1750 \end{bmatrix}.$$

Then at this time there are 5000 trees in class 1, 2000 in class 2, 1250 in class 3, and 1750 in class 4. ☐

Now suppose that just after a harvest the profile of the forest is

$$\mathbf{P} = \begin{bmatrix} p_1 \\ p_2 \\ p_3 \\ p_4 \end{bmatrix}.$$

Over the next growing season some of the trees in classes 1, 2, and 3 will grow into the next higher class. (We assume that a tree can only grow into a class one higher than the one it started in.) For $i = 1, 2,$ and 3, let

$g_i =$ Proportion of trees in class i that grow into class $i + 1$ during a growing season.

Then

$1 - g_i =$ Proportion of trees in class i that remain in class i during a growing season.

Therefore, at the time of the next harvest, there will be

$(1 - g_1)p_1$	trees in class 1,
$g_1 p_1 + (1 - g_2)p_2$	trees in class 2,
$g_2 p_2 + (1 - g_3)p_3$	trees in class 3,
$g_3 p_3 + p_4$	trees in class 4.

Hence, the profile of the forest at the time of the next harvest will be

$$\begin{bmatrix} (1 - g_1)p_1 \\ g_1 p_1 + (1 - g_2)p_2 \\ g_2 p_2 + (1 - g_3)p_3 \\ g_3 p_3 + p_4 \end{bmatrix} = \begin{bmatrix} 1 - g_1 & 0 & 0 & 0 \\ g_1 & 1 - g_2 & 0 & 0 \\ 0 & g_2 & 1 - g_3 & 0 \\ 0 & 0 & g_3 & 1 \end{bmatrix} \begin{bmatrix} p_1 \\ p_2 \\ p_3 \\ p_4 \end{bmatrix}.$$

The matrix

$$\mathbf{G} = \begin{bmatrix} 1 - g_1 & 0 & 0 & 0 \\ g_1 & 1 - g_2 & 0 & 0 \\ 0 & g_2 & 1 - g_3 & 0 \\ 0 & 0 & g_3 & 1 \end{bmatrix}$$

is called the **growth matrix.** We have shown that

MATRICES

> If **P** is the profile of the forest just after one harvest, then **GP** is the profile of the forest at the time of the next harvest.

EXAMPLE 2 Suppose the profile of the forest just after a harvest is as in Example 1, and that during the growing season

- 10% of the class 1 trees grow into class 2;
- 20% of the class 2 trees grow into class 3;
- 30% of the class 3 trees grow into class 4.

Then

$$\mathbf{G} = \begin{bmatrix} 0.9 & 0 & 0 & 0 \\ 0.1 & 0.8 & 0 & 0 \\ 0 & 0.2 & 0.7 & 0 \\ 0 & 0 & 0.3 & 1 \end{bmatrix},$$

and the profile of the forest at the time of the next harvest will be

$$\mathbf{GP} = \begin{bmatrix} 0.9 & 0 & 0 & 0 \\ 0.1 & 0.8 & 0 & 0 \\ 0 & 0.2 & 0.7 & 0 \\ 0 & 0 & 0.3 & 1 \end{bmatrix} \begin{bmatrix} 5000 \\ 2000 \\ 1250 \\ 1750 \end{bmatrix} = \begin{bmatrix} 4500 \\ 2100 \\ 1275 \\ 2125 \end{bmatrix}. \quad \square$$

For $i = 1, 2, 3, 4$, let

$$h_i = \text{Number of trees harvested from class } i.$$

The vector

$$\mathbf{H} = \begin{bmatrix} h_1 \\ h_2 \\ h_3 \\ h_4 \end{bmatrix}$$

is called the **harvest vector**. Thus, the total number of trees harvested is $h_1 + h_2 + h_3 + h_4$, and since every tree cut down is immediately replaced by a seedling, this is also the number of seedlings planted at harvest time. Since seedlings are class 1 trees, the **replacement vector R** is given by

$$\mathbf{R} = \begin{bmatrix} h_1 + h_2 + h_3 + h_4 \\ 0 \\ 0 \\ 0 \end{bmatrix}.$$

Now, if the yield of the forest is to be sustainable, then **H** (and hence **R**) will be the same every year; also, since the forest is to be completely replenished between harvests, the forest profile at the time of one harvest must be the same as the forest profile at the time of the next harvest. These conditions imply that

> If the yield is to be sustainable, then the forest profile just after one harvest must equal the forest profile just after the next harvest.

Now suppose that **P** is the profile just after one harvest; then **GP** is the profile at the time of the next harvest, and, hence, **GP** − **H** + **R** is the profile just after the next harvest. Thus, if the yield is to be sustainable, then

$$\mathbf{P} = \mathbf{GP} - \mathbf{H} + \mathbf{R},$$

which we may rewrite as

$$\mathbf{H} - \mathbf{R} = (\mathbf{G} - \mathbf{I})\mathbf{P},$$

where **I** is the 4 × 4 identity matrix. But

$$\mathbf{H} - \mathbf{R} = \begin{bmatrix} h_1 \\ h_2 \\ h_3 \\ h_4 \end{bmatrix} - \begin{bmatrix} h_1 + h_2 + h_3 + h_4 \\ 0 \\ 0 \\ 0 \end{bmatrix} = \begin{bmatrix} 0 & -1 & -1 & -1 \\ 0 & 1 & 0 & 0 \\ 0 & 0 & 1 & 0 \\ 0 & 0 & 0 & 1 \end{bmatrix} \begin{bmatrix} h_1 \\ h_2 \\ h_3 \\ h_4 \end{bmatrix},$$

so if we set

$$\mathbf{S} = \begin{bmatrix} 0 & -1 & -1 & -1 \\ 0 & 1 & 0 & 0 \\ 0 & 0 & 1 & 0 \\ 0 & 0 & 0 & 1 \end{bmatrix},$$

we can write **H** − **R** = **SH**. Therefore, if the yield is to be sustainable, we must have

$$\mathbf{SH} = (\mathbf{G} - \mathbf{I})\mathbf{P}.$$

In other words,

> Given the profile **P** just after a harvest and the growth matrix **G,** if we can find a harvest vector **H** that satisfies
>
> $$\mathbf{SH} = (\mathbf{G} - \mathbf{I})\mathbf{P},$$
>
> then **H** will define a sustainable harvest.

EXAMPLE 3   Let **P** and **G** be as in Examples 1 and 2. Then

$$\mathbf{SH} = (\mathbf{G} - \mathbf{I})\mathbf{P}$$

becomes

$$\begin{bmatrix} 0 & -1 & -1 & -1 \\ 0 & 1 & 0 & 0 \\ 0 & 0 & 1 & 0 \\ 0 & 0 & 0 & 1 \end{bmatrix} \begin{bmatrix} h_1 \\ h_2 \\ h_3 \\ h_4 \end{bmatrix}$$

$$= \left(\begin{bmatrix} 0.9 & 0 & 0 & 0 \\ 0.1 & 0.8 & 0 & 0 \\ 0 & 0.2 & 0.7 & 0 \\ 0 & 0 & 0.3 & 1 \end{bmatrix} - \begin{bmatrix} 1 & 0 & 0 & 0 \\ 0 & 1 & 0 & 0 \\ 0 & 0 & 1 & 0 \\ 0 & 0 & 0 & 1 \end{bmatrix} \right) \begin{bmatrix} 5000 \\ 2000 \\ 1250 \\ 1750 \end{bmatrix},$$

which reduces to

$$\begin{bmatrix} -h_2 - h_3 - h_4 \\ h_2 \\ h_3 \\ h_4 \end{bmatrix} = \begin{bmatrix} -500 \\ 100 \\ 25 \\ 375 \end{bmatrix},$$

(Check this.) Therefore $h_2 = 100$, $h_3 = 25$, $h_4 = 375$, and there are no restrictions on h_1 other than $0 \le h_1 \le 4500$. (The number of trees in class 1 at harvest time is 4500; see Example 1.) Hence, each year we may harvest 100 trees from class 2, 25 from class 3, 375 from class 4, and h_1 from class 1, $0 \le h_1 \le 4500$, and this will be a sustainable harvest. (Note that if in fact we take $h_1 > 0$, then we will be harvesting seedlings and replacing them with seedlings; since this seems to make little sense economically, in practice we would probably take $h_1 = 0$.) ☐

To conclude, we remark that sustainable harvests are not always possible: a given profile vector **P** and growth matrix **G** might not allow a sustainable harvest. (See Exercise 2.) However, for a given growth matrix **G**, of all profiles that *do* allow sustainable harvests, there will be one that yields the maximum possible profit. (See Exercise 5.)

SUPPLEMENTARY EXERCISES

1. Suppose a forest contains 10,000 trees partitioned into four height classes. Let the forest profile just after a harvest be

$$\mathbf{P} = \begin{bmatrix} 4000 \\ 3000 \\ 2000 \\ 1000 \end{bmatrix}$$

and assume that during each growing season 25% of the trees in class i grow into class $i + 1$, $i = 1, 2, 3$. Find the sustainable harvest for the forest.

2. Repeat Exercise 1 if

$$P = \begin{bmatrix} 2000 \\ 3000 \\ 4000 \\ 1000 \end{bmatrix}.$$

3. Suppose the trees in a forest are partitioned by age as follows:

		Age (years)
	1	[0, 1)
	2	[1, 3)
Class	3	[3, 6)
	4	[6, 10)
	5	[10, +∞)

If the forest consists of 10,000 trees and its profile after a harvest is

$$P = \begin{bmatrix} 1000 \\ 2000 \\ 2700 \\ 2000 \\ 2300 \end{bmatrix},$$

find its sustainable harvest. (*Hint:* The proportion of trees in a class that grows into the next class depends only on the age range of the class. Thus, 100% of the class 1 trees grow into class 2, 50% of the class 2 trees grow into class 3, and so on.)

4. A farmer raises pigs, classifying them into four weight classes, and selling some of them once each year. Those sold are replaced by piglets, so the farmer always has a total of 1100 pigs. Between the annual sales, 90% of the pigs in each of the first three weight classes grow into the next class. If the profile vector just after a sale is

$$P = \begin{bmatrix} 500 \\ 300 \\ 200 \\ 100 \end{bmatrix},$$

find the sustainable yield of pigs.

*5. Assuming that no crops are ever harvested from the first class (i.e., that $h_1 = 0$), we can find for a given growth matrix G the profile P that will yield the sustainable harvest of maximum profit. Thus, suppose the crop is partitioned into n classes. Let

$$g_i = \text{Growth constant for class } i, \qquad i = 1, 2, \ldots, n - 1.$$

(Thus g_i is the proportion of class i that grows into class $i + 1$.) Let

$$y_i = \text{Profit per unit of class } i, \qquad i = 2, 3, \ldots, n.$$

Let

$$T = \text{Total number of crop members.}$$

For $k = 2, 3, \ldots, n$, find

$$Y_k = \frac{y_k T}{1/g_1 + \cdots + 1/g_{k-1}}.$$

Choose the value of k that gives the largest Y_k, and let

$$\mathbf{P} = \frac{T}{1/g_1 + \cdots + 1/g_{k-1}} \begin{bmatrix} 1/g_1 \\ \vdots \\ 1/g_{k-1} \\ 0 \\ \vdots \\ 0 \end{bmatrix}.$$

Then $\mathbf{P}$ is the profile that yields the sustainable harvest of maximum profit. This harvest will always involve harvesting the entire kth class and doing no harvesting from any of the other classes, and the maximum profit will be Y_k.

Suppose that for the forest of Examples 1, 2, and 3 of this supplement, the profits on the trees in classes 2, 3, and 4 are as follows:

Class	2	3	4
Profit per tree	$20	$50	$60

Assuming that no class 1 trees are ever harvested, find the profile that yields the sustainable harvest of maximum profit, the sustainable harvest of maximum profit, and the maximum profit.

*6. Suppose the profits on the pigs of Exercise 4 are as follows:

Class	2	3	4
Profit per pig	$50	$120	$200

Assuming that no class 1 pigs are ever sold, find the profile that gives the sustainable harvest of maximum profit, the sustainable harvest of maximum profit, and the maximum profit.

5

Linear Programming

In this chapter we discuss an important mathematical technique known as **linear programming.** Linear programming is one of the most useful techniques of applied mathematics; it is particularly helpful in solving problems concerning resource allocation and scheduling that are very common in the business world. For instance, consider the following problem:

> A company has budgeted $55,000 to purchase a combination of television and newspaper advertisements. Each television ad will cost $5000 and reach an audience of 80,000 people, and each newspaper ad will cost $2500 and reach an audience of 50,000 people. The advertising director has recommended purchasing a minimum of four television ads and eight newspaper ads. Given that the company wants the ads to reach the maximum possible audience, how should it allocate its advertising budget?

In an attempt to solve this problem, suppose we let

x = Number of television ads the company should purchase

and y = Number of newspaper ads the company should purchase.

Then the advertising director has recommended

$$x \geq 4 \quad \text{and} \quad y \geq 8.$$

The total cost of running x television ads and y newspaper ads will be $\$5000x + \$2500y$, and this must remain within budget, so we must have

$$5000x + 2500y \leq 55{,}000.$$

Finally, the company seeks to maximize the size of the audience for the ads, which will be $A = 80{,}000x + 50{,}000y$. Thus we must

Maximize $\quad A = 80{,}000x + 50{,}000y$

subject to the conditions

$$5000x + 2500y \leq 55{,}000,$$
$$x \geq 4,$$
$$y \geq 8.$$

This is a typical linear programming problem in two variables. Notice that it requires the optimization of some quantity (here, the audience size A) subject to a system of linear inequalities in the variables.

Since linear programming problems frequently involve systems of linear inequalities, we begin this chapter with a consideration of linear inequalities in two variables. We then start our discussion of linear programming proper with a section devoted to modeling linear programming problems. This is followed by a discussion of linear programming problems in two variables; such problems can be solved by graphical methods. The next two sections consider linear programming problems in more than two variables, and present a general method for solving them known as the **simplex method.** The chapter concludes with a brief discussion of duality.

5.1 LINEAR INEQUALITIES IN TWO VARIABLES

In this section we introduce the notion of a linear inequality in two variables, show how such an inequality can be solved, and discuss the solution of systems of such inequalities.

Solving Linear Inequalities in Two Variables

Linear inequalities in two variables are inequalities of the forms

$$Ax + By < C,$$
$$Ax + By \leq C,$$
$$Ax + By > C,$$

and
$$Ax + By \geq C,$$

LINEAR PROGRAMMING

where A, B, and C are real numbers. The solution set for a linear inequality in two variables is that subset of the plane consisting of all points (x, y) whose coordinates satisfy the inequality. For example, the point $(0, 0)$ is an element of the solution set of the inequality

$$2x + 3y \leq 12$$

because

$$2(0) + 3(0) = 0 \leq 12.$$

Similarly,

$$2(1) + 3(1) = 5 \leq 12$$

and

$$2(-2) + 3(-1) = -7 \leq 12,$$

so the points $(1, 1)$ and $(-2, -1)$ are also elements of the solution set. However,

$$2(5) + 3(1) = 13 > 12$$

and

$$2(3) + 3(4) = 18 > 12,$$

so the points $(5, 1)$ and $(3, 4)$ are not elements of the solution set for this inequality.

Since the line having equation

$$Ax + By = C$$

consists of those points whose coordinates satisfy the equation, all points lying on this line are elements of both the solution set of the linear inequality

$$Ax + By \leq C$$

and the solution set of the linear inequality

$$Ax + By \geq C.$$

However, no point lying on this line is an element of the solution set of the linear inequality

$$Ax + By < C$$

or the solution set of the linear inequality

$$Ax + By > C.$$

Here is a method for finding the solution set of a linear inequality in two variables:

The Solution Set of a Linear Inequality in Two Variables

To find the solution set of a linear inequality in two variables:

1. Replace the inequality symbol in the given linear inequality by an equal sign, thus obtaining the equation of a line.

2. Graph the line. If the given linear inequality is of the form

$$Ax + By < C \quad \text{or} \quad Ax + By > C,$$

   use a dashed line to indicate that the points on the line are not in the solution set.

3. Choose any point that does not lie on the line and substitute its coordinates in the given inequality.
   a. If the coordinates of the chosen point satisy the inequality, then that point and all points on the same side of the line as that point are elements of the solution set. All points on the other side of the line are not elements of the solution set.
   b. If the coordinates of the chosen point do not satisfy the inequality, then that point and all points on the same side of the line as that point are not elements of the solution set. All points on the other side of the line are elements of the solution set.

4. If the linear inequality is of the form

$$Ax + By < C \quad \text{or} \quad Ax + By > C,$$

the solution set consists only of those points described in step 3.
   If the linear inequality is of the form

$$Ax + By \leq C \quad \text{or} \quad Ax + By \geq C,$$

the solution set consists of those points described in step 3 and all points on the line.

EXAMPLE 1

Find the solution set of the linear inequality $2x + 3y \leq 12$.

1. Replace $\leq$ with $=$ to obtain

$$2x + 3y = 12.$$

2. Graph the line having this equation. See Figure 5–1.

3. Since the line does not pass through the origin, we choose the point $(0, 0)$. The coordinates of the point $(0, 0)$ satisfy the inequality, since

$$2(0) + 3(0) = 0 \leq 12.$$

Hence, all points on the same side of the line as the point $(0, 0)$ also satisfy the inequality.

4. Since $2x + 3y \leq 12$ is of the form $Ax + By \leq C$, the solution set consists of all points that lie on the same side of the line as the point $(0, 0)$ and all points on the line. The solution set is the shaded region and the line of Figure 5–1. ◻

EXAMPLE 2 Find the solution set of the linear inequality $4x - 3y > 24$.

1. We replace $>$ with $=$ to obtain the line
$$4x - 3y = 24.$$

2. Since the points on this line are not in the solution set of
$$4x - 3y > 24,$$
we graph the line as a dashed line. See Figure 5–2.

3. Since the line does not pass through the origin, and since the origin does not satisfy the inequality, all points lying on the other side of the line from $(0, 0)$ are in the solution set.

4. Therefore, the solution set of $4x - 3y > 24$ is the shaded region in Figure 5–2. ◻

EXAMPLE 3 Find the solution set of $2x + y \geq 0$.

Since the line $2x + y = 0$ goes through $(0, 0)$, we cannot use $(0, 0)$ as our test point. However, the point $(1, 0)$ is not on the line, and since
$$2(1) + 0 \geq 0,$$
the point $(1, 0)$ is in the solution set of the inequality. Therefore, the solution set of this inequality consists of the line
$$2x + y = 0$$
and all points on the same side of the line as $(1, 0)$. See Figure 5–3. ◻

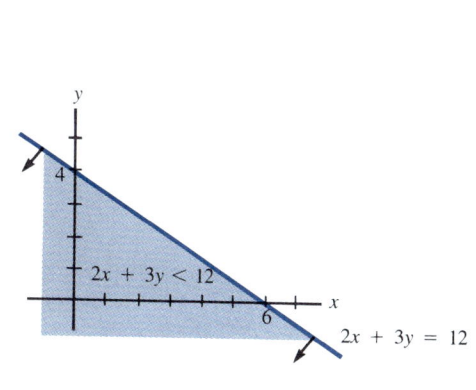

FIGURE 5–1

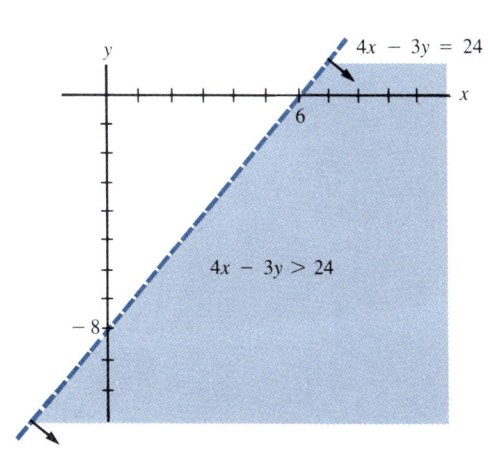

FIGURE 5–2

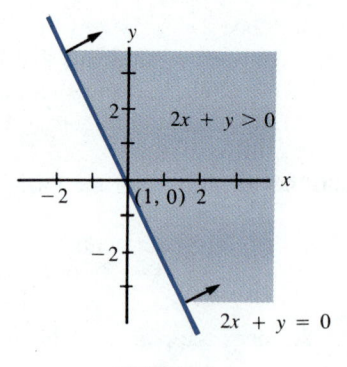

FIGURE 5–3

EXAMPLE 4 Find the solution sets of $x \geq 3$ and $y < 4$.
The solution sets for these inequalities are shown in Figure 5–4. ☐

Systems of Linear Inequalities in Two Variables

A **system** of linear inequalities in two variables is a collection of inequalities such as the following:

$$2x + 3y \leq 12$$
$$x + y \geq 4$$
$$x \geq 0$$
$$y \geq 0.$$

The **solution set** for such a system of inequalities is the set of all points in the plane that satisfy all the inequalities simultaneously. Hence, the solution set for the system is the intersection of the solution sets of the individual inequalities that make up the system. Thus, for example, the solution set for the preceding system is the intersection of the solution sets of the four inequalities

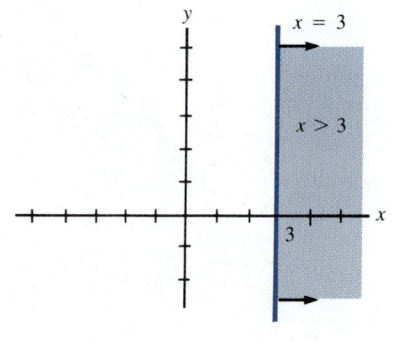

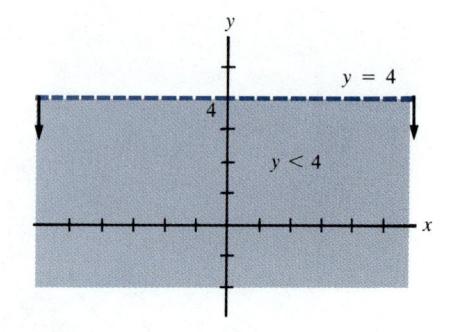

$$2x + 3y \leq 12, \qquad x + y \geq 4, \qquad x \geq 0, \qquad \text{and} \qquad y \geq 0.$$

FIGURE 5–4

LINEAR PROGRAMMING   273

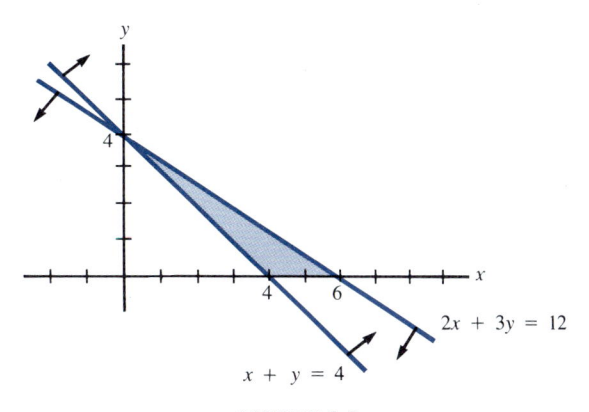

FIGURE 5–5

Figure 5–5 depicts the solution set of the system.

The solution set for a system of linear inequalities is also referred to as the **feasibility region** for the system, since any point that lies within the feasibility region is a **feasible combination** for the system in the sense that the coordinates of such a point satisfy all the inequalities of the system simultaneously. The corners formed by the intersections of boundary lines of a feasibility region are called the **vertices** (singular: **vertex**) of the region. In Figure 5–5, the vertices of the feasibility region are the points (0, 4), (6, 0), and (4, 0).

EXAMPLE 5 Let us find the feasibility region and its vertices for the following system:

$$3x + 4y \leq 24$$
$$3x + 2y \leq 18$$
$$x \geq 0$$
$$y \geq 0.$$

The feasibility region is the shaded area, with its boundaries, shown in Figure 5–6. Its vertices are the points (0, 0), (0, 6), (4, 3), and (6, 0). The vertex (4, 3) is the point of intersection of the lines

$$3x + 4y = 24 \quad \text{and} \quad 3x + 2y = 18,$$

and hence was found by solving the equations of these lines simultaneously for x and y, thus:

$$\begin{aligned} 3x + 4y &= 24 \\ -3x - 2y &= -18 \\ \hline 2y &= 6 \\ y &= 3 \\ 3x + 4(3) &= 24 \\ x &= 4. \quad \square \end{aligned}$$

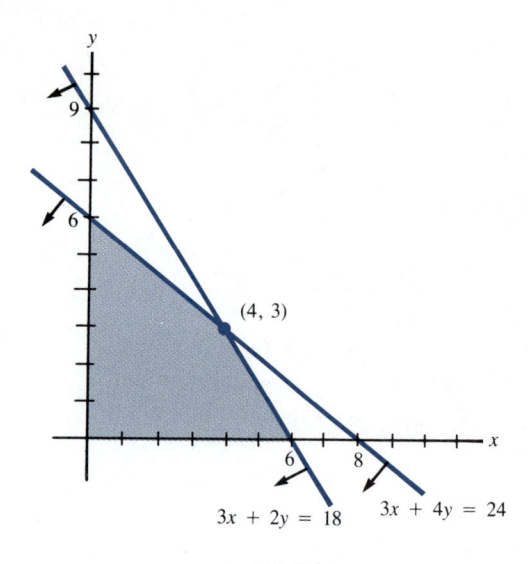

FIGURE 5–6

FIGURE 5–7

EXAMPLE 6 Find the feasibility region and its vertices for the following system:

$$x + y \geq 12$$
$$2x + 5y \geq 30$$
$$x \geq 0$$
$$y \geq 0.$$

The feasibility region is shown in Figure 5–7. Its vertices are (0, 12), (10, 2), and (15, 0). Note that in this case the feasibility region is unbounded. ☐

Finding the feasibility region for a system of linear inequalities can be of practical importance.

EXAMPLE 7 Suppose Alpha Company makes small and large watercoolers, and that each small watercooler requires 2 kg of plastic and 2 kg of steel and each large one requires 9 kg of plastic and 3 kg of steel. Suppose also that Alpha has 2700 kg of plastic and 1200 kg of steel available.

Alpha's production is constrained by its limited resources. For instance, Alpha can make 100 small and 200 large watercoolers, because this will use up only

$$2(100) + 9(200) = 2000 \leq 2700$$

kg of plastic and

$$2(100) + 3(200) = 800 \leq 1200$$

kg of steel. However, the company cannot make 100 small and 300 large watercoolers, because it would then use up

$$2(100) + 9(300) = 2900 \not\leq 2700$$

kg of plastic. Therefore the combination of 100 small and 200 large watercoolers is feasible for Alpha, but the combination of 100 small ones and 300 large ones is not feasible.

We can find all feasible combinations for Alpha by expressing the constraints as a system of linear inequalities and then finding the feasibility region for the system. We let

$x =$ Number of small watercoolers to be produced,

$y =$ Number of large watercoolers to be produced.

Then surely $x \geq 0$ and $y \geq 0$. We find the other constraints with the aid of a table. We write down the resource requirements and the amounts of the resources available:

	Watercoolers		
	Small x	Large y	
Requirements			Amount
Plastic	2	9	2700
Steel	2	3	1200

Thus Alpha will use $2x + 9y$ kg of plastic, and since this cannot total more than the amount available, we must have

$$2x + 9y \leq 2700.$$

Similarly,

$$2x + 3y \leq 1200.$$

Thus Alpha's production of watercoolers is constrained by the system of linear inequalities

$$2x + 9y \leq 2700$$
$$2x + 3y \leq 1200$$
$$x \geq 0$$
$$y \geq 0.$$

The feasibility region for this system is shown in Figure 5–8. Any point (x, y) that lies within the feasibility region (the shaded region and its boundaries in Figure 5–8) represents a combination of small and large watercoolers that Alpha can produce; any point outside the feasibility region represents a combination that Alpha cannot produce. ☐

Notice how we set up the constraining inequalities in Example 7. The method used there can be followed in general:

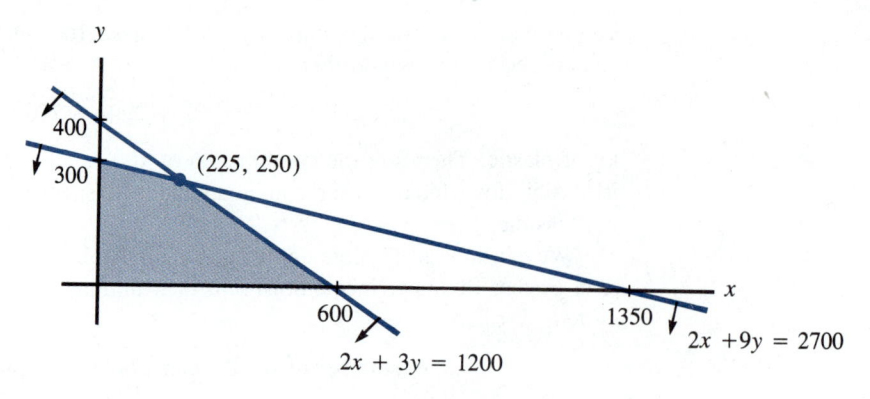

[Figure 5-8: Graph showing region bounded by $2x + 3y = 1200$ and $2x + 9y = 2700$, intersecting at $(225, 250)$, with intercepts at $y=400$, $y=300$, $x=600$, $x=1350$.]

FIGURE 5–8

Setting Up Constraints

To set up a system of constraints:

1. Assign a variable to each of the quantities that are constrained by the available resources.

2. Make a table, with one line for each constraining resource. On each line of the table write
   a. The amount of the resource used by each unit of each variable.
   b. The total amount of the resource available.

3. Each line of the table then translates into a constraint on the variables.

5.1 EXERCISES

Solving Linear Inequalities in Two Variables

In Exercises 1 through 21, find the solution set for the given inequality.

1. $x + y < 1$
2. $2x + 3y \leq 6$
3. $3x - 2y \geq 6$
4. $-x + 2y > 4$
5. $-5x + 6y \leq 9$
6. $-3x - 4y > 12$
7. $3x + 8y \leq 12$
8. $3x + 8y < 0$
9. $x - 4y \geq 0$
10. $3x - 6y \leq -12$
11. $-4x - 2y > 10$
12. $5y < 10$
13. $2y \geq 3$
14. $3x > 15$
15. $-x \leq 3$
16. $y \leq x$
17. $y \geq 3x$
18. $x < 2y$
19. $x \geq 0$
20. $y \geq 0$
21. $x + y \geq 0$

Systems of Linear Inequalities in Two Variables

In Exercises 22 through 36, find the feasibility region and its vertices for the given system of inequalities.

22. $x + y \leq 6$
    $x + y \geq 2$
    $x \geq 0$
    $y \geq 0$

23. $2x + 5y \leq 20$
    $y \geq 2$
    $x \geq 0$

24. $x \geq 4$
    $y \geq 2$
    $x \leq 8$
    $y \leq 10$

25. $2x + 4y \leq 40$
    $3x + y \leq 15$
    $x \geq 0$
    $y \geq 0$

26. $3x + 8y \leq 48$
    $3x + 2y \leq 24$
    $x \geq 0$
    $y \geq 0$

27. $3x + 8y \leq 48$
    $3x + 2y \leq 24$
    $x - y \geq 0$
    $x \geq 0$
    $y \geq 0$

28. $x + y \geq 8$
    $x \geq 4$
    $y \geq 3$

29. $2x + 6y \geq 36$
    $3x + 3y \geq 36$
    $x + 6y \geq 24$
    $x \geq 0$
    $y \geq 0$

30. $3x + y \geq 15$
    $x + 2y \leq 30$
    $2x + y \geq 12$
    $y \geq 1$
    $x \geq 0$

31. $4x + 5y \leq 80$
    $2x + 3y \leq 36$
    $x + y \geq 8$
    $x \geq 0$
    $y \geq 0$

32. $5x + 6y \geq 30$
    $2x + y \geq 8$
    $x \geq 6$
    $y \geq 5$

33. $x + y \leq 20$
    $x + 3y \geq 36$
    $x \geq 14$
    $y \geq 0$

34. $-x + y \leq 4$
    $x - y \leq 4$
    $x + y \geq 4$
    $x + y \leq 12$
    $x \geq 0$
    $y \geq 0$

35. $x + y \geq 4$
    $x + y \leq 12$
    $x - y \geq 4$
    $x \geq 0$
    $y \geq 0$

36. $-2x + y \leq 5$
    $x + 2y \leq 20$
    $3x - y \leq 27$
    $x + y \geq 4$
    $x \leq 12$
    $y \leq 12$
    $x \geq 0$
    $y \geq 0$

Management

37. A company has budgeted $55,000 to purchase a combination of television and newspaper advertisements. Each television ad will cost $5000, and each newspaper ad will cost $2500. The advertising director has recommended purchasing a minimum of four television ads and eight newspaper ads. Find the feasibility region for the number of each type of ad the company should run.

38. A company makes two products: sockets and wickets. Each socket takes 6 hours for fabrication and 6 hours for testing. Each wicket takes 12 hours for fabrication and 2 hours for testing. The company has 120 hours available for fabrication and 30 hours available for testing. Find the feasibility region for the number of sockets and wickets the company can produce.

39. Repeat Exercise 38 if the firm has only 20 hours available for testing.

40. A firm ships parts from a warehouse to two plants. Plant A requires at least 1000 parts each week, and plant B needs at least 1200 per week. It costs $5 to ship a part from the warehouse to plant A, and $7.50 to ship one from the warehouse to plant B. The firm has budgeted $20,000 per week for shipping costs. Find the feasibility region for the number of parts to be shipped to each plant.

41. Repeat Exercise 40 if the weekly demand for parts is at least 1500 for plant A and at least 1700 for plant B.

42. Repeat Exercise 40 if the weekly demand for parts is at least 1300 for plant A and at least 1800 for plant B.

43. An oil company mixes high octane and low octane gasoline to create two intermediate grades of gas, grade A and grade B. Grade A consists of 60% high octane and 40% low octane gas; grade B consists of 30% high octane and 70% low octane. The company has 120,000 gallons of high octane and 112,000 gallons of low octane gas available and must produce at least 50,000 gallons of grade A and 80,000 gallons of grade B. Find the feasibility region for the amounts of grade A and grade B gas the company can produce.

44. A firm makes both a regular and a deluxe model battery. Each battery has one plastic casing. The regular model can be assembled in 10 minutes, but the deluxe model takes 15 minutes. The firm must produce at least 120 regular and 150 deluxe batteries. It has 300 plastic cases on hand and eight employees in the assembly department, each of whom works an 8-hour day. Find the feasibility region for the number of regular and deluxe batteries the firm can produce.

45. A job shop mills, drills, and grinds two types of parts. Each unit of the first type requires 2 machine-hours of milling, 1 machine-hour of drilling, and 2 machine-hours of grinding. Each unit of the second requires 2.5 machine-hours of milling, 0.75 machine-hours of drilling, and 1.25 machine-hours of grinding. Each day the shop has available 20 machine-hours of milling, 15 of drilling, and 12 of grinding. Find the feasibility region for the number of parts of each type that can be worked per day.

Life Science

46. A patient needs at least 60 units of diet supplement A and at least 40 units of diet supplement B. Brand X capsules contain 8 units of A and 4 units of B, and Brand Y capsules contain 5 units of A and 5 units of B. Find the feasibility region for the number of capsules of each brand the patient should take.

47. Repeat Exercise 46 if each Brand X capsule contains 6 units of A and none of B.

48. In order to feed his cattle, a farmer must harvest at least 100 bales of hay and 300 bushels of corn. He must weed the hay at least 1/30 of an hour per bale per week and the corn at least 1/20 hour per bushel per week. Each bale of hay requires the application of 0.5 lb of fertilizer and each bushel of corn needs 0.2 lb of fertilizer. Each bale of hay requires 0.125 acres of ground and each bushel of corn requires 0.025 acres. The farmer can devote 40 hours per week to weeding his crops, can afford 200 lb of fertilizer, and has 40 acres of land. Find the feasibility region for the amount of hay and corn the farmer can grow.

49. Repeat Exercise 48 if the farmer can devote only 28 hours per week to weeding.

50. A nutritionist is planning a diet for a patient who can eat only grain and milk. Suppose that each ounce of grain contains 0.04 oz of protein, 0.001 oz of vitamin A, 0.04 oz of fat, and no cholesterol, and each ounce of milk contains 0.16 oz of protein, 0.02 oz of vitamin A, 0.1 oz of fat, and 0.02 oz of cholesterol. The patient must receive at least 1 oz of protein and 0.1 oz of vitamin A per meal, and not more than 0.8 oz of fat and 0.15 oz of cholesterol per meal. Find the feasibility region for the amount of grain and milk the patient can eat at each meal.

Social Science

51. A political scientist is designing an opinion poll. She does not wish to poll more than 1200 voters. Of those she does poll, at least 300 must be men and at least 300 must be women. Also, she wishes to poll at least twice as many women as men. Find the feasibility region for her poll design.

5.2 THE LINEAR PROGRAMMING MODEL

Many problems in business, economics, and the sciences require that a function be maximized or minimized subject to a set of constraints. If the function to be maximized or minimized is linear and the constraints are linear equations or inequalities, the problem is called a **linear programming problem.** In this section we discuss this linear programming model and demonstrate how to set up linear programming problems. We postpone discussion of the solution of such problems to the following sections.

Let us begin by considering again the Alpha Company problem of Example 7, Section 5.1. Recall that Alpha's production of x small and y large watercoolers was constrained by the amounts of plastic and steel available, and that we embodied the constraining information in the following table:

	Watercoolers		
	Small x	Large y	
Requirements			Amount
Plastic	2	9	2700
Steel	2	3	1200

Now suppose that Alpha makes a profit of $60 on each small watercooler and $100 on each large one. We add this information to the table:

	Watercoolers		
	Small x	Large y	
Profit	60	100	
Requirements			Amount
Plastic	2	9	2700
Steel	2	3	1200

Obviously, Alpha would like to maximize its profit subject to its constraints, so we have the following problem:

$$\text{Maximize } P = 60x + 100y$$
$$\text{subject to } 2x + 9y \le 2700$$
$$2x + 3y \le 1200.$$

Of course, $x \geq 0$ and $y \geq 0$, so the complete problem is

$$\text{Maximize} \quad P = 60x + 100y$$
$$\text{subject to} \quad 2x + 9y \leq 2700$$
$$2x + 3y \leq 1200$$
$$x \geq 0$$
$$y \geq 0.$$

This is an example of a typical linear programming problem.

The Linear Programming Model

A **linear programming problem** consists of

1. A linear **objective function** that is to be optimized (maximized or minimized).
2. A set of **constraints,** each of which is either a linear equality or a linear inequality.

(Note that in the Alpha Company problem, the objective function is the profit function $P = 60x + 100y$.)

The Alpha Company problem is an example of a **product mix** problem. Here is an example of an **investment problem.**

EXAMPLE 1   You intend to invest a total of $50,000 in either a money market fund, a mutual fund, or both. The money market fund pays 8% per year and the mutual fund pays 9%. Because the mutual fund is riskier than the money market fund, you do not wish to place more than $30,000 in it. Also, you wish to have some of the money readily available if needed, so you intend to put at least $10,000 into the money market fund. How much should you allot to each investment in order to maximize your return?

We must determine the amount that should be allocated to each fund in order to maximize the return on investment. Let

$$x = \text{Amount invested in the money market fund,}$$
$$y = \text{Amount invested in the mutual fund.}$$

Clearly $x \geq 0$ and $y \geq 0$, and we have

LINEAR PROGRAMMING

	Fund		
	Money Market x	Mutual y	
Return	0.08	0.09	
Requirements			Amount
Total invested	1	1	$50,000
Money market total	1		$10,000
Mutual fund total		1	$30,000

Since the total invested must be exactly $50,000, we have

$$x + y = 50{,}000;$$

since the money market total must be at least $10,000 and the mutual fund total no more than $30,000, we have

$$x \geq 10{,}000 \quad \text{and} \quad y \leq 30{,}000.$$

Therefore the linear programming model for this problem is

$$\text{Maximize} \quad R = 0.08x + 0.09y$$

$$\begin{aligned}
\text{subject to} \quad & x + y = 50{,}000 \\
& x \geq 10{,}000 \\
& y \leq 30{,}000 \\
& x \geq 0 \\
& y \geq 0. \quad \square
\end{aligned}$$

Our next example is a **blending problem;** notice that it requires the minimization of the objective function.

EXAMPLE 2 Two types of oil, low-sulfur and high-sulfur, are to be mixed. The low-sulfur oil has a sulfur content of 2%, while the high-sulfur oil has a sulfur content of 6%; the mixture is to have a sulfur content of 4%, and at least 1000 barrels of it are needed. The low-sulfur oil costs $20 per barrel, and the high-sulfur oil costs $16 per barrel. How much of each type of oil is needed if the cost is to be minimized?

We must find the number of barrels of each type of oil needed. Let

$$x = \text{Number of barrels of low-sulfur oil needed,}$$

$$y = \text{Number of barrels of high-sulfur oil needed.}$$

Then surely $x \geq 0$ and $y \geq 0$. We have

	Oil		
	Low-Sulfur	High-Sulfur	
	x	y	
Cost	20	16	
Requirements			Amount
Total	1	1	1000
Sulfur content	0.02	0.06	0.04(total) = 0.04(x + y)

(The last row of the table follows from the fact that combining x barrels of oil having a 2% sulfur content with y barrels of oil having a 6% sulfur content results in $x + y$ barrels of oil that must have a 4% sulfur content.) Therefore the linear programming model is

$$\text{Minimize} \quad C = 20x + 16y$$
$$\text{subject to} \quad x + y \geq 1000$$
$$0.02x + 0.06y = 0.04(x + y)$$
$$x \geq 0$$
$$y \geq 0.$$

If we rewrite the second constraint to put it into standard form, with all its variables on the left side of the inequality, we have

$$\text{Minimize} \quad C = 20x + 16y$$
$$\text{subject to} \quad x + y \geq 1000$$
$$-0.02x + 0.02y = 0$$
$$x \geq 0$$
$$y \geq 0. \quad \square$$

Here is another example of a product mix problem:

EXAMPLE 3   A manufacturer makes a low-priced, a medium-priced, and a high-priced radio. Each low-priced radio requires 20 minutes for assembly, each medium-priced one requires 30 minutes, and each high-priced one requires 45 minutes. Each low-priced radio uses ½ lb of plastic, each medium-priced one uses ¾ lb, and each high-priced one uses 1 lb. Today there are 8 hours of assembly time and 100 lb of plastic available. The firm makes a profit of $10 on each low-priced radio, $25 on each medium-priced one, and $50 on each high-priced one. At least 10 low-priced, 10 medium-priced, and 5 high-priced radios must be made to fill orders on hand. Maximize today's profit.

We seek to find the number of low-, medium-, and high-priced radios that the company must make in order to maximize its profit. Let

x_1 = Number of low-priced radios to be made,

x_2 = Number of medium-priced radios to be made,

x_3 = Number of high-priced radios to be made.

LINEAR PROGRAMMING

Then we have

	Radios			
	Low-Priced x_1	Medium-Priced x_2	High-Priced x_3	
Profit	10	25	50	
Requirement				Amount
Assembly time	20	30	45	480
Plastic	0.5	0.75	1	100
Low-priced total	1			10
Medium-priced total		1		10
High-priced total			1	5

(Note that since the assembly times for the radios were given in minutes, we converted the 8 hours of time available to $8(60) = 480$ minutes.) Thus the linear programming model is

$$\text{Maximize } P = 10x_1 + 25x_2 + 50x_3$$

$$\text{subject to } \quad 20x_1 + 30x_2 + 45x_3 \leq 480$$

$$0.5x_1 + 0.75x_2 + x_3 \leq 100$$

$$x_1 \geq 10$$

$$x_2 \geq 10$$

$$x_3 \geq 5. \quad \square$$

Another type of linear programming problem that is frequently encountered is the **machine assignment problem**.

EXAMPLE 4 A furniture company makes coffee tables, each of which has a top and four legs. The legs and tops must be sanded on a sander. The company has three different sanders available; Table 5–1 gives the time, in hours, that each part requires if it is sanded by a given sander. It costs $25 per hour to operate sander number 1, $27 per hour to operate sander number 2, and $32 per hour to operate sander number 3. The company works a 40-hour week, so each sander is available for no more than 40 hours per week. The company intends to make 100 coffee tables this week, so it must sand 100 tops and 400 legs. Minimize the cost of sanding the legs and tops.

TABLE 5–1
Sanding Times (hours)

Sander	Top	Leg
1	0.50	0.20
2	0.45	0.18
3	0.40	0.12

Let

$$x_1 = \text{Number of tops to sand on sander 1},$$
$$y_1 = \text{Number of legs to sand on sander 1},$$
$$x_2 = \text{Number of tops to sand on sander 2},$$
$$y_2 = \text{Number of legs to sand on sander 2},$$
$$x_3 = \text{Number of tops to sand on sander 3},$$
$$y_3 = \text{Number of legs to sand on sander 3}.$$

Of course, all six of these variables must be nonnegative. Then we have

	Sander 1		Sander 2		Sander 3		
	x_1	y_1	x_2	y_2	x_3	y_3	
Cost	25(.5)	25(.2)	27(.45)	27(.18)	32(.4)	32(.12)	
Requirements							Amount
#1 time	0.5	0.2					40
#2 time			0.45	0.18			40
#3 time					0.4	0.12	40
Total tops	1		1		1		100
Total legs		1		1		1	400

Therefore the linear programming model is

Minimize $C = 12.5x_1 + 5y_1 + 12.15x_2 + 4.86y_2 + 12.8x_3 + 3.84y_3$
subject to
$$0.5x_1 + 0.2y_1 \le 40$$
$$0.45x_2 + 0.18y_2 \le 40$$
$$0.4x_3 + 0.12y_3 \le 40$$
$$x_1 + x_2 + x_3 = 100$$
$$y_1 + y_2 + y_3 = 400$$
$$x_1 \ge 0, \; y_1 \ge 0, \; x_2 \ge 0, \; y_2 \ge 0, \; x_3 \ge 0, \; y_3 \ge 0. \quad \square$$

5.2 EXERCISES

For each of the following exercises, you are to set up a linear programming problem whose solution will answer the question asked in the exercise. Do not attempt to solve the linear programming problem.

1. You intend to invest a total of at least $10,000, but not more than $25,000, in either a money market fund, a bond fund, or a combination of both. You do not want to invest more than $16,000 in the money market fund. If the bond fund pays 8.2% per year and the money market fund pays 9.5% per year, how much should you invest in each in order to maximize your annual return?

2. You need at least 24 quarts of a solution that is 6% alcohol. How many quarts of a 4% solution, a 10% solution, and a 12% solution should you mix in order to meet

your needs at minimum cost if the 4% solution costs $0.69, the 10% solution costs $0.99, and the 12% solution costs $1.19 per quart?

3. Refer to Exercise 2: How many quarts of each should you mix if only 4 quarts of the 10% solution are available?

4. A high-school class earned $4500 working on a project and wishes to invest it in three local banks: Amity Bank, Bank & Trust, and Community Bank. Because the father of one of the students worked on the project and is president of Amity Bank, they must invest at least half the amount in Amity. They also wish to invest at least $1000 in Bank & Trust and no more than a total of $3000 in Amity Bank and Community Bank. How much should they invest in each bank in order to maximize their return if the interest rates are 6%, 6.5%, and 8% for Amity Bank, Bank & Trust, and Community Bank, respectively?

5. A woman plans to invest $30,000 in municipal bonds, savings bonds, and Treasury bills. She wishes to invest a minimum of $5,000 in each of the three. If the interest rates are 12% for municipal bonds, 8% for savings bonds, and 10% for Treasury bills, how much should she invest in each?

6. How much should the woman of Exercise 5 allot to each investment if she does not wish to invest a total of more than $15,000 in the two bonds?

7. A party mix consists of miniature pretzels, peanuts, and wheat-a-bits. Standards require that at least 25% of the mixture be pretzels, at least 40% be peanuts, and at least 20% be wheat-a-bits. During a given week, availability is 80 lb of pretzels, 100 lb of peanuts, and 60 lb of wheat-a-bits. If a minimum of 200 lb of mix is required and the costs are $1.50 per pound of pretzels, $1.00 per pound of peanuts, and $1.25 per pound of wheat-a-bits, how many pounds of each should be mixed to minimize the cost of the mix?

8. A library has been granted funds amounting to $6000 for the purchase of new books in the categories of adventure stories, mysteries, and science fiction, provided that they buy no more than twice as many books in any one category as in any other category. The library has a source from which it can purchase all books in any one category at a standard price. The prices are $15 for adventure stories, $20 for mysteries, and $20 for science fiction. In a recent survey the library found that 30% of their patrons liked adventure stories, 50% liked mysteries, and 35% liked science fiction. How many books in each category should the library purchase in order to maximize patron satisfaction?

Management

9. The XY Corporation manufactures product X and product Y. Product X requires 8 lb of resource A and 6 lb of resource B. Product Y requires 4 lb of resource A and 6 lb of resource B. On a certain day, the XY Corporation has on hand 560 lb of resource A and 630 lb of resource B. If the profit on product X is $12.25 per unit and the profit on product Y is $8.75 per unit, how many units of product X and how many of product Y should the XY Corporation manufacture on that day?

10. If the XY Corporation of Exercise 9 must manufacture at least 40 units of product X, how many units of product X and how many of product Y should they manufacture on that day in order to maximize their profit?

11. Suppose the XY Corporation of Exercise 9 must manufacture at least 40 units of product X and wishes to minimize its cost. If it costs $8.00 to manufacture each

unit of product X and $10.00 to manufacture each unit of product Y, how many of each should they manufacture?

12. Snick and Snee Candy Co. manufactures Blo bubble gum and Choco candy bars. Snick and Snee is under contracts with distributors to supply a total of at least 60,000 units of bubble gum and candy bars, of which at least 40,000 must be candy bars. If their costs are $0.01 for each package of bubble gum and $0.02 for each candy bar, how many of each should they produce in order to minimize their cost?

13. VitaBeer Company is a small brewery that produces both light and dark nonalcoholic beer. A case of light beer requires 2 lb of malt and 3 lb of hops, while a case of dark beer requires 1 lb of malt and 4 lb of hops. During a given week, VitaBeer has 160 lb of malt and 340 lb of hops available. If the profit for each of light and dark beer is $1.00 per case, how many cases of light and how many cases of dark beer should VitaBeer produce that week?

14. Refer to Exercise 13: If VitaBeer has an order for 70 cases of light beer, how many cases of light and how many cases of dark beer should VitaBeer produce?

15. General Eclectic has plants in Belpre, Ohio, and Berwyn, Illinois. Each of these plants is capable of producing two types of cloth: psychedelic patterns and plainjane prints. General Eclectic has orders for 24,000 bolts of psychedelic patterns and 12,000 bolts of plainjane prints. The Belpre plant can produce 400 bolts of psychedelic patterns and 400 bolts of plainjane prints each day, whereas the Berwyn plant can produce 1200 bolts of psychedelic patterns and 400 bolts of plainjane prints. If the daily costs are $10,000 for the Belpre plant and $15,000 for the Berwyn plant, how many days should each plant devote to manufacturing these cloths in order to minimize costs and still meet or exceed the quantities ordered?

16. Dutch Bulbs, Inc., grows bulbs of tulips, jonquils, and hyacinths for sale in lots of one dozen bulbs of a kind. Each dozen tulip bulbs require 6 ounces (oz) of fertilizer and 3 square feet (ft^2) of land. Each dozen jonquil bulbs require 4 oz of fertilizer, 2 ft^2 of land, and 6 oz of X-tra Gro. Each dozen hyacinth bulbs require 5 oz of fertilizer, 4 ft^2 of land, and 5 oz of X-tra Gro. During a certain week Dutch Bulbs has available 13 lb 2 oz of fertilizer, 180 ft^2 of land, and 15 lb of X-tra Gro. If their profits are $7.00 per dozen tulip bulbs, $8.00 per dozen jonquil bulbs, and $9.00 per dozen hyacinth bulbs, how many bulbs of each variety should they set out that week?

17. A hardware store mixes three kinds of paint to obtain a special blend. The first of the three paints has a secret ingredient content of 2%, the second 4%, and the third 10%. The special blend is to have a secret ingredient content of 6%. How many gallons of each of the three kinds of paint should be mixed to minimize costs if the first costs $4.00 per gallon, the second $2.00 per gallon, the third $6.00 per gallon, and the hardware store has orders for 400 gallons?

18. Refer to Exercise 17: Suppose the hardware store has only 100 gallons of the second paint. How many gallons of each of the three paints should they mix?

19. (This problem is taken from Stevenson: *Introduction to Management Science*, Irwin, 1989.) An advertising firm often uses linear programming to determine an optimal allocation of advertising budgets. Recently, a client asked for a plan that would allocate a $12,000 budget among radio, TV, and newspaper advertisements, with the stipulation that no more than 40% of the budget be allocated to any one

medium. The client wanted a plan that would maximize the effectiveness of the advertisements. The advertising firm did some research that yielded an effectiveness index for each medium and also determined the cost for an ad in each medium as shown:

Medium	Effectiveness	Cost per Ad
Radio	2.4	$200
TV	3.2	400
Newspaper	1.6	300

How many ads should run in each medium in order to maximize effectiveness?

20. A lumber company uses some of its lumber to produce mouldings which they sell at discount prices. Their budget is $2000 for the production of three types of moulding: apron, band, and chair rail. Experience has shown that they normally sell at least 200 board feet of each type and at least twice as many board feet of chair rail moulding as apron moulding and band moulding combined. If their production costs and profits are as shown in the table, how many board feet of each type should they produce in order to maximize their profit?

Moulding	Cost (per board ft)	Profit (per board ft)
Apron	$2.00	$0.25
Band	1.00	0.50
Chair rail	2.00	0.25

21. A computer parts manufacturer produces CPU housings and monitor shells, each of which must be finished on one of three finishing machines. It costs $20 per hour to operate the first of the three finishing machines, $35 per hour to operate the second, and $30 per hour to operate the third. The company works a 40-hour week, so it has a maximum of 40 hours available for using each finishing machine. The company plans to produce 300 CPU housings and 100 monitor shells during a certain week. If the finishing times for the CPU housings and monitor shells are as shown in the table below, minimize the cost for finishing.

	Finishing Time (in hours)	
Finisher	CPU Housing	Monitor Shell
1	0.2	0.4
2	0.3	0.5
3	0.2	0.3

22. A computer vendor wishes to add software to his line of products and decides to start with three types: spreadsheet, database, and word processor. He plans to order from software houses EasySoft and HardSoft, both of which sell all three types, and he wants to inventory at least two of each type from each house. His fairness policy mandates that his total purchases from either house must be between 40 and 60% of the total of all purchases. If the costs are as indicated in the table, how many software packages of each type of software should he purchase from each software house in order to minimize his cost?

	EasySoft	HardSoft
Spreadsheet	$150.00	$145.00
Database	300.00	350.00
Wordprocessor	250.00	200.00

23. Royal China, Ltd., manufactures fine bone china in three patterns: Aster, Baby's Breath, and Candytuft. The 10-inch plates in all three of these patterns can be made on each of three machines, designated Mach1, Mach2, and Mach3, each of which is available 40 hours per week. During a given week, Royal China has orders scheduled for 1200 Aster, 800 Baby's Breath, and 900 Candytuft 10-inch plates. If the number of each plate pattern produced by each machine per hour of operation and the machine operating costs are as shown in the following tables, how should Royal China schedule the machines so as to minimize cost?

Plates Produced per Hour

	Aster	Baby's Breath	Candytuft
Mach1	30	20	25
Mach2	25	20	30
Mach3	20	25	20

Operating Cost ($ per hour)

	Aster	Baby's Breath	Candytuft
Mach1	0.24	0.20	0.28
Mach2	0.32	0.24	0.32
Mach3	0.36	0.20	0.32

*24. (This problem is taken from Stevenson: *Introduction to Management Science*, Irwin, 1989.) A classic linear programming problem involves minimizing trim loss. Here is one version of the problem.

A mill cuts 20-foot pieces of wood into several different lengths: 8-foot, 10-foot, and 12-foot. The mill has a certain amount of 20-foot stock on hand and orders for the various sizes. The objective is to fill the orders with as little waste as possible. For example, if two 8-foot lengths are cut from a 20-foot piece, there will be a loss of 4 feet, the leftover amount. Currently, the mill has 350 twenty-foot pieces of wood on hand and the following orders, which must be filled from stock on hand:

Size (ft)	Number Ordered
8	275
10	100
12	250

Formulate a linear programming model that will enable the mill operator to satisfy the orders with minimum trim loss. (*Hint:* List the different ways the 20-foot pieces could be cut into the desired sizes.)

Life Science

25. A patient requires daily dosages of between 250 and 400 mg of vitamin A, between 100 and 200 mg of vitamin B, and between 350 and 500 mg of vitamin C.

The vitamins can be obtained from three prescription medicines, Med-1, Med-2, and Med-3, which are in liquid form, but the patient must take at least one teaspoon of Med-2 and Med-3 combined. The patient's doctor has computed that the costs per teaspoon of the three medicines are $0.18 for Med-1, $0.22 for Med-2, and $0.20 for Med-3. If the vitamin contents of the three medicines are as shown in the table, how many teaspoons per day of each of the three medicines should the doctor prescribe in order to minimize the cost to the patient?

Vitamin	Med-1 (mg)	Med-2 (mg)	Med-3 (mg)
A	100	50	100
B	50	100	50
C	100	100	50

5.3 TWO-DIMENSIONAL LINEAR PROGRAMMING

In this section we show how to solve linear programming problems in two variables. We begin by again considering the Alpha Company example of Sections 5.1 and 5.2. In Section 5.2 we constructed the following linear programming model for Alpha:

$$\text{Maximize} \quad P = 60x + 100y$$
$$\text{subject to} \quad 2x + 9y \leq 2700$$
$$2x + 3y \leq 1200$$
$$x \geq 0$$
$$y \geq 0.$$

Here x is the number of small and y the number of large watercoolers Alpha should make in order to maximize its profit P. We will now solve this linear programming problem.

We obtained the feasibility region for Alpha's constraints in Example 7 of Section 5.1; it is depicted in Figure 5–9. Thus we already know what combinations of small and large watercoolers are feasible for Alpha to produce. We seek a feasible combination that will maximize Alpha's profit. But its profit is

$$P = 60x + 100y$$

dollars, and this equation represents a family of parallel lines: each choice of P yields a different line, but all the lines have slope equal to $-60/100 = -0.6$. Some of these parallel lines are shown in Figure 5–10. We want P to be as large as possible, subject to the condition that the point (x, y) lie within the feasibility region; as Figure 5–11 shows, this occurs when the line

$$60x + 100y = P$$

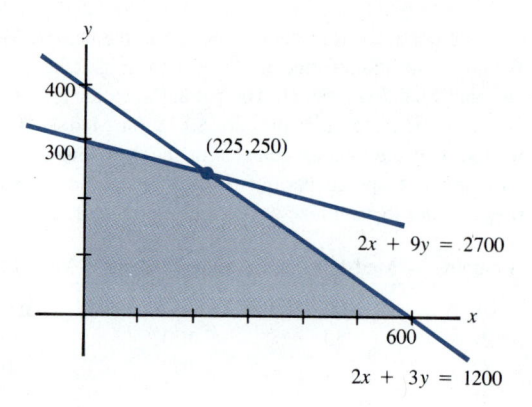

FIGURE 5–9

intersects the feasibility region at the vertex (225, 250). Hence Alpha should make $x = 225$ small and $y = 250$ large watercoolers: this combination is feasible and results in the maximum profit of

$$P = \$60(225) + \$100(250) = \$38,500.$$

Note that in the Alpha Company problem the optimum value of the objective function occurred at a vertex of the feasibility region. The Fundamental Theorem of Linear Programming says that this is always the case.

The Fundamental Theorem of Linear Programming

In a linear programming problem, the optimum value of the objective function, if it exists, occurs at a vertex of the feasibility region.

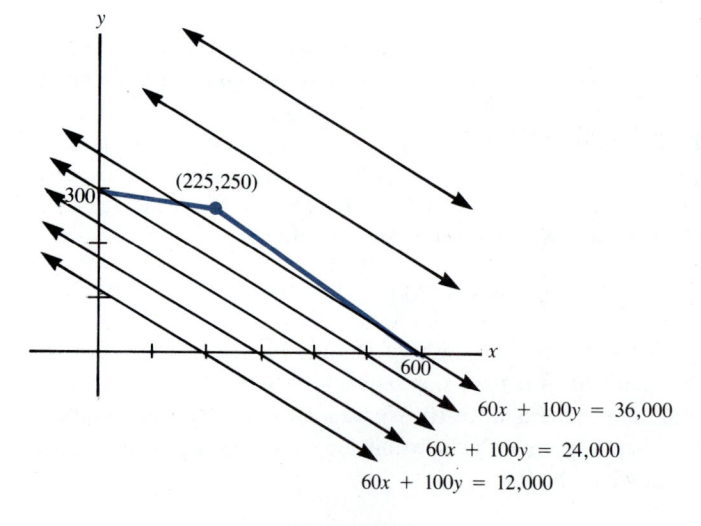

FIGURE 5–10

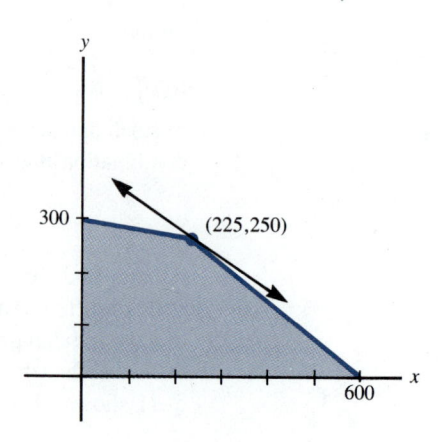

FIGURE 5–11

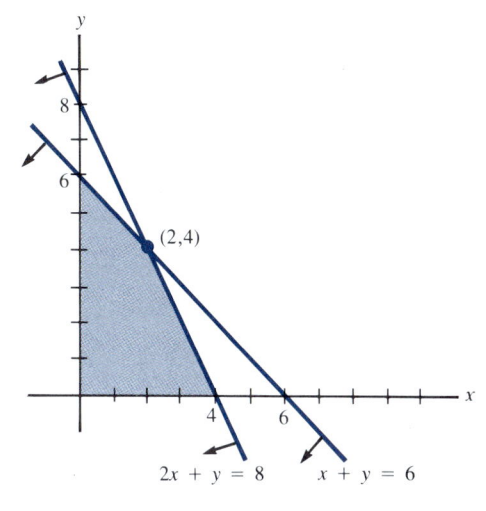

LINEAR PROGRAMMING   291

[Figure 5-12: Feasibility region bounded by lines 2x + y = 8 and x + y = 6, with vertex marked at (2,4)]

FIGURE 5–12

Thus, according to the Fundamental Theorem, in order to solve a linear programming problem it is sufficient to find the vertices of the feasibility region and then evaluate the objective function at each vertex to find the one that yields the optimum value. (If two vertices both yield the optimum value, then so do all points on the line segment joining them.) We also note that, as our statement of the Fundamental Theorem suggests, a linear programming problem need not have a solution.

EXAMPLE 1 Let us solve the following linear programming problem:

$$\text{Maximize} \quad F = 3x + 4y$$
$$\text{subject to} \quad x + y \leq 6$$
$$2x + y \leq 8$$
$$x \geq 0$$
$$y \geq 0.$$

The feasibility region for the constraints is shown in Figure 5–12. The vertex at (2, 4) was obtained by finding the intersection of the lines $x + y = 6$ and $2x + y = 8$:

$$\begin{aligned} -x - y &= -6 \\ 2x + y &= 8 \\ \hline x &= 2 \end{aligned}$$

$$2 + y = 6$$
$$y = 4.$$

Now we find the maximum value of F by evaluating it at each vertex:

Vertex	$F = 3x + 4y$
(0, 0)	$F = 3(0) + 4(0) = 0$
(0, 6)	$F = 3(0) + 4(6) = 24$
(2, 4)	$F = 3(2) + 4(4) = 22$
(4, 0)	$F = 3(4) + 4(0) = 12.$

Hence, the maximum value of the objective function f is 24 when $x = 0$ and $y = 6$. ☐

Before we continue with examples, let us summarize our method for solving linear programming problems in two variables.

The Linear Programming Algorithm

To solve a linear programming problem in two variables:

1. Write the objective function and constraints for the problem.

2. Find the feasibility region for the constraints and identify its vertices.

3. Evaluate the objective function at each vertex. The solution occurs at the vertex that yields the optimum value of the objective function.

EXAMPLE 2 Beta Company manufactures two products: A and B. Each unit of product A contains 2 lb of steel and 1 lb of tin. Each unit of B contains 14 lb of steel and 2 lb of tin. Beta has available 70 lb of steel and 20 lb of tin. Beta makes a profit of $2 on each unit of A and $5 on each unit of B. Find Beta Company's maximum profit and the number of units of A and B it must manufacture in order to obtain its maximum profit.

1. Let

$$x = \text{Number of units of product A that Beta can manufacture,}$$

$$y = \text{Number of units of product B that Beta can manufacture.}$$

Clearly, $x \geq 0$ and $y \geq 0$.

We tabulate the information given in the problem:

	Product A (x)	Product B (y)	Amount
Profit	2	5	
Requirements			
Steel	2	14	70
Tin	1	2	20

LINEAR PROGRAMMING   293

Therefore we may state the problem as follows:

$$\text{Maximize profit} \quad P = 2x + 5y$$
$$\text{subject to} \quad 2x + 14y \le 70$$
$$x + 2y \le 20$$
$$x \ge 0$$
$$y \ge 0.$$

2. We find the feasibility region and its vertices. See Figure 5–13. Note that the vertices are (0, 0), (0, 5), (14, 3), and (20, 0). The vertex (14, 3) was found by solving the equations $2x + 14y = 70$ and $x + 2y = 20$ simultaneously.

3. We evaluate the profit function at each vertex:

Vertex	$P = 2x + 5y$
(0, 0)	$P = 2(0) + 5(0) = 0$
(0, 5)	$P = 2(0) + 5(5) = 25$
(14, 3)	$P = 2(14) + 5(3) = 43$
(20, 0)	$P = 2(20) + 5(0) = 40.$

Therefore, Beta should produce 14 units of product A and 3 of product B for a maximum profit of $43. ☐

EXAMPLE 3 Suppose we repeat Example 2, with the only change being that Beta makes a profit of $1 on each unit of A and $7 on each unit of B. The constraints and therefore the feasibility region remain the same, but the objective function is $P = x + 7y$. Evaluating the objective function at the vertices of the feasibility region, we have

Vertex	$P = x + 7y$
(0, 0)	$P = 0 + 7(0) = 0$
(0, 5)	$P = 0 + 7(5) = 35$
(14, 3)	$P = 14 + 7(3) = 35$
(20, 0)	$P = 20 + 7(0) = 20.$

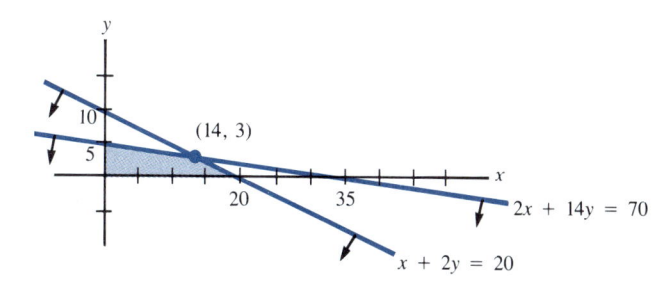

FIGURE 5–13

Therefore, in this case the problem has more than one solution: Beta Company can obtain its maximum profit of $35 by producing 0 units of A and 5 units of B or by producing 14 units of A and 3 units of B. In fact, any point (x, y) that lies between $(0, 5)$ and $(14, 3)$ on the line

$$2x + 14y = 70$$

represents a feasible combination that yields the maximum profit of $35. Thus, for example, Beta can also obtain its maximum profit of $35 by producing 7 units of A and 4 units of B, since $(7, 4)$ lies on the line segment joining $(0, 5)$ and $(14, 3)$. ☐

EXAMPLE 4

Theta Corporation makes industrial cleaner in two strengths: weak and strong. Each barrel of the weak cleaner contains 2 gallons of acid solution and 10 lb of emulsifiers. Each barrel of the strong cleaner contains 4 gallons of acid solution and 30 lb of emulsifiers. According to contracts it has signed with its suppliers, Theta must use at least 80 gallons of acid solution and 450 lb of emulsifiers each week. It costs Theta $2.50 to make each barrel of weak cleaner and $3.25 to make each barrel of strong cleaner. How many barrels of each should the firm make each week in order to minimize its cost?

If we let x denote the number of barrels of weak cleaner and y the number of barrels of strong cleaner to be made each week, we can tabulate the information in the problem as follows:

	Cleaner		
	Weak	Strong	
	x	y	
Cost	2.50	3.25	
Requirements			Amount
Acid	2	4	80
Emulsifiers	10	30	450

Hence the problem may be stated as follows:

$$\text{Minimize cost} \quad C = 2.5x + 3.25y$$

$$\text{subject to} \quad 2x + 4y \geq 80$$

$$10x + 30y \geq 450$$

$$x \geq 0$$

$$y \geq 0.$$

The feasibility region and its vertices are shown in Figure 5–14.

We evaluate the objective function at the vertices of the feasibility region:

Vertex	$C = 2.5x + 3.25y$
(0, 20)	$C = 2.5(0) + 3.25(20) = 65.00$
(30, 5)	$C = 2.5(30) + 3.25(5) = 91.25$
(45, 0)	$C = 2.5(45) + 3.25(0) = 112.50$

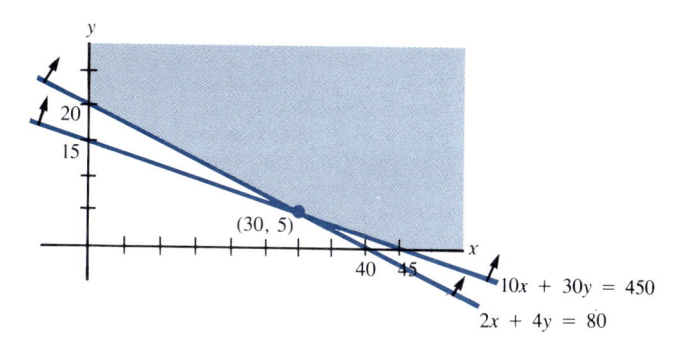

LINEAR PROGRAMMING

FIGURE 5-14

Therefore the minimum cost of $65 occurs when Theta makes 0 barrels of the weak cleaner and 20 barrels of the strong cleaner. □

EXAMPLE 5 A patient needs at least 40 units of vitamin A, at least 30 of vitamin C, and at least 30 of vitamin E each day. Each Brand X multivitamin pill contains 4 units of A, 6 of C, and 2 of E; each Brand Y pill contains 5 units of A, 3 of C, and 5 of E. If each Brand X pill costs $0.22 and each Brand Y pill costs $0.27, how many pills should the patient take each day in order to minimize the cost?

Let

$$x = \text{Number of Brand X pills to take each day},$$
$$y = \text{Number of Brand Y pills to take each day}.$$

Then

	Pills		
	Brand X	Brand Y	
	x	y	
Cost	22	27	
Requirements			Amount
Vitamin A	4	5	40
Vitamin C	6	3	30
Vitamin E	2	5	30

Hence, the problem is

Minimize $C = 22x + 27y$

subject to
$$4x + 5y \geq 40$$
$$6x + 3y \geq 30$$
$$2x + 5y \geq 30$$
$$x \qquad \geq 0$$
$$y \geq 0.$$

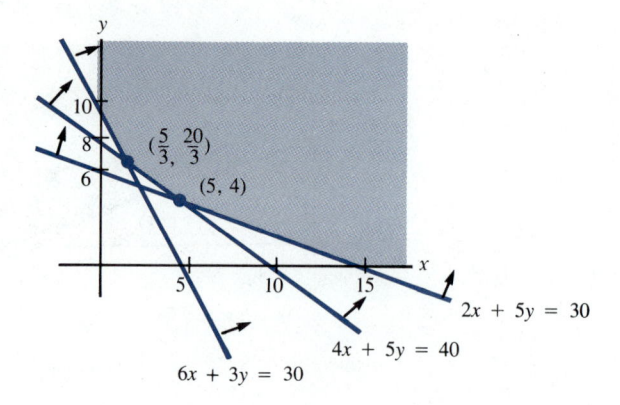

FIGURE 5–15

The feasibility region is shown in Figure 5–15. Note that the vertex (5/3, 20/3) was obtained as the point of intersection of the lines $4x + 5y = 40$ and $6x + 3y = 30$, and that the vertex (5, 4) was obtained as the point of intersection of the lines $2x + 5y = 30$ and $4x + 5y = 40$.

Evaluating the objective function at the vertices yields the following:

Vertex	$C = 22x + 27y$
(0, 10)	$C = 22(0) + 27(10) = 270$
(5/3, 20/3)	$C = 22(5/3) + 27(20/3) = 650/3$
(5, 4)	$C = 22(5) + 27(4) = 218$
(15, 0)	$C = 22(15) + 27(0) = 330$

Since $650/3 = 216\frac{2}{3}$, the minimum cost of aproximately $2.17 per day occurs if the patient takes $1\frac{2}{3}$ Brand X and $6\frac{2}{3}$ Brand Y pills per day. □

5.3 EXERCISES

In Exercises 1 through 10, solve the given linear programming problem.

1. Maximize $F = 2x + y$
    subject to
    $x + y \leq 6$
    $x + y \geq 2$
    $x \geq 0$
    $y \geq 0$
    (See Exercise 22, Exercises 5.1)

2. Maximize $F = 3x + 2y$
    subject to
    $2x + 4y \leq 40$
    $3x + y \leq 15$
    $x \geq 0$
    $y \geq 0$
    (See Exercise 25, Exercises 5.1)

3. Maximize $F = 6x + 4y$
    subject to
    $3x + 8y \leq 48$
    $3x + 2y \leq 24$
    $x \geq 0$
    $y \geq 0$
    (See Exercise 26, Exercises 5.1)

4. Maximize $F = x + 2y$
    subject to
    $3x + 8y \leq 48$
    $3x + 2y \leq 24$
    $x - y \geq 0$
    $x \geq 0$
    $y \geq 0$
    (See Exercise 27, Exercises 5.1)

5. Minimize $F = 4x + 5y$
   subject to
   $$x \geq 4$$
   $$y \geq 2$$
   $$x \leq 8$$
   $$y \leq 10$$
   (See Exercise 24, Exercises 5.1)

6. Minimize $F = 2x - y$
   subject to
   $$x + y \geq 8$$
   $$x \geq 4$$
   $$y \geq 3$$
   (See Exercise 28, Exercises 5.1)

7. Minimize $F = 3x + 5y$
   subject to
   $$2x + 6y \geq 36$$
   $$3x + 3y \geq 36$$
   $$x + 6y \geq 24$$
   $$x \geq 0$$
   $$y \geq 0$$
   (See Exercise 29, Exercises 5.1)

8. Minimize $F = x + 0.5y$
   subject to
   $$3x + y \geq 15$$
   $$x + 2y \leq 30$$
   $$2x + y \geq 12$$
   $$y \geq 1$$
   $$x \geq 0$$
   (See Exercise 30, Exercises 5.1)

9. Maximize $F = 5x + 12y$
   subject to
   $$-x + y \leq 4$$
   $$x - y \leq 4$$
   $$x + y \geq 4$$
   $$x + y \leq 12$$
   $$x \geq 0$$
   $$y \geq 0$$
   (See Exercise 34, Exercises 5.1)

10. Maximize $F = -12x + 4y$
    subject to
    $$-2x + y \leq 5$$
    $$x + 2y \leq 20$$
    $$3x - y \leq 27$$
    $$x + y \geq 4$$
    $$x \leq 12$$
    $$y \leq 12$$
    $$x \geq 0$$
    $$y \geq 0$$
    (See Exercise 36, Exercises 5.1)

*11. A linear programming problem may not have a solution because its constraints are **inconsistent,** in which case there are no feasible combinations for the problem. Show that this is so for the following problem:

   Maximize $F = 2x + 3y$
   subject to
   $$10x + 10y \leq 1000$$
   $$x \geq 150$$
   $$y \geq 100.$$

*12. A linear programming problem may not have a solution because unboundedness of the feasibility region results in no optimum value for the objective function. Show that this is so for the following problem:

   Maximize $F = 2x + 3y$
   subject to
   $$x + y \geq 400$$
   $$x \geq 100$$
   $$y \geq 200.$$

13. Solve Exercise 1 of Exercises 5.2

14. You have $10,000 to invest, and you are considering investing in one or both of two stocks: X and Y. Stock X pays an annual dividend of 10% but is risky. Stock Y is safer but pays an annual dividend of only 6%. You decide on the following guidelines for your investments:
    (a) You will not invest more than $8000 in Stock X.
    (b) Your investment in Stock X will not exceed four times your investment in Stock Y.
    (c) You will invest at least $1000 in Stock X.
    Find the maximum return on your investment and how much you should invest in each stock in order to receive it.

Management

15. Solve Exercise 9 of Exercises 5.2.

16. Solve Exercise 13 of Exercises 5.2.

17. A company has budgeted $55,000 to purchase a combination of television and newspaper advertisements. Each television ad will cost $5000 and reach an audience of 80,000 people, and each newspaper ad will cost $2500 and reach an audience of 50,000 people. The advertising director has recommended purchasing a minimum of four television ads and eight newspaper ads. Find the number of television and newspaper ads and the company should run in order to maximize its audience. (See Exercise 37, Exercises 5.1.)

18. A company makes two types of electronic calculators: a business model and a scientific model. Each business model has one plastic case and contains 10 microcircuits. Each scientific model has one plastic case and contains 20 microcircuits. If the company has 24 plastic cases and 320 microcircuits on hand, and if it makes a profit of $3 on each business model and $4 on each scientific model, how many of each must it make in order to maximize its profit?

19. Repeat Exercise 18 if the profit is $3 on each business model and $10 on each scientific model.

20. Repeat Exercise 18 if the profit is $3 on both models.

21. Suppose that in addition to the restrictions concerning plastic cases and microcircuits in Exercise 18, each business model requires 5 worker-hours and each scientific model 6 worker-hours to assemble and the company has 120 worker-hours available for assembly. If the company makes a profit of $2 on each business model and $2.50 on each scientific model, find its maximum profit and the number of units of each model that must be made in order to attain it.

22. Repeat Exercise 21 if, in addition, the company must produce at least 16 scientific models.

23. A company makes two products: sockets and wickets. Each socket requires 6 hours for fabrication and 6 hours for testing. Each wicket requires 12 hours for fabrication and 2 hours for testing. The company has 120 hours available for fabrication and 30 hours available for testing, and makes a profit of $4 on each socket and $3 on each wicket. Find the company's maximum profit and the number of sockets and wickets it must make in order to attain its maximum profit. (See Exercise 38, Exercises 5.1.)

24. Solve Exercise 12 of Exercises 5.2.

25. Solve Exercise 15 of Exercises 5.2.

26. Alpha Company makes two kinds of tires: bias-ply and radial. Alpha must produce at least 4000 tires per day, of which at least 1600 must be radials. Each bias-ply tire takes 2 minutes and costs $20 to produce. Each radial tire takes 6 minutes and costs $30 to produce. According to its union contract, Alpha must use at least 320 worker-hours (19,200 worker-minutes) per day. Find Alpha's minimum cost and how many of each type of tire the company must make in order to attain it.

27. Repeat Exercise 26 if, in addition, Alpha must produce at least 600 bias-ply tires per day.

28. Delta Company produces two products, A and B, and stores them in inventory in its warehouse. Each unit of A occupies 6 ft^3 of space, and each unit of B occupies 4 ft^3 of space. The warehouse has 1440 ft^3 of space. Furthermore, Delta must produce a total of at least 100 units of product, of which at least 50 must be Product B. Delta cannot possibly produce a total of more than 300 units of product and cannot produce more than 150 units of Product A. If each unit of A is worth $5 and each unit of B is worth $7, find the maximum value of Delta Company's inventory.

29. An oil company mixes high octane and low octane gasoline to create two intermediate grades of gas, grade A and grade B. Grade A consists of 60% high octane and 40% low octane gas, and grade B consists of 30% high octane and 70% low octane. The company has 120,000 gallons of high octane and 112,000 gallons of low octane gas available and must produce at least 50,000 gallons of grade A and 80,000 gallons of grade B. If the company sells each gallon of A for $1.20 and each gallon of B for $1.12, find the amount of grade A and grade B the company should produce in order to maximize its revenue. (See Exercise 43, Exercises 5.1.)

30. Suppose the high octane gasoline of Exercise 29 costs the company $0.98 per gallon and the low octane gas costs them $0.86 per gallon. How many gallons of grade A and grade B must the company produce in order to minimize its cost? What is the minimum cost?

31. A firm makes both a regular and a deluxe model battery. Each battery has one plastic case. Each regular battery can be assembled in 10 minutes, and each deluxe model in 15 minutes. The firm must produce at least 120 regular and 150 deluxe batteries, has 300 plastic cases on hand, and has eight employees in the assembly department, each working an 8-hour day. If the firm makes a profit of $25 on each regular battery and $35 on each deluxe battery, how many of each should it make in order to maximize its daily profit? (See Exercise 44 of Exercises 5.1.)

32. A job shop mills, drills, and grinds two types of parts. Each unit of the first type requires 2 machine-hours of milling, 1 machine-hour of drilling, and 2 machine-hours of grinding. Each unit of the second requires 2.5 machine-hours of milling, 0.75 machine-hours of drilling, and 1.25 machine-hours of grinding. Each day the shop has available 20 machine-hours of milling, 15 of drilling, and 12 of grinding. If the shop makes a profit of $12 on each part of the first type and $15 on each part of the second type, find the maximum profit and the number of each type of part that must be worked in order to attain it. (See Exercise 45, Exercises 5.1.)

33. You must develop next month's advertising budget for your company. You can spend the money on either TV commercials or radio commercials or a combination of both. Each TV commercial costs $1000 and reaches 50,000 people. Each radio commercial costs $200 and reaches 15,000 people. Your boss has told you that
    (a) your commercials must reach at least 300,000 people;
    (b) you must run at least three TV commercials and at least eight radio commercials during the month;
    (c) the number of radio commercials you run cannot exceed twice the number of TV commercials you run.
    Minimize the cost of reaching at least 300,000 people. How many of each type commercial will you run?

34. An advertising manager must allocate the budget for an advertising campaign between two monthly magazines, A and B. Magazine A has a circulation of

1,200,000, and magazine B has a circulation of 750,000. An ad in A will cost $25,000, and an ad in B will cost $18,000. The ad campaign will cover 12 months, and at least one ad must appear each month. At least two ads will be placed in each magazine. If the budget is $536,000, how many ads should be placed in each magazine in order to maximize the number of reader exposures?

35. Two plants, X and Y, both ship the parts they make to a common distribution center. Each week the distribution center needs to receive at least 1000 parts from X and at least 1500 from Y; however, it cannot handle more than 6000 parts total. In addition, the center would always like to have on hand at least twice as many parts from X as from Y. If it costs $0.10 to ship a part to the distribution center from X and $0.14 to ship one to the center from Y, how many should be shipped each week in order to minimize the cost?

36. A company blends two types of oil, heavy and light, to make its special all-purpose oil. Each gallon of heavy oil has a sulfur content of 3%, while each gallon of light oil has a sulfur content of 1%. The company needs at least 10,000 gallons of the all-purpose oil, which must have a sulfur content of at least 1.5% but not more than 2%. Heavy oil costs $0.60 per gallon, and light oil costs $0.75 per gallon. How many gallons of each oil, heavy and light, must the company blend in order to minimize its cost?

37. Knifty-Knit Sweaters is about to schedule a week's production run of ski sweaters. The sweaters are made in two sizes, medium and large, and Knifty-Knit already has orders for 2000 medium and 1000 large sweaters. Each medium sweater is made of 8 oz of wool, and each large one is made of 10 oz. It takes 10 minutes to knit a sweater, regardless of its size, and each finished sweater is packed in a plastic envelope for shipping. Knifty-Knit has purchased 72,000 oz of wool and 8000 plastic envelopes. Also, the firm's contract with the Knitters Union guarantees each of 15 knitters at least 40 hours of work each week. If Knifty-Knit makes a profit of $4 on each medium sweater and $6 on each large sweater, how many of each size must it produce in order to maximize profit?

*38. Solve the linear programming problem of Example 1, Section 5.2.

*39. Solve the linear programming problem of Example 2, Section 5.2.

*40. A firm produces its product at two plants. One plant must produce at least 800 units per week, the other at least 600 per week, and next week the total production of both plants is to be exactly 2000 units. Suppose that it costs $5 to produce a unit at the first plant and $6 to produce one at the second plant. Minimize the firm's cost.

Life Science

41. A patient needs at least 60 units of diet supplement A and at least 40 units of diet supplement B. Brand X capsules contain 8 units of A and 4 units of B, and Brand Y capsules contain 5 units of A and 5 units of B. If each Brand X capsule costs $0.32 and each Brand Y capsule $0.18, how many of each should the patient take in order to minimize the cost? (See Exercise 46, Exercises 5.1.)

42. Repeat Exercise 41 if each Brand X capsule costs $1 and each Brand Y capsule costs $2.20.

43. Each Brand X vitamin pill contains 50 mg of vitamin A, 10 mg of vitamin B, and 20 mg of vitamin C. Each Brand Y pill contains 40 mg of vitamin A, 50 mg of

vitamin B, and 60 mg of vitamin C. Each Brand X pill costs $0.035 and each Brand Y pill costs $0.042. How many of each brand should you take if you must supplement your daily diet with at least 440 mg of vitamin A, 320 mg of vitamin B, and 440 mg of vitamin C and you wish to minimize your daily cost?

44. Two species of birds inhabit the same ecological environment and compete for the same sources of food. Each nesting pair of species A needs 33 kg of food during the nesting season and a territory of 100 m². Each nesting pair of species B needs 58 kg of food during the nesting season and a territory of 75 m². A biologist estimates that the particular region he is studying can supply 60,000 kg of food for birds during a nesting season and has an area of 150,000 m². What is the maximum combined number of nesting pairs that the region can support?

45. In order to feed his cattle, a farmer must harvest at least 100 bales of hay and 300 bushels of corn. He must weed the hay at least $1/30$ of an hour per bale per week and the corn at least $1/20$ hour per bushel per week. Each bale of hay requires the application of 0.5 lb of fertilizer and each bushel of corn needs 0.2 lb of fertilizer. Each bale of hay requires 0.125 acres of ground and each bushel of corn requires 0.025 acres. The farmer can devote 40 hours per week to weeding his crops, can afford 200 lb of fertilizer, and has 40 acres of land. Suppose that a bushel of corn has 1.8 times as much nutritional value as a bale of hay. How many bales of hay and bushels of corn should the farmer grow in order to maximize the nutritional value of his feed? (See Exercise 48, Exercises 5.1.)

46. A nutritionist is planning a diet for a patient who can eat only grain and milk. Suppose that each ounce of grain contains 0.04 oz of protein, 0.001 oz of vitamin A, 0.04 oz of fat, and no cholesterol, and each ounce of milk contains 0.16 oz of protein, 0.02 oz of vitamin A, 0.1 oz of fat, and 0.02 oz of cholesterol. The patient must receive at least 1 oz of protein and 0.1 oz of vitamin A per meal, and not more than 0.8 oz of fat and 0.15 oz of cholesterol per meal. If each ounce of grain costs $0.11 and each ounce of milk costs $0.16, how much of each should the patient receive in order to minimize the cost of the diet? (See Exercise 50, Exercises 5.1.)

Social Science

*47. A political scientist is designing an opinion poll. She does not wish to poll more than 1200 voters. Of those she does poll, at least 300 must be men and at least 300 must be women. Also, she wishes to poll twice as many women as men. If it costs $12 to interview each voter, how many men and how many women should be polled in order to minimize the cost? (See Exercise 51, Exercises 5.1.)

48. A judge has ordered a town to implement busing in order to desegregate its two high schools. Currently, East School has 1800 minority students, 1000 nonminority students, and a capacity of 3000 students. West School has 1200 minority students, 2200 nonminority students, and a capacity of 3500 students. The judge has ruled that East School can have at most 60% minority enrollment and West School can have at most 60% nonminority enrollment. Only minority students will be transferred from East School to West School, and only nonminority students will be transferred from West to East. Minimize the total number of students who must be transferred.

5.4 THE SIMPLEX METHOD: MAXIMIZATION WITH ≤ CONSTRAINTS

In Section 5.3 we introduced a graphical method for solving linear programming problems in two variables. However, many linear programming problems have more than two variables (indeed, hundreds of variables are not uncommon), and such problems cannot be solved by graphical methods. Fortunately, in 1947 George Dantzig invented the **simplex method** for solving linear programming problems having any number of variables. In this section and the next, we describe and illustrate the simplex method.

Before beginning our discussion of the simplex method we should remark that it is not the only general method for solving linear programming problems. Recently two other methods have been discovered; the newest of these, known as **Karmarkar's algorithm**, was found in 1984 by the mathematician Narendra Karmarker. One of the properties of Karmarkar's algorithm is that it requires fewer iterations than the simplex method, and hence has the potential of being much faster. Karmarkar's algorithm has undergone extensive testing and it does appear to be faster than the simplex method, at least for some types of problems. For instance, it is now being used to solve problems that have to do with the routing of telephone calls. Even if Karmarkar's algorithm comes into general use, however, the simplex method will remain important, for it is easy to use, exceptionally well-suited to many types of problems, and there is an enormous investment in it.

The Initial Tableau

Now we are ready to begin our explanation of the simplex method. As we have already noted, every linear programming problem consists of two parts: a linear objective function that is to be either maximized or minimized and a set of linear constraints. The constraints may be less-than-or-equal-to inequalities, greater-than-or-equal-to inequalities, or equalities. For example, the problem

$$\text{Maximize} \quad F = 2x_1 + 3x_2 + x_3$$
$$\text{subject to} \quad x_1 + 2x_2 + 2x_3 \leq 100$$
$$2x_1 + x_2 \geq 20$$
$$x_2 - x_3 = 0$$
$$x_1 \geq 0$$
$$x_2 \geq 0$$
$$x_3 \geq 0$$

is a maximization problem in three variables with six constraints. Note that the constant term on the right-hand side of each constraint is nonnegative and that the variables are restricted to nonnegative values. Most linear programming problems

LINEAR PROGRAMMING

of a practical nature satisfy these two conditions, and we shall confine our attention to those that do.

We will illustrate the simplex method by using it to solve the following problem: Acme Company makes three products, A, B, and C. Table 5–2 shows the amounts of raw materials needed to produce these products, the amount of each raw material that Acme has on hand, and the profit Acme makes on each unit of each product. If we let

$$x_1 = \text{Number of units of product A to be made,}$$
$$x_2 = \text{Number of units of product B to be made,}$$

and

$$x_3 = \text{Number of units of product C to be made,}$$

TABLE 5–2

	A	B	C	Available
Steel (lb)	2	10	14	70
Tin (lb)	1	0	2	20
Chrome (lb)	0	1	1	10
Profit ($)	2	1	5	

then Acme Company's problem is

$$\text{Maximize} \quad F = 2x_1 + x_2 + 5x_3$$
$$\text{subject to} \quad 2x_1 + 10x_2 + 14x_3 \leq 70$$
$$x_1 \qquad\qquad + 2x_3 \leq 20$$
$$x_2 + x_3 \leq 10$$
$$x_1 \geq 0, x_2 \geq 0, x_3 \geq 0.$$

Now we are ready to illustrate the simplex method. The amount of a resource that is not used is referred to as its **slack.** For example, in the Acme Company problem the combination $x_1 = 1$, $x_2 = 1$, and $x_3 = 4$ is a feasible one because it satisfies all of the constraints. However, this combination uses only

$$2(1) + 10(1) + 14(4) = 68 \text{ lb of steel.}$$

Therefore, there is a slack of 2 lb of steel when $x_1 = 1$, $x_2 = 1$, and $x_3 = 4$. The amount of slack can vary, depending on the feasible combination chosen, but it must always be either positive (if the resource is not used up) or zero (if the resource is used up). For each resource we use a nonnegative **slack variable** s to stand for the slack that is present. Adding the slack variable to the less-than-or-equal-to constraint for the resource converts the inequality into an equality, since the slack variable represents exactly the portion of the resource that remains unused.

The Acme Company problem has three less-than-or-equal-to constraints, so we must add a slack variable to each of them. Thus we add a nonnegative slack variable s_1 to the first constraint, a nonnegative slack variable s_2 to the second constraint, and a nonnegative slack variable s_3 to the third constraint. Adding these slack variables to the constraints converts the constraints from inequalities into equalities. We also add the three slack variables to the objective function F, assigning each of them a coefficient of zero so that F is not actually changed. Hence the original problem is transformed into the equivalent problem

$$\text{Maximize} \quad F = 2x_1 + x_2 + 5x_3 + 0s_1 + 0s_2 + 0s_3$$

$$\text{subject to} \quad 2x_1 + 10x_2 + 14x_3 + s_1 = 70$$

$$x_1 + 2x_3 + s_2 = 20$$

$$x_2 + x_3 + s_3 = 10$$

$$x_1 \geq 0, x_2 \geq 0, x_3 \geq 0, s_1 \geq 0, s_2 \geq 0, s_3 \geq 0.$$

One solution for the three constraint equations may be found by setting

$$x_1 = x_2 = x_3 = 0$$

and solving for the remaining variables to obtain $s_1 = 70$, $s_2 = 20$, and $s_3 = 10$. This gives us our initial vertex of the feasibility region: it is the vertex where $x_1 = x_2 = x_3 = 0$. The value of F at this vertex is 0. The variables we have set equal to zero to obtain a vertex are called **nonbasic variables,** whereas the variables for which we solve are called **basic variables, or the basis.** In this case, x_1, x_2, and x_3 are our initial nonbasic variables, and s_1, s_2, and s_3 are the initial basic variables.

Let us form the augmented matrix of the system of three equations in the unknowns $x_1, x_2, x_3, s_1, s_2, s_3$:

$$\begin{array}{c} \\ s_1 \\ s_2 \\ s_3 \end{array} \begin{array}{c} x_1 \ \ x_2 \ \ x_3 \ \ s_1 \ \ s_2 \ \ s_3 \ \ \ \text{sol} \\ \left[\begin{array}{cccccc|c} 2 & 10 & 14 & 1 & 0 & 0 & 70 \\ 1 & 0 & 2 & 0 & 1 & 0 & 20 \\ 0 & 1 & 1 & 0 & 0 & 1 & 10 \end{array} \right] \end{array}.$$

Note that we have labeled each column to the left of the vertical bar with its variable and the column to the right of the vertical bar with "sol" (for solution). Note also that each row of the matrix is labeled with the basic variable that appears in that row.

Next we incorporate the objective function F into our matrix. We do this by adding a new row, called the **F-row,** at the bottom of the matrix. The entries of the F-row to the left of the vertical bar are the negatives of the coefficients of F; the entry of the F-row to the right of the vertical bar is the value of F at the basis. Thus, since

$$F = 2x_1 + x_2 + 5x_3 + 0s_1 + 0s_2 + 0s_3,$$

LINEAR PROGRAMMING 305

and since F has value 0 at the initial basis s_1, s_2, s_3, we have

$$\begin{array}{c c} & \begin{array}{c c c c c c c} x_1 & x_2 & x_3 & s_1 & s_2 & s_3 & \text{sol} \end{array} \\ \begin{array}{c} s_1 \\ s_2 \\ s_3 \\ F \end{array} & \left[\begin{array}{c c c c c c | c} 2 & 10 & 14 & 1 & 0 & 0 & 70 \\ 1 & 0 & 2 & 0 & 1 & 0 & 20 \\ 0 & 1 & 1 & 0 & 0 & 1 & 10 \\ \hline -2 & -1 & -5 & 0 & 0 & 0 & 0 \end{array} \right]. \end{array}$$

This matrix is the **initial simplex tableau** for the Acme Company problem. Every linear programming problem that is one of maximization and in which all resource constraints are of the less-than-or-equal-to variety has an initial simplex tableau constructed in this manner.

EXAMPLE 1 Set up the initial simplex tableau for the following problem:

$$\text{Maximize} \quad F = 2x_1 - x_2 + 2x_3 + x_4$$

$$\text{subject to} \quad x_1 + x_2 + 2x_3 + x_4 \le 100$$

$$2x_1 \qquad\qquad + x_3 + 2x_4 \le 50$$

$$x_1 + x_2 \qquad\qquad\qquad \le 20$$

$$x_1 \ge 0,\ x_2 \ge 0,\ x_3 \ge 0,\ x_4 \ge 0.$$

Adding slack variables to each of the resource constraints, we transform the problem into

$$\text{Maximize} \quad F = 2x_1 - x_2 + 2x_3 + x_4 + 0s_1 + 0s_2 + 0s_3$$

subject to

$$x_1 + x_2 + 2x_3 + x_4 + s_1 \qquad\qquad\qquad = 100$$

$$2x_1 \qquad\qquad + x_3 + 2x_4 \qquad + s_2 \qquad\quad = 50$$

$$x_1 + x_2 \qquad\qquad\qquad\qquad\qquad\quad + s_3 = 20$$

$$x_1 \ge 0,\ x_2 \ge 0,\ x_3 \ge 0,\ x_4 \ge 0,$$

$$s_1 \ge 0,\ s_2 \ge 0,\ s_3 \ge 0.$$

Therefore the initial tableau is as follows:

$$\begin{array}{c c} & \begin{array}{c c c c c c c c} x_1 & x_2 & x_3 & x_4 & s_1 & s_2 & s_3 & \text{sol} \end{array} \\ \begin{array}{c} s_1 \\ s_2 \\ s_3 \\ F \end{array} & \left[\begin{array}{c c c c c c c | c} 1 & 1 & 2 & 1 & 1 & 0 & 0 & 100 \\ 2 & 0 & 1 & 2 & 0 & 1 & 0 & 50 \\ 1 & 1 & 0 & 0 & 0 & 0 & 1 & 20 \\ \hline -2 & 1 & -2 & -1 & 0 & 0 & 0 & 0 \end{array} \right]. \end{array}$$ ∎

Transforming the Simplex Tableau

We continue with the Acme Company problem. As we have seen, this problem has the following initital simplex tableau:

$$\begin{array}{c} \\ s_1 \\ s_2 \\ s_3 \\ F \end{array} \left[\begin{array}{cccccc|c} x_1 & x_2 & x_3 & s_1 & s_2 & s_3 & \text{sol} \\ 2 & 10 & 14 & 1 & 0 & 0 & 70 \\ 1 & 0 & 2 & 0 & 1 & 0 & 20 \\ 0 & 1 & 1 & 0 & 0 & 1 & 10 \\ \hline -2 & -1 & -5 & 0 & 0 & 0 & 0 \end{array} \right].$$

Now we are going to obtain vertices of the feasibility region that will produce successively larger values for the objective function F. We shall do this by transforming the initial simplex tableau into a second simplex tableau, the second tableau into a third tableau, and so on, until we find the solution to the problem. As we do this, it is important to realize that at any stage of the process the current simplex tableau will tell us the values currently assigned to the variables: nonbasic variables always have value 0, and the values currently assigned to the basic variables may be read from the tableau by ignoring the nonbasic columns. For example, the initial tableau for the Acme Company problem tells us that at this stage s_1, s_2, and s_3 form the basis, Hence,

$$x_1 = x_2 = x_3 = 0,$$

since x_1, x_2, and x_3 are nonbasic, and by ignoring the nonbasic columns of the tableau we see that $s_1 = 70$, $s_2 = 20$, and $s_3 = 10$. Furthermore, the entry in the F-row and the solution column always tells us the value of the objective function F at the current basis.

For the Acme Company problem, we have

$$F = 2x_1 + x_2 + 5x_3 + 0s_1 + 0s_2 + 0s_3.$$

Since nonbasic variables are always equal to 0, it follows that in order to increase the value of F we must let one of x_1, x_2, or x_3 become a basic variable so that it can take on a value greater than 0. Note that

- If x_1 becomes basic while x_2 and x_3 remain nonbasic, then $F = 2x_1$.
- If x_2 becomes basic while x_1 and x_3 remain nonbasic, then $F = x_2$.
- If x_3 becomes basic while x_1 and x_2 remain nonbasic, then $F = 5x_3$.

Therefore, it is clear that if we let x_3 enter the basis we stand to increase the value of F by a greater amount than if we let either x_1 or x_2 enter. Hence, we choose x_3 to be the **entering variable.**

We could also have reached the conclusion that x_3 should enter the basis by examining the simplex tableau. Since the entries to the left of the vertical bar in the F-row of the tableau are the negatives of the coefficients in the expression for F, the entry that has the largest absolute value makes the greatest contribution to F. Note that, of the negative entries in the F-row of our initial simplex tableau, the one having the largest absolute value is -5, which is in the x_3-column. This tells us that x_3 must be the entering variable. This reasoning may be applied to any simplex tableau.

> ### The Entering Variable
>
> To determine which variable should enter the basis, proceed as follows:
>
> **1.** Examine the negative entries to the left of the vertical bar in the F-row, and choose the one having the largest absolute value. This entry will be in the column of the entering variable.
>
> **2.** If two or more negative entries tie for the largest absolute value, choose any one of them.
>
> **3.** If none of the entries in the F-row are negative, the value of F cannot be increased.

Now that we know that x_3 must enter the basis, one of s_1, s_2, or s_3 must leave the basis to make room for x_3. We cannot chose the variable that is to leave the basis arbitrarily, for removing a particular variable from the basis may force the entering variable to come into the basis at a value that is not feasible. For example, suppose we try to let s_3 leave as x_3 enters. Then x_1, x_2, and s_3 would be nonbasic and hence have value 0, and the constraint

$$x_2 + x_3 + s_3 = 10$$

would force $x_3 = 10$. But 10 is not a feasible value for x_3 because $x_1 = x_2 = 0$ and $x_3 = 10$ violate the first constraint of the original problem (Check this.)

Similarly, if we try to let s_2 leave the basis as x_3 enters, we would have

$$x_1 = x_2 = s_2 = 0,$$

and the constraint

$$x_1 + 2x_3 + s_2 = 20$$

would again force $x_3 = 10$. However, if we let s_1 leave as x_3 enters, then we would have

$$x_1 = x_2 = s_1 = 0,$$

and the constraint

$$2x_1 + 10x_2 + 14x_3 + s_1 = 70$$

would force $x_3 = 5$. It is easy to check that $x_1 = x_2 = 0$ and $x_3 = 5$ is a feasible combination for the problem. Hence we have no choice but to let s_1 leave the basis as x_3 enters. We refer to s_1 as the **departing variable.**

We can also use the simplex tableau to identify the departing variable. Since we known that x_3 is to enter our basis, let us divide each entry (except the last one) in the solution column by the corresponding entry in the x_3-column:

	x_3	sol	quotient
s_1	14	70	$70/14 = 5$
s_2	2	20	$20/2 = 10$
s_3	1	10	$10/1 = 10.$

Each quotient tells us the value at which x_3 will enter the basis if the basic variable that labels its row leaves the basis. If we choose the smallest nonnegative quotient, we will ensure that x_3 enters the basis at a feasible value. Thus, since the smallest nonnegative quotient is 5, which is in the s_1-row, s_1 must be the departing variable. Similar reasoning may be applied to any simplex tableau.

The Departing Variable

To determine which variable should leave the basis, proceed as follows:

1. Determine the entering variable.

2. For each row of the tableau except the last, form the quotient

$$\frac{\text{Entry in solution column}}{\text{Entry in column of entering variable}}.$$

Do not form this quotient if the entry in the column of the entering variable is not positive.

3. Of the quotients obtained in step 2, choose the one having the smallest nonnegative value; the row that gave rise to this quotient is the row of the departing variable.

4. If two or more quotients tie for the smallest nonnegative value, choose any one of them.

5. If there are no nonnegative quotients, the linear programming problem has no solution.

We have determined that x_3 must enter the basis and s_1 must leave the basis. Therefore, we must transform the tableau in such a way as to make the x_3-column look exactly like the current s_1-column. We accomplish this by using Type II and Type III row operations and the pivoting process of Gauss-Jordan elimination (as described in Section 4.1) to change the x_3-column from

$$\begin{bmatrix} 14 \\ 2 \\ 1 \\ -5 \end{bmatrix} \quad \text{to} \quad \begin{bmatrix} 1 \\ 0 \\ 0 \\ 0 \end{bmatrix}.$$

LINEAR PROGRAMMING

This will result in a new simplex tableau.

To accomplish this transformation, we first multiply the s_1-row of the tableau by $1/14$ to produce a 1 in the pivot spot, then use this 1 to produce 0 in the other positions of the x_3-column:

$$\begin{array}{c} \\ s_1 \\ s_2 \\ s_3 \\ \hline F \end{array} \left[\begin{array}{cccccc|c} x_1 & x_2 & x_3 & s_1 & s_2 & s_3 & \text{sol} \\ 2 & 10 & \boxed{14} & 1 & 0 & 0 & 70 \\ 1 & 0 & 2 & 0 & 1 & 0 & 20 \\ 0 & 1 & 1 & 0 & 0 & 1 & 10 \\ \hline -2 & -1 & -5 & 0 & 0 & 0 & 0 \end{array} \right] \xrightarrow{(1/14)R_1}$$

$$\left[\begin{array}{cccccc|c} x_1 & x_2 & x_3 & s_1 & s_2 & s_3 & \text{sol} \\ 1/7 & 5/7 & \boxed{1} & 1/14 & 0 & 0 & 5 \\ 1 & 0 & 2 & 0 & 1 & 0 & 20 \\ 0 & 1 & 1 & 0 & 0 & 1 & 10 \\ \hline -2 & -1 & -5 & 0 & 0 & 0 & 0 \end{array} \right] \begin{array}{l} \xrightarrow{-2R_1 + R_2} \\ \xrightarrow{-1R_1 + R_3} \\ \xrightarrow{5R_1 + R_4} \end{array}$$

$$\begin{array}{c} \\ x_3 \\ s_2 \\ s_3 \\ \hline F \end{array} \left[\begin{array}{cccccc|c} x_1 & x_2 & x_3 & s_1 & s_2 & s_3 & \text{sol} \\ 1/7 & 5/7 & 1 & 1/14 & 0 & 0 & 5 \\ 5/7 & -10/7 & 0 & -1/7 & 1 & 0 & 10 \\ -1/7 & 2/7 & 0 & -1/14 & 0 & 1 & 5 \\ \hline -9/7 & 18/7 & 0 & 5/14 & 0 & 0 & 25 \end{array} \right].$$

Thus we have a new simplex tableau. Note that x_3 has replaced s_1 in the basis, and hence the first row of the new tableau is labeled x_3. The values of the basic variables are $x_3 = 5$, $s_2 = 10$, and $s_3 = 5$. Since x_1, x_2, and s_1 are nonbasic, each of them has value 0. The value of the objective function F at this basis is 25.

The procedure used above may be used to transform any simplex tableau into a new one. Let us summarize it.

Obtaining a New Simplex Tableau

To obtain a new simplex tableau from the old one, proceed as follows:

1. First determine the entering and departing variables.

2. Then multiply the row labeled by the departing variable by a number that will produce a 1 in the column of the entering variable.

3. Use this 1 and Type III row operations to produce zeros in the rest of the entering variable's column.

Now that we have obtained a new simplex tableau for the Acme Company problem, we repeat the entire procedure with the new tableau as our starting point. First, we determine the entering and departing variables:

	x_1	x_2	x_3	s_1	s_2	s_3	sol	quotient
x_3	$1/7$	$5/7$	1	$1/14$	0	0	5	$5 \div 1/7 = 35$
s_2	$5/7$	$-10/7$	0	$-1/7$	1	0	10	$10 \div 5/7 = 14 \longrightarrow$
s_3	$-1/7$	$2/7$	0	$-1/14$	0	1	5	
F	$-9/7$	$18/7$	0	$5/14$	0	0	25	

↑

The negative entry in the F-row that has the largest absolute value (indeed, the only negative entry in the F-row) is $-9/7$, which is in the x_1-column. Hence x_1 is the entering variable. The smallest nonnegative quotient of the form

$$\frac{\text{Solution column entry}}{x_1\text{-column entry}}$$

is 14, which is in the s_2-row. (Note that since $-1/7$ is not positive, we did not form the quotient $5 \div (-1/7)$). Hence, s_2 is the departing variable. Now we use row operations to transform the tableau into one in which the x_1-column will have the following form:

$$\begin{bmatrix} 0 \\ 1 \\ 0 \\ 0 \end{bmatrix}.$$

We have

	x_1	x_2	x_3	s_1	s_2	s_3	sol
x_3	$1/7$	$5/7$	1	$1/14$	0	0	5
s_2	⑤/7	$-10/7$	0	$-1/7$	1	0	10
s_3	$-1/7$	$2/7$	0	$-1/14$	0	1	5
F	$-9/7$	$18/7$	0	$5/14$	0	0	25

$\xrightarrow{(7/5)R_2}$

	x_1	x_2	x_3	s_1	s_2	s_3	sol
	$1/7$	$5/7$	1	$1/14$	0	0	5
	①	-2	0	$-1/5$	$7/5$	0	14
	$-1/7$	$2/7$	0	$-1/14$	0	1	5
	$-9/7$	$18/7$	0	$5/14$	0	0	25

$\xrightarrow{(-1/7)R_2 + R_1}$
$\xrightarrow{(1/7)R_2 + R_3}$
$\xrightarrow{(2/7)R_2 + R_4}$

	x_1	x_2	x_3	s_1	s_2	s_3	sol
x_3	0	0	1	$1/10$	$-1/5$	0	3
x_1	1	-2	0	$-1/5$	$1/5$	0	14
s_3	0	0	0	$-1/10$	$1/5$	1	7
F	0	0	0	$1/10$	$9/5$	0	43

Note that every entry in the F-row is nonnegative. This means that further change of the basis will not increase the value of F. Hence we have obtained the maximum

LINEAR PROGRAMMING

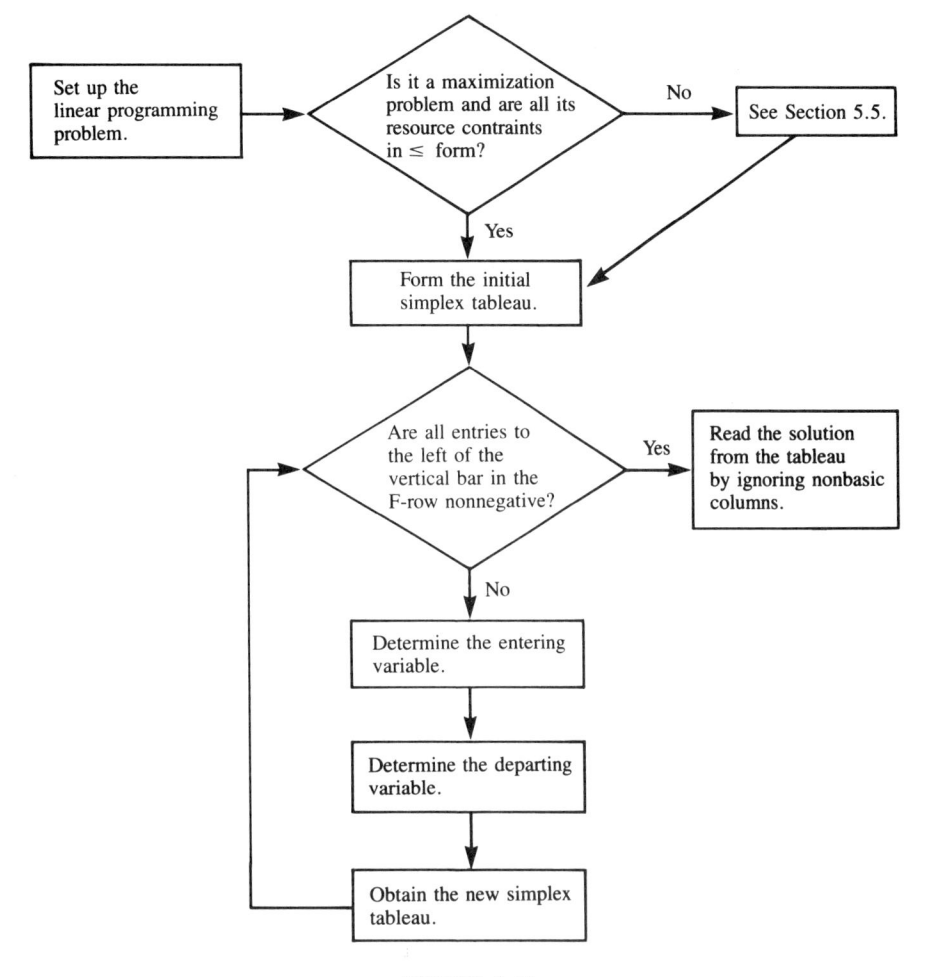

FIGURE 5–16
Simplex Method Flow Chart

value of F and thus have solved the problem. The solution is $x_1 = 14$, $x_2 = 0$ (because x_2 is nonbasic), $x_3 = 3$, with the maximum value of F equal to 43. Therefore Acme should produce 14 units of product A, 0 units of product B, and 3 units of product C, for a maximum profit of $43.

The simplex method is summarized in flowchart form in Figure 5–16. The method as we have developed it so far may be applied to all linear programming problems that are maximization problems and in which all resource constraints (those that are not of the form $x_i \geq 0$) are of the less-than-or-equal-to variety. In the next section we will show how the method may be generalized to apply to minimization problems and those that have equal-to or greater-than-or-equal-to resource constraints.

EXAMPLE 2 Let us use the simplex method to solve the following problem:

$$\text{Maximize} \quad F = 2x_1 + 3x_2 - x_3$$
$$\text{subject to} \quad x_1 + 2x_2 + x_3 \leq 10$$
$$x_1 - x_2 \leq 6$$
$$3x_2 + 2x_3 \leq 12$$
$$x_1 \geq 0, x_2 \geq 0, x_3 \geq 0.$$

1. The problem is given as a maximization problem all of whose resource constraints are of the less-than-or-equal-to variety. We proceed to form the initial simplex tableau, then determine the entering and departing variables:

	x_1	x_2	x_3	s_1	s_2	s_3	sol	quotient
s_1	1	2	1	1	0	0	10	$10/2 = 5$
s_2	1	-1	0	0	1	0	6	
s_3	0	3	2	0	0	1	12	$12/3 = 4 \longrightarrow s_3$ departs
F	-2	-3	1	0	0	0	0	

x_2 enters

2. Multiplying through the s_3-row by $\frac{1}{3}$ and then clearing the x_2-column gives the new tableau. (Check the computations.)

	x_1	x_2	x_3	s_1	s_2	s_3	sol	quotient
s_1	1	0	$-\frac{1}{3}$	1	0	$-\frac{2}{3}$	2	$2/1 = 2 \longrightarrow s_1$ departs
s_2	1	0	$\frac{2}{3}$	0	1	$\frac{1}{3}$	10	$10/1 = 10$
x_2	0	1	$\frac{2}{3}$	0	0	$\frac{1}{3}$	4	
F	-2	0	3	0	0	1	12	

x_1 enters

Note that we cannot form the third quotient because doing so would involve division by 0.

3. The s_1-row already has a 1 in the x_1-column, so we do not need to multiply through the row to produce it. We next clear the x_1-column to obtain the new tableau. (Check the computations.)

	x_1	x_2	x_3	s_1	s_2	s_3	sol	quotient
x_1	1	0	$-\frac{1}{3}$	1	0	$-\frac{2}{3}$	2	
s_2	0	0	1	-1	1	1	8	$8 \div 1 = 8 \longrightarrow s_2$ departs
x_2	0	1	$\frac{2}{3}$	0	0	$\frac{1}{3}$	4	$4 \div \frac{1}{3} = 12$
F	0	0	$\frac{7}{3}$	2	0	$-\frac{1}{3}$	16	

s_3 enters

4. Finally, clearing the s_3-column results in

$$\begin{array}{c} x_1 \\ s_3 \\ x_2 \\ F \end{array} \begin{array}{c} \begin{array}{cccccc} x_1 & x_2 & x_3 & s_1 & s_2 & s_3 \end{array} \quad \text{sol} \\ \left[\begin{array}{cccccc|c} 1 & 0 & 1/3 & 1/3 & 2/3 & 0 & 22/3 \\ 0 & 0 & 1 & -1 & 1 & 1 & 8 \\ 0 & 1 & 1/3 & 1/3 & -1/3 & 0 & 4/3 \\ \hline 0 & 0 & 8/3 & 5/3 & 1/3 & 0 & 56/3 \end{array} \right]. \end{array}$$

5. Since none of the entries in the F-row is negative, we are through. The solution is $x_1 = 22/3$, $x_2 = 4/3$, and $x_3 = 0$; the maximum value of F is $56/3$. ◻

■ **Computer Manual: Program SIMPLEX**

5.4 EXERCISES

The Initial Tableau

In Exercises 1 through 6, form the initial simplex tableau.

1. Maximize $F = 2x_1 + 5x_2$
   subject to
   $$4x_1 + 5x_2 \leq 20$$
   $$3x_1 + x_2 \leq 10$$
   $$x_1 \geq 0, x_2 \geq 0$$

2. Maximize $F = 3x_1 + x_2$
   subject to
   $$5x_1 + x_2 \leq 30$$
   $$4x_1 + 3x_2 \leq 24$$
   $$x_1 \geq 0, x_2 \geq 0$$

3. Maximize $F = 3x_1 + x_2 + 4x_3$
   subject to
   $$5x_1 + x_2 + x_3 \leq 100$$
   $$x_1 \geq 0, x_2 \geq 0, x_3 \geq 0$$

4. Maximize $F = 2x_1 + 3x_2 + 7x_3$
   subject to
   $$x_1 + x_2 + x_3 \leq 40$$
   $$3x_1 + x_2 \leq 20$$
   $$2x_2 + 3x_3 \leq 30$$
   $$x_1 \geq 0, x_2 \geq 0, x_3 \geq 0$$

5. Maximize $F = 2x_1 + x_2 + 3x_3 - x_4$
   subject to
   $$3x_2 + x_3 + 2x_4 \leq 8$$
   $$5x_1 + x_2 \leq 10$$
   $$x_1 \geq 0, x_2 \geq 0, x_3 \geq 0, x_4 \geq 0$$

6. Maximize $F = -x_1 + 3x_2 + 15x_3 + 5x_4$
   subject to
   $$x_1 + x_2 + x_3 + x_4 \leq 50$$
   $$2x_1 - x_3 \leq 30$$
   $$5x_1 + x_2 + 3x_4 \leq 80$$
   $$x_2 + 3x_3 \leq 20$$
   $$x_4 \leq 10$$
   $$x_1 \leq 10$$
   $$x_1 \geq 0, x_2 \geq 0, x_3 \geq 0, x_4 \geq 0$$

Exercises 7 through 10 present initial simplex tableaux for linear programming problems in which the objective function is to be maximized, the resource constraints are all of the less-than-or-equal-to variety, and all variables are to be nonnegative. In each case, reconstruct the linear programming problem from the tableau.

7. $$\begin{array}{c} s_1 \\ s_2 \\ F \end{array} \begin{array}{c} \begin{array}{cccc} x_1 & x_2 & s_1 & s_2 \end{array} \quad \text{sol} \\ \left[\begin{array}{cccc|c} 2 & 1 & 1 & 0 & 10 \\ 3 & 0 & 0 & 1 & 5 \\ \hline -2 & -5 & 0 & 0 & 0 \end{array} \right] \end{array}$$

8. $$\begin{array}{c} s_1 \\ s_2 \\ s_3 \\ F \end{array} \begin{array}{c} \begin{array}{ccccc} x_1 & x_2 & s_1 & s_2 & s_3 \end{array} \quad \text{sol} \\ \left[\begin{array}{ccccc|c} 5 & 1 & 1 & 0 & 0 & 8 \\ 3 & 2 & 0 & 1 & 0 & 12 \\ 2 & -1 & 0 & 0 & 1 & 16 \\ \hline -3 & -1 & 0 & 0 & 0 & 0 \end{array} \right] \end{array}$$

9.

$$\begin{array}{c} \\ s_1 \\ s_2 \\ s_3 \\ F \end{array} \begin{bmatrix} x_1 & x_2 & x_3 & s_1 & s_2 & s_3 & \text{sol} \\ 1 & 1 & -1 & 1 & 0 & 0 & 20 \\ 2 & 0 & 3 & 0 & 1 & 0 & 30 \\ 0 & 5 & 0 & 0 & 0 & 1 & 40 \\ -3 & 2 & -4 & 0 & 0 & 0 & 0 \end{bmatrix}$$

10.

$$\begin{array}{c} \\ s_1 \\ s_2 \\ F \end{array} \begin{bmatrix} x_1 & x_2 & x_3 & x_4 & s_1 & s_2 & \text{sol} \\ 2 & 0 & 0 & 6 & 1 & 0 & 15 \\ 3 & 2 & 1 & 0 & 0 & 1 & 15 \\ -2 & -8 & -5 & -1 & 0 & 0 & 0 \end{bmatrix}$$

Transforming the Simplex Tableau

In Exercises 11 through 20, find the entering and departing variable for the tableau of the given exercise.

11. Exercise 1 **12.** Exercise 2 **13.** Exercise 3 **14.** Exercise 4
15. Exercise 5 **16.** Exercise 6 **17.** Exercise 7 **18.** Exercise 8
19. Exercise 9 **20.** Exercise 10

In Exercises 21 through 28, solve the maximization problem using the simplex method.

21. Maximize $F = 2x_1 + x_2$
subject to $x_1 + 2x_2 \leq 16$
$x_1 + x_2 \leq 10$
$x_1 \geq 0, x_2 \geq 0$

22. Maximize $F = 2x_1 + 3x_2$
subject to $x_1 - 2x_2 \leq 16$
$x_1 - x_2 \leq 10$
$x_1 \geq 0, x_2 \geq 0$

23. Maximize $F = 3x_1 + 4x_2$
subject to $x_1 + 2x_2 \leq 16$
$x_1 + x_2 \leq 10$
$x_1 \geq 0, x_2 \geq 0$

24. Maximize $F = 3x_1 + x_2 + 2x_3$
subject to $2x_1 + 6x_2 + 4x_3 \leq 24$
$x_1 + 5x_3 \leq 20$
$x_1 \geq 0, x_2 \geq 0, x_3 \geq 0$

25. Maximize $F = 2x_1 + 4x_2 - x_3$
subject to $x_1 + 2x_2 + x_3 \leq 20$
$2x_1 + x_2 + 3x_3 \leq 30$
$3x_1 + 2x_2 + 2x_3 \leq 15$
$x_1 \geq 0, x_2 \geq 0, x_3 \geq 0$

26. Maximize $F = 3x_1 + 5x_2 + 4x_3$
subject to $2x_1 + 2x_2 + 5x_3 \leq 50$
$3x_1 + 2x_2 + x_3 \leq 12$
$2x_1 + x_2 + 4x_3 \leq 18$
$x_1 \geq 0, x_2 \geq 0, x_3 \geq 0$

27. Maximize $F = 8x_1 + 5x_2 + 6x_3$
subject to $2x_1 + x_2 + 2x_3 \leq 8$
$2x_1 + x_2 + 9x_3 \leq 10$
$3x_1 + 2x_2 + x_3 \leq 15$
$x_1 \geq 0, x_2 \geq 0, x_3 \geq 0$

28. Maximize $F = 3x_1 + 10x_2 + 7x_3 + 2x_4$
subject to $2x_1 + 6x_2 + 3x_3 + x_4 \leq 120$
$5x_1 + 4x_4 \leq 22$
$3x_1 + x_2 + x_4 \leq 15$
$2x_2 + 2x_3 + 5x_4 \leq 45$
$x_1 \geq 0, x_2 \geq 0, x_3 \geq 0, x_4 \geq 0$

Management

29. Limex Company makes both a small and a large alarm clock. Each small clock takes 2 hours to assemble and each large one takes 3 hours to assemble. Each clock, of either model, has one clock face, which Limex purchases from a supplier. The firm makes a profit of $4 on each small clock and $5 on each large one. On a given day, Limex has 120 assembly-hours and 50 clock faces available. Use the simplex method to find the number of each model of clock Limex should produce in order to maximize profit.

30. A plant that makes refrigerators must plan its daily production. The plant assembles, tests, and paints both 9-ft^3 and 12-ft^3 refrigerators. The worker-hours required to carry out these operations are as follows:

	Worker-hours		
	For Assembly	For Testing	For Painting
9 ft^3	6	1.25	1.0
12 ft^3	5	1.25	1.4

Each day the plant has 480 worker-hours available for assembly, 120 for testing, and 112 for painting. Each 9-ft^3 refrigerator contributes $70 to profit and each 12-ft^3 one contributes $85 to profit. Use the simplex method to find out how the plant should schedule its daily production.

*31. A sales manager must hire new salespeople and then assign them to three new territories, A, B, and C. Territory A has a population of 8 million, territory B a population of 7 million, and territory C a population of 5 million. Travel expenses for salespeople are $900 per week in territory A, $700 per week in territory B, and $600 per week in territory C. If sales in a territory are proportional to the population of the territory, if the sales manager's travel budget for the three territories will be $7000 per week, and if she has been authorized to hire no more than 10 new salespeople, how many should she hire and how many should she assign to each territory?

32. Wheeler Corporation makes wheelbarrows in three sizes: small, medium, and large. Each wheelbarrow has one rubber wheel. Each small wheelbarrow takes 2 hours to assemble, each medium wheelbarrow takes 3 hours to assemble, and each large wheelbarrow takes 4 hours to assemble. Wheeler makes a profit of $3 on each small wheelbarrow, $5 on each medium one, and $7 on each large one. On a given day Wheeler has 200 rubber wheels in stock and 660 assembly hours available. Maximize Wheeler's profit.

33. A local radio station has offered Mike's Men's Shop a package deal on commercials at a very attractive price. The station will sell the store a package of 15-second, 30-second, and 45-second commercials subject to the following conditions:

- The total number of commercials cannot exceed 30.
- The number of 45-second commercials cannot exceed the number of 30-second commercials by more than 2.
- The number of 15-second commercials cannot exceed the number of 30-second commercials by more than 4.

The manager must select the number of each type of commercial he wants, subject to these conditions. He has estimated that each 45-second commercial has an effectiveness rating of 2, each 30-second commercial an effectiveness rating of 1, and each 15-second commercial an effectiveness rating of $1/2$. How many of each type of commercial should he buy in order to maximize their total effectiveness?

34. A bakery makes three types of bread: raisin bread (R), cinnamon bread (C), and cinnamon-raisin bread (CR). The bakery makes a profit of $0.40 on each loaf of raisin bread, and a profit of $0.30 on each loaf of cinnamon bread. However, because of the extra time required to produce the cinnamon-raisin bread, it loses $0.20 on each loaf. Each loaf of bread has certain requirements as to its ingredi-

ents and the bakery has only a certain amount of each ingredient on hand, as the following table shows. Maximize the bakery's profit.

	Loaf R	Loaf C	Loaf CR	On Hand
Salt (oz)	1	2	3	400
Raisins (oz)	3		2	120
Cinnamon (oz)		3	1	150

35. A gasoline company blends high and low octane gasoline into three intermediate grades: regular, premium, and super premium. The regular grade consists of 40% high octane and 60% low octane, the premium consists of 50% of each, and the super premium consists of 70% high octane and 30% low octane. The company has 120,000 gallons of high octane and 112,000 gallons of low octane gas on hand. Regular gas sells for $1.10 per gallon, premium for $1.22 per gallon, and super premium for $1.34 per gallon. How many gallons of each grade should the company produce in order to maximize its revenue?

36. The owner of a garage is trying to schedule tomorrow's work. The garage does tune-ups, wheel-balancing, lubrications, and oil changes. The owner does all tune-ups and all wheel-balancing himself; his assistant does all lubrications and all oil changes. It takes 3 hours to do a tune-up, 2 hours to do a wheel-balancing, 1 hour to do a lubrication, and $\frac{1}{2}$ hour to do an oil change. Both the owner and his assistant work an 8-hour day. The owner knows from experience that they should not attempt to handle more than 10 jobs per day. If he charges $30 for a tune-up, $20 for a wheel-balancing, $10 for a lubrication, and $8 for an oil change, how many of each type of job should he schedule in order to maximize tomorrow's revenue?

37. Solve Exercise 16 of Exercises 5.2.
38. Solve Exercise 20 of Exercises 5.2.

Life Science

39. A patient is to receive three painkillers, X, Y, and Z, singly or in combination. The relative effectiveness of the painkillers is 1 for each unit of X, 2 for each unit of Y, and 2.5 for each unit of Z. Because of possible side effects, the patient cannot receive more than 50 units of X and Y combined, more than 40 units of X and Z combined, or more than 80 units of Y and Z combined. How many units of each should the patient be given in order to maximize the painkilling effect?

40. A chemist is attempting to blend four ingredients to make a new drug. The ingredients A, B, C, and D have potency ratings of 1, 2, $\frac{1}{4}$, and $\frac{1}{2}$, respectively. Because of possible side effects, the chemist knows that she cannot use more than 2 units of B, more than 4 units of A and B combined, more than 6 units of A, C, and D combined, or more than 5 units of C and D combined. How many units of each ingredient must she blend together in order to maximize the potency of the drug?

Social Science

41. A psychologist is attempting to put together a group of people who will be efficient at problem solving. He has divided the candidates into three classes: extremely

competitive persons, normally competitive persons, and extremely noncompetitive persons. He believes that too many extremely competitive persons would be bad for the harmony of the group, as would too many extremely noncompetitive persons, so he has set a limit of three extremely competitive people and two extremely noncompetitive people for the group. The group will consist of no more than eight people. The psychologist estimates that each extremely competitive person will contribute one and one-half times as much to the group's ability as each normally competitive person will and each extremely noncompetitive person will contribute three-fourths as much as each normally competitive person will. How many of each type must be in the group in order to maximize its problem-solving ability?

42. The state's Department of Health and Human Services has three projects it can undertake: a rent subsidy program, an income supplement program, and a child care program:

Project	Cost	Benefit-to-Cost Ratio
Rent subsidy	$200	0.70
Income supplement	$250	0.75
Child care	$300	0.80

(Cost figures in millions of dollars.) Suppose HHS's total budget is $600 million. It can use part of this amount to carry out some proportion x_1 of the rent subsidy program, part to carry out some proportion x_2 of the income supplement program, and part to carry out some proportion x_3 of the child care program. The total benefit obtained by doing this will be $0.70x_1 + 0.75x_2 + 0.80x_3$. Find the values of x_1, x_2, and x_3 that maximize the total benefit.

5.5 THE SIMPLEX METHOD: $\geq$, = CONSTRAINTS AND MINIMIZATION

The simplex method as we have developed it thus far can only be applied to maximization problems in which the resource constraints are of the less-than-or-equal-to variety. In this section we generalize the method so that it can be used to solve problems having greater-than-or-equal-to or equal-to resource constraints, as well as those that require the minimization of the objective function.

Greater-than-or-Equal-to and Equal-to Constraints

Suppose we wish to solve the following linear programming problem:

$$\text{Maximize} \quad F = 2x_1 + 3x_2 + x_3$$
$$\text{subject to} \quad 2x_1 + x_2 + x_3 \leq 70$$
$$x_1 + x_2 + x_3 \geq 5$$
$$x_1 \geq 0, x_2 \geq 0, x_3 \geq 0.$$

The first constraint may be converted to an equality by adding a slack variable s_1 to its left-hand side. The second constraint, however, can lead to a condition of surplus. For example, if we let

$$x_1 = x_2 = x_3 = 2$$

in the second constraint, we have a surplus of 1 on the left-hand side, and if we wish to rewrite the constraint as an equality, we must subtract this surplus. It follows that in order to convert the second constraint to an equality, we must subtract a nonnegative **surplus variable** s_2 from its left-hand side. If we do this, we may rewrite the problem in the following equivalent form:

$$\text{Maximize} \quad F = 2x_1 + 3x_2 + x_3 + 0s_1 + 0s_2$$

$$\text{subject to} \quad 2x_1 + x_2 + x_3 + s_1 = 10$$

$$\phantom{\text{subject to} \quad} x_1 + x_2 + x_3 - s_2 = 5$$

$$x_1 \geq 0, x_2 \geq 0, x_3 \geq 0, s_1 \geq 0, s_2 \geq 0.$$

Now suppose that we attempt to find an initial basis by setting

$$x_1 = x_2 = x_3 = 0$$

and solving for s_1 and s_2, thus obtaining $s_1 = 10$ and $s_2 = -5$. But $s_2 = -5$ contradicts the constraint $s_2 \geq 0$, so this cannot be done. In order to find an initial basis, we must add a nonnegative **artificial variable** A_2 to the left-hand side of the second constraint, obtaining

$$x_1 + x_2 + x_3 + A_2 - s_2 = 5.$$

Now we can find an initial basis by setting

$$x_1 = x_2 = x_3 = s_2 = 0$$

and solving for s_1 and A_2 to obtain $s_1 = 10$ and $A_2 = 5$.

As noted, the artificial variable A_2 is needed to find an initial basis, and hence it is an initial basic variable. Ultimately, however, the artificial variable A_2 must become nonbasic (and therefore have value 0), for if it does not, the equality

$$x_1 + x_2 + x_3 - s_2 = 5$$

cannot be satisfied. In order to ensure that A_2 will eventually leave the basis, we assign it a coefficient in the objective function of $-K$, where K is a very large positive number. Thus, we rewrite the problem in the following form:

$$\text{Maximize} \quad F = 2x_1 + 3x_2 + x_3 + 0s_1 - KA_2 + 0s_2$$

$$\text{subject to} \quad 2x_1 + x_2 + x_3 + s_1 = 10$$

$$\phantom{\text{subject to} \quad} x_1 + x_2 + x_3 + A_2 - s_2 = 5$$

$$x_1 \geq 0, x_2 \geq 0, x_3 \geq 0, s_1 \geq 0, s_2 \geq 0, A_2 \geq 0, K > 0, K \text{ large}.$$

Clearly, if F is to be maximized, the term $-KA_2$ must disappear from the expression for F. As the simplex method proceeds to maximize F, it will cause this to happen by forcing A_2 out of the basis, thus assigning it a value of 0.

LINEAR PROGRAMMING

Now we are ready to form the **pretableau** for our problem:

$$\begin{array}{c} \\ s_1 \\ A_2 \\ F \end{array} \begin{array}{c} x_1 \quad x_2 \quad x_3 \quad s_1 \quad A_2 \quad s_2 \quad \text{sol} \\ \left[\begin{array}{cccccc|c} 2 & 1 & 1 & 1 & 0 & 0 & 10 \\ 1 & 1 & 1 & 0 & 1 & -1 & 5 \\ \hline -2 & -3 & -1 & 0 & K & 0 & 0 \end{array}\right] \end{array}.$$

This pretableau is *not* the initital simplex tableau; the entry in the lower right-hand corner of the matrix is not the value of F at the initial basis $s_1 = 10, A_2 = 5$. To convert the pretableau into the initial simplex tableau, we perform Type III row operations on it in such a way as to replace each K that presently appears in the F-row with 0. Thus, if we multiply the A_2-row by $-K$ and add it to the F-row, we obtain the initial simplex tableau

$$\begin{array}{c} \\ s_1 \\ A_2 \\ F \end{array} \begin{array}{c} x_1 \quad\quad x_2 \quad\quad x_3 \quad s_1 \quad A_2 \quad s_2 \quad \text{sol} \\ \left[\begin{array}{cccccc|c} 2 & 1 & 1 & 1 & 0 & 0 & 10 \\ 1 & 1 & 1 & 0 & 1 & -1 & 5 \\ \hline -2-K & -3-K & -1-K & 0 & 0 & K & -5K \end{array}\right] \end{array}.$$

Now we apply the simplex method in the usual fashion. Since K is large and positive, the negative entry to the left of the vertical bar in the F-row that has the largest absolute value is $-3 - K$; hence x_2 enters the basis and A_2 leaves the basis:

$$\begin{array}{c} \\ s_1 \\ A_2 \\ F \end{array} \begin{array}{c} x_1 \quad\quad x_2 \quad\quad x_3 \quad s_1 \quad A_2 \quad s_2 \quad \text{sol} \quad \text{quotient} \\ \left[\begin{array}{cccccc|c} 2 & 1 & 1 & 1 & 0 & 0 & 10 \\ 1 & 1 & 1 & 0 & 1 & -1 & 5 \\ \hline -2-K & -3-K & -1-K & 0 & 0 & K & -5K \end{array}\right] \begin{array}{l} 10/1 = 10 \\ 5/1 = 5 \longrightarrow \end{array} \\ \quad\quad\quad\quad\quad \uparrow \end{array}.$$

Next we obtain the new simplex tableau:

$$\left[\begin{array}{cccccc|c} 2 & 1 & 1 & 1 & 0 & 0 & 10 \\ 1 & \circled{1} & 1 & 0 & 1 & -1 & 5 \\ \hline -2-K & -3-K & -1-K & 0 & 0 & K & -5K \end{array}\right] \begin{array}{l} \xrightarrow{-R_2 + R_1} \\ \\ \xrightarrow{(3+K)R_2 + R_3} \end{array}$$

$$\begin{array}{c} \\ s_1 \\ x_2 \\ F \end{array} \begin{array}{c} x_1 \quad x_2 \quad x_3 \quad s_1 \quad A_2 \quad s_2 \quad \text{sol} \\ \left[\begin{array}{cccccc|c} 1 & 0 & 0 & 1 & -1 & 1 & 5 \\ 1 & 1 & 1 & 0 & 1 & -1 & 5 \\ \hline 1 & 0 & 2 & 0 & 3+K & -3 & 15 \end{array}\right] \end{array}.$$

Now s_2 enters the basis and s_1 leaves the basis, so multiplying the s_1-row by 3 and adding it to the F-row, then adding the s_1-row to the x_2-row, gives the third simplex tableau:

	x_1	x_2	x_3	s_1	A_2	s_2	sol
s_1	1	0	0	1	-1	1	5
x_2	2	1	1	1	0	0	10
F	4	0	2	3	K	0	30

Therefore, the solution is $x_1 = 0$ and $x_2 = 10$, with the maximum value of F equal to 30.

Equal-to-constraints may be handled in a manner similar to that just discussed for greater-than-or-equal-to constraints. Of course, it is not necessary to subtract a surplus variable from an equal-to constraint, but it is necessary to add to it an artificial variable in order to find an initial basis.

EXAMPLE 1   Let us use the simplex method to solve the following problem:

$$\text{Maximize} \quad F = 8x_1 + 5x_2$$
$$\text{subject to} \quad x_1 + x_2 \leq 20$$
$$2x_1 + 3x_2 \geq 12$$
$$2x_1 - x_2 = 4$$
$$x_1 \geq 0, \, x_2 \geq 0.$$

We add a slack variable s_1 to the first constraint, add an artificial variable A_2 to and subtract a surplus variable s_2 from the second constraint, and add an artificial variable A_3 to the third constraint. We add the new variables to the objective function, assigning the slack and surplus variables coefficients of 0 and the artificial variables coefficients of $-K$, where K is large and positive. Thus, the problem becomes the following:

$$\text{Maximize} \quad F = 8x_1 + 5x_2 + 0s_1 - KA_2 + 0s_2 - KA_3$$
$$\text{subject to} \quad x_1 + x_2 + s_1 = 20$$
$$2x_1 + 3x_2 + A_2 - s_2 = 12$$
$$2x_1 - x_2 + A_3 = 4$$
$$x_1 \geq 0, \, x_2 \geq 0, \, s_1 \geq 0, \, A_2 \geq 0, \, s_2 \geq 0, \, A_3 \geq 0, \, K > 0, \, K \text{ large}.$$

Now we form the pretableau:

	x_1	x_2	s_1	A_2	s_2	A_3	sol
s_1	1	1	1	0	0	0	20
A_2	2	3	0	1	-1	0	12
A_3	2	-1	0	0	0	1	4
F	-8	-5	0	K	0	K	0

LINEAR PROGRAMMING

To convert the pretableau into the initial simplex tableau, we use Type III row operations on the pretableau to replace the K's in the A_2- and A_3-columns with zeros. Thus, multiplying the A_2-row by $-K$ and adding the result to the F-row, then multiplying the A_3-row by $-K$ and adding to the F-row, we obtain the initial simplex tableau:

$$\begin{array}{c} \\ s_1 \\ A_2 \\ A_3 \\ F \end{array} \begin{array}{c} x_1 \quad\; x_2 \quad s_1 \;\; A_2 \;\; s_2 \;\; A_3 \quad \text{sol} \\ \left[\begin{array}{cccccc|c} 1 & 1 & 1 & 0 & 0 & 0 & 20 \\ 2 & 3 & 0 & 1 & -1 & 0 & 12 \\ 2 & -1 & 0 & 0 & 0 & 1 & 4 \\ \hline -8-4K & -5-2K & 0 & 0 & K & 0 & -16K \end{array}\right]. \end{array}$$

(Check this.) Now we see that x_1 enters the basis and A_3 leaves the basis. (Why?) Performing the appropriate row operations on the tableau yields the new tableau:

$$\begin{array}{c} \\ s_1 \\ A_2 \\ x_1 \\ F \end{array} \begin{array}{c} x_1 \;\; x_2 \;\; s_1 \;\; A_2 \;\; s_2 \quad A_3 \quad \text{sol} \\ \left[\begin{array}{cccccc|c} 0 & 3/2 & 1 & 0 & 0 & -1/2 & 18 \\ 0 & 4 & 0 & 1 & -1 & -1 & 8 \\ 1 & -1/2 & 0 & 0 & 0 & 1/2 & 2 \\ \hline 0 & -9-4K & 0 & 0 & K & 4+2K & 16-8K \end{array}\right]. \end{array}$$

Now x_2 enters and A_2 leaves the basis, and this leads in turn to the third tableau:

$$\begin{array}{c} \\ s_1 \\ x_2 \\ x_1 \\ F \end{array} \begin{array}{c} x_1 \; x_2 \; s_1 \quad A_2 \quad s_2 \quad A_3 \quad \text{sol} \\ \left[\begin{array}{cccccc|c} 0 & 0 & 1 & -3/8 & 3/8 & -1/8 & 15 \\ 0 & 1 & 0 & 1/4 & -1/4 & -1/4 & 2 \\ 1 & 0 & 0 & 1/8 & -1/8 & 3/8 & 3 \\ \hline 0 & 0 & 0 & 9/4+K & -9/4 & 7/4+K & 34 \end{array}\right]. \end{array}$$

Now s_2 enters and s_1 leaves the basis, and we obtain the fourth tableau:

$$\begin{array}{c} \\ s_2 \\ x_2 \\ x_1 \\ F \end{array} \begin{array}{c} x_1 \; x_2 \;\; s_1 \;\; A_2 \; s_2 \quad A_3 \quad \text{sol} \\ \left[\begin{array}{cccccc|c} 0 & 0 & 8/3 & -1 & 1 & -1/3 & 40 \\ 0 & 1 & 2/3 & 0 & 0 & -1/3 & 12 \\ 1 & 0 & 1/3 & 0 & 0 & 1/3 & 8 \\ \hline 0 & 0 & 6 & K & 0 & 1+K & 124 \end{array}\right]. \end{array}$$

(Check these calculations.) Hence, the solution is $x_1 = 8$ and $x_2 = 12$, and the maximum value of F is 124. $\square$

Let us summarize our results.

> **Greater-than-or-Equal-to and Equal-to Constraints**
>
> To form the initial simplex tableau for a linear programming problem with greater-than-or-equal-to or equal-to resource constraints, complete the following steps.
>
> 1. a. Add a slack variable to the left-hand side of each $\leq$ constraint.
>   b. Add an artificial variable to and subtract a surplus variable from the left-hand side of each $\geq$ constraint.
>   c. Add an artificial variable to the left-hand side of each $=$ constraint. Each slack variable and each artificial variable will appear in the initial basis; surplus variables do not appear in the initial basis.
> 2. Add the slack, surplus, and artificial variables to the objective function, assigning slack and surplus variables coefficients of 0 and artificial variables coefficients of $-K$, where $K > 0$, K large.
> 3. Use the results of steps 1 and 2 to form the pretableau.
> 4. Obtain the initial simplex tableau by performing Type III row operations on the pretableau in such a way as to replace the K's that appear in the columns of the artificial variables with zeros.

After the initial simplex tableau has been obtained, the simplex method continues exactly as outlined in Section 5.4.

Minimization

Finally, in order to solve a linear programming problem that requires the minimization of the objective function F, we simply maximize $-1F$: the values of the variables that maximize $-1F$ will minimize F, and the minimum value of F will be -1 times the maximum value of $-1F$. We illustrate with an example.

EXAMPLE 2   Let us use the simplex method to solve the following linear programming problem:

$$\text{Minimize} \quad F = 2x_1 + 3x_2$$
$$\text{subject to} \quad x_1 + x_2 \geq 6$$
$$2x_1 - x_2 = 3$$
$$x_1 \geq 0, x_2 \geq 0.$$

We first recast the problem as a maximization problem for $-1F$:

$$\text{Maximize} \quad -1F = -2x_1 - 3x_2$$
$$\text{subject to} \quad x_1 + x_2 \geq 6$$
$$2x_1 - x_2 = 3$$
$$x_1 \geq 0, x_2 \geq 0.$$

LINEAR PROGRAMMING

Next we introduce the necessary artificial and surplus variables:

$$\text{Maximize} \quad -1F = -2x_1 - 3x_2 - KA_1 + 0s_1 - KA_2$$

$$\text{subject to} \quad x_1 + x_2 + A_1 - s_1 = 6$$

$$2x_1 - x_2 + A_2 = 3$$

$$x_1 \geq 0, x_2 \geq 0, A_1 \geq 0, s_1 \geq 0, A_2 \geq 0, K > 0, K \text{ large}.$$

We form the pretableau:

$$\begin{array}{c} \\ A_1 \\ A_2 \\ -1F \end{array} \begin{array}{c} x_1 \quad x_2 \quad A_1 \quad s_1 \quad A_2 \quad \text{sol} \\ \left[\begin{array}{ccccc|c} 1 & 1 & 1 & -1 & 0 & 6 \\ 2 & -1 & 0 & 0 & 1 & 3 \\ 2 & 3 & K & 0 & K & 0 \end{array} \right] \end{array}.$$

Using Type III row operations to replace the K's in the A_1-column and A_2-column with zeros, we obtain the initial simplex tableau:

$$\begin{array}{c} \\ A_1 \\ A_2 \\ -1F \end{array} \begin{array}{c} x_1 \quad x_2 \quad A_1 \quad s_1 \quad A_2 \quad \text{sol} \\ \left[\begin{array}{ccccc|c} 1 & 1 & 1 & -1 & 0 & 6 \\ 2 & -1 & 0 & 0 & 1 & 3 \\ 2-3K & 3 & 0 & K & 0 & -9K \end{array} \right] \end{array}.$$

Thus, x_1 enters the basis and A_2 leaves the basis, and we obtain the second simplex tableau:

$$\begin{array}{c} \\ A_1 \\ x_1 \\ -1F \end{array} \begin{array}{c} x_1 \quad x_2 \quad A_1 \quad s_1 \quad A_2 \quad \text{sol} \\ \left[\begin{array}{ccccc|c} 0 & 3/2 & 1 & -1 & -1/2 & 9/2 \\ 1 & -1/2 & 0 & 0 & 1/2 & 3/2 \\ 0 & 4-3/2 K & 0 & K & -1+3/2 K & -3-9/2 K \end{array} \right] \end{array}.$$

Now x_2 enters and A_1 departs the basis, leading to the third tableau:

$$\begin{array}{c} \\ x_2 \\ x_1 \\ -1F \end{array} \begin{array}{c} x_1 \quad x_2 \quad A_1 \quad s_1 \quad A_2 \quad \text{sol} \\ \left[\begin{array}{ccccc|c} 0 & 1 & 2/3 & -2/3 & -1/3 & 3 \\ 1 & 0 & 1/3 & -1/3 & 1/3 & 3 \\ 0 & 0 & -8/3+K & 8/3 & 1/3+K & -15 \end{array} \right] \end{array}.$$

Since K is large and positive, all entries in the $-1F$-row to the left of the vertical bar are nonnegative, and hence we are through. The solution to the maximization problem is $x_1 = x_2 = 3$, with the maximum value of $-1F$ equal to -15. Therefore the solution to the original minimization problem is $x_1 = x_2 = 3$, with the minimum value of F equal to 15. □

In the next section we will present a method for solving minimization problems that does not require the use of artificial and surplus variables.

■ *Computer Manual: Program SIMPLEX*

5.5 EXERCISES

Greater-than-or-Equal-to and Equal-to Constraints

Solve Exercises 1 through 8 by the simplex method. (All x_i are ≥ 0.)

1. Maximize $F = 3x_1 + x_2$
   subject to
   $x_1 + x_2 \leq 20$
   $x_1 + x_2 \geq 10$

2. Maximize $F = x_1 + 3x_2$
   subject to
   $2x_1 + 4x_2 \geq 12$
   $x_1 + x_2 \leq 8$
   $x_2 \geq 2$

3. Maximize $F = x_1 + 2x_2$
   subject to
   $5x_1 + x_2 \leq 10$
   $2x_1 - x_2 = 1$

4. Maximize $F = x_1 - 3x_2 + 4x_3$
   subject to
   $6x_1 + x_2 - x_3 \leq 12$
   $4x_1 \quad\quad + 2x_3 \leq 8$
   $x_2 + x_3 = 10$

5. Maximize $F = 2x_1 + 4x_2 + 3x_3$
   subject to
   $x_1 \leq 20$
   $x_2 \leq 30$
   $x_3 \leq 40$
   $x_1 + x_2 + x_3 \geq 10$

6. Maximize $F = 5x_1 - 3x_2 + 7x_3$
   subject to
   $4x_1 + 3x_2 + x_3 \leq 50$
   $2x_1 + x_2 + x_3 = 20$
   $x_1 + x_2 + x_3 = 15$

7. Maximize $F = 3x_1 + x_2 - x_3$
   subject to
   $2x_1 + 3x_2 + x_3 \geq 24$
   $x_1 - x_2 + 2x_3 \geq 20$
   $x_1 + x_2 \quad\quad = 8$

8. Maximize $F = x_1 + 2x_2 + 2x_3 + x_4$
   subject to
   $x_1 + x_2 \quad\quad\quad \leq 20$
   $x_1 - x_2 \quad\quad\quad \geq 10$
   $\quad\quad 2x_3 + 6x_4 \leq 80$
   $\quad\quad x_3 - 2x_4 \geq 20$

9. An investor has $40,000 to divide between a bond fund and a money market fund. The money market fund is paying 8% per year, and the bond fund is paying 10% per year. The investor intends to have at least $5000 invested in each fund, but also wants the amount invested in the bond fund to be no more than three times the amount invested in the money market fund. Use the simplex method to find the amount that must be invested in each fund in order to maximize the investor's return.

10. An investor is considering investing in a combination of the three stocks A, B, and C. Stock A pays an annual dividend of 10%, stock B one of 8%, and stock C one of 12%. The investor can invest up to $10,000 in the stocks, but, because stock C is riskier than the other two, she does not want to put any more than $5000 into stock C. Furthermore, she wishes to invest a total of at least $6000 in the two higher-yielding stocks, A and C. Maximize her return on investment.

11. Solve Exercise 4 of Exercises 5.2.

12. Solve Exercise 5 of Exercises 5.2.

*13. You are considering investing in four stocks, and have collected the following information on them:

Stock	Price per Share	Dividend per Share	Risk Coefficient
A	$10	$0.05	2.5
B	$20	$0.25	1.5
C	$30	$0.10	1.0
D	$40	$0.20	2.0

You have $100,000 to invest. How much should you invest in each stock in order to maximize the total dividends you will receive if

- You do not want to invest more than one-half the money in stock D.
- You want to have at least 100 shares of each stock.
- The total weighted risk of the investment must be no more than 1.75. (Weighted risk for a stock is the amount invested in the stock times its risk coefficient, divided by the total invested in all stocks.)

Management

14. A manufacturer of trash compactors makes a profit of $100 on the ½-horsepower (hp) model and $130 on the ¾-hp model. If the manufacturer must make at least 1000 of the ½-hp model and at least 1400 of the ¾-hp model, but does not have the capacity to make more than 4000 units total, how many of each will maximize profit? Use the simplex method.

15. A sales manager must allocate 50 salespeople among three territories, A, B, and C. Travel expenses for salespeople are $1000 per week in territory A, $800 per week in territory B, and $1200 per week in territory C. Salespeople in territory A average $3500 per week in sales, and those in B and C average $3000 and $3800 per week respectively. The sales manager's travel budget for the three territories is $70,000 per week. If no territory is to be left without any salespeople, how should the sales manager assign the salespeople?

16. A gasoline company blends high and low octane gasoline into three intermediate grades: regular, premium, and super premium. The regular grade consists of 40% high octane and 60% low octane, the premium consists of 50% of each, and the super premium consists of 70% high octane and 30% low octane. The company has 120,000 gallons of high octane and 112,000 gallons of low octane gas on hand, and it must produce at least 40,000 gallons of the regular gas and at least 100,000 gallons of the two premium grades combined. Regular gas sells for $1.10 per gallon, premium for $1.22 per gallon, and super premium for $1.34 per gallon. How many gallons of each grade should the company produce in order to maximize its revenue? (See Exercise 35, Exercises 5.4.)

17. A mining company owns three mines: Paydirt, Richstrike, and Goldstar. Each mine yields gold, silver, and copper in the amounts given below:

	Yields per Ton of Ore Mined		
	Gold (oz)	Silver (oz)	Copper (lb)
Paydirt	0.2	30	50
Richstrike	0.3	20	20
Goldstar	0.4	30	25

At the Paydirt mine it takes 1 worker-hour to extract a ton of ore, whereas at both Richstrike and Goldstar it takes 2 worker-hours per ton extracted. The company has 300 worker-hours available. At least 100 tons of ore must be mined at Richstrike so the mine can be shored up. The company has a contract for 5000 lb of copper. Gold sells for $400 per ounce, silver for $10 per ounce, and copper for $5 per pound. Maximize the company's revenue.

18. A company's media budget is $480,000, which is to be spent on a combination of television commercials, radio commercials, newspaper ads, and magazine ads. Each airing of a television commercial costs $24,000, each airing of a radio commercial costs $8000, each appearance of a newspaper ad costs $6000, and each appearance of a magazine ad costs $12,000. The television commercials reach 400,000 viewers, the radio commercials reach 100,000 listeners, the newspaper ads reach 120,000 readers, and the magazine ads reach 200,000 readers. The company has already decided that it wants to run a combined total of at least 10 commercials in the electronic media and a combined total of at least 12 ads in the print media. How should the company allocate its media budget if it wishes to maximize the total audience for its advertising?

19. Solve Exercise 19 of Exercises 5.2.

20. Solve Exercise 21 of Exercises 5.2.

Life Science

21. A patient must take two drugs, the first of which has an effectiveness rating of 1 and the second of which has an effectiveness rating of 2.2. The patient must take at least 10 units of the first and 20 units of the second, but cannot take a total of more than 80 units of the two combined. Use the simplex method to determine how many units of each drug the patient must take for maximum effectiveness.

22. A medical researcher is testing a combination of three drugs. He intends to mix the drugs and give a total of 15 units of the mixture to a patient. Information about the drugs is given below. How many units of each drug should go into the mixture if its effectiveness rating is to be maximized?

	Drugs		
	Number 1	Number 2	Number 3
Maximum dose allowed	10 units	—	—
Minimum effective dose	—	3 units	5 units
Effectiveness rating	8	4	5

Social Science

23. A psychologist is conducting an experiment using college students as participants. She pays the freshmen who participate $1.00, the sophomores $2.00, the juniors $2.50, and the seniors $3.50. If she has just $450 to pay the students but must have at least 100 underclass students (freshmen and sophomores) and at least 120 upperclass students (juniors and seniors) taking part in the experiment, maximize the total number of students who can take part.

Minimization

Solve Exercises 24 through 31 by the simplex method. (All x_i are ≥ 0.)

24. Minimize $F = x_1 + x_2$
    subject to
    $3x_1 + 4x_2 \geq 12$
    $4x_1 + 2x_2 \geq 8$

25. Minimize $F = 3x_1 + 6x_2$
    subject to
    $x_1 + x_2 \geq 10$
    $2x_1 + x_2 \leq 20$
    $x_1 \geq 5$

26. Minimize $F = 3x_1 + 2x_2 + x_3$
    subject to
    $x_1 + x_2 + x_3 \geq 100$
    $2x_1 + x_3 \geq 50$

27. Minimize $F = 5x_1 + x_2 + x_3$
    subject to
    $x_1 + x_2 \geq 10$
    $x_2 + x_3 \geq 20$
    $x_1 + x_2 + x_3 \leq 100$

28. Minimize $F = 2x_1 + x_2 + 2x_3$
    subject to
    $x_1 + 2x_2 \geq 40$
    $x_1 + 2x_2 + 3x_3 \geq 60$
    $x_2 + x_3 \geq 30$

29. Minimize $F = 5x_1 + x_2 + 4x_3$
    subject to
    $x_1 + x_2 + x_3 = 60$
    $x_1 \geq 10$
    $x_2 + x_3 \geq 20$

30. Minimize $F = 2x_1 + 3x_2 + x_3 + 2x_4$
    subject to
    $x_1 + x_2 + x_3 = 80$
    $x_2 + x_3 \geq 40$
    $x_3 - x_4 \geq 60$

31. Minimize $F = x_1 + 2x_2 + 3x_3 + 4x_4$
    subject to
    $x_1 + x_3 = 1$
    $x_2 + x_4 = 1$
    $x_1 + x_2 + x_3 + x_4 \leq 3$

32. Solve Exercise 2 of Exercises 5.2.

Management

33. An oil company mixes high-sulfur oil with low-sulfur oil to form a mixture with an intermediate-sulfur content. Suppose the high-sulfur oil is 8% sulfur, the low-sulfur oil is 2% sulfur, and the mixture must contain no more than 4% sulfur. Suppose also that the company must produce at least 100,000 barrels of the mixture, and that the high-sulfur oil costs $16.50 per barrel and the low-sulfur costs $22.40 per barrel. Use the simplex method to determine how many barrels of each type of oil the company should use in order to minimize its cost.

34. Solve Exercise 17 of Exercises 5.2.

35. Beta Company makes three products: A, B, and C. Each unit of A costs $2, each unit of B costs $1, and each unit of C costs $3 to produce. Beta must produce at least 10 A's, 20 B's, and 30 C's, and cannot produce more than 100 total units of A's, B's, and C's combined. Minimize Beta's cost.

36. Peanuts cost $0.80 per pound, cashews cost $1.30 per pound, and brazil nuts cost $1.40 per pound. Each day a store mixes these three types of nuts and sells the mixture by the pound. Suppose that daily demand for the mixture is 50 lb and that it is store policy that the cashews plus the brazil nuts must make up at least 50% of the mixture. How many pounds of each variety of nut should the store use if it wishes to minimize its cost?

37. Repeat Exercise 36 if in addition it is store policy that at least 20% of the mixture must be brazil nuts.

*38. A firm makes two products, A and B. Its fixed cost is $50,000, and its variable cost per unit is $20 for A and $30 for B. It sells each unit of A for $25 and each unit of B for $36, and it has an order for 100 units of A. The firm's break-even quantity is the total number of units of A plus the total number of units of B it must make in order to have cost equal to revenue. Find the firm's break-even quantity. (*Hint:* Minimize cost subject to the conditions that total cost equal total revenue and that the number of units of A produced must be at least 100.)

*39. A firm makes low-, medium-, and high-priced modems. Its fixed cost is $250,000, and its variable cost per unit is $100 for each low-priced modem, $180 for each medium-priced one, and $250 for each high-priced one. It sells the low-priced modems for $125 each, the medium-priced ones for $225 each, and the high-priced ones for $350 each. It has orders for 4000 of the low-priced modems and 2000 of the medium-priced ones, and it knows that demand for the high-priced ones will not exceed 5000 units. Find the firm's break-even quantity.

40. It costs a store $200 for each commercial it runs on the local TV station, $150 for each ad in the local daily newspaper, and $180 for each ad in the local Sunday newspaper. The store owner does not want to purchase more than 10 TV commercials or fewer than 4 ads in the daily newspaper. He estimates that each TV commercial is seen by 10,000 people and each newspaper ad (daily or Sunday) is seen by 5000 people. He wants his advertising to reach at least 50,000 people. How many of each type of ad should he run in order to minimize his cost?

41. Solve Exercise 24 of Exercises 5.2.

Life Science

42. A nutritionist is preparing a supplement to the diet of a patient. The patient needs at least 100 additional mg of iron and 120 mg of niacin per day, but not more than 150 mg of niacin. The nutritionist has Brand X capsules, which cost $0.20 each and contain 10 mg of iron and 2 mg of niacin; Brand Y capsules, which cost $0.40 each and contain 8 mg of iron and 12 mg of niacin; and Brand Z capsules, which cost $0.30 each and contain no iron and 4 mg of niacin. Minimize the cost of supplementing the patient's diet.

43. A patient can take three cold remedies, A, B, and C, singly or in combination. Each tablet of A contains 20 mg of cough suppressant and 12 mg of antihistamines, each tablet of B contains 25 mg of cough suppressant and 25 mg of antihistamines, and each tablet of C contains 15 mg of cough suppressant and 30 mg of antihistamines. The patient wants at least 200 mg of cough suppressant and 200 mg of antihistamines. Find the minimum number of tablets the patient must take.

44. Solve Exercise 25 of Exercises 5.2.

Social Science

45. A state political party is holding a convention. Fifty party members will be invited to attend. Those from the southern part of the state will receive $200 in travel expenses, and those from the northern part of the state, who have a shorter distance to travel, will receive only $100. The party chairperson wants to make sure that the total number of northerners, male and female, who attend the convention will not be fewer than 15 or more than 30. The chairperson also wants to make sure that at least 20 female delegates attend the convention. Minimize the cost of meeting these requirements.

5.6 DUALITY

Associated with every linear programming problem is another problem called its **dual.** (In this context the original problem is known as the **primal problem.**) As we shall see, solving the dual problem by means of the simplex method automatically solves the primal problem also. Thus, if we wish we can solve a given linear programming problem by solving its dual problem, and it is often useful to proceed in this manner. In this section we show how to form the dual problem from the primal problem and how to use the final simplex tableau of the dual solution to find the primal solution.

Our first task is to show how to obtain the dual of a given linear programming problem. In this regard, it will be helpful if we have available the notion of the **transpose** of a matrix.

> ### The Transpose of a Matrix
> If $\mathbf{A}$ is an $m \times n$ matrix, its **transpose** $\mathbf{A}^{\mathbf{T}}$ is the $n \times m$ matrix whose
>
> 1st row is the 1st column of $\mathbf{A}$,
> 2nd row is the 2nd column of $\mathbf{A}$,
> $\vdots$
> nth row is the nth column of $\mathbf{A}$.

EXAMPLE 1   If

$$\mathbf{A} = \begin{bmatrix} 2 & 4 \\ -1 & 3 \end{bmatrix}, \quad \text{then} \quad \mathbf{A}^{\mathbf{T}} = \begin{bmatrix} 2 & -1 \\ 4 & 3 \end{bmatrix}.$$

If

$$\mathbf{B} = \begin{bmatrix} 3 & 2 & 5 & 1 \\ 4 & 9 & -1 & 0 \\ 6 & 0 & 3 & 1 \end{bmatrix}, \quad \text{then} \quad \mathbf{B}^{\mathbf{T}} = \begin{bmatrix} 3 & 4 & 6 \\ 2 & 9 & 0 \\ 5 & -1 & 3 \\ 1 & 0 & 1 \end{bmatrix}. \quad \square$$

Now we are ready to show how to form the dual of a linear programming problem. Consider the problem

$$\text{Minimize} \quad F = 2x_1 + 3x_2$$
$$\text{subject to} \quad x_1 + x_2 \geq 8$$
$$x_1 + 2x_2 \geq 12$$
$$x_1 \geq 2.$$

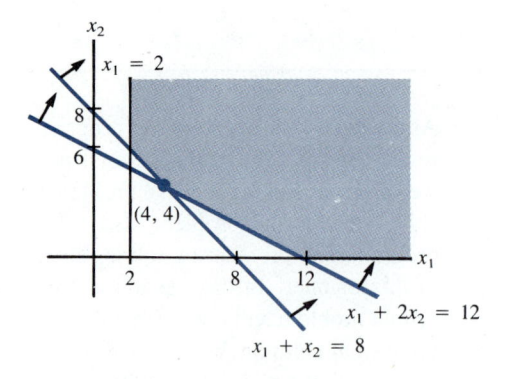

FIGURE 5–17

(As usual, we assume that all variables are nonnegative.) It is easy to check that the solution of this problem is $x_1 = x_2 = 4$, minimum value of $F = 20$. (See Figure 5–17.) Now we form the **matrix M** of the problem:

$$\mathbf{M} = \left[\begin{array}{cc|c} 1 & 1 & 8 \\ 1 & 2 & 12 \\ 1 & 0 & 2 \\ \hline 2 & 3 & 0 \end{array}\right].$$

Note that **M** is *not* a simplex tableau: it is just the matrix of the coefficients of the constraints, augmented on the right by the constant terms of the constraints and on the bottom by the coefficients of the objective function. (The 0 in the lower right-hand corner is just a spaceholder.)

Let us form the transpose of the matrix **M**:

$$\mathbf{M}^T = \left[\begin{array}{ccc|c} 1 & 1 & 1 & 2 \\ 1 & 2 & 0 & 3 \\ \hline 8 & 12 & 2 & 0 \end{array}\right].$$

Now we construct a maximization problem with nonnegative variables whose matrix is $\mathbf{M}^T$ and all of whose constraints are of the $\leq$ variety. We have

$$\text{Maximize} \quad G = 8y_1 + 12y_2 + 2y_3$$

subject to

$$y_1 + y_2 + y_3 \leq 2$$
$$y_1 + 2y_2 \qquad\quad \leq 3.$$

This is the **dual problem** of the original minimization problem. Notice that

- The primal problem was a minimization problem and the dual is a maximization problem,
- All of the primal constraints were of the $\geq$ variety, and all of the dual constraints are of the $\leq$ variety,

LINEAR PROGRAMMING

- The primal had three constraints and two variables, and the dual has two constraints and three variables.

These conclusions hold in general:

- The dual of a minimization problem is a maximization problem,
- If the primal problem has only $\geq$ constraints, the dual problem will have only $\leq$ constraints,
- If the primal problem has m constraints and n variables, the dual problem will have n constraints and m variables.

If we use the simplex method to solve the dual problem, the final simplex tableau will be

$$\begin{array}{c} \\ y_1 \\ y_2 \\ G \end{array} \begin{array}{c} \begin{array}{cccccc} y_1 & y_2 & y_3 & s_1 & s_2 & \text{sol} \end{array} \\ \left[\begin{array}{cccccc|c} 1 & 0 & 2 & 2 & -1 & 1 \\ 0 & 1 & -1 & -1 & 1 & 1 \\ 0 & 0 & 2 & 4 & 4 & 20 \end{array} \right] . \end{array}$$

(Check this.) Notice that

- The maximum value of the dual objective function is the same as the minimum value of the primal objective function (both are equal to 20),
- The solutions for the primal problem ($x_1 = x_2 = 4$) appear in the final simplex tableau for the dual problem at the bottom of the slack variable columns.

The Duality Theorem, which we will quote shortly, shows that these conclusions also hold for any minimization problem and its dual.

Let us outline the procedure for forming the dual problem.

Forming the Dual Problem

Let the primal problem be a minimization problem all of whose constraints are of the $\geq$ variety. To form the dual problem, proceed as follows:

1. Form the matrix **M** of the primal problem: this is the matrix whose entries are the coefficients of the constraints, augmented on the right by the constant terms of the constraints and at the bottom by the coefficients of the objective function.

2. Write a maximization problem whose matrix is $\mathbf{M}^T$ and all of whose constraints are of the $\leq$ variety.

EXAMPLE 2

Let us find the dual problem of the following primal:

$$\text{Minimize} \quad F = 2x_1 + 3x_2 + 4x_3$$

subject to
$$x_1 + x_2 + x_3 \geq 100$$
$$2x_1 \quad\quad + 3x_3 \geq 120$$
$$4x_2 + 5x_3 \geq 200$$
$$x_1 \quad\quad\quad\quad \geq 20.$$

We have

$$\mathbf{M} = \left[\begin{array}{ccc|c} 1 & 1 & 1 & 100 \\ 2 & 0 & 3 & 120 \\ 0 & 4 & 5 & 200 \\ 1 & 0 & 0 & 20 \\ \hline 2 & 3 & 4 & 0 \end{array}\right].$$

Therefore

$$\mathbf{M}^T = \left[\begin{array}{cccc|c} 1 & 2 & 0 & 1 & 2 \\ 1 & 0 & 4 & 0 & 3 \\ 1 & 3 & 5 & 0 & 4 \\ \hline 100 & 120 & 200 & 20 & 0 \end{array}\right],$$

and hence the dual problem is

Maximize $G = 100y_1 + 120y_2 + 200y_3 + 20y_4$

subject to
$$y_1 + 2y_2 \quad\quad + y_4 \leq 2$$
$$y_1 \quad\quad + 4y_3 \quad\quad \leq 3$$
$$y_1 + 3y_2 + 5y_3 \quad\quad \leq 4. \quad\square$$

Having found the dual, the following theorem shows that we can use it to solve the primal:

The Duality Theorem

Let the primal problem be a minimization problem, and form and solve its dual problem by the simplex method. Then

1. The entries in the slack variable columns and objective function row of the final simplex tableau of the dual are the solutions of the primal.

2. The maximum value of the dual objective function is the minimum value of the primal objective function.

This affords us a new way to solve minimization problems, one that does not require the use of surplus or artificial variables: we merely form and solve the dual of the minimization problem. We illustrate with an example.

EXAMPLE 3   Let us use the dual to solve

$$\text{Minimize} \quad F = 5x_1 + x_2 + 4x_3$$

$$\text{subject to} \quad x_1 + x_2 + x_3 \geq 60$$

$$x_1 \geq 10$$

$$x_2 \geq 20.$$

The dual problem is

$$\text{Maximize} \quad G = 60y_1 + 10y_2 + 20y_3$$

$$\text{subject to} \quad y_1 + y_2 \leq 5$$

$$y_1 + y_3 \leq 1$$

$$y_1 \leq 4.$$

(Check this.) The simplex solution of the dual proceeds as follows:

$$\begin{array}{c} \\ s_1 \\ s_2 \\ s_3 \\ G \end{array} \begin{array}{c} y_1 \\ \begin{bmatrix} 1 \\ \textcircled{1} \\ 1 \\ -60 \end{bmatrix} \end{array} \begin{array}{c} y_2 \\ 1 \\ 0 \\ 0 \\ -10 \end{array} \begin{array}{c} y_3 \\ 0 \\ 1 \\ 0 \\ -20 \end{array} \begin{array}{c} s_1 \\ 1 \\ 0 \\ 0 \\ 0 \end{array} \begin{array}{c} s_2 \\ 0 \\ 1 \\ 0 \\ 0 \end{array} \begin{array}{c} s_3 \\ 0 \\ 0 \\ 1 \\ 0 \end{array} \begin{array}{c} \text{sol} \\ 5 \\ 1 \\ 4 \\ 0 \end{array} \begin{array}{l} \xrightarrow{-R_2 + R_1} \\ \\ \xrightarrow{-R_2 + R_3} \\ \xrightarrow{60R_2 + R_4} \end{array}$$

$$\begin{array}{c} \\ s_1 \\ y_1 \\ s_3 \\ G \end{array} \begin{array}{c} y_1 \\ \begin{bmatrix} 0 \\ 1 \\ 0 \\ 0 \end{bmatrix} \end{array} \begin{array}{c} y_2 \\ \textcircled{1} \\ 0 \\ 0 \\ -10 \end{array} \begin{array}{c} y_3 \\ -1 \\ 1 \\ -1 \\ 40 \end{array} \begin{array}{c} s_1 \\ 1 \\ 0 \\ 0 \\ 0 \end{array} \begin{array}{c} s_2 \\ -1 \\ 1 \\ -1 \\ 60 \end{array} \begin{array}{c} s_3 \\ 0 \\ 0 \\ 1 \\ 0 \end{array} \begin{array}{c} \text{sol} \\ 4 \\ 1 \\ 3 \\ 60 \end{array} \xrightarrow{10R_1 + R_4}$$

$$\begin{array}{c} \\ y_2 \\ y_1 \\ s_3 \\ G \end{array} \begin{array}{c} y_1 \\ \begin{bmatrix} 0 \\ 1 \\ 0 \\ 0 \end{bmatrix} \end{array} \begin{array}{c} y_2 \\ 1 \\ 0 \\ 0 \\ 0 \end{array} \begin{array}{c} y_3 \\ -1 \\ 1 \\ -1 \\ 30 \end{array} \begin{array}{c} s_1 \\ 1 \\ 0 \\ 0 \\ 10 \end{array} \begin{array}{c} s_2 \\ -1 \\ 1 \\ -1 \\ 50 \end{array} \begin{array}{c} s_3 \\ 0 \\ 0 \\ 1 \\ 0 \end{array} \begin{array}{c} \text{sol} \\ 4 \\ 1 \\ 3 \\ 100 \end{array} .$$

Therefore, the solution of the primal problem is $x_1 = 10$, $x_2 = 50$, $x_3 = 0$, and the minimum value of its objective function F is 100. ☐

We have thus far said nothing about how to form the dual of a minimization problem if it contains $\leq$ or $=$ constraints. However, such constraints can always be replaced by one or more $\geq$ constraints. To see this, suppose a minimization problem contains the constraints

$$2x_1 + 3x_2 + x_3 \leq 50 \quad \text{and} \quad x_1 + 2x_2 + 4x_3 = 20.$$

Any $\leq$ constraint can be converted to a $\geq$ constraint by multiplying it by -1; thus

$$2x_1 + 3x_2 + x_3 \leq 50$$

can be replaced by

$$-2x_1 - 3x_2 - x_3 \geq -50.$$

Any $=$ constraint can be replaced by a pair of inequality constraints:

$$x_1 + 2x_2 + 4x_3 = 20$$

can be replaced by the pair

$$x_1 + 2x_2 + 4x_3 \geq 20,$$

$$x_1 + 2x_2 + 4x_3 \leq 20.$$

But then

$$x_1 + 2x_2 + 4x_3 \leq 20$$

can be replaced by

$$-x_1 - 2x_2 - 4x_3 \geq -20.$$

Hence, every minimization problem can be written using only $\geq$ constraints, and then our procedure for forming its dual applies.

Finally, what about the dual of a maximization problem? In a manner similar to that just discussed, every maximization problem can be written by using only $\leq$ constraints. Once this is done, we can form its matrix $\mathbf{M}$ and then use the transpose $\mathbf{M}^T$ to create the dual problem. The dual of a maximization problem all of whose constraints are of the $\leq$ variety is a minimization problem all of whose constraints are of the $\geq$ variety; if the dual is solved by the simplex method, the solutions of the primal will appear in the final tableau at the bottom of the surplus variable columns, and the minimum value of the dual objective function will be the same as the maximum value of the primal objective function. It is sometimes advantageous to solve a maximization problem with many constraints and relatively few variables by solving its dual. This is so because the dual will have relatively few constraints and many variables, and in general it is the number of constraints rather than the number of variables that makes it difficult to solve a linear programming problem; thus in a case such as this, the dual may well be easier to solve than the primal.

■ *Computer Manual: Programs MATRICES, SIMPLEX*

5.6 EXERCISES

In Exercises 1 through 4, find the transpose of the given matrix.

1. $\begin{bmatrix} 1 & 3 \\ -2 & 9 \end{bmatrix}$
2. $\begin{bmatrix} 6 & 2 & 1 \\ 1 & 3 & -1 \end{bmatrix}$
3. $\begin{bmatrix} 4 & 2 & 3 \\ -1 & 3 & 2 \\ 0 & 1 & 0 \end{bmatrix}$
4. $\begin{bmatrix} 2 & 3 & 4 \\ 3 & 1 & 7 \\ 4 & 7 & 0 \end{bmatrix}$

In Exercises 5 through 12, form the dual of the given problem. (As usual, all variables are nonnegative.)

5. Minimize $F = x_1 + 2x_2$
   subject to
   $2x_1 + x_2 \geq 20$
   $3x_1 + 4x_2 \geq 60$

6. Minimize $F = 2x_1 + x_2 + x_3$
   subject to
   $x_1 + x_2 + x_3 \geq 10$
   $x_1 \geq 5$
   $x_2 \geq 5$
   $x_2 + 2x_3 \geq 12$
   $2x_1 + x_3 \geq 12$

7. Minimize $F = 4x_1 + 2x_2 + x_3 + 4x_4$
   subject to
   $2x_1 + x_2 + x_3 \geq 100$
   $x_1 + x_3 + 2x_4 \geq 80$
   $2x_2 + 3x_3 + 5x_4 \geq 220$

8. Minimize $F = 5x_1 + x_2 + x_3$
   subject to
   $x_1 + x_2 + x_3 \geq 50$
   $x_1 + x_2 \geq 20$
   $x_2 + x_3 \geq 20$
   $x_1 \geq 8$
   $x_2 \geq 8$
   $x_3 \geq 8$

9. Minimize $F = 3x_1 + x_2 + 2x_3$
   subject to
   $4x_1 + 2x_2 \geq 25$
   $3x_1 + 5x_3 \leq 125$
   $6x_2 + 2x_3 \geq 40$

10. Minimize $F = 4x_1 + x_2$
    subject to
    $x_1 + x_2 \geq 10$
    $x_1 + x_2 \leq 30$
    $x_1 + 5x_2 = 45$

11. Maximize $F = 5x_1 + 2x_2$
    subject to
    $3x_1 + 2x_2 \leq 50$
    $4x_1 + 3x_2 \leq 90$
    $x_1 \leq 10$

12. Maximize $F = 3x_1 + x_2 + 2x_3$
    subject to
    $x_1 + x_2 + x_3 \geq 12$
    $x_1 \leq 50$
    $x_2 \leq 50$
    $x_2 + x_3 = 30$

In Exercises 13 through 20, solve the given problem by solving its dual problem. All variables are nonnegative.

13. Minimize $F = 3x_1 + x_2$
    subject to
    $x_1 + x_2 \geq 10$
    $2x_1 + 3x_2 \geq 24$

14. Minimize $F = 5x_1 + 2x_2$
    subject to
    $x_1 + x_2 \geq 10$
    $x_1 \geq 5$
    $x_2 \geq 5$
    $2x_1 + x_2 \geq 16$
    $x_1 + 2x_2 \geq 16$

15. Minimize $F = 3x_1 + x_2 + x_3$
    subject to $x_1 + x_2 + x_3 \geq 200$
    $2x_1 + x_3 \geq 100$

16. Minimize $F = 2x_1 + 3x_2 + 2x_3$
    subject to $x_1 + x_2 \geq 20$
    $x_2 + x_3 \geq 30$
    $x_1 + x_2 + x_3 \geq 40$

17. Minimize $F = 2x_1 + x_2 - x_3 + 3x_4$
    subject to $x_1 + x_2 + x_3 + x_4 \geq 80$
    $x_1 + 2x_2 + 3x_4 \geq 100$

18. Minimize $F = x_1 + 3x_2 + x_3 + x_4$
    subject to $2x_1 + x_4 \geq 10$
    $x_1 + x_2 + x_3 + x_4 \leq cf1100$
    $x_2 + x_3 = 40$

19. Maximize $F = x_1 + 2x_2$
    subject to $x_1 + 3x_2 \leq 60$
    $4x_1 + 3x_2 \leq 150$

20. Maximize $F = x_1 + 2x_2 + 3x_3$
    subject to $2x_1 + x_2 + x_3 \leq 100$
    $x_1 + x_2 \geq 10$

In Exercises 21 through 27, solve the linear programming problem by solving its dual.

Management

21. A firm makes two products, A and B. The variable cost per unit for A is $40, and for B it is $25. The firm must make at least 75 units of A and at least 100 units of B. If each unit of A sells for $75 and each unit of B for $30, and if the revenue from sales of A's and B's must be at least $2250, find the number of units of each product it must make in order to minimize its cost.

22. Each TV commercial costs $200,000 to run and reaches 10 million people. Each magazine ad costs $80,000 and reaches 3.5 million people. A direct mail campaign costs $120,000 and reaches 8 million people. A campaign of radio ads costs $50,000 and reaches 2.8 million people. An advertiser wishes to reach a total audience of at least 100 million people and has already committed to four TV commercials, two magazine ads, and one direct mail campaign. Minimize the advertiser's cost.

23. A company makes six products, A, B, C, D, E, and F. Each unit of A costs $5 to produce; each unit of B, $5; each unit of C, $8; each unit of D, $11; each unit of E, $9; and each unit of F, $7. The company must make at least 400 total units, of which the A's, B's, and C's must total at least 60 and the D's, E's, and F's must total at least 50. Minimize the cost of production.

24. A print shop must blend inks that are 1%, 2%, and 4% acid to obtain an ink that is at least 3.5% acid. It needs at least 100 gallons of the 3.5% ink. If the 1% ink costs $8 per gallon, the 2% ink $10 per gallon, and the 4% ink $13 per gallon, minimize the cost of blending the inks.

25. A cereal is blended from three grains, which contain 20% protein and 6% fiber, 15% protein and 8% fiber, and 10% protein and 8% fiber, and cost $1.50 per pound, $1.20 per pound, and $1.10 per pound, respectively. The cereal must contain at least 12% protein and at least 7% fiber. Minimize the cost of blending at least 120 lb of the cereal.

26. Blapple Juice is a mixture of apple juice, blueberry juice, and water. At least 10% of the mixture must be blueberry juice; at least 40% of it must be juice (not water). There must be at least two and one-half times as much apple juice in the mixture as there is blueberry juice. Apple juice costs $0.60 per gallon, blueberry juice $0.80

per gallon, and water $0.05 per gallon. Minimize the cost of mixing at least 1000 gallons of Blapple Juice.

27. A company makes three products, A, B, and C. It can make at most 200 total units, at most 120 units of the A's and B's combined, at most 130 units of the A's and C's combined, at most 160 units of the B's and C's combined, at most 100 units of the A's, at most 100 units of the B's, and at most 90 units of the C's. In addition, each unit of A uses 1 kg of steel; each unit of B, 2 kg; and each unit of C, 3 kg. The company has 450 kg of steel on hand. If the firm makes a profit of $2 on each unit of A and B and $2.50 on each unit of C, maximize its profit.

SUMMARY

This summary consists of terms and symbols whose meaning you should know, facts you should know, and techniques and methods you should be able to employ.

Section 5.1

Linear inequality in two variables. Solution set of a linear inequality in two variables. Finding the solution set of a linear inequality in two variables. System of linear inequalities in two variables. Feasibility region, feasible combination. Vertex of a feasibility region. Finding the feasibility region. Setting up systems of linear inequalities for applied problems and finding their feasibility regions.

Section 5.2

Linear programming problem. The linear programming model. Objective function. Constraints. Setting up linear programming models for applied problems.

Section 5.3

Linear programming problems with two variables. The Fundamental Theorem of Linear Programming. The Linear Programming Algorithm. Solving two-dimensional linear programming problems. Setting up and solving two-dimensional linear programming models for applied problems.

Section 5.4

Simplex method. Karmarkar's algorithm. Slack, slack variable. Basic, nonbasic variables. F-row. Initial simplex tableau. Setting up the initial simplex tableau. Entering variable. Determining the entering variable. Departing variable. Determining the departing variable. Obtaining a new simplex tableau using row operations. The simplex algorithm. Using the simplex method to solve maximization problems with $\leq$ constraints.

Section 5.5

Surplus, surplus variable. Artificial variable. Pretableau. Forming the pretableau. Transforming the pretableau into the initial tableau. Using the simplex method to solve maximization problems having $\geq$, $=$ constraints. Using the simplex method to solve minimization problems.

Section 5.6

Dual problem, primal problem. Transpose of a matrix, $\mathbf{A}^T$. Forming the dual of a minimization problem with $\geq$ constraints. The Duality Theorem. Using the solution of the dual to find the solution of the primal. Using the dual to solve minimization problems. Forming the dual in the presence of $\leq$, $=$ constraints. Forming the dual of a maximization problem.

REVIEW EXERCISES

In Exercises 1 through 4, find the solution set for the given inequality.

1. $2x + 6y < 18$
2. $5x - 4y \geq 40$
3. $-3x + 7y \leq 42$
4. $5x - 9y > 0$

In Exercises 5 through 8, find the feasibility region and its vertices for the given system of inequalities.

5. $x + y \leq 10$
   $2x + y \geq 4$
   $x \geq 0$
   $y \geq 0$

6. $4x + 6y \leq 96$
   $2x + y \leq 20$
   $x \geq 0$
   $y \geq 0$

7. $2x + 5y \geq 20$
   $3x + 2y \geq 12$
   $x \geq 0$
   $y \geq 2$

8. $x + y \geq 5$
   $3x + 5y \leq 45$
   $2x + y \leq 23$
   $x \geq 3$
   $y \geq 2$

9. An optical company is considering the purchase of additional lens-grinding and lens-polishing machines. Each grinding machine occupies 100 ft² of floor space, and four operators must be available to tend it; each polishing machine occupies 125 ft² and three operators must be available. The company must buy at least five grinding machines and has 2000 ft² of floor space and 72 operators available for the new machines. Find the company's feasibility region and the vertices of the feasibility region.

10. Maximize $F = 3x + 4y$ subject to the constraints of Exercise 5.
11. Maximize $F = 5x + 3y$ subject to the constraints of Exercise 6.
12. Minimize $F = 5x + 4y$ subject to the constraints of Exercise 7.
13. Maximize and minimize $F = 6x + 10y$ subject to the constraints of Exercise 8.
14. Referring to Exercise 9, suppose each grinding machine can grind 4200 lenses per day and each polishing machine can polish 3600 lenses per day. How many of each type of machine should the company purchase if it wishes to maximize the total number of lenses that can be worked each day?
15. A city has two newspapers, *The Bugle* and *The Clarion*, each of which publishes daily and puts out a suburban supplement once a week. Any full-page ad run in

either daily paper is also run in that paper's suburban supplement at no extra charge. Full-page ads in *The Bugle* reach 60,000 readers, and those in its suburban supplement reach 20,000 readers. Full-page ads in *The Clarion* reach 90,000 readers, and those in its supplement reach 10,000 readers. Suppose you wish to purchase enough full-page ads to reach at least 2,700,000 daily newspaper readers and at least 500,000 readers of the suburban supplements. Suppose also that you have already agreed to buy at least 12 ads in *The Clarion* and that full-page ads will cost you $5000 each in *The Bugle* and $6000 each in *The Clarion*. How many ads should you buy in each paper in order to minimize your cost?

16. Use the simplex method to solve the following problem:

$$\text{Maximize } F = 0.5x_1 + 2x_2 + x_3$$
$$\text{subject to } \quad 2x_1 + 3x_2 + 4x_3 \le 72$$
$$x_1 \le 18$$
$$x_2 \le 6$$
$$x_3 \le 12$$
$$x_1 \ge 0, x_2 \ge 0, x_3 \ge 0.$$

17. A widget company makes small, medium, and large widgets. Space restrictions limit total production to 200 widgets per day. Also, each small widget takes 2 worker-hours to assemble, each medium one 4 worker-hours, and each large one 8 worker-hours. The firm has 1240 worker-hours available in the assembly department each day. If each small widget contributes $20 to profit, each medium one $22, and each large one $24, maximize the firm's daily profit.

In Exercises 18 and 19, use the simplex method to solve.

18. Maximize $F = 2x_1 + 3x_2$
    subject to
    $x_1 + 2x_2 \le 8$
    $2x_1 + x_2 \le 8$
    $x_1 + x_2 \ge 2$
    $x_1 \ge 0, x_2 \ge 0$

19. Minimize $F = 3x_1 + x_2 - x_3$
    subject to
    $2x_1 + x_2 \ge 10$
    $x_1 + x_2 - x_3 = 4$
    $x_1 \ge 0, x_2 \ge 0, x_3 \ge 0$

20. A store stocks three brands of a certain product: A, B, and C. Each unit of Brand A occupies 2 ft² of shelf space; each unit of Brand B, 3 ft²; and each unit of Brand C, 1 ft². The store has 80 ft² of shelf space to allocate to this product. In addition, the store must have on hand at least 10 units of Brand A, and at least 20 units of Brands B and C combined. Each unit of A costs the store $5; each unit of B, $4; and each unit of C, $7. Minimize the store's cost.

In Exercises 21 through 23, solve the given linear programming problem by solving its dual problem. All variables are ≥ 0.

21. Minimize $F = 40y_1 + 15y_2$
    subject to
    $2y_1 + 3y_2 \ge 3$
    $4y_1 + y_2 \ge 2$

22. Maximize $F = 3x_1 + 3x_2 + 2x_3$
    subject to
    $2x_1 + 3x_2 + x_3 \le 3$
    $6x_1 + 3x_2 + 6x_3 \le 5$

23. Minimize $F = 50y_1 + 20y_2 - 20y_3 + 15y_4 - 15y_5$
    subject to
    $4y_1 + 2y_2 - 2y_3 + y_4 - y_5 \ge 5$
    $3y_1 + y_2 - y_3 + y_4 - y_5 \ge 3$
    $y_1 + y_2 - y_3 + y_4 - y_5 \ge 7$

ADDITIONAL TOPICS

Here are some suggestions for topics not covered in the text that you might want to investigate on your own.

1. A **transportation problem** is a particular type of linear programming problem. Transportation problems are concerned with minimizing the cost of shipping goods from one or more sources to one or more destinations, and hence are very common in the business world. They have their own characteristic form and are usually solved not by the simplex method but by a method designed especially for them, known as the **transportation algorithm.** Find out about transportation problems and the transportation algorithm.

2. An **assignment problem** is another particular type of linear programming problem. Assignment problems are scheduling problems: they are concerned with assigning people or machines to tasks in such a way as to minimize cost or maximize efficiency. Assignment problems also have their own characteristic form, as well as a special method of solution known as the **Hungarian method.** Find out about assignment problems and the Hungarian method.

3. Today linear programming problems are usually solved by computerized versions of the simplex method. See if your school has available a computerized version of the simplex method, and if so, find out how to use it.

4. At the beginning of Section 5.4 we mentioned that recently two new methods for solving linear programming problems have been discovered. One of these is Karmarkar's algorithm; the other is **Khachian's algorithm,** which was announced in 1979 by the Soviet mathematician L. G. Khachian. Find out about these algorithms: how they work, their strengths and weaknesses, and whether they are currently being used to solve practical problems.

SUPPLEMENT: SHADOW PRICES AND RIGHT-HAND-SIDE RANGING

The final simplex tableau for a linear programming problem not only gives us the solution to the problem, but also contains much other interesting and valuable information. In this supplement we discuss two topics, **shadow prices** and **right-hand-side ranging,** that make use of some of this information. As an illustrative example, we will utilize the Acme Company problem of Section 5.4:

$$\text{Maximize profit} \quad F = 2x_1 + x_2 + 5x_3$$
subject to
$$\begin{aligned} \text{steel:} \quad & 2x_1 + 10x_2 + 14x_3 \leq 70 \\ \text{tin:} \quad & x_1 + 2x_3 \leq 20 \\ \text{chrome:} \quad & x_2 + x_3 \leq 10 \\ & x_1 \geq 0 \\ & x_2 \geq 0 \\ & x_3 \geq 0. \end{aligned}$$

Linear Programming

Here profit is in dollars, the resources are in pounds, and x_1, x_2, and x_3 are the number of units of products A, B, and C, respectively, that Acme Company must make. The final simplex tableau for the Acme Company problem is

	x_1	x_2	x_3	s_1	s_2	s_3	sol
x_3	0	1	1	0.1	-0.2	0	3
x_1	1	-2	0	-0.2	1.4	0	14
s_3	0	0	0	-0.1	0.2	1	7
F	0	0	0	0.1	1.8	0	43

Thus $x_1 = 14$ units, $x_2 = 0$ units, $x_3 = 3$ units, and the maximum profit is \$43.

Shadow Prices

The entries in the F-row and the slack variable columns of the final simplex tableau (i.e., the solutions of the dual problem) are called the **shadow prices** for the resources associated with the slack variables. Thus, for the Acme Company problem, 0.1 is the shadow price for steel, because 0.1 is in the F-row and s_1-column and s_1 is the slack variable for the steel resource. Similarly, 1.8 is the shadow price for tin and 0 is the shadow price for chrome. Each shadow price represents the gross per-unit contribution its resource makes to the objective function, provided that the variables in the basis do not change. In other words:

> ### Shadow Prices
> Let resource i have shadow price p_i. Then as long as the same variables remain in the basis, an additional unit of resource i made available at no cost will change the value of the objective function by the amount p_i.

For instance, in the Acme Company example, we have seen that the shadow price of steel is 0.1. Hence an additional pound of steel made available at no cost would increase the company's profit by \$0.10. Similarly, since the shadow price of tin is 1.8, an additional pound of tin made available at no cost would increase profit by \$1.80. An additional pound of chrome made available at no cost would increase profit by \$0, that is, not at all. (This makes sense, because the solution $x_1 = 14$, $x_2 = 0$, $x_3 = 3$ leaves 7 lb of chrome unused, and therefore increasing the amount of chrome available will not change the profit.) These facts can be helpful in analyzing the effect that additional resources will have on Acme Company's profit. For instance, if it costs more than \$0.10 to obtain an additional pound of steel, it will not pay to do so. Also, suppose it will actually cost \$0.08 per pound to obtain additional steel and \$1.40 per pound to obtain additional tin. Then for \$1.40 Acme Company can get an additional 1 lb of tin, which will increase net profit by \$1.80 − \$1.40 = \$0.40, or it can get an additional 17.5 lb of steel, which will increase net profit by (17.5)(\$0.10 − \$0.08) = \$0.35. Therefore under these circumstances it would be wise for Acme to spend its money on additional tin rather than on additional steel. It must be emphasized that the conclusions reached in this paragraph will hold only as long as the shadow prices remain valid, that is, only as long as the variables x_1, x_3, and s_3 remain in the basis.

Right-Hand-Side Ranging

Changing the amount of resources available may change the variables in the basis and hence change the shadow prices. Therefore in order to determine when the shadow prices obtained from a final simplex tableau remain valid, we must determine the effect that changing the amount of each resource available will have on the basis. Since the amount of each resource available is the number on the right-hand-side of its constraint, the technique for doing this is known as **right-hand-side ranging.**

> **Right-Hand-Side Ranging**
>
> Let s_i be the slack variable of constraint i. Using each row of the final simplex tableau except the F-row, form the ratios
>
> $$\frac{\text{Solution column entry}}{s_i\text{-column entry}}.$$
>
> Let a be the value of the smallest nonnegative ratio, and let b be the absolute value of the nonpositive ratio having the smallest absolute value. Then the right-hand-side of the ith constraint can be decreased by a and increased by b without changing the variables appearing in the basis.

To illustrate, suppose we begin with the s_1-column of the final Acme Company tableau. We have

s_1 (steel)	Sol.	Ratio
0.1	3	3/0.1 = 30
−0.2	14	14/−0.2 = −70
−0.1	7	7/−0.1 = −70

The smallest nonnegative ratio is 30, so $a = 30$; the nonpositive ratio with the smallest absolute value is -70, so $b = |-70| = 70$. Therefore Acme can decrease the right-hand-side of the steel constraint from 70 to $70 - 30 = 40$ and increase it from 70 to $70 + 70 = 140$ and still have x_1, x_3, and s_3 in the basis.

Now let us perform right-hand-side ranging for the tin and chrome constraints. We have

s_2 (tin)	Sol	Ratio	s_3 (chrome)	Sol	Ratio
−0.2	3	3/−0.2 = −15	0	3	—
1.4	14	14/1.4 = 10	0	14	—
0.2	7	7/0.2 = 35	1	7	7/1 = 7

LINEAR PROGRAMMING

We see that for tin, $a = 10$ and $b = 15$; therefore Acme can decrease the right-hand-side of the tin constraint from 20 to $20 - 10 = 10$ and increase it from 20 to $20 + 15 = 35$ and still have x_1, x_3, and s_3 in the basis. For chrome $a = 7$ and there are no restrictions on b; therefore Acme can decrease the right-hand-side of the chrome constraint from 10 to $10 - 7 = 3$ and increase it without bound and still have x_1, x_3, and s_3 in the basis.

Suppose we display our results for Acme Company in a table:

Basic Variables x_1, x_3, s_3

Resource	Shadow Price	Right-Hand-Side Range
Steel	$0.10	[40, 140]
Tin	$1.80	[10, 35]
Chrome	$0.00	[3, $+\infty$]

Thus, as long as the amount of steel available is in the interval [40, 140], the amount of tin available is in the interval [10, 35], and the amount of chrome available is in the interval [3, $+\infty$], the variables x_1, x_3, and s_3 will remain in the basis. Therefore as long as the resources available are in these ranges, Acme will make products A and C but not B, and the chrome will not be used up but the steel and tin will be.

■ *Computer Manual: Program SIMPLEX*

SUPPLEMENTARY EXERCISES

Find the shadow prices and right-hand-side ranges, and interpret your results, for the given linear programming problem.

1. The problem of Exercise 29, Exercises 5.4.
2. The problem of Exercise 30, Exercises 5.4.
3. The problem of Exercise 31, Exercises 5.4.
4. The problem of Exercise 33, Exercises 5.4.
5. The problem of Exercise 35, Exercises 5.4.
6. The problem of Exercise 39, Exercises 5.4.
7. The problem of Exercise 40, Exercises 5.4.
8. The problem of Exercise 9, Exercises 5.5.
9. The problem of Exercise 10, Exercises 5.5.
10. The problem of Exercise 13, Exercises 5.5.
11. The problem of Exercise 18, Exercises 5.5.
12. The problem of Exercise 21, Exercises 5.5.
13. The problem of Exercise 36, Exercises 5.5.
14. The problem of Exercise 39, Exercises 5.5.
15. The problem of Exercise 43, Exercises 5.5.

6

Probability

The **probability** of an event is a measure of the likelihood that the event will occur. For instance, a weather forecaster who says "there is a 30 percent chance of rain tomorrow" could just as well have said "the probability is 0.30 that it will rain tomorrow"; both statements express the same estimate of the likelihood of rain tomorrow. Similarly, when asked what the chances are of getting heads if we toss a fair coin once, we might reply "50–50" or "1 chance in 2"; another way of saying the same thing would be to state that the probability of getting heads is 0.50 or $\frac{1}{2}$.

In virtually every sphere of life it is important to consider the likelihood that certain events will occur, so probabilities are used in nearly every field. They are important in particle physics, genetics, quality control, forecasting, marketing, political polling, gambling and gaming, and a host of other areas. For instance, probabilities can be used to answer questions such as the following:

- If there is a 10% chance that a machine will turn out a defective part, what is the likelihood that of the next 20 parts it turns out none will be defective?
- What is the likelihood that two brown-eyed parents will have a blue-eyed child?
- If we play the game of roulette over and over, betting on red each time, how much can we expect to win or lose in the long run?

In this chapter we introduce the basic notions of probability theory and demonstrate some of their applications. The first three sections are devoted to the definitions of the various types of probability and the properties of probabilities. The fourth section is concerned with permutations and combinations, which are methods of counting, and their application to probability theory. This is followed by a section concerned with the important topic of binomial probabilities. The chapter concludes with a discussion of random variables, probability distributions, and expected value.

6.1 PROBABILITIES

In this section we introduce the notion of an event and the probability of an event, explain the connection between probabilities and odds, and discuss relative frequency and subjective probabilities.

Events

Suppose we conduct an **experiment,** such as tossing a coin or rolling a pair of dice. The experiment will have several possible results, called its **outcomes.** The set consisting of all possible outcomes of the experiment is called the **sample space** for the experiment. Any subset of the sample space is referred to as an **event** of the experiment.

EXAMPLE 1 Let an experiment consist of tossing a coin. There are two possible outcomes: heads and tails. Therefore, the sample space $\mathcal{U}$ for this experiment is

$$\mathcal{U} = \{\text{heads, tails}\}.$$

There are four subsets of the sample space $\mathcal{U}$ and hence four events for this experiment:

- {heads}, the event {the coin comes up heads}.
- {tails}, the event {the coin comes up tails}.
- $\emptyset$, the event {the coin comes up neither heads nor tails}.
- $\mathcal{U}$, the event {the coin comes up either heads or tails}.

Note that event $\emptyset$ cannot occur, while event $\mathcal{U}$ is certain to occur. ∎

EXAMPLE 2 Let an experiment consist of rolling a pair of dice. We think of rolling one die first, then rolling the second, and we record the results as ordered pairs (n, m), where n is the number that comes up on the first die and m is the number that comes up on the second die. The sample space $\mathcal{U}$ for this experiment is thus

$$\mathcal{U} = \{(1, 1), (1, 2), (1, 3), (1, 4), (1, 5), (1, 6),$$
$$(2, 1), (2, 2), (2, 3), (2, 4), (2, 5), (2, 6),$$
$$(3, 1), (3, 2), (3, 3), (3, 4), (3, 5), (3, 6),$$
$$(4, 1), (4, 2), (4, 3), (4, 4), (4, 5), (4, 6),$$
$$(5, 1), (5, 2), (5, 3), (5, 4), (5, 5), (5, 6),$$
$$(6, 1), (6, 2), (6, 3), (6, 4), (6, 5), (6, 6)\}.$$

Any subset of $\mathcal{U}$ is an event. For instance,

- $\{(1, 1)\}$ is the event {the numbers total 2}.
- $\{(3, 6), (4, 5), (5, 4), (6, 3)\}$ is the event {the numbers total 9}.
- $\{(1, 2), (2, 1), (5, 6), (6, 5)\}$ is the event {the numbers total 3 or 11}. ☐

Let $\mathcal{U}$ be the sample space for an experiment, and let A be an event of the experiment. Denote by $n(\mathcal{U})$ and $n(A)$ the number of elements in the sets $\mathcal{U}$ and A, respectively.

The Probability of an Event

If all outcomes of the experiment are equally likely, and A is an event of the experiment, the **probability $P(A)$ of A** is defined to be

$$P(A) = \frac{n(A)}{n(\mathcal{U})}.$$

Note that $P(A)$ is the ratio of the number of ways event A can happen to the total number of ways the experiment can turn out. Hence $P(A)$ measures the likelihood that event A will occur. Note also that $P(A)$ is a nonnegative fraction less than or equal to 1 or, if divided out, a nonnegative decimal less than or equal to 1. Therefore,

for any event A,
$$0 \leq P(A) \leq 1.$$

EXAMPLE 3 Consider again the coin-tossing experiment of Example 1. If we assume that the coin is fair, so that heads and tails are equally likely to occur, we may apply our definition of probability to any of the four possible events of the experiment. For example,

$$P(\{\text{heads}\}) = \frac{n(\{\text{heads}\})}{n(\{\text{heads, tails}\})} = \frac{1}{2} = 0.50.$$

Thus, there is one chance out of two that the coin will come up heads, that is, the coin will come up heads 50% of the time. Similarly,

$$P(\{\text{tails}\}) = \frac{n(\{\text{tails}\})}{n(\mathcal{U})} = \frac{1}{2},$$

$$P(\{\text{neither heads nor tails}\}) = \frac{n(\emptyset)}{n(\mathcal{U})} = \frac{0}{2} = 0,$$

$$P(\{\text{either heads or tails}\}) = \frac{n(\mathcal{U})}{n(\mathcal{U})} = 1. \quad \square$$

EXAMPLE 4 Consider again the experiment of rolling a pair of dice introduced in Example 2. If we assume that the dice are fair, so that all outcomes are equally likely, then

$$P(\{\text{numbers total 2}\}) = \frac{n(\{(1, 1)\})}{n(\mathcal{U})} = \frac{1}{36},$$

$$P(\{\text{numbers total 9}\}) = \frac{n(\{(3, 6), (4, 5), (5, 4), (6, 3)\})}{n(\mathcal{U})}$$

$$= \frac{4}{36} = \frac{1}{9},$$

and

$$P(\{\text{numbers total 3 or 11}\}) = \frac{4}{36} = \frac{1}{9}. \quad \square$$

In order to simplify the notation, from now on we will omit the set braces around events when writing their probabilities, writing $P(\text{heads})$ rather than $P(\{\text{heads}\})$, for instance. This should cause no difficulty.

According to our definitions of sample space $\mathcal{U}$ and event A, A is certain to occur if and only if $A = \mathcal{U}$. Hence A is certain to occur if and only if $P(A) = 1$. On the other hand, it is impossible for A to occur if and only if $A = \emptyset$, hence if and only if $P(A) = 0$. Therefore

---

For any event A:

$P(A) = 1$ if and only if A is certain to occur.
$P(A) = 0$ if and only if it is impossible for A to occur.

Odds

Odds are an alternative way of stating probabilities. If A is an event and we know that $P(A) = 2/5$, say, then there are two ways A can occur out of a total of five possible outcomes. Therefore, there are three ways A cannot occur, compared with two ways A can occur, and we say that the **odds against** A are 3 to 2. On the other hand, suppose B is an event and we are told that the odds against B are 7 to 5. Then there are 7 ways B cannot occur, compared with 5 ways B can occur, or 5 ways B can occur out of 12 possible outcomes. Hence $P(B) = 5/12$.

Odds and Probabilities

Let A be an event.

1. If $P(A) = n/m$, the odds against A are $m - n$ to n.

2. If the odds against A are r to s, then

$$P(A) = \frac{s}{r + s}.$$

EXAMPLE 5 The probability that a horse will win a certain race is $4/9$. Hence the odds against the horse winning the race are 5 to 4. ☐

EXAMPLE 6 The odds against a football team winning its next game are 10 to 11. Hence the probability that the team will win the game is $11/21$. ☐

Relative Frequency and Subjective Probabilities

The definition of the probability of an event that we have used thus far in this section is known as the **classical definition of probability.** Classical probabilities are defined in terms of the number of possible outcomes of events and experiments and thus in some sense are inherent in the situation. But this is not always the case. For example, suppose we wish to assign a probability to the event

{it will rain tomorrow}

or to the event

{I will get the job I have applied for}.

We cannot use the classical definition of probability to assign probabilities to these events, for there is no way to count outcomes for events such as these. When probabilities cannot be assigned by means of the classical definition of probability, two other definitions may be used: the **relative frequency definition** and the **subjective definition.**

The relative frequency definition of probability assigns **relative frequency probabilities** to events on the basis of past history. To illustrate, suppose a bakery owner wishes to assign a probability to the event {sell 5 cakes tomorrow}. He consults his records and finds the following:

Number of Cakes Sold in a Day	Number of Days This Occurred
0	20
1	30
2	60
3	40
4	40
5	25
6	15
7	5
	235

Thus, to the event {sell 5 cakes tomorrow} he assigns the relative frequency probability $25/235$, since this is the relative frequency with which sales of five cakes per day occurred in the past. Similarly,

$$P(\text{sell 2 cakes tomorrow}) = 60/235,$$

and so on.

The subjective definition of probability assigns **subjective probabilities** to events based on personal opinion as to the likelihood that the events will occur. A subjective probability is an educated guess based on experience, hunch, intuition, and so on. For example, you may believe that you have a good chance of getting the job you have applied for and may therefore assign the event {I get the job} a large probability, say,

$$P(\text{I get the job}) = 0.80.$$

This would reflect your personal opinion that there is an 80% chance that you will get the job and only a 20% chance that you will not.

No matter what method is used to assign probabilities to events, the rules of probability must be obeyed.

Rules of Probability

Let A be an event and let $P(A)$ denote the probability of A. Then

1. $0 \leq P(A) \leq 1$.
2. $P(A) = 1$ if and only if A is certain to occur.
3. $P(A) = 0$ if and only if it is impossible for A to occur.
4. The larger $P(A)$ is, the more likely A is to occur; the smaller $P(A)$ is, the less likely A is to occur. Therefore, if B is an event that is less likely than A, then it must be the case that

$$P(B) < P(A).$$

EXAMPLE 7 If A and B are events, with $P(A) + P(B) = 1$ and B three times as likely to occur as A, then we have

$$P(A) + P(B) = 1$$

and

$$P(B) = 3\,P(A).$$

Thus

$$P(A) + 3\,P(A) = 1,$$

which implies that $P(A) = \frac{1}{4}$, and hence that $P(B) = \frac{3}{4}$. ☐

6.1 EXERCISES

Events and The Probability of an Event

Consider the experiment that consists of rolling a single die whose sides are numbered 1 through 6. Exercises 1 through 6 refer to this experiment.

1. Write out the event {the number that comes up is 2} as a subset of the sample space and find its probability.
2. Write out the event {the number that comes up is even} as a subset of the sample space and find its probability.
3. Write out the event {the number that comes up is either 3 or 5} as a subset of the sample space and find its probability.
4. Write out the event {the number that comes up is greater than 4} as a subset of the sample space and find its probability.
5. Write out the event {the number that comes up is less than 8} as a subset of the sample space and find its probability.
6. Write out the event {the number that comes up is greater than 8} as a subset of the sample space and find its probability.

Consider the experiment that consists of rolling a pair of dice. Exercises 7 through 12 refer to this experiment.

7. Write the event {the numbers that come up total 2} as a subset of the sample space and find its probability.
8. Write the event {the numbers that come up total 3} as a subset of the sample space and find its probability.
9. Write the event {the numbers that come up total 4} as a subset of the sample space and find its probability.
10. Write the event {the numbers that come up total either 7 or 11} as a subset of the sample space and find its probability.
11. Write the event {the numbers that come up total less than 7} as a subset of the sample space and find its probability.

12. Write the event {the numbers that come up total more than 7} as a subset of the sample space and find its probability.

Consider the experiment of tossing a fair coin twice. Exercises 13 through 20 refer to this experiment.

13. Write the event {get 2 heads} as a subset of the sample space and find its probability.

14. Write the event {get 1 head and 1 tail} as a subset of the sample space and find its probability.

15. Write the event {get 2 tails} as a subset of the sample space and find its probability.

16. Write the event {get at least 1 head} as a subset of the sample space and find its probability.

17. Write the event {get at least 1 tail} as a subset of the sample space and find its probability.

18. Write the event {get no heads} as a subset of the sample space and find its probability.

19. Write the event {get at most 1 head} as a subset of the sample space and find its probability.

20. Write the event {get either 1 head or 1 tail} as a subset of the sample space and find its probability.

Consider the experiment that consists of drawing a single card from a deck of 52 cards. (Such a deck contains 13 spades, 13 hearts, 13 diamonds, and 13 clubs; each suit consists of the cards

$$1 \text{ (Ace)}, 2, 3, 4, 5, 6, 7, 8, 9, 10, \text{ jack, queen, king.}$$

Spades and clubs are black suits, hearts and diamonds are red suits.) Exercises 21 through 30 refer to this experiment.

21. Write the event {draw a king} as a subset of the sample space and find its probability.

22. Write the event {draw a red king} as a subset of the sample space and find its probability.

23. Write the event {draw the ace of hearts} as a subset of the sample space and find its probability.

24. Write the event {draw a heart} as a subset of the sample space and find its probability.

25. Write the event {draw a king or a queen} as a subset of the sample space and find its probability.

26. Write the event {draw a heart or a king} as a subset of the sample space and find its probability.

27. Write the event {draw a heart and a king} as a subset of the sample space and find its probability.

28. Write the event {draw a black card less than 4} as a subset of the sample space and find its probability.

29. Write the event {draw a king and an ace} as a subset of the sample space and find its probability.

30. Write the event {draw a king, a queen, or a black jack} as a subset of the sample space and find its probability.

Odds

In Exercises 31 through 38 state the odds against A if the probability of A is as given.

31. $P(A) = 1/3$
32. $P(A) = 2/7$
33. $P(A) = 1/2$
34. $P(A) = 9/10$
35. $P(A) = 0.1$
36. $P(A) = 0.3$
37. $P(A) = 0.45$
38. $P(A) = 0.35$

In Exercises 39 through 46 find $P(A)$ if the odds against A are as stated.

39. 3 to 2
40. 5 to 4
41. 7 to 1
42. 1 to 2
43. 2 to 5
44. 20 to 1
45. 11 to 8
46. 99 to 1

47. If the probability of rain is 0.4, what are the odds against rain?

48. If the probability that you will get a job you have applied for is 0.37, what are the odds against your getting the job?

49. You write an integer n, $1 \le n \le 10$, on a piece of paper. Your opponent tries to guess the integer.
    (a) What are the odds against her guessing the integer?
    (b) What is the probability that she will guess the integer?

50. A roulette wheel has 38 numbers on it:
    $$00, \ 0, \ 1, \ 2, \ 3, \ \ldots, \ 34, \ 35, \ 36.$$
    (a) If you bet on a single one of the numbers, what is the probability that you will win? What are the odds against your winning?
    (b) If you bet on "evens," you win if any one of the numbers 2, 4, 6, . . . , 36 comes up. If you bet on "evens," what is the probability that you will win? What are the odds against your winning?

*51. According to the *1989 World Almanac* (Pharos books, New York), in 1987 in the United States there were 48,700 deaths caused by motor vehicle accidents, 11,300 caused by falls, 5300 caused by drowning, 4800 caused by fire, and 357 caused by commercial aviation accidents. Also according to the *World Almanac,* the population of the United States in 1987 was approximately 243,400,000. Estimate the odds against a U.S. resident dying by a motor vehicle accident, by a fall, by drowning, by fire, or by a commercial airline accident in 1987.

Relative Frequency and Subjective Probabilities

52. Let A, B, C, D be events, with $P(A) = 1/2$, $P(B) = 1$, $P(C) = 1/4$, and $P(D) = 0$. Rank the events in the order of their likelihood, from least likely to occur to most likely to occur.

53. Repeat Exercise 52 if $P(A) = 1/3$, $P(B) = 3/10$, $P(C) = 2/5$, and $P(D) = 2/3$.

54. Repeat Exercise 52 if $P(A) = 0.521$, $P(B) = 0.519$, $P(C) = 0.51$, and $P(D) = 0.524$.

55. If event A is twice as likely to occur as event B and $P(A) + P(B) = 1$, what probabilities must be assigned to A and B?

56. If event A is three times as likely to occur as event B and $P(A) + P(B) = 1$, what probabilities must be assigned to A and B?

*57. If event A is 2.5 times as likely to occur as event B, what probabilities could be assigned to B?

*58. If event A is twice as likely to occur as event B, B is three times as likely to occur as event C, and $P(A) + P(B) + P(C) = 1$, what probabilities must be assigned to A, B, and C?

*59. If event A is 3.5 times as likely to occur as event B, and event C is 4.5 times as likely to occur as B, what probabilities could be assigned to B?

60. Records show that it has rained on 12 of the last 20 graduation days at Dimes College. What is the probability that it will rain on the next graduation day at Dimes College?

61. Surveys show that 75% of all students who graduate from college with business degrees immediately obtain jobs that pay at least $26,000 per year. If your brother is graduating from college with a business degree, what is the probability that he will immediately obtain a job that pays at least $26,000 per year?

*62. (This exercise was suggested by one in Aczel, *Complete Business Statistics*, Irwin, 1989.) Suppose an average of 10 flights per day arrive at an airport, the flights averaging 200 passengers each and each passenger on them averaging two pieces of checked luggage. Suppose also that the airport averages three pieces of luggage lost every two days. If you fly into the airport with one piece of checked luggage, estimate the probability that it will be lost.

One thousand light bulbs were tested to see how long they would last before burning out. The results were as follows:

Hours Lasted	< 500	500–699	700–899	≥ 900
Number of Bulbs	5	50	550	395

Exercises 63 through 67 refer to this situation.

63. Find P(a bulb will last fewer than 500 hours).
64. Find P(a bulb will last 900 or more hours).
65. Find P(a bulb will last fewer than 700 hours).
66. Find P(a bulb will last 700 or more hours).
67. Find P(a bulb will last between 500 and 899 hours, inclusive).

Management

68. Experts estimate that 68% of new businesses fail within five years. What is the probability that a new business will fail within five years?

69. It is estimated that 7 out of 12 middle managers will be fired at least once in their careers. What is the probability that a middle manager will be fired at least once?

Records show that the number of hot dogs sold by a vendor at a baseball stadium were as follows:

PROBABILITY

Dozens of Hot Dogs Sold	1	2	3	4	5	6
Number of Days This Occurred	2	5	15	13	2	1

Exercises 70 through 73 refer to this situation.

70. Find *P*(the vendor will sell one dozen hot dogs on a given day).
71. Find *P*(the vendor will sell fewer than four dozen hot dogs on a given day).
72. Find *P*(the vendor will sell more than four dozen hot dogs on a given day).
73. Find *P*(the vendor will sell either four or five dozen hot dogs on a given day).
74. A firm's sales department feels that there are 2 chances out of 10 that next year's sales will increase, 4 chances out of 10 that they will decrease, and 4 chances out of 10 that they will stay the same. Find the probabilities of the events {sales will increase}, {sales will decrease}, and {sales will stay the same}.

Alpha Company is about to introduce a new product. Sanchez, Bernstein, and Johnson are discussing whether the product will be successful. Exercises 75 through 79 refer to this situation.

75. Sanchez is certain the product will be successful. What probability should Sanchez assign to the event {the product will be successful}?
76. Johnson is certain that the product will fail. What probability should Johnson assign to the event {the product will be successful}?
77. Bernstein thinks there is a 40% chance that the product will be successful. What probability should Bernstein assign to the event {the product will be successful}?
78. Suppose Chang enters the discussion and states that the odds against the product being successful are 5 to 4. What probability should Chang assign to the event {the product will be successful}?
79. Suppose Avanian joins the discussion and states that the chances for the product's success are better than Bernstein thinks but not as good as Chang thinks. What probabilities could Avanian assign to the event {the product will be successful}?

Life Science

80. Medical research shows that 80% of all patients diagnosed with liver cancer will survive for at least three years. Find the probability that a patient diagnosed with liver cancer will survive for at least three years.
81. In a drug-testing experiment, 632 out of 948 patients treated with the drug experienced improvement in their condition. Find the probability that a patient treated with the drug will experience improvement.
82. If it is true that three out of four doctors surveyed recommend Rylenol for pain relief, what is the probability that a doctor selected at random will recommend Rylenol?

Of 200 fledgling robins hatched during a study of bird mortality, 35 lived for fewer than 2 weeks, 44 lived at least 2 weeks but fewer than 4 weeks, 59 lived at least 4 weeks but fewer than 8 weeks, 34 lived at least 8 weeks but fewer than 12 weeks, and the rest lived for 12 or more weeks. Exercises 83 through 85 refer to this study.

83. Find the probability that a fledgling robin will live for at least four weeks.

356 CHAPTER 6

84. Find the probability that a fledgling robin will not live for at least 12 weeks.
85. Find the probability that a fledgling robin will live for fewer than 2 or for 12 or more weeks.
*86. A doctor estimates that the chance that a new drug will help a patient is equal to the chance that it will have no effect and is 1.5 times the chance that it will do harm. Assign probabilities to the events {the drug will help}, {the drug will have no effect}, and {the drug will do harm}.

Social Science

87. If 44.5% of the adults in a state have a high-school diploma, find the probability that a person chosen at random from the state will have a high-school diploma.
88. If 37.4% of the persons receiving welfare will be off the welfare rolls within two years, what is the probability that a person now on the welfare rolls will be off them within two years?

A poll of voters surveyed before an election revealed their party affiliations to be as follows:

Party	Democrat	Republican	Conservative	Liberal	None
Number	247	268	84	44	357

Exercises 89 through 92 refer to this poll.

89. Find P(a voter is a Democrat).
90. Find P(a voter is not a Republican).
91. Find P(a voter is neither a Democrat nor a Republican).
92. Find P(a voter is either a Liberal or a Conservative).

6.2 PROBABILITIES OF COMPOUND EVENTS

Since an event is just a set of outcomes, we can create new events from old ones by taking complements and by forming unions and intersections. In this section and the next we show how the probabilities of such newly created events are related to the probabilities of the events from which they are formed.

Combining Events

If $\mathcal{U}$ is a sample space for an experiment and A and B are events of the experiment, then A and B are subsets of $\mathcal{U}$ and we may consider the **compound events**

$$A', \quad A \cup B, \quad \text{and} \quad A \cap B.$$

The event A' is the **complementary event** to A and consists of all outcomes of the experiment that are not elements of A. Thus A' is the event {A does not occur}. The event A' is shaded in Figure 6–1.

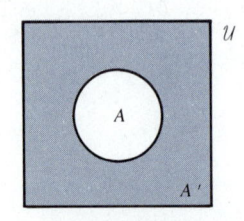

FIGURE 6–1
The Complement of Event A

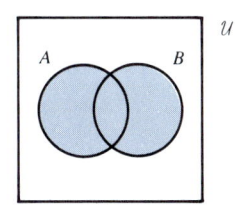

FIGURE 6–2
The Union of Events A and B

The event $A \cup B$, the **union** of A and B, consists of all outcomes that are elements of either A or B. Thus $A \cup B$ is the event {either A occurs or B occurs}. The event $A \cup B$ is shaded in Figure 6–2.

The event $A \cap B$, the **intersection** of A and B, consists of all outcomes that are elements of both A and B. Thus $A \cap B$ is the event {A occurs and B occurs}. The event $A \cap B$ is shaded in Figure 6–3.

TABLE 6–1

| | | Party Registration | | |
Age	Democrat (D)	Republican (R)	Independent (I)	Total
Young (Y)	7	3	10	20
Middle-aged (M)	12	18	6	36
Elderly (E)	6	9	2	17
Total	25	30	18	73

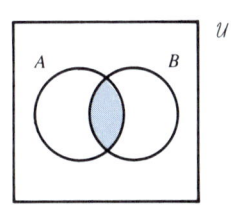

FIGURE 6–3
The Intersection of Events A and B

To illustrate these ideas in a concrete way, we present Table 6–1, which cross-classifies the registered voters in a small town according to their ages and party registrations. Suppose we choose a voter at random from among the 73 registered. Then

$$P(\text{voter chosen is not a Democrat}) = P(D') = \frac{73 - 25}{73} = \frac{48}{73},$$

since 48 of the voters are not Democrats. Similarly,

$$P(M') = P(\text{voter chosen is not middle-aged}) = \frac{73 - 36}{73} = \frac{37}{73}.$$

Also,

$$P(\text{voter chosen is middle-aged or elderly}) = P(M \cup E) = \frac{36 + 17}{73} = \frac{53}{73};$$

$P(M \cup R) = P(\text{voter chosen is middle-aged or a Republican})$

$$= \frac{12 + 18 + 6 + 3 + 9}{73} = \frac{48}{73};$$

$$P(\text{voter chosen is a young Democrat}) = P(Y \cap D) = \frac{7}{73};$$

and

$$P(E \cap I) = P(\text{voter chosen is an elderly Independent}) = \frac{2}{73}.$$

In the remainder of this section we consider complementary events and unions of events separately.

Complementary Events

Let $\mathcal{U}$ be a sample space and A an event in $\mathcal{U}$. As Figure 6–4 indicates, the number of outcomes in A plus the number in A' equals the number in $\mathcal{U}$:

$$n(A) + n(A') = n(\mathcal{U}).$$

(The dots in the figure represent outcomes.) Therefore,

$$P(A') = \frac{n(A')}{n(\mathcal{U})} = \frac{n(\mathcal{U}) - n(A)}{n(\mathcal{U})} = 1 - \frac{n(A)}{n(\mathcal{U})} = 1 - P(A).$$

Thus,

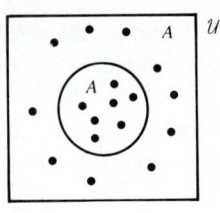

FIGURE 6–4

> For any event A,
> $$P(A') = 1 - P(A).$$

EXAMPLE 1 If $P(A) = 2/5$, then $P(A') = 1 - 2/5 = 3/5$. ☐

EXAMPLE 2 Suppose we draw a card from an ordinary deck of 52 cards. Then

$$P(\text{we draw a heart}) = 13/52,$$

since there are 13 hearts in the deck. Therefore,

$$P(\text{we do not draw a heart}) = 1 - 13/52 = 39/52 = 3/4. \quad \square$$

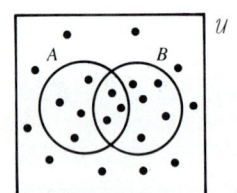

Union of Events

Let $\mathcal{U}$ be a sample space, and let A and B be events in $\mathcal{U}$. If we wish to find $n(A \cup B)$, we may count the elements of A and the elements of B separately; but then we will have counted the elements of $A \cap B$ twice, so to obtain $n(A \cup B)$ we must subtract $n(A \cap B)$ once. Thus

$$n(A \cup B) = n(A) + n(B) - n(A \cap B).$$

FIGURE 6–5 See Figure 6–5. It follows that

> For any events A and B,
> $$P(A \cup B) = P(A) + P(B) - P(A \cap B).$$

EXAMPLE 3 Suppose $P(A) = 0.30$, $P(B) = 0.50$, and $P(A \cap B) = 0.10$. Then

$$P(A \cup B) = P(A) + P(B) - P(A \cap B) = 0.30 + 0.50 - 0.10 = 0.70. \quad \square$$

EXAMPLE 4 Suppose we draw a card from an ordinary deck and we wish to find

$$P(\text{draw either a heart or a king}).$$

Since the deck has 13 hearts, 4 kings, and 1 king of hearts, we have

$P(\text{draw either a heart or a king})$

$$= P(\text{draw a heart}) + P(\text{draw a king}) - P(\text{draw a heart and a king})$$

$$= P(\text{draw a heart}) + P(\text{draw a king}) - P(\text{draw king of hearts})$$

$$= {}^{13}/_{52} + {}^{4}/_{52} - {}^{1}/_{52}$$

$$= {}^{16}/_{52} = {}^{4}/_{13}. \quad \square$$

In the special case where A and B are **mutually exclusive events,** also called **disjoint events,** we have

$$A \cap B = \emptyset.$$

This means that $A \cap B$ is an impossible event and hence that

$$P(A \cap B) = 0.$$

Therefore

If A and B are mutually exclusive events, then
$$P(A \cup B) = P(A) + P(B).$$

EXAMPLE 5 If $P(A) = 0.50$, $P(B) = 0.30$, and A and B are mutually exclusive events, then

$$P(A \cup B) = P(A) + P(B) = 0.80. \quad \square$$

EXAMPLE 6 Suppose we draw a card from an ordinary deck and wish to find

$$P(\text{draw either a king or an ace}).$$

Since it is impossible to draw a single card which is both a king and an ace, the events {draw a king} and {draw an ace} are mutually exclusive. Thus,

$$P(\text{draw either a king or an ace}) = P(\text{draw a king}) + P(\text{draw an ace})$$

$$= {}^{4}/_{52} + {}^{4}/_{52}$$

$$= {}^{8}/_{52} = {}^{2}/_{13}. \quad \square$$

6.2 EXERCISES

Combining Events

Table 6–2 cross-classifies the students at a certain college by class and gender. In Exercises 1 through 12, find the probability that a student chosen at random from this college will be of the given category.

TABLE 6–2

	Class			
Gender	Freshman	Sophomore	Junior	Senior
Female	200	150	100	75
Male	200	180	70	90

1. a female
2. not a female
3. a junior
4. not a freshman
5. not a sophomore
6. a senior
7. a male freshman
8. a female sophomore
9. a female junior
10. a male senior
11. a sophomore or a junior
12. a female or a senior

Table 6–3 cross-classifies all pizza restaurants in the United States by the region in which they are located and whether they are outlets for one of the national chains or independent operators. Suppose a pizza restaurant is selected at random. In Exercises 13 through 24, find the probability that the restaurant chosen will belong to the given category, and state the category in words.

13. N
14. H
15. S'
16. E'
17. $E \cap H$
18. $N \cap Q$
19. $Q' \cap W$
20. $I \cap E'$
21. $E \cup W$
22. $S \cup Q$
23. $H \cup S'$
24. $I' \cup N$

TABLE 6–3

	Region			
Affiliation	North (N)	East (E)	South (S)	West (W)
Pizza House (H)	300	300	800	1200
Quakey's (Q)	500	700	200	200
Independent (I)	600	1000	500	900

Complementary Events

25. If $P(A) = 1/3$, find $P(A')$.
26. If $P(A) = 3/7$, find $P(A')$.
27. If $P(A) = 0.623$, find $P(A')$.
28. If $P(A) = 0$, find $P(A')$. Interpret.
29. If $P(A) = 1$, find $P(A')$. Interpret.

Suppose you draw a card from an ordinary deck. In Exercises 30 through 34 find the given probability.

30. P(it is not a spade)

31. P(it is not a king)

32. P(it is not a face card) (Face cards are jacks, queens, kings, and aces.)

33. P(it is neither a spade nor a king)

34. P(it is not the king of spades)

35. If P(you will get the job you have applied for) $= 0.70$, find P(you won't get the job you have applied for).

Union of Events

36. If $P(A) = 1/5$, $P(B) = 4/5$, and $P(A \cap B) = 2/5$, find $P(A \cup B)$.

37. If $P(A) = 4/7$, $P(B) = 5/7$, and $P(A \cap B) = 3/7$, find $P(A \cup B)$.

38. If $P(A) = 0.39$, $P(B) = 0.66$, and $P(A \cap B) = 0.12$, find $P(A \cup B)$.

39. Is it possible to have $P(A) = 0.62$, $P(B) = 0.58$, and $P(A \cap B) = 0.10$? Why or why not?

40. If $P(A) = 0.44$, $P(B) = 0.68$, and $A \cup B$ is certain to occur, find $P(A \cap B)$.

41. If $P(A) = 4/11$, $P(B) = 3/11$, and A and B are mutually exclusive, find $P(A \cup B)$.

42. If $P(A) = 0.8$, $P(B) = 0.2$, and A and B are mutually exclusive, find $P(A \cup B)$. Interpret.

43. If $P(A) = 0.25$, $P(B) = 0.31$, and A and B are mutually exclusive, find $P(A \cup B)$.

44. If $P(A) = 0.25$, $P(B) = 0.37$, and $P(A \cup B) = 0.62$, are A and B mutually exclusive?

45. If $P(A) = 3/5$, $P(B) = 4/5$, and $P(A \cup B) = 4/5$, are A and B mutually exclusive?

46. If $P(A) = 3/11$, $P(B) = 4/11$, and $P(A \cup B) = 20/33$, are A and B mutually exclusive?

Suppose you draw a card from an ordinary deck. In Exercises 47 through 51 find the given probability.

47. P(it is either a 10 or a jack)

48. P(it is either a spade or a jack)

49. P(it is either a spade or a club)

50. P(it is either a spade or a king)

51. P(it is either a heart or a face card)

A vase has four marbles in it: two red ones, one white one, and one blue one. You draw a marble from the vase. In Exercises 52 through 55 find the given probability.

52. P(marble drawn is red)

53. P(marble drawn is white)

54. P(marble drawn is either red or white)

55. P(marble drawn is either white or blue)

Management

Table 6–4 cross-classifies the accounts receivable of a business according to amount and when they were paid. Exercises 56 through 63 refer to this table.

TABLE 6-4

Amount of Account	Account Paid		
	On time (D)	Late (E)	Never (F)
Less then $100 (A)	30	6	1
$100 through $500 (B)	22	12	2
More than $500 (C)	16	7	4

56. Find P(an account will be paid on time).
57. Find P(an account for more than $500 will never be paid).
58. Find P(an account will not be paid late).
59. Find P(an account will either be for less than $100 or it will be paid on time).
60. Find $P(B')$, and state the event B' in words.
61. Find $P(B \cap E)$, and state the event $B \cap E$ in words.
62. Find $P(C \cup F)$, and state the event $C \cup F$ in words.
63. Find $P(A \cup C)$, and state the event $A \cup C$ in words.

Social Science

A psychologist is studying the effects of sleep deprivation on memory. She asks her subjects to memorize a list of words and then to repeat the list the next day. Some of the subjects are allowed 8 hours of sleep between memorization and recall, others are allowed only 4 hours, and some are not allowed any sleep at all. The results of the experiment are as shown in Table 6–5. Exercises 64 through 71 refer to this table. In each case, find the given probability, and state the event in words.

TABLE 6-5

Hours of Sleep	Number Who Recall		
	Entire List (D)	Part of List (E)	None of List (F)
8 (A)	26	42	8
4 (B)	13	52	20
0 (C)	7	23	44

64. $P(B')$
65. $P(A \cap F)$
66. $P(A \cup C)$
67. $P(B \cup E)$
68. $P(B \cap E')$
69. $P(C \cup E')$
70. $P((C \cup E)')$
71. $P((A \cap C)')$

6.3 CONDITIONAL PROBABILITIES

The likelihood of an event will often depend on the occurrence of some other event. For example, on a given summer day, the probability of going on a picnic if it is raining is likely to be different from the probability of going on a picnic if it is not raining. Events such as {going on a picnic if it is raining} are known as **conditional**

Conditional Events

events, and their probabilities are **conditional probabilities.** In this section we discuss conditional probabilities and their applications.

Conditional Events

Let A and B be events. We define the **conditional events** $A \mid B$ and $B \mid A$ as follows:

> **Conditional Events**
>
> $A \mid B$ is the event $\{A$ occurs given that B has occurred$\}$;
> $B \mid A$ is the event $\{B$ occurs given that A has occurred$\}$.

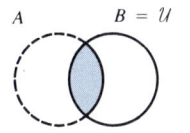

FIGURE 6-6
The Conditional Event $A|B$

The conditional event $A \mid B$ is depicted in Figure 6–6. Note that in the figure we have indicated that B is playing the role of the sample space $\mathcal{U}$. This is so because, when considering the event $A \mid B$, event B has already occurred, and hence we are interested only in those outcomes belonging to B that also belong to A.

To illustrate some conditional events and their probabilities, consider Table 6–6, which cross-classifies the students at a certain college according to their gender and whether they reside in the college dormitories or commute to school from their homes. Suppose we choose a student at random. Then

P(student chosen is female given that she lives in the dormitory)

$$= P(F \mid D) = 1400/2400 = 7/12.$$

This is so because we know to begin with that the student chosen is one of the 2400 who live in a dormitory; hence we have only those 2400 students to select from, and of them, 1400 are females. Similarly,

P(student chosen is a commuter given that he is male)

$$= P(C \mid M) = 500/1500 = 1/3.$$

Also,

$P(D \mid F) = P$(student chosen is a dormitory resident given that she is female)

$$= 1400/1600 = 7/8,$$

TABLE 6-6

Gender	Dormitory Residents (D)	Commuters (C)	Total
Male (M)	1000	500	1500
Female (F)	1400	200	1600
Total	2400	700	3100

and

$P(M \mid C) = P$(student chosen is male given that he is a commuter)

$= {}^{500}/_{700} = {}^{5}/_{7}.$

Note that $P(F \mid D) \neq P(D \mid F)$ and $P(C \mid M) \neq P(M \mid C)$.

EXAMPLE 1 Suppose we draw a card from an ordinary deck, then draw a second card without replacing the first one. Then

P(2nd card is an ace given that 1st card was an ace) $= {}^{3}/_{51},$

since there are 3 aces and 51 cards left on the second draw. Similarly,

P(2nd card is an ace given that 1st card was not an ace) $= {}^{4}/_{51},$

P(2nd card is a king given that 1st card was an ace) $= {}^{4}/_{51},$

P(2nd card is a spade given that 1st card was a spade) $= {}^{12}/_{51},$

and so on. ◻

EXAMPLE 2 Suppose we again draw two cards from a deck, but that we replace the first card before we draw the second one. In this case,

P(2nd card is an ace given that 1st card was an ace) $= {}^{4}/_{52},$

since on the second draw there are still 4 aces and 52 cards to draw from. Similarly,

P(2nd card is an ace given that 1st card was not an ace) $= {}^{4}/_{52},$

P(2nd card is a king given that 1st card was an ace) $= {}^{4}/_{52},$

P(2nd card is a spade given that 1st card was a spade) $= {}^{13}/_{52},$

and so on. ◻

Examples 1 and 2 illustrate the difference between drawing *with replacement* (Example 2) and drawing *without replacement* (Example 1).

We have already noted that when considering a conditional event $A \mid B$, the event B plays the role of the sample space because we are concerned only with elements of B to begin with. Therefore

$$P(A \mid B) = \frac{n(A \mid B)}{n(B)}.$$

But, as Figure 6–6 shows, those elements of A that can occur when B has already occurred are exactly the elements of $A \cap B$. Therefore

$$P(A \mid B) = \frac{n(A \cap B)}{n(B)},$$

and it follows that

$$P(A \mid B) = \frac{P(A \cap B)}{P(B)}.$$

A similar formula holds for $P(B \mid A)$.

Conditional Probabilities

For any events A and B,

$$P(A \mid B) = \frac{P(A \cap B)}{P(B)}, \qquad \text{provided that } P(B) \neq 0,$$

and

$$P(B \mid A) = \frac{P(A \cap B)}{P(A)}, \qquad \text{provided that } P(A) \neq 0.$$

EXAMPLE 3   Suppose $P(A \cap B) = 0.20$, $P(A) = 0.30$, and $P(B) = 0.40$. Then

$$P(A \mid B) = \frac{0.20}{0.40} = 0.50 \qquad \text{and} \qquad P(B \mid A) = \frac{0.20}{0.30} \approx 0.67. \quad \square$$

Events are said to be **independent** if they do not affect one another. Events that do affect one another are said to be **dependent**. For example, suppose we draw a card from a deck, replace it, and then draw another card. The two draws are independent, because what happens on either draw does not depend on what happens on the other. (See Example 2.) However, if we draw a card from a deck and then draw another without replacing the first, the draws are dependent, because clearly the result of the second draw will depend on what happens on the first draw. (See Example 1.) If two events A and B are independent, then

$$P(A \mid B) = P(A),$$

because, since event B does not affect event A, the probability that A will occur must be the same whether or not B has already occurred. Similarly, if A and B are independent, then

$$P(B \mid A) = P(B).$$

On the other hand, if

$$P(A \mid B) = P(A) \qquad \text{and} \qquad P(B \mid A) = P(B),$$

then B does not affect A and A does not affect B, so A and B must be independent. Therefore:

> **Independent Events**
>
> Events A and B are independent if and only if
> $$P(A \mid B) = P(A) \quad \text{and} \quad P(B \mid A) = P(B).$$

EXAMPLE 4 Suppose $P(A) = 0.20$, $P(B) = 0.30$, and $P(A \cap B) = 0.10$. Are A and B independent events? No, because
$$P(A \mid B) = \frac{P(A \cap B)}{P(B)} = \frac{0.10}{0.30} \neq P(A). \quad \square$$

EXAMPLE 5 Suppose $P(A) = 0.20$, $P(B) = 0.30$, and $P(A \cap B) = 0.06$. Are A and B independent events? Yes, because
$$P(A \mid B) = \frac{P(A \cap B)}{P(B)} = \frac{0.06}{0.30} = 0.20 = P(A)$$

and

$$P(B \mid A) = \frac{P(A \cap B)}{P(A)} = \frac{0.06}{0.20} = 0.30 = P(B). \quad \square$$

Intersection of Events

Let $\mathcal{U}$ be a sample space, and let A and B be events in $\mathcal{U}$. If $P(B) \neq 0$ and we solve
$$P(A \mid B) = \frac{P(A \cap B)}{P(B)}$$
for $P(A \cap B)$, we obtain the following formula for the intersection of two events:

> **The Probability of an Intersection of Events**
>
> For any events A and B,
> $$P(A \cap B) = P(A \mid B) \cdot P(B).$$

(This formula also holds if $P(B) = 0$. Why?)

In the special case where A and B are known to be independent events, $P(A \mid B) = P(A)$, so the intersection formula reduces to the following:

The Probability of an Intersection of Independent Events

If A, B are independent events,

$$P(A \cap B) = P(A) \cdot P(B).$$

EXAMPLE 6 Suppose $P(A) = 0.20$, $P(B) = 0.35$, and A and B are known to be independent events. Then

$$P(A \cap B) = P(A) \cdot P(B) = (0.20)(0.35) = 0.07. \quad \square$$

EXAMPLE 7 Suppose $P(A) = 0.20$, $P(B) = 0.35$, and $P(A \mid B) = 0.10$. Note that A and B are not independent, since $P(A \mid B) \neq P(A)$. Thus

$$P(A \cap B) = P(A \mid B) \cdot P(B) = (0.10)(0.35) = 0.035. \quad \square$$

When dealing with a sequence of events, we sometimes find it convenient to depict the events as branches in a **tree diagram.** Each path through the tree diagram corresponds to an outcome of the experiment, and the probability of an outcome is the product of the probabilities of the events along its path.

EXAMPLE 8 A vase has 10 marbles in it, 6 red *(R)* and 4 white *(W)*. Suppose we draw a marble from the vase, replace it, and then draw another. (Note that since we are drawing with replacement, the draws are independent.)

A tree diagram for this experiment is shown in Figure 6–7. We have placed the probability of each event on its branch. The probability of an outcome is just

First draw	Second draw	Outcome	Probability
Red $\frac{6}{10}$	Red $\frac{6}{10}$	R_1R_2	$\frac{6}{10} \cdot \frac{6}{10} = \frac{36}{100}$
	White $\frac{4}{10}$	R_1W_2	$\frac{6}{10} \cdot \frac{4}{10} = \frac{24}{100}$
White $\frac{4}{10}$	Red $\frac{6}{10}$	W_1R_2	$\frac{4}{10} \cdot \frac{6}{10} = \frac{24}{100}$
	White $\frac{4}{10}$	W_1W_2	$\frac{4}{10} \cdot \frac{4}{10} = \frac{16}{100}$

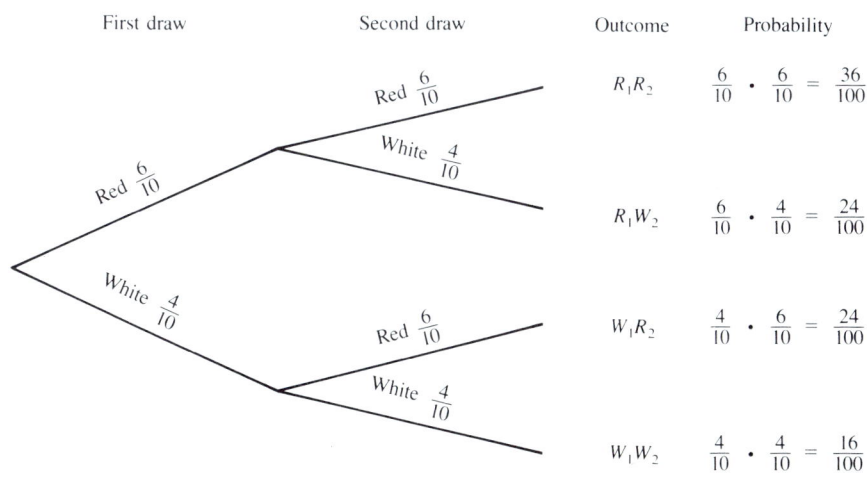

FIGURE 6–7

the product of the probabilities along the branches leading to the outcome, by virtue of the formula

$$P(A \cap B) = P(A) \cdot P(B)$$

for independent events.

Now suppose that we are asked to find

$$P(\text{both marbles drawn are red}) = P(R_1 \cap R_2).$$

We have

$$P(R_1 \cap R_2) = P(R_1)P(R_2) = \frac{6}{10} \cdot \frac{6}{10} = \frac{36}{100} = \frac{9}{25}.$$

Similarly, the probability that just one of the two marbles drawn is red is

$$P((R_1 \cap W_2) \cup (W_1 \cap R_2)) = P(R_1 \cap W_2) + P(W_1 \cap R_2)$$

$$= P(R_1)P(W_2) + P(W_1)P(R_2)$$

$$= \frac{6}{10} \cdot \frac{4}{10} + \frac{4}{10} \cdot \frac{6}{10} = \frac{48}{100} = \frac{12}{25},$$

and

$$P(\text{1st marble drawn is white}) = P((W_1 \cap R_2) \cup (W_1 \cap W_2))$$

$$= P(W_1 \cap R_2) + P(W_1 \cap W_2)$$

$$= P(W_1)P(R_2) + P(W_1)P(R_2)$$

$$= \frac{4}{10} \cdot \frac{6}{10} + \frac{4}{10} \cdot \frac{4}{10} = \frac{40}{100} = \frac{2}{5}. \quad \square$$

EXAMPLE 9 A vase contains 3 red (R), 5 white (W), and 2 blue (B) marbles. Suppose we draw 2 marbles from the vase, drawing the second without replacing the first. (Note that since we are drawing without replacement, the draws are not independent.)

The tree diagram for this experiment is shown in Figure 6–8. The probabilities on the branches for the second draw are conditional probabilities, since the results of the second draw depend on those of the first draw. Note also that the probability of an outcome is again the product of the probabilities of the branches that lead to it, this time by virtue of the formula

$$P(A \cap B) = P(A) \cdot P(B \mid A).$$

Thus the probability that neither marble drawn is blue is

$$P((R_1 \cap R_2) \cup (R_1 \cap W_2) \cup (W_1 \cap R_2) \cup (W_1 \cap W_2))$$

$$= P(R_1 \cap R_2) + P(R_1 \cap W_2) + P(W_1 \cap R_2) + P(W_1 \cap W_2)$$

$$= P(R_1)P(R_2 \mid R_1) + P(R_1)P(W_2 \mid R_1) + P(W_1)P(R_2 \mid W_1) + P(W_1)P(W_2 \mid W_1)$$

$$= \frac{3}{10} \cdot \frac{2}{9} + \frac{3}{10} \cdot \frac{5}{9} + \frac{5}{10} \cdot \frac{3}{9} + \frac{5}{10} \cdot \frac{4}{9} = \frac{56}{90} = \frac{28}{45}.$$

PROBABILITY

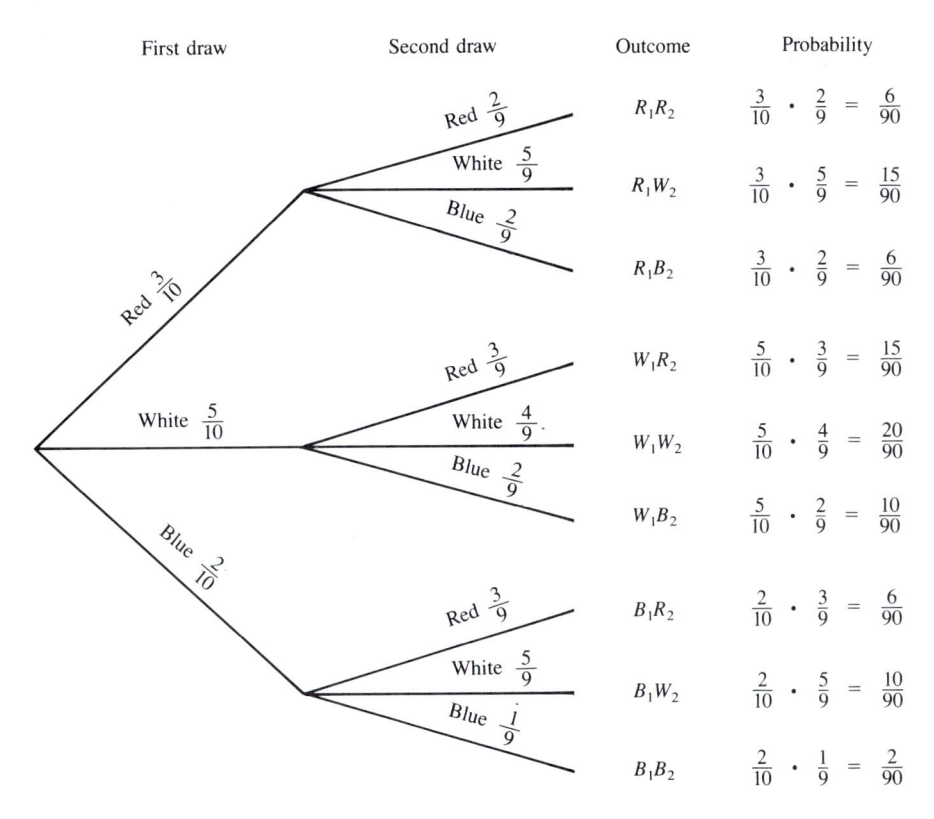

FIGURE 6–8

Similarly,

P(just one of the marbles drawn is white)

$= P((R_1 \cap W_2) \cup (W_1 \cap R_2) \cup (B_1 \cap W_2) \cup (W_1 \cap B_2))$

$= P(R_1 \cap W_2) + P(W_1 \cap R_2) + P(B_1 \cap W_2) + P(W_1 \cap B_2)$

$= P(R_1)P(W_2 \mid R_1) + P(W_1)P(R_2 \mid W_1) + P(B_1)P(W_2 \mid B_1) + P(W_1)P(B_2 \mid W_1)$

$= \dfrac{3}{10} \cdot \dfrac{5}{9} + \dfrac{5}{10} \cdot \dfrac{3}{9} + \dfrac{2}{10} \cdot \dfrac{5}{9} + \dfrac{5}{10} \cdot \dfrac{2}{9} = \dfrac{50}{90} = \dfrac{5}{9}.$ ◻

EXAMPLE 10 You are writing a four-line computer program that will run only if each line has no errors. Suppose the probability that you make no errors writing a line is 0.9, and suppose that each line's status is independent of that of the other lines. Let A_i denote the event {line i has no errors}; then

P(the program will run)

$= P$(1st and 2nd and 3rd and 4th lines have no errors)

$= P(A_1 \cap A_2 \cap A_3 \cap A_4)$

$$= P(A_1)P(A_2)P(A_3)P(A_4)$$
$$= (0.9)(0.9)(0.9)(0.9)$$
$$= 0.6561. \quad \square$$

EXAMPLE 11 During the winter, television weatherman Al Sleet predicts snow 40% of the time. The probability that it will snow when Al predicts snow is 0.1, while the probability that it will snow when he does not predict it is 0.5. Find the probability that it will snow on a given day.

Let S denote the event {it snows} and F the event {Al forecasts snow}. Then F' is the event {Al does not forecast snow}, and we have

$$P(\text{it snows}) = P(S) = P((F \cap S) \cup (F' \cap S))$$
$$= P(F \cap S) + P(F' \cap S)$$
$$= P(F)P(S \mid F) + P(F')P(S \mid F')$$
$$= (0.4)(0.1) + (0.6)(0.5)$$
$$= 0.34. \quad \square$$

We conclude this section with a summary of our probability formulas:

Probability Formulas

Let A and B be events. Then

1. $P(A') = 1 - P(A)$.
2. $P(A \cup B) = P(A) + P(B) - P(A \cap B)$.
3. If A and B are mutually exclusive events,
$$P(A \cup B) = P(A) + P(B).$$
If $A_1, \ldots, A_n$ are mutually exclusive events,
$$P(A_1 \cup \cdots \cup A_n) = P(A_1) + \cdots + P(A_n).$$
4. $P(A \mid B) = \dfrac{P(A \cap B)}{P(B)}$ if $P(B) \neq 0$.
5. $P(A \cap B) = P(A \mid B)P(B)$.
6. If A and B are independent events,
$$P(A \cap B) = P(A)P(B).$$
If $A_1, \ldots, A_n$ are independent events,
$$P(A_1 \cap \cdots \cap A_n) = P(A_1) \cdots P(A_n).$$

6.3 EXERCISES

Conditional Events

Table 6–2 cross-classified the students at a certain college by gender and class. We repeat the table here:

TABLE 6–2

	Class			
Gender	Freshman	Sophomore	Junior	Senior
Female	200	150	100	75
Male	200	180	70	90

Exercises 1 through 8 refer to this table. Find the probability that a student chosen at random will be of the given category.

1. A freshman given that he is male.
2. A male given that he is a freshman.
3. A female given that she is either a junior or a senior.
4. A junior or a senior given that she is female.
5. A freshman given that he is a male.
6. A male given that he is either a junior or a sophomore.
7. A freshman given that she is not a male.
8. A female given that she is not a senior.

Table 6–3 cross-classified pizza restaurants in the United States by region and affiliation. We repeat the table here:

TABLE 6–3

	Region			
Affiliation	North (N)	East (E)	South (S)	West (W)
Pizza House (H)	300	300	800	1200
Quakey's (Q)	500	700	200	200
Independent (I)	600	1000	500	900

Exercises 9 through 16 refer to this table. Find the given probability and also state the given conditional event in words.

9. $P(H \mid W)$
10. $P(W \mid H)$
11. $P(Q \mid E')$
12. $P(S \mid H \cup Q)$
13. $P(H \mid E \cup S)$
14. $P(S \cup W \mid Q)$
15. $P(N' \mid Q \cup I)$
16. $P(Q \mid (N \cup E)')$
17. If $P(A) = 0.30$, $P(B) = 0.20$, and $P(A \cap B) = 0.15$, find $P(A \mid B)$ and $P(B \mid A)$.

18. If $P(A) = 0.62$, $P(B) = 0.45$, and $P(A \cap B) = 0.23$, find $P(A \mid B)$ and $P(B \mid A)$.
19. If $P(A) = 2/3$, $P(B) = 1/5$, and $P(A \cap B) = 1/10$, are A and B independent?
20. If $P(A) = 1/2$, $P(B) = 1/5$, and $P(A \cap B) = 1/10$, are A and B independent?

A vase contains 10 marbles, 6 red and 4 white. You draw two marbles from the vase, replacing the first one before you draw the second one. Exercises 21 through 26 refer to this situation. Find the given probability.

21. P(1st marble drawn is red)
22. P(2nd marble drawn is red given that 1st is red)
23. P(1st marble drawn is white)
24. P(2nd marble drawn is white given that 1st is white)
25. P(2nd marble drawn is white given that 1st is red)
26. P(2nd marble drawn is red given that 1st is white)

A vase contains 10 marbles, 6 red and 4 white. You draw one marble from the vase, then draw a second without replacing the first. Exercises 27 through 32 refer to this situation. Find the given probability.

27. P(1st marble drawn is red)
28. P(2nd marble drawn is red given that 1st is red)
29. P(1st marble drawn is white)
30. P(2nd marble drawn is white given that 1st is white)
31. P(2nd marble drawn is white given that 1st is red)
32. P(2nd marble drawn is red given that 1st is white)

Management

Table 6–4 cross-classified the accounts receivable of a business by amount and when they were paid. We repeat the table here:

TABLE 6–4

Amount of Account	On Time (D)	Late (E)	Never (F)
Less than $100 (A)	30	6	1
$100 through $500 (B)	22	12	2
More than $500 (C)	16	7	4

Exercises 33 through 40 refer to this table. Find the given probability.

33. P(account will be paid late given that it is for less than $100)
34. P(account will be for more than $500 given that it is paid on time)
35. P(account will be paid given that it is for less than $100)
36. P(account is for more than $500 $\mid$ it will be paid)
37. $P(E \mid B)$; what is the event $E \mid B$?

38. $P(A \mid D')$; what is the event $A \mid D'$?

39. $P(C' \mid F)$; what is the event $C' \mid F$?

40. $P(A \cup B \mid D)$; what is the event $A \cup B \mid D$?

Social Science

Table 6–5 cross-classified the subjects of an experiment according to how much sleep they got and how well they remembered a list of words. We repeat the table here:

TABLE 6–5

	Number Who Recall		
Hours of Sleep	Entire List (D)	Part of List (E)	None of List (F)
8 (A)	26	42	8
4 (B)	13	52	20
0 (C)	7	23	44

Exercises 41 through 48 refer to this table. Find the given probability and state the given conditional event in words.

41. $P(A \mid E)$ **42.** $P(E \mid A)$ **43.** $P(B \mid E')$ **44.** $P(B \mid E \cup F)$

45. $P(C \mid D')$ **46.** $P(D \mid C')$ **47.** $P(A \cup B \mid F)$ **48.** $P(E \mid (A \cup B)')$

Intersection of Events

49. If A and B are independent events with $P(A) = 2/5$ and $P(B) = 4/5$, find $P(A \cap B)$.

50. If A and B are independent events with $P(A) = 0.36$ and $P(B) = 0.47$, find $P(A \cap B)$.

51. If A_1, A_2, and A_3 are independent events with $P(A_1) = 0.30$, $P(A_2) = 0.40$, and $P(A_3) = 0.50$, find $P(A_1 \cap A_2 \cap A_3)$.

52. If $A_1, A_2, \ldots, A_6$ are independent events that are equally likely to occur, find $P(A_1 \cap A_2 \cap \cdots \cap A_6)$.

53. If $P(B) = 2/3$ and $P(A \mid B) = 1/6$, find $P(A \cap B)$.

54. If $P(A) = 0.45$ and $P(B \mid A) = 0.27$, find $P(A \cap B)$.

A vase has eight marbles in it, five red and three blue. You draw two marbles from the vase without replacement. Exercises 55 through 58 refer to this situation. Find the given probability.

55. P(both marbles drawn are blue)

56. P(just 1 of the marbles drawn is blue)

57. P(the 2nd marble drawn is blue)

58. P(at least 1 of the marbles drawn is blue)

A vase has eight marbles in it, five red and three blue. You draw two marbles from the vase with replacement. Exercises 59 through 62 refer to this situation. Find the given probability.

59. *P*(both marbles drawn are blue)

60. *P*(just 1 of the marbles drawn is blue)

61. *P*(the 2nd marble drawn is blue)

62. *P*(at least 1 of the marbles drawn is blue)

A vase contains 12 marbles: 4 red, 7 white, and 1 blue. You draw three marbles with replacement. Exercises 63 through 66 refer to this situation. Find the given probability.

63. *P*(all 3 are red)

64. *P*(1 is red, 1 white, and 1 blue)

65. *P*(2nd marble drawn is either red or blue)

66. *P*(either all 3 are red or none are red)

A vase has 12 marbles in it: 4 red, 7 white, and 1 blue. You draw three marbles from the vase without replacement. Exercises 67 through 70 refer to this situation. Find the given probability.

67. *P*(all 3 are red)

68. *P*(1 is red, 1 white, and 1 blue)

69. *P*(2nd marble drawn is either red or blue)

70. *P*(either all 3 are red or none are red)

71. An automatic teller machine has two cardreaders, one of which serves as a backup for the other. The teller will work if either of its cardreaders works. Suppose that *P*(a cardreader works) = 0.95. Assuming that the two cardreaders work independently of one another, find *P*(the automatic teller will work).

*72. Refer to Exercise 71: How many backup cardreaders would the automatic teller need in order to have *P*(the teller will work) $\geq$ 0.9999?

Management

73. A widget machine has five components, each of which must work properly if the machine is to produce a perfect widget. Suppose each component is 99% reliable; that is,

    P(a component works properly) = 0.99.

    Assuming that the components work independently of one another, find

    P(the machine makes a perfect widget).

74. Repeat Exercise 73 if the machine has 10 components; 20 components.

75. An oil company intends to do a geological survey on a tract of land and then drill for oil on it. They have estimated the following probabilities:

    P(survey will indicate oil is present) = 0.40,

    P(survey will indicate oil is not present) = 0.60,

    P(will strike oil given that survey indicates oil present) = 0.60,

    P(will strike oil given that survey indicates oil not present) = 0.05.

    Find *P*(will strike oil).

Life Science

Suppose that

$$P(\text{a child will be a boy}) = P(\text{a child will be a girl}) = 0.5.$$

Exercises 76 through 80 refer to a couple who intend to have several children. Find the given probability.

76. $P(\text{their 2nd child will be a boy} \mid \text{their 1st child is a girl})$
77. $P(\text{their 2nd child will be a boy})$
78. $P(\text{their 1st 3 children will be girls})$
79. $P(\text{their 3rd child will be a girl})$
80. $P(\text{at least 2 of their first 4 children will be girls})$

Suppose the probability that two brown-eyed parents will have a blue-eyed child is $1/3$. Exercises 81 through 84 refer to a family of two brown-eyed parents and four children. Find the given probability.

81. $P(\text{none of the 4 children are blue-eyed})$
82. $P(\text{1 of the 4 children is blue-eyed})$
83. $P(\text{2 of the 4 children are blue-eyed})$
84. $P(\text{all of the 4 children are blue-eyed})$

Sickle-cell anemia is a genetic disease controlled by two genes: the dominant S-gene and the recessive s-gene. A child receives either gene S or gene s from each parent. If the child receives two s-genes (ss), it will develop the disease. If it receives one s and one S (sS), it will not develop the disease, but can pass it on to its offspring and hence is a carrier of the disease. If the child receives two S genes (SS), it cannot develop the disease and cannot be a carrier. A parent can only pass on to a child a gene that the parent has; if the parent is an sS-person, the probability that the child will receive the s-gene is $1/2$. Prospective parents can now be tested to see whether or not they are carriers of the disease and thus be advised as to the chances that their children will develop sickle cell anemia or be carriers of it. Two prospective parents, neither of whom has sickle cell anemia, are tested, and it is found that one is an SS-person while the other is an sS-person. Exercises 85 through 90 refer to this situation. Find the given probability.

85. $P(\text{their first child will be an ss-person and thus develop the disease})$
86. $P(\text{their first child will be a carrier of the disease (an sS-person)})$
87. $P(\text{their first child will neither develop the disease nor be a carrier (an SS-person)})$
88. $P(\text{at least one of their first two children will develop the disease})$
89. $P(\text{at least one of their first two children will be a carrier (an sS-person)})$
90. $P(\text{both of their first two children will be SS-persons})$

Referring to the situation of Exercises 85 through 90, suppose both parents are sS-persons. In Exercises 91 through 96 find the given probability.

91. $P(\text{their first child will be an ss-person and thus develop the disease})$
92. $P(\text{their first child will be a carrier of the disease (an sS-person)})$
93. $P(\text{their first child will neither develop the disease nor be a carrier (an SS-person)})$

94. *P*(at least one of their first two children will develop the disease)
95. *P*(at least one of their first two children will be a carrier (an sS-person))
96. *P*(both of their first two children will be SS-persons)

6.4 COUNTING

In order to find probabilities we must be able to count outcomes. Up until now, this has been easy for us to do, since we have essentially been able to list all the outcomes of the experiments we have considered. However, in many common situations it is simply not possible to list all outcomes of an experiment. For instance, suppose we deal a hand of 13 cards from an ordinary deck of 52 cards. How many different such hands are there? The answer is approximately 635,000,000,000, so we cannot possibly list the outcomes of this experiment. Thus we must develop methods of counting that do not depend on making lists. This section is devoted to a discussion of several such methods, namely, the **Multiplication Principle, permutations,** and **combinations,** and their application to probability theory. Before we begin our discussion of these methods, however, we must introduce the notion of the **factorial** of a nonnegative integer.

Factorials

If *n* is a nonnegative integer, the symbol *n*! is called **n factorial.** By definition,

$$0! = 1,$$
$$1! = 1.$$

Then for $n > 1$, we define

$$n! = n(n - 1)(n - 2) \cdots 2 \cdot 1.$$

Hence

$$2! = 2 \cdot 1 = 2, \quad 3! = 3 \cdot 2 \cdot 1 = 6, \quad 4! = 4 \cdot 3 \cdot 2 \cdot 1 = 24,$$
$$5! = 5 \cdot 4 \cdot 3 \cdot 2 \cdot 1 = 120, \quad 6! = 6 \cdot 5 \cdot 4 \cdot 3 \cdot 2 \cdot 1 = 720,$$

and so on.

Factorials build up very rapidly to large numbers. For instance,

$$11! = 39,916,800$$

and

$$20! \approx 2,432,902,000,000,000,000.$$

As we shall see later in this section, factorials are useful in counting outcomes.

PROBABILITY

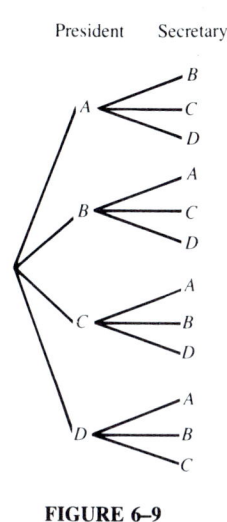

The Multiplication Principle and Permutations

Suppose a club has four members, whom we will designate as A, B, C, and D, and we wish to pick a president and a secretary at random from among the club members. In how many different ways can we do this, assuming that no member of the club can hold more than one office?

Suppose we choose the president first, then the secretary. Figure 6–9 is a tree diagram that shows all the ways of selecting a president and secretary for the club. Note that we have four ways to choose the president, then three ways to choose the secretary, for a total of

$$4 \cdot 3 = 12$$

different choices of officers. In solving this problem, we have made use of the following principle:

FIGURE 6–9

The Multiplication Principle

If event A_1 can occur in n_1 ways, event A_2 can occur in n_2 ways, . . . , and event A_k can occur in n_k ways, then, provided the events do not influence one another, A_1 followed by A_2 followed by . . . followed by A_k can occur in

$$n_1 \cdot n_2 \cdots n_k$$

ways.

EXAMPLE 1 A supermarket sells eight different brands of bread. Each brand comes in small, medium, and large sizes, and each size is available in rye, white, whole wheat, and bran. How many different choices of bread does the market offer?

We apply the multiplication principle, with the aid of a table:

	Brand	Size	Type	
Number of Choices	8 ·	3 ·	4	= 96

Therefore, there are 96 different choices available. ☐

EXAMPLE 2 A compact disk has 10 musical selections on it, and we wish to play 4 of them. In how many ways can we play 4 selections chosen from 10 selections in order to obtain different sequences of music?

We have

	1st sel	2nd sel	3rd sel	4th sel	
Number of Choices	10 ·	9 ·	8 ·	7	= 5040

Therefore, there are 5040 different choices. ☐

EXAMPLE 3 Massachusetts license plates consist of three digits followed by three letters, such as 123 ABC, for instance. How many different Massachusetts license plates are possible if no plate number can begin with a zero and the letters O and Q cannot be used?

Since the plate number cannot begin with a zero, there are 9 choices for its first digit and 10 for each of the other two digits. Since the letters O and Q cannot be used, there are 24 choices for each of the three letters. Hence by the Multiplication Principle we have

Digit			Letter		
1st	2nd	3rd	1st	2nd	3rd
9 ·	10 ·	10 ·	24 ·	24 ·	24 = 12,441,600

different plates possible. ∎

A **permutation** of the elements of a set is an ordering of the elements of the set. In Example 2 we found all permutations of 4 things chosen from a set of 10 things. The multiplication principle told us there were 5040 such permutations.

EXAMPLE 4 How many permutations are there of three things chosen from a set of seven things? In other words, in how many different orders can we choose three elements from a set of seven elements?

The Multiplication Principle tells us that the answer is

$$7 \cdot 6 \cdot 5 = 210,$$

since there are seven choices for the first element, then six choices for the second, and finally five choices for the third. ∎

In general, if we seek to find all permutations of k things chosen from n things, where $k \leq n$, then we have

	1st Thing	2nd Thing	3rd Thing	$\cdots$	kth Thing
Number of Choices	n ·	$(n-1)$ ·	$(n-2)$	$\cdots$	$(n-k+1)$

Hence the number of such permutations is

$$n(n-1)(n-2)\cdots(n-k+1)$$
$$= \frac{n(n-1)(n-2)\cdots(n-k+1)(n-k)(n-k-1)\cdots 2 \cdot 1}{(n-k)(n-k-1)\cdots 2 \cdot 1} = \frac{n!}{(n-k)!}.$$

Therefore, as a special case of the Multiplication Principle we have the following formula.

The Permutation Formula

The number of permutations $P(n, k)$ of k things chosen from n things, $0 \leq k \leq n$, is

$$P(n, k) = \frac{n!}{(n-k)!}.$$

EXAMPLE 5 Find the number of permutations of 5 things chosen from 10 things.
Here $n = 10$ and $k = 5$, so the number of permutations is

$$P(10, 5) = \frac{10!}{(10 - 5)!} = \frac{10!}{5!} = \frac{10 \cdot 9 \cdot 8 \cdot 7 \cdot 6 \cdot 5 \cdot 4 \cdot 3 \cdot 2 \cdot 1}{5 \cdot 4 \cdot 3 \cdot 2 \cdot 1}$$

$$= 10 \cdot 9 \cdot 8 \cdot 7 \cdot 6 = 30,240.$$

The same result can be obtained directly from the Multiplication Principle. Since there are 10 choices for the first thing, 9 for the second, . . . , 6 for the fifth, the Multiplication Principle tells us that there are

$$10 \cdot 9 \cdot 8 \cdot 7 \cdot 6 = 30,240$$

different choices.

EXAMPLE 6 There are eight candidates running for town council in the town of Dudley. In how many different orders can their names be listed on the ballot? In other words, how many permutations are there of eight names chosen from eight names? The answer is

$$8 \cdot 7 \cdot 6 \cdot 5 \cdot 4 \cdot 3 \cdot 2 \cdot 1 = 40,320.$$

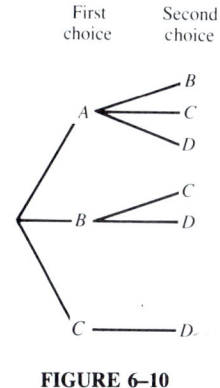

Combinations

Again consider a club with four members, designated by A, B, C, and D. This time we intend to choose two club members to serve on a committee. How many different committees are possible? Here we are not interested in the order in which the club members are chosen, but only in which of them end up on the committee. The tree diagram of Figure 6–10 shows that there are six different committees possible. We have thus found that there are six **combinations** of two things chosen from four things: a **combination** of the elements of a set is a subset of the set without regard to the order in which the elements of the subset were chosen.

FIGURE 6–10

EXAMPLE 7 Find the number of combinations of four things chosen from six things. In other words, find the number of different subsets of four elements that can be chosen from a set of six elements.

First, there are $6 \cdot 5 \cdot 4 \cdot 3$ different permutations of four things chosen from six things. However, some of these permutations yield the same combination: two permutations yield the same combination if they consist of the same four elements arranged in different orders. The same four elements can be arranged in $4 \cdot 3 \cdot 2 \cdot 1$ different ways, so there are $4 \cdot 3 \cdot 2 \cdot 1$ permutations that consist of exactly the same four elements. Hence there are

$$\frac{6 \cdot 5 \cdot 4 \cdot 3}{4 \cdot 3 \cdot 2 \cdot 1} = 15$$

combinations of four things chosen from six things.

The reasoning of Example 7 can be used to show that the number of combinations of k things chosen from n things is

$$\frac{P(n, k)}{k!} = \frac{n!}{k!(n - k)!}.$$

This number is denoted by the **binomial coefficient** $\binom{n}{k}$ or by $C(n, k)$.

The Combination Formula

The number of combinations of k things chosen from n things, $0 \leq k \leq n$, is

$$C(n, k) = \binom{n}{k} = \frac{n!}{k!(n - k)!}.$$

EXAMPLE 8 How many combinations are there of 6 things chosen from 10 things? In other words, how many different subsets of 6 elements can be chosen from a set of 10 elements?

Here, $n = 10$ and $k = 6$, so the answer is

$$C(10, 6) = \binom{10}{6} = \frac{10!}{6!(10 - 6)!} = \frac{10!}{6!4!} = \frac{10 \cdot 9 \cdot 8 \cdot 7 \cdot 6 \cdot 5 \cdot 4 \cdot 3 \cdot 2 \cdot 1}{6 \cdot 5 \cdot 4 \cdot 3 \cdot 2 \cdot 1 \cdot 4 \cdot 3 \cdot 2 \cdot 1}$$

$$= \frac{10 \cdot \overset{3}{\cancel{9}} \cdot \cancel{8} \cdot 7}{\cancel{4} \cdot \cancel{3} \cdot \cancel{2} \cdot 1} = 10 \cdot 3 \cdot 7 = 210. \quad \square$$

EXAMPLE 9 There are seven candidates for town council in Dudley. Three will be elected. How many different town councils are possible? In other words, how many combinations are there of three things chosen from seven things? The answer is

$$C(7, 3) = \binom{7}{3} = \frac{7!}{3!(7 - 3)!} = \frac{7!}{3!4!} = \frac{7 \cdot 6 \cdot 5 \cdot 4 \cdot 3 \cdot 2 \cdot 1}{3 \cdot 2 \cdot 1 \cdot 4 \cdot 3 \cdot 2 \cdot 1} = \frac{7 \cdot \cancel{6} \cdot 5}{\cancel{3} \cdot \cancel{2} \cdot 1}$$

$$= 7 \cdot 5 = 35. \quad \square$$

Applications to Probability

The Multiplication Principle, permutations, and combinations are useful in probability problems because they provide efficient ways of counting the number of outcomes of experiments. As we proceed, keep the following in mind:

Keys to Counting Outcomes

If the order of events is important, think of the Multiplication Principle and/or permutations.

If the order of events is not important, think of combinations.

EXAMPLE 10 A club has 15 members. The club is to choose a president, vice president, secretary, and treasurer. If Schwartz and O'Brien are both club members, what is the probability that Schwartz will be president and O'Brien will be secretary, assuming that the selection is made randomly and that no member can hold more than one office?

We think of choosing the president first, then the vice president, then the secretary, then the treasurer; thus, the order of events is important here, so we think of permutations. There are $P(15, 4)$ ways to choose 4 people from 15 people when order is important. If Schwartz is to be president, and O'Brien secretary, there are $P(13, 2)$ ways to choose the remaining 2 officers from the remaining 13 people. Hence,

$$P(\text{Schwartz President and O'Brien Secretary}) = \frac{P(13, 2)}{P(15, 4)} = \frac{13 \cdot 12}{15 \cdot 14 \cdot 13 \cdot 12}$$

$$= \frac{1}{210}. \quad \square$$

EXAMPLE 11 A firm has four promotions available for its employees. Fifteen employees are competing for the promotions. If D'Amore and Yee are among the 15, and if the promotions will be awarded randomly, what is the probability that both D'Amore and Yee will receive a promotion?

Here order is not important: we merely wish to choose 4 people to be promoted from among 15 people, without regard to the order in which they are chosen. Therefore, we think of combinations. There are

$$C(15, 4) = \binom{15}{4} = \frac{15!}{4!11!} = 15 \cdot 7 \cdot 13$$

ways to choose 4 people from 15 people without regard to order. If D'Amore and Yee are both to receive promotions, the remaining 2 people promoted must be chosen from the remaining 13 people eligible, without regard to order. There are

$$C(13, 2) = \binom{13}{2} = \frac{13!}{2!11!} = 13 \cdot 6$$

ways to do this. Hence

$$P(\text{D'Amore and Yee both promoted}) = \frac{C(13, 2)}{C(15, 4)} = \frac{13 \cdot 6}{15 \cdot 7 \cdot 13} = \frac{2}{35}. \quad \square$$

EXAMPLE 12 A club has 12 members, 6 men and 6 women. A committee of five is to be chosen at random from among the club members. What is the probability that the committee will consist of three women and two men?

There are

- $\binom{6}{3}$ ways to choose the three women from the six women.

- $\binom{6}{2}$ ways to choose the two men from the six men.

- $\binom{12}{5}$ ways to choose the committee.

Therefore

$$P(3 \text{ women and 2 men on committee}) = \frac{\binom{6}{3} \cdot \binom{6}{2}}{\binom{12}{5}} = \frac{25}{66}.$$

EXAMPLE 13 In the game of bridge, each of four players is dealt 13 cards from an ordinary 52-card deck. Find the probability that a player will be dealt a hand that contains exactly seven spades or exactly nine hearts.

There are

- $\binom{13}{7}\binom{39}{6}$ ways to be dealt 7 spades from the 13 spades *and* 6 nonspades from the 39 nonspades.

- $\binom{13}{9}\binom{39}{4}$ ways to be dealt 9 hearts from the 13 hearts *and* 4 nonhearts from the 39 nonhearts.

- $\binom{52}{13}$ ways to be dealt 13 cards from 52 cards.

Therefore, since the events {dealt exactly 7 spades} and {dealt exactly 9 hearts} are mutually exclusive,

$P(\text{dealt exactly 7 spades or exactly 9 hearts})$

$$= \frac{\binom{13}{7}\binom{39}{6} + \binom{13}{9}\binom{39}{4}}{\binom{52}{13}} \approx 0.0089.$$

■ *Computer Manual: Program PERMCOMB*

6.4 EXERCISES

Factorials

In Exercises 1 through 10, evaluate the given expression.

1. $7!$
2. $8!$
3. $9!$
4. $\dfrac{10!}{3!5!}$
5. $\dfrac{12!}{7!2!}$
6. $\dfrac{6!}{3!3!}$
7. $\dfrac{10!}{(4-2)!7!}$
8. $\dfrac{9!}{5!(9-5)!}$
9. $\dfrac{n!}{(n-0)!0!}$
10. $\dfrac{n!}{(n-n)!n!}$

The Multiplication Principle and Permutations

11. Find the number of permutations of three things chosen from five things.
12. Find the number of permutations of two things chosen from four things.
13. Find the number of permutations of three things chosen from 10 things.
14. In how many ways can you choose four books from eight books and arrange them in order on a shelf?
15. In how many ways can you arrange five people in a line?
16. In how many ways can six different candidates for office be listed on the ballot?
17. Ten people hold raffle tickets, and first, second, and third prizes will be given. If no person can receive more than one prize, in how many different ways can the prizes be given out?
18. In how many different ways can you be dealt five cards from a deck?
19. Every telephone number in the town of New Boston is of the form 487-XXXX. How many telephone numbers are possible in the town?
20. Refer to Exercise 19: How many numbers are possible if no number can have four identical digits after the hyphen?

In July of 1988 the New England Telephone Company created a new area code in eastern Massachusetts because, it said, it was running out of numbers in the old area code. Using the fact that all telephone numbers in an area code are of the form XXX-XXXX, answer the questions of Exercises 21 through 24.

21. How many possible telephone numbers are there in an area code?
22. How many possible telephone numbers are there in an area code if no number can begin with a 0?
23. How many possible telephone numbers are there in an area code if no number can begin with the same digit repeated three times? (Thus 555-XXXX would not be allowed, for example.)
24. How many possible telephone numbers are there in an area code if no number can begin with a 0 and no number can begin with the same digit repeated three times?

384   CHAPTER 6

License plates in New Hampshire now consist of three letters followed by a three-digit number, such as ABC 123, for instance; they used to consist of a six-digit number, such as 123456, for instance. Exercises 25 through 28 refer to these facts.

25. How many different license plates are possible in New Hampshire now?
26. How many different license plates were possible in New Hampshire under the old system?
27. How many different license plates are possible in New Hampshire now if the letter O cannot be used?
28. How many different license plates were possible in New Hampshire under the old system if no number could begin with a 0?

Management

29. A computer manufacturer sells three different computers. With each computer the customer has a choice of three different operating systems, seven different software packages, and four different service plans. How many different options does a customer face when buying a computer from this manufacturer?

Social Science

30. A sociologist identifies nine different socioeconomic classes in a community. If she selects one person from each class to interview, how many different sequences of interviews are possible?

Combinations

In Exercises 31 through 39, evaluate the given expression.

31. $\binom{6}{3}$ 32. $\binom{3}{2}$ 33. $\binom{5}{2}$ 34. $\binom{10}{7}$ 35. $\binom{12}{9}$ 36. $\binom{7}{7}$ 37. $\binom{8}{0}$

38. $\binom{n}{0}$, any $n \geq 0$, $n \in \mathbf{Z}$

39. $\binom{n}{n}$, any $n \geq 0$, $n \in \mathbf{Z}$

40. How many combinations are there of three things chosen from nine things?
41. How many combinations are there of four things chosen from seven things?
42. How many combinations are there of 4 things chosen from 16 things?
43. How many combinations are there of 98 things chosen from 100 things?
44. How many combinations are there of four things chosen from 50 things?
45. How many combinations are there of 0 things chosen from n things?
46. How many combinations are there of n things chosen from n things?
47. A set consists of 10 elements. How many distinct subsets of four elements does it have?
48. A set consists of 10 elements. How many distinct subsets of seven elements does it have?
49. How many different committees of five people can be chosen from eight people?

50. How many different committees of two people can be chosen from eight people?
51. If 12 people are running for school committee, and 4 will be elected, how many different school committees are possible?
52. How many different seven-card hands can be dealt from an ordinary deck?
53. How many different 13-card hands can be dealt from an ordinary deck? (*Hint:* $52! \approx 8.066(10^{67})$, $39! \approx 2.04(10^{46})$, $13! \approx 6.227(10^9)$.)

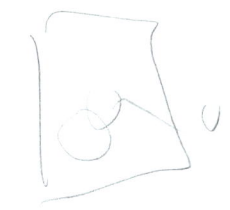

Applications to Probability

A club has 10 members, including Lopez and Connors. A president, vice president, and secretary are selected at random from among the club members, and no member can hold more than one office. Exercises 54 through 58 refer to this situation. Find the given probability.

54. *P*(Lopez will be president)
55. *P*(Lopez will be president and Connors will be secretary)
56. *P*(Lopez will be an officer)
57. *P*(Lopez and Connors will both be officers)
58. *P*(either Lopez or Connors will be an officer)

A club has 10 members, including Lewis and Morelli. A committee of three members is chosen at random. Exercises 59 through 63 refer to this situation. Find the given probability.

59. *P*(Lewis will be on the committee)
60. *P*(Lewis and Morelli will both be on the committee)
61. *P*(either Lewis or Morelli will be on the committee)
62. *P*(exactly one of Lewis or Morelli will be on the committee)
63. *P*(neither Lewis nor Morelli will be on the committee)

You are dealt three cards from an ordinary deck. Exercises 64 through 67 refer to this situation. Find the given probability.

64. *P*(all 3 are diamonds)
65. *P*(all 3 are of the same suit)
66. *P*(2 are aces and 1 is a king)
67. *P*(all 3 are face cards)

A state lottery sells tickets having a three-digit red number, a two-digit white number, and a one-digit blue number; for example,

Red	White	Blue
123	45	6

Every week the state lottery commission draws a red number, a white number, and a blue number. If the red, white, and blue numbers on your ticket match those drawn by the commission, you win $50,000. If only the red and white numbers match, you win $1000. If only the red and blue numbers match, you win $100. If only the red numbers match, you win $10. If only the white and blue numbers match, you win $5. If only the blue number matches, you win $1. Exercises 68 through 73 assume that you buy one ticket. Find the given probability.

68. *P*(you will win $50,000)

69. *P*(you will win $1000)

70. *P*(you will win $100)

71. *P*(you will win $10)

72. *P*(you will win $5)

73. *P*(you will win nothing)

In the Massachusetts Megabucks lottery, a player pays $1 to purchase a ticket containing six numbers chosen from the integers 1, 2, . . . , 36. If the six numbers on the ticket match the six drawn at random by the lottery commission, the player wins the grand prize; if exactly five of the six match, the player wins $400; if exactly four of the six match, the player wins $40. Exercises 74 through 77 refer to this situation. Find the given probability.

74. *P*(a player who buys 1 ticket will win the grand prize)

75. *P*(a player who buys 1 ticket will win $400)

76. *P*(a player who buys 1 ticket will win $40)

77. *P*(a player who buys 1 ticket will not win any money)

*78. If there are n people at a party, $n > 1$, what is the probability that at least two of them have the same birthday?

*79. If there are 23 people at a party, what are the odds that at least 2 of them have the same birthday?

Management

A restaurant offers 8 appetizers, 12 entrees, and 6 desserts. Of the appetizers, 3 are seafood; of the entrees, 5 are seafood. Assume that every customer orders an entree, but that some will not order an appetizer or a dessert. Assume also that every order is as likely as every other order. Exercises 80 through 83 refer to this situation. Find the given probability.

80. *P*(a customer will order a seafood appetizer, a seafood entree, and no dessert)

81. *P*(a customer will order no appetizer, a nonseafood entree, and a dessert)

82. *P*(a customer will order either a seafood appetizer or a seafood entree)

83. *P*(a customer will order a seafood appetizer and a seafood entree)

The union at Ajax Company must choose a six-person committee to conduct contract negotiations with management. The committee will be chosen at random from among the 14 members of the union's executive council, 6 of whom are leaning toward a strike and 8 of whom are leaning toward settling without a strike. Exercises 84 through 87 refer to this situation. Find the given probability.

84. *P*(committee will consist entirely of those who are leaning toward a strike)

85. *P*(committee will be divided equally between those leaning toward a strike and those leaning toward settling without a strike)

86. *P*(committee will have a majority who are leaning toward a strike)

87. *P*(committee will have a majority who are leaning toward settling without a strike)

A firm intends to promote three employees to new positions. There are 11 employees eligible for these promotions, among whom are Chang and Walker. Assume that the promotions will be made at random. Exercises 88 through 90 refer to this situation. Find the given probability.

88. *P*(Chang will be promoted)

89. *P*(both Chang and Walker will be promoted)

90. *P*(either Chang or Walker will be promoted)

Life Science

A medical researcher chooses 7 patients from a group of 13 patients and tests a new drug on the 7 chosen. Burke and Hare are among the 13 patients. Exercises 91 through 93 refer to this situation. Find the given probability.

91. *P*(neither Burke nor Hare receives the new drug)

92. *P*(both Burke and Hare receive the new drug)

93. *P*(one of Burke or Hare receives the new drug while the other does not)

Social Science

94. Suppose the sociologist of Exercise 30 interviews the nine people she has chosen in random order. What is the probability that she will interview the person from the highest class first and the one from the lowest class last?

*95. Referring to Exercise 94: what is the probability that the interviews of the person from the highest class and the one from the lowest class will follow one another?

6.5 THE BINOMIAL FORMULA

An experiment often consists of repetitions of a single event, such as repeated tossing of a coin. Problems involving this type of experiment can frequently be solved with the aid of binomial probabilities. In this section we discuss binomial probabilities and their uses.

Bernoulli Processes

An experiment is said to be a **Bernoulli process** if it satisfies the following four requirements:

1. It consists of repetitions of the same action. Each repetition is called a **trial**. The number of trials is denoted by *n*.
2. Each trial has only two possible outcomes. We label one of the outcomes ''success'' and the other ''failure.''

3. The probability of a success on each trial is always the same. We set
$$P(\text{success on any trial}) = p;$$
thus
$$P(\text{failure on any trial}) = 1 - p.$$
4. The trials are independent of one another.

The prototypical Bernoulli process is a coin-tossing experiment. Suppose we toss a fair coin 3 times. Then:

1. The experiment consists of repetitions of the same action. In this case, the action being repeated is the coin toss. Each toss is a trial, so we have $n = 3$ trials.
2. Each trial has only two outcomes: heads or tails. We may label heads "success" and tails "failure."
3. The probability of a success on each trial is always the same, since
$$P(\text{success on any trial}) = P(\text{heads}) = \tfrac{1}{2}.$$
Therefore $p = \tfrac{1}{2}$, and
$$P(\text{failure}) = P(\text{tails}) = 1 - p = \tfrac{1}{2}.$$
4. The trials are independent of one another, because the outcome of a toss does not depend on the outcome of any other toss.

Therefore, our coin-tossing experiment is a Bernoulli process. As we shall see, many situations may be modeled as Bernoulli processes.

Binomial Probabilities

Suppose an experiment is a Bernoulli process having n trials with
$$P(\text{success on any trial}) = p,$$
and we wish to find
$$P(\text{exactly } k \text{ successes in } n \text{ trials}),$$
where k is one of $0, 1, 2, \ldots, n$. One way to have exactly k successes in n trials is to have the first k trials result in successes (S) and the remaining $n - k$ trials result in failures (F):

$$\underbrace{S \cdot S \cdots S}_{k} \cdot \underbrace{F \cdot F \cdots F}_{n - k}$$

Since $P(\text{success}) = p$ and therefore $P(\text{failure}) = 1 - p$, and since the trials are independent, it follows that

$$P(\underbrace{S \cdot S \cdots S}_{k} \cdot \underbrace{F \cdot F \cdots F}_{n-k}) = \underbrace{P(S) \cdot P(S) \cdots P(S)}_{k} \cdot \underbrace{P(F) \cdot P(F) \cdots P(F)}_{n-k}$$

$$= p^k (1 - p)^{n-k}.$$

Of course, the sequence of k successes and $n - k$ failures we have considered is not the only such sequence, but because of the independence of the trials it will always be the case that

$$P(\text{a sequence of } k \text{ successes and } n - k \text{ failures}) = p^k(1 - p)^{n-k}.$$

Thus, the question becomes this: how many sequences of k successes and $n - k$ failures are there? Since each such sequence is the result of a particular choice of k successes from n possible successes, the answer is that there are

$$\binom{n}{k}$$

such sequences. Hence we have the **binomial formula.**

The Binomial Formula

If a Bernoulli process consists of n trials with

$$P(\text{success on any trial}) = p,$$

then

$$P(\text{exactly } k \text{ successes in the } n \text{ trials}) = \binom{n}{k} p^k (1 - p)^{n-k}.$$

EXAMPLE 1

Suppose we toss a fair coin $n = 3$ times. Defining heads to be a success, we have $P(\text{success}) = \frac{1}{2}$. Hence

$$P(0 \text{ heads in 3 tosses}) = \binom{3}{0}\left(\frac{1}{2}\right)^0\left(1 - \frac{1}{2}\right)^3 = \frac{1}{8};$$

$$P(1 \text{ heads in 3 tosses}) = \binom{3}{1}\left(\frac{1}{2}\right)^1\left(1 - \frac{1}{2}\right)^2 = \frac{3}{8};$$

$$P(2 \text{ heads in 3 tosses}) = \binom{3}{2}\left(\frac{1}{2}\right)^2\left(1 - \frac{1}{2}\right)^1 = \frac{3}{8};$$

and

$$P(3 \text{ heads in 3 tosses}) = \binom{3}{3}\left(\frac{1}{2}\right)^3\left(1 - \frac{1}{2}\right)^0 = \frac{1}{8}. \quad \square$$

EXAMPLE 2

A machine stamps out medallions and sometimes stamps out a defective one. Whether or not a particular medallion is defective is independent of the outcome for any other medallion, and the probability that any single medallion is defective is 0.10. Suppose we are interested in the number of defective medallions in the next five stamped out by the machine.

This is a Bernoulli process with $n = 5$ trials. We define a defective medallion to be a success, with

$$P(\text{success}) = p = 0.10.$$

390 CHAPTER 6

Hence

$$P(\text{none of the 5 are defective}) = \binom{5}{0}(0.10)^0(0.90)^5 = 0.59049;$$

$$P(1 \text{ of the 5 is defective}) = \binom{5}{1}(0.10)^1(0.90)^4 = 0.32805;$$

$$P(\text{at least 1 of the 5 is defective}) = 1 - P(\text{none of the 5 are defective})$$
$$= 1 - 0.59049$$
$$= 0.40951;$$

$$P(\text{at most 1 of the 5 is defective}) = P(1 \text{ is defective}) + P(\text{none are defective})$$
$$= 0.32805 + 0.59049$$
$$= 0.91854. \quad \square$$

It is often tedious to use the binomial formula to calculate binomial probabilities. Therefore we often use binomial tables, which give the binomial probabilities

$$\binom{n}{k}p^k(1-p)^{n-k}$$

for various values of n, p, and k. Table 5 in the appendix is such a binomial table.

EXAMPLE 3 A multiple choice test has five questions, each of which has five possible answers to choose from. A student selects his answers at random. What is the probability that he gets three of five correct? at least one correct?
This is a Bernoulli process with $n = 5$ trials and

$$P(\text{choose correct answer on any question}) = p = \tfrac{1}{5} = 0.20.$$

We look in Table 5 and find the portion of the table for which $n = 5$; then we locate the column for which $p = 0.20$. Our answers will be found in this column. From the table, we see that

$$P(3 \text{ of 5 correct}) = 0.0512;$$

$$P(\text{at least 1 of 5 correct}) = 1 - P(0 \text{ of 5 correct})$$
$$= 1 - 0.3277$$
$$= 0.6723. \quad \square$$

EXAMPLE 4 An article by Alfred Rappaport ("Stock Market Signals to Managers," *Harvard Business Review,* November–December, 1987, quoted in Aczel, *Complete Business Statistics,* Irwin, 1989) reports that 60% of the top executives of American companies think that their company's stock is undervalued. If 10 top executives are selected at random, what is the probability that at least 7 of them will think that their company's stock is undervalued?

Selecting the executives is a Bernoulli process with $n = 10$ trials. If we define a success to be the selection of an executive who thinks his or her company's stock is undervalued, then $P(\text{success}) = 0.6$, and we must find

$$P(\text{at least 7 successes in 10 trials}).$$

But if we now attempt to use Table 5, we find that it does not contain any entries for $n = 10$, $p = 0.6$. The table does have entries for $n = 10$, $p = 0.4$, however, so if we define a success to be the selection of an executive who does *not* think his or her company's stock is undervalued and recast the problem in terms of this definition, we will be able to use the table. Thus,

$$P(\text{at least 7 of 10 think stock is undervalued})$$
$$= P(\text{at most 3 of 10 do not think stock is undervalued})$$
$$= P(\text{3 or fewer successes in 10 trials when } p = 0.4)$$
$$= P(0 \text{ successes}) + P(1 \text{ success}) + P(2 \text{ successes})$$
$$\quad + P(3 \text{ successes})$$
$$= 0.0060 + 0.0403 + 0.1209 + 0.2150$$
$$= 0.3822. \quad \square$$

We must point out that it is not possible to use the binomial table to solve all binomial problems, for the table gives binomial probabilities only for a few values of n and p. If the table cannot be used, binomial probabilities should be calculated directly from the binomial formula.

■ **Computer Manual: Program BINOMDST**

6.5 EXERCISES

Bernoulli Processes

In Exercises 1 through 8, if the given experiment is a Bernoulli process, identify n and p; if it is not, state why it is not.

1. Toss a fair coin 10 times.
2. Toss a coin three times; the coin is loaded so that it comes up heads two thirds of the time.
3. Roll a die four times, if you are interested in the number that comes up each time.
4. Roll a die four times, if you are interested only in whether you get an even number or an odd number each time.
5. Draw five cards from an ordinary deck, with replacement, if you are interested only in whether you get an ace each time.

6. Draw five cards from an ordinary deck, without replacement, if you are interested only in whether you get an ace each time.

7. Select at random a president, secretary, and treasurer for a club consisting of six men and four women, if you are interested only in whether an officer is male or female and if no member of the club can hold more than one office.

8. Repeat Exercise 7, except that members of the club may hold more than one office.

Binomial Probabilities

In Exercises 9 through 12, a fair coin is tossed four times. Find the probability of getting

9. 0 heads 10. 1 head 11. 2 heads 12. at least 1 head

You take a quiz consisting of 10 true-or-false questions, choosing your answers by guessing. In Exercises 13 through 16, find the probability of getting:

13. none right 14. exactly 2 right 15. at least 1 right 16. fewer than 4 right

You take a multiple-choice quiz having 15 questions, each of which has 4 choices, and you select your answers by guessing. In Exercises 17 through 20, find the probability of getting

17. exactly 3 right 18. 5 or more right 19. fewer than 10 right 20. 4, 5, or 6 right

A professional golfer can make a putt of 6 feet or less two thirds of the time. In Exercises 21 through 24, find the probability that, of the next three such putts he attempts, he will

21. make at least 1 22. make all 3 23. make none 24. make more than 1

Surveys show that two out of every five people who buy ice cream buy vanilla. Suppose that 15 people buy ice cream at a store. Exercises 25 through 27 refer to this situation. Find the given probability. (You may assume that the trials are independent.)

25. P(6 of them will buy vanilla) 26. P(more than 6 of them will buy vanilla)

27. P(8 or fewer will not buy vanilla)

Management

A machine stamps out printed circuits. The probability that it will stamp out a defective circuit is 0.05. Exercises 28 through 30 refer to this situation. Find the given probability. (You may assume that the trials are independent.)

28. P(4 of the next 5 circuits will be good) 29. P(at least 14 of the next 15 will be good)

30. P(9 or fewer of the next 10 will be good)

A computer company gives new employees a training course in computer operation and maintenance. It has been the experience of the company that 8% of those who start the course fail to finish it. Suppose four new employees are hired this week. Exercises 31 through 33 refer to this situation. Find the given probability. (You may assume that the trials are independent.)

31. P(all of the 4 will finish the course) 32. P(not all of the 4 will finish the course)

33. P(at least 3 of the 4 will finish the course)

Oil company records show that 13% of wildcat wells strike oil. Exercises 34 through 36 refer to this situation. Find the given probability. (You may assume that the trials are independent.)

34. P(2 of the next 3 wildcat wells drilled will strike oil)

35. P(at least 1 of the next 5 wildcat wells drilled will strike oil)

36. P(1, 2, or 6 of the next 6 wildcat wells drilled will strike oil)

(Exercises 37 through 40 were suggested by material in Niebel, *Motion and Time Study,* 8th edition, Irwin, 1988.) A machine is said to be "down" if it is not working. Suppose a shop has four identical machines and that there is a 5% chance that a machine will be down sometime during the day. Exercises 37 through 40 refer to this situation. Find the given probability. (You may assume that the trials are independent.)

37. P(none of the 4 machines will be down during the day)

38. P(at least one of the 4 machines will be down during the day)

39. P(3 of the 4 machines will be down during the day)

40. P(all 4 of the machines will be down during the day)

Life Science

A new drug improves the condition of 75% of the patients who take it. Suppose 10 patients take the drug. Exercises 41 through 44 refer to this situation. Find the given probability. (You may assume that the trials are independent.)

41. P(all 10 will improve) **42.** P(at least 8 will improve)

43. P(fewer than 4 will improve) **44.** P(5, 6, 7, or 8 will improve)

Suppose one third of all newly hatched robins will live to maturity, and suppose that a nest contains three newly hatched robins. Exercises 45 through 48 refer to this situation. Find the given probability. (You may assume that the trials are independent.)

45. P(none will live to maturity) **46.** P(at least 2 will live to maturity)

47. P(fewer than 3 will live to maturity) **48.** P(at least 2 will not live to maturity)

Suppose the probability that parents who are not red-haired will have a child who has red hair is 0.10. Two parents who are not red-haired have five children. Exercises 49 through 51 refer to this situation. Find the given probability. (You may assume that the trials are independent.)

49. P(at least 1 of the children will have red hair)

50. P(none of the children will have red hair)

51. P(fewer than 2 of the children will have red hair)

Social Science

Suppose that 60% of the voters in a town are Democrats and you take a poll by calling 15 registered voters chosen at random. Exercises 52 through 55 refer to this situation. Find the given probability. (You may assume that the trials are independent.)

52. P(at least 10 of those you call are Democrats) **53.** P(all 15 are Democrats)

54. P(at least 8 are not Democrats) **55.** P(fewer than 7 are Democrats)

Suppose that 12% of the people in a community are college students and you stop four people at random on the street for a survey. Exercises 56 through 58 refer to this situation. Find the given probability. (You may assume that the trials are independent.)

56. P(none of those you stop are college students)   **57.** P(at least 1 is a college student)

58. P(not all of them are college students)

6.6 RANDOM VARIABLES, PROBABILITY DISTRIBUTIONS, AND EXPECTED VALUE

In this section we continue our study of probability theory by introducing the concepts of a **random variable,** the **probability distribution** of a random variable, and the **expected value** of a random variable. These notions are important in both probability and statistics, and expected values are widely used in decision making.

Random Variables

A **random variable** is a variable that takes on numerical values assigned to the outcomes of an experiment. Random variables are usually designated by the letter X.

EXAMPLE 1 Let us toss a fair coin. The possible outcomes of this experiment are heads and tails. Define the random variable X by

$$X = \text{number of heads obtained.}$$

Then X can take on the values 0 and 1:

$$X = 0 \text{ if the coin comes up tails;}$$

$$X = 1 \text{ if the coin comes up heads.} \quad \square$$

EXAMPLE 2 Let us toss a fair coin twice. Define the random variable X by

$$X = \text{number of heads obtained.}$$

Then X can take on the values 0, 1, and 2, since we can obtain either zero heads, one head, or two heads in two tosses of the coin. $\quad \square$

EXAMPLE 3 Let us roll a single die. Define the random variable X by

$$X = \text{number that comes up.}$$

Then X can take on the values 1, 2, 3, 4, 5, and 6. $\quad \square$

Probability Distributions

A **probability distribution** for a random variable X is a function that gives the probabilities that X takes on any of its values. Probability distributions are often given in tabular and pictorial form as well as in functional form.

PROBABILITY

EXAMPLE 4 Consider the coin-tossing experiment of Example 1, where X is the number of heads obtained in one toss of a fair coin. Since

$$P(X = 0) = P(\text{tails}) = \tfrac{1}{2} \quad \text{and} \quad P(X = 1) = P(\text{heads}) = \tfrac{1}{2},$$

the probability distribution for X is

$$P(X = k) = \tfrac{1}{2}, \quad \text{for } k = 0, 1.$$

This probability distribution may be written in tabular form as follows:

X	$P(X)$
0	$\tfrac{1}{2}$
1	$\tfrac{1}{2}$

Note that the probability column of the table sums to 1. This must be so for every probability distribution. (Why?)

Finally, the probability distribution may be given in pictorial form, as in Figure 6–11. ◻

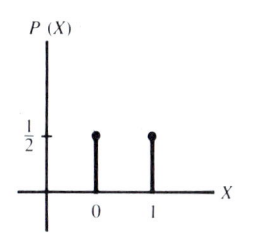

FIGURE 6–11

EXAMPLE 5 Consider the experiment of Example 2, where X is the number of heads obtained in two tosses of a fair coin.

The outcomes and their probabilities are as shown in Table 6–7. Therefore, the probability distribution for X is

$$P(X = k) = \begin{cases} \tfrac{1}{4}, & \text{for } k = 0, 2, \\ \tfrac{1}{2}, & \text{for } k = 1. \end{cases}$$

TABLE 6–7

	1st Toss	2nd Toss	Probability
2 heads	Heads	Heads	$\tfrac{1}{4}$
1 head	{ Heads	Tails	$\tfrac{1}{4}$
	{ Tails	Heads	$\tfrac{1}{4}$
0 heads	Tails	Tails	$\tfrac{1}{4}$

In tabular form, the distribution is as follows:

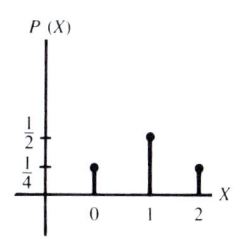

X	$P(X)$
0	$\tfrac{1}{4}$
1	$\tfrac{1}{2}$
2	$\tfrac{1}{4}$

FIGURE 6–12

In pictorial form, it is shown in Figure 6–12. ◻

EXAMPLE 6

The probability distribution for the random variable X of Example 3 is

X	$P(X)$
1	$1/6$
2	$1/6$
3	$1/6$
4	$1/6$
5	$1/6$
6	$1/6$

or

$$P(X = k) = 1/6 \quad \text{for } k = 1, 2, \ldots, 6. \quad \square$$

It is sometimes possible to describe an entire class of experiments by means of a general probability distribution that applies to all of them. For example, as we saw in Section 6.5, every Bernoulli process is described by its number of trials n, the probability p of a success on any trial, and the binomial formula

$$P(k \text{ successes in } n \text{ trials}) = \binom{n}{k} p^k (1-p)^{n-k}.$$

For any Bernoulli process we may define a random variable X by

$$X = \text{number of successes in } n \text{ trials}.$$

Then the Bernoulli process is described by the **binomial probability distribution**

$$P(X = k) = \binom{n}{k} p^k (1-p)^{n-k},$$

where $k = 0, 1, \ldots, n$.

EXAMPLE 7

Suppose we roll a die four times and are interested only in whether we get a number less than 3 each time. Then this experiment is a Bernoulli process with $n = 4$ trials and

$$p = P(\text{success}) = P(\text{get 1 or 2}) = 1/3.$$

If we let

$$X = \text{number of successes in 4 trials},$$

then the probability distribution for X is given in tabular form in Table 6–8. $\square$

Expected Value

Suppose we toss a fair coin once and define the random variable X, as usual, by

$$X = \text{number of heads obtained}.$$

What is the average value of X in the long run? That is, if we do this experiment over and over, what can we expect to happen, on average? Since $X = 0$ and

TABLE 6–8

X	P(X)
0	$\binom{4}{0}\left(\frac{1}{3}\right)^0\left(\frac{2}{3}\right)^4 = \frac{16}{81}$
1	$\binom{4}{1}\left(\frac{1}{3}\right)^1\left(\frac{2}{3}\right)^3 = \frac{32}{81}$
2	$\binom{4}{2}\left(\frac{1}{3}\right)^2\left(\frac{2}{3}\right)^2 = \frac{24}{81}$
3	$\binom{4}{3}\left(\frac{1}{3}\right)^3\left(\frac{2}{3}\right)^1 = \frac{8}{81}$
4	$\binom{4}{4}\left(\frac{1}{3}\right)^4\left(\frac{2}{3}\right)^0 = \frac{1}{81}$

$X = 1$ both occur with probability $\frac{1}{2}$, in the long run we can expect the value of X to average

$$0(\tfrac{1}{2}) + 1(\tfrac{1}{2}) = \tfrac{1}{2}.$$

We express this fact by saying that the **expected value** $E(X)$ of X is $\frac{1}{2}$.

The Expected Value of a Random Variable

Let X be a random variable that can take on the values $k_1, k_2, \ldots, k_n$. Then the **expected value of X** is

$$E(X) = k_1 \cdot P(X = k_1) + k_2 \cdot P(X = k_2) + \cdots + k_n \cdot P(X = k_n).$$

The expected value $E(X)$ is the average value of X in the long run.

Note that the expected value of X need not be one of the values that X can actually take on.

EXAMPLE 8 Suppose we toss a fair coin twice, and let $X =$ number of heads. Then

$$\begin{aligned} E(X) &= 0 \cdot P(X = 0) + 1 \cdot P(X = 1) + 2 \cdot P(X = 2) \\ &= 0(\tfrac{1}{4}) + 1(\tfrac{1}{2}) + 2(\tfrac{1}{4}) \\ &= 1. \end{aligned}$$

Thus, in the long run, we will average one head for every time the experiment is performed. ◻

EXAMPLE 9 A baker's records show the number of wedding cakes she has sold each week for the past 60 weeks:

Wedding Cakes Sold	Number of Weeks This Occurred
0	10
1	25
2	20
3	5
	60

If we let

$$X = \text{number of wedding cakes sold per week,}$$

then X can take on the values 0, 1, 2, and 3, and we have

$$E(X) = 0 \cdot P(X = 0) + 1 \cdot P(X = 1) + 2 \cdot P(X = 2) + 3 \cdot P(X = 3)$$
$$= 0(^{10}/_{60}) + 1(^{25}/_{60}) + 2(^{20}/_{60}) + 3(^{5}/_{60})$$
$$= ^{4}/_{3}.$$

Hence in the long run the baker can expect to average $1\frac{1}{3}$ wedding cakes sold per week. ☐

The concept of expected value is of considerable use when analyzing gambling games. In such cases we usually let the random variable X be defined as the monetary payoff of the game. Then the game is a **fair game** if the expected value $E(X) = 0$, for this signifies that in the long run the players will neither win nor lose money.

EXAMPLE 10 Suppose you and an opponent toss a fair coin twice and if the coin comes up

- Heads both times, you win $1 from your opponent;
- Heads once and tails once, neither of you wins or loses;
- Tails both times, your opponent wins $1 from you.

Let

$$X = \text{the amount of your winnings.}$$

Then X can take on the values 1, 0, and -1, and the expected value of your winnings is

$$E(X) = 1(\tfrac{1}{4}) + 0(\tfrac{1}{2}) + (-1)(\tfrac{1}{4}) = 0.$$

Hence in the long run you will break even playing this game, so it is a fair game. ☐

EXAMPLE 11 A roulette wheel has 38 numbers on it:

$$00, \ 0, \ 1, \ 2, \ 3, \ \ldots, \ 35, \ 36.$$

Suppose you bet on a specific number between 1 and 36, inclusive. If the number comes up, you will win $36; if not, you will lose $1.

If X = the amount of your winnings, then

$$E(X) = 36 \cdot P(X = \$36) + (-1) \cdot P(X = -\$1)$$
$$= 36(\tfrac{1}{38}) + (-1)(\tfrac{37}{38})$$
$$= -\tfrac{1}{38}$$
$$\approx -0.026.$$

Thus, roulette is not a fair game: in the long run you can expect to average a loss of approximately 2.6 cents for every dollar you bet on a specific number between 1 and 36. ☐

Games played in gambling casinos are not fair games: the expected winnings for the player are always negative, and those for the casino are always positive. Hence in the long run the casino must win. Gambling games can be made more (or less) fair by adjusting the payoffs. For example, the single-number bet on roulette discussed in Example 11 could be made fair by setting the payoff for winning at $37 rather than $36.

The concept of expected value is often useful when considering decision-making problems.

EXAMPLE 12 An architect is trying to decide whether or not to enter a design competition. If he does enter, it will cost him $2000 to prepare and submit his design. The prize for winning the competition is $30,000. He knows there will be nine other entrants in the competition, and he feels that each of them will have same chance of winning as he will. Should he enter the competition?

Let

$$X = \text{amount of architect's net winnings}$$

if he does enter. Then X can take on the values

$$30,000 - 2000 = 28,000 \quad \text{and} \quad -2,000.$$

Therefore,

$$E(X) = 28,000\, P(X = 28,000) + (-2000)\, P(X = -2000)$$
$$= 28,000(\tfrac{1}{10}) + (-2000)(\tfrac{9}{10})$$
$$= 1,000.$$

Since his expected winnings are positive, he should enter the competition. ☐

We will have much more to say about the use of expected values in decision making in Chapter 7.

■ *Computer Manual: Programs HISTGRAM, EXPCTVAL*

6.6 EXERCISES

Random Variables and Probability Distributions

In Exercises 1 through 8, give the probability distribution for the given random variable X.

1. Experiment: toss a fair coin three times
   X = number of heads obtained

2. Experiment: toss a fair coin four times
   X = number of heads obtained

3. Experiment: draw one card from an ordinary deck
   X = number of hearts drawn

4. Experiment: draw two cards from an ordinary deck, with replacement
   X = number of hearts drawn

5. Experiment: draw two cards from an ordinary deck, without replacement
   X = number of hearts drawn

6. Experiment: choose a four-person committee at random from a group of four men and six women
   X = number of women on the committee

7. Experiment: choose a president, secretary, and treasurer at random from the members of a club consisting of four men and six women if no member can hold more than one office
   X = number of women chosen to be officers

8. Experiment: roll a pair of dice
   X = the total shown on the dice

Expected Value

9. A random variable X has the following probability distribution:

X	$P(X)$
1	$1/6$
2	$2/3$
3	$1/6$

   Find its expected value $E(X)$.

10. Repeat Exercise 9 for the following probability distribution:

X	$P(X)$
0	0.20
2	0.40
4	0.10
6	0.30

11. Find and interpret $E(X)$ for the random variable X of Exercise 1.

12. Find and interpret $E(X)$ for the random variable X of Exercise 2.

PROBABILITY   401

13. Find and interpret $E(X)$ for the random variable X of Exercise 3.
14. Find and interpret $E(X)$ for the random variable X of Exercise 4.
15. Find and interpret $E(X)$ for the random variable X of Exercise 5.
16. Find and interpret $E(X)$ for the random variable X of Exercise 6.
17. Find and interpret $E(X)$ for the random variable X of Exercise 7.
18. Find and interpret $E(X)$ for the random variable X of Exercise 8.
19. You buy a raffle ticket. The prize is a TV set worth $500. If 1000 tickets were sold at $1 each, find your expected net winnings.
20. You buy a raffle ticket on a new car worth $8500. If 12,500 tickets are sold at $1 each, find your expected net winnings.
21. A vase has seven red and four white marbles in it. If you draw three marbles from the vase (with replacement), what is the expected number of red marbles?
22. Repeat Exercise 21 if you draw without replacement.
23. If you deal three cards from an ordinary deck, find the expected number of spades.
24. If you deal three cards from an ordinary deck, find the expected number of aces.
25. If you deal three cards from an ordinary deck, find the expected number of face cards (jacks, queens, kings, and aces).
26. If you deal three cards from an ordinary deck, find the expected number of red cards.
27. You draw one card from an ordinary deck. If you draw a spade, you win $1, but if you do not draw a spade, you lose $2. Find your expected winnings. Is this a fair game?
28. Repeat Exercise 27 if when you win, you win $4, but when you lose, you lose $1.
29. A vase has six red and four white marbles in it. For a $1 bet, you are allowed to reach in and draw one marble from the vase. If it is white, you win; but if it is red, you lose your $1. What amount must you be paid for winning if this is to be a fair game?
30. In the game of numbers, a $1 bet allows you to pick any three-digit number from 000 through 999. If the number you have picked comes up, you win $600. Is this a fair game? What payoff for winning would make it a fair game?
31. A roulette wheel has the numbers

    00, 0, 1, 2, 3, . . . , 35, 36

    on it. If you bet on "odds" and one of the numbers 1, 3, 5, . . . , 35 comes up, you win $1; if not, you lose $1. Find the expected winnings of betting this way on roulette. What payoff for winning would make this a fair bet?
32. In Europe roulette wheels have only the numbers 0, 1, . . . , 36. If you bet on a number between 1 and 36, inclusive, and it comes up, you win $35; otherwise you lose $1. Find the expected value of such a bet.
33. If you bet on "odds" in European roulette (see Exercise 32) and one of the numbers 1, 3, . . . , 35 comes up, you win $1; otherwise you lose $1. Find the expected value of such a bet.

34. A lottery ticket has a three-digit red number and a two-digit white number. If only the white number on your ticket matches that drawn by the lottery commission, you win $10. If only the red number matches, you win $100. If both numbers match, you win $1000. Find your expected net winnings if you buy 1 ticket for $1.

35. In the Massachusetts Megabucks lottery, a player pays $1 to purchase a ticket containing six numbers chosen from the integers 1, 2, . . . , 36. If the six numbers on his ticket match the six drawn at random by the lottery commission, the player wins the grand prize; if exactly five of the six match, the player wins $400; if exactly four of the six match, he wins $40. Find the expected winnings on a $1 bet if the grand prize is worth $2 million.

36. The Illinois state lottery requires that each player choose six numbers from the integers 1, 2, . . . , 54. Suppose the payoff for matching six of six numbers is $5 million, for matching five of six is $2000, and for matching four of six is $100. Find the expected winnings on a $1 bet.

37. A survey showed the number of defective light bulbs per carton sampled to be as follows:

Defective Bulbs in Carton:	0	1	2	3	4	5
Number of Cartons:	42	36	14	3	1	1

Find the expected number of defective bulbs per carton.

38. You are about to graduate from college and have been offered two jobs. The first pays $15,000 per year to start, but if you are promoted, it will pay you $25,000 by your third year; if not promoted, your salary will remain $15,000. The second job pays $18,000 to start, guaranteed to rise to $20,000 by the third year. You estimate that you have a 70% chance of being promoted if you take the first job. Based on expected salary in the third year, which job should you take?

39. Repeat Exercise 38 if you estimate that you have a 50% chance of being promoted if you take the first job.

40. Repeat Exercise 38 if you estimate that you have a 40% chance of being promoted if you take the first job.

Management

41. A hot dog vendor sells 4 dozen hot dogs per day 40% of the time, 5 dozen per day 30% of the time, 6 dozen per day 20% of the time, and 7 dozen per day 10% of the time. Find the expected number of hot dogs sold per day.

42. A firm's records show that its annual return on investment (ROI) has been as follows:

Annual ROI:	8.8%	9.0%	9.2%	9.4%	9.6%
Number of Years This Occurred:	2	6	9	7	8

What is the firm's expected annual ROI? If the firm intends to invest $10 million next year, estimate its ROI.

43. A restaurant's daily records show the following:

PROBABILITY 403

Percent of Capacity:	70%	75%	80%	85%	90%	95%
Probability:	0.1	0.2	0.2	0.25	0.2	0.05
Revenue ($1000s):	2.0	2.5	3.2	3.8	4.2	4.4

Find the restaurant's expected daily revenue.

44. Ace Company knows that 20% of its salespersons earn commissions of $10,000 per year, 45% earn commissions of $15,000 per year, 30% earn commissions of $20,000 per year, and 5% earn commissions of $30,000 per year. Ace has 200 salespersons, and the company is attempting to estimate the amount that should be budgeted next year for payment of commissions. What should the estimate be?

Life Science

45. Analysis of water samples from 25 wells in a town shows the following concentrations of pollutants, in parts per million:

Pollutants:	2	5	6	8	10	11	12
No. of Wells:	8	4	7	2	2	1	1

Find the expected concentration of pollutants in the town's wells.

46. The probabilities for the size of a dog's litter are as follows:

Number of Pups:	1	2	3	4	5	6
Probability:	0.01	0.11	0.13	0.44	0.22	0.09

Find the expected size of the litter.

47. Suppose the probability that two brown-eyed parents will have a blue-eyed child is $\frac{1}{3}$. If two such parents have six children, what is the expected number who will have blue eyes?

48. Repeat Exercise 47 if the probability that two brown-eyed parents will have a blue-eyed child is $\frac{1}{5}$.

Sickle-cell anemia is a genetic disease controlled by two genes: the dominant S-gene and the recessive s-gene. A child receives either gene S or gene s from each parent. If the child receives two s-genes (ss), it will develop the disease. If it receives one s and one S (sS), it will not develop the disease, but can pass it on to its offspring and hence is a carrier of the disease. If the child receives two S genes (SS), it cannot develop the disease and cannot be a carrier. A parent can only pass on to a child a gene that the parent has; if the parent is an sS-person, the probability that the child will receive the s-gene is $\frac{1}{2}$. Prospective parents can now be tested to see whether or ot they are carriers of the disease and thus be advised as to the chances that their children will develop sickle-cell anemia or be carriers of it. Suppose two parents are tested. Exercises 49 through 54 refer to this situation.

49. If neither of the parents has sickle-cell anemia, but both are carriers, how many of their four children can be expected to develop the disease?

50. If neither of the parents has sickle-cell anemia, but both are carriers, how many of their four children can be expected to be carriers?

51. If one parent is a carrier and the other does not have sickle-cell anemia and is not a carrier, how many of their four children can be expected to be carriers?

52. If one parent is a carrier and the other has the disease, how many of their four children can be expected to develop the disease?

53. If one parent is a carrier and the other has the disease, how many of their four children can be expected to be carriers?

54. If one parent has the disease and the other does not and is not a carrier, how many of their four children can be expected to be carriers?

Social Science

According to information on page 22 of the May 1986 issue of *Scientific American*, if

$X =$ number of males in a family who will marry,

the probability distribution for X is as follows:

X:	0	1	2	3	4	5	6 or more
P(X):	0.317	0.364	0.209	0.080	0.023	0.005	0.002

Exercises 55 through 61 refer to this random variable.

55. Find P(a family will have at least 1 male who will marry).

56. Find P(a family will have at most 1 male who will marry).

57. Find P(a family will have 5 or more males who will marry).

58. Find and interpret $P(X = 3)$.

59. Find and interpret $P(X < 3)$.

60. Find and interpret $P(X > 3)$.

61. Replace the entry "6 or more" with "6," and then find and interpret $E(X)$.

62. The length of time it took 21 rats to make their way through a maze on their first and second tries was as follows:

First Try

Time (Seconds):	30	45	60	75	90	105
Number of Rats:	1	5	6	7	1	1

Second Try

Time (Seconds):	30	45	60	75	90	105
Number of Rats:	4	10	4	2	1	0

What is the expected time for a rat to get through the maze on its first attempt? on its second attempt?

63. Political candidate A estimates that if she can get her opponent, B, to debate, she can win 52% of the vote in the coming election, whereas if she cannot entice B into a debate, she will receive only 46% of the vote. Candidate A also estimates that the probability that she will be able to goad B into a debate is 0.45. What is A's expected percentage of the vote?

Decision Matrices

A **decision matrix** is a matrix that gives the payoffs for alternative strategies under different states of nature. (See Exercises 4.1.) If probabilities are assigned to the states of na-

*64. Suppose you can invest in either a money market fund, a stock fund, or a bond fund, and the decision matrix is as follows (payoffs in dollars):

$$\text{Strategies:} \begin{array}{r} \\ \text{money market} \\ \text{stock fund} \\ \text{bond fund} \end{array} \begin{array}{c} \text{States of Nature:} \\ \text{Interest Rates} \\ \begin{array}{ccc} \text{fall} & \text{stay same} & \text{rise} \\ 0.2 & 0.5 & 0.3 \end{array} \\ \begin{bmatrix} -100 & 20 & 150 \\ -200 & -100 & 200 \\ 100 & 0 & -85 \end{bmatrix} \end{array}$$

The number at the top of each column is the probability of its state of nature. Find the expected value of each strategy. Which strategy is best by the expected value criterion?

*65. A manufacturer must decide whether to build a small new plant, build a large new plant, expand the old plant, or do nothing. Suppose the payoff matrix is as follows (payoffs in millions of dollars):

$$\text{Strategies:} \begin{array}{r} \\ \text{small new plant} \\ \text{large new plant} \\ \text{expand old plant} \\ \text{do nothing} \end{array} \begin{array}{c} \text{States of Nature:} \\ \begin{array}{ccc} \text{demand} & \text{demand} & \text{demand} \\ \text{high} & \text{medium} & \text{low} \\ 0.22 & 0.42 & 0.36 \end{array} \\ \begin{bmatrix} 20 & 15 & -10 \\ 40 & -5 & -30 \\ 20 & 5 & -15 \\ 5 & 0 & -5 \end{bmatrix} \end{array}$$

Find the expected value of each strategy. Which strategy is the best by the expected value criterion?

Let $S_1, \ldots, S_n$ be the states of nature, and define the **expected value under certainty (EVC)** as follows:

$$\text{EVC} = P(S_1)(\text{best payoff for } S_1) + \cdots + P(S_n)(\text{best payoff for } S_n).$$

The expected value under certainty is the decision maker's expected value if he or she knows with certainty which state of nature will occur. Now define the **expected value under perfect information (EVPI)** by

$$\text{EVPI} = \text{EVC} - \text{expected value of best strategy}.$$

The expected value under perfect information is the maximum amount that the decision maker will pay for information about which state of nature will occur.

*66. Find and interpret EVPI for the decision matrix of Exercise 64.

*67. Find and interpret EVPI for the decision matrix of Exercise 65.

SUMMARY

This summary consists of terms and symbols whose meaning you should know, facts you should know, and techniques and methods you should be able to employ.

Section 6.1

Experiment, outcome of an experiment. Sample space. Event. The probability of an event, $P(A)$. Finding probabilities by counting outcomes. The probability of a certain event. The probability of an impossible event. The probabilities of two events when one is more likely than the other. The rules of probability. Odds against an event. Finding the odds against an event. The relationship between probability and odds. Converting from probabilities to odds and from odds to probabilities. Relative frequency probabilities. Finding relative frequency probabilities of events. Subjective probabilities. Assigning subjective probabilities to events.

Section 6.2

Complementary event A' (not A). Union of events $A \cup B$ (A or B). Intersection of events $A \cap B$ (A and B). Using complementation, union, intersection to find probabilities. Formula for the probability of A'. Using the complementation formula to find probabilities. Formula for the probability of $A \cup B$. Mutually exclusive events. Formula for the probability of $A \cup B$ when A, B are mutually exclusive. Using the formulas for the probability of a union to find probabilities.

Section 6.3

Conditional events $A \mid B$, $B \mid A$. Finding probabilities of conditional events. Drawing with and without replacement and conditional events. Conditional probability formula. Using the conditional probability formula to find probabilities. Dependent and independent events. Events A, B independent if and only if $P(A \mid B) = P(A)$ and $P(B \mid A) = P(B)$. Showing that events are dependent or independent. Formula for the probability of the intersection of two events. Formula for the probability of the intersection of two independent events. Using the formulas for the probability of the intersection of events to find probabilities. The probability formulas.

Section 6.4

Factorial of a nonnegative integer, $0!$, $n!$ for $n > 0$. The Multiplication Principle. Using the Multiplication Principle to determine the number of ways an event can occur. Permutation of k things chosen from n things. Number of permutations of k things chosen from n things, $P(n,k)$. Using permutations to determine the number of ways an event can occur. Combination of k things chosen from n things. Binomial coefficient $\binom{n}{k}$. Number of combinations of k things chosen from n things, $C(n, k) = \binom{n}{k}$. Using combinations to find the number of ways an event can occur. Using the Multiplication Principle, permutations, and combinations to find probabilities.

Section 6.5

Bernoulli process. Trial, success, failure, P(success) = p. Binomial formula. Using the binomial formula to find probabilities.

Section 6.6

Random variable X. Probability distribution for a random variable. Finding the probability distribution for a random variable. The expected value $E(X)$ of a random variable X. The expected value formula. Finding expected values for random variables. Fair game. Using expected values to determine whether a game is fair. Using expected values to make decisions.

REVIEW EXERCISES

Consider the experiment of tossing a fair coin three times. Exercises 1 through 11 refer to this experiment.

1. Write the sample space for the experiment.
2. Write the event {get 3 heads} as a subset of the sample space.
3. Write the event {get at least 2 heads} as a subset of the sample space.
4. Write the event {get at least 2 tails} as a subset of the sample space.
5. Write the event {get no heads} as a subset of the sample space.
6. Write the event {get either 1 head or 1 tail} as a subset of the sample space.
7. Find P(get 3 heads).
8. Find P(get at least 2 heads).
9. Find P(get at least 2 tails).
10. Find P(get no heads).
11. Find P(get either 1 head or 1 tail).

Four hundred towns had their drinking water tested for particulate matter and the results were as follows:

Particulate Matter (ppm):	0–4	5–9	10–14	15–19	20 or more
Number of Towns:	220	110	40	20	10

Exercises 12 through 15 refer to this situation.

12. Find P(particulate matter for a town is 9 ppm or less).
13. Find P(particulate matter for a town is between 10 and 19 ppm, inclusive).
14. Find P(particulate matter for a town is more than 9 ppm).
15. Find P(particulate matter for a town is either less than 5 ppm or greater than 14 ppm).

16. If events A, B, C, and D are such that $P(A) + P(B) + P(C) + P(D) = 1$, A and C are equally likely, B is twice as likely as A, and D is one-third as likely as A, assign probabilities to A, B, C, and D.

TABLE 6–9

	Size		
Style	Subcompact (D)	Compact (E)	Full-Sized (F)
2-Door sedan *(A)*	80	50	20
4-Door sedan *(B)*	60	40	30
Station wagon *(C)*	10	20	5

Table 6–9 cross-classifies the cars available at a large dealership according to style and size. You choose a car at random to test-drive. Exercises 17 through 31 refer to this situation.

17. Find *P*(car you choose will be a subcompact 4-door sedan).
18. Find *P*(the car you choose will not be a 4-door sedan).
19. Find *P*(the car you choose will be either a compact or a full-sized car).
20. Find *P*(the car you choose will be either a 2-door sedan or a compact).
21. Find *P*(the car you choose will be neither a 4-door sedan nor a subcompact).
22. Find $P(E')$ and state the event E' in words.
23. Find $P(A \cap F)$ and state the event $A \cap F$ in words.
24. Find $P((A \cap F)')$ and state the event $(A \cap F)'$ in words.
25. Find $P(A \cup C)$ and state the event $A \cup C$ in words.
26. Find $P((A \cup C)')$ and state the event $(A \cup C)'$ in words.
27. Find $P(C \cup E)$ and state the event $C \cup E$ in words.
28. Find *P*(the car you choose will be a compact if you are choosing from among the station wagons).
29. Find *P*(the car you choose will be a station wagon if you are choosing from among the compacts).
30. Find $P(B \mid F)$. State the event $B \mid F$ in words.
31. Find $P(D \mid B')$. State the event $D \mid B'$ in words.

A vase contains four large and six small marbles. Of the large ones, two are red and two are white; of the small ones, one is red and five are white. You draw a single marble from the vase. Exercises 32 through 37 refer to this experiment.

32. Find *P*(marble drawn is small).
33. Find *P*(marble drawn is white).
34. Find *P*(marble drawn is large and white).
35. Find *P*(marble drawn is small and red).
36. Find *P*(marble drawn is large or white).
37. Find *P*(marble drawn is small or red).

You draw two cards with replacement from an ordinary deck. Exercises 38 through 41 refer to this experiment.

38. Find P(both cards drawn are aces).

39. Find P(just 1 of the cards drawn is an ace).

40. Find P(the 2nd card drawn is an ace).

41. Find P(at least 1 of the 2 cards drawn is an ace).

You draw two cards without replacement from an ordinary deck. Exercises 42 through 45 refer to this experiment.

42. Find P(both cards drawn are aces).

43. Find P(just 1 of the cards drawn is an ace).

44. Find P(the 2nd card drawn is an ace).

45. Find P(at least 1 of the 2 cards drawn is an ace).

46. An organization is planning to print raffle tickets. Each ticket will have a red three-digit number and a green two-digit number on it. If no two tickets are to be identical, how many can the organization print? Assuming that all tickets are printed and sold, find the probability that you will win the raffle if you buy 10 tickets.

47. A club with 15 members is choosing a slate of officers (president, vice president, secretary, and treasurer). No member can hold more than one office.
  (a) How many possible slates of officers are there?
  (b) What is the probability that you will be an officer if you are a member of the club?

48. Suppose the club of Exercise 47 is choosing a steering committee consisting of four members.
  (a) How many possible steering committees are there?
  (b) What is the probability that you will be on the steering committee if you are a member of the club?

49. A plumber has a dozen different wrenches hanging on the wall in his shop. He is called out on an emergency job and grabs three of the wrenches at random to take with him. Suppose only one of the dozen wrenches is appropriate for the job. What is the probability that when he gets to the job he will have the correct wrench?

50. Repeat Exercise 49 if four of the dozen wrenches could be used to do the job.

Baseball player Wade Swamps bats .300 against pitcher Frank Cello. In Exercises 51 through 55 find the probability that Swamps will get the given number of hits in his next five at bats against Cello. You may assume independence of trials.

51. 0 hits   **52.** at least 1 hit   **53.** fewer than 5 hits

54. more than 3 hits   **55.** 1, 2, or 3 hits

56. Refer to Exercises 51 through 55: What is Swamps's expected number of hits in five at bats against Cello?

If he comes to bat with a man on first base and fewer than two outs, two thirds of the time baseball player Jim Dice will hit into a double play. Suppose Dice comes up four times in a game with a man on first and fewer than two outs. In Exercises 57 through 61 find the given probability. You may assume independence of trials.

57. P(Dice will not hit into any double plays).
58. P(Dice will hit into at least 1 double play).
59. P(Dice will hit into more than 1 double play).
60. P(Dice will hit into fewer than 4 double plays).
61. P(Dice will hit into 4 double plays).
62. Refer to Exercises 57 through 61: What is the expected number of double plays that Dice will hit into during the game?
63. You deal two cards from an ordinary deck. Let X denote the number of face cards dealt (face cards are jacks, queens, kings, and aces). Write the probability distribution for the random variable X, and find the expected number of face cards dealt.
64. In the game of Keno, 80 balls, numbered 1 through 80, are shaken in a cage and then 20 of them are selected. If the number between 1 and 80 which you have chosen matches that on any of the 20 balls, you win $2.20; if not, you lose $1. Find your expected winnings. Is this a fair game? If not, how should the payoff for winning be changed to make it a fair game?

An auto dealership's records show the following:

Number of Pickup Trucks Sold per Year:	200	210	220	230	240	250
Number of Years This Occurred:	2	6	10	4	2	1

Exercises 65 through 67 refer to this situation.

65. Find the dealership's expected number of pickup trucks sold per year.
66. If the dealership makes a profit of $400 on each pickup truck sold, find the expected annual profit from selling pickups.
67. The dealership is considering dropping its line of pickups for a line of recreational vehicles (RVs). It estimates that it can make a profit of $800 on each RV sold and that the probabilities of annual RV sales are as follows:

Annual RV Sales:	120	130	140	150
Probability:	0.10	0.50	0.30	0.10

Should they drop the line of pickup trucks in favor of a line of RVs?

ADDITIONAL TOPICS

Here are some suggestions for topics not covered in the text that you might want to investigate on your own.

1. The game of **craps** is played as follows: you throw a pair of dice; if you get a 7 or 11 on the first throw, you win; if you get a 2, 3, or 12 on the first throw, you lose; if you get a 4, 5, 6, 8, 9, or 10 on the first throw, the number you get is called your **point,** and you continue to throw the dice until you either get your point again, in which case you win, or you get a 7 or 11, in which case you lose. Find the odds of winning at craps.

2. In the game of **chuck-a-luck,** three dice are tossed. The bettor bets $1 on a number between 1 and 6, inclusive, and wins $1 for each appearance of the number bet on. For instance, if the three dice come up 4–4–5 and the bettor has bet $1 on the number 4, then he will win $2 (and get his $1 bet back). Calculate the expected value of chuck-a-luck; is it a fair game?

3. Find out how probability applies to poker hands and how a knowledge of probability can be of advantage to a poker player.

4. Find out about the game of **blackjack.** (Also known as **twenty-one.**) In certain circumstances it is actually possible for a blackjack player to gain an advantage over the casino by a method known as **card counting.** Find out how card counting works and how it gives the player using it an advantage.

5. Show that if a Bernoulli process has n trials, with the probability of a success on any trial equal to p, then the expected number of successes in the n trials is np.

6. A **Poisson process** is one that involves random arrivals, such as telephone calls at a switchboard, for instance. Every Poisson process can be described by a Poisson-distributed random variable, in much the same way that a Bernoulli process is described by a binomially-distributed random variable. Find out about the Poisson distribution and how it is used.

7. Among the earliest contributors to the theory of probability were the French mathematician Pierre Fermat (1601?–1665) and the French mathematician and philosopher Blaise Pascal (1623–1662). Their investigation of probability was instigated by a question about gambling posed to Pascal by a gentleman gambler, and carried out through a series of letters between them in 1654. Find out about this series of letters and the contributions of Fermat and Pascal to probability theory.

SUPPLEMENT: QUALITY CONTROL[1]

Suppose a producer of computer chips sends them to a consumer in large lots. The producer can inspect the lots for defective chips before they are shipped, or the consumer can inspect them when they are received. In either case, the usual procedure will be to take a sample of n chips from each lot and test them; the lot will then be accepted if the sample contains c or fewer defective chips and rejected if it contains more than c defective ones. This is a **quality control procedure** that requires knowledge of two numbers: the **sample size n** and the **acceptance number c.** Once the sample size and acceptance number are known, a **sampling plan** can be created.

EXAMPLE 1 Suppose $n = 20$ and $c = 3$. Then the sampling plan for the lots of computer chips would be as follows:

Randomly select a sample of 20 chips from each lot and test them. If three or fewer of the chips tested are defective, accept the lot; otherwise, reject it. ◻

---

[1] The material in this supplement follows that in Section 16.4 of Chase and Aquilino, *Production and Operations Management,* fourth edition, Irwin, 1985.

Both the producer and the consumer are usually willing to tolerate some defective parts in a lot. The producer will consider a lot to be good if its percentage of defectives is less than or equal to some benchmark level called the **acceptance quality level (AQL)**. Similarly, the consumer will consider a lot to be bad if its percentage of defectives is greater than another benchmark level called the **lot tolerance percentage defective (LTPD)**.

EXAMPLE 2

Suppose the producer of computer chips sets an AQL of 1% and the consumer sets an LTPD of 6%. Then the producer will consider lots having 1% or fewer defective chips to be good and hence will want the sampling plan to accept such lots. The consumer will consider lots having more than 6% defective chips to be bad and hence will want the sampling plan to reject such lots. ☐

Now, a lot may be a good one from the producer's point of view (because its percentage of defectives is less than or equal to the AQL), but just by bad luck the sample taken from the lot may contain a disproportionate number of defectives, so many that the sampling plan will call for its rejection. Thus a good lot would be rejected; the probability that this will happen is called **producer's risk**. Similarly, a lot may be a bad one from the consumer's point of view (because its percentage of defectives is greater than the LTPD), but just by bad luck the sample taken from the lot may fail to contain enough defectives to cause its rejection. Thus a bad lot would be accepted; the probability that this will occur is called **consumer's risk**.

Producer's Risk and Consumer's Risk

For a given sampling plan,

α = producer's risk
= P(the plan will reject a lot whose percentage of defectives is less than or equal to the AQL)

and

β = consumer's risk
= P(the plan will accept a lot whose percentage of defectives is greater than the LTPD)

(α and β are the Greek letters alpha and beta, respectively.)

Naturally, the producer wishes to make α small so that fewer good lots will be rejected, and the consumer wishes to make β small so that fewer bad lots will be accepted. It turns out that the smaller we make these risks, the larger we must make the sample size n. But the larger n is, the more it costs to carry out the sampling plan. Hence α and β are usually chosen in such a way as to make the sample size reasonably small. It is very common to take $\alpha = 0.05$ and $\beta = 0.10$.

Once AQL, LTPD, α, and β have been decided upon, the sample size n and acceptance number c can be determined and a sampling plan can be created. The procedure for finding n and c directly is too complex to be presented here, but fortunately they can be found from tables in reference works on quality control. Table 6–10 is a portion of such a table for $\alpha = 0.05$ and $\beta = 0.10$. We will show how to use Table 6–10 with an example.

TABLE 6–10

LTPD/AQL	c	n(AQL)
44.89	0	0.052
10.95	1	0.355
6.51	2	0.818
4.89	3	1.366
4.06	4	1.970
3.55	5	2.613
3.21	6	3.286
2.96	7	3.981

EXAMPLE 3 Suppose the producer of computer chips normally makes them in lots with 1% or fewer defectives, but is willing to take a 5% risk of having the sampling plan reject such a lot. Suppose also that the consumer of the chips considers lots with more than 6% defectives to be unacceptable, but is willing to take a 10% risk of having the sampling plan accept such a lot. Then

$$\text{AQL} = 0.01, \qquad \alpha = \text{producer's risk} = 0.05,$$

and

$$\text{LTPD} = 0.06, \qquad \beta = \text{consumer's risk} = 0.10.$$

Hence

$$\frac{\text{LTPD}}{\text{AQL}} = \frac{0.06}{0.01} = 6.$$

In Table 6–10 find the smallest value in the LTPD/AQL column that is greater than or equal to 6; it is 6.51, which corresponds to $c = 2$. In the row for $c = 2$ and the n(AQL) column is the value 0.818. Therefore

$$n(\text{AQL}) = n(0.01) = 0.818,$$

so $n = 81.8$. Since a sample size must be a natural number, we take $n = 82$. Thus the sampling plan is as follows:

Randomly select a sample of $n = 82$ chips from each lot and test them. If $c = 2$ or fewer of the chips tested are defective, accept the lot; otherwise, reject it. ☐

Every sampling plan has risks. In fact, given a lot with $x\%$ defectives, there is always a risk that the plan will reject the lot when it should accept it or accept it when it should reject it. A graph of

$$P(\text{the plan will accept a lot having } x\% \text{ defectives})$$

against x is called an **operating characteristic (OC) curve** for the plan.

EXAMPLE 4 Consider the sampling plan of Example 3. Let us find P(the plan will accept a lot having 3% defectives). If we sample with replacement, then

P(the plan will accept a lot having 3% defectives)

= P(getting 2 or fewer defectives out of the 82 sampled when the probability of a defective is always 0.03).

But if we let "success" be "get a defective chip," this probability is a binomial probability with $n = 82$, $k \leq 2$, and

$$p = P(\text{success}) = P(\text{get a defective}) = 0.03.$$

Thus,

P(the plan will accept a lot having 3% defectives)

$$= \binom{82}{0}(0.03)^0(0.97)^{82} + \binom{82}{1}(0.03)^1(0.97)^{81} + \binom{82}{2}(0.03)^2(0.97)^{80}$$

$\approx 0.55.$

We can similarly find the probability that the sampling plan will accept a lot with $x\%$ defectives for any x between 0 and 100. See Table 6–11 for the results for some values of x. If we plot x on the horizontal axis and the probabilities on the vertical axis, we obtain the OC curve for the sampling plan. See Figure 6–13. (Note that the value for producer's risk is 0.12 rather than the 0.10 specified prior to the creation of the sampling plan. The discrepancy is due to rounding off in our calculations and to the fact that, for the risks α and β to be *exactly* 0.05 and 0.10, we would need $n = 81.8$.) □

TABLE 6–11

x	P(the plan will accept a lot with $x\%$ defectives)
0.0	1.00
0.5	0.99
1.0	0.95 (= 1 − producer's risk)
2.0	0.77
3.0	0.55
4.0	0.36
5.0	0.22
6.0	0.12 (= consumer's risk)
7.0	0.07

The operating characteristic curve for a sampling plan allows us to estimate at a glance the probability that the plan will accept a lot with $x\%$ defectives. All OC curves have the shape of the one in Figure 6–13: They start at 1, decrease slowly for a time, suddenly drop to a lower level, and then decrease slowly again. The steeper the drop, the better the sampling plan is at discriminating between good and bad lots. The plan can be made more discriminating by keeping c fixed and increasing the sample size n.

P (accepting a lot with *x* percent defectives)

FIGURE 6–13
The OC Curve

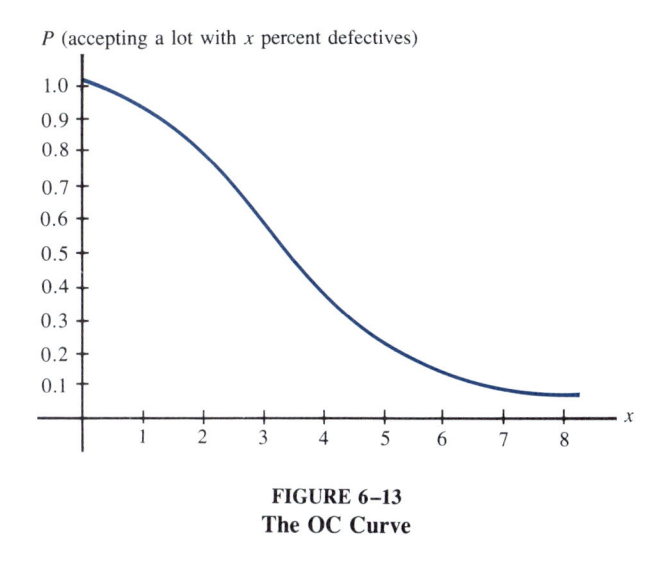

We conclude by remarking that in Example 4 we assumed sampling with replacement. In practice, we might find it more convenient to sample without replacement; in this case, the probability of getting a defective chip when we draw one from the lot would *not* remain the same each time (why not?); hence the probabilities involved would not be binomial. However, if the sample size is small compared with the lot size, then the probability of getting a defective chip would remain *approximately* the same from draw to draw and, hence, the probabilities would be *approximately* binomial. Thus if we sample without replacement, the OC curve constructed by the method of Example 4 will give an approximate picture of the probability of accepting a lot having $x\%$ defectives; the smaller the sample size is as a proportion of the lot size, the better this approximation will be.

SUPPLEMENTARY EXERCISES

1. Create a sampling plan if AQL = 0.02, LTPD = 0.08, α = 0.05, and β = 0.10.
2. Repeat Exercise 1 if AQL = 0.005 and LTPD = 0.05.
3. Sketch the OC curve for the sampling plan of Exercise 1.
4. Sketch the OC curve for the sampling plan of Exercise 2.
5. We have remarked that if sampling is without replacement and the sample size is small as a proportion of the lot size, then the OC curve can be constructed as if the sampling were with replacement, and it will give a good approximation to the true probability of accepting a lot with $x\%$ defectives. This is not the case when sampling is without replacement and the sample size is a large proportion of the lot size, however. To illustrate, suppose a sampling plan says

take a sample of 3 parts from each lot of 8 parts, sampling without replacement, test them, and accept the lot if 0 of them are defective.

(a) Find an approximate OC curve for this sampling plan. (That is, use binomial probabilities to construct the OC curve.)
(b) Find the actual OC curve for this sampling plan.
(c) Compare the OC curves of (a) and (b).

Suppose a producer samples each lot using a sampling plan. Also suppose that if a lot is rejected it is inspected completely, with all defective parts being replaced by good ones. Under such a system, the **average outgoing quality (AOQ)** of the lots is given by

$$\text{AOQ} = P(\text{the plan will accept a lot having } x\% \text{ defectives})(0.01x).$$

(This assumes sampling with replacement or sampling without replacement when the sample size is a small proportion of the lot size.) If AOQ is plotted against x, the result is an AOQ curve. The AOQ curve will show that the average percentage of defectives in the outgoing lots versus the percentage of defectives in incoming lots is small when the percentage of defectives is small, and it is also small when the percentage of defectives is large. (Why?) The largest value of AOQ is called the **average outgoing quality limit (AOQL)**.

*6. Sketch the AOQ curve for the sampling plan of Exercise 1 and estimate AOQL.

*7. Sketch the AOQ curve for the sampling plan of Exercise 2 and estimate AOQL.

7

Applications of Probability

In this chapter we continue our study of probability theory by considering several of its most important and extensive applications. Our work here will allow us to solve problems such as the following:

- It will cost $100 million to make and distribute a flu vaccine in preparation for a possible epidemic. If the epidemic occurs and the vaccine is available, the epidemic will cost the nation $3 billion; if the epidemic occurs and the vaccine is not available, it will cost the nation $20 billion. If the probability that the epidemic will occur is 0.7, should the vaccine be made and distributed?

- Three brands of soap, X, Y, and Z, are fighting for market share in a region. Each month X loses 10% of its customers to Y and 5% to Z, Y loses 5% of its customers to X and 10% to Z, and Z loses 20% of its customers to X and 15% to Y. What share of the market will each firm have in the long run?

- Two candidates are running for the same political office. Each has the option of attacking the other or discussing the issues. If both candidates attack one another, or if both discuss the issues, there will be no change in the support of either. However, if either of them attacks the other while the other discusses the issues, the attacker will gain 1% of the nonattacker's support each week for as many weeks as the situation continues. What will each candidate do?

The first of these problems can be handled with the aid of a probabilistic **decision tree,** the second by means of a **Markov chain,** and the third is a problem in **game theory.** We begin this chapter with a section devoted to decision trees, then discuss Markov chains, and follow with a consideration of game theory. The final two sections of the chapter are devoted to **Bayes' Law** and **Bayesian decision making.**

7.1 DECISION TREES

At the end of Section 6.6 we mentioned that expected values are useful in decision making. In this section, and again in Section 7.5, we consider the topic of decision making using expected value criteria in more depth. We begin with a discussion of decision trees.

Suppose that every year a rancher must decide whether to take his cattle to market in the spring or in the fall. The profit he will make per head of cattle will depend on whether demand is high or low at the particular time he goes to market. The profit per head in dollars is displayed in a **payoff matrix:**

$$\text{Go to Market} \quad \begin{array}{c} \\ \text{spring} \\ \text{fall} \end{array} \begin{array}{c} \text{Profit per Head} \\ \begin{array}{cc} \text{demand high} & \text{demand low} \end{array} \\ \begin{bmatrix} 100 & 40 \\ 160 & -20 \end{bmatrix} \end{array}.$$

Now we make a **decision tree** for this situation (see Figure 7–1). The decision tree is constructed by representing each decision as a line segment radiating from a square. In our example there are two possible decisions: "go to market spring" and "go to market fall." The square represents a point where a choice between the decisions must be made. **States of nature** are events over which the decision maker has no control. States of nature are represented by line segments radiating from circles. In our example, the states of nature are "demand high" and "demand low." A circle represents a point where one or more states of nature occur.

As it stands, the decision tree of Figure 7–1 is of limited use to the rancher. If he is optimistic, he may wish to go to market in the fall, in the hope of getting

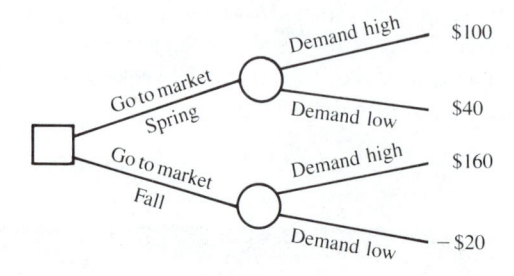

FIGURE 7–1

APPLICATIONS OF PROBABILITY 421

$160 per head for his cattle. Of course, this entails a risk of losing $20 per head, so if he is pessimistic he may choose to go to market in the spring, for if he does so at least he cannot lose money. In order to make the decision tree useful, the rancher must assign probabilities to each of the states of nature. Suppose the rancher assigns the following probabilities to each of the states of nature:

$$P(\text{demand will be high in spring}) = 0.50,$$

$$P(\text{demand will be low in spring}) = 0.50,$$

$$P(\text{demand will be high in fall}) = 0.30,$$

and

$$P(\text{demand will be low in fall}) = 0.70.$$

These are subjective probabilities, based on the rancher's best guess as to the likelihood of the events.

We now place these probabilities on the proper branches of the decision tree (see Figure 7–2). Wherever states of nature emanate from a circle, we find the expected value and place it in the circle. In each square we place the best of the expected values in the circles radiating from it. The decision with the largest expected value, in this case "go to market spring," is the one that pays off best in the long run. Therefore, on the basis of expected value, the rancher should choose to go to market every spring.

Note that the rancher cannot actually achieve the expected value of $70 per head if he goes to market in the spring; he will actually receive either $100 or $40 per head. However, in the long run, as he faces this situation again and again, he can expect to average $70 per head by going to market in the spring, as opposed to $34 per head by going in the fall.

Decision trees with expected values are often used to help make decisions even when the situations are not repetitive ones. In such cases the expected values of the alternatives may never be realizable, even in the long run, but they do serve to give a picture of the comparative worth of the alternatives.

EXAMPLE 1 Meryl Launch has just received $10,000 worth of stock as an inheritance and is trying to decide whether to keep the stock or cash it in and use the proceeds to buy

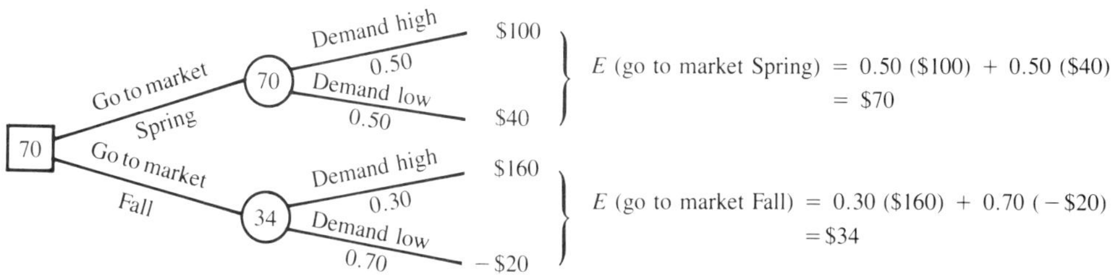

FIGURE 7–2

gold. In either case, she plans to liquidate her investment a year from now in order to make a down payment on a house. Therefore the situation is certainly not a repetitive one, but Meryl wishes to use a decision tree with expected value in order to determine the comparative worth of the investment alternatives.

Meryl fees that the value of her investment will depend on the average inflation rate over the next year according to the following payoff matrix:

$$\text{Investment} \begin{array}{c} \text{stock} \\ \text{gold} \end{array} \begin{array}{c} \text{Average Inflation Rate} \\ \begin{array}{ccc} 5\% & 10\% & 15\% \end{array} \\ \begin{bmatrix} 12{,}000 & 10{,}000 & 9{,}000 \\ 7{,}000 & 11{,}000 & 14{,}000 \end{bmatrix} \end{array}.$$

She also estimates the probabilities of inflation over the next year as follows:

$$P(\text{inflation will average } 5\%) = 0.20;$$

$$P(\text{inflation will average } 10\%) = 0.70;$$

$$P(\text{inflation will average } 15\%) = 0.10.$$

Since

$$E(\text{keep stock}) = 0.20(\$12{,}000) + 0.70(\$10{,}000) + 0.10(\$9000)$$

$$= \$10{,}300$$

and

$$E(\text{buy gold}) = 0.20(\$7000) + 0.70(\$11{,}000) + 0.10(\$14{,}000)$$

$$= \$10{,}500,$$

Meryl's decision tree is as shown in Figure 7–3. Thus, based on the comparative worth of the alternatives as measured by their expected values, Meryl should cash in the stock and buy gold. ☐

EXAMPLE 2   New Boston, N.H., has no high school and currently pays tuition to send its students to high school in a neighboring town. The town is debating whether to build its

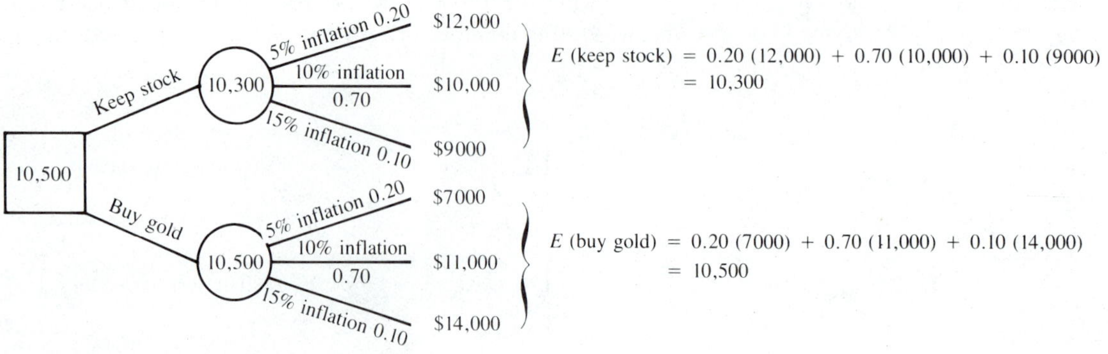

FIGURE 7–3

own high school, and, if the decision is to build, whether the new school should be small or large. It is estimated that a small school (capacity 400 students) will cost an average of $600,000 per year over the next 10 years, while a large one (capacity 800 students) will cost an average of $1,200,000 per year over the same period. It is also estimated that under conditions of slow growth the town will be responsible for the education of an average of 250 students per year, under conditions of moderate growth the figure will be 350 students per year, and under conditions of high growth it will be 600 students per year. If the town decides not to build a school, sending students to the neighboring town will cost an average of $3000 per year in tuition. Also, if the town builds a school that is not large enough to accommodate the student population, the excess students will have to be sent to the neighboring town at a cost of $3000 per year. Suppose that

$$P(\text{slow growth}) = 0.05,$$

$$P(\text{moderate growth}) = 0.35,$$

and

$$P(\text{high growth}) = 0.60.$$

On the basis of expected cost over the next 10 years, what should the town do?

Figure 7–4 shows the decision tree for this situation. The costs for the top branch of the tree were obtained by multiplying the number of students under each

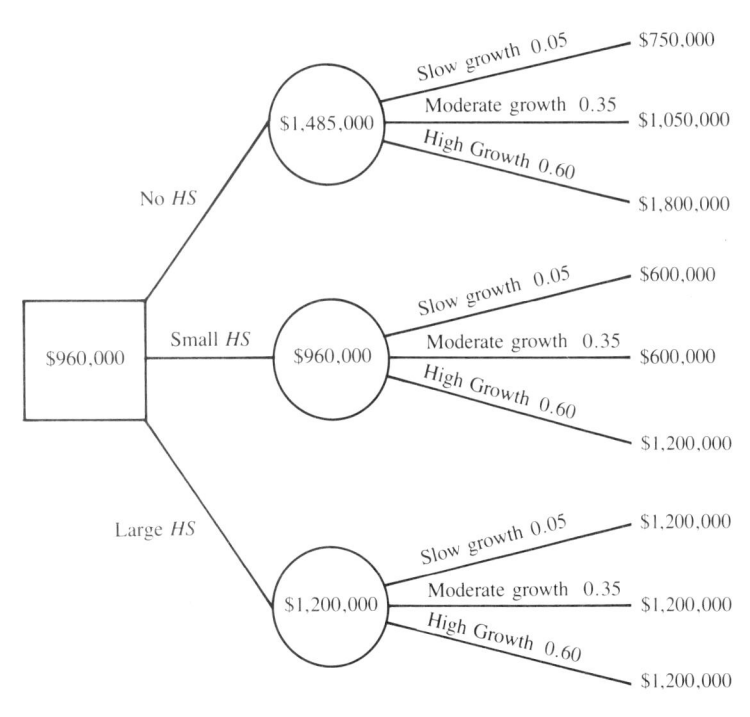

FIGURE 7–4

state of nature by $3000. The costs for the middle branch were obtained as follows: for slow and moderate growth, the cost is just that of the small high school; for high growth, it is the cost of the small high school plus the cost of tuition for $600 - 400 = 200$ excess students. The costs for the bottom branch of the tree are all equal to the cost of the large high school. Since the expected cost is smallest for the middle branch of the tree, the town should build the small high school. □

7.1 EXERCISES

1. An investor has $10,000 to invest for one year. He can either put the money into a savings account that pays 6% per year compounded quarterly or invest it all in a stock that will pay $500 in dividends during the year. Also, if he invests his money in the stock, he estimates that he will be able to sell the stock at the end of the year for either $11,000 or $9500, depending on whether the stock market is up or down at the time. His stockbroker has told him that

$$P(\text{market will be up at end of year}) = 0.40,$$

and

$$P(\text{market will be down at end of year}) = 0.60.$$

On the basis of expected return on investment, what should the investor do?

2. Repeat Exercise 1 if

$$P(\text{market will be up at the end of the year}) = 0.60$$

and

$$P(\text{market will be down at the end of the year}) = 0.40.$$

*3. Refer to Exercises 1 and 2: Let

$$P(\text{the market will be up at the end of the year}) = p.$$

and

$$P(\text{market will be down at the end of the year}) = 1 - p.$$

Find the values of p such that
   (a) the investor will choose the savings account;
   (b) the investor will choose stocks;
   (c) the investor will be indifferent between the savings account and stocks.

Management

4. A company is trying to decide which of two new products, A or B, it should market. The payoff matrix for the company is as follows:

$$\text{Market} \begin{array}{c} \\ \text{product A} \\ \text{product B} \end{array} \overset{\begin{array}{cc} \text{Product Is} & \\ \text{successful} & \text{unsuccessful} \end{array}}{\begin{bmatrix} 1 & -0.2 \\ 6 & -4 \end{bmatrix}}.$$

The payoffs are in millions of dollars. The company estimates that

$$P(\text{product A will be successful}) = 0.5$$

APPLICATIONS OF PROBABILITY

and
$$P(\text{product B will be successful}) = 0.5.$$

Draw a decision tree. On the basis of expected payoff, what should the company do?

5. Repeat Exercise 4 if
$$P(\text{product A will be successful}) = 0.5$$
and
$$P(\text{product B will be successful}) = 0.4.$$

6. An oil company must decide whether or not to drill a well on a certain tract of land it has leased. The lease cost the company $10,000. To drill will cost an additional $60,000. If the well strikes oil, the revenue from it will be $200,000. The oil company's geologist estimates that the probability a well on the land will strike oil is 0.20. Draw a decision tree. On the basis of expected profit, what should the company do?

7. Repeat Exercise 6 if
$$P(\text{well on the land will strike oil}) = 0.50.$$

*8. Refer to Exercises 6 and 7: Let
$$P(\text{well will strike oil}) = p.$$

Find the values of p such that
(a) the company will drill for oil;
(b) the company will not drill for oil;
(c) the company will be indifferent between drilling and not drilling.

9. Tattle Toy Company has invested $100,000 in the development of a new electronic robot toy. Each unit of the toy will cost $20 to make and will sell for $35. Tattle is trying to decide how many units of the toy to produce for the Christmas season. Its facilities will allow it to produce 40,000, 60,000, or 80,000 units of the toy. If the toy turns out to be a success, Tattle will be able to sell as many units as it can produce, but it is not a success, it will be able to sell only 50,000 units. Tattle's marketing department estimates that
$$P(\text{toy will be a success}) = 0.40.$$

On the basis of expected profit, what should Tattle do?

10. An oil company is considering how to meet its demands for fuel oil next winter. It can refine either 25%, 50%, or 80% of its needs, in each case purchasing the rest of its needs from other refiners. It will cost the company $0.48 per gallon to refine its own fuel oil and $0.55 per gallon to purchase it from other refiners. The demand for the company's fuel oil, and thus the price it will bring, will depend on the severity of the winter:

		Demand for Oil (Million Gal)	Price per Gallon
	mild	100	$0.82
Winter is:	normal	140	$0.88
	harsh	200	$0.95

Also,
$$P(\text{winter will be mild}) = 0.20,$$
$$P(\text{winter will be normal}) = 0.75,$$
and
$$P(\text{winter will be harsh}) = 0.05.$$

On the basis of expected profit, what should the company do?

11. Repeat Exercise 10 if the company can sell the oil for $0.82 per gallon if the winter is mild, $0.95 per gallon if it is normal, and $1.00 per gallon if it is harsh.

12. A buyer for a department store must purchase women's swimsuits in January for sale the following summer. She has already decided to buy either string bikinis or maillots or a combination of both. The bikinis cost $16 each and the maillots $24 each, and she has enough money to buy either 3000 bikinis, or 2000 maillots, or a combination of 1500 bikinis and 1000 maillots. Sales next summer will depend on the popularity of each type of suit at the time. There are four possibilities:

- both types of suit will be popular;
- bikinis will be popular but maillots will not be;
- maillots will be popular but bikinis will not be;
- neither type of suit will be popular.

The buyer estimates that these four events are of equal probability. She also knows that if either type of suit is popular, the store will be able to sell all of it, whereas if it is not popular, the store will be able to sell only two-thirds of the amount in stock. Each bikini will sell for $30 and each maillot for $36. On the basis of expected profit, what should the buyer do?

13. Repeat Exercise 12 if
$$P(\text{bikinis will be popular but maillots will not}) = 0.15;$$
$$P(\text{maillots will be popular but bikinis will not}) = 0.15;$$
$$P(\text{both bikinis and maillots will be popular}) = 0.65;$$
and
$$P(\text{neither bikinis nor maillots will be popular}) = 0.05.$$

Life Science

14. It will cost $100 million to make and distribute a flu vaccine in preparation for a possible epidemic next year. If the epidemic occurs and the vaccine is available, the epidemic will cost the nation $3 billion; if the epidemic occurs and the vaccine is not available, the epidemic will cost the nation $20 billion. Suppose that $P(\text{the epidemic will occur}) = 0.70$. On the basis of expected cost, should the vaccine be made and distributed?

15. Repeat Exercise 14 if the cost of making and distributing the vaccine is $8 billion and the cost to the economy will be $20 billion if the vaccine is needed and unavailable, $10 billion if it is needed and available.

Social Science

16. A congresswoman who is running for reelection estimates that if the turnout of voters is low on election day, her party organization will be able to deliver 55% of the

vote to her, but if the turnout is high, she will only be able to get 48% of the vote. The congresswoman has two options for her campaign: attack or ignore her opponent. She has further estimated the probability of either a high or a low turnout under each option:

$$\text{Option} \begin{array}{c} \text{attack opponent} \\ \text{ignore opponent} \end{array} \begin{array}{c} \text{Probability of Turnout} \\ \begin{array}{cc} \text{high} & \text{low} \end{array} \\ \begin{bmatrix} 0.75 & 0.25 \\ 0.50 & 0.50 \end{bmatrix} \end{array}.$$

Based on her expected percentage of the vote, what should the congresswoman do?

17. A town is trying to decide whether to contract for a mosquito control program next summer. If the mosquito population next summer turns out to be large, the town definitely will need such a program, but if it turns out to be small, the town can get by without the program. If the town contracts for the program now, it will cost them $50,000, but if they wait until next summer to contract, it will cost them $60,000. A scientist from the state environmental agency has estimated that

P(mosquito population will be large next summer) $= 0.55$.

On the basis of expected cost, what should the town do?

18. Repeat Exercise 17 if

P(mosquito population will be large next summer) $= 0.85$.

*19. Refer to Exercises 17 and 18: Let

P(mosquito population will be large next summer) $= p$.

Find the values of p such that, on the basis of expected cost, the town will
(a) contract for the program now;
(b) wait until next year to decide;
(c) be indifferent between contracting now and waiting until next year.

Utilities

A **utility** is a number that measures satisfaction or dissatisfaction. A positive utility represents satisfaction, and the larger the utility, the greater the satisfaction. A utility of zero represents indifference. A negative utility represents dissatisfaction, and the larger the absolute value of the utility, the greater the degree of dissatisfaction. Exercises 20 through 25 use utilities.

*20. If a gambler bets $10, he can either win $10, break even, or lose $10. If he bets $100, he can either win $100, break even, or lose $100. His utilities in each case are as follows:

$$\text{Bet} \begin{array}{c} \$10 \\ \$100 \end{array} \begin{array}{c} \text{Utilities} \\ \begin{array}{ccc} \text{win} & \text{break even} & \text{lose} \end{array} \\ \begin{bmatrix} 2 & -\frac{1}{2} & -1 \\ 50 & -5 & -20 \end{bmatrix} \end{array}.$$

(Note that the gambler gets more satisfaction from winning a bet of a given size that he suffers dissatisfaction from losing the same bet; also, he suffers dissatisfaction if he merely breaks even on a bet.) Suppose that

P(winning a bet) $= 0.4$,
P(breaking even) $= 0.2$,

and
$$P(\text{losing a bet}) = 0.4.$$

Draw a decision tree for the gambler using his utilities. If he will only make a bet when his expected utility is nonnegative, which bet ($10 or $100) will he make?

*21. Repeat Exercise 20 if the gambler's utilities are as follows:

$$\text{Bet} \begin{array}{c} \$10 \\ \$100 \end{array} \begin{bmatrix} \overset{\text{win}}{5} & \overset{\text{break even}}{0} & \overset{\text{lose}}{-5} \\ 50 & 0 & -50 \end{bmatrix}.$$

*22. Repeat Exercise 20 if the gambler's utilities are as follows:

$$\text{Bet} \begin{array}{c} \$10 \\ \$100 \end{array} \begin{bmatrix} \overset{\text{win}}{5} & \overset{\text{break even}}{0} & \overset{\text{lose}}{-10} \\ 50 & 0 & -200 \end{bmatrix}.$$

*23. A doctor is considering two possible treatments, A and B, for a patient with a chronic illness. Each treatment could improve the patient's condition, leave it unchanged, or worsen it. The patient's utilities for these possibilities are as follows:

a treatment improves his condition: $+5$;

a treatment leaves his condition unchanged: 0;

a treatment worsens his condition: -10.

The doctor estimates the following probabilities for each of the treatments:

$$\text{Treatment} \begin{array}{c} A \\ B \end{array} \begin{bmatrix} \overset{\text{improve condition}}{0.45} & \overset{\text{leave condition unchanged}}{0.30} & \overset{\text{worsen condition}}{0.25} \\ 0.35 & 0.45 & 0.20 \end{bmatrix}.$$

Draw a decision tree for the patient. On the basis of expected utility, what treatment should the doctor recommend?

*24. Repeat Exercise 23 if the patient's utilities are as follows:

a treatment improves his condition: $+10$;

a treatment leaves his condition unchanged: 0;

a treatment worsens his condition: -10.

*25. Repeat Exercise 23 if the patient's utilities are as follows:

a treatment improves his condition: $+20$;

a treatment leaves his condition unchanged: $+15$;

a treatment worsens his condition: -10.

7.2 Markov Chains

A **Markov chain** is a sequence of stages, each of which depends probabilistically only on the preceding stage. As we shall demonstrate, Markov chains are frequently

APPLICATIONS OF PROBABILITY 429

useful in modeling situations in which the distribution of some commodity among several sites changes over time.

Suppose we own two stores that rent garden tools, one in the town of Auburn (A) and the other in the nearby town of Boxboro (B). Each week we take an inventory of the tools available for rental at each location. We know from experience that 90% of the customers who rent tools at a particular store return the tools to the same store, while the remaining 10% return the tools to the other store. We may embody this information in a **transition matrix T:**

$$\text{Rent at} \begin{array}{c} \\ A \\ B \end{array} \begin{array}{c} \text{Return to} \\ \begin{array}{cc} A & B \end{array} \\ \begin{bmatrix} 0.90 & 0.10 \\ 0.10 & 0.90 \end{bmatrix} = \mathbf{T}.\end{array}$$

The transition matrix **T** is a matrix of probabilities. The first row of the matrix tells us that if a customer rents tools at the Auburn store, the probability that he will return them there is 0.90, while the probability that he will return them to the Boxboro store is 0.10. We interpret the second row of the matrix similarly. Note that each row of the transition matrix must sum to 1. (Why is this so?)

Let us suppose that as of week 1, we have 2000 tools available for rental at the Auburn store and 1000 available at the Boxboro store. We embody this information in a row vector:

$$\begin{array}{c} \quad A \quad B \\ \text{Week 1:} \quad [2000, \ 1000]. \end{array}$$

Now let us see what happens to the garden tools, week by week, assuming that all tools available are rented each week. We have a sequence:

$$\text{Week 1} \xrightarrow{\mathbf{T}} \text{week 2} \xrightarrow{\mathbf{T}} \text{week 3} \xrightarrow{\mathbf{T}} \text{week 4} \longrightarrow \cdots$$
$$[2000, \ 1000]$$

The transition matrix **T** tells us how the tools get redistributed among the stores each week:

$$\text{Week 1:} \quad \begin{array}{cc} A & B \\ [2000, & 1000] \end{array} \begin{bmatrix} 0.90 & 0.10 \\ 0.10 & 0.90 \end{bmatrix} = \begin{array}{cc} A & B \\ [1900, & 1100]. \end{array}$$

Therefore, as of week 2 we will have 1900 tools at the Auburn store and 1100 at the Boxboro store. Similarly, since

$$\begin{array}{cc} A & B \\ [1900, & 1100] \end{array} \begin{bmatrix} 0.90 & 0.10 \\ 0.10 & 0.90 \end{bmatrix} = \begin{array}{cc} A & B \\ [1820, & 1180], \end{array}$$

as of week 3 we will have 1820 tools available at Auburn and 1180 at Boxboro. Continuing in this manner, we can find the tools available for rental at each location in each week; if $[a_i, b_i]$ is the vector that gives the number a_i available at Auburn and the number b_i available at Boxboro as of week i, then

$$[a_i, b_i]\mathbf{T} = [a_{i+1}, b_{i+1}]$$

gives the number available at each location as of week $i + 1$.

Note that since

$$[a_i, b_i]\mathbf{T} = [a_{i+1}, b_{i+1}]$$

and

$$[a_{i+1}, b_{i+1}]\mathbf{T} = [a_{i+2}, b_{i+2}],$$

we have

$$[a_i, b_i]\mathbf{T}^2 = [a_{i+1}, b_{i+1}]\mathbf{T} = [a_{i+2}, b_{i+2}].$$

Similarly, for any positive integer n,

$$[a_i, b_i]\mathbf{T}^n = [a_{i+n}, b_{i+n}],$$

and, in particular,

$$[a_1, b_1]\mathbf{T}^n = [a_{1+n}, b_{1+n}].$$

In other words, the distribution of the tools in week $n + 1$ may be found by multiplying the distribution of the tools in week 1 by the nth power of the transition matrix $\mathbf{T}$. For example, as of week 4 we will have

$$\begin{matrix} \text{A} & \text{B} \\ [2000, & 1000] \end{matrix} \mathbf{T}^3 = [2000, 1000] \begin{bmatrix} 0.756 & 0.244 \\ 0.244 & 0.756 \end{bmatrix} = \begin{matrix} \text{A} & \text{B} \\ [1756, & 1244], \end{matrix}$$

and hence will have 1756 tools at Auburn and 1244 at Boxboro.

We have just illustrated the concept of a Markov chain. In general, a Markov chain is a sequence of stages, with a transition matrix $\mathbf{T}$ that tells us how to get from one stage to the next:

$$\text{stage 1} \xrightarrow{\mathbf{T}} \text{stage 2} \xrightarrow{\mathbf{T}} \text{stage 3} \xrightarrow{\mathbf{T}} \text{stage 4} \xrightarrow{\mathbf{T}} \cdots$$

The entries of $\mathbf{T}$ are probabilities, and each row of $\mathbf{T}$ must sum to 1. If $\mathbf{V}_i$ is the row vector that describes stage i of the chain, then

$$\mathbf{V}_i\mathbf{T} = \mathbf{V}_{i+1},$$

and in fact,

$$\mathbf{V}_1\mathbf{T}^n = \mathbf{V}_{n+1}, \quad \text{for every positive integer } n.$$

EXAMPLE 1 Three brands of soap, Brand X, Brand Y, and Brand Z, currently have 60%, 30%, and 10% of the market, respectively. Market research shows that each month:

- 85% of Brand X customers remain loyal to Brand X;
- 10% of Brand X customers switch to Brand Y;

- 5% of Brand X customers switch to Brand Z;
- 85% of Brand Y customers remain loyal to Brand Y;
- 5% of Brand Y customers switch to Brand X;
- 10% of Brand Y customers switch to Brand Z;
- 65% of Brand Z customers remain loyal to Brand Z;
- 20% of Brand Z customers switch to Brand X;
- 15% of Brand Z customers switch to Brand Y.

The transition matrix **T** that describes the monthly changes in market share is thus

$$\text{Switch from Brand} \begin{array}{c} X \\ Y \\ Z \end{array} \begin{bmatrix} 0.85 & 0.10 & 0.05 \\ 0.05 & 0.85 & 0.10 \\ 0.20 & 0.15 & 0.65 \end{bmatrix} = \mathbf{T}.$$

(columns labeled: Switch to Brand X, Y, Z)

The Markov chain is

$$\text{month 1} \xrightarrow{\mathbf{T}} \text{month 2} \xrightarrow{\mathbf{T}} \text{month 3} \xrightarrow{\mathbf{T}} \cdots$$

In month 1 the market shares are given by

$$\mathbf{V}_1 = [0.60,\ 0.30,\ 0.10].$$

(columns X, Y, Z)

Therefore, in month 2 the market shares will be

$$\mathbf{V}_2 = \mathbf{V}_1 \mathbf{T} = [0.60, 0.30, 0.10]\mathbf{T} = [0.545, 0.33, 0.125].$$

Thus, in month 2 Brand X will have 54.5% of the market, Brand Y will have 33%, and Brand Z will have 12.5%. Let us find the market shares in month 5. Since

$$\mathbf{V}_1 \mathbf{T}^4 = \mathbf{V}_5$$

for any Markov chain, the market shares in month 5 are given by

$$[0.60, 0.30, 0.10]\mathbf{T}^4 = \mathbf{V}_5.$$

Since

$$\mathbf{T}^4 = \begin{bmatrix} 0.5887 & 0.2835 & 0.1278 \\ 0.2023 & 0.6073 & 0.1904 \\ 0.3901 & 0.3461 & 0.2638 \end{bmatrix}$$

(rounded to four decimal places), we have

$$[0.60, 0.30, 0.10]\mathbf{T}^4 = [0.453, 0.387, 0.160],$$

approximately. Therefore, by month 5, Brand X will have approximately 45.3%, Brand Y approximately 38.7%, and Brand Z approximately 16.0% of the market. □

Consider again the garden tool rentals at Auburn and Boxboro with which we introduced Markov chains. We started with 2000 tools at Auburn and 1000 at Boxboro. Note that, as the weeks passed, the numbers of tools available at each location came closer and closer together. Will the Markov chain reach steady state? That is, will the number of tools available at each location eventually remain unchanged week after week? This is equivalent to asking if there is some vector $[a_{i+1}, b_{i+1}]$ giving the number of tools available at each store as of week $i + 1$ that is the same as the vector $[a_i, b_i]$ of the previous week; for if

$$[a_{i+1}, b_{i+1}] = [a_i, b_i],$$

then

$$[a_{i+2}, b_{i+2}] = [a_{i+1}, b_{i+1}]\mathbf{T} = [a_i, b_i]\mathbf{T} = [a_{i+1}, b_{i+1}] = [a_i, b_i],$$

$$[a_{i+3}, b_{i+3}] = [a_{i+2}, b_{i+2}]\mathbf{T} = [a_i, b_i]\mathbf{T} = [a_{i+1}, b_{i+1}] = [a_i, b_i],$$

and so on, so that all vectors from the ith one on are identical. For convenience let us drop the subscript i. What we have shown is that the Markov chain will reach steady state if there is some stage vector $[a, b]$ such that

$$[a, b]\begin{bmatrix} 0.90 & 0.10 \\ 0.10 & 0.90 \end{bmatrix} = [a, b],$$

or

$$0.90a + 0.10b = a$$
$$0.10a + 0.90b = b,$$

which we rewrite as

$$-0.10a + 0.10b = 0$$
$$0.10a - 0.10b = 0.$$

These last two equations are really the same (the second is the negative of the first); however, we must also have

$$a + b = 3000,$$

since the total tools available equal 3000. Hence we can find a and b by solving the system

$$-0.10a + 0.10b = 0$$
$$a + b = 3000.$$

The solution is $a = b = 1500$. Therefore, in the long run the Markov chain will reach steady state with 1500 garden tools available at each location, and this will not change thereafter.

In general, a Markov chain will reach steady state if the transition matrix $\mathbf{T}$ or some power of $\mathbf{T}$ has no zero entry. (A Markov chain with this property is called

APPLICATIONS OF PROBABILITY 433

a **regular** chain.) If this is the case, steady state may be found by solving the matrix equation

$$VT = V$$

for the row vector **V**. In order to do so, it is necessary to use the fact that the entries of **V** must sum to some predetermined value.

EXAMPLE 2 Consider again Example 1 concerning the market shares of three brands of soap: X, Y, and Z. The transition matrix is

$$T = \begin{bmatrix} 0.85 & 0.10 & 0.05 \\ 0.05 & 0.85 & 0.10 \\ 0.20 & 0.15 & 0.65 \end{bmatrix}.$$

Since **T** has no zero entry, the Markov chain will reach steady state. To find steady state, we solve

$$[x, y, z]T = [x, y, z]$$

for x, y, and z. We have

$$0.85x + 0.50y + 0.20z = x$$
$$0.10x + 0.85y + 0.15z = y$$
$$0.05x + 0.10y + 0.65z = z$$

or

$$-0.15x + 0.05y + 0.20z = 0$$
$$0.10x - 0.15y + 0.15z = 0$$
$$0.05x + 0.10y - 0.35z = 0.$$

The last of these equations is just the sum of the two preceding it with the signs changed and hence does not tell anything that the first two do not. However, since x, y, and z are market shares which must sum to 100%, we also have

$$x + y + z = 1.$$

Therefore, we solve the system

$$-0.15x + 0.05y + 0.20z = 0$$
$$0.10x - 0.15y + 0.15z = 0$$
$$x + y + z = 1$$

to obtain $x = {}^{15}\!/_{39}$, $y = {}^{17}\!/_{39}$, and $z = {}^{7}\!/_{39}$. Hence the Markov chain will eventually reach a steady state in which Brand X has ${}^{15}\!/_{39} \approx 38.5\%$ of the market, Brand Y has ${}^{17}\!/_{39} \approx 43.6\%$ of the market, and Brand Z has ${}^{7}\!/_{39} \approx 17.9\%$ of the market, and this will not change thereafter. ∎

■ *Computer Manual: Program MARKOV*

7.2 EXERCISES

In Exercises 1 through 6, the given matrix T is the transistion matrix and the vector V_1 is the initial vector for a Markov chain. Find V_2, V_3, and V_4 in each case.

1. $T = \begin{bmatrix} 0.4 & 0.6 \\ 0.6 & 0.4 \end{bmatrix}$, $V_1 = [10, 20]$

2. $T = \begin{bmatrix} 0.5 & 0.5 \\ 0.5 & 0.5 \end{bmatrix}$, $V_1 = [20, 10]$

3. $T = \begin{bmatrix} 0.4 & 0.2 & 0.4 \\ 0.2 & 0.8 & 0.0 \\ 0.0 & 0.6 & 0.4 \end{bmatrix}$, $V_1 = [100, 200, 100]$

4. $T = \begin{bmatrix} 0.3 & 0.6 & 0.1 \\ 0.2 & 0.7 & 0.1 \\ 0.4 & 0.5 & 0.1 \end{bmatrix}$, $V_1 = [1000, 500, 2000]$

5. $T = \begin{bmatrix} 0 & 1 & 0 \\ 0 & 0 & 1 \\ 1 & 0 & 0 \end{bmatrix}$, $V_1 = [1000, 2000, 3000]$

6. $T = \begin{bmatrix} 0 & 1 & 0 & 0 \\ 1 & 0 & 0 & 0 \\ 0 & 0 & 0 & 1 \\ 0 & 0 & 1 & 0 \end{bmatrix}$, $V_1 = [1000, 2000, 3000, 4000]$

In Exercises 7 through 12, find the steady-state vector (if there is one) for the Markov chain whose transition matrix and initial vector are those of the given exercise.

7. Exercise 1
8. Exercise 2
9. Exercise 3
10. Exercise 4
11. Exercise 5
12. Exercise 6

Management

A car rental agency has outlets at Kennedy and La Guardia airports. Experience shows that, of the cars available for rental at Kennedy, 80% of them either remain there or are rented and returned there each week; the remaining 20% are rented there but turned in at La Guardia. Of the cars available for rental at La Guardia, 70% either remain there or are rented and returned there; the remaining 30% are rented there but turned in at Kennedy. As of week 1, the agency has 200 cars at Kennedy and 150 at La Guardia. Exercises 13 through 15 refer to this situation.

13. Write the transition matrix.
14. How many cars will the agency have at each airport as of week 2? as of week 3? as of week 5?
15. Find the number of cars at each airport when the Markov chain reaches steady state.

Suppose the car rental agency of Exercises 13 through 15 also has an outlet at Newark Airport and that

- Of the cars available at Kennedy each week:
  20% are rented and turned in at La Guardia;
  20% are rented and turned in at Newark;
  the rest either remain at Kennedy or are rented and returned there.

- Of the cars available at La Guardia each week:
  40% are rented and turned in at Kennedy;
  10% are rented and turned in at Newark;
  the rest either remain at La Guardia or are rented and returned there.

- Of the cars available at Newark each week:
  60% are rented and turned in at Kennedy;
  20% are rented and turned in at La Guardia;
  the rest either remain at Newark or are rented and returned there.

- The agency starts with 500 cars at each location.

Exercises 16 through 18 refer to this situation.

16. Write the transition matrix.

17. How many cars will the agency have at each airport as of week 2? as of week 3? as of week 5?

18. Find the number of cars at each airport when the Markov chain reaches steady state.

Admiral Motors, Sword Autos, and Toyolla Motors share the automobile market. Currently Admiral has 40% of the market, Sword 35%, and Toyolla 25%. Each year 4% of Admiral owners switch to Sword and 6% switch to Toyolla, 3% of Sword owners switch to Admiral and 3% switch to Toyolla, and 1% of Toyolla owners switch to Admiral and 2% switch to Sword. Exercises 19 through 21 refer to this situation.

19. Write the transition matrix.

20. What percentage of the market will each manufacturer have in one year? in two years?

21. What percentage of the market will each manufacturer have when steady state is reached?

According to an article in *The Wall Street Journal* of October 10, 1988, market shares for colas in the Phoenix area in June of 1988 were as follows:

Coca-Cola: 29.5%   Pepsi-Cola: 35.1%   Others: 35.4%

At that time, Coca-Cola and Pepsi-Cola were locked in a "cola war" characterized by price cutting to gain market share. Suppose that because of the cola war, each month

- 4% of Coca-Cola drinkers switched to Pepsi-Cola and 1% switched to other brands;
- 3% of Pepsi-Cola drinkers switched to Coca-Cola and 1% switched to other brands;
- 5% of those who drank other brands switched to Coca-Cola and 5% switched to Pepsi-Cola.

Exercises 22 through 25 refer to this situation.

22. Write the transition matrix.

23. What were the market shares in July 1988?

24. What were the market shares in September 1988?

25. What were the market shares in the long run?

Each year managers in a firm must rate the performance of their subordinates as being either superior, average, or poor. Experience shows that

- of those rated superior one year, the next year 72% will be rated superior, 18% will be rated average, and the rest will be rated poor;
- of those rated average one year, the next year 12% will be rated superior, 66% will be rated average, and the rest will be rated poor;
- of those rated poor one year, the next year 2% will be rated superior, 30% will be rated average, and the rest will be rated poor.

Exercises 26 and 27 refer to this situation.

26. Write the transition matrix.

27. Find the percentage of employees in each rating category in the long run.

A credit bureau updates its ratings each month using the following transition matrix:

$$\text{Credit Rating:} \quad \begin{array}{c} \\ \text{good} \\ \text{fair} \\ \text{poor} \end{array} \begin{array}{c} \text{Credit Rating:} \\ \begin{array}{ccc} \text{good} & \text{fair} & \text{poor} \end{array} \\ \begin{bmatrix} 0.90 & 0.05 & 0.05 \\ 0.20 & 0.50 & 0.30 \\ 0.00 & 0.25 & 0.75 \end{bmatrix} \end{array}.$$

(This was suggested by material in Budnick, McLeavey, and Mojena, *Principles of Operations Research for Management,* second edition, Irwin, 1988.) Exercises 28 through 30 refer to this situation.

28. What does the transition matrix tell you about how the credit ratings change from month to month?

29. If the bureau has 30,000 accounts rated good, 40,000 rated fair, and 20,000 rated poor as of one month, how many will it have in each category two months later, assuming no new accounts?

30. Find and interpret steady state.

The following transition matrix describes the movement of hospital patients on a day-to-day basis:

$$\text{From} \quad \begin{array}{c} \\ \text{IC} \\ \text{EM} \\ \text{ward} \\ \text{out} \end{array} \begin{array}{c} \text{To} \\ \begin{array}{cccc} \text{IC} & \text{EM} & \text{ward} & \text{out} \end{array} \\ \begin{bmatrix} 0.40 & 0.00 & 0.60 & 0.00 \\ 0.20 & 0.00 & 0.30 & 0.50 \\ 0.10 & 0.00 & 0.50 & 0.40 \\ 0.00 & 0.30 & 0.70 & 0.00 \end{bmatrix} \end{array}.$$

Here IC denotes intensive care, EM denotes emergency room, and "out" stands for "outside the hospital." (This was suggested by material in Budnick, McLeavey, and Mojena, *Principles of Operations Research for Management,* second edition, Irwin, 1988.) Exercises 31 and 32 refer to this situation.

31. What does the transition matrix tell you about the day-to-day movement of patients in the hospital?

32. If the hospital starts with 80 patients in IC, 120 in EM, 500 in the wards, and 231 admittances, find and interpret steady state.

Social Science

A study of a metropolitan area consisting of a central city and its suburbs finds that each year the city loses 6% of its inhabitants to the suburbs, while the suburbs lose 2% of their inhabitants to the city. Also, population gains and losses due to movement into and out of the metropolitan area balance out each year, for both city and suburbs. The city currently has a population of 120,000; the suburbs a population of 200,000. Exercises 33 through 35 refer to this situation.

33. Write the transition matrix.

34. Find the population of the city and the suburbs one year from now; two years from now.

35. At what values will the populations stabilize?

Every election year 80% of the voters in a community who are registered Democrats remain so, while 20% of them re-register as Republicans. Similarly, 85% of the registered Republicans remain so, while 15% re-register as Democrats. Exercises 36 and 37 refer to this situation.

36. Write the transition matrix.

37. Find the percentages of Democratic and Republican voters in the community in the long run.

Students are classified as overachievers, achievers, and underachievers. Each year

- 10% of the achievers are reclassified as overachievers and 10% as underachievers;
- 12% of the overachievers are reclassified as achievers and 3% as underachievers;
- 20% of the underachievers are reclassified as achievers and 15% as overachievers.

Exercises 38 and 39 refer to this situation.

38. Write the transition matrix.

39. Find the percentages of students in each category in the long run.

A charity's records have led it to the following transition matrix:

$$\text{This Year} \begin{array}{c} \\ \text{contribute} \\ \text{do not contribute} \end{array} \overset{\begin{array}{cc} \text{Next Year} & \\ \text{contribute} & \text{do not contribute} \end{array}}{\begin{bmatrix} 0.75 & 0.25 \\ 0.40 & 0.60 \end{bmatrix}}.$$

(This was suggested by material in Rorres and Anton, *Applications of Linear Algebra*, Wiley, 1977.) Exercises 40 through 42 refer to this situation.

40. What does the transition matrix tell you about the percentages of people who contribute year to year?

41. What is the probability that a person who did not contribute last year will do so this year? will do so next year?

42. What is the long run probability that a person will contribute in a given year?

A university currently has 200 faculty members who are nontenured and 80 who are tenured, and it must hire 20 new faculty members to replace those who are leaving. It is known that each year 50% of nontenured faculty will remain nontenured, 40% will be granted tenure, and 10% will leave. Also, none of the tenured faculty will become nontenured, 90% will remain at the university as tenured faculty, and 10% will leave. Finally, of those who enter, 90% enter as nontenured faculty and 10% as tenured faculty. Exercises 43 and 44 refer to this situation.

43. Write the transition matrix.

44. What will the composition of the faculty be in the long run, assuming that its size remains constant at 300?

7.3 TWO-PERSON ZERO-SUM GAMES

A **game** is a competitive situation. A **zero-sum game** is one in which winning players win, in total, what losing players lose, in total. A **two-person game** is a game with two players. Therefore, a **two-person zero-sum game** is a competitive situation with two players, in which whatever one wins the other must lose. Many competitive situations in areas such as marketing, war, and politics can be analyzed as two-person zero-sum games. In this section we study such games and their applications.

Strictly Determined Games

Each player in a game has possible plays that result in wins or losses. For a two-person zero-sum game, the wins and losses for both players may be exhibited in a single **payoff matrix.** For example, suppose we have a two-person zero-sum game with players A and B and the following payoff matrix:

$$\text{A's Plays} \begin{array}{c} \\ A_1 \\ A_2 \\ A_3 \\ A_4 \end{array} \overset{\overset{\text{B's Plays}}{B_1 \quad B_2 \quad B_3 \quad B_4}}{\begin{bmatrix} -5 & 8 & -6 & 9 \\ 6 & 7 & 9 & 10 \\ 3 & 2 & 10 & 3 \\ -7 & 4 & 8 & 7 \end{bmatrix}}.$$

This payoff matrix is that of the **row player A,** but it also tells us the payoffs for the **column player B** because the game is zero-sum. Thus, if A plays A_1 and B plays B_1, then A will lose 5 and hence B will gain 5; if A plays A_2 and B plays B_3, then A will gain 9 and hence B will lose 9.

A play is said to **dominate** another play if every payoff of the first is at least as good as every payoff of the second. In our example, play A_2 dominates play A_4, because in every case the payoff for A_2 is at least as good as that for A_4:

6 is better for A than -7;
7 is better for A than 4;
9 is better for A than 8;
10 is better for A than 7.

Note that there is no other row dominance present; for instance, A_2 does not dominate A_3 because while 6 is better than 3, 7 is better than 2, and 10 is better than 3 for A, 9 is not as good as 10.

Now suppose our game is played under uncertainty so that neither player knows which play the other will choose. Then player A will never choose A_4, because no matter how B plays, A_2 always results in a better return for A than A_4 does. Since A will never choose A_4, it may be crossed out of the matrix:

APPLICATIONS OF PROBABILITY

$$\text{A's Plays} \quad \begin{array}{c} \\ A_1 \\ A_2 \\ A_3 \\ A_4 \end{array} \overset{\overset{\displaystyle \text{B's Plays}}{B_1 \quad B_2 \quad B_3 \quad B_4}}{\begin{bmatrix} -5 & 8 & -6 & 9 \\ 6 & 7 & 9 & 10 \\ 3 & 2 & 10 & 3 \\ \text{-7} & \text{-4} & \text{-8} & \text{-7} \end{bmatrix}}.$$

Working with the remaining entries of the matrix and adopting the viewpoint of the column player B, we see that play B_1 dominates play B_4 because winning 5 is better than losing 9, losing 6 is better than losing 10, and losing 3 is the same as losing 3. (Remember that the entries in the payoff matrix are the row player's payoffs, so a positive number represents a loss and a negative number a win for the column player.) Since player B will never choose B_4, we may remove it from the payoff matrix:

$$\text{A's Plays} \quad \begin{array}{c} \\ A_1 \\ A_2 \\ A_3 \\ A_4 \end{array} \overset{\overset{\displaystyle \text{B's Plays}}{B_1 \quad B_2 \quad B_3 \quad B_4}}{\begin{bmatrix} -5 & 8 & -6 & 9 \\ 6 & 7 & 9 & 10 \\ 3 & 2 & 10 & 3 \\ \text{-7} & \text{-4} & \text{-8} & \text{-7} \end{bmatrix}}.$$

Thus the game has been reduced to one with the following payoff matrix:

$$\text{A's Plays} \quad \begin{array}{c} \\ A_1 \\ A_2 \\ A_3 \end{array} \overset{\overset{\displaystyle \text{B's Plays}}{B_1 \quad B_2 \quad B_3}}{\begin{bmatrix} -5 & 8 & -6 \\ 6 & 7 & 9 \\ 3 & 2 & 10 \end{bmatrix}}.$$

If there were any dominated rows and/or columns in the reduced payoff matrix, we could remove them and reduce the game still further. Since there are no such rows or columns here, we have reduced the game as far as possible. Now let us analyze the reduced game and see what plays the players will choose.

Notice that

if A plays A_1, the worst A can do is lose 6;
if A plays A_2, the worst A can do is gain 6;
if A plays A_3, the worst A can do is gain 2.

Therefore A's best guaranteed result occurs when A plays A_2. Similarly,

if B plays B_1, the worst B can do is lose 6;
if B plays B_2, the worst B can do is lose 8;
if B plays B_3, the worst B can do is lose 10.

Therefore B's best guaranteed result occurs when B plays B_1. Hence A will play A_2 and B will play B_1, with the result being a gain of 6 for A and a loss of 6 for B.

The pair of plays (A_2, B_1) is called a **saddle point** of the game. A game that has a saddle point is a **strictly determined game,** because the plays each player will make and the outcome of the game are indeed strictly determined by the structure of the game. As we shall see, not all two-person zero-sum games are strictly determined.

How did we find the saddle point of our game? We did so by looking for the worst outcome for the row player under each of his or her plays and the worst outcome for the column player under each of his or her possible plays. But the worst outcome for a row play is the minimum entry in its row, and the worst outcome for a column play is the maximum entry in its column. If an entry is the minimum in its row and the maximum in its column (as 6 was for the game we considered), it will define a saddle point.

Finding Saddle Points

To find the saddle points of a two-person zero-sum game, find the minimum entry in each row and the maximum entry in each column of the payoff matrix. If an entry is both the minimum in its row and the maximum in its column, it defines a saddle point.

(This method of finding saddle points assumes that the payoff matrix gives the payoffs for the row player.)

EXAMPLE 1

Consider the two-person zero-sum game whose payoff matrix (payoffs for the row player) is

$$\begin{array}{c} & B_1 & B_2 & B_3 \\ A_1 \\ A_2 \\ A_3 \end{array} \begin{bmatrix} 4 & -2 & 1 \\ 6 & 3 & 4 \\ 5 & 2 & 1 \end{bmatrix}.$$

We find the minimum entry in each row and the maximum entry in each column:

$$\begin{array}{c} & B_1 & B_2 & B_3 & \text{Min} \\ A_1 \\ A_2 \\ A_3 \end{array} \begin{bmatrix} 4 & -2 & 1 \\ 6 & 3 & 4 \\ 5 & 2 & 1 \end{bmatrix} \begin{array}{c} -2 \\ 3 \\ 1 \end{array}$$
$$\text{Max} \quad\; 6 \quad\;\; 3 \quad\;\; 4$$

Therefore since 3 is the minimum entry in row 2 and the maximum entry in column 2, (A_2, B_2) is a saddle point for the game: the row player will play A_2 for a return of 3 and the column player will play B_2 for a return of -3. ☐

A game can have more than one saddle point.

EXAMPLE 2

The two-person zero-sum game with payoff matrix

$$\begin{array}{c} & B_1 & B_2 & B_3 & B_4 \\ \begin{array}{c} A_1 \\ A_2 \end{array} & \left[\begin{array}{cccc} -2 & 1 & 2 & -2 \\ -3 & 4 & 3 & -4 \end{array} \right] \end{array}$$

has saddle points (A_1, B_1) and (A_1, B_4), since -2 is the minimum entry in the first row and the maximum entry in both the first and fourth columns. Each of these saddle points returns a loss of 2 for the row player and a gain of 2 for the column player. ☐

EXAMPLE 3

Alpha Company and Omega Company share a market. Alpha has prepared four new advertising campaigns, A_1, A_2, A_3, and A_4; it will choose one of them. Omega has also prepared four ad campaigns, B_1, B_2, B_3, and B_4; it will choose one of them. Since whatever share of the market either company gains from its new campaign must come at the expense of the other, the situation can be treated as a two-person zero-sum game. Suppose the payoff matrix for Alpha is

$$\begin{array}{c} & B_1 & B_2 & B_3 & B_4 \\ \begin{array}{c} A_1 \\ A_2 \\ A_3 \\ A_4 \end{array} & \left[\begin{array}{cccc} 0.05 & 0.10 & 0.02 & 0.07 \\ 0.12 & 0.05 & 0.05 & 0.11 \\ 0.07 & 0.02 & -0.02 & 0.12 \\ -0.05 & 0.05 & 0.00 & 0.14 \end{array} \right] \end{array},$$

where the entries are the proportions of the market that Alpha will take from or lose to Omega.

Since the only entry that is the minimum in its row and the maximum in its column is the entry 0.05 in row 2, column 3 (check this), this game has only one saddle point, namely, (A_2, B_3). Hence Alpha will choose campaign A_2, Omega will choose campaign B_3, and Alpha will gain 5% of the market from Omega. ☐

Not all two-person zero-sum games have saddle points. Consider, for instance, the game with payoff matrix

$$\begin{array}{c} & B_1 & B_2 \\ \begin{array}{c} A_1 \\ A_2 \end{array} & \left[\begin{array}{cc} 3 & 1 \\ -1 & 2 \end{array} \right] \end{array}.$$

No entry of this matrix is both the minimum in its row and the maximum in its column, so the game has no saddle point and thus is not strictly determined: there is no single best play for either player, and the outcome of the game cannot be determined before playing it.

Mixed Strategies

In a game that is not strictly determined, each player must choose a **mixed strategy**, according to which he or she will alternate plays so as to maximize expected return.

CHAPTER 7

We illustrate how such mixed strategies are determined with the following two-person zero-sum game:

$$\begin{array}{c} & \text{Player B} \\ & \begin{array}{cc} B_1 & B_2 \end{array} \\ \text{Player A} & \begin{array}{c} A_1 \\ A_2 \end{array} \begin{bmatrix} 3 & 1 \\ -1 & 2 \end{bmatrix}. \end{array}$$

Suppose player A chooses to play A_1 with probability p, and therefore A_2 with probability $1 - p$. Let

$$E_A(B \text{ plays } B_1) = E_A(B_1)$$

denote the expected return to player A if player B plays B_1. Then

$$E_A(B_1) = 3p + (-1)(1 - p) = 4p - 1.$$

Similarly,

$$E_A(B_2) = 1p + 2(1 - p) = -p + 2.$$

Since p is a probability, we have $0 \leq p \leq 1$. If we graph the line segments

$$E_A(B_1) = 4p - 1, \quad 0 \leq p \leq 1$$

and

$$E_A(B_2) = -p + 2, \quad 0 \leq p \leq 1$$

on the same set of axes, we obtain the diagram shown in Figure 7–5. Note that the line segments intersect at the point $(3/5, 7/5)$. Now, player A's expected return can lie on either line segment, depending on whether B plays B_1 or B_2. In particular, for $0 \leq p < 3/5$ or for $3/5 < p \leq 1$, A's expected return might lie on the lower of

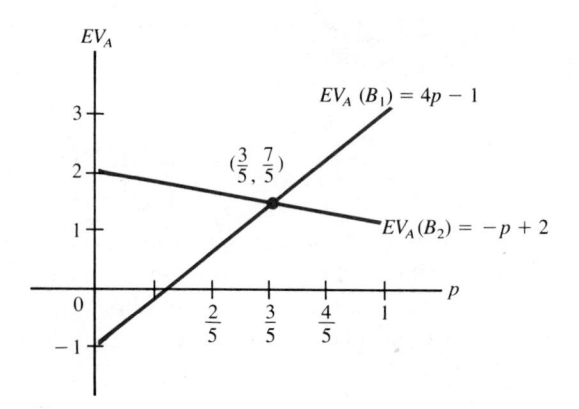

FIGURE 7–5

the two line segments, and hence be less than $7/5$. But at the point $(3/5, 7/5)$, player A is guaranteed an expected return of $7/5$, regardless of whether B plays B_1 or B_2. Therefore the best guaranteed return for A is $7/5$, which occurs when $p = 3/5$. Thus, we have

$$P(A \text{ plays } A_1) = p = 3/5;$$

$$P(A \text{ plays } A_2) = 1 - p = 2/5,$$

and hence player A's optimal strategy should be to play A_1 three-fifths of the time and A_2 two-fifths of the time, for this will maximize A's expected return. In the long run, this strategy will lead to an average win of $7/5$ for A.

(Of course, if A does play A_1 three-fifths of the time and A_2 two-fifths of the time, he cannot do so according to any pattern, for if he does follow a pattern, B could discover it and take advantage of it. Player A must arrange things so that he switches back and forth between A_1 and A_2 in a random manner, while still managing to play A_1 three-fifths of the time and A_2 two-fifths of the time. One way to accomplish this would be for him to draw a marble from a vase containing three red and two white marbles each time he is ready to play. If he draws a red marble, he plays A_1; if he draws a white one, he plays A_2.)

Let us complete this illustration by finding the optimal strategy for player B. We already know that A's expected return is $7/5$, and therefore, since the game is zero-sum, B's expected return must be $-7/5$. In the long run, this is the best that B can do. We seek the strategy that will allow B to accomplish this.

Let us set

$$P(B \text{ plays } B_1) = q$$

and

$$P(B \text{ plays } B_2) = 1 - q, \quad 0 \leq q \leq 1.$$

Then B's expected returns are

$$E_B(A_1) = -3q - 1(1 - q) = -2q - 1$$

and

$$E_B(A_2) = 1q - 2(1 - q) = 3q - 2.$$

If we set

$$-2q - 1 = 3q - 2$$

and solve for q, we obtain $q = 1/5$. Therefore, B's strategy should be to play B_1 one-fifth of the time and B_2 four-fifths of the time, switching between B_1 and B_2 in a random manner.

■ **Computer Manual: Program ZEROSUM**

7.3 EXERCISES

Strictly Determined Games

For each of the two-person zero-sum games in Exercises 1 through 12, remove all dominated rows and columns, find all saddle points, and determine the outcome of the game, if possible. If a game is not strictly determined, state that fact. As usual, each payoff matrix gives the row player's payoffs.

1. $\begin{array}{c} & B_1 & B_2 \\ A_1 & \begin{bmatrix} 4 & 2 \\ 3 & 1 \end{bmatrix} \\ A_2 & \end{array}$

2. $\begin{array}{c} & B_1 & B_2 \\ A_1 & \begin{bmatrix} 4 & 1 \\ 3 & 2 \end{bmatrix} \\ A_2 & \end{array}$

3. $\begin{array}{c} & B_1 & B_2 \\ A_1 & \begin{bmatrix} 4 & -2 \\ -2 & 4 \end{bmatrix} \\ A_2 & \end{array}$

4. $\begin{array}{c} & B_1 & B_2 \\ A_1 & \begin{bmatrix} 5 & 1 \\ 6 & -1 \\ 2 & -2 \end{bmatrix} \\ A_2 & \\ A_3 & \end{array}$

5. $\begin{array}{c} & B_1 & B_2 \\ A_1 & \begin{bmatrix} 4 & 1 \\ 2 & 3 \\ 5 & -1 \\ 3 & 4 \end{bmatrix} \\ A_2 & \\ A_3 & \\ A_4 & \end{array}$

6. $\begin{array}{c} & B_1 & B_2 \\ A_1 & \begin{bmatrix} 2 & 0 \\ 4 & -2 \\ 3 & -1 \end{bmatrix} \\ A_2 & \\ A_3 & \end{array}$

7. $\begin{array}{c} & B_1 & B_2 & B_3 \\ A_1 & \begin{bmatrix} 2 & -3 & 4 \\ -1 & 2 & 0 \end{bmatrix} \\ A_2 & \end{array}$

8. $\begin{array}{c} & B_1 & B_2 & B_3 \\ A_1 & \begin{bmatrix} 2 & -1 & 3 \\ 5 & -2 & -3 \\ -3 & -1 & 4 \end{bmatrix} \\ A_2 & \\ A_3 & \end{array}$

9. $\begin{array}{c} & B_1 & B_2 & B_3 \\ A_1 & \begin{bmatrix} 2 & 4 & -3 \\ 6 & 1 & -2 \\ 5 & -4 & -1 \end{bmatrix} \\ A_2 & \\ A_3 & \end{array}$

10. $\begin{array}{c} & B_1 & B_2 & B_3 \\ A_1 & \begin{bmatrix} -4 & -6 & 3 \\ 2 & 0 & 5 \\ -1 & -3 & 1 \end{bmatrix} \\ A_2 & \\ A_3 & \end{array}$

11. $\begin{array}{c} & B_1 & B_2 & B_3 & B_4 \\ A_1 & \begin{bmatrix} -2 & 6 & 2 & 1 \\ 4 & 2 & -1 & -2 \\ 7 & -2 & -4 & 0 \\ 3 & 6 & 2 & 2 \end{bmatrix} \\ A_2 & \\ A_3 & \\ A_4 & \end{array}$

12. $\begin{array}{c} & B_1 & B_2 & B_3 & B_4 \\ A_1 & \begin{bmatrix} -2 & 6 & 2 & 2 \\ 4 & 6 & -1 & 1 \\ 7 & 3 & 2 & 2 \\ 8 & 0 & 2 & 1 \end{bmatrix} \\ A_2 & \\ A_3 & \\ A_4 & \end{array}$

Management

13. Gamma Company is considering changing the price of its product, either by raising it or lowering it. Its only competitor, Delta Company, can respond by raising or lowering its price or by leaving it unchanged. The following zero-sum payoff matrix gives Gamma's payoffs in millions of dollars:

		Delta	
	raises price	lowers price	leaves price same
Gamma raises price	20	10	40
Gamma lowers price	60	−10	10
Gamma leaves price same	20	−20	−5

    What should each company do, and what will the outcome be for each?

14. Repeat Exercise 13 if the payoff matrix is

		Delta	
	raises price	lowers price	leaves price same
Gamma raises price	0	30	20
Gamma lowers price	−20	−10	30
Gamma leaves price same	0	−20	−5

APPLICATIONS OF PROBABILITY   445

15. Repeat Exercise 13 if the payoff matrix is

$$\text{Gamma} \begin{array}{c} \text{raises price} \\ \text{lowers price} \\ \text{leaves price same} \end{array} \begin{array}{c} \text{Delta} \\ \begin{array}{ccc} \text{raises price} & \text{lowers price} & \text{leaves price same} \end{array} \\ \begin{bmatrix} 10 & 20 & 40 \\ 60 & -10 & 10 \\ 30 & 0 & 40 \end{bmatrix} \end{array}.$$

16. Firm A has planned five marketing strategies, and its competitor B has planned four marketing strategies. The payoff matrix, with entries equal to the percentage of the market that A will gain or lose under each possibility, is as follows:

$$A \begin{array}{c} A_1 \\ A_2 \\ A_3 \\ A_4 \\ A_5 \end{array} \begin{array}{c} B \\ \begin{array}{cccc} B_1 & B_2 & B_3 & B_4 \end{array} \\ \begin{bmatrix} 5 & 4 & -2 & 3 \\ 6 & 2 & 1 & 1 \\ -4 & 1 & 1 & 2 \\ 3 & 5 & 2 & 3 \\ -2 & 3 & -2 & 2 \end{bmatrix} \end{array}.$$

What will happen?

17. Repeat Exercise 16 if the payoff matrix is

$$A \begin{array}{c} A_1 \\ A_2 \\ A_3 \\ A_4 \\ A_5 \end{array} \begin{array}{c} B \\ \begin{array}{cccc} B_1 & B_2 & B_3 & B_4 \end{array} \\ \begin{bmatrix} 5 & 4 & 2 & 3 \\ 6 & 2 & 1 & -1 \\ -4 & 1 & 1 & 4 \\ 3 & 5 & 2 & 3 \\ -2 & 3 & -2 & 2 \end{bmatrix} \end{array}.$$

Social Science

18. Polls show that two opposing political candidates each have 50% of the vote with eight weeks remaining before the election. Each week each candidate must decide whether to spend the week attacking his or her opponent or discussing the issues. The following zero-sum payoff matrix shows the percentage of the vote gained or lost each week by candidate A:

$$\text{Candidate A} \begin{array}{c} \text{attacks B} \\ \text{discusses issues} \end{array} \begin{array}{c} \text{Candidate B} \\ \begin{array}{cc} \text{attacks A} & \text{discusses issues} \end{array} \\ \begin{bmatrix} 0 & 1 \\ -1 & 0 \end{bmatrix} \end{array}.$$

What will each candidate do? What will the outcome of the election be?

19. Repeat Exercise 18 if the payoff matrix is

$$\text{Candidate A} \begin{array}{c} \text{attacks B} \\ \text{discusses issues} \end{array} \begin{array}{c} \text{Candidate B} \\ \begin{array}{cc} \text{attacks A} & \text{discusses issues} \end{array} \\ \begin{bmatrix} 0 & -1 \\ 2 & -1 \end{bmatrix} \end{array}.$$

20. Repeat Exercise 18 if the payoff matrix is

$$\begin{array}{c} \\ \text{Candidate A} \end{array} \begin{array}{c} \\ \text{attacks B} \\ \text{discusses issues} \end{array} \begin{bmatrix} \begin{array}{cc} \text{Candidate B} & \\ \text{attacks A} & \text{discusses issues} \end{array} \\ \begin{array}{cc} 2 & 1 \\ 0 & -1 \end{array} \end{bmatrix}.$$

Military

(Adapted from "Military Decision and Game Theory," O. G. Haywood, Jr., *Journal of Operations Research Society of America,* November 1954.) During World War II, a Japanese convoy was about to sail from Rabaul to New Guinea. The convoy could choose a northern route or a southern route. The weather on the northern route was bad, but that on the southern route was good. The commander of the Allied Air Forces in the region would have to search for the convoy. He had sufficient planes to concentrate heavy reconnaissance on one of the routes, but only light reconnaissance on the other. There were thus four possibilities:

- The convoy could choose the northern route and the Allied commander could send the heavy reconnaissance on the northern route. In this case the reconnaissance, hampered by the poor weather, would find the convoy on the second day, leaving two days for bombing it.

- The convoy could choose the northern route and the Allied commander could send the heavy reconnaissance on the southern route. In this case the lighter reconnaissance on the northern route, hampered by the bad weather, would not find the convoy until the third day, leaving only one day for bombing.

- The convoy could choose the southern route and the Allied commander could send the heavy reconnaissance on the northern route. In this case the lighter reconnaissance on the southern route, aided by the good weather, would find the convoy on the second day, leaving two days for bombing.

- The convoy could choose the southern route and the Allied commander could send the heavy reconnaissance on the southern route. In this case the reconnaissance, aided by the good weather, would find the convoy on the first day, leaving three days for bombing.

The Japanese and Allied commanders were aware of these facts. Exercises 21 and 22 refer to this situation.

21. Analyze the situation as a two-person zero-sum game and find its outcome. (This is actually what did happen.)

22. What would the outcome have been if the weather on the northern route had been so bad that it would have taken the heavy reconnaissance three days to find the convoy on that route, and the light reconnaissance would not have found it at all?

Mixed Strategies

For each of the two-person zero-sum games in Exercises 23 through 28, remove all dominated rows and columns and find each player's mixed strategy and expected return.

23.
$$\begin{array}{c}\begin{array}{cc}B_1 & B_2\end{array}\\\begin{array}{c}A_1\\A_2\end{array}\left[\begin{array}{rr}2 & -3\\-1 & 4\end{array}\right]\end{array}$$

24.
$$\begin{array}{c}\begin{array}{cc}B_1 & B_2\end{array}\\\begin{array}{c}A_1\\A_2\end{array}\left[\begin{array}{rr}0 & -2\\-2 & 1\end{array}\right]\end{array}$$

25.
$$\begin{array}{c}\begin{array}{cc}B_1 & B_2\end{array}\\\begin{array}{c}A_1\\A_2\\A_3\end{array}\left[\begin{array}{rr}4 & 0\\1 & 2\\5 & 1\end{array}\right]\end{array}$$

26.
$$\begin{array}{c}\begin{array}{ccc}B_1 & B_2 & B_3\end{array}\\\begin{array}{c}A_1\\A_2\\A_3\end{array}\left[\begin{array}{rrr}2 & 1 & 2\\-3 & 3 & 4\\-1 & 7 & 10\end{array}\right]\end{array}$$

27.
$$\begin{array}{c}\begin{array}{ccc}B_1 & B_2 & B_3\end{array}\\\begin{array}{c}A_1\\A_2\\A_3\end{array}\left[\begin{array}{rrr}-3 & -2 & 0\\4 & 3 & 1\\-3 & 5 & 2\end{array}\right]\end{array}$$

28.
$$\begin{array}{c}\begin{array}{cccc}B_1 & B_2 & B_3 & B_4\end{array}\\\begin{array}{c}A_1\\A_2\\A_3\\A_4\end{array}\left[\begin{array}{rrrr}3 & 6 & 2 & 4\\3 & 5 & -3 & 2\\2 & 2 & -3 & 0\\-4 & -2 & 4 & 6\end{array}\right]\end{array}$$

Management

29. Alpha Company and Beta Company share the market for designer bluejeans for toddlers. Every autumn, when each company introduces its new line of jeans, each must decide how best to promote the line: through magazine ads, through sales promotion at the retail level, or through a combination of these strategies. The following zero-sum payoff matrix shows the percentage of the market gained or lost by Alpha:

$$\begin{array}{c}\text{Beta promotes through}\\\begin{array}{ccc}\text{magazines} & \text{retail outlets} & \text{combination}\end{array}\\\begin{array}{c}\text{Alpha}\text{magazines}\\\text{promotes}\text{retail outlets}\\\text{through}\text{combination}\end{array}\left[\begin{array}{ccc}5 & -3 & 4\\3 & 1 & -3\\4 & -3 & -3\end{array}\right].\end{array}$$

Find the optimal strategy for each company and each company's expected market share.

30. Repeat Exercise 13 if the payoff matrix is as follows:

$$\begin{array}{c}\text{Delta}\\\begin{array}{ccc}\text{raises price} & \text{lowers price} & \text{leaves price same}\end{array}\\\begin{array}{c}\phantom{\text{Gamma}}\text{raises price}\\\text{Gamma}\text{lowers price}\\\phantom{\text{Gamma}}\text{leaves price same}\end{array}\left[\begin{array}{ccc}-20 & 10 & 40\\60 & -10 & 10\\20 & -20 & 5\end{array}\right].\end{array}$$

31. Repeat Exercise 16 if the payoff matrix is

$$\begin{array}{c}\text{B}\\\begin{array}{cccc}B_1 & B_2 & B_3 & B_4\end{array}\\\text{A}\begin{array}{c}A_1\\A_2\\A_3\\A_4\\A_5\end{array}\left[\begin{array}{rrrr}5 & 4 & -2 & 3\\-1 & 5 & -1 & 1\\6 & 4 & 6 & 6\\3 & 5 & 2 & 3\\-2 & 3 & -2 & 2\end{array}\right].\end{array}$$

Life Science

32. (Suggested by material in Rorres and Anton, *Applications of Linear Algebra*, Wiley, 1977.) Two strains of influenza, A and B, are rampant. There are two vaccines available: vaccine 1 and vaccine 2. The following payoff matrix has as its entries the percent effectiveness of the vaccines against each strain of flu:

Flu Strain

$$\text{Vaccine} \begin{array}{c} 1 \\ 2 \end{array} \begin{bmatrix} 0.75 & 0.80 \\ 0.90 & 0.60 \end{bmatrix}.$$

(Thus, for instance, vaccine 1 is 75% effective against A and 80% effective against B.) What should the public health strategy be, and what is its payoff?

Social Science

33. Repeat Exercise 18 if the payoff matrix is as follows:

$$\text{Candidate A} \begin{array}{c} \text{attacks B} \\ \text{discusses issues} \end{array} \begin{bmatrix} 1 & 0 \\ 0 & 1 \end{bmatrix}.$$

(columns: Candidate B — attacks A, discusses issues)

34. Repeat Exercise 18 if the payoff matrix is as follows:

$$\text{Candidate A} \begin{array}{c} \text{attacks B} \\ \text{discusses issues} \end{array} \begin{bmatrix} 1 & 0 \\ -2 & 1 \end{bmatrix}.$$

(columns: Candidate B — attacks A, discusses issues)

Non-Zero-Sum Games

Any two-person zero-sum game with only two plays for each player has a solution: either the game will have a saddle point or each player will have an optimal mixed strategy. The situation is different for non-zero-sum games, however, because selfishness can enter the picture. For instance, consider the **prisoner's dilemma** game: A and B are partners in crime who are being questioned separately by the police about the crime they have just committed. The following payoff matrix describes the possibilities:

Prisoner B

	confesses	does not confess
Prisoner A confesses	2 \ 2	5 \ 0
Prisoner A does not confess	0 \ 5	1 \ 1

The entries in the matrix are the prison sentences, in years, that each player will receive, with A's sentences in the lower left corner and B's in the upper right corner of each box. (Thus, for instance, if A confesses and B does not, A will go free while B will receive a five-year sentence.)

*35. Analyze the prisoner's dilemma. What will happen if both prisoners confess? if one confesses and the other does not? if neither confesses? What is the "rational" thing for each prisoner to do?

*36. Two stores compete. If either lowers its prices and the other does not respond, the one lowering its prices obtains a competitive advantage. The payoff matrix is as follows:

	Store B leaves prices as is	Store B lowers prices
Store A leaves prices as is	50 / 50	75 / 25
Store A lowers prices	25 / 75	40 / 40

The entries in the matrix represent revenue in thousands of dollars. Analyze the game. Do you see why "price wars" can occur?

*37. Two nations are rivals. The following payoff matrix describes the possibilities if they increase or decrease military spending; the entries represent relative advantages or disadvantages for each nation in terms of economic growth, prestige, and so forth.

	Nation B increases military spending	Nation B decreases military spending
Nation A increases military spending	−10 / −10	−50 / 100
Nation A decreases military spending	100 / −50	10 / 10

Analyze the game. Do you see why arms races can occur?

7.4 BAYES' LAW

In this section we develop an important probability formula known as **Bayes' Law.** Not only is this important for its own sake but, as we shall see in Section 7.5, it may be used together with decision trees to incorporate conditional information into the decision-making process.

Consider the probability formula

$$P(A \cap B) = P(A \mid B) \cdot P(B),$$

developed in Section 6.3. By interchanging A and B in the formula, we also have

$$P(B \cap A) = P(B \mid A) \cdot P(A).$$

But since $P(A \cap B) = P(B \cap A)$, it follows that

$$P(A \mid B) \cdot P(B) = P(B \mid A) \cdot P(A).$$

Therefore, if $P(B) \neq 0$,

$$P(A \mid B) = \frac{P(B \mid A) \cdot P(A)}{P(B)}.$$

Let us now write

$$B = (B \cap A) \cup (B \cap A')$$

and observe that $B \cap A$ and $B \cap A'$ are mutually exclusive, so that (see Figure 7–6)

$$P(B) = P(B \cap A) + P(B \cap A') = P(B \mid A) \cdot P(A) + P(B \mid A') \cdot P(A').$$

Substituting this expression for $P(B)$ in

$$P(A \mid B) = \frac{P(B \mid A) \cdot P(A)}{P(B)},$$

we obtain Bayes' Law:

Bayes' Law

For any two events A and B of a sample space,

$$P(A \mid B) = \frac{P(B \mid A) \cdot P(A)}{P(B \mid A) \cdot P(A) + P(B \mid A') \cdot P(A')}.$$

Bayes' Law is useful when it is necessary to reason backwards and find the probability that a given event occurred if a subsequent event is known to have occurred.

EXAMPLE 1 Suppose $P(A) = 0.40$, $P(B \mid A) = 0.20$, and $P(B \mid A') = 0.30$. Then

$$P(A') = 1 - P(A) = 0.60,$$

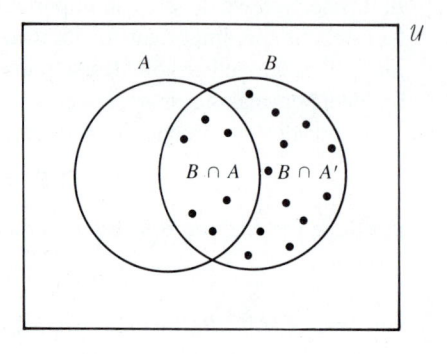

FIGURE 7–6

APPLICATIONS OF PROBABILITY

and by Bayes' Law,

$$P(A \mid B) = \frac{(0.20)(0.40)}{(0.20)(0.40) + (0.30)(0.60)} \approx 0.31.$$

See Figure 7–7. ☐

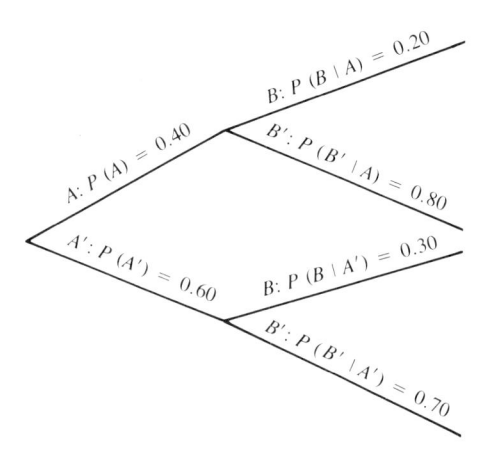

FIGURE 7–7

EXAMPLE 2 Suppose the probability of rain today is 0.25, the probability that John will go on a picnic if it rains is 0.05, and the probability that John will go on a picnic if it does not rain is 0.65. John did go on a picnic today; what is the probability that it rained?

Let

$$J = \text{John goes on a picnic}$$

and

$$R = \text{rain today}.$$

Then we know that

$$P(\text{rain today}) = P(R) = 0.25,$$
$$P(\text{John goes on picnic} \mid \text{rain today}) = P(J \mid R) = 0.05,$$
$$P(\text{John goes on picnic} \mid \text{no rain today}) = P(J \mid R') = 0.65,$$

from the information given. Also,

$$P(\text{no rain today}) = P(R') = 1 - P(R) = 0.75.$$

Therefore by Bayes' Law,

P(it rained today given that John went on a picnic)
$$= P(R \mid J)$$
$$= \frac{P(J \mid R) \cdot P(R)}{P(J \mid R) \cdot P(R) + P(J \mid R') \cdot P(R')}$$
$$= \frac{(0.05)(0.25)}{(0.05)(0.25) + (0.65)(0.75)} = 0.025. \quad \square$$

EXAMPLE 3 Sky Fy and Stereo Village are the only two stores in town that sell compact disks. Sky Fy has 60% of the CD market, and Stereo Village has the rest. The probability that a disk purchased from Sky Fy will be defective is 0.02, while the probability that one bought at Stereo Village will be defective is 0.06. If a CD purchased in town is found to be defective, what is the probability that it was purchased at Sky Fy? at Stereo Village?

Let
$$D = \text{defective disk}$$
$$F = \text{disk purchased at Sky Fy,}$$
and
$$V = \text{disk purchased at Stereo Village.}$$

Then we know that
$$P(D \mid F) = 0.02,$$
$$P(D \mid V) = 0.06,$$
$$P(F) = 0.60,$$
and
$$P(V) = P(F') = 0.40.$$

We wish to find $P(F \mid D)$ and $P(V \mid D)$. But
$$P(F \mid D) = \frac{P(D \mid F) \cdot P(F)}{P(D \mid F) \cdot P(F) + P(D \mid V) \cdot P(V)}$$
$$= \frac{(0.02)(0.60)}{(0.02)(0.60) + (0.06)(0.40)} = \frac{1}{3}.$$

Hence,
$$P(V \mid D) = 1 - P(F \mid D) = 1 - \frac{1}{3} = \frac{2}{3}. \quad \square$$

It should be noted that Bayes' Law may be generalized to more than two events.

Bayes' Law Generalized

If $A_1, A_2, \ldots, A_n$ are mutually exclusive events whose union is the sample space and B is an event of the sample space, then for each k, $1 \leq k \leq n$,

$$P(A_k \mid B) = \frac{P(B \mid A_k) \cdot P(A_k)}{P(B \mid A_1) \cdot P(A_1) + P(B \mid A_2) \cdot P(A_2) + \cdots + P(B \mid A_n) \cdot P(A_n)}.$$

EXAMPLE 4 An automobile dealer has three mechanics, Adams, Bernstein, and Chen, each of whom handles dealer preparation for one-third of the new cars. Among the tasks they perform is that of filling up the windshield washer container with washer fluid. It is known that Adams forgets to do this 6% of the time, Bernstein forgets 9% of the time, and Chen forgets 15% of the time. Suppose you pick up your new car at the dealership and find that it has no washer fluid. What is the probability that Adams prepared your car?

If we let W denote "no washer fluid" and A, B, and C that Adams, Bernstein, and Chen, respectively, prepared the car, then we wish to find $P(A \mid W)$ given that

$$P(W \mid A) = 0.06, \quad P(W \mid B) = 0.09, \quad P(W \mid C) = 0.15,$$

and

$$P(A) = P(B) = P(C) = \tfrac{1}{3}.$$

Hence

$$P(A \mid W) = \frac{P(W \mid A)P(A)}{P(W \mid A)P(A) + P(W \mid B)P(B) + P(W \mid C)P(C)}$$

$$= \frac{(0.06)(\tfrac{1}{3})}{(0.06)(\tfrac{1}{3}) + (0.09)(\tfrac{1}{3}) + (0.15)(\tfrac{1}{3})}$$

$$= 0.2. \quad \square$$

7.4 EXERCISES

1. If $P(A) = 0.25$, $P(B \mid A) = 0.30$, and $P(B \mid A') = 0.22$, find $P(A \mid B)$.
2. If $P(A) = 0.06$, $P(B \mid A) = 0.08$, and $P(B \mid A') = 0.35$, find $P(A \mid B)$.
3. If $P(A) = \tfrac{2}{3}$, $P(B \mid A) = \tfrac{1}{6}$, and $P(B \mid A') = \tfrac{1}{3}$, find $P(A \mid B)$.
4. If $P(A) = \tfrac{4}{5}$, $P(B \mid A) = \tfrac{3}{5}$, and $P(B \mid A') = \tfrac{1}{5}$, find $P(A \mid B)$.
5. If $P(A) = 0.85$, $P(B \mid A) = 0.27$, and $P(B \mid A') = 0.44$, find $P(A \mid B)$ and $P(A' \mid B)$.
6. If $P(B) = 0.44$, $P(A \mid B) = 0.05$, and $P(A \mid B') = 0.13$, find $P(B \mid A)$ and $P(B' \mid A)$.

Let A, B, and C be mutually exclusive events whose union is the sample space and suppose that $P(D \mid A) = 0.4$, $P(D \mid B) = 0.2$, $P(D \mid C) = 0.1$, $P(A) = 0.3$, $P(B) = 0.5$, and $P(C) = 0.2$. Exercises 7 through 9 refer to this situation.

7. Find $P(A \mid D)$ **8.** Find $P(B \mid D)$ **9.** Find $P(C' \mid D)$

Let A, B, C, and D be mutually exclusive events whose union is the sample space, and suppose that $P(X \mid A) = 1/9$, $P(X \mid B) = 1/3$, $P(X \mid C) = 1/4$, $P(X \mid D) = 1/6$, $P(A) = 1/3$, $P(B) = 1/6$, $P(C) = 3/8$, and $P(D) = 1/8$. Exercises 10 through 13 refer to this situation.

10. Find $P(A \mid X)$ **11.** Find $P(B \mid X)$ **12.** Find $P(C \mid X)$

13. Find $P(D' \mid X)$

You have two vases: the first contains seven red and three white marbles, the second six red and four white marbles. You select a vase at random and draw a marble from it. Exercises 14 through 17 refer to this experiment.

14. Find P(the marble you drew came from the first vase|it was red).

15. Find P(the marble you drew came from the second vase|it was red).

16. If the marble you drew was white, find the probability that it came from the first vase.

17. If the marble you drew was white, find the probability that it came from the second vase.

You have three vases: the first contains seven red and three white marbles, the second contains six red and four white marbles, the third contains eight red and two white marbles. You select a vase at random and draw a marble from it. Exercises 18 through 21 refer to this experiment.

18. Find P(the marble you drew came from the first vase|it was red).

19. Find P(the marble you drew came from the second vase|it was white).

20. If the marble you drew was white, find the probability that it came from the third vase.

21. If the marble you drew was red, find the probability that it came from the third vase.

You have two vases: the first contains three red, five white, and two blue marbles, the second contains two red, three white, and five blue marbles. You select a vase at random and draw a marble from it. Exercises 22 through 25 refer to this experiment.

22. If the marble you drew was red, find the probability that it came from the first vase.

23. If the marble you drew was blue, find the probability that it came from the second vase.

24. If the marble you drew was not red, find the probability that it came from the first vase.

25. If the marble you drew was not white, find the probability that it came from the second vase.

26. Elena estimates that the probability that she passed her final exam in history is 0.60. She also estimates that the probability that she will pass the history course if she passed the final exam is 0.80, while the probability that she will pass the course if she did not pass the final is 0.20. It turns out that Elena passes the history course. What is the probability that she passed the final exam?

27. Repeat Exercise 26 if Elena estimates that P(she passed the final exam) $= 0.10$.

28. The only things that can cause a computer to go down (stop working) are problems with the power supply and problems with the disk drive. Suppose that if there is a power supply problem, the probability that the computer will go down is 0.95, whereas if there is a problem with the disk drive, the probability that the computer will go down is 0.5. Suppose also that the probability of a power supply problem occurring is 0.02, and the probability of a disk drive problem occurring is 0.01. If the computer goes down, find the probability that this was caused by a power supply problem; by a disk drive problem.

29. Refer to Exercise 28: Suppose trouble with the CPU (central processing unit) can also cause the computer to go down, that the probability of CPU trouble is 0.005, and that the probability that the computer will go down if there is CPU trouble is 0.8. If the computer goes down, find the probability that this was caused by a power supply problem; by a disk drive problem; by a CPU problem.

Management

30. Al and Betty work on an assembly line putting backplates on calculators. Each puts backplates on 50% of the calculators. The probability that Al puts a backplate on upside down is 0.02, while the probability that Betty does so is 0.03. If a calculator comes off the assembly line with its backplate on upside down, what is the probability that Al did it? that Betty did it?

31. Suppose the probability that a firm's property tax assessment will be more than $1 million this year is 0.20, the probability that the firm will close its local plant if the assessment is more than $1 million is 0.60, and the probability that the firm will close the plant if the assessment is $1 million or less is 0.05. The firm announces that it will close the plant. What is the probability that the tax assessment was more than $1 million?

32. Alpha Company advertises its products in magazines. It is known that 20% of all potential customers buy Alpha's products. Of those who do, 35% have seen the magazine ads; of those who do not, 40% have seen the ads. Find the probability that someone who has seen the ads buys the products.

33. Alpha Company of Exercise 32 also advertises in newspapers. Of those who buy Alpha's products, 45% have seen the newspaper ads; of those who do not, 25% have seen the newspaper ads. Find the probability that someone who has seen the newspaper ads buys Alpha's products. Which type of advertising, magazine or newspaper, seems more effective for Alpha?

34. A department store knows that 8.5% of its charge accounts will not be paid on time. The probability that an account that is not paid on time will have a balance over $500 is 0.65; the probability that an account that is paid on time will have a balance over $500 is 0.32. If an account has a balance over $500, what is the probability that the account will not be paid on time?

(Suggested by material in Duncan, *Quality Control and Industrial Statistics,* fifth edition, Irwin, 1986.) Suppose the quality of cotter pins produced by a manufacturer varies:

> 40% of the time, 5% of each lot of pins is defective;
>
> 50% of the time, 4% of each lot of pins is defective;
>
> 10% of the time, 3% of each lot of pins is defective.

A customer receives a lot of cotter pins from the manufacturer, takes 100 pins at random from the lot, and finds 5 of them to be defective. Exercises 35 through 37 refer to this situation.

35. Find the probability that 5% of the lot is defective.
36. Find the probability that 4% of the lot is defective.
37. Find the probability that 3% of the lot is defective.

Life Science

38. The probability that a smoker will develop lung cancer is 0.12; the probability that a nonsmoker will develop the disease is 0.01. Suppose one-third of the population smokes. If a person develops lung cancer, find the probability that the person is a smoker.

39. Human immunodeficiency virus (HIV) is the virus that causes AIDS. According to the March–April, 1989, issue of *Focus,* the newsletter of the Mathematical Association of America, the standard diagnostic test for HIV gives a false positive reading 7% of the time and a false negative reading 2.3% of the time. (A false positive is a finding that the virus is present when in fact it is not; a false negative is a finding that the virus is not present when in fact it is.) Thus:

   if a person carries HIV, the diagnostic test will detect it with a probability of 0.977 and will fail to detect it with a probability of 0.023;

   if a person does not carry HIV, the test will correctly report its absence with a probability of 0.93 and falsely report its presence with a probability of 0.07.

   Suppose 1% of the population carries HIV.
   (a) If a person selected at random is given the diagnostic test and the test indicates the presence of HIV, what is the probability that the person really does carry HIV?
   (b) A person selected at random donates blood, and the blood is then tested for the presence of HIV using the diagnostic test. Suppose the test indicates that the blood is not infected with HIV. What is the probability that it really is infected?

40. A biostatistician has collected data that show that persons whose occupations are stressful have a probability of 0.225 of developing heart disease, whereas those whose occupations are not stressful have a probability of 0.143 of doing so. If 20% of the population have stressful occupations, find the probability that a person who develops heart disease has a stressful occupation.

Social Science

41. In the 1988 presidential election, the Democrat Dukakis received 46% of the vote and the Republican Bush received 54% of the vote. Of those who voted for Dukakis, 80% were Democrats; of those who voted for Bush, 20% were Democrats. Find the percentage of Democrats who voted for Bush.

42. Twelve percent of the United States population have incomes below the poverty level; of these, 25% are age 65 or older. Six percent of those who have incomes above the poverty level are 65 or older. What is the probability that a person who is 65 or older has an income below the poverty level?

A poll shows that 22% of voters 18 through 30 years of age are conservatives, 28% of those who are 31 through 55 years old are conservatives, and 33% of those who are 56 or

APPLICATIONS OF PROBABILITY

older are conservatives. Also, 18- through 30-year-olds make up 20% of the electorate and 31- through 55-year-olds make up 51% of the electorate. Exercises 43 through 45 refer to these facts.

43. Find the probability that a voter who is a conservative is 18 through 30 years of age.

44. Find the probability that a voter who is a conservative is 31 through 55 years of age.

45. Find the probability that a voter who is not a conservative is 56 years old or older.

7.5 Bayesian Decision Trees

As we mentioned in Section 7.4, Bayes' Law may be used to incorporate conditional information into the decision-making process. In this section we will illustrate how this technique can be used in conjunction with decision trees to refine the decision-making process.

Consider a farmer whose crop is worth $10,000 but is currently being attacked by insects. He must decide whether or not to spray the crop with an insecticide. Spraying will cost him $1000. If he decides to spray the crop, he will be able to save all of it, provided it does not rain during the week after the spraying; if it does rain, he will be able to save only 50% of the crop. On the other hand, if he decides not to spray the crop, then he will lose it all if it does not rain during the next week but will be able to save 60% of it if it does rain. The farmer's payoff matrix therefore is as follows:

$$\begin{array}{c} \\ \text{spray} \\ \text{don't spray} \end{array} \begin{array}{cc} \text{no rain} & \text{rain} \\ \begin{bmatrix} 10{,}000 - 1000 & 0.5(10{,}000) - 1000 \\ 0 & 0.6(10{,}000) \end{bmatrix}, \end{array}$$

or

$$\begin{array}{c} \\ \text{spray} \\ \text{don't spray} \end{array} \begin{array}{cc} \text{no rain} & \text{rain} \\ \begin{bmatrix} 9000 & 4000 \\ 0 & 6000 \end{bmatrix}. \end{array}$$

Suppose the farmer estimates the probability of rain next week to be 0.40. Then his decision tree is that shown in Figure 7–8. The symbols S and S' stand for spray and don't spray, respectively, while R and R' stand for rain and no rain, respectively. Based on expected value, the farmer's decision will be to spray the crop.

Now suppose we incorporate more information into the decision-making process by assuming that the farmer can purchase a weather forecast from a professional meteorologist for $100 and that the meteorologist is known to be 90% accurate in her forecasts. Now the farmer must first decide whether to purchase the forecast, denoted by F, or not to purchase the forecast, denoted by F'. If he does not purchase

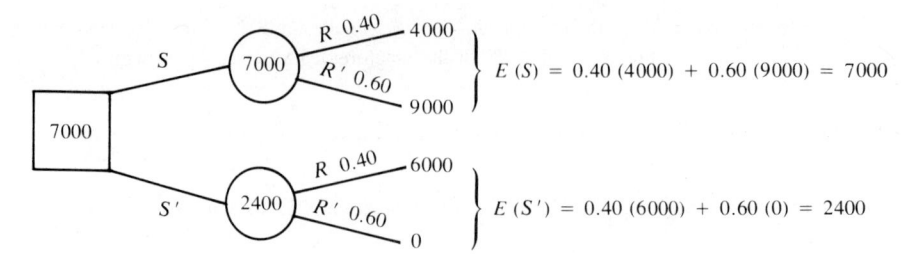

```
              R  0.40
                      4000
        S  ┌─────┐R'
    ┌───┤7000 │ 0.60
    │    └─────┘
    │                 9000      }  E(S) = 0.40(4000) + 0.60(9000) = 7000
┌───┤
│7000│
    │         R  0.40
    │                  6000
        S' ┌─────┐R'
        ┌──┤2400 │ 0.60
           └─────┘
                         0       }  E(S') = 0.40(6000) + 0.60(0) = 2400
```

FIGURE 7–8

the forecast, everything is as before, but if he does, then his decision as to whether or not to spray the crop will depend on the meteorologist's forecast. If we let

$$FR = \text{meteorologist forecasts rain}$$

and

$$FR' = \text{meteorologist forecasts no rain},$$

then the farmer's decision tree is that shown in Figure 7–9.

Note that the payoffs for the top half of the tree have been adjusted to incorporate the cost of purchasing the forecast. Note also that we have not yet assigned probabilities to the states of nature S, S', FR, and FR' in the top half of the tree. Since every state of nature must be assigned a probability, we must find the probabilities for these branches of the tree. To do so we will use Bayes' Law, since the likelihood of the events occurring must reflect the meteorologist's forecast and its accuracy.

Consider the R (or rain) branch at the very top of the tree. This branch comes after an FR (or forecast rain) branch, so its probability is

$$P(\text{rain given that rain was forecast}) = P(R \mid FR).$$

By Bayes' Law,

$$P(R \mid FR) = \frac{P(FR \mid R) \cdot P(R)}{P(FR \mid R) \cdot P(R) + P(FR \mid R') \cdot P(R')}.$$

But

$$P(FR \mid R) = P(\text{forecast rain given that it rains}) = 0.90,$$

because the meteorologist is known to be 90% accurate in her forecasts. Also,

$$P(FR \mid R') = P(\text{forecast rain given that it does not rain}) = 0.10,$$

because this represents an inaccurate forecast by the meteorologist, and we know that the probability that she is inaccurate is 10%. Furthermore, the farmer's prior estimate of the probability of rain was 0.40, so we have

$$P(R) = 0.40 \quad \text{and thus} \quad P(R') = 0.60.$$

APPLICATIONS OF PROBABILITY

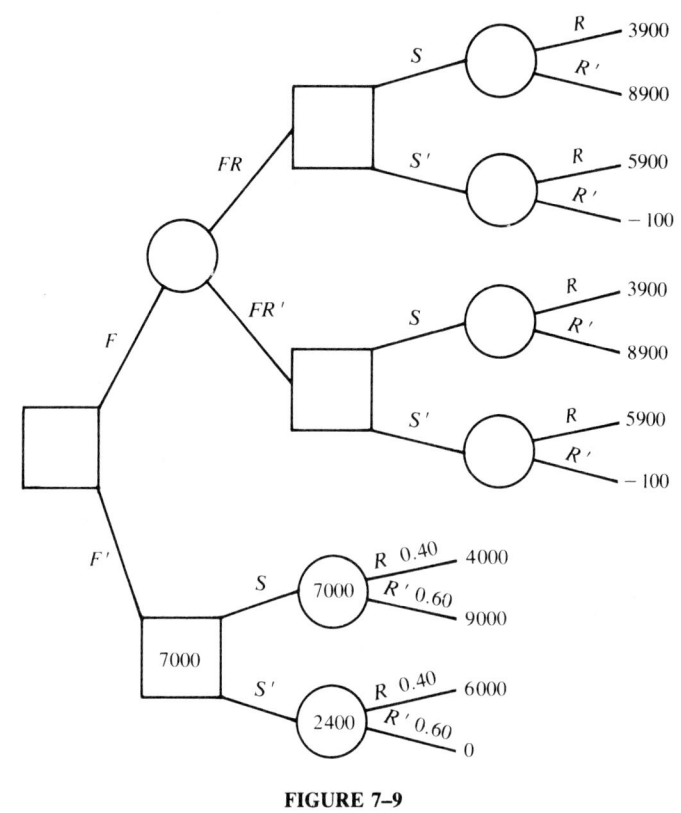

[Figure 7-9: Decision tree with branches F and F'. Under F: FR and FR'; each leads to S/S' then R/R' with payoffs. FR-S: R 3900, R' 8900. FR-S': R 5900, R' -100. FR'-S: R 3900, R' 8900. FR'-S': R 5900, R' -100. Under F': S (7000) → R 0.40 → 4000, R' 0.60 → 9000; S' (2400) → R 0.40 → 6000, R' 0.60 → 0. F' node shows 7000.]

FIGURE 7-9

Substituting these probabilities into the Bayes' Law formula, we obtain

$$P(R \mid FR) = \frac{(0.90)(0.40)}{(0.90)(0.40) + (0.10)(0.60)} \approx 0.86.$$

It follows that

$$P(R' \mid FR) = 1 - P(R \mid FR) \approx 1 - 0.86 = 0.14.$$

These probabilities may now be placed on the R and R' branches that follow the FR branch. (See Figure 7–10.)

We must next find the probabilities for the R and R' branches that follow the FR' branch. If R follows FR', then by Bayes' Law,

$$P(R \mid FR') = \frac{P(FR' \mid R) \cdot P(R)}{P(FR' \mid R) \cdot P(R) + P(FR' \mid R') \cdot P(R')}.$$

But, reasoning as we did above, we have

$$P(FR' \mid R) = 0.10,$$
$$P(FR' \mid R') = 0.90,$$
$$P(R) = 0.40,$$

[Figure 7–10: decision tree diagram]

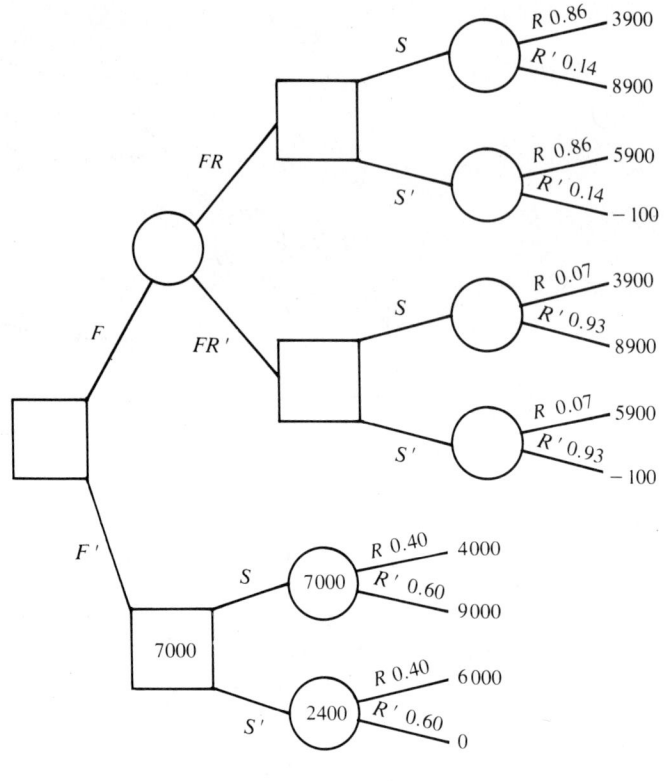

FIGURE 7–10

and
$$P(R') = 0.60,$$

so
$$P(R \mid FR') = \frac{(0.10)(0.40)}{(0.10)(0.40) + (0.90)(0.60)} \approx 0.07.$$

Hence
$$P(R' \mid FR') = 1 - P(R \mid FR') \approx 1 - 0.07 = 0.93.$$

These probabilities may now be placed on the R and R' branches that follow FR'. See Figure 7–10.

Finally, we must assign probabilities to the states of nature FR and FR'. We have

$$\begin{aligned} P(FR) &= P((FR \cap R) \cup (FR \cap R')) \\ &= P(FR \cap R) + P(FR \cap R') \\ &= P(FR \mid R) \cdot P(R) + P(FR \mid R') \cdot P(R') \\ &= (0.90)(0.40) + (0.10)(0.60) = 0.42. \end{aligned}$$

Applications of Probability

Then

$$P(FR') = 1 - P(FR) = 1 - 0.42 = 0.58.$$

These probabilities may now be placed on the FR and FR' branches of the tree. The complete tree is shown in Figure 7–11.

Now we are ready to do the expected value calculations that will tell the farmer what decisions to make. We proceed through the tree from right to left. Whenever we encounter a circle, we find the expected value of the branches that radiate from it and place that expected value in the circle; whenever we encounter a square, we place in it the best of the expected values at the end of the lines that radiate from it. For example, at the top of the tree we have

$$0.86(3900) + 0.14(8900) = 4600$$

and

$$0.86(5900) + 0.14(-100) = 5060$$

for the first two circles at the right. The square that precedes these two circles is assigned the best of these expected values, or 5060.

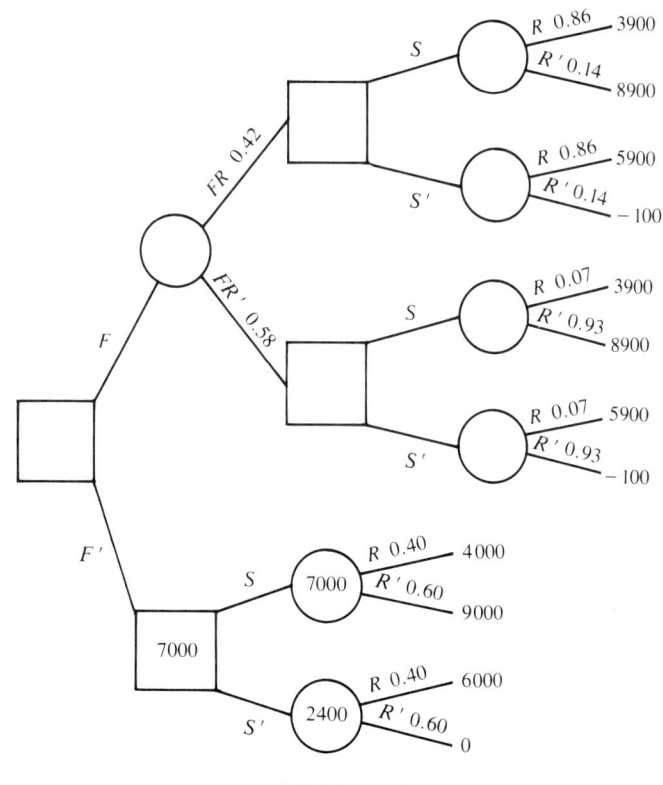

FIGURE 7–11

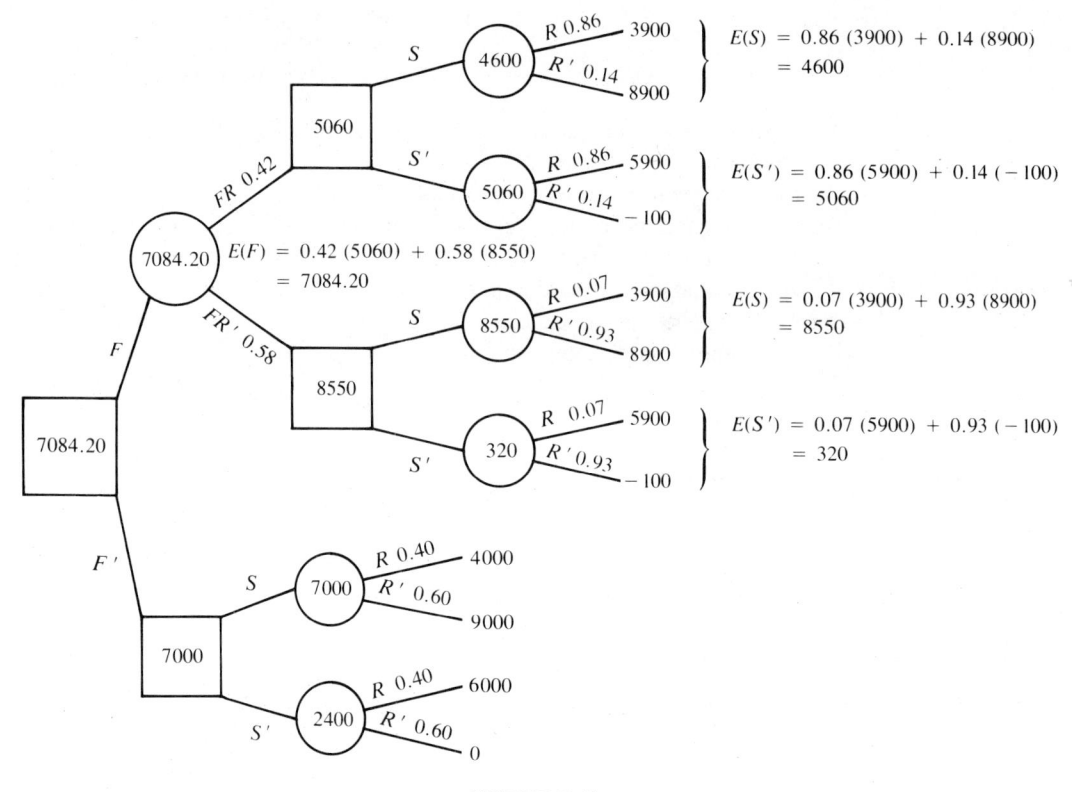

FIGURE 7–12

When we work our way back to the circle from which the *FR* and *FR'* branches radiate, its expected value is

$$0.42(5060) + 0.58(8550) = 7084.20.$$

The completed decision tree, with all expected values included, is shown in Figure 7–12. The tree tells us that, on the basis of expected value, the farmer should purchase the forecast from the meteorologist. Then, if the forecast is for rain *(FR)*, he should not spray, while if the forecast is for no rain *(FR')*, he should spray.

7.5 EXERCISES

1. An investor has $10,000 to invest for one year in either a savings account that pays 6% per year compounded quarterly or a stock that will pay $500 in dividends during the year and which the investor will be able to sell at the end of the year for either $11,000 or $9500, depending on whether the market is up or down at the time. The investor's broker has told him that

   P(market will be up at end of year) $= 0.40$

and
$$P(\text{market will be down at end of year}) = 0.60.$$
For $200 the investor can purchase a market forecast that will predict whether the market will be up or down at the end of the year; this forecast has a record of being 80% accurate in its predictions. Based on expected return on investment, what should the investor do? (See Exercise 1, Exercises 7.1.)

2. Repeat Exercise 1 if the market forecast has a record of 60% accuracy.

Management

3. A company is trying to decide whether to market a new product. If demand for the product turns out to be high, the profit from the product will be $17 million; if demand is low, the company will lose $7 million. The company estimates that it is twice as likely that demand will be as low as it is that demand will be high. The company can do a market survey, at a cost of $1 million, that will be 70% accurate in predicting the demand for the product. Draw a decision tree. What should the company do?

4. Repeat Exercise 3 if it is twice as likely that demand will be high as it is that demand will be low.

5. An oil company must decide whether to drill a well on a certain tract of land it has leased. The lease cost the company $10,000, and the well will cost an additional $60,000. If the well strikes oil, the revenue from it will be $200,000. The oil company's geologist estimates that the probability a well on the land will strike oil is 0.2. The company can conduct a seismic test, at a cost of $40,000, that will correctly predict the presence or absence of oil 75% of the time. On the basis of expected profit, what should the company do? (See Exercise 6, Exercises 7.1.)

6. Repeat Exercise 5 if the seismic test will cost $10,000 and be 90% accurate.

7. Tattle Toy Company has invested $100,000 in the development of a new electronic robot toy. Each unit of the toy will cost $20 to make and will sell for $35. Tattle is trying to decide how many units of the toy to produce for the Christmas season. Its facilities will allow it to produce either 20,000, 60,000, or 80,000 units. If the toy turns out to be a success, Tattle will be able to sell as many units as it can produce, but if the toy is not a success, the company will be able to sell only 50,000 units. Tattle's marketing department estimates that
$$P(\text{toy will be a success}) = 0.4.$$
Tattle can carry out a market survey, at a cost of $80,000, that will predict whether the toy will be a success with 70% accuracy. On the basis of expected profit, what should Tattle do? (See Exercise 9, Exercises 7.1.)

8. Repeat Exercise 7 if the market survey will cost $50,000 and be 90% accurate.

9. A retail oil company purchases heating oil in the summer from a wholesaler at a price of $0.60 per gallon. The amount of oil the company will need and the price it can charge for the oil depend on the severity of the winter:

Winter	Company Will Need	Company Can Charge
Mild	100,000 gallons	$0.80 per gallon
Normal	120,000 gallons	$0.90 per gallon
Harsh	150,000 gallons	$0.95 per gallon

The company's records show that in the past, 20% of the winters have been mild, 50% have been normal, and 30% have been harsh. If the company has oil left over at the end of a winter, it must be carried in inventory over the summer at a cost of $0.10 per gallon; if the company runs out of oil during a winter, it must purchase an emergency supply at $1.00 per gallon. In the summer the company can purchase a long-range forecast that will predict whether the coming winter will be mild, normal, or harsh. This forecast makes the correct prediction 80% of the time, makes each of the two possible incorrect predictions 10% of the time, and costs $2000. On the basis of expected profit, what should the company do?

10. Repeat Exercise 9 if the forecast costs $5000 and makes the correct prediction 60% of the time and each of the two possible incorrect predictions 20% of the time.

Life Science

11. It will cost $100 million to make and distribute a flu vaccine in preparation for a possible epidemic next year. If the epidemic occurs and the vaccine is available, the epidemic will cost the nation $3 billion; if the epidemic occurs and the vaccine is not available, the epidemic will cost the nation $20 billion. Suppose that a preliminary estimate puts

$$P(\text{the epidemic will occur}) = 0.70.$$

A more accurate estimate will require the efforts of a team of experts: the team's work will cost $10 million and the team's prediction of whether the epidemic will occur will be 65% accurate. On the basis of expected cost, what should be done? (See Exercise 14, Exercises 7.1.)

*12. How accurate must the prediction of the team in Exercise 11 be if it is to be worthwhile to use the team?

Exercises 13 and 14 make use of utilities. For an explanation of utilities, see Exercises 7.1.

*13. A doctor is considering whether to operate on a patient. The doctor estimates that the probability that the patient needs the operation is 0.65 and that the patient's utilities are as follows:

$$\text{Patient} \begin{array}{c} \text{needs operation} \\ \text{doesn't need operation} \end{array} \begin{array}{cc} \overset{\text{Doctor}}{\overset{\text{operates doesn't operate}}{}} \\ \begin{bmatrix} 10 & -20 \\ -5 & 10 \end{bmatrix} \end{array}.$$

The doctor can call in a specialist who is 90% accurate in his diagnoses. This will result in a delay of a month in the decision, in which case the patient's utilities will be as follows:

$$\text{Patient} \begin{array}{c} \text{needs operation} \\ \text{doesn't need operation} \end{array} \begin{array}{cc} \overset{\text{Doctor}}{\overset{\text{operates doesn't operate}}{}} \\ \begin{bmatrix} 10 & -30 \\ -10 & 10 \end{bmatrix} \end{array}.$$

What should the doctor do?

*14. How accurate must the specialist of Exercise 13 be in order to justify following his advice?

Social Science

15. A mayoral candidate has enough money to conduct a voter registration drive or run a series of TV spots, but not both. She estimates that the registration drive will net her 30,000 votes, while the TV spots will net her 100,000 votes if they are successful but only 10,000 if they are not. She also estimates that there is 1 chance in 4 that the spots will be successful. The candidate could hire a consultant who would predict, with 85% accuracy, whether or not the spots would be successful. However, doing this would leave less money available for the campaign, and thereby reduce the votes gained by the registration drive to 25,000 and those gained by the TV spots to 80,000 if they are successful, 0 if they are not. What should the candidate do?

16. A congressman running for reelection can either attack or ignore his opponent, and election day turnout can be either high or low. The percentage of the vote the congressman can expect under each alternative is as follows:

$$\text{Strategy} \begin{array}{c} \text{attack} \\ \text{ignore} \end{array} \begin{bmatrix} \overset{\text{Turnout}}{\overset{\text{high} \quad \text{low}}{49\% \quad 52\%}} \\ 54\% \quad 44\% \end{bmatrix}.$$

The congressman estimates that the probability of a high turnout is 0.75. His campaign can invest money in polling to determine whether turnout will in fact be high or low; this polling will be 80% accurate in predicting the turnout, but its cost will leave less money available for other campaign activities, thus reducing the percentage of the vote the congressman can expect to receive to the following:

$$\text{Strategy} \begin{array}{c} \text{attack} \\ \text{ignore} \end{array} \begin{bmatrix} \overset{\text{Turnout}}{\overset{\text{high} \quad \text{low}}{48\% \quad 50\%}} \\ 52\% \quad 45\% \end{bmatrix}.$$

On the basis of expected percentage of the vote, what should the congressman do?

17. A town is trying to decide whether to contract for a mosquito control program next summer. If the mosquito population next summer turns out to be large, the town will need the program, but if the population turns out to be small, the town can get by without it. If the town contracts for the program now it will cost $50,000, but if it waits until next summer, the program will cost $60,000. A scientist from the state environmental agency has estimated that

P(mosquito population will be large next summer) = 0.75.

The town can make a more accurate estimate of whether the mosquito population will be large or small next summer by examining the local places where mosquitoes breed. This will cost $4000, and the result will be 90% accurate. On the basis of expected cost, what should the town do?

18. Repeat Exercise 17 if the cost of the estimate will be $8000 and it will be 75% accurate.

SUMMARY

This summary consists of terms and symbols whose meaning you should know, facts you should know, and techniques and methods you should be able to employ.

Section 7.1

Payoff matrix. Decision tree. States of nature. Constructing decision trees. Using decision trees and expected value to make decisions.

Section 7.2

Markov chain. Transition matrix. Using the transition matrix to find stage $i + 1$ from stage i; to find stage $i + 1$ from stage 1. Steady state. Regular Markov chain. Every regular Markov chain reaches steady state. Finding steady state for a Markov chain.

Section 7.3

Game. Zero-sum game. Two-person game. Two-person zero-sum game. Plays of a game. Payoff matrix for a two-person zero-sum game. Row player, column player. Row and column dominance. Reducing a payoff matrix by removing dominated rows and columns. Saddle point. Strictly determined game. Finding the saddle points of a game. Not all games are strictly determined. Mixed strategy. Find the optimal mixed strategies for the row and column players of a 2 × 2 zero-sum game.

Section 7.4

Bayes' Law. Using Bayes' Law to find the probability of an event when a subsequent event is known to have occurred. Generalized form of Bayes' Law. Using the generalized form of Bayes' Law to find the probability of an event when a subsequent event is known to have occurred.

Section 7.5

Using Bayes' Law to incorporate conditional information into decision trees.

REVIEW EXERCISES

In Exercises 1 through 3, find V_2, V_3, and V_4 for the given transition matrix T and initial vector V_1.

1. $T = \begin{bmatrix} 0.25 & 0.35 & 0.40 \\ 0.50 & 0 & 0.50 \\ 0.20 & 0.40 & 0.40 \end{bmatrix}$, $V_1 = [1000, 2000, 10{,}000]$

2. $T = \begin{bmatrix} 0 & 1 \\ 1 & 0 \end{bmatrix}$, $V_1 = [100, 400]$

3. $T = \begin{bmatrix} 0 & 1 & 0 & 0 \\ 0 & 0 & 1 & 0 \\ 0 & \frac{1}{2} & 0 & \frac{1}{2} \\ \frac{1}{2} & 0 & 0 & \frac{1}{2} \end{bmatrix}$, $V_1 = [1000, 1000, 2000, 2000]$

4. Find the steady-state vector for the transition matrix of Exercise 1.

5. Find the steady-state vector for the transition matrix of Exercise 2.

6. Find the steady-state vector for the transition matrix of Exercise 3.

A cable television company has an exclusive franchise in a town where 20% of the homes currently subscribe to cable TV. Each month 10% of those who did not previously have cable add it, while 5% of those who did have it cancel it. Exercises 7 through 9 refer to this situation.

7. Write a transition matrix that describes this situation.

8. Find the percentage of homes in the town that will have cable TV two months from now.

9. Find the percentage of homes that will eventually have cable TV.

In Exercises 10 and 11, analyze the two-person zero-sum games. If a game is strictly determined, find its outcome; if not, find the mixed strategies and expected returns for each player.

10.
$\begin{array}{c} \\ A_1 \\ A_2 \\ A_3 \end{array} \begin{array}{ccc} B_1 & B_2 & B_3 \end{array} \\ \begin{bmatrix} 2 & 3 & -2 \\ 5 & 1 & 1 \\ -3 & 4 & -1 \end{bmatrix}$

11.
$\begin{array}{c} \\ A_1 \\ A_2 \\ A_3 \end{array} \begin{array}{ccc} B_1 & B_2 & B_3 \end{array} \\ \begin{bmatrix} 2 & 3 & -2 \\ 5 & 1 & -2 \\ -3 & 4 & -1 \end{bmatrix}$

12. Two nations are competing to gain influence over other nations. Each of the competitors can rely primarily on diplomatic, economic, or military pressure. The competition is zero-sum, with payoff matrix

$$\text{Nation A} \begin{array}{c} \text{diplomatic} \\ \text{economic} \\ \text{military} \end{array} \begin{array}{c} \text{Nation B} \\ \begin{array}{ccc} \text{diplomatic} & \text{economic} & \text{military} \end{array} \\ \begin{bmatrix} 2 & -3 & 2 \\ -3 & 4 & 4 \\ -1 & -4 & 0 \end{bmatrix} \end{array}.$$

The entries in the matrix measure the amount of influence gained or lost. What strategies will the two nations adopt?

13. A paperback book publisher is trying to decide whether to ship 125,000 or 200,000 copies of a new novel to bookstores. The publisher makes a profit of $1.20 on each copy sold, but the bookstores will return all unsold copies for a credit of $2 each. If the novel becomes a bestseller, all copies shipped will sell; if not, 75,000 copies will sell. What should the publisher do if the probability that the book will become a bestseller is 0.40?

14. Refer to Exercise 13: Let

$$P(\text{the book will be a best seller}) = p.$$

For what values of p will the publisher
- **(a)** ship 125,000 copies of the book;
- **(b)** ship 250,000 copies of the book.

15. A manufacturer is about to introduce a new high-definition television set to the market, and must decide on a marketing strategy. There are three possibilities:

Skimming: set a high price initially, quickly sell relatively few units at this price, then when sales at the high price cease, cut the price drastically to stimulate sales; this strategy will cost $25 million to implement.

Moderation: set the price at a medium level initially, steadily sell a moderate number of units, cutting the price by small amounts when necessary to stimulate sales; this strategy will cost $15 million to implement.

Lowballing: set the price as low as possible initially, steadily sell many units; this strategy will cost $35 million to implement.

The demand for the television sets will determine which of these strategies will perform the best. The payoff matrix for the strategies under various demands is as follows:

$$\begin{array}{c c} & \text{Demand} \\ & \begin{array}{ccc} \text{high} & \text{medium} & \text{low} \end{array} \\ \begin{array}{c} \text{skimming} \\ \text{moderation} \\ \text{lowballing} \end{array} & \left[\begin{array}{ccc} 150 & 80 & 20 \\ 130 & 110 & 10 \\ 180 & 100 & 40 \end{array} \right] \end{array}.$$

The entries in the matrix are revenues in millions of dollars. The probabilities for the three demand levels are as follows:

$$P(\text{demand will be high}) = 0.25,$$
$$P(\text{demand will be medium}) = 0.45,$$

and

$$P(\text{demand will be low}) = 0.30.$$

Draw a decision tree for the manufacturer. On the basis of expected profit, what should the manufacturer do?

16. Repeat Exercise 15 if

$$P(\text{demand will be high}) = 0.6,$$
$$P(\text{demand will be medium}) = 0.3,$$

and

$$P(\text{demand will be low}) = 0.1.$$

17. If $P(A) = 0.48$, $P(B \mid A) = 0.45$, and $P(B \mid A') = 0.44$, find $P(A \mid B)$.

18. The probability that a new car will be a lemon if it was built on a Monday is 0.01; the probability that it will be a lemon if it was not built on a Monday is 0.001. Twenty percent of all cars are built on Mondays. If your car is a lemon, find the probability that it was built on a Monday.

19. If $P(A_1) = 0.22$, $P(A_2) = 0.43$, $P(A_3) = 0.08$, $P(A_4) = 0.27$, $P(B \mid A_1) = 0.33$, $P(B \mid A_2) = 0.25$, $P(B \mid A_3) = 0.12$, and $P(B \mid A_4) = 0.04$, find $P(A_3 \mid B)$.

20. It is estimated that there is a 50-50 chance that the greenhouse effect (the warming of the earth due to a buildup of carbon dioxide in the atmosphere) has started to occur. However, five of the hottest summers ever recorded have occurred during the 1980s, and it is also estimated that the probability that this would happen if there were no greenhouse effect is $\frac{1}{3}$. Use these facts and Bayes' Law to find the revised probability that the greenhouse effect is occurring.

21. The New Boston town meeting is considering the question of whether to float a bond issue to build a new school. There are 150 Republicans, 120 Democrats, and 230 independents at the meeting. An informal poll taken prior to the meeting indicated that 38% of the Republicans, 42% of the Democrats, and 63% of the independents were opposed to the bond issue. If a person speaks in favor of the bond issue at the town meeting, what is the probability that he or she is
    (a) a Republican   (b) a Democrat   (c) an independent.

22. Repeat Exercise 13 if, in addition to the information given there, the publisher can commission a market survey that will cost $40,000 and will be 90% reliable in predicting if the novel will be a bestseller.

23. The manufacturer of Exercise 15 can commission a market survey that will predict whether demand for the new high-definition television set will be high, medium, or low. The survey will cost $5 million, and its accuracy will be as follows:

$$P(\text{survey predicts demand high} \mid \text{demand high}) = 0.80,$$
$$P(\text{survey predicts demand high} \mid \text{demand medium}) = 0.05,$$
$$P(\text{survey predicts demand high} \mid \text{demand low}) = 0.10,$$
$$P(\text{survey predicts demand medium} \mid \text{demand high}) = 0.05,$$
$$P(\text{survey predicts demand medium} \mid \text{demand medium}) = 0.90,$$
$$P(\text{survey predicts demand medium} \mid \text{demand low}) = 0.10,$$
$$P(\text{survey predicts demand low} \mid \text{demand high}) = 0.15,$$
$$P(\text{survey predicts demand low} \mid \text{demand medium}) = 0.05,$$
$$P(\text{survey predicts demand low} \mid \text{demand low}) = 0.80.$$

What should the manufacturer do?

ADDITIONAL TOPICS

Here are some suggestions for topics not covered in the text that you might want to investigate on your own.

1. In this chapter we have considered only *two-person* zero-sum games. Many zero-sum games have more than two players, however. Find out about *n*-person zero-sum games, $n > 2$. (A good reference for game theory is Luce and Raiffa, *Games and Decisions*, Wiley, 1964.)

2. In Exercises 7.3 we briefly discussed non-zero-sum games. Find out about such games.

3. The application of game theory to economics was pioneered in a famous book by Von Neumann and Morgenstern entitled *Theory of Games and Economic Behavior* (Princeton, 1947). Find out about some of the applications of game theory to economics, as discussed in this book.

4. Techniques for using decision trees and expected value in making decisions form part of a body of knowledge that is often referred to as "decision theory." There is an extensive literature on decision theory, and many operations research and management science textbooks discuss decision theory in more detail than we have been able to. Find out about some applications of decision trees and expected value to business and public-sector decision making by consulting the decision theory literature.

5. Game theory as a discipline within mathematics was virtually created by the Hungarian-American mathematician John Von Neumann (1903–1957). Von Neumann made important contributions to many areas of mathematics and had a very colorful career; for instance, he was "present at the creation" of both computers and atomic weapons. Find out about Von Neumann's life and his contributions to mathematics.

SUPPLEMENT: GAME THEORY AND LINEAR PROGRAMMING

In Section 7.3 we showed how to determine optimal mixed strategies for certain two-person zero-sum games that were not completely determined. The method we introduced there is not a general one, however, for it depends on assigning probabilities p and $1 - p$ to the rows and q and $1 - q$ to the columns of the reduced payoff matrix, and hence it can be used only if the reduced payoff matrix is 2×2 or smaller. But many two-person zero-sum games have payoff matrices that cannot be reduced to a 2×2 or smaller matrix. Consider, for example, the game whose payoff matrix is

$$\text{A's Plays} \begin{array}{c} \\ A_1 \\ A_2 \\ A_3 \end{array} \begin{array}{c} \text{B's Plays} \\ \begin{array}{ccc} B_1 & B_2 & B_3 \end{array} \\ \begin{bmatrix} 3 & 1 & 2 \\ 1 & 2 & 3 \\ 1 & 3 & 1 \end{bmatrix} \end{array}.$$

This matrix has no dominated rows or columns (check this) and hence cannot be reduced at all. In this supplement we will show how to find the optimal mixed strategies for such a game by using linear programming.

Consider the two-person zero-sum game introduced above. By a **mixed strategy** for the row player A, we mean a set $\{p_1, p_2, p_3\}$ of probabilities that will make A's expected return as large as possible when A plays A_1 with probability p_1, A_2 with probability p_2, and A_3 with probability p_3. Note that since p_1, p_2, and p_3 are probabilities,

$$p_1 \geq 0, \qquad p_2 \geq 0, \qquad \text{and} \qquad p_3 \geq 0.$$

APPLICATIONS OF PROBABILITY   471

Also, since A must play either A_1, A_2, or A_3,

$$p_1 + p_2 + p_3 = 1.$$

Now let w be any number such that A's expected return is always greater than or equal to w, no matter how the column player B plays. Thus, if $E_A(\text{B plays } B_j)$ denotes A's expected return if B plays B_j, then we have

$$E_A(\text{B plays } B_1) = 3p_1 + p_2 + p_3 \geq w,$$

$$E_A(\text{B plays } B_2) = p_1 + 2p_2 + 3p_3 \geq w,$$

and

$$E_A(\text{B plays } B_3) = 2p_1 + 3p_2 + p_3 \geq w.$$

Since the entries in the payoff matrix are all positive, A's expected payoffs will all be positive. Hence, we may choose $w > 0$, which allows us to rewrite the preceding inequalities as

$$3\frac{p_1}{w} + \frac{p_2}{w} + \frac{p_3}{w} \geq 1,$$

$$\frac{p_1}{w} + 2\frac{p_2}{w} + 3\frac{p_3}{w} \geq 1,$$

and

$$2\frac{p_1}{w} + 3\frac{p_2}{w} + \frac{p_3}{w} \geq 1.$$

Similarly,

$$\frac{p_1}{w} + \frac{p_2}{w} + \frac{p_3}{w} = \frac{1}{w}$$

and

$$\frac{p_1}{w} \geq 0, \qquad \frac{p_2}{w} \geq 0, \qquad \frac{p_3}{w} \geq 0.$$

If we let $x_1 = p_1/w$, $x_2 = p_2/w$, and $x_3 = p_3/w$, we can rewrite these conditions as

$$3x_1 + x_2 + x_3 \geq 1,$$

$$x_1 + 2x_2 + 3x_3 \geq 1,$$

$$2x_1 + 3x_2 + x_3 \geq 1,$$

$$x_1 + x_3 + x_3 = 1/w,$$

and

$$x_1 \geq 0, \qquad x_2 \geq 0, \qquad x_3 \geq 0.$$

Now, the row player wants his expected return to be as large as possible, and since his expected return is greater than or equal to w, he would like w to be as large as possible. But since $w > 0$, this is the same as saying that $1/w$ should be as small as possible. Thus the row player wants to minimize

$$x_1 + x_2 + x_3 = 1/w.$$

Therefore if we can solve the linear programming problem

$$\text{Minimize} \quad F = x_1 + x_2 + x_3$$

$$\text{subject to} \quad 3x_1 + x_2 + x_3 \geq 1$$

$$x_1 + 2x_2 + 3x_3 \geq 1$$

$$2x_1 + 3x_2 + x_3 \geq 1$$

$$x_1 \geq 0, \quad x_2 \geq 0, \quad x_3 \geq 0$$

for x_1, x_2, and x_3, we will be able to find the largest possible value of w and the optimal strategy probabilities $p_1 = x_1 w$, $p_2 = x_2 w$, and $p_3 = x_3 w$. The maximum value of w will then be the row player's expected return under the optimal strategy.

(Notice that the resource constraints of this linear programming problem may be expressed as the "matrix inequality"

$$\mathbf{P}^T \begin{bmatrix} x_1 \\ x_2 \\ x_3 \end{bmatrix} \geq \begin{bmatrix} 1 \\ 1 \\ 1 \end{bmatrix},$$

where $\mathbf{P}$ is the payoff matrix for the game and $\mathbf{P}^T$ denotes the transpose of $\mathbf{P}$.)

The solution of the linear programming problem, obtained by the simplex method, is $x_1 = 4/17$, $x_2 = 2/17$, and $x_3 = 3/17$, and the minimum value of $F = 1/w$ is $9/17$. Hence the maximum value of w is $17/9$, and since $p_i = x_i w$, we find that

$$p_1 = (4/17)(17/9) = 4/9,$$

$$p_2 = (2/17)(17/9) = 2/9,$$

and

$$p_3 = (3/17)(17/9) = 1/3.$$

Therefore the row player A should play A_1 four-ninths of the time, A_2 two-ninths of the time, and A_3 one-third of the time (in a random manner) for an expected return of $17/9$.

The method for finding the optimal mixed strategy of the row player that we have just illustrated is a general one.

APPLICATIONS OF PROBABILITY   **473**

> **Determining the Optimal Mixed Strategy for the Row Player**
>
> Let **P** be an $m \times n$ payoff matrix for a two-person zero-sum game. Assume that the entries of **P** give the payoffs for the row player and that all the entries are positive. To find the optimal mixed strategy for the row player, proceed as follows:
>
> 1. Solve the linear programming problem
>
> $$\text{Minimize} \quad F = x_1 + \cdots + x_m$$
>
> subject to
>
> $$\mathbf{P}^{\mathrm{T}} \begin{bmatrix} x_1 \\ \vdots \\ x_m \end{bmatrix} \geq \begin{bmatrix} 1 \\ \vdots \\ 1 \end{bmatrix},$$
>
> $x_1, \ldots, x_m$ all nonnegative.
>
> 2. Let
>
> $$w = \frac{1}{\text{minimum value of } F}.$$
>
> The optimal mixed strategy for the row player is to play row i of **P** with probability $p_i = x_i w$, for $i = 1, \ldots, m$. This strategy will result in an expected return of w for the row player.

Note that the method requires that all entries in the payoff matrix **P** be positive. If this is not the case, we can create a new payoff matrix **P**′ that does have all its entries positive by adding a suitable constant c to every entry of **P**. The optimal mixed strategy for **P**′ will be the same as that for **P**, and the expected return for **P**′ will be equal to that of **P** plus the number c. For example, suppose

$$\mathbf{P} = \begin{bmatrix} 1 & -1 & 0 \\ -1 & 0 & 1 \\ -1 & 1 & -1 \end{bmatrix}.$$

Then adding 2 to each entry of **P** results in a new matrix

$$\mathbf{P}' = \begin{bmatrix} 3 & 1 & 2 \\ 1 & 2 & 3 \\ 1 & 3 & 1 \end{bmatrix}$$

all of whose entries are positive. Since **P**′ is the matrix we used in our illustrative example, we know that the optimal mixed strategy for **P**′ is $p_1 = 4/9$, $p_2 = 2/9$, and

$p_3 = \frac{1}{3}$, with an expected return of $\frac{17}{9}$. Therefore the optimal mixed strategy for the matrix **P** is also $p_1 = \frac{4}{9}$, $p_2 = \frac{2}{9}$, and $p_3 = \frac{1}{3}$, with an expected return of $\frac{17}{9} - 2 = -\frac{1}{9}$.

Finally, we note that our method for finding optimal mixed strategies applies only to the row player. To find the optimal mixed strategy for the column player, we could rewrite the payoff matrix so as to turn the row player into the column player and use our method. However, if the same sort of analysis as we performed earlier to convert the row player's returns into a linear programming problem is done for the column player, the result will be the dual problem of the one we obtained for the row player.

Determining the Optimal Mixed Strategy for the Column Player

Let **P** be an $m \times n$ payoff matrix for a two-person zero-sum game. Assume that the entries of **P** give the payoffs for the row player and that all the entries are positive. To find the optimal mixed strategy for the column player, proceed as follows:

1. Solve the linear programming problem

$$\text{Maximize} \quad G = y_1 + \cdots + y_n$$

subject to

$$\mathbf{P} \begin{bmatrix} y_1 \\ \vdots \\ y_n \end{bmatrix} \leq \begin{bmatrix} 1 \\ \vdots \\ 1 \end{bmatrix},$$

$y_1, \ldots, y_n$ all nonnegative.

2. Let

$$w = \frac{1}{\text{maximum value of } G}.$$

The optimal mixed strategy for the column player is to play column j of **P** with probability $q_j = y_j w$, for $j = 1, \ldots, n$. This strategy will result in an expected return of $-w$ for the column player.

But since the linear programming problems for the row and column player are duals of one another, we can solve the linear programming problem for either player and the solution for the other will appear at the bottom of the final simplex tableau.

EXAMPLE 1   Consider again the game with payoff matrix

$$\mathbf{P} = \begin{bmatrix} 3 & 1 & 2 \\ 1 & 2 & 3 \\ 1 & 3 & 1 \end{bmatrix},$$

and suppose that we did not know the optimal mixed strategy for the row player. The linear programming problem whose solution will give us the optimal mixed strategy for the column player is

$$\text{Maximize} \quad G = y_1 + y_2 + y_3$$

$$\text{subject to} \quad 3y_1 + y_2 + 2y_3 \leq 1$$

$$y_1 + 2y_2 + 3y_3 \leq 1$$

$$y_1 + 3y_2 + y_3 \leq 1$$

$$y_1 \geq 0, \quad y_2 \geq 0, \quad y_3 \geq 0.$$

The final simplex tableau for this problem is

	y_1	y_2	y_3	s_1	s_2	s_3	soln
y_1	1	0	0	$-7/17$	$-5/17$	$1/17$	$3/17$
y_3	0	0	1	$-1/17$	$8/17$	$-5/17$	$2/17$
y_2	0	1	0	$-2/17$	$-1/17$	$7/17$	$4/17$
G	0	0	0	$4/17$	$2/17$	$3/17$	$9/17$

Hence the solution for the column player is $y_1 = 3/17$, $y_2 = 4/17$, and $y_3 = 2/17$, maximum value of $G = 9/17$; it follows that the column player's optimal mixed strategy is to play column j with probability q_j, where

$$q_1 = (3/17)(17/9) = 1/3,$$

$$q_2 = (4/17)(17/9) = 4/9,$$

and

$$q_3 = (2/17)(17/9) = 2/9.$$

The solution for the row player is the dual solution, which the final simplex tableau tells us is $x_1 = 4/17$, $x_2 = 2/17$, and $x_3 = 3/17$. Thus the row player's optimal mixed strategy is to play row i with probability p_i, where

$$p_1 = (4/17)(17/9) = 4/9,$$

$$p_2 = (2/17)(17/9) = 2/9,$$

and

$$p_3 = (3/17)(17/9) = 1/3.$$

The expected value of the game for the column player is $-17/9$, and for the row player it is $17/9$. □

■ **Computer Manual: Program SIMPLEX**

SUPPLEMENTARY EXERCISES

In Exercises 1 through 5, use the technique demonstrated in this supplement to determine the optimal mixed strategies for the players of the given zero-sum game.

1. $\begin{array}{c} \\ A_1 \\ A_2 \end{array} \begin{array}{cc} B_1 & B_2 \\ \begin{bmatrix} 2 & 4 \\ 4 & 2 \end{bmatrix} \end{array}$

2. $\begin{array}{c} \\ A_1 \\ A_2 \end{array} \begin{array}{cc} B_1 & B_2 \\ \begin{bmatrix} 1 & -3 \\ -3 & 4 \end{bmatrix} \end{array}$

3. $\begin{array}{c} \\ A_1 \\ A_2 \\ A_3 \end{array} \begin{array}{ccc} B_1 & B_2 & B_3 \\ \begin{bmatrix} 1 & 2 & 1 \\ 0 & 1 & 2 \\ 3 & 1 & 1 \end{bmatrix} \end{array}$

4. $\begin{array}{c} \\ A_1 \\ A_2 \\ A_3 \end{array} \begin{array}{ccc} B_1 & B_2 & B_3 \\ \begin{bmatrix} 2 & -1 & 5 \\ 4 & -2 & -1 \\ -2 & 1 & -1 \end{bmatrix} \end{array}$

5. $\begin{array}{c} \\ A_1 \\ A_2 \\ A_3 \\ A_4 \end{array} \begin{array}{ccccc} B_1 & B_2 & B_3 & B_4 & B_5 \\ \begin{bmatrix} 4 & 3 & -1 & -2 & 2 \\ 3 & 1 & 0 & 2 & 3 \\ -2 & -1 & 2 & 1 & -2 \\ 1 & -1 & -2 & -2 & -2 \end{bmatrix} \end{array}$

6. Two competing supermarkets can each advertise on radio, in the newspaper, and by direct mail. The payoffs, in percentage of the market gained or lost, are as follows:

$$\text{Supermarket A} \begin{array}{c} \text{radio} \\ \text{newspaper} \\ \text{mail} \end{array} \overset{\displaystyle\text{Supermarket B}}{\begin{array}{ccc} \text{radio} & \text{newspaper} & \text{mail} \\ \begin{bmatrix} 1.0 & -1.0 & 0.0 \\ -1.5 & 1.0 & 0.5 \\ 0.5 & 1.0 & -1.0 \end{bmatrix} \end{array}}$$

Find each supermarket's optimal strategy.

7. Mal DeMer is running for political office. He can either attack his opponent on a personal basis *(A)*, differ with his opponent's stand on the issues *(D)*, praise his own personal qualities *(P)*, or explain his own stands on the issues *(E)*. His opponent Sal Monella has the same four options. The payoff matrix of this zero-sum game, with entries of the percentage of the vote lost or gained each week, is as follows:

$$\text{Mal} \begin{array}{c} A \\ D \\ P \\ E \end{array} \overset{\displaystyle\text{Sal}}{\begin{array}{cccc} A & D & P & E \\ \begin{bmatrix} 0.0 & -0.4 & 0.4 & -0.1 \\ -0.4 & 0.0 & 0.3 & 0.4 \\ 0.1 & -0.2 & 0.0 & -0.3 \\ 0.4 & -0.1 & -0.2 & 0.0 \end{bmatrix} \end{array}}$$

Find the optimal strategy for each candidate.

8. The game of paper-scissors-rock is played like this: each of two players simultaneously puts out either a flat hand (paper), two fingers extended (scissors), or a fist (rock).

If one player shows paper and the other rock, paper wins (paper covers rock).

If one player shows scissors and the other rock, rock wins (rock blunts scissors).

If one player shows scissors and the other paper, scissors wins (scissors cuts paper).

If both players show the same thing, neither wins.

(a) Suppose each player bets $1 on the outcome of the game. Write a payoff matrix for the amount won or lost.

(b) Find each player's optimal strategy.

8

Statistics

A **population** is a set of data consisting of *all* the data that are of interest; a **sample** is a set of data consisting of only *some* of the data of interest. For example, suppose an advertising executive is interested in how many hours per week children watch television. If he conducts a survey that asks *all* children about the amount of television they watch per week, their responses form a population; if he surveys only *some* children about the amount they watch per week, their responses form a sample. **Statistics** is the study of populations and samples.

Statistics has become important in virtually every area of human endeavor. Its techniques are employed in business in such areas as market research, advertising, and quality control; in the life and social sciences in the design, control, and analysis of experiments; in politics in the creation and interpretation of political polls; in sports in the measurement of team and individual performance; and in a host of other disciplines.

The field of statistics may be divided into two subfields: **descriptive statistics** and **inferential statistics.** Descriptive statistics deals with the collection, arrangement, and summarization of data, while inferential statistics deals with methods of using samples to draw conclusions about populations. To illustrate, consider again the advertising executive who wants to find out how much television children watch. Since he cannot possibly ask all children how much television they watch, he must take a survey of some children and ask them how much they watch. In this way he will obtain a sample of responses. The techniques of descriptive statistics will allow him to arrange and summarize his sample data, and thus to answer questions such as: what percentage of children in the sample watch between 10 and 15 hours of television per week? how many hours of television does the typical child in the sample watch per week? The techniques of inferential statistics will then allow him to use his knowledge of the sample to generalize about the population, and thus to make estimates about the percentage of *all* children who watch between 10 and 15 hours per week, or the hours watched by a typical child in the population, for instance.

In this chapter we will mainly be concerned with descriptive statistics. We begin with a section devoted to the arrangement of data into frequency distributions. This is followed by sections that discuss the summarization of data using measures of central tendency and dispersion. The fourth section introduces the normal distribution, which is an important probability distribution that occurs often in statistics. The chapter concludes with a brief discussion of a topic from inferential statistics known as regression and correlation analysis.

8.1 FREQUENCY DISTRIBUTIONS

A set of numerical data, whether it forms a population or a sample, usually must be arranged and organized in order for it to be comprehensible. This often entails putting the data into a table so that they are sorted into **classes.** Such a table is called a **frequency distribution** for the data; the number of pieces of data in a class is called the **frequency** of the class. To illustrate, suppose a comparison shopper who is interested in the prices of digital watches goes to 25 stores and records the price of each store's most popular digital watch, thus obtaining the following sample of prices:

$22.50	$29.00	$20.95	$18.00	$29.65
$25.00	$27.00	$30.50	$22.00	$24.05
$22.00	$23.00	$32.15	$24.00	$21.49
$28.00	$25.00	$26.50	$21.99	$26.00
$31.49	$24.00	$23.75	$30.00	$29.99

One possible frequency distribution for this set of data is that of Table 8–1. Table 8–1 shows that there are two stores whose most popular watch sells for $18.00 through $20.99, seven whose most popular watch sells for $21.00 through $23.99, and so on. Note that the data have been arranged in five classes; the frequency of the first class is 2, the frequency of the second class is 7, and so on.

If we wish, we may expand the frequency distribution of Table 8–1 by including

TABLE 8–1

Price ($)	Number of Stores
$18.00–20.99	2
21.00–23.99	7
24.00–26.99	7
27.00–29.99	5
30.00–32.99	4
	25

columns for relative frequency and cumulative frequency. The **relative frequency** of a class is the frequency of the class divided by the total of all frequencies; it tells us the percentage of the data that fall into the class. The **cumulative frequency** of a class is the sum of the frequency of the class and the frequencies of all preceding classes. The frequency distribution of Table 8–1 is expanded to include relative and cumulative frequencies in Table 8–2. Thus, the relative frequency column tells us that 8% of the data fall into the first class, 28% into the second class, and so on. The cumulative frequency column tells us that 2 pieces of data fall into the first class, 9 into the first two classes, 16 into the first three classes, and so on.

TABLE 8–2

Price ($)	Frequency	Relative Frequency	Cumulative Frequency
$18.00–20.99	2	$2/25 = 0.08$	2
21.00–23.99	7	$7/25 = 0.28$	9
24.00–26.99	7	$7/25 = 0.28$	16
27.00–29.99	5	$5/25 = 0.20$	21
30.00–32.99	4	$4/25 = 0.16$	25
	25	1.00	

We may depict the information contained in frequency distributions such as Table 8–2 by bar graphs known as **histograms**. A **frequency histogram** is constructed by marking off the classes on the horizontal axis and erecting over each class a rectangle whose height is the frequency of the class. **Relative frequency histograms** and **cumulative frequency histograms** are constructed in a similar manner. Figure 8–1 shows a frequency histogram, a relative frequency histogram, and a cumulative frequency histogram for the frequency distribution of Table 8–2.

An alternative way to display the information in a frequency distribution pictorially is to use polygons. A **frequency polygon** is constructed by plotting for each class the point (m,f), where m is the **midpoint** (or **class mark**) of the class and f its frequency, and then connecting the plotted points with line segments. **Relative frequency polygons** and **cumulative frequency polygons** are constructed in a similar manner. Figure 8–2 shows a frequency polygon, a relative frequency polygon, and a cumulative frequency polygon for the frequency distribution of Table 8–2.

How does one go about constructing a frequency distribution from a set of data? Suppose, for instance, that we wish to construct a frequency distribution for the following set of data:

13	7	10	19	17	12
11	13	14	15	15	18
21	8	4	15	14	11
9	16	7	20	18	6
13	15	14	16	14	15

We begin by deciding how many classes to use. In general, the larger the set of data, the greater the number of classes, but we should never use fewer than 5 or more than 15 classes. Since we have only 30 pieces of data here, which is not really very many, let us try six classes.

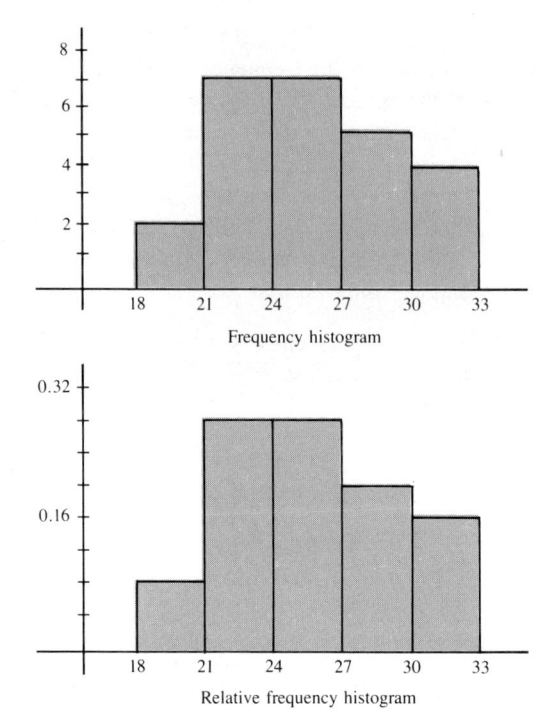

FIGURE 8–1

Next we must determine the size of the classes. It is a good idea to make all classes of equal width, when possible, so let us try for six classes of equal width. We first find the **range** of the data: the range is the largest value minus the smallest value, in this case,

$$21 - 4 = 17.$$

Therefore, if we wish to have six classes of equal width, each class must be of width

$${}^{17}\!/_6 = 2{}^5\!/_6.$$

[Frequency polygon]

[Relative frequency polygon]

[Cumulative frequency polygon]

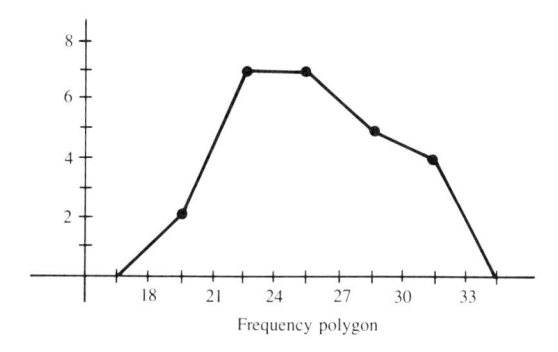

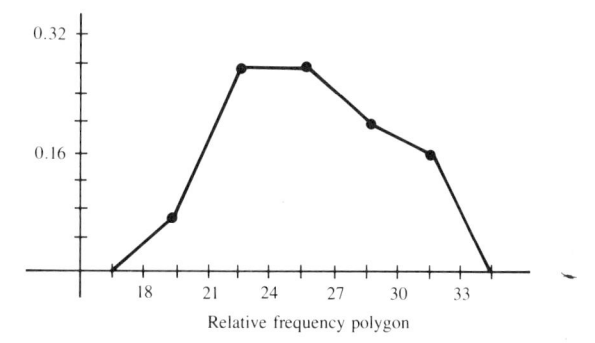

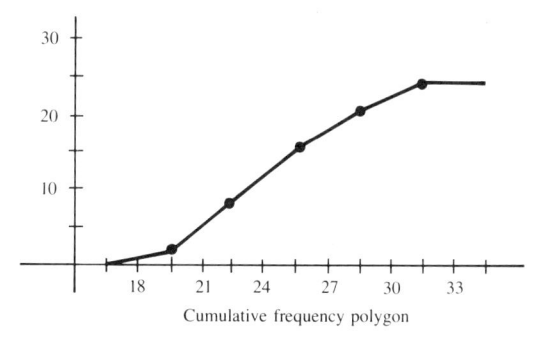

FIGURE 8–2

Since classes of width $2\frac{5}{6}$ would be awkward to handle, we round this up to 3 and decide to use six classes of width 3.

Our first class can start at 4, our smallest value. Since each class is to be of width 3, we might say that our first class would go from 4 to 7, the second from 7 to 10, and so on. However, if we write the classes in this way, we create an overlap at the number 7, which is not a good idea, for it means that we would not know to which class the number 7 belongs, the first or the second. To avoid this, we must make our first class go from 4 up to but not including 7, the second go from 7 up to but not including 10, and so on. Since our data are whole numbers, the class from 4 up to but not including 7 actually covers the numbers 4, 5, and 6, so we may write it as the class 4–6. Similarly, the class from 7 up to but not including 10 actually covers 7, 8, and 9, so we may write it as the class 7–9, and so on.

Now we are ready to set up our classes and tally the number of pieces of data that fall into each class (see Table 8–3).

TABLE 8–3

Class	Tally
4–6	||
7–9	||||
10–12	||||
13–15	|||| |||| ||
16–18	||||
19–21	|||

Thus the frequency distribution, with relative and cumulative frequencies included, is as shown in Table 8–4.

TABLE 8–4

Class	Frequency	Relative Frequency	Cumulative Frequency
4–6	2	$1/15$	2
7–9	4	$2/15$	6
10–12	4	$2/15$	10
13–15	12	$2/5$	22
16–18	5	$1/6$	27
19–21	3	$1/10$	30
	30	1	

EXAMPLE 1 Construct a frequency distribution for the following set of data:

16.0	20.2	21.1	24.8	27.2
25.9	30.5	21.8	23.9	17.6
14.6	19.3	22.7	23.5	12.2
16.5	22.8	25.3	18.2	25.8
23.4	17.0	22.3	24.8	17.1

Since we have only 25 pieces of data here, let us use five classes. Our range is

$$30.5 - 12.2 = 18.3,$$

and since

$$\frac{18.3}{5} = 3.66,$$

we will use five classes of width 4. Our first class could go from 12.2 up to but not including 16.2, our second from 16.2 up to but not including 20.2, and so on.

However, the classes will be easier to handle if we start at 12 rather than 12.2; so, for the sake of simplicity, let us set up our first class to go from 12 up to but not including 16, our second to go from 16 up to but not including 20, and so on. Since our data are given to one decimal place, the class from 12 up to but not including 16 actually covers the numbers from 12.0 through 15.9 and hence may be written as 12.0–15.9. Similarly, the second class may be written as 16.0–19.9, and so on. Thus, we have Table 8–5. ◻

TABLE 8–5

Class	Tally	Frequency
12.0–15.9	||	2
16.0–19.9	ⅢⅠ ||	7
20.0–23.9	ⅢⅠ ||||	9
24.0–27.9	ⅢⅠ |	6
28.0–31.9	|	1
		25

Finally, we must mention that the purpose of a frequency distribution is to organize and display the pattern of the data. A distribution in which all classes have nearly the same frequency is usually not displaying the pattern of the data properly, because either too many or too few classes have been used. Furthermore, a frequency distribution should never contain a class whose frequency is zero; if this occurs, too many classes have been used.

■ *Computer Manual: Programs FREQDIST, HISTGRAM*

8.1 EXERCISES

Expand each of the tables in Exercises 1 through 6 to include relative and cumulative frequencies, and for each construct a frequency histogram, a frequency polygon, a relative frequency histogram, and a cumulative frequency polygon.

1.

Class	Frequency
0–4	8
5–9	12
10–14	22
15–19	13
20–24	7

2.

Class	Frequency
100–119	22
120–139	16
140–159	60
160–179	9
180–199	52
200–219	18
220–239	23

3.

Class	Frequency
250–279.9	162
280–309.9	128
310–349.9	96
350–389.9	66
390–469.9	43
470–549.9	11

4.

Class	Frequency
32.0–35.9	2
36.0–40.9	8
41.0–45.9	12
46.0–50.9	17
51.0–55.9	26
56.0–59.9	35

486 CHAPTER 8

5.

Class	Frequency
10.00–12.99	52
13.00–15.99	34
16.00–18.99	16
19.00–21.99	9
22.00–24.99	20
25.00–27.99	42
28.00–30.99	55

6.

Class	Frequency
0.000–0.249	2
0.250–0.499	4
0.500–0.749	12
0.750–0.999	48
1.000–1.249	3
1.250–1.499	6
1.500–1.749	18
1.750–1.999	72

Management

7. The following data represent the number of employees in each of Ajax Company's 30 departments. Make a frequency distribution using 5 classes of equal width. Include relative and cumulative frequencies.

12	19	21	23	23	28	6	24	13	30
25	29	18	25	25	29	19	25	24	16
35	19	18	14	30	8	33	25	26	20

8. A firm has calculated the average number of sick days taken per year by all its employees who have more than one year of service with the company. The results are given below. Make a frequency distribution using seven classes of equal width, and draw a frequency histogram.

2.0	2.1	4.1	7.0	3.2	3.9	1.8	2.7	1.0	8.6
4.3	0.0	1.6	5.2	3.1	3.3	5.8	2.4	5.2	3.6
1.5	4.1	6.2	9.8	0.5	4.5	2.1	8.1	10.5	6.8
0.7	2.9	3.7	3.3	1.0	5.3	3.8	0.9	7.4	4.1

9. The payroll department of Apex Company has calculated the average hourly wage, including overtime, for each of its 50 employees who are paid by the hour. The results are given below. Make a frequency distribution using eight classes of equal width, and draw a frequency polygon.

$6.22	$7.81	$6.45	$8.03	$4.39	$6.53	$7.45	$8.00
$7.77	$5.51	$3.02	$4.78	$5.82	$6.37	$7.14	$8.54
$3.31	$9.25	$9.61	$4.44	$3.99	$9.03	$10.22	$6.75
$10.80	$5.43	$6.82	$7.00	$7.36	$6.55	$5.14	$8.15
$8.96	$7.47	$4.20	$9.43	$5.01	$9.17	$8.62	$8.01
$10.00	$6.50	$8.88	$5.44	$6.12	$7.17	$7.28	$6.35
$4.11	$6.72

Life Science

10. The weights, in pounds, of 25 people participating in a diet experiment are given below. Make a frequency distribution using five classes of equal width. Include relative and cumulative frequencies, and draw a frequency histogram.

423	385	295	244	271	236	302	297	255
402	396	377	326	211	312	309	283	434
356	317	333	452	342	350	358

11. The yield of peaches, in bushels per acre, for an orchard over the past 32 years is given below. Make a frequency distribution using six classes of equal width, and draw a relative frequency polygon.

10.4	14.4	18.4	18.6	29.8	6.3	17.8	20.8
24.8	24.7	19.3	8.2	14.6	11.1	24.5	22.1
27.0	22.4	26.3	21.5	13.6	19.5	23.2	23.6
9.1	15.3	21.9	23.8	6.1	16.1	21.2	19.4

Social Science

12. A psychologist administers an intelligence quotient (IQ) test to 40 engineers. The test scores are given below. Make a frequency distribution using seven classes of equal width. Include relative and cumulative frequencies, and draw a frequency polygon.

126	130	128	119	133	144	141	135	106
111	118	124	128	120	119	129	136	102
96	114	109	125	119	116	124	116	105
99	132	140	126	121	115	125	129	117
124	108	113	127					

13. The spatial perceptions of 50 kindergarten children were tested by recording the time it took each of them to complete a simple jigsaw puzzle. The times, in minutes, are recorded below. Make a frequency distribution using eight classes of equal width, and draw a frequency histogram.

2.0	2.5	3.0	2.1	2.7	2.7	3.4	2.9	2.3	1.8
3.0	1.1	2.9	1.4	3.7	2.5	2.6	2.2	1.3	2.9
2.9	2.1	1.6	2.7	2.2	2.4	1.5	4.0	2.8	3.2
1.9	2.2	2.4	3.3	3.8	4.3	1.6	4.8	2.7	2.3
3.5	1.8	2.0	2.6	3.1	2.6	2.5	2.6	2.8	2.8

14. The high-school dropout rates in 1986 for the 50 states and the District of Columbia are given below. (Data are from the *1989 World Almanac*, Pharos Books, 1989.) The figures are the percentages of high-school students who dropped out during the year. Make a frequency distribution for the data using eight classes of equal width. Include relative frequencies and draw frequency and relative frequency histograms.

32.7	31.7	37.0	22.0	33.3	26.9	10.2	29.3	43.2
38.0	37.3	29.2	21.0	24.2	28.3	12.5	18.5	31.4
37.3	23.5	23.4	23.3	32.2	8.6	36.7	24.4	12.8
11.9	34.8	26.7	22.4	27.7	35.8	30.0	10.3	19.6
28.4	25.9	21.5	32.7	35.5	18.5	32.6	35.7	19.7
22.4	26.1	24.8	24.8	13.7	18.8			

*15. The fiscal year 1986 per capita tax burdens of the 50 states, in dollars, are given below. (Data are from the *1989 World Almanac*, Pharos Books, 1989.) Make a frequency distribution for the data. (You will probably find it useful to use classes of unequal width.)

740	3490	975	770	1144	718	1202	1343	780	806
1400	743	848	810	863	777	863	807	940	1047
1314	1019	1163	731	712	755	700	1084	472	1096
989	1278	881	907	843	895	715	898	908	863
570	682	667	820	923	836	1169	964	1148	1569

8.2 MEASURES OF CENTRAL TENDENCY

Given a set of data, we often find it convenient to consider some number as being typical of the data. For example, suppose the data are the test scores of 25 students. We could find the average of the 25 test scores and consider this to be the typical test score, the one which best represents all of them. Alternatively, we could arrange the scores in order, from lowest to highest, and pick out the one in the middle as the typical one. Or, if one score occurs more often than any other, we could consider it to be the typical score. These measures of typicality are called **measures of central tendency.** In this section we discuss measures of central tendency and their uses. Before we begin, however, we must discuss the use of the summation symbol Σ.

Summation Notation

The **summation symbol** Σ is the Greek capital letter sigma. It is used in mathematics as shorthand for "sum up what follows." Thus, if $x_1, x_2, \ldots, x_n$ are numbers, then by definition,

$$\sum_{i=1}^{n} x_i = x_1 + x_2 + \cdots + x_n.$$

The left-hand side of the preceding identity is read as "the sum of x sub i from i equal 1 to n." Note that the sum is formed by setting i equal to 1, then 2, then 3, and so on, up to n, and adding the resulting terms. Where there is no possibility of confusion, we shall write $\Sigma\ x$ for $\Sigma_{i=1}^{n} x_i$.

EXAMPLE 1 Let $x_1 = 2$, $x_2 = 5$, $x_3 = 0$, $x_4 = 6$, and $x_5 = -1$. Then

$$\sum_{i=1}^{5} x_i = x_1 + x_2 + x_3 + x_4 + x_5 = 2 + 5 + 0 + 6 + (-1) = 12,$$

$$\sum_{i=1}^{4} x_i = x_1 + x_2 + x_3 + x_4 = 2 + 5 + 0 + 6 = 13,$$

$$\sum_{i=1}^{3} x_i = x_1 + x_2 + x_3 = 2 + 5 + 0 = 7,$$

$$\sum_{i=1}^{2} x_i = x_1 + x_2 = 2 + 5 = 7,$$

and

$$\sum_{i=1}^{1} x_i = x_1 = 2. \quad \square$$

EXAMPLE 2 Consider the summation

$$\sum_{i=1}^{4} i^2.$$

We have

$$\sum_{i=1}^{4} i^2 = 1^2 + 2^2 + 3^2 + 4^2 = 1 + 4 + 9 + 16 = 30.$$

Similarly,

$$\sum_{j=0}^{3} 2(j+1) = 2(0+1) + 2(1+1) + 2(2+1) + 2(3+1) = 20.$$

(Note that we could have factored the 2 out of this summation before evaluating it:

$$\sum_{j=0}^{3} 2(j+1) = 2\sum_{j=0}^{3} (j+1) = 2[(0+1) + (1+1) + (2+1) + (3+1)]$$

$$= 2[10] = 20.)$$

Also,

$$\sum_{k=3}^{5} \frac{k+1}{k} = \frac{3+1}{3} + \frac{4+1}{4} + \frac{5+1}{5} = \frac{4}{3} + \frac{5}{4} + \frac{6}{5} = \frac{227}{60}. \quad \square$$

EXAMPLE 3   Note that

$$\sum_{i=1}^{6} 3 = \underbrace{3 + 3 + 3 + 3 + 3 + 3}_{6 \text{ times}} = 6(3) = 18.$$

In general, if a is any number and n is a positive integer, then

$$\sum_{i=1}^{n} a = na. \quad \square$$

The Mean

The most useful measure of central tendency for most purposes is the **arithmetic mean,** or **arithmetic average.** As the second of these names implies, the arithmetic mean is just the common average with which we are familiar. However, it is necessary here and throughout the remainder of this chapter to make a distinction between data that form a population and data that form a sample. Recall that a population consists of everything we are interested in at the moment, while a sample is only part of a population.

> **The Mean**
>
> Let a set of data consist of the numbers $x_1, x_2, \ldots, x_n$.
>
> 1. If the data form a population, the **arithmetic mean** of the population is
>
> $$\mu = \frac{\sum_{i=1}^{n} x_i}{n}.$$
>
> (μ is the Greek letter mu, pronounced "mew.")
>
> 2. If the data form a sample, the **arithmetic mean** of the sample is
>
> $$\bar{x} = \frac{\sum_{i=1}^{n} x_i}{n}.$$
>
> (The symbol $\bar{x}$ is read "x bar.")

Whether the data form a population or a sample, note that the mean is found the same way: by adding the values of the data and dividing the sum by the number of pieces of data. It is only the notation that differs for populations and samples.

Before we present an example, we must add a word about rounding answers. In general, it is a good idea to report answers to one more significant digit than that of the original data. This means that if the data are given to the nearest whole number, then a figure calculated from them, such as the mean, should be reported to the nearest tenth (0.1); if the data are given to the nearest tenth, then the answer should be reported to the nearest hundredth (0.01), and so on. In the rounding-off process, digits that are 5 or higher are rounded up, and those from 0 through 4 are rounded down.

EXAMPLE 4 A sample of eight grades received on a mathematics test is as follows:

$$71, \quad 86, \quad 38, \quad 67, \quad 92, \quad 84, \quad 77, \quad 78.$$

Find the mean grade.

Since the data form a sample, the mean is denoted by $\bar{x}$. Hence the mean grade is

$$\bar{x} = \frac{71 + 86 + 38 + 67 + 92 + 84 + 77 + 78}{8}$$

$$= \frac{593}{8} = 74.125.$$

The data were given to the nearest whole number, so we report the mean to the nearest tenth. Therefore, we report the mean grade as

$$\bar{x} = 74.1. \quad \square$$

STATISTICS

EXAMPLE 5

A village in New Hampshire has a population of six voters. Their ages in years are as follows:

$$32.7, \quad 42.6, \quad 55.9, \quad 48.6, \quad 72.4, \quad 66.5.$$

Find the mean age.

Since the data form a population, the mean is denoted by μ. Hence the mean age is

$$\mu = \frac{32.7 + 42.6 + 55.9 + 48.6 + 72.4 + 66.5}{6}$$

$$= \frac{318.7}{6} = 53.116667.$$

The data were given to the nearest tenth, so we report the mean to the nearest hundredth. Therefore, we report the mean age in years as

$$\mu = 53.12. \quad \square$$

Sometimes it is necessary to estimate the mean of data that have been grouped into a frequency distribution.

The Mean for Grouped Data

To estimate the mean of data that have been grouped into a frequency distribution, let m denote the class mark (midpoint) of each class and f the frequency of the class. Then the mean is approximately equal to

$$\frac{\Sigma \, fm}{n},$$

where n is the number of pieces of data and the sum is taken over all classes of the frequency distribution.

Note that the mean estimated from a frequency distribution is only an approximation of the true mean since it is obtained by letting the midpoint m of each class represent all the data in the class. The class mark m of a class may be found by using the formula

$$m = (\tfrac{1}{2})(\text{lower limit of class} + \text{upper limit of class}).$$

EXAMPLE 6

A population of 300 students entering a certain college had the following scores on the mathematics portion of their Scholastic Aptitude Tests (SATs):

Math SAT Score	Number of Students
250–299	10
300–399	74
400–499	96
500–599	102
600–649	18
	300

Estimate the mean score.

Since the data form a population, the mean is denoted by μ. In Table 8–6 we expand the frequency distribution table to include the midpoint m of each class and the product fm for each class. Hence the mean score is given by

$$\mu \approx \frac{\Sigma fm}{n} = \frac{139{,}050}{300} = 463.5. \quad \square$$

TABLE 8–6

Class	Frequency f	Midpoint m	fm
250–299	10	274.5	2,745
300–399	74	349.5	25,863
400–499	96	449.5	43,152
500–599	102	549.5	56,049
600–649	18	624.5	11,241
	300		139,050

The Median

The **median** is another measure of central tendency; it locates the middle of the data when they are arranged in increasing order.

The Median

Let a set of data have n entries, arranged in increasing order.

1. If n is odd, the median is the middle entry.

2. If n is even, the median is the mean of the two entries nearest the middle.

EXAMPLE 7   Find the median of the following data:

$$26, \quad 82, \quad 43, \quad 19, \quad 84, \quad 17, \quad 3.$$

Here $n = 7$ is odd; arranging the data in increasing order, we have

$$3, \quad 17, \quad 19, \quad \textcircled{26}, \quad 43, \quad 82, \quad 84.$$

The middle entry is circled. Hence the median is 26. $\square$

EXAMPLE 8   Find the median of the following data:

$$114, \quad 426, \quad 897, \quad 36, \quad 520, \quad 0, \quad 91, \quad 44.$$

Here $n = 8$ is even; arranging the data in increasing order, we have

$$0, \quad 36, \quad 44, \quad \boxed{91, \quad 114}, \quad 426, \quad 520, \quad 897.$$

The two entries nearest the middle are boxed. Hence the median is

$$\frac{91 + 114}{2} = 102.5. \quad \square$$

The Mode

The **mode** of a set of data is the entry that occurs most often. A set of data can have several modes if several of its entries each occurs most often. On the other hand, if no entry occurs more often than any other, the set of data has no mode.

EXAMPLE 9   Find the mode for the following data:

    2,   8,   3,   5,   3,   7,   2,   3,   8,   11,   3.

The mode is 3, because this is the entry that occurs most often.

EXAMPLE 10   The set of data

    14,   23,   9,   −3,   17,   13,   6,   19

has no mode.

EXAMPLE 11   The set of data

    102,   86,   14,   29,   102,   31,   76,   14,   99

is bimodal; that is, it has two modes, 14 and 102.

Comparison of the Mean, Median, and Mode

The mean, the median, and the mode are all measures of central tendency, that is, measures of what is typical of a set of data. However, they measure what is typical in different ways: the mean measures the average value, the median measures the middle value, and the mode measures the most frequently occurring value. Each has its advantages and disadvantages, and each is useful in certain circumstances.

The mean is the measure of central tendency that is most often used. It has several advantages: it is familiar, it is relatively easy to find, it always exists for numerical data, and it is unique. The major disadvantage of the mean is that it is influenced by **extreme values,** that is, by values that are very large or very small compared to the rest of the data.

EXAMPLE 12   Five people have the following annual incomes:

    $20,000, $20,000, $20,000, $20,000, $500,000.

The mean income is $116,000, which does not give a true picture of the typical income of the five people. The reason this happens is that the value of the mean is pulled toward the extreme value of $500,000.

The median is often used as a measure of central tendency. Its advantages are that it always exists for numerical data, it is unique, and it is not much influenced by extreme values. For instance, the median income of Example 12 is $20,000. If the mean is an inappropriate measure to use because of the influence of extreme values, the median may be more appropriate. This is the case in Example 12. The median also has the property that it can be found for nonnumerical data.

EXAMPLE 13

Five consumers rank a product. The rankings are

poor, fair, good, good, excellent.

There is no mean ranking, but the median ranking is good. ☐

The major disadvantage of the median is that, if the set of data is a large one, the median will be difficult to find, for it is quite difficult to arrange a large collection of data in increasing order.

The mode is the measure of central tendency that is least often used. Its advantages are that it is easy to find and that it can be applied to nonnumerical data. For instance, in Example 13 the modal ranking is good. The mode may be the appropriate measure to use if both the mean and median are inappropriate.

EXAMPLE 14

Nine students take a test. Their scores are

70, 70, 70, 70, 85, 90, 90, 90, 90.

The mean score is 80.6, the median score 85. Neither of these is very typical of the data. There are two modal scores, 70 and 90, and it probably would be best to consider these to be the typical scores on the test. ☐

The disadvantages of the mode are that it may not exist and that it may not be unique if it does exist.

■ *Computer Manual: Programs DATUGRPD, DATAGRPD*

8.2 EXERCISES

Summation Notation

In Exercises 1 through 4, evaluate each sum given that $x_1 = 8$, $x_2 = 4$, $x_3 = -3$, $x_4 = 10$, $x_5 = 0$, and $x_6 = 9$.

1. $\sum_{i=1}^{6} x_i$
2. $\sum_{i=1}^{3} x_i$
3. $\sum_{i=1}^{5} 2x_i$
4. $\sum_{i=1}^{4} 3x_i$

In Exercises 5 through 13, evaluate the given sum.

5. $\sum_{i=1}^{5} i$
6. $\sum_{i=2}^{5} 3i$
7. $\sum_{j=0}^{6} 4(j + 3)$
8. $\sum_{k=2}^{6} k(k + 2)$
9. $\sum_{i=1}^{3} i(2i + 1)$
10. $\sum_{i=1}^{5} \frac{i}{i + 1}$
11. $\sum_{j=0}^{3} \frac{j - 1}{j + 1}$
12. $\sum_{i=1}^{3} 4$
13. $\sum_{i=1}^{100} \frac{1}{2}$

The Mean

In Exercises 14 through 19, find the mean of the given data.

14. the sample 2, 18, 12, 4, 6
15. the sample 2.8, 3.2, 7.1, 0.2, 8.1, 4.6, 5.2

STATISTICS

16. the sample 243, 246, 197, 222, 261, 219, 225, 250
17. the population 42.4, 56.2, 58.0, 72.1, 68.3, 73.5
18. the population 500, 1700, 600, 300, 2200, 1200, 1400, 900
19. the population 1.05, 1.08, 2.97, 3.62, 1.45, 2.62

In Exercises 20 through 23, estimate the mean from the given frequency distribution.

20. The data form a sample.

Class	Frequency
0–4	2
5–9	5
10–14	12
15–19	6
20–24	1

21. The data form a sample.

Class	Frequency
10–24	10
25–49	30
50–74	50
75–99	40
100–114	20
115–124	10

22. The data form a population.

Class	Frequency
100.0–101.4	8
101.5–102.9	11
103.0–104.4	23
104.5–105.9	2
106.0–107.4	15

23. The data form a population.

Class	Frequency
0.00–4.99	8
5.00–9.99	14
10.00–12.49	62
12.50–14.99	102
15.00–17.49	105
17.50–19.99	47
20.00–24.99	11
25.00–29.99	4

Management

24. A firm's annual sales, in thousands of dollars, for the past 10 years are as follows:
105, 210, 322, 330, 362, 368, 420, 425, 435, 440.
Find the firm's mean annual sales over the past 10 years.

25. Eleven small businesses were sampled to find the annual salary drawn by their owners. The results, in thousands of dollars per year, were as follows:
38.1, 42.4, 26.6, 19.0, 40.2, 32.5, 24.7, 20.1, 28.5, 25.0, 27.2.
Find the mean annual salary of the owners.

26. A survey of 50 retailers who sell by direct mail revealed the following (the amounts spent are in millions of dollars):

Amount Spent on Direct Mail	Number of Retailers
0.00–0.99	2
1.00–2.49	6
2.50–4.99	21
5.00–9.99	5
10.00–19.99	12
20.00–29.99	3
30.00–39.99	1

Estimate the mean amount spent on direct mail by the retailers.

27. According to an article in the October 4, 1988, edition of *The Wall Street Journal*, for companies having fewer than 5000 employees the average sales per employee was as follows:

Size of Company (Number of Employees)	Sales per Employee (Thousands of Dollars)
1–4	112
5–19	128
20–99	127
100–499	118
500–4999	120

Estimate the average sales per employee for all firms having fewer than 5000 employees.

28. A sample of 200 employees of Beta Corporation revealed the following:

Annual Salary ($)	Number of Employees
10,000–14,999	86
15,000–19,999	73
20,000–24,999	22
25,000–29,999	12
30,000–39,999	6
40,000–59,999	1

(a) Estimate the mean annual salary of the 200 employees.
(b) Assuming that this sample of 200 is typical of all employees of Beta Corporation, estimate the total annual amount paid in salary by the corporation if it has 5000 employees.

Life Science

29. The annual yield from an apple tree over the past eight years, in bushels of apples, has been as follows:
    16.5,   12.4,   12.8,   13.2,   11.7,   11.2,   10.9,   10.5.
Find the mean annual yield of the tree over the last eight years.

30. The size of the gypsy moth population, in millions, in a certain region for the past 15 years has been as follows:

    22.85,   6.02,   3.51,   2.22,   8.41,
    18.68,   4.33,   4.75,   6.13,   5.05,
    3.29,   2.42,   20.95,   8.45,   2.72.

Find the mean size of the gypsy moth population in the region over the past 15 years.

31. The following table shows the number of years 200 patients survived after their first positive test for AIDS:

Years Survived	Number of Patients
0.0–0.9	12
1.0–1.9	21
2.0–2.9	28
3.0–3.9	65
4.0–4.9	42
5.0–5.9	20
6.0–6.9	12

Estimate the mean survival time for the patients.

Social Science

32. A sample of 10 people showed their years of formal education to be as follows:

15, 12, 16, 9, 17, 16, 12, 12, 12, 11.

Find the mean years of formal education for this sample.

33. A sample of 20 people ranked a governor's performance in office on a scale from 1 (very poor) to 5 (excellent). The results were the following set of rankings:

2, 1, 2, 2, 4, 2, 1, 3, 5, 3,
3, 2, 2, 2, 3, 5, 4, 2, 1, 2.

Find the mean ranking.

34. The following table gives the number of weeks of unemployment suffered by 100 wage earners who were unemployed last year:

Weeks of Unemployment	Number of Unemployed
0–2	82
3–5	7
6–8	5
9–11	3
12–14	2
15–17	1

Estimate the mean number of weeks of unemployment suffered by the wage earners.

35. A pyschologist has developed a new intelligence test. In order for the new test to be comparable to existing IQ tests, the mean score on the test should be 100. The new test has been tried on some volunteers with the following results:

Score	Number Who Had This Score
80–89	6
90–99	18
100–109	34
110–119	22
120–129	22
130–139	10

What does this indicate about the new test?

An article by Lester Thurow ("A Surge in Inequality," *Scientific American,* May 1987) presented the following data about the net worth of U.S. families:

	Percentage of Families		
Net Worth ($)	1970	1977	1983
0–4,999	38	35	33
5,000–9,999	6	5	5
10,000–24,999	14	13	12
25,000–49,999	17	17	16
50,000–99,999	13	15	17
100,000–249,999	9	12	12
250,000–499,999	2	2	3
500,000 or more	1	1	2

Exercises 36 and 37 refer to this information.

36. Estimate the mean net worth of U.S. families in each year, using a class mark of $1 million for the last class.
37. Repeat Exercise 36 using a class mark of $10 million for the last class.

The Median

In Exercises 38 through 43, find the median.
38. 13, 9, 6, 23, 18, 4, 7, 11, 6
39. 202, 461, 196, 327, 487, 306
40. 1.35, 2.07, 8.65, 4.21, 0.09, 2.77, 8.21
41. 22, 14, 16, 22, 18, 17, 16, 22, 18, 13, 25
42. 42, 119, 36, 48, 97, 102, 77, 55
43. 2412, 6319, 4200, 8635, 6000, 3398, 5417, 7750

Management

44. Find the median sales for the firm of Exercise 24.
45. Find the median salary for the owners of Exercise 25.

Life Science

46. Find the median yield for the apple tree of Exercise 29.
47. Find the median size of the gypsy moth population of Exercise 30.

Social Science

48. Find the median years of education for the sample of Exercise 32.
49. Find the median ranking for the sample of Exercise 33.

The Mode

In Exercises 50 through 53 find the mode.
50. 2, 5, 3, 7, 5, 8, 3, 5, 4
51. 12, 16, 5, 3, 9, 6, 2, 18, 14
52. 400, 700, 600, 400, 800, 700, 400, 500, 700, 900
53. 50, 73, 62, 50, 86, 50, 44, 73, 62, 73, 49, 62
54. According to *The Boston Globe,* at a recent time the 10 largest money market funds were paying the following annualized interest rates:

    7.2%, 7.4%, 7.0%, 7.2%, 7.0%, 6.9%, 7.1%, 7.2%, 7.0%, 7.3%.

    Find the modal interest rate of the funds.

Social Science

55. Find the modal number of years of formal education for the sample of Exercise 32.
56. Find the modal ranking for the sample of Exercise 33.

Comparison of the Mean, Median, and Mode

In Exercises 57 through 62, state which measure of central tendency you think should be used.

STATISTICS

57. Annual salaries of six professors: $39,100, $39,700, $38,600, $40,000, $39,500, $58,400.

58. Annual salaries of six executives of a firm: $165,000, $165,000, $163,000, $167,000, $164,500, $166,000.

59. Annual salaries of seven Acme company employees: $21,000, $16,000, $42,000, $16,000, $42,000, $42,000, $16,000.

60. Eight student evaluations of an instructor: poor, average, average, excellent, poor, average, excellent, average.

61. Preferences of nine consumers as to brand of soap: Brand X, Brand Y, Brand X, Brand Z, Brand Y, Brand Z, Brand X, Brand Y, Brand X.

62. Eleven test scores: 74, 80, 83, 90, 80, 77, 70, 86, 83, 80, 77.

Management

63. For the past six years, Alpha Company's annual sales, in millions of dollars, have been

    46.2,   48.8,   55.7,   59.3,   62.4,   66.0.

For the same six years, Alpha's profits, as a percentage of sales, have been

    10.2,   10.1,   6.3,   10.4,   10.3,   10.2.

(a) Find Alpha's mean and median annual sales for the past six years. Which of these measures of central tendency would you prefer to use to describe Alpha's sales over the past six years?

(b) Find Alpha's mean and median profit as a percentage of sales over the past six years. Which of these measures of central tendency would you prefer to use to describe Alpha's profits?

64. In an article in the September 26, 1988, edition of *The Wall Street Journal*, 16 economists gave their forecasts for the percentage growth in the U.S. Gross National Product for the fourth quarter of 1988. The forecasts were as follows:

    3.1%, 5.7%, 3.0%, 3.0%, 2.5%, 2.6%, 2.6%, 3.0%,
    3.0%, 3.0%, 2.5%, 3.0%, 2.0%, 3.3%, 2.5%, 1.4%.

Find the mean, median, and modal forecasts. Which would you prefer to use as the most typical forecast?

Life Science

65. A nutritionist is conducting weight-loss experiments using two different diets. Using the first diet, eight subjects had weight losses (in kilograms) of

    10.2,   13.4,   17.6,   15.5,   7.6,   20.3,   44.7,   and   22.3.

Using the second diet, seven subjects had weight losses (in kilograms) of

    22.7,   12.3,   18.8,   8.7,   22.6,   22.0,   and   22.4.

(a) Find the mean and median weight loss for those on each of the diets.
(b) Which diet is more effective, in your opinion?

Social Science

66. Voters are asked to rate the performance of the governor of their state on a 5-point scale. The results are as follows:

Verbal Rating:	Very Poor	Poor	Average	Good	Excellent
Numerical Rating:	1	2	3	4	5
Number of Voters:	60	130	80	50	40

(a) Find the mean numerical rating.
(b) Find the median and modal verbal ratings.
(c) What rating best characterizes the response of the voters?

Estimating the Median and Mode

It is possible to estimate the median and the mode for data that have been grouped into a frequency distribution. Define the **modal class** of a frequency distribution to be the class with the largest frequency. The estimate for the mode of the data is then the midpoint of the modal class. For the median, suppose there are n pieces of data in the table (that is, that the sum of the frequency column is n) and define the **median class** to be the class in which the $(n/2)$nd datum lies. Then the median is approximately equal to

$$L + \frac{0.5n - C}{f} w,$$

where L is the lower boundary of the median class, w is the width of the median class, f is the frequency of the median class, and C is the cumulative frequency of the class prior to the median class. Exercises 67 through 69 make use of these facts.

*67. Estimate the median and modal amount spent by retailers on direct mail by using the data of Exercise 26.

*68. Estimate the median and modal years of survival for the AIDS patients of Exercise 31.

*69. Estimate the median and modal scores for the IQ test of Exercise 35.

Likert Scales

(Suggested by material in Emory, *Business Research Methods,* third edition, Irwin, 1985.) In an attempt to find out how strongly voters feel about a political issue, they are read a quote concerning the issue and asked to respond to it using the following Likert scale:

1	2	3	4	5
strongly disagree	disagree	undecided, don't care	agree	strongly agree

The results are as follows:

Response	Number of Voters
Strongly Disagree	140
Disagree	230
Undecided, Don't Care	300
Agree	210
Strongly Agree	120
	1000

When dealing with responses such as these, where many of the respondents are undecided or do not care, the analyst may be more interested in the relative strengths of the extreme opinions than in the overall opinion. One way to examine the strengths of the extremes is to select, say, the top 25% and bottom 25% of the responses, and treat them separately:

Top 25%:

Response	Number of Voters
Agree	130
Strongly Agree	120
	250

Bottom 25%:

Response	Number of Voters
Strongly Disagree	140
Disagree	110
	250

Now let $\bar{x}$ be the mean response of all the voters, $\bar{x}_T$ the mean response of the top 25% of the voters, and $\bar{x}_B$ the mean response of the bottom 25% of the voters. If these means are relatively far apart, there are many who feel strongly about the issue, and polarization is possible; if the means are relatively close together, few feel strongly about the issue, and polarization is unlikely.

*70. Estimate $\bar{x}$, $\bar{x}_T$, and $\bar{x}_B$ for the sample of 1000 voters. (*Hint:* The class marks are the numbers 1, 2, 3, 4, and 5.) Do many people feel strongly about the issue addressed by the quotation?

*71. Suppose the responses of 500 voters to another question are as follows:

Response	Number of Voters
Strongly Disagree	5
Disagree	25
Undecided, Don't Care	435
Agree	28
Strongly Agree	7
	500

Using the top 25% and bottom 25% of the sample, estimate $\bar{x}$, $\bar{x}_T$, and $\bar{x}_B$. Do people feel as strongly about this issue as they do about the one addressed by Exercise 70?

8.3 MEASURES OF DISPERSION

Measures of central tendency do not tell us all we need to know about a collection of data. Consider, for example, the two collections

$$2, \quad 4, \quad 6, \quad 6, \quad 6, \quad 8, \quad 10$$

and

$$5, \quad 6, \quad 6, \quad 6, \quad 6, \quad 6, \quad 7.$$

Both have a mean of 6, a median of 6, and a mode of 6. Yet the two collections are clearly different in that the first is more "spread out" than the second. We need some measures that will tell us how spread out a collection of data is, or, as it is more commonly stated, how **dispersed** a collection of data is. In this section we study **measures of dispersion.**

The Range

We have already encountered one measure of dispersion, and that is the **range.** Recall that the range of a collection of data is defined as follows:

The Range

The **range** of a set of data is defined by

Range = Largest value − Smallest value.

EXAMPLE 1 Find the range of the following data:

18, 6, 14, 27, 12, 44, 38, 33.

The range is 44 − 6 = 38. ◻

For numerical data, the range always exists and is unique, but it only tells us how far apart the extreme values of the data are and does not tell us anything about how the values in between the extremes are dispersed. For example, the collections of data

2, 7, 7, 7, 12

and

2, 3, 4, 5, 6, 7, 8, 9, 10, 11, 12

both have range 12 − 2 = 10, but clearly, the values between the extremes of 2 and 12 are dispersed differently in the two cases. This demonstrates the necessity for developing better measures of dispersion than the range.

The Variance and Standard Deviation

Consider the population

2, 5, 9, 13, 16.

We seek to measure the dispersion of this population by comparing its elements with its mean, which is $\mu = 9$. Suppose we begin by asking ourselves how far from the mean each element is. If we use x to label the data, then $x - \mu$ measures the distance from element x to mean μ:

x:	2	5	9	13	16
$x - \mu = x - 9$:	−7	−4	0	4	7

Note that the sum of the values of $x - \mu$ is zero:

$$\Sigma(x - \mu) = 0.$$

This is true for any collection of data: if we take the sum of the distances from the mean to each entry, the result will be zero. Therefore, we cannot use $\Sigma(x - \mu)$ as a measure of dispersion. However, if we square each distance $x - \mu$, we eliminate the minus signs, and the sum $\Sigma(x - \mu)^2$ will not be zero unless all the data are equal:

x	$x - \mu = x - 9$	$(x - \mu)^2 = (x - 9)^2$
2	-7	49
5	-4	16
9	0	0
13	4	16
16	7	49
45		130

Note that the sum $\Sigma(x - \mu)^2 = 130 \neq 0$.

Obviously, the larger the value of $\Sigma(x - \mu)^2$, the further from the mean the data are. Therefore, the sum $\Sigma(x - \mu)^2$ measures the dispersion of the data from the mean. If we divide $\Sigma(x - \mu)^2$ by the number n of pieces of data, we produce an average that measures the distance of a typical piece of data from the mean. This average

$$\frac{\Sigma(x - \mu)^2}{n}$$

is called the **variance** of the population and is denoted by σ^2 (read "sigma squared"). For our example, the variance is

$$\sigma^2 = \frac{\Sigma(x - \mu)^2}{n} = \frac{130}{5} = 26.$$

Unfortunately, the variance is in squared units because of the presence in it of the terms $(x - \mu)^2$. Thus, for example, if the data are in dollars, then the variance σ^2 is in square dollars, which makes no sense. To return to the original units of the data, we must take the square root of the variance, thus obtaining the **standard deviation** σ (sigma). For our example, the standard deviation is

$$\sigma = \sqrt{\sigma^2} = \sqrt{26} \approx 5.1.$$

The standard deviation of a population measures the dispersion of the population from the mean by measuring the average distance from the mean of a typical element of the population, in the original units of the data

We have illustrated the concept of the standard deviation of a population. The standard deviation of a sample is not arrived at in quite the same way.

The Variance and Standard Deviation

Let a collection of data consist of the numbers

$$x_1, x_2, \ldots, x_n.$$

1. If the data form a population, its variance σ^2 is given by

$$\sigma^2 = \frac{\sum_{i=1}^{n} (x_i - \mu)^2}{n},$$

where μ is the mean of the population, and its standard deviation σ is given by

$$\sigma = \sqrt{\sigma^2}.$$

2. If the data form a sample, its variance s^2 is given by

$$s^2 = \frac{\sum_{i=1}^{n} (x_i - \bar{x})^2}{n-1},$$

where $\bar{x}$ is the mean of the sample, and its standard deviation s is given by

$$s = \sqrt{s^2}.$$

Note that when finding the variance of a population, we divide by n, the number of pieces of data in the population, whereas when finding the variance of a sample, we divide by $n - 1$, one less than the number of pieces of data in the sample. Therefore we must know whether the data form a population or a sample. (The discrepancy is due to the fact that the sample variance is often used as an estimate for the variance of the population from which the sample comes, and dividing by $n - 1$ rather than n makes this estimate more accurate.)

EXAMPLE 2 Consider the following data:

$$15, \quad 12, \quad 19, \quad 13, \quad 16, \quad 12.$$

Suppose the data form a population. Let us find its variance and standard deviation.

We set up our calculations in a table:

x	$x - \mu$	$(x - \mu)^2$
15	0.5	0.25
12	−2.5	6.25
19	4.5	20.25
13	−1.5	2.25
16	1.5	2.25
12	−2.5	6.25
87		37.50

STATISTICS

The mean of the population is

$$\mu = \frac{87}{6} = 14.5$$

Therefore,

$$x - \mu = x - 14.5$$

for each x. We have

$$\Sigma(x - \mu)^2 = 37.5,$$

so the variance of the population is

$$\sigma^2 = \frac{\Sigma(x - \mu)^2}{n} = \frac{37.5}{6} = 6.25.$$

Its standard deviation is

$$\sigma = \sqrt{\sigma^2} = \sqrt{6.25} = 2.5. \quad \square$$

EXAMPLE 3   Let us now find the variance and standard deviation of the data of Example 2, assuming that they form a sample.

We have

x	$x - \bar{x}$	$(x - \bar{x})^2$
15	0.5	0.25
12	−2.5	6.25
19	4.5	20.25
13	−1.5	2.25
16	1.5	2.25
12	−2.5	6.25
87		37.50

since the sample mean is

$$\bar{x} = \frac{87}{6} = 14.5$$

and

$$x - \bar{x} = x - 14.5$$

for each x. The variance of the sample is thus

$$s^2 = \frac{\Sigma(x - \bar{x})^2}{n - 1} = \frac{37.5}{5} = 7.5.$$

Its standard deviation is

$$s = \sqrt{s^2} = \sqrt{7.5} \approx 2.74. \quad \square$$

The formulas we have given for the variance and standard deviation require that we first find the mean, μ or $\bar{x}$. However, there are shortcut formulas that allow us to find the variance without finding the mean first.

> **Shortcut Formulas for the Variance**
>
> For a population:
>
> $$\sigma^2 = \frac{\Sigma x^2 - \frac{(\Sigma x)^2}{n}}{n}.$$
>
> For a sample:
>
> $$s^2 = \frac{\Sigma x^2 - \frac{(\Sigma x)^2}{n}}{n - 1}.$$
>
> Here n is the number of pieces of data.

EXAMPLE 4 Consider again the data of Examples 2 and 3:

$$15, \quad 12, \quad 19, \quad 13, \quad 16, \quad 12.$$

We have

x	x^2
15	255
12	144
19	361
13	169
16	256
12	144
87	1299

Therefore,

$$\Sigma x^2 = 1299 \quad \text{and} \quad (\Sigma x)^2 = (87)^2 = 7569.$$

Hence if the data form a population,

$$\sigma^2 = \frac{1299 - \frac{7569}{6}}{6} = 6.25.$$

If the data form a sample,

$$s^2 = \frac{1299 - \frac{7569}{6}}{5} = 7.5. \quad \square$$

Sometimes it is necessary to estimate the variance and standard deviation for data grouped into a frequency distribution.

The Variance and Standard Deviation for Grouped Data

To estimate the variance and standard deviation for data grouped into a frequency distribution, let m denote the midpoint of each class, f the frequency of the class, and n the number of pieces of data.

1. If the data form a population,

$$\sigma^2 \approx \frac{\Sigma f(m - \mu)^2}{n} \quad \text{and} \quad \sigma \approx \sqrt{\sigma^2}.$$

2. If the data form a sample,

$$s^2 \approx \frac{\Sigma f(m - \bar{x})^2}{n - 1} \quad \text{and} \quad s \approx \sqrt{s^2}.$$

EXAMPLE 5 Estimate the variance and standard deviation for the population data grouped as follows:

Class	Frequency
0–9	5
10–19	10
20–29	15
30–39	7
40–49	3
	40

We must first estimate the mean μ, then use it to estimate the variance σ^2. We make Table 8–7.

TABLE 8–7

Class	f	m	fm	$m - \mu$	$(m - \mu)^2$	$f(m - \mu)^2$
0–9	5	4.5	22.5	−18.3	334.89	1674.45
10–19	10	14.5	145.0	−8.3	68.89	688.90
20–29	15	24.5	367.5	1.7	2.89	43.35
30–39	7	34.5	241.5	11.7	136.89	958.23
40–49	3	44.5	133.5	21.7	470.89	1412.67
	40		910.0			4777.60

Since

$$\mu \approx \frac{\Sigma fm}{n} = \frac{910.0}{40} = 22.8,$$

every

$$m - \mu = m - 22.8.$$

Therefore,

$$\sigma^2 \approx \frac{\Sigma f(m - \mu)^2}{n} = \frac{4777.60}{40} = 119.44$$

and

$$\sigma \approx \sqrt{119.44} = 10.93. \quad \square$$

Suppose we have two sets of data, one with mean $\mu = 100$ and standard deviation $\sigma = 20$ and the other with mean $\mu = 10$ and standard deviation $\sigma = 3$. Which set has the more variable data? This question can be answered with the aid of the **coefficient of variation.**

The Coefficient of Variation

The **coefficient of variation** V of a data set is defined by

$$V = \frac{\text{Standard deviation}}{\text{Mean}} 100.$$

The coefficient of variation V expresses dispersion a a percentage of the mean, and thus measures the variability of the data; the larger V is, the more variable the data. The coefficient of variation is useful for comparing data sets to see which one is more variable.

EXAMPLE 6   Suppose population A has mean $\mu_A = 100$ and standard deviation $\sigma_A = 20$, while population B has mean $\mu_B = 10$ and standard deviation $\sigma_B = 3$. Then their coefficients of variation are

$$V_A = \frac{20}{100} 100 = 20 \quad \text{and} \quad V_B = \frac{3}{10} 100 = 30,$$

and hence population B exhibits more variability than population A. $\quad \square$

■ *Computer Manual: Programs DATUGRPD, DATAGRPD*

8.3 EXERCISES

The Range

In Exercises 1 through 4, find the range for each collection of data.

1. 12, 6, 3, 15, 27, 19, 82, 13
2. 8.6, 4.2, 9.7, 13.1, 5.2, 6.8, 14.0, 9.3, 8.1, 5.7
3. 9.2, 8.7, 9.8, 9.2, 9.0, 9.8, 9.1, 9.8, 9.0, 9.8, 9.2
4. 60.4, 66.3, 58.7, 58.8, 61.1, 58.7, 58.9, 62.4, 101.8, 58.4

The Variance and Standard Deviation

In Exercises 5 through 10, find the variance and standard deviation.

5. the population 4, 3, 0, 1, 5, 7, 2, 9
6. the population 26, 41, 13, 24, 37, 39
7. the sample 12, 12, 12, 12
8. the sample 18.2, 16.5, 21.3, 17.4, 18.6, 16.5
9. the population 18.2, 16.5, 21.3, 17.4, 18.6, 16.5
10. the sample 20, 33, 14, 17, 25, 19, 24, 30

In Exercises 11 through 14, estimate the variance and standard deviation from the given frequency distribution.

11. The data form a sample.

Class	Frequency
0–4	3
5–9	6
10–14	9
15–19	6
20–24	3

12. The data form a sample.

Class	Frequency
10–19	1
20–29	4
30–39	10
40–59	12
60–79	6
80–99	2

13. The data form a population.

Class	Frequency
5–7	3
8–10	5
11–15	8
16–20	6
21–30	4
31–40	2

14. The data form a population.

Class	Frequency
0.0–0.9	10
1.0–1.9	20
2.0–2.9	40
3.0–3.9	30
4.0–4.9	5

In Exercises 15 through 18, find the coefficient of variation for the given data sets and state which set is more variable.

15. population A: 20, 36, 29, 42, 37, 48
    population B: 14, 26, 19, 37, 41, 13
16. population A: 2, 4, 6, 8, 10, 12, 14, 16
    population B: 2, 5, 9, 9, 9, 9, 13, 16
17. sample A: 202, 186, 193, 244, 226, 210, 225
    sample B: −6, 18, 14, −9, −12, 26, 40, 8
18. sample A: 4, 1, 3, 7, 7, 5, 9, 2, −1, 0, 4, 8, 6
    sample B: 12, 12, 12, 12, 9, 10, 50
19. Estimate the coefficient of variation for the data of Exercise 11.
20. Estimate the coefficient of variation for the data of Exercise 13.

Management

21. A firm's annual sales for the past eight years, in millions of dollars, were as follows:

8.6, 7.9, 8.8, 9.2, 9.3, 9.8, 10.4, 9.6.

Find the variance and standard deviation of this sample of sales figures.

22. Repeat Exercise 21 if the firm was founded eight years ago, so that the sales figures form a population.

23. Ace Appliance Company was founded six years ago, and since then the firm's annual sales, in thousands of dollars, have been as follows:

20.3, 16.4, 22.8, 27.9, 32.2.

Find the variance and standard deviation of Ace's sales.

24. The annual profits of Ace Appliance Company since the firm's founding, in thousands of dollars, have been as follows:

−14.4, 6.2, 6.8, 7.5, 7.9.

Find the variance and standard deviation of Ace's profits.

25. In an article in the September 26, 1988, edition of *The Wall Street Journal*, 16 economists gave their forecasts for the percentage growth in the U.S. Gross National Product for the fourth quarter of 1988. The forecasts were as follows:

3.1%, 5.7%, 3.0%, 3.0%, 2.5%, 2.6%, 2.6%, 3.0%,
3.0%, 3.0%, 2.5%, 3.0%, 2.0%, 3.3%, 2.5%, 1.4%.

Find the variance and standard deviation of this sample of forecasts.

26. Find the coefficient of variation for the firm of Exercise 21.

27. Find the coefficient of variation for both the sales and profits of the Ace Appliance Company of Exercises 23 and 24. Have Ace's sales or its profits been more variable?

28. A store takes a sample of 100 of its customers' charge accounts and finds the following:

Outstanding Balance ($)	Number of Accounts
0–19	14
20–39	20
40–59	40
60–79	20
80–99	6

Estimate the variance and standard deviation of the outstanding balances.

29. According to an article in the October 4, 1988, edition of *The Wall Street Journal*, for companies having fewer than 5000 employees the average sales per employee was as follows:

Size of Company (Number of Employees)	Sales per Employee (Thousands of Dollars)
1–4	112
5–19	128
20–99	127
100–499	118
500–4999	120

Estimate the variance and standard deviation of average sales per employee for all firms having fewer than 5000 employees.

STATISTICS 511

30. A chain of stores has 50 outlets. Their annual returns on investment (ROIs) for last year were as follows:

ROI	Number of Stores
0.0%–4.9%	6
5.0%–7.9%	14
8.0%–9.9%	20
10.0%–11.9%	4
12.0%–14.9%	6

Estimate the variance and standard deviation ROI for the chain.

31. Two stores owned by the same company have the following sales records, by month:

Store A		Store B	
Sales	Number of Months	Sales	Number of Months
150–153	2	130–134	1
154–157	5	135–139	3
158–161	2	140–144	7
162–165	2	145–149	1
166–169	1		

The sales figures are in thousands of dollars. Estimate the coefficient of variation for each store's sales. Which store exhibits more variability in sales?

The hourly wages of employees at Alpha Company and at Beta Corporation were recorded as follows:

Alpha		Beta	
Hourly Wages	Number of Employees	Hourly Wages	Number of Employees
$5.00–$5.99	20	$5.00–$5.99	83
$6.00–$6.99	35	$6.00–$6.99	42
$7.00–$7.99	48	$7.00–$7.99	59
$8.00–$8.99	167	$8.00–$8.99	38
$9.00–$9.99	55	$9.00–$9.99	79
$10.00–$10.99	25	$10.00–$10.99	49

Exercises 32 through 34 refer to these facts.

32. Estimate the variance and standard deviation of Alpha's hourly wages.

33. Estimate the variance and standard deviation of Beta's hourly wages.

34. Which firm has the more variable wage structure?

Life Science

35. A tribe of rare monkeys was introduced into a new region of the Amazon rain forest 12 years ago. The size of the tribe at the end of each year from then until now has been as follows:

20, 26, 38, 42, 54, 76, 80, 88, 96, 110, 118, 130.

Find the variance and standard deviation of the monkey population.

36. A sample of the air collected from 10 different locations in a city showed ozone concentrations of

48, 16, 82, 54, 58, 21, 73, 72, 42, and 35

parts per million. Find the variance and standard deviation of the ozone concentrations.

37. Find the coefficient of variation for the data of Exercise 36.

38. Two heart patients undergo weekly tests of the level of the drug digitalis in their bloodstream. The results for the past seven weeks were as follows:

| Patient 1: | 9.2 | 10.5 | 11.2 | 11.8 | 10.4 | 11.0 | 11.2 |
| Patient 2: | 9.0 | 11.4 | 10.6 | 11.9 | 13.2 | 12.4 | 10.1 |

Find the coefficient of variation for each patient's digitalis level. Which patient suffers more variability in the digitalis level?

39. The number of trout caught in each of two streams over the past 20 years is as follows:

| Stream A | | Stream B | |
Number Caught	Years	Number Caught	Years
1000–1999	10	700–899	1
2000–2499	6	900–1099	2
2500–2999	1	1100–1299	1
3000–3499	2	1300–1499	7
3500–3999	1	1500–1699	5
		1700–1899	4

Which stream shows more variability in the number of trout caught per year?

Social Science

40. A psychologist ran the same rat through nine different mazes. The times it took the rat to complete the mazes, in seconds, were as follows:

142, 116, 109, 107, 101, 95, 92, 98, 96.

Find the variance and standard deviation of this population of times.

41. A psychologist studying learning administered a test to a sample of 15 subjects. The times they needed to complete the test, in minutes, were as follows:

18, 22, 16, 14, 17, 20, 9, 15, 14, 17, 18, 19, 15, 14, 16.

The test was then modified and given to another sample of 15 subjects, and their times were as follows:

19, 12, 17, 13, 14, 12, 20, 11, 8, 9, 10, 12, 14, 15, 23.

Find the coefficients of variation of the two samples. Which version of the test appears to result in more variability in the times needed to complete it?

42. Students at a school took a standardized reading test. The results were as follows:

STATISTICS

Score	Number of Students
40–44	3
45–49	1
50–54	6
55–59	5
60–64	8
65–69	7
70–74	4
75–79	5
80–84	1

Estimate the variance, standard deviation, and coefficient of variation for the test scores.

43. An article by Lester Thurow ("A Surge in Inequality," *Scientific American,* May 1987) presented the following data about the net worth of U.S. families:

	Percentage of Families		
Net Worth ($)	1970	1977	1983
0–4,999	38	35	33
5,000–9,999	6	5	5
10,000–24,999	14	13	12
25,000–49,999	17	17	16
50,000–99,999	13	15	17
100,000–249,999	9	12	12
250,000–499,999	2	2	3
500,000 or more	1	1	2

Estimate the coefficient of variation for the populations of 1970, 1977, and 1983, and interpret your results. Use a class mark of $1 million for the last class.

Chebyshev's Theorem

A remarkable result known as Chebyshev's Theorem says the following:

> If a population has mean μ and standard deviation σ and an element X is selected at random from the population, then the probability that X lies between $\mu - k\sigma$ and $\mu + k\sigma$ is at least $1 - 1/k^2$:
>
> $$P(\mu - k\sigma \leq X \leq \mu + k\sigma) \geq 1 - \frac{1}{k^2}.$$

Exercises 44 through 67 make use of Chebyshev's Theorem.

Let X be an element selected at random from a population with mean $\mu = 25$ and standard deviation $\sigma = 4$. Exercises 44 through 63 refer to this experiment. In Exercises 44 through 47 find the minimum value of the given probability.

*44. $P(17 \leq X \leq 33)$ ***45.** $P(13 \leq X \leq 37)$

*46. $P(19 \leq X \leq 31)$ ***47.** $P(16 \leq X \leq 34)$

In Exercises 48 through 51 find the maximum value of the given probability.

*48. $P(X < 21 \text{ or } X > 29)$ ***49.** $P(X < 13 \text{ or } X > 37)$

*50. $P(X < 15 \text{ or } X > 35)$ ***51.** $P(X < 12 \text{ or } X > 38)$

In Exercises 52 through 55 find the minimum percentage of the population that must lie in the given interval.

*52. [9, 41] ***53.** [18, 32] ***54.** [20, 30] ***55.** [15, 35]

In Exercises 56 through 59 find the maximum percentage of the population that can lie *outside* the given interval.

*56. [9, 41] ***57.** [16, 34] ***58.** [11, 39] ***59.** [14, 36]

*60. Find the smallest interval in which at least 75% of the population must lie.

*61. Find the smallest interval in which at least 80% of the population must lie.

*62. Find the smallest interval in which at least 90% of the population must lie.

*63. Find the smallest interval in which at least 99% of the population must lie.

Management

*64. The average sale at a discount store is $47.75 with a standard deviation of $7.45. Find the smallest interval such that 90% of the store's sales fall within the interval.

*65. Refer to Exercise 64: Find the smallest interval such that 95% of the store's sales fall within the interval.

Life Science

*66. Suppose condor eggs hatch in an average of 14.5 days with a standard deviation of 2.1 days. Find the smallest time interval such that 80% of all condor eggs will hatch within the interval.

Social Science

*67. Suppose that wage earners who become unemployed remain so for an average of 12.5 weeks with a standard deviation of 3.5 weeks. If a state wants to make sure that its unemployment benefits will last long enough to cover at least 85% of unemployed wage earners, for how long after the beginning of unemployment must the benefits continue?

Investment Risk

Exercises 68 through 71 were suggested by material in Higgins, *Analysis for Financial Management,* Irwin, 1984. They concern investment risk. Investment risk can be measured by dispersion: If all possible returns on an investment are equally likely, its risk is measured by the standard deviation of the returns. Thus, given two such investments, the one whose returns have the largest standard deviation is the riskiest one.

*68. Let investment A have possible returns

$$-\$1000, \quad -\$500, \quad \$0, \quad \$500, \quad \text{and} \quad \$1000,$$

all equally likely. Let investment B have possible returns

$$-\$100, \quad -\$50, \quad \$0, \quad \$50, \quad \text{and} \quad \$100,$$

all equally likely. Find the standard deviation of each investment's returns. Which investment carries the most risk?

If the returns on an investment are not all equally likely, they must be weighted by their probabilities before the mean and standard deviation are calculated. This is done as follows: if the investment has possible returns

STATISTICS

$$r_1, r_2, \ldots, r_n,$$

with probabilities

$$p_1, p_2, \ldots, p_n,$$

then its **mean return** μ is defined to be its expected value:

$$\mu = \sum_{i=1}^{n} p_i r_i.$$

Its standard deviation then is

$$\sigma = \sqrt{\sum_{i=1}^{n} p_i (r_i - \mu)^2}.$$

*69. Find the mean return and standard deviation return for the investment whose returns and their probabilities are as follows:

Return:	−$10,000	−$5000	−$2000	$0	$3000	$8000
Probability:	0.05	0.10	0.20	0.25	0.25	0.15

*70. Repeat Exercise 69 for the investment with

Return:	−$30,000	−$10,000	−$2000	$3000	$15,000	$50,000
Probability:	0.20	0.10	0.10	0.20	0.30	0.10

*71. Which of the investments of Exercises 69 and 70 carries the most risk?

8.4 THE NORMAL DISTRIBUTION

The probability distributions we studied in Sections 6.5 and 6.6 were **discrete distributions,** that is, distributions for random variables that take on **discrete values,** such as 0, 1, 2, 3, ... However, not all probability problems can be handled with the aid of discrete distributions. Measurement problems, for example, might require that the random variable take on infinitely many values that can be arbitrarily close to one another. Suppose, for instance, that we are interested in the weight X of adult male Americans. The variable X can take on infinitely many values, for there are infinitely many possible weights for adult male Americans; furthermore, these weights can be arbitrarily close to one another if the weighing is done accurately enough. Thus, weight X cannot be a discrete random variable. Instead it is a **continuous random variable,** and its probability distribution is a **continuous distribution.**

Continuous probability distributions are described by continuous curves. For example, if X is the weight (in pounds) of adult male Americans, the distribution for X might be that shown in Figure 8–3. The area between the curve and the X-axis tells us the proportion of adult male Americans whose weights lie within any given range. For example, the curve indicates that most adult male Americans have weights near 170 lb. The further away from 170 lb a weight is, the fewer the men who have that weight. Relatively few men weigh more than 230 lb, and even fewer weigh less than 110 lb.

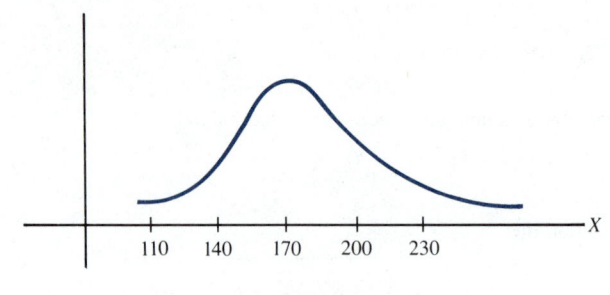

FIGURE 8-3

The Normal Curve

The most commonly used continuous distribution is the **normal distribution.** A random variable X is **normally distributed with mean μ and standard deviation σ** if its probability distribution is a **symmetric, bell-shaped curve** such as that shown in Figure 8–4. This curve is known as the **normal curve.** The normal curve has certain properties.

Properties of the Normal Curve

1. Its peak is located opposite the mean μ.

2. It is symmetric about the mean μ. Thus, the two halves of the curve on either side of μ are mirror images of one another.

3. The total area between the curve and the X-axis is 1. Hence the area under the curve on either side of μ is 0.5.

4. Approximately:
   - 68% of the area under the curve lies between

   $$\mu - \sigma \quad \text{and} \quad \mu + \sigma;$$

   - 95% of the area under the curve lies between

   $$\mu - 2\sigma \quad \text{and} \quad \mu + 2\sigma;$$

   - 99.7% of the area under the curve lies between

   $$\mu - 3\sigma \quad \text{and} \quad \mu + 3\sigma.$$

The symmetry of the normal curve and the percentages of property 4 produce the percentage breakdown of the area under the curve shown in Figure 8–5. The figure gives the approximate percentage areas under the curve over the given subdivision of the X-axis. (See Exercise 49.)

STATISTICS

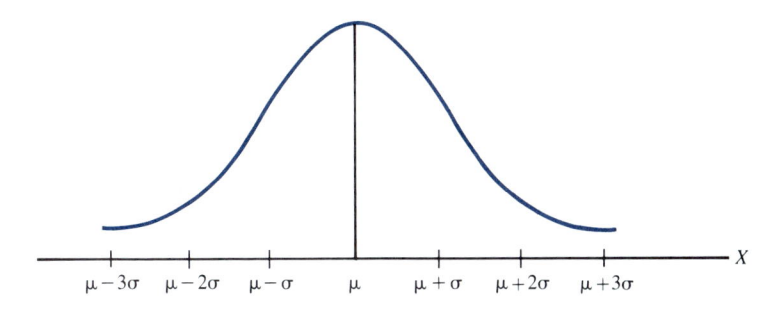

FIGURE 8–4
The Normal Curve

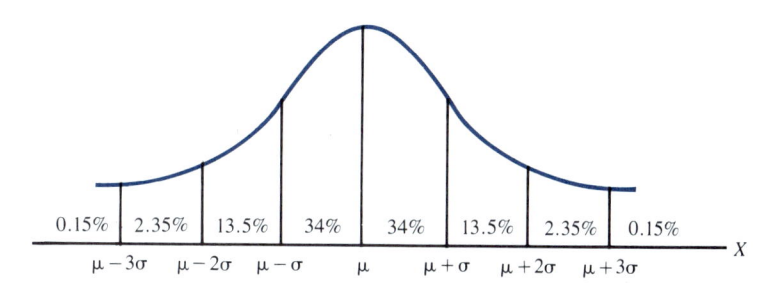

FIGURE 8–5
Percentages under the Normal Curve

EXAMPLE 1 Suppose scores on intelligence tests (IQ scores) are normally distributed with mean score $\mu = 100$ and standard deviation score $\sigma = 15$.

Then the distribution of IQ scores is that shown in Figure 8–6. Therefore, of all IQ scores, approximately:

- 50% will be greater than or equal to 100;
- 34% will be between 100 and 115;
- 13.5% will be between 70 and 85;
- 15.85% will be between 115 and 145;

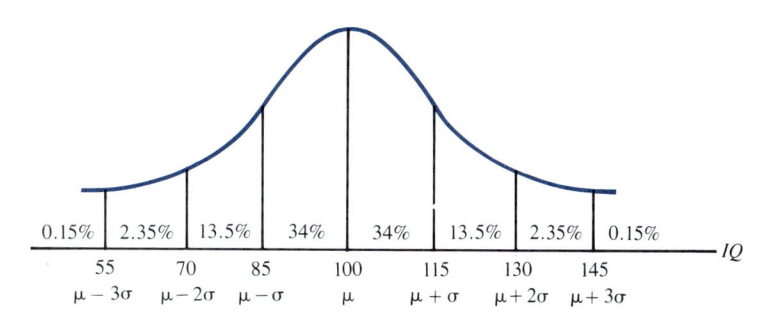

FIGURE 8–6

- 0.15% will be greater than 145;
- 2.50% will be less than 70;

and so on. Stated another way, if a person is selected at random and given an IQ test,

- the probability that the person will score 100 or more is 0.5000;
- the probability that the person will score between 100 and 115 is 0.3400;
- the probability that the person will score more than 145 is 0.0015;

and so on. ◻

Since a normally distributed random variable can take on infinitely many values, the probability that it takes on a specific single value must be zero. However, the probability that it takes on a value *between* two other values will be positive. Similarly, the probability that it takes on a value *greater than* or *less than* a given value will be positive. Such probabilities are represented by areas under the normal curve. Thus, if X is normally distributed with mean μ and standard deviation σ, and if a, b, and c are given numbers, with $a < b$, then

$$P(a \leq X \leq b), \qquad P(X \geq c), \qquad \text{and} \qquad P(X \leq c)$$

are represented by the shaded areas in Figure 8–7. We know from Example 1 how to calculate such probabilities if the numbers a, b, and c happen to be of the forms

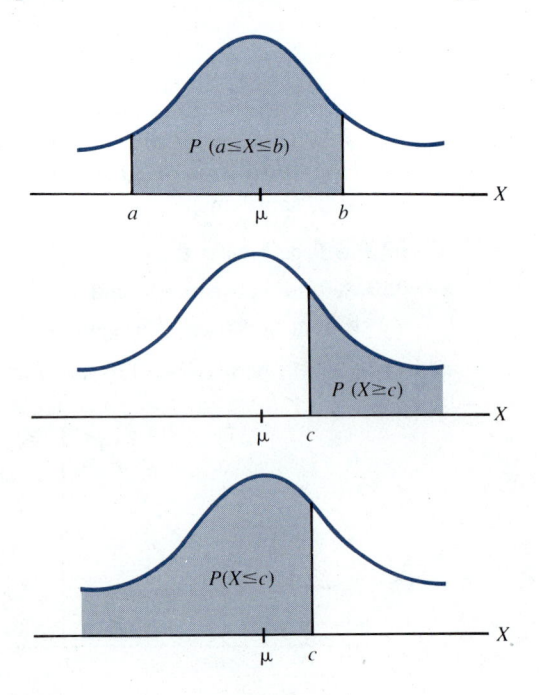

FIGURE 8–7
Probability as Area

STATISTICS

$\mu + \sigma$, $\mu - \sigma$, $\mu + 2\sigma$, and so on. However, if a, b, and c are not of such forms, then we must resort to the **standard normal distribution** to find the probabilities.

The Standard Normal Distribution

The **standard normal distribution** is the normal distribution with mean $\mu = 0$ and standard deviation $\sigma = 1$. The standard normal curve is shown in Figure 8–8. The standard normal random variable is traditionally designated by z.

Table 6 of the appendix is a table for the standard normal distribution. The table gives the proportion of the area under the standard normal curve between $z = 0$ and $z = a$ for positive values of the number a. In other words, Table 6 gives the probability $P(0 \leq z \leq a)$ that z is between 0 and a, for positive values of a (a to two decimal places). We illustrate the use of Table 6 with examples. Throughout z denotes the standard normal random variable.

EXAMPLE 2 Find $P(0 \leq z \leq 1.25)$.

This probability corresponds to the shaded area in Figure 8–9. We look up this probability in Table 6 in the row for $z = 1.2$ and the column for $z = 0.05$. Thus,

$$P(0 \leq z \leq 1.25) = 0.3944.$$

EXAMPLE 3 Find $P(-2.33 \leq z \leq 0)$.

As Figure 8–10 shows,

$$P(-2.33 \leq z \leq 0) = P(0 \leq z \leq 2.33),$$

because of the symmetry of the normal curve. Therefore, from Table 6,

$$P(-2.33 \leq z \leq 0) = P(0 \leq z \leq 2.33) = 0.4901.$$

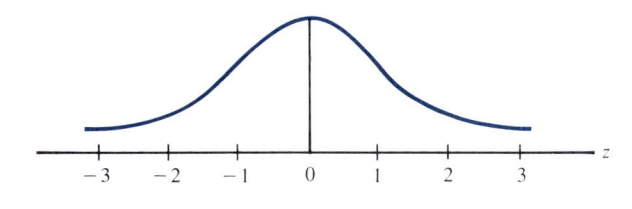

FIGURE 8–8
The Standard Normal Curve

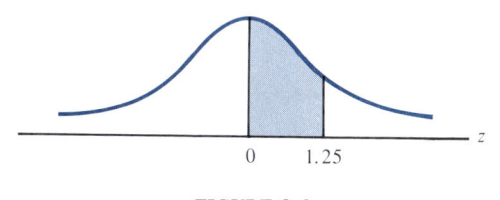

FIGURE 8–9

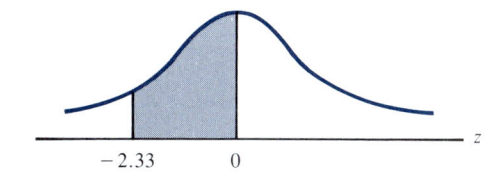

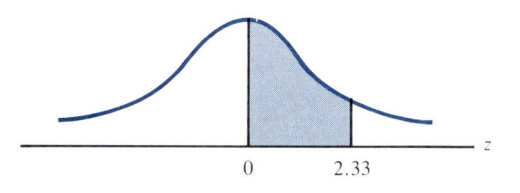

FIGURE 8–10

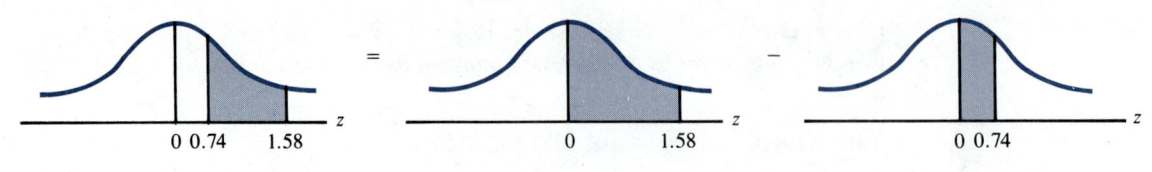

FIGURE 8-11

EXAMPLE 4 Find $P(0.74 \leq z \leq 1.58)$.
Figure 8–11 shows that

$$P(0.74 \leq z \leq 1.58) = P(0 \leq z \leq 1.58) - P(0 \leq z \leq 0.74).$$

Since

$$P(0 \leq z \leq 1.58) = 0.4429 \quad \text{and} \quad P(0 \leq z \leq 0.74) = 0.2704$$

from Table 6, we have

$$P(0.74 \leq z \leq 1.58) = 0.4429 - 0.2704 = 0.1725. \quad \square$$

EXAMPLE 5 Find $P(-1.58 \leq z \leq -0.74)$.
As Figure 8–12 shows,

$$P(-1.58 \leq z \leq -0.74) = P(0.74 \leq z \leq 1.58),$$

by the symmetry of the normal curve. Hence

$$P(-1.58 \leq z \leq -0.74) = P(0.74 \leq z \leq 1.58) = 0.1725. \quad \square$$

EXAMPLE 6 Find $P(-1.50 \leq z \leq 0.75)$.
Figure 8–13 shows that

$$P(-1.50 \leq z \leq 0.75) = P(-1.50 \leq z \leq 0) + P(0 \leq z \leq 0.75).$$

But since

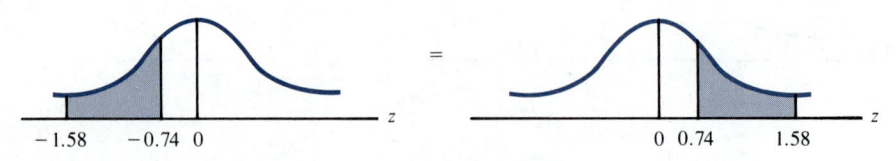

FIGURE 8-12

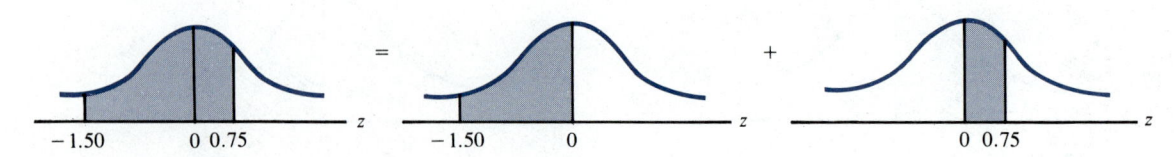

FIGURE 8-13

$$P(-1.50 \le z \le 0) = P(0 \le z \le 1.50)$$

by the symmetry of the normal curve, we have

$$P(-1.50 \le z \le 0.75) = P(0 \le z \le 1.50) + P(0 \le z \le 0.75)$$
$$= 0.4332 + 0.2734 = 0.7066. \quad \square$$

EXAMPLE 7 Find $P(z \ge 1.80)$.

Figure 8–14 shows that

$$P(z \ge 1.80) = P(z \ge 0) - P(0 \le z \le 1.80).$$

Since $P(z \ge 0) = 0.5000$ (because the area under any normal curve on either side of the mean is $\tfrac{1}{2}$), we have

$$P(z \ge 1.80) = 0.5000 - 0.4641 = 0.0359. \quad \square$$

EXAMPLE 8 Find $P(z \le 2.44)$.

Figure 8–15 shows that

$$P(z \le 2.44) = P(z \le 0) + P(0 \le z \le 2.44).$$

Thus,

$$P(z \le 2.44) = 0.5000 + 0.4927 = 0.9927. \quad \square$$

EXAMPLE 9 Find the number a such that $P(0 \le z \le a) = 0.3051$.

From Table 6,

$$P(0 \le z \le 0.86) = 0.3051.$$

Hence $a = 0.86$. $\quad \square$

Converting to the Standard Normal Random Variable

Any normally distributed random variable X may be converted to the standard normal random variable z by means of the **conversion formula:**

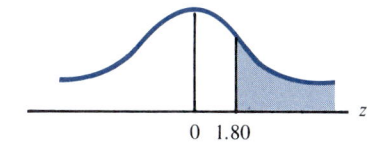

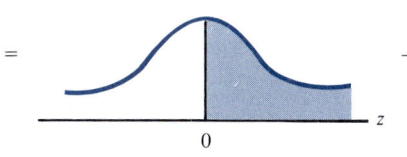

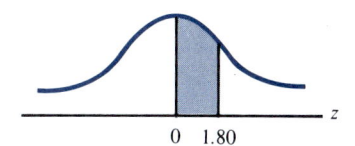

FIGURE 8–14

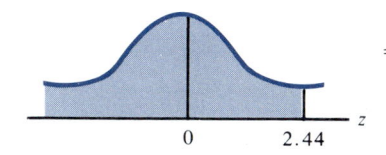

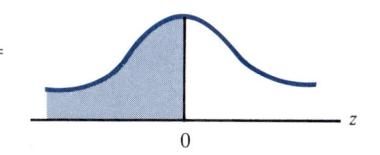

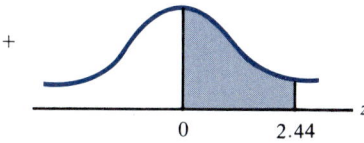

FIGURE 8–15

> **The Conversion Formula**
>
> If the random variable X is normally distributed with mean μ and standard deviation σ and z denotes the standard normal random variable, then
>
> $$z = \frac{X - \mu}{\sigma}.$$

With the aid of this formula, any problem involving a normal distribution may be converted to an equivalent problem involving the standard normal distribution, which can in turn be solved with the aid of Table 6.

We illustrate with examples of IQ test scores. As before, we let X = IQ test scores and suppose that X is normally distributed with mean $\mu = 100$ and standard deviation $\sigma = 15$.

EXAMPLE 10 Find the probability that a randomly selected person will score between 100 and 120 on an IQ test.

We seek $P(100 \leq X \leq 120)$. Using the conversion formula

$$z = \frac{X - \mu}{\sigma} = \frac{X - 100}{15},$$

$P(100 \leq X \leq 120)$ converts to

$$P\left(\frac{100 - 100}{15} \leq z \leq \frac{120 - 100}{15}\right) = P(0 \leq z \leq 1.33) = 0.4082,$$

from Table 6. □

EXAMPLE 11 Find the probability that a randomly selected person will score between 120 and 140 on an IQ test.

We seek $P(120 \leq X \leq 140)$. Using the conversion formula

$$z = \frac{X - 100}{15},$$

$P(120 \leq X \leq 140)$ converts to

$$P\left(\frac{120 - 100}{15} \leq z \leq \frac{140 - 100}{15}\right) = P(1.33 \leq z \leq 2.67)$$

$$= P(0 \leq z \leq 2.67) + P(0 \leq z \leq 1.33)$$

$$= 0.4962 - 0.4082 = 0.0880. \quad \square$$

EXAMPLE 12 Find the percentage of IQ test scores that fall between 90 and 130.

We seek $P(90 \leq X \leq 130)$, which converts to

STATISTICS

$$P\left(\frac{90-100}{15} \leq z \leq \frac{130-100}{15}\right) = P(-0.67 \leq z \leq 2.00)$$
$$= P(0 \leq z \leq 0.67) + P(0 \leq z \leq 2.00)$$
$$= 0.2486 - 0.4772 = 0.7258.$$

Therefore, 72.58% of the scores will fall between 90 and 130. □

EXAMPLE 13 Find the percentage of IQ test scores that are less than 70: $P(X \leq 70)$ converts to

$$P\left(z \leq \frac{70-100}{15}\right) = P(z \leq -2.00)$$
$$= P(z \geq 2.00)$$
$$= P(z \geq 0) - P(0 \leq z \leq 2.00)$$
$$= 0.5000 - 0.4772 = 0.0228.$$

Thus, 2.28% of the scores will be less than 70. □

EXAMPLE 14 Find the IQ test score that 10% of all scores will be greater than.

We seek to find the number a such that $P(X \geq a) = 0.10$. Converting to z, we have

$$P\left(z \geq \frac{a-100}{15}\right) = 0.10.$$

But

$$P\left(z \geq \frac{a-100}{15}\right) = 0.5000 - P\left(0 \leq z \leq \frac{a-100}{15}\right),$$

so we have

$$0.5000 - P\left(0 \leq z \leq \frac{a-100}{15}\right) = 0.10,$$

or

$$P\left(0 \leq z \leq \frac{a-100}{15}\right) = 0.4000.$$

From Table 6, $P(0 \leq z \leq 1.28) = 0.3997$, and this is as close as we can come to 0.4000. Hence

$$\frac{a-100}{15} \approx 1.28,$$

and

$$a \approx 119.2.$$

Therefore, 10% of all IQ scores will be greater than 119.2, approximately. □

Note that in the preceding examples when we converted from X to z we rounded off to the nearest hundredth, since Table 6 gives values of z only to the nearest hundredth.

■ *Computer Manual: Program NORMDIST*

8.4 EXERCISES

The Normal Curve

The weight of the cereal in boxes of Crunchy Cereal is normally distributed with a mean of 12 ounces and a standard deviation of 0.5 ounce. In Exercises 1 through 4, find the approximate probability that the cereal in a randomly selected box will weigh the given amount.

1. 12 oz or less
2. between 11.5 and 13.0 oz
3. between 10.5 and 13.0 oz
4. less than 13.0 oz
5. If the Crunchy Cereal Company of Exercises 1 through 4 cannot sell boxes that contain less than 11.5 ounces because they are underweight, what percentage of the company's production cannot be marketed?

The lifetime of a certain type of television picture tube is normally distributed with a mean of 4.5 years and a standard deviation of 1.2 years. In Exercises 6 through 9, find the percentage of picture tubes that will fail in the given time period.

6. 0.9 years or less
7. 2.1 years or less
8. 6.9 years or less
9. 8.1 years or less
10. Suppose the picture tubes of Exercises 6 through 9 carry a warranty that says they will be replaced without charge if they fail within the first x years of their life. What must x be if the warranty is to be applicable to only 2.5% of the tubes?

Management

The lengths of the pieces of wire produced by a wire-cutting machine are normally distributed with a mean of 2.54 centimeters (cm) and a standard deviation of 0.02 cm. In Exercises 11 through 14, find the approximate percentage of wires that have the given length.

11. 2.54 cm or more.
12. Between 2.52 and 2.56 cm.
13. Between 2.50 and 2.56 cm.
14. Less than 2.50 cm.
15. Refer to Exercises 11 through 14:
    (a) If pieces of wire less than 2.48 or more than 2.58 cm long must be discarded as defective, what percentage of the machine's production will be discarded?
    (b) Approximately 97.5% of the machine's production of wires will be less than what length?

Social Science

The number of roll-call votes missed by representatives during a session of the House of Representatives is normally distributed with a mean of 42 votes missed and a standard deviation of 12 votes missed. In Exercises 16 through 19, find the approximate percentage of representatives who missed the given number of votes.

16. 6 or fewer votes
17. Between 18 and 30 votes
18. Between 30 and 66 votes
19. 78 or more votes
20. Refer to Exercises 16 through 19: The 16% of the representatives who missed the most votes missed approximately how many votes?

The Standard Normal Distribution

Let z denote the standard normal random variable. In Exercises 21 through 31 find the given probability.

21. (a) $P(0 \leq z)$ (b) $P(z \leq 0)$
22. (a) $P(0 \leq z \leq 0.37)$ (b) $P(0 \leq z \leq 1.26)$ (c) $P(0 \leq z \leq 2.59)$
23. (a) $P(-0.44 \leq z \leq 0)$ (b) $P(-1.22 \leq z \leq 0)$ (c) $P(-2.80 \leq z \leq 0)$
24. (a) $P(0.03 \leq z \leq 1.21)$ (b) $P(1.35 \leq z \leq 2.40)$ (c) $P(2.66 \leq z \leq 3.01)$
25. (a) $P(-0.55 \leq z \leq -0.50)$ (b) $P(-1.32 \leq z \leq -0.27)$ (c) $P(-2.33 \leq z \leq -1.65)$
26. (a) $P(-0.69 \leq z \leq 0.69)$ (b) $P(-2.02 \leq z \leq 2.02)$
27. (a) $P(-1.35 \leq z \leq 0.86)$ (b) $P(-2.44 \leq z \leq 2.91)$ (c) $P(-0.04 \leq z \leq 1.32)$
28. (a) $P(z \geq 0.78)$ (b) $P(z \geq 2.56)$ (c) $P(z \geq 1.84)$
29. (a) $P(z \geq -0.52)$ (b) $P(z \geq -1.01)$ (c) $P(z \geq -2.47)$
30. (a) $P(z \leq -1.83)$ (b) $P(z \leq -0.99)$ (c) $P(z \leq -2.45)$
31. (a) $P(z \leq 2.88)$ (b) $P(z \leq 1.71)$ (c) $P(z \leq 0.64)$

In Exercises 32 through 39, z again denotes the standard normal random variable. In each case find the number a that yields the given probability.

32. $P(0 \leq z \leq a) = 0.4625$
33. $P(0 \leq z \leq a) = 0.4500$
34. $P(z \geq a) = 0.0630$
35. $P(z \geq a) = 0.7500$
36. $P(z \geq a) = 0.9800$
37. $P(z \leq a) = 0.2054$
38. $P(z \leq a) = 0.1000$
39. $P(z \leq a) = 0.9000$

Converting to the Standard Normal Random Variable

40. Let X be normally distributed with mean 14 and standard deviation 6. Find $P(14 \leq X \leq 22)$.
41. Let X be normally distributed with mean 12 and standard deviation 0.5. Find $P(11.6 \leq X \leq 12.6)$.
42. Let X be normally distributed with mean 28 and standard deviation 4.2. Find $P(30 \leq X \leq 40)$.
43. Let X be normally distributed with mean 3 and standard deviation 0.08. Find $P(X \leq 2.86)$.
44. Let X be normally distributed with mean 40 and standard deviation 14. Find $P(X \geq 62)$.
45. Let X be normally distributed with mean 8 and standard deviation 2. Find a such that $P(X \geq a) = 0.25$.
46. Let X be normally distributed with mean 75 and standard deviation 16. Find a such that $P(X \geq a) = 0.90$.

47. Let X be normally distributed with mean 102 and standard deviation 30. Find a such that $P(X \leq a) = 0.12$.

48. Let X be normally distributed with mean 24 and standard deviation 6.5. Find a such that $P(X \leq a) = 0.85$.

49. Use the conversion formula and Table 6 to show that the percentages of Figure 8–5 are approximately correct.

50. The lifetime of a certain brand of light bulb is normally distributed with a mean of 1000 hours and a standard deviation of 250 hours. Find the probability that a light bulb of this brand will last:

(a) more than 1000 hours
(b) between 1000 and 1500 hours
(c) between 1100 and 1200 hours
(d) between 900 and 1100 hours

51. Refer to Exercise 50: Find the percentage of bulbs that will last

(a) more than 1350 hours
(b) less than 825 hours
(c) more than 900 hours
(d) less than 1450 hours

52. Refer to Exercise 50: The company that makes the light bulbs intends to offer a money-back guarantee on any bulb that does not last for a certain length of time. The guarantee will be stated: "If this light bulb does not last for x hours, we will refund your money." What should x be in this statement if the company wants no more than 1% of the bulbs to qualify for the refund?

Management

53. The diameters of washers produced by a machine are normally distributed with a mean of 2.00 cm and a standard deviation of 0.01 cm. Find the percentage of washers produced by the machine that have the given diameter.

(a) Between 2.00 and 2.02 cm
(b) Between 2.00 and 2.015 cm
(c) Between 2.015 and 2.025 cm
(d) Between 1.995 and 2.015 cm

54. Refer to Exercise 53: Find the probability that a randomly selected washer produced by the machine will have a diameter that is

(a) more than 2.015 cm
(b) more than 1.983 cm
(c) less than 1.985 cm
(d) less than 2.018 cm

55. Refer to Exercise 53:

(a) The 10% of the washers with the largest diameters will have diameters greater than what number?
(b) The 70% of the washers with the largest diameters will have diameters greater than what number?

Life Science

56. The growing times for a certain strain of corn are normally distributed with a mean of 112 days and a standard deviation of three days. If you plant this strain of corn, find the probability that your corn crop will be ready for harvesting

(a) between 112 and 115 days after planting
(b) between 110 and 114 days after planting

57. Refer to Exercise 56: Find the probability that your corn crop will be ready for harvesting

(a) in 102 or fewer days after planting
(b) in 108 or more days after planting
(c) in 116 or more days after planting

58. Refer to Exercise 56: Suppose you intend to sign a contract that states that you will deliver your corn crop to a buyer x days after planting. What should the value of x

be if you want to have a 95% chance of honoring the contract and you know that you can deliver the crop to the buyer two days after it is ready to be harvested?

Social Science

59. A sociology professor gives a test. The scores are normally distributed with a mean of 70 and a standard deviation of 8. If a score of 90 or above is an A, 80 up to 90 is a B, 60 up to 80 is a C, 50 up to 60 is a D, and below 50 is an F, find the percentage of students who will receive a grade of
    (a) A (b) B (c) C (d) D (e) F (f) D or above (g) C or below

60. Refer to Exercise 59: Suppose the professor does not like the grade breakdown given and instead decides to assign the top 5% of the scores a grade of A, the next 20% a grade of B, the next 50% a grade of C, the next 20% a grade of D, and the bottom 5% a grade of F. Under these circumstances, what test scores will correspond to a grade of
    (a) A? (b) B? (c) C? (d) D? (e) F?

The Normal Approximation to the Binomial

We introduced the binomial distribution in Section 6.5. Finding binomial probabilities can be quite tedious when the number of trials n is large. However, when n is large, it is often possible to approximate binomial probabilities by normal probabilities. This is done as follows: suppose we wish to find the binomial probability

$$P(k \text{ successes in } n \text{ trials}),$$

where $p = P(\text{a success on any one trial})$. If we set $\mu = np$ and $\sigma = \sqrt{np(1-p)}$ and let X be normally distributed with mean μ and standard deviation σ, then

$$P(k \text{ successes in } n \text{ trials}) \approx P(k - 0.5 \leq X \leq k + 0.5).$$

This approximation is generally held to be valid if $n \geq 20$ and $np \geq 5$, $n(1-p) \geq 5$. In Exercises 61 through 66, you are to use this approximation technique.

*61. Approximate the binomial probability
$$P(20 \text{ successes in } 50 \text{ trials})$$
if $p = 0.5$.

*62. Approximate the binomial probability
$$P(15 \text{ or } 16 \text{ successes in } 25 \text{ trials})$$
if $p = 0.4$.

*63. If you toss a fair coin 100 times, find the probability that you will obtain at least 45 but not more than 55 heads.

*64. If you toss a fair coin 100 times, find the probability that you will get fewer than 40 heads.

*65. If a basketball player makes 90% of his free throws and shoots 200 free throws in a season, find the probability that he will miss fewer than 10 free throws during the season.

*66. If the probability that a machine turns out a defective part is 0.001 and a lot consists of 10,000 parts, find the probability that there will be at least five defective parts in a lot.

Inventory Control

Exercises 67 through 70 were suggested by material in Stevenson, *Production/Operations Management*, second edition, Irwin, 1986. They are concerned with finding the reorder point for an item. Thus, suppose that when the stock of an item declines to r units, an order for new stock is generated. The number r is called the **reorder point** for the item. Let L denote the time that elapses between the generation of an order for new stock and the arrival of the stock; L is called the **lead time** for an order. **Safety stock** is extra stock that is on hand to lessen the risk of a **stockout,** that is, of running out of stock before the next order arrives. The reorder point r depends on the lead time and the safety stock according to the following equation:

$$r = \text{Expected demand for stock during lead time} + \text{safety stock.}$$

Now suppose the lead time L for an item is always the same and that daily demand for the item is normally distributed with a mean of μ_d units per day and a standard deviation of σ_d items per day. If the probability of a stockout is p, then

$$\text{Expected demand during lead time} = \mu_d L$$

and

$$\text{Safety stock} = z_p \sigma_d \sqrt{L},$$

where z_p is determined from the following standard normal diagram:

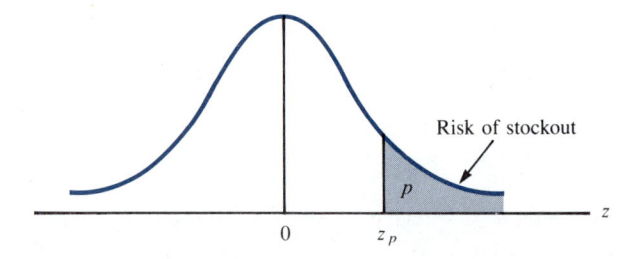

*67. A supermarket's sales of bar soap are normally distributed with average sales of 200 bars per day and a standard deviation of 30 bars per day. The reorder time for bar soap is five days. If the store is willing to have

$$P(\text{stockout of bar soap}) = 0.01,$$

find the expected demand during lead time, the safety stock, and the reorder point for bar soap.

*68. Repeat Exercise 67 if the store is willing to run a 5% risk of a stockout.

If the lead time L is not always the same for an item, it may itself be normally distributed with a mean time of μ_L and a standard deviation time of σ_L. Then

$$\text{Expected demand during lead time} = \mu_d \mu_L$$

and

$$\text{Safety stock} = z_p \sqrt{\mu_L \sigma_d^2 + \mu_d^2 \sigma_L^2},$$

where z_p is determined as before.

*69. Suppose a service station's gasoline sales are normally distributed with a mean of 400 gallons per day and a standard deviation of 80 gallons per day. Suppose also that the lead time for a delivery of gasoline is normally distributed with a mean of 4.5 days and a standard deviation of 0.4 days. If the station is willing to run a 2% risk of a stockout of gas, find the expected demand during lead lead time, the safety stock, and the reorder point for gasoline.

*70. Repeat Exercise 69 if the lead time changes so that its standard deviation is 1.1 days.

STATISTICS

8.5 LINEAR REGRESSION AND CORRELATION

Students often ask how the functions used to model practical situations are "made up." We have seen throughout this book that such functions can often be constructed from an analysis of the situation. For instance, as we saw in Section 3.2, if it costs a firm v dollars to make each unit of its product and if its fixed cost is F dollars, then the cost C of making x units is given by $C = vx + F$. Often, however, functional models cannot be developed through analysis of a situation, but instead must be created by fitting curves to experimental data. In this section we show how to fit a linear function to experimental data and how to measure the "goodness of fit" of the line to the data. The linear function that best fits a set of data is called the **least-squares regression line** for the data, and the goodness-of-fit measurement is referred to as the **correlation coefficient** of the data.

Linear Regression

Suppose we want to find a function that models annual family savings as a function of annual family income. To begin with we have no information on the relationship, and thus no reason to prefer any particular type of function (e.g., linear, quadratic, etc.) as a model. If we collected some data on the relationship between income and savings, perhaps we would be able to decide on the type of function to use as a model. Let us assume we have somehow obtained the following data:

X Annual Family Income (in 1000s of $)	Y Annual Family Savings (in 1000s of $)
42	5.0
48	6.0
37	4.5
52	6.1
60	7.1
54	6.1
44	4.9
48	5.6

Let us plot the data as points (x_i, y_i). The result is the diagram of Figure 8–16, which is called a **scatter diagram** for the data. We now guess the type of curve that would best fit the scatter diagram in the sense that it would come as close as possible to all the points of the diagram. The type of curve that best fits the diagram depends on the pattern formed by the points of the diagram: it might be a line, a parabola, or some other curve. In this case it looks as if a line would fit the scatter diagram fairly well. See Figure 8–17. Thus our data lead us to choose a linear function to model the relationship between income and savings.

Our next task is to determine the equation $Y = mX + b$ of our linear model by finding the values of m and b that make the line come as close as possible to all the points of the scatter diagram simultaneously. For any scatter diagram and

FIGURE 8–16

FIGURE 8–17

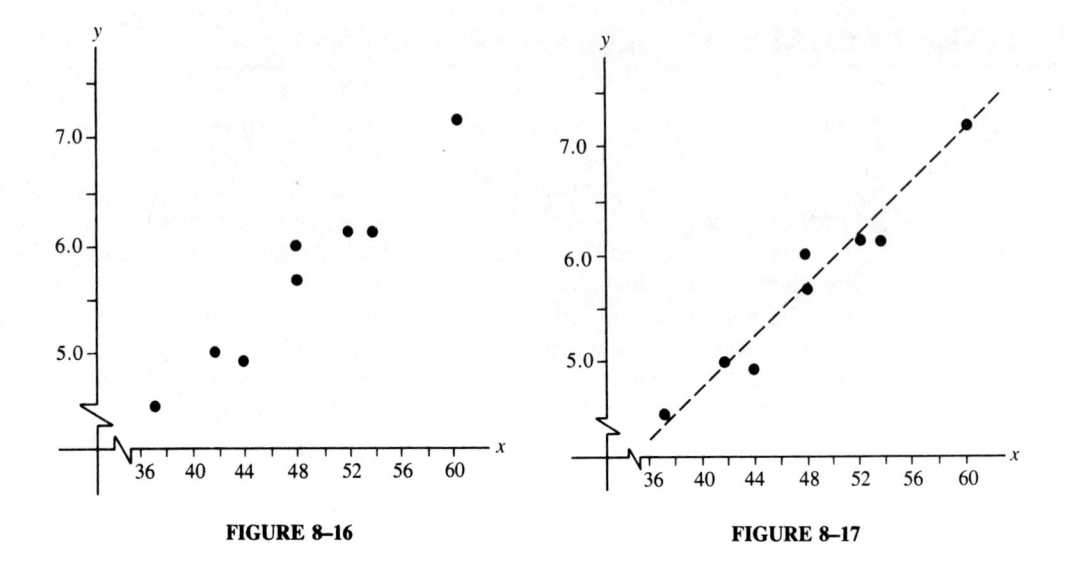

any line, the closeness of the line to the points could be measured by the sum of the deviations of the points from the line: see Figure 8–18, which is a picture of a general scatter diagram and line $Y = mX + b$ with the deviations $d_1, d_2, \ldots, d_n$, where

$$d_i = (mx_i + b) - y_i$$

for each i. However, since some of the d_i's (those corresponding to points that lie below the line) will be positive while others (those corresponding to points that lie above the line) will be negative, there will be some "canceling out" in the sum $d_1 + d_2 + \cdots + d_n$, and this could make the line appear to be closer to the points

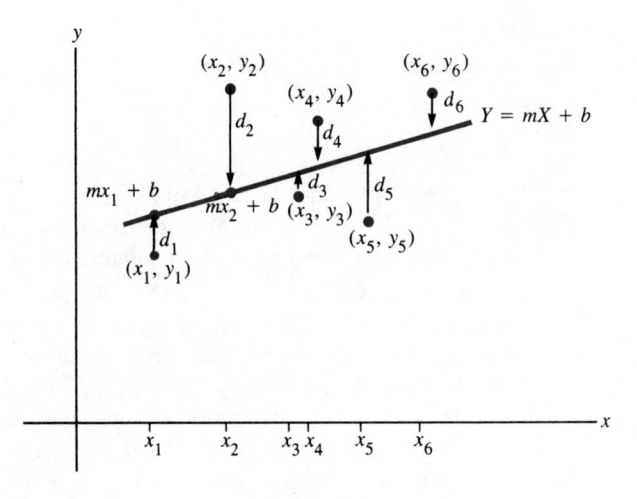

FIGURE 8–18

than it really is. Therefore instead of using the d_i's to measure deviation from the line, we use their squares d_i^2. The line for which

$$\sum_{i=1}^{n} d_i^2 = \sum_{i=1}^{n} (mx_i + b - y_i)^2$$

is minimized is called the **least-squares regression line** for the data. Thus, given a set $\{(x_1, y_1), \ldots, (x_n, y_n)\}$ of n data points, the least-squares regression line for the data is the line

$$Y = mX + b$$

that best fits the data in the sense that m and b are chosen in such a way as to minimize the sum

$$\sum_{i=1}^{n} (mx_i + b - y_i)^2.$$

The methods of calculus may be used to develop the following formulas for m and b:

The Least-Squares Line Formulas

Given a set $\{(x_1, y_1), \ldots, (x_n, y_n)\}$ of n data points, the **least-squares regression line** for the data is the line

$$Y = mX + b,$$

where

$$m = \frac{n \sum_{i=1}^{n} x_i y_i - \left(\sum_{i=1}^{n} x_i\right)\left(\sum_{i=1}^{n} y_i\right)}{n \left(\sum_{i=1}^{n} x_i^2\right) - \left(\sum_{i=1}^{n} x_i\right)^2}$$

and

$$b = \frac{\sum_{i=1}^{n} y_i - m \sum_{i=1}^{n} x_i}{n}.$$

Let us use these formulas to find the least-squares line for the data on family savings and family income that we presented at the beginning of this section. We need the sum of the x_i's, the sum of the y_i's, the sum of the $x_i y_i$'s, and the sum of the x_i^2's. We have

X	Y	XY	X^2
42	5.0	210.0	1764
48	6.0	288.0	2304
37	4.5	166.5	1369
52	6.1	317.2	2704
60	7.1	426.0	3600
54	6.1	329.4	2916
44	4.9	215.6	1936
48	5.6	268.8	2304
385	45.3	2221.5	18,897

Thus

$$m = \frac{8(2221.5) - (385)(45.3)}{8(18{,}897) - (385)^2} \approx 0.112$$

and

$$b = \frac{1}{8}[(45.3) - 0.112(385)] \approx 0.273.$$

Hence the least-squares line that expresses family savings Y as a function of family income X is

$$Y = 0.112X + 0.273.$$

This equation may be used for prediction. When this is done, what is being predicted is the *average* value of Y for a given value of X. For instance, if $X = 50$, then

$$Y = 0.112(50) + 0.273 = 5.873.$$

Thus we predict that the *average* annual family savings for all families whose annual income is $50,000 is $5873. Note also that since $m = 0.112$ is positive, average annual family savings Y increases as annual family income X increases; in fact, we can say that every increase of $1000 in annual family income will result in an increase of $0.112(\$1000) = \$112$ in average annual family savings.

There is another point to keep in mind when using a least-squares line to make predictions: The equation is valid *only* over the range of the original data, so we must not **extrapolate** it. This means that we cannot use the equation to make predictions for values of X that are outside the range of the data from which the equation was derived. For instance, attempting to use the least-squares equation

$$Y = 0.112X + 0.273$$

to predict the average value of Y when $X = 70$ would constitute extrapolation, since $X = 70$ is not in the range covered by our data.

Correlation

The least-squares regression line is the line that best fits the data, but it is possible that even the best-fitting line does not provide a very good fit. To measure the goodness of fit of the regression line to the data, we make use of the **correlation coefficient.**

The Correlation Coefficient

If $\{(x_i, y_i) \mid i = 1, 2, \ldots, n\}$ is a set of n data points, their **correlation coefficient** r is defined to be

$$r = \frac{n \sum x_i y_i - (\sum x_i)(\sum y_i)}{\sqrt{[n \sum x_i^2 - (\sum x_i)^2][n \sum y_i^2 - (\sum y_i)^2]}},$$

with all sums being taken from $i = 1$ to n. The correlation coefficient r has the following properties (see Figure 8–19):

1. $-1 \leq r \leq 1$.

2. If $r > 0$, then y tends to increase as x increases; if $r < 0$, then y tends to decrease as x increases.

3. r^2 gives the proportion of the variation among the y-values in the data that is explained by the variation among the x-values according to the regression line.

4. It follows from **3.** that the closer $|r|$ is to 1, the better the regression line fits the data; if $|r| = 1$, the fit is perfect and all data points lie on the line.

5. It follows from **3.** that the closer $|r|$ is to 0, the worse the fit of the regression line to the data; if $|r| = 0$, there is no linear relationship between x and y.

For our example relating family savings to family income, we have

$$n \sum_{i=1}^{n} x_i y_i - \left(\sum_{i=1}^{n} x_i\right)\left(\sum_{i=1}^{n} y_i\right) = 8(2221.5) - (385)(45.3) = 331.5,$$

$$n \sum_{i=1}^{n} x_i^2 - \left(\sum_{i=1}^{n} x_i\right)^2 = 8(18{,}897) - (385)^2 = 2951,$$

and

$$n \sum_{i=1}^{n} y_i^2 - \left(\sum_{i=1}^{n} y_i\right)^2 = 8(261.45) - (45.3)^2 = 39.51.$$

Hence,

$$r = \frac{331.5}{\sqrt{(2951)(39.51)}} \approx 0.971.$$

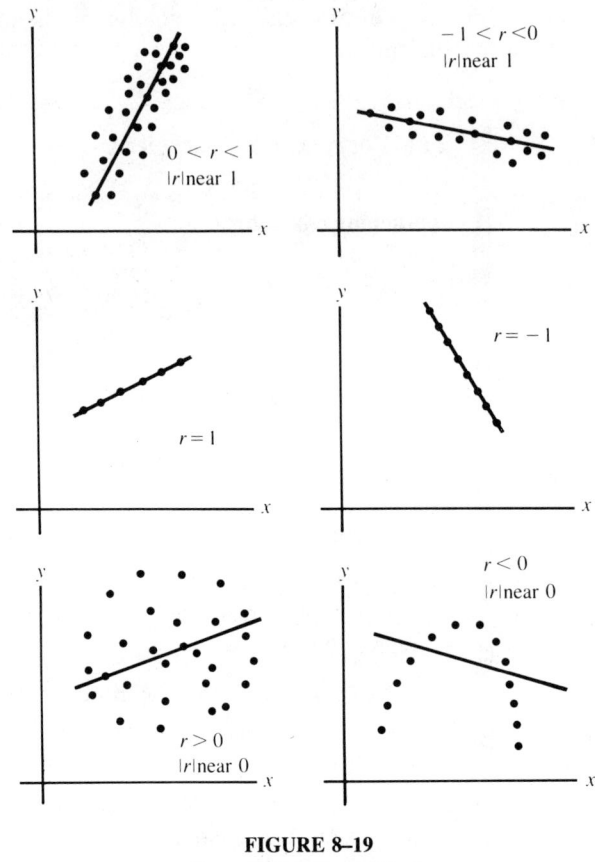

**FIGURE 8–19
Correlation Coefficients**

Thus, since $r > 0$, annual savings Y tends to increase as annual income X increases. Since r is close to 1, the least-squares line fits the data quite closely, and hence the linear relation

$$Y = 0.112X + 0.273$$

between X and Y is a close one. In fact, since $r^2 \approx 0.943$, we can say that approximately 94.3% of the variation in the family savings data is explained by the variation in the family income data according to the least-squares line.

Finally, we must note that just because there is a strong linear relationship between X and Y, it does not follow that X causes Y or has a direct effect on Y. Both variables may be affected by more fundamental factors that have not been considered, or the apparent relationship between them may even be coincidental. (See Exercise 11.) The fact that X and Y are highly correlated only means that most of the variation in the Y data is accounted for by variation in the X data.

■ *Computer Manual: Program REGRCORR*

8.5 EXERCISES

In Exercises 1 through 8, make a scatter diagram, find the least-squares regression line and correlation coefficient, and interpret your results.

1. | X: | 1 | 2 | 3 | 4 |
    |---|---|---|---|---|
    | Y: | 10 | 13 | 19 | 22 |

2. | X: | 2 | 4 | 5 | 7 |
    |---|---|---|---|---|
    | Y: | 45 | 42 | 36 | 27 |

3. | X: | 2 | 3 | 4 | 5 | 6 |
    |---|---|---|---|---|---|
    | Y: | 8 | 10 | 7 | 11 | 8 |

4. | X: | 1 | 2 | 3 | 4 | 5 |
    |---|---|---|---|---|---|
    | Y: | 1 | 4 | 5 | 3 | 0 |

5. | X: | 12 | 18 | 23 | 24 | 30 | 32 |
    |---|---|---|---|---|---|---|
    | Y: | 63 | 46 | 38 | 25 | 35 | 16 |

6. | X: | 4.1 | 5.4 | 6.9 | 8.2 | 9.1 | 9.9 |
    |---|---|---|---|---|---|---|
    | Y: | 42 | 44 | 48 | 61 | 75 | 90 |

7. | X: | 20.2 | 31.4 | 45.6 | 54.8 | 66.2 | 78.0 | 85.7 |
    |---|---|---|---|---|---|---|---|
    | Y: | 34.4 | 34.0 | 33.9 | 33.2 | 32.8 | 31.1 | 30.5 |

8. | X: | 6.6 | 8.4 | 10.5 | 12.3 | 14.6 | 17.1 | 19.2 | 21.4 |
    |---|---|---|---|---|---|---|---|---|
    | Y: | 68 | 42 | 36 | 30 | 28 | 34 | 48 | 70 |

9. An instructor collected the following information concerning five students:

Student:	A	B	C	D	E
Absences from Class:	1	2	4	6	7
Grade on Final Exam:	91	86	85	74	73

   (a) Find the linear model that best describes grade on the final exam as a function of the number of absences from class.
   (b) Describe the goodness of fit of the model to the data.
   (c) What is the significance of the slope of the model?
   (d) Predict the average grade on the final exam for all students who have five absences from class.

10. According to the *1989 World Almanac* (Pharos Books, 1989) world gold production over the period 1982–87 was as follows:

Production:	43.1	45.2	46.8	49.2	51.6	52.5
Year t:	0	1	2	3	4	5

    The production figures are in millions of troy ounces, and the years are coded so that $t = 0$ represents 1982. Make a scatter diagram of production versus time and then find the linear least-squares model for gold production as a function of time. What is the significance of the slope of the model? Describe the goodness of fit of the model to the data.

11. According to data on page 34 of the June 1988 issue of *Scientific American*, West Germany's birth rate and its population of brooding storks have both been declining in recent years:

536 CHAPTER 8

Year:	1965	1970	1975	1980
Pairs of Brooding Storks:	1900	1400	1050	900
Millions of Newborn Babies:	1.1	0.88	0.65	0.65

Find a linear least-squares model for births as a function of the number of pairs of brooding storks, find the correlation coefficient, and interpret your results.

Management

12. A cost accountant is attempting to find a linear model that describes the cost of producing a firm's product. She has the following data:

Units Produced (in Thousands):	2	3	4	5
Total Cost (in Thousands of $):	110	135	155	180

    (a) Find the linear model that best describes cost as a function of production.
    (b) Describe the goodness of fit of the model to the data.
    (c) What is the significance of the slope of the linear model?
    (d) Predict the average cost of producing 3500 units.

13. The following table compares weekly sales of gas at a certain gas station with the price per gallon:

Price per Gallon ($):	1.10	1.25	1.35	1.50
Weekly Sales (Hundreds of Gallons):	16.4	14.3	13.7	12.5

    (a) Find the linear model that best describes weekly sales as a function of price.
    (b) Describe the goodness of fit of the model to the data.
    (c) What is the significance of the slope of the linear model?
    (d) Predict the average weekly sales if gas costs $1.45 per gallon.

14. The following table presents historical data relating the price of a commodity to the quantity of the commodity supplied to the market:

Quantity Supplied (in Millions of Units):	2.5	3.4	3.8	4.6	5.2	6.0
Price per Unit ($):	15.82	13.44	13.15	12.35	11.85	11.20

    (a) Find a linear supply function for the commodity, writing price as a function of quantity supplied.
    (b) Discuss the meaning of the model.
    (c) Predict the average price per unit when 5 million units are supplied to the market.

15. According to an article in the October 3, 1988, edition of *The Wall Street Journal*, the profits of U.S. oil companies and the price of Mideast light crude oil in 1987 and 1988 were as follows:

	Profits (Billions of Dollars)	Price of Light Crude (Dollars per Barrel)
1987 Qtr I	2.936	17.25
II	3.507	17.38
III	3.429	17.65
IV	3.600	17.02
1988 Qtr I	4.395	15.05
II	4.257	14.68

(a) Find the least-squares line for profit as a function of price per barrel of oil. What is the significance of the slope of this line?

(b) Find the correlation coefficient for profit as a function of price per barrel, and interpret your result.

(c) Predict the oil companies' profit if the price per barrel is $16.00.

16. A firm has the following data on its profit from making and selling x units of product:

X Units Made and Sold	Y Profit ($)
5	−155
10	−114
15	−77
20	−40
25	−2
30	44
35	83
40	120
45	162
50	194

Make a scatter diagram and find the equation of the least-squares line for the data. Can you estimate the firm's fixed cost and the contribution to profit made by each unit?

17. An article by Richard Lester ("Rethinking Nuclear Power," *Scientific American*, March 1986) gives the following data concerning the average construction time, in months, for nuclear power plants in the United States during the period 1971–86:

Year	Average Construction Time (Months)
1971	46
1972	73
1973	71
1974	83
1975	78
1976	87
1977	100
1978	112
1979	125

538 CHAPTER 8

Year	Average Construction Time (Months)
1980	120
1981	140
1982	158
1983	106
1984	134
1985	150
1986	125

Find a linear least-squares model for construction time as a function of time, find the correlation coefficient, and interpret your results.

Life Science

18. A researcher has gathered the following data regarding the rate of lung cancer among ex-smokers:

Years Since Stopped Smoking:	2	4	5	8	10
Cases of Lung Cancer per 1000 Ex-smokers:	42	37	33	24	20

   (a) Find the linear model that best describes lung cancer cases as a function of time since smoking was stopped.
   (b) Discuss the meaning of the model.
   (c) Predict the average number of lung cancer cases for those who quit smoking seven years ago.

19. A new drug that shrinks tumors is being tested in various concentrations. The results of the tests are as follows:

X Drug Dosage (Units)	Y Time until Tumor Disappears (Days)
1	19.0
3	17.5
4	16.6
6	15.4
7	14.4
8	13.7
10	12.1
12	10.8
13	9.6
14	9.0

Make a scatter diagram and find the least-squares line for the data. Can you estimate the effect that increasing the dosage by 1 unit has on the tumor?

20. An article by Richard Stolarski ("The Antarctic Ozone Hole," *Scientific American*, June 1988) indicates that a NASA satellite collected the following data on ozone levels over Antarctica from 1979 through 1986:

STATISTICS

Year	Ozone Level (in Dobson Units)
1979	283
1980	237
1981	239
1982	216
1983	196
1984	192
1985	208
1986	168

Find a linear least-squares model for the ozone level as a function of time, find the correlation coefficient, and interpret your results.

Social Science

An article in the July 3, 1988, edition of *The New York Times* reported that the percentage of eligible voters who voted in the seven presidential elections from 1960 through 1984 was as follows:

Year:	1960	1964	1968	1972	1976	1980	1984
% Voting:	63.1	61.8	60.7	55.1	53.6	52.6	53.1

The same article also estimated the average hours of television watched per day by U.S. viewers to be as follows:

Year:	1960	1964	1968	1972	1976	1980	1984
Avg. Hours per Day:	5.0	5.4	5.9	6.5	6.3	6.8	7.2

Exercises 21 through 25 refer to these data sets.

21. Find the linear model that best describes voter participation as a function of time.

22. Use the model of Exercise 21 to "predict" voter participation in the election of 1988. (This constitutes extrapolation. Why?) Actual voter participation in 1988 was 50%; how does this compare with your "prediction"?

23. Find the linear model that best describes average hours of television watched per day as a function of time.

24. Find the linear model that best describes voter participation as a function of hours spent watching television. What is the significance of the slope of this line?

25. Find the correlation coefficient for voter participation as a function of hours spent watching television, and interpret your result.

Stock Volatility

Exercises 26 and 27 were suggested by material in Higgins, *Analysis for Financial Management,* Irwin, 1984. They are concerned with measuring the volatility of a stock. Thus, let s be a stock, and let σ_s be the standard deviation of its monthly return to investors. Let p denote a large benchmark portfolio of stocks (such as Standard & Poor's 500 stocks, for instance), and let σ_p be the standard deviation of the portfolio's monthly return to investors. Let r be the correlation coefficient between the monthly returns of stock s and the monthly returns of portfolio p. Define the **volatility** or **β-risk** of stock s to be β_s, where

$$\beta_s = \frac{r\sigma_s}{\sigma_p}.$$

(β is the Greek letter beta.) The larger the value of β_s, the more volatile, and therefore presumably the riskier, stock s is relative to the portfolio p. Here are monthly prices for a stock s_1, a stock s_2, and a benchmark portfolio p:

Monthly Price of s_1:	$20	$21	$20	$22	$24
Monthly Price of s_2:	$55	$58	$56	$58	$59
Monthly Price of p:	$120	$121	$122	$122	$123

*26. Find the volatility of stock s_1.

*27. Find the volatility of stock s_2. Is s_2 more or less volatile than s_1?

SUMMARY

This summary consists of terms and symbols whose meaning you should know, facts you should know, and techniques and methods you should be able to employ.

Section 8.1

Frequency distribution. Class of a frequency distribution. Frequency, relative frequency, cumulative frequency of a class. Histogram. Frequency histogram, relative frequency histogram, cumulative frequency histogram. Class mark or midpoint of a class. Polygon. Frequency polygon, relative frequency polygon, cumulative frequency polygon. Range of data. Constructing a frequency distribution for a collection of raw data and drawing its histograms and polygons.

Section 8.2

Measure of central tendency. Summation notation, summation symbol. Mean or arithmetic average. Population mean μ, sample mean $\bar{x}$. Finding the mean of a population or a sample. Formula for the mean of grouped data. Finding the mean of grouped data. Median. Finding the median. Mode. Finding the mode. Advantages and disadvantages of the mean, median, and mode. Choosing the best measure of central tendency for a data set.

Section 8.3

Dispersion of a set of data. Measures of dispersion. Range. Variance. Standard deviation. Variance of a population σ^2, standard deviation of a population σ. Variance of a sample s^2, standard deviation of a sample s. Finding the variance and standard deviation of a population or a sample. Shortcut formulas for the

STATISTICS

variance. Finding variances using the shortcut formulas. Formulas for the variance of grouped data. Finding the variance and standard deviation of grouped data. Coefficient of variation V. Significance of the coefficient of variation. Using the coefficient of variation to compare the variability of data sets.

Section 8.4

Discrete distribution. Continuous random variable. Continuous distribution. Normally distributed random variable with mean μ and standard deviation σ. Normal curve. Properties of the normal curve. The percentage breakdown of areas under the normal curve that are delimited by values of the form $\mu \pm k\sigma$. Using the properties of the normal curve and the percentage breakdowns of areas under the curve to find probabilities and percentages. Standard normal distribution. Standard normal random variable z. Finding standard normal probabilities using the properties of the normal curve and the standard normal table. Finding the value of z that yields a given probability. The conversion formula. Finding probabilities and percentages for a normal random variable by converting it to standard normal. Finding the value of a normal random variable that yields a given probability or percentage.

Section 8.5

Least-squares regression line. Correlation coefficient. Scatter diagram. Least-squares formulas. Finding the least-squares regression line. Significance of the slope of the least-squares line. Using the least-squares line to predict the average value of Y. Extrapolation. The danger of extrapolation. Correlation coefficient r. Properties of the correlation coefficient. Significance of the correlation coefficient. Significance of r^2. Finding and interpreting the correlation coefficient. A high degree of correlation does not imply causality.

REVIEW EXERCISES

1. Thirty students took a statistics test. Their grades were as follows:

87	64	91	80	72	63
64	85	89	66	57	69
60	82	88	98	82	76
61	52	65	85	80	63
74	65	85	67	78	69

    Make a frequency distribution for this set of data. Use five classes of equal width, and include relative and cumulative frequencies. Also, draw a frequency histogram and a cumulative frequency polygon.

2. Find the mean, median, and mode for the following data:

    7, 16, 12, 9, 12, 10, 14, 13, 10, 8, 12, 7,
    8, 12, 10, 13, 15, 6, 8, 5, 11, 11, 8, 9

3. Twelve employees of a large firm have the following years of service with the company:

12, 6, 8, 9, 7, 8, 10, 11, 12, 8, 4, 9.

Find the mean, median, and modal years of service for this sample.

4. A firm's quarterly revenues for the past four years, in thousands of dollars, have been as follows:

	Quarter			
Year	I	II	III	IV
1	232.4	245.6	242.9	255.6
2	255.8	262.9	264.9	273.0
3	274.5	280.1	278.3	288.4
4	286.0	292.3	290.1	300.2

Find the firm's average quarterly revenue over the last four years.

5. For the firm of Exercise 4, find the average revenue for quarter I, for quarter II, for quarter III, and for quarter IV over the last four years.

6. A restaurant franchiser forecasts that its franchises will have the following gross revenues next year:

Gross Revenues	Number of Franchises
$100,000–$249,999	22
$250,000–$399,999	28
$400,000–$599,999	34
$600,000–$749,999	12
$750,000–$900,000	4

Each franchise pays the franchiser a fee equal to 3% of its gross revenues. Estimate the franchiser's income from this fee next year.

7. Patients participating in a study of the effectiveness of a painkiller report on the length of time between injection of the painkiller and disappearance of the pain:

Time (Minutes)	Number of Patients
0.0–1.0	1
1.1–2.0	5
2.1–3.0	12
3.1–4.0	44
4.1–5.0	36
5.1–6.0	8
6.1–7.0	2

Find the mean time of effectiveness for the painkiller.

In Exercises 8 through 10, find the mean, median, and mode for the given set of data, and state which of these is preferable as a measure of the central tendency of the data.

8. Number of sick days taken last year by eight employees:

5, 8, 3, 10, 2, 3, 7, 6.

9. Time, in minutes, that it took to check out 12 successive customers of a supermarket:

3, 1, 2, 2, $1\frac{1}{2}$, 3, 4, 2, 3, $\frac{1}{2}$, 3, 2.

10. Percentage returns on six investments:

$$-7\%, \quad 10\%, \quad 8\%, \quad 12\%, \quad 9\%, \quad 10\%.$$

11. Find the range, variance, and standard deviation of the following set of population data:

$$14, \quad 9, \quad 15, \quad 10, \quad 18, \quad 8, \quad 12, \quad 10, \quad 16.$$

12. Repeat Exercise 11 if the data form a sample.

13. Find the variance and standard deviation for the years-of-service data of Exercise 3.

14. A sample of 10 retailers in a city made the following prediction for Christmas retail sales (figures in millions of dollars):

$$200, \quad 220, \quad 185, \quad 205, \quad 210, \quad 220, \quad 225, \quad 245, \quad 190, \quad 200.$$

A sample of eight economists made the following predictions for Christmas retail sales:

$$210, \quad 208, \quad 215, \quad 220, \quad 215, \quad 175, \quad 205, \quad 212.$$

Which sample shows more variability, as measured by its coefficient of variation?

15. Estimate the variance and standard deviation for the data of Exercise 7.

16. A population has a mean of 42 and a standard deviation of 6.
    (a) Find the minimum percentage of the population that lies in the interval [18, 66].
    (b) Find the smallest interval in which at least 90% of the population lies.

17. Using the data of Exercise 6 and Chebyshev's Theorem, estimate an interval in which the gross revenue of at least 96% of the franchises will fall.

Let X be a normally distributed random variable with a mean of 20 and a standard deviation of 5. In Exercises 18 through 25 find the given probability.

18. $P(20 \leq X \leq 34)$
19. $P(17.5 \leq X \leq 28.5)$
20. $P(14.2 \leq X \leq 18)$
21. $P(13.9 \leq X \leq 33.6)$
22. $P(X \geq 5)$
23. $P(X \geq 30.5)$
24. $P(X \leq 12.4)$
25. $P(X \leq 27.5)$

Again let X be a normally distributed random variable with mean 20 and standard deviation 5. In Exercises 26 through 33, find the number a.

26. $P(20 \leq X \leq a) = 0.4505$
27. $P(a \leq X \leq 20) = 0.3975$
28. $P(X \geq a) = 0.4$
29. $P(X \leq a) = 0.25$
30. $P(X \geq a) = 0.72$
31. $P(X \leq a) = 0.09$
32. $P(X \leq a) = 0.55$
33. $P(15 \leq X \leq a) = 0.1$

34. Suppose the time customers in a fast-food restaurant must wait to be served is normally distributed with a mean of 2.8 minutes and a standard deviation of 0.6 minute. Find
    (a) the percentage of customers who will wait less than 1 minute to be served
    (b) the probability that a customer will be served in 3.5 minutes or less
    (c) the probability that a customer will wait longer than 2.2 minutes to be served
    (d) the percentage of customers who will wait between 1 and 2 minutes before they are served
    (e) the probability that a customer will wait longer than 4.25 minutes before being served
    (f) the percentage of customers who will wait between 2.5 and 3 minutes before being served

35. Refer to Exercise 34.
   (a) The 6% of the customers who must wait the longest will wait longer than how many minutes?
   (b) The 20% of the customers who are served quickest will be served within how many minutes?

In Exercises 36 and 37, make a scatter diagram for the given data set, find its least-squares regression line and correlation coefficient, and interpret your results.

36.

X:	2	8	12	20	25
Y:	25	65	85	135	165

37.

X:	2	8	12	20	25
Y:	25	65	85	60	20

38. Five farmers have tested a new fertilizer by applying it to their corn crops in different concentrations. The results were as follows:

Farmer:	A	B	C	D	E
Fertilizer Concentration Used (Tons Per Acre):	1.50	1.00	1.75	1.25	2.00
Corn Yield (Bushels Per Acre):	382	330	406	348	444

   (a) Find the linear model that best describes corn yield as a function of fertilizer concentration.
   (b) Describe the goodness of fit of the model to the data.
   (c) What is the significance of the slope of the linear model?
   (d) Predict average corn yield for farmers who use 1.30 tons of the new fertilizer per acre.

ADDITIONAL TOPICS

Here are some suggestions for topics not covered in the text that you might want to investigate on your own.

1. The only continuous distribution we have studied is the normal distribution, but there are other continuous distributions that are discussed in statistics textbooks. The simplest of these is the **uniform** or **rectangular** distribution. This distribution is used to describe populations in which the percentage of the population that lies within an interval is directly proportional to the length of the interval. Find out about the uniform distribution and some of its applications.

2. Another important continuous distribution is the **exponential distribution.** This distribution is often used when dealing with the probability that a product will fail within a given period of time. Find out about the exponential distribution and some of its applications.

3. A third continuous distribution that is of great importance is the **chi-square** (χ^2) distribution. The χ^2 distribution is used for goodness-of-fit tests, which use sample data to test whether the population from which the sample was drawn has a specified distribution. It can also be used to test two variables to see if they are dependent or independent. Find out about the χ^2 distribution and some of its applications.

4. The linear regression introduced in Section 8.5 is often referred to as "simple" regression, because it allows the variable Y to depend only on one independent variable X. Multiple linear regression, on the other hand, allows Y to depend on several independent variables $X_1, \ldots, X_n$ according to an equation of the form

$$Y = m_1 X_1 + \cdots + m_n X_n + b.$$

Find out about multiple linear regression and some of its uses.

5. Statistics as an organized quantitative discipline is of relatively recent origin. The founder of statistics is generally considered to be Karl Pearson (1857–1936) of Great Britain, who invented many of the statistical techniques and measurements used today (including the correlation coefficient of Section 8.5). Find out about Pearson and his contributions to statistics.

SUPPLEMENT: CONFIDENCE INTERVALS

One of the common problems in the applications of statistics is that of estimating a population mean from a sample mean. To illustrate, we consider a production process that fills boxes with cereal. Each box is supposed to receive exactly 12 ounces of cereal, but if the process is not working correctly this may not be the case. We want to monitor the process by checking the average weight of cereal per box for each day's production. Thus our population will consist of all the boxes filled during a day, our random variable X will be defined by

$$X = \text{Weight of cereal in each box, in ounces,}$$

and we want to find

$$\mu = \text{Mean weight of cereal in the day's boxes.}$$

It would probably be inconvenient, if not impossible, to find μ by weighing the cereal in all the boxes filled during a day, so instead each day we will take a sample of n boxes, weigh the cereal in them, and find the mean weight $\bar{x}$ for the sample. Then we can use $\bar{x}$ as a **point estimate** for μ. For instance, if it turns out that $\bar{x} = 11.9$ ounces, then we can conclude that $\mu \approx 11.9$ ounces.

By itself, a point estimate $\bar{x}$ for μ is not very useful, for we do not know how good the estimate is: The actual value of μ might be close to $\bar{x}$, but it also might be quite far from $\bar{x}$. What we need is a way to use the sample mean $\bar{x}$ to "pin down" the population mean μ. In this supplement we show how this can be done by constructing **confidence intervals** for μ.

Suppose that instead of taking just one sample of size n from our population we take *all possible* samples of size n. Each of these samples will have its own mean, so if there are m such samples we will obtain m sample means

$$\bar{x}_1, \ldots, \bar{x}_m.$$

The set of numbers $\{\bar{x}_1, \ldots, \bar{x}_m\}$ is a population, and we define a random variable $\bar{X}$ on it by letting $\bar{X}$ take on the values $\bar{x}_1, \ldots, \bar{x}_m$. The random variable $\bar{X}$ is called the **sampling distribution of the mean** for samples of size n.

EXAMPLE 1

Suppose our population of cereal boxes consisted of just four boxes, designated A, B, C, and D, with weights as follows:

Box:	A	B	C	D
X = weight of cereal:	11.9	12.2	12.0	11.8

Let us find the sampling distribution of the mean for samples of size 2. There are six samples of size 2 that can be chosen from this population: {A, B}, {A, C}, {A, D}, {B, C}, {B, D}, and {C, D}. Hence we have

Sample	X	$\overline{X}$
{A, B}	{11.9, 12.2}	12.05
{A, C}	{11.9, 12.0}	11.95
{A, D}	{11.9, 11.8}	11.85
{B, C}	{12.2, 12.0}	12.10
{B, D}	{12.2, 11.8}	12.00
{C, D}	{12.0, 11.8}	11.90

Thus the sampling distribution of the mean for samples of size 2 is the random variable $\overline{X}$ of the table. □

Since the sampling distribution of the mean for samples of size n is a random variable, it has a mean and a standard deviation, which we denote by $\mu_{\overline{x}}$ and $\sigma_{\overline{x}}$, respectively. The following result is extremely important in statistics. We present it without proof.

The Sampling Distribution of the Mean

If a random variable X defined on a population is normally distributed with mean μ and standard deviation σ, and if $\overline{X}$ is the sampling distribution of the mean for samples of size n chosen from the population, then $\overline{X}$ is normally distributed with mean $\mu_{\overline{x}} = \mu$ and standard deviation $\sigma_{\overline{x}} = \sigma/\sqrt{n}$.

This result allows us to find probabilities involving $\overline{X}$, and thus to make probabilistic statements about sample means $\overline{x}$, when we know μ, σ, and n. We illustrate with an example.

EXAMPLE 2

Suppose we knew that our random variable

$$X = \text{Weight of cereal boxes}$$

was normally distributed with mean $\mu = 12$ ounces and standard deviation $\sigma = 0.1$ ounce. Then the sampling distribution of the mean for samples of size $n = 25$ would be normally distributed with mean $\mu_{\overline{x}} = 12$ ounces and standard deviation $\sigma_{\overline{x}} = 0.1/\sqrt{25} = 0.02$ ounce. See Figure 8–20.

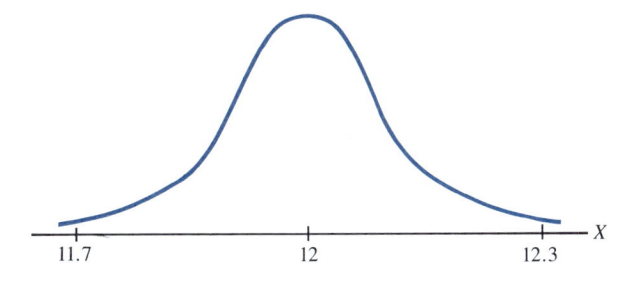

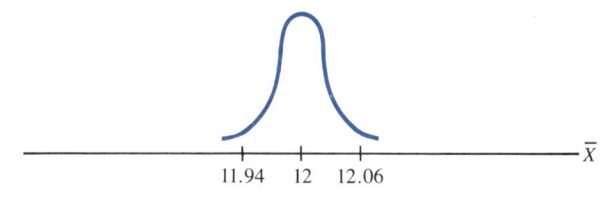

FIGURE 8–20

Now we can use the conversion formula

$$z = \frac{\overline{X} - \mu_{\overline{x}}}{\sigma_{\overline{x}}} = \frac{\overline{X} - 12}{0.02}$$

and the standard normal table to find probabilities for $\overline{X}$. For instance,

$$P(11.9608 \leq \overline{X} \leq 12.0392) = P\left(\frac{11.9608 - 12}{0.02} \leq z \leq \frac{12.0392 - 12}{0.02}\right)$$

$$= P(-1.96 \leq z \leq 1.96)$$

$$= 0.95.$$

Thus if we knew that X was normally distributed with mean 12 and standard deviation 0.1 and we took a sample of size 25 from the population, the probability that the interval [11.9608, 12.0392] would contain the sample mean $\overline{x}$ would be 0.95. Hence, we could be fairly certain that under these circumstances the mean of any sample of size 25 would lie in this interval. ☐

Of course, the result of Example 2 goes in the wrong direction for our purposes. We want to use the sample mean $\overline{x}$ to pin down the population mean μ; we do not want to use the population mean to pin down the sample mean (which is what we did in the example). However, the process used in Example 2 can be reversed, as we now demonstrate.

Let us suppose that our random variable X is normally distributed with an unknown mean μ and a standard deviation of 0.1 ounce, and that we intend to take a sample of size 25 from the population. We find from Table 6 that

$$P(-1.96 \leq z \leq 1.96) = 0.95.$$

But since $\overline{X}$ is normally distributed with mean $\mu_{\overline{x}} = \mu$ and standard deviation $\sigma_{\overline{x}} = \sigma/\sqrt{n}$,

$$z = \frac{\overline{X} - \mu_{\overline{x}}}{\sigma_{\overline{x}}} = \frac{\overline{X} - \mu}{\sigma/\sqrt{n}} = \frac{\overline{X} - \mu}{0.1/\sqrt{25}} = \frac{\overline{X} - \mu}{0.02}.$$

Thus

$$P\left(-1.96 \leq \frac{\overline{X} - \mu}{0.02} \leq 1.96\right) = 0.95,$$

or

$$P(-0.0392 \leq \overline{X} - \mu \leq 0.0392) = 0.95,$$

or

$$P(-\overline{X} - 0.0392 \leq -\mu \leq -\overline{X} + 0.0392) = 0.95,$$

or

$$P(\overline{X} + 0.0392 \geq \mu \geq \overline{X} - 0.0392) = 0.95,$$

or

$$P(\overline{X} - 0.0392 \leq \mu \leq \overline{X} + 0.0392) = 0.95.$$

Therefore, whatever $\overline{X}$ is, the probability is 0.95 that the interval

$$[\overline{X} - 0.0392, \overline{X} + 0.0392]$$

contains μ. For instance, if we now take a sample of size 25 and calculate its mean $\bar{x}$ to be 11.98 ounces, the probability is 0.95 that the interval

$$[11.98 - 0.0392, 11.98 + 0.0392] = [11.9408, 12.0192]$$

contains μ. The interval [11.9408, 12.0192] is called a **95% confidence interval for μ**.

The preceding analysis can be applied using any confidence level $100r\%$, $0 < r < 1$.

Confidence Intervals

Let X be a normally distributed random variable with mean μ and standard deviation σ, let $\bar{x}$ be the sample mean of a sample of size n, and let $0 < r < 1$. If z_r is such that

$$P(-z_r \leq z \leq z_r) = r,$$

then

$$\left[\bar{x} - z_r \frac{\sigma}{\sqrt{n}}, \bar{x} + z_r \frac{\sigma}{\sqrt{n}}\right]$$

is a **$100r\%$ confidence interval for μ**: the probability is r that the interval contains μ.

STATISTICS

EXAMPLE 3 Assume once again that

$$X = \text{weight of cereal in boxes}$$

is normally distributed with an unknown mean μ and a known standard deviation $\sigma = 0.1$ ounce. Suppose we take a sample of size 25 and find that its mean is $\bar{x} = 11.98$ ounces. If we want a 99% confidence interval for μ, we must find z_r such that

$$P(-z_r < z < z_r) = 0.99;$$

from Table 6, $z_r \approx 2.58$. Hence the 99% confidence interval for μ is (approximately)

$$\left[11.98 - 2.58\,\frac{0.1}{\sqrt{25}},\ 11.98 + 2.58\,\frac{0.1}{\sqrt{25}}\right] = [11.9284,\ 12.0316].$$

Comparing this with the 95% confidence interval obtained previously using the same values for σ, n, and $\bar{x}$, we see that increasing the confidence level makes the confidence interval larger. □

EXAMPLE 4 Let X, σ, and $\bar{x}$ be as in Example 3, but suppose now that our sample is of size 400. Then a 99% confidence interval for μ will be (approximately)

$$\left[11.98 - 2.58\,\frac{0.1}{\sqrt{400}},\ 11.98 + 2.58\,\frac{0.1}{\sqrt{400}}\right] \approx [11.9671,\ 11.9929].$$

Comparing this result with that of Example 3, we see that increasing the sample size makes the interval smaller. □

SUPPLEMENTARY EXERCISES

1. A random variable X is normally distributed with a standard deviation of 10. A sample of size 100 is taken, and it is found that the sample mean $\bar{x} = 42$. Find 90%, 95%, and 99% confidence intervals for the population mean μ.

2. A random variable X is normally distributed with a standard deviation of 0.5. A sample of size 400 is taken, and it is found that the sample mean $\bar{x} = 3$. Find 80%, 92%, and 97% confidence intervals for the population mean μ.

*3. A random variable X is normally distributed with a standard deviation of 12. How large a sample must be taken in order to have 90% confidence that the sample mean will be within ± 1 of the population mean?

*4. Repeat Exercise 3 using a confidence level of 99%.

Management

5. A machine fills cans with cola. The amount of fill is normally distributed with a standard deviation of 0.20 ounce. A sample of 25 filled cans is checked, and it is found that on average these contain 11.98 ounces of cola. Find 90%, 95%, and 99% confidence intervals for the average amount of cola in all cans filled by the machine.

6. An independent testing laboratory tests new cars. It is known from experience that mileage is normally distributed with a standard deviation of 2.4 miles per gallon (mpg). Suppose the lab tests 16 new X-cars and finds that they average 27 mpg. Find a 95% confidence interval for the average miles per gallon for all X-cars.

*7. Refer to Exercise 6: How large a sample should be taken if the lab wants to have 95% confidence that its average miles per gallon is within ± 0.5 of the true value?

*8. (Suggested by material in Meigs, Whittington, and Meigs, *Principles of Auditing*, eighth edition, Irwin, 1985.) Auditors wish to estimate the value of a large firm's accounts receivable by taking a sample of the accounts receivable and finding its mean value. The values of the accounts receivable are normally distributed. Suppose the auditors wish their result to be within $\pm$ $1 of the true mean with 95% confidence.

 (a) How large a sample should they take if the estimated standard deviation for the value of the accounts receivable is $25?
 (b) Suppose the sample mean turns out to be $\bar{x}$ = $422. If the firm has 20,000 accounts receivable and shows their value on its books as $9,500,000, should the auditors ask for an explanation?

*9. (Suggested by material in Stevenson, *Production/Operations Management*, second edition, Irwin, 1986.) A **control chart,** such as that pictured in the figure, is a graphical depiction of a confidence interval.

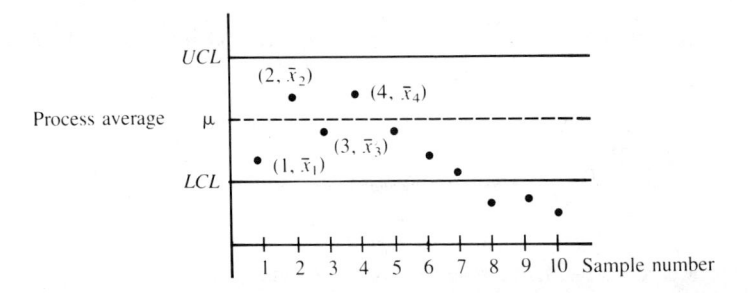

Control charts are used to check whether a production process is under control: If the sample means remain within the band defined by the lower control limit (LCL) and the upper control limit (UCL), the process is considered to be under control, but if they stray outside this band, the process is considered to be out of control. The interval [LCL, UCL] is just a confidence interval for the process average μ. In practice, the control limits are often determined by setting

$$\text{LCL} = \mu - \frac{3\sigma}{\sqrt{n}},$$

and

$$\text{UCL} = \mu + \frac{3\sigma}{\sqrt{n}},$$

where n is the sample size. When this is done, what confidence level is being utilized?

Life Science

10. Blood was taken from 36 patients who were on anticoagulant medication, and it was found that, on average, it took their blood 22 minutes to coagulate. It is known that coagulation time is normally distributed with a standard deviation of 2.5 minutes. Find a 90% confidence interval for the average coagulation time for all patients who are on the medication.

*11. Refer to Exercise 10: How large a sample should be taken if the researcher demands 98% confidence that the estimate for average coagulation time be correct to within ± 15 seconds?

Social Science

12. One hundred ninety-six college students are given an IQ test and their average score is 112. It is known that IQ scores are normally distributed with a standard deviation of 15. Find a 99% confidence interval for the average IQ score of all college students.

*13. Refer to Exercise 12: How large a sample must be taken in order to have 99.5% confidence that the average score of the sample is within ± 1 point of the average score for all college students?

9

The Mathematics of Finance

The major portion of this chapter is devoted to a discussion of **compound interest, present value,** and **annuities.** These financial topics are of great importance in the business world; a knowledge of them will allow us to answer questions such as the following:

- If $100,000 is invested at 8% interest per year compounded quarterly, what will the investment be worth at the end of five years?

- If a company can borrow money from one lender at 10% interest per year compounded quarterly and from another at 10.05% per year compounded semiannually, which lender should it choose?

- If at the end of each month a family deposits $100 in a savings account that pays 6% interest per year compounded monthly, how much will there be in the account at the end of 10 years?

- If a firm can invest $1 million in a product line that will return $200,000 annually for 10 years, or $1.2 million in another product line that will return $175,000 annually for 15 years, and if the firm will not make any investment that does not return at least 8% per year compounded annually, which of the product lines should it invest in?

We begin the chapter with a section devoted to arithmetic and geometric sequences, in which we develop some results we will need later and also briefly discuss simple interest. This is followed by a section that concerns compound interest and the present value of money. We conclude the chapter with a discussion of annuities.

9.1 ARITHMETIC AND GEOMETRIC SEQUENCES

A **sequence** is a function from the set of natural numbers **N** to the set of real numbers **R**. In this section we introduce two special types of sequences, **arithmetic sequences** and **geometric sequences,** and discuss their properties. We will need some of the results we develop here in the rest of this chapter as we discuss compound interest and annuities.

We have said that a sequence is a function from **N** to **R**. To illustrate, suppose we define a function a from **N** to **R** by setting

$$a(n) = n^2 + 1, \quad n \in \mathbf{N}.$$

This function is a sequence. The notation will be simpler if we write a_n for $a(n)$; if we do this, the sequence is defined by the equation

$$a_n = n^2 + 1, \quad n \in \mathbf{N}.$$

Thus,

$$a_1 = 1^2 + 1 = 2, \quad a_2 = 2^2 + 1 = 5, \quad a_3 = 3^2 + 1 = 10,$$

and so on. We may therefore consider the sequence to be the list of numbers

$$2, 5, 10, 17, 26, \ldots.$$

The numbers $a_1, a_2, a_3, \ldots$ are called the **terms** of the sequence; a_n is called the **nth term,** or **general term,** of the sequence. If the general term is given by an equation, as is the case here, we can use the equation to find any term of the sequence. For instance, since here $a_n = n^2 + 1$,

$$a_{12} = 12^2 + 1 = 145 \quad \text{and} \quad a_{200} = 200^2 + 1 = 40{,}001.$$

EXAMPLE 1 Let us write the first four terms and the tenth term of the sequences whose general terms are

$$a_n = 2^{-n} \quad \text{and} \quad b_n = \frac{3n}{n+1}.$$

We have

$$a_1 = 2^{-1} = \frac{1}{2}, \quad a_2 = 2^{-2} = \frac{1}{4}, \quad a_3 = 2^{-3} = \frac{1}{8}, \quad \text{and} \quad a_4 = 2^{-4} = \frac{1}{16}.$$

Therefore the sequence whose general term is a_n is

$$\tfrac{1}{2}, \tfrac{1}{4}, \tfrac{1}{8}, \tfrac{1}{16}, \ldots.$$

The tenth term of this sequence is

$$a_{10} = 2^{-10} = \frac{1}{1024}.$$

Similarly,

$$b_1 = \frac{3 \cdot 1}{1+1} = \frac{3}{2}, \quad b_2 = \frac{3 \cdot 2}{2+1} = 2,$$

$$b_3 = \frac{3 \cdot 3}{3+1} = \frac{9}{4}, \quad \text{and} \quad b_4 = \frac{3 \cdot 4}{4+1} = \frac{12}{5},$$

so the sequence is

$$\tfrac{3}{2}, 2, \tfrac{9}{4}, \tfrac{12}{5}, \ldots$$

and $b_{10} = \frac{30}{11}$. ◻

A sequence in which each term after the first is obtained by adding the same number to the previous term is called an **arithmetic sequence.** For instance, the sequence

$$2, 6, 10, 14, 18, \ldots$$

is an arithmetic sequence because each term after the first is obtained by adding 4 to the previous term. In general, if the first term of an arithmetic sequence is $a_1 = a$ and the number that is added to each term to obtain the next one is d, then the sequence has the form

$$a, a+d, (a+d)+d, ((a+d)+d)+d, \ldots,$$

or

$$a, a+d, a+2d, a+3d, \ldots.$$

Hence its nth term is

$$a_n = a + (n-1)d.$$

Arithmetic Sequence

A sequence defined by an equation of the form

$$a_n = a + (n-1)d$$

is an **arithmetic sequence** with **first term** a and **common difference** d.

EXAMPLE 2 The sequence defined by

$$a_n = 6 + (n-1)3$$

is arithmetic with first term $a = 6$ and common difference $d = 3$; it is the sequence

$$6, 9, 12, 15, 18, \ldots.$$

Now consider the sequence

$$17, 12, 7, 2, -3, -8, \ldots$$

The difference between any term and the preceding term is -5, so this sequence is arithmetic with first term $a = 17$ and common difference $d = -5$. Hence its general term is

$$a_n = 17 + (n - 1)(-5),$$

or

$$a_n = 17 - 5(n - 1).$$

Finally, the sequence

$$2, 8, 15, 23, \ldots$$

is not arithmetic because we do not obtain the same value every time we subtract a term from the next term. □

If an amount of money is invested, it is usual for the investor to receive a fee for allowing the borrower to use his or her money. The amount invested (or borrowed) is called the **principal,** and the fee is the **interest** on the principal. Interest is calculated as a percentage of the principal over some period of time.

If the interest earned at any time depends only on the principal and not on any interest previously accumulated, it is called **simple interest.** Thus, if we invest \$10,000 at 8% simple interest per year, each year we will earn 8% of \$10,000, or \$800, in interest. We will use arithmetic sequences to develop a formula for simple interest.

Suppose a principal of P dollars is invested at the simple interest rate of $i\%$ per year. This means that if $r = i/100$, then at the end of each year the investment earns rP dollars in interest. Therefore we have the following:

Year	Value of Investment
0	P
1	$P + rP$
2	$(P + rP) + rP = P + 2rP$
3	$(P + 2rP) + rP = P + 3rP$
.	.
.	.
.	.

Hence the values of the investment form an arithmetic sequence

$$P, P + rP, P + 2rP, P + 3rP, \ldots$$

and if A denotes the value of the investment at the end of t years, then

$$A = P + trP = P(1 + rt).$$

The Simple Interest Formula

If P dollars is invested at $i\%$ simple interest per year, $r = i/100$, and A is the value of the investment at the end of t years, then

$$A = P(1 + rt).$$

The value A is called the **future value** of the principal P at the end of t years.

EXAMPLE 3   If $10,000 is invested at 8% simple interest per year, then at the end of three years the value of the investment will be

$$A = \$10{,}000(1 + (0.08)3) = \$10{,}000(1.24) = \$12{,}400. \quad \square$$

It is sometimes necessary to add the first n terms of an arithmetic sequence. If the sequence is

$$a, a + d, a + 2d, \ldots,$$

the sum of its first n terms is

$$S_n = a + (a + d) + \cdots + (a + (n - 2)d) + (a + (n - 1)d).$$

If we reverse the order of the addition, we may write

$$S_n = (a + (n - 1)d) + (a + (n - 2)d) + \cdots + (a + d) + a;$$

adding these two expressions for S_n, we obtain

$$
\begin{aligned}
S_n &= a + (a + d) + \cdots + (a + (n - 1)d) \\
S_n &= (a + (n - 1)d) + (a + (n - 2)d) + \cdots + a \\
\hline
2S_n &= 2a + (n - 1)d + 2a + (n - 1)d + \cdots + 2a + (n - 1)d
\end{aligned}
$$

or

$$2S_n = n(2a + (n - 1)d).$$

The Sum of the First n Terms of an Arithmetic Sequence

The sum of the first n terms of an arithmetic sequence whose nth term is given by $a_n = a + (n - 1)d$ is

$$S_n = \frac{n}{2}(2a + (n - 1)d).$$

EXAMPLE 4 Consider the sequence

$$5, 9, 13, 17, \ldots,$$

that is, the arithmetic sequence defined by

$$a_n = 5 + 4(n - 1).$$

The sum of the first four terms of this sequence is

$$S_4 = \frac{4}{2}(2(5) + (4 - 1)4) = 44;$$

the sum of the first 100 terms of this sequence is

$$S_{100} = \frac{100}{2}(2(5) + (100 - 1)4) = 20{,}300. \quad \square$$

EXAMPLE 5 Suppose you receive a sequence of monthly payments from an investment, with the first payment being $400 and the payments increasing by $25 each month. How much will you receive over the first 12 months of payments?

To answer the question, we must sum the first 12 terms of the arithmetic sequence

$$\$400, \$425, \$450, \ldots.$$

Thus the answer is

$$S_{12} = \frac{12}{2}(2(\$400) + (12 - 1)(\$25)) = \$6450. \quad \square$$

Another important kind of sequence is the **geometric sequence:** a sequence is geometric if each of its terms after the first is obtained by multiplying the preceding term by the same number r, called the **common ratio** of the sequence. For instance, the sequence

$$3, 6, 12, 24, \ldots$$

is geometric with common ratio $r = 2$, since each term is two times the preceding one.

If a sequence is geometric with first term a and common ratio r, then it must have the form

$$a, ar, (ar)r, ((ar)r)r, \ldots,$$

or

$$a, ar, ar^2, ar^3, \ldots.$$

Hence its nth term will be ar^{n-1}.

THE MATHEMATICS OF FINANCE 559

> **Geometric Sequence**
>
> A sequence defined by an equation of the form
>
> $$a_n = ar^{n-1}$$
>
> is a **geometric sequence** with **first term** *a* and **common ratio** *r*.

EXAMPLE 6 The sequence defined by

$$a_n = 2 \cdot 3^{n-1}$$

is geometric with first term $a = 2$ and common ratio $r = 3$; it is the sequence

$$2, 6, 18, 54, 162, \ldots .$$

Now consider the sequence

$$6, -3, \tfrac{3}{2}, -\tfrac{3}{4}, \tfrac{3}{8}, -\tfrac{3}{16}, \ldots ;$$

If we divide any term of this sequence by the preceding one, we obtain $-\tfrac{1}{2}$, and thus the sequence is geometric with first term 6 and common ratio $-\tfrac{1}{2}$. Therefore it is defined by

$$a_n = 6\left(-\frac{1}{2}\right)^{n-1}.$$

Finally, the sequence

$$4, 8, 24, 96, \ldots$$

is not geometric because we do not obtain the same value every time we divide a term by the preceding term. □

When we discuss annuities later in this chapter, we will need the formula for the sum of the first *n* terms of a geometric sequence

$$a, ar, ar^2, \ldots .$$

This sum will be of the form

$$S_n = a + ar + ar^2 + \cdots + ar^{n-2} + ar^{n-1}.$$

If we multiply both sides of the preceding equation by $-r$, we obtain

$$-rS_n = -ar - ar^2 - ar^3 - \cdots - ar^{n-1} - ar^n.$$

Now we add these two equations:

$$\begin{array}{l} S_n = a + ar + ar^2 + \cdots + ar^{n-2} + ar^{n-1} \\ -rS_n = - ar - ar^2 - \cdots - ar^{n-2} - ar^{n-1} - ar^n \\ \hline (1-r)S_n = a \phantom{+ ar + ar^2 + \cdots + ar^{n-2} + ar^{n-1}} - ar^n \end{array}$$

Therefore, if $r \neq 1$,

$$S_n = \frac{a - ar^n}{1 - r}.$$

The Sum of the First n Terms of a Geometric Sequence

The sum of the first n terms of a geometric sequence whose nth term is given by $a_n = ar^{n-1}$, $r \neq 1$, is

$$S_n = \frac{a(1 - r^n)}{1 - r}.$$

EXAMPLE 7 The sum of the first four terms of the geometric sequence

$$2, 6, 18, 54, 162, \ldots,$$

that is, the sequence defined by

$$a_n = 2 \cdot 3^{n-1},$$

is

$$S_4 = \frac{2(1 - 3^4)}{1 - 3} = \frac{2(1 - 81)}{-2} = 80;$$

the sum of the first 15 terms of this sequence is

$$S_{15} = \frac{2(1 - 3^{15})}{1 - 3} = 3^{15} - 1 = 14{,}348{,}906. \quad \square$$

EXAMPLE 8 There is an ancient folktale concerning a peasant who saves the life of a king. In gratitude, the king offers to let the peasant name his own reward. The peasant, professing to be a simple man who knows nothing of the value of gold or jewels, asks only that he be rewarded with the value of all the wheat that can be accumulated by placing one grain on the first square of a chessboard, two grains on the second square, four grains on the third square, and so on. The king thinks he has gotten away cheaply until he calculates the number of grains whose value the peasant will receive: since a chessboard has 64 squares, this number is the sum of the first 64 terms of the geometric sequence

$$1, 2, 4, 8, 16, \ldots,$$

which is

$$S_{64} = \frac{1(1 - 2^{64})}{1 - 2} = 2^{64} - 1.$$

This is such a huge number ($\approx 18{,}450{,}000{,}000{,}000{,}000{,}000$) that even if a grain of wheat were worth only one one-millionth of a cent, the peasant would be due an amount far greater than the value of everything in the kingdom. $\quad \square$

9.1 EXERCISES

In Exercises 1 through 10 the general term of a sequence is given. Write out the first four terms of the sequence and state whether it is arithmetic, geometric, or neither. If it is arithmetic, identify its common difference; if it is geometric, identify its common ratio.

1. $a_n = 5 + 7n$
2. $a_n = 2 - \frac{3n}{2}$
3. $a_n = 2 + n^2$
4. $a_n = 3^{n-1}$
5. $a_n = 3 \cdot 4^{n-1}$
6. $a_n = \frac{1}{n}$
7. $a_n = \frac{n+1}{n}$
8. $a_n = -2 + 1.3n$
9. $a_n = 4^{n/2}$
10. $a_n = \left(\frac{2}{3}\right)^{n+1}$

In Exercises 11 through 22, state whether the given sequence is arithmetic, geometric, or neither. If it is arithmetic or geometric, find the equation for its nth term.

11. $2, 3, 5, 8, 12, \ldots$
12. $19, 13, 7, 1, -5, \ldots$
13. $\frac{2}{3}, 3, \frac{16}{3}, \frac{23}{3}, 10, \ldots$
14. $\frac{1}{2}, -\frac{1}{4}, \frac{1}{8}, -\frac{1}{16}, \frac{1}{32}, \ldots$
15. $\frac{1}{2}, \frac{1}{4}, \frac{1}{12}, \frac{1}{48}, \ldots$
16. $1, -\frac{1}{7}, \frac{1}{49}, -\frac{1}{343}, \ldots$
17. $3, 9, 27, 81, \ldots$
18. $-42, -25, -8, 9, \ldots$
19. $\frac{2}{3}, \frac{3}{4}, \frac{4}{5}, \frac{5}{6}, \ldots$
20. $\frac{2}{2}, \frac{3}{4}, \frac{5}{6}, \frac{11}{12}, 1, \ldots$
21. $68, 16, -36, -88, \ldots$
22. $4, 4, 4, 4, \ldots$

In Exercises 23 through 32, find the sum of the first n terms of the given sequence for the given value of n.

23. $2, 11, 20, 29, \ldots, n = 12$
24. $15, 8, 1, -6, \ldots, n = 8$
25. $1, \frac{1}{2}, \frac{1}{4}, \frac{1}{8}, \ldots, n = 7$
26. $5, 10, 20, 40, \ldots, n = 6$
27. $6, -4, \frac{8}{3}, -\frac{16}{9}, \ldots, n = 5$
28. $11, \frac{17}{2}, 6, \frac{7}{2}, \ldots, n = 40$
29. $1, 2, 3, 4, \ldots, n = 100$
30. $1, 2, 4, 8, \ldots, n = 10$
31. $2.6, 6.8, 11.0, 15.2, \ldots, n = 1000$
32. $20, 24, 28.8, 34.56, \ldots, n = 12$

*33. Find a formula for the sum $1 + 2 + 3 + \cdots + n$.

In Exercises 34 through 41, use the simple interest formula.

34. An investment of $12,000 earns 8% simple interest per year. How much is it worth at the end of 10 years?

35. A principal of $5000 is borrowed at 24% simple interest per year for 18 months. How much will the borrower owe at the end of this time?

36. How much has to be invested now at 6% simple interest per year in order to have $1000 seven years from now?

37. A borrower paid 12% simple interest per year on a loan, and at the end of 10 months owed $50,000. What was the amount of the loan?

38. How long will it take for a principal of $10,000 to grow to a future value of $19,900 at a simple interest rate of 9% per year?

39. If an investment earns simple interest at 10% per year, how long will it take for an investor to triple his or her investment?

40. An investment of $6000 earned simple interest each year for 11 years and at the end of this time was worth $21,000. What was the simple interest rate?

41. A borrower paid simple interest on a loan each week, and at the end of one year (52 weeks) the amount the borrower owed was 27 times the amount he borrowed. What was the interest rate he paid?

42. If you receive a sequence of payments from a trust fund, with the first payment being $1000 and the payments increasing by $100 each time, how much will you receive in the first 20 payments?

43. If you take a job for $25,000 per year and you get a raise of $1800 each year, how much will you be making during your tenth year on the job? How much will you make during your first 10 years on the job?

44. Repeat Exercise 43 if instead of an $1800 raise each year you get a 5% raise each year.

45. If you save $1 today, $2 tomorrow, $3 the next day, and so on, how much will you be saving 365 days from now? How much will you save over the course of the year?

46. Suppose you save 1 cent today, 2 cents tomorrow, 4 cents the next day, 8 cents the next, and so on. How much will you be saving 30 days from now? How much will you save over the course of the month?

*47. Consider the savings plan of Exercise 45: How long would it take you to save $88,410?

Management

48. A machine depreciates by 25% of its book value each year; its original book value was equal to its purchase price of $80,000.
    (a) The book values of the machine in year 1, year 2, . . . of its life form a sequence. Write the general term of this sequence.
    (b) What will the book value of the machine be in the seventh year of its life?

49. Consider again the machine of Exercise 48: The amounts by which it is depreciated each year form a sequence.
    (a) Write the general term of this sequence.
    (b) By how much will the machine be depreciated in the seventh year of its life?
    (c) What will the total amount of depreciation be over the first seven years of the machine's life?

Life Science

50. A certain strain of bacteria increases its population by 12% every hour. If a colony of this strain is started with 100 bacteria, how many will there be in the colony 24 hours later?

9.2 COMPOUND INTEREST AND PRESENT VALUE

We discussed simple interest in Section 9.1. In the business world, simple interest is usually encountered only when dealing with short-term loans and investments; longer-term financial dealings generally involve the use of **compound interest.** In this section we consider compound interest and introduce the concept of present value.

Compound Interest

If the interest earned on an investment depends not only on the principal but also on any interest that has previously accumulated, we say that the interest is **compounded.** For example, if we invest $10,000 at an interest rate of 8% per year compounded annually, then at the end of each year, we will earn 8% interest not only on the principal of $10,000 but also on any interest that has accumulated up to that time. Thus,

- at the end of one year we will have accumulated

$$\$10{,}000 + 0.08(\$10{,}000) = \$10{,}800;$$

- at the end of two years we will have accumulated

$$\$10{,}800 + 0.08(\$10{,}800) = \$11{,}664;$$

- at the end of three years we will have accumulated

$$\$11{,}664 + 0.08(\$11{,}664) = \$12{,}597.12;$$

and so on. (Contrast this with the investment of $10,000 at 8% simple interest discussed in Example 3 of Section 9.1.)

Interest may be compounded over any specified period of time. In the preceding illustration we compounded the interest annually (once each year), but interest is often compounded **semiannually** (at the end of every six months), **quarterly** (at the end of every three months), **monthly, daily,** and so on. Suppose, for example, that we invest $10,000 at an interest rate of 8% per year compounded quarterly. The stated yearly rate, in this case 8%, is called the **nominal interest rate.** Since the interest is compounded quarterly, it is compounded four times each year and hence is

$$\frac{8\%}{4} = 2\% \text{ per quarter.}$$

Thus,

- at the end of the first quarter we will have

$$\$10{,}000 + 0.02(\$10{,}000) = \$10{,}200;$$

- at the end of the second quarter we will have

$$\$10{,}200 + 0.02(\$10{,}200) = \$10{,}404;$$

- at the end of the third quarter we will have

$$\$10{,}404 + 0.02(\$10{,}404) = \$10{,}612.08;$$

- at the end of the fourth quarter we will have

$$\$10{,}612.08 + 0.02(\$10{,}612.08) = \$10{,}824.32;$$

and so on.

We develop a general formula for calculating compound interest. Suppose we invest principal P at an interest rate of $i\%$ per year compounded m times per year. Then the nominal interest rate is $i\%$ per year, and the interest rate for each compounding period is therefore $i\%/m$, or, in decimal form, $i/100m$. Let $r = i/100m$:

- at the end of the first compounding period we will have accumulated

$$P + rP = P(1 + r);$$

- at the end of the second compounding period we will have accumulated

$$P(1 + r) + rP(1 + r) = P(1 + r)(1 + r) = P(1 + r)^2;$$

- at the end of the third compounding period we will have accumulated

$$P(1 + r)^2 + rP(1 + r)^2 = P(1 + r)^2(1 + r) = P(1 + r)^3.$$

Thus the values of the investment at the ends of the compounding periods form the sequence

$$P(1 + r),\ P(1 + r)^2,\ P(1 + r)^3,\ \ldots\ .$$

But this is a geometric sequence with first term $P(1 + r)$ and common ratio $1 + r$, so its nth term is

$$P(1 + r)(1 + r)^{n-1} = P(1 + r)^n.$$

If the interest is left to accumulate for t years, then, since it is compounded m times per year, it will have been compounded $n = mt$ times. Hence at the end of t years the value of the investment will be $P(1 + r)^{mt}$ dollars.

The Compound Interest Formula

Let P be the principal invested at an annual interest rate of $i\%$ per year compounded m times per year. If A is the amount accumulated at the end of t years, then

$$A = P(1 + r)^{mt},$$

where $r = i/100m$ is the interest rate per compounding period, expressed as a decimal.

As with simple interest, the amount A is called the **future value** of the principal P.

EXAMPLE 1   We invest $1000 at the rate of 6% per year compounded annually. How much will we have at the end of three years?

Here $P = \$1000$, $i\% = 6\%$, $m = 1$, and $t = 3$. Hence

$$r = \frac{6}{100(1)} = 0.06$$

and

$$A = \$1000(1 + 0.06)^{1 \cdot 3} = \$1000(1.06)^3.$$

Now we must calculate $(1.06)^3$. We can do this using a calculator or with the aid of Table 1 at the back of the book. (Table 1 gives the value of $(1 + r)^k$ for various values of r and k.) Using a calculator with an x^y key, we enter 1.06, hit the x^y key, then enter 3, and find that $(1.06)^3 = 1.191016$. Therefore

$$A = \$1000(1.191016)$$

or, rounded to the nearest cent,

$$A = \$1191.02.$$

Note that of the total of $1191.02, $1000 is principal and hence the interest earned over the three-year period is

$$\$1191.02 - \$1000 = \$191.02. \quad \square$$

EXAMPLE 2   We invest $1000 at the rate of 6% per year compounded semiannually. How much will we have at the end of three years?

Here $P = \$1000$, $i\% = 6\%$, $m = 2$, and $t = 3$. Hence

$$r = \frac{6}{100(2)} = 0.03$$

and

$$A = \$1000(1 + 0.03)^{2 \cdot 3} = \$1000(1.03)^6.$$

Since $(1.03)^6 = 1.194052$, to six decimal places, the amount A, rounded to the nearest cent, is

$$A = \$1000(1.194052) = \$1194.05.$$

The interest earned over the three-year period is thus

$$\$1194.05 - \$1000 = \$194.05. \quad \square$$

Note that P, i, and t in Example 2 were the same as they were in Example 1, but that the interest earned in Example 2 was greater than that earned in Example 1. This occurred because we compounded more frequently in Example 2. In general, for a given nominal interest rate $i\%$, the more often we compound, the faster the interest accumulates.

Effective Interest Rates

Suppose we invest money at a rate of 6% per year compounded quarterly. Let us find the amount an investment of $1.00 will yield in one year.

We take $P = \$1$, $i = 6$, $m = 4$, and $t = 1$. Then $r = 0.015$ and

$$A = \$1(1.015)^4 = \$1(1.0613635) = \$1.0613635.$$

Of the amount $1.0613635, $1.00 is principal; hence the interest earned is $0.0613635, which is 6.13635% of the $1.00 invested. Thus, a nominal rate of 6% compounded quarterly is equivalent to an **effective rate** of 6.13635% compounded annually. If we invest principal P at the nominal rate of 6% compounded quarterly, at the end of one year we will have the same amount as if we had invested P at the effective rate of 6.13635% compounded annually.

The concept of effective rate per year is important because it enables us to compare nominal rates that have different compounding periods.

EXAMPLE 3 Acme Company wishes to borrow money to finance a new plant. Three banks have quoted Acme Company different interest rates. Bank X will lend Acme the money at 8% per year compounded quarterly. Bank Y will lend it the money at 8.05% compounded semiannually. Bank Z will lend it the money at 8.25% compounded annually. From which bank should Acme borrow the money?

We cannot compare the three interest rates directly since the periods over which they are compounded are different. To obtain comparable interest rates, we must convert each of the given nominal rates to its equivalent effective rate per year.

- For each dollar borrowed from Bank X, Acme would have to pay

$$\$1\left(1 + \frac{8}{100(4)}\right)^4 = \$1(1.02)^4 = \$1.0824322$$

in principal and interest at the end of a year. Thus, the interest would be $0.0824322 for an effective rate of 8.24322%.

- For each dollar borrowed from Bank Y, Acme would have to pay

$$\$1\left(1 + \frac{8.05}{100(2)}\right)^2 = \$1(1.04025)^2 = \$1.0821201$$

in principal and interest at the end of a year. Thus, the interest would be $0.0821201 for an effective rate of 8.21201%.

- Since the interest at Bank Z is compounded annually, the nominal rate and the effective rate per year are equal. Thus, the effective rate is 8.25%.

Comparing the effective rates, we see that the one quoted by Bank Y is the lowest, so Acme should borrow the money from Bank Y. ☐

Present Value

The amount P that must be invested now in order to have a specified amount A in the future is called the **present value** of A.

EXAMPLE 4

How much must we invest now at 8% per year compounded quarterly in order to have exactly $10,000 in 2.5 years?

Here $i\% = 8\%$, $m = 4$, $t = 2.5$, and the future value $A = \$10,000$. We must find P. We have

$$r = \frac{8}{400} = 0.02,$$

and thus

$$\$10,000 = P(1.02)^{10}.$$

Since $(1.02)^{10} = 1.218994$, this becomes

$$\$10,000 = P(1.218994),$$

so

$$P = \frac{\$10,000}{1.218994} = \$8203.49.$$

Thus we say that $8203.49 is the present value of $10,000 due 2.5 years from now if money is worth 8% per year compounded quarterly. ☐

Note that the present value of an amount due in the future is less than the future amount; the difference between the future amount and its present value measures the cost of waiting for the future amount.

Since the compound interest formula relates the future value A to the present value P, we need only solve the formula for P to obtain the present value formula.

The Present Value Formula

Let A be an amount that is due t years in the future. Suppose money is worth $i\%$ per year compounded m times per year, and let $r = i/100m$. Then the present value P of amount A is given by

$$P = A(1 + r)^{-mt}$$

The process of finding the present value of a future amount is often referred to as **discounting** the future amount. The quantity $(1 + r)^{-mt}$ may be found with the aid of a calculator or by looking it up in Table 2 in the back of the book.

EXAMPLE 5

Let us find the present value of $10,000 due five years from now if money is worth 6% compounded quarterly.

Here the future value $A = \$10,000$, $i = 6$, $m = 4$, and $t = 5$, so $r = 0.015$ and

$$P = \$10,000(1.015)^{-20}.$$

Using a calculator with an x^y key, we find that $(1.015)^{20} = 1.346855$, so

$$(1.015)^{-20} = \frac{1}{(1.015)^{20}} = \frac{1}{1.346855} = 0.742470.$$

Hence

$$P = \$10,000(0.742470) = \$7424.70.$$

Thus if money is worth 6% per year compounded quarterly, the present value of $10,000 due five years from now is $7424.70; another way to express this is to say that $10,000 discounted for five years at 6% per year compounded quarterly is $7424.70. In either case, what is meant is that $7424.70 invested now at 6% per year compounded quarterly will amount to $10,000 five years from now, or, said another way, if a return on investment of 6% per year compounded quarterly is required, then $10,000 due five years from now is worth $7424.70 today. ◻

It is important to realize that the longer we have to wait for a future amount, the smaller its present value will be.

EXAMPLE 6 If money is worth 12% per year compounded quarterly, $10,000 due five years from today has a present value of

$$P = \$10,000(1.03)^{-20}$$
$$= \$10,000(0.553676)$$
$$= \$5536.76.$$

The same amount at the same interest rate due 10 years from today has a present value of only

$$P = \$10,000(1.03)^{-40}$$
$$= \$10,000(0.306557)$$
$$= \$3065.57. \quad ◻$$

Present values are important because they allow us to compare future amounts with present amounts. They are widely used, for instance, in making decisions about investments. The usual method of doing this is to consider the **net present value** (NPV) of the investment; this is the present value of its future payoff minus its cost. If the NPV of an investment is positive, its present value is greater than its cost, and it is desirable; if its NPV is negative, its present value is less than its cost, and it is not desirable. We illustrate with an example.

EXAMPLE 7 Gamma Corporation is considering the purchase of a new press. A Brand X press will cost $60,000 and will save the company $85,000 three years from now; a Brand Y press will cost $50,000 and will save $64,000 two years from now. Gamma has a policy that no investment will be undertaken unless it will earn at least 12% per year compounded monthly. Which press, if either, should Gamma purchase?

In order to answer this question, we must find the net present value of the future savings due to each press. Since Gamma evaluates investments at 12% per year compounded monthly, we have

$$r = \frac{12}{12(100)} = 0.01.$$

Thus the present value of the future savings due to press X is

$$P = \$85{,}000(1.01)^{-36} = \$85{,}000(0.698925) = \$59{,}408.63.$$

The NPV of press X is its present value minus its cost, or

$$\$59{,}408.63 - \$60{,}000 = -\$591.37.$$

Since the NPV of press X is negative, investment in the press will not return the required 12% per year compounded monthly to Gamma; hence Gamma will not purchase press X.

On the other hand, the present value of the savings due to press Y is

$$P = \$64{,}000(1.01)^{-24} = \$64{,}000(0.787566) = \$50{,}404.23,$$

and therefore its net present value is

$$\$50{,}404.23 - \$50{,}000 = \$404.23.$$

Since this NPV is positive, investment in press Y will return more than the required 12% per year compounded monthly to Gamma; hence Gamma will purchase press Y. ☐

■ *Computer Manual: Programs INTEREST, INTTABLE, PRSNTVAL*

9.2 EXERCISES

Compound Interest

1. How much will you have at the end of one year if you invest $15,000 at 8% per year compounded annually?
2. Repeat Exercise 1 if the interest is compounded semiannually.
3. Repeat Exercise 1 if the interest is compounded quarterly.
4. How much will you owe at the end of one year if you borrow $20,000 at 12% per year compounded annually?
5. Repeat Exercise 4 if the interest is compounded semiannually.
6. Repeat Exercise 4 if the interest is compounded quarterly.
7. Repeat Exercise 4 if the interest is compounded monthly.
8. How much will you owe at the end of two years if you borrow $8000 at 6% per year compounded semiannually?

9. Repeat Exercise 8 if the interest is compounded quarterly.
10. Repeat Exercise 8 if the interest is compounded monthly.
11. How much will you have at the end of five years if you invest $50,000 at 8% per year compounded annually?
12. Repeat Exercise 11 if the interest is compounded semiannually.
13. Repeat Exercise 11 if the interest is compounded quarterly.
14. You borrow $7000 at 10% per year compounded quarterly. If you pay back the loan at the end of six years, how much will you pay?
15. Repeat Exercise 14 if you pay back the loan at the end of three years.
16. Repeat Exercise 14 if you pay back the loan at the end of two years and nine months.
17. You invest $22,000 at 12% per year compounded monthly. How much will you have at the end of three years?
18. Repeat Exercise 17 if the question is how much you will have at the end of three years and two months.
19. Repeat Exercise 17 if the question is how much you will have at the end of three years and five months.
20. Repeat Exercise 17 if the question is how much you will have at the end of three years and 10 months.

Effective Interest Rates

21. What is the effective interest rate paid if money is borrowed at 12% per year compounded annually?
22. Repeat Exercise 21 if the nominal rate is 12% per year compounded semiannually.
23. Repeat Exercise 21 if the nominal rate is 12% per year compounded quarterly.
24. What is the effective interest rate paid if money is borrowed at 6% per year compounded semiannually?
25. Repeat Exercise 24 if the nominal rate is 6% per year compounded quarterly.
26. Repeat Exercise 24 if the nominal rate is 6% per year compounded monthly.
27. What effective interest rate is equivalent to a nominal rate of 9.95% per year compounded semiannually?
28. Repeat Exercise 27 if the nominal rate is 9.90% per year compounded quarterly.

Management

29. Alpha Company intends to borrow money to purchase new equipment. Bank A will lend Alpha the money at 9.10% per year compounded annually. Bank B will lend the company the money at 9% per year compounded semiannually. Bank C will lend them the money at 8.96% per year compounded monthly. From which bank should Alpha borrow the money?
30. Beta Company can invest its earnings at 6% per year compounded monthly, at 6.15% per year compounded semiannually, or at 6.20% per year compounded annually. Which rate should Beta choose?

*31. Gamma Corporation can invest its earnings in investment A, which will pay 8% per year compounded monthly, or in investment B, which will pay i% per year compounded annually. What must i be in order for investment B to be preferable to investment A?

*32. Repeat Exercise 31 if investment A will pay 8% per year compounded quarterly.

Present Value

33. How much should you invest now at 8% per year compounded semiannually in order to have $10,000 five years from now?

34. Repeat Exercise 33 if the interest is compounded quarterly.

35. How much should you invest now at 6% per year compounded monthly in order to have $14,000 three years and four months from now?

36. Repeat Exercise 35 if you want to have $14,000 three and one-half years from now.

37. Find the present value of $12,000 due four years from now if money is worth 6% per year compounded annually.

38. Repeat Exercise 37 if the compounding is done semiannually.

39. Repeat Exercise 37 if the compounding is done quarterly.

40. Repeat Exercise 37 if the compounding is done monthly.

41. Find the present value of $30,000 due three years from now if money is worth 8% per year compounded semiannually.

42. Repeat Exercise 41 if the $30,000 is due 4 years from now.

43. Repeat Exercise 41 if the $30,000 is due 5 years from now.

44. Repeat Exercise 41 if the $30,000 is due 10 years from now.

45. Discount $50,000 for three years at 12% per year compounded monthly.

46. Repeat Exercise 45 if the discount period is 3.5 years.

47. Repeat Exercise 45 if the discount period is 3.75 years.

48. Repeat Exercise 45 if the discount period is 4 years.

49. You are considering buying 500 shares of a stock that sells for $10 per share. The stock pays no dividends but you are certain that you will be able to sell it three years from now at $15 per share. You will not make any investment unless it returns at least 6% per year compounded quarterly. Should you buy the stock?

50. Repeat Exercise 49 if you require a return of at least 16% per year compounded quarterly.

51. Repeat Exercise 49 if you require a return of at least 14% per year compounded annually.

Management

52. Alpha Corporation is considering the purchase of a new computer. The firm has three models to choose from: X, Y, and Z. Model X will cost $50,000 now and will save Alpha $70,000 two years from now. Model Y will cost $75,000 now and save $100,000 three years from now. Model Z will cost $100,000 now and will save $150,000 four years from now. Alpha will not make any investment unless it returns at least 10% per year compounded quarterly. What should Alpha do?

53. Repeat Exercise 52 if Alpha will not make any investment unless it returns at least 6% per year compounded quarterly.

54. Repeat Exercise 52 if Alpha will not make any investment unless it returns at least 20% per year compounded semiannually.

55. Beta Company can invest in one or more of three new product lines. Product line A will cost Beta $1.5 million and will return $2.8 million four years later. Product line B will cost $1.8 million and will return $3.3 million six years later. Product line C will cost $2.4 million and will return $4.3 million eight years later. If Beta requires a return of at least 10% per year compounded annually on any investment, which, if any, of the product lines should it invest in?

56. Repeat Exercise 55 if Beta requires a return of at least 14% per year compounded annually.

57. Repeat Exercise 55 if Beta requires a return of at least 14% per year compounded annually and product line C will cost only $1.5 million.

Short-term loans are often written on a **discounted basis.** This means that the borrower receives the present value of the loan and pays back the amount of the loan. Exercises 58 through 61 are concerned with discounted loans.

*58. Suppose Gamma Company borrows $100,000 discounted at 6% per year compounded quarterly for six months. Then Gamma will receive the present value of the $100,000, and six months later will pay back $100,000. How much will Gamma receive?

*59. Repeat Exercise 58 if the discount rate is 12% per year compounded annually and the loan is for eight months.

*60. Suppose Epsilon Company can borrow $1 million discounted at 12% per year compounded quarterly for six months from Bank X, and can also borrow $1 million discounted at 12.24% per year compounded semiannually for six months from Bank Y. On which loan would Epsilon pay more interest?

*61. Repeat Exercise 60 if Bank Y uses a discount rate of 12.18% per year compounded semiannually.

Businesses sometimes **factor** their accounts receivable: This means that they sell them at a discount to another business that then collects the money from the customer at the time it is due. The business buying the accounts receivable usually pays the business selling them an amount equal to their present value, calculated at some agreed-upon discount rate. Exercises 62 through 65 are concerned with factoring.

*62. Aardvark Appliances sells its goods on terms of 30% down with the balance payable at the end of six months. Aardvark factors its accounts receivable with the Filbert Finance Company, which discounts Aardvark's accounts receivable at 12% per year compounded monthly. If Aardvark sells $5000 worth of appliances today, how much will it receive from Filbert? (*Hint:* Aardvark's accounts receivable will be $5000 − 0.3($5000).)

*63. Repeat Exercise 62 if Aardvark sells $15,000 worth of appliances today.

*64. Repeat Exercise 62 if Aardvark sells $20,000 worth of appliances today and Filbert discounts Aardvark's accounts receivable at 6% per year compounded quarterly.

*65. Repeat Exercise 62 if Aardvark sells $25,000 worth of appliances today and Filbert discounts Aardvark's accounts receivables at 9% per year compounded semiannually.

9.3 ANNUITIES

In this section we discuss annuities. An **annuity** is a sequence of payments made at regular intervals. For example, if you deposit $100 in a savings account at the end of each month, you are creating a monthly annuity. We will consider annuities in which payments are made at the end of each compounding period and for which all payments are equal.

Ordinary Annuities

Suppose we create an annuity by depositing $100 at the end of each month in a savings account that pays 6% per year compounded monthly. An annuity such as this is called an **ordinary annuity** because the payments are all equal and are made at the *end* of each compounding period. What will be the value of this annuity at the end of three months? That is, how much will be in the savings account at the end of three months?

At the end of the first month we will have just made the first payment of $100 into the account. This payment will earn interest for two months (until the end of the third month), so the compound interest formula tells us that it will accumulate to

$$100(1.005)^2 = \$101.0025.$$

The $100 we deposit at the end of the second month will earn interest for only one month, so it will accumulate to

$$100(1.005)^1 = \$100.50.$$

The $100 we deposit at the end of the third month will not earn any interest before the end of the third month, so it will "accumulate" to $100. Therefore, the total amount in the account at the end of three months will be

$$\$101.0025 + \$100.50 + \$100 = \$301.5025.$$

This is the value of this three-month annuity.

We develop a general formula for the value of an annuity. Assume that we make equal payments of amount R at the end of each compounding period and that each payment earns interest at the rate of $i\%$ per year compounded m times per year. As usual we let $r = i/100m$, and we let S be the value of the annuity at the end of t years. The first payment of amount R will earn interest for $mt - 1$ compounding periods and so will accumulate to

$$R(1 + r)^{mt-1}.$$

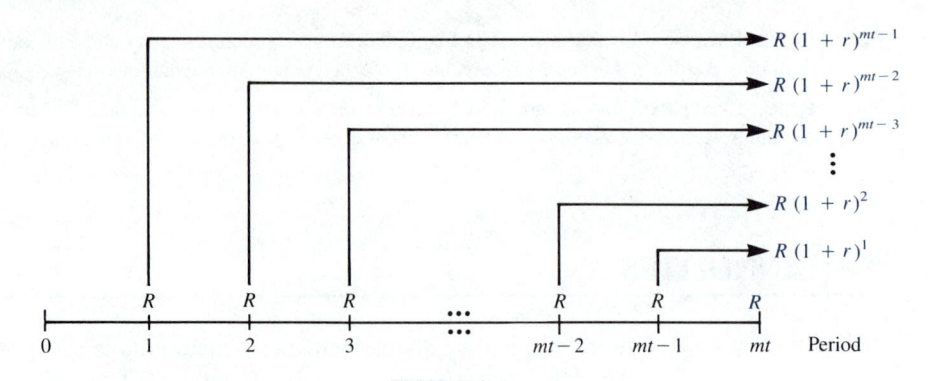

FIGURE 9–1
The Value of the Annuity Is the Sum of the Terms in Color

See Figure 9–1. The second payment will earn interest for $mt - 2$ compounding periods and so will accumulate to

$$R(1 + r)^{mt-2},$$

and so on. The next-to-last payment will earn interest for only one compounding period, so it will accumulate to

$$R(1 + r)^1,$$

and the last payment will earn no interest at all and so will "accumulate" to R. Therefore, the value of the annuity will be

$$S = R(1 + r)^{mt-1} + R(1 + r)^{mt-2} + \cdots + R(1 + r)^1 + R.$$

But this is the sum of the first mt terms of the geometric sequence with first term R and common ratio $1 + r$. Hence,

$$S = \frac{R[1 - (1 + r)^{mt}]}{1 - (1 + r)},$$

or

$$S = \frac{R[(1 + r)^{mt} - 1]}{r}.$$

The Ordinary Annuity Formula

Let an annuity be created with equal payments of amount R at the *end* of each compounding period, where each payment earns interest at the rate of $i\%$ per year compounded m times per year. Let $r = i/100m$, and let S denote the value of the annuity at the end of t years. Then

$$S = \frac{R[(1 + r)^{mt} - 1]}{r}.$$

THE MATHEMATICS OF FINANCE 575

It is common practice to set

$$s_{\overline{k}|r} = \frac{(1+r)^k - 1}{r}$$

for every positive integer k and hence to write the annuity formula as

$$S = R s_{\overline{mt}|r}.$$

Table 3 in the back of the book gives the value of $s_{\overline{k}|r}$ for various values of k and r.

EXAMPLE 1 An annuity is created by depositing $100 at the end of each month in a savings account that pays 6% per year compounded monthly. What is the value of the annuity at the end of four years?

Here we have $R = \$100$, $i = 6$, $m = 12$, and $t = 4$. Thus $r = 0.005$, and

$$S = \frac{\$100[(1.005)^{48} - 1]}{0.005}.$$

Now we could use a calculator (or Table 1) to find the value of $(1.005)^{48}$, then substitute this value in the formula and find S. However, since we have an annuity table available, it is easier to write

$$S = \$100 s_{\overline{48}|0.005}$$

and find the value of $s_{\overline{48}|0.005}$ from Table 3. We find that

$$s_{\overline{48}|0.005} = 54.097832,$$

so

$$S = \$100(54.097832) = \$5409.78,$$

to the nearest cent. ☐

EXAMPLE 2 Find the value of an annuity that is created by investing $1000 at the end of each quarter at 8% per year compounded quarterly for 10 years.

The answer is

$$S = \$1000 s_{\overline{40}|0.02} = \$1000(60.401983) = \$60{,}401.98. \quad \square$$

EXAMPLE 3 The value of an annuity can exceed by far the total amount paid in if the interest is compounded frequently over a long period of time. For example, suppose that you begin at the age of 20 to deposit $100 at the end of each month in an account that pays 9% per year compounded monthly. Then when you are 65, the value of the annuity will be

$$S = \frac{\$100[(1.0075)^{540} - 1]}{0.0075}.$$

Using a calculator with an x^y key, we find that

$$(1.0075)^{540} = 56.536588.$$

Hence

$$S = \frac{\$100[56.536588 - 1]}{0.0075},$$

or

$$S = \$740{,}487.85,$$

of which the amount deposited is only $54,000. ☐

A **sinking fund** is an annuity created in order to have a specified amount of money at some future time. The following examples illustrate the uses of sinking funds.

EXAMPLE 4   The Smiths will need $40,000 fifteen years from now to finance their child's college education. They wish to create a sinking fund by making payments at the end of each quarter into a savings account that pays 6% per year compounded quarterly. How much should each payment be if they want to have $40,000 in the sinking fund 15 years from now?

Here we have an annuity whose value 15 years from now should be $40,000. Since $i = 6$, $m = 4$, $t = 15$, and $S = \$40{,}000$, we have

$$\$40{,}000 = \frac{R[(1.015)^{60} - 1]}{0.015} = Rs_{\overline{60}|0.015}$$

or

$$\$40{,}000 = R(96.214652).$$

Thus,

$$R = \$415.74,$$

so the Smiths must pay $415.74 into the sinking fund at the end of each quarter. ☐

EXAMPLE 5   It is your goal to have a nest egg of $1 million when you retire at the age of 60. If you are now 20 and estimate that you can earn 8% per year compounded annually on your investments, how much must you invest each year in order to reach your goal?

This is equivalent to asking what the annual payment must be to a sinking fund that pays 8% per year compounded annually if the fund is to amount to $1 million 40 years from now. Thus we have

$$\$1{,}000{,}000 = \frac{R[(1.08)^{40} - 1]}{0.08} = \frac{R[21.724521 - 1]}{0.08},$$

or

$$\$1{,}000{,}000 = R[259.05652].$$

THE MATHEMATICS OF FINANCE 577

Hence

$$R = \frac{\$1{,}000{,}000}{259.05652} = \$3860.16. \quad \square$$

We have thus far considered only ordinary annuities, those to which equal payments R are made at the *end* of each compounding period. An annuity to which equal payments are made at the *beginning* of each compounding period is called an **annuity due.** A formula for annuities due is given in Exercise 28 of Exercises 9.3.

The Present Value of an Annuity

We turn now to the concept of the **present value of an annuity,** or, as it is sometimes called, the **present value of a future stream of payments.** Suppose a creditor owes us $300 and agrees to pay us $100 at the end of each month for three months. What is the present value of this three-month annuity if money is worth 6% per year compounded monthly?

The present value of each payment is $\$100(1.005)^{-k}$, where k is the number of months we must wait for the payment. Thus, since we must wait one month for the first payment, two months for the second, and three months for the third, their present values are, respectively,

$$\$100(1.005)^{-1} = \$99.5025,$$

$$\$100(1.005)^{-2} = \$99.0075,$$

and

$$\$100(1.005)^{-3} = \$98.5149.$$

Therefore, the present value V of the annuity, to the nearest cent, is

$$V = \$99.5025 + \$99.0075 + \$98.5149 = \$297.02.$$

Hence the future stream of the three $100 payments is worth $297.02 to us now.

We develop a general formula for the present value V of an annuity. We assume that equal payments of amount R are to be received at the end of each compounding period for t years and that money is worth $i\%$ per year compounded m times per year. Then, letting $r = i/100m$, we have

$$V = R(1 + r)^{-1} + R(1 + r)^{-2} + \cdots + R(1 + r)^{-mt},$$

since we must wait one compounding period for the first payment, two compounding periods for the second payment, . . . , and mt compounding periods for the last payment. See Figure 9–2. But then V is the sum of the first mt terms of the geometric sequence with first term $R(1 + r)^{-1}$ and common ratio $(1 + r)^{-1}$. Therefore

$$V = \frac{R(1 + r)^{-1}[1 - (1 + r)^{-mt}]}{1 - (1 + r)^{-1}},$$

$$R(1+r)^{-mt}$$
$$R(1+r)^{-mt+1}$$
$$R(1+r)^{-mt+2}$$
$$\vdots$$
$$R(1+r)^{-3}$$
$$R(1+r)^{-2}$$
$$R(1+r)^{-1}$$

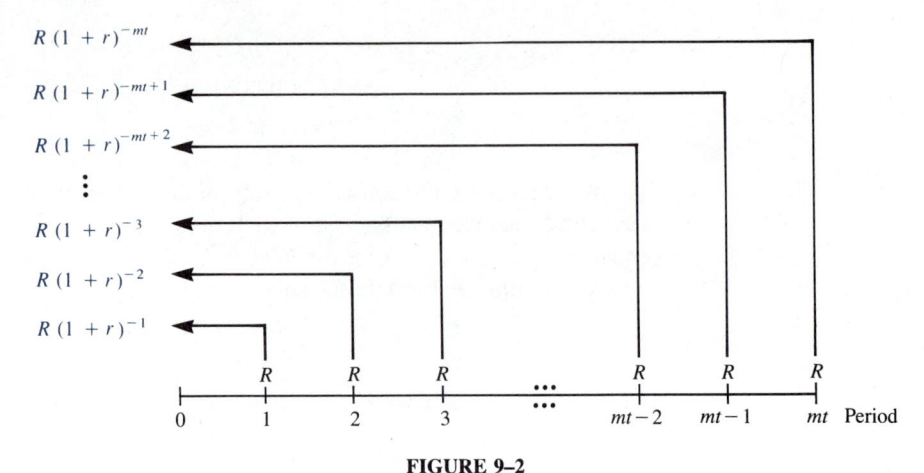

FIGURE 9–2
The Present Value of the Annuity Is the Sum of the Terms in Color

which simplifies to

$$V = \frac{R[1 - (1+r)^{-mt}]}{r}.$$

The Present Value Formula for an Ordinary Annuity

Let equal payments of amount R be due at the end of each compounding period for t years, and suppose that money is worth $i\%$ per year, compounded m times per year. Let $r = i/100m$, and let V be the present value of the annuity (that is, of the future stream of payments). Then

$$V = \frac{R[1 - (1+r)^{-mt}]}{r}.$$

It is common practice to set

$$a_{\overline{k}|r} = \frac{1 - (1+r)^{-k}}{r}$$

for every positive integer k and thus to write the present value formula in the form

$$V = R a_{\overline{mt}|r}.$$

Table 4 in the back of the book gives the value of $a_{\overline{k}|r}$ for various values of k and r.

EXAMPLE 6 We are to receive payments of $100 at the end of each month for the next four years. Let us find the present value of this annuity if money is worth 6% per year compounded monthly. We have $R = 100$, $m = 12$, and $t = 4$, so $r = 0.005$ and

THE MATHEMATICS OF FINANCE

$$V = \frac{\$100[1 - (1.005)^{-48}]}{0.005}.$$

We could use a calculator (or Table 2) to find the value of $(1.005)^{-48}$, then substitute this value in the formula and find V. However, since we have Table 4 available, it is easier to write

$$V = \$100 a_{\overline{48}|0.005}$$

and find

$$a_{\overline{48}|0.005} = 42.580318$$

from Table 4. Hence

$$V = \$100(42.580318) = \$4258.03. \quad \square$$

EXAMPLE 7 Find the present value of a future stream of payments received at the end of each quarter for 10 years if each payment is $1000 and money is worth 8% compounded quarterly. The answer is

$$V = \$1000 a_{\overline{40}|0.02} = \$1000(27.355479) = \$27{,}355.48. \quad \square$$

EXAMPLE 8 An investor has just purchased a bond that will pay her $500 at the end of every six months for the next 20 years. If she discounts this stream of payments at 10% per year compounded semiannually, what is its present value? The answer is

$$V = \frac{\$500[1 - (1.05)^{-40}]}{0.05} = \frac{\$500[1 - 0.1420456]}{0.05} = \$8579.54.$$

Notice that the investor will actually receive payments totalling

$$40(\$500) = \$20{,}000. \quad \square$$

Business decisions often require finding the net present value (NPV) of annuities in order to compare the payoffs of different investments. We illustrate with an example.

EXAMPLE 9 Alpha Company is considering buying a new drill press. There are three presses from which to choose. Press X costs $100,000 and will save Alpha $20,000 per year over the next 10 years. Press Y costs $160,000 and will save Alpha $25,000 per year over the next 12 years. Press Z costs $240,000 and will save Alpha $28,000 per year over the next 15 years. It is the policy of Alpha Company that an investment will not be made unless it returns at least 8% per year compounded annually. Which of the presses, if any, will the firm buy?

If we think of the savings generated by a press as occurring at the end of a year, then the savings form an annuity for Alpha Company. We must find the net present value (= present value − cost) of each annuity. For press X, we have

$$V = \$20{,}000 a_{\overline{10}|0.08} = \$20{,}000(6.710081) = \$134{,}201.62.$$

Hence the NPV for press X is

$$\$134{,}201.62 - \$100{,}000 = \$34{,}201.62.$$

Since this NPV is positive, investment in press X will return more than the required 8% per year to Alpha. For press Y, we have

$$V = \$25{,}000 a_{\overline{12}|0.08} = \$25{,}000(7.536078) = \$188{,}401.94.$$

Hence the NPV for press Y is

$$\$188{,}401.94 - \$160{,}000 = \$28{,}401.94.$$

Since this NPV is positive, investment in press Y will return more than the required 8% per year to Alpha. For press Z, we have

$$V = \$28{,}000 a_{\overline{15}|0.08} = \$28{,}000(8.559479) = \$239{,}665.40.$$

Hence the NPV for press Z is

$$\$239{,}665.40 - \$240{,}000 = -\$334.60.$$

Since this NPV is negative, investment in press Z will not return the required 8% per year to Alpha. Therefore Alpha will not buy press Z. To decide between press X and press Y, Alpha will choose the one having the larger NPV, since this is the one that offers the larger return on investment. Thus Alpha will purchase press X. □

We have considered the present value of ordinary annuities here. Similar results may be obtained for annuities due. Exercise 54 of Exercises 9.3 gives a formula for the present value of an annuity due.

■ *Computer Manual: Programs ANNUITY, ANNTABLE, SNKNGFND*

9.3 EXERCISES

Ordinary Annuities

1. Find the value of an annuity created by depositing $200 at the end of each month for one year in an account that pays 6% per year compounded monthly.
2. Repeat Exercise 1 if the annuity lasts for two years.
3. Repeat Exercise 1 if the annuity lasts for three years.
4. Repeat Exercise 1 if the annuity lasts for four years.
5. If you invest $500 at the end of each quarter in an investment that pays 8% per year compounded quarterly, how much will you have at the end of two years?
6. Repeat Exercise 5 if the question is how much you will have at the end of five years.
7. Repeat Exercise 5 if the question is how much you will have at the end of eight years.

8. Repeat Exercise 5 if the question is how much you will have at the end of 10 years.

9. Suppose you invest $50 at the end of each month in a mutual fund that pays 12% per year compounded monthly. How much will your investment be worth at the end of two years?

10. Repeat Exercise 9 if the question is how much the investment will be worth at the end of 3.5 years.

11. Repeat Exercise 9 if the question is how much the investment will be worth at the end of 4.25 years.

12. Repeat Exercise 9 if the question is how much the investment will be worth at the end of five years.

13. Suppose that at the end of each day you deposit $10 in an account that pays 8% per year compounded daily. How much will there be in the account at the end of 10 years? Use a 365-day year.

14. Repeat Exercise 13 if the question is how much will be in the account at the end of 20 years.

15. Repeat Exercise 13 if the question is how much will be in the account at the end of 40 years.

16. An insurance policy requires the payment of a $200 premium at the end of each six months. The premiums accumulate at 4% per year compounded semiannually. If a person purchases this policy at the age of 25, how much will it be worth when the person turns 65?

17. Repeat Exercise 16 if the premiums accumulate at 6% per year compunded semiannually.

18. Repeat Exercise 16 if the premiums accumulate at 12% per year compounded semiannually.

19. A sinking fund is created by making equal payments at the end of each quarter into an account that pays 6% per year compounded quarterly. What must the payment be if the fund is to contain $50,000 five years after it was started?

20. Repeat Exercise 19 if the account pays 8% per year compounded quarterly.

21. Repeat Exercise 19 if the account pays 12% per year compounded quarterly.

22. Repeat Exercise 19 if the account pays 16% per year compounded quarterly.

23. You wish to create a sinking fund so that three years from now you will have $20,000 to buy a new car. If you make equal payments at the end of each month into an account that pays 6% per year compounded monthly, what must each payment be?

24. Repeat Exercise 23 if you require the $20,000 four years from now.

25. Suppose you create a sinking fund by making equal payments at the end of each day into an account that pays 8% per year compounded daily. What must each payment be if at the end of 30 years the fund is to contain $100,000? Use 365-day years.

26. Repeat Exercise 25 if the fund is to contain $500,000.

27. Repeat Exercise 25 if the fund is to contain $1,000,000.

*28. Let an annuity be created with equal payments of $R at the *beginning* of each compounding period. Such an annuity is called an **annuity due.** Show that the formula for the future value S of an annuity due is

$$S = \frac{R(1 + r)[(1 + r)^{mt} - 1]}{r} = (1 + r)s_{\overline{mt}|r}.$$

*29. Repeat Exercise 1, assuming the annuity is an annuity due.

*30. Repeat Exercise 5, assuming the annuity is an annuity due.

*31. Repeat Exercise 9, assuming the annuity is an annuity due.

*32. Repeat Exercise 13, assuming the annuity is an annuity due.

*33. An annuity is created by making equal payments of $600 at the beginning of each month into an account that pays 12% per year compounded monthly. How much will the annuity be worth at the end of three years?

*34. Repeat Exercise 33 if the account earns 9% per year compounded monthly.

Management

35. A firm's new product will produce revenue of $100,000 per year for the next 20 years before it becomes obsolete. If the revenue is invested at the end of each year and earns 10% per year compounded annually, what is the value of this investment?

36. Repeat Exercise 35 if the revenue is invested at 20% per year compounded annually.

37. A business will need a new computer system eight years from now. The new system will cost $250,000 and will be paid for with the proceeds of a sinking fund. Equal payments will be made to the fund at the end of each quarter and will earn 8% per year compounded quarterly. How much should the payments be?

38. A firm has just floated a $25 million issue of 20-year bonds. Thus at the end of the 20 years the firm must redeem the bonds for $25 million; it will do so with the proceeds of a sinking fund that earns 10% per year compounded annually. If equal payments will be made to the fund at the end of each year, how much should each payment be?

39. Repeat Exercise 38 if the bonds are 15-year bonds.

40. Repeat Exercise 38 if the bonds are 10-year bonds.

The Present Value of an Annuity

41. Find the present value of an annuity if the payments of $500 are received at the end of each quarter for five years and money is worth 8% per year compounded quarterly.

42. Repeat Exercise 41 if money is worth 12% per year compounded quarterly.

43. Repeat Exercise 41 if money is worth 6% per year compounded quarterly.

44. Suppose a payment of $1000 is received at the end of each month for two years. Find the present value of this annuity using a discount rate of 6% per year compounded monthly.

THE MATHEMATICS OF FINANCE 583

45. Repeat Exercise 44 if the annuity lasts for four years.
46. Repeat Exercise 44 if the annuity lasts for 10 years.
47. Payments of $12,000 are received at the end of each year for five years. Discount this annuity using a rate of 12% per year compounded annually.
48. Payments of $6000 are received at the end of each six months for five years. Discount this annuity using a rate of 12% per year compounded semiannually.
49. Payments of $3000 are received at the end of each quarter for five years. Discount this annuity using a rate of 12% per year compounded quarterly.
50. Payments of $1000 are received at the end of each month for five years. Discount this annuity using a rate of 12% per year compounded monthly.
51. Jones holds an IOU from Smith for $2000. Smith has agreed to pay Jones $500 at the end of each quarter for one year. What is the present value of this annuity to Jones if he must earn 8% per year compounded quarterly on his investment?
52. Repeat Exercise 51 if Jones must earn 10% per year compounded quarterly.
53. A lottery winner wins $5,000,000, payable in annual installments over the next 20 years. Find the present value of this annuity if money is worth 8% per year compounded annually.
*54. Show that the formula for the present value of an annuity due (see Exercise 28) is

$$V = \frac{R(1 + r)[1 - (1 + r)^{-mt}]}{r} = (1 + r)Ra_{\overline{mt}|r}.$$

*55. Repeat Exercise 41 if the annuity is an annuity due.
*56. Repeat Exercise 44 if the annuity is an annuity due.
*57. Repeat Exercise 47 if the annuity is an annuity due.
*58. Repeat Exercise 49 if the annuity is an annuity due.
*59. An annuity pays $1000 at the beginning of each quarter. Find the present value of this annuity if it lasts for two years and money is worth 6% per year compounded quarterly.
*60. Repeat Exercise 59 if the annuity lasts for 3.75 years.

Management

61. A company intends to produce a new product. The product will earn revenues of $200,000 per year at the end of each of the next 10 years. Find the present value of this stream of revenues if money is worth 10% per year compounded annually.
62. Repeat Exercise 61 if money is worth 16% per year compounded annually.
63. It will cost a firm $400,000 to produce and market a new product. The product will bring in revenues of $120,000 per year at the end of each year for five years. If the firm must earn at least 8% per year compounded annually on any product, will it go ahead with the new one?
64. Repeat Exercise 63 if the firm must earn at least 16% per year compounded annually on any product.
65. Alpha Company is considering installing a new telephone system. For $140,000 vendor X will sell them a system that will save them $18,000 per year for 20 years

before becoming obsolete. For $120,000, vendor Y will sell them a system that will save them $27,000 per year for 10 years before becoming obsolete. For $100,000, vendor Z will sell them a system that will save them $30,000 per year for eight years before becoming obsolete. Alpha will not invest in any new equipment unless it returns at least 10% per year compounded annually. Which of the telephone systems, if any, should Alpha purchase?

66. Repeat Exercise 65 if Alpha must earn at least 4% per year compounded annually on any investment.

67. Repeat Exercise 65 if Alpha must earn at least 25% per year compounded annually on any investment.

68. Beta Corporation will not undertake any investment unless it returns at least 18% per year compounded monthly. Beta is considering the development of three new products, X, Y, and Z.

Product	Development Costs	Monthly Profits
X	$450,000	$17,500 per month for three years
Y	$530,000	$12,000 per month for six years
Z	$750,000	$20,000 per month for five years

If Beta can afford to develop all three of the products, what should it do? What if it can only afford to develop one of them?

*69. Refer to Exercise 68. Using trial and error, can you find a discount rate that will result in all three new products being developed, assuming that Beta can afford to do so? Can you find a discount rate that will result in none of the three being developed?

Bond Yields

Exercises 70 through 75 were suggested by material in Stigum, *Money Market Calculations,* Dow-Jones Irwin, 1981. They are concerned with bond yields. Every bond has both a **face value** F and a **coupon rate** c. The face value F is the amount for which the bond can be redeemed at its maturity. The coupon rate is an interest rate: Every six months, the bondholder receives a coupon payment equal to $0.005cF$. For instance, a bond with a face value of $10,000 and a coupon rate of 8% can be cashed in at maturity for $10,000, and until then each bondholder will receive a coupon payment of $0.005(8)(\$10,000) = \$400$ every six months. Bonds normally sell at a discount: A bond with a $10,000 face value will sell for less than $10,000. Thus the bondholder's return comes from two sources: the annuity formed by the coupon payments and the difference between the purchase price of the bond and its face value F. Each of these sources has a present value. The return on a bond is measured by its **yield to maturity,** which is the interest rate i such that

the annuity of coupon payments discounted from
the date of maturity to the date of purchase at
$i\%$ per year compounded semiannually

plus

> the face value discounted from the date of maturity to the date of purchase at $i\%$ per year compounded semiannually

equals

> the purchase price of the bond.

*70. Suppose you pay $9227.82 for a bond with a face value of $10,000, a coupon rate of 8%, and a maturity date exactly five years from the date of purchase. Assume that you will receive coupon payments starting exactly six months after the date of purchase and continuing to the maturity date. Show that
   (a) at the time of purchase, the annuity of coupon payments, discounted at 10% per year compounded semiannually, is $3088.69;
   (b) at the time of purchase, the present value of the face value of the bond, discounted at 10% per year compounded semiannually, is $6139.13.
   Therefore, since $3088.69 + $6139.13 = $9227.82, the yield to maturity for this bond is 10%.

*71. A bond has a face value of $5000 and coupon rate of 7.25%. Its maturity date is exactly 10 years from date of purchase and coupon payments will be received starting exactly six months after date of purchase and continuing until the date of maturity. If the bond's yield to maturity is 9.15%, find its purchase price.

*72. Suppose a bond has face value F, coupon rate c, and yield to maturity $i\%$. Suppose also that its maturity date is exactly n years from the date of purchase and that coupon payments will be received starting exactly six months after date of purchase and continuing until date of maturity. Find a formula for the purchase price P of the bond.

*73. Use the result of Exercise 72 to find the purchase price of a bond with face value $10,000, coupon rate 8.36%, and yield to maturity 12.45%, if its date of maturity is exactly 4.5 years from the date of purchase and coupon payments will be received starting exactly six months after date of purchase and continuing until date of maturity.

*74. A bond purchased for $8632.18 has a face value of $10,000, a yield to maturity of 9.5%, and a maturity date exactly five years from the date of purchase. If coupon payments are received starting exactly six months after the date of purchase and continuing until the date of maturity, find the bond's coupon rate.

If the face value F, coupon rate c, time to maturity n (in years), and purchase price P of a bond are known, it is possible to find its yield to maturity i. This can be done by trial and error or by more sophisticated methods that we cannot present here. When trial and error is used to find the yield to maturity, a good first estimate for i is

$$i_0 = \frac{1 + 0.01cn - P/F}{n - (1 - P/F)(0.5n + 0.25)}.$$

*75. A bond's face value is $10,000, its coupon rate is 8%, its time to maturity is four years, and its purchase price is $9447.28. Find i_0, the first estimate of its yield to maturity.

*76. Use trial and error to find the yield to maturity of the bond of Exercise 75.

SUMMARY

This summary consists of terms and symbols whose meaning you should know, facts you should know, and techniques and methods you should be able to employ.

Section 9.1

Sequence. Terms of a sequence, nth term a_n, general term. Arithmetic sequence. Common difference d. Formula for the nth term of an arithmetic sequence. Finding the nth term of an arithmetic sequence. Simple interest, principal, future value. Simple interest formula. Using the simple interest formula. Formula for the sum of the first n terms of an arithmetic sequence. Finding the sum of the first n terms of an arithmetic sequence. Geometric sequence. Common ratio r. Formula for the nth term of a geometric sequence. Finding the nth term of a geometric sequence. Formula for the sum of the first n terms of a geometric sequence. Finding the sum of the first n terms of a geometric sequence.

Section 9.2

Compound interest. Principal, future value, compounding period. Compound interest formula. Finding the future value. Nominal interest rate. Effective interest rate. Finding effective interest rates. Using effective interest rates to make business decisions. Present value. Discounting, discount rate. Present value formula. Finding the present value. Net present value (NPV). Finding net present value. Using net present value to make business decisions.

Section 9.3

Ordinary annuity. Ordinary annuity formula. Finding the future value of an ordinary annuity. Sinking fund. Finding the value of a sinking fund. Finding the payment for a sinking fund. Annuity due. Present value of an ordinary annuity. Present value formula for an ordinary annuity. Finding the present value of an ordinary annuity. Net present value of an ordinary annuity. Using the net present value of an ordinary annuity to make business decisions. Present value of an annuity due.

REVIEW EXERCISES

1. How much will you owe at the end of five years if you borrow $8000 at 6% per year compounded quarterly? at 9% per year compounded monthly?

2. How much will you have at the end of 10 years if you invest $50,000 at 10% per year compounded semiannually? at 10.5% per year compounded monthly?

THE MATHEMATICS OF FINANCE   587

3. What is the effective annual interest rate if money is borrowed at 12% per year compounded monthly?

4. If a firm can invest its earnings at either 13.1% per year compounded quarterly, 13.3% per year compounded semiannually, or 12.96% per year compounded monthly, which rate should it choose?

5. A firm can reinvest its profits at 14% per year compounded quarterly, or it can invest them in an outside investment that pays i% per year compounded annually. What must i be in order for the outside investment to be preferable?

6. How much should you invest now at 9% per year compounded monthly in order to have $10,000 two and one-half years from now?

7. Find the present value of $5000 due eight years from now if money is worth 10% per year compounded quarterly.

8. Omega Corporation is considering the purchase of Alpha Company or Beta Associates. Purchasing Alpha will cost $5.5 million and will return $7.2 million four years from now. Purchasing Beta will cost $1.3 million and will return $4.4 million eight years from now. Omega will not undertake any investment that does not return at least 10% per year compounded annually. What will Omega do?

9. Suppose Bank X will lend Zeta Company $2 million discounted at 7.92% per year compounded quarterly, while Bank Y will lend Zeta the same amount discounted at 7.88% per year compounded monthly. Which loan will Zeta take if it wishes to minimize the interest paid?

10. A furniture store sells sofas for 5% down and the balance payable in two months. It factors its accounts receivable with a finance company that discounts them at 24% per year compounded monthly. If the store sells $10,000 worth of sofas, how much will it receive from the finance company?

11. An annuity is created by making equal payments of $500 at the end of each year into an account that pays 8% per year compounded annually. What is the value of the annuity at the end of 20 years?

12. A company wishes to create a sinking fund so that six years from now it can purchase $250,000 worth of new equipment. If equal payments are to be made at the end of each month into an account that pays 12% per year compounded monthly, how much should each payment be?

13. A firm's earnings from a product will be $2.5 million per year, received at the end of each year. If the product will sell for 12 years before becoming obsolete, and if the firm reinvests its earnings at 12% per year compounded annually, find the future value of the product's earnings.

14. A corporation issues $50 million worth of 20-year bonds and simultaneously creates a sinking fund to redeem them. If the fund earns 10% per year compounded annually and if equal payments are made to the fund at the end of each year, how much should each payment be?

15. Find the present value of an annuity of $400 received at the end of every six months for the next five years if money is worth 6% per year compounded semiannually.

16. Gamma Corporation is considering the purchase of one of three small genetic engineering firms: D, N, or A. Firm D can be bought for $800,000, firm N for $1,000,000, and firm A for $1,250,000. Gamma estimates that firm D would provide annual profits of $100,000 for the next 15 years, firm N annual profits of $160,000 for the next 12 years, and firm A annual profits of $200,000 for the next

10 years. Gamma evaluates all investments using a discount rate of 10% per year compounded annually. Which of the firms, if any, will Gamma purchase?

17. Repeat Exercise 16 if Gamma evaluates all investments using a discount rate of 5% per year compounded annually.

18. Repeat Exercise 16 if Gamma evaluates all investments using a discount rate of 20% per year compounded annually.

19. An annuity is created with payments of $1000 at the beginning of each quarter. If the annuity earns 6% per year compounded quarterly, find its value at the end of 5.25 years.

20. A stockholder received dividends of $250 per quarter at the beginning of each quarter for eight years. Find the present value of this annuity, using a discount rate of 12% per year compounded quarterly.

ADDITIONAL TOPICS

Here are some suggestions for topics not covered in the text that you might want to investigate on your own.

1. At the end of Exercises 9.3 we discussed bond yields. There we assumed that bonds were purchased at such a time that the next coupon payment was due exactly six months later. In reality, bonds may be purchased at any time; if a bond is purchased at a time that is between coupon payments, part of the next coupon payment belongs to the person who sold the bond and part belongs to the person who bought it. Find out how the coupon payment is allocated in such a case and how this affects the yield to maturity calculation.

2. Many consumer loans, such as auto loans, for instance, are written on a discounted basis: the borrower makes equal monthly payments of amount R for n periods and receives the present value V of this stream of payments as the amount of the loan. The **true annual interest rate** is defined to be the annual rate i such that

$$V = \frac{12R[1 - (1 + 0.01i/12)^{-n}]}{0.01i}.$$

Show that this formula follows from that developed in Section 9.3 for the present value of an annuity. Find out how true annual interest rates are calculated and what the federal Truth in Lending Act has to say about disclosure of true annual interest rates to borrowers.

3. If an installment loan is paid off earlier than scheduled, the borrower receives an **interest rebate**. The interest rebate can be calculated with the aid of the so-called **Rule of 78**. Find out about the Rule of 78 and how it is used to calculate interest rebates.

SUPPLEMENT: AMORTIZATION

Many loans are **amortized**, that is, paid back by means of a sequence of payments, with part of each payment allocated to repayment of the principal and the rest to payment of interest. Auto loans and home mortgage loans, for instance, are usually

THE MATHEMATICS OF FINANCE

amortized on a monthly basis. In this supplement we discuss amortization and amortization schedules.

Suppose we borrow $10,000 at 8% per year compounded quarterly, agreeing to pay back the loan is equal quarterly installments over a two-year period. Thus we will make eight equal payments on the loan. The first thing we must do is find the amount of each payment. To do this, we consider the payments to form an ordinary annuity with the principal of the loan as its present value. Therefore we will use the formula

$$V = Ra_{\overline{mt}|r}.$$

Here the principal $V = \$10{,}000$, $m = 4$, $t = 2$, and $r = 0.02$. Hence

$$\$10{,}000 = Ra_{\overline{8}|0.02},$$

or

$$\$10{,}000 = R(7.325481),$$

so

$$R = \$1365.10.$$

Thus we will make eight quarterly payments of $1365.10 each. Note that this will result in a total payment of 8($1365.10) = $10,920.80. Therefore we will be paying a total of

$$\$10{,}920.80 - \$10{,}000 = \$920.80$$

in interest on the $10,000 loan.

Of each payment of $1365.10, a portion will be allocated to repayment of the principal and the rest will be allocated to payment of the interest. The way this works is as follows: At any given time, the **unpaid balance** is the amount still owed on the principal. Since the interest rate is 2% per quarter, that is, 2% per payment, at the time of each payment we owe 2% of the unpaid balance in interest. Thus an amount equal to 2% of the unpaid balance will be deducted from each payment and allocated to payment of interest; the remainder of the payment will be allocated to repayment of the principal. To illustrate, before the first payment is made the unpaid balance will be the full $10,000. Since 2% of $10,000 is $200, $200 of the first payment of $1365.10 will be allocated to interest and $1365.10 - $200 = $1165.10 will be allocated to repayment of the principal.

Now we can make an **amortization schedule** for our loan. This is a table that shows

- The unpaid balance before each payment
- The amount of each payment
- The portion of each payment allocated to payment of interest
- The portion of each payment allocated to repayment of the principal
- The unpaid balance after each payment

Here is the amortization schedule for our loan:

Payment Number	Unpaid Balance before Payment	Payment	To Interest	To Principal	Unpaid Balance after Payment
1	$10,000.00	$1,365.10	$200.00	$1,165.10	$8,834.90
2	8,834.90	1,365.10	176.70	1,188.40	7,646.50
3	7,646.50	1,365.10	152.93	1,212.17	6,434.33
4	6,434.33	1,365.10	128.69	1,236.41	5,197.92
5	5,197.92	1,365.10	103.96	1,261.14	3,936.78
6	3,936.78	1,365.10	78.74	1,286.36	2,650.42
7	2,650.42	1,365.10	53.01	1,312.09	1,338.33
8	1,338.33	1,365.10	26.77	1,338.33	0.00
		$10,920.80	$920.80	$10,000.00	

The amortization schedule was constructed as follows:

- The unpaid balance before the first payment is the amount of the loan.
- The amount allocated to interest for each payment is 2% of the unpaid balance before the payment.
- The amount allocated to principal for each payment is equal to the payment minus the amount allocated to interest for the payment.
- The unpaid balance after the payment is the unpaid balance before the payment minus the amount allocated to principal for the payment.
- After the first payment, the unpaid balance before the payment is equal to the unpaid balance after the previous payment.

Notice that as the amortization process continues, the amount of each payment allocated to interest decreases while the amount allocated to repayment of the principal increases. This is typical of amortization schedules.

Now let us make an amortization schedule for a home mortgage loan. Suppose we borrow $100,000 at 10.5% per year to buy a house. Mortgage loans are compounded and paid monthly, and they usually run for 20, 25, or 30 years. We will assume that ours runs for 25 years. Then the $100,000 principal is the present value of an annuity of $25(12) = 300$ monthly payments at $10.5\%/12 = 0.875\%$ per month. Therefore the monthly payment (exclusive of taxes, insurance, or any assessments) will be R, where

$$\$100,000 = \frac{R[1 - (1.00875)^{-300}]}{0.00875} = R[105.91182].$$

Hence, the monthly mortgage payment is

$$R = \frac{\$100,000}{105.91182} = \$944.18.$$

Here is the amortization schedule for the mortgage:

THE MATHEMATICS OF FINANCE   591

Month Number	Unpaid Balance before Payment	Monthly Payment	To Interest	To Principal	Unpaid Balance after Payment
1	$100,000.00	$ 944.18	$ 875.00	$ 69.18	$99,930.82
2	99,930.82	944.18	874.39	69.79	99,861.03
3	99,861.03	944.18	873.78	70.40	99,790.63
4	99,790.63	944.18	873.17	71.01	99,719.62
.	.	.	.	.	.
.	.	.	.	.	.
.	.	.	.	.	.
149	79,202.61	944.18	693.02	251.16	78,951.45
150	78,951.45	944.18	690.83	253.35	78,698.10
151	78,698.10	944.18	688.61	255.57	78,435.53
.	.	.	.	.	.
.	.	.	.	.	.
.	.	.	.	.	.
297	3,697.91	944.18	32.36	911.82	2,786.09
298	2,786.09	944.18	24.38	919.80	1,866.29
299	1,866.29	944.18	16.33	927.85	938.44
300	938.44	946.65	8.21	938.44	0.00
		$283,256.47	$183,256.47	$100,000.00	

Notice how the last payment had to be adjusted from $944.18 to $946.65 in order to make the unpaid balance after the last payment equal to zero. This sort of adjustment in the final payment often occurs when making amortization schedules. It is the result of round-off error in the calculations; as the amortization schedule is constructed, round-off error accumulates and must be compensated for at the time of the last payment.

The amortization schedule for the mortgage shows that at the beginning of the amortization process, nearly all of the monthly mortgage payment goes for interest, with very little being applied to reducing the principal. It is only toward the end of the life of the mortgage that most of the monthly payment is applied to reduction of the principal. Also, notice that the total interest paid is

$$\$283{,}256.47 - \$100{,}000 = \$183{,}256.47.$$

This is typical of home mortgages: the interest paid over the life of the mortgage is far more than the amount of the loan.

We conclude with the remark that when dealing with long-term amortization schedules (such as mortgage schedules) it is frequently useful to have a method for finding the unpaid balance before payment k, for any k. If u_k denotes the unpaid balance before payment k, then it can be shown that

$$u_k = V(1 + r)^{k-1} - R s_{\overline{k-1}|r}.$$

Here, as usual, V is the present value of the stream of payments ($=$ the amount of the loan), R is the amount of each payment, r is the interest rate per payment period, in decimal form, and

$$s_{\overline{k-1}|r} = \frac{(1+r)^{k-1} - 1}{r}.$$

(See Exercise 9.) This formula will be useful in some of the following exercises.

■ *Computer Manual: Programs AMORTSC1, AMORTSC2*

SUPPLEMENTARY EXERCISES

1. You borrow $5000 at 8% per year compounded semiannually, paying back the loan in equal semiannual installments over a three-year period. Make an amortization schedule for the loan.

2. You borrow $6000 at 12% per year compounded monthly, paying back the loan in eight equal monthly installments. Make an amortization schedule for the loan.

3. You borrow $10,000 at 10% per year compounded quarterly, paying back the loan in equal quarterly installments over 1.75 years. Make an amortization schedule for the loan.

4. You borrow $100,000 at 12% per year compounded quarterly, paying back the loan in equal quarterly installments over a six-year period. Write the portion of the amortization schedule for the loan that covers the twelfth, thirteenth, and fourteenth payments.

5. A homeowner has a 20-year mortgage for $100,000 at 12% per year. Write the amortization schedule for months 1, 2, 3, 4, 100, 101, 102, 238, 239, and 240 of this mortgage.

6. Repeat Exercise 5 if the interest rate is 6%.

Amortization schedules can be made even when the payments are not all equal. The idea is the same: A portion of each payment equal to a certain percentage of the unpaid balance is applied to payment of the interest, with the rest of the payment being allocated to repayment of the principal. Exercises 7 and 8 require the construction of such schedules.

*7. A credit card charges interest of 1.5% per month on the unpaid balance. Suppose you have a $500 unpaid balance on this card, and decide to pay it off as follows: each month you will pay either 25% of the unpaid balance or $50, whichever is greater, until the unpaid balance falls below $100, at which time you will pay it in full. Make an amortization schedule for this payment plan.

*8. Repeat Exercise 7 if the credit card charges interest of 2% per month on the unpaid balance, your unpaid balance is $1000, and each month you will pay either 15% of the unpaid balance or $100, whichever is greater, until the unpaid balance falls below $500, at which time you will pay it in full.

*9. Show that if V is the present value of the stream of payments (= the amount of the loan), R is the amount of each payment, and r is the interest rate per payment period, in decimal form, then the unpaid balance before payment k is

$$u_k = V(1 + r)^{k-1} - Rs_{\overline{k-1}|r}.$$

10

The Derivative

The remainder of this book is devoted to calculus. Calculus is a method for studying functions. For instance, suppose a demographer concludes that a certain function will describe a nation's population size as a function of time over the next few years. The techniques of calculus can be applied to this function to answer questions such as: At what rate will the population be changing at any given time? When will the population be increasing and when will it be decreasing? What will be the maximum and minimum sizes of the population over the next several years? The particular concept of calculus needed to answer questions like these is called the **derivative.**

In this chapter we study the derivative. The derivative is defined in terms of another concept of calculus called the **limit,** so we will begin with a brief intuitive introduction to limits. We will then define the derivative, develop some rules for finding derivatives of functions, show how to use the derivative to find rates of change, exhibit some other useful applications of the derivative, and conclude with a discussion of the various notations for the derivative and for derivatives of the derivative.

10.1 LIMITS

The **limit** of a function f as its independent variable x approaches a number a is a number L with the property that, as x gets closer and closer to a (without ever being equal to a), the functional values $f(x)$ get closer and closer to L. For a given function f and limiting value a such a number L may or may not exist. The existence or nonexistence of the limit, and its value if it does exist, thus tells us something about the behavior of the function for values of x near a. Hence limits are important in analyzing functions. In this section we illustrate the notion of the limit of a function geometrically, state the properties of limits, and show how to use these properties to find limits of polynomial and rational functions.

The Limit of a Function

Consider the linear function f whose defining equation is $f(x) = 2x + 1$. We ask the following question: if we let x approach the number 1 on the x-axis, from either side of 1 and without ever letting $x = 1$, what can we say about the behavior of the corresponding functional values $f(x)$? For instance, Table 10–1 shows some functional values $f(x) = 2x + 1$ as x approaches 1 from the left and from the right:

TABLE 10–1

	\multicolumn{5}{c}{x approaches 1 from the left}
x	0.8	0.9	0.99	0.999	0.9999
$f(x) = 2x + 1$	2.6	2.8	2.98	2.998	2.9998
	\multicolumn{5}{c}{x approaches 1 from the right}
x	1.2	1.1	1.01	1.001	1.0001
$f(x) = 2x + 1$	3.4	3.2	3.02	3.002	3.0002

Notice that as x gets closer and closer to 1, from either side of 1, the values $f(x)$ get closer and closer to the single number 3. Hence we say that the **limit of $f(x)$ as x approaches 1 is 3,** written

$$\text{Lim}_{x \to 1} f(x) = 3,$$

or

$$\text{Lim}_{x \to 1} (2x + 1) = 3.$$

We can also tell that $\text{Lim}_{x \to 1} (2x + 1) = 3$ from the graph of the function. Figure 10–1 shows the graph of f. As the figure indicates, as x approaches 1 from either side on the x-axis the functional values $f(x)$ approach 3 on the y-axis.

Having illustrated the idea of the limit of a function, we now present an informal definition of the concept.

THE DERIVATIVE

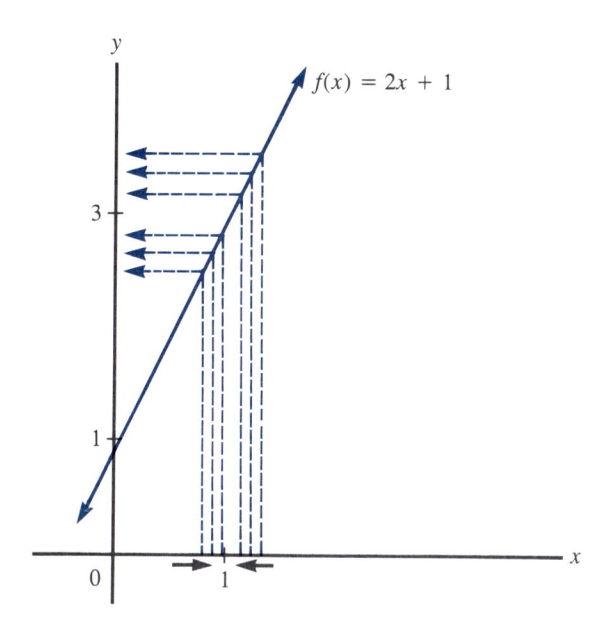

FIGURE 10-1
$\text{Lim}_{x \to 1} (2x + 1) = 3$

The Limit of a Function

If f is a function and a is a number, the **limit of f as x approaches a is L,** written

$$\text{Lim}_{x \to a} f(x) = L,$$

provided that as x comes arbitrarily close to a from either side, without ever being equal to a, the corresponding functional values $f(x)$ come arbitrarily close to the single number L. If there is no such single number L, the limit does not exist.

EXAMPLE 1   If $f(x) = x^2 + 1$ and $a = 0$, we claim that $\text{Lim}_{x \to 0} f(x) = 1$. Figure 10–2 verifies this: it shows that as x approaches 0 on the x-axis, from either side and without ever being equal to 0, the corresponding functional values approach (that is, come arbitrarily close to) the single number 1 on the y-axis. ☐

EXAMPLE 2   If f is defined by

$$f(x) = \begin{cases} x - 1 & \text{if } 0 \leq x \leq 3, \\ x + 1 & \text{if } x > 3, \end{cases}$$

then $\text{Lim}_{x \to 3} f(x)$ does not exist. To see this, look at Figure 10–3, which depicts the graph of this function. Notice that as x approaches 3 on the x-axis from the left,

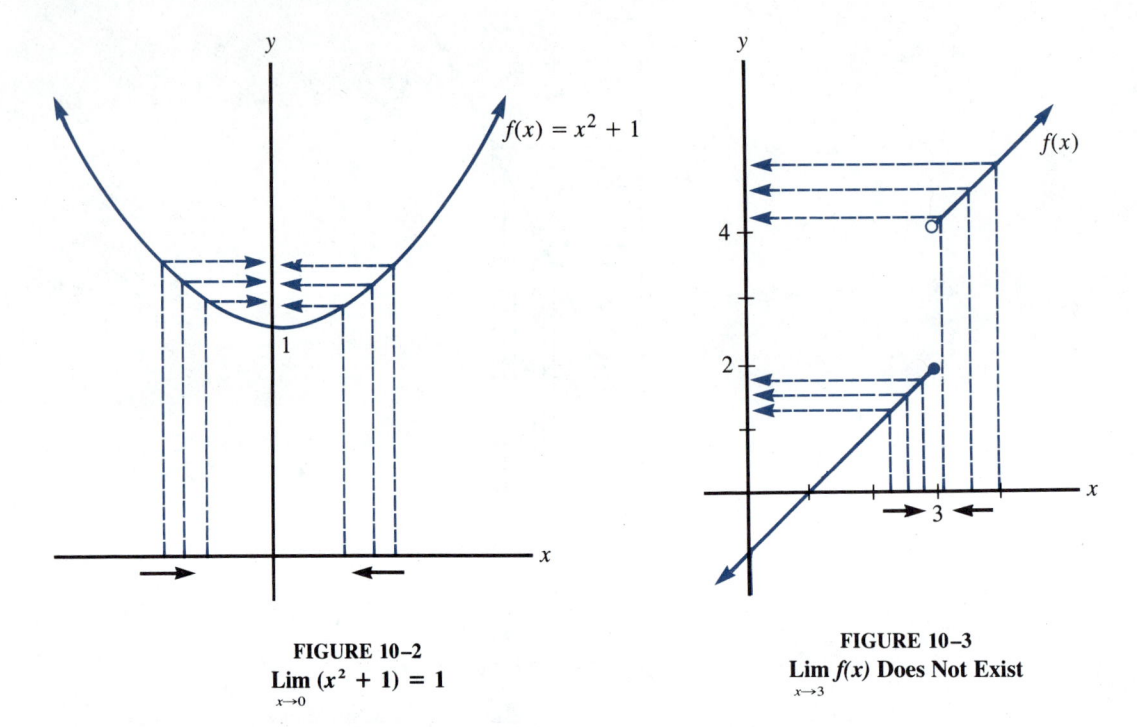

FIGURE 10–2
$\operatorname*{Lim}_{x \to 0} (x^2 + 1) = 1$

FIGURE 10–3
$\operatorname*{Lim}_{x \to 3} f(x)$ Does Not Exist

the corresponding functional values approach 2 on the y-axis, but as x approaches 3 from the right, the corresponding functional values approach 4 on the y-axis. Therefore there is no single number L that the functional values approach as x approaches 3, so the limit does not exist.

EXAMPLE 3 Let h be defined by

$$h(x) = \frac{x+1}{x-1}.$$

Figure 10–4 shows the graph of h. From the figure, we see that as x approaches -1 on the x-axis, from either the right or the left, the functional values $h(x)$ approach 0. Hence

$$\operatorname*{Lim}_{x \to -1} h(x) = \operatorname*{Lim}_{x \to -1} \frac{x+1}{x-1} = 0.$$

On the other hand, as x approaches 1 from the right, the functional values $h(x)$ get larger and larger without bound, and thus cannot approach any number L. Hence

$$\operatorname*{Lim}_{x \to 1} h(x) = \operatorname*{Lim}_{x \to 1} \frac{x+1}{x-1} \quad \text{does not exist.}$$

EXAMPLE 4 Figure 10–5 shows the graph of a function g. From the graph, we see that

a. $\operatorname*{Lim}_{x \to -1} g(x) = 0$, because as x approaches -1 from either side, without ever being equal to -1, the functional values $g(x)$ approach 0;

THE DERIVATIVE

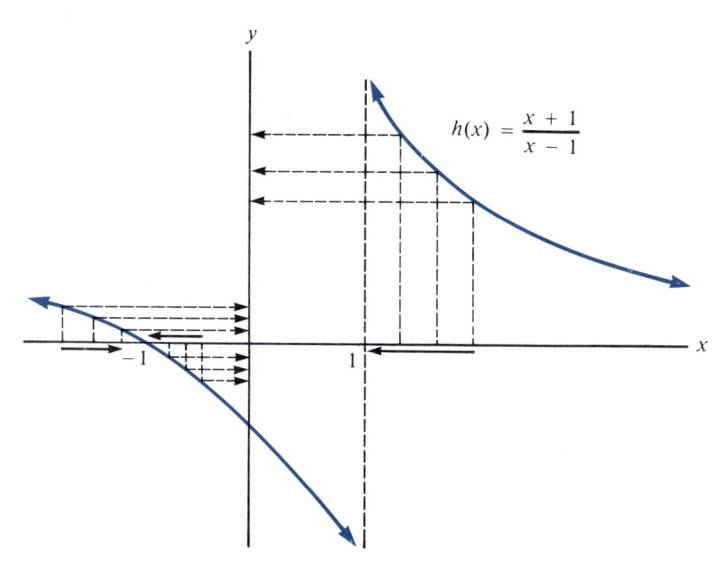

FIGURE 10-4

$h(x) = \dfrac{x+1}{x-1}$

b. $\lim_{x \to 0} g(x) = 3$, because as x approaches 0 from either side, without ever being equal to 0, the functional values actually become equal to 3 (and hence are certainly arbitrarily close to 3);

c. $\lim_{x \to 1} g(x) = 5$, because as x approaches 1 from either side, without ever being equal to 1, the functional values approach 5; note that this is so even though the functional value $g(1) \neq 5$;

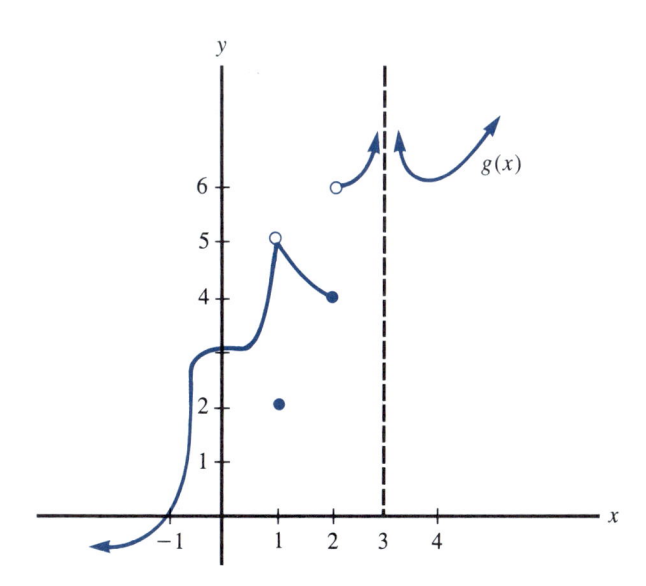

FIGURE 10-5

d. $\text{Lim}_{x \to 2}\, g(x)$ does not exist, because as x approaches 2 from the left the functional values approach 4, but as x approaches 2 from the right they approach 6;

e. $\text{Lim}_{x \to 3}\, g(x)$ does not exist, because as x approaches 3 from the left (and also from the right), the functional values grow larger and larger without bound and hence do not approach a single number L. ☐

We should remark that our definition of limit is highly intuitive and not mathematically precise. A more rigorous definition of the limit of a function is presented in suggestion 1 of the additional topics at the end of this chapter.

Properties of Limits

The following properties are useful when it is necessary to evaluate limits. They may be proved using the rigorous definition of limit referred to in the preceding paragraph.

Properties of Limits

Let a be any number and assume that both $\text{Lim}_{x \to a}\, f(x)$ and $\text{Lim}_{x \to a}\, g(x)$ exist.

1. If $f(x) = g(x)$ whenever $x \neq a$, then $\text{Lim}_{x \to a}\, f(x) = \text{Lim}_{x \to a}\, g(x)$.

2. For any number c, $\text{Lim}_{x \to a}\, c = c$.

3. For any number r, $\text{Lim}_{x \to a}\, x^r = a^r$, provided that a^r is defined; if a^r is not defined as a real number, the limit does not exist.

4. For any real number c, $\text{Lim}_{x \to a}\, cf(x) = c \cdot \text{Lim}_{x \to a}\, f(x)$.

5. $\text{Lim}_{x \to a}\, [f(x) \pm g(x)] = \text{Lim}_{x \to a}\, f(x) \pm \text{Lim}_{x \to a}\, g(x)$.

6. $\text{Lim}_{x \to a}\, [f(x)g(x)] = [\text{Lim}_{x \to a}\, f(x)][\text{Lim}_{x \to a}\, g(x)]$.

7. If $\text{Lim}_{x \to a}\, g(x) \neq 0$, then

$$\text{Lim}_{x \to a}\, \frac{f(x)}{g(x)} = \frac{\text{Lim}_{x \to a}\, f(x)}{\text{Lim}_{x \to a}\, g(x)}.$$

8. If $\text{Lim}_{x \to a}\, g(x) = 0$ and $\text{Lim}_{x \to a}\, f(x) \neq 0$, then $\text{Lim}_{x \to a}\, \frac{f(x)}{g(x)}$ does not exist.

EXAMPLE 5   The functions f and g defined by

$$f(x) = x^2 + 1 \quad \text{and} \quad g(x) = \frac{x^3 + x}{x}$$

THE DERIVATIVE

are equal except when $x = 0$, and we know from Example 1 that
$$\lim_{x \to 0} f(x) = 1.$$

Therefore by Property 1 we have
$$\lim_{x \to 0} g(x) = \lim_{x \to 0} f(x) = 1. \quad \square$$

EXAMPLE 6 By Property 2, we have $\lim_{x \to a} 4 = 4$ for any number a. $\quad \square$

EXAMPLE 7 Property 3 shows that $\lim_{x \to -2} x = -2$, $\lim_{x \to 3} x^2 = 3^2 = 9$, and
$$\lim_{x \to 4} \sqrt{x} = \lim_{x \to 4} x^{1/2} = 4^{1/2} = 2.$$

But also by Property 3,
$$\lim_{x \to 0} x^{-1} \text{ does not exist, since } 0^{-1} = 1/0 \text{ is not defined,}$$

and
$$\lim_{x \to -1} \sqrt[4]{x} \text{ does not exist, since } \sqrt[4]{-1} \text{ is not defined as a real number.} \quad \square$$

EXAMPLE 8 By Properties 4 and 3,
$$\lim_{x \to 2} 3x^2 = 3 \cdot \lim_{x \to 2} x^2 = 3(2^2) = 12. \quad \square$$

EXAMPLE 9 By Properties 2 through 5, we have
$$\lim_{x \to 3} (2x^3 + 4x^2 - x + 5) = \lim_{x \to 3} (2x^3) + \lim_{x \to 3} (4x^2) - \lim_{x \to 3} x + \lim_{x \to 3} 5$$
$$= 2 \cdot \lim_{x \to 3} x^3 + 4 \cdot \lim_{x \to 3} x^2 - \lim_{x \to 3} x + 5$$
$$= 2(3^3) + 4(3^2) - 3 + 5$$
$$= 92. \quad \square$$

EXAMPLE 10 Properties 2 through 6 imply that
$$\lim_{x \to 3} [(2x^{1/2})(3x - 1)] = \left[\lim_{x \to 3} 2\sqrt{x} \right] \left[\lim_{x \to 3} (3x - 1) \right]$$
$$= [2\sqrt{3}][3(3) - 1]$$
$$= 16\sqrt{3}. \quad \square$$

EXAMPLE 11 Properties 2 through 7 show that

$$\text{Lim}_{x \to 0} \frac{x^2 + 2}{2x + 3} = \frac{\text{Lim}_{x \to 0} (x^2 + 2)}{\text{Lim}_{x \to 0} (2x + 3)}$$

$$= \frac{0^2 + 2}{2(0) + 3}$$

$$= \frac{2}{3}. \quad \square$$

EXAMPLE 12 By Property 8,

$$\text{Lim}_{x \to 1} \frac{x^2 + 1}{x - 1}$$

does not exist because

$$\text{Lim}_{x \to 1} (x^2 + 1) = 2$$

while

$$\text{Lim}_{x \to 1} (x - 1) = 0. \quad \square$$

Recall that a **polynomial function** is one whose defining equation is of the form

$$p(x) = a_n x^n + a_{n-1} x^{n-1} + \cdots + a_1 x + a_0.$$

As Example 9 suggests, Properties 2 through 5 imply that we can find the limit of any polynomial function as x approaches a simply by substituting a for x in the defining equation of the function.

Limit of a Polynomial Function

If p is a polynomial function, then $\text{Lim}_{x \to a} p(x) = p(a)$, for any a.

EXAMPLE 13 We have

$$\text{Lim}_{x \to 2} (x^3 - 5x^2 + 3x - 12) = 2^3 - 5(2^2) + 3(2) - 12 = -18. \quad \square$$

A function that is a quotient of polynomials is called a **rational function.** For example, the function h of Example 4 of this section defined by

$$h(x) = \frac{x + 1}{x - 1}$$

is a rational function. If $f = p/q$ is a rational function, where the polynomials p and q have a common factor, Property 1 implies that we can cancel this common

THE DERIVATIVE

factor from p/q without affecting the limit of f. Therefore to find the limit of a rational function, we first reduce the function to its lowest terms by canceling common factors from its numerator and denominator. We then use our result for the limit of polynomials and apply either Property 7 or Property 8. We illustrate with examples.

EXAMPLE 14   Consider

$$\lim_{x \to 3} \frac{x^2 + x - 12}{x^2 - x - 6} = \lim_{x \to 3} \frac{(x + 4)(x - 3)}{(x + 2)(x - 3)}.$$

We cancel the common factor $x - 3$ from the rational expression. Then by Property 7 we have

$$\lim_{x \to 3} \frac{x^2 + x - 12}{x^2 - x - 6} = \lim_{x \to 3} \frac{x + 4}{x + 2} = \frac{\lim_{x \to 3}(x + 4)}{\lim_{x \to 3}(x + 2)}$$

$$= \frac{3 + 4}{3 + 2}$$

$$= \frac{7}{5}. \quad \square$$

EXAMPLE 15   We use the same rational function as in Example 14, but this time take its limit as x approaches -2. Now,

$$\lim_{x \to -2} \frac{x^2 + x - 12}{x^2 - x - 6} = \frac{\lim_{x \to -2}(x + 4)}{\lim_{x \to -2}(x + 2)},$$

and

$$\lim_{x \to -2}(x + 4) = -2 + 4 = 2,$$

while

$$\lim_{x \to -2}(x + 2) = 0.$$

Hence by Property 8,

$$\lim_{x \to -2} \frac{x^2 + x - 12}{x^2 - x - 6} = \frac{\lim_{x \to -2}(x + 4)}{\lim_{x \to -2}(x + 2)} = \frac{2}{0}$$

does not exist. $\quad \square$

EXAMPLE 16   Since

$$\lim_{x \to 3} \frac{x^3 - 9x}{x^3 - 6x^2 + 9x} = \lim_{x \to 3} \frac{x(x - 3)(x + 3)}{x(x - 3)^2}$$

$$= \lim_{x \to 3} \frac{x + 3}{x - 3} = \frac{\lim_{x \to 3}(x + 3)}{\lim_{x \to 3}(x - 3)} = \frac{6}{0},$$

this limit does not exist. However,

$$\operatorname*{Lim}_{x\to 3}\frac{x^3-6x^2+9x}{x^3-9x}=\operatorname*{Lim}_{x\to 3}\frac{x-3}{x+3}=\frac{\operatorname*{Lim}_{x\to 3}(x-3)}{\operatorname*{Lim}_{x\to 3}(x+3)}=\frac{0}{6}=0. \quad \square$$

Sometimes functions are defined in "pieces," with each piece being a polynomial or rational expression. (The function of Example 2 is such a function.) Our techniques for finding limits of polynomial and rational functions can be applied to each piece of such a function.

EXAMPLE 17 Let f be defined by

$$f(x) = \begin{cases} 2 & \text{if } 0 \le x < 3, \\ x-1 & \text{if } 3 \le x \le 5, \\ 16-2x & \text{if } x > 5. \end{cases}$$

Figure 10–6 shows the graph of f. Clearly,

$$\operatorname*{Lim}_{x\to 3} f(x) = 2 \quad \text{and} \quad \operatorname*{Lim}_{x\to 5} f(x) \text{ does not exist.}$$

If $0 < a < 3$, then $f(x) = 2$ when x is near a, so

$$\operatorname*{Lim}_{x\to a} f(x) = \operatorname*{Lim}_{x\to a} 2 = 2;$$

if $3 < a < 5$, then $f(x) = x - 1$ when x is near a, so

$$\operatorname*{Lim}_{x\to a} f(x) = \operatorname*{Lim}_{x\to a} (x-1) = a - 1;$$

if $a > 5$, then $f(x) = 16 - 2x$ when x is near a, so

$$\operatorname*{Lim}_{x\to a} f(x) = \operatorname*{Lim}_{x\to a} (16-2x) = 16 - 2a. \quad \square$$

EXAMPLE 18 The cost of cleaning up an oil spill of x gallons in a harbor is $C(x)$ dollars, where

$$C(x) = \begin{cases} 250{,}000 + 2x, & 0 < x \le 10{,}000, \\ 210{,}000 + 6x, & x > 10{,}000. \end{cases}$$

See Figure 10–7. Thus if $0 < a < 10{,}000$,

$$\operatorname*{Lim}_{x\to a} C(x) = \operatorname*{Lim}_{x\to a} (250{,}000 + 2x) = 250{,}000 + 2a.$$

Similarly, if $a > 10{,}000$, then

$$\operatorname*{Lim}_{x\to a} C(x) = \operatorname*{Lim}_{x\to a} (210{,}000 + 6x) = 210{,}000 + 6a.$$

Note also from Figure 10–7 that

$$\operatorname*{Lim}_{x\to 10{,}000} C(x) = 270{,}000. \quad \square$$

■ *Computer Manual: Program LIMIT*

THE DERIVATIVE

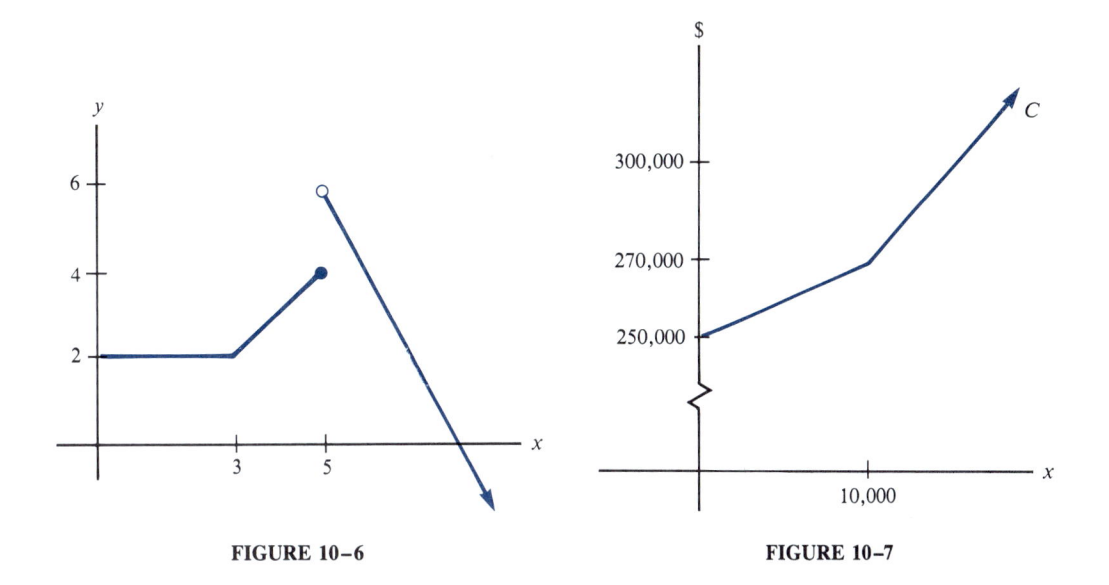

FIGURE 10–6

FIGURE 10–7

10.1 EXERCISES

The Limit of a Function

In Exercises 1 through 4, complete the table and find the limit if it exists; if the limit does not exist, so state.

1. $\lim_{x \to 3} f(x)$, where $f(x) = x^2 - 9$.

	x approaches 3 from the left				
x	2.8	2.9	2.99	2.999	2.9999
f(x)					

	x approaches 3 from the right				
x	3.2	3.1	3.01	3.001	3.0001
f(x)					

2. $\lim_{x \to 3} f(x)$, where $f(x) = \dfrac{x^2 - 9}{x - 3}$.

	x approaches 3 from the left				
x	2.8	2.9	2.99	2.999	2.9999
f(x)					

	x approaches 3 from the right				
x	3.2	3.1	3.01	3.001	3.0001
f(x)					

3. $\lim_{x \to 0} f(x)$, where $f(x) = \dfrac{1}{x}$.

x approaches 0 from the left

x	−0.1	−0.01	−0.001	−0.0001	−0.00001
f(x)					

x approaches 0 from the right

x	0.1	0.01	0.001	0.0001	0.00001
f(x)					

4. $\lim_{x \to 1} f(x)$, where $f(x) = \begin{cases} x & \text{if } 0 \leq x < 1, \\ 2x & \text{if } 1 \leq x < 2. \end{cases}$

x approaches 1 from the left

x	1/2	3/4	7/8	15/16	31/32	63/64
f(x)						

x approaches 1 from the right

x	3/2	5/4	9/8	17/16	33/32	65/64
f(x)						

The following is the graph of a function f:

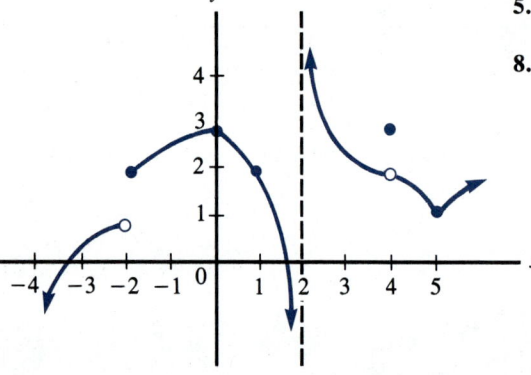

In Exercises 5 through 10, find the indicated limit if it exists; if it does not exist, so state.

5. $\lim_{x \to -2} f(x)$ 6. $\lim_{x \to 0} f(x)$ 7. $\lim_{x \to 1} f(x)$

8. $\lim_{x \to 2} f(x)$ 9. $\lim_{x \to 4} f(x)$ 10. $\lim_{x \to 5} f(x)$

The following is the graph of a function g:

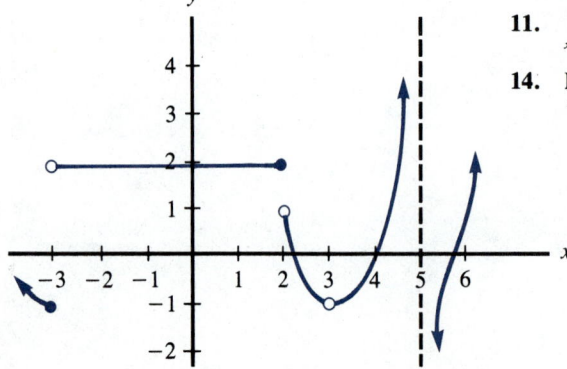

In Exercises 11 through 16, find the indicated limit if it exists; if it does not exist, so state.

11. $\lim_{x \to -3} g(x)$ 12. $\lim_{x \to -1} g(x)$ 13. $\lim_{x \to 0} g(x)$

14. $\lim_{x \to 2} g(x)$ 15. $\lim_{x \to 3} g(x)$ 16. $\lim_{x \to 5} g(x)$

The following is the graph of a function h:

In Exercises 17 through 22, find the indicated limit if it exists; if it does not exist, so state.

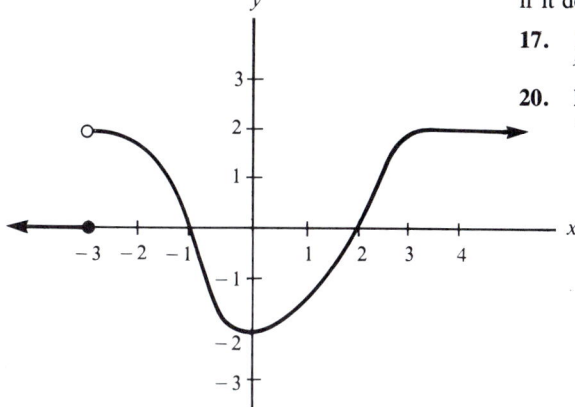

17. $\lim\limits_{x \to -3} h(x)$ 18. $\lim\limits_{x \to -1} h(x)$ 19. $\lim\limits_{x \to 0} h(x)$

20. $\lim\limits_{x \to 2} h(x)$ 21. $\lim\limits_{x \to 3} h(x)$ 22. $\lim\limits_{x \to 4} h(x)$

Properties of Limits

In Exercises 23 through 64, find the indicated limit if it exists; if it does not exist, so state.

23. $\lim\limits_{x \to 2} 5$
24. $\lim\limits_{x \to -2} 3$
25. $\lim\limits_{x \to -4} \sqrt{3}$
26. $\lim\limits_{x \to a} \pi$
27. $\lim\limits_{x \to 2} x^3$

28. $\lim\limits_{x \to 9} \sqrt{x}$
29. $\lim\limits_{x \to 3} 1/x$
30. $\lim\limits_{x \to 4} x^{-2}$
31. $\lim\limits_{x \to 16} x^{3/4}$
32. $\lim\limits_{x \to 2} 5x$

33. $\lim\limits_{x \to -3} 3x^4$
34. $\lim\limits_{x \to 4} -5x^{3/2}$
35. $\lim\limits_{x \to 0} 4x^{-1/2}$
36. $\lim\limits_{x \to 8} 3x^{-1/3}$
37. $\lim\limits_{x \to 4} (3x + 2)$

38. $\lim\limits_{x \to 0} (5x - 7)$
39. $\lim\limits_{x \to 5} (5 - x)$
40. $\lim\limits_{x \to 2} (x^2 + x)$
41. $\lim\limits_{x \to 3} (x^2 + 3x)$
42. $\lim\limits_{x \to 1} (x^2 - 1)$

43. $\lim\limits_{x \to 0} (2x^3 - 5x^2 + 3x - 17)$
44. $\lim\limits_{x \to 2} (x^3 + 2x^2 - 6x + 1)$
45. $\lim\limits_{x \to -2} (-3x^3 + 2x^2 + 5x + 3)$

46. $\lim\limits_{x \to 0} (x^{99} + 2x^{98} + 3x^{97} + \cdots + 99x + c)$
47. $\lim\limits_{x \to 4} (x^{-2} + 5x)$
48. $\lim\limits_{x \to 2} \left(\dfrac{2}{x} - \dfrac{3}{x^2} + 1\right)$

49. $\lim\limits_{x \to 0} (2x + 4)(3x - 2)$
50. $\lim\limits_{x \to 2} (5x - 3)(x^2 + 5)$
51. $\lim\limits_{x \to 1} (x^2 - 4x - 6)(x^3 + x + 2)$

52. $\lim\limits_{x \to 1} \dfrac{2}{x + 1}$
53. $\lim\limits_{x \to 1} \dfrac{2}{x - 1}$
54. $\lim\limits_{x \to 0} \dfrac{x + 1}{2x + 3}$

55. $\lim\limits_{x \to 1} \dfrac{4x + 7}{5x - 2}$
56. $\lim\limits_{x \to 3} \dfrac{x^2 - 9}{x + 2}$
57. $\lim\limits_{x \to 1} \dfrac{x^2 + 4x + 3}{x - 2}$

58. $\lim\limits_{x \to 2} \dfrac{x^2 + 4x + 3}{x - 2}$
59. $\lim\limits_{x \to 3} \dfrac{x^2 - 5x + 6}{x^2 + x - 12}$
60. $\lim\limits_{x \to 3} \dfrac{x^2 - 5x + 6}{x^2 + 10x + 24}$

61. $\lim\limits_{x \to 2} \dfrac{x^2 - 5x - 14}{x^2 - 4}$
62. $\lim\limits_{x \to 7} \dfrac{x^2 - 5x - 14}{x^2 - 4}$
63. $\lim\limits_{x \to 1} \dfrac{x^3 - 2x^2 - 15x}{x^4 - 4x^3 - 12x^2}$

64. $\lim\limits_{x \to 0} \dfrac{x^3 - 2x^2 - 15x}{x^4 - 4x^3 - 12x^2}$

65. Let f be defined by
$$f(x) = \begin{cases} 0, & 0 \le x < 2, \\ 1, & x \ge 2. \end{cases}$$

Graph f and find
(a) $\lim\limits_{x \to 2} f(x)$ (b) $\lim\limits_{x \to a} f(x)$ if $0 < a < 2$ (c) $\lim\limits_{x \to a} f(x)$ if $a > 2$

66. Let f be defined by
$$f(x) = \begin{cases} -1, & 0 \le x \le 2, \\ 1, & 2 < x \le 3, \\ 2, & x > 3. \end{cases}$$

Graph f and find
(a) $\lim\limits_{x \to 2} f(x)$, $\lim\limits_{x \to 3} f(x)$ (b) $\lim\limits_{x \to a} f(x)$ if $0 < a < 2$, $2 < a < 3$, and $a > 3$

67. Let f be defined by
$$f(x) = \begin{cases} x, & 0 \le x \le 3, \\ 6 - x, & 3 < x \le 6. \end{cases}$$

Graph f and find
(a) $\lim\limits_{x \to 3} f(x)$ (b) $\lim\limits_{x \to a} f(x)$ if $0 < a < 3$ and $3 < a < 6$

68. Let f be defined by
$$f(x) = \begin{cases} 1, & -1 \le x \le 1, \\ 2x, & 1 < x < 2, \\ 3, & x \ge 2. \end{cases}$$

Graph f and find
(a) $\lim\limits_{x \to 1} f(x)$, $\lim\limits_{x \to 2} f(x)$ (b) $\lim\limits_{x \to a} f(x)$ if $-1 < a < 1$, $1 < a < 2$, and $a > 2$

69. Let f be defined by
$$f(x) = \begin{cases} 1 + x, & 0 \le x \le 3, \\ 4, & 3 < x \le 4, \\ 12 - 2x, & 4 < x < 5, \\ 0, & x \ge 5. \end{cases}$$

Graph f and find
(a) $\lim\limits_{x \to 3} f(x)$, $\lim\limits_{x \to 4} f(x)$, $\lim\limits_{x \to 5} f(x)$
(b) $\lim\limits_{x \to a} f(x)$ if $0 < a < 3$, $3 < a < 4$, $4 < a < 5$, and $a > 5$

70. Let f be defined by
$$f(x) = \begin{cases} x^2, & 0 \le x \le 1, \\ 2x, & 1 < x \le 2, \\ 10 - 3x, & 2 < x < 4, \\ 16 - x^2, & x \ge 4. \end{cases}$$

Graph f and find
(a) $\lim\limits_{x \to 1} f(x)$, $\lim\limits_{x \to 2} f(x)$, $\lim\limits_{x \to 4} f(x)$
(b) $\lim\limits_{x \to a} f(x)$ if $0 < a < 1$, $1 < a < 2$, $2 < a < 4$, and $a > 4$

Management

71. If a biochemical company produces x grams of interferon, its variable cost per gram is $v(x)$ dollars, where

$$v(x) = \begin{cases} 50, & 0 < x \leq 1000, \\ 40, & 1000 < x \leq 5000, \\ 35, & x > 5000. \end{cases}$$

Graph the function v and find
(a) $\lim_{x \to 1000} v(x)$, $\lim_{x \to 5000} v(x)$
(b) $\lim_{x \to a} v$ if $0 < a < 1000$, $1000 < a < 5000$, and $a > 5000$

72. A software company sold its word-processing package for $s(t)$ dollars t years after its introduction, where

$$s(t) = \begin{cases} 120, & 0 < t < 2, \\ 99, & 2 \leq t < 4, \\ 79, & t \geq 4. \end{cases}$$

Graph the function s and find
(a) $\lim_{t \to 2} s(t)$, $\lim_{t \to 4} s(t)$
(b) $\lim_{t \to a} s(t)$ if $0 < a < 2$, $2 < a < 4$, and $a > 4$

73. Suppose the cost of recovering x barrels of oil from a stripper well is C dollars, where

$$C(x) = \begin{cases} 40{,}000 + 18x, & 0 < x < 5000, \\ 50{,}000 + 20x, & x \geq 5000. \end{cases}$$

Graph this function and find $\lim_{x \to a} C(x)$ if $0 < a < 5000$, $a = 5000$, and $a > 5000$.

74. If the price of a gallon of gasoline is p dollars, gasoline consumption will be y billion gallons per year, where

$$y = \begin{cases} 500 - 10p, & 0 \leq p \leq 1.50, \\ 560 - 50p, & p > 1.50. \end{cases}$$

Graph this function, and find $\lim_{p \to a} y$ if $0 < a < 1.50$, $a = 1.50$, and $a > 1.50$.

75. Repeat Exercise 74 if

$$y = \begin{cases} 500 - 10p, & 0 \leq p \leq 1.50, \\ 400 - 40p, & p > 1.50. \end{cases}$$

10.2 ONE-SIDED LIMITS, INFINITE LIMITS, AND LIMITS AT INFINITY

In the previous section we introduced the notion of the limit of a function f as its independent variable x approaches a limiting value a, and we saw how the existence or nonexistence of $\lim_{x \to a} f(x)$ describes the behavior of f when x is near a. There is more to be said about limits, however: There are several refinements of the notion of limit that can be used to describe the behavior of functions with greater precision.

FIGURE 10–8
$\operatorname*{Lim}_{x \to 3^+} f(x) = 4$

FIGURE 10–9
$\operatorname*{Lim}_{x \to 3^-} f(x) = 2$

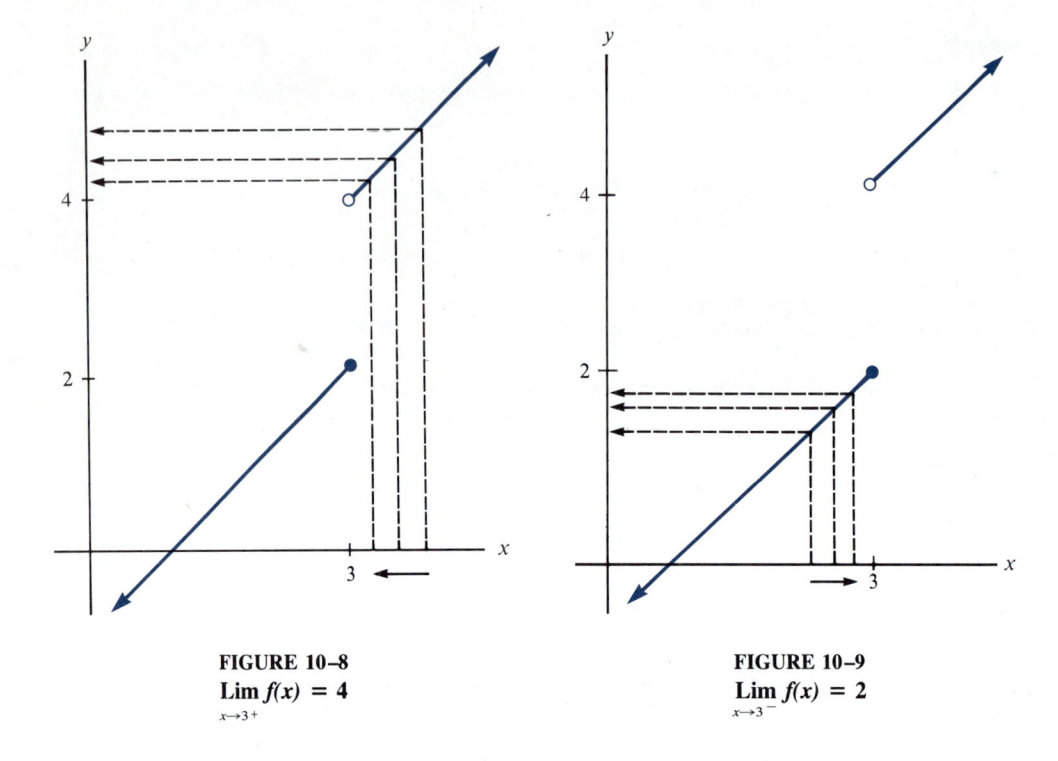

In this section we consider three of these refinements: one-sided limits, infinite limits, and limits at infinity.

One-Sided Limits

Consider the function f defined by

$$f(x) = \begin{cases} x - 1, & 0 \le x \le 3, \\ x + 1, & x > 3. \end{cases}$$

See Figure 10–8. Note that if we let x approach 3 *only* from the right, then the corresponding functional values $f(x)$ approach the number 4. We express this fact by means of the **one-sided limit**

$$\operatorname*{Lim}_{x \to 3^+} f(x) = 4.$$

Notice the notation $x \to 3^+$, which indicates that x approaches 3 only from the right. Similarly, letting $x \to 3^-$ indicate that x approaches 3 only from the left, we have the one-sided limit

$$\operatorname*{Lim}_{x \to 3^-} f(x) = 2.$$

This is so because, as Figure 10–9 shows, as x approaches 3 from the left, the functional values $f(x)$ approach 2. One-sided limits such as these are often convenient in describing the behavior of functions.

THE DERIVATIVE   611

EXAMPLE 1   Let f be defined by

$$f(x) = \begin{cases} -1, & x < 0, \\ 1, & x \geq 0. \end{cases}$$

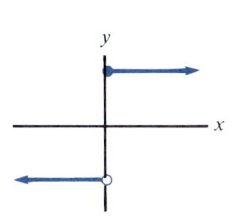

See Figure 10–10. We have

$$\operatorname*{Lim}_{x \to 0^-} f(x) = -1 \quad \text{and} \quad \operatorname*{Lim}_{x \to 0^+} f(x) = 1.$$

Note that $\operatorname*{Lim}_{x \to 0^-} f(x) \neq \operatorname*{Lim}_{x \to 0^+} f(x)$ and that $\operatorname*{Lim}_{x \to 0} f(x)$ does not exist. ☐

FIGURE 10–10

EXAMPLE 2   Let f be defined by

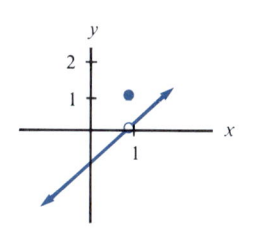

$$f(x) = \begin{cases} x, & x < 1, \\ 2, & x = 1, \\ x, & x > 1. \end{cases}$$

See Figure 10–11. Then

$$\operatorname*{Lim}_{x \to 1^-} f(x) = 1 \quad \text{and} \quad \operatorname*{Lim}_{x \to 1^+} f(x) = 1.$$

Note that $\operatorname*{Lim}_{x \to 1^-} f(x) = \operatorname*{Lim}_{x \to 1^+} f(x) = 1$ and that $\operatorname*{Lim}_{x \to 1} f(x) = 1$ also. ☐

FIGURE 10–11

EXAMPLE 3   Let g be given by $g(x) = \sqrt{x}$. Because $g(x)$ is not defined for $x < 0$, we can only consider the one-sided limit as $x \to 0^+$. From Figure 10–12 we see that

$$\operatorname*{Lim}_{x \to 0^+} g(x) = \operatorname*{Lim}_{x \to 0^+} \sqrt{x} = 0. \quad \square$$

It should be clear that if the one-sided limits $\operatorname*{Lim}_{x \to a^-} f(x)$ and $\operatorname*{Lim}_{x \to a^+} f(x)$ exist and are equal (as in Example 2), then $\operatorname*{Lim}_{x \to a} f(x)$ exists and is equal to them. On the other hand, if either of the one-sided limits does not exist, or if they both exist but are not equal (as in Example 1), then $\operatorname*{Lim}_{x \to a} f(x)$ does not exist.

FIGURE 10–12

EXAMPLE 4   Let f be the function whose graph is shown in Figure 10–13. Then

$$\operatorname*{Lim}_{x \to 1^-} f(x) = -2$$

and $\operatorname*{Lim}_{x \to 1^+} f(x)$ does not exist, so $\operatorname*{Lim}_{x \to 1} f(x)$ does not exist. ☐

EXAMPLE 5   If a firm makes and sells 6000 or fewer units, its profit is $P(x) = 250x - 400{,}000$ dollars. However, in order to make more than 6000 units, it must put on a second shift, which causes its profit to be $P(x) = 200x - 300{,}000$ dollars. Thus

$$P(x) = \begin{cases} 250x - 400{,}000, & 0 \leq x \leq 6000, \\ 200x - 300{,}000, & x > 6000. \end{cases}$$

Thus $\operatorname*{Lim}_{x \to 6000^-} P(x) = 1{,}100{,}000$ and $\operatorname*{Lim}_{x \to 6000^+} P(x) = 900{,}000$. (See Figure 10–14.) Hence as the firm's production approaches 6000 units from the left, its

FIGURE 10–13

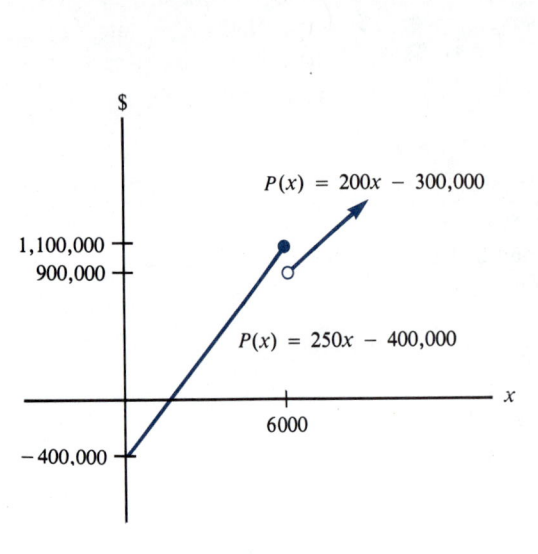

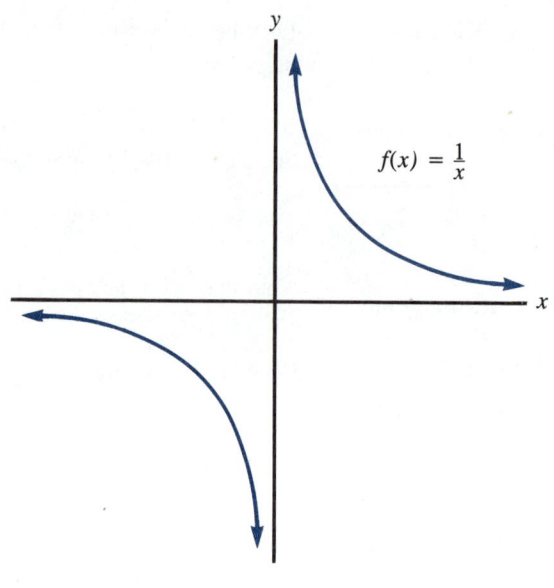

FIGURE 10–14

FIGURE 10–15

profit approaches $1,100,000, and as its production approaches 6000 units from the right, its profit approaches $900,000. ☐

Infinite Limits

Consider the function f defined by $f(x) = 1/x$. See Figure 10–15. The one-sided limit $\text{Lim}_{x \to 0^+} f(x)$ does not exist because as $x \to 0^+$ the corresponding functional values $f(x)$ increase without bound. When this happens, we say that the limit is **positive infinity,** written $+\infty$. Thus we write

$$\text{Lim}_{x \to 0^+} f(x) = \text{Lim}_{x \to 0^+} \frac{1}{x} = +\infty.$$

This does *not* mean that the one-sided limit exists: $+\infty$ is not a number but a symbol for a process of increasing without bound. The statement $\text{Lim}_{x \to 0^+} f(x) = +\infty$ is thus a way of saying that the limit does not exist *because* the functional values increase without bound.

Returning to Figure 10–15, we see also that $\text{Lim}_{x \to 0^-} 1/x$ does not exist because as $x \to 0^-$ the corresponding functional values decrease without bound. In such a case we say that the limit is **negative infinity,** written $-\infty$. Hence we have

$$\text{Lim}_{x \to 0^-} \frac{1}{x} = -\infty.$$

Again, this does not mean that the limit exists but that it does *not* exist because the functional values decrease without bound.

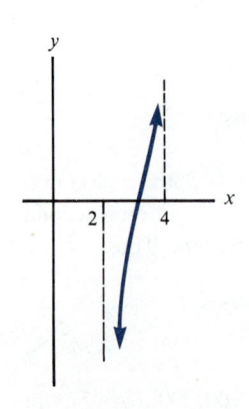

FIGURE 10–16

If $\text{Lim}_{x \to a^+} f(x) = \pm\infty$ or $\text{Lim}_{x \to a^-} f(x) = \pm\infty$, the line $x = a$ is called a

vertical asymptote for f. For instance, the line $x = 0$ (the y-axis) is a vertical asymptote for $f(x) = 1/x$ as $x \to 0^+$ and also as $x \to 0^-$. Notice in Figure 10–15 how the graph of f approaches the y-axis more and more closely as x approaches 0 from either side. Finding asymptotes will be important to us when we discuss curve sketching in Chapter 12.

EXAMPLE 6 The function g whose graph is shown in Figure 10–16 has $\text{Lim}_{x \to 2^+} g(x) = -\infty$ and $\text{Lim}_{x \to 4^-} g(x) = +\infty$. Therefore $x = 2$ is a vertical asymptote for g as $x \to 2^+$, and $x = 4$ is a vertical asymptote for g as $x \to 4^-$. Notice how the graph of g approaches these lines more and more closely as $x \to 2^+$ and as $x \to 4^-$. ☐

Polynomial functions cannot have vertical asymptotes, because if p is such a function and a is any real number, then $\text{Lim}_{x \to a} p(x) = p(a)$, and $p(a) \neq \pm\infty$. However, rational functions can have vertical asymptotes: these will occur at those values of x for which the denominator of the function is zero while its numerator is nonzero.

EXAMPLE 7 Consider the rational function f defined by

$$f(x) = \frac{x}{x - 1}.$$

Because $x = 1$ makes the denominator of f zero and its numerator nonzero, the line $x = 1$ is a vertical asymptote for the function, and hence $\text{Lim}_{x \to 1^+} f(x) = \pm\infty$. We can decide between the alternatives $+\infty$ and $-\infty$ for the limit by examining the sign of the fraction $x/(x - 1)$ as $x \to 1^+$. Note that as x approaches 1 from the right, x is greater than 1, so both x and $x - 1$ are positive. Therefore as $x \to 1^+$, the fraction $x/(x - 1)$ is always positive, and thus

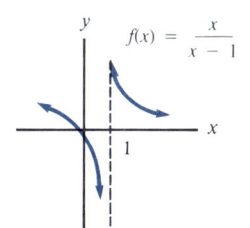

$$\text{Lim}_{x \to 1^+} \frac{x}{x - 1} = +\infty.$$

Similarly, since $x/(x - 1)$ is negative when x is near 1 but less than 1, it follows that

$$\text{Lim}_{x \to 1^-} \frac{x}{x - 1} = -\infty.$$

See Figure 10–17. ☐

FIGURE 10–17

EXAMPLE 8 It is estimated that the cost of removing $x\%$ of the pollutants from the air in the Los Angeles basin would be

$$C(x) = \frac{60x}{100 - x}$$

billion dollars. (This model implies that it would be impossible to remove all pollutants from the air. Why?) Since $C(x)$ is positive when x is near 100 and less than 100, we have $\text{Lim}_{x \to 100^-} C(x) = +\infty$. Thus the cost of removing $x\%$ of the pollutants would become prohibitively large as x approached 100. ☐

Limits at Infinity

The infinity symbols $\pm\infty$ are also useful when considering how a function behaves as its independent variable is allowed to increase or decrease without bound. We write

$x \to +\infty$ to indicate that x is allowed to increase without bound (move to the right along the x-axis without bound).

$x \to -\infty$ to indicate that x is allowed to decrease without bound (move to the left along the x-axis without bound).

Now consider the function f whose graph is shown in Figure 10-18. Notice that as $x \to +\infty$, the corresponding functional values $f(x)$ approach the number L; as $x \to -\infty$, the functional values approach the number K. We express these facts by writing the **limits at infinity**

$$\lim_{x \to +\infty} f(x) = L \quad \text{and} \quad \lim_{x \to -\infty} f(x) = K,$$

respectively, and referring to the lines $y = L$ and $y = K$ as **horizontal asymptotes** for f.

EXAMPLE 9   The function f whose graph is depicted in Figure 10-19 has $\lim_{x \to +\infty} f(x) = 3$ and $\lim_{x \to -\infty} f(x) = 0$. Hence the line $y = 3$ is a horizontal asymptote for f as x approaches $+\infty$ and the line $y = 0$ (the x-axis) is a horizontal asymptote for f as x approaches $-\infty$. ☐

If the functional values $f(x)$ increase without bound as $x \to +\infty$, we write $\lim_{x \to +\infty} f(x) = +\infty$; if they decrease without bound, we write $\lim_{x \to +\infty} f(x) = -\infty$. Similar remarks apply to $\lim_{x \to -\infty} f(x)$.

EXAMPLE 10   The function f of Figure 10-20 has $\lim_{x \to +\infty} f(x) = +\infty$ and $\lim_{x \to -\infty} f(x) = -\infty$. ☐

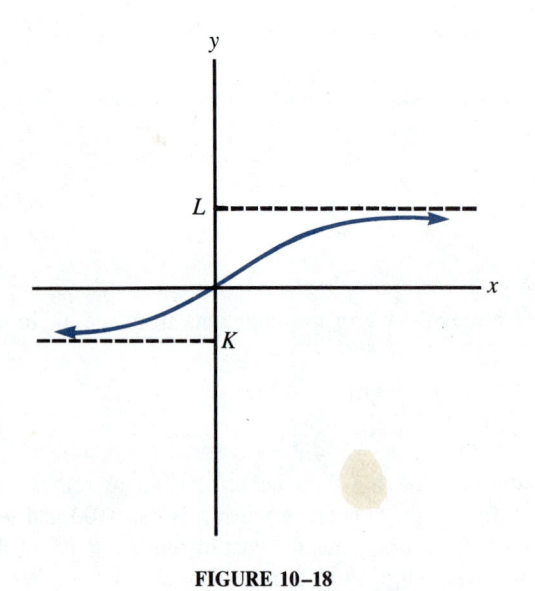

FIGURE 10-18

FIGURE 10-19
$\lim_{x \to +\infty} f(x) = 3, \quad \lim_{x \to -\infty} f(x) = 0$

THE DERIVATIVE	615

EXAMPLE 11 Let us find the limits at infinity of the function f defined by $f(x) = 1/x$. When x is a large positive number, $1/x$ is a small positive fraction; we can make $1/x$ as close to 0 as we want by taking x sufficiently large. Therefore

$$\lim_{x \to +\infty} \frac{1}{x} = 0.$$

This shows that the x-axis is a horizontal asymptote for the function f as $x \to +\infty$. Similar reasoning as $x \to -\infty$ shows that the x-axis is also a horizontal asymptote for f as $x \to -\infty$. The graph of f is shown in Figure 10–21. ☐

In Example 11 we showed that $\lim_{x \to \pm\infty} (1/x) = 0$. The reasoning used there holds also for $1/x^n$, where n is any positive integer. On the other hand, as x increases without bound, so does x^n for any positive integer n, so $\lim_{x \to +\infty} x^n = +\infty$. Similarly, as $x \to -\infty$, x^n approaches $+\infty$ if n is even and approaches $-\infty$ if n is odd.

Limits at Infinity for Powers of x

For any positive integer n:

1. $\lim\limits_{x \to +\infty} \dfrac{1}{x^n} = 0$ and $\lim\limits_{x \to -\infty} \dfrac{1}{x^n} = 0$.

2. $\lim\limits_{x \to +\infty} x^n = +\infty$, $\lim\limits_{x \to -\infty} x^n = +\infty$ if n is even, and

   $\lim\limits_{x \to -\infty} x^n = -\infty$ if n is odd.

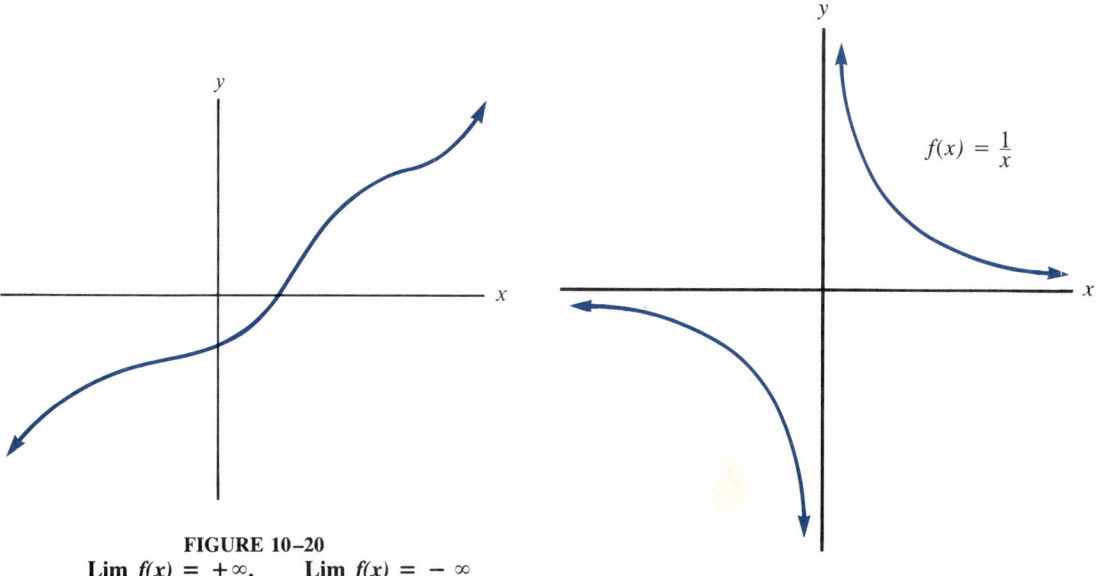

FIGURE 10–20
$\lim\limits_{x \to +\infty} f(x) = +\infty,\qquad \lim\limits_{x \to -\infty} f(x) = -\infty$

FIGURE 10–21

EXAMPLE 12

$$\operatorname*{Lim}_{x \to +\infty} -\frac{2}{x^2} = -2 \cdot \operatorname*{Lim}_{x \to +\infty} \frac{1}{x^2} = -2(0) = 0,$$

$$\operatorname*{Lim}_{x \to +\infty} 4x^3 = 4 \cdot \operatorname*{Lim}_{x \to +\infty} x^3 = 4(+\infty) = +\infty,$$

$$\operatorname*{Lim}_{x \to -\infty} 4x^3 = 4 \cdot \operatorname*{Lim}_{x \to -\infty} x^3 = 4(-\infty) = -\infty. \quad \square$$

Because $\operatorname{Lim}_{x \to \pm\infty} x^n = \pm\infty$, polynomial functions do not have horizontal asymptotes. Rational functions can have horizontal asymptotes, however; to find them, we must be able to evaluate the limits at infinity of such functions. This can be accomplished by dividing each term of the numerator and denominator by the largest power of x that appears in the denominator and using our results about limits at infinity for powers of x. We illustrate with examples.

EXAMPLE 13 Let us evaluate

$$\operatorname*{Lim}_{x \to +\infty} \frac{2x^2 + 1}{x^2 - 5x + 4}.$$

The largest power of x in the denominator is x^2, so we rewrite the limit as

$$\operatorname*{Lim}_{x \to +\infty} \frac{2x^2/x^2 + 1/x^2}{x^2/x^2 - 5x/x^2 + 4/x^2} = \operatorname*{Lim}_{x \to +\infty} \frac{2 + 1/x^2}{1 - 5(1/x) + 4(1/x^2)}.$$

But $\operatorname{Lim}_{x \to +\infty} (1/x) = 0$ and $\operatorname{Lim}_{x \to +\infty} (1/x^2) = 0$ by our results for limits at infinity for powers of x. Hence

$$\operatorname*{Lim}_{x \to +\infty} \frac{2x^2 + 1}{x^2 - 5x + 4} = \frac{2 + 0}{1 - 5(0) + 4(0)} = \frac{2}{1} = 2.$$

Thus $y = 2$ is a horizontal asymptote for f as $x \to +\infty$. $\quad \square$

EXAMPLE 14 Let us evaluate

$$\operatorname*{Lim}_{x \to -\infty} \frac{x^3 + 8}{x^2 - 5}.$$

The highest power of x present in the denominator is x^2, so

$$\operatorname*{Lim}_{x \to -\infty} \frac{x^3/x^2 + 8/x^2}{x^2/x^2 - 5/x^2} = \operatorname*{Lim}_{x \to -\infty} \frac{x + 8/x^2}{1 - 5/x^2}.$$

But $\operatorname*{Lim}_{x \to -\infty} x = -\infty$ and $\operatorname*{Lim}_{x \to -\infty} (1/x^2) = 0$, so

$$\operatorname*{Lim}_{x \to -\infty} \frac{x^3 + 8}{x^2 - 5} = \operatorname*{Lim}_{x \to -\infty} \frac{x + 8(1/x^2)}{1 - 5(1/x^2)} = \frac{-\infty + 8(0)}{1 - 5(0)} = -\infty.$$

Therefore this function has no horizontal asymptotes. $\quad \square$

EXAMPLE 15 If it makes x units of its product, a firm's average cost per unit is $A(x)$ dollars, where

$$A(x) = \frac{12x + 200}{x}.$$

THE DERIVATIVE

Since

$$\lim_{x \to +\infty} A(x) = \lim_{x \to +\infty} \frac{12x + 200}{x} = \lim_{x \to +\infty} \frac{12 + 200/x}{1} = \frac{12 + 0}{1} = 12,$$

we see that as the firm makes more and more units, its average cost per unit approaches $12. ☐

In addition to vertical and horizontal asymptotes, functions can also possess **oblique asymptotes.** An oblique asymptote is a line $y = mx + b$, $m \neq 0$, such that as $x \to \pm\infty$, the functional values $f(x)$ approach $mx + b$. This means that as $x \to \pm\infty$, the graph of f will approach the graph of $y = mx + b$ more and more closely. We illustrate with an example.

EXAMPLE 16 Consider the rational function g defined by

$$g(x) = \frac{x^2 + 1}{x}.$$

Notice that we can write g in the form

$$g(x) = x + \frac{1}{x}.$$

Then

$$\lim_{x \to 0^+} g(x) = \lim_{x \to 0^+} \left(x + \frac{1}{x} \right) = 0 + (+\infty) = +\infty$$

and

$$\lim_{x \to 0^-} g(x) = \lim_{x \to 0^-} \left(x + \frac{1}{x} \right) = 0 + (-\infty) = -\infty.$$

Therefore the y-axis is a vertical asymptote for g as $x \to 0^+$ and as $x \to 0^-$. Also,

$$\lim_{x \to +\infty} g(x) = \lim_{x \to +\infty} \frac{x^2 + 1}{x} = \lim_{x \to +\infty} \frac{x + 1/x}{1} = \frac{+\infty + 0}{1} = +\infty$$

and

$$\lim_{x \to -\infty} g(x) = \lim_{x \to -\infty} \frac{x^2 + 1}{x} = \lim_{x \to -\infty} \frac{x + 1/x}{1} = \frac{-\infty + 0}{1} = -\infty,$$

so g has no horizontal asymptotes. However, closer examination of the behavior of $g(x)$ as $x \to \pm\infty$ shows that the line $y = x$ is an oblique asymptote for g. This is so because as $x \to \pm\infty$, $1/x$ approaches 0, and hence

$$g(x) = x + \frac{1}{x} \approx x + 0 = x.$$

Therefore as $x \to \pm\infty$, the functional values $g(x)$ approach x. Figure 10–22 shows the graph of g: notice how the graph approaches the line $y = x$ as an asymptote. ☐

■ *Computer Manual: Programs LIMIT, INFLIMIT*

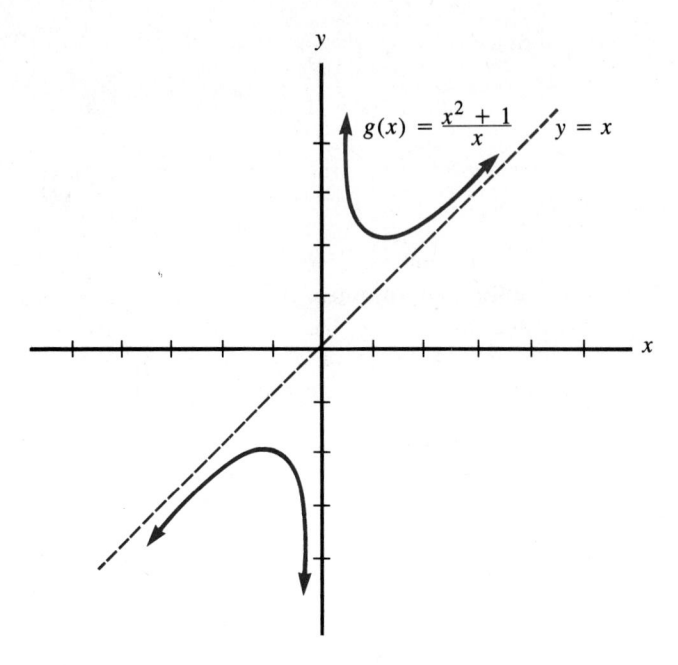

FIGURE 10–22

10.2 EXERCISES

One-Sided Limits

1. The following is the graph of a function f. Find $\lim_{x \to 0^+} f(x)$, $\lim_{x \to 0^-} f(x)$, and $\lim_{x \to 0} f(x)$.

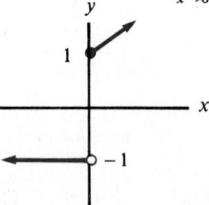

2. The following is the graph of a function f. Find $\lim_{x \to 2^+} f(x)$, $\lim_{x \to 2^-} f(x)$, and $\lim_{x \to 2} f(x)$.

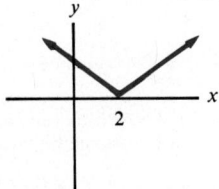

The Derivative

3. The following is the graph of a function f. Find $\lim_{x \to 1^+} f(x)$, $\lim_{x \to 1^-} f(x)$, and $\lim_{x \to 1} f(x)$.

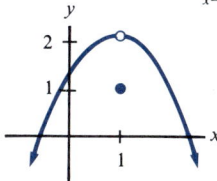

4. The following is the graph of a function f. Find $\lim_{x \to -3^+} f(x)$, $\lim_{x \to -3^-} f(x)$, $\lim_{x \to -3} f(x)$, $\lim_{x \to 2^+} f(x)$, $\lim_{x \to 2^-} f(x)$, and $\lim_{x \to 2} f(x)$.

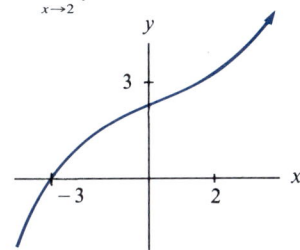

5. The following is the graph of a function f. Find $\lim_{x \to -1^+} f(x)$, $\lim_{x \to -1^-} f(x)$, $\lim_{x \to -1} f(x)$, $\lim_{x \to 1^+} f(x)$, $\lim_{x \to 1^-} f(x)$, and $\lim_{x \to 1} f(x)$.

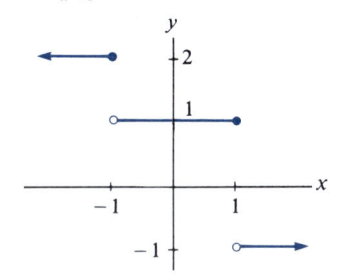

In Exercises 6 through 21, find the indicated limit.

6. $\lim_{x \to 5^+} 2$

7. $\lim_{x \to 2^-} x$

8. $\lim_{x \to 2^+} 3x$

9. $\lim_{x \to 3^+} (5x + 4)$

10. $\lim_{x \to 4^-} (x - 2)$

11. $\lim_{x \to 0^+} |x|$

12. $\lim_{x \to 0^-} |x|$

13. $\lim_{x \to 1^+} \sqrt{x - 1}$

14. $\lim_{x \to -5^+} \sqrt{x + 5}$

15. $\lim_{x \to (1/2)^-} \sqrt{1 - 2x}$

16. $\lim_{x \to 3^-} 2\sqrt{3 - x}$

17. $\lim_{x \to 0^+} f(x)$, $\lim_{x \to 0^-} f(x)$, $\lim_{x \to 0} f(x)$ if $f(x) = \begin{cases} -1, & x < 0, \\ x, & x > 0. \end{cases}$

18. $\lim_{x \to 2^+} f(x)$, $\lim_{x \to 2^-} f(x)$, $\lim_{x \to 2} f(x)$ if $f(x) = \begin{cases} x + 1, & x \leq 2, \\ 5 - x, & x > 2. \end{cases}$

19. $\lim_{x \to 0^+} f(x)$, $\lim_{x \to 0^-} f(x)$, $\lim_{x \to 0} f(x)$ if $f(x) = \begin{cases} x, & x < 0, \\ 2, & x = 0, \\ x, & x > 0. \end{cases}$

20. $\lim_{x \to -2^+} f(x)$, $\lim_{x \to -2^-} f(x)$, $\lim_{x \to -2} f(x)$, $\lim_{x \to 1^+} f(x)$, $\lim_{x \to 1^-} f(x)$, and $\lim_{x \to 1} f(x)$ if

$$f(x) = \begin{cases} 2x + 1, & x < -2, \\ x - 1, & -2 \leq x \leq 1, \\ x + 2, & x > 1. \end{cases}$$

*21. $\lim_{x \to 0^+} f(x)$, $\lim_{x \to 0^-} f(x)$, $\lim_{x \to 0} f(x)$ if $f(x) = \begin{cases} |x|/x, & x \neq 0, \\ 0, & x = 0. \end{cases}$

Management

22. A company's cost function is defined by

$$C(x) = \begin{cases} 20x + 15{,}000, & 0 \leq x \leq 1000, \\ 24x + 16{,}000, & x > 1000. \end{cases}$$

Find and interpret $\lim_{x \to 1000^+} C(x)$ and $\lim_{x \to 1000^-} C(x)$.

23. A firm's profit function is given by

$$P(x) = \begin{cases} 10x - 5000, & 0 \leq x \leq 2500, \\ 6x + 5000, & 2500 < x \leq 5000, \\ 5x + 6000, & 5000 < x \leq 7500. \end{cases}$$

Find and interpret $\lim_{x \to 2500^+} P(x)$, $\lim_{x \to 2500^-} P(x)$, $\lim_{x \to 5000^+} P(x)$, and $\lim_{x \to 5000^-} P(x)$.

24. If a printing company purchases x gallons of ink it costs them $C(x)$ dollars, where

$$C(x) = \begin{cases} 5.50x, & 0 \leq x < 25, \\ 5.00x, & 25 \leq x < 100, \\ 4.75x, & x \geq 100. \end{cases}$$

Find and interpret $\lim_{x \to 25^+} C(x)$, $\lim_{x \to 25^-} C(x)$, $\lim_{x \to 100^+} C(x)$, and $\lim_{x \to 100^-} C(x)$.

25. It costs a mail-order company y dollars to ship a package that weighs x pounds, where

$$y = \begin{cases} 0.5 + 5.0x, & 0 < x < 2, \\ 1.5 + 4.5x, & 2 \leq x < 10, \\ 6.5 + 4.0x, & x \geq 10. \end{cases}$$

Find and interpret $\lim_{x \to 2^+} y$, $\lim_{x \to 2^-} y$, $\lim_{x \to 10^+} y$, and $\lim_{x \to 10^-} y$.

Infinite Limits

In Exercises 26 through 30 use the symbols $\pm\infty$.

26. The following is the graph of a function f. Find $\lim_{x \to 2^+} f(x)$, $\lim_{x \to 2^-} f(x)$, and the vertical asymptotes of f.

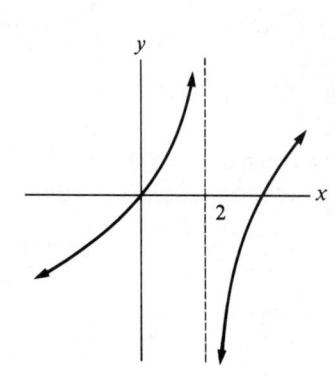

THE DERIVATIVE

27. The following is the graph of a function f. Find $\lim\limits_{x\to 1^+} f(x)$, $\lim\limits_{x\to 1^-} f(x)$, and the vertical asymptotes of f.

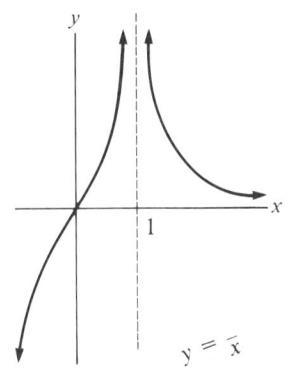

28. The following is the graph of a function f. Find $\lim\limits_{x\to 2^+} f(x)$, $\lim\limits_{x\to 2^-} f(x)$, $\lim\limits_{x\to -2^+} f(x)$, $\lim\limits_{x\to -2^-} f(x)$, and the vertical asymptotes of f.

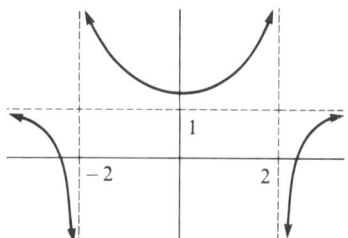

29. The following is the graph of a function f. Find $\lim\limits_{x\to 3^+} f(x)$, $\lim\limits_{x\to 3^-} f(x)$, and the vertical asymptotes of f.

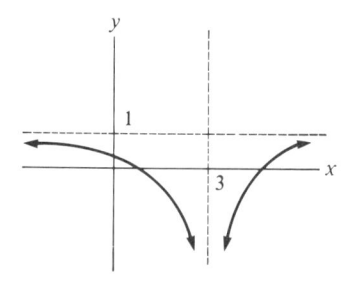

30. The following is the graph of a function f. Find $\lim\limits_{x\to -2^+} f(x)$, $\lim\limits_{x\to 2^+} f(x)$, $\lim\limits_{x\to 2^-} f(x)$, $\lim\limits_{x\to 4^-} f(x)$, and the vertical asymptotes of f.

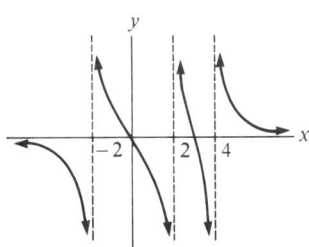

In Exercises 31 through 42, find the indicated limits. Use the symbols $\pm\infty$ as appropriate.

31. $\lim\limits_{x\to 0^+} \dfrac{1}{x^3}$, $\lim\limits_{x\to 0^-} \dfrac{1}{x^3}$

32. $\lim\limits_{x\to 0^+} \dfrac{2}{x^4}$, $\lim\limits_{x\to 0^-} \dfrac{2}{x^4}$

33. $\lim\limits_{x\to 0^+} \dfrac{-2}{x}$, $\lim\limits_{x\to 0^-} \dfrac{-2}{x}$

34. $\lim\limits_{x\to 2^+} \dfrac{1}{x-2}$, $\lim\limits_{x\to 2^-} \dfrac{1}{x-2}$

35. $\lim\limits_{x \to -1^+} \dfrac{1}{1-x}$, $\lim\limits_{x \to -1^-} \dfrac{1}{1-x}$

36. $\lim\limits_{x \to 1^+} \dfrac{x}{x-1}$, $\lim\limits_{x \to 1^-} \dfrac{x}{x-1}$

37. $\lim\limits_{x \to -3^+} \dfrac{x}{x+3}$, $\lim\limits_{x \to -3^-} \dfrac{x}{x+3}$

38. $\lim\limits_{x \to 5^+} \dfrac{|x|}{x-5}$, $\lim\limits_{x \to 5^-} \dfrac{|x|}{x-5}$

39. $\lim\limits_{x \to -1^+} \dfrac{-2x}{x+1}$, $\lim\limits_{x \to -1^-} \dfrac{-2x}{x+1}$

40. $\lim\limits_{x \to 1^+} \dfrac{x^2}{x-1}$, $\lim\limits_{x \to 1^-} \dfrac{x^2}{x-1}$

41. $\lim\limits_{x \to 2^+} \dfrac{x}{x^2-4}$, $\lim\limits_{x \to 2^-} \dfrac{x}{x^2-4}$

42. $\lim\limits_{x \to -1^+} \dfrac{x^2}{x^2-1}$, $\lim\limits_{x \to -1^-} \dfrac{x^2}{x^2-1}$

43. Suppose the cost of eliminating all air pollution in the Los Angeles basin is $C(x)$ billion dollars, where

$$C(x) = \dfrac{60x}{120 - x}.$$

Find and interpret $\lim\limits_{x \to 100^-} C(x)$ (cf. Example 8).

Management

44. The demand function for a commodity is given by

$$q = 20\sqrt{100 - p}.$$

Find and interpret $\lim\limits_{x \to 100^-} q$.

45. Reducing the pollution caused by auto emissions by $x\%$ over the next five years would cost the auto manufacturers y billion dollars, where

$$y = \dfrac{10x}{20 - 0.4x}.$$

Find and interpret $\lim\limits_{x \to 50^-} y$.

46. The cost of decommissioning a nuclear power plant that has operated for t years is estimated to be y million dollars, where

$$y = \dfrac{40t + 30}{30 - t}.$$

Find and interpret $\lim\limits_{t \to 30^-} y$.

47. An oil field is estimated to contain 100 million barrels of oil. The cost of obtaining $x\%$ of this oil will be

$$C(x) = \dfrac{400x - 3.5x^2}{95 - x}$$

million dollars. Find and interpret $\lim\limits_{x \to 95^-} C(x)$.

48. The supply function for a commodity is given by

$$q = \dfrac{p^2 + 100p - 50{,}000}{500 - p}.$$

Find and interpret $\lim\limits_{p \to 500^-} q$.

Limits at Infinity

In Exercises 49 through 53, find $\lim_{x \to \pm \infty} f(x)$ and the horizontal asymptotes for the function f whose graph is shown. Use the symbols $\pm \infty$ as appropriate.

49.

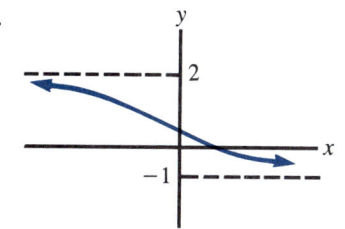

50.

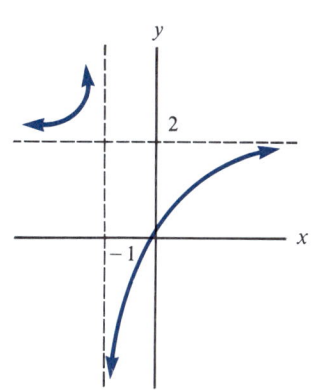

51.

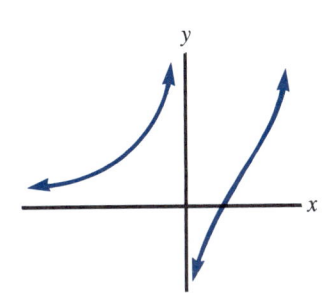

52.

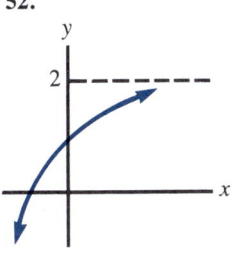

53.

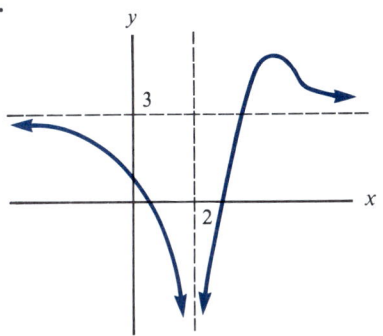

In Exercises 54 through 71, find the indicated limit at infinity. Use the symbols $\pm \infty$ as appropriate.

54. $\lim\limits_{x \to +\infty} (2x + 3)$

55. $\lim\limits_{x \to -\infty} (x^2 - 2)$

56. $\lim\limits_{x \to -\infty} (x^3 + 4)$

57. $\lim\limits_{x \to +\infty} \dfrac{1}{x^2}$

58. $\lim\limits_{x \to -\infty} \dfrac{2}{x^3}$

59. $\lim\limits_{x \to -\infty} \dfrac{3}{x^4}$

60. $\lim\limits_{x \to +\infty} -\dfrac{2}{x^2}$

61. $\lim\limits_{x \to +\infty} \dfrac{x}{x + 5}$

62. $\lim\limits_{x \to -\infty} \dfrac{3x}{2x + 4}$

63. $\lim\limits_{x \to +\infty} \dfrac{2x + 1}{2 - 3x}$

64. $\lim\limits_{x \to -\infty} \dfrac{3x}{x^2 + 1}$

65. $\lim\limits_{x \to +\infty} \dfrac{2x^2 + 1}{3x - 2}$

66. $\lim\limits_{x \to +\infty} \dfrac{3x^2}{5x^2 + 3}$

67. $\lim\limits_{x \to +\infty} \dfrac{x^2 + 5x - 7}{x^3 - 2x + 1}$

68. $\lim\limits_{x \to +\infty} \dfrac{2x^2 + 5x - 7}{4x^2 + 8x + 2}$

69. $\lim\limits_{x \to -\infty} \dfrac{5x^3 + 6x + 12}{11x^3 - 2x + 3}$

70. $\lim\limits_{x \to +\infty} \dfrac{4x^2 + 6x + 9}{2x^4 + 3x^3 + 2}$

71. $\lim\limits_{x \to +\infty} \dfrac{6x^4 - 2x^3 + 8x + 1}{x^2 - 3x + 2}$

In Exercises 72 through 81, find all asymptotes, vertical, horizontal, and oblique, of the given function f.

72. $f(x) = \dfrac{x}{x-3}$

73. $f(x) = \dfrac{6x}{3x-2}$

74. $f(x) = \dfrac{x+1}{x-3}$

75. $f(x) = \dfrac{5x+2}{2x-6}$

76. $f(x) = \dfrac{2-2x}{1+x}$

77. $f(x) = \dfrac{2+3x}{x}$

78. $f(x) = \dfrac{3x}{x^2-1}$

79. $f(x) = \dfrac{x-1}{x^2-4}$

80. $f(x) = \dfrac{x^2+1}{x}$

81. $f(x) = \dfrac{2x^2-2}{x}$

Management

82. If x units of a product are made, its average cost per unit is $A(x)$ dollars, where
$$A(x) = 22 + \dfrac{50{,}000}{x}.$$
Find and interpret $\lim\limits_{x \to +\infty} A(x)$.

83. Acme Company's average cost per unit produced is given by
$$A(x) = \dfrac{5x+80}{x+1},$$
where x is the number of units produced and $A(x)$ is the average cost per unit in dollars. Find and interpret $\lim\limits_{x \to +\infty} A(x)$.

84. A firm's average cost per unit if it makes x units is $A(x)$ dollars, where
$$A(x) = \dfrac{2x^2+50}{x}.$$
Find and interpret $\lim\limits_{x \to +\infty} A(x)$.

85. A computer company is introducing a new model of computer. The firm estimates that it will take $T(x)$ worker-hours to build the xth unit of the new computer, where
$$T(x) = \dfrac{300x^3 + 50x + 100}{x^3}.$$
Find and interpret $\lim\limits_{x \to +\infty} T(x)$.

Life Science

86. If a person who suffers a bite by a poisonous snake receives an immediate shot of antivenin, then t seconds after the shot is given there will be
$$y = \dfrac{0.5t + 2000}{2t + 3}$$
ppm of poison in the victim's blood. Find and interpret $\lim\limits_{t \to +\infty} y$.

Social Science

87. Subjects of a memory experiment are asked to memorize a list of numbers, and it is found that t days later the typical subject can recall

$$y = 2 + \frac{20}{t+2}$$

of the numbers. Find and interpret $\lim\limits_{t \to +\infty} y$.

Computer Science

*88. According to an article by Peter Denning ("The Science of Computing: Speeding Up Parallel Processing," *American Scientist*, July–August 1988), if the size of a problem is held constant while the number of processors in a computer increases, then the speedup in processing is

$$F = \frac{aN^2 p}{aN^2 + pb}.$$

Here p is the number of processors, $p \geq 1$, N is a measure of the size of the problem, aN^2 is the total computation time, and b is the communication time between processors. Find and interpret $\lim\limits_{p \to +\infty} F$.

10.3 THE DERIVATIVE

The concept of the derivative is the basis of calculus. The derivative of a function tells us many things about the function. For instance, as we shall see later in this chapter, we can use the derivative to examine the rate of change of a function. Thus, suppose a certain function describes the concentration of an antibiotic in a patient's blood as a function of time; then the derivative of the function will tell us the rate at which the concentration of the antibiotic is changing with respect to time. Derivatives can also help us graph functions, find their maximum and minimum values, and in general analyze them in many useful ways. Before we can do any of these things, however, we must first develop a thorough understanding of what the derivative is. In this section we introduce the derivative, discuss its geometric significance, and briefly consider the concepts of differentiability and continuity of a function.

Definition of the Derivative

In order to make our initial discussion of the derivative specific, we choose to consider the function f defined by $f(x) = x^2$ and the point (1, 1) on its graph. We may construct a unique line that just touches the graph at the point (1, 1). See Figure 10–23. This line is called the **line tangent to f at (1, 1).** If we knew the slope of this tangent line, we could write its equation; we will now try to find its slope.

Suppose we add a number h to 1. The point

$$(1 + h, f(1 + h)) = (1 + h, (1 + h)^2)$$

will lie on the graph of f, and we can draw a line through the points (1, 1) and $(1 + h, (1 + h)^2)$. This line is called a **secant line.** Figure 10–24 depicts the situation if h is positive. The secant line has

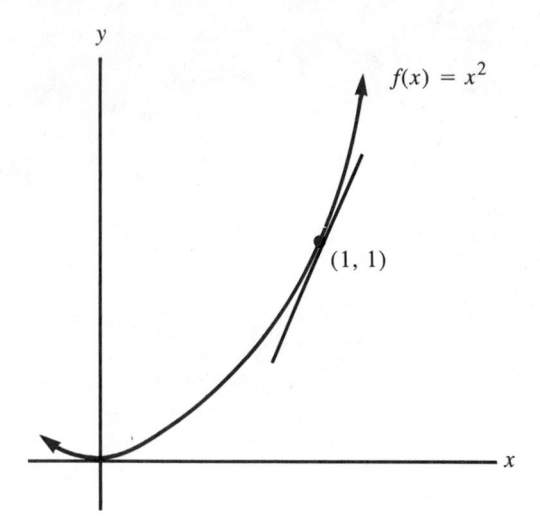

FIGURE 10–23

FIGURE 10–24

$$\text{Slope} = \frac{(1+h)^2 - 1}{1+h-1} = \frac{1+2h+h^2-1}{h} = \frac{h(2+h)}{h} = 2+h.$$

Now let h approach 0. As h gets smaller and smaller, $1+h$ approaches 1 on the x-axis, the point $(1+h, (1+h)^2)$ "slides" along the graph toward the point $(1, 1)$, and the secant line approaches the tangent line at $(1, 1)$. Figure 10–25 shows how as h approaches 0 the secant line "pivots" at $(1, 1)$ and approaches the tangent line. Since the secant line approaches the tangent line, it follows that the slope of the secant line must approach the slope of the tangent line. In other words,

$$\lim_{h \to 0}(\text{slope of secant line}) = \text{Slope of tangent line},$$

or

$$\lim_{h \to 0}(2+h) = 2 + 0 = 2.$$

Thus the slope of the line tangent to f at $(1, 1)$ is equal to 2. This slope is called the **derivative of f at $x = 1$.**

The preceding analysis can be applied to any function f at any number in its domain as long as the function has a unique nonvertical tangent line at the point $(x, f(x))$ on its graph. Figure 10–26, p. 628, depicts the process. We begin by adding h to x to obtain the point $(x+h, f(x+h))$ on the graph. We then form the secant line through $(x, f(x))$ and $(x+h, f(x+h))$. The secant line will have

$$\text{Slope} = \frac{f(x+h) - f(x)}{x+h-x} = \frac{f(x+h) - f(x)}{h},$$

and thus

$$\text{Slope of tangent line} = \lim_{h \to 0} \frac{f(x+h) - f(x)}{h}.$$

The limit on the right side of this equation is called the **derivative of f at x.**

THE DERIVATIVE

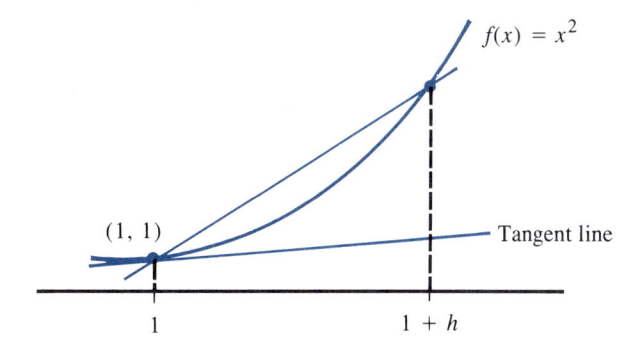

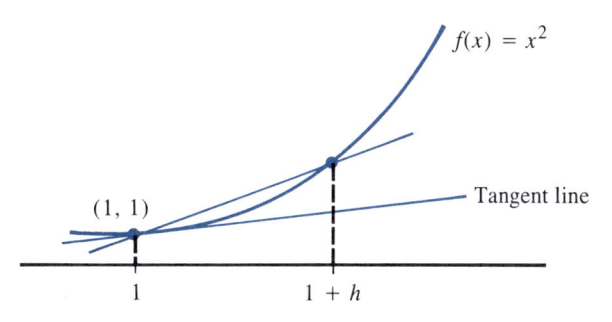

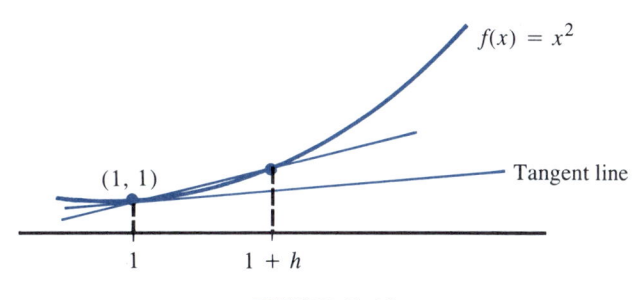

[Figure: Three graphs of $f(x) = x^2$ showing secant lines through points $(1, 1)$ and $(1+h, (1+h)^2)$ approaching the tangent line as $h \to 0$.]

FIGURE 10–25

The Derivative of a Function

Let f be a function. The **derivative of f at x,** denoted by $f'(x)$, is defined to be

$$f'(x) = \lim_{h \to 0} \frac{f(x+h) - f(x)}{h},$$

provided this limit exists. The notation $f'(x)$ is read "f prime of x."

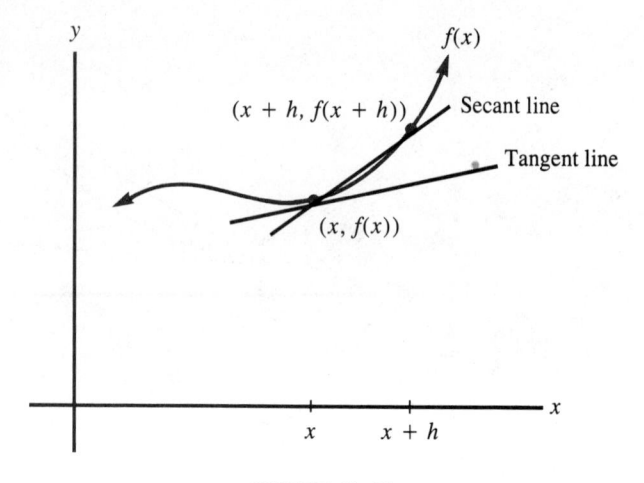

FIGURE 10–26

As we have seen, the derivative $f'(x)$ is equal to the slope of the line tangent to f at the point $(x, f(x))$.

EXAMPLE 1   Let $f(x) = x^2$. Let us find the derivative $f'(x)$. We begin by forming the quotient

$$\frac{f(x + h) - f(x)}{h},$$

which is called the **difference quotient of f at x**. Since $f(x) = x^2$, we have

$$f(x + h) = (x + h)^2,$$

so

$$\frac{f(x + h) - f(x)}{h} = \frac{(x + h)^2 - x^2}{h}.$$

Next we simplify the difference quotient, our goal being the removal of the factor h from the denominator:

$$\frac{f(x + h) - f(x)}{h} = \frac{(x + h)^2 - x^2}{h}$$

$$= \frac{x^2 + 2xh + h^2 - x^2}{h}$$

$$= \frac{h(2x + h)}{h}$$

$$= 2x + h.$$

Therefore the derivative of $f(x) = x^2$ at x is

$$f'(x) = \lim_{h \to 0} \frac{f(x + h) - f(x)}{h} = \lim_{h \to 0} (2x + h) = 2x + 0 = 2x. \quad \square$$

The Derivative

Notice that the derivative of the function f defined by $f(x) = x^2$ is another function f' defined by $f'(x) = 2x$. If we let $x = a$ in the equation $f'(x) = 2x$, we obtain the slope of the line tangent to f at the point $(a, f(a))$. For instance, if we take $x = 1$, then $f'(1) = 2(1) = 2$, so the slope of the line tangent to f at $(1, f(1)) = (1, 1)$ is equal to 2. This is the result we obtained at the beginning of this section.

The procedure we used in Example 1 to find the derivative is known as the **Three-Step Rule.**

The Three-Step Rule

To find the derivative $f'(x)$ of $f(x)$:

1. Form the difference quotient $\dfrac{f(x + h) - f(x)}{h}$.

2. Simplify the difference quotient to eliminate the factor h from the denominator.

3. Then
$$f'(x) = \lim_{h \to 0} \frac{f(x + h) - f(x)}{h}$$

if this limit exists.

EXAMPLE 2 Let f be given by $f(x) = x^2 + 2x + 4$. We will use the Three-Step Rule to find the defining equation for the derivative f'. We have

$$\frac{f(x + h) - f(x)}{h} = \frac{[(x + h)^2 + 2(x + h) + 4] - [x^2 + 2x + 4]}{h}$$

$$= \frac{x^2 + 2xh + h^2 + 2x + 2h + 4 - x^2 - 2x - 4}{h}$$

$$= \frac{h(2x + h + 2)}{h}$$

$$= 2x + h + 2.$$

Therefore

$$f'(x) = \lim_{h \to 0} \frac{f(x + h) - f(x)}{h} = \lim_{h \to 0} (2x + h + 2) = 2x + 2. \quad \square$$

EXAMPLE 3 If

$$g(x) = \frac{1}{x},$$

find $g'(x)$ and then find the equation of the line tangent to g at $(2, g(2))$.

We have

$$\frac{g(x+h)-g(x)}{h} = \frac{1/(x+h) - 1/x}{h}$$

$$= \frac{\frac{x-(x+h)}{x(x+h)}}{h}$$

$$= \frac{-h}{hx(x+h)}$$

$$= \frac{-1}{x(x+h)}.$$

Thus

$$g'(x) = \operatorname*{Lim}_{h\to 0} \frac{g(x+h)-g(x)}{h} = \operatorname*{Lim}_{h\to 0} \frac{-1}{x(x+h)} = \frac{-1}{x(x+0)} = \frac{-1}{x^2}.$$

Hence if $g(x) = 1/x$, then $g'(x) = -1/x^2$. Thus the line tangent to g at $(2, g(2)) = (2, \frac{1}{2})$ has slope equal to

$$g'(2) = -\frac{1}{2^2} = -\frac{1}{4},$$

and the equation of the tangent line is

$$y = -\frac{1}{4}x + b.$$

But the point $(2, \frac{1}{2})$ is on the tangent line, so

$$\frac{1}{2} = -\frac{1}{4}(2) + b,$$

which implies that $b = 1$. Hence the equation of the tangent line is

$$y = -\frac{1}{4}x + 1.$$

See Figure 10–27. □

Differentiability and Continuity

If a function f has a derivative $f'(a)$ at $x = a$, we say that the function is **differentiable at a;** if it does not have a derivative at a, we say it is **not differentiable at a**.

EXAMPLE 4   The function f defined by $f(x) = \sqrt{x}$ is differentiable at a if $a > 0$, but it is not differentiable at 0. To see this, suppose $a \geq 0$ and note that

$$\operatorname*{Lim}_{h\to 0} \frac{\sqrt{a+h}-\sqrt{a}}{h} = \operatorname*{Lim}_{h\to 0} \frac{(\sqrt{a+h}-\sqrt{a})(\sqrt{a+h}+\sqrt{a})}{h(\sqrt{a+h}+\sqrt{a})}$$

$$= \operatorname*{Lim}_{h\to 0} \frac{a+h-a}{h(\sqrt{a+h}+\sqrt{a})}$$

$$= \operatorname*{Lim}_{h\to 0} \frac{1}{\sqrt{a+h}+\sqrt{a}}.$$

THE DERIVATIVE 631

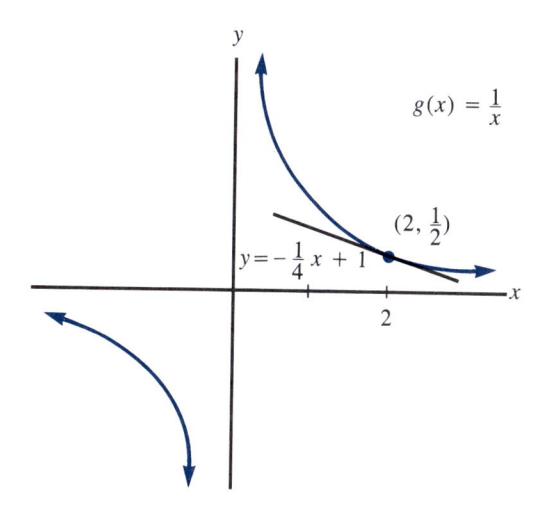

FIGURE 10-27

If $a > 0$, this limit is equal to

$$\frac{1}{\sqrt{a+0}+\sqrt{a}} = \frac{1}{2\sqrt{a}}.$$

Thus $f'(a) = \dfrac{1}{2\sqrt{a}}$ exists if $a > 0$. However, if $a = 0$, the limit becomes

$$\lim_{h \to 0} \frac{1}{\sqrt{0+h}+\sqrt{0}} = \lim_{h \to 0} \frac{1}{\sqrt{h}},$$

which does not exist, and therefore f is not differentiable at 0. ☐

If the derivative $f'(a)$ exists, it is the slope of the unique nonvertical line tangent to f at $(a, f(a))$. Therefore if f does not have a unique tangent line at $(a, f(a))$, as in Figures 10-28 and 10-29, or if the tangent line is vertical, as in Figure 10-30, then the derivative $f'(a)$ cannot exist. For instance, the square-root function of Example 4 has a vertical tangent line at (0, 0) (see Figure 10-31), so it is not differentiable at $x = 0$.

Closely allied to the notion of differentiability is the notion of continuity. A function f is **continuous at $x = a$** if we can trace its graph through the point $(a, f(a))$ without lifting our pencil from the paper. This means that f is continuous at $x = a$ if there is neither a jump nor a gap in its graph at $x = a$. Figure 10-32 shows the graph of a function that is continuous at every real number; such a function is said to be **continuous everywhere.** If f is not continuous at $x = a$, we say it is **discontinuous at $x = a$.** Figures 10-33, 10-34, and 10-35 show different ways in which a function can be discontinuous at $x = a$: in Figure 10-33 the function has a jump in its graph at $x = a$; in Figures 10-34 and 10-35 it has a gap in its graph at $x = a$.

We can state the definition of continuity of a function at $x = a$ in terms of limits.

632 CHAPTER 10

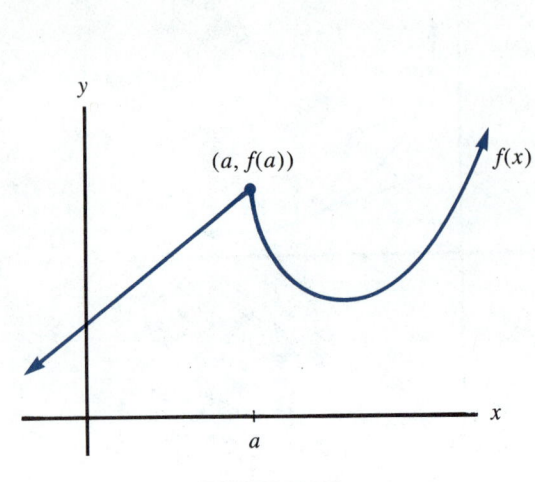

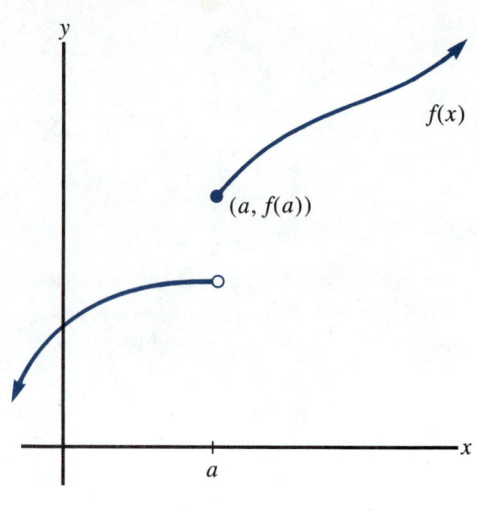

FIGURE 10–28

FIGURE 10–29

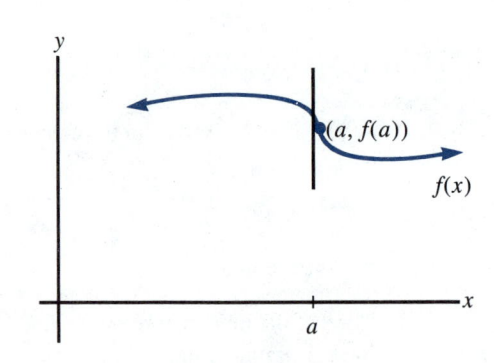

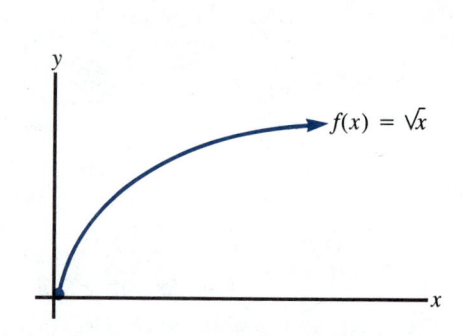

FIGURE 10–30

FIGURE 10–31

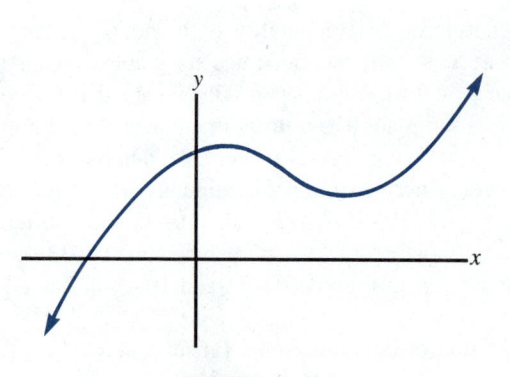

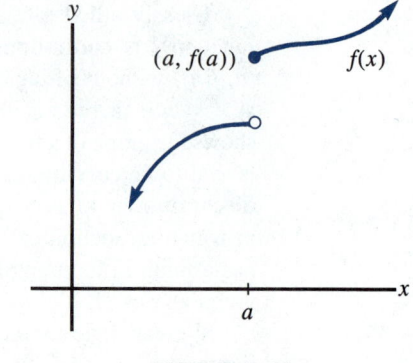

FIGURE 10–32

FIGURE 10–33

The Derivative

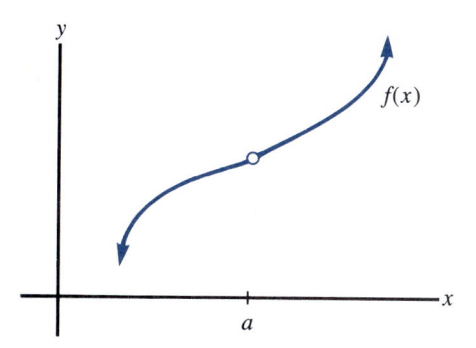

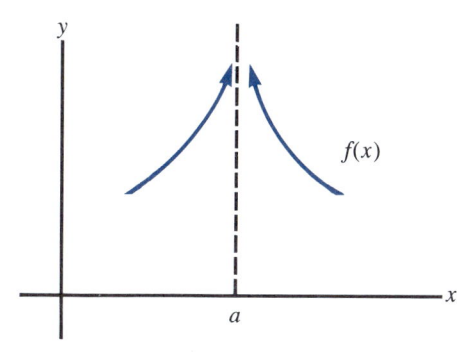

FIGURE 10-34

FIGURE 10-35

> ### The First Limit Definition of Continuity
> A function f is continuous at $x = a$ if $f(a)$ exists and
> $$\lim_{h \to 0} f(x) = f(a).$$

This definition of continuity agrees with our previous geometric one: if $f(a)$ exists, then the point $(a, f(a))$ is on the graph, so there can be no gap in the graph of f at $x = a$; if $\lim_{x \to a} f(x) = f(a)$, then there can be no jump in the graph at $x = a$. (Why?)

Note that if we set $x = a + h$, then we may write

$$\lim_{x \to a} f(x) \quad \text{as} \quad \lim_{h \to 0} f(a + h).$$

Hence we may rewrite the first limit definition of continuity as

> ### The Second Limit Definition of Continuity
> A function f is continuous at $x = a$ if $f(a)$ exists and
> $$\lim_{h \to 0} f(a + h) = f(a).$$

This second version of the limit definition of continuity will be useful to us later.

We saw in Section 10.1 that if p is a polynomial function, then

$$\lim_{x \to a} p(x) = p(a).$$

FIGURE 10–36
The Postal Function

FIGURE 10–37

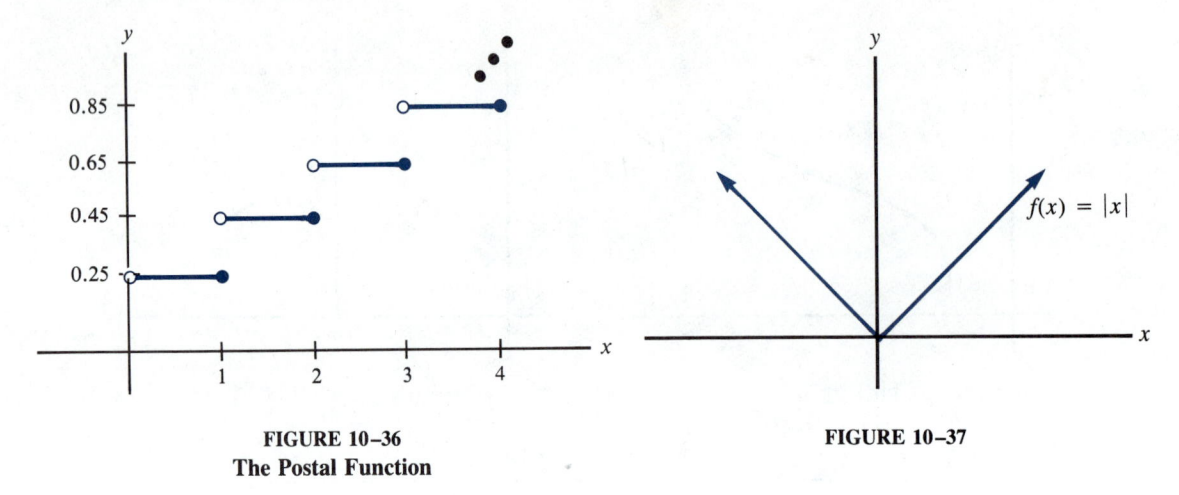

Thus polynomial functions are continuous at every real number. Rational functions that have been reduced to lowest terms are continuous at every real number for which they are defined. However, there are functions that are *not* continuous at every real number for which they are defined.

EXAMPLE 5 Let P be the postal function defined by

$$P(x) = \text{Cost of mailing a letter that weighs } x \text{ ounces.}$$

Currently postage is $0.25 for a letter weighing 1 ounce or less, and an additional $0.20 for each additional ounce or part of an ounce. Therefore

$$P(x) = \begin{cases} \$0.25, & 0 < x \leq 1, \\ \$0.45, & 1 < x \leq 2, \\ \$0.65, & 2 < x \leq 3, \\ \vdots & \end{cases}$$

The graph of P is shown in Figure 10–36. Note that P is discontinuous at $x = 1, 2, 3, \ldots$ and continuous in the intervals $(0, 1)$, $(1, 2)$, $(2, 3)$, ... ◻

The relationship between continuity and differentiability is this: a function differentiable at $x = a$ is continuous at $x = a$. To see this, suppose the function f has a derivative at $x = a$. If f were not continuous at $x = a$, it would have either a jump or a gap in its graph at $x = a$. But if this were the case, it could not have a unique tangent line at $(a, f(a))$, and hence it could not have a derivative at $x = a$. Thus f must be continuous at $x = a$.

> **Differentiability Implies Continuity**
>
> If the function f is differentiable at $x = a$, then it is continuous at $x = a$.

The converse of the preceding statement is *not* true; continuity does *not* imply differentiability. In other words, it is possible for a function to be continuous at

THE DERIVATIVE   635

$x = a$ but not differentiable at $x = a$. For example, the absolute value function defined by $f(x) = |x|$ has the graph shown in Figure 10–37; from the graph we see that this function is continuous at $x = 0$, but it is not differentiable at $x = 0$ because it does not have a unique tangent line there.

■ **Computer Manual: Program DFQUOTNT**

10.3 EXERCISES

Definition of the Derivative

In Exercises 1 through 16, use the Three-Step Rule to find $f'(x)$.

1. $f(x) = x$
2. $f(x) = -2x$
3. $f(x) = 3x + 2$
4. $f(x) = 3x^2$
5. $f(x) = 1 - x^2$
6. $f(x) = x - x^2$
7. $f(x) = x^3$
8. $f(x) = 2x^3 + x^2$
9. $f(x) = 3/x$
10. $f(x) = x^{-2}$
11. $f(x) = x + \dfrac{1}{x}$
12. $f(x) = \dfrac{x}{x-1}$
13. $f(x) = \dfrac{x-1}{x+1}$
14. $f(x) = \sqrt{2x}$
15. $f(x) = \sqrt{x+1}$
16. $f(x) = x^{-1/2}$

In Exercises 17 through 32, find the equation of the line tangent to f at $(a, f(a))$. You may wish to use the results of Exercises 1 through 16.

17. $f(x) = x$, $a = 0$
18. $f(x) = -2x$, $a = 4$
19. $f(x) = 3x + 2$, $a = -\frac{1}{2}$
20. $f(x) = 3x^2$, $a = 1$
21. $f(x) = 1 - x^2$, $a = 3$
22. $f(x) = x - x^2$, $a = -2$
23. $f(x) = x^3$, $a = -1$
24. $f(x) = 2x^3 + x^2$, $a = 0$
25. $f(x) = 3/x$, $a = 2$
26. $f(x) = x^{-2}$, $a = \frac{1}{2}$
27. $f(x) = x + \dfrac{1}{x}$, $a = 1$
28. $f(x) = \dfrac{x}{x-1}$, $a = -1$
29. $f(x) = \dfrac{x-1}{x+1}$, $a = -1$
30. $f(x) = \sqrt{2x}$, $a = 2$
31. $f(x) = \sqrt{x+1}$, $a = -1$
32. $f(x) = x^{-1/2}$, $a = 9$

*33. If $f(x) = x^{1/3}$, use the Three-Step Rule to find $f'(x)$. (*Hint:* Multiply numerator and denominator of the difference quotient by $(x + h)^{2/3} + x^{1/3}(x + h)^{1/3} + x^{2/3}$.)

*34. If $f(x) = |x|$, use the Three-Step Rule to find $f'(x)$. (*Hint:* Consider the cases $x > 0$, $x = 0$, and $x < 0$ separately.)

Differentiability and Continuity

35. Find the values of x at which the functions of Exercises 1 through 8 are not differentiable.

36. Find the values at x at which the functions of Exercises 9 through 16 are not differentiable.

In Exercises 37 through 46, use the given graph to find
(a) the intervals on which the function is continuous;
(b) the values of x at which the function is discontinuous;
(c) the values of x at which the function is not differentiable.

37.
38.
39.
40.
41.
42.
43.
44.
45.
46.

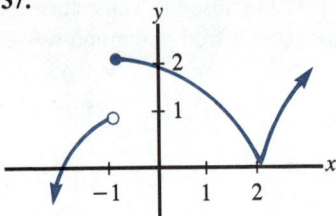

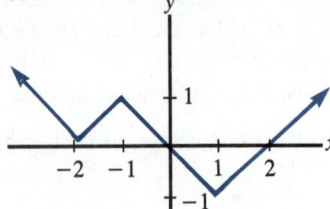

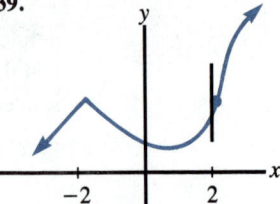

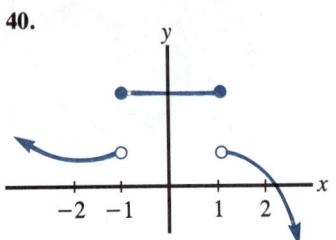

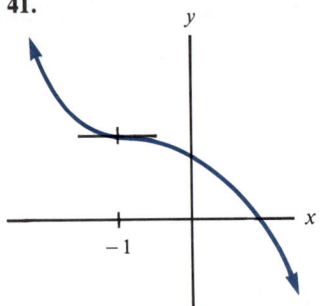

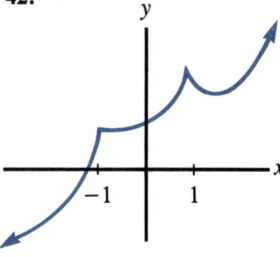

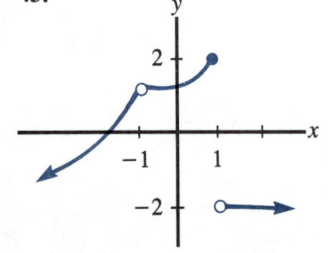

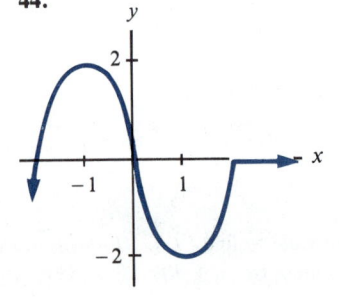

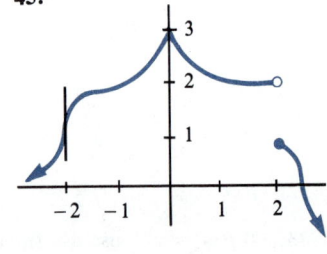

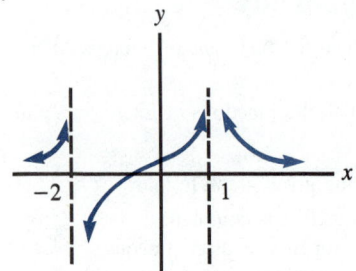

In Exercises 47 through 58, find the intervals on which the function f is continuous and the values of x at which it is discontinuous.

47. $f(x) = x^2 - 3x + 2$

48. $f(x) = x^{15} - 19x^{13} + 27x^3 + 1$

49. $f(x) = \dfrac{x}{x - 2}$

50. $f(x) = \dfrac{2x - 4}{x^2 + 1}$

51. $f(x) = \dfrac{x^2 - 6x + 8}{x^2 - 4}$

52. $f(x) = \dfrac{x^3 - 1}{x^2 + 2x - 3}$

53. $f(x) = \begin{cases} 2, & 0 \leq x \leq 1, \\ 4, & 1 < x \leq 2, \\ 6, & 2 < x \leq 3 \end{cases}$

54. $f(x) = \begin{cases} 1, & 0 \leq x \leq 1, \\ 2 - x, & 1 < x < 2, \\ 0, & x > 2 \end{cases}$

55. $f(x) = \begin{cases} 2x + 1, & 0 \leq x < 2, \\ 5, & 2 \leq x \leq 3, \\ 5 - x, & x > 3 \end{cases}$

56. $f(x) = \begin{cases} x^2, & -1 \leq x < 0, \\ x^3, & 0 \leq x \leq 1, \\ 4 - 2x, & 1 < x \leq 2 \end{cases}$

57. $f(x) = \begin{cases} 2/x, & x < 2, \\ x/2, & x \geq 2 \end{cases}$

*58. $f(x) = [x]$, where $[x]$ denotes the greatest integer less than or equal to x.

59. The cost of renting a VCR is $12 per week or any part of a week. Find and graph a function that describes the cost of renting a VCR for x weeks, $0 \leq x \leq 4$. Where is this function continuous? Where is it discontinuous? Where is it not differentiable?

60. The cost of an international telephone call is $3 per minute for the first minute and another $2 per minute for each additional minute or part of a minute. Find and graph a function that describes the cost of making a call that lasts for t minutes, $0 \leq t \leq 5$. Where is this function continuous? Where is it discontinuous? Where is it not differentiable?

Management

61. (See Exercise 71, Exercises 10.1.) If a biochemical company makes x grams of interferon, its variable cost per gram is $v(x)$ dollars, where

$$v(x) = \begin{cases} 50, & 0 < x \leq 1000, \\ 40, & 1000 < x \leq 5000, \\ 35, & x > 5000. \end{cases}$$

Where is this function continuous? Where is it discontinuous? Where is it not differentiable?

62. (See Exercise 72, Exercises 10.1.) A software company sold its word-processing package for $s(t)$ dollars t years after its introduction, where

$$s(t) = \begin{cases} 120, & 0 < t < 2, \\ 99, & 2 \leq t < 4, \\ 79, & t \geq 4. \end{cases}$$

Where is this function continuous? Where is it discontinuous? Where is it not differentiable?

63. (See Exercise 74, Exercises 10.1.) If the price of a gallon of gasoline is p dollars, gasoline consumption will be y billion gallons per year, where

$$y = \begin{cases} 500 - 10p, & 0 \leq p \leq 1.50, \\ 560 - 50p, & p > 1.50. \end{cases}$$

Where is this function continuous? Where is it discontinuous? Where is it not differentiable?

64. (See Exercise 75, Exercises 10.1.) Repeat Exercise 63 if
$$y = \begin{cases} 500 - 10p, & 0 \leq p \leq 1.50, \\ 400 - 40p, & p > 1.50. \end{cases}$$

65. In order for a firm to produce more than 10,000 units of its product, it must put on a second shift, and this increases its variable cost per unit. Thus its cost function is given by
$$C(x) = \begin{cases} 100,000 + 8x, & 0 \leq x \leq 10,000, \\ 80,000 + 10x, & x > 10,000. \end{cases}$$
Graph this function. Where is it continuous? Where is it discontinuous? Where is it not differentiable?

66. Repeat Exercise 65 if
$$C(x) = \begin{cases} 100,000 + 8x, & 0 \leq x \leq 10,000, \\ 80,000 + 12x, & x > 10,000. \end{cases}$$

67. A firm that sells computer supplies charges $9.00 per ribbon for printer ribbons on orders of fewer than 10 ribbons, $8.60 per ribbon on orders of 10 through 24 ribbons, and $8.20 per ribbon on orders of 25 or more ribbons. Find and graph a function that expresses the cost of an order of x ribbons. Where is the function continuous? Where is it discontinuous? Where is it not differentiable?

68. A candy company has three machines that can make chocolate bonbons. Machine A, which is new, can make 2000 bonbons per hour at a cost of 1.8 cents each. Machine B, which is older, can make 1500 bonbons per hour at a cost of 2.2 cents each. Machine C, which is ancient, can make 800 bonbons per hour at a cost of 3.5 cents each. Find and graph a function that expresses the cost of making x bonbons per hour. (*Hint:* If $x \leq 2000$, the company will use only machine A; if $2000 < x \leq 3500$, they will use A and B, etc.) Where is this function continuous? Where is it discontinuous? Where is it not differentiable?

Life Science

69. The number of bacteria present t hours after the start of an experiment is y thousand, where
$$y = \begin{cases} 2t^2 + 100, & 0 \leq t < 2, \\ 160 - 26t, & 2 \leq t \leq 6. \end{cases}$$
Graph this function. Where is it continuous? Where is it discontinuous? Where is it not differentiable?

10.4 RULES OF DIFFERENTIATION

The process of finding the derivative of a function is called **differentiation**. As of now, we can only differentiate functions by means of the Three-Step Rule, which is often both time-consuming and tedious to apply. Fortunately, it is possible to use the Three-Step Rule to prove some very general rules of differentiation. Once we

THE DERIVATIVE

have these rules of differentiation available and understand how to use them, we will be able to differentiate entire classes of functions (such as the polynomial functions) almost automatically. In this section we prove some of the most important rules of differentiation and illustrate their usage.

Let c be any number and consider the **constant function** f defined by

$$f(x) = c \quad \text{for all numbers } x.$$

By the Three-Step Rule,

$$f'(x) = \lim_{h \to 0} \frac{f(x + h) - f(x)}{h} = \lim_{h \to 0} \frac{c - c}{h} = \lim_{h \to 0} 0 = 0.$$

Thus we have proved the **constant rule**.

The Constant Rule

If c is a number and $f(x) = c$ for all x, then

$$f'(x) = 0.$$

That is, the derivative of a constant function is zero.

EXAMPLE 1 Let $f(x) = 2$ and $g(x) = \sqrt{2}$ for all x. Then $f'(x) = 0$ and $g'(x) = 0$ for all x; thus $f'(5) = 0$, $f'(\pi) = 0$, $g'(0) = 0$, and so on. ☐

Now let n be a positive integer and consider the **power function** defined by

$$f(x) = x^n \quad \text{for all numbers } x.$$

We have

$$f'(x) = \lim_{h \to 0} \frac{f(x + h) - f(x)}{h} = \lim_{h \to 0} \frac{(x + h)^n - x^n}{h}.$$

But by the Binomial Theorem,

$$(x + h)^n = x^n + nx^{n-1}h + \frac{n(n - 1)}{2} x^{n-2}h^2 + \cdots + h^n,$$

where all the terms after the first two on the right side of the equality have h^2 as a factor. Therefore if we factor h out of all terms after the first one, we will obtain

$$(x + h)^n = x^n + h\left[nx^{n-1} + \frac{n(n - 1)}{2} x^{n-2}h + \cdots + h^{n-1}\right],$$

where now all the terms inside the brackets after the first one still have h as a factor, and hence will approach 0 as $h \to 0$. Thus

$$f'(x) = \lim_{h \to 0} \frac{x^n + h[nx^{n-1} + (n(n-1)/2)x^{n-2}h + \cdots + h^{n-1}] - x^n}{h}$$

$$= \lim_{h \to 0} \left[nx^{n-1} + \frac{n(n-1)}{2} x^{n-2}h + \cdots + h^{n-1} \right]$$

$$= nx^{n-1} + 0 + \cdots + 0$$

$$= nx^{n-1}.$$

We have shown that if $f(x) = x^n$, then $f'(x) = nx^{n-1}$. This fact, which is known as the **power rule,** remains true if we replace n by any nonzero number (though we shall not prove this).

The Power Rule

If r is any number except 0 and $f(x) = x^r$, then

$$f'(x) = rx^{r-1}.$$

That is, the derivative of x raised to a nonzero power is the power times x raised to the power minus 1.

EXAMPLE 2 If $f(x) = x$, then $f(x) = x^1$, and $f'(x) = 1x^{1-1} = 1x^0 = 1 \cdot 1 = 1$. Thus if $f(x) = x$, then $f'(x) = 1$. ☐

EXAMPLE 3 If $g(x) = x^2$, then $g'(x) = 2x^{2-1} = 2x^1 = 2x$. ☐

EXAMPLE 4 If $h(x) = x^5$, then $h'(x) = 5x^{5-1} = 5x^4$. ☐

EXAMPLE 5 If $k(x) = \sqrt{x}$, then $k(x) = x^{1/2}$, so

$$k'(x) = \frac{1}{2} x^{(1/2)-1} = \frac{1}{2x^{1/2}} = \frac{1}{2\sqrt{x}}. \quad \square$$

EXAMPLE 6 If $s(x) = 1/x = x^{-1}$, then $s'(x) = (-1)x^{-1-1} = -x^{-2} = -1/x^2$. ☐

The power rule shows how to differentiate $f(x) = x^r$ when r is any nonzero number. We would like to be able to differentiate functions whose defining equations are of the form $f(x) = cx^r$, where c is a number. But if g is any function for which $g'(x)$ exists, and $f(x) = c \cdot g(x)$, then

$$f'(x) = \lim_{h \to 0} \frac{f(x+h) - f(x)}{h} = \lim_{h \to 0} \frac{c \cdot g(x+h) - c \cdot g(x)}{h}$$

$$= c \cdot \lim_{h \to 0} \frac{g(x+h) - g(x)}{h} = c \cdot g'(x).$$

and we have proved the **constant-multiple rule.**

The Derivative

> **The Constant-Multiple Rule**
>
> If c is a number and g is a function such that $f(x) = c \cdot g(x)$ and $g'(x)$ exists, then
> $$f'(x) = c \cdot g'(x).$$
> That is, the derivative of a constant times a function is the constant times the derivative of the function.

EXAMPLE 7   If $f(x) = 2x$, then $f'(x) = (2x)' = 2(x)' = 2 \cdot 1 = 2.$ ☐

EXAMPLE 8   If $h(x) = 3x^2$, then $h'(x) = (3x^2)' = 3(x^2)' = 3(2x) = 6x.$ ☐

EXAMPLE 9   If $k(x) = \frac{1}{3}x^{-3}$, then $k'(x) = -x^{-4}.$ ☐

Now we can differentiate functions whose defining equations are of the form $f(x) = cx^r$. Polynomials are formed by taking sums and differences of such functions, and we would like to be able to differentiate polynomials. But if $k(x) = f(x) + g(x)$, where $f'(x)$ and $g'(x)$ exist, then

$$k'(x) = \lim_{h \to 0} \frac{k(x+h) - k(x)}{h}$$

$$= \lim_{h \to 0} \frac{[f(x+h) + g(x+h)] - [f(x) + g(x)]}{h}$$

$$= \lim_{h \to 0} \frac{f(x+h) - f(x)}{h} + \lim_{h \to 0} \frac{g(x+h) - g(x)}{h}$$

$$= f'(x) + g'(x).$$

Similarly, if $k(x) = f(x) - g(x)$, then $k'(x) = f'(x) - g'(x)$. Thus we have the **sum rule.**

> **The Sum Rule**
>
> If $f'(x)$ and $g'(x)$ both exist and $k(x) = f(x) \pm g(x)$, then
> $$k'(x) = f'(x) \pm g'(x).$$
> That is, the derivative of a sum (difference) is the sum (difference) of the derivatives.

EXAMPLE 10   If $k(x) = -2x^2 + 3$, then
$$k'(x) = (-2x^2 + 3)' = -2(x^2)' + (3)' = -2(2x) + 0 = -4x.$$ ☐

CHAPTER 10

EXAMPLE 11   If $s(x) = x^4 - 5x^3$, then
$$s'(x) = 4x^3 - 5(3x^2) = 4x^3 - 15x^2. \quad \square$$

We note that the sum rule also holds if there are more than two summands. Thus if
$$k(x) = f_1(x) \pm f_2(x) \pm \cdots \pm f_n(x),$$
then
$$k'(x) = f_1'(x) \pm f_2'(x) \pm \cdots \pm f_n'(x).$$

The four rules of differentiation we have proved in this section allow us to differentiate polynomials and also functions that are formed by taking sums and differences of nonintegral powers of the variable.

EXAMPLE 12   If $f(x) = 2x^3 + 5x^2 - 12x + 6$, then
$$f'(x) = 2(3x^2) + 5(2x) - 12(1) + 0 = 6x^2 + 10x - 12. \quad \square$$

EXAMPLE 13   If $g(x) = 2\sqrt{x} - 4x^{6/5} + 6x - 10/x^2 = 2x^{1/2} - 4x^{6/5} + 6x - 10x^{-2}$, then
$$g'(x) = 2\left(\frac{1}{2}x^{-1/2}\right) - 4\left(\frac{6}{5}x^{1/5}\right) + 6(1) - 10(-2x^{-3})$$
$$= x^{-1/2} - \frac{24}{5}x^{1/5} + 6 + 20x^{-3}$$
$$= \frac{1}{\sqrt{x}} - \frac{24}{5}\sqrt[5]{x} + \frac{20}{x^3} + 6. \quad \square$$

EXAMPLE 14   Let $f(x) = x^5 - 2x^3 + 7x + 2$. Find the equation of the line tangent to f at the point $(1, f(1)) = (1, 8)$.

We find the derivative
$$f'(x) = 5x^4 - 6x^2 + 7$$
and evaluate it at $x = 1$ to obtain
$$f'(1) = 5(1^4) - 6(1^2) + 7 = 6.$$
Since $f'(1) = 6$, the tangent line has slope equal to 6. Hence its equation is
$$y = 6x + b.$$
But since the point $(1, 8)$ is on the line, we have
$$8 = 6(1) + b = 6 + b.$$
Thus $b = 2$. Therefore the equation of the tangent line is
$$y = 6x + 2. \quad \square$$

EXAMPLE 15   A patient's white blood cell count t hours after being admitted to a hospital is
$$f(t) = 5t^2 - 80t + 500$$

THE DERIVATIVE

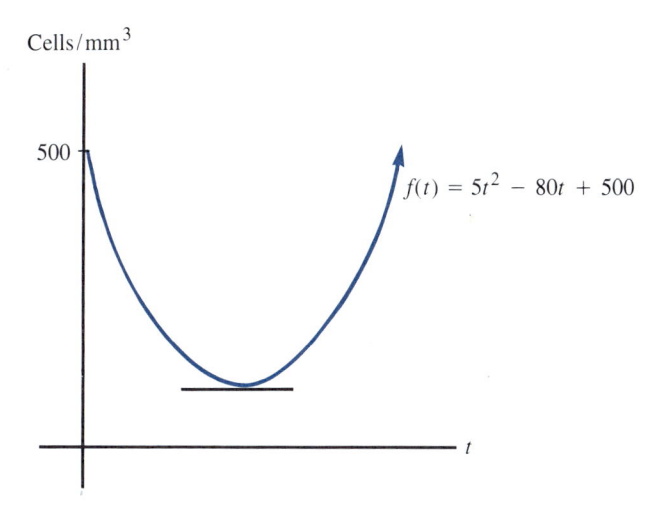

FIGURE 10-38

cells per cubic millimeter. The graph of this quadratic function is the parabola of Figure 10–38. Note that the line tangent to the graph at the vertex of the parabola is horizontal and therefore has slope zero. But the slope of the line tangent to f at $(t, f(t))$ is $f'(t)$. Thus if

$$f'(t) = 10t - 80 = 0,$$

then $t = 8$. Since

$$f(8) = 5(8)^2 - 80(8) + 500 = 180,$$

the vertex of the parabola is $(8, f(8)) = (8, 180)$. Therefore the patient's minimum white blood cell count occurs 8 hours after admission and is equal to 180 cells per cubic millimeter. ◻

■ *Computer Manual: Program DIFFPOLY*

10.4 EXERCISES

In Exercises 1 through 50, use the rules of differentiation to find $f'(x)$.

1. $f(x) = 2$
2. $f(x) = \sqrt{3}$
3. $f(x) = -4$
4. $f(x) = x^3$
5. $f(x) = x^7$
6. $f(x) = x^{100}$
7. $f(x) = x^{-2}$
8. $f(x) = 1/x^3$
9. $f(x) = 1/x^5$
10. $f(x) = x^{3/2}$
11. $f(x) = x^{7/5}$
12. $f(x) = x^{-2/3}$
13. $f(x) = x^{-2.7}$
14. $f(x) = \sqrt[3]{x}$
15. $f(x) = \sqrt[5]{x}$
16. $f(x) = 1/\sqrt{x}$
17. $f(x) = 3x$
18. $f(x) = 5x^2$
19. $f(x) = -2x^3$
20. $f(x) = 3x^9$
21. $f(x) = 2x^{3/2}$

22. $f(x) = -3x^{-2/3}$
23. $f(x) = -4\sqrt{x}$
24. $f(x) = 5/x$
25. $f(x) = -3/x^2$
26. $f(x) = 4/x^5$
27. $f(x) = 3/\sqrt[3]{x}$
28. $f(x) = 6x + 2$
29. $f(x) = 3 - 8x$
30. $f(x) = 15x - 9$
31. $f(x) = x^2 - 5x$
32. $f(x) = 2x^2 - 8x + 12$
33. $f(x) = 8x^2 - 12x + 3$
34. $f(x) = -1.5x^2 + 2.2x + 11.3$
35. $f(x) = x^3 - 7x^2 + 9x$
36. $f(x) = -x^3 + 8x + 2$
37. $f(x) = x^4 - 3x^3 - 3x^2 + 5x + 1$
38. $f(x) = -\frac{3}{4}x^4 + \frac{7}{3}x^3 - \frac{7}{2}x^2 + 1$
39. $f(x) = -2x^{100} + 3x^{99} + 10$
40. $f(x) = 4\sqrt{x} + 3x^{2/3}$
41. $f(x) = 3\sqrt{x} + 6x^{-3/2} + 1$
42. $f(x) = \frac{1}{x} - \frac{2}{x^2} + \frac{3}{x^3} - \frac{4}{x^4}$
43. $f(x) = \frac{4}{x} + \frac{1}{x^{2/5}}$
44. $f(x) = 2x^{-2} + 5x^{0.6} + 10x^{-0.4}$
45. $f(x) = 6\sqrt[3]{x} + 5\sqrt{x} - \frac{2}{\sqrt{x}}$
46. $f(x) = (x + 1)(x - 4)$
47. $f(x) = (2x - 5)(3x + 6)$
48. $f(x) = (x^2 - 2x)(3x + 9)$
49. $f(x) = \frac{x + 2x^2}{x^2}$
50. $f(x) = \frac{2x^{3/4} - 4}{x^{5/4}}$

In Exercises 51 through 60, find the equation of the line tangent to f at $(a, f(a))$.

51. $f(x) = 5x - 3$, $a = 2$
52. $f(x) = 2x^2 - 3$, $a = 1$
53. $f(x) = 1 - x - x^2$, $a = 3$
54. $f(x) = x^2 - 3x + 1$, $a = 0$
55. $f(x) = x^3 - 1$, $a = -2$
56. $f(x) = -x^3 + 2x^2 - x + 1$, $a = -1$
57. $f(x) = 4\sqrt{x} - 5x$, $a = 9$
58. $f(x) = 3x^{4/3} - 2x^{2/3}$, $a = 27$
59. $f(x) = \frac{2}{x} - \frac{4}{x^2}$, $a = 2$
60. $f(x) = 6x^{-2} + x^{-3/4}$, $a = 16$

Management

61. A firm's profit function is given by $P(x) = -x^2 + 40x - 350$, where x is in units and profit is in thousands of dollars.
    (a) Graph this function.
    (b) The firm's maximum profit occurs at the point on the graph where the tangent line is horizontal. Use this fact to find the firm's maximum profit and the value of x at which it occurs.

62. Repeat Exercise 61 if $P(x) = -x^2 + 800x - 42{,}000$, where x is in units and profit is in dollars.

63. Over the past five years the average productivity of a company's plant has been $y = 0.4t^2 - 30t + 1400$ units per day in month t, where $t = 0$ represents the beginning of the five-year period.
    (a) Graph this function.
    (b) The plant's minimum average productivity occurred at the point on the graph where the tangent line is horizontal. Find the plant's minimum average productivity over the five-year period and the time when it occurred.

64. Repeat Exercise 63 if $y = -0.02t^2 + 1.6t + 12$. (In this case you will be finding the plant's *maximum* average productivity.)

65. A firm's average cost per unit is given by $A(x) = 10x + 1000/x$. As x increases, the average cost per unit decreases, then increases again; the firm's minimum average cost per unit occurs at the value of x for which the slope of the tangent line is 0. Find this value of x and the firm's minimum average cost per unit.

66. A firm's monthly sales over the past year are given by
$$S(t) = 2t^3 - 39t^2 + 240t + 100,$$
where t is in months, with $t = 0$ representing the beginning of the year, and sales are in thousands of dollars. During the year monthly sales increased for a time, then decreased, and finally increased again. The points on the graph where sales changed direction have horizontal tangent lines. Find the time when sales stopped increasing and began to decrease and the time when they stopped decreasing and began to increase.

67. Over the past decade, the productivity of a firm's employees decreased, then increased, and finally decreased again. Suppose employee productivity in year t was y units produced per week, where $y = -t^3 + 18t^2 - 96t + 300$. Here $t = 0$ represents the beginning of the decade. The points on the graph where productivity changed direction have horizontal tangent lines. Find the time when productivity stopped decreasing and began to increase and the time when it stopped increasing and began to decrease.

Life Science

68. The number of black bear cubs born in New England in year t was estimated to be y, where $y = -0.5t^2 + 5t + 100$ and $t = 0$ represents 1980.
    (a) Graph this function.
    (b) Find the maximum number of cubs born in New England and the year in which this occurred by finding the value of t for which the tangent line is horizontal.

69. Repeat Exercise 68 if $y = t^2 - 8t + 100$. (In this case you will be finding the *minimum* number of cubs born.)

70. Biologists estimate that over the next several years the number of elephants in a certain region of East Africa will decrease somewhat, then increase temporarily, and finally begin a pronounced decline. Suppose the number of elephants alive t years from now will be y, where $y = -2t^3 + 45t^2 - 300t + 200$. The points on the graph where the temporary increase and then the final decline occur have horizontal tangent lines. Find the time when the temporary increase and the final decline will occur.

71. A new antibiotic is tested by applying it to a culture containing bacteria, with the result that t days later there are y thousand bacteria alive in the culture, where $y = t^5/5 - 29t^3/3 + 100t + 140$. The graph of this function shows that after the application of the antibiotic the number of bacteria grows for a time, then declines, and then begins to grow again. The points on the graph where growth turns into decline and decline into growth have horizontal tangent lines. Find these points.

Social Science

72. A politician's approval rating is given by $y = 0.0002t^2 - 0.01t + 0.6$, where t is the number of months since the last election and y is the proportion of voters who approve of the politician's performance in office.

(a) Graph this function.
(b) Find the politician's minimum approval rating and the time that it occurred by finding the value of t for which the slope of the tangent line is 0.

73. Over the past four years the unemployment rate in a state has been given by $u(t) = 0.0002t^3 - 0.0147t^2 + 0.216t + 5.9$. Here $u(t)$ is the unemployment rate, as a percentage of the work force, in month t, with $t = 0$ being the beginning of the four-year period. The graph of this function shows that the unemployment rate rose, then fell, then rose again, and the points on the graph where the rate changed from rising to falling and from falling to rising occurred where the tangent lines are horizontal. Find these points.

Bond Duration

If a bond is purchased for $p and its future stream of earnings discounted at $y\%$ per year compounded semiannually equals p, then y is called its **yield to maturity**; thus, the lower the price p, the higher the yield to maturity y. Changes in prevailing interest rates affect bond prices: if interest rates rise, a bond's yield to maturity must also rise, and the only way this can happen is if the price of the bond falls. According to an article by Martin Leibowitz ("Analytical Measures in Today's Bond Market," an advertisement in *Scientific American*, June 1987), if $p = f(y)$ describes bond price as a function of yield to maturity, then the **duration** D of the bond is defined by

$$D = \frac{f'(y)}{f(y)} 100.$$

Duration measures the short-term risk to the bondholder of a change in interest rates: D is the approximate percentage change in price that will result from a 1% rise in interest rates. (We will see why this is so in Section 11.3.) Exercises 74 and 75 refer to bond duration.

*74. Suppose a bond's price as a function of its yield to maturity is given by $p = 8000/y$, $0 < y \le 100$. Find and interpret the bond's duration if its current yield to maturity is 8%; 10%; 16%.

*75. Repeat Exercise 74 if $p = y^2 - 120y + 2000$.

10.5 RATES OF CHANGE

We have several times referred to the fact that the derivative of a function describes the rate of change of the function with respect to its independent variable. In this section we see why this is true, and we demonstrate how to use this property of the derivative in some practical applications.

Suppose we make a journey in a small airplane. We fly in a straight line, and after t hours in the air we have traveled a distance of

$$s(t) = 4t^2 + 150t$$

miles from our starting point, where $0 \le t \le 4$. What is the average velocity of our plane during the second hour of the trip?

THE DERIVATIVE

Average velocity is distance traveled divided by elapsed time. During the second hour of the trip the time changes from $t = 1$ to $t = 2$, so the distance traveled during the second hour is $s(2) - s(1)$ miles, and therefore the average velocity during the second hour is

$$\frac{s(2) - s(1)}{2 - 1} = \frac{316 - 154}{2 - 1} = 162$$

miles per hour. The same reasoning holds for any time interval: the average velocity of the plane from time t to time $t + h$ is

$$\frac{s(t + h) - s(t)}{t + h - t} = \frac{s(t + h) - s(t)}{h}$$

miles per hour. Notice that the right side of the equation is the difference quotient of the **distance function** $s = s(t)$. If we take the limit of this difference quotient as h approaches 0, we obtain the derivative $s'(t)$. On the other hand, as h approaches 0 the difference quotient gives the average velocity over shorter and shorter intervals of time. Hence, if we take the limit as h approaches 0, we obtain the instantaneous velocity at time t. Thus the derivative

$$s'(t) = 8t + 150$$

of $s(t)$ gives the instantaneous velocity of the airplane at time t. For example, when $t = 2$ we have

$$s'(2) = 8(2) + 150 = 16 + 150 = 166.$$

Therefore at the instant when $t = 2$ the velocity of the airplane is 166 miles per hour.

A distance function such as the one we have just considered is simply a means of specifying the position of an object at time t relative to its starting point. The analysis we performed for $s(t) = 4t^2 + 150t$ can be repeated for any function $s = s(t)$ that gives the position of an object relative to some reference point, as long as the motion is in a straight line.

> ## Velocity
>
> If an object moves in a straight line so that its position at time t relative to some reference point is given by the function $s = s(t)$, then its instantaneous velocity at time t is
>
> $$v(t) = s'(t).$$

If we take the reference point to be the zero point on the line of motion, then one direction along the line is positive and the other is negative. Thus if the velocity is positive at time $t = a$, the object is moving along the line in the positive direction; if the velocity is negative, it is moving in the negative direction.

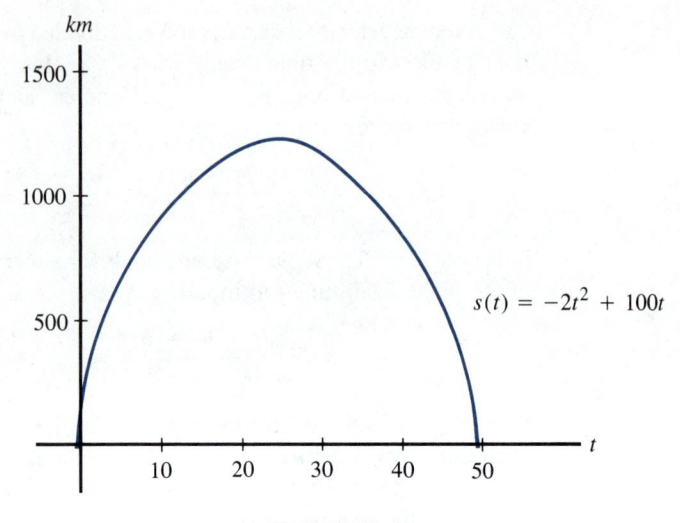

FIGURE 10–39

EXAMPLE 1   A rocket is launched straight up into the atmosphere. The altitude of the rocket t minutes after it is launched is

$$s(t) = -2t^2 + 100t$$

kilometers. We have chosen the rocket's zero point to be its launch pad and its upward motion to be positive motion. Figure 10–39 shows the graph of the position function $s = s(t)$. Note that this is the graph of the rocket's distance from its launch pad (that is, its altitude) at time t, and not a picture of its motion; the rocket moves in a straight line.

The instantaneous velocity of the rocket at time t is

$$v(t) = s'(t) = -4t + 100$$

kilometers per minute. For instance, exactly 10 minutes after launching, the rocket is at an altitude of

$$s(10) = 800$$

kilometers and its velocity at the instant when $t = 10$ is

$$v(10) = 60$$

kilometers per minute. Therefore at the instant when $t = 10$, the rocket's altitude is increasing at a rate of 60 kilometers per minute; we can tell that the altitude is increasing because $v(10)$ is positive. See Figure 10–40.

Similarly, 45 minutes after launching, the rocket is at an altitude of

$$s(45) = 450$$

kilometers and has a velocity of

$$v(45) = -80$$

THE DERIVATIVE

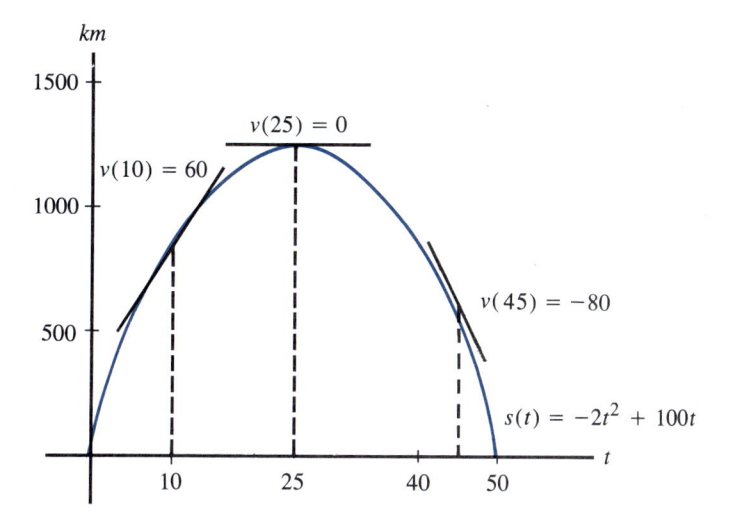

FIGURE 10-40

kilometers per minute. Hence when $t = 45$ the rocket's altitude is decreasing at the rate of 80 kilometers per minute; the fact that the velocity is negative tells us that the altitude is decreasing. See Figure 10-40.

Notice that $v(25) = 0$. Thus at the instant when $t = 25$, the rocket's altitude is neither increasing nor decreasing. Therefore the rocket reaches its peak altitude 25 minutes after launching. Again, see Figure 10-40. □

We have seen that the derivative of a position function is a velocity function. But velocity is just the name we give to the rate of change of position with respect to time. Thus the derivative of a position function is the rate of change of the position function with respect to time. We can apply the same reasoning to any function to conclude that its derivative represents the rate of change of the function with respect to the independent variable.

The Derivative as Rate of Change

For any function f, the derivative f' gives the instantaneous rate of change of f with respect to its independent variable.

EXAMPLE 2 A stone dropped from a height falls .

$$s(t) = 16t^2$$

feet in t seconds. The velocity of the stone at time t is thus

$$v(t) = s'(t) = 32t$$

feet per second. For instance, the velocity of the stone exactly 2 seconds after it is dropped is

$$v(2) = 64$$

feet per second. The derivative of $v(t)$ will give the stone's rate of change of velocity with respect to time; this is known as its **acceleration.** If we let $a(t)$ denote the stone's acceleration at time t, then

$$a(t) = v'(t) = 32$$

feet per second per second. Note that the acceleration of the stone is constant. ☐

In the preceding example we said that the acceleration of the stone was the derivative of its velocity. This is true in general.

Acceleration

If $v(t)$ is the velocity of an object at time t and $a(t)$ its acceleration at time t, then

$$a(t) = v'(t).$$

The fact that the derivative of a function can be viewed as its rate of change with respect to its independent variable has many applications in addition to finding velocity and acceleration.

EXAMPLE 3 Alpha Company has studied the productivity of its employees and has found that after t years of experience an employee's monthly productivity can be expected to be

$$f(t) = -2t^2 + 120t + 100$$

units. The derivative

$$f'(t) = -4t + 120$$

gives the rate of change of monthly productivity with respect to years of experience. For instance, when $t = 10$ we have $f'(10) = 80$. Thus employees with 10 years of experience are increasing their monthly productivity at a rate of 80 units per year. Similarly, since $f'(20) = 40$, employees with exactly 20 years of experience are also increasing their monthly productivity, but only at the rate of 40 units per year.

Note that $f'(30) = 0$, and hence employees with exactly 30 years of experience will be neither increasing nor decreasing their productivity. Since

$$f'(t) = -4t + 120$$

is negative for $t > 30$, employee productivity decreases when $t > 30$. It follows that greatest productivity is reached when $t = 30$, and hence that maximum monthly productivity is $f(30) = 1900$ units. See Figure 10–41 for a graph of the productivity function. ☐

Units

1900

$f(t) = -2t^2 + 120t + 100$

100

10 20 30 40 50 60   t

FIGURE 10–41

EXAMPLE 4 If Beta Corporation makes x units of product, its average cost per unit is

$$A(x) = 0.10x + \frac{40}{x}$$

dollars. Thus, if Beta produces 10 units, its average cost per unit is

$$A(10) = 0.10(10) + \frac{40}{10} = 5$$

dollars. The derivative

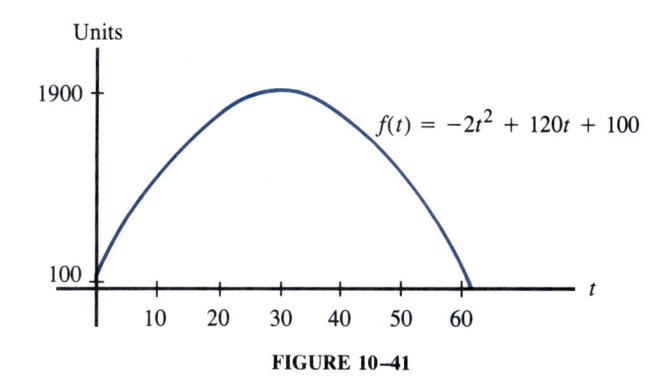

$$A'(x) = 0.10 - \frac{40}{x^2}$$

gives the rate of change of Beta's average cost per unit. For instance, since

$$A'(10) = 0.10 - \frac{40}{10^2} = -0.30,$$

when Beta produces 10 units its average cost per unit is decreasing by $0.30 per unit. ☐

■ *Computer Manual: Program DIFFPOLY*

10.5 EXERCISES

1. In t hours an automobile travels $s(t) = 40t$ miles in a straight line from its starting point. Find its velocity and acceleration t hours after it started.

2. In t minutes a rocket travels $s(t) = 250t$ kilometers in a straight line. Find its velocity and acceleration t minutes after it started.

3. A particle moves in a straight line in such a manner that t seconds after its start it has traveled $s(t) = 0.2t^{5/2}$ meters.
   (a) Find the distance it has traveled 1, 4, and 9 seconds after its start.
   (b) Find its velocity 1, 4, and 9 seconds after its start.
   (c) Find its acceleration 1, 4, and 9 seconds after its start.

4. A vehicle travels $s(t) = 40t^2 - 3t^3$ kilometers in a straight line in t hours, where $0 \le t \le 4$. Find the vehicle's
   - (a) velocity 1, 2, and 3.5 hours after it started;
   - (b) acceleration 1, 2, and 3.5 hours after it started.

5. A drag racer's car has traveled $s(t) = 12t^2 + 110t$ feet from the starting line t seconds after the start of a race.
   - (a) What is the car's velocity 1 second after the start of the race? 4 seconds after the start?
   - (b) What is the car's acceleration 1 second after the start of the race? 4 seconds after the start?

You throw a ball straight up. The altitude of the ball t seconds after you release it is $s(t) = -2t^2 + 20t$ feet. Exercises 6 through 10 refer to this situation.

6. What is the altitude of the ball 2 seconds after you release it? What is its velocity at this time? Is it going up or coming down? What is its acceleration?

7. Repeat Exercise 6 for $t = 5$.

8. Repeat Exercise 6 for $t = 7$.

9. Repeat Exercise 6 for $t = 10$.

10. What is the maximum altitude of the ball? When will it attain its maximum altitude?

A mortar is fired, and t seconds later the shell is at an altitude of
$$s(t) = -t^2 + 8t + 220$$
feet. Exercises 11 through 15 refer to this situation.

11. What is the altitude of the mortar shell 2.5 seconds after firing? What is its velocity at this time? Is it going up or coming down? What is its acceleration?

12. Answer the questions of Exercise 11 for $t = 4$.

13. Answer the questions of Exercise 11 for $t = 5.5$.

14. Answer the questions of Exercise 11 for $t = 8$.

15. What is the maximum altitude of the shell? When does it attain its maximum altitude?

A stone dropped from rest falls in a straight line. Suppose that its velocity t seconds after it was dropped is $v(t) = 32t$ feet per second. Exercises 16 through 20 refer to this situation.

16. Find the stone's acceleration t seconds after it was dropped.

*17. Find a function that describes the distance the stone has fallen as a function of the time since it was dropped. (*Hint:* Find a function $s(t)$ whose derivative is $v(t)$, using the fact that $s(0) = 0$.)

18. Find the distance the stone fell, its velocity, and its acceleration 1 second after it was dropped.

19. Repeat Exercise 18 for $t = 2$.

20. Repeat Exercise 18 for $t = 4$.

On the moon the velocity of a stone dropped from a height is $v(t) = 1.6t$ meters per second. Exercises 21 through 25 refer to this situation.

The Derivative

21. Find the stone's acceleration at time t.

*22. Find a function that describes the distance the stone falls in the first t seconds after it is dropped.

23. Find the distance the stone falls, its velocity, and its acceleration 1.1 seconds after it is dropped.

24. Repeat Exercise 23 for $t = 2.2$.

25. Repeat Exercise 23 for $t = 4.4$.

26. Stefan's Law states that the amount of energy radiated by a body per unit of surface area is directly proportional to the fourth power of its temperature in degrees Kelvin: $y = kT^4$, where T is the temperature of the body in °K, y is the amount of energy radiated by the body per unit area of its surface, and k is a constant that depends on the body. The surface temperature of the sun is approximately 6000° K. Find the rate of change of the energy radiated by the sun per unit of surface area.

Management

27. A firm's cost function is given by $C(x) = 0.025x^2 + 25{,}000$. Find the rate at which its cost is changing when production is 10 units; 100 units; 1000 units.

28. Delta Company's cost function is given by $C(x) = 5x^2 + 2000$, where x is in thousands of units produced. Delta's average cost per thousand units is therefore given by $A(x) = C(x)/x$. At what rate is Delta's average cost per thousand units changing when production is 5000 units? 10,000 units? 20,000 units? 30,000 units? At each of these production levels, is average cost per thousand units increasing or decreasing?

29. Total monthly sales of Gamma Corporation t months after the founding of the company are

$$S(t) = t^3 - 7t^2 + 10t + 100$$

thousand dollars.

(a) What were total monthly sales one month after the founding of the corporation? At what rate were sales changing then? Were sales increasing or decreasing?

(b) Answer the questions of part (a) for $t =$ three months.

30. A firm's profit function is given by

$$P(x) = -0.02x^3 + 4.5x^2 - 300x - 5000.$$

At what rate is profit changing when 40 units are being produced? when 80 units are being produced? when 150 units are being produced? At each of these production levels, is profit per unit increasing or decreasing?

31. If the supply function for a product is given by

$$q = p^2 + 2p,$$

where q is in thousands of units, find the rate at which the quantity of the product supplied to the market is changing when the price is $5 per unit; $100 per unit.

32. If the demand function for a product is given by $q = -p^2 - 30p + 4000$, where q is in thousands of units, find the rate at which the quantity of the product demanded by consumers is changing when the price is $10 per unit; when the price is $50 per unit.

33. The supply function for a commodity is $q = 12p - 2p^{1/2}$, where q is in millions of units and p is in dollars. Find the rate of change of quantity supplied with respect to price when the price per unit is
   (a) $1   (b) $4   (c) $9   (d) $16   (e) $25

34. The demand function for a product is given by $q = 10 - 0.5\sqrt{p}$, where q is in thousands of units.
   (a) Find the rate of change of quantity demanded with respect to price when the price is $100 per unit; $256 per unit.
   (b) Find the rate of change of price with respect to quantity demanded when quantity demanded is 5000 units; when quantity demanded is 2000 units. (*Hint:* Solve the demand equation for p in terms of q.)

35. A firm's revenue function is $R(x) = x^3/3 - 15x^2 + 216x$. Find the rate of change of revenue with respect to sales when the number of units sold is
   (a) 0   (b) 6   (c) 12   (d) 14   (e) 16   (f) 18   (g) 21   (h) 24

36. A firm's monthly sales over the past year are given by
$$S(t) = 2t^3 - 39t^2 + 240t + 100,$$
   where t is in months, with $t = 0$ representing the beginning of the year, and sales are in thousands of dollars. Find the rate at which sales are changing when $t = 2$, 5, 6, 8, and 10. In each case state whether sales are increasing or decreasing at the given time. (See Exercise 66 of Exercises 10.4.)

37. Employee productivity in a firm in year t was y units produced per week, where
$$y = -t^3 + 18t^2 - 96t + 300.$$
   Here $t = 0$ represents the beginning of the decade. Find the rate at which employee productivity was changing when $t = 1, 4, 6, 8,$ and 9. In each case state whether productivity was increasing or decreasing at the given time. (See Exercise 67 of Exercises 10.4.)

Life Science

38. The volume of an artery of length 10 centimeters (cm) is $V = 10\pi r^2$ cm^3, where r is its radius. If a drug increases the radius of the artery, find the rate at which its volume is changing when
   (a) $r = 0.2$ cm   (b) $r = 0.3$ cm   (c) $r = 0.4$ cm

39. A reagent is added to a chemical solution containing an enzyme and the enzyme precipitates out of solution. The amount of enzyme precipitated out t seconds after the reagent has been added is $y = -0.04t^2 + 0.20t$ grams, where $0 \le t \le 2.5$. Find the rate at which the enzyme is precipitating out of solution 0.5 second, 1 second, 1.5 seconds, and 2 seconds after the reagent is added.

40. The number of bacteria present in a culture t hours after the start of an experiment is given by $f(t) = t^3 + 6t^2 + 100$. Find the rate at which the size of the bacteria population in the culture is changing when $t = 2, t = 5,$ and $t = 10$.

41. If there are x million cars on the road in a city, there will be y parts per million of carbon monoxide in its air, where
$$y = 0.2x^3 + 4.$$
   Find the rate at which the carbon monoxide concentration is changing when the number of cars on the road, in millions, is
   (a) 1.0   (b) 2.0   (c) 3.0   (d) 3.25   (e) 3.5

42. A field is sprayed with weed killer and t days later, $0 \leq t \leq 15$, it contains an average of
$$W = t^3 - 3t^2 - 144t + 900$$
weeds per acre. Find the rate at which the average number of weeds per acre is changing 1 day after spraying; 5 days after spraying; 10 days after spraying; 12 days after spraying. Is the average number of weeds per acre increasing or decreasing at these times?

43. The number of bacteria alive in a culture t days after an antibiotic is applied to it is y thousand, where
$$y = \frac{t^5}{5} - \frac{29t^3}{3} + 100t + 140.$$
Find the rate at which the bacteria population is changing when $t = 1, 2, 3, 4, 5,$ and 6. In each case state whether the bacteria population is increasing or decreasing. (See Exercise 71 of Exercises 10.4.)

44. The proportion of a scar that remains visible t days after an operation is y, where
$$y = 0.2 + 0.8t^{-1/2}, \quad t \geq 1.$$
Find the rate at which the visibility of the scar is changing 16 days after the operation; 30 days after the operation.

Social Science

45. In a memory experiment, subjects memorize a list of 20 numbers and then watch rock videos for various lengths of time. After t minutes of watching the videos, the typical subject can remember approximately
$$y = -0.02t^2 + 20$$
of the numbers. Find the rate at which the typical subject forgets the numbers when $t = 5, t = 10, t = 15,$ and $t = 25$ minutes.

46. Students who take an intensive foreign language course in which only the language being taught is spoken to them typically understand
$$P = 0.005t^3 - 0.05t^2 + 2t$$
percent of what is being said to them by the end of day t of the course, $0 \leq t \leq 25$. Find the rate at which the comprehension of a typical student is changing 5 days, 10 days, and 15 days into the course.

47. The unemployment rate in a state over a four-year period has been $u(t)$, where
$$u(t) = 0.0002t^3 - 0.0147t^2 + 0.216t + 5.9.$$
Here $u(t)$ is the unemployment rate as a percentage of the work force and t is in months, with $t = 0$ representing the beginning of the four-year period. Find the rate at which unemployment was changing when $t = 5, 9, 30,$ and 40. In each case state whether the unemployment rate was increasing or decreasing at the given time. (See Exercise 72, Exercises 10.4.)

48. In a certain state expenditures on welfare y depend on the unemployment rate x according to the equation
$$y = 0.005x^4 + 0.1x^2 + 15,$$
where x is the percentage unemployment rate and y is in millions of dollars. Find the rate of change of welfare expenditures when the unemployment rate is
(a) 4% (b) 6% (c) 8% (d) 10%

49. A politician's campaign manager estimates that the proportion of voters who support the politician t days before an election is y, where

$$y = 0.45 + \frac{0.10}{t}$$

and $1 \leq t \leq 10$. Find the rate at which the politician's support is changing 10 days, 5 days, and 1 day before the election.

10.6 THE THEORY OF THE FIRM

The **theory of the firm** is concerned with questions such as

- How many units of its product must a firm make and sell in order to maximize its profit?
- What price should the firm charge for its product in order to attain its maximum profit?
- How will changing the product's price affect the firm's revenue?

In this section we show how to use calculus to answer these questions. We do this by exploiting both viewpoints of the derivative: that it is the slope of the tangent line, and that it gives the rate of change with respect to the independent variable.

Marginality

If a firm produces and sells x units of its product, then, as we have seen, it has a **cost function** C, a **revenue function** R, and a **profit function**

$$P = R - C.$$

These functions need not be linear, but as in the linear case we can find the firm's **break-even quantities** by setting $R(x) = C(x)$ and solving for x, or, alternatively, by setting $P(x) = 0$ and solving for x.

EXAMPLE 1 Suppose Acme Company's cost and revenue functions are given by

$$C(x) = x^2 + 500{,}000 \quad \text{and} \quad R(x) = 1500x,$$

respectively. Then Acme's profit function is given by

$$P(x) = R(x) - C(x) = -x^2 + 1500x - 500{,}000.$$

Since

$$P(x) = -(x - 500)(x - 1000),$$

setting $P(x) = 0$ and solving for x yields $x = 500$ and $x = 1000$. Therefore the firm's break-even quantities are 500 units and 1000 units. Figures 10–42 and 10–43 show the graphs of C, R, and P. Note that Acme will make a profit if it makes and sells between 500 and 1000 units, but will suffer a loss if it makes and sells fewer than 500 or more than 1000 units. ∎

The Derivative

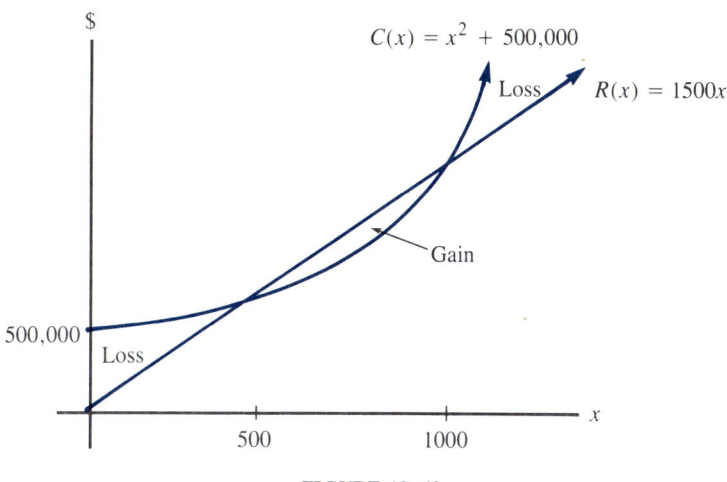

FIGURE 10-42

The derivatives of the cost, revenue, and profit functions are called **marginal functions.** Thus C' is the **marginal cost function,** R' the **marginal revenue function,** and P' the **marginal profit function.** The economic interpretation of the marginal cost function is as follows: since the derivative $C'(a)$ is the instantaneous rate of change of the cost when $x = a$, $C'(a)$ represents the approximate additional cost of producing one more unit. In other words, $C'(a)$ is the approximate additional cost of producing the $(a + 1)$st unit. Similarly, $R'(a)$ is the approximate additional revenue obtained from selling the $(a + 1)$st unit, and $P'(a)$ is the approximate additional contribution to profit made by the $(a + 1)$st unit.

EXAMPLE 2 Let Acme company's cost, revenue, and profit functions be as in Example 1. Then Acme's marginal cost, revenue, and profit functions are given by

$$C'(x) = 2x, \quad R'(x) = 1500, \quad \text{and} \quad P'(x) = -2x + 1500,$$

respectively. Since $C'(600) = 2(600) = 1200$, it will cost Acme approximately $1200 to make the 601st unit. Similarly, since $R'(600) = 1500$, Acme's revenue from selling the 601st unit will be approximately $1500. (Actually, in this example Acme's revenue from selling the 601st unit will be *exactly* $1500, since every unit

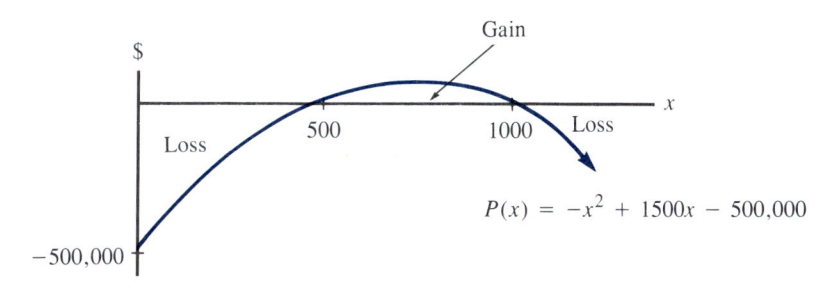

FIGURE 10-43

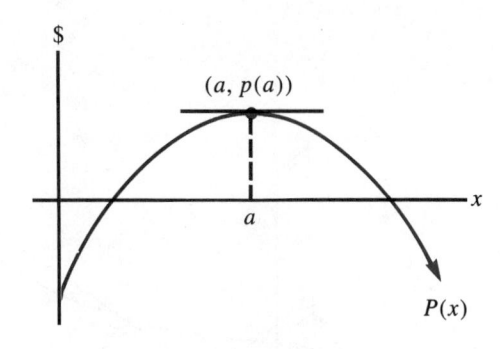

FIGURE 10-44

sells for $1500.) Also, the approximate additional contribution to profit made by the 601st unit will be $P'(600) = \$300$. ∎

Figure 10-44 shows the graph of a typical profit function P. Notice that the maximum profit occurs when $x = a$ units are made and sold. Since the tangent line to P at $(a, P(a))$ is horizontal, its slope will be zero; hence we must have $P'(a) = 0$. (Why?) But since $P(x) = R(x) - C(x)$,

$$0 = P'(a) = R'(a) - C'(a),$$

and thus $R'(a) = C'(a)$. Therefore profit is maximized at the value of x for which

$$R'(x) = C'(x),$$

or, in the language of economics, at the quantity where marginal revenue equals marginal cost.

Profit Maximization

Profit is maximized at the quantity for which marginal revenue equals marginal cost.

EXAMPLE 3 Consider Acme Company of Examples 1 and 2 again. If we set marginal revenue equal to marginal cost and solve for x, we have

$$R'(x) = C'(x)$$
$$1500 = 2x$$
$$750 = x.$$

Therefore, as Figure 10-45 indicates, Acme's profit is maximized when it makes and sells 750 units, and its maximum profit is $P(750) = \$62,500$. ∎

What if a firm has not yet determined the price it should charge for its product? For a single firm, the price that must be charged in order to obtain a given revenue

THE DERIVATIVE

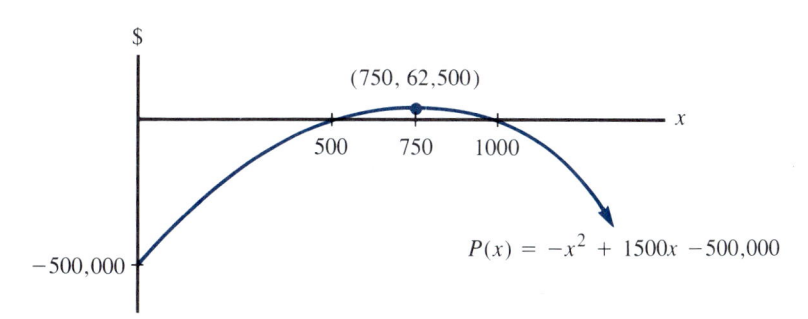

Figure 10–45: Graph of $P(x) = -x^2 + 1500x - 500{,}000$ with maximum point $(750, 62{,}500)$; x-intercepts near 500 and 1000; y-intercept at $-500{,}000$.

FIGURE 10–45

depends on the quantity of the product demanded: if the quantity demanded is large, the firm can sell many units at a low price; if the quantity demanded is small, it cannot sell very many units, and therefore must charge a higher price per unit in order to obtain the same revenue. If we let p denote the unit price of the product and x the quantity demanded, then p depends on x by means of a demand function $p = d(x)$. Since revenue is quantity sold times unit price, it follows that the firm's revenue is given by

$$R(x) = xp = xd(x).$$

EXAMPLE 4   As in the previous examples, suppose Acme Company's cost function is given by

$$C(x) = x^2 + 500{,}000,$$

but now suppose that the firm has not yet established a price for its product. Let the demand function for the product be defined by

$$p = 2100 - 0.5x.$$

Then Acme's revenue function is given by

$$R(x) = xp = x(2100 - 0.5x) = 2100x - 0.5x^2.$$

As before, we maximize profit by setting $R'(x) = C'(x)$ and solving for x: we have

$$R'(x) = C'(x)$$
$$2100 - x = 2x$$
$$700 = x.$$

Therefore Acme will maximize its profit by making and selling 700 units. In order to sell 700 units, the firm must charge

$$p = 2100 - 0.5(700) = \$1750$$

per unit. We have thus determined that if Acme prices its product at $1750 per unit, it will sell 700 units for a maximum profit of $P(700) = \$235{,}000$. □

Compare Example 4 with Example 3. In Example 3 we assumed that Acme had set a price of $1500 per unit and found that under this condition the firm should make and sell 750 units for a maximum profit of $62,500. In Example 4 it turned out that if Acme's demand function is defined by $p = 2100 - 0.5x$, then the firm should make fewer units, sell them at a higher price, and make a larger profit. A different choice of demand function could have led to the opposite conclusion: that Acme would be better off selling more units at a lower price. The property of the demand function that leads to such results is called **elasticity of demand.**

Elasticity of Demand

A demand function for a product relates quantity demanded q to unit price p. Here we consider the price p to be the independent variable so that the demand function is defined by an equation of the form $q = d(p)$. Figure 10–46 shows the graph of a typical demand function. Note that as price p increases, quantity demanded q decreases.

The derivative $q' = d'(p)$ gives the rate of change of quantity demanded with respect to price. Since the line tangent to the demand curve at any point will have negative slope, as indicated in Figure 10–47, $d'(p)$ will always be negative. If $|d'(p)|$ is large, then quantity demanded is decreasing quite rapidly as price increases; if $|d'(p)|$ is small, then quantity demanded is decreasing quite slowly as price increases. In other words, if $|d'(p)|$ is large, a small change in price results in a large change in quantity demanded; if $|d'(p)|$ is small, a small change in price results in a small change in quantity demanded.

EXAMPLE 5 Suppose

$$d(p) = \frac{10,000}{p}.$$

Then

$$d'(p) = -\frac{10,000}{p^2},$$

so $d'(2) = -2500$ and $d'(100) = -1$. Thus when the price is $2, a small increase in price will cause a large decrease in quantity demanded (in fact, we can say that

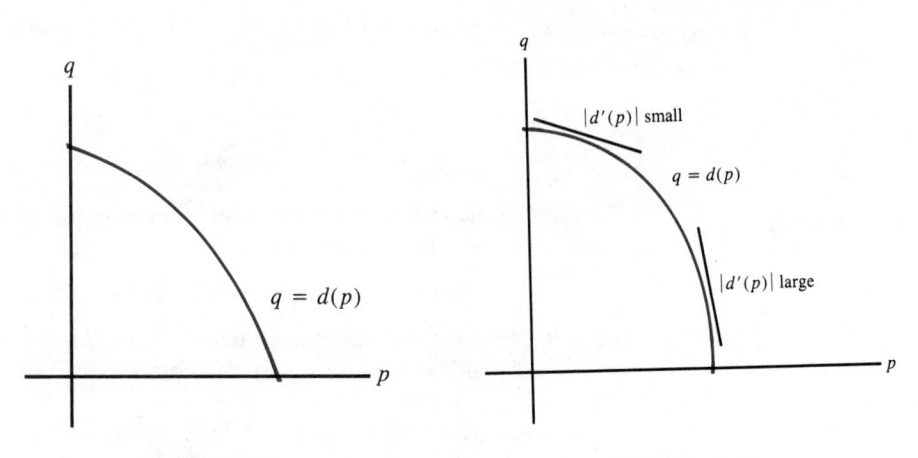

FIGURE 10–46

FIGURE 10–47

THE DERIVATIVE

when $p = \$2$, an increase of $1 in the price will cause quantity demanded to decrease by approximately 2500 units). On the other hand, when the price is $100, a small increase in price will cause only a small decrease in quantity demanded. See Figure 10–48. ☐

When a small change in price results in a large change in quantity demanded, we say that demand is **elastic.** When a small change in price results in only a small change in quantity demanded, we say that demand is **inelastic.** In order to make these notions precise, we define the **elasticity of demand** for a commodity.

The Elasticity of Demand Formula

If the demand function for a commodity is defined by an equation of the form $q = d(p)$, the commodity's **elasticity of demand at price p** is $E(p)$, where

$$E(p) = \frac{p[d'(p)]}{d(p)}.$$

Exercise 74 at the end of this section outlines the derivation of the elasticity formula.

The elasticity of demand $E(p)$ measures the proportional change in quantity demanded relative to price. If $|E(p)| > 1$, demand is elastic; if $|E(p)| < 1$, demand is inelastic. A product can have elastic demand at one price and inelastic demand at another.

EXAMPLE 6 If the demand function for a product is defined by

$$q = -1000p + 10{,}000,$$

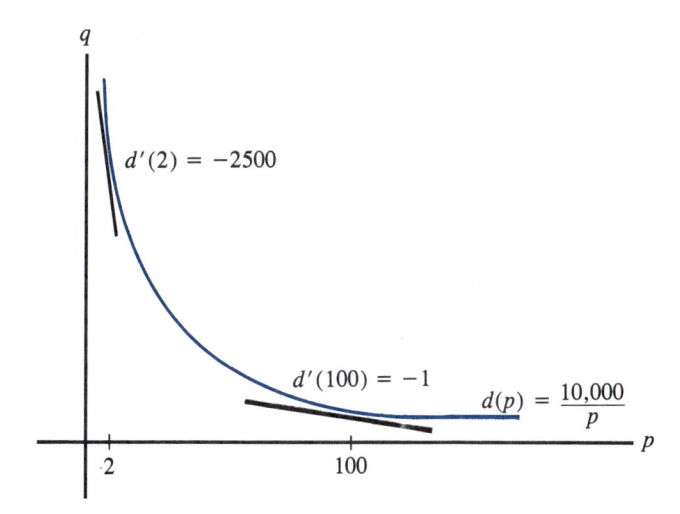

FIGURE 10–48

then the elasticity of demand for the product is

$$E(p) = \frac{p[d'(p)]}{d(p)} = \frac{-1000p}{-1000p + 10{,}000}.$$

Suppose that $p = \$2$. Then

$$|E(p)| = |E(2)| = |-0.25| = 0.25 < 1,$$

so demand for the product is inelastic at a price of \$2. Note that when $p = \$2$, quantity demanded $q = 8000$ units and revenue is thus $\$2(8000) = \$16{,}000$. If the price is increased by 10% to \$2.20, the quantity demanded decreases to 7800 and the revenue increases to $\$2.20(7800) = \$17{,}160$. Hence, when $p = \$2$, a 10% increase in price results in a small decrease in quantity demanded and an increase in revenue.

Now suppose that $p = \$8$. Then

$$|E(p)| = |E(8)| = |-4| = 4 > 1,$$

so demand for the product is elastic at a price of \$8. When $p = \$8$, quantity demanded $q = 2000$ units and revenue is $\$8(2000) = \$16{,}000$ again. But now a 10% increase in the price, from \$8 to \$8.80, results in a decrease in quantity demanded to 1200 units, which in turn yields revenue of \$10,560. Hence when $p = \$8$, a 10% increase in price leads to a large decrease in quantity demanded and a large decrease in revenue.

When $|E(p)| = 1$, demand is said to be of **unit elasticity.** For instance, in this example demand for the product is of unit elasticity when $p = \$5$. Notice also that $|E(p)| < 1$ when $p < \$5$ and $|E(p)| > 1$ when $p > \$5$. Hence demand is inelastic for $p < \$5$, of unit elasticity when $p = \$5$, and elastic when $p > \$5$. ☐

Note that in Example 6, when demand was inelastic at a price of \$2, increasing the price by a small amount led to an increase in revenue; when demand was elastic at a price of \$8, increasing the price by a small amount led to a decrease in revenue. These facts are examples of general truths:

Elasticity and Revenue

If demand is inelastic at a price,

- increasing the price by a small amount will result in an increase in revenue;
- decreasing the price by a small amount will result in a decrease in revenue.

If demand is elastic at a price,

- increasing the price by a small amount will result in a decrease in revenue;
- decreasing the price by a small amount will result in an increase in revenue.

If demand is of unit elasticity at a price, increasing or decreasing the price by a small amount will leave revenue approximately constant.

We will ask for a proof of these facts in Exercises 12.1.

EXAMPLE 7 Let $q = -1000p + 10{,}000$ be a demand function. Then, as we saw in Example 6, demand is inelastic for $p < \$5$; hence when $p = \$3$, for instance, increasing the price by a small amount will result in an increase in revenue and decreasing it by a small amount will result in a decrease in revenue. Similarly, when $p = \$7.50$ demand is elastic, so increasing the price by a small amount will result in a decrease in revenue and decreasing it by a small amount will lead to an increase in revenue. ☐

We conclude this section by remarking that instead of considering quantity demanded to be a function of price, as we have done here, economists usually consider price to be a function of quantity demanded. Therefore they define demand functions by equations of the form $p = d(q)$. When price is thus considered to be a function of quantity demanded, the formula for elasticity of demand has the alternative form

$$E(q) = \frac{d(q)}{q[d'(q)]}.$$

■ *Computer Manual: Program DIFFPOLY*

10.6 EXERCISES

Marginality

Zeta Company's cost function is given by $C(x) = 0.5x^2 + 100{,}000$ and Zeta sells each unit it makes for $600. Exercises 1 through 5 refer to this situation.

1. Find Zeta's total cost, revenue, and profit if it makes and sells 300 units.
2. Find Zeta's approximate additional cost from making the 301st unit.
3. Find Zeta's approximate additional revenue from selling the 301st unit.
4. Find Zeta's approximate additional profit from making and selling the 301st unit.
5. Find Zeta's maximum profit and the number of units it must make and sell in order to attain its maximum profit.

Eta Corporation's cost and revenue functions are given by

$$C(x) = 5x^2 + 500{,}000 \quad \text{and} \quad R(x) = -x^3 - 25x^2 + 14{,}400x,$$

respectively. Exercises 6 through 9 refer to this situation.

6. Find Eta's cost if it makes 25 units, and find the approximate additional cost of making the twenty-sixth unit.
7. Find Eta's revenue if it sells 50 units, and find the approximate additional revenue from selling the fifty-first unit.
8. Find Eta's profit if it makes and sells 75 units, and find the approximate additional profit from making and selling the seventy-sixth unit.
9. Find Eta's maximum profit and the number of units it must make and sell in order to attain its maximum profit.

Zeta Company's cost function is given by $C(x) = 0.5x^2 + 100{,}000$ and the demand function for its product by $p = 1200 - 2x$. Exercises 10 through 14 refer to this situation.

10. Find Zeta's total cost, revenue, and profit if it makes and sells 100 units.
11. Find Zeta's approximate additional cost from making the 101st unit.
12. Find Zeta's approximate additional revenue from selling the 301st unit.
13. Find Zeta's approximate additional profit from making and selling the 301st unit.
14. Find Zeta's maximum profit, the number of units it must make and sell, and the price per unit it must charge in order to attain its maximum profit.

Zeta Company's cost function is given by $C(x) = 0.5x^2 + 100{,}000$, and the demand function for its product by $p = 700 - 0.5x$. Exercises 15 through 19 refer to this situation.

15. Find Zeta's total cost, revenue, and profit if it makes and sells 200 units.
16. Find Zeta's approximate additional cost from making the 201st unit.
17. Find Zeta's approximate additional revenue from selling the 301st unit.
18. Find Zeta's approximate additional profit from making and selling the 301st unit.
19. Find Zeta's maximum profit, the number of units it must make and sell, and the price per unit it must charge in order to attain its maximum profit.

Eta Corporation's cost function is given by $C(x) = 5x^2 + 500{,}000$, and the demand function for its product by $p = -x^2 - 130x + 66{,}000$. Exercises 20 through 23 refer to this situation.

20. Find Eta's cost if it makes 50 units, and find the approximate additional cost of making the 51st unit.
21. Find Eta's revenue if it sells 50 units, and find the approximate additional revenue from selling the 51st unit.
22. Find Eta's profit if it makes and sells 75 units, and find the approximate additional profit from making and selling the 76th unit.
23. Find Eta's maximum profit, the number of units it must make and sell, and the price per unit it must charge in order to attain its maximum profit.

Eta Corporation's cost function is given by $C(x) = 5x^2 + 500{,}000$, and the demand function for its product by $p = -x^2 - 10x + 9000$. Exercises 24 through 27 refer to this situation.

24. Find Eta's cost if it makes 75 units, and find the approximate additional cost of making the 76th unit.
25. Find Eta's revenue if it sells 50 units, and find the approximate additional revenue from selling the 51st unit.
26. Find Eta's profit if it makes and sells 75 units, and find the approximate additional profit from making and selling the 76th unit.
27. Find Eta's maximum profit, the number of units it must make and sell, and the price per unit it must charge in order to attain its maximum profit.
28. A Corporation's cost function is given by $C(x) = 20x^2 + 40{,}000$, and the demand function for its product by $p = 6000 - 100x$. Find its maximum profit, the num-

ber of units the company must make and sell, and the price per unit it must charge in order to attain its maximum profit.

29. Repeat Exercise 28 if $C(x) = 2.3x^2 + 4000x$, $p = 10,000 - 0.10x$, and x is in thousands of units.

30. A firm's fixed cost is $100,000, its marginal cost function is given by $C'(x) = 500x$, and its demand function is linear, with quantity demanded being 0 units when price is equal to $214,500 and price being $0 when quantity demanded is equal to 8580 units. Find the firm's maximum profit and the number of units and price per unit required to attain it.

31. Theta, Inc. knows the following:

—Its fixed cost is $20,000.

—Its marginal cost function is given by $C'(x) = 10x$.

—Its demand function is linear.

—At a price of $2650 per unit, it sold 90 units.

—At a price of $2800 per unit, it sold 80 units.

Find Theta's maximum profit, the number of units the company must make and sell, and the price per unit it must charge in order to attain its maximum profit.

The average cost function is the cost function divided by the number of units produced; the **marginal average cost function** is the derivative of the average cost function. Similarly, marginal average revenue is the derivative of average revenue, and marginal average profit is the derivative of average profit. Exercises 32 through 35 are concerned with marginal average cost, revenue, and profit.

*32. Suppose Zeta Company's cost function is given by $C(x) = 0.5x^2 + 100,000$ and that Zeta sells each unit it produces for $600. Find and interpret Zeta's marginal average cost, marginal average revenue, and marginal average profit when it makes and sells 300 units.

*33. Repeat Exercise 32 if Zeta makes and sells 200 units.

*34. Suppose Eta Corporation's cost function is given by $C(x) = 5x^2 + 500,000$ and its revenue function by $R(x) = -x^3 - 25x^2 + 14,400x$, respectively. Find and interpret Eta's marginal average cost, marginal average revenue, and marginal average profit when it makes and sells 50 units.

*35. Repeat Exercise 34 if Eta's cost function is given by $C(x) = 5x^2 + 500,000$ and the demand function for its product by $p = -x^2 - 130x + 66,000$.

Elasticity of Demand

The demand function for a product is given by $q = -3000p + 120,000$. Exercises 36 through 38 refer to this product.

36. Find the elasticity of demand for the product when $p = \$15$. Find quantity demanded and revenue when $p = \$15$, $p = \$13.50$, and $p = \$16.50$.

37. Find the elasticity of demand for the product when $p = \$20$. Find quantity demanded and revenue when $p = \$20$, $p = \$19$, and $p = \$21$.

38. Find the elasticity of demand for the product when $p = \$30$. Find quantity demanded and revenue when $p = \$30$, $p = \$27$, and $p = \$33$.

The demand function for a commodity is given by $q = -100p + 25{,}000$. Exercises 39 through 41 refer to this commodity.

39. Find the elasticity of demand for the commodity when $p = \$150$. Find quantity demanded and revenue when $p = \$150$, $p = \$153$, and $p = \$147$.

40. Find the elasticity of demand for the commodity when $p = \$125$. Find quantity demanded and revenue when $p = \$125$, $p = \$127.50$, and $p = \$122.50$.

41. Find the elasticity of demand for the commodity when $p = \$100$. Find quantity demanded and revenue when $p = \$100$, $p = \$102$, and $p = \$98$.

The demand function for a commodity is given by $q = -50p^2 + 60{,}000$. Exercises 42 through 44 refer to this commodity.

42. Find the elasticity of demand for the commodity when $p = \$25$. Find quantity demanded and revenue when $p = \$25$, $p = \$25.25$, and $p = \$24.75$.

43. Find the elasticity of demand for the commodity when $p = \$20$. Find quantity demanded and revenue when $p = \$20$, $p = \$20.20$, and $p = \$19.80$.

44. Find the elasticity of demand for the commodity when $p = \$15$. Find quantity demanded and revenue when $p = \$15$, $p = \$15.15$, and $p = \$14.85$.

The demand function for a product is defined by $q = -5p + 3000$. Exercises 45 through 47 refer to this product.

45. Find the elasticity of demand for the product when $p = \$100$ and when $p = \$500$.

46. Find quantity demanded and revenue when $p = \$100$ and when $p = \$500$, and also when the price is increased by 10% in each case.

47. Find the price at which the product has unit elasticity of demand. Find quantity demanded and revenue when the price is increased or decreased by 5% from this value.

The demand function for a firm's product is given by $q = -2250p + 65{,}250$. Exercises 48 through 50 refer to this product.

48. Find an expression for the elasticity of demand for the product.

49. Over what range of prices is demand for the product elastic? inelastic? At what price is demand of unit elasticity?

50. Suppose the firm is currently selling the product for $10 per unit. Will increasing the price by a small amount increase or decrease revenue? Answer the same question if the firm is currently selling the product for $20 per unit.

The demand function for a commodity is given by $q = -200p + 180{,}000$. Exercises 51 through 54 refer to this commodity.

51. Find an expression for the elasticity of demand of the commodity as a function of p.

52. Find the price at which demand for the commodity is of unit elasticity. Find the range of prices over which demand for the commodity is elastic and the range over which it is inelastic.

53. If the commodity is currently selling for $500 per unit, what will happen to revenue if the price is increased by a small amount? if it is decreased by a small amount?

THE DERIVATIVE

54. Answer the questions of Exercise 53 if the commodity is currently selling for $450; for $400.

The demand function for a commodity is given by $q = -160p^2 + 1,200,000$. Exercises 55 through 58 refer to this commodity.

55. Find an expression for the elasticity of demand of the commodity as a function of p.

56. Find the price at which demand for the commodity is of unit elasticity. Find the range of prices over which demand for the commodity is elastic and the range over which it is inelastic.

57. If the commodity is currently selling for $45 per unit, what will happen to revenue if the price is increased by a small amount? If it is decreased by a small amount?

58. Answer the questions of Exercise 57 if the commodity is currently selling for $50; for $55.

59. The daily demand for ice cream bars at the beach is given by

$$q = 10,000 - 2500p.$$

Find the price at which demand is of unit elasticity, the price range where it is elastic, and the price range where it is inelastic. If ice cream bars currently sell for $1.50 each, what will happen to sales revenue if the price is increased to $1.75?

60. The monthly demand for a certain brand of automobile is given by

$$q = 12,000 - 0.75p.$$

Find the price at which demand is of unit elasticity, the price range where it is elastic, and the price range where it is inelastic. If the automobile currently sells for $10,000, will raising the price by $500 increase or decrease sales revenue?

61. Cutter Cotter Pin Company needs to increase its revenue. The marketing department has recommended a 5% decrease in the price the company charges for its cotter pins, arguing that this would increase the quantity demanded so much that revenue would actually increase. The accounting department has recommended a 5% increase in the price, arguing that this would increase revenue because the resulting decrease in quantity demanded would be more than offset by the higher price. Suppose the demand function for cotter pins is given by $q = -1250p + 25,000$, where p is price per dozen cotter pins and q is in dozens of cotter pins. Whose recommendation should be taken if the current price per dozen of cotter pins is
(a) $6?   (b) $10?   (c) $12?

62. The demand for gasoline in a certain country is given by

$$q = \begin{cases} -10p + 50, & 0 \le p < 1.20, \\ -35p + 80, & 1.20 \le p \le 2.00, \end{cases}$$

where p is the price in dollars per gallon and q is in millions of gallons per day.
(a) Find the elasticity of demand for gasoline when $p = 0.80$, $p = 1.15$, $p = 1.25$, and $p = 1.50$.
(b) Graph the demand function. Can you explain why the price per gallon must rise to at least $1.20 before significant conservation of gasoline occurs?

The demand function for a commodity is given by $p = -0.01q + 1600$. Exercises 63 through 67 refer to this commodity.

63. Use the elasticity of demand formula in the form $E(q) = d(q)/q[d'(q)]$ to find an expression for the elasticity of demand for the commodity as a function of q.

668   CHAPTER 10

64. Use the expression obtained in Exercise 63 to find the elasticity of demand for the commodity when $q = 20{,}000$ units.

65. Use the expression obtained in Exercise 63 to find the elasticity of demand for the commodity when $p = \$400$.

66. Solve the demand equation for p in terms of q and then use the elasticity of demand formula in the form $E(p) = p[d'(p)]/d(p)$ to find an expression for the elasticity of demand for the commodity as a function of p.

67. Use the expression obtained in Exercise 66 to find the elasticity of demand for the commodity when $p = \$1400$ and when $q = 120{,}000$.

The demand function for a commodity is given by $p = -0.005q^2 + 80$. Exercises 68 through 70 refer to this commodity.

68. Use the elasticity of demand formula in the form $E(q) = d(q)/q[d'(q)]$ to find an expression for the elasticity of demand for the commodity as a function of q.

69. Use the expression obtained in Exercise 68 to find the elasticity of demand for the commodity when $q = 100$ units.

70. Use the expression obtained in Exercise 68 to find the elasticity of demand for the commodity when $q = 50$ units.

*71. Show that a commodity whose demand function is given by an equation of the form $q = a/p$, where a is a positive number, will have unit elasticity of demand at every price and that revenue from sales of the commodity will therefore remain constant no matter what the price.

*72. If a commodity has a demand function whose equation is of the form $q = c$, where c is a constant, show that its elasticity of demand is 0. This is called **perfect inelasticity**; when it occurs, revenue depends only on price. (Why?)

*73. If the demand function for a commodity has an equation of the form $p = c$, where c is a constant, show that its elasticity of demand does not exist. This is known as **perfect elasticity**; when it occurs, revenue depends only on quantity demanded. (Why?)

*74. If the price of a commodity changes by some percentage, then in response the quantity of the commodity demanded by consumers will change by some percentage. Elasticity of demand is defined to be the ratio of the percentage change in quantity demanded to the percentage change in price. If $d(p)$ is a demand function and the price changes from p to $p + h$, then the quantity demanded changes from $d(p)$ to $d(p + h)$, and hence

$$\text{Percentage change in price} = \frac{(p + h) - p}{p} 100 = \frac{h}{p} 100$$

and

$$\text{Percentage change in quantity demanded} = \frac{d(p + h) - d(p)}{d(p)} 100.$$

Take the limit as h approaches 0 of the ratio

$$\frac{\text{Percentage change in quantity demanded}}{\text{Percentage change in price}}$$

and interpret the result.

10.7 DIFFERENTIAL NOTATION AND HIGHER DERIVATIVES

Thus far we have written the derivative of a function f as f'. This is known as the **prime notation** for derivatives. Another common way of writing derivatives, known as **differential notation,** is sometimes more convenient and useful than prime notation. In this section we discuss differential notation for derivatives. We also consider the higher derivatives of a function: higher derivatives are derivatives of derivatives. For instance, as we saw in Section 10.5, velocity is the derivative of position, and acceleration is the derivative of velocity; thus acceleration is a higher derivative of position.

Differential Notation

Consider the function f defined by the equation

$$f(x) = x^3 + 4x + 1.$$

Using prime notation, we write the derivative of this function as

$$f'(x) = 3x^2 + 4.$$

If the function is given in the form $y = x^3 + 4x + 1$, we can still use prime notation and write the derivative as $y' = 3x^2 + 4$.

Another notation often used for the derivative of a function is **differential notation:** if f is a function of x, its derivative is written in differential notation as

$$\frac{df}{dx}$$

and referred to as the **derivative of f with respect to x.** The notation

$$\left.\frac{df}{dx}\right|_{x=a}$$

indicates that the derivative is to be evaluated at $x = a$.

EXAMPLE 1   Let $f(x) = x^3 + 2x^2 + 3x$, $y = \dfrac{2}{x} + 3x$, and $g(t) = 3\sqrt{t}$. Then the derivative of f with respect to x is

$$\frac{df}{dx} = 3x^2 + 4x + 3$$

and

$$\left.\frac{df}{dx}\right|_{x=1} = 3(1)^2 + 4(1) + 3 = 10.$$

Similarly, the derivative of y with respect to x is

$$\frac{dy}{dx} = -\frac{2}{x^2} + 3$$

and
$$\left.\frac{dy}{dx}\right|_{x=2} = \frac{5}{2}.$$

Finally,
$$\frac{dg}{dt} = \frac{3}{2\sqrt{t}}$$

and
$$\left.\frac{dg}{dt}\right|_{t=4} = \frac{3}{2\sqrt{4}} = \frac{3}{4}. \quad \square$$

At this time we will not consider df/dx and dy/dx to be ratios, but merely symbols for the derivative. Later we will show how these symbols can indeed be interpreted as ratios.

Differential notation is especially convenient when we must make explicit which variable is to be regarded as the independent one.

EXAMPLE 2   Let $y = x^3$. Then dy/dx is the derivative of y with respect to x. As usual, x is the independent variable and y is the dependent one. Thus
$$\frac{dy}{dx} = \frac{d}{dx}(x^3) = 3x^2.$$

On the other hand, dx/dy is the derivative of x with respect to y, so here y is to be regarded as the independent variable and x as the dependent one. If we solve the equation $y = x^3$ for x in terms of y to obtain $x = y^{1/3}$, we find that
$$\frac{dx}{dy} = \frac{d}{dy}(y^{1/3}) = \frac{1}{3}y^{-2/3}. \quad \square$$

Higher Derivatives

If f' is the derivative of f, then f' may itself have a derivative $(f')'$, which we write as f''. Similarly, f'' may have a derivative $(f'')' = f'''$, and so on. The derivative f' is the **first derivative of f**, f'' is the **second derivative of f**, and f''' is the **third derivative of f**. If $n > 3$, we denote the nth derivative of f by $f^{(n)}$.

EXAMPLE 3   Let $f(x) = x^4 + 5x^3 + 3x^2 + 4x - 1$. Then we have

first derivative: $\quad f'(x) = 4x^3 + 15x^2 + 6x + 4$
second derivative: $\quad f''(x) = 12x^2 + 30x + 6$
third derivative: $\quad f'''(x) = 24x + 30$
fourth derivative: $\quad f^{(4)}(x) = 24$
fifth derivative: $\quad f^{(5)}(x) = 0$

Note that $f^{(n)}(x) = 0$ for $n \geq 5$. $\quad \square$

The Derivative

When using differential notation, we write the second derivative as

$$\frac{d^2f}{dx^2},$$

with the obvious extension to higher derivatives.

EXAMPLE 4 Let $y = 1/x$. Then

$$\frac{dy}{dx} = -\frac{1}{x^2}, \quad \frac{d^2y}{dx^2} = \frac{2}{x^3},$$

$$\frac{d^3y}{dx^3} = -\frac{6}{x^4}, \quad \frac{d^4y}{dx^4} = \frac{24}{x^5}. \quad \square$$

■ *Computer Manual: Program HIGHPOLY*

10.7 EXERCISES

Differential Notation

In Exercises 1 through 22, find the indicated derivatives.

1. $\dfrac{dy}{dx}$, $y = x^2 - 5x + 2$

2. $\dfrac{dy}{dx}$, $y = 2x^3 - 3x + 12$

3. $\dfrac{df}{dx}$, $f(x) = x^3 - 5x^2 + 6x - 2$

4. $\dfrac{df}{dt}$, $f(t) = t^4 - 3t^2 + 9$

5. $\dfrac{dy}{dx}$, $y = 5\sqrt{x} - 2x^{2/3}$

6. $\dfrac{df}{dx}$, $f(x) = 7x^{-2} + 3x + 1$

7. $\dfrac{df}{dt}$, $f(t) = 3t^7 + t^{-2} + 12$

8. $\dfrac{dW}{dq}$, $W = 14q^2 + 10q - q^{-1}$

9. $\dfrac{dg}{dx}$, $g(x) = \dfrac{2}{x} - \dfrac{4}{x^3}$

10. $\dfrac{ds}{dt}$, $s = \sqrt{t} + 3t^2 + 2$

11. $\dfrac{df}{dx}\bigg|_{x=2}$, $f(x) = x^2 + 5x - 2$

12. $\dfrac{dy}{dx}\bigg|_{x=1}$, $y = 5x^4 - 20x + 2$

13. $\dfrac{dp}{dq}\bigg|_{q=9}$, $p = q^{3/2} + 5q$

14. $\dfrac{dy}{dx}\bigg|_{x=4}$, $y = x^5 - 5x^{3/2} + 2$

15. $\dfrac{dh}{dv}\bigg|_{v=8}$, $h(v) = 2v^{-3} + 2v^{-1/3}$

16. $\dfrac{df}{dy}\bigg|_{y=-1}$, $f(y) = 2y^{2/3} + 4$

17. $\dfrac{df}{dx}\bigg|_{x=a}$, $f(x) = x^3 + 2x^2$

18. $\dfrac{dq}{dp}\bigg|_{p=b}$, $q = 400p^2 - 1200p + 100$

19. $\dfrac{dy}{dx}$, $\dfrac{dx}{dy}$, $3x + 2y = 5$

20. $\dfrac{dp}{dq}$, $\dfrac{dq}{dp}$, $40p - 35q = 1250$

21. $\dfrac{dy}{dt}$, $\dfrac{dt}{dy}$, $4ty = 1$

22. $\dfrac{du}{dv}$, $\dfrac{dv}{du}$, $u^3 = 27v$

Higher Derivatives

In Exercises 23 through 44, find the indicated higher derivatives.

23. $f''(x)$, $f'''(x)$, $f(x) = x^3 + x^2 + x + 1$

24. $f''(x)$, $f^{(4)}(x)$, $f(x) = x^5 - 2x^3 + 4x$

25. $f''(x)$, $f(x) = 3x^5 - 12x^3 + 1$

26. $g'''(x)$, $g(x) = x^4 + 1$

27. $f^{(5)}(x)$, $f(x) = 6x^5 + 20x^3 + 3$

28. $f^{(100)}(x)$, $f(x) = 2x^{76} + 36x^{48} + 99x^{25} - 455x^7 + 123$

29. $\dfrac{d^2f}{dx^2}$, $\dfrac{d^3f}{dx^3}$, $f(x) = x^4 - 4x^3 + 3x + 12$

30. $\dfrac{d^2f}{dx^2}$, $f(x) = x^3 + 3x^2 - 9x + 10$

31. $\dfrac{d^2y}{dx^2}$, $\dfrac{d^4y}{dx^4}$, $y = \dfrac{2}{x^2} - \dfrac{1}{x}$

32. $\dfrac{d^2g}{dt^2}$, $g(t) = \dfrac{5}{t^2} + \dfrac{1}{t} + 2$

33. $\dfrac{d^2y}{dx^2}$, $\dfrac{d^3y}{dx^3}$, $y = x^{-3}$

34. $\dfrac{d^3y}{dx^3}$, $\dfrac{d^4y}{dx^4}$, $y = 2\sqrt{x}$

35. $\dfrac{d^5y}{dx^5}$, $y = x^3 + x^{-3}$

36. $f''(3)$, $f(u) = u^4 + 3u + 2$

37. $f'''(0)$, $f(t) = 2t^5 + t^4$

38. $g''(4)$, $g^{(4)}(4)$, $g(p) = 2p^{5/2}$

39. $\left.\dfrac{d^2y}{dx^2}\right|_{x=0}$, $y = 3x^3 + 2x - 7$

40. $\left.\dfrac{d^3y}{dx^3}\right|_{x=1}$, $y = x^2 + 2\sqrt{x}$

41. $\left.\dfrac{d^2p}{dq^2}\right|_{q=1}$, $\left.\dfrac{d^3p}{dq^3}\right|_{q=1}$, $p = 3q^{-1} + q^2$

42. $\left.\dfrac{d^2f}{dx^2}\right|_{x=4}$, $\left.\dfrac{d^3f}{dx^3}\right|_{x=-4}$, $f(x) = \dfrac{5}{x} - \dfrac{1}{x^2}$

43. $f^{(4)}(a)$, $f(x) = x^8 + x^4 + x$

44. $g^{(4)}(b)$, $g(x) = x^3 + 2x^2 + 3x + 15$

SUMMARY

This summary consists of terms and symbols whose meaning you should know, facts you should know, and techniques and methods you should be able to employ.

Section 10.1

Meaning of "x approaches a": $x \to a$. Limit of a function f as x approaches a: $\mathrm{Lim}_{x \to a} f(x)$. Finding limits using tables of functional values. Finding limits using graphs. Properties of limits. Finding limits using the properties of limits. Limits of polynomial functions. Finding limits of polynomial functions. Rational function. Finding limits of rational functions. Finding limits of functions having different defining equations on different intervals.

Section 10.2

Meaning of "x approaches a from the right," "x approaches a from the left": $x \to a^+$, $x \to a^-$. One-sided limits: $\mathrm{Lim}_{x \to a^+} f(x)$, $\mathrm{Lim}_{x \to a^-} f(x)$. Finding one-sided limits from graphs. $\mathrm{Lim}_{x \to a} f(x) = L$ if and only if $\mathrm{Lim}_{x \to a^+} f(x) = \mathrm{Lim}_{x \to a^-} f(x) = L$. Infinite limits: $\mathrm{Lim}_{x \to a^+} f(x) = \pm\infty$, $\mathrm{Lim}_{x \to a^-} f(x) = \pm\infty$.

THE DERIVATIVE

Meaning of $\pm\infty$ as limits. Vertical asymptote. Polynomials have no vertical asymptotes. A rational function has vertical asymptotes at those values that make its denominator zero and its numerator nonzero. When $x = a$ is an asymptote of a rational function, determining whether a one-sided limit as x approaches a is $+\infty$ or $-\infty$. Meaning of "x approaches $+\infty$," "x approaches $-\infty$": $x \to +\infty$, $x \to -\infty$. Limits at infinity: $\lim_{x \to +\infty} f(x)$, $\lim_{x \to -\infty} f(x)$. Finding limits at infinity from graphs. Horizontal asymptote. Limits at infinity for powers of x. Polynomial functions do not have horizontal asymptotes. Finding limits at infinity for rational functions. Finding horizontal asymptotes for rational functions. Oblique asymptote. Finding oblique asymptotes for rational functions.

Section 10.3

Line tangent to f at $(x, f(x))$. Secant line. The derivative of f at x:

$$f'(x) = \lim_{h \to 0} \frac{f(x+h) - f(x)}{h}$$

The derivative $f'(x)$ as the slope of the line tangent to f at the point $(x, f(x))$. The difference quotient. The Three-Step Rule. Using the Three-Step Rule to find derivatives. Using the derivative to find the equation of the line tangent to f at $(x, f(x))$. Function differentiable at $x = a$. Function not differentiable at $x = a$. Determining whether a function is differentiable or not differentiable at $x = a$ from its graph. Function continuous at $x = a$. Function discontinuous at $x = a$. Function continuous everywhere. Determining whether a function is continuous or discontinuous at $x = a$ from its graph. First limit definition of continuity. Second limit definition of continuity. Polynomial functions are continuous everywhere. Rational functions are continuous everywhere they are defined. Differentiability implies continuity. Continuity does not imply differentiability.

Section 10.4

Constant function. The constant rule. Using the constant rule. Power function. The power rule. Using the power rule. The constant-multiple rule. Using the constant-multiple rule. The sum rule. The sum rule for more than two summands. Using the sum rule.

Section 10.5

Distance function. Average velocity. Instantaneous velocity. Velocity as the derivative of position. Finding velocities. The derivative as rate of change. Acceleration as the derivative of velocity. Finding accelerations. Finding rates of change.

Section 10.6

The theory of the firm. Cost, revenue, profit functions. Break-even quantities. Finding break-even quantities. Marginal cost, marginal revenue, marginal profit functions. The meaning of marginality: the meaning of $C'(a)$, $R'(a)$, $P'(a)$. Finding marginal cost, revenue, profit. Maximum profit occurs when marginal revenue equals marginal cost. Finding maximum profit and when it occurs. Revenue equals price per unit times quantity demanded. Finding the price that must be charged in order to attain maximum profit. Quantity demanded q as a function of price p. The rate of change of quantity demanded with respect to price. Elastic demand, inelastic demand. The elasticity of demand formula. Demand is elastic when $|E(p)| > 1$, inelastic when $|E(p)| < 1$, of unit elasticity when $|E(p)| = 1$. Finding elasticities. Finding changes in revenue when price is changed. Relationships between elasticity, price changes, and revenue changes. Elasticity formula $E(q)$ when price p is considered to be a function of quantity demanded.

Section 10.7

Prime notation f'. Differential notation df/dx. The derivative evaluated at $x = a$: $df/dx|_{x=a}$. Using differential notation. First derivative f', second derivative f'', third derivative f''', nth derivative $f^{(n)}$. Finding higher derivatives. Differential notation for higher derivatives: $d^n f/dx^n$, $d^n f/dx^n|_{x=a}$. Using differential notation for higher derivatives.

REVIEW EXERCISES

Let the function f have the following graph:

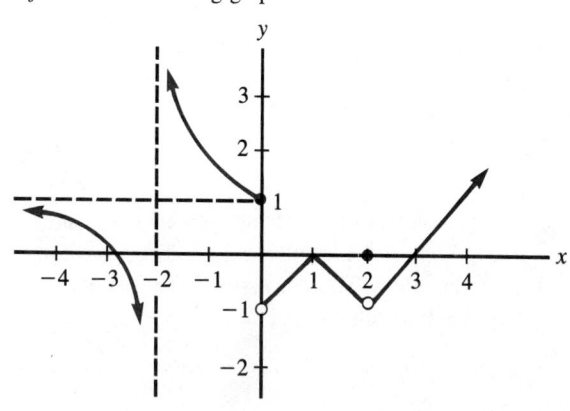

In Exercises 1 through 20, find the indicated limit. Use the symbols $\pm\infty$ as appropriate.

1. $\lim\limits_{x \to -2} f(x)$
2. $\lim\limits_{x \to -2^+} f(x)$
3. $\lim\limits_{x \to -2^-} f(x)$
4. $\lim\limits_{x \to -1} f(x)$
5. $\lim\limits_{x \to -1^+} f(x)$
6. $\lim\limits_{x \to -1^-} f(x)$
7. $\lim\limits_{x \to 0^+} f(x)$
8. $\lim\limits_{x \to 0^-} f(x)$
9. $\lim\limits_{x \to 0} f(x)$
10. $\lim\limits_{x \to 1^+} f(x)$
11. $\lim\limits_{x \to 1^-} f(x)$
12. $\lim\limits_{x \to 1} f(x)$
13. $\lim\limits_{x \to 2^+} f(x)$
14. $\lim\limits_{x \to 2^-} f(x)$
15. $\lim\limits_{x \to 2} f(x)$
16. $\lim\limits_{x \to 3} f(x)$
17. $\lim\limits_{x \to 3^+} f(x)$
18. $\lim\limits_{x \to 3^-} f(x)$
19. $\lim\limits_{x \to +\infty} f(x)$
20. $\lim\limits_{x \to -\infty} f(x)$

In Exercises 21 through 42, find the indicated limit. Use the symbols $\pm\infty$ as appropriate.

21. $\lim\limits_{x \to 2} (3x^2 - 5x - 2)$

22. $\lim\limits_{x \to 1} (x^4 + x^3 - 2x^2 + 5x - 1)$

23. $\lim\limits_{x \to 2^+} (5x^2 + 3x - 2)$

24. $\lim\limits_{x \to 0^-} (3x^3 + 5x^2 - 7x + 13)$

25. $\lim\limits_{x \to 9} \left(4\sqrt{x} - \dfrac{2}{x}\right)$

26. $\lim\limits_{x \to 1} \left(\dfrac{1}{x} - \dfrac{2}{x^2}\right)$

27. $\lim\limits_{x \to -1} (x^4 + x + 1)(x^2 - x - 1)$

28. $\lim\limits_{x \to 1} \dfrac{2}{x - 1}$

29. $\lim\limits_{x \to 2} \dfrac{x + 3}{x + 2}$

30. $\lim\limits_{x \to 4^+} \dfrac{x + 3}{x - 4}$

31. $\lim\limits_{x \to -2^-} \dfrac{12x + 5}{3x + 6}$

32. $\lim\limits_{x \to 2} \dfrac{x + 3}{x - 2}$

33. $\lim\limits_{x \to 0^+} \dfrac{2x + 5}{x^2 + 2}$

34. $\lim\limits_{x \to 0^+} \dfrac{2x + 5}{x^2 + 2x}$

35. $\lim\limits_{x \to 0} \dfrac{3x + 1}{2x^2 + 5}$

36. $\lim\limits_{x \to 5} \dfrac{x^2 - 7x + 10}{x^2 + 2x - 15}$

37. $\lim\limits_{x \to 5} \dfrac{x^2 + 7x + 10}{x^2 + 2x - 15}$

38. $\lim\limits_{x \to +\infty} \dfrac{4x^2 - 3x}{x + 2}$

39. $\lim\limits_{x \to 2^-} \dfrac{x^2 + x - 6}{x^2 + 3x - 10}$

40. $\lim\limits_{x \to 3^+} \dfrac{x^3 + x}{x^2 - 2x - 3}$

41. $\lim\limits_{x \to -\infty} \dfrac{x^2 + 2x}{4x^2 - 3x}$

42. $\lim\limits_{x \to +\infty} \dfrac{x + 2}{4x^2 - 3x}$

In Exercises 43 through 45, find all asymptotes of the given function.

43. $f(x) = \dfrac{3x - 7}{4 - 2x}$

44. $y = \dfrac{x^2 + 1}{x - 4}$

45. $g(x) = \dfrac{x^2 - 3x + 2}{x}$

46. Use the Three-Step Rule to find $k'(x)$ if $k(x) = 5x^3$. Find the equation of the line tangent to k at $(-3, k(-3))$.

47. Use the Three-Step Rule to find $g'(x)$ if $g(x) = \sqrt[3]{3x}$. For what values of x does $g'(x)$ not exist?

48. The following is the graph of a function h:

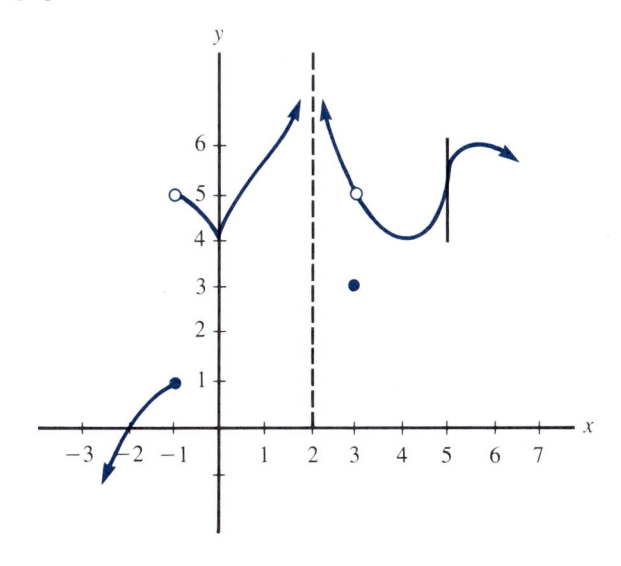

(a) At what values of x is h defined but not differentiable?
(b) At what values of x is h defined and discontinuous?

In Exercises 49 through 51, find the intervals over which the function f is continuous, the values of x at which f is defined but discontinuous, and the values of x at which f is defined but $f'(x)$ does not exist.

49. $f(x) = 2x^2 - 8x + 12$

50. $f(x) = \dfrac{x^2 - 2x - 8}{x^2 + 9x + 20}$

51. $f(x) = \begin{cases} x + 4, & x < -1, \\ -2x + 1, & -1 \le x \le 0, \\ 2, & 0 < x < 1, \\ x^2 + 1, & 1 \le x < 2, \\ 2, & x = 2, \\ x^2 - 8x + 17, & x > 2 \end{cases}$

In Exercises 52 through 59, find $f'(x)$.

52. $f(x) = -402$

53. $f(x) = \sqrt{3}x$

54. $f(x) = 3x^2 + 6x + 12$

55. $f(x) = -x^6 + 4x^5 - 12x^3 + 16x^2 - 8x - 3$

56. $f(x) = \dfrac{2}{x^3} - \dfrac{1}{x} + 3$

57. $f(x) = 4x^{-2} + 2x^{-3}$

58. $f(x) = 4\sqrt{x}$

59. $f(x) = 5x^{4/5} - 2x^{-1/2} + \dfrac{3}{x^{2/3}}$

In Exercises 60 through 65, find the equation of the line tangent to f at the point $(a, f(a))$.

60. $f(x) = 2/\sqrt{3}$, $a = -5$

61. $f(x) = 4x^2 - 6x + 2$, $a = 3$

62. $f(x) = x^5 + 2x^4 - 9x^2 + 3x + 12$, $a = 0$

63. $f(x) = 3\sqrt{x} + \dfrac{2}{x} - 2$, $a = 1$

64. $f(x) = 2x^{6/5}$, $a = 32$

65. $f(x) = 4x^{-3/4} + 1$, $a = 81$

66. A mortar shell is fired straight up into the air. The altitude of the shell t seconds after firing is
$$s(t) = -5t^2 + 150t$$
meters. Find the shell's velocity and acceleration 3 seconds and 6 seconds after firing. Find the shell's maximum altitude. How long is the shell in the air?

67. A firm estimates that the value of a computer t years after it is purchased new will be
$$V(t) = 576 - 192\sqrt{t}$$
thousand dollars. Here $0 \le t \le 9$. Find the rate at which the value of the computer is changing at the end of one year; at the end of four years; at the end of eight years.

68. A firm's cost function is given by $C(x) = x^2 + 320$, and the demand function for its product by $p = 56 - x$. Find the firm's maximum profit, the number of units that must be made and sold in order to attain the maximum profit, and the price per unit the firm must charge in order to attain the maximum profit.

69. The demand function for a commodity is given by $q = -6000p + 240{,}000$. Find
    (a) the elasticity of demand for the commodity when $p = \$10$, $p = \$20$, and $p = \$30$;
    (b) the revenue when $p = \$10$, $p = \$20$, and $p = \$30$;
    (c) the revenue when $p = \$11$, $p = \$22$, and $p = \$33$.

70. For the commodity of Exercise 69, find the price at which unit elasticity occurs, the range of prices where demand for the commodity is elastic, and the range where it is inelastic. If the commodity sells for $25 per unit, what will happen to revenue if the price is increased to $26? if it is decreased to $24?

In Exercises 71 through 75, find the indicated derivatives.

71. $f(x) = 2x^4 - 5x^2 + 12$, $\dfrac{df}{dx}, \dfrac{d^2f}{dx^2}, \dfrac{d^3f}{dx^3}\bigg|_{x=2}$

72. $y = 2\sqrt{x} + \dfrac{2}{\sqrt{x}}$, $\dfrac{dy}{dx}, \dfrac{d^2y}{dx^2}$

73. $g(x) = 3x^3 - 4x^{4/5} + 2$, $g'(x), g''(x), g'''(1)$

74. $w = 2u^2 + 3bu + 2\sqrt{u} + 1$, b a constant, $\dfrac{dw}{du}, \dfrac{d^2w}{du^2}$

75. $\dfrac{8x}{y} = 11$, $\dfrac{dy}{dx}, \dfrac{dx}{dy}$

ADDITIONAL TOPICS

Here are some additional topics not covered in the text that you might wish to investigate on your own.

1. A more rigorous definition of the limit of a function than the one presented in Section 10.1 is the following:

   $\text{Lim}_{x \to a} f(x) = L$ if for every positive real number ϵ there exists a positive real number δ such that $|f(x) - L| < \epsilon$ whenever $0 < |x - a| < \delta$.

   Explain how this definition of limit "translates" into the definition we gave in Section 10.1.

2. Show how the rigorous definition of limit given in suggestion 1 is used to prove the following properties of limits:

   $$\text{Lim}_{x \to a} c = c, \quad \text{Lim}_{x \to a} cf(x) = c \cdot \text{Lim}_{x \to a} f(x),$$

   $$\text{Lim}_{x \to a} [f(x) + g(x)] = \text{Lim}_{x \to a} f(x) + \text{Lim}_{x \to a} g(x),$$

   $$\text{Lim}_{x \to a} f(x)g(x) = [\text{Lim}_{x \to a} f(x)][\text{Lim}_{x \to a} g(x)].$$

3. In Section 10.3 we gave an intuitive argument to show that if a function is differentiable at a point then it must be continuous there. Use the Second Limit Definition of Continuity to prove this.

4. Given a function f, it is often important to be able to find numbers x_r such that $f(x_r) = 0$. Such numbers are called **roots** of f, and are often difficult to find. **Newton's method** (also called the **Newton-Raphson method**) uses the derivative of f to find approximate roots of f. Find out how Newton's method works, why it works, why it sometimes fails to work, and some of its applications.

5. The derivative was introduced to mathematics in the seventeenth century by Isaac Newton and Gottfried Wilhelm Leibniz. Find out about the early development and use of the derivative: why the concept of the derivative was needed by Newton and Leibniz, the different ways in which each defined and used his discovery, and how

each of them introduced his version of the derivative to the mathematical world. A good place to begin research on this topic would be with the supplement to this chapter.

SUPPLEMENT: NEWTON AND LEIBNIZ

No one person "invented" calculus. The development of calculus was a lengthy process contributed to by many mathematicians over many centuries. Nevertheless, there was a short period of time during which the partial results and incomplete methods previously discovered were greatly extended and then organized into a coherent mathematical theory that was thereafter known as "the calculus." This rapid development of the subject occurred in the latter part of the seventeenth century and was the independent work of Isaac Newton (1643–1727) of England and Gottfried Wilhelm Leibniz (1646–1716) of Germany.

Isaac Newton was one of the greatest of human intellects. He made monumental contributions to physics by enunciating the law of universal gravitation and the three laws of motion that bear his name, working out many of the consequences of these laws as they apply to astronomy and mechanics, experimenting in optics, and proposing a particle theory of light. His development of calculus went hand-in-hand with his work in physics as he created new mathematical tools to solve problems in celestial and earthly mechanics. In addition to these activities, Newton also experimented in chemistry and alchemy, studied and wrote on theology, and served as Master of the Royal Mint.

Newton apparently developed what he called his "fluxional calculus" during the period 1665–1666. Quantities that changed with time he referred to as **fluents;** he called their rates of change with respect to time **fluxions.** For instance, if a particle moves along a curve its x- and y-coordinates change with time and hence are fluents, and therefore their rates of change with respect to time, which Newton denoted by $\dot{x}$ and $\dot{y}$, respectively, are fluxions. Thus the fluxions $\dot{x}$ and $\dot{y}$ are what we would now write as dx/dt and dy/dt, the derivatives of x and y with respect to time. The notation $\dot{x}$ for the derivative with respect to time survives today in physics and engineering.

The method by which Newton found derivatives was rather cumbersome, proceeding much in the manner of our Three-Step Rule but using the idea of infinitesimal quantities rather than limits. His reasoning was as follows: if h is the duration of an infinitely short time period, then $\dot{x}h$ and $\dot{y}h$ are infinitesimals, that is, infinitely small nonzero amounts by which x and y change during the time period. If we substitute $x + \dot{x}h$ for x and $y + \dot{y}h$ for y in the relationship between x and y, subtract the original equation from the result, and then drop all terms containing powers of h that are 2 or greater because they are infinitesimally small compared with the remaining terms, we can obtain an expression for $\dot{y}$ in terms of $\dot{x}$. To illustrate, suppose

$$y = x^2.$$

Then

$$y + \dot{y}h = (x + \dot{x}h)^2,$$

or

$$y + \dot{y}h = x^2 + 2x\dot{x}h + (\dot{x})^2h^2.$$

Subtracting $y = x^2$ from this last equation, and dropping the term containing h^2 because it is infinitesimal compared with the rest of the terms, we have

$$\dot{y}h = 2x\dot{x}h,$$

or

$$\dot{y} = 2x\dot{x}.$$

This last equation can be rewritten in the form

$$\frac{\dot{y}}{\dot{x}} = 2x,$$

which expresses what we today call the derivative of y with respect to x as a ratio of the time derivatives $\dot{y}$ and $\dot{x}$.

Newton seems to have been a secretive man who was not anxious to share his discoveries with others. For many years he did not communicate his discoveries in calculus and physics to the world at large, but instead revealed them in piecemeal fashion to a few selected colleagues. Finally, at the instigation of his friend Edmund Halley, he published in 1687 a book entitled *Philosophiae Naturalis Principia Mathematica*, known today as the *Principia*. In the *Principia* Newton presented for the first time his work on gravitation and celestial mechanics, as well as the fluxional calculus that made it possible. Unfortunately, Newton recast all the calculus into geometrical form, in approved classical style, and this served to obscure the power and utility of his method of fluxions.

During the years 1673–1676, after Newton had developed his version of the calculus but before he had published it, Leibniz independently created a slightly different version. He, too, utilized infinitesimals, but he called his infinitesimals **differentials:** the differentials for x and y he denoted by dx and dy, respectively. These were defined geometrically as shown in Figure 10–49: dx is an arbitrary infinitesimal increment to x, and dy is determined by dx and the tangent line at (x, y). Leibniz defined the derivative of y with respect to x to be the ratio dy/dx of the differentials dy and dx.

Leibniz's terminology and notation have survived to the present day. This is why we use the differential notation dy/dx for the derivative. We will discuss differentials in more detail in Section 11.3.

The method by which Leibniz found derivatives was much like Newton's. In the relationship between x and y, x was replaced by $x + dx$, y was replaced by $y + dy$, and the original equation was subtracted from the result. Then all terms containing powers of dx or dy that were 2 or greater, or all terms with a factor of the form $dx\, dy$, were eliminated on the grounds that they were infinitesimally small

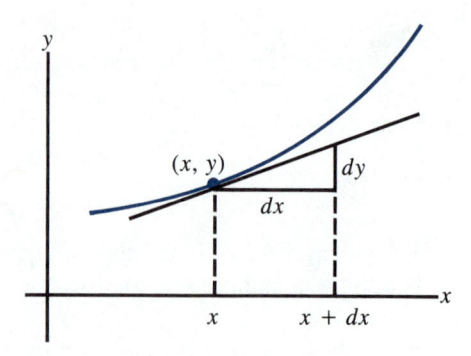

FIGURE 10-49

compared with the remaining terms. The end result was an expression relating dy and dx. However, Leibniz also worked out and used many of the rules of differentiation, such as (in his notation)

$$d(cx) = c \cdot dx$$

and

$$d(x^n) = nx^{n-1}\, dx.$$

Unlike Newton, Leibniz was eager to let others know of his discoveries. Even before the formal publication of his results in 1684, knowledge of his methods was quickly spread through the European mathematical community by personal correspondence and word of mouth.

In 1676 Newton learned that Leibniz was working on a version of calculus; in an attempt to assert his priority, he sent Leibniz a letter in which he mentioned his earlier work in the field. Leibniz replied with a letter outlining his own methods. When Leibniz published his results in 1684, he did not mention Newton's claim of priority; in response Newton, in the *Principia,* told of this exchange of letters. This started a quarrel, not so much between Newton and Leibniz but between their followers, over who had priority of discovery. Accusations of plagiarism were made by both sides, and the situation became, particularly in England, one of national pride. As a result, English mathematicians for many years insisted on Newton as the ultimate mathematical authority and adopted his terminology and notation exclusively. Since the terminology and notation used by Leibniz were generally superior to those of Newton, this had an unfortunate stultifying effect on English mathematics.

Newton and Leibniz both used infinitesimals in their development of calculus. Such "infinitely small" but nonzero quantities seemed illogical to mathematicians and made them uneasy. Newton himself was not satisfied with his reliance on infinitesimals, and in his later work he tried to get by without them; in so doing, he nearly arrived at the notion of the limit of a function. However, it was not until 1821, when the French mathematician Augustin Louis Cauchy introduced the idea of limit in its modern form, that it was possible for calculus to be put on a firm logical foundation.

Calculus is one of the greatest tools ever discovered for the investigation of nature. It was developed as a method for solving problems about the movement of objects, particularly astronomical bodies, but its applications soon began to expand, first throughout the physical sciences and engineering, and then to the biological and social sciences, including management and economics. The usefulness of calculus continues to grow today as it is used to investigate new problems as they arise in these and other fields.

Suggestion

Many books describe the development of calculus in much greater detail than we have been able to do here. If you are interested in this topic and would like to read more about it, here are some places to start:

Andrade, E. N. Da C. "Isaac Newton." In *The World of Mathematics,* vol. 1, ed. James R. Newman. New York: Simon and Schuster, 1956.

Bell, E. T. *Men of Mathematics.* New York: Simon and Schuster, 1937.

Boyer, Carl B. *History of the Calculus and Its Conceptual Development.* New York: Dover, 1959.

Boyer, Carl B. *A History of Mathematics.* New York: John Wiley & Sons, 1968.

Eves, Howard. *An Introduction to the History of Mathematics.* New York: Holt, Rinehart, and Winston, 1961.

Eves, Howard. *Great Moments in Mathematics After 1650.* Dolciani Mathematical Expositions No. 7, The Mathematical Association of America, 1983.

11

More about Differentiation

In Chapter 10 we introduced the derivative and presented some of the basic rules of differentiation: the constant rule, the power rule, the constant-multiple rule, and the sum rule. These rules are not sufficient to handle all functions of interest, however. For instance, they do not tell us how to differentiate functions that are products or quotients, such as those defined by

$$f(x) = (x^2 + 2x + 3)(x^5 + 4x^3 - 12)$$

and

$$g(x) = \frac{3x^2 + 2x - 5}{5x + 3},$$

or those that are defined as generalized powers, such as

$$h(x) = (2x^3 - 5x + 7)^{12}.$$

In this chapter we continue our study of differentiation by developing the product rule, quotient rule, and chain rule for derivatives. These rules will allow us to differentiate functions such as those referred to above. We also discuss the topic of differentials, showing how the differential notation dy/dx can indeed be regarded as a ratio, and demonstrate the useful technique of implicit differentiation. We conclude the chapter with a section devoted to the solution of related rates problems.

11.1 THE PRODUCT RULE AND THE QUOTIENT RULE

In this section we develop and illustrate rules of differentiation that allow us to find the derivatives of functions that are products and quotients of other functions.

The Product Rule

Let f and g be functions and suppose that $f'(x)$ and $g'(x)$ both exist. We would like to find the derivative of the product function $k = fg$ at x. Using the Three-Step Rule, we form the difference quotient of k at x, then rewrite it in terms of the difference quotients of f and g by adding and subtracting $f(x)g(x + h)$ in its numerator:

$$\frac{k(x + h) - k(x)}{h} = \frac{f(x + h)g(x + h) - f(x)g(x)}{h}$$

$$= \frac{f(x + h)g(x + h) - f(x)g(x + h) + f(x)g(x + h) - f(x)g(x)}{h}$$

$$= \frac{[f(x + h) - f(x)]g(x + h) + f(x)[g(x + h) - g(x)]}{h}$$

$$= \frac{f(x + h) - f(x)}{h} g(x + h) + f(x) \frac{g(x + h) - g(x)}{h}.$$

Because $f'(x)$ and $g'(x)$ exist, we have

$$\lim_{h \to 0} \frac{f(x + h) - f(x)}{h} = f'(x) \quad \text{and} \quad \lim_{h \to 0} \frac{g(x + h) - g(x)}{h} = g'(x).$$

Also, since $g'(x)$ exists $g(x)$ is differentiable and hence continuous at x; therefore by the second limit definition of continuity (Section 10.3),

$$\lim_{h \to 0} g(x + h) = g(x).$$

Thus

$$k'(x) = \lim_{h \to 0} \frac{f(x + h) - f(x)}{h} \lim_{h \to 0} g(x + h)$$

$$+ \lim_{h \to 0} f(x) \lim_{h \to 0} \frac{g(x + h) - g(x)}{h}$$

$$= f'(x)g(x) + f(x)g'(x).$$

This is the **product rule** for derivatives.

MORE ABOUT DIFFERENTIATION   685

> **The Product Rule**
>
> If $k = fg$ and $f'(x)$ and $g'(x)$ both exist, then
>
> $$k'(x) = f(x)g'(x) + f'(x)g(x).$$

In words: the derivative of a product is the first factor times the derivative of the second factor plus the derivative of the first factor times the second factor. Notice that the derivative of a product is *not* the product of the derivatives.

EXAMPLE 1   If $k(x) = (6x + 7)(3x - 2)$, then by the product rule,

$$\begin{aligned}k'(x) &= (6x + 7)(3x - 2)' + (6x + 7)'(3x - 2)\\ &= (6x + 7)3 + 6(3x - 2)\\ &= 18x + 21 + 18x - 12\\ &= 36x + 9.\end{aligned}$$

(Of course, we could have found $k'(x)$ without using the product rule by multiplying out its right-hand side and then taking the derivative:

$$k(x) = 18x^2 + 9x - 14, \quad \text{so} \quad k'(x) = 36x + 9.)\ \square$$

EXAMPLE 2   If $k(x) = (x^2 - 3x + 4)(x^3 + x)$, then

$$\begin{aligned}k'(x) &= (x^2 - 3x + 4)(x^3 + x)' + (x^2 - 3x + 4)'(x^3 + x)\\ &= (x^2 - 3x + 4)(3x^2 + 1) + (2x - 3)(x^3 + x)\\ &= 5x^4 - 12x^3 + 15x^2 - 6x + 4.\ \square\end{aligned}$$

EXAMPLE 3   Let $f(x) = (x^2 + 1)(x^3 + 2)(x^4 + 3)$. We can find $f'(x)$ by applying the product rule twice. Since $f(x)$ is the product of $x^2 + 1$ and $(x^3 + 2)(x^4 + 3)$, a first application of the product rule gives

$$f'(x) = (x^2 + 1)[(x^3 + 2)(x^4 + 3)]' + (x^2 + 1)'(x^3 + 2)(x^4 + 3).$$

A second application of the product rule now allows us to find the derivative of $(x^3 + 2)(x^4 + 3)$:

$$\begin{aligned}f'(x) =\ & (x^2 + 1)[(x^3 + 2)(x^4 + 3)' + (x^3 + 2)'(x^4 + 3)]\\ & + (x^2 + 1)'(x^3 + 2)(x^4 + 3)\\ =\ & (x^2 + 1)[(x^3 + 2)4x^3 + 3x^2(x^4 + 3)] + 2x(x^3 + 2)(x^4 + 3),\end{aligned}$$

so

$$\begin{aligned}f'(x) &= (x^2 + 1)[7x^6 + 8x^3 + 9x^2] + 2x(x^7 + 2x^4 + 3x^3 + 6)\\ &= 9x^8 + 7x^6 + 12x^5 + 15x^4 + 8x^3 + 9x^2 + 12x.\ \square\end{aligned}$$

The Quotient Rule

For functions f and g, suppose that $f'(x)$ and $g'(x)$ both exist and that $g(x) \neq 0$. We would like to find the derivative of the quotient function $k = f/g$ at x. Using a trick similar to that employed in deriving the product rule, we write

$$\frac{k(x+h) - k(x)}{h} = \frac{f(x+h)/g(x+h) - f(x)/g(x)}{h}$$

$$= \frac{f(x+h)g(x) - f(x)g(x+h)}{hg(x)g(x+h)}$$

$$= \frac{f(x+h)g(x) - f(x)g(x) + f(x)g(x) - f(x)g(x+h)}{hg(x)g(x+h)}$$

$$= \frac{f(x+h) - f(x)}{h} \cdot \frac{g(x)}{g(x)g(x+h)}$$

$$- \frac{f(x)}{g(x)g(x+h)} \cdot \frac{g(x+h) - g(x)}{h}.$$

As was the case for the product rule, the existence of $f'(x)$ and $g'(x)$ implies that

$$\lim_{h \to 0} \frac{f(x+h) - f(x)}{h} = f'(x) \quad \text{and} \quad \lim_{h \to 0} \frac{g(x+h) - g(x)}{h} = g'(x).$$

Also, the existence of $g'(x)$ implies that g is differentiable and hence continuous at x; therefore by the second limit definition of continuity (Section 10.3),

$$\lim_{h \to 0} g(x+h) = g(x).$$

Thus

$$k'(x) = \lim_{h \to 0} \frac{f(x+h) - f(x)}{h} \lim_{h \to 0} \frac{g(x)}{g(x)g(x+h)}$$

$$- \lim_{h \to 0} \frac{f(x)}{g(x)g(x+h)} \lim_{h \to 0} \frac{g(x+h) - g(x)}{h}$$

$$= f'(x) \frac{g(x)}{g(x)g(x)} - \frac{f(x)}{g(x)g(x)} g'(x)$$

$$= \frac{g(x)f'(x) - f(x)g'(x)}{[g(x)]^2}.$$

This is the **quotient rule** for derivatives.

The Quotient Rule

If $k = f/g$, $f'(x)$ and $g'(x)$ both exist, and $g(x) \neq 0$, then

$$k'(x) = \frac{g(x)f'(x) - f(x)g'(x)}{[g(x)]^2}.$$

MORE ABOUT DIFFERENTIATION

In words: the derivative of a quotient is the denominator times the derivative of the numerator minus the numerator times the derivative of the denominator, all divided by the square of the denominator. Notice that the derivative of a quotient is *not* the quotient of the derivatives.

EXAMPLE 4   If
$$k(x) = \frac{2x + 3}{5x + 4},$$
then by the quotient rule,
$$\begin{aligned} k'(x) &= \frac{(5x + 4)(2x + 3)' - (2x + 3)(5x + 4)'}{[5x + 4]^2} \\ &= \frac{(5x + 4)2 - (2x + 3)5}{[5x + 4]^2} \\ &= \frac{10x + 8 - 10x - 15}{[5x + 4]^2} \\ &= -\frac{7}{25x^2 + 40x + 16}. \quad \square \end{aligned}$$

EXAMPLE 5   If
$$f(x) = \frac{x^2 + x + 1}{2x^3 + 1},$$
then
$$\begin{aligned} f'(x) &= \frac{(2x^3 + 1)(x^2 + x + 1)' - (x^2 + x + 1)(2x^3 + 1)'}{[2x^3 + 1]^2} \\ &= \frac{(2x^3 + 1)(2x + 1) - (x^2 + x + 1)(6x^2)}{[2x^3 + 1]^2} \\ &= \frac{4x^4 + 2x^3 + 2x + 1 - 6x^4 - 6x^3 - 6x^2}{[2x^3 + 1]^2} \\ &= \frac{-2x^4 - 4x^3 - 6x^2 + 2x + 1}{4x^6 + 4x^3 + 1}. \quad \square \end{aligned}$$

Sometimes it is necessary to use the product and quotient rules together.

EXAMPLE 6   Let
$$f(x) = (x^2 + 1)\left(\frac{2x}{4x + 9}\right).$$
To find $f'(x)$, we apply the product rule followed by the quotient rule:

$$f'(x) = (x^2 + 1)\left(\frac{2x}{4x + 9}\right)' + (x^2 + 1)'\left(\frac{2x}{4x + 9}\right)$$

$$= (x^2 + 1)\left(\frac{(4x + 9)(2x)' - 2x(4x + 9)'}{(4x + 9)^2}\right) + 2x\left(\frac{2x}{4x + 9}\right)$$

$$= (x^2 + 1)\frac{(4x + 9)2 - 2x(4)}{[4x + 9]^2} + 2x\frac{2x}{4x + 9}$$

$$= (x^2 + 1)\frac{18}{[4x + 9]^2} + \frac{4x^2}{4x + 9}$$

$$= \frac{16x^3 + 54x^2 + 18}{[4x + 9]^2}. \quad \square$$

EXAMPLE 7 A patient is placed on a regimen that includes taking a daily dose of a certain drug. The amount of the drug concentrated in the patient's liver t days after the start of the regimen is y grams, where

$$y = \frac{20t}{t + 15}.$$

At what rate is the amount concentrated in the liver changing after 30 days on the regimen?

By the quotient rule,

$$\frac{dy}{dt} = \frac{(t + 15)20 - 20t(1)}{[t + 15]^2},$$

or

$$\frac{dy}{dt} = \frac{300}{[t + 15]^2}.$$

Therefore at the end of 30 days the amount of the drug concentrated in the liver is increasing at the rate of

$$\left.\frac{dy}{dt}\right|_{t=30} = \frac{300}{[30 + 15]^2} = \frac{4}{27}$$

gram per day. $\square$

■ *Computer Manual: Programs PRODPOLY, QUOPOLY*

11.1 EXERCISES

In Exercises 1 through 40, use the product rule or the quotient rule to find the derivative of the given function.

1. $k(x) = (3x + 5)(2x - 1)$
2. $k(x) = (2x + 6)(4x - 15)$
3. $f(x) = (5x + 1)(x^2 - 2)$
4. $f(x) = (x^2 + 3)(2x + 5)$

5. $f(x) = (x^2 + 5)(x^2 - 3)$
6. $k(x) = (3x^5 + 2x)(x^2 + 4x - 1)$
7. $k(x) = \dfrac{4}{x + 2}$
8. $k(x) = -\dfrac{2}{3x - 1}$
9. $f(x) = \dfrac{3}{\sqrt{x} - 2}$
10. $f(x) = -\dfrac{2x}{3x + 1}$
11. $f(x) = \dfrac{x}{2x - 1}$
12. $f(x) = \dfrac{x + 1}{2x + 3}$
13. $k(x) = \dfrac{2x + 4}{5x - 1}$
14. $k(x) = \dfrac{4 - 3x}{7 - x}$
15. $f(x) = \dfrac{x^2 + 3x}{x + 2}$
16. $f(x) = \dfrac{x^2 + 3x}{3x + 4}$
17. $k(x) = \dfrac{x^2 + 3x - 2}{9 - 3x}$
18. $k(x) = (5x + 3)^2$
19. $k(x) = (x^2 + 4x)^2$
20. $k(x) = (x^4 + 2)x^{-3}$
21. $k(x) = 2x^{-3}(x^3 + 4)$
22. $f(x) = (2 + \sqrt{x})(x^3 + x^2)$
23. $f(x) = 2(x^{-3} + x^{-2})(x^2 + 4x)$
24. $f(x) = \dfrac{5x^2 + 3}{5x - 2}$
25. $f(x) = \dfrac{x^2 + 3x}{x^2 - 2}$
26. $k(x) = (3x^{1/3} - 2\sqrt{x})(x^2 - 3x + 1)$
27. $k(x) = \dfrac{2x^{-1} + 3x^{-2}}{x^{-2} - 1}$
28. $k(x) = x(2x + 1)(3x + 2)$
29. $f(x) = (x + 1)(x^2 + 1)(x^3 + 1)$
30. $k(x) = \dfrac{(3x + 5)(2x - 7)}{6x + 1}$
31. $k(x) = \dfrac{5x + 6}{(2x - 1)(3x + 9)}$
32. $k(x) = \dfrac{8x + 3}{(x^2 + 1)(7 - 2x)}$
33. $k(x) = \dfrac{(x^2 - 3)(4 - 3x)}{5x + 11}$
34. $k(x) = \dfrac{(x^2 + 1)(x^3 + 2)}{x^2 - 2x}$
35. $k(x) = \dfrac{x^2 - 2x}{(x^2 + 1)(x^3 + 2)}$
36. $k(x) = (1 - 2x)(x^2 - 1)(3 + 4x)(x^3 + 2x)$
37. $k(x) = x(x^2 + 1)(x^3 + x^2 + 1)$
38. $k(x) = (x^2 + 3)(x^2 + 8)(x^3 + 2x^2 + 1)(x^4 - 1)$
39. $k(x) = \dfrac{5x + 3}{7x + 2} \cdot \dfrac{6x + 1}{x - 1}$
40. $k(x) = \dfrac{4x + 3}{5 - 2x} \cdot \dfrac{x^2 + 1}{x^2 - 1}$

Management

41. Weekly revenue from box office receipts of a movie t weeks after the movie is released is y million dollars, where

$$y = \dfrac{20t}{0.5t + 5}.$$

Find the rate of change of revenue when $t = 1, 2,$ and 5 weeks. Also find and interpret $\lim\limits_{t \to +\infty} y$.

42. A firm's cost function is given by

$$C(x) = \frac{x^2}{x+1} - 1, \quad x \geq 1.$$

Find the firm's marginal cost when $x = 1$, $x = 10$, and $x = 20$, and interpret your results.

43. If Acme Company makes $x \geq 1$ units of its product, its average cost per unit is given by

$$A(x) = \frac{20x - 4}{5x - 0.5}.$$

Find the rate at which Acme's average cost per unit is changing when $x = 1$, $x = 10$, and $x = 100$. Also find $\lim\limits_{x \to +\infty} A(x)$ and interpret your result.

44. If a firm produces x units of its product, its average cost per unit is $A(x)$ dollars, where

$$A(x) = \frac{6x + 53}{2x + 1}, \quad x \geq 1.$$

(a) Find and interpret $A'(1)$, $A'(2)$, and $A'(10)$.
(b) Find and interpret $\lim\limits_{x \to +\infty} A(x)$.

45. Repeat Exercise 44 if

$$A(x) = \frac{0.2x^2 + 100}{3x + 2}.$$

46. The demand function for a commodity is given by

$$p = \frac{1000}{q + 1},$$

where q is the quantity demanded in thousands of units and p is the unit price in dollars. Find the marginal revenue when the quantity demanded is 9000, 99,000 and 999,000 units. Also find $\lim\limits_{q \to +\infty} p$ and interpret your result.

47. The demand function for a product is given by

$$p = \frac{2q + 200}{8q + 10}.$$

Here q is in millions of units, and p is in dollars. Find the marginal revenue when $q = 5$, 25, and 100 million units. Also find $\lim\limits_{q \to +\infty} p$ and interpret your result.

Life Science

48. Suppose that t hours after the introduction of a single bacterium into a culture there are

$$y = \frac{200t}{t + 10}$$

bacteria in the culture. Find the rate at which the number of bacteria is growing when $t = 1$, $t = 5$, and $t = 10$. Also find $\lim\limits_{t \to +\infty} y$ and interpret your result.

49. The number of signs a chimpanzee knows after t months of being trained in sign language is y, where

$$y = \frac{540t}{2t + 7}.$$

Find the rate at which the chimpanzee is learning signs when $t = 0, 6,$ and 12 months. Also find and interpret $\lim_{t \to +\infty} y$.

50. A drug is given to reduce a benign tumor. A dosage of x units will shrink the tumor by y cm, where

$$y = \frac{x^2 + 1}{2 + 3x}.$$

Find the rate at which the tumor will shrink if the dosage is 5 units; if the dosage is 10 units.

Social Science

51. A psychologist asks the subjects of an experiment to perform a series of complex tasks, and finds that after x repetitions of the series they can, on average, complete

$$y = \frac{100x + 30}{x + 1}$$

percent of the tasks without mistakes. Find the rate at which learning takes place as a function of the number of repetitions if the number of repetitions is 0, 2, and 5.

52. A municipality estimates that t years from now the proportion of its budget that will be devoted to debt service is y, where

$$y = (2t + 1)(7t + 8)^{-1}.$$

(a) Find and interpret $\left.\dfrac{dy}{dt}\right|_{t=0}$, $\left.\dfrac{dy}{dt}\right|_{t=1}$, and $\left.\dfrac{dy}{dt}\right|_{t=2}$.

(b) Find and interpret $\lim_{t \to +\infty} y$.

11.2 THE CHAIN RULE

We can create new functions from old ones by addition, subtraction, multiplication, and division, and the sum rule, product rule, and quotient rule for derivatives show us how to differentiate such new functions in terms of the old ones. However, there is another method of creating new functions from old ones: this method combines two functions to form a new one known as their **composite function.** In this section we discuss composite functions and introduce the **chain rule,** which tells how to differentiate them.

Composite Functions

To illustrate composite functions, suppose f and g are defined by

$$f(x) = 3x + 7 \quad \text{and} \quad g(x) = -2x.$$

We create a new function k from f and g by substituting $g(x)$ for x in the formula for f:

$$k(x) = f(g(x)) = f(-2x) = 3(-2x) + 7 = -6x + 7,$$

or

$$k(x) = -6x + 7.$$

The function k is called the **composite of g with f**.

Composite Functions

If f and g are functions, the **composite of g with f** is the function k defined by

$$k(x) = f(g(x)).$$

EXAMPLE 1   Let

$$f(x) = x^2 - 2x + 2 \quad \text{and} \quad g(x) = 2x - 1.$$

The composite of g with f is the function k defined by

$$\begin{aligned} k(x) &= f(g(x)) \\ &= (g(x))^2 - 2(g(x)) + 2 \\ &= (2x - 1)^2 - 2(2x - 1) + 2 \\ &= 4x^2 - 4x + 1 - 4x + 2 + 2 \\ &= 4x^2 - 8x + 5. \end{aligned}$$

We can also form the composite of f with g, which is the function h defined by $h(x) = g(f(x))$:

$$\begin{aligned} h(x) &= g(f(x)) \\ &= 2(f(x)) - 1 \\ &= 2(x^2 - 2x + 2) - 1 \\ &= 2x^2 - 4x + 3. \end{aligned}$$

Note that $f(g(x)) \neq g(f(x))$. □

EXAMPLE 2   If $f(x) = \sqrt{3x - 2}$ and $g(x) = x^2 + 4$, then

$$f(g(x)) = f(x^2 + 4) = \sqrt{3(x^2 + 4) - 2} = \sqrt{3x^2 + 10}$$

and

$$g(f(x)) = g(\sqrt{3x - 2}) = (\sqrt{3x - 2})^2 + 4 = 3x - 2 + 4 = 3x + 2. \quad \square$$

Composite functions are important because it is often the case that one quantity is a function of a second, and the second is a function of a third. In such a case, the first quantity may be described as a function of the third by means of a composite function. We illustrate with an example.

EXAMPLE 3 Suppose the demand function for a commodity is given by

$$q = 100{,}000 - 0.5p^2$$

and that the price is changing with time according to the equation

$$p = 100 + 0.1t,$$

where t is in years with $t = 0$ representing the current year. Then

$$q = 100{,}000 - 0.5(100 + 0.1t)^2$$

or

$$q = 95{,}000 - 10t - 0.005t^2.$$

The equation $q = 95{,}000 - 10t - 0.005t^2$ thus describes quantity demanded as a function of time. □

Many functions can be written as composites of simpler ones. For instance, the function k defined by $k(x) = (x + 1)^2$ may be written as the composite of $g(x) = x + 1$ with $f(x) = x^2$, because

$$f(g(x)) = f(x + 1) = (x + 1)^2 = k(x).$$

It will be important for us to be able to write functions as composites when appropriate. We give an example to show how this is done.

EXAMPLE 4 Let $k(x) = (2x + 3)^{20}$. We seek to write the function k as a composite function by finding f and g such that $k(x) = f(g(x))$. The way to do this is to choose $g(x)$ to be the "inside" expression in $k(x)$ and $f(x)$ to be the "outside" expression in $k(x)$. Thus if we take $g(x) = 2x + 3$ and $f(x) = x^{20}$, then

$$f(g(x)) = f(2x + 3) = (2x + 3)^{20} = k(x).$$

Similarly, if

$$h(x) = \sqrt{2x^3 + 4x},$$

and we take

$$g(x) = 2x^3 + 4x \quad \text{and} \quad f(x) = \sqrt{x},$$

then

$$f(g(x)) = f(2x^3 + 4x) = \sqrt{2x^3 + 4x} = h(x). \quad □$$

The Chain Rule

Now we develop the rule for differentiating composite functions. Suppose $k(x) = f(g(x))$, where $f'(g(x))$ and $g'(x)$ both exist. We form the difference quotient

$$\frac{k(x+h) - k(x)}{h} = \frac{f(g(x+h)) - f(g(x))}{h}.$$

If $g(x+h) - g(x) \neq 0$, we may rewrite this difference quotient as

$$\frac{k(x+h) - k(x)}{h} = \frac{f(g(x+h)) - f(g(x))}{g(x+h) - g(x)} \frac{g(x+h) - g(x)}{h}.$$

Since $g'(x)$ exists, g is continuous at x. Therefore

$$\lim_{h \to 0} g(x+h) = g(x),$$

or

$$\lim_{h \to 0} [g(x+h) - g(x)] = 0.$$

Let us set $r = g(x+h) - g(x)$. As h approaches 0, r approaches 0, and

$$g(x+h) = g(x) + r.$$

Thus

$$k'(x) = \lim_{h \to 0} \frac{f(g(x+h)) - f(g(x))}{g(x+h) - g(x)} \lim_{h \to 0} \frac{g(x+h) - g(x)}{h}$$

$$= \lim_{r \to 0} \frac{f(g(x) + r) - f(g(x))}{r} g'(x)$$

$$= f'(g(x))g'(x).$$

We have thus shown that if $k(x) = f(g(x))$, then $k'(x) = f'(g(x))g'(x)$, a result known as the **chain rule**. We have proved the chain rule under the assumption that $g(x+h) - g(x) \neq 0$, but it is true in general.

The Chain Rule: First Formulation

If $k(x) = f(g(x))$ and $f'(g(x))$ and $g'(x)$ both exist, then

$$k'(x) = f'(g(x))g'(x).$$

EXAMPLE 5 Suppose $k(x) = (2x + 3)^{20}$. We will use the chain rule to find $k'(x)$. From Example 4,

$$k(x) = f(g(x)),$$

where

MORE ABOUT DIFFERENTIATION

$$f(x) = x^{20} \quad \text{and} \quad g(x) = 2x + 3.$$

Since

$$f'(x) = 20x^{19} \quad \text{and} \quad g'(x) = 2,$$

the chain rule tells us that

$$\begin{aligned} k'(x) &= f'(g(x))g'(x) \\ &= f'(2x+3)g'(x) \\ &= 20(2x+3)^{19}(2) \\ &= 40(2x+3)^{19}. \quad \square \end{aligned}$$

EXAMPLE 6 Let $h(x) = \sqrt{2x^3 + 4x}$. In Example 4 we showed that

$$h(x) = f(g(x)),$$

where

$$f(x) = \sqrt{x} \quad \text{and} \quad g(x) = 2x^3 + 4x.$$

Since

$$f'(x) = \frac{1}{2\sqrt{x}} \quad \text{and} \quad g'(x) = 6x^2 + 4,$$

we have

$$\begin{aligned} h'(x) &= f'(g(x))g'(x) \\ &= f'(2x^3 + 4x)g'(x) \\ &= \frac{1}{2\sqrt{2x^3 + 4x}}(6x^2 + 4) \\ &= \frac{3x^2 + 2}{\sqrt{2x^3 + 4x}}. \quad \square \end{aligned}$$

We do not really need to write out $f(x)$, $g(x)$, $f'(x)$, and $g'(x)$ every time we differentiate a composite function. Roughly speaking, what the chain rule says is that to differentiate a composite expression $f(g(x))$ we must differentiate from the "outside in." In other words, we first differentiate the "outside" expression $f(x)$, then multiply the result by the derivative of the "inside" expression $g(x)$. With a little practice this becomes easy to do.

EXAMPLE 7 Let us differentiate $k(x) = (2x + 3)^{20}$ again. This time we will not write out $f(x)$, $g(x)$, $f'(x)$, and $g'(x)$, but instead will differentiate from the outside in as follows:

$$\begin{aligned} k'(x) &= [(2x+3)^{20}]' = 20(2x+3)^{19}(2x+3)' \\ &= 20(2x+3)^{19}(2) \\ &= 40(2x+3)^{19}. \quad \square \end{aligned}$$

EXAMPLE 8 Let us differentiate $h(x) = \sqrt{2x^3 + 4x}$ again. This time we will differentiate from the outside in. We have

$$h'(x) = (\sqrt{2x^3 + 4x})'$$
$$= [(2x^3 + 4x)^{1/2}]'$$
$$= \frac{1}{2}(2x^3 + 4x)^{-1/2}(2x^3 + 4x)'$$
$$= \frac{1}{2\sqrt{2x^3 + 4x}}(6x^2 + 4)$$
$$= \frac{3x^2 + 2}{\sqrt{2x^3 + 4x}}.$$

The same differentiation using differential notation would be written as follows:

$$\frac{dh}{dx} = \frac{d}{dx}(\sqrt{2x^3 + 4x}) = \frac{1}{2\sqrt{2x^3 + 4x}} \frac{d}{dx}(2x^3 + 4x)$$
$$= \frac{1}{2\sqrt{2x^3 + 4x}}(6x^2 + 4)$$
$$= \frac{3x^2 + 2}{\sqrt{2x^3 + 4x}}. \quad \square$$

The chain rule may be used with the other rules of differentiation.

EXAMPLE 9 Let

$$p(x) = (x^2 + 1)^3(x^4 - 1)^5.$$

Then by the product rule we have

$$p'(x) = (x^2 + 1)^3[(x^4 - 1)^5]' + [(x^2 + 1)^3]'(x^4 - 1)^5.$$

Now we apply the chain rule to find

$$[(x^4 - 1)^5]' = 5(x^4 - 1)^4(x^4 - 1)' = 5(x^4 - 1)^4(4x^3)$$
$$= 20x^3(x^4 - 1)^4$$

and

$$[(x^2 + 1)^3]' = 3(x^2 + 1)^2(x^2 + 1)' = 3(x^2 + 1)^2(2x)$$
$$= 6x(x^2 + 1)^2.$$

Therefore

$$p'(x) = (x^2 + 1)^3[20x^3(x^4 - 1)^4] + [6x(x^2 + 1)^2](x^4 - 1)^5$$
$$= 2x(x^2 + 1)^2(x^4 - 1)^4[10x^2(x^2 + 1) + 3(x^4 - 1)]$$
$$= 2x(x^2 + 1)^2(x^4 - 1)^4[13x^4 + 10x^2 - 3]. \quad \square$$

EXAMPLE 10 Let $q(x) = [(x + 1)/(x - 1)]^{2/3}$. To differentiate $q(x)$, we use the chain rule followed by the quotient rule:

$$q'(x) = \left[\left(\frac{x + 1}{x - 1}\right)^{2/3}\right]' = \frac{2}{3}\left(\frac{x + 1}{x - 1}\right)^{-1/3}\left(\frac{x + 1}{x - 1}\right)'$$

$$= \frac{2}{3}\left(\frac{x + 1}{x - 1}\right)^{-1/3}\frac{(x - 1)(x + 1)' - (x + 1)(x - 1)'}{(x - 1)^2}$$

$$= \frac{2}{3}\left(\frac{x + 1}{x - 1}\right)^{-1/3}\frac{(x - 1)(1) - (x + 1)(1)}{(x - 1)^2}$$

$$= -\frac{4}{3(x - 1)^2}\left(\frac{x + 1}{x - 1}\right)^{-1/3} \quad \square$$

The functions to which we have thus far applied the chain rule have all been defined by equations of the form

$$k(x) = [g(x)]^r.$$

But the chain rule tells us that the derivative of such a function is

$$k'(x) = r[g(x)]^{r-1}g'(x).$$

(Why?) This special case of the chain rule is known as the **general power rule.**

The General Power Rule

If

$$k(x) = [g(x)]^r, \quad r \neq 0,$$

then

$$k'(x) = r[g(x)]^{r-1}g'(x).$$

EXAMPLE 11 If

$$k(x) = (x^3 - 2x^2 + 17)^{12},$$

then

$$k'(x) = 12(x^3 - 2x^2 + 17)^{11}(x^3 - 2x^2 + 17)'$$
$$= 12(3x^2 - 4x)(x^3 - 2x^2 + 17)^{11}. \quad \square$$

There is another formulation of the chain rule that is often useful. If $y = f(x)$ defines y as a function of x and $x = g(t)$ defines x as a function of t, then

$$y = f(g(t))$$

defines y as a function of t. Differentiating with respect to t by means of the chain rule, we have

$$\frac{dy}{dt} = \frac{df}{dx}\frac{dg}{dt},$$

which we may write as

$$\frac{dy}{dt} = \frac{dy}{dx}\frac{dx}{dt}.$$

This is the second formulation of the chain rule.

The Chain Rule: Second Formulation

If $y = f(x)$ and $x = g(t)$ and dy/dx and dx/dt both exist, then

$$\frac{dy}{dt} = \frac{dy}{dx}\frac{dx}{dt}.$$

EXAMPLE 12 Suppose $y = x^2 + 1$ and $x = 3t - 2$. Then by the second formulation of the chain rule,

$$\frac{dy}{dt} = \frac{dy}{dx}\frac{dx}{dt} = \frac{d}{dx}(x^2 + 1)\frac{d}{dt}(3t - 2) = 2x \cdot 3 = 6x.$$

If we wish, we may write dy/dt as a function of t by substituting $3t - 2$ for x:

$$\frac{dy}{dt} = 6x = 6(3t - 2) = 18t - 12. \quad \square$$

In the previous example it would have been easy to substitute $3t - 2$ for x in the equation that defines y and then find dy/dt directly:

$$y = x^2 + 1 = (3t - 2)^2 + 1 = 9t^2 - 12t + 5,$$

so

$$\frac{dy}{dt} = \frac{d}{dt}(9t^2 - 12t + 5) = 18t - 12.$$

Nevertheless, as we shall see later in this chapter, we often need to find dy/dt without writing y explicitly as a function of t. We then must use the second formulation of the chain rule.

The second formulation of the chain rule may be used with variables other than x, y, and t.

EXAMPLE 13 Let $u = 3v^2 - 2v + 2$ and $v = 2\sqrt{w}$. Then

$$\frac{du}{dw} = \frac{du}{dv}\frac{dv}{dw}$$

$$= \frac{d}{dv}(3v^2 - 2v + 2)\frac{d}{dw}(2\sqrt{w})$$

$$= \frac{6v - 2}{\sqrt{w}}. \quad \square$$

It is important to realize that the chain rule is a statement about rates of change: it says that if y depends on x and x depends on t, then the rate of change of y with respect to t is the product of the rate of change of y with respect to x and the rate of change of x with respect to t.

EXAMPLE 14 Suppose that the demand function for a product is given by

$$q = 100{,}000 - 200p^2$$

and that the price p depends on the time t according to the equation

$$p = 2\sqrt{t} + 100,$$

where t is time in years with $t = 0$ representing the current year. Then the rate of change of quantity demanded with respect to time is dq/dt, and by the chain rule,

$$\frac{dq}{dt} = \frac{dq}{dp}\frac{dp}{dt} = (-400p)\left(\frac{1}{\sqrt{t}}\right) = -\frac{400p}{\sqrt{t}}.$$

Four years from now the price will be

$$p = 2\sqrt{4} + 100 = 104$$

dollars, and thus

$$\left.\frac{dq}{dt}\right|_{t=4} = -\frac{400(104)}{\sqrt{4}} = -20{,}800.$$

Hence four years from now quantity demanded will be decreasing at a rate of 20,800 units per year. $\square$

■ *Computer Manual: Programs COMPOSIT, POWRPOLY*

11.2 EXERCISES

Composite Functions

In Exercises 1 through 10, find $f(g(x))$ and $g(f(x))$.

1. $f(x) = 4x + 3$, $g(x) = 5x - 2$
2. $f(x) = 2x + 1$, $g(x) = \frac{1}{2}x - \frac{1}{2}$
3. $f(x) = x^2 + 2x$, $g(x) = 3x + 1$
4. $f(x) = 1 - x$, $g(x) = x^2 - 3x + 4$

5. $f(x) = 2x^2$, $g(x) = 2x^3 + 1$

6. $f(x) = x^2$, $g(x) = \sqrt{x}$

7. $f(x) = \sqrt{5x + 4}$, $g(x) = x + 2$

8. $f(x) = \dfrac{1}{x}$, $g(x) = 7x + 3$

9. $f(x) = \dfrac{2}{x^2}$, $g(x) = \dfrac{3}{x^2}$

10. $f(x) = \dfrac{x}{x + 1}$, $g(x) = \dfrac{x + 1}{2x}$

11. If $y = 5x + 3$ and $x = t^2 + 4$, express y as a function of t.
12. If $y = 4x - 16$ and $x = t + 4$, express y as a function of t.
13. If $y = 3u^3 - 1$ and $u = 1 - x$, express y as a function of x.
14. If $w = 8t^2 + 3t + 1$ and $t = \sqrt{x + 1}$, express w as a function of x.
15. If $u = 17v^2 + 2\sqrt{v} + 3$ and $v = w^{3/4}$, express u as a function of w.
16. If $y = 4x - 16$ and $x = t + 4$, express t as a function of y.

In Exercises 17 through 28, write k as a composite function.

17. $k(x) = (3x + 5)^{10}$
18. $k(x) = (2 - 3x)^{12}$
19. $k(x) = (5 - 7x)^{-4}$
20. $k(x) = (x^2 + 4)^{11}$
21. $k(x) = (x^2 - 2x)^5$
22. $k(x) = (2x^2 + 3x - 1)^{-5/2}$
23. $k(x) = \sqrt{4x - 1}$
24. $k(x) = \sqrt{1 - 3x}$
25. $k(x) = (x^2 + 7)^{3/2}$
26. $k(x) = \sqrt[3]{x^2 + 1}$
27. $k(x) = \sqrt[5]{x^2 + 3}$
28. $k(x) = (1 - 2x^{-1})^{-3/2}$

29. Suppose the cats eat the rats and the rats eat the bats, and that if c is the number of cats, r the number of rats, and b the number of bats, then $c = 0.1r^{2/3}$ and $r = \sqrt{b}$. Find an expression for the number of cats as a function of the number of bats, and use it to determine how many cats 1000 bats can support.

Management

30. A firm's daily profit is P dollars, where $P(x) = 1.2x - 10{,}000$ and production x is a function of the number of employees n according to the equation $x = 125n$. Express the firm's daily profit as a function of n and find the daily profit when $n = 100$.

31. Suppose the demand function for a product is given by $q = 10{,}000/p$ and that the price has been increasing according to the equation $p = 0.02t + 5$, where t is in months, with $t = 0$ representing the current month. Find an expression for quantity demanded as a function of time, and use it to forecast quantity demanded six months from now.

32. Repeat Exercise 31 if $q = 500p^2 + 3000p + 10{,}000$ and $p = 60(t + 10)^{-1}$.

33. For the product of Exercise 31, find an expression for revenue as a function of time.

34. For the product of Exercise 32, find an expression for revenue as a function of time.

Life Science

35. It is estimated that the percentage of ozone remaining in the ozone layer when a total of x tons of chlorofluorocarbons (CFCs) have been released into the atmosphere is

MORE ABOUT DIFFERENTIATION

$$f(x) = \frac{100}{0.16x + 1}.$$

It is also estimated that the total amount of CFCs released into the atmosphere is $x = 12t + 22$, where t is in years with $t = 0$ representing 1980. Find an expression for the percentage of ozone remaining in the ozone layer as a function of time.

36. The carbon monoxide concentration in the air of a large city depends on the number of cars driven in the city each day. The number of cars driven in the city is expected to increase with time. Suppose that if x cars are driven in the city each day, then the concentration of carbon monoxide in the atmosphere is

$$c(x) = 10^{-9}x^2 + 10^{-5}x + 1$$

parts per million (ppm); suppose also that the number of cars driven in the city per day is forecast to be

$$x = 10,000t + 100,000$$

in year t, where $t = 0$ represents 1987. Find a function that describes the carbon monoxide concentration as a function of time, and forecast this concentration in the year 1995.

The Chain Rule

In Exercises 37 through 67, use the chain rule to find $k'(x)$.

37. $k(x) = (3x + 5)^{10}$
38. $k(x) = (2 - 3x)^{12}$
39. $k(x) = (5 - 7x)^{-4}$
40. $k(x) = (x^2 + 4)^{11}$
41. $k(x) = (x^2 - 2x)^5$
42. $k(x) = (2x^2 + 3x - 1)^{-5/2}$
43. $k(x) = (5x^2 - 3x + 1)^{100}$
44. $k(x) = \sqrt{4x - 1}$
45. $k(x) = \sqrt{1 - 3x}$
46. $k(x) = (x^2 + 7)^{3/2}$
47. $k(x) = \sqrt[3]{x^2 + 1}$
48. $k(x) = \sqrt[5]{x^2 + 3}$
49. $k(x) = (1 - 2x^{-1})^{-3/2}$
50. $k(x) = \sqrt[3]{x^5 - 3x^2}$
51. $k(x) = x^3(5x^2 - 1)^2$
52. $k(x) = (2x + 1)^4(3x - 2)^5$
53. $k(x) = (2x^2 + 1)^{11}(x^3 - 1)^9$
54. $k(x) = 2x^2\sqrt{2x + 3}$
55. $k(x) = x^3\sqrt[3]{4x + 3}$
56. $k(x) = 2x\sqrt[3]{x^2 + 8}$
57. $k(x) = \sqrt{x^2 - 1}\sqrt[3]{x^3 - 1}$
58. $k(x) = [(3 - 2x - x^4)(x^2 - 4)]^7$
59. $k(x) = [(4\sqrt{x} - 3x^{-2})(4x^{4/3} + 10)]^{-3}$
60. $k(x) = \sqrt{(5x + 1)(x^2 + 8)}$
61. $k(x) = \sqrt{(x^2 + 1)(x^2 - 3)}$
62. $k(x) = \left(\dfrac{2x}{x - 3}\right)^5$
63. $k(x) = \left(\dfrac{x + 2}{x - 2}\right)^4$
64. $k(x) = \left(\dfrac{5x - 1}{x + 2}\right)^{-5}$
65. $k(x) = \dfrac{3x + 1}{\sqrt{x^2 + 3}}$
66. $k(x) = \sqrt{\dfrac{x^2 + 2}{x^2 - 1}}$
67. $k(x) = \left(\dfrac{x^{-1} + \sqrt{x}}{x^{-1} - \sqrt{x}}\right)^3$

68. Use the second formulation of the chain rule to find $\dfrac{dy}{dt}$ if $y = x^2$ and $x = 4 - 2t$. Check your answer by expressing y as a function of t and finding $\dfrac{dy}{dt}$ directly.

69. Repeat Exercise 68 if $y = 2x^2 - 3x + 1$ and $x = 5t + 1$.

70. Use the second formulation of the chain rule to find $\dfrac{du}{dv}$ if $u = \sqrt{x+1}$ and $x = v^2 - 1$. Check your answer by expressing u as a function of v and finding $\dfrac{du}{dv}$ directly.

71. Use the second formulation of the chain rule to find $\dfrac{dy}{dt}$ if

   (a) $y = \sqrt{x^2 + 1}$ and $x = t^2$
   (b) $y = x^2$ and $x = t^2 + 1$

72. Use the second formulation of the chain rule to find $\dfrac{dy}{dt}$ if

   (a) $y = \dfrac{1}{x}$ and $x = 4t - 5$
   (b) $y = (x^3 + 3x - 1)^5$, $x = u^2$ and $u = 5t$

73. Use the second formulation of the chain rule to find $\dfrac{dz}{dw}$ if

   (a) $z = 3\sqrt{v}$, $v = \dfrac{1}{q}$, and $q = w^2 + 1$
   (b) $z = \dfrac{p-1}{2-p}$, $p = 5s + 1$, and $s = w^{-3}$

74. Find $\dfrac{dy}{dt}$ if $y = 3x^2 + 5x - 7$ and $x = 2t + 11$.

75. Find $\dfrac{dx}{ds}$ if $x = \dfrac{t}{8} + 4$ and $t = 4s^2 - 12$.

76. Find $\dfrac{dp}{dr}$ if $p = \sqrt{5q^2 + 2}$ and $q = \dfrac{8}{r}$.

77. Find $\dfrac{dy}{dv}$ if $y = 2(x-1)^{-1}$, $x = 3t + 5$, and $t = v^2 + 8$.

Management

78. A firm's revenue function is given by
$$R(x) = x\left(1 - \dfrac{x^2}{100}\right)^2$$
and its cost function by
$$C(x) = \dfrac{x^2}{x+1} + 1,$$
both valid for $x \geq 1$. Find the firm's marginal revenue and marginal profit when $x = 1$, $x = 10$, and $x = 20$, and interpret your results.

79. Suppose a firm's cost and revenue functions are given by
$$C(x) = \sqrt{x + 100} \quad \text{and} \quad R(x) = \sqrt{x^2 + 2},$$
respectively, for $x \geq 1$. Find the firm's marginal cost, marginal revenue, and marginal profit when $x = 10$, and interpret your results.

80. A firm's cost and revenue functions are given by
$$C(x) = (x + 0.25)^{1/2} \quad \text{and} \quad R(x) = (x^2 + 32)^{1/2},$$

respectively, where x is in thousands of units and cost and revenue are in millions of dollars. Find the firm's maximum profit and the number of units required to attain it.

81. For the firm of Exercise 30, find an expression for the rate of change of profit with respect to the number of employees.

82. The supply function for a commodity is given by $q = \sqrt{100p - 1000}$, $p \geq 10$. Find the rate of change of quantity supplied with respect to price when $p = \$20$, $\$30$, and $\$50$.

83. For the product of Exercise 31, find the rate of change of quantity demanded with respect to time six months from now.

84. Find the marginal revenue function for the revenue function of Exercise 33.

Life Science

85. A rare species of seal was placed on the endangered species list in 1984. It is estimated that the size of the population of this species in year t is given by

$$y = \sqrt{\frac{8t + 2}{2t + 9}},$$

where $t = 0$ represents 1984 and y is in thousands. Find the rate at which the population was changing in 1986 and 1990. Also find and interpret $\lim\limits_{t \to +\infty} y$.

86. The number of bacteria in a culture t hours after the start of an experiment is y thousand, where

$$y = \left(\frac{t + 2}{5t + 3}\right)^{3/4}.$$

Find the rate of change of the bacteria population when $t = 0$ and when $t = 1$.

87. The number of bacteria in a culture t hours after the start of an experiment is y, where $y = 1000\sqrt{300t^2 + 10{,}000}$. Find the rate of change of the bacteria population when $t = 1$, 2, and 10.

88. It is estimated that t years from now $y\%$ of the population of a country will be afflicted with the disease bilharzia, where $y = 100(0.3t + 4)^{-1/2}$. Find both the incidence of the disease and the rate at which it will be declining **(a)** now **(b)** 40 years from now.

89. Refer to Exercise 35: Find the rate at which the ozone concentration was changing with respect to time in 1988.

90. Refer to Exercise 36: Forecast the rate of change of the carbon monoxide concentration in 1995.

Social Science

91. Suppose the proportion of eligible voters who participated in presidential elections in the United States since World War II is y, where

$$y = \frac{1}{\sqrt{0.5t + 2}}$$

and $t = 0$ represents 1948, $t = 1$ represents 1952, and so on. Find the rate at which voter participation was changing in the election of 1980 and in the election of 1984.

Marginal Revenue Product

If $p = f(q)$ is the demand function for a firm's product, and output q is a function of the number of employees n, then revenue R is a function of n. The derivative of R with respect to n is called the **marginal revenue product**. Exercises 92 through 94 refer to the marginal revenue product.

*92. Show that the marginal revenue product $\dfrac{dR}{dn}$ is given by

$$\frac{dR}{dn} = \left(p + q\frac{dp}{dq}\right)\frac{dq}{dn}.$$

*93. Find and interpret the marginal revenue product when $p = 100 - 0.1q$, $q = 10n - 0.02n^2$, and
(a) $n = 10$ (b) $n = 90$.

*94. If p and q are as in Exercise 93, set $\dfrac{dR}{dn} = 0$ and solve for n, and then examine the sign of the marginal revenue product on either side of your solution. What does this tell you about the relationship between the firm's number of employees and its revenue?

11.3 DIFFERENTIALS

In Section 10.7 we discussed the use of differential notation dy/dx for the derivative and stated that for the time being we would regard this expression not as a ratio but merely as another symbol for the derivative. Now it is time to demonstrate how the symbol dy/dx can indeed be regarded as the ratio of the **differentials** dy and dx. This view of the derivative as a ratio will be extremely useful to us in later chapters when we discuss the process of integration. We also show in this section how differentials can be used to approximate the change in a function's value brought about by a small change in its independent variable.

The Differentials dy and dx

Let f be a function and let $(x, f(x))$ be a point on its graph. Let Δx be a number added to x. (The symbol Δ is the Greek capital delta, and Δx is read "delta x.") The number Δx is called an **increment** and is referred to as the **change in x**; Δx may be positive or negative. Figure 11–1 depicts the situation if Δx is positive.

Now suppose we set $dx = \Delta x$ and dy equal to the vertical distance from the point $(x + \Delta x, f(x))$ to the tangent line to f at $(x, f(x))$. The number dx is the **differential of x**, and dy is the **differential of y**. See Figure 11–2. Since dy/dx is the slope of the tangent line, and since this slope is the derivative $f'(x)$, we have

$$\frac{dy}{dx} = f'(x).$$

This shows how the derivative may be regarded as a ratio of the differentials dy and dx.

MORE ABOUT DIFFERENTIATION

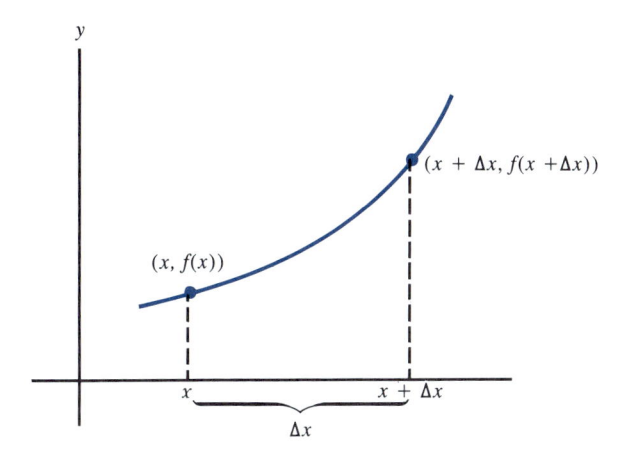

FIGURE 11–1

Now that we can regard dy and dx as individual quantities, we can solve the equation

$$\frac{dy}{dx} = f'(x)$$

for dy to obtain the differential formula:

The Differential Formula

If f is differentiable at x, then

$$dy = f'(x)\, dx.$$

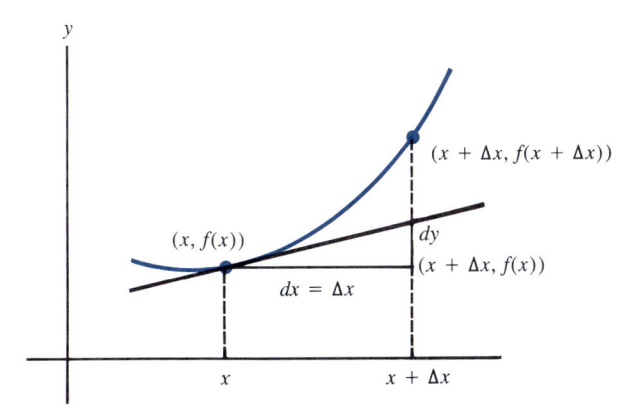

FIGURE 11–2

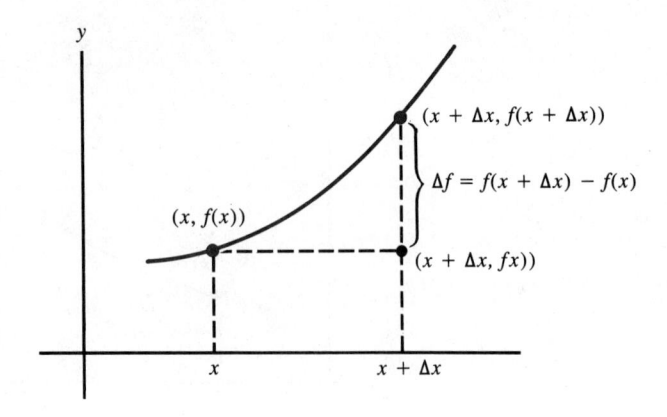

FIGURE 11–3

EXAMPLE 1 If $f(x) = x^2 - 5x + 3$, then $dy = (2x - 5)\,dx$. ☐

Approximation by Differentials

As before, we let Δx be an increment added to x. The point $(x + \Delta x, f(x + \Delta x))$ is on the graph of f, and the quantity

$$\Delta f = f(x + \Delta x) - f(x)$$

is the **change in** f brought about by the change Δx in x. See Figure 11–3.

EXAMPLE 2 Let $f(x) = x^2$, $x = 1$, and take $\Delta x = 0.1$. Then

$$x + \Delta x = 1.1,$$

and thus

$$\Delta f = f(1.1) - f(1) = (1.1)^2 - 1^2 = 0.21.$$

See Figure 11–4. ☐

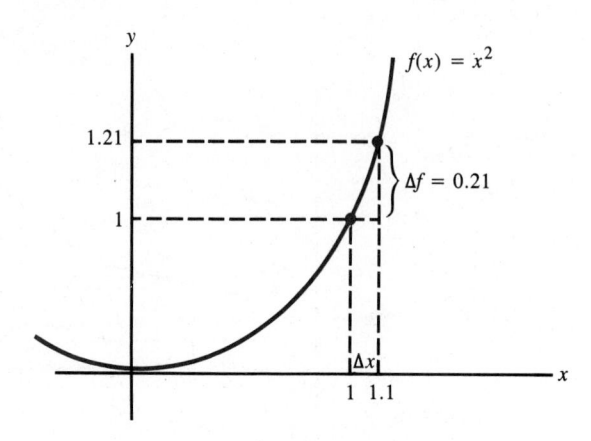

FIGURE 11–4

More about Differentiation

What happens if the increment Δx is allowed to approach zero? Figure 11-5 shows that as Δx approaches 0, the point
$$(x + \Delta x, f(x + \Delta x)) = (x + \Delta x, f(x) + \Delta f)$$
"slides" down the graph toward the point (x, y) and hence Δf approaches dy. Therefore for values of Δx that are close to (but not equal to) zero, we have
$$\Delta f \approx dy.$$

Approximation by Differentials

If f is differentiable at x and Δx is close to but not equal to zero, then
$$\Delta f \approx f'(x)\,\Delta x.$$

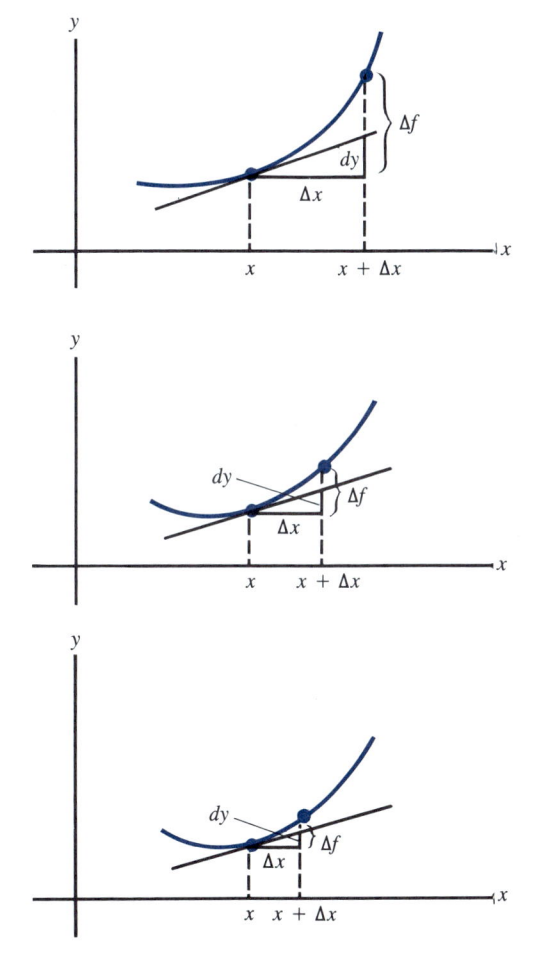

FIGURE 11-5
As $\Delta x \to 0$, $\Delta f \to dy$

Approximation by differentials is often used to produce formulas for the approximate amount of change in the dependent variable as a function of the amount of change in the independent variable. We illustrate with examples.

EXAMPLE 3 If $f(x) = x^2$, then

$$\Delta f \approx 2x \, \Delta x$$

when Δx is close to zero. Thus, for instance, if x is changed from $x = 1$ to $x = 1.1$, so that $\Delta x = 0.1$, then the change Δf in f will be

$$\Delta f \approx 2(1)(0.1) = 0.2.$$

Compare this result with that of Example 2, where we found the *exact* change in f as x went from 1 to 1.1 to be 0.21. ☐

EXAMPLE 4 Suppose the supply function for a commodity is $p = 3q^{2/3}$. The change in price Δp brought about by a small change Δq in quantity supplied is given by

$$\Delta p \approx 2q^{-1/3} \, \Delta q$$

or

$$\Delta p \approx \frac{2}{\sqrt[3]{q}} \Delta q.$$

Thus, for instance, if quantity supplied q changes from 1000 units to 990 units, Δq will be -10, and the corresponding change in price will be

$$\Delta p \approx \frac{2}{\sqrt[3]{1000}}(-10) = -\$2. \quad \square$$

EXAMPLE 5 Suppose the cost of making x units of a product is $C(x)$ dollars. If production is increased from x units to $x + 1$ units, then $\Delta x = 1$ and hence the change ΔC in cost is given by

$$\Delta C \approx C'(x) \, \Delta x = C'(x) \cdot 1 = C'(x).$$

In other words, the additional cost of making one more unit is approximately equal to the marginal cost, a fact we stated without proof in Section 10.6. ☐

As Figure 11–3 shows, when x changes by an amount Δx, the functional value $f(x)$ changes by an amount Δf. The **proportional change in $f(x)$** is thus defined to be the ratio

$$\frac{\Delta f}{f(x)}.$$

Since $\Delta f \approx f'(x) \, \Delta x$ when Δx is close to zero, we have

MORE ABOUT DIFFERENTIATION

> **The Proportional Change in $f(x)$**
>
> If f is differentiable at x and Δx is close to but not equal to zero, then
>
> $$\frac{\Delta f}{f(x)} \approx \frac{f'(x)\,\Delta x}{f(x)}.$$

If we multiply the proportional change in $f(x)$ by 100, we obtain the percentage change in $f(x)$.

EXAMPLE 6 If one side of a square plot of land is measured with a possible measurement error of Δx units, estimate the proportional and percentage errors that might occur in the value for the area of the plot.

If the side of the plot is x units, its area is $A(x) = x^2$ square units. Hence, the proportional change in area that could be introduced by a change of Δx in the length of a side is

$$\frac{\Delta A}{A(x)} \approx \frac{2x\,\Delta x}{x^2},$$

or

$$\frac{\Delta A}{A(x)} \approx \frac{2\,\Delta x}{x}.$$

The percentage error is 100 times the proportional error. For instance, if the side were measured to be 10 meters long, with a possible measurement error of ± 0.1 meter, then the proportional error in the area would be

$$\frac{\Delta A}{A(10)} \approx \frac{2(\pm 0.1)}{10} = \pm 0.02.$$

The percentage error would thus be approximately $\pm 2\%$. □

11.3 EXERCISES

The Differentials dy and dx

In Exercises 1 through 8, find dy.

1. $f(x) = 3x^2 - 2x + 5$
2. $y = -5x^3 + 4x^2 - 2x + 11$
3. $y = (x^3 - 5x + 2)^8$
4. $f(x) = \sqrt{x^2 - 2x + 5}$
5. $y = (x^2 + 2x)(x^3 + 4)$
6. $y = (3x^2 - 2x + 1)(x^2 + 4)^4$
7. $f(x) = \dfrac{x^2 - 1}{x^3 + 2x}$
8. $y = \sqrt{\dfrac{3x + 2}{5x - 1}}$

Approximation by Differentials

In Exercises 9 through 16, find an expression for the approximate change in f brought about by a change of Δx in x. Then use this expression to find the approximate change in f brought about by the stated change in x.

9. $f(x) = x^3$, x changes from 1 to 1.2
10. $f(x) = x^3$, x changes from 1 to 1.02
11. $f(x) = \dfrac{1}{x}$, x changes from 10 to 10.5
12. $f(x) = \dfrac{1}{x}$, x changes from 100 to 100.5
13. $f(x) = \sqrt{x}$, x changes from 4 to 3.9
14. $f(x) = \sqrt{x}$, x changes from 100 to 99
15. $f(x) = \sqrt[3]{x}$, x changes from 64 to 63.7
16. $f(x) = x^{2/3}$, x changes from 27 to 27.1

*17. Show that when t is near 0, $\sqrt{1+t} \approx 1 + \dfrac{1}{2}t$. (Hint: Let $f(x) = \sqrt{x}$, $x = 1$, $\Delta x = t$.)

*18. Show that when t is near 0, $(1 + t)^r \approx 1 + rt$.

In Exercises 19 through 24, find an expression for the approximate proportional change in f brought about by a change of Δx in x. Then use this expression to find the approximate proportional and percentage changes in f brought about by the stated change in x.

19. $f(x) = x$, x changes from 2 to 2.1
20. $f(x) = x^2$, x changes from 10 to 9.5
21. $f(x) = x^3$, x changes from 4.2 to 4.1
22. $f(x) = x^4$, x changes from 5 to 5.02
23. $f(x) = \sqrt{x}$, x changes from 4 to 4.1
24. $f(x) = \dfrac{1}{x}$, x changes from 100 to 98.5

25. The side of a square is measured with a possible measurement error of Δx units.
    (a) Find an expression for the approximate proportional error in the perimeter of the square due to measurement error.
    (b) Use the expression of part (a) to find the approximate proportional and percentage errors in the perimeter of the square due to measurement error if its side measures 10 cm with a possible error of ± 0.1 cm.

26. The radius of a circle is measured with a possible measurement error of Δx units.
    (a) Find an expression for the approximate proportional error in the circumference of the circle due to measurement error.
    (b) Use the expression of part (a) to find the approximate proportional and percentage errors in the circumference of the circle due to measurement error if its radius measures 10 cm with a possible error of ± 0.1 cm.

27. Repeat Exercise 26, substituting the word "area" for "circumference."

28. The edge of a cube is measured to be 4 feet with a possible measurement error of ± 0.5 inch. Find the approximate proportional and percentage errors in
    (a) the surface area   (b) the volume
    of the cube due to measurement error.

29. The radius of a sphere is measured to be 12 meters with a possible measurement error of ± 2 millimeters. Find the approximate proportional and percentage errors in
    (a) the surface area $(4\pi r^2)$ (b) the volume $\left(\dfrac{4\pi r^3}{3}\right)$
    of the sphere due to measurement error.

30. Find the approximate proportional and percentage changes in the area of a square if the length of each of its sides is increased by 1%.

More about Differentiation

31. Find the approximate proportional and percentage changes in the volume of a cube if the length of each of its sides is increased by 1%.

32. Find the approximate proportional and percentage changes in the area of a circle if the length of its radius is decreased by 2%.

33. Find the approximate proportional and percentage changes in the volume of a sphere if the length of its radius is decreased by 1%.

*34. How accurately must the side of a square be measured in order that
  (a) its area   (b) its perimeter
  be determined to within $\pm 5\%$ of the true value? (*Hint:* For (a) you want $-0.05\, A(x) \leq \Delta A(x) \leq 0.05\, A(x)$.)

*35. How accurately must the radius of a circle be measured in order that
  (a) its area   (b) its circumference
  be determined to within $\pm 2\%$ of the true value?

*36. How accurately must the edge of a cube be measured in order that
  (a) its volume   (b) its surface area
  be determined to within $\pm 1\%$ of the true value?

*37. If the earth is a sphere, how accurately must its radius be measured if
  (a) its surface area   (b) its volume
  is to be determined to within $\pm 0.1\%$?

Management

38. The supply function for a product is given by $p = 0.02\sqrt{q}$.
  (a) Find an expression for the approximate change in price brought about by a small change in quantity supplied.
  (b) Use the expression of part (a) to find the approximate change in price when quantity supplied changes from 400 units to 405 units.

39. The demand function for a product is given by $p = 5000q^{-1/2}$.
  (a) Find an expression for the approximate change in price brought about by a small change in quantity demanded.
  (b) Use the expression of part (a) to find the approximate change in price when quantity demanded changes from 3600 units to 3580 units.

40. The supply function for a commodity is given by $q = 200(p - 1)^{4/3}$.
  (a) Find an expression for the approximate change in quantity supplied brought about by a small change in price.
  (b) Use the expression of part (a) to find the approximate change in quantity supplied when the price changes from \$9 per unit to \$9.50 per unit.

41. A firm's weekly output of goods is worth $y = 20x^{1/5}$ thousand dollars, where x is the amount (in thousands of dollars) that the firm spends on labor during the week. If the firm is currently spending \$32,000 per week on labor, find the approximate change in the weekly output if new hires increase labor costs to \$35,000 per week.

42. With an input of x thousand dollars of capital, a firm can produce y units of product, where $y = 200(x + 10)^{4/5}$. If the firm's capital input is currently \$22,000 per week, find the approximate change in output if capital input is decreased to \$20,000 per week.

43. A firm's profit function is given by $P(x) = -x^2 + 80x - 200$.
   (a) Find an expression for the approximate proportional change in profit brought about by a small change in the number of units made and sold.
   (b) Use the expression of part (a) to find the approximate proportional and percentage change in profit if the number of units made and sold changes from 20 to 21.

44. A company's cost function is given by $C(x) = 0.5x^2 + 5x + 2000$.
   (a) Find an expression for the approximate proportional change in the company's cost brought about by a change in the number of units made.
   (b) Use the expression of part (a) to find the approximate proportional and percentage change in cost if the number of units sold changes from 25 to 27.

45. A firm's revenue function is given by $R(x) = x^2 + 4x$. Find the approximate proportional and percentage change in revenue if sales fall from 100 units to 96 units.

Life Science

46. The number of bacteria in a culture t hours after an experiment has begun is $f(t) = 200t^{3/2} + 4000$.
   (a) Find an expression for the approximate change in the number of bacteria during a small period of time.
   (b) Use the expression of part (a) to find the approximate change in the number of bacteria during the period from $t = 4$ hours to $t = 4$ hours and 5 minutes.

47. The number of fish remaining in a lake t months after it was stocked is $y = 20(4000 - 120t)^{1/2}$.
   (a) Find an expression for the approximate change in the number of fish in the lake during a small period of time.
   (b) Use the expression of part (a) to find the approximate change in the number of fish during the period from $t = 30$ to $t = 32$.

48. The concentration of pollutants in a reservoir as of year t is given by
$$y = 10 + 2t - 0.5\sqrt{t + 3}.$$
Here y is in parts per million and $t = 0$ corresponds to January 1, 1989. Find the approximate change in the concentration of pollutants during the first three months of 1990.

49. A patient is injected with a drug, and t minutes later the concentration of the drug in the patient's bloodstream is y milligrams per cubic centimeter, where $y = 10(1 + 2t)^{-1/2}$. Find the approximate change in the concentration of the drug from time $t = 0$ to time $t = 1$.

50. An artery 100 millimeters (mm) long has a circular cross section. As plaques build up in the artery, its radius decreases. Suppose that the radius decreases from its original value of 2.0 mm to 1.75 mm. Find the proportional and percentage change in
   (a) the cross-sectional area of the artery;
   (b) the volume of the artery. (*Hint:* The volume of a circular cylinder of radius r and height h is $\pi r^2 h$.)

Social Science

51. A rumor is started and t days later, $t \geq 0$, $y = 200\sqrt{t} + 1$ people have heard it.
   (a) Find an expression for the approximate change in the number of people who hear the rumor during a short period of time.

(b) Use the expression of part (a) to find the approximate number of people who hear the rumor during the tenth day after it was started.

52. If a person learns a foreign language but does not use it, it is forgotten. Suppose that t years after learning a language, a person who has never used it recalls $f(t)\%$ of it, where $f(t) = 100(t + 1)^{-3/4}$. Find the approximate change in the percentage of the language retained in the first year after it is learned.

Sustainable Growth Rates

Exercises 53 through 61 were suggested by material in Higgins, *Analysis for Financial Management*, Irwin, 1984. They are concerned with the sustainable growth rate of a corporation: if S is the annual sales of a corporation and ΔS is its change in annual sales, then its **sustainable growth rate** is the relative change in sales $\Delta S/S$. Sales growth requires new assets, which a corporation can finance by increasing its debt (i.e., by borrowing), by retaining its profits, or by selling new stock. Suppose we let

P = earnings/sales (= profit margin).
D = dividends paid/earnings (= dividend payout ratio).
T = asset value/sales (= asset-to-sales ratio).
L = debt/equity (= debt-to-equity ratio).

(Equity is the total ownership value of the corporation; it is equal to the corporation's assets minus its liabilities and can be increased by selling stock or by retaining profits.) In what follows we will assume that the corporation wants its debt-to-equity ratio to remain constant and that it cannot or will not sell more stock.

*53. Show that over the next year the corporation's retained profits will be equal to $P(S + \Delta S)(1 - D)$. (*Hint:* Over the next year sales will be $S + \Delta S$ and retained profits = earnings minus dividends paid.)

Each dollar of retained profit increases equity by \$1, and since L is to remain constant, this means that the firm's debt must also increase by \$1. Thus over the next year,

$$\text{New debt} = P(S + \Delta S)(1 - D)L.$$

The retained profits and new debt will be used to finance new assets. Hence the change in the value of the corporation's assets over the next year will be

$$\Delta T = \text{Retained profits plus new debt}.$$

*54. Show that $\Delta T = P(S + \Delta S)(1 - D)(1 + L)$.

But a change in sales ΔS requires a proportional change in assets ΔT. Therefore the asset-to-sales ratio will remain constant, so $\Delta T/\Delta S = T$, or $(\Delta S)T = \Delta T$.

*55. Show that

$$\frac{\Delta S}{S} = \frac{P(1 - D)(1 + L)}{T - P(1 - D)(1 + L)}.$$

(*Hint:* Write out $(\Delta S)T = \Delta T$ and solve for ΔS.)

The equation of Exercise 55 is called the **sustainable growth equation.** It gives the relative rate of change in sales that will occur for given values of P, D, T, and L.

*56. Suppose $P = 0.1$, $D = 0.4$, $T = 0.8$, and $L = 0.5$. Find the sustainable growth rate $\Delta S/S$.

*57. Find the sustainable growth rate of a corporation if its sales this year were $50 million, its earnings were $10 million, its dividend payout was $500,000, its debt was $30 million, its equity was $50 million, and its assets were worth $80 million.

*58. How can a firm increase its sustainable growth rate? (Remember: L is to remain fixed.)

*59. Many new firms pay no dividends, find they cannot use their assets more efficiently, and refuse to dilute their ownership by issuing new stock. How must such firms finance their growth? Are there dangers to such firms in trying to grow too fast?

If sales S are a function of time t, so that $S = S(t)$, then

$$\frac{\Delta S}{S} \approx \frac{S'(t) \Delta t}{S(t)}.$$

(Why?) Hence we can estimate the sustainable growth rate of a corporation from its sales function.

*60. Suppose a corporation's sales function is given by $S = 20t + 200$, where sales S is in millions of dollars and t is time in years, with $t = 0$ representing the present. Estimate the firm's sustainable growth rate over the next year.

*61. Suppose a corporation's sales function is given by $S = 0.02t^2 + 1.5t + 10$, where S is sales in millions of dollars and t is time in years with $t = 0$ representing the present. Estimate the firm's sustainable growth rate over the next six months.

11.4 IMPLICIT DIFFERENTIATION

A defining equation of the form $y = f(x)$ is said to define y **explicitly** as a function of x. For instance, the equation $y = x^2 + 3x - 5$ defines y explicitly as a function of x. Sometimes, however, an equation involving x and y will not be solved for y in terms of x; examples are

$$4x + 2y = 12 \quad \text{and} \quad x^2 + y^2 = 1.$$

Such equations are said to define y **implicitly** as a function of x. It is sometimes necessary to be able to find the derivative of y with respect to x when y is defined implicitly as a function of x. In this section we show how to use **implicit differentiation** to accomplish this.

Equations that define y implicitly as a function of x can sometimes be solved to yield one or more explicitly defined functions. For example, the equation

$$4x + 2y = 12$$

can be solved for y to obtain the explicitly defined function

$$y = -2x + 6.$$

More about Differentiation

Similarly, the equation
$$x^2 + y^2 = 1$$
can be solved for y to yield the two explicitly defined functions
$$y = +\sqrt{1 - x^2} \quad \text{and} \quad y = -\sqrt{1 - x^2}.$$
However, not all implicitly defined functions can be converted to explicit ones, and even when such conversion is possible, it is often far too tedious to carry out. Therefore we determine the derivatives of implicitly defined functions by the method of **implicit differentiation,** which requires that we apply the rules of differentiation to all terms of the defining equation while keeping in mind the fact that y is actually a function of x. We illustrate with examples.

EXAMPLE 1

Let us find the derivative dy/dx if y is defined implicitly as a function of x by the equation
$$xy + y^2 = 1.$$

To remind ourselves that y is indeed a function of x, we will replace y by $f(x)$ in the equation:
$$xf(x) + [f(x)]^2 = 1.$$
Now we differentiate each term of this equation, using the product rule to differentiate the term $xf(x)$ and the chain rule to differentiate the term $[f(x)]^2$. We have
$$\frac{d}{dx}(xf(x)) + \frac{d}{dx}([f(x)]^2) = \frac{d}{dx}(1),$$
or
$$x\frac{d}{dx}(f(x)) + \frac{d}{dx}(x) \cdot f(x) + 2[f(x)]\frac{d}{dx}(f(x)) = 0,$$
or
$$x\frac{d}{dx}(f(x)) + 1 \cdot f(x) + 2f(x)\frac{d}{dx}(f(x)) = 0,$$
or
$$x\frac{dy}{dx} + y + 2y\frac{dy}{dx} = 0,$$
or
$$(x + 2y)\frac{dy}{dx} = -y.$$

Solving the last equation for dy/dx, we find that
$$\frac{dy}{dx} = -\frac{y}{x + 2y}. \quad \square$$

EXAMPLE 2 Let us find dy/dx if
$$\frac{2x^2}{y} = x + y^3.$$

This time we will not replace y by $f(x)$, but will simply remember as we differentiate that y is defined implicitly as a function of x. It will also be convenient to use prime notation in this example. We have
$$\left(\frac{2x^2}{y}\right)' = (x + y^3)'.$$

Now we differentiate the left side of the equation by the quotient rule and the right side by the sum rule to obtain
$$\frac{y(2x^2)' - 2x^2 y'}{y^2} = (x)' + (y^3)',$$

or
$$\frac{4xy - 2x^2 y'}{y^2} = 1 + 3y^2 y',$$

or
$$4xy - 2x^2 y' = y^2 + 3y^4 y',$$

so
$$(3y^4 + 2x^2)y' = 4xy - y^2,$$

and thus
$$\frac{dy}{dx} = y' = \frac{4xy - y^2}{3y^4 + 2x^2}. \quad \square$$

EXAMPLE 3 We find the equation of the line tangent to the circle
$$x^2 + y^2 = 5$$

at the point $(1, 2)$. We have
$$\frac{d}{dx}(x^2) + \frac{d}{dx}(y^2) = \frac{d}{dx}(5),$$

or
$$2x + 2y \frac{dy}{dx} = 0.$$

Hence
$$\frac{dy}{dx} = -\frac{x}{y},$$

so at (1, 2) we have

$$\frac{dy}{dx} = -\frac{1}{2}.$$

Thus the slope of the line tangent to the circle at (1, 2) is $-\frac{1}{2}$, and the equation of the tangent line is therefore

$$y = -\frac{1}{2}x + b.$$

Since (1, 2) is on the line, $b = \frac{5}{2}$, and the equation is

$$y = -\frac{1}{2}x + \frac{5}{2}. \quad \square$$

If y is defined implicitly as a function of x, and x is a function of a third variable t, then y is a function of t, and we can use implicit differentiation and the chain rule to find dy/dt.

EXAMPLE 4 Suppose

$$xy + y^2 + x = 1$$

and that x is a function of t. Let us find dy/dt.

Differentiating with respect to t, we have

$$\frac{d}{dt}(xy) + \frac{d}{dt}(y^2) + \frac{d}{dt}(x) = \frac{d}{dt}(1).$$

or

$$x\frac{dy}{dt} + y\frac{dx}{dt} + 2y\frac{dy}{dt} + \frac{dx}{dt} = 0$$

or

$$(x + 2y)\frac{dy}{dt} + (1 + y)\frac{dx}{dt} = 0.$$

Thus

$$\frac{dy}{dt} = -\frac{1 + y}{x + 2y} \cdot \frac{dx}{dt}. \quad \square$$

11.4 EXERCISES

In Exercises 1 through 16, use implicit differentiation to find $\frac{dy}{dx}$.

1. $y^2 = x$
2. $xy^2 - 2x = 4$
3. $x^2 - y^3 = 2$
4. $x^2 - 2y^2 = 1$
5. $y^2 + 4x^3 - 2y = 0$
6. $x^2y^2 + x = 6$

7. $y^3 = x^2$

8. $x^3y^2 = 4$

9. $\dfrac{5x}{y^2} = 1$

10. $\dfrac{x}{y^3} - x^2 = 0$

11. $\dfrac{x}{y} - \dfrac{y^2}{x} = 1$

12. $(x^2 + y^2)^5 = 100$

13. $(x^2 + 2y)^3 = 1$

14. $\sqrt{xy} + y^2 = 0$

15. $\dfrac{y}{x} - \sqrt{xy} + 2x - 3y = 8$

16. $\sqrt{\dfrac{x}{y}} + 2x + y^{-2} = 6$

In Exercises 17 through 26, find the equation of the line tangent to the graph of the function at the given point.

17. $2x - 5y^2 = 0$, $(10, 2)$

18. $x^2 + y^2 = 10$, $(1, 3)$

19. $x^3 - y^3 = 0$, $(1, 1)$

20. $3xy^2 = 1$, $(3, \frac{1}{3})$

21. $2x^3y^2 - 8x = 0$, $(1, -2)$

22. $xy^2 - yx^2 + 6 = 0$, $(-1, -3)$

23. $\dfrac{x}{y^2} + 2y = 3$, $(1, 1)$

24. $\dfrac{y}{x^2} + \dfrac{y^2}{x} = 0$, $(1, -1)$

25. $(5x - 1)^2(y^2 - 1)^2 = 25$, $(1, \frac{3}{2})$

26. $(x + 2)^2 + (y - 2)^2 = 4$, $(0, 2)$

In Exercises 27 through 32, x and y are functions of t. Find dy/dt.

27. $x^2 + y^2 = 4$

28. $x^3 + y^3 = 1$

29. $\dfrac{x^2}{2} - \dfrac{y^2}{4} = 1$

30. $xy^2 + yx^2 = 1$

31. $x^2 - y^2x + 2x - 3y = 0$

32. $(xy + 1)^{1/2} = 9$

In Exercises 33 through 38, find $\dfrac{d^2y}{dx^2}$.

*33. $2x + y^2 = 1$

*34. $x^2 - 3y^2 = 0$

*35. $xy^3 - 3x = 5$

*36. $2xy^2 + y^3 = 8$

*37. $2x^2 - \dfrac{y^2}{x} = 2$

*38. $\dfrac{x}{y} - \sqrt{y} = 1$

Management

39. A demand function is given by the equation
$$p = 10{,}000 - 0.05q.$$
   (a) Solve the equation for q in terms of p and find $\dfrac{dq}{dp}$.

   (b) Use implicit differentiation to find $\dfrac{dq}{dp}$ without solving the equation for q.

40. A demand function is given in the form
$$p = 200 - 0.002q^2.$$

Use implicit differentiation to find $\frac{dq}{dp}$ and then find the rate at which demand is changing when
(a) $q = 200$ (b) $p = \$195$

41. A demand function is given in the form
$$p = -0.25q^2 - 0.5q + 100.$$
Use implicit differentiation to find $\frac{dq}{dp}$ and then find the rate at which demand is changing when
(a) $q = 9$ (b) $p = \$10$

*42. (a) Suppose quantity demanded q is written as a function of price p, say $q = f(p)$. Show that the elasticity of demand formula of Section 10.6 may be written in the form
$$E(p) = \frac{p(dq/dp)}{q}.$$

(b) Suppose a demand function is written in the form $p = g(q)$. Show that the elasticity of demand formula may be written in the form
$$E(q) = \frac{p}{q(dp/dq)}.$$

11.5 RELATED RATES

There is a common type of problem in which information about the rate of change of one variable with respect to another is known and it is required to find the rate of change of the second variable with respect to a third one. Such a problem is known as a **related rates** problem. Some examples of related rates problems are

- Finding the rate at which the quantity demanded of a product is changing if the rate at which its price is changing is known.
- Finding the rate at which the area of a circle is changing if the rate at which its radius is changing is known.
- Finding the rate at which the top of a ladder is sliding down a wall if the rate at which the bottom of the ladder is being pulled away from the base of the wall is known.

Related rates problems are solved with the aid of the chain rule. The key to solving related rates problems is to find an equation that relates the variable whose rate of change we want to the one whose rate of change we know; we can then differentiate this equation with the aid of the chain rule and solve for the unknown rate of change. Sometimes implicit differentiation is helpful. We illustrate with examples.

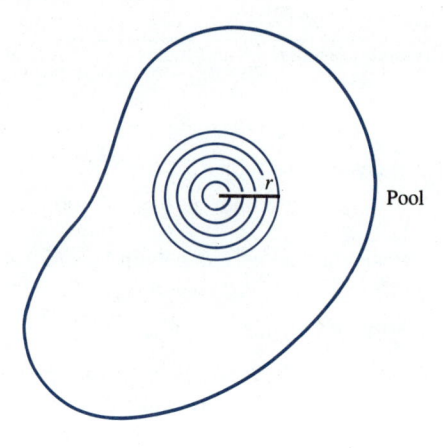

FIGURE 11–6

EXAMPLE 1 A stone is dropped into a pool and the ripples created move outward in a circular pattern from the point where the stone entered the water. The radius of the outermost ripple increases at a rate of 0.2 meter per second. Find the rate at which the area covered by the ripples is increasing when the radius of the outermost ripple is 1 meter.

The radius r of the outermost ripple and hence the area A of the circle covered by the ripples are both functions of time t. See Figure 11–6 for a picture of the situation. We wish to find the rate at which A is changing with respect to time when $r = 1$; that is, we wish to find dA/dt when $r = 1$. Since the area A is the area of a circle of radius r, we have

$$A = \pi r^2.$$

Because we want to find dA/dt, we use the chain rule to differentiate this expression with respect to t and thus obtain an equation involving dA/dt:

$$\frac{dA}{dt} = 2\pi r \frac{dr}{dt}.$$

But the radius r of the outermost ripple is changing at the rate of 0.2 meter per second (mps); that is, $dr/dt = 0.2$ mps. Also, we want dA/dt when $r = 1$. Therefore when $r = 1$,

$$\frac{dA}{dt} = 2\pi(1)(0.2) = 0.4\pi$$

m²ps. ◻

The method we used to solve the related rates problem of Example 1 can be applied to any related rates problem. We summarize it here:

MORE ABOUT DIFFERENTIATION

> **Keys to Solving Related Rates Problems**
>
> 1. Assign variables to the quantities that vary with time and draw a picture of the situation if possible.
>
> 2. Identify the desired rate of change as a derivative.
>
> 3. Find a mathematical relationship between the variables (the picture may be helpful here) and differentiate it with respect to time to obtain an equation involving the desired derivative.
>
> 4. The equation of step 3 will contain derivatives and/or variables. Find the values of the derivatives and the values of the variables at the desired time, substitute these values into the equation of step 3, and solve for the desired derivative.

In the examples that follow, notice how we use this method to solve related rates problems.

EXAMPLE 2 Suppose the demand function for a product is given by

$$q = 10{,}000 - 400p.$$

where q is quantity demanded and p is the unit price of the product. Suppose also that the price is given as a function of time by the equation

$$p = 5 + 2\sqrt{t},$$

where t is time in months, with $t = 0$ representing the present. Let us find the rate at which quantity demanded will be changing with respect to time four months from now.

The variables p and q have already been assigned to the quantities that vary with time. No picture is possible here. We want to find the rate of change of quantity demanded q with respect to time t when $t = 4$; that is, we want to find dq/dt when $t = 4$. The relationship between p and q is given as

$$q = 10{,}000 - 400p,$$

so we differentiate it with respect to t to obtain

$$\frac{dq}{dt} = -400 \frac{dp}{dt}.$$

Now,

$$p = 5 + 2\sqrt{t},$$

so

$$\frac{dp}{dt} = \frac{1}{\sqrt{t}}.$$

CHAPTER 11

Hence

$$\frac{dq}{dt} = -\frac{400}{\sqrt{t}},$$

and therefore

$$\left.\frac{dq}{dt}\right|_{t=4} = -\frac{400}{\sqrt{4}} = -200.$$

Thus four months from now quantity demanded will be decreasing at the rate of 200 units per month. ☐

EXAMPLE 3   A 10-foot ladder is leaning against a wall. The bottom of the ladder is pulled away from the wall at a rate of 2 feet per second. Find the rate at which the top of the ladder is sliding down the wall when the bottom is 6 feet from the base of the wall.

We begin by sketching a picture of the situation. See Figure 11–7. Here x is the distance from the base of the wall to the bottom of the ladder, and y is the distance from the base of the wall to the top of the ladder. Both x and y change with time, so they are functions of time t. We must find dy/dt when x equals 6 feet. We need an equation relating x and y that we can differentiate to find dy/dt. But the triangle in Figure 11–7 is a right triangle, so by the Theorem of Pythagoras we have

$$x^2 + y^2 = 10^2,$$

or

$$x^2 + y^2 = 100.$$

Differentiating this last equation implicitly with respect to t, we find that

$$2x\frac{dx}{dt} + 2y\frac{dy}{dt} = 0,$$

or

$$\frac{dy}{dt} = -\frac{x}{y}\frac{dx}{dt}.$$

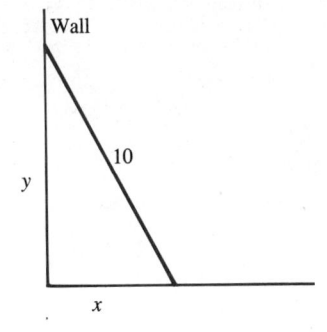

FIGURE 11–7

More about Differentiation 723

We know that x is increasing at the rate of 2 feet per second (fps). That is,

$$\frac{dx}{dt} = 2$$

feet per second. Substituting 6 for x in the equation $x^2 + y^2 = 100$, we obtain

$$6^2 + y^2 = 100,$$

which yields

$$y = \sqrt{64} = 8.$$

Hence when $x = 6$,

$$\frac{dy}{dt} = -\frac{6}{8}(2) = -1.5$$

fps. ☐

EXAMPLE 4   A person who is 5 feet tall is walking toward a building at a rate of 3 feet per second. The building is 100 feet high and has a light on its top. At what rate is the length of the person's shadow changing?

Figure 11–8 depicts the situation: x is the distance from the building to the person, and s is the length of the person's shadow. Both x and s are functions of time t. We must find ds/dt.

We need to find an equation relating s and x that we can differentiate to obtain ds/dt. From Figure 11–8 we see that the triangles ABE and CDE are similar, so their corresponding sides are in proportion. Thus

$$\frac{100}{x+s} = \frac{5}{s},$$

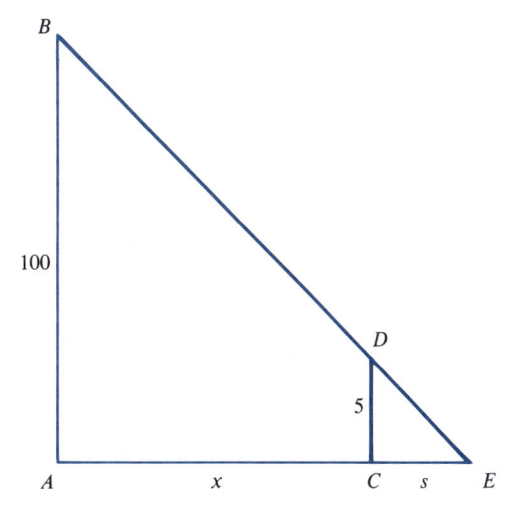

FIGURE 11–8

which yields

$$s = \frac{x}{19}.$$

Therefore

$$\frac{ds}{dt} = \frac{1}{19}\frac{dx}{dt}.$$

But the person is approaching the building at a rate of 3 feet per second; that is, $dx/dt = -3$ fps. (Note that dx/dt is *negative:* this is so because the distance x is *decreasing* as the person walks toward the building.) Hence

$$\frac{ds}{dt} = \frac{1}{19}(-3) = -\frac{3}{19}$$

fps. ☐

11.5 EXERCISES

1. If you blow up a balloon so that its radius increases at a rate of 2 centimeters per second (cps), find the rate at which its surface area is changing 4 seconds after you begin to blow it up. (The surface area of a sphere of radius r is $4\pi r^2$.)

2. Referring to Exercise 1, find the rate at which the volume of the balloon is changing 4 seconds after you begin to blow it up. (The volume of a sphere of radius r is $\frac{4}{3}\pi r^3$.)

3. The length of the edge of a cube is increasing at a rate of 2 cps. How fast is the volume of the cube changing at the instant when the edge is 20 cm long?

4. Boyle's Gas Law states that $V = T/P$, where V denotes the volume of a gas in cubic inches, P the pressure in pounds per square inch, and T the temperature in degrees Kelvin. Suppose the temperature is constant at 100°K and the pressure is increasing at 2 pounds per square inch (psi) per second. Find the rate at which the volume of the gas is changing with respect to time when the pressure is 20 psi.

5. Refer to Exercise 4: Suppose P is fixed at atmospheric pressure (14.7 psi) and the temperature is increasing at 2°K per minute. Find the rate of change of volume with respect to time.

6. A person who is 6 feet tall walks toward an 18-foot high street light at a rate of 2 fps. How fast is the length of the person's shadow changing?

7. A person who is 6 feet tall walks away from a 30-foot-high street light at a rate of 5 fps. How fast is the length of the person's shadow changing?

8. Two airplanes take off from the same airport at the same time. One flies due south at 300 mph, the other due west at 400 mph. At what rate is the distance between the planes changing 3 hours after takeoff?

9. Two cars are approaching an intersection, one from due north and the other from due east. The car from the north has a constant speed of 30 mph. When the cars are

each 1 mile from the intersection, the distance between them is decreasing at a rate of 40 mph. How fast is the car coming from the east traveling at the instant when both cars are 1 mile from the intersection?

10. A baseball player runs from first base to second base at a rate of 20 fps. How fast is the player's distance from third base changing when the player is 10 feet from second base? (*Hint:* A baseball diamond is a square with a side of 90 feet.)

11. A side street is perpendicular to a straight main highway. A police cruiser is situated on the side street 40 feet from the highway, and its radar is locked onto a car that is traveling along the highway toward the side street. The radar indicates that at the instant when the straight-line distance from the cruiser to the car is 200 feet, the distance between them is decreasing at a rate of 100 fps. How fast is the car traveling along the highway at this instant?

12. A helicopter is ascending straight up at a rate of 10 fps. An observer is 100 feet from the point where the helicopter took off. Find the rate at which the distance between the helicopter and the observer is changing when the helicopter is 50 feet above the ground.

13. A water tank is cylindrical with a circular base of radius 2 feet. If water is drawn from the tank at a rate of 1 gallon per minute, how fast is the water level in the tank changing? (*Hint:* The volume of a circular cylinder is $V = \pi r^2 h$, where r is the radius of its base and h is its height.)

14. A water tank has the shape of a cone 30 meters tall with a circular base whose radius is 3 meters. Water is being drawn out of the tank at a rate of 5 cubic meters per minute. How fast is the water level falling when the water is 20 meters deep? (The volume of a cone of height h whose base is a circle of radius r is $\frac{1}{3}\pi r^2 h$.)

15. An observer on the shore is watching a ship steam through a canal. The ship heads down the middle of the canal, which is one-quarter mile wide, at a rate of 20 mph. Find the rate at which the distance between the ship and the observer is changing when the ship is 1 mile from the observer.

Management

16. A firm's weekly cost function is given by $C(x) = 40x + 50{,}000$, where x is units produced per week and the number of units produced is increasing at the rate of 100 per week. Find the rate of change of cost with respect to time measured in weeks.

17. A company's daily revenue function is given by
$$R(x) = 0.03x^2 + 0.9x,$$
where x is the number of units sold per day. If the number of units sold is decreasing at a rate of 15 per day, find the rate of change of the company's revenue when $x = 100$.

18. The supply function for a product is given by
$$q = p^2 + 10p,$$
where q is quantity supplied and p is unit price. The price as of month t is given by
$$p = 25 - 0.5t,$$
where $t = 0$ represents the present. Find the rate at which quantity supplied will be changing three months from now.

19. The demand function for a commodity is given by
$$q = 100 - 0.02p^2,$$
with q in thousands of units and p in dollars. The price t months from now is forecast to be
$$p = 1.1t^{1/2} + 5$$
dollars. Find the rate at which quantity demanded will be changing with respect to time four months from now.

20. A firm's monthly cost function is given by
$$C(x) = 0.02x^2 + 100x + 20,000.$$
If production t months from now is expected to be
$$x = 1000 - 40t$$
units, find the rate at which cost will be changing six months from now.

21. The demand function for a company's product is given by
$$q = 100,000 - 5p$$
and the price t years from now is expected to be
$$p = 12 + 0.2t^{3/2}.$$
Find the rate at which the company's revenue will be changing one year from now.

22. A firm plans to build a new manufacturing plant. It is estimated that if the plant produces x units per month, its contribution to the firm's profits will be P dollars, where
$$P = 0.5x^2 - 3x - 50,000.$$
It is also estimated that t months after it opens, the plant will produce
$$x = 200\sqrt{t} + 50$$
units per month. Find the rate at which the plant's contribution to profits will be changing 16 months after it opens.

23. A store's monthly sales depend on the number of newspaper ads it runs, according to the equation
$$S = 20\sqrt{x + 100}.$$
Here S is monthly sales in thousands of dollars, and x is the number of ads run during a month. If the store is increasing the number of ads it runs at the rate of five per month, find the rate at which sales are increasing when 96 ads are run during a month.

24. If a firm has access to x thousand dollars of capital it can produce $y = 30(x + 10)^{2/3}$ units of its product. If the firm will have access to $x = 1.5t^{3/4} + 5$ units of capital t weeks from now, find the rate at which its production will be changing with respect to time 16 weeks from now.

Life Science

25. A circular oil slick is created as the result of the blowout of an offshore oil well. The radius of the slick increases at the rate of 400 feet per hour. Find the rate at which the area polluted by the slick is increasing 12 hours after the blowout.

26. The daily amount of smog produced by cars coming into the city is given by
$$y = 2 + 0.01x^{3/2},$$

where y is smog concentration in parts per million and x is the number of cars in thousands. The number of cars coming into the city is increasing at the rate of 250 per day. Find the rate at which the smog concentration will be changing when 225,000 cars are coming into the city.

27. Cases of a new disease are increasing at a rate of 10,000 per year, and the number of deaths y depends on the number of cases x according to the equation $y = 0.1\sqrt{x}$. Find an expression that gives the annual death rate due to the disease.

28. A patient's systolic blood pressure S depends on the concentration C of medication in the patient's blood stream according to the equation $S = 180 - 12(C + 10)^{1/2}$, and the concentration of the medication depends on the time t since the patient began taking the medication according to the equation $C = 20t^{1/2}$. Find the rate at which the patient's blood pressure is changing with respect to time nine days after medication was begun.

29. If a blood vessel has a circular cross section of radius r mm, Poiseuille's Law states that the velocity V of blood that is a mm from the central axis of the blood vessel is given by $V = k(r^2 - a^2)$, where k is a constant. Suppose a drug that dilates blood vessels causes an artery whose radius is 2 mm to expand at a rate of 0.1 mm per minute. Find the rate at which the velocity of the blood in the artery is changing. Treat a as a constant.

Social Science

30. If the population of a state is x million people, the state will need y thousand social workers in its department of social services, where

$$y = 3\sqrt{x} + 20.$$

Suppose the population of the state is increasing at a rate of 50,000 people per year. Find the annual rate at which the number of social workers needed is changing when the state's population is 5 million people.

SUMMARY

This summary consists of terms and symbols whose meaning you should know, facts you should know, and techniques and methods you should be able to employ.

Section 11.1

The product rule. Using the product rule to find derivatives. The quotient rule. Using the quotient rule to find derivatives.

Section 11.2

Composite function. The composite of f with g: $f(g(x))$. The composite of g with f: $g(f(x))$. Finding $f(g(x))$, $g(f(x))$ given $f(x)$ and $g(x)$. Writing a given function k as a composite function. The first formulation of the chain rule. Using the first formulation of the chain rule to find derivatives: differentiating "from the outside

CHAPTER 11

in." The general power rule. Using the general power rule to find derivatives. The second formulation of the chain rule. Using the second formulation of the chain rule to find derivatives.

Section 11.3

Increment Δx, change in x. The differentials dx, dy. The derivative dy/dx as a ratio of the differentials dy, dx. The differential formula. Using the differential formula to find dy. The change in f: Δf. Approximation by differentials. Using the approximation by differentials formula to find expressions for the approximate change in f. The proportional change in f. The percentage change in f. Approximating the proportional and percentage change in f.

Section 11.4

Explicitly defined function. Implicitly defined function. The technique of implicit differentiation. Using implicit differentiation to find derivatives.

Section 11.5

Related rates problem. Method for solving related rates problems. Solving related rates problems.

REVIEW EXERCISES

In Exercises 1 through 16, find $f'(x)$.

1. $f(x) = (6x^2 - 2x)(x^3 + 4)$
2. $f(x) = \dfrac{3x + 7}{5x - 3}$
3. $f(x) = \dfrac{2x^2 - x}{4x + 6}$
4. $f(x) = (2\sqrt{x} + 3)(x^3 - 1)$
5. $f(x) = \dfrac{(5x + 2)(x^2 + 3)}{2x^2 + 1}$
6. $f(x) = \dfrac{x^2 - 4}{(5x + 2)(x^2 + 1)}$
7. $f(x) = (2x^2 + x)\dfrac{x^3 - 1}{x^2 + 1}$
8. $f(x) = \left(x^2 + \dfrac{x}{x - 1}\right)\left(x^2 - \dfrac{x}{x + 1}\right)$
9. $f(x) = (7x - 9)^{10}$
10. $f(x) = (3x^2 + 5x + 1)^{12}$
11. $f(x) = \sqrt{6x^3 - 2x + 3}$
12. $f(x) = \sqrt{\dfrac{5x + 2}{x^2 - 2}}$
13. $f(x) = (2x + 3)^6(x^2 - 5x)^8$
14. $f(x) = [(2x - 2)^4(x^2 + 1)]^5$
15. $f(x) = \left(\dfrac{3x - 1}{x^2 + 2}\right)^7$
16. $f(x) = \dfrac{(3x + 4)^5}{(5x^2 - 1)^4}$

17. If $y = 6x^2 - 2x + 1$ and $x = 4t^2 - 9$, find $\dfrac{dy}{dt}$.

18. If $y = x^2 - 5x + 2$ and $x = 8t - 3$, find $\dfrac{dy}{dt}$.

MORE ABOUT DIFFERENTIATION

19. If $v = 7w - 3\sqrt{w}$ and $u = (v^4 - 2v + 11)^{11}$, find $\dfrac{du}{dw}$.

20. The supply function for a commodity is given by
$$q = 5000(0.5p^2 - 2\sqrt{p})^{3/4}.$$
Find the rate of change of quantity supplied q when $p = \$16$.

21. A demand function is given by $p = 80{,}000(10q)^{-1/2}$.
    (a) Find an expression for the approximate change in price brought about by a change of Δq in quantity demanded.
    (b) Find the approximate change in price brought about by a change in quantity demanded from 1000 units to 1005 units.

22. For the demand function of Exercise 21,
    (a) Find an expression for the approximate proportional change in price brought about by a change of Δq in quantity demanded.
    (b) Find the approximate proportional and percentage change in price if quantity demanded changes from 250 units to 248 units.

23. A conical water tank whose base is a circle with a radius of 20 feet has its height measured as 50 feet with an error of ± 3 inches. Find the approximate proportional and percentage errors in its volume $\left(\dfrac{\pi r^2 h}{3}\right)$ due to measurement error.

24. How accurately must the radius of the base of a conical water tank whose height is 50 feet be measured if its volume is to be determined to within $\pm 1\%$?

In Exercises 25 through 28, find $\dfrac{dy}{dx}$.

25. $8y^2 - 2x = 5$ 26. $xy^3 - x^2y = 10$ 27. $x^2 + y^2 + \dfrac{x}{y} = 1$ 28. $x^2 + y^2 - \dfrac{y}{x} = 1$

In Exercises 29 and 30, find the equation of the line tangent to the graph of the function at the given point.

29. $-x^2 + y^2 + 3x = 9$, $(0, 3)$ 30. $x^2 + 3\dfrac{x}{y^2} = 8$, $(-3, 3)$

31. Find $\dfrac{d^2y}{dx^2}$ if $2x^2y^2 = 4$.   32. Find $\dfrac{d^2y}{dx^2}$ if $\dfrac{2x}{\sqrt{y}} - x^2 = 0$.

33. If x and y are functions of u and $3xy^2 + \dfrac{2y}{x} = 1$, find $\dfrac{dy}{du}$.

34. If price p and quantity demanded q for a certain commodity are related by the equation
$$q = -p^2 - 50p + 5000$$
and the price is increasing at a rate of \$0.50 per month, find the rate at which quantity demanded is changing with respect to time when the price is \$25.

35. If one person drives due north from point A at 55 mph and another person leaves point A two hours after the first and drives due east at 50 mph, find the rate at which the distance separating them is changing 4 hours after the second person left.

ADDITIONAL TOPICS

Here are some additional topics not covered in the text that you might wish to investigate on your own.

1. In Section 11.2 we proved the chain rule under the assumption that $g(x + h) - g(x) \neq 0$. Show how the chain rule is proved when this is not assumed to be the case.

2. There is one general rule of differentiation that we have not presented. It involves inverse functions: Recall that if f is a one-to-one function, then f^{-1} is the function defined by $f^{-1}(y) = x$ if and only if $f(x) = y$. (See Section 2.4) There is a relationship between the derivatives df/dx and df^{-1}/dx; this relationship is known as the **inverse function rule.** Find out about the inverse function rule, how it is proved, and how it is used.

3. In Section 10.4 we introduced the power rule
$$\frac{d}{dx}(x^r) = rx^{r-1}, \qquad r \neq 0,$$
and outlined a proof of the rule in the case where r is a positive integer. Now prove the power rule when r is a negative integer.

4. Prove the power rule when r is a rational number p/q.

5. A pair of equations of the form $x = f(t)$, $y = g(t)$ are called **parametric equations** for x and y with **parameter** t. Find out how parametric equations are used to describe curves. Given parametric equations for x and y, can you find expressions for dy/dx and dx/dy?

SUPPLEMENT: CONSUMPTION AND THE MULTIPLIER

It is not uncommon to find news items such as this: "Today XYZ Corporation announced that it will build a new plant in Smallville. The plant will provide employment for 250 workers at an operating cost of some $1 million per year. It is estimated that the presence of the plant will add $4 million per year to the Smallville economy."

How was the figure of $4 million per year determined? It must have taken into account the fact that some of the income generated by an investment (such as the $1 million per year that XYZ is investing in Smallville) will be saved and invested by those who receive it; this new round of investment will in turn produce more income, some of which will again be saved and invested to produce still more income, and so on. This "chain reaction" of income generation is called the **multiplier effect.** Let us see if we can quantify the multiplier effect.

Evidently the strength of the multiplier effect depends on the proportion of income saved and invested at each stage of the process. Therefore we need an expression that relates income to savings and investment. For simplicity we assume that all income that is saved is invested, so savings equals investment. Thus we need a relationship between income and savings.

Economists divide income into the portion that is saved and the portion that is not saved; the latter is called **consumption.** Thus,

$$\text{Income} = \text{Savings} + \text{Consumption},$$

or

$$\text{Consumption} = \text{Income} - \text{Savings}.$$

If we denote income by Y, consumption by C, and savings by S, then consumption is given as a function of income by a **consumption function**

$$C = Y - S.$$

EXAMPLE 1 Let

$$C = 5 + 8\sqrt{Y}$$

be a consumption function, with Y and C in millions of dollars. Thus when $Y = 100$ consumption is

$$C(100) = 5 + 8\sqrt{100} = 85$$

million dollars. The remaining $100 - 85 = 15$ million dollars of income is saved. ☐

The derivative of C with respect to Y is the rate of change of consumption with respect to income, which is known as the **marginal propensity to consume.** The marginal propensity to consume is the proportion of an additional $1 of income that would be spent rather than saved.

EXAMPLE 2 Let

$$C = 5 + 8\sqrt{Y}$$

as in Example 1. Then the marginal propensity to consume is

$$\frac{dC}{dY} = \frac{4}{\sqrt{Y}}.$$

For instance, when $Y = 100$, we have

$$\text{Marginal propensity to consume} = \frac{4}{\sqrt{100}} = 0.40.$$

Hence at an income level of $100 million, $0.40 of each additional dollar of income would be spent, and the remaining $0.60 would be saved. Note that the marginal propensity to consume $dC/dY = 4/\sqrt{Y}$ is a decreasing function of Y. Therefore in this example, as income increases a smaller proportion of it is spent and a larger proportion is saved. ☐

The equation $C = Y - S$ shows that savings S is a function of income Y also. As we remarked previously, the strength of the multiplier effect depends on the proportion of income invested at each stage of the income-generation process. Thus, since we are assuming that savings equals investment, we can say that the multiplier effect is measured by the rate at which *income* changes with respect to *savings*, or dY/dS. The derivative dY/dS is called the **multiplier.**

From $C = Y - S$ we have

$$\frac{dC}{dS} = \frac{dY}{dS} - \frac{dS}{dS}.$$

Applying the chain rule to the left side gives

$$\frac{dC}{dY}\frac{dY}{dS} = \frac{dY}{dS} - 1.$$

Solving this last equation for dY/dS, we obtain

$$\frac{dY}{dS} = \frac{1}{1 - dC/dY} = \frac{1}{1 - \text{marginal propensity to consume}}.$$

EXAMPLE 3 Again let $C = 5 + 8\sqrt{Y}$. Then the multiplier is

$$\frac{dY}{dS} = \frac{1}{1 - 4/\sqrt{Y}} = \frac{\sqrt{Y}}{\sqrt{Y} - 4}.$$

Thus, at an income level of $100 million,

$$\frac{dY}{dS} = \frac{\sqrt{100}}{\sqrt{100} - 4} = \frac{5}{3}.$$

Hence each additional $1 of savings generates an additional $1.67 in income. ☐

EXAMPLE 4 Suppose the consumption function for Smallville is

$$C = 20 + 0.75Y,$$

where Y and C are in millions of dollars. The multiplier is then

$$\frac{1}{1 - dC/dY} = \frac{1}{1 - 0.75} = 4,$$

so if XYZ Corporation invests $1 million per year in Smallville it will generate approximately $4 million in additional income for the town. ☐

The multiplier effect is what is behind the idea of "pump priming" in the economy: if there are many opportunities for investment, the multiplier is large; therefore a relatively small investment (the "priming of the pump") can lead to a large increase in income. Unfortunately, the reverse is also true: a relatively small decrease in investment can lead to a large decrease in income.

SUPPLEMENTARY EXERCISES

1. Let $C = 20 + 0.6Y$ be a consumption function, with Y and C in millions of dollars, $Y \geq 0$. Find and interpret the marginal propensity to consume when the income level is $10 million; when it is $100 million.

2. Let $C = 350 + 0.32Y$ be a consumption function, with Y and C in billions of dollars, $Y \geq 0$. Find and interpret the marginal propensity to consume at every income level.

3. Let $C = 200 + 20\sqrt{Y}$ be a consumption function, with Y and C in billions of dollars, $Y \geq 100$. Find and interpret the marginal propensity to consume when the income level is
   (a) $400 billion
   (b) $900 billion

4. Let $C = 50 + 10\sqrt{Y - 10}$ be a consumption function, with Y and C in billions of dollars, $Y \geq 35$. Find and interpret the marginal propensity to consume when the income level is
   (a) $410 billion
   (b) $2510 billion

5. Let $C = 245 + 75\sqrt[3]{Y}$ be a consumption function, with Y and C in millions of dollars, $Y \geq 125$. Find and interpret the marginal propensity to consume when the income level is
   (a) $125 million
   (b) $1000 million
   (c) $8000 million

6. Find and interpret the multiplier for the consumption function of Exercise 1.

7. Find and interpret the multiplier for the consumption function of Exercise 2.

8. Find and interpret the multiplier for the consumption function of Exercise 3 when the income level is
   (a) $400 billion
   (b) $900 billion

9. Find and interpret the multiplier for the consumption function of Exercise 4 when the income level is
   (a) $410 billion
   (b) $2510 billion

10. Find and interpret the multiplier for the consumption function of Exercise 5 when the income level is
    (a) $8000 million
    (b) $1000 million
    (c) $125 million

Since $C = Y - S$, we have $S = Y - C$. The derivative dS/dY is called the **marginal propensity to save**. In Exercises 11 through 15 find and interpret the marginal propensity to save for the given consumption function at the given level of income.

11. The consumption function of Exercise 1, at income levels of $10 million and $100 million.

12. The consumption function of Exercise 2, at every income level.

13. The consumption function of Exercise 3, at income levels of $400 billion and $900 billion.

14. The consumption function of Exercise 4, at income levels of $410 billion and $2510 billion.

15. The consumption function of Exercise 5, at income levels of $125 million, $1000 million, and $8000 million.

12

Applications of Differentiation

The graph of a function is usually of considerable help in analyzing the behavior of the function. For instance, Figure 12–1 shows the graph of a function that predicts high-school enrollment in a large city for the years 1990 through 2020. From the graph we see the following:

- From 1990 to 1996 enrollment is predicted to decline, with the rate of decline decreasing as 1996 approaches.
- The minimum enrollment will be 200,000 students in 1996.
- From 1996 to 2004 enrollment will grow at an increasing rate.
- From 2004 to 2010 enrollment will continue to grow, but at a decreasing rate.
- The maximum enrollment will be 250,000 students in 2010.
- From 2010 to 2020 enrollment will decline at an increasing rate.

To analyze the behavior of a function as we have just done, we must be able to draw the graph accurately enough to show

- Where the function is increasing and where it is decreasing.
- What its maximum and minimum values are and where they occur.
- The general shape of the curve.

In this chapter we show how to use the first and second derivatives to determine these things. The first two sections of the chapter are devoted to finding relative maxima and minima of functions, the third to finding absolute maxima and minima, and the last demonstrates how calculus may be used to sketch the graph of a function.

FIGURE 12–1

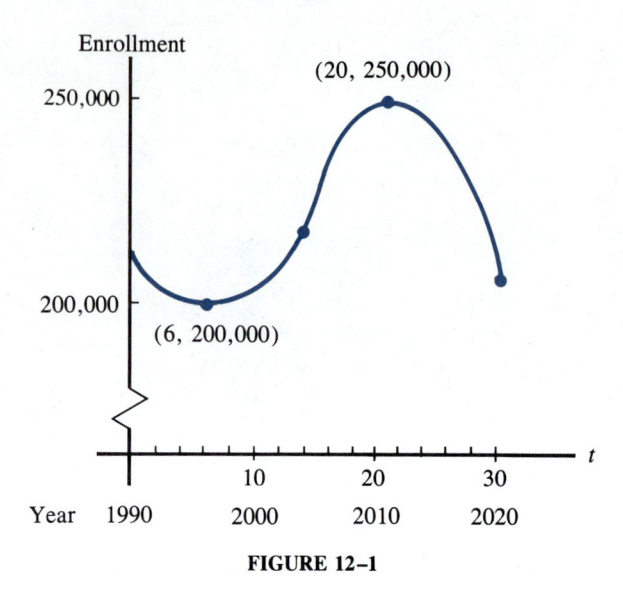

12.1 RELATIVE MAXIMA AND MINIMA: THE FIRST DERIVATIVE TEST

Suppose a function has the graph shown in Figure 12–2. Notice that the graph comes to a peak at the point $(-1, 10)$ and forms a trough at the point $(2, 1)$. The point $(-1, 10)$ is an example of a **relative maximum point**; the point $(2, 1)$ is an example of a **relative minimum point**. When graphing functions we must find their relative maximum and relative minimum points. In this section we show how to use the first derivative to do this.

Figure 12–3 shows the graph of a function f which has a relative maximum point at $(a, f(a))$ and a relative minimum point at $(b, f(b))$. Consider the point $(a, f(a))$: from the figure it is clear that for all values of x near a, on either side of a, $f(x)$ exists and $f(x) \leq f(a)$. Similarly, for all values of x near b, on either side of b, $f(x)$ exists and $f(x) \geq f(b)$. These properties define the concepts of relative maximum and relative minimum:

Relative Maximum and Relative Minimum

Let f be a function.

1. The point $(a, f(a))$ is a **relative maximum point** for f if for all x near a, on either side of a, $f(x)$ exists and $f(x) \leq f(a)$. In such a case we say that $f(a)$ is a **relative maximum for f**.

2. The point $(b, f(b))$ is a **relative minimum point** for f if for all x near b, on either side of b, $f(x)$ exists and $f(x) \geq f(b)$. In such a case we say that $f(b)$ is a **relative minimum for f**.

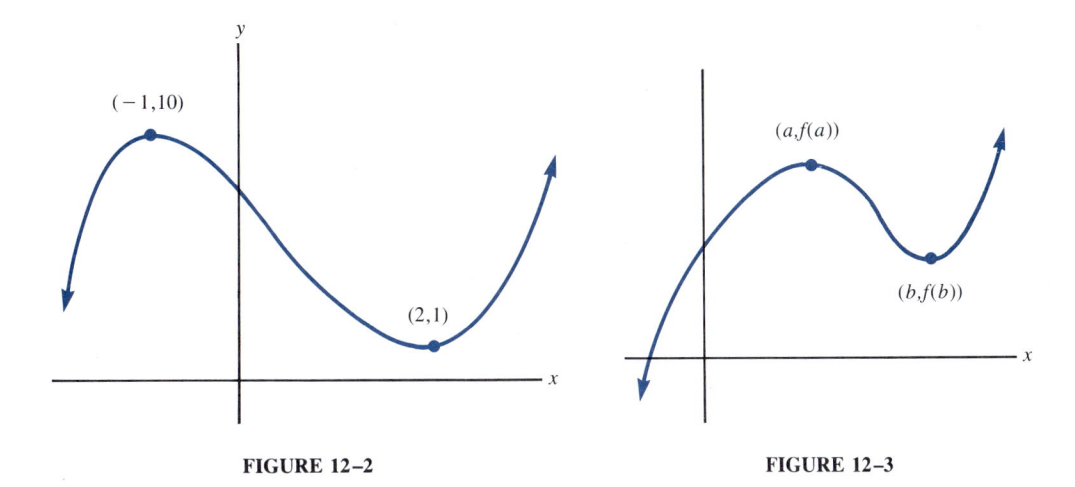

APPLICATIONS OF DIFFERENTIATION

FIGURE 12–2

FIGURE 12–3

EXAMPLE 1 The function whose graph is shown in Figure 12–4 has relative maxima at the points (4, 8) and (9, 7) and relative minima at the points (3, 2) and (7, 4). ☐

How can we find the relative maxima and relative minima for a function f? As Figure 12–5 shows, if $(a, f(a))$ is a relative maximum point *and* there is a unique tangent line at $(a, f(a))$, then the tangent line must be horizontal; in other words, if $f(a)$ is a relative maximum and $f'(a)$ exists, then $f'(a) = 0$. The same reasoning holds for relative minima. Therefore we can find at least *some* of the relative maxima and relative minima of f by finding the values of x for which $f'(x) = 0$. We might not find *all* relative maxima and relative minima for f by doing this, because f might have relative maxima or relative minima at points where the derivative $f'(x)$ does not exist. As an example of such behavior, the function of Figure 12–4 has a relative maximum point at (9, 7) and a relative minimum point at (3, 2), but $f'(9)$ and $f'(3)$ do not exist. (Why not?)

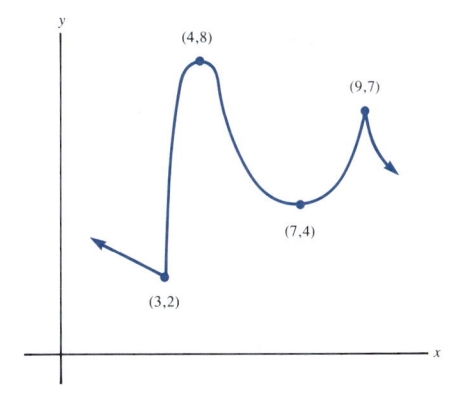

FIGURE 12–4

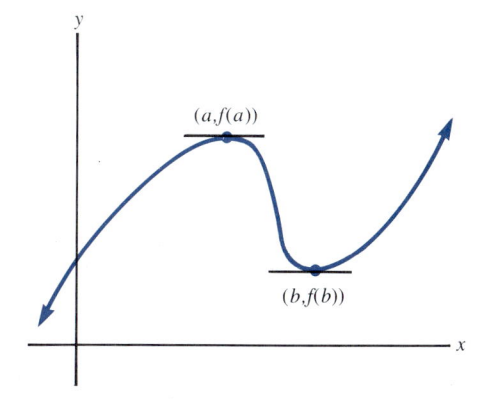

Our conclusion, then, is that if we are looking for the relative maxima and relative minima of a function f, we must find all values of x in the domain of f such that $f'(x) = 0$ or $f'(x)$ does not exist. Such values are called **critical values** of the function.

> **Critical Values**
>
> Let f be a function. The value $x = a$ is a **critical value** for f if a is in the domain of f and either $f'(a) = 0$ or $f'(a)$ does not exist.

EXAMPLE 2 Let us find the critical values for the functions f and g defined by

$$f(x) = 2x^3 - 12x^2 + 18x + 1 \quad \text{and} \quad g(x) = x^{2/3}.$$

We begin by finding the derivative of f:

$$f'(x) = 6x^2 - 24x + 18 = 6(x - 1)(x - 3).$$

Now we set $f'(x) = 0$ and solve for x to obtain $x = 1$ and $x = 3$. Since $f'(1) = 0$ and $f'(3) = 0$ and both 1 and 3 are in the domain of f, $x = 1$ and $x = 3$ are critical values for f. Because $f'(x)$ exists for all x, these are the only critical values for f.

Now we find the critical values for g. We have

$$g'(x) = \frac{2}{3x^{1/3}},$$

so $g'(x)$ is never zero, but $g'(x)$ does not exist at $x = 0$. Since $g(0) = 0$, 0 is in the domain of g, and hence $x = 0$ is the only critical value for g. ☐

Having found the critical values for a function, we must somehow test them to find out if they yield relative maxima or relative minima. In this connection, it is important to realize that if $x = a$ is a critical value for f, it is not the case that $f(a)$ must be either a relative maximum or a relative minimum: it might be neither. For instance, Figure 12–6 shows the graph of a function f that has a critical value

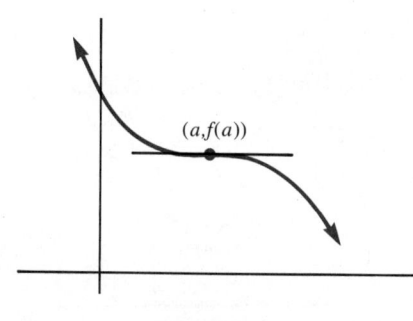

FIGURE 12–6

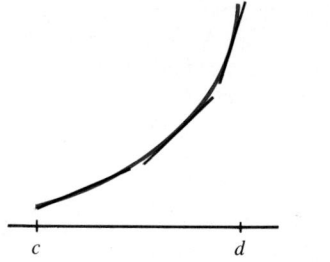

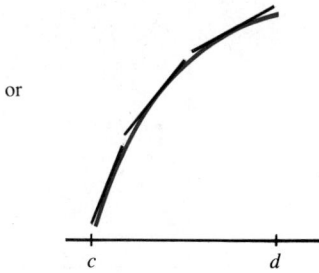

FIGURE 12–7
$f'(x) > 0$ for x in (c, d)

APPLICATIONS OF DIFFERENTIATION

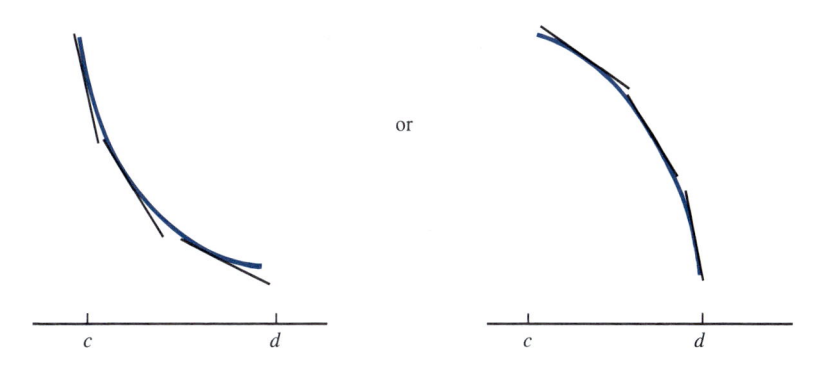

FIGURE 12–8
$f'(x) < 0$ for x in (c, d)

at $x = a$ (because $f'(a)$ is the slope of tangent line at $(a, f(a))$ and this slope is 0), but the point $(a, f(a))$ is neither a relative maximum point nor a relative minimum point for f.

How can we test a critical value to determine whether it yields a relative maximum, a relative minimum, or neither? The answer lies in the sign of the derivative. Since $f'(x)$ is the slope of the line tangent to f at $(x, f(x))$, it follows that if $f'(x) > 0$ for all x in an interval (c, d), all the tangent lines to f over the interval have positive slope, as in Figure 12–7. However, as Figure 12–7 shows, this means that the functional values $f(x)$ increase as x increases from c to d; in such a case, we say that the function f is **increasing on the interval.** Similarly, if $f'(x) < 0$ for all x in an interval, then as Figure 12–8 demonstrates, the functional values $f(x)$ decrease as x increases over the interval, and we say that f is **decreasing on the interval.**

The Sign of the Derivative

1. If $f'(x) > 0$ for all x in an interval (c, d), then f is increasing on the interval.

2. If $f'(x) < 0$ for all x in an interval (c, d), then f is decreasing on the interval.

EXAMPLE 3 Let f be defined by $f(x) = x^2 - 4x + 2$. Then
$$f'(x) = 2x - 4 = 2(x - 2).$$
Thus:

if $x < 2$, $x - 2 < 0$, so $f'(x) = 2(x - 2) < 0$ and therefore f is decreasing on the interval $(-\infty, 2)$;

if $x > 2$, $x - 2 > 0$, so $f'(x) = 2(x - 2) > 0$ and therefore f is increasing on the interval $(2, +\infty)$.

FIGURE 12-9

FIGURE 12-10

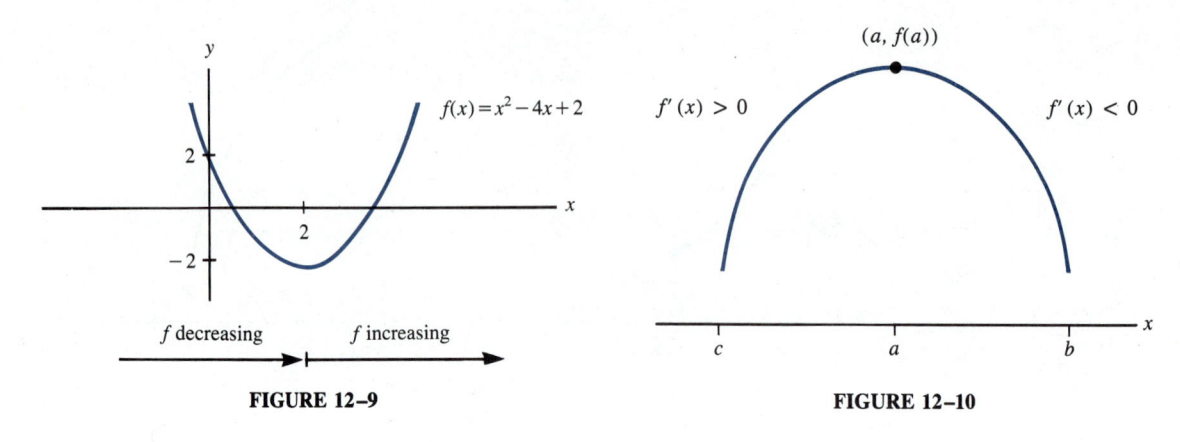

The graph of f is shown in Figure 12–9. Note that the graph of f is indeed decreasing over $(-\infty, 2)$ and increasing over $(2, +\infty)$. □

Now suppose that $x = a$ is a critical value for f. If $f'(x) > 0$ for all x in an interval (c, a) and $f'(x) < 0$ for all x in an interval (a, b) then f is increasing to the left of the point $(a, f(a))$ and decreasing to the right of $(a, f(a))$, as in Figure 12–10; hence $(a, f(a))$ must be a relative maximum point for f. Similarly, if $f'(x) < 0$ on an interval (c, a) and $f'(x) > 0$ on an interval (a, b), then $(a, f(a))$ must be a relative minimum point for f. See Figure 12–11. Thus we can test a critical value $x = a$ to see whether it yields a relative maximum or a relative minimum (or neither) by examining the sign of $f'(x)$ for x near a. We illustrate with an example.

EXAMPLE 4 Let f be defined by $f(x) = 2x^3 - 12x^2 + 18x + 1$. Then

$$f'(x) = 6x^2 - 24x + 18 = 6(x - 1)(x - 3),$$

and as we have seen, the critical values for f are $x = 1$ and $x = 3$. Note that if $x < 1$, $x - 1 < 0$ and $x - 3 < 0$, so

$$f'(x) = 6(x - 1)(x - 3) > 0;$$

if $1 < x < 3$, $x - 1 > 0$ and $x - 3 < 0$, so

$$f'(x) = 6(x - 1)(x - 3) < 0;$$

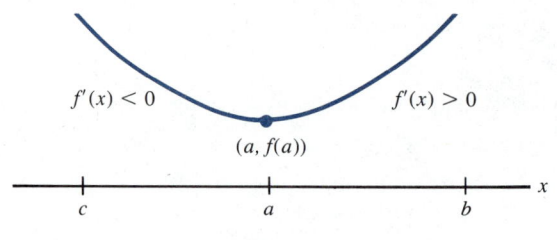

FIGURE 12-11

Applications of Differentiation

if $x > 3$, $x - 1 > 0$ and $x - 3 > 0$, so

$$f'(x) = 6(x - 1)(x - 3) > 0.$$

We record this information on a diagram of the x-axis:

```
      +           -            +
  ----+-----------+-----------------  x
      1           3
```

Here we have used the critical values to subdivide the x-axis into intervals; a + sign over an interval means that f' is positive on the interval, and a − sign over an interval means that f' is negative on the interval. Therefore f is increasing on $(-\infty, 1)$ and $(3, +\infty)$, and decreasing on $(1, 3)$. We add this information to the diagram:

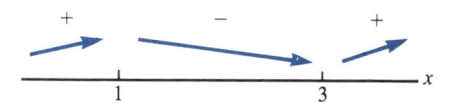

```
      +  ↗        -  ↘         +  ↗
  ----+-----------+-----------------  x
      1           3
```

Here an upward-slanting arrow ↗ indicates that f is increasing and a downward-slanting arrow ↘ indicates that f is decreasing. This last diagram shows which critical values yield relative maxima and which yield relative minima: the pattern of arrows ↗↘ at $x = 1$ shows that f has a relative maximum at $(1, f(1)) = (1, 9)$; the pattern of arrows ↘↗ at $x = 3$ shows that f has a relative minimum at $(3, f(3)) = (3, 1)$. Figure 12–12 shows the graph of f: note the relative maximum at $(1, 9)$ and the relative minimum at $(3, 1)$, and note also that f is increasing over the intervals $(-\infty, 1)$ and $(3, +\infty)$ and decreasing over the interval $(1, 3)$. ☐

The procedure we used in Example 4 to find relative maxima and minima is called the **first derivative test.**

$(1,9)$ $f(x) = 2x^3 - 12x^2 + 18x + 1$

$(3,1)$

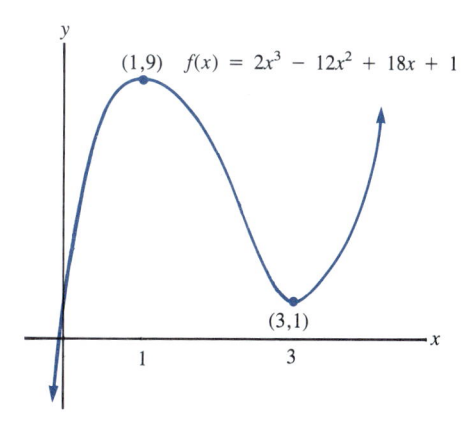

FIGURE 12–12

> **The First Derivative Test**
>
> To find the relative maxima and relative minima of a function f, proceed as follows:
>
> 1. Find all values of x such that $f'(x) = 0$ or $f'(x)$ does not exist.
>
> 2. Plot the values obtained in step 1 on the x-axis, thereby dividing the x-axis into intervals.
>
> 3. a. On each interval of the diagram of step 2, determine whether the derivative f' is positive or negative; record this information on the diagram using $+$ and $-$ signs.
>
>   b. f is increasing on those intervals where f' is positive and decreasing on those intervals where f' is negative; record this information on the diagram using arrows.
>
> 4. The diagram of step 3b will show which critical values yield relative maxima, which yield relative minima, and which yield neither.

(Notice that in step 1 of the first derivative test we will certainly obtain all the critical values of f, but we may also obtain numbers x that are not critical values because neither $f(x)$ nor $f'(x)$ exists. Since such numbers are not critical values, relative maxima and minima cannot occur at them; however, the derivative can have different signs on either side of such values, and when we discuss curve sketching in Section 12.4 it will be convenient to have the behavior of the function around them analyzed as part of the first derivative test.)

We illustrate the use of the first derivative test with examples.

EXAMPLE 5 Let us use the first derivative test to find the relative maxima and relative minima of the function f defined by $f(x) = -x^2 + 8x - 12$.

1. We have
$$f'(x) = -2x + 8 = -2(x - 4).$$
Setting $f'(x) = 0$ and solving for x yields $x = 4$, so $x = 4$ is a critical value for f. There are no values of x for which $f'(x)$ does not exist.

2. We plot the critical value $x = 4$ on the x-axis:

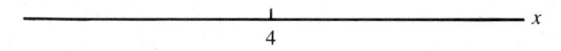

$$\xrightarrow{\qquad\qquad\qquad\underset{4}{|}\qquad\qquad\qquad} x$$

3. a. We determine the sign of $f'(x) = -2(x - 4)$ on the intervals $(-\infty, 4)$ and $(4, +\infty)$:

   if $x < 4$, $x - 4$ is negative, so $f'(x)$ is positive;

   if $x > 4$, $x - 4$ is positive, so $f'(x)$ is negative.

APPLICATIONS OF DIFFERENTIATION

We record this information on the diagram:

$$\begin{array}{c|c} + & - \\ \hline & 4 \end{array} \longrightarrow x$$

b. Since f' is positive on $(-\infty, 4)$ and negative on $(4, +\infty)$, f is increasing on $(-\infty, 4)$ and decreasing on $(4, +\infty)$. We add this information to the diagram:

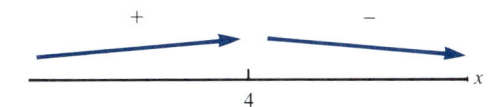

4. The diagram of step 3b shows that f has a relative maximum at $(4, f(4)) = (4, 4)$. There are no relative minima. The graph of f is shown in Figure 12–13. ☐

EXAMPLE 6 We find the relative maxima and relative minima for the function g defined by $g(x) = x^{2/3}$.

1. We have

$$g'(x) = \frac{2}{3x^{1/3}}.$$

Thus $g'(x)$ is never 0, but it does not exist at $x = 0$. (We saw in Example 2 that since 0 is in the domain of g, $x = 0$ is a critical value for g.)

2. We plot $x = 0$ on the x-axis:

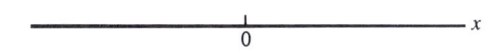

3. **a.** We determine the sign of

$$g'(x) = \frac{2}{3x^{1/3}}$$

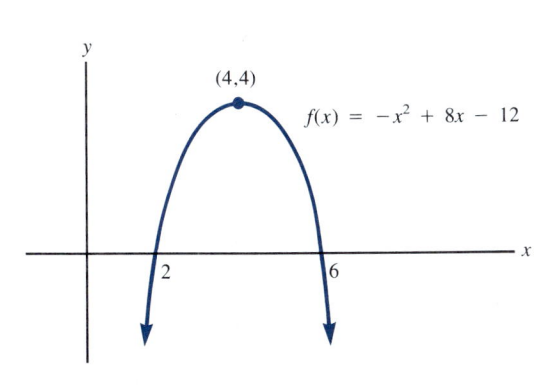

$f(x) = -x^2 + 8x - 12$

$g(x) = x^{2/3}$

FIGURE 12–13

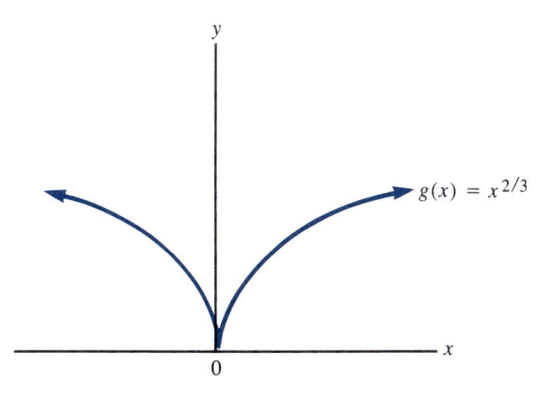

FIGURE 12–14

on the intervals $(-\infty, 0)$ and $(0, +\infty)$. Since $x^{1/3} = \sqrt[3]{x}$ is negative when $x < 0$ and positive when $x > 0$, $g'(x)$ is negative over $(-\infty, 0)$ and positive over $(0, +\infty)$. Hence we have the diagram

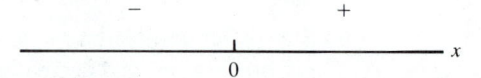

        −   +
———————|———————— x
          0

b. Since g' is negative over $(-\infty, 0)$ and positive over $(0, +\infty)$, g is decreasing over $(-\infty, 0)$ and increasing over $(0, +\infty)$. Hence the diagram becomes

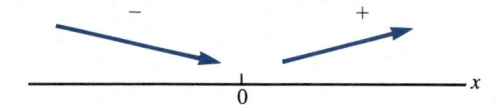

        −   +
———————|———————— x
          0

4. The diagram of step 3b shows that g has a relative minimum at $(0, g(0)) = (0, 0)$. There are no relative maxima. The graph of g is shown in Figure 12–14. ☐

Step 3a of the first derivative test requires that we determine the sign of f' on each of the intervals between critical points of f. If f' is continuous on such an interval, we can determine its sign by choosing a number from the interval and substituting it for x in $f'(x)$. This fact can often be used to simplify the work in step 3a.

EXAMPLE 7   Let us find the relative maxima and relative minima of the function h given by

$$h(x) = -x^3 + 6x^2 + 15x.$$

1. Since

$$h'(x) = -3x^2 + 12x + 15 = -3(x + 1)(x - 5),$$

setting $h'(x) = 0$ and solving for x yields $x = -1$ and $x = 5$. There are no values of x for which $h'(x)$ does not exist.

2. We have

———————|———————|———————— x
       −1   5

3. a. Since h' is a polynomial, it is continuous on the intervals $(-\infty, -1)$, $(-1, 5)$, and $(5, +\infty)$. Therefore we can determine its sign on each of these intervals by selecting a number from the interval and substituting it into $h'(x)$. For instance, if we select -2 from $(-\infty, -1)$, 0 from $(-1, 5)$, and 6 from $(5, +\infty)$, we have

$$h'(-2) = -3(-2 + 1)(-2 - 5) < 0,$$
$$h'(0) = -3(0 + 1)(0 - 5) > 0,$$
$$h'(6) = -3(6 + 1)(6 - 5) < 0.$$

APPLICATIONS OF DIFFERENTIATION

Thus $h'(x)$ is negative on $(-\infty, -1)$ and $(5, +\infty)$, and positive on $(-1, 5)$. Hence we have

```
        -           +           -
      ──┼───────────────────────┼──── x
       -1                       5
```

b. Thus h is increasing over $(-1, 5)$, and decreasing over $(-\infty, -1)$ and $(5, +\infty)$. Therefore the diagram becomes

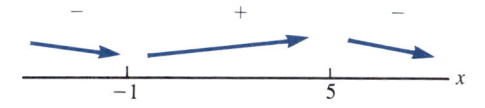

```
        -           +           -
      ──┼───────────────────────┼──── x
       -1                       5
```

4. The diagram of step 3b shows that $(-1, h(-1)) = (-1, -8)$ is a relative minimum point for h and that $(5, h(5)) = (5, 100)$ is a relative maximum point for h. Figure 12–15 shows the graph of h. ☐

EXAMPLE 8   We find the relative maxima and relative minima of the function k defined by

$$k(x) = -x^4 + 8x^3 - 18x^2.$$

1. Since

$$k'(x) = -4x^3 + 24x^2 - 36x = -4x(x - 3)^2,$$

setting $k'(x) = 0$ and solving for x yields $x = 0$ and $x = 3$. There are no values for which $k'(x)$ does not exist.

2. We have

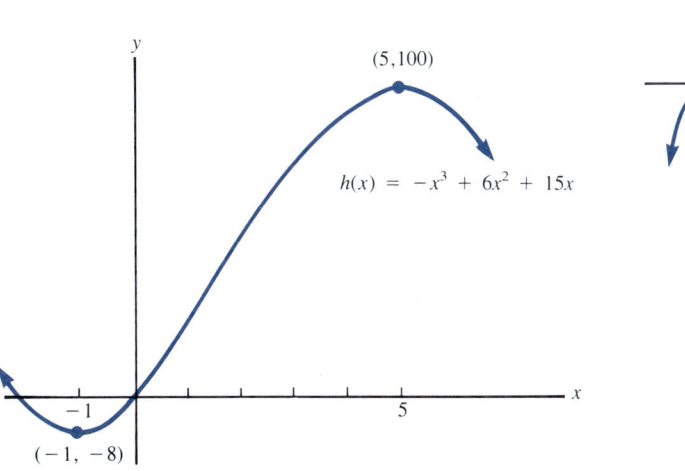

```
      ──┼───────────────────────┼──── x
        0                       3
```

FIGURE 12–15

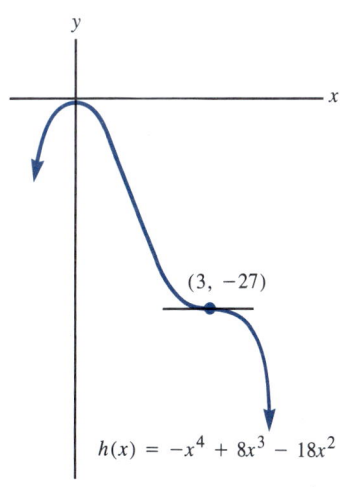

FIGURE 12–16

3. Note that k' is continuous on $(-\infty, 0)$, $(0, 3)$, and $(3, +\infty)$. (Why?) If we choose $x = -1$ from $(-\infty, 0)$, $x = 1$ from $(0, 3)$, and $x = 4$ from $(3, +\infty)$, we find that $k'(-1) = 64$ and $k'(1) = k'(4) = -16$.

   a. Hence k' is positive on $(-\infty, 0)$, and negative on $(0, 3)$ and on $(3, +\infty)$:

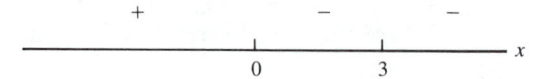

   b. Thus k is increasing on $(-\infty, 0)$, and decreasing on $(0, 3)$ and on $(3, +\infty)$:

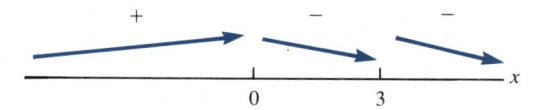

4. The diagram of step 3b shows that there is a relative maximum at $(0, k(0)) = (0, 0)$. Note that since k is decreasing on both sides of $x = 3$, there is neither a relative maximum nor a relative minimum for k at $x = 3$. In fact, the graph of k "flattens out" at $(3, k(3)) = (3, -27)$, because $k'(3) = 0$ implies that the graph has a horizontal tangent line at $(3, -27)$. See Figure 12–16. ∎

EXAMPLE 9 Let us find the relative maxima and minima of the rational function f defined by

$$f(x) = \frac{x^2}{x^2 - 1}.$$

Note that $f(1)$ and $f(-1)$ do not exist.

Since

$$f'(x) = -\frac{2x}{(x^2 - 1)^2},$$

setting $f'(x) = 0$ and solving for x yields $x = 0$. Thus 0 is a critical value for f. The derivative f' does not exist at $x = \pm 1$, but ± 1 are not critical values because they are not in the domain of f. Thus our step 1 diagram is as follows:

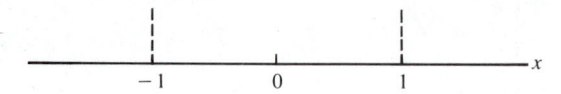

Here we have employed dashed lines at $x = \pm 1$ to remind us that f cannot have relative maxima or minima at these values because they are not critical values.

Now,

if $x < -1$, $f'(x) > 0$;
if $-1 < x < 0$, $f'(x) > 0$;

if $0 < x < 1$, $f'(x) < 0$;
if $x > 1$, $f'(x) < 0$.

Therefore we obtain the diagram

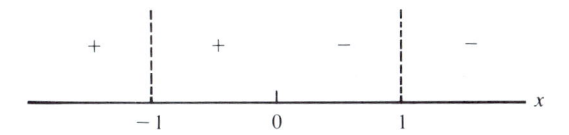

```
    +   |   +   |   -      -
  ──────┼───────┼──────────── x
       -1      0       1
```

which leads in turn to

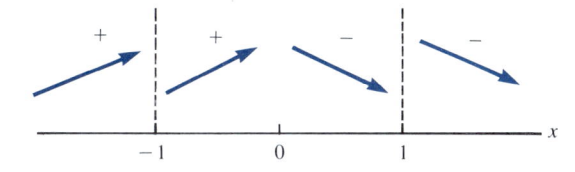

```
    +   |   +   |   -      -
   ↗    |  ↗    |   ↘       ↘
  ──────┼───────┼──────────── x
       -1      0       1
```

This last diagram shows that f has a relative maximum at $(0, f(0)) = (0, 0)$. The graph of f is shown in Figure 12–17.

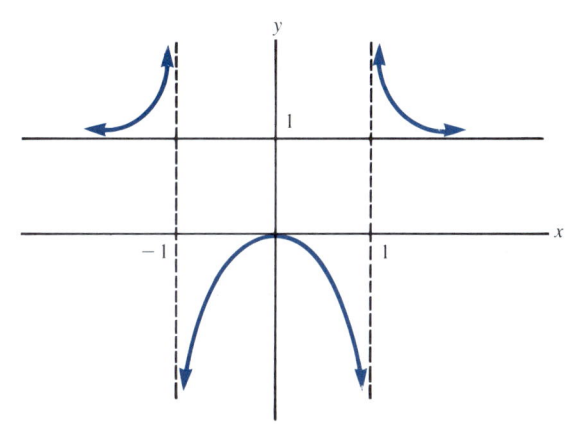

FIGURE 12–17

■ *Computer Manual: Program DERTEST1*

12.1 EXERCISES

In Exercises 1 through 6, find all critical values, relative maxima, and relative minima of the function whose graph is shown.

1.

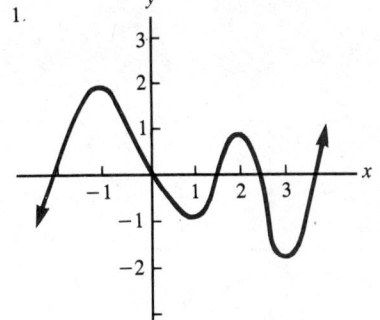

2.

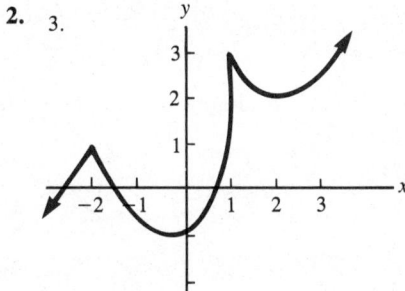

3.

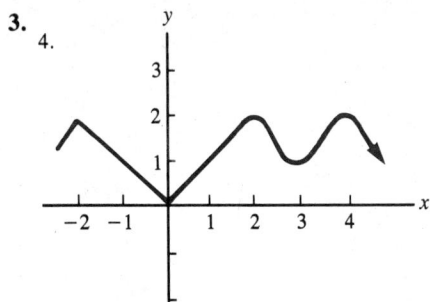

4.

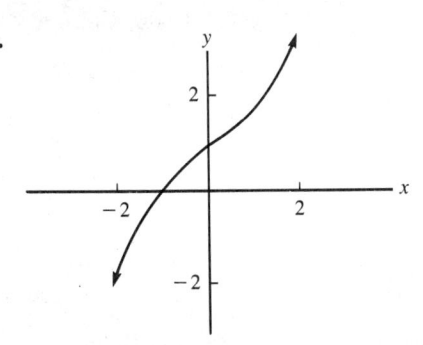

5.

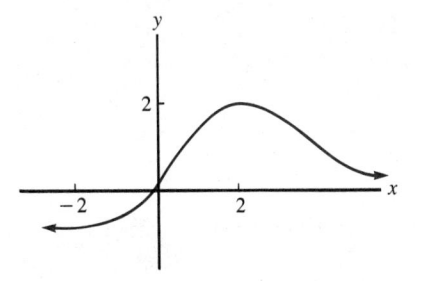

6.

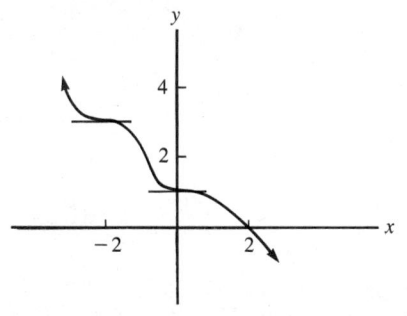

In Exercises 7 through 22, find the intervals on which f is increasing and those on which it is decreasing.

7. $f(x) = x^2 - 10x + 24$
8. $f(x) = -x^2 + 2x + 3$
9. $f(x) = x^3$
10. $f(x) = -x^3 + 3x^2 - 3x$
11. $f(x) = -2x^3 - 9x^2 + 60x + 10$
12. $f(x) = 2x^3 - 27x^2 - 120x$
13. $f(x) = \dfrac{x^3}{3} - \dfrac{3x^2}{2} + \dfrac{5x}{4} + 1$
14. $f(x) = x^4 - 1$
15. $f(x) = x^4 - 8x^2 + 1$
16. $f(x) = 3x^4 + 4x^3 - 12x^2$
17. $f(x) = \sqrt{x}$
18. $f(x) = x^{1/3}$
19. $f(x) = \dfrac{1}{x}$
20. $f(x) = \dfrac{x}{x+1}$
21. $f(x) = \dfrac{x+1}{2x-4}$
22. $f(x) = \dfrac{x^2}{x+1}$

In Exercises 23 through 53, use the first derivative test to find all relative maxima and relative minima for f.

23. $f(x) = x^2 - 4x + 3$
24. $f(x) = -2x^2 - 8x + 14$
25. $f(x) = 0.25x^2 + 3x - 9$
26. $f(x) = x^3 + 4$
27. $f(x) = x^3 - 2x^2$
28. $f(x) = 3x^3 - 9x$

29. $f(x) = -x^3 + 12x$
30. $f(x) = x^3 + 6x^2 + 12x + 8$
31. $f(x) = x^3 + 3x^2 - 9x$
32. $f(x) = -x^3 + 21x^2 - 120x + 20$
33. $f(x) = x^3 - 6x^2 + 12x + 2$
34. $f(x) = -2x^3 + 15x^2 - 24x + 8$
35. $f(x) = 2x^3 + 9x^2 - 108x + 20$
36. $f(x) = x^4$
37. $f(x) = x^4 - 2x^2$
38. $f(x) = 3x^4 + 4x^3$
39. $f(x) = 3x^4 + 4x^3 - 12x^2$
40. $f(x) = -3x^4 + 4x^3 + 6x^2 - 12x$ [Hint: $f'(x) = -12(x - 1)(x^2 - 1)$.]
41. $f(x) = 3x^5 - 20x^3 + 10$
42. $f(x) = 3x^5 - 10x^3 + 15x + 1$ [Hint: $f'(x) = 15(x^2 - 1)^2$.]
43. $f(x) = 2x^6 - 27x^4$
44. $f(x) = x^{4/3}$
45. $f(x) = \dfrac{2x - 1}{x + 2}$
46. $f(x) = \dfrac{3x + 2}{x - 4}$
47. $f(x) = \dfrac{x + 2}{x^2 - 3}$
48. $f(x) = \dfrac{x + 1}{-3 - x^2}$
49. $f(x) = \dfrac{x^2}{x + 1}$
50. $f(x) = \dfrac{10x^2 + 4}{20x + 3}$
51. $f(x) = x + \dfrac{1}{x}$
52. $f(x) = x^2 + \dfrac{2}{x}$
53. $f(x) = \dfrac{1}{3}x^3 + \dfrac{16}{x}$

54. During a superconductivity experiment, a solution is cooled rapidly and then heated more slowly. If the temperature of the solution t minutes after the start of the experiment is T degrees Kelvin, where $T = 2t + 4/t$, find the intervals on which T is increasing and those on which it is decreasing, and find the minimum temperature of the solution and the time when it occurs.

Management

55. A worker's productivity is given by
$$f(t) = -0.05t^2 + 1.8t + 150,$$
where t is the number of years the worker has been on the job and $f(t)$ is the dollar value of the worker's average daily production in year t. Find the intervals on which the worker's productivity is increasing and those on which it is decreasing. Find the worker's maximum productivity and the year in which it occurs.

56. A firm's profit function is given by
$$P(x) = -x^3 + 135x^2 - 2400x - 75,000.$$
Find the intervals on which profit is increasing and those on which it is decreasing, and find the maximum profit and the number of units required to attain it.

57. If Beatific Corporation spends x thousand dollars on advertising, its sales will be $f(x)$ dollars, where
$$f(x) = 2x^3 - 210x^2 + 6000x + 100,000, \quad 0 \le x \le 40.$$
Find the intervals on which sales are increasing and those on which they are decreasing. Find the firm's maximum sales and the amount that must be spent on advertising to attain maximum sales.

58. Amity Company's average cost function is given by
$$A(x) = 10x + \dfrac{4000}{x},$$

where x is the number of units produced, $x \geq 1$. Find the intervals on which the average cost is increasing and those on which it is decreasing, and find the minimum average cost per unit.

59. A radio station's share of the audience during 1989 was given by

$$s = \frac{t-2}{t^2+12} + \frac{3}{10},$$

where t is in months, $1 \leq t \leq 12$, with $t = 1$ representing January 1989. Find the intervals on which the audience share was increasing and those on which it was decreasing, and find the maximum audience share and the time during the year when it occurred.

*60. Show that
  (a) if marginal revenue is less than marginal cost on an interval, then profit is decreasing on the interval;
  (b) if marginal revenue is greater than marginal cost on an interval, then profit is increasing on the interval.

61. Suppose the demand function for a product is given by

$$q = -1000p + 10,000.$$

  (a) Find the intervals on which revenue from sales of the product is increasing and those on which it is decreasing.
  (b) Find the maximum revenue and the price at which it occurs.
  (c) Find the intervals on which demand for the product is elastic and those on which it is inelastic, and find the price at which demand is of unit elasticity. Compare your answers to those of parts *(a)* and *(b)*.

62. Repeat Exercise 61 if the demand function is $q = 216,000 - 20p^2$.

*63. Show the following:
  (a) if demand for a product is elastic on a price interval, then revenue from sales of the product decreases as the price increases within the interval;
  (b) if demand for a product is inelastic on a price interval, then revenue from sales of the product increases as the price increases within the interval;
  (c) maximum revenue occurs at the price for which demand is of unit elasticity.

Life Science

64. The number of birds in a wildlife sanctuary in month t of the year is estimated to be

$$y = -t^3 + 5.25t^2 + 33t + 1000.$$

The maximum bird population occurs at a relative maximum of this function. Find the intervals on which the size of the bird population is increasing and those on which it is decreasing, and find the maximum bird population and the time during the year when it occurs.

65. The number of gypsy moths at time t is given by

$$y = \frac{10t}{t^2+4}, \quad t \geq 0,$$

where t is in months, with $t = 0$ representing the present, and y is in millions. Find the intervals on which the gypsy moth population is increasing and those on which it is decreasing, and find the time at which the population will peak.

66. During an epidemic the number of people inoculated during month t was given by

$$y = -24t^5 + 165t^4 + 800t^3, \quad 0 \le t \le 10.$$

Find the intervals on which the number inoculated was increasing and those on which it was decreasing, and find the maximum number inoculated during a month and when it occurred.

Social Science

67. Enrollment in the higher education system of a nation for the period 1990–2010 is forecast to be

$$y = 0.20t^5 - 7.5t^4 + 99t^3 - 594t^2 + 1620t + 5000$$

thousand students in year t, where $t = 0$ represents 1990. Find the intervals on which enrollment is increasing and those on which it is decreasing. Find the maximum and minimum enrollments and the years in which they will occur. [*Hint:* $y' = (t - 3)(t - 6)^2(t - 15)$.]

68. Political advisors for a certain candidate estimate that if x thousand people vote in the next election, their candidate will receive y thousand votes, where

$$y = 10\sqrt{x} - 0.5x - 5.$$

Find the intervals on which the candidate's vote is increasing and those on which it is decreasing, and find the candidate's maximum number of votes.

12.2 RELATIVE MAXIMA AND MINIMA: THE SECOND DERIVATIVE TEST

In this section we present another method for testing the critical values of a function to find out whether they yield relative maxima or relative minima. The test we will develop, which is called the **second derivative test,** involves substituting critical values into the second derivative of the function.

In the previous section we noted that if f' is positive on an interval, then f is increasing over the interval. Since this is true for any function and its derivative, let us apply it to the function f' and its derivative f'': if $f''(x) > 0$ for all x in an interval, then f' must be increasing over the interval. But if f' is increasing over the interval, then the slopes of the tangent lines to f are increasing over the interval, as shown in Figure 12–18, for instance. Therefore in order for the graph of f to fit the tangent lines, it must lie above them and bend upward when $f''(x) > 0$.

Similarly, if the derivative of a function is negative over an interval, then the function is decreasing over the interval. Thus if $f''(x) < 0$ for all x in an interval, then f' is decreasing over the interval. But then the slopes of the tangent lines to f are decreasing, as in Figure 12–19. Thus, in order to fit the tangent lines, the graph of f must lie below them and bend downward over the interval.

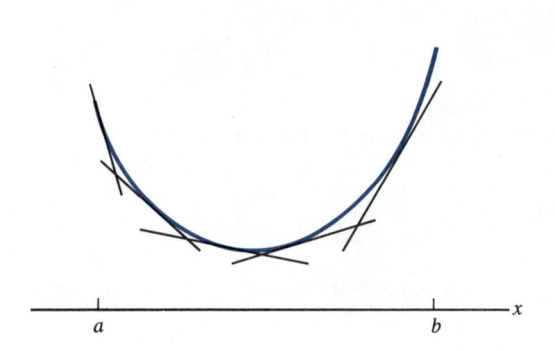

FIGURE 12–18
Slopes of Tangent Lines Increasing on *(a, b)*

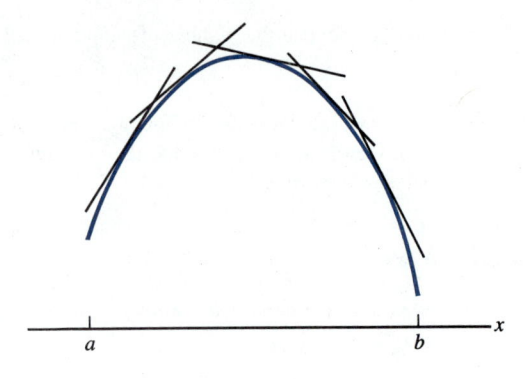

FIGURE 12–19
Slopes of Tangent Lines Decreasing on *(a, b)*

Concavity

1. If $f''(x) > 0$ for all x in an interval, the graph of f bends upward over the interval, as shown in Figure 12–20, and f is said to be **concave upward** over the interval.

2. If $f''(x) < 0$ over an interval, the graph of f bends downward over the interval, as shown in Figure 12–21, and f is said to be **concave downward** over the interval.

EXAMPLE 1 If $f(x) = x^2$, then $f''(x) = 2$. Hence $f''(x)$ is positive for all x, and therefore f is concave upward for all x. See Figure 12–22. ◻

EXAMPLE 2 If $g(x) = \frac{1}{3}x^3 - x^2 - 3x + 2$, then $g''(x) = 2(x - 1)$. Note that if $x < 1$, then $g''(x) < 0$, while if $x > 1$, then $g''(x) > 0$. Thus $g''(x) < 0$ over the interval $(-\infty, 1)$, and $g''(x) > 0$ over the interval $(1, +\infty)$. Therefore g is concave downward over $(-\infty, 1)$ and concave upward over $(1, +\infty)$. See Figure 12–23. ◻

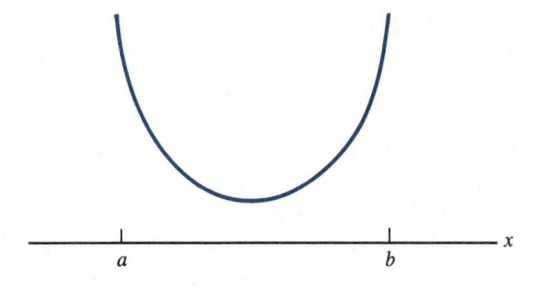

FIGURE 12–20
f Concave Upward over *(a, b)*

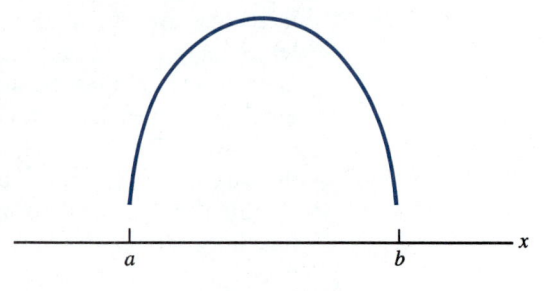

FIGURE 12–21
f Concave Downward over *(a, b)*

APPLICATIONS OF DIFFERENTIATION 753

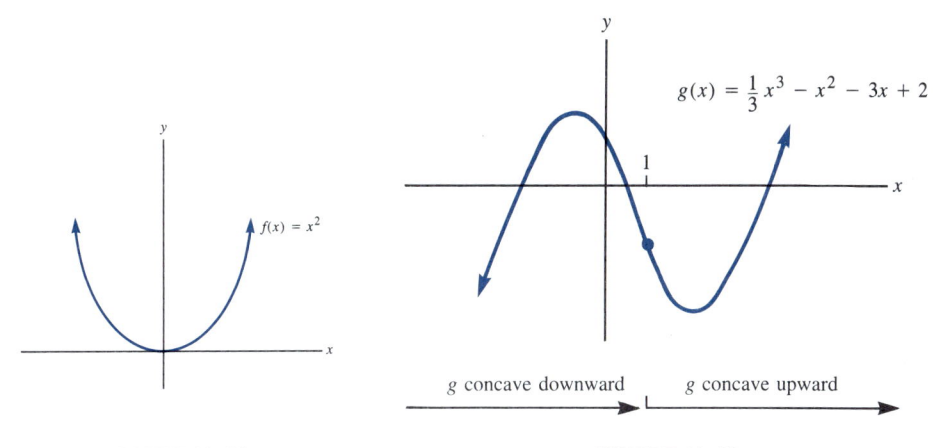

FIGURE 12–22

FIGURE 12–23

Now suppose that $f'(a) = 0$ and $f''(a) < 0$. Then f is concave downward at $(a, f(a))$ and also has a horizontal tangent there (see Figure 12–24), and thus $(a, f(a))$ must be a relative maximum point for f. Similarly, if $f'(a) = 0$ and $f''(a) > 0$, f is concave upward and has a horizontal tangent at $(a, f(a))$ (see Figure 12–25), so $(a, f(a))$ must be a relative minimum point for f. This shows that we can sometimes test critical values by substituting them into the second derivative.

The Second Derivative Test

1. If $f'(a) = 0$ and $f''(a) > 0$, then $(a, f(a))$ is a relative minimum point for f.

2. If $f'(a) = 0$ and $f''(a) < 0$, then $(a, f(a))$ is a relative maximum point for f.

3. If $f'(a) = 0$ and $f''(a) = 0$ or $f''(a)$ does not exist, the test fails.

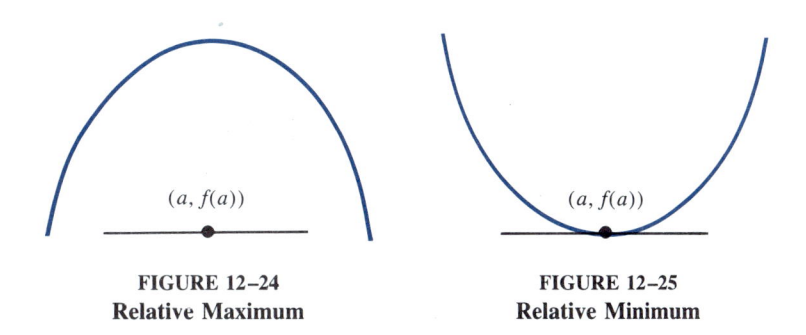

FIGURE 12–24
Relative Maximum

FIGURE 12–25
Relative Minimum

EXAMPLE 3 Let $f(x) = -2x^3 - 3x^2 + 36x + 5$. We have
$$f'(x) = -6x^2 - 6x + 36 = -6(x+3)(x-2)$$
and
$$f''(x) = -12x - 6 = -6(2x+1).$$
Since
$$f'(-3) = 0 \quad \text{and} \quad f''(-3) = 30 > 0,$$
f has a relative minimum at $(-3, f(-3)) = (-3, -76)$. Since
$$f'(2) = 0 \quad \text{and} \quad f''(2) = -30 < 0,$$
f has a relative maximum at $(2, f(2)) = (2, 49)$. □

The second derivative test is often easier to apply than the first derivative test, but it is not as general as the latter. First of all, if $x = a$ is a critical value and $f'(a)$ does not exist, then the second derivative test does not apply. Second, even if $f'(a) = 0$, the test tells us nothing about whether $(a, f(a))$ is a relative maximum or a relative minimum if $f''(a) = 0$ or if $f''(a)$ does not exist. In such cases we must use some other means (for example, the first derivative test) to analyze the behavior of f at $(a, f(a))$.

EXAMPLE 4 Let g be defined by $g(x) = x^{2/3}$. Then $x = 0$ is a critical value for g (see Example 2, Section 12.1), but we cannot use the second derivative test to determine whether $(0, g(0)) = (0, 0)$ is a relative maximum or relative minimum point for g because $g'(0)$ does not exist. The first derivative test (see Example 6, Section 12.1) shows that in fact g has a relative minimum at $(0, 0)$. □

EXAMPLE 5 Let $f(x) = -x^3 + 6x^2 - 12x + 8$. Then
$$f'(x) = -3(x-2)^2 \quad \text{and} \quad f''(x) = -6(x-2),$$
so $f'(2) = 0$ and $f''(2) = 0$. Thus the second derivative test cannot be used to classify the point $(2, f(2)) = (2, 0)$ as a relative maximum or a relative minimum point for f. The first derivative test leads to the diagram

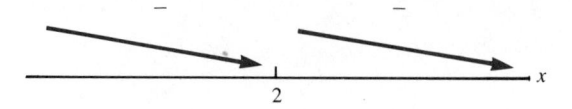

(Check this.) Hence $(2, 0)$ is neither a relative maximum nor a relative minimum point for f, but rather a point where the graph "flattens out." □

EXAMPLE 6 Let $g(x) = x^{4/3}$. We have
$$g'(x) = \frac{4}{3} x^{1/3} \quad \text{and} \quad g''(x) = \frac{4}{9} x^{-2/3}.$$

Therefore $g'(0) = 0$, but $g''(0)$ does not exist, and again the second derivative test cannot be used. The first derivative test leads to the diagram

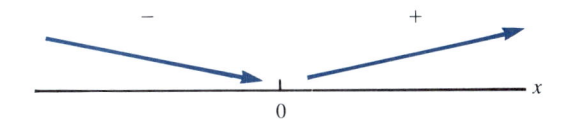

Hence $(0, g(0)) = (0, 0)$ is a relative minimum for g. ☐

■ *Computer Manual: Program DERTEST2*

12.2 EXERCISES

In Exercises 1 through 12, find the intervals over which the given function is concave upward and those over which it is concave downward.

1. $f(x) = x^2 - 4x$
2. $f(x) = -x^2 + 8x - 9$
3. $f(x) = x^3$
4. $f(x) = -x^3 + 2x$
5. $f(x) = 2x^3 - 12x^2 + 6x - 9$
6. $f(x) = -2x^3 + 15x^2 - 12x + 4$
7. $f(x) = x^4 - 16$
8. $f(x) = x^4 - 24x^2$
9. $f(x) = -x^4 + 2x^3 + 36x^2 + 4x$
10. $f(x) = x^5 - 10x$
11. $f(x) = -x^5 + 80x^2$
12. $f(x) = 3x^5 - 5x^4$

In Exercises 13 through 32, use the second derivative test to find all relative maxima and minima for the given function. If the second derivative test cannot be applied or if it fails, use the first derivative test.

13. $f(x) = -x^2 + 6x + 5$
14. $f(x) = 3x^2 - 30x + 1$
15. $f(x) = x^3 + 9x^2$
16. $f(x) = x^3 + 12x^2 + 48x + 3$
17. $f(x) = 2x^3 - 15x^2 - 216x + 25$
18. $f(x) = -4x^3 + 6x^2 - 3x + 5$
19. $f(x) = -x^3 + 6x^2 + 63x$
20. $f(x) = -x^4 + 4x^3 + 30$
21. $f(x) = x^4 - 4x^2 + 10$
22. $f(x) = x^4/4 + 2x^3 + 5x^2/2 - 10$
23. $f(x) = -x^4 + 8x^3 - 22x^2 + 24x$ [Hint: $f'(x) = -4(x - 1)(x^2 - 5x + 6)$]
24. $f(x) = 3x^4 - 40x^3 + 150x^2 - 10$
25. $f(x) = 3x^5 + 5x^3$
26. $f(x) = 6x^5 - 15x^4 + 10x^3 + 25$
27. $f(x) = -6x^5 + 45x^4 - 110x^3 + 90x^2 - 50$ [Hint: $f'(x) = -30x(x - 1)(x^2 - 5x + 6)$]
28. $f(x) = 4x^5 + 15x^4 - 40x^2$ [Hint: $f'(x) = 20x(x - 1)(x^2 + 4x + 4)$]
29. $f(x) = x^{8/5}$
30. $f(x) = x^{4/5}$
31. $f(x) = (x - 1)^{11/5}$
32. $f(x) = -2(x - 1)^{7/5}$

*33. The second derivative test is not always easier to apply than the first derivative test. To see this, use the first derivative test and then the second derivative test to find the relative maxima and relative minima of f if

$$f(x) = \frac{x + 1}{x^2 + 3}.$$

Management

34. Caritas, Inc., knows that the demand function for its product is given by
$$p = 10{,}000 - 8x.$$
The company's maximum revenue occurs at a relative maximum of its revenue function. Use the second derivative test to find the maximum revenue.

35. Dismal Company's cost function is given by $C(x) = x^2 + 12x + 900$. The firm's minimum average cost per unit occurs at a relative minimum of its average cost function. Use the second derivative test to find the minimum average cost per unit.

36. If a firm spends x thousand dollars on advertising, its net profit will be $P(x)$ dollars, where
$$P(x) = x^3 - 240x^2 + 18{,}900x - 100{,}000, \qquad 0 \le x \le 100.$$
The firm's maximum profit occurs at a relative maximum of this function. Use the second derivative test to find the firm's maximum profit and the amount that must be spent on advertising in order to attain it.

37. A firm's profit function is given by
$$P(x) = -0.25x^4 + 20x^3 - 800{,}000.$$
The firm's maximum profit occurs at a relative maximum of P. Use the second derivative test to find the maximum profit.

*38. Show that average cost per unit is minimized at the value of x for which average cost per unit equals marginal cost. You may assume that the cost function is concave upward for all $x \ge 0$.

*39. Exercise 38 shows that the average cost per unit is minimized at the value $x = a$ for which average cost equals marginal cost. Show also that
   (a) average cost is greater than marginal cost on the interval $(0, a)$;
   (b) average cost is less than marginal cost on the interval $(a, +\infty)$.

Life Science

40. Suppose that t minutes after the injection of a temperature-reducing drug a patient's temperature is $y°F$, where
$$y = -0.006t^2 + 0.36t + 102.8.$$
The maximum temperature occurs at a relative maximum of this function. How long after the injection of the drug does the maximum temperature occur? What is the maximum temperature? Use the second derivative test to answer these questions.

41. The number of bacteria in a culture at time t was y thousand, where
$$y = 2\sqrt{t+1} + \frac{8}{\sqrt{t+1}}.$$
Here t is in hours since the start of an experiment. The minimum number of bacteria present occurred at a relative minimum of this function. Use the second derivative test to find the minimum number of bacteria present and when this minimum occurred.

Social Science

42. Demographics suggest that over the next 20 years, the cohort of 18- to 25-year-olds in a city will contain approximately

$$y = 0.2t^5 - 6.625t^4 + 51t^3 + 40{,}000$$

individuals. Use the second derivative test to find the maximum and minimum number in the cohort, assuming that these occur at a relative maximum and relative minimum of the function.

43. Police records for a large city suggest that the city's crime rate may be described by the equation

$$y = 12.5\sqrt{t} + \frac{400}{t}.$$

Here y is the number of crimes per thousand population in year t, and $t = 1$ represents 1980. The minimum crime rate occurs at a relative minimum of this function. Use the second derivative test to find the minimum crime rate and the year in which it will occur.

12.3 ABSOLUTE MAXIMA AND MINIMA

The largest value a function can assume on an interval is called its **absolute maximum** on the interval; the smallest value it can assume is its **absolute minimum** on the interval. For instance, Figure 12–26 shows the graph of a function which describes

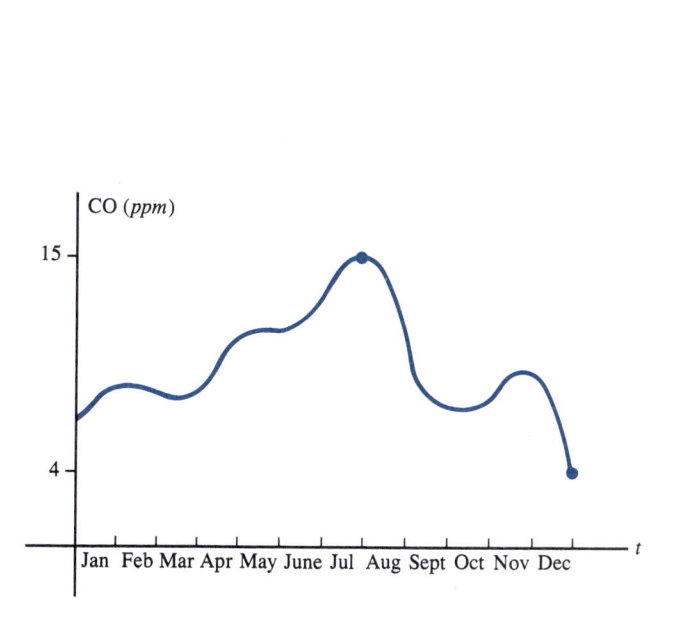

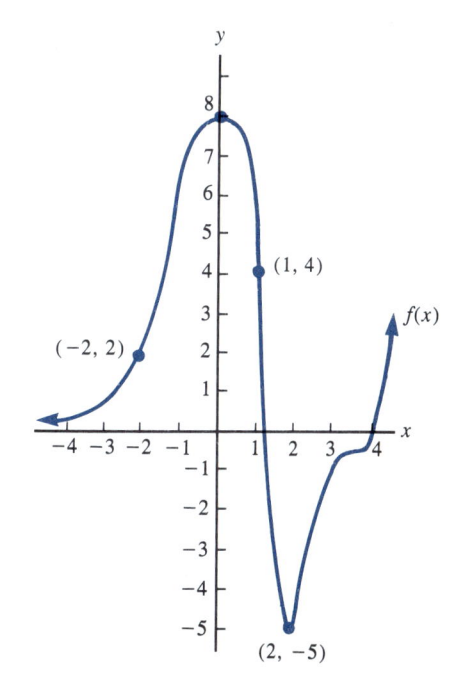

FIGURE 12–26

FIGURE 12–27

the concentration in parts per million of carbon monoxide in the air of a large city during 1989. The absolute maximum of this function is 15 ppm; it occurred on August 1. The absolute minimum of the function is 4 ppm, which occurred on December 31. It is often necessary to find the absolute maximum or absolute minimum of a function. In this section we show how to use the first derivative test to do this.

The existence and value of the absolute maximum and absolute minimum of a function depend not only on the function itself but also on the interval over which it is considered. For instance, consider the function f whose graph is pictured in Figure 12–27:

- On the entire x-axis f has an absolute minimum of -5 at $x = 2$; however, f has no absolute maximum on the entire x-axis because as x approaches $+\infty$, the functional values $f(x)$ increase without bound.
- On the interval $[-1, 4]$, f has an absolute minimum of -5 at $x = 2$ and an absolute maximum of 8 at $x = 0$.
- On the interval $[-2, 1]$, f has an absolute minimum of 2 at $x = -2$ and an absolute maximum of 8 at $x = 0$.
- On the interval $(-\infty, -2]$, f has an absolute maximum of 2 at $x = -2$; however, f has no absolute minimum on $(-\infty, -2]$ because as x approaches $-\infty$, the functional values $f(x)$ approach but never equal zero.

It should now be clear that the absolute maximum or absolute minimum of a function on an interval may or may not exist; if it does exist, it may occur at either a relative maximum or relative minimum within the interval or at an endpoint of the interval. Therefore in order to find the absolute maximum or absolute minimum of a function on an interval, we must examine its relative maxima or minima and the values of the function at the endpoints of the interval.

Finding the Absolute Maximum and Absolute Minimum

Let f be a continuous function defined on an interval $[a, b]$.

1. The absolute maximum of f on $[a, b]$ is the largest of $f(a)$, $f(b)$, and all the relative maxima that lie within $[a, b]$.

2. The absolute minimum of f on $[a, b]$ is the smallest of $f(a)$, $f(b)$, and all the relative minima that lie within $[a, b]$.

The first derivative test is very useful when we must find absolute maxima and absolute minima, because the diagram of step 3b of the test tells us where to look for them. We illustrate with examples.

EXAMPLE 1   Let

$$f(x) = x^3 - 3x^2 - 45x + 1.$$

APPLICATIONS OF DIFFERENTIATION

Then
$$f'(x) = 3x^2 - 6x - 45 = 3(x + 3)(x - 5),$$
and the diagram of step 3b of the first derivative test looks like this:

```
      +           -          +
  ─────▶     ─────▶     ─────▶
─────────┼──────────┼──────────── x
        -3          5
```

(Check this.) Suppose we want to find the absolute maximum and absolute minimum of f on the interval $[0, 8]$. Then we can only consider the function over this interval, so our diagram becomes

```
            -              +
     ───────────▶    ─────▶
     ┼──────────────┼──────┼
     0              5      8
```

The diagram shows that the absolute minimum must occur at $x = 5$; therefore the absolute minimum is equal to $f(5) = -174$. It also shows that the absolute maximum must occur at either $x = 0$ or $x = 8$. Since $f(0) = 1$ and $f(8) = -39$, the absolute maximum occurs at $x = 0$ and is equal to 1. □

EXAMPLE 2 Let f be as in Example 1, but this time let us find its absolute maximum and absolute minimum on the interval $[-5, 6]$. The diagram of step 3b of the first derivative test then becomes

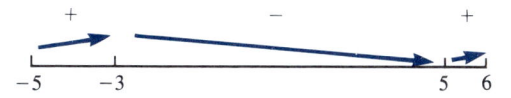

```
      +                -            +
  ─────▶       ───────────▶    ───▶
  ┼──────┼──────────────────┼───┼
 -5     -3                  5   6
```

Therefore the absolute maximum of f on the interval $[-5, 6]$ must occur at either $x = -3$ or $x = 6$. Since $f(-3) = 82$ and $f(6) = -161$, the absolute maximum is 82 at $x = 3$. Similarly, the absolute minimum must occur at either $x = -5$ or $x = 5$. Since $f(-5) = 26$ and $f(5) = -174$, the absolute minimum is -174 at $x = 5$. □

EXAMPLE 3 Again let f be as in Example 1.

a. On the interval $[0, +\infty)$ the diagram becomes

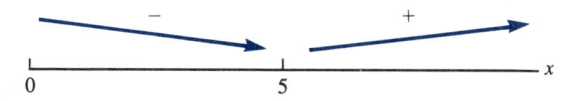

```
            -              +
     ───────────▶    ─────▶
     ┼──────────────┼────────── x
     0              5
```

Thus on $[0, +\infty)$ the function has an absolute minimum of -174 at $x = 5$. It has no absolute maximum, since its functional values increase without bound as $x \to +\infty$.

b. On the entire x-axis the diagram becomes

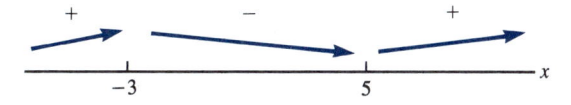

```
      +           -          +
  ─────▶     ─────▶     ─────▶
─────────┼──────────┼──────────── x
        -3          5
```

so on the x-axis the function has neither an absolute maximum nor an absolute minimum. (Why not?) ☐

Now let us turn to some practical problems involving the absolute maximum and absolute minimum. Our first problem concerns profit maximization.

EXAMPLE 4 The profit function of Altruism, Inc., is given by

$$P(x) = -0.01x^3 + 300x - 10,000.$$

Let us find Altruism's maximum daily profit if the firm can make at most 400 units of product per day.

We seek to maximize

$$P(x) = -0.01x^3 + 300x - 10,000$$

for $0 \le x \le 400$, that is, on the interval $[0, 400]$.
Setting $P'(x) = 0$, we have

$$P'(x) = -0.03x^2 + 300 = 0,$$

which has solutions $x = 100$ and $x = -100$. Since -100 is not in our interval, we discard it, and our diagram is

Therefore the maximum profit is $P(100) = \$10,000$. Note that if we had said that Altruism could make at most 50 units per day, then the maximum profit would have been $P(50) = \$3750$. (Why?) ☐

EXAMPLE 5 Suppose we wish to minimize the amount of fencing that will be required to enclose a rectangular field of 900 square meters. The field will look like this:

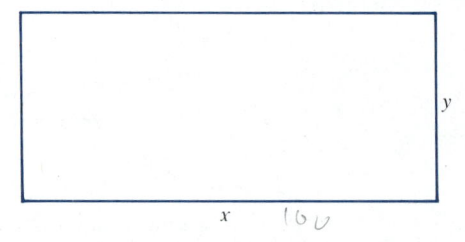

Here x and y are in meters. We must minimize the amount of fencing

$$F = 2x + 2y$$

subject to the constraint that the area of the field is 900 square meters, that is, that

$$xy = 900$$

APPLICATIONS OF DIFFERENTIATION

square meters. In order to apply our methods of minimization to F, we must write it as a function of a single variable. The constraint equation $xy = 900$ allows us to do this, for we may solve it for y to obtain

$$y = \frac{900}{x}.$$

Hence

$$F = 2x + 2\left(\frac{900}{x}\right) = 2x + \frac{1800}{x}.$$

Thus we must minimize

$$F = 2x + \frac{1800}{x}, \qquad x > 0.$$

But

$$\frac{dF}{dx} = 2 - \frac{1800}{x^2} = 0$$

implies that

$$2x^2 - 1800 = 0,$$

or

$$x = \pm 30.$$

Since we require $x > 0$, we discard the negative critical value. To test the value $x = 30$, we use the first derivative test: the diagram is

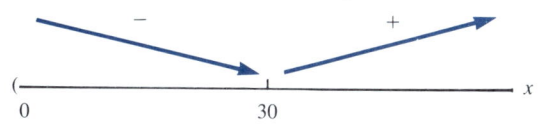

(Check this, and note that we have placed a parenthesis "(" at $x = 0$ to show that 0 is not included in our interval.) Hence $x = 30$ yields a minimum. Therefore $y = 900/30 = 30$, and the minimum amount of fencing is $F = 2(30) + 2(30) = 120$ meters. ☐

EXAMPLE 6   We wish to construct a box with a square bottom and top and rectangular sides. The material for the sides costs $2.00 per square foot and that for the top and bottom costs $0.25 per square foot. We intend to spend $24.00 on materials. Let us find the dimensions that will give us the box of maximum volume.

The box will look like this:

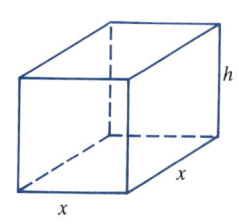

Here x and h are in feet. The volume V of the box is its length times its width times its height, or
$$V = x^2 h$$
cubic feet. The cost of the material used to make the box is
$$\$0.25 \text{ (area of top + area of bottom)} + \$2.00 \text{(area of four sides)}$$
$$= \$0.25(2x^2) + \$2.00(4xh)$$
$$= \$0.50x^2 + \$8.00xh,$$
and this must equal \$24.00. Thus we seek to maximize
$$V = x^2 h$$
subject to the constraint
$$0.5x^2 + 8xh = 24.$$
Now we use the constraint equation to express V in terms of a single variable: we solve
$$0.5x^2 + 8xh = 24$$
for h to obtain
$$h = \frac{24 - 0.5x^2}{8x} = \frac{3}{x} - \frac{x}{16}$$
and then substitute this into the expression for V to get
$$V = x^2 \left(\frac{3}{x} - \frac{x}{16} \right) = 3x - \frac{x^3}{16}.$$
Thus we seek to maximize
$$V = 3x - \frac{x^3}{16}, \qquad x > 0.$$
But
$$\frac{dV}{dx} = 3 - \frac{3x^2}{16} = 0$$
implies that
$$x^2 = 16,$$
so $x = \pm 4$. We may discard the negative critical value. Since
$$\frac{dV}{dx} > 0 \quad \text{if} \quad 0 < x < 4 \quad \text{and} \quad \frac{dV}{dx} < 0 \quad \text{if} \quad x > 4,$$
we have

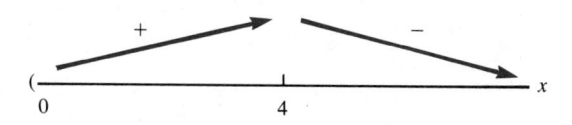

APPLICATIONS OF DIFFERENTIATION   763

Hence $x = 4$ yields a maximum. Therefore the dimensions that give the box of maximum volume are $x = 4$ feet and

$$h = \frac{3}{4} - \frac{4}{16} = \frac{1}{2}$$

foot, and the maximum volume is

$$V = 4^2 \left(\frac{1}{2}\right) = 8$$

cubic feet. ∎

An **inventory problem** is one concerned with minimizing the total cost of purchasing, ordering, and carrying a product in inventory. Total inventory cost is thus the sum of purchasing cost, ordering cost, and carrying cost, where

Purchasing cost = (Number of units purchased) (Price per unit),
Ordering cost = (Number of orders placed) (Cost of placing an order),

and

Carrying cost = (Average number of units in inventory) ×
                 (Cost of carrying a unit in inventory).

We solve an inventory problem by finding the **optimal reorder quantity,** which is the number of units that should be ordered each time to minimize total inventory cost.

EXAMPLE 7 Suppose the Shoe Shop sells 2000 pairs of Lazy Loafers each year. Each pair of Lazy Loafers is purchased for $10, and it costs $100 to place an order for them. Also, it costs $1.60 to carry a pair of shoes in inventory for one year. Let us find the optimal reorder quantity for this problem.

Let x denote the number of pairs of Lazy Loafers that should be ordered each time an order is made. Since 2000 pairs of Lazy Loafers are required during the course of a year, they must be ordered $2000/x$ times per year. Suppose we assume that pairs of Lazy Loafers are sold at a constant rate and that they sell out just as a new order arrives; then since each order consists of x pairs, the average number of pairs in inventory at any time will be $x/2$. Therefore

Annual purchasing cost = (2000 pairs)($10 per pair) = $20,000,

Annual ordering cost   = $\left(\dfrac{2000}{x} \text{ orders}\right)$ ($100 per order) = $\dfrac{\$200,000}{x}$,

Annual carrying cost   = $\left(\dfrac{x}{2} \text{ pairs}\right)$ ($1.60 per pair) = $0.80x$.

Thus if $I(x)$ is the total annual inventory cost that results from ordering x pairs of Lazy Loafers per order, then

$$I(x) = 20{,}000 + \frac{200{,}000}{x} + 0.80x,$$

where $x > 0$. We must minimize the inventory cost function I on the interval $(0, +\infty)$.

Since
$$I'(x) = -\frac{200{,}000}{x^2} + 0.80,$$

setting $I'(x) = 0$ and solving for x yields $x = \pm 500$. Because x must be positive, we may discard the critical value -500, and the diagram of the first derivative test looks like this:

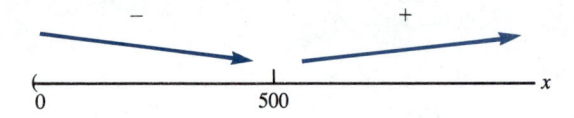

(Check this.) It is clear that the minimum inventory cost occurs at $x = 500$. Therefore the optimal reorder quantity is 500 pairs of Lazy Loafers per order. This reorder quantity will result in $2000/500 = 4$ orders per year, and the store's minimum annual inventory cost will be

$$I(500) = \$20{,}000 + \frac{\$200{,}000}{500} + \$0.80(500) = \$20{,}800. \quad \square$$

■ *Computer Manual: Program ABMAXMIN*

12.3 EXERCISES

1. Let f have the following graph:

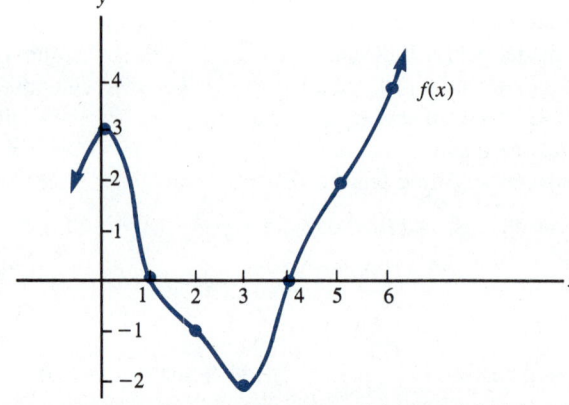

Find the absolute maximum and absolute minimum of f, if they exist, on
(a) the interval $[0, 4]$
(b) the interval $[2, 6]$
(c) the interval $(-\infty, 0]$
(d) the interval $[3, +\infty)$

2. Let g have the following graph:

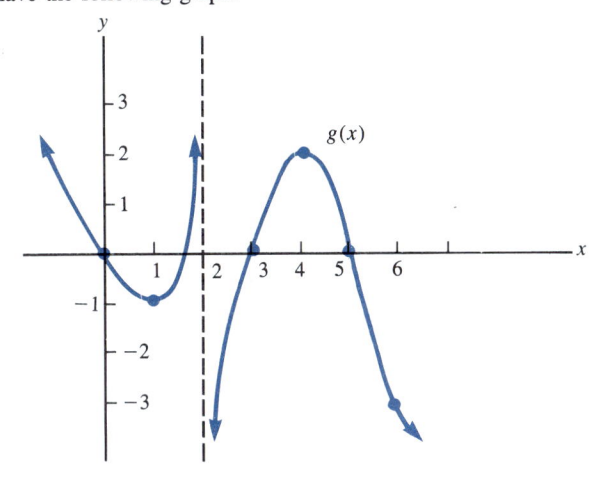

Find the absolute maximum and absolute minimum of g, if they exist, on
(a) the interval [0, 2)   (b) the interval (2, 6]
(c) the interval [3, 6]   (d) the x-axis

In Exercises 3 through 14, find the absolute maximum and absolute minimum of the given function on the given intervals.

3. $f(x) = 3x^2 - 18x - 5$, [0, 5], [6, 8], [0, $+\infty$), the x-axis.
4. $f(x) = -2x^2 + 8x + 14$, [-1, 1], [1, $+\infty$), ($-\infty$, -1], the x-axis.
5. $f(x) = x^3 - 12x$, [0, 6], [-1, 1], [-6, 6], the x-axis.
6. $f(x) = -x^3 + x^2$, ($-\infty$, 1], [1, $+\infty$), [0, 1], the x-axis.
7. $f(x) = -x^3 + 21x^2 - 120x + 200$, [0, 12], [0, 5], [5, 12], the x-axis.
8. $f(x) = 2x^3 + 9x^2 - 60x + 1$, [$-6$, 6], [$-6$, 3], [1, $+\infty$), ($-\infty$, 0].
9. $f(x) = x^4 - 32x$, ($-\infty$, 0], [-3, 3], [0, 4], [0, $+\infty$).
10. $f(x) = -x^4 + 2x^2 + 6$, [-2, 2], [-1, 0], [0, 1], [1, 2].
11. $f(x) = \dfrac{x-3}{x-1}$, ($-\infty$, 1), (1, $+\infty$), [0, 2], [2, 4].
12. $f(x) = \dfrac{x+1}{2x-3}$, [-1, $\frac{3}{2}$), ($\frac{3}{2}$, 2], [-1, 2], [-1, 1].
13. $f(x) = \dfrac{x^2}{x^2-9}$, ($-3$, 3), (3, $+\infty$), [-2, 2], x-axis.
14. $f(x) = \dfrac{x^2-1}{2x-4}$, [0, $+\infty$), ($-\infty$, 0], [0, 2), (2, 4].

15. A boat that travels at a speed of x miles per hour uses $y = 6x^{-1} + x/150$ gallons of fuel per mile. What speed gives the minimum fuel cost?

16. What are the dimensions of the rectangular field of maximum area that can be fenced in with exactly 800 feet of fencing?

*17. Prove that the rectangle of maximum area having a fixed perimeter P is actually a square with side $P/4$.

18. What are the dimensions of the rectangular field whose area is 10,000 square meters if the fencing around the field is to be a minimum?

*19. Prove that the rectangle of minimum perimeter having a fixed area A is actually a square with side $\sqrt{A}$.

20. You wish to fence in a rectangular field. One side of the field will be bordered by a stone wall and therefore will not require any fencing. The area of the field must be 12,800 square feet, and fencing costs $2 per foot. What should the dimensions of the field be in order to minimize the cost of the fencing if the stone wall is 200 feet long?

21. Do Exercise 20 if the stone wall is 100 feet long.

22. A square plot of land has sides 30 meters long. You fence off a corner of the plot with a straight fence 20 meters long. Find the dimensions that will maximize the area of the fenced-off triangle.

23. A rectangular plot of land with an area of 600 square meters is fenced in on all four sides and then divided into two plots of equal area by another fence parallel to one of the outside fences. Find the minimum fencing required.

24. A window is in the shape of a rectangle surmounted by a semicircle. If the perimeter of the window is 16 feet, find the dimensions that maximize its area.

25. A wire 100 centimeters (cm) long is cut into two pieces and one of the pieces is formed into a square while the other is bent into an equilateral triangle. Where should the wire be cut and which part of it should be bent into the square if the total area of the two figures is to be maximized? (*Hint:* The area of an equilateral triangle of side s is $s^2\sqrt{3}/4$.)

26. Repeat Exercise 25 if the total area of the two figures is to be minimized.

27. A road is built from point A to point B in the figure:

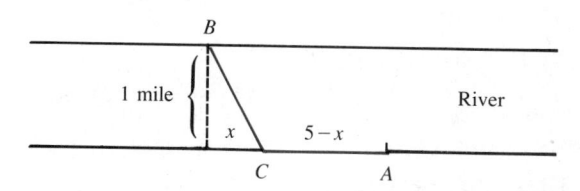

Suppose it costs $10,000 per mile to build the portion of the road along the river from A to C and $20,000 per mile to build the bridge from C to B. Where should C be located in order to minimize the cost of the road?

28. Repeat Exercise 27 if it costs $9000 per mile to build the bridge from C to B.

29. The base of a toy chest is to be a rectangle twice as long as it is wide. The material for the top and bottom of the chest costs $5 per square foot, and that for the sides costs $3 per square foot. The materials budget is $240. Find the dimensions that maximize the volume of the toy chest.

30. A box with an open top is to have a square base and a volume of 500 cubic centimeters. If the material for the box costs $0.50 per square centimeter, find the dimensions of the box that minimize its cost.

31. A cylindrical container with a circular base and lid is to have a volume of 54 cubic inches. Find the dimensions that minimize the amount of material from which the container is made.

Management

32. A firm's profit function is given by
$$P(x) = -2x^3 + 270x^2 - 4800x - 6000,$$
where x is in thousands of units and profit is in dollars. Find the firm's maximum profit if it can produce at most 100,000 units; if it can produce at most 50,000 units.

33. A firm's profit function is given by
$$P(x) = -2x^3 + 216x^2 - 4320x - 20,000.$$
Find the firm's maximum profit if it can make and sell at most
(a) 100 units (b) 50 units

34. Derma Corporation's cost function is given by
$$C(x) = 0.01x^2 + 10x + 40,000.$$
Find Derma's minimum average cost per unit if it can produce at most 5000 units; if it can produce at most 1500 units.

35. The demand function for a company's product is defined by
$$p = 1000 - \frac{q}{200}.$$
Find the company's maximum revenue and the price per unit that the company must charge to attain it if the firm can produce at most 200,000 units; if it can produce at most 80,000 units.

36. A chain of furniture stores sells 10,000 sofas each year. Each sofa costs the chain $500, and it costs $200 to order sofas from the supplier. It also costs the chain $100 to carry a sofa in inventory for a year. Assuming an average inventory equal to one half the optimal reorder quantity, find the optimal reorder quantity, the number of orders the chain must make each year, and the minimum annual inventory cost.

37. The Drive N' Shop supermarket sells 100 boxes of soap each week. Each box costs the store $0.50, each order of soap costs $2.00, and it costs $0.01 to carry a box of soap in inventory for one week. Assuming an average inventory equal to one half the optimal reorder quantity, find the optimal reorder quantity, the number of orders required each week, and the minimum weekly inventory cost.

*38. Octopus Industries makes radios. The annual demand for its radios is 50,000 units. It costs Octopus Industries $70 to make a radio, $2000 to set up a production run of radios, and $2 to carry a radio in inventory for a year. Assuming an average inventory equal to one half the number of radios produced per production run, find the number of radios that Octopus should produce in each production run, the number of production runs required each year, and the minimum inventory cost for the radios. (*Hint:* Let x denote the number of production runs the company should make each year.)

*39. Suppose that the quantity of a product demanded is Q units per time period, that it costs K dollars to place each order for the product, and that it costs A dollars to carry 1 unit of the product in inventory for the time period. Assuming that the average inventory is one half the optimal reorder quantity, show that the optimal reorder quantity is

$$x = \sqrt{\frac{2QK}{A}}.$$

40. A health club offers to give employees of a certain corporation a discount on its membership fee. If 20 employees (the minimum number the club will accept) join, the fee will be $100 each. For every additional employee who joins, the club will reduce the fee for each employee by $1. Find the number of employees that will maximize the club's revenue if the club can accept at most
    (a) 80 new members   (b) 50 new members

41. A promoter can sell 20,000 tickets to a rock concert at $10 each. For every dollar the price of a ticket is raised, it is estimated that 200 fewer tickets will be sold. Find the promoter's maximum revenue.

42. Slicendice Company has $120,000 to spend on advertising. Each radio ad will cost $3000, and each newspaper ad $6000. If Slicendice runs x radio ads and y newspaper ads, it will obtain

$$E = 10{,}000x - 30{,}000xy + 20{,}000y + 6{,}000{,}000$$

effective exposures to its ads. How many of each type of ad should the company run in order to maximize effective exposures?

43. Gruel World Corporation makes regular and deluxe model globes. Each regular globe costs $4 to make, and each deluxe one costs $8 to make. The corporation has budgeted $800 for daily production costs. If it makes x regular and y deluxe globes per day, its daily profit is $P = -x^2 + 3xy - 400$ dollars. How many of each type of globe must the company produce in order to maximize daily profit?

44. An advertising handbill is to have 24 square inches of printed material, with margins of 1 inch along each of the two sides, 2 inches at the top, and 1 inch at the bottom. Find the dimensions of the handbill that will minimize the cost of the paper used.

45. A florist sells 160 bouquets a day at $12 each. For each $1 rise in the price of a bouquet, sales will decrease by 10 bouquets per day. Find the florist's maximum revenue from sales of bouquets and the price that must be charged in order to attain it.

46. Repeat Exercise 45 if each $1 rise in price causes sales to decrease by 15 bouquets per day.

*47. (This exercise was suggested by material in Budnick, McLeavey, and Mojena, *Principles of Operations Research for Management*, second edition, Irwin, 1988.)
    A drill press costs $500,000 new and depreciates until its scrap value is $50,000. Its average operating cost t years after purchase is given by

$$\overline{C}(t) = 40{,}000 + 4500t.$$

Find the time when the drill press should be replaced. (*Hint:* Minimize the average cost per time period, which is given by

$$\frac{500{,}000 - 50{,}000}{t} + \overline{C}(t).)$$

*48. (This exercise was suggested by material in Budnick, McLeavey, and Mojena, *Principles of Operations Research for Management*, second edition, Irwin, 1988.)

Let p denote the purchase price of an asset and s its salvage value. Assume that s is independent of the asset's age. Then the total capital cost of the asset is $p - s$, and its average cost at time t is therefore

$$A(t) = \frac{p - s}{t}.$$

Let the average operating cost of the asset at time t be linear,

$$\overline{C}(t) = mt + b.$$

Show that the asset should be replaced t_0 years after purchase, where

$$t_0 = \sqrt{\frac{p - s}{m}}.$$

Life Science

49. An egg ranch has 120 chickens, each of which produces 250 eggs per year. If more chickens are squeezed into the chicken coop, the resulting overcrowding will reduce egg production by one egg per chicken per year for each additional chicken squeezed in. Find the maximum egg production and the number of chickens that will yield this maximum production.

50. Do Exercise 49 if each additional chicken squeezed in reduces egg production by three eggs per chicken per year.

51. A lake was stocked with trout in 1965, and t years later there were

$$y = -\frac{1}{3}t^3 + 12.5t^2 - 100t + 1000$$

trout in the lake. Find the maximum and minimum numbers of trout in the lake from 1965 through 1985.

52. If a farmer sows x seeds per square foot, the resulting yield will be y bushels of grain, where

$$y = -2x^3 - 39x^2 + 4620x.$$

Find the farmer's maximum yield if there are enough seeds on hand to sow 30 seeds per square foot; if there are enough to sow 20 seeds per square foot.

53. Poiseuille's law states that (if the units of measurement are properly chosen) the velocity of blood x units from the central axis of an artery of radius r is

$$V(x) = r^2 - x^2.$$

Find the position within the artery where blood flows the fastest.

54. If a blood vessel contracts, the velocity at which blood flows through it is affected. If the units of measurement are properly chosen, the average velocity of the blood flowing through an artery of radius x, where x is less than or equal to the normal radius r, is

$$V(x) = x^2(r - x).$$

Find the amount of contraction that maximizes the average velocity of the blood.

12.4 CURVE SKETCHING

Sometimes a knowledge of the maxima and minima of a function, relative or absolute, is all we need to know in order to answer a question. However, if we want to analyze a function in the manner described earlier, we need to know

- The intervals on which the function is increasing and those on which it is decreasing.
- The location of its relative maxima and minima.
- The intervals on which the function is concave upward and those on which it is concave downward.

It is also useful to know the intercepts of the function and its asymptotes, if any. Given all this information, we can then make an accurate analysis of the behavior of the function. In this section we collect our results of Sections 12.1 and 12.2 in a procedure that will allow us to analyze functions with relative ease.

Recall that the first derivative test locates relative maxima and minima. The diagram used in the test also tells us the intervals on which the function is increasing and those on which it is decreasing. The second derivative tells us about the concavity of the function: if the second derivative is positive on an interval, the function is concave upward there; if the second derivative is negative, the function is concave downward on the interval. A point on its graph where the function changes its concavity is called **a point of inflection.** The function f whose graph is shown in Figure 12–28 has points of inflection at $(a, f(a))$ and $(b, f(b))$. (Note that in Figure 12–28, $x = b$ is a critical value for f, because the tangent line at $(b, f(b))$ is horizontal, so that $f'(b) = 0$; hence critical values can yield points of inflection.)

If $(x, f(x))$ is a point of inflection for a function f, the function changes concavity at $(x, f(x))$, and therefore f'' has different signs on either side of x. But this can happen only if $f''(x) = 0$ or $f''(x)$ does not exist. Thus we can find the points of inflection for f by finding those values of x in its domain such that $f''(x) = 0$ or $f''(x)$ does not exist.

Suppose we use the first and second derivatives of a function to find its relative maxima, its relative minima, its points of inflection, and to determine where the function is increasing, where it is decreasing, where it is concave upward, and where it is concave downward. Suppose we also find the intercepts and asymptotes of the function. All this information will enable us to sketch an accurate graph of the function. Let us outline a procedure for this.

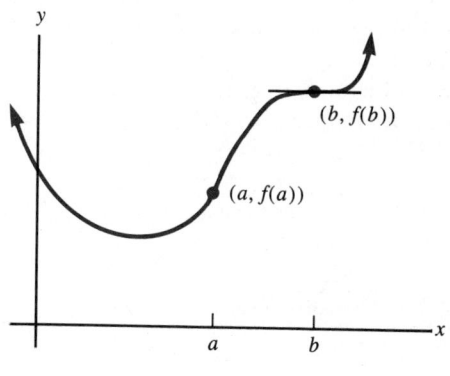

FIGURE 12–28

The Curve-Sketching Procedure

To sketch the graph of a function f, proceed as follows:

1. Use the first derivative test to locate all relative maxima and relative minima of the function, and to find the intervals on which it is increasing and those on which it is decreasing.

2. a. Find all solutions of the equation $f''(x) = 0$ and also all values of x for which $f''(x)$ does not exist, and plot these values on the x-axis.
   b. On each interval of the diagram of step 2a, determine whether f'' is positive or negative, and add this information to the diagram.
   c. On intervals where f'' is positive, f is concave upward; on intervals where f'' is negative, f is concave downward. Add this information to the diagram. The diagram now shows the intervals on which f is concave upward and those on which it is concave downward and indicates the values of x that yield points of inflection.

3. Find the intercepts of f.
   a. If $f(0)$ is defined, the y-intercept is $(0, f(0))$.
   b. To find the x-intercepts, set $f(x) = 0$ and solve for x. (In practice, this step is often omitted when the equation $f(x) = 0$ is not easily solved.)

4. Find the asymptotes for f.
   a. If the denominator of $f(x)$ is zero and the numerator is nonzero when $x = c$, then the vertical line $x = c$ is an asymptote for f. To determine how the graph approaches the vertical asymptote as x approaches c, find

   $$\lim_{x \to c^+} f(x) \quad \text{and} \quad \lim_{x \to c^-} f(x).$$

   b. If

   $$\lim_{x \to +\infty} f(x) = K \quad \text{or} \quad \lim_{x \to -\infty} f(x) = K,$$

   then $y = K$ is an asymptote for f as $x \to +\infty$ or $x \to -\infty$, respectively. (Here K can be either a number or an expression involving x.)

5. Sketch the graph of f using the information obtained in steps 1 through 4.

EXAMPLE 1   Let us use the curve-sketching procedure to draw the graph of the quadratic function defined by

$$f(x) = x^2 - 8x + 15.$$

1. Since $f'(x) = 2x - 8$, the first derivative test yields the following diagram:

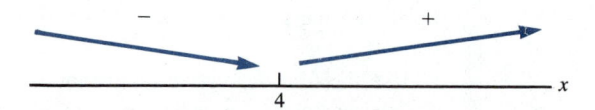

```
         −              +
─────────────────┼──────────────── x
                 4
```

Hence the function has a relative minimum at $(4, f(4)) = (4, -1)$ and is decreasing on the interval $(-\infty, 4)$ and increasing on the interval $(4, +\infty)$.

2. **a.** Since $f''(x) = 2$, the equation $f''(x) = 0$ has no solution. Also, $f''(x)$ exists for all x. Therefore the diagram is

```
──────────────────────────────── x
```

b. Since $f''(x) = 2$ is positive for all x, we have

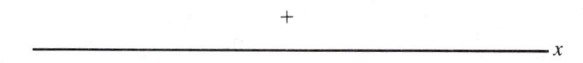

```
                 +
──────────────────────────────── x
```

c. Since $f''(x)$ is positive for all x, f is concave upward for all x, so we have

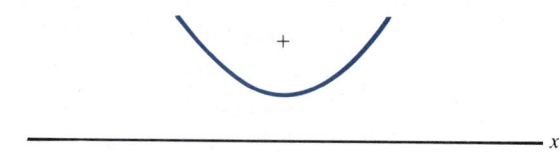

```
           \      +     /
            _____/
──────────────────────────────── x
```

Since the function never changes concavity, it has no points of inflection.

3. **a.** The y-intercept is $(0, f(0)) = (0, 15)$.
   b. The x-intercepts occur at the solutions of
   $$x^2 - 8x + 15 = 0,$$
   or
   $$(x - 3)(x - 5) = 0,$$
   and thus are $(3, 0)$ and $(5, 0)$.

4. The function is a polynomial, and polynomials have no asymptotes.

5. The graph of the function is shown in Figure 12–29. Notice how each piece of information determined in steps 1 through 4 has been used to sketch the graph. □

EXAMPLE 2 Let us sketch the graph of the polynomial function defined by
$$f(x) = x^3 - 3x^2 - 24x + 3.$$

1. Since
$$f'(x) = 3x^2 - 6x - 24 = 3(x + 2)(x - 4),$$

APPLICATIONS OF DIFFERENTIATION 773

$$f(x) = x^2 - 8x + 15$$

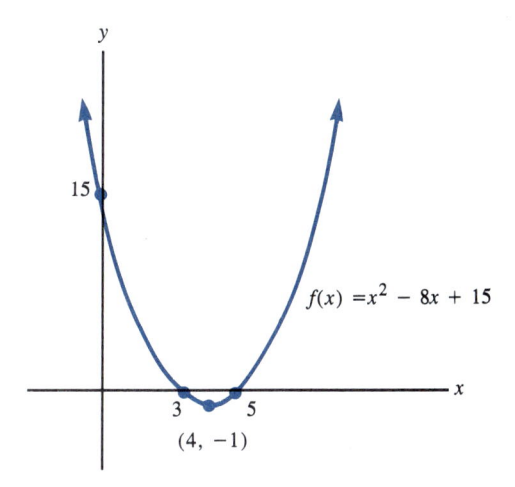

FIGURE 12-29

the diagram of the first derivative test is

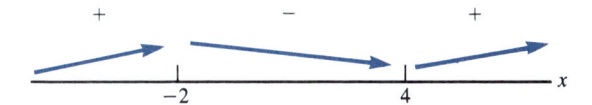

```
      +           -           +
  ─────────┼───────────┼─────────── x
          -2           4
```

Hence the function

- has a relative maximum at $(-2, f(-2)) = (-2, 31)$;
- has a relative minimum at $(4, f(4)) = (4, -77)$;
- is increasing on the intervals $(-\infty, -2)$ and $(4, +\infty)$;
- is decreasing on the interval $(-2, 4)$.

2. a. Since

$$f''(x) = 6x - 6 = 6(x - 1),$$

setting $f''(x) = 0$ and solving for x yields $x = 1$. The second derivative exists for all x. Hence we have the diagram

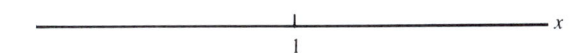

```
  ──────────────┼────────────── x
                1
```

b. Since $f''(x) < 0$ if $x < 1$ and $f''(x) > 0$ if $x > 1$, we have

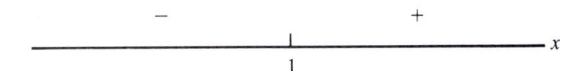

```
         -              +
  ──────────────┼────────────── x
                1
```

c. Therefore f is concave downward on $(-\infty, 1)$ and concave upward on $(1, +\infty)$, and we have the diagram

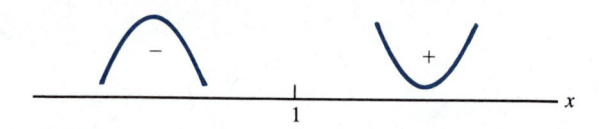

Since f changes concavity at $(1, f(1)) = (1, -23)$, this is a point of inflection for the function.

3. **a.** The y-intercept is $(0, f(0)) = (0, 3)$.
   b. The x-intercepts occur at the values of x for which
   $$x^3 - 3x^2 - 24x + 3 = 0.$$
   Since this equation is not easily solved, we omit this step.
4. The function is a polynomial and hence has no asymptotes.
5. The function is graphed in Figure 12–30. ☐

FIGURE 12–30

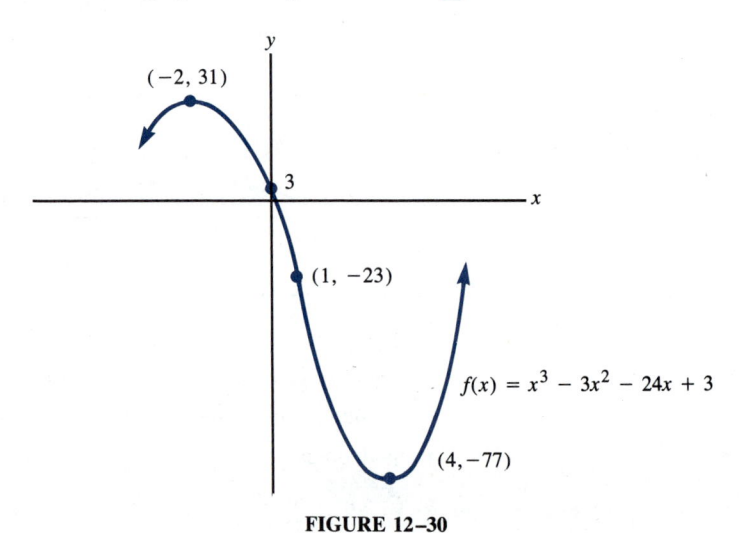

EXAMPLE 3 We sketch the graph of the rational function whose defining equation is
$$f(x) = \frac{4x - 3}{x - 2}.$$

Note that $f(2)$ is not defined.

1. Since
$$f'(x) = -\frac{5}{(x - 2)^2},$$
the equation $f'(x) = 0$ has no solution. Since $f'(2)$ does not exist and $x = 2$ is not a critical value for f, we have the diagram

APPLICATIONS OF DIFFERENTIATION 775

Note the vertical dashed line at $x = 2$: this reminds us that $x = 2$ is not a critical value and thus cannot yield a relative maximum or minimum. Since $f'(x) < 0$ for all values of x, the function is decreasing on the intervals $(-\infty, 2)$ and $(2, +\infty)$:

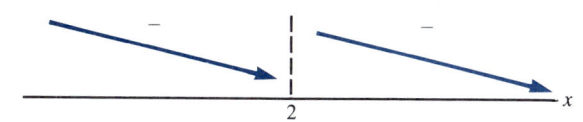

2. a. Since
$$f''(x) = \frac{10}{(x-2)^3},$$
the equation $f''(x) = 0$ has no solution. The second derivative does not exist at $x = 2$. Our diagram is

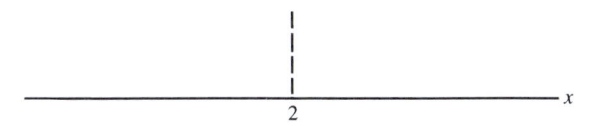

Again, note the dashed line at $x = 2$ to remind us that $f(2)$ is not defined, and hence that $x = 2$ cannot yield a point of inflection.

b. Since $f''(x) < 0$ if $x < 2$ and $f''(x) > 0$ if $x > 2$, we have

c. The diagram of step 2b shows that f is concave downward on $(-\infty, 2)$ and concave upward on $(2, +\infty)$. Hence we have

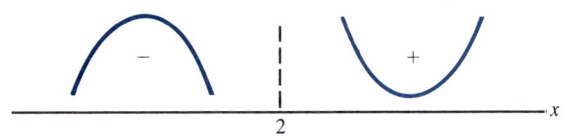

The function has no points of inflection.

3. a. The y-intercept is $(0, f(0)) = (0, 3/2)$.
   b. The x-intercepts occur at the values of x for which
   $$\frac{4x - 3}{x - 2} = 0.$$

The only solution of this equation is $x = 3/4$. Thus the function's single x-intercept is $(3/4, 0)$.

4. **a.** When $x = 2$, the denominator of $f(x)$ is zero but the numerator is not zero. Therefore the line $x = 2$ is a vertical asymptote for f. As $x \to 2^+$, $4x - 3$ and $x - 2$ are both positive, so

$$\lim_{x \to 2^+} f(x) = \lim_{x \to 2^+} \frac{4x - 3}{x - 2} = +\infty.$$

Similarly, as $x \to 2^-$,

$$\lim_{x \to 2^-} f(x) = -\infty.$$

b. Since

$$\lim_{x \to \pm\infty} \frac{4x - 3}{x - 2} = \lim_{x \to \pm\infty} \frac{4 - 3(1/x)}{1 - 2(1/x)} = \frac{4 - 3(0)}{1 + 2(0)} = 4,$$

the line $y = 4$ is a horizontal asymptote for f as $x \to +\infty$ and as $x \to -\infty$.

5. The graph of the function is shown in Figure 12–31. ∎

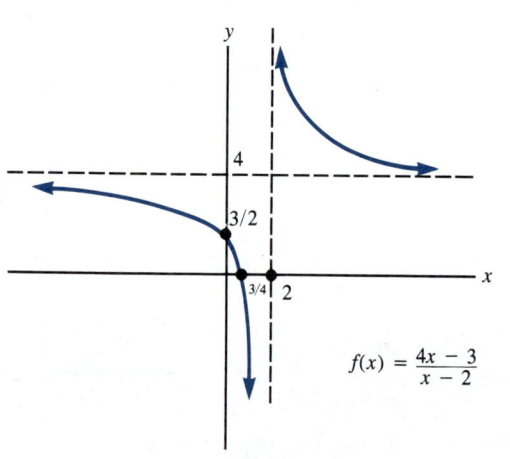

FIGURE 12–31

EXAMPLE 4

Let us sketch the graph of the rational function f defined by

$$f(x) = x + \frac{1}{x}.$$

Note that $f(0)$ is not defined.

1. Since

$$f'(x) = 1 - \frac{1}{x^2},$$

APPLICATIONS OF DIFFERENTIATION   777

$f'(0)$ does not exist and the equation $f'(x) = 0$ has solutions $x = \pm 1$. Therefore we have

Since $f'(x) > 0$ for $x < -1$ and $x > 1$, and $f'(x) < 0$ for $-1 < x < 0$ and $0 < x < 1$, the diagram becomes

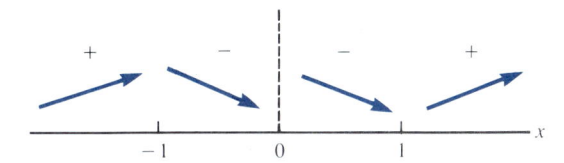

Therefore f is increasing on $(-\infty, -1)$ and $(1, +\infty)$, decreasing on $(-1, 0)$ and $(0, 1)$, and has relative maxima at $(-1, -2)$ and $(1, 2)$.

2. Since

$$f''(x) = \frac{2}{x^3},$$

$f''(0)$ does not exist and the equation $f''(x) = 0$ has no solution. Therefore we have

Since $f''(x) < 0$ for $x < 0$ and $f''(x) > 0$ for $x > 0$, the diagram becomes

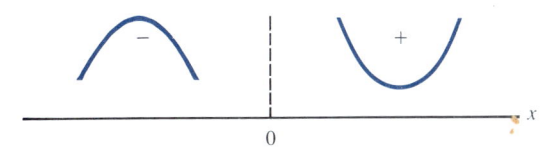

Hence f is concave downward on $(-\infty, 0)$, concave upward on $(0, +\infty)$, and has no points of inflection.

3. Since f is not defined at 0, the function has no y-intercept. It has no x-intercepts because the equation $f(x) = 0$ has no solution.

4. The line $x = 0$ (the y-axis) is a vertical asymptote for f. Note that

$$\lim_{x \to 0^+} f(x) = \lim_{x \to 0^+} \left(x + \frac{1}{x} \right) = 0 + (+\infty) = +\infty$$

and

$$\lim_{x \to 0^-} f(x) = \lim_{x \to 0^-} \left(x + \frac{1}{x}\right) = 0 + (-\infty) = -\infty.$$

Also,

$$\lim_{x \to +\infty} f(x) = \lim_{x \to +\infty} \left(x + \frac{1}{x}\right) = +\infty + 0 = +\infty$$

and

$$\lim_{x \to -\infty} f(x) = \lim_{x \to -\infty} \left(x + \frac{1}{x}\right) = -\infty + 0 = -\infty.$$

Therefore f has no horizontal asymptotes. However, as $x \to \pm\infty$,

$$f(x) = x + \frac{1}{x} \to x + 0 = x.$$

Hence the line $y = x$ is an oblique asymptote for f.

5. The graph of f is shown in Figure 12–32. ◻

EXAMPLE 5   Suppose that Acme Company's profit function is given by

$$P(x) = -x^3 + 36x^2 - 240x - 100,$$

where x is in thousands of units and profit is in thousands of dollars. Let us analyze Acme's profits. We begin by sketching the graph of the profit function. Since x

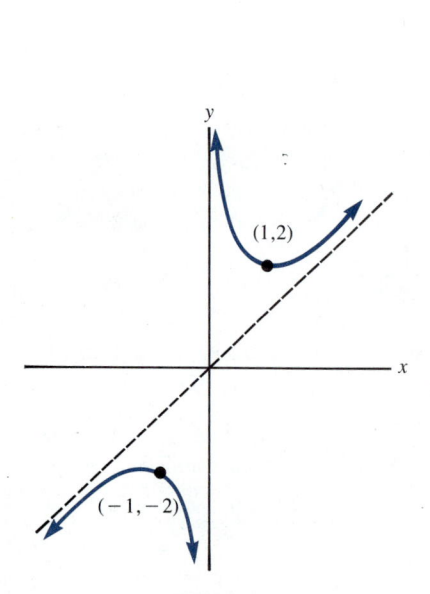

FIGURE 12–32

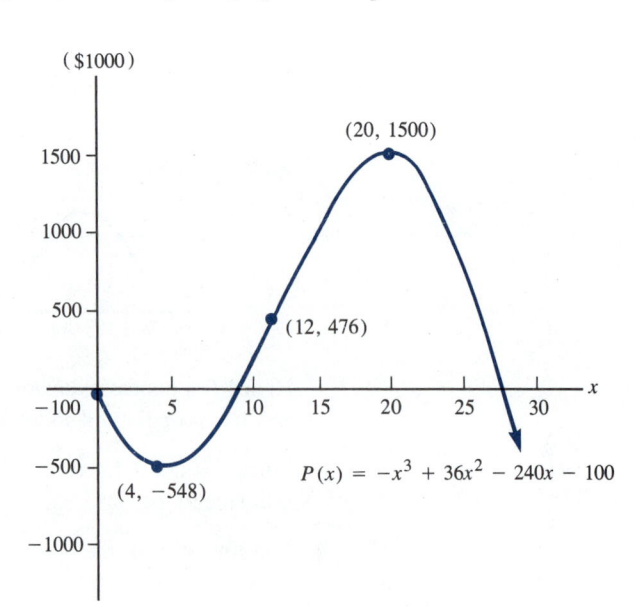

FIGURE 12–33

APPLICATIONS OF DIFFERENTIATION 779

must be nonnegative, we can restrict our diagrams to the interval $[0, +\infty)$. We will leave out some of the details of the curve-sketching procedure: you should check these details.

1. The diagram of the first derivative test is

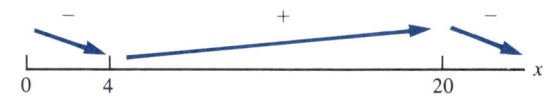

    [Number line with signs: − on $[0, 4)$, + on $(4, 20)$, − on $(20, \infty)$; marks at 0, 4, 20]

    The relative maximum occurs at $(20, 1500)$, and the relative minimum occurs at $(4, -548)$.

2. The concavity diagram is

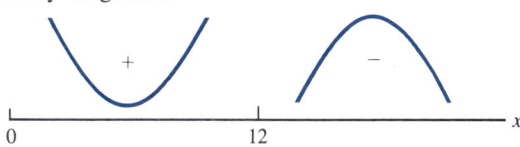

    [Number line showing concave up (+) on $[0, 12)$ and concave down (−) on $(12, \infty)$; marks at 0, 12]

    The point of inflection is $(12, 476)$.

3. The y-intercept is $(0, -100)$.
4. There are no asymptotes.
5. The graph of Acme's profit function is shown in Figure 12–33.

Now let us use the graph to analyze Acme's profit. Starting at $x = 0$ and moving to the right along the x-axis, we observe the following:

  a. Since $P(0) = -100$, when Acme produces 0 units of its product it suffers a loss of $100,000. Thus Acme's fixed cost is $100,000.
  b. As production increases from 0 units to 4000 units, Acme's profit decreases, but at a slower and slower rate, until it reaches a minimum of $-$$548,000 at 4000 units.
  c. As production increases from 4000 units to 12,000 units, profit increases, and it does so at a faster and faster rate as the point of inflection is approached.
  d. At a production level of 12,000 units, Acme's profit is $476,000. As production increases from 12,000 units to 20,000 units, profit continues to increase, but at a slower and slower rate, until it reaches a maximum of $1.5 million at 20,000 units.
  e. If production increases beyond 20,000 units, profit will decrease, and it will do so at a faster and faster rate as production goes up; eventually losses will begin to accrue at an ever-increasing rate.

Note also that careful drawing of the graph shows us that Acme will break even somewhere between 9000 and 10,000 units, and again somewhere near 27,000 units. (The exact break-even quantities are the solutions of the equation

$$-x^3 + 36x^2 - 240x - 100 = 0.)$$ ☐

■ *Computer Manual: Program CURSKECH*

12.4 EXERCISES

In Exercises 1 through 38, use the curve-sketching procedure to sketch the graph of the given function.

1. $f(x) = x^2 - 2x - 15$
2. $f(x) = -2x^2 + 18x - 28$
3. $f(x) = 2x^3 + 3x^2 - 12x + 2$
4. $f(x) = -2x^3 - 12x^2 - 24x + 30$
5. $f(x) = -x^3 + 21x^2 - 120x + 20$
6. $f(x) = -3x^3 + 6x^2$
7. $f(x) = x^3 + 3x^2 - 9x + 10$
8. $f(x) = -2x^3 + 27x^2 - 108x + 12$
9. $f(x) = x^4 - 32x$
10. $f(x) = x^4 - x^2$
11. $f(x) = 3x^4 + 4x^3$
12. $f(x) = 6x^5 + 15x^4 + 10x^3$
13. $f(x) = x^4 - 8x^2 + 1$
14. $f(x) = 3x^5 + 15x^4 - 10x^3$
15. $f(x) = \frac{1}{5}x^5 - \frac{8}{3}x^3 + 7x$
16. $f(x) = x^5 - 20x^2 + 1$
17. $f(x) = 2x^6 - 27x^4$
18. $f(x) = 5x^6 - 18x^5 + 30x^4 - 40x^3 + 45x^2 - 30x$
19. $f(x) = x^{2/3}$
20. $f(x) = x^{4/3}$
21. $f(x) = (x - 3)^{11/5}$
22. $f(x) = (x - 3)^{6/5}$
23. $f(x) = (x - 3)^{1/5}$
24. $f(x) = \dfrac{2}{x^2}$
25. $f(x) = \dfrac{3}{x^3}$
26. $f(x) = \dfrac{1}{x - 1}$
27. $f(x) = \dfrac{2}{(x + 1)^2}$
28. $f(x) = \dfrac{x - 3}{x + 2}$
29. $f(x) = \dfrac{2x - 1}{3x + 2}$
30. $f(x) = \dfrac{3x + 2}{x - 4}$
31. $f(x) = \dfrac{x}{x^2 - 1}$
32. $f(x) = \dfrac{x^2}{x + 1}$
33. $f(x) = \dfrac{x^2 - 1}{x}$
34. $f(x) = \dfrac{x^2}{x^2 - 3}$
35. $f(x) = x + \dfrac{4}{x}$
36. $f(x) = 0.5x^2 - 1/x$
37. $f(x) = \dfrac{x^2 - 3}{x^2}$
38. $f(x) = \dfrac{x^3 + 1}{x^2}$

*39. Use the curve-sketching procedure to graph
   (a) $f(x) = ax^2 + bx + c$, $\quad a > 0$
   (b) $f(x) = ax^2 + bx + c$, $\quad a < 0$

40. An economist estimates that if inflation is q percent per year, then families will save approximately s percent of their income, where
$$s = -0.001q^3 + 0.06q^2 - 1.2q + 12.$$
Draw the graph and analyze the effect of inflation on savings.

Management

41. Brawny Corporation's profit function is given by
$$P(x) = -x^3 + 90x^2 - 1500x - 9,$$
where x is in thousands of units and profit is in thousands of dollars. Sketch the graph of the profit function and analyze Brawny's profit in the manner of Example 5.

APPLICATIONS OF DIFFERENTIATION

42. Since it was founded 15 years ago, a firm's annual sales have been S million dollars per year, where

$$S = 0.025t^3 - 1.2t^2 + 18t + 12$$

and $t = 0$ represents the year the firm was founded. Sketch the graph of S and analyze annual sales since the firm's founding.

43. Suppose the cost of removing $x\%$ of the pollutants in a chemical plant's waste water is

$$y = \frac{8x}{100 - x}$$

million dollars, $0 \leq x < 100$. Sketch the graph and analyze the cost.

44. Repeat Exercise 43 if

$$y = \frac{6x}{110 - x}, \qquad 0 \leq x \leq 100.$$

45. The cost of decommissioning a nuclear power plant that produces electricity for t years, $0 \leq t \leq 25$, is estimated to be y million dollars, where

$$y = \frac{60t + 10}{30 - t}.$$

Graph this function, and analyze the cost.

46. Each weekday *The Wall Street Journal* publishes a **yield curve** for U.S. government securities. The yield curve depicts the percentage yields y on government securities as a function of their maturities, which run from 3 months to 30 years. Suppose a yield curve is given by

$$y = \frac{5t + 0.5}{0.5t + 0.15}, \qquad 0.25 \leq t \leq 30,$$

t in years. Graph this function and analyze yield as a function of maturity.

47. The average cost function of Gormless, Inc., is given by

$$A(x) = 2x + \frac{50}{x}, \qquad x > 0.$$

Here x is in hundreds of units and $A(x)$ is in dollars. Draw the graph of A and discuss Gormless's average cost per hundred units produced.

48. Delta Company's average cost function is given by

$$A(x) = \frac{300}{x} + 0.6x, \qquad x > 0.$$

Graph the function A and analyze Delta's average cost per unit.

49. A firm's average cost per thousand units is given by

$$A(x) = 0.02x + 600/x^2, \qquad x > 0.$$

Sketch the graph and analyze average cost per thousand.

Life Science

50. The number of deer in a region is

$$y = -t^3 + 18t^2 - 96t + 2000,$$

781

where $t = 0$ represents 1976, t in years. Sketch the graph and discuss the changes in the deer population from 1976 through 1986. At one time deer hunting was banned in the region for a four-year period. Can you tell when this occurred?

51. The cost of removing $x\%$ of the pollutants from a lake is estimated to be
$$C(x) = \frac{3x}{100 - x}, \quad 0 \leq x < 100.$$
Draw the graph of C and analyze the cost.

52. An immunization campaign is described by the equation
$$y = (9t + 40)/(10t + 125),$$
where t is time in months since the campaign began and y is the proportion of the population immunized at time t. Sketch the graph and analyze the campaign.

53. Suppose that in the t years since they have been declared an endangered species, the population of whooping cranes has grown to y birds, where
$$y = \frac{20t^2 + 25}{t + 1}, \quad t \geq 0.$$
Graph this function and analyze the recovery of the whooping crane.

54. It is estimated that t years after the start of an international effort to save the whales the number of blue whales alive was y hundred, where
$$y = \frac{(t + 1)^3 + 128}{t + 1}, \quad t \geq 0.$$
Graph this function and discuss the changes in the blue whale population.

Social Science

55. Housing starts in a city are described by the equation
$$y = -3t^4 + 80t^3 - 750t^2 + 3000t + 500,$$
where t is in years, with $t = 0$ representing 1978, and y is the number of new houses started in the city in year t. Draw the graph and analyze housing starts in the city from 1978 through 1990. (*Hint:* $t - 10$ and $(t - 5)^2$ are factors of the derivative of y with respect to t.)

56. The number of children attending public school in a large city in year t is
$$y = -\frac{t^4}{4} + \frac{14t^3}{3} - 20t^2 + 200,$$
where $t = 0$ represents 1989 and y is in thousands. Draw the graph and analyze enrollment from 1989 through 2001. Do you think it would make sense to sell school buildings in 1992 because of declining enrollment?

57. The cost of providing x million elementary school children with a free glass of milk at lunch is estimated to be
$$y = \frac{25x}{10 + x}$$
million dollars, $x \geq 0$. Graph this function and analyze the cost.

58. It is estimated that if a state's economy continues to grow at its current pace, the unemployment rate in the state t months from now will be $y\%$, where

$$y = \frac{2.4t + 1.3}{t + 0.25}, \qquad 0 \le t \le 12.$$

Graph this function and discuss the estimated unemployment rate over the next year.

Computer Science

An article by Peter Deming ("The Science of Computing: Speeding Up Parallel Processing," *American Scientist*, July–August, 1988) states that the best possible speed-up in running a program on a computer with $p > 1$ processors rather than 1 processor is given by

$$f(p) = \frac{np}{n + s(p - 1)}.$$

Here n is the number of operations required by the program and s is the number of operations in the longest sequential path in the program. Exercises 59 through 61 refer to the function f. (Note that p is not a continuous variable, since $p = 2, 3, 4, \ldots$; however, for the purposes of these exercises, you may treat p as if it were continuous.)

*59. Graph f for $p > 1$ when $n = 1000$ and
   (a) $s = 10$ (b) $s = 100$ (c) $s = 1000$

*60. Graph f for $p > 1$ and any n and s if
   (a) $s < n$ (b) $s = n$

*61. Show that for all p, n, and s, $f(p) \le n/s$. This inequality is known as Amdahl's Law, and it was long taken as showing that there is a fundamental limit to how much a program could be speeded up by the use of multiprocessors. Can you explain why the inequality implies such a limit? (In fact, it has recently been demonstrated that Amdahl's Law does not hold under all circumstances.)

SUMMARY

This summary consists of terms and symbols whose meaning you should know, facts you should know, and techniques and methods you should be able to employ.

Section 12.1

Relative maximum, relative minimum. If $(a, f(a))$ is a relative maximum or relative minimum point and $f'(a)$ exists, then $f'(a) = 0$. Critical values. Relative maxima and relative minima occur at critical values. Not all critical values yield relative maxima or relative minima. Finding critical values. Function increasing on an interval, decreasing on an interval. Increasing and decreasing functions and the first derivative: if $f'(x) > 0$ on an interval, f is increasing on the interval; if $f'(x) < 0$ on an interval, f is decreasing on the interval. Determining the intervals over which a function is increasing and those over which it is decreasing. The first derivative test. Using the first derivative test to find relative maxima and relative minima.

Section 12.2

Concavity: concave upward, concave downward. Concavity and the second derivative: if $f''(x) > 0$ on an interval, f is concave upward over the interval; if $f''(x) < 0$ on an interval, f is concave downward over the interval. Determining the intervals over which a function is concave upward and those over which it is concave downward. The second derivative test. Using the second derivative test to find relative maxima and relative minima. The second derivative test fails when the second derivative is zero or does not exist.

Section 12.3

Absolute maximum, absolute minimum. Finding the absolute maximum and absolute minimum of a continuous function on an interval. Setting up and solving maximum/minimum problems. Inventory problems.

Section 12.4

Point of inflection. The curve-sketching procedure. Using the curve-sketching procedure to sketch the graphs of functions. Analyzing the graphs of models.

REVIEW EXERCISES

Exercises 1 through 5 refer to the function f whose graph is as follows:

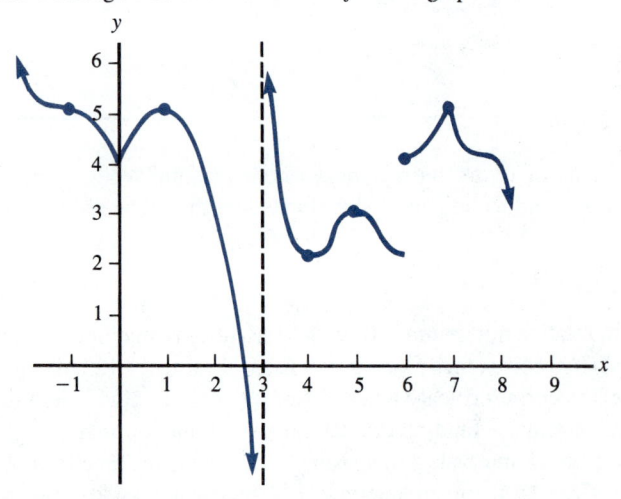

1. Find all relative maxima and relative minima for f and where they occur.
2. Find all critical values for f.
3. Find all intervals where f is increasing and all those where it is decreasing.

4. Find all intervals where f is concave upward and all those where it is concave downward.

5. Find all points of inflection of f.

In Exercises 6 through 11, use the first derivative test to find all relative maxima and relative minima for the given function.

6. $f(x) = -x^2 + 7x - 4$

7. $f(x) = x^3 + 6x^2 - 180x - 1200$

8. $f(x) = x^4 + 4x^3$

9. $f(x) = 3x^5 - 20x^3 - 480x + 200$

10. $f(x) = \dfrac{x^2 - 8}{2x + 1}$

11. $f(x) = \dfrac{2x + 1}{x^2 - 8}$

12. A firm's average cost per unit is given by
$$A(x) = 0.5x^2 - 54\sqrt{x} + 200.$$
Find the firm's minimum average cost per unit.

13. A new disease that affects birch trees is introduced into a locality, and t years later the proportion of birches that have the disease is
$$y = \dfrac{t^2 + 8t + 2}{4t^2 + 24}.$$
What was the maximum proportion of birches that contracted the disease?

In Exercises 14 through 19, use the second derivative test to find all relative maxima and relative minima for the given function. If the second derivative test fails, use the first derivative test.

14. $f(x) = -2x^2 + 9x - 3$

15. $f(x) = x^3 + 6x^2 - 15x$

16. $f(x) = x^4 - 4.5x^2 + 1$

17. $f(x) = \dfrac{1}{5}x^5 - 4x^2 + 1$

18. $f(x) = x^{7/6}$

19. $f(x) = x^{5/6}$

20. The demand function for a firm's product is given by
$$p = 6000 - 2\sqrt{q}.$$
Use the second derivative test to find the firm's maximum revenue.

A function has the following graph:

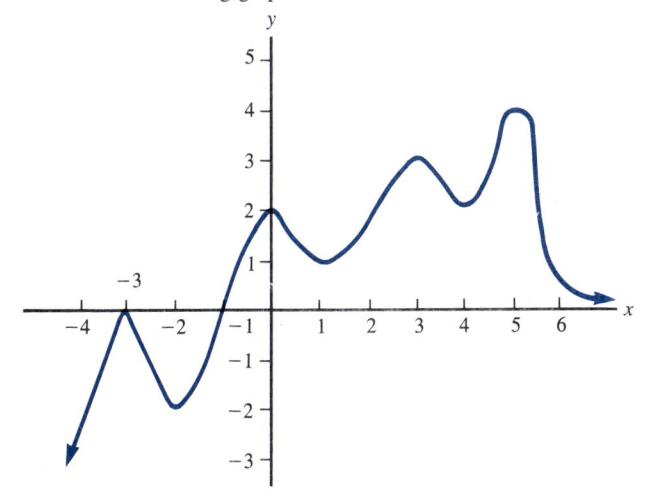

In Exercises 21 through 28, find the absolute maximum and absolute minimum of the function on the given interval.

21. [0, 5]   22. [0, 3]   23. [−2, 5]   24. [−2, 3]
25. [0, +∞) 26. [5, +∞) 27. (−∞, −2] 28. (−∞, 0]

29. Find the absolute maximum and absolute minimum of the function defined by
$$f(x) = x^4 - 32x^2$$
on

    (a) the x-axis   (b) the interval $[-1, 1]$
    (c) the interval $[-5, 6]$ (d) the interval $[1, 6]$

30. A firm's profit function is given by
$$P(x) = -x^3 + 150x^2 - 4800x.$$
Find its maximum profit if it can produce and sell at most 50 units; an unlimited number of units.

31. A car dealership sells 400 new cars per year. Each new car costs $12,000, it costs the dealership $1000 to order cars, and is costs $2000 to carry a car in inventory for a year. Assuming that the average inventory of cars is one half the number ordered, find the optimal reorder quantity and the minimum annual inventory cost for the dealership.

32. Find the dimensions of a cylindrical container with a circular base and top if its volume is to be maximized and its surface area is to be 6 square feet. ($V = \pi r^2 h$ and $S = 2\pi r^2 + 2\pi rh$.)

In Exercises 33 through 36, use the curve-sketching procedure to sketch the graph of the function.

33. $f(x) = -2x^2 + 11x + 14$ 34. $f(x) = -x^3 + 6x^2 + 96x$

35. $f(x) = \dfrac{x^2 - 4}{x^2 - 9}$ 36. $f(x) = \dfrac{x^3 + 1}{2x^2}$

37. A pollster finds that t months after inauguration y percent of the voters approve of the job that a governor is doing, where
$$y = 0.01\left(-\frac{t^3}{3} + 21t^2 - 360t + 6000\right), \quad 0 \le t \le 46.$$
Sketch the graph of this function and analyze the governor's popularity.

38. The cost of desalinating sea water to produce fresh water depends on the proportion of salt removed from the sea water. Suppose that the cost of removing proportion p of the salt from sea water is y dollars per gallon, where
$$y = \frac{0.20p}{1 - p}, \quad 0 \le p < 1.$$
Graph and analyze this function.

39. The neutron radiation given off by a cold-fusion experiment was measured to be y times the normal background radiation, where
$$y = \frac{261t - 30t^2}{30t + 100} + 1, \quad 0 \le t \le 10,$$
t in hours since the start of the experiment. Graph this function and analyze the radiation.

ADDITIONAL TOPICS

Here are some additional topics not covered in the text that you might wish to investigate on your own.

1. Rolle's Theorem says the following:

    If f is continuous on $[a, b]$ and f' exists on (a, b), and if $f(a) = f(b) = 0$, then there is some number c in (a, b) such that $f'(c) = 0$.

    Find out what Rolle's Theorem says about the graph of f and find out how it is proved.

2. The Mean Value Theorem for derivatives says the following:

    If f is continuous on $[a, b]$ and f' exists on (a, b), there is some number c in (a, b) such that $f(b) = f(a) + f'(c)(b - a)$.

    Find out what the Mean Value Theorem says about the graph of f and find out how it is proved.

3. The Extended Mean Value Theorem says the following:

    If f and f' are continuous on $[a, b]$ and f'' exists on (a, b), then there is some number c in (a, b) such that

    $$f(b) = f(a) + f'(a)(b - a) + \frac{f''(c)}{2}(b - a)^2.$$

    Find out how the Extended Mean Value Theorem is proved.

4. If f' exists in some interval containing a, then the **linear approximation of f** for x near a is

    $$f(x) \approx f(a) + f'(a)(x - a).$$

    Show how this approximation can be derived. If f'' also exists and is continuous in the interval, then the error of the linear approximation is $E(x)$, where

    $$E(x) = f(a) + f'(a)(x - a) - f(x).$$

    Show that

    $$|E(x)| \leq 0.5 \text{ Max } \{|f''(c)| \mid c \text{ between } a \text{ and } x\}(x - a)^2.$$

    Demonstrate how the linear approximation of f and the bound on its error are used.

5. Cauchy's form of the Mean Value Theorem says the following:

    If f and g are continuous on $[a, b]$ and f' and g' exist on (a, b), and if $g'(x) \neq 0$ for any x in (a, b), then there is some number c in (a, b) such that

    $$\frac{f(b) - f(a)}{g(b) - g(a)} = \frac{f'(c)}{g'(c)}.$$

    Find out the geometric meaning of Cauchy's form of the Mean Value Theorem and find out how it is proved.

6. L'Hôpital's Rule says the following:

    If f and g are continuous on $[a, b]$ and f' and g' exist on (a, b), except possibly at r, and if

    $$\lim_{x \to r} f(x) = \lim_{x \to r} g(x) = 0,$$

then

$$\operatorname*{Lim}_{x \to r} \frac{f(x)}{g(x)} = \operatorname*{Lim}_{x \to r} \frac{f'(x)}{g'(x)},$$

provided the limit on the right-hand side exists.

Find out how L'Hôpital's Rule is proved. L'Hôpital's Rule has several extensions: its conclusion holds when

$$\operatorname*{Lim}_{x \to r} f(x) = \pm \infty \quad \text{and} \quad \operatorname*{Lim}_{x \to r} g(x) = \pm \infty$$

and when the number r is replaced by $\pm \infty$. Show how L'Hôpital's Rule and its extensions are used to evaluate limits.

SUPPLEMENT: QUALITATIVE SOLUTIONS OF DIFFERENTIAL EQUATIONS

A **differential equation** is an equation involving derivatives. For instance, suppose y is a function of t; then

$$\frac{dy}{dt} = 2t,$$

or

$$y' = 2t,$$

is a differential equation for y. A **solution** of a differential equation is a function that satisfies the equation. For example, the function defined by $y = t^2 + c$, where c is any number, is a solution of the differential equation $y' = 2t$. This is so because if $y = t^2 + c$, then

$$y' = \frac{d}{dt}(t^2 + c) = 2t + 0 = 2t.$$

Differential equations are important because in many applications the rate of change (= derivative) of a function is known, but the function itself is not known. To see how this can occur, suppose a biologist studying an animal population finds that the size of the population is increasing at a constant rate of 2% per year. If $y = y(t)$ denotes the size of the population at time t, t in years, then y' is the rate of change of the population, and hence (as we noted in Section 11.3), y'/y is the *proportional* rate of change of the population. But to say that the size of the population is increasing at 2% per year is the same as saying that its proportional rate of change is 0.02, so

$$\frac{y'}{y} = 0.02,$$

or

$$y' = 0.02y.$$

APPLICATIONS OF DIFFERENTIATION 789

Thus the biologist's knowledge of the growth rate of the population translates into a differential equation concerning the function that models population size.

We will consider differential equations in more detail in Section 15.6, and there we will see how to solve an equation such as $y' = 0.02y$ for y. The point we wish to make here, however, is that it is often possible to sketch the graph of a solution to a differential equation without actually finding the solution explicitly. This procedure is referred to as finding a **qualitative solution** to the differential equation. We illustrate with an example.

EXAMPLE 1

Suppose that a biologist's knowledge of the growth rate of a population translates into the differential equation

$$y' = 0.02y,$$

as before. Suppose also that the biologist knows that the population size will be positive for all $t \geq 0$, that is, that $y > 0$ on the interval $[0, +\infty)$. Then $y > 0$ on $[0, +\infty)$ implies that

$$y' = 0.02y > 0$$

on $[0, +\infty)$. But if a function has a positive derivative on an interval, it is increasing there. Therefore we conclude that

y is increasing on $[0, +\infty)$.

We can also look at the second derivative of y:

$$y' = 0.02y$$

implies that

$$y'' = 0.02y'.$$

Since we already know that $y' > 0$ on $[0, +\infty)$, it follows that $y'' > 0$ on $[0, +\infty)$. But if a function has a positive second derivative on an interval, it is concave upward there. Hence

y is concave upward on $[0, +\infty)$.

Thus we know that $y(0) > 0$ (because $y > 0$ for all $t \geq 0$) and that y is increasing and concave upward on $[0, +\infty)$. Therefore the graph of y must be as shown in Figure 12–34. This graph is the qualitative solution of the differential equation. ☐

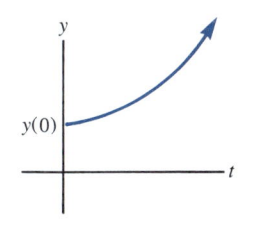

FIGURE 12–34

Qualitative solutions of differential equations are important because often we do not require an explicit solution, but only need to know how the solution behaves in a general way: is it increasing or decreasing, concave upward or downward, does it have maxima, minima, or points of inflection, does it approach a limit, and so on.

EXAMPLE 2

Again let $y = y(t)$ be the size of a population at time t. Suppose a biologist's knowledge of the population results in the differential equation

$$y' = ky(L - y),$$

where k and L are positive numbers. Assume also that
$$0 < y < L$$
for all $t \geq 0$, and that $y(0) < L/2$. Let us find the qualitative solution of the differential equation on $[0, +\infty)$.

Since k, y, and $L - y$ are all positive on $[0, +\infty)$, so is y'. Hence
$$y \text{ is increasing on } [0, +\infty).$$
By the product rule,
$$y'' = ky(-y') + ky'(L - y),$$
or
$$y'' = ky'(L - 2y).$$
Since k and y' are positive on $[0, +\infty)$, we see that
$$y'' > 0 \quad \text{if } y < L/2$$
and
$$y'' < 0 \quad \text{if } y > L/2.$$
Hence,
$$y \text{ is concave upward when } y < L/2;$$
$$y \text{ is concave downward when } y > L/2.$$

Putting all this information together, we obtain the qualitative solution shown in Figure 12–35. Notice the point of inflection: this tells us that population size increases at an increasing rate until it reaches $L/2$, and thereafter it increases at a decreasing rate. Note also that y increasing, concave downward, and less than L implies that y approaches L from below. Thus in the long run the population size grows toward the limiting value L. ☐

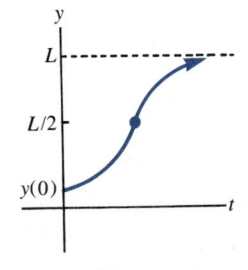

FIGURE 12–35

We conclude this supplement with a predator-prey model. To be specific, we will take our predators to be wolves and their prey to be rabbits. The sizes of the wolf and rabbit populations are interrelated: a large number of rabbits can support a relatively large number of wolves, but if the wolves are too successful in hunting the rabbits, the rabbit population will "crash" to a low level which can support only a small number of wolves. We are interested in how the wolf and rabbit populations behave in the long run.

Let $w = w(t)$ be the size of the wolf population at time t and $r = r(t)$ the size of the rabbit population at time t. Biologists describe the behavior of these populations by means of the **Lotka-Volterra equations**
$$r' = p(a - w)r$$
$$w' = q(r - b)w,$$
where a, b, p, and q are positive numbers. We will also assume that both populations are positive on $[0, +\infty)$: $r > 0$ and $w > 0$ for $t \geq 0$. Then it follows from the Lotka-Volterra equations that

if $w < a$, $r' > 0$, so r is increasing;
if $w > a$, $r' < 0$, so r is decreasing;
if $r < b$, $w' < 0$, so w is decreasing;
if $r > b$, $w' > 0$, so w is increasing.

We will not consider the concavity of r and w, as the equations for the second derivatives are rather complex.

Given a configuration for the initial population sizes, we can now determine how the populations behave in the long run by finding qualitative solutions of the Lotka-Volterra equations. For instance, suppose we assume that

$$w(0) < a < b < r(0).$$

This implies that initially both r and w are increasing. As w increases, it will eventually become greater than a, at which time r will begin to decrease; w will continue to increase while r decreases until r becomes less than b, at which time w will begin to decrease; and so on. Figure 12–36 shows the result of this analysis. Note that the sizes of the wolf and rabbit populations oscillate about the values a and b.

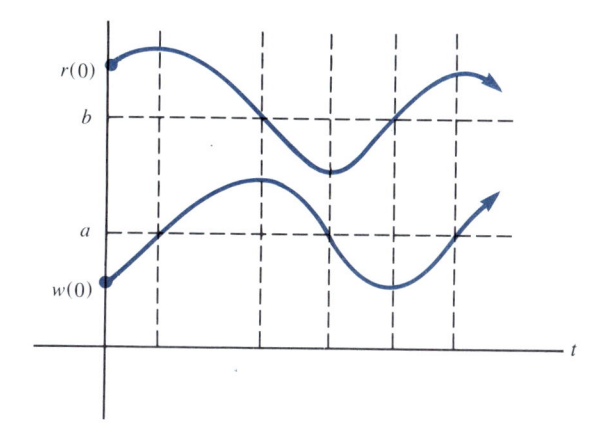

FIGURE 12–36

SUPPLEMENTARY EXERCISES

1. A biologist models the growth of an animal population as a function of time t by means of the differential equation
$$y' = ky, \quad k > 0, \quad y > 0, \quad t \geq 0.$$
Sketch the qualitative solution of the differential equation and analyze the growth of the population.

2. Repeat Exercise 1 if the model is
$$y' = -ky, \quad k > 0, \quad y > 0, \quad t \geq 0,$$
and $\lim_{t \to +\infty} y = L$.

3. The rate at which patents in the area of superconductivity are being applied for is modeled as a function of time t by the differential equation
$$y' = k(L - y), \quad k > 0, \quad L > 0, \quad 0 < y < L, \quad t \geq 0.$$
Sketch the qualitative solution of this differential equation and analyze the growth in patent applications.

4. The rate at which timber is harvested from a forest t years after the start of logging is given by the differential equation
$$y' = k(L - y), \quad k > 0, \quad L > 0, \quad y > L, \quad t \geq 0.$$
Sketch the qualitative solution of this differential equation and analyze the amount harvested as a function of time.

5. An endemic disease responds to treatment in such a way that the rate of change of new cases per year t is given by the differential equation
$$y' = ky(L - y), \quad k > 0, \quad L > 0, \quad y > L, \quad t \geq 0.$$
Sketch the qualitative solution of this differential equation and analyze the progress of the disease.

*6. Repeat Exercise 5 if $0 < y < L$ and

   (a) $y(0) \geq L/2$ (b) $0 < y(0) < \dfrac{L}{2}$.

Exercises 7 through 9 refer to the Lotka-Volterra wolf-rabbit model of this supplement.

7. Sketch the behavior of the wolf and rabbit populations if
$$a < w(0) < b < r(0).$$

8. Repeat Exercise 7 if $r(0) < b < w(0) < a$.

9. Repeat Exercise 7 if $r(0) < b < a < w(0)$.

10. The following system of differential equations can be used to model the interaction of two competing species X and Y:

$$\frac{dx}{dt} = p(a - y)x$$

$$\frac{dy}{dt} = q(b - x)y.$$

Here $x = x(t)$ and $y = y(t)$ are the sizes of the populations of X and Y, respectively, at time t, and a, b, p, and q are positive constants. Sketch the qualitative solutions of this system if $x(0) > 0$, $y(0) > 0$, and $x(0) < a < b < y(0)$. What happens to the two species?

11. Repeat Exercise 10 for other configurations of $x(0)$, $y(0)$, a, and b.

13

Exponential and Logarithmic Functions

In this chapter we introduce **exponential functions** and **logarithmic functions.** These functions are important because they can be used to model many situations of growth and decline. For instance, suppose a firm's annual sales were $5 million in the year it was founded and since then have been doubling every 10 years. If S denotes annual sales and t the time in years since the firm's founding, the equation

$$S = 5 \cdot 2^{0.1t}$$

expresses annual sales as a function of time. Notice that the independent variable t appears in the exponent of this equation. The function defined by the equation is thus called an **exponential function.** Exponential functions and their inverses, which are called **logarithmic functions,** are often used in the study of population growth, the growth of investments, radioactive decay, the spread of epidemics, and similar topics.

We begin this chapter with introductions to exponential and logarithmic functions and discussions of their properties. We then consider the differentiation of these functions, and finally conclude with a section devoted to some of their applications.

13.1 EXPONENTIAL FUNCTIONS

In this section we define exponential functions, discuss their characteristics, show how to graph them, and demonstrate a few of their applications. We repeatedly make use of the material on exponents presented in Section 1.4.

Exponentials

Let x be a real number. If x is an integer or a rational number, then we can calculate 2^x using the results of Section 1.4. For instance,

$$2^0 = 1, \qquad 2^1 = 2, \qquad 2^2 = 2 \cdot 2 = 4,$$

$$2^{-1} = \frac{1}{2^1} = \frac{1}{2}, \qquad 2^{-2} = \frac{1}{2^2} = \frac{1}{4}, \qquad 2^{1/2} = \sqrt{2},$$

$$2^{2/3} = (\sqrt[3]{2})^2, \qquad 2^{-1/2} = \frac{1}{\sqrt{2}}, \qquad 2^{-2/3} = \frac{1}{(\sqrt[3]{2})^2}.$$

If x is an irrational number, we can approximate x to any desired degree of accuracy by a rational number, and this allows us to approximate 2^x to any desired degree of accuracy. For example, if $x = \sqrt{3}$, successively better approximations of x are

$$\sqrt{3} \approx 1.73 \quad \text{and} \quad \sqrt{3} \approx 1.732,$$

so successively better approximations of $2^{\sqrt{3}}$ are

$$2^{1.73} = 2^{173/100} = 3.317278 \quad \text{and} \quad 2^{1.732} = 2^{1732/1000} = 3.321880.$$

In this way we can calculate $2^{\sqrt{3}}$ to any required degree of accuracy. Similar reasoning allows us to calculate 2^x when x is any irrational number.

Now that we know how to calculate 2^x for any number x, suppose we define a function f by setting

$$f(x) = 2^x$$

for all x. This function is called an **exponential function with base 2.** Its graph is shown in Figure 13–1. Note that

- The domain of the function is the set of all real numbers, and its range is the set of all positive real numbers.
- The function is one-to-one: for every real number $y > 0$, there is exactly one real number x such that $2^x = y$.
- The y-intercept of the function is $(0, f(0)) = (0, 2^0) = (0, 1)$.
- The x-axis is an asymptote for the function as $x \to -\infty$; that is,

$$\lim_{x \to -\infty} 2^x = 0.$$

Also, $\lim_{x \to +\infty} 2^x = +\infty$.

EXPONENTIAL AND LOGARITHMIC FUNCTIONS

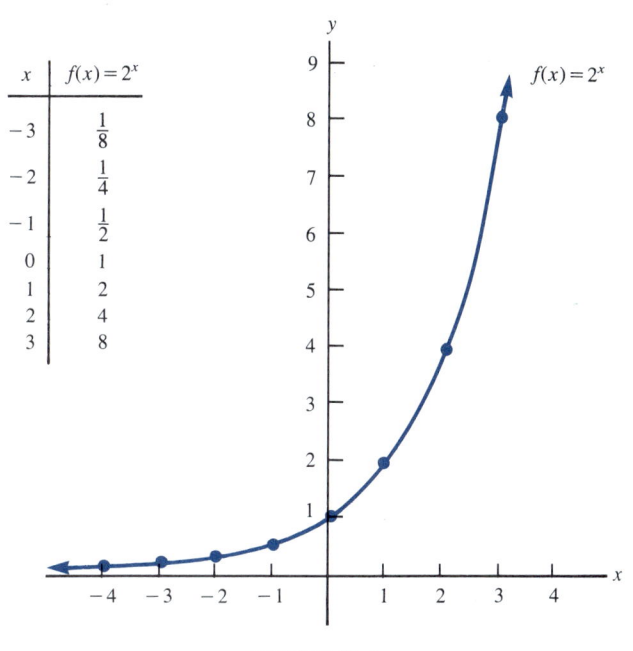

x	$f(x) = 2^x$
-3	$\frac{1}{8}$
-2	$\frac{1}{4}$
-1	$\frac{1}{2}$
0	1
1	2
2	4
3	8

FIGURE 13-1

- The function is continuous at every value of x.

Now suppose we define a function g by setting

$$g(x) = 3 \cdot 2^{0.5x}.$$

Since we can calculate $2^{0.5x}$ for any real number x, the function g is defined for every real number x. Its graph is shown in Figure 13-2. The function g is also referred to as an exponential function with base 2.

It should be clear that what we did above using 2 as a base can be done using any positive number as a base. (We cannot use a negative number as a base because if b is negative, b^x will not be defined for all x; for instance, when b is negative, $b^{1/2} = \sqrt{b}$ is not defined.) Therefore given any number $b > 0$ and any numbers $c \neq 0$, $r \neq 0$, we can define an exponential function f by setting

$$f(x) = cb^{rx}.$$

Because $1^x = 1$ for all x, we omit the case $b = 1$.

Exponential Functions

Let b be any positive real number, $b \neq 1$, and let c and r be real numbers, with $c \neq 0$, $r \neq 0$. The function f defined by

$$f(x) = cb^{rx}$$

is an **exponential function with base b**.

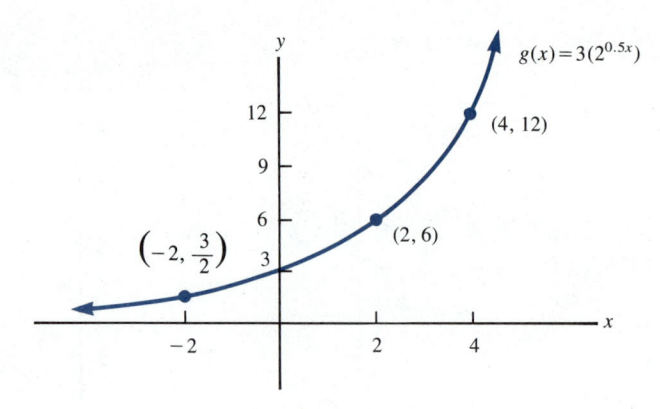

FIGURE 13–2

EXAMPLE 1 The functions f and g defined by $f(x) = 3^{2x}$ and $g(x) = 2(3^{2x})$ are both exponential with base 3. Their graphs are shown in Figure 13–3. ◻

EXAMPLE 2 The functions f and g defined by $f(x) = 2^{-x/2}$ and $g(x) = (½)^{-x/2}$ are exponential with base 2 and base ½, respectively. Their graphs are shown in Figure 13–4. ◻

Consider the exponential function defined by

$$y = (½)^{3x}.$$

Since $½ = 2^{-1}$, we may rewrite the equation as

$$y = (2^{-1})^{3x}$$

or

$$y = 2^{-3x}.$$

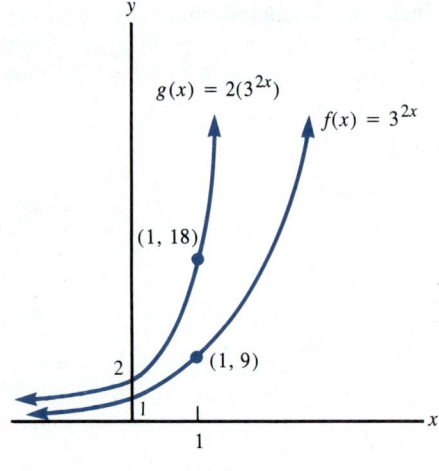

FIGURE 13–3

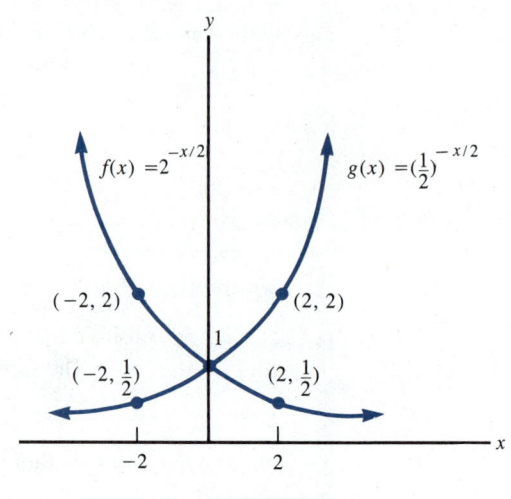

FIGURE 13–4

EXPONENTIAL AND LOGARITHMIC FUNCTIONS

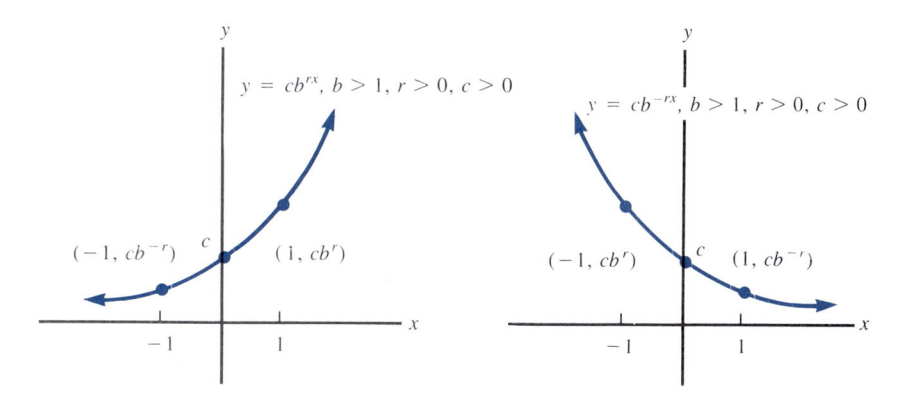

FIGURE 13–5
The Graphs of Exponential Functions

This same change of base can be carried out for any exponential function whose base is less than 1: if $y = c(1/b)^{rx}$, where $b > 1$, then $y = c(b^{-1})^{rx} = cb^{-rx}$. Hence if we wish we can always write our exponential functions using bases greater than 1. Figure 13–5 shows the graphs of $y = cb^{rx}$ and $y = cb^{-rx}$ when $b > 1$, $r > 0$, and $c > 0$. Note the following:

- The functions defined by $y = cb^{rx}$ and $y = cb^{-rx}$ are continuous and one-to-one.
- Their y-intercepts are $(0, c)$.
- $\lim\limits_{x \to +\infty} cb^{rx} = +\infty$, $\lim\limits_{x \to -\infty} cb^{rx} = 0$.
- $\lim\limits_{x \to +\infty} cb^{-rx} = 0$, $\lim\limits_{x \to -\infty} cb^{-rx} = +\infty$.

As we have previously remarked, exponential functions are often used to model situations of growth or decline. We illustrate with examples.

EXAMPLE 3   The average number of people employed by a state's Department of Motor Vehicles in year t is y, where

$$y = 1200 \cdot 2^{0.1t}.$$

Here $t = 0$ represents the year 1989. The average number of employees in 1989 was therefore

$$y = 1200 \cdot 2^{0.1(0)} = 1200 \cdot 1 = 1200.$$

Figure 13–6 shows the graph of this exponential function. Notice that the average number of employees is growing at a greater and greater rate.

Suppose we wish to predict the average number of employees the Department of Motor Vehicles will have in 1992. In 1992, $t = 3$, so the average number of employees y will be

$$y = 1200 \cdot 2^{0.1(3)} = 1200 \cdot 2^{0.3}.$$

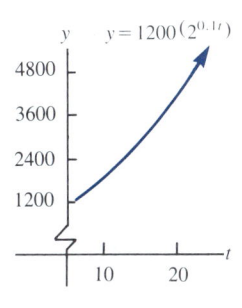

FIGURE 13–6

Using a calculator with an x^y key, we find that $2^{0.3} \approx 1.2311$. Therefore the average number of employees in 1992 will be approximately

$$y = 1200(1.2311) \approx 1477.$$

When will the average number of employees be 2400? If y is to be equal to 2400, we must have

$$2400 = 1200 \cdot 2^{0.1t}.$$

Dividing both sides of this equation by 1200, we obtain

$$2 = 2^{0.1t}$$

or

$$2^1 = 2^{0.1t}.$$

But because the exponential function defined by $y = 1200 \cdot 2^{0.1t}$ is one-to-one, $2^1 = 2^{0.1t}$ implies that

$$1 = 0.1t,$$

Hence $t = 10$ and thus the average number of employees will be 2400 in 1999. ☐

EXAMPLE 4 The number of the sea otters still alive t months after a marine oil spill has contaminated their environment is y, where

$$y = 450 \cdot 3^{-0.6t}.$$

Thus the number of sea otters alive when the oil spill occurred was

$$y = 450 \cdot 3^{-0.6(0)} = 450 \cdot 1 = 450.$$

Figure 13–7 shows the graph of this exponential function. Notice that the number of otters remaining alive declines toward zero.

Five months after the oil spill, the number of otters still alive will be

$$y = 450 \cdot 3^{-0.6(5)} = 450 \cdot 3^{-3} = 450(1/27) \approx 17.$$

How long will it take for two-thirds of the otters to die? Since the number of otters alive at the time of the spill was 450, this is equivalent to asking how long it will take until $(1/3)450 = 150$ remain alive. But if $y = 150$, we have

$$150 = 450 \cdot 3^{-0.6t}.$$

Dividing both sides of this equation by 450, we obtain

$$\frac{1}{3} = 3^{-0.6t}$$

or

$$3^{-1} = 3^{-0.6t}.$$

Therefore, since the exponential function is one-to-one,

$$-1 = -0.6t,$$

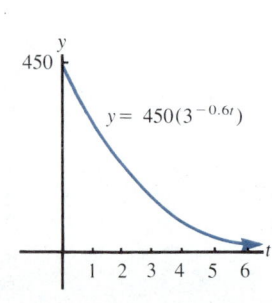

FIGURE 13–7

so
$$t = \frac{1}{0.6} = \frac{5}{3}.$$

Hence it will take $1\tfrac{2}{3}$ months from the time of the spill until two-thirds of the otters have died. ☐

How were the exponential models of Examples 3 and 4 created? Notice that if $f(t) = c \cdot 2^{t/m}$, then

$$f(0) = c \cdot 2^{0/m} = c \cdot 2^0 = c \cdot 1 = c,$$
$$f(m) = c \cdot 2^{m/m} = 2c = 2f(0),$$
$$f(2m) = c \cdot 2^{2m/m} = 2 \cdot 2c = 2f(m),$$
$$f(3m) = c \cdot 2^{3m/m} = 2 \cdot 4c = 2f(2m),$$

and so on. Hence $f(t)$ doubles every m periods. Similarly, if $g(t) = c \cdot 3^{t/m}$, $g(t)$ will triple every m periods. Thus if we know that a quantity experiences a b-fold change every m periods, we can model its growth or decline using an exponential equation of the form $y = cb^{t/m}$.

EXAMPLE 5 Suppose a company's average cost per unit is increasing in such a way that it doubles every six years. Then if $A(t)$ denotes the company's average cost per unit in year t we can model this situation by means of an equation of the form

$$A(t) = c \cdot 2^{t/6}.$$

If we know the value of $A(t)$ for some value of t, we can find the value of c. Thus, suppose that in 1989 the company's average cost per unit was $5; then if we let $t = 0$ represent 1989, we have

$$5 = A(0) = c \cdot 2^{0/6} = c \cdot 1 = c.$$

Hence the model is

$$A(t) = 5 \cdot 2^{t/6}.$$

We can now use the model to predict the average cost per unit in 1992: it will be

$$A(3) = 5 \cdot 2^{3/6} = 5\sqrt{2} \approx \$7.07$$

per unit. ☐

EXAMPLE 6 A radioactive element decays in such a way that no matter how much of the element is present to begin with, 10 years later only one-half of it is left. (Thus we say that the **half-life** of this element is 10 years.) Therefore if y is the amount of the element left after t years, then

$$y = c(\tfrac{1}{2})^{t/10},$$

or, since $\tfrac{1}{2} = 2^{-1}$,

$$y = c \cdot 2^{-0.1t}.$$

Note that when $t = 0$, $y = c$; therefore c is the amount of the element originally present. ☐

There is a special number denoted by e that is very important in the study of exponential functions and their applications. The number e is defined as a limit at infinity.

The Number e

$$e = \lim_{x \to +\infty} \left(1 + \frac{1}{x}\right)^x$$

To verify that this limit exists and to get an idea of the value of e, suppose we use a calculator with an x^y key to calculate $(x + 1/x)^x$ for some values of x:

x	$\left(1 + \dfrac{1}{x}\right)^x$
10	2.5937425
100	2.7048138
1000	2.7169239
10000	2.7181459
100000	2.7182682
1000000	2.7182805
10000000	2.7182817

Thus it appears that the limit does exist and is approximately equal to 2.71828. In fact, the number e is irrational and to 11 significant digits,

$$e = 2.7182818285.$$

Since e is positive, we may use it to construct exponential functions with base e. Any function whose defining equation is of the form

$$y = ce^{rx},$$

where c and r are nonzero real numbers, is thus an **exponential function with base e.**

EXAMPLE 7 Let us graph the exponential function defined by $y = e^x$. To do this, we must be able to evaluate e^x for any real number x. Many calculators have an exp or e^x key by means of which such evaluation may be carried out. Also, Table 7 in the back of the book gives values of e^x for selected values of x. Thus, using either a calculator or Table 7, we find that

$$e^{-2.0} \approx 0.1353, \quad e^{-1.5} \approx 0.2231,$$
$$e^{-1.0} \approx 0.3679, \quad e^{-0.5} \approx 0.6065,$$
$$e^{0.5} \approx 1.6487, \quad e^{1.0} \approx 2.7183,$$
$$e^{1.5} \approx 4.4817, \quad e^{2.0} \approx 7.3891,$$

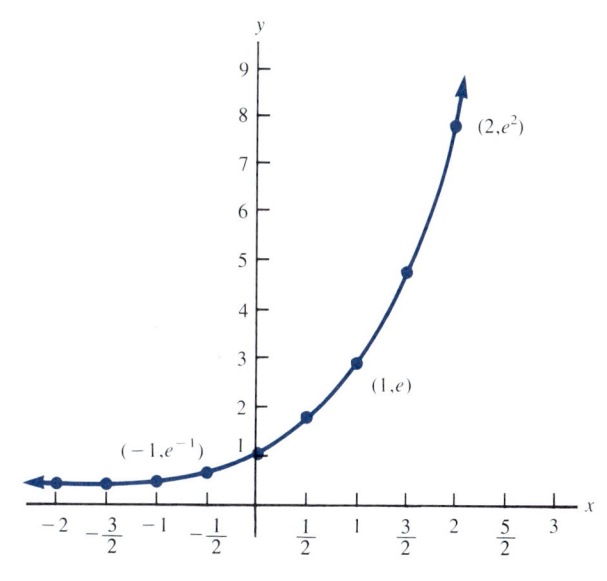

FIGURE 13–8
The Graph of $y = e^x$

and so on. (Check these.) Also, of course, $e^0 = 1$. We use these values of e^x to graph $y = e^x$. The graph is shown in Figure 13–8.

EXAMPLE 8   A radioactive isotope decays exponentially according to the equation

$$y = 100e^{-0.25t}.$$

Here t is time in years and y is the amount of the isotope present, in grams, at time t. Note that the amount of the isotope present initially is

$$y = 100e^{-0.25(0)} = 100e^0 = 100$$

grams, and that the amount present declines toward zero. See Figure 13–9.

Let us find the amount of the isotope present at the end of 10 years. It is

$$y = 100e^{-0.25(10)} = 100e^{-2.5}$$

grams. Using either a calculator with an e^x key or Table 7, we find that $e^{-2.5} \approx 0.08208$. Hence the amount of the isotope present at the end of 10 years is approximately

$$y = 100(0.08208) = 8.208$$

grams.

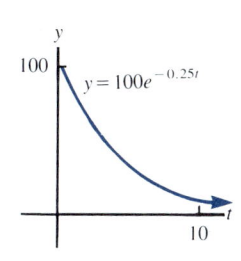

FIGURE 13–9

Why do we want to use e as a base for exponential functions? The answer is that e occurs naturally when many phenomena are modeled by exponential functions; we shall see an example of this in a moment when we discuss continuous compounding of interest. Also, as we will discover later in this chapter, using e as the base for exponential functions simplifies the process of differentiating such functions. Because of these facts, there is a tendency to use base e in preference to other bases. This is not a limitation, for if b is any base, the fact that $y = e^x$ has the positive real numbers as its range and is one-to-one implies that there is a unique number a such that $b = e^a$. (See Figure 13–10.) Hence the exponential function $y = cb^{rx}$ may be written as $y = ce^{arx}$.

EXAMPLE 9 Let us rewrite the exponential equation

$$y = 3 \cdot 2^{0.3t}$$

using e as a base. Since

$$e^{0.6931} \approx 2$$

(check this with a calculator), we have (approximately)

$$y = 3(e^{0.6931})^{0.3t}$$

or

$$y = 3e^{0.20793t}. \quad \square$$

FIGURE 13–10

Continuous Compounding of Interest

If the principal, interest rate, and number of compounding periods per year are fixed, the compound interest formula

$$A = P(1 + r)^{mt}$$

of Section 9.1 expresses the future value A as an exponential function of time t with base $1 + r$.

EXAMPLE 10 An investment of $10,000 earns 8% per year compounded quarterly. Hence at the end of t years it will be worth

$$A = 10{,}000(1.02)^{4t}$$

dollars. $\quad \square$

In Section 9.1 we remarked that for a fixed principal, nominal interest rate, and time, more frequent compounding results in a larger future value. For instance, suppose we invest $100,000 at 12% per year for one year, but vary the number of times that interest is compounded during the year:

Compounding	m	$A = \$100{,}000(1 + 0.12/m)^m$
Annually	1	$112,000.00
Semiannually	2	$112,360.00
Quarterly	4	$112,550.88
Monthly	12	$112,682.50
Daily	365	$112,747.46
Hourly	8760	$112,749.60

EXPONENTIAL AND LOGARITHMIC FUNCTIONS

The table shows that more frequent compounding does indeed increase the interest earned, but it also seems to show that as the number of compounding periods per year increases, the additional interest earned gets smaller and smaller. What happens if we compound interest at every instant of time? This is the most frequent compounding possible, and is known as **continuous compounding.** Under continuous compounding, m, the number of compounding periods per year, becomes infinite, and we have

$$A = \lim_{m \to +\infty} P(1 + r)^{mt} = P \cdot \lim_{m \to +\infty} \left(1 + \frac{i/100}{m}\right)^{mt}.$$

For instance, if $P = \$1$, $i\% = 100\%$, and $t = 1$ year, then

$$A = \lim_{m \to +\infty} \left(1 + \frac{1}{m}\right)^m.$$

But by definition,

$$\lim_{m \to +\infty} \left(1 + \frac{1}{m}\right)^m = e.$$

More generally, it can be shown (though we shall not do so) that for any number a,

$$\lim_{m \to +\infty} \left(1 + \frac{a}{m}\right)^{mt} = e^{at}.$$

Therefore if we take $a = i/100$, we have

$$A = P \cdot \lim_{m \to +\infty} \left(1 + \frac{i/100}{m}\right)^{mt} = Pe^{(i/100)t},$$

and we have developed the formula for continuous compounding of interest.

The Formula for Continuous Compounding of Interest

If principal P is invested at $i\%$ per year compounded continuously, then $r = i/100$ is the annual interest rate expressed in decimal form, and the future value A of the investment at the end of t years is

$$A = Pe^{rt}.$$

EXAMPLE 11 If $1000 is invested at 6% per year compounded continuously, then at the end of three years the investment will be worth

$$A = \$1000 e^{(0.06)3} = \$1000 e^{0.18}.$$

Using a calculator with an e^x key, we find that

$$e^{0.18} = 1.1972174.$$

Thus
$$A = \$1000(1.1972174) = \$1197.22,$$
to the nearest cent. ∎

It is important to be aware of the fact that for a fixed nominal interest rate, compounding continuously makes the future value grow as fast as possible. Thus, in Example 11, if the interest was compounded m times per year for any finite value of m, the future value of the $1000 at the end of 3 years would be less than $\$1000e^{0.18}$.

We can also apply the concept of present value when interest is compounded continuously:

> **The Present Value Formula for Continuous Compounding**
> If future value A is discounted at $i\%$ per year compounded continuously for t years and $r = i/100$, then the present value P of A is
> $$P = Ae^{-rt}.$$

EXAMPLE 12 The present value of $25,000 discounted at 10% per year compounded continuously for five years is
$$P = \$25{,}000 e^{-0.1(5)}$$
$$= \$25{,}000 e^{-0.5}$$
$$= \$25{,}000(0.6065307)$$
$$= \$15{,}163.27. \quad \blacksquare$$

■ *Computer Manual: Programs LIMIT, INFLIMIT, FNCTNTBL, INTEREST, PRSNTVAL*

13.1 Exercises

Exponentials

In Exercises 1 through 16, graph the given exponential function.

1. $y = 3^x$
2. $y = 3^{-x}$
3. $y = 2(5)^x$
4. $y = 3(5)^{-x}$
5. $y = 3^{x/2}$
6. $y = 5^{-x/2}$
7. $y = 4^{2x/3}$
8. $y = 5^{-3x/2}$
9. $y = 2(3)^{-x/3}$
10. $y = \left(\dfrac{1}{2}\right)^x$
11. $y = -5\left(\dfrac{2}{5}\right)^{x/2}$
12. $y = 3\left(\dfrac{4}{3}\right)^{-2x}$
13. $y = 2^{0.2x}$
14. $y = 5^{-0.8x}$
15. $y = 2(3)^{-0.3x}$
16. $y = 4(0.6)^{0.8x}$

In Exercises 17 through 24, find the given power of e.

17. $e^{0.06}$
18. e^4
19. $e^{0.75}$
20. $e^{-0.14}$
21. e^{-1}
22. $e^{-5.6}$
23. $e^{-3.5}$
24. $e^{9.9}$

In Exercises 25 through 34, graph the given exponential function.

25. $y = e^{-x}$
26. $y = e^{2x}$
27. $y = e^{-2x}$
28. $y = 2e^{2x}$
29. $y = e^{x/2}$
30. $y = -5e^{3x/2}$
31. $y = 1.4e^{0.5x}$
32. $y = 2e^{-0.8x}$
33. $y = 3e^{-0.05x}$
34. $y = 5.2e^{1.25x}$

In Exercises 35 through 40, rewrite the exponential function using base e. You may use the facts that

$$e^{0.6931} \approx 2 \quad \text{and} \quad e^{1.0986} \approx 3.$$

35. $y = 3 \cdot 2^{0.4t}$
36. $y = 5 \cdot 3^{-2x}$
37. $y = (½)^{3x}$
38. $y = -4(⅓)^{-1.2t}$
39. $y = 2(⅑)^{4x}$
40. $y = 6^{-3t}$

41. A radioactive isotope decays according to the equation

$$y = 300 \cdot 2^{-t/25}.$$

   Here y is the amount of the isotope, in grams, present at time t, t in years.
   (a) Graph the exponential function defined by this equation.
   (b) What amount of the isotope is present at the end of 12.5 years?
   (c) What is the half-life of the isotope?
   (d) Find and interpret $\lim_{t \to +\infty} y$.

42. A radioactive isotope decays according to the equation

$$y = 200e^{-0.05t},$$

   where t is in years and y is the amount of the isotope remaining, in grams, at time t. Answer the questions of Exercise 41 for this isotope. (*Hint:* $e^{-0.6931} \approx ½$.)

43. Suppose the half-life of a radioactive isotope is 12 years and you start with 100 grams of the isotope. Find a model that describes the amount of the isotope present at the end of t years.

Management

44. A firm's sales are given by $S = 5 \cdot 2^{0.1t}$, where S is annual sales in millions of dollars and t is time in years since the firm was founded.
    (a) Graph this exponential function.
    (b) Find the firm's annual sales 20 years after it was founded; 25 years after it was founded.
    (c) How long did it take the firm to double its annual sales?

45. Use the fact that $e^{0.6931} \approx 2$ to express the sales of the firm of Exercise 44 as an exponential function with base e.

46. A company's average cost per unit was $22.50 in 1985 and has been doubling every eight years. Find a model that describes the company's average cost per unit as a function of time, and use it to predict the average cost per unit in 1993.

47. Repeat Exercise 46 if the company's average cost per unit has been tripling every 10 years.

*48. Repeat Exercise 46 if the company's average cost per unit has been decreasing by 20% every five years.

49. A firm's cost function is given by $C(x) = 300e^{0.04x}$, where cost C is in thousands of dollars and x is in thousands of units.
    (a) Graph the firm's cost function. What is its fixed cost?
    (b) What is the firm's cost if it makes 10,000 units?

50. An oil field will produce oil at a rate of $y = 40e^{-0.3t}$ million barrels per year t years after it is opened.
    (a) Graph the exponential function defined by this equation. At what rate will the field produce oil when it is opened?
    (b) At what rate will the field produce oil five years after it is opened?
    (c) Find and interpret $\lim\limits_{t \to +\infty} y$.

The **double-declining balance** method of depreciation depreciates an asset in year t of its useful life by the amount

$$D = \frac{2C}{n}\left(1 - \frac{2}{n}\right)^{t-1}.$$

Here n is the number of years of useful life the asset has when new, C is its original cost, and D is the amount of its depreciation in year t, $1 \leq t \leq n$. Exercises 51 through 53 make use of this depreciation formula.

51. A car that cost $25,000 new has a useful life of five years. Write an expression for the amount of its depreciation in year t of its life using the double-declining balance method, and find the amount of its depreciation during the third and fifth years of its life.

52. Fill in the following depreciation schedule for the car of Exercise 51:

Year	Depreciation	Value of Car at End of Year
1	$10,000	$15,000
2		
3		
4		
5		

53. An apartment building cost $1.2 million to construct and has a useful life of 30 years. Find an expression for the amount of its depreciation in year t of its useful life using the double-declining balance method, and find the amount of its depreciation in the first, tenth, and twenty-fifth years of its life.

Life Science

54. The number of bacteria in a culture t hours after the start of an experiment was y thousand, where $y = 10 \cdot 2^{0.2t}$.
    (a) Graph the exponential function defined by $y = 10 \cdot 2^{0.2t}$. How many bacteria were in the culture at the start of the experiment?
    (b) How many bacteria were in the culture 5 hours after the start of the experiment?

55. An article by Simons, Sherrod, Collopy, and Jenkins ("Restoring the Bald Eagle," *American Scientist*, May–June, 1988) indicates that y percent of bald eagle colonies that are founded by x individual eagles will become extinct, where $y = 107e^{-0.082x}$.
    (a) Graph the function defined by $y = 107e^{-0.082x}$.
    (b) What percentage of colonies founded by two eagles become extinct? What percentage of those founded by 20 eagles become extinct?

56. In the early years of the AIDS epidemic, the number of diagnosed cases was approximately y, where $y = 100e^{0.46t}$, with t being time in years since 1976.

(a) Graph the function defined by $y = 100e^{0.46t}$.
(b) If the epidemic had continued to be described by this exponential function, how many cases would there have been by 1991? by 1996?

Continuous Compounding of Interest

57. How much will you have at the end of one year if you invest $10,000 at 8% per year compounded continuously?

58. How much will you have at the end of 3.5 years if you invest $45,000 at 10.5% per year compounded continuously?

59. If you borrow $15,000 at 6% per year compounded continuously, how much will you owe at the end of five years?

60. If you borrow $8,000 at 12.2% per year compounded continuously, how much will you owe at the end of 6.25 years?

61. Make a table showing the future value of $10,000 invested for one year at 6% per year compounded annually, semiannually, quarterly, monthly, and continuously.

62. Make a table showing the future value of $1 million invested for four years at 12% per year compounded annually, semiannually, quarterly, monthly, and continuously.

63. How much must be invested now at 10% per year compounded continuously in order to have $10,000 three years from now?

64. How much must be invested now at 6.5% per year compounded continuously in order to have $25,000 six years from now?

65. Find the present value of $100,000 discounted for eight years at 5% per year compounded continuously.

66. Find the present value of $40,000 discounted for 6.4 years at 8.5% per year compounded continuously.

67. Make a table showing the present value of $10,000 discounted for one year at 9% per year compounded annually, semiannually, quarterly, monthly, and continuously.

68. Make a table showing the present value of $1 million discounted for five years at 6% per year compounded annually, semiannually, quarterly, monthly, and continuously.

Management

69. If Alpha Company borrows $100,000 for three months discounted at 15% per year compounded continuously, how much will Alpha actually receive?

70. Beta Corporation has $250,000 worth of accounts receivable that are due six months from now. If Beta factors these accounts receivable with a finance company that discounts their value by 18% per year compounded continuously, how much will Beta receive?

71. A company can purchase either (or neither) of two new telephone systems. System X will cost $100,000 and save $200,000 three years from now, while system Y will cost $150,000 and save $375,000 five years from now. If the company has a policy that no investment will be made unless it returns at least 12% per year compounded continuously, which system will it purchase?

72. Repeat Exercise 71 if the company requires a return of at least 20% per year compounded continuously.

73. Repeat Exercise 71 if system X will save $200,000 three and one-half years from now and the company requires a return of 20% per year compounded continuously.

13.2 LOGARITHMIC FUNCTIONS

Since every exponential function defined by an equation of the form $y = b^x$ is one-to-one, it has an inverse function, called the **logarithmic function to the base b**. Logarithmic functions are important: as we shall see later in this chapter, they can be used to model some growth situations, and they possess several properties that make them useful in computation. In this section we introduce logarithmic functions, state the properties of logarithms, and show how to use the properties of logarithms to solve exponential equations.

Logarithms

In Section 2.4 we introduced the notion of the **inverse function** of a one-to-one function. Recall that if f is a one-to-one function we define its inverse function f^{-1} by setting $f^{-1}(u) = v$ if and only if $u = f(v)$. See Figure 13–11. Thus f^{-1} is a function whose domain is the range of f and whose range is the domain of f.

Since the exponential function defined by $f(v) = b^v$ is one-to-one, with domain the set of real numbers and range the set of positive real numbers, its inverse function f^{-1} exists and has domain the set of positive real numbers and range the set of real numbers; furthermore, f^{-1} is defined by setting $f^{-1}(u) = v$ if and only if $u = b^v$. This inverse function f^{-1} is called the **logarithmic function to the base b,** and is denoted by $\log_b$: $f^{-1}(u) = \log_b u$. Thus $\log_b u = v$ if and only if $u = b^v$. Replacing u and v by x and y, we have the following definition of logarithmic functions.

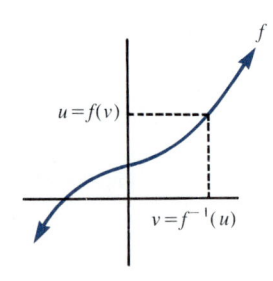

FIGURE 13–11

The Logarithmic Function to the Base b

Let b be a positive real number, $b \neq 1$. The inverse function of the exponential function defined by $y = b^x$ is called the **logarithmic function to the base b,** and is denoted by $\log_b$ (read as "log to the base b"). The logarithmic function to the base b is defined by

$$y = \log_b x \quad \text{if and only if} \quad x = b^y.$$

The domain of the logarithmic function to the base b is the set of positive real numbers and its range is the set of real numbers.

In Section 2.4 we noted that the graph of f^{-1} is that of f reflected about the line $y = x$. Therefore the graph of $y = \log_b x$ is that of $y = b^x$ reflected about the line

EXPONENTIAL AND LOGARITHMIC FUNCTIONS

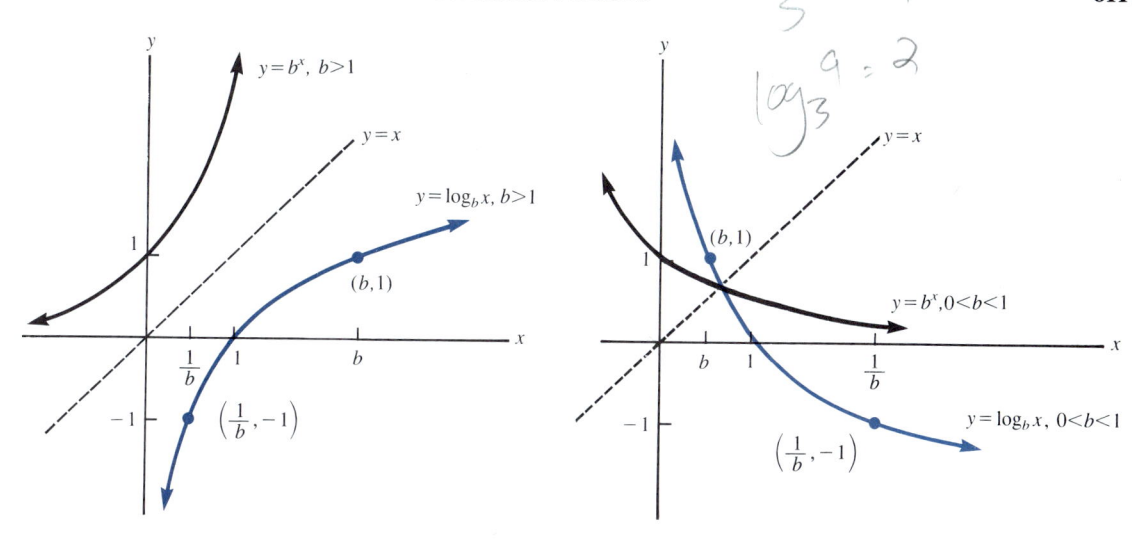

FIGURE 13–12
The Graphs of Logarithmic Functions

$y = x$. Figure 13–12 shows the graph of the logarithmic function to the base b when $b > 1$ and when $0 < b < 1$. Notice that

- The function defined by $y = \log_b x$ is continuous and one-to-one.
- Its x-intercept is $(1, 0)$.
- If $b > 1$, $\lim\limits_{x \to +\infty} \log_b x = +\infty$, $\lim\limits_{x \to 0^+} \log_b x = -\infty$.
- If $0 < b < 1$, $\lim\limits_{x \to +\infty} \log_b x = -\infty$, $\lim\limits_{x \to 0^+} \log_b x = +\infty$.

Before we proceed to some examples, we wish to emphasize that the key to understanding logarithms lies in the definition of $y = \log_b x$. Remember,

$$y = \log_b x \quad \text{means} \quad x = b^y.$$

EXAMPLE 1

$\log_2 8 = 3$ because $2^3 = 8$,

$\log_3 81 = 4$ because $3^4 = 81$,

$\log_2(1/2) = -1$ because $2^{-1} = 1/2$,

$\log_3(1/27) = -3$ because $3^{-3} = 1/27$,

$\log_{1/5} 25 = -2$ because $(1/5)^{-2} = 25$. ∎

EXAMPLE 2

$\log_{10} 10 = 1$, $\log_{10} 100 = 2$, $\log_{10} 1000 = 3$, $\log_{10} 10{,}000 = 4$,

$\log_{10} 0.1 = -1$, $\log_{10} 0.01 = -2$, $\log_{10} 0.001 = -3$. ∎

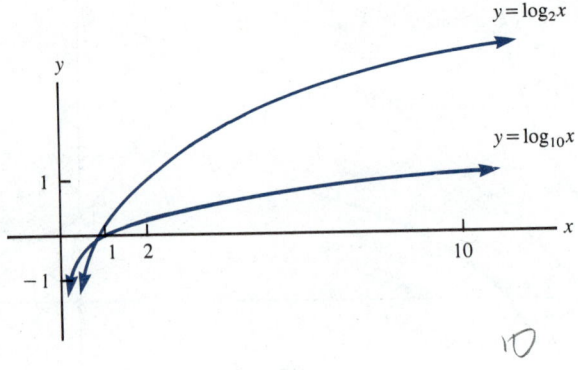

FIGURE 13-13

EXAMPLE 3 The graphs of $y = \log_2 x$ and $y = \log_{10} x$ are shown on the same set of axes in Figure 13-13. □

Let b be a positive real number, $b \neq 1$. Then

- $\log_b 1 = 0$ because $1 = b^0$;
- $\log_b b = 1$ because $b = b^1$;
- $\log_b b^r = r$ because $b^r = b^r$;
- $b^{\log_b x} = x$ because if $\log_b x = y$, then $x = b^y$, so $b^{\log_b x} = b^y = x$.

The preceding identities are just some of the **properties of logarithms.**

The Properties of Logarithms

Let b be a positive real number, $b \neq 1$. Let r be any real number and x, x_1, and x_2 be positive numbers. Then

1. $\log_b 1 = 0$.
2. $\log_b b^r = r$; in particular, $\log_b b = 1$.
3. $b^{\log_b x} = x$.
4. $\log_b(x_1 x_2) = \log_b x_1 + \log_b x_2$.
5. $\log_b(x_1/x_2) = \log_b x_1 - \log_b x_2$.
6. $\log_b x^r = r \cdot \log_b x$.
7. if $\log_b x_1 = \log_b x_2$, then $x_1 = x_2$.

Properties 1, 2, and 3 were proved above; property 7 follows from the fact that $y = \log_b x$ is a one-to-one function; and properties 4, 5, and 6 follow from the

Exponential and Logarithmic Functions

properties of exponents of Section 1.4 and the fact that $y = \log_b x$ means that $x = b^y$.

EXAMPLE 4

$$\log_2 1 = 0, \quad \log_2 2 = 1;$$

$$\log_3 9 = \log_3 3^2 = 2, \quad \log_5(1/625) = \log_5 5^{-4} = -4;$$

$$7^{\log_7 38} = 38;$$

if $\log_{10} 2x = \log_{10} 44$, then $2x = 44$, so $x = 22$. ☐

EXAMPLE 5

Given that $10^{0.3010} = 2$, $10^{0.4771} = 3$, and $10^{0.6990} = 5$, so that $\log_{10} 2 = 0.3010$, $\log_{10} 3 = 0.4771$, and $\log_{10} 5 = 0.6990$, we use the properties of logarithms to find:

$$\log_{10} 6 = \log_{10}(2 \cdot 3) = \log_{10} 2 + \log_{10} 3 = 0.3010 + 0.4771 = 0.7781;$$

$$\log_{10} 8 = \log_{10} 2^3 = 3 \log_{10} 2 = 3(0.3010) = 0.9030;$$

$$\log_{10} 45 = \log_{10}(3^2 \cdot 5) = 2 \log_{10} 3 + \log_{10} 5 = 2(0.4771) + 0.6990$$

$$= 1.6532;$$

$$\log_{10}(\tfrac{2}{3}) = \log_{10} 2 - \log_{10} 3 = 0.3010 - 0.4771 = -0.1761;$$

$$\log_{10}(\tfrac{75}{4}) = \log_{10}(3 \cdot 5^2) - \log_{10} 2^2 = \log_{10} 3 + 2 \log_{10} 5 - 2 \log_{10} 2$$

$$= 0.4771 + 2(0.6990) - 2(0.3010)$$

$$= 1.2731. \quad ☐$$

Just as there is a tendency when working with exponential functions to use base e, so there is a tendency when working with logarithms to use base e. The logarithm to the base e is called the **natural logarithm** and is written ln. Thus, ln 2 means $\log_e 2$, ln 5 means $\log_e 5$, and so on. Many calculators have an ln key, which allows calculation of ln x for positive values of x. Also, Table 8 of the appendix is a table of values of ln x for selected values of x.

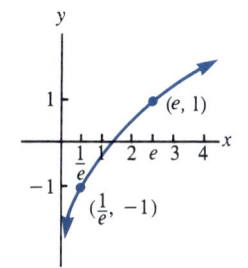

FIGURE 13–14
The Graph of $y = \ln x$

EXAMPLE 6

We find some natural logs (check these using a calculator or Table 8):
ln 0.5 = −0.6931,   ln 2 = 0.6931,   ln 5 = 1.6094,
ln 10 = 2.3026,   ln ($\tfrac{8}{25}$) = −1.1394. ☐

EXAMPLE 7

Figure 13–14 shows the graph of $y = \ln x$. ☐

In mathematics the natural logarithm is far more common than logarithms to other bases; hence from now on we will be using natural logarithms almost exclusively. However, logarithms to bases other than e are sometimes encountered in business and the sciences. Logarithms to the base 10 are called **common logarithms**; they are well adapted to computation and are therefore frequently encountered in applications. Also, logarithms to the base 2 often appear in computer work. In a moment we will demonstrate how to change from one base to another.

In Section 13.1 we observed that it is possible to write any exponential function using base e because for any positive number b, $b \neq 1$, it is always possible to find a number r such that $e^r = b$. More generally, given any two positive numbers a and b, neither of them equal to 1, we can always find a number r such that $a^r = b$: since $a^{\log_a b} = b$, we may take $r = \log_a b$.

> **The Change-of-Base Formula for Exponentials**
>
> Let a, b be positive numbers not equal to 1. Then
> $$a^{\log_a b} = b.$$

EXAMPLE 8 Let us rewrite the exponential function defined by $y = 2 \cdot 5^{3x}$ in terms of base e. To do so, we must find a number r such that $e^r = 5$; using the change-of-base formula with $a = e$ and $b = 5$, we see that $r = \log_e 5 = \ln 5 = 1.6094$. Hence
$$y = 2 \cdot 5^{3x} = 2(e^{1.6094})^{3x} = 2e^{4.8282x}. \quad \square$$

Just as it is sometimes convenient to change bases in exponential functions, so too it is sometimes useful to change bases in logarithmic functions. Thus let a and b be positive numbers, neither of which is equal to 1, and suppose we wish to express $\log_b x$ in terms of $\log_a x$. If
$$\log_b x = y$$
so that
$$x = b^y,$$
then the change-of-bases formula for exponentials tells us that
$$x = (a^{\log_a b})^y = a^{y(\log_a b)}.$$
But
$$x = a^{y(\log_a b)}$$
implies that
$$\log_a x = y \log_a b,$$
and hence that
$$y = \frac{\log_a x}{\log_a b}.$$

> **The Change-of-Base Formula for Logarithms**
>
> If a, b are positive numbers not equal to 1, then
> $$\log_b x = \frac{\log_a x}{\log_a b}.$$

EXAMPLE 9 Suppose we wish to rewrite the logarithmic function $y = \log_{10} x$ using base e. Taking $a = e$ and $b = 10$ in the change-of-base formula, we have

$$y = \log_{10} x = \frac{\log_e x}{\log_e 10} = \frac{\ln x}{\ln 10} \approx \frac{\ln x}{2.3026}. \quad \square$$

Solving Exponential Equations

One of the most important uses of logarithms occurs in the solving of **exponential equations:** these are equations in which the quantity to be determined appears as an exponent. The method of solution is to take the logarithm of both sides of the equation and then use the fact that $\log_b A^x = x \cdot \log_b A$ to "get x out of the exponent." In doing this we could use any base b, but as a matter of convenience we will use the natural logarithm. We illustrate with examples.

EXAMPLE 10 Let us solve the exponential equation

$$2^{5x} = 12$$

for x. Taking the natural logarithm of both sides of the equation, we have

$$\ln 2^{5x} = \ln 12,$$

so

$$5x(\ln 2) = \ln 12,$$

or

$$x = \frac{\ln 12}{5(\ln 2)}.$$

Since $\ln 12 = 2.4849$ and $\ln 2 = 0.6931$,

$$x = \frac{2.4849}{5(0.6931)} = 0.717. \quad \square$$

EXAMPLE 11 The decay of a radioactive isotope is described by the exponential function

$$y = 200 e^{-0.06t},$$

where y is the amount of the isotope, in grams, present at time t, with t in years. Let us find the half-life of the isotope.

To do so, we must determine how long it will take for one-half of the original amount present to decay. Since 200 grams are present initially (why?), we must find out how long it will take until there are only 100 grams left; in other words, we must solve the exponential equation

$$100 = 200\, e^{-0.06t}$$

for t. Simplifying the equation by dividing both sides by 200, we have

$$\tfrac{1}{2} = e^{-0.06t}.$$

Taking the natural logarithm of both sides, we obtain

$$\ln 0.5 = \ln e^{-0.06t},$$

or

$$-0.6931 = -0.06t(\ln e) = -0.06t(1) = -0.06t.$$

Hence

$$t = \frac{0.6931}{0.06} = 11.55 \text{ years.} \quad \square$$

EXAMPLE 12 If $10,000 is invested at 8% per year compounded continuously, how long will it be until the investment is worth $30,000?

We use the formula $A = Pe^{rt}$ with $P = \$10{,}000$, $r = 0.08$, and $A = \$30{,}000$. Thus we have

$$30{,}000 = 10{,}000 e^{0.08t},$$

or

$$3 = e^{0.08t}.$$

Taking the natural logarithm of both sides of this last equation, we find that

$$\ln 3 = 0.08t(\ln e) = 0.08t.$$

Therefore

$$t = \frac{\ln 3}{0.08} = \frac{1.0986}{0.08} = 13.73$$

years. $\quad \square$

EXAMPLE 13 What interest rate, compounded quarterly, must you obtain if an investment of $5000 is to grow to $20,000 in 12 years?

We use the compound interest formula $A = P(1 + r)^{mt}$ with $P = \$5000$, $m = 4$, $t = 12$, and $A = \$20{,}000$. Thus we have

$$20{,}000 = 5000(1 + r)^{4(12)},$$

or

$$4 = (1 + r)^{48}.$$

Taking the natural logarithm of both sides of this last equation, we find that

$$\ln 4 = 48 \cdot \ln(1 + r).$$

Hence

$$\ln(1 + r) = \frac{\ln 4}{48} = \frac{1.3863}{48} = 0.0289.$$

Therefore

$$e^{\ln(1 + r)} = e^{0.0289},$$

Exponential and Logarithmic Functions

or
$$1 + r = 1.0293,$$
and
$$r = 0.0293.$$

Since r is the quarterly interest rate, the annual rate is
$$4r = 4(0.0293) = 0.1172,$$
or 11.72% per year compounded quarterly. □

■ **Computer Manual: Programs LIMIT, INFLIMIT, FNCTNTBL, NTRSTYRS, NTRSTRAT**

13.2 EXERCISES

Logarithms

In Exercises 1 through 4, evaluate the given expressions.

1. $\log_2 16$, $\log_2 128$, $\log_2(1/4)$, $\log_{1/2}(1/4)$, $\log_{1/2} 8$
2. $\log_3 81$, $\log_3 243$, $\log_3(1/243)$, $\log_{1/3}(1/9)$, $\log_{1/3} 27$
3. $\log_5 1$, $\log_5 5$, $\log_{1/5} 1$, $\log_{1/5}(1/5)$, $\log_{1/5} 5$
4. $\log_{10} 100{,}000$, $\log_{10} 1{,}000{,}000$, $\log_{10} 0.0000001$, $\log_{10} 0.000001$, $\log_{0.1} 100{,}000$

In Exercises 5 through 24, use the properties of logarithms to evaluate the given expression.

5. $\log_2 2^{16}$
6. $\log_5 (625 \cdot 125)$
7. $2^{\log_2 6}$
8. $\ln e^{-2}$
9. $\log_2 (64 \cdot 128)$
10. $\ln \sqrt{e}$
11. $\log_5 \dfrac{5^{14}}{5^{17}}$
12. $\log_3 \dfrac{81}{729}$
13. $3^{\log_3 5.1}$
14. $\log_3 81^{15}$
15. $\ln e^{4.25}$
16. $e^{\ln 5}$
17. $2^{3 \cdot \log_2 5}$
18. $5^{4 \cdot \log_5 3}$
19. $e^{\log_2 8}$
20. $\log_{10} 100^{100}$
21. $e^{-2 \cdot \ln 1.5}$
22. $\dfrac{\log_2 64}{\log_2 8}$
23. $\dfrac{\log_5 125}{\ln (1/e)}$
24. $e^{\ln e^2}$

In Exercises 25 through 28, sketch the graph of the given function.

25. $y = \log_3 x$
26. $y = 2 \cdot \log_5 x$
27. $y = \log_{1/3} x$
28. $y = \ln x^2$

In Exercises 29 through 32, change the base of the given exponential function to base e.

29. $y = 10^x$
30. $y = 2^{-x}$
31. $y = 2 \cdot 7^{4x}$
32. $y = 3(1.25)^{-2x}$

In Exercises 33 through 36, change the base of the given logarithmic function to base e.

33. $y = \log_2 x$
34. $y = 2 \cdot \log_3 x$
35. $y = 5 \cdot \log_5 x^{-2}$
36. $y = 4 \cdot \log_{11} x^3$

The intensity of earthquakes is measured by the **Richter scale r,** where
$$r = \log_{10} \dfrac{x}{x_0}.$$

Here x_0 represents the intensity of a benchmark quake and x is the intensity of a given quake in multiples of x_0. Thus, if a quake is 10 times as intense as the benchmark quake, then $x = 10x_0$. Exercises 37 through 39 refer to the Richter scale.

37. Find the Richter scale measurement r for an earthquake that is
    (a) 100 (b) 10,000 (c) 10,000,000
    times as intense as the benchmark quake.

38. Suppose an earthquake measures 5 on the Richter scale. How much more intense is it than the benchmark quake?

39. The most destructive earthquakes measure between 8.0 and 9.0 on the Richter scale. How much more intense than the benchmark quake is such a quake?

Management

Exercises 40 through 45 were suggested by material in Coleman, *Introduction to Mathematical Sociology*, Free Press of Glencoe, 1964. They are concerned with modeling consumer purchases as a function of consumer income. For any particular consumer good, let $p(x)$ denote the proportion of consumers whose income is x who do *not* purchase the good within one year's time. One possible model for the relationship between $p(x)$ and x is

$$\ln p(x) = \begin{cases} 0, & 0 \le x < c, \\ -a(x - c), & x \ge c. \end{cases}$$

Here $a > 0$, $c \ge 0$, and a and c depend on the consumer good.

*40. Solve

$$\ln p(x) = \begin{cases} 0, & 0 \le x < c, \\ -a(x - c), & x \ge c, \end{cases}$$

for $p(x)$ in terms of x.

*41. Use the result of Exercise 40 to graph p as a function of x. Interpret your graph in terms of the proportion of consumers who will or will not purchase the good at a particular income.

*42. Find the percentage of consumers who have incomes of $25,000 who will not purchase a VCR within one year if $c = \$20,000$ and $a = 0.05$ for VCRs. (Measure income x in thousands of dollars.)

*43. Repeat Exercise 42 for consumers who have incomes of $15,000, $30,000, and $50,000.

*44. Suppose that for a particular consumer good, the constant a remains fixed but the constant c increases. What can you say about consumer purchases of the good and their relation to consumer income?

*45. Answer the question of Exercise 44 if c remains fixed but a increases.

Solving Exponential Equations

In Exercises 46 through 55, use logarithms to solve the given exponential equation.

46. $e^{2x} = 5$ 47. $e^{-3x} = 11$ 48. $e^{-2.25x} = 1.9$ 49. $e^{0.4x} = 2$
50. $e^{-2.4x} = 0.06$ 51. $2^x = 3$ 52. $3^{-x/2} = 12$ 53. $5^{x/3} = 9$
54. $(1.3)^{2x} = 11.5$ 55. $(2.2)^{0.5x} = 10.2$

In Exercises 56 through 61, change the base of the given function as indicated.

56. $y = e^x$ to base 10   **57.** $y = 10^{-2x}$ to base 2   **58.** $y = 4 \cdot 3^{1.5x}$ to base 5

59. $y = \ln x$ to base 3   **60.** $y = 3 \cdot \log_2 x$ to base 10   **61.** $y = 4 \cdot \log_5 x^2$ to base 11

62. If inflation persists at 10% per year, in t years the purchasing power of \$1, in terms of today's dollar, will be

$$y = e^{-0.1t}$$

dollars. Under these conditions, how long will it take for money to lose one-half of its value? 99% of its value?

63. The decay of a radioactive isotope is described by the equation

$$y = 1000e^{-0.02t},$$

t in days, y in grams. What is the half-life of the isotope? How long will it take for 90% of the isotope to decay?

64. The decay of a radioactive isotope is described by the equation

$$y = ce^{-0.15t},$$

t in years. What is the half-life of the isotope? How long will it take for 30% of it to decay? How long will it take for 99% of it to decay?

65. How long will it take for an investment of \$12,000 to double if it earns 9% per year compounded continuously?

66. How long will it take for an investment of \$25,000 to grow to \$40,000 if it earns 11.5% per year compounded continuously?

67. How long will it take for an investment of \$7000 to triple if it earns 6% per year compounded quarterly?

68. How long will it take for a debt of \$20,000 to grow to \$35,000 if interest is added to the debt at 12% per year compounded monthly?

69. What annual interest rate must be earned if \$8000 is to grow to \$20,000 in five years under continuous compounding?

70. What annual interest rate must be earned if \$8000 is to grow to \$16,000 in five years under continuous compounding?

71. What annual interest rate i must be earned if \$22,000 is to increase 10-fold in 30 years at i% per year compounded quarterly?

72. If a future value of \$10,000 is discounted for four years at a rate compounded annually and its present value is \$6500, what is the rate?

73. A sinking fund is created by making payments of \$5000 at the end of each year into an account that pays 10% per year compounded annually. How long will it be until the fund contains \$100,000?

74. A sinking fund is created by making payments of \$100 at the end of each month into an account that pays 12% per year compounded monthly. How long will it be until the fund contains \$50,000?

75. If you can deposit \$200 at the end of each month into an account that pays 6% per year compounded quarterly, how long will it be until you have the \$20,000 you need for a new car?

Management

76. If a firm can invest $500,000 in a new product and receive $750,000 in return three years later, what interest rate is it earning on its investment, assuming continuous compounding?

77. Eta Associates takes a six-month loan of $100,000 discounted at a continuously compounded rate. If Eta receives $95,000, what was the rate?

78. Pluperfect Pools sells above-ground swimming pools on terms of no money down and the balance payable in three months. Suppose that Pluperfect sells $10,000 worth of pools and after factoring the accounts receivable from these sales with a finance company receives $9000 for them. If the discount rate the finance company uses is compounded monthly, what is it?

79. A firm's annual sales are given by the equation $S = 200e^{0.05t}$, where t is in years with $t = 0$ representing 1985 and annual sales S is in millions of dollars. When will annual sales attain a level of $1 billion?

80. If an airplane manufacturer has x thousand employees, it can build a plane in y worker-hours, where $y = x^{2.78}$. How many employees would be needed to build a plane in exactly 10,000 worker-hours?

Life Science

81. The number of bacteria in a culture t hours after the start of an experiment is y thousand, where $y = 120 \cdot 2^{-0.25t}$. How long will it take until the number of bacteria in the culture is 10% of the number present at the start of the experiment?

82. A patient is injected with a drug and t hours later $y = 100b^{-t/12}$ percent of the drug remains in the patient's body. If 50% of the drug is eliminated in 9 hours, what is b?

Social Science

83. The population of the earth t years after 1988 is forecast to be y billion people, where $y = 5.0562e^{0.015t}$. How long will it take for the population to reach 100 billion? 1 trillion (1000 billion)?

Internal Rate of Return

Exercises 84 through 88 were suggested by material in Higgins, *Analysis for Financial Management*, Irwin, 1984. They are concerned with the **internal rate of return** of an investment. The internal rate of return (IRR) of an investment is defined as follows:

> the IRR is the annually compounded discount rate that makes the net present value of the investment zero.

*84. Find the IRR of an investment that will pay $100,000 four years from now and costs $75,000 now.

*85. Find a formula for calculating the IRR of an investment that costs P dollars now and will pay A dollars t years from now.

*86. If a firm invests $400,000 now in a new product, it will receive $500,000 four years from now. Use the formula of Exercise 85 to find the IRR of this investment. Suppose the firm must borrow the $400,000 at i% per year compounded continuously; should they do so if $i = 5$? if $i = 7$?

***87.** If a firm can invest $250,000 now in a new billing system it will save $340,000 three years from now. If the firm must borrow the $250,000 at $i\%$ per year compounded continuously, for what values of i will it make sense for the company to borrow the money?

***88.** A firm has three investment opportunities, A, B, and C. If A costs $350,000 and will pay $500,000 in three and one-half years, B costs $400,000 and will pay $600,000 in four years, and C costs $460,000 and will pay $730,000 in five and one-quarter years, use their IRRs to rank the investments in order of desirability.

Average Daily Yield

Exercises 89 through 92 were suggested by material in Stigum, *Money Market Calculations*, Dow Jones-Irwin, 1981. They are concerned with the **average daily yield** of an investment. Short-term securities are usually purchased at a discount and redeemed for the face value of the security. For instance, an investor might purchase a 90-day note with a face value of $1000 for $950; then 90 days after the note was issued it matures, at which time the investor can redeem it for $1000. If $P is the price paid by the investor for a security and $F is its face value, then $P is considered to be the present value of $F discounted at a rate that is compounded daily: thus

$$P = F(1 + r)^{-k},$$

where k is the number of days from the purchase of the security to its redemption and $100r\%$ is the **average daily yield** of the security.

***89.** An investor pays $950 for a note that will be redeemed 90 days later for $1000. Find the average daily yield of the note.

***90.** Find a formula for calculating the average daily yield of a security that is purchased for $P, has a face value of $F, and will be redeemed in k days.

***91.** Use the result of Exercise 90 to find the average daily yield of a security whose face value is $1000 and which is purchased for $982 for redemption 60 days later.

***92.** Average daily yields are usually annualized to present a truer picture of the interest rate earned on the investment. This is done as follows: the average daily yield is a daily interest rate; the annualized average daily yield is the annual effective rate equivalent to this daily rate. Suppose a security has a face value of $1000 and is purchased for $968. Find its annualized daily yield if it is redeemed 30 days after purchase; 60 days after purchase. Use a 365-day year.

13.3 DIFFERENTIATION OF LOGARITHMIC FUNCTIONS

In Section 13.2 we introduced logarithmic functions and discussed some of their properties and uses. The graph of $y = \ln x$ suggests that there is a unique tangent line at each point on the curve (Figure 13–15), and hence that the derivative of the natural logarithm function must exist for all $x > 0$. In this section we show how to differentiate the natural logarithm and all other logarithmic functions as well.

Differentiation of Logarithmic Functions

Let $f(x) = \ln x$. To find $f'(x)$, we must take the limit as h approaches 0 of the difference quotient

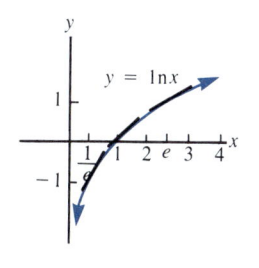

FIGURE 13–15

CHAPTER 13

$$\frac{f(x+h) - f(x)}{h} = \frac{\ln(x+h) - \ln x}{h}.$$

We rewrite the different quotient using the properties of logarithms:

$$\frac{\ln(x+h) - \ln x}{h} = \frac{1}{h}[\ln(x+h) - \ln x]$$

$$= \frac{1}{h} \ln \frac{x+h}{x}$$

$$= \ln\left(1 + \frac{h}{x}\right)^{1/h}.$$

Now let $m = x/h$. Then

$$\frac{\ln(x+h) - \ln x}{h} = \ln\left(1 + \frac{1}{m}\right)^{m/x} = \ln\left[\left(1 + \frac{1}{m}\right)^m\right]^{1/x}.$$

But by definition,

$$\lim_{m \to +\infty} \left(1 + \frac{1}{m}\right)^m = e,$$

and it is possible to use the continuity of the natural logarithm function to show that

$$\lim_{m \to +\infty} \ln\left[\left(1 + \frac{1}{m}\right)^m\right]^{1/x} = \ln\left[\lim_{m \to +\infty} \left(1 + \frac{1}{m}\right)^m\right]^{1/x} = \ln e^{1/x}.$$

(The details of the argument are rather technical, and we omit them.) Thus, since $m = x/h \to +\infty$ as $h \to 0$, we have

$$f'(x) = \lim_{h \to 0} \frac{\ln(x+h) - \ln x}{h}$$

$$= \lim_{m \to +\infty} \ln\left[\left(1 + \frac{1}{m}\right)^m\right]^{1/x}$$

$$= \ln e^{1/x}$$

$$= \frac{1}{x} \ln e$$

$$= \frac{1}{x}.$$

The Derivative of ln x

If $f(x) = \ln x$, then

$$f'(x) = \frac{1}{x}.$$

EXPONENTIAL AND LOGARITHMIC FUNCTIONS 823

EXAMPLE 1   If $f(x) = 2 \cdot \ln x$, then $f'(x) = 2(1/x) = 2/x$. □

EXAMPLE 2   If $f(x) = x \ln x$, then by the product rule,

$$f'(x) = x(\ln x)' + (x)' \ln x = x \frac{1}{x} + 1 \cdot \ln x = 1 + \ln x. \quad \square$$

Now suppose that $u = g(x) > 0$ and consider $\frac{d}{dx}(\ln u)$. By the chain rule

$$\frac{d}{dx}(\ln u) = \frac{d}{du}(\ln u) \frac{du}{dx} = \frac{1}{u}\frac{du}{dx} = \frac{1}{g(x)}\frac{d}{dx}(g(x)).$$

Therefore we have

The Derivative of ln $g(x)$

If $f(x) = \ln g(x)$, then

$$f'(x) = \frac{g'(x)}{g(x)}.$$

EXAMPLE 3   If $f(x) = \ln(x^2 + 3x + 1)$, then

$$f'(x) = \frac{(x^2 + 3x + 1)'}{x^2 + 3x + 1} = \frac{2x + 3}{x^2 + 3x + 1}. \quad \square$$

EXAMPLE 4   Let us use the curve-sketching procedure to draw the graph of the function defined by

$$f(x) = x \ln x$$

over the interval $[e^{-2}, +\infty)$.

From Example 2, $f'(x) = 1 + \ln x$. Setting $f'(x) = 0$ and solving for x yield

$$1 + \ln x = 0,$$

or

$$\ln x = -1,$$

and hence

$$e^{\ln x} = e^{-1}.$$

Since $e^{\ln x} = x$, we have

$$x = e^{-1}.$$

Thus $x = e^{-1}$ is the only critical value. Since e^{-2} is the interval $[e^{-2}, e^{-1})$, we may use it as a test value. Then

$$f'(e^{-2}) = 1 + \ln e^{-2} = 1 - 2 = -1$$

shows that $f'(x) < 0$ on the interval $[e^{-2}, e^{-1})$. Similarly, since e is in the interval $(e^{-1}, +\infty)$, we may use it as a test value. Then

$$f'(e) = 1 + \ln e = 1 + 1 = 2$$

implies that $f'(x) > 0$ on the interval $(e^{-1}, +\infty)$. Hence we have the diagram

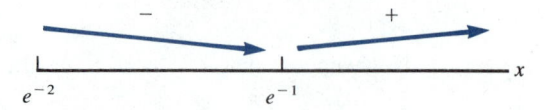

and there is a relative minimum at $(e^{-1}, f(e^{-1})) = (e^{-1}, -e^{-1})$.

The second derivative is

$$f''(x) = \frac{1}{x},$$

which is always positive for $x \geq e^{-2}$. Therefore the function is concave upward on the interval $[e^{-2}, +\infty)$. Since 0 is not in the interval $[e^{-2}, +\infty)$, the function has no y-intercept. If

$$x \ln x = 0,$$

then $x \neq 0$ implies that

$$\ln x = 0,$$

and hence that

$$x = 1.$$

Therefore the only x-intercept is the point $(1, 0)$. The function has no asymptotes. Its graph is shown in Figure 13–16. □

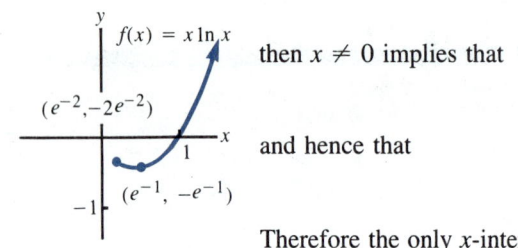

FIGURE 13–16

What about the derivative of $f(x) = \log_b g(x)$, where $b \neq e$? Using the change-of-base formula for logarithms (Section 13.2) with $a = e$, we see that

$$f(x) = \frac{\ln g(x)}{\ln b} = \frac{1}{\ln b} \ln g(x).$$

Hence

$$f'(x) = \frac{1}{\ln b} (\ln g(x))' = \frac{1}{\ln b} \frac{g'(x)}{g(x)}.$$

The Derivative of $\log_b g(x)$

If $f(x) = \log_b g(x)$, then

$$f'(x) = \frac{1}{\ln b} \frac{g'(x)}{g(x)}.$$

Compare the formula for the derivative of ln g(x) with that for the derivative of $\log_b g(x)$, $b \neq e$; the simplicity of the former is one reason why base e is preferred to other bases. Also note that if $b = e$, the formula for the derivative of $\log_b g(x)$ reduces to that for the derivative of ln g(x).

EXAMPLE 5 If $f(x) = \log_2(x^2 + 1)$, then

$$f'(x) = \frac{1}{\ln 2} \frac{(x^2+1)'}{x^2+1} = \frac{2x}{(\ln 2)(x^2+1)}. \quad \square$$

Logarithmic Differentiation

Sometimes it is convenient, or even necessary, to take the logarithm of both sides of an expression before differentiating it. We illustrate with an example.

EXAMPLE 6 Let us differentiate $f(x) = x^2(x^2+1)^2(x^4+1)^4$. We could do this by repeatedly using the product rule and the chain rule, but we prefer to take the natural logarithm of both sides of the equation, differentiate the result, and solve for $f'(x)$. We have

$$\ln f(x) = \ln x^2(x^2+1)^2(x^4+1)^4,$$

or

$$\ln f(x) = \ln x^2 + \ln(x^2+1)^2 + \ln(x^4+1)^4,$$

which simplifies to

$$\ln f(x) = 2 \cdot \ln x + 2 \cdot \ln(x^2+1) + 4 \cdot \ln(x^4+1).$$

Now we differentiate both sides of the last equation to obtain

$$\frac{f'(x)}{f(x)} = \frac{2}{x} + 2\left(\frac{2x}{x^2+1}\right) + 4\left(\frac{4x^3}{x^4+1}\right),$$

so that

$$f'(x) = f(x)\left(\frac{2}{x} + \frac{4x}{x^2+1} + \frac{16x^3}{x^4+1}\right),$$

or

$$f'(x) = x^2(x^2+1)^2(x^4+1)^4 \left(\frac{2}{x} + \frac{4x}{x^2+1} + \frac{16x^3}{x^4+1}\right). \quad \square$$

The technique used in Example 6 is called **logarithmic differentiation**. It is often useful when the function to be differentiated is the product or quotient of several factors.

Sometimes logarithmic differentiation is the only way to proceed.

EXAMPLE 7 Let f be given by $f(x) = x^x$. None of our rules of differentiation apply to this function: it is neither of the form x^r, where r is a fixed real number, nor of the form b^x, where b is a fixed real number. In order to find $f'(x)$, we must use logarithmic differentiation:

$$\ln f(x) = \ln x^x,$$

so

$$\ln f(x) = x \ln x.$$

Differentiating both sides of this last equation, we obtain

$$\frac{f'(x)}{f(x)} = x \frac{1}{x} + 1 \cdot \ln x.$$

Therefore

$$f'(x) = f(x)[1 + \ln x] = x^x[1 + \ln x]. \quad \square$$

■ *Computer Manual: Program DFQUOTNT*

13.3 EXERCISES

Differentiation of Logarithmic Functions

In Exercises 1 through 22, find $f'(x)$.

1. $f(x) = 3 \ln x$
2. $f(x) = -4 \ln x$
3. $f(x) = \ln(7x + 2)$
4. $f(x) = \ln(4 - 2x)$
5. $f(x) = \ln(x^2 + 5x)$
6. $f(x) = \ln(x^2 + 5x - 3)$
7. $f(x) = \ln(x^4 + 2x^2 + 200)^4$
8. $f(x) = 2 \ln \sqrt{x^2 + 1}$
9. $f(x) = 2x \ln 3x$
10. $f(x) = x^2 \ln 2x$
11. $f(x) = x^3 \ln x^2$
12. $f(x) = (x^2 + 1) \ln (x^2 + 1)$
13. $f(x) = \ln \dfrac{2x + 1}{3x + 2}$
14. $f(x) = \ln \dfrac{x^2 - 1}{x^2 + 1}$
15. $f(x) = \dfrac{\ln 5x}{\ln 6x}$
16. $f(x) = \dfrac{\ln(x^2 + 4)}{x^2 + 4}$
17. $f(x) = (\ln x)^3$
18. $f(x) = [\ln(x^5 + 1)]^4$
19. $f(x) = \sqrt{\ln(x^3 + 3x^2 + 4x + 3)}$
20. $f(x) = \ln(\ln x)$
21. $f(x) = \ln(\ln(\ln x))$
22. $f(x) = \ln|x|$

In Exercises 23 through 34, use the curve-sketching procedure to draw the graph of the given function.

23. $y = \ln(2x + 3)$
24. $y = \ln(4x - 12)$
25. $y = \ln(ax + b), a > 0, b > 0$
26. $y = x - \ln x$
27. $f(x) = x \ln 2x$ on $[e^{-3}, +\infty)$
28. $f(x) = x \ln x^2$ on $[e^{-4}, +\infty)$
29. $y = x^2 \ln x$ on $[e^{-2}, +\infty)$
30. $y = x^3 \ln x$ on $[e^{-5}, +\infty)$
31. $f(x) = (\ln x)^2$
32. $y = \dfrac{\ln x}{x}$
33. $y = \dfrac{\ln x}{x^2}$
34. $y = \dfrac{x}{\ln x}$

In Exercises 35 through 42, find $f'(x)$.

35. $f(x) = \log_2 x$
36. $f(x) = \log_3 2x$
37. $f(x) = \log_7 2x^3$
38. $f(x) = \log_{11}(3x + 5)$

39. $f(x) = 4 \log_3(x^2 + 4x)$

40. $f(x) = x \log_2 x$

41. $f(x) = x^2 \log_{10}(2x - 1)$

42. $f(x) = \log_2(2x + 3) \cdot \log_3 x^2$

Management

43. A firm's annual sales are given by the equation
$$S = \ln(10t + 8),$$
where t is in years, with $t = 0$ representing 1988, and sales S are in millions of dollars. Forecast the rate at which sales will be changing in 1992; in 2010.

44. A firm's revenue function is given by the equation
$$R = \ln(25x + 1),$$
where x is in thousands of units and R is in millions of dollars. Find and interpret the firm's marginal revenue when x is 10 thousand and when x is 100 thousand. Then sketch the graph of the revenue function and analyze revenue as x increases.

45. If steel production in year t was y million tons, where
$$y = 96 \ln(\sqrt{t} + 3)$$
and $t = 0$ represents 1955, find the rate at which production was changing in 1970, 1980, and 1990. Then sketch the graph of the function and analyze steel production as t increases.

46. A firm's cost function is
$$C(x) = \ln\sqrt{x^2 + 2}$$
and its revenue function is
$$R(x) = \ln(100x + 1),$$
both valid for $x \geq 1$. Here cost and revenue are in millions of dollars. Find the firm's maximum profit.

47. The value of all goods produced by an industry in year t is y million dollars, where $y = 1000[\ln(0.1t + 0.1)/(t + 1)] + 20$, with $t = 0$ representing the year 1950. Find the maximum value of the industry's production and the year in which it occurred.

Life Science

48. The concentration of radioactive iodine in seawater increased during the period 1945–1965 due to aboveground testing of atomic weapons. Suppose the concentration of such iodine in year t, where $t = 0$ represents 1945, was
$$y = 3 \ln(5 + t)$$
parts per billion. Find the rate at which the iodine concentration was changing in 1955 and 1960.

49. A lake that is being deacidified has its pH y described by
$$y = 2 \ln(3t + 6).$$
Here t is in years, with $t = 0$ representing the start of the deacidification process. Find the rate at which the deacidification process is proceeding 5 and 10 years after its start. Then sketch the graph of the function and analyze the deacidification process as a function of time. (Deacidification ceases when pH-7.)

Social Science

50. The amount spent on social services by a state is y million dollars, where y depends on the unemployment rate $x\%$ according to the equation

$$y = 90 \ln(10x + 50).$$

Find the rate at which spending will be changing when the unemployment rate is 5%.

51. The number of people working in a state's bureaucracy is modelled by

$$y = 1.5 \ln(60t^{2/5} + 20),$$

where y is the number of people in thousands and t is time in years, with $t = 0$ representing 1985. Find the rate at which the bureaucracy will be growing in 1995. Then sketch the graph of the function and analyze the growth of the bureaucracy as a function of time.

Nuclear Testing

52. According to Lynn Sykes and Dan Davis ("The Yields of Soviet Strategic Warheads," *Scientific American*, January, 1987), if x is the yield of an underground nuclear explosion, in kilotons, and y is the magnitude of the surface seismic waves generated by the explosion, then

$$y = \log_{10}(114.4x + 1144.5).$$

Sketch the graph of this function and analyze the magnitude of the seismic waves as a function of the yield of the explosion.

Geology

53. The intensity of earthquakes is measured by the Richter scale r, where

$$r = \log_{10} \frac{x}{x_0}.$$

Here x_0 represents the intensity of a benchmark quake and x is the intensity of a given quake in multiples of x_0. Find the rate of change of the Richter scale measurement as a function of the intensity of a quake.

Logarithmic Differentiation

In Exercises 54 through 66, use logarithmic differentiation to find $f'(x)$.

54. $f(x) = x^2(x^4 + 2)^4$

55. $f(x) = x(x^3 + 1)^3(x^5 + 1)^5$

56. $f(x) = \dfrac{(x^2 + 3)^4}{(x^4 - 2)^7}$

57. $f(x) = \dfrac{(x^3 + 1)^2(x^2 + 1)^3}{(x^2 + 4x + 4)^4}$

58. $f(x) = \dfrac{(x^6 + 1)^3(x^9 - 2x + 1)^4}{(x^5 + 2)^5(x^8 - 7x + 1)^3}$

59. $f(x) = x^{2x}$

60. $f(x) = (2x)^x$

61. $f(x) = x^{-x}$

62. $f(x) = (2x)^{-2x}$

63. $f(x) = (x^2)^{x+1}$

64. $f(x) = (x + 1)^x$

65. $f(x) = (\ln x)^x$

66. $f(x) = x^{\ln x}$

13.4 DIFFERENTIATION OF EXPONENTIAL FUNCTIONS

In the previous section we showed how to differentiate logarithmic functions. Now we are ready to consider exponential functions. The graph of $y = e^x$ suggests that there is a unique tangent line at each point on the curve (Figure 13–17), and hence that the derivative of the exponential function with base e must exist for all x. In this section we discuss the differentiation of this and other exponential functions.

We can use our knowledge about the derivative of the natural logarithm to develop a rule for the derivative of e^x: if we differentiate both sides of the identity

$$x = \ln e^x$$

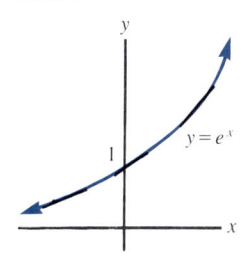

FIGURE 13–17

with respect to x, we have

$$1 = \frac{(e^x)'}{e^x}.$$

Thus

$$(e^x)' = e^x.$$

The Derivative of e^x
If $f(x) = e^x$, then
$$f'(x) = e^x.$$

EXAMPLE 1 If $f(x) = 3e^x$, then $f'(x) = 3(e^x)' = 3e^x$. □

If $f(x) = e^{g(x)}$, let $u = g(x)$. Then by the chain rule,

$$\frac{df}{dx} = \frac{d}{dx}(e^u) = \frac{d}{du}(e^u)\frac{du}{dx} = e^u \frac{du}{dx} = e^{g(x)}g'(x).$$

The Derivative of $e^{g(x)}$
If $f(x) = e^{g(x)}$, then
$$f'(x) = e^{g(x)}g'(x).$$

EXAMPLE 2 If $f(x) = e^{-3x}$, then
$$f'(x) = e^{-3x}(-3x)' = e^{-3x}(-3) = -3e^{-3x}. \quad \square$$

EXAMPLE 3 If $f(x) = 2e^{x^2+x}$, then
$$f'(x) = 2e^{x^2+x}(x^2 + x)' = 2(2x + 1)e^{x^2+x}. \quad \square$$

EXAMPLE 4 If $f(x) = xe^{-x}$, then by the product rule,
$$f'(x) = x(e^{-x})' + (x)'e^{-x} = xe^{-x}(-1) + 1 \cdot e^{-x} = e^{-x}(1-x). \quad \square$$

EXAMPLE 5 Find the equation of the line tangent to the graph of
$$y = \frac{e^x - e^{-x}}{e^x + e^{-x}}$$
at $x = 0$. By the quotient rule,
$$\begin{aligned}\frac{dy}{dx} &= \frac{(e^x + e^{-x})\dfrac{d}{dx}(e^x - e^{-x}) - (e^x - e^{-x})\dfrac{d}{dx}(e^x + e^{-x})}{(e^x + e^{-x})^2} \\ &= \frac{(e^x + e^{-x})(e^x + e^{-x}) - (e^x - e^{-x})(e^x - e^{-x})}{(e^x + e^{-x})^2} \\ &= \frac{(e^{2x} + 2 + e^{-2x}) - (e^{2x} - 2 + e^{-2x})}{(e^x + e^{-x})^2} \\ &= \frac{4}{(e^x + e^{-x})^2}.\end{aligned}$$

Therefore
$$\left.\frac{dy}{dx}\right|_{x=0} = \frac{4}{(1+1)^2} = 1.$$

Thus, since $y = 0$ when $x = 0$ (check this), the equation of the tangent line at $x = 0$ is $y = x$. $\square$

EXAMPLE 6 Let us sketch the graph of the function defined by
$$f(x) = xe^{-x}.$$
Since $f'(x) = e^{-x}(1-x)$ by Example 4, setting $f'(x) = 0$ yields
$$e^{-x}(1-x) = 0.$$
Since e to any power is positive, this implies that $x = 1$. Furthermore, when $x < 1$, $f'(x) = e^{-x}(1-x)$ is positive, and when $x > 1$, $f'(x) = e^{-x}(1-x)$ is negative. Hence the diagram of the first derivative test looks like this:

Therefore there is a relative maximum at $(1, f(1)) = (1, e^{-1})$.
The second derivative is
$$\begin{aligned}f''(x) &= e^{-x}(1-x)' + (e^{-x})'(1-x) \\ &= -e^{-x} - e^{-x}(1-x) \\ &= e^{-x}(x-2).\end{aligned}$$

EXPONENTIAL AND LOGARITHMIC FUNCTIONS   831

Hence, setting the second derivative equal to zero and solving for x will yield $x = 2$. Since $f''(x) < 0$ when $x < 2$ and $f''(x) > 0$ when $x > 2$, the concavity diagram looks like this:

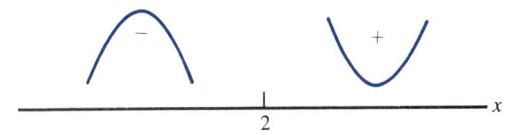

Therefore there is a point of inflection at $(2, f(2)) = (2, 2e^{-2})$.

The point $(0, 0)$ is the only intercept. It is easy to check that as $x \to +\infty$, the value of xe^{-x} approaches 0, and that as $x \to -\infty$, the value of xe^{-x} approaches $-\infty$. Thus

$$\lim_{x \to +\infty} xe^{-x} = 0 \quad \text{and} \quad \lim_{x \to -\infty} xe^{-x} = -\infty.$$

Hence the x-axis is an asymptote as $x \to +\infty$. The graph of the function is shown in Figure 13–18.

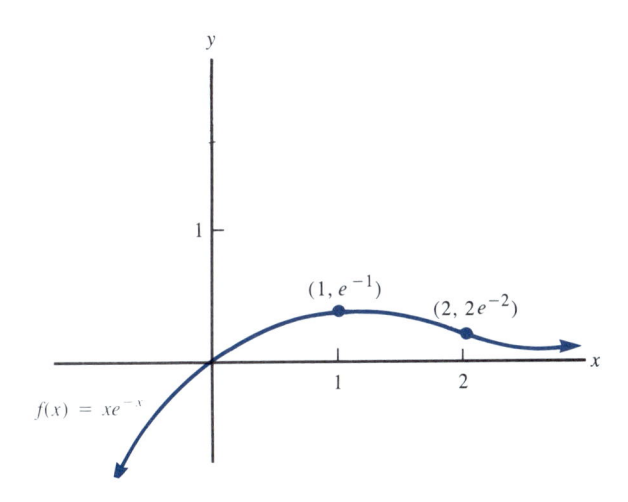

FIGURE 13–18

EXAMPLE 7 Let us sketch the graph of $y = e^{-x^2}$. We have

$$\frac{dy}{dx} = -2xe^{-x^2}.$$

Setting the derivative equal to zero and solving for x yield $x = 0$. The diagram of the first derivative test looks like this:

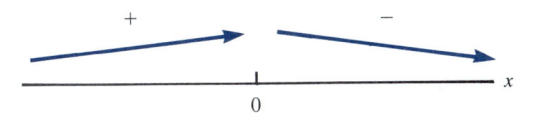

(Check this.) Hence the function has a relative maximum at (0, 1).
The second derivative is

$$\frac{d^2y}{dx^2} = -2e^{-x^2} + 4x^2e^{-x^2} = -2e^{-x^2}(1 - 2x^2),$$

so setting it equal to 0 and solving for x yield $x = \pm 1/\sqrt{2}$. The diagram of the second derivative test looks like this:

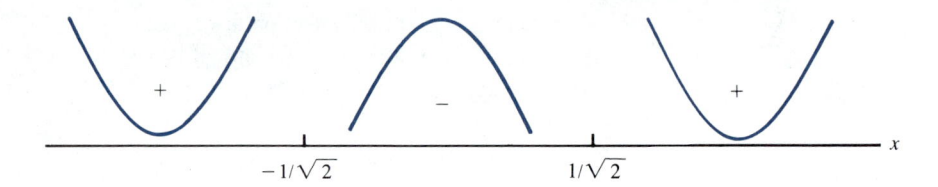

(Check this.) Hence there are points of inflection at $(\pm 1/\sqrt{2}, e^{-1/2})$. It is easy to check that $\text{Lim}_{x \to \pm \infty} e^{-x^2} = 0$. The graph is shown in Figure 13–19. Notice that this graph resembles that of the normal curve (Section 8.4); in fact, the standard normal curve is the graph of $y = (2\pi)^{-1/2}e^{-x^2/2}$. (See Exercises 41, 42.)

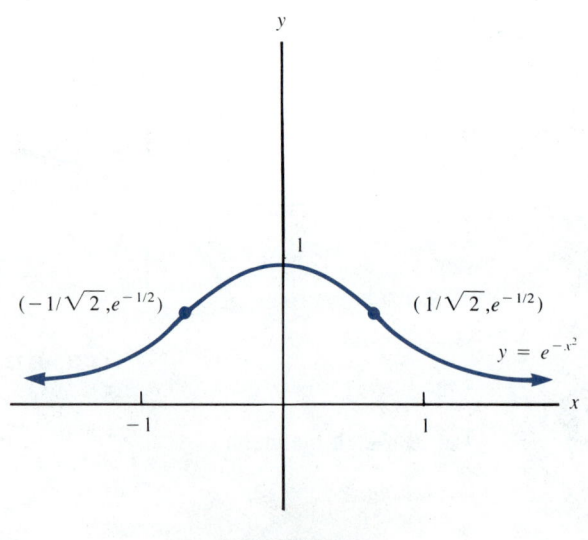

FIGURE 13–19

EXAMPLE 8 Suppose sales of electronic calculators t years after their introduction were y million units, where

$$y = \frac{100}{1 + 49e^{-0.4t}}.$$

Let us find the rate at which sales were changing when $t = 10$.
If we rewrite the equation in the form

$$y = 100(1 + 49e^{-0.4t})^{-1},$$

the general power rule shows that

$$\frac{dy}{dt} = -100(1 + 49e^{-0.4t})^{-2}(49(-0.4)e^{-0.4t}),$$

or

$$\frac{dy}{dt} = \frac{1960e^{-0.4t}}{(1 + 49e^{-0.4t})^2}.$$

Hence

$$\left.\frac{dy}{dt}\right|_{t=10} = \frac{1960e^{-4}}{(1 + 49e^{-4})^2} = 9.971.$$

Therefore when $t = 10$ sales were increasing at a rate of 9,971,000 units per year. □

What about the derivative of $f(x) = b^{g(x)}$, $b \neq e$? Since $e^{\ln b} = b$, we have

$$f(x) = (e^{\ln b})^{g(x)} = e^{(\ln b)g(x)}.$$

Therefore

$$f'(x) = e^{(\ln b)g(x)}((\ln b)g(x))' = e^{(\ln b)g(x)}(\ln b)g'(x),$$

which we may rewrite as

$$f'(x) = b^{g(x)}(\ln b)g'(x).$$

The Derivative of $b^{g(x)}$

If $f(x) = b^{g(x)}$, then

$$f'(x) = (\ln b)b^{g(x)}g'(x).$$

The simplicity of the formula for the derivative of $e^{g(x)}$, as compared with the formula for the derivative of $b^{g(x)}$, $b \neq e$, is another reason why base e is preferred to other bases. Note also that if $b = e$, the formula for the derivative of $b^{g(x)}$ reduces to that for the derivative of $e^{g(x)}$.

EXAMPLE 9 If $f(x) = 3^{2x}$, then

$$f'(x) = (\ln 3)3^{2x}(2x)' = 2(\ln 3)3^{2x} \quad □$$

■ *Computer Manual: Program DFQUOTNT*

13.4 EXERCISES

In Exercises 1 through 24, find $f'(x)$.

1. $f(x) = e^{2x}$
2. $f(x) = 3e^x$
3. $f(x) = 2e^{-x}$
4. $f(x) = -12e^{-x/2}$
5. $f(x) = \frac{1}{2}(e^x - e^{-x})$
6. $f(x) = \frac{1}{2}(e^x + e^{-x})$
7. $f(x) = 5e^{x^2+3x+4}$
8. $f(x) = 2e^{x^2-2x+7}$
9. $f(x) = 2e^{x^3}$
10. $f(x) = 4e^{\sqrt{x^2+1}}$
11. $f(x) = e^{3(\ln x)}$
12. $f(x) = e^{e^x}$
13. $f(x) = 2xe^{-4x}$
14. $f(x) = 3x^2 e^{2x}$
15. $f(x) = \frac{e^x}{x}$
16. $f(x) = \frac{2e^{-2x}}{x^2+1}$
17. $f(x) = e^x \ln x$
18. $f(x) = e^{-x} \ln x$
19. $f(x) = e^{2x} \ln 3x$
20. $f(x) = \frac{e^x}{\ln x}$
21. $f(x) = e^{e^x}$
22. $f(x) = (e^x)^{e^x}$
23. $f(x) = x^e$
24. $f(x) = x^{e^x}$

In Exercises 25 through 40, use the curve-sketching procedure to draw the graph of the function.

25. $f(x) = e^{4x}$
26. $y = 2e^{-3x}$
27. $f(x) = xe^x$
28. $f(x) = 2xe^{3x}$
29. $y = x^2 e^x$
30. $y = x^2 e^{-x}$
31. $y = x^3 e^{-3x}$
32. $y = \frac{e^x}{x}$
33. $y = e^{-x}\sqrt{x}$
34. $y = \frac{1}{2}(e^x - e^{-x})$
35. $y = \frac{1}{2}(e^x + e^{-x})$
36. $y = \frac{e^x - e^{-x}}{e^x + e^{-x}}$
37. $y = 1 - e^{-x}$
38. $y = 1 + e^{-x}$
39. $y = \frac{1}{1+e^{-x}}$
40. $y = xe^{-x^2}$

41. In Section 8.4 we introduced the standard normal distribution. The bell curve that describes the standard normal distribution is the graph of the function f defined by

$$f(x) = \frac{1}{\sqrt{2\pi}} e^{-x^2/2}.$$

Use the curve-sketching procedure to graph this function.

42. In Section 8.4 we introduced the normal distribution with mean μ and standard deviation σ. The bell curve that describes such a normal distribution is the graph of the function f defined by

$$f(x) = \frac{1}{\sigma\sqrt{2\pi}} e^{-(x-\mu)^2/2\sigma^2}.$$

Use the curve-sketching procedure to graph this function.

In Exercises 43 through 50, find $f'(x)$.

43. $f(x) = 2^x$
44. $f(x) = 3^{-5x}$
45. $f(x) = 3(5^{-2x})$
46. $f(x) = 10^{2x^2+3}$
47. $f(x) = x^2 5^x$
48. $f(x) = 2^{-3x}(7^{x^2+2x})$
49. $f(x) = 3^{2x}/x$
50. $f(x) = 4^{-x}/x^2$

51. If $10,000 is invested at 8% per year compounded continuously, find the rate at which the investment is increasing at the end of 5 years; at the end of 10 years.

52. If $25,000 is invested at 6% per year compounded quarterly, find the rate at which the investment is increasing at the end of four years; at the end of eight years.

53. If amount P is invested at i% per year compounded continuously, find the rate at which the investment is changing at the end of t years.

54. If amount P is invested at i% per year compounded m times per year, find the rate at which the investment is changing at the end of t years.

55. The amount of a radioactive isotope present t minutes after 10 grams of it are created in a reactor is $y = 100e^{-0.2t}$ grams. Find the rate at which the amount of isotope present is changing when $t = 0$ and when $t = 5$.

56. A radioactive isotope has a half-life of 30 days.
    (a) Find an equation that describes the amount of the isotope present as an exponential function of time. Use base 2.
    (b) Use the equation of part (a) to find the rate at which the amount of isotope present is changing when $t = 0$ and when $t = 60$.

Management

57. A firm's profit in year t is given by $P(t) = 100{,}000e^{-0.2t}$, where $t = 0$ represents 1985. Find the rate at which the firm's profit will be changing in 1990 and in 1995.

58. The supply function for a commodity is given by $q = e^{0.025p} - 1$ and the demand function by $q = 1000 - 2e^{-0.06p}$. Graph the supply and demand functions on the same set of axes and use the graph to estimate the equilibrium price.

59. A company's average cost per unit is given by
$$f(x) = 2 + 3e^{-0.1x},$$
and its average revenue per unit by
$$g(x) = 5 - e^{-0.2x}.$$
Find the rate of change of average cost per unit, average revenue per unit, and average profit per unit when $x = 10$. Then graph these functions and analyze average cost per unit and average revenue per unit.

60. The worker-hours needed to produce the xth unit of a product are y, where
$$y = 2000(2^{-0.2x}) + 1000, \quad x \geq 1.$$
Find the rate at which production time is changing when $x = 1$, 5, and 25.

61. (This exercise and the next were suggested by material in Budnick, McLeavey, and Mojena, *Principles of Operations Research for Management,* second edition, Irwin, 1988.) Slice-n-Dice, Inc., advertises its "slicer-dicer" kitchen implement only on television. The firm comes into a city, advertises the slicer-dicer extensively on local television for a few weeks, and then, when the market is exhausted, moves on to another city. Suppose that a city contains 300,000 potential customers for the slicer-dicer and that the proportion of them who will buy it in response to the ad campaign is $y = 1 - e^{-0.14t}$, where t is the length of the campaign in days. Suppose also that it costs Slice-n-Dice $200,000 to set up the ad campaign and $10,000 per day to run it, and that the firm makes a profit of $5 on each slicer-

dicer sold. How long should the ad campaign run in order to maximize Slice-n-Dice's profits?

*62. Refer to Exercise 61: suppose that a city contains c potential customers, that it costs Slice-n-Dice $\$F$ to set up the ad campaign and $\$v$ per day to run it, that the firm's profit is $\$m$ per slicer-dicer sold, and that the proportion of customers in the city who will buy the slicer-dicer in response to the ad campaign is $y = 1 - e^{-rt}$, $r > 0$, t the length of the ad campaign in days. Find the value of t that will maximize Slice-n-Dice's profits.

Life Science

63. The concentration of medication in a patient's blood t days after going off the medication is y micrograms per milliliter, where $y = 10(3^{-0.4t})$. Find the concentration and the rate at which it is changing five days after going off the medication.

64. The pollutants in the atmosphere near a power plant will be y parts per million t years from now, where $y = 30 + 125e^{-0.06t}$. Find the rate of change of the pollutants now, and also 4 and 12 years from now. Then graph this function and analyze the plant's pollution as a function of time.

65. The number of bacteria in a culture t hours after an experiment begins will be y thousand, where $y = 100 - 10(5^{-0.1t})$. Find the rate of change of the bacteria population when the experiment begins and also 12 and 24 hours later.

66. A flu epidemic hits a town and t days later y of the inhabitants have contracted the disease, where
$$y = \frac{5000}{1 + 4999e^{-0.5t}}.$$
Find the rate at which the epidemic is proceeding when $t = 0$, 17, and 25.

67. The concentration of a drug in a patient's bloodstream t minutes after injection is $y = 10te^{-0.1t}$ milligrams per milliliter. Graph this function for $t \geq 0$, and find the maximum concentration of the drug and when this maximum occurs.

Social Science

68. The growth of a state bureaucracy is described by the equation $y = 225e^{0.005t}$. Here y is the size of the bureaucracy, in thousands, and t is time in years with $t = 0$ representing 1985. Find the rate at which the size of the bureaucracy was changing in 1985 and the rate at which it will be changing in 1995.

69. A college student starts a rumor about another student, and t hours later y of the students at the college have heard it, where
$$y = \frac{2000}{1 + 1999e^{-1.1t}}.$$
Find the rate at which the rumor spreads when $t = 0$, 5, and 10 hours.

70. A state's aid to education t years from now is forecast to be $y = 200te^{-0.05t}$ million dollars. Find the maximum aid to education and when it will occur. Then graph the function and analyze aid to education as a function of time.

Optimum Holding Period

Exercises 71 through 73 were suggested by material in Budnick, McLeavey, and Mojena, *Principles of Operations Research for Management*, second edition, Irwin, 1988. They are concerned with the optimum holding period for an investment.

*71. Suppose you are considering investing $10,000 in real estate and you estimate that the investment will be worth

$$I(t) = 10,000(1 + t)$$

dollars t years from now. You eventually intend to sell the real estate for a gain, and you would like to determine how long you should hold it before selling. It is your policy to discount future values at 10% per year compounded continuously. Should you buy the real estate? If so, how long should you hold onto it before selling? (*Hint:* Maximize the net present value of $I(t)$.)

*72. Repeat Exercise 71 if t years from now the investment will be worth $I(t) = 10,000\sqrt{t}$ and you discount future values at 12% per year compounded continuously.

*73. Suppose an investor pays C dollars for an asset that will be worth $I(t)$ dollars t years from now, where $I(t)$ is a continuous and increasing function of t. Suppose also that the investor discounts the future worth of the investment at $100r\%$ per year compounded continuously. Find the optimum holding period for the asset.

13.5 EXPONENTIAL GROWTH AND DECLINE

We have seen throughout this chapter that exponential functions can be used to model many situations involving growth and decline. In this section we discuss some useful exponential models and also briefly consider a logarithmic growth model.

Exponential Growth and Decline

A function defined by an equation of the form $y = ce^{rt}$ has the property that its rate of change is directly proportional to y:

$$\frac{dy}{dt} = rce^{rt} = ry.$$

As we shall see in Section 15.6, functions of this form are the only ones having this property. It therefore follows that if the rate of change of y is directly proportional to y, then $y = ce^{rt}$, for some numbers c and r.

Models of Growth and Decline

If a quantity changes in such a way that its rate of change at time t is directly proportional to the amount of it present at time t, then it can be modeled by means of an equation of the form

$$y = ce^{rt}.$$

If the quantity increases with time, the model is said to be one of **exponential growth**; if the quantity decreases with time, the model is said to be one of **exponential decline** or **exponential decay**.

Figure 13–20 shows the graph of an exponential growth model when $c > 0$, and Figure 13–21 shows the graph of an exponential decline model when $c > 0$.

We illustrate the construction of exponential growth and decline models with examples.

EXAMPLE 1 Suppose we have reason to believe that the rate of population growth for a caribou herd depends directly on the size of the herd. Suppose also that there were 10,000 caribou in the herd in 1987 and 11,000 in 1989, so that the size of the herd is increasing. We may then model the size of the herd by an exponential growth function whose defining equation is of the form

$$y = ce^{rt}$$

To complete the model, we must find numbers c and r such that the resulting function describes the size of the herd.

If we measure time in years, with $t = 0$ representing 1987, we must have

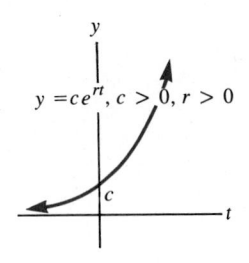

$$y = 10{,}000 \quad \text{when } t = 0 \quad \text{and} \quad y = 11{,}000 \quad \text{when } t = 2.$$

FIGURE 13–20 Exponential Growth

Substituting 0 for t and 10,000 for y in the model yields

$$10{,}000 = ce^{r(0)} = ce^0 = c \cdot 1 = c.$$

Therefore $c = 10{,}000$, so

$$y = 10{,}000 e^{rt}.$$

To find r, we substitute 2 for t and 11,000 for y to obtain

$$11{,}000 = 10{,}000 e^{r(2)} = 10{,}000 e^{2r}.$$

Solving the exponential equation

$$11{,}000 = 10{,}000 e^{2r}$$

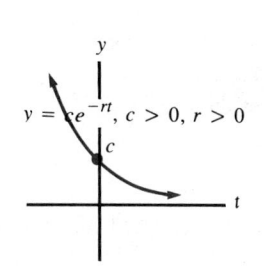

FIGURE 13–21 Exponential Decline

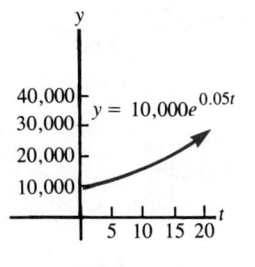

for r, we have

$$\frac{11{,}000}{10{,}000} = e^{2r},$$

or

$$1.1 = e^{2r},$$

so

$$\ln 1.1 = \ln e^{2r} = 2r,$$

or

$$0.095310 = 2r.$$

Thus $r = 0.05$, rounded to two decimal places, and our model is

$$y = 10{,}000 e^{0.05t}.$$

FIGURE 13–22 The graph of this function is shown in Figure 13–22.

EXPONENTIAL AND LOGARITHMIC FUNCTIONS 839

Now we can use the model for prediction. For instance, if we wish to predict the size of the herd in 1997, we need only substitute 10 for t in the model:

$$y = 10{,}000 e^{0.05(10)} = 10{,}000 e^{0.5} = 10{,}000(1.6487) = 16{,}487.$$

Similarly, if we wish to predict how long after 1987 it will take for the size of the herd to grow to 20,000 caribou, then we must substitute 20,000 for y and solve the exponential equation

$$20{,}000 = 10{,}000 e^{0.05t}$$

for t. Dividing both sides of this last equation by 10,000 and then taking the natural logarithm of both sides, we obtain

$$\ln 2 = 0.05t(\ln e),$$

or

$$\ln 2 = 0.05t.$$

Therefore

$$t = \ln 2/0.05 \approx 13.9 \text{ years.} \quad \square$$

EXAMPLE 2   Radioactive decay of an isotope is known to proceed at a rate that is directly proportional to the amount of the isotope present, and the amount of the isotope present decreases with time. Therefore radioactive decay can be modeled by exponential decline functions defined by equations of the form

$$y = ce^{rt}.$$

For instance, the radioactive isotope of uranium known as U^{235} has a half-life of 12 million years. (Thus any amount of U^{235} will decrease by one half in 12 million years.) Let us measure time in millions of years. When $t = 0$,

$$y = ce^{r(0)} = ce^0 = c \cdot 1 = c,$$

so U^{235} is initially present in amount c; but since its half-life is 12 million years, when $t = 12$ we must have $y = c/2$:

$$\frac{1}{2}c = ce^{r(12)} = ce^{12r}.$$

Solving this exponential equation for r, we find that

$$\frac{1}{2} = e^{12r},$$

or

$$\ln \frac{1}{2} = \ln e^{12r},$$

or

$$-0.693147 = 12r,$$

so $r = -0.06$, rounded to two decimal places. Therefore the model is
$$y = ce^{-0.06t}.$$
The graph is shown in Figure 13–23.

We see that at the end of 50 million years approximately 5% of the original amount of U^{235} will be left, because when $t = 50$ we have
$$y = ce^{-0.06(50)} = ce^{-3} = c(0.049787) \approx 0.05c.$$

How long will it be until only 1 percent of the original amount of U^{235} remains? One percent of the original amount is $0.01c$, and if we set $y = 0.01c$ in the model and solve
$$0.01c = ce^{-0.06t}$$
for t, we find that $t \approx 76.8$ million years. (Check this.) ☐

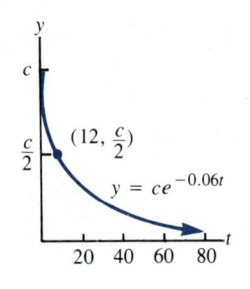

FIGURE 13–23

Limited Growth and Decline

Exponential growth models of the type discussed above are often unrealistic because they assume unlimited growth. For instance, the size of the caribou herd of Example 1 certainly cannot grow without limit, since, for one thing, there will be only a certain amount of food available for the caribou. Thus, if we wish to make our models of exponential growth realistic, it will sometimes be necessary to allow them to reflect limited growth. Similar remarks apply to situations of limited decline.

To see what a model for limited exponential growth might be like, consider a function defined by an equation of the form
$$y = a - ce^{-rt}, \qquad a > c > 0, \quad r > 0$$
on $[0, +\infty)$. Figure 13–24 shows the graph of such a function. (See Exercise 23.) Note that

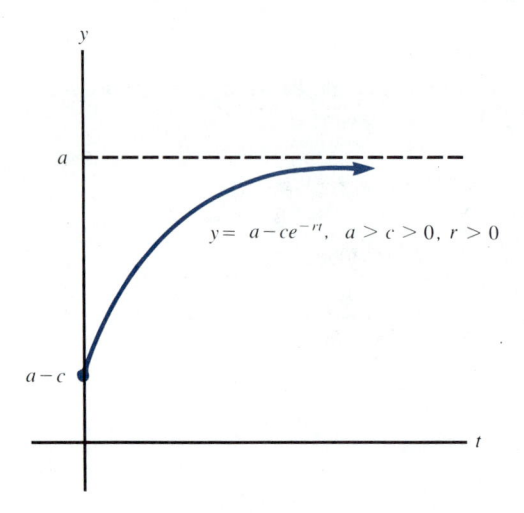

FIGURE 13–24
Limited Growth

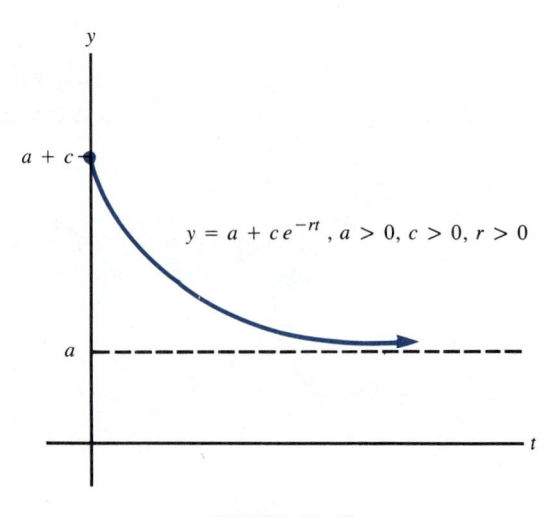

FIGURE 13–25
Limited Decline

EXPONENTIAL AND LOGARITHMIC FUNCTIONS 841

- As t increases, y grows.
- As t increases, the rate at which y grows decreases.
- As $t \to +\infty$, growth levels off and y approaches the limiting value a from below.

Functions of this type are often used to model situations of limited growth when the rate of growth decreases with time and some limiting value is approached from below. Similarly, functions defined by equations of the form

$$y = a + ce^{-rt}, \quad a > 0, \quad c > 0, \quad r > 0$$

on $[0, +\infty)$ are often used to model situations of limited decline. Figure 13-25 shows the graph of such a function. (See Exercise 24.)

Limited Growth and Decline

1. Suppose that as time increases, a quantity increases, its rate of growth decreases, and its growth levels off and approaches a limiting value $a > 0$ from below. Then the amount of the quantity present at time t can be modeled by an equation of the form

$$y = a - ce^{-rt}, \quad a > c > 0, \quad r > 0.$$

2. Suppose that as time increases, a quantity decreases, its rate of decline decreases, and its decline levels off and approaches a limiting value $a > 0$ from above. Then the amount of the quantity present at time t can be modeled by an equation of the form

$$y = a + ce^{-rt}, \quad c > 0, \quad r > 0.$$

We illustrate limited growth and decline with examples.

EXAMPLE 3 Consider again the caribou herd of Example 1. As before, suppose the size of the herd is 10,000 in 1987 and 11,000 in 1989, but now we will assume that as time goes on the growth of the herd will slow down and its size will be limited by the food supply to no more than 15,000. If we model the size of the herd by a function whose defining equation is of the form

$$y = a - ce^{-rt},$$

where $t = 0$ represents 1987, then the limiting value $a = 15,000$, so the model is

$$y = 15,000 - ce^{-rt}.$$

Since $y = 10,000$ when $t = 0$, we have

$$10,000 = 15,000 - ce^{-r(0)} = 15,000 - c.$$

Therefore $c = 5000$, and the model is
$$y = 15{,}000 - 5000e^{-rt}.$$
Finally, since $y = 11{,}000$ when $t = 2$, we have
$$11{,}000 = 15{,}000 - 5000e^{-r(2)},$$
or
$$-4000 = -5000e^{-2r},$$
or
$$0.80 = e^{-2r}.$$
Taking the natural logarithm of both sides of this last equation and solving for r give $r = 0.11$, rounded to two decimal places. (Check this.) Thus the final model is
$$y = 15{,}000 - 5000e^{-0.11t}.$$
The graph is shown in Figure 13–26. With this model our prediction for the size of the herd in 1997 is
$$y = 15{,}000 - 5000e^{-0.11(10)} = 15{,}000 - 5000e^{-1.1} = 13{,}336. \quad \square$$

An important application of limited growth is the **learning curve.** Learning curves relate productivity to experience. Typically, productivity increases as employees become more experienced, since with experience they become more efficient. Productivity cannot grow without limit, however, for even perfect efficiency would result in a certain finite productivity. Thus the rate of productivity growth slows down as productivity increases toward some limit, and hence productivity can be described by a model of the form
$$y = a - ce^{-rt}.$$

EXAMPLE 4 Data collected at a plant that makes microwave carts show that a worker who has no experience can assemble carts at a rate of approximately 200 per week, whereas a worker who has five weeks of experience can assemble them at a rate of approximately 233 per week. It is also known that the most experienced and efficient workers can approach an assembly rate of 300 carts per week. Let us find the learning curve for a typical worker at the plant.

We use the model
$$y = a - ce^{-rt},$$
where t is worker experience, measured in weeks, and y is carts assembled per week. We may take the limiting value a to be 300, so our model is
$$y = 300 - ce^{-rt}.$$
We know that $y = 200$ when $t = 0$, and $y = 233$ when $t = 5$; hence
$$200 = 300 - ce^{-r(0)} = 300 - c,$$

so $c = 100$. But then
$$y = 300 - 100e^{-rt},$$
and the fact that $y = 233$ when $t = 5$ tells us that
$$233 = 300 - 100e^{-5r},$$
or
$$-67 = -100e^{-5r},$$
or
$$0.67 = e^{-5r}.$$

Taking the natural logarithm of both sides of this last equation and solving for r, we find that $r = 0.08$, to two decimal places. Hence the learning curve is
$$y = 300 - 100e^{-0.08t}.$$

Figure 13–27 shows the graph of this learning curve.

Now suppose that a worker must assemble at least 250 carts per week in order to cover the cost of his or her salary. Since $y \approx 247$ when $t = 8$ and $y \approx 251$ when $t = 9$, a worker cannot be expected to cover the cost of his or her salary until the ninth week on the job. ∎

To illustrate a limited decline model, let us consider learning curves from a different viewpoint. In Example 4 we were interested in productivity as a function of experience, but sometimes it is more important to consider production time as a function of the number of units produced. What typically occurs is that each unit can be produced in slightly less time than the preceding one because production becomes more efficient with experience. Even with perfect efficiency, however, there must be some minimum time necessary to produce a unit. Hence the time required to produce a unit will approach some limiting value from above.

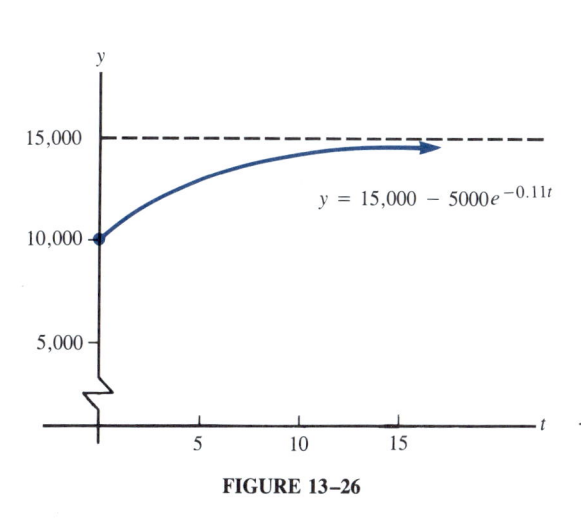

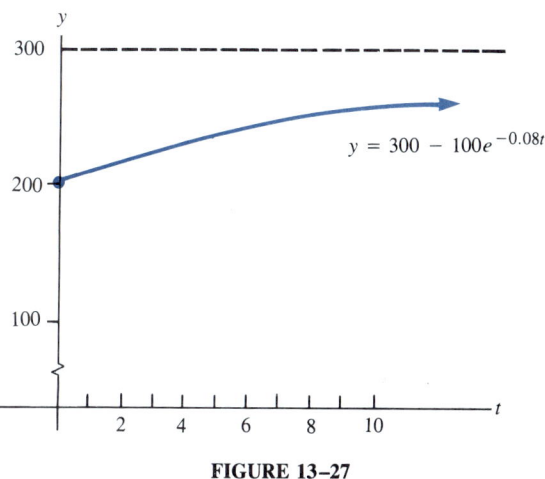

FIGURE 13–26

FIGURE 13–27

EXAMPLE 5 An airplane manufacturer about to begin production of a new plane estimates that it will take 5000 worker-hours to build the first plane, 4600 worker-hours to build the second one, and that it would take 2800 worker-hours to build a plane under conditions of perfect efficiency. Let us find the learning curve that describes the number of worker-hours needed to build the xth plane, $x \geq 1$.

Our model will be of the form

$$y = a + ce^{-rx}.$$

Since the limiting value is 2800 worker-hours, we may take $a = 2800$, so that

$$y = 2800 + ce^{-rx}.$$

When $x = 1$, $y = 5000$, and when $x = 2$, $y = 4600$. Therefore

$$5000 = 2800 + ce^{-r}$$

and

$$4600 = 2800 + ce^{-2r},$$

or

$$2200 = ce^{-r}$$

and

$$1800 = ce^{-2r}.$$

We divide the first equation by the second to obtain

$$\frac{2200}{1800} = \frac{ce^{-r}}{ce^{-2r}} = e^r.$$

Solving for r yields $r = 0.20$, and then

$$2200 = ce^{-0.20}$$

implies that $c = 2687$. Therefore our learning curve is

$$y = 2800 + 2687e^{-0.20x}.$$

The graph is shown in Figure 13–28. If we wished, we could estimate the total number of worker-hours required to build the first 25 planes, say, by substituting $x = 1, 2, \ldots, 25$ into the model and summing the resulting y-values. ☐

Logarithmic Growth

Functions defined by equations of the form

$$y = \ln(at + b), \quad a > 0, \quad b > 1$$

are sometimes used to model situations of growth in which growth is unlimited but the rate of growth is decreasing. Figure 13–29 shows the graph of such a function. (See Exercise 53.) Note that as t increases, y increases at a decreasing rate and $\lim_{t \to +\infty} y = +\infty$. A situation that can be modeled by such a function is said to be one of **logarithmic growth**.

Exponential and Logarithmic Functions

FIGURE 13–28

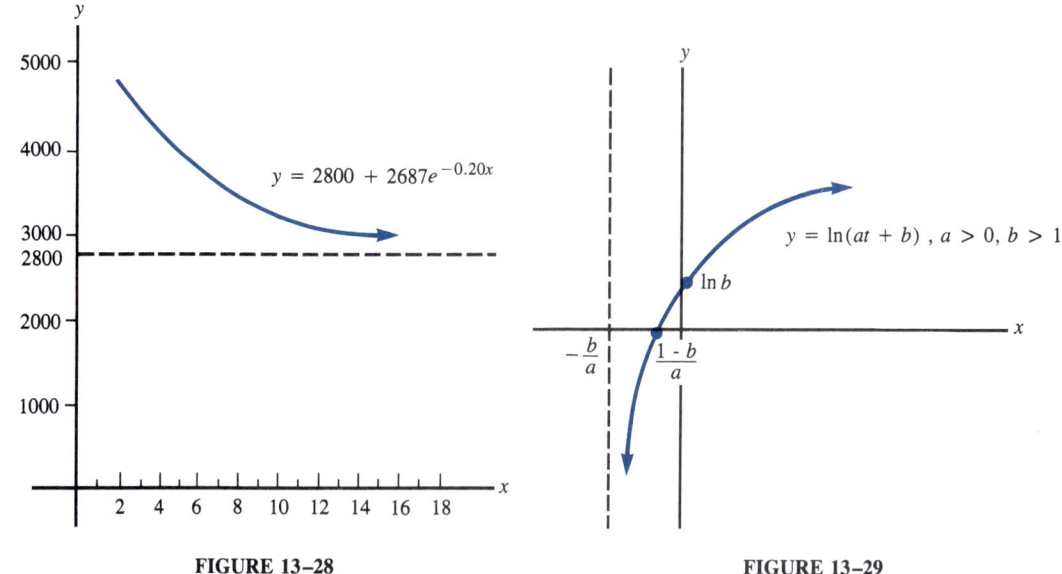

[Graph showing $y = 2800 + 2687e^{-0.20x}$ decreasing toward asymptote at 2800, with x-axis from 2 to 18 and y-axis from 1000 to 5000]

FIGURE 13–29
Logarithmic Growth

[Graph showing $y = \ln(at + b)$, $a > 0$, $b > 1$, with point at $\ln b$ on y-axis, x-intercept at $\frac{1-b}{a}$, and vertical asymptote at $-\frac{b}{a}$]

EXAMPLE 6   The amount of steel produced in a certain country has been increasing at a decreasing rate. The amount produced was 4 million tons in 1986 and 6 million tons in 1988. Let us describe the country's steel production with a model of the form

$$y = \ln(at + b), \quad a > 0, \quad b > 1,$$

where y is in millions of tons of steel and t is in years, with $t = 0$ representing 1986.

When $t = 0$, $y = 4$, so

$$4 = \ln(a(0) + b) = \ln b.$$

But

$$4 = \ln b$$

implies that

$$e^4 = e^{\ln b} = b.$$

Similarly, since $y = 6$ when $t = 2$, we have

$$6 = \ln(2a + b) = \ln(2a + e^4),$$

so

$$e^6 = e^{\ln(a + e^4)} = 2a + e^4,$$

which shows that

$$a = \frac{1}{2}(e^6 - e^4).$$

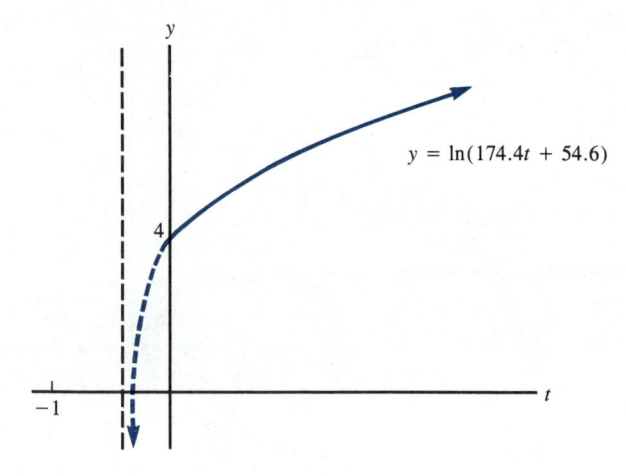

FIGURE 13–30

Converting a and b to decimal form, we have

$$a = \frac{1}{2}(e^6 - e^4) \approx 174.4, \qquad b = e^4 \approx 54.6.$$

Therefore our model is

$$y = \ln(174.4t + 54.6).$$

The graph is shown in Figure 13–30. ☐

■ *Computer Manual: Programs EXPGRWTH, LIMGRWTH, LOGGRWTH*

13.5 EXERCISES

Exponential Growth and Decline

1. The rate of growth of an investment is directly proportional to the value of the investment. Suppose the original amount of the investment was $10,000 and that five years later the investment had a value of $14,918. Find a model that describes the value of the investment as a function of time.

2. The rate of growth of an investment is directly proportional to the value of the investment. Suppose that the original amount invested was $5000 and that eight years later the value of the investment had doubled. Find
   (a) a model that describes the value of the investment as a function of time;
   (b) the value of the investment 12 years after its inception;
   (c) how long it will be until the investment is worth $30,000.

*3. The rate of growth of an investment is directly proportional to the value of the investment. Suppose the original amount invested was P_o dollars and that n years later the investment was worth P_n dollars. Find a model that describes the worth of the investment as a function of time.

4. A radioactive isotope has a half-life of 500 years. Find a model that describes the amount of the isotope present as a function of time.

5. A radioactive isotope has a half-life of 16 days. Find
   (a) a model that describes the amount of the isotope present as a function of time;
   (b) the percentage of the original amount remaining at the end of 10 days;
   (c) how long it will take for 90% of the isotope to decay.

6. Ten grams of a radioactive isotope were created in a reactor and 4 minutes later 7.5 grams were left. Find a model that describes the amount of the isotope present as a function of time, and then find the half-life of the isotope.

*7. If a radioactive isotope has a half-life of T years, find a model that describes the amount of the isotope present as a function of time.

8. Inflation causes the purchasing power of money to decline at a rate proportional to the purchasing power. Suppose inflation persists at an annual rate of 10%. Find a function that describes the purchasing power of a dollar as a function of time. Then find the purchasing power of a dollar after 10 years of 10% inflation; after 15 years of such inflation.

Radiocarbon Dating

The element carbon has two isotopes: C^{14}, which is radioactive with a half-life of 5730 years, and C^{12}, which is not radioactive. In living organisms, the C^{14} incorporated into tissue decays but is replaced by C^{14} from the atmosphere in such a way that the ratio of C^{14} to C^{12} in the tissue remains constant. In dead tissue, however, the C^{14} decays and is not replaced, so the C^{14} to C^{12} ratio decreases at a rate proportional to the ratio. By comparing the C^{14} to C^{12} ratio in dead tissue to that in living tissue, it is thus possible to determine how long ago an organism died. Exercises 9 and 10 concern this technique of **radiocarbon dating.**

*9. Let y be the value of the C^{14} to C^{12} ratio in tissue t years after the tissue has died.
   (a) Find an exponential model that describes y as a function of time. (*Hint:* The ratio will decrease by one-half in 5730 years.)
   (b) Find the age of wheat grains found in an Egyptian tomb if the C^{14} to C^{12} ratio for the grains is 75% of today's ratio.
   (c) Date an archaeological site if charcoal fragments found at the site have a C^{14} to C^{12} ratio that is 20% of today's ratio.

*10. (a) Find an equation that describes the time T since an organism died as a function of the C^{14} to C^{12} ratio of its tissue.
    (b) Use the result of part (a) to date a papyrus whose C^{14} to C^{12} ratio is 82% of today's ratio.

Management

11. The annual sales of Ampersand, Inc. have been increasing at a rate proportional to sales. They were $100,000 in 1985 and $739,000 in 1990. Find
    (a) a function that describes Ampersand's annual sales;
    (b) the firm's sales in 1992;
    (c) when Ampersand's sales will be $15 million per year.

12. Battledore Company stopped advertising its product on January 1, and since then sales have been declining at a rate proportional to sales. January sales were 50,000 units, and February sales were 45,000 units. Find

(a) a function that describes Battledore's monthly sales;
(b) the company's sales in March;
(c) when sales will be 20,000 units per month.

13. The annual profits of Grimditch Associates were $2.5 million in 1980 and $1.4 million in 1983. If profits decrease at a rate proportional to profits, find
    (a) a function that describes Grimditch's annual profits;
    (b) the firm's profits in 1990;
    (c) when Grimditch's annual profits will be $250,000.

14. Dynamic Corporation's revenues have been increasing at a rate proportional to revenue. Suppose they have been doubling every 10 years and were $1 million in 1985. Predict Dynamic's 1994 revenues. When will Dynamic's annual revenues attain the level of $5.5 million per year?

15. The average cost per unit for Elwin Company was $80 in 1980, and it has been decreasing at a rate proportional to itself, declining by 50% every 20 years. When will Elwin's average cost per unit be $30?

16. The quantity demanded of a commodity was 95,123 units when the price was $1 per unit, and 90,484 units when the price was $2 per unit. Assuming that the rate of change of quantity demanded is proportional to quantity demanded, find the demand function for the commodity and also find the quantity demanded when the price is $5 per unit.

17. The quantity supplied of the commodity of Exercise 16 was 20,816 units when the price was $1 per unit, and 23,470 units when the price was $4 per unit. Assuming that the rate of change of quantity supplied is proportional to quantity supplied, find the supply function for the commodity and then find its equilibrium price.

Life Science

18. Suppose that 2 hours after the beginning of an experiment there are 40,000 bacteria in a culture and that 5 hours after the beginning of the experiment there are 50,000 bacteria in the culture. Assuming that the rate of change of the bacteria population is proportional to its size, find a function that describes the number of bacteria present as a function of the time since the experiment began.

19. Biologists are attempting to revive a lake that has suffered from acid rain. It is estimated that there are currently 80,000 fish in the lake, and that over the next 5 years this number will increase to 120,000 as the lake continues to revive. If the fish population increases at a rate proportional to its size, find the number of fish the lake will support when the revivification effort is complete 12 years from now.

20. The number of cases of a new disease was 200 in month 0 and has been increasing at a rate proportional to the number of cases, tripling every 20 months. Find a model for the number of cases as a function of time.

21. The *Physician's Desk Reference* (Medical Economics Co., 1984) indicates that when administering a certain anesthetic to a patient, the vapor concentration should depend on the elapsed time since anesthesia began according to the following table:

Elapsed time (minutes)	2	4	12	24
Vapor concentration	2%	1.5%	0.475%	0.085%

Assuming that the rate of change of the vapor concentration is proportional to the vapor concentration, find a function that fits this set of data and use it to estimate the correct vapor concentration 10 minutes after anesthesia begins.

Social Science

22. A psychologist teaches subjects a complex sequence of tasks. Upon testing the subjects 10 days later, it is found that the typical subject can perform only 20% of the tasks correctly. Assuming that the rate of change of performance is proportional to performance, find a model that describes the percentage of the tasks a typical subject can perform as a function of time.

Limited Growth and Decline

23. Use the curve-sketching procedure to sketch the graph of
$$y = a - ce^{-rt}, \quad a > c > 0, \quad r > 0.$$

24. Use the curve-sketching procedure to sketch the graph of
$$y = a + ce^{-rt}, \quad a > 0, \quad c > 0, \quad r > 0.$$

Management

25. A worker with no experience can produce camshafts at a rate of 60 per week; with 1 week of experience he can produce them at a rate of 95 per week. The most experienced workers can approach a production rate of 200 camshafts per week. Find the learning curve and sketch its graph. How long will it be until a worker can produce camshafts at a rate of 140 per week?

26. Repeat Exercise 25 if a worker with one week of experience can produce camshafts at the rate of only 75 per week.

27. When it has just opened, a new plant will produce refrigerators at a rate of 20,000 per year; two months after it has opened, it will produce them at a rate of 40,000 per year. Its designed maximum rate of production is 100,000 refrigerators per year. Find the learning curve for the plant.

28. This exercise was suggested by material in McGann and Russell, *Advertising Media,* second edition, Irwin, 1988. It concerns advertising learning curves. Consumers who are exposed to an advertisement repeatedly will absorb more and more of the information content of the ad, and the amount they absorb follows an advertising learning curve of the form
$$y = 1 - ce^{-rx}, \quad r > 0.$$
Here x is the number of exposures to the advertisement and y is the proportion of the total information the ad contains that has been absorbed by the consumer. Suppose that a consumer who has seen an ad once absorbs 50% of its information content. (Of course, a consumer who has seen an ad 0 times absorbs 0% of its information content.) Find and graph the advertising learning curve for the ad.

*29. A manufacturing plant can fabricate units at the rate of 70,000 per year by the end of its first month of operation, 80,000 per year by the end of its second month of operation, and 88,000 per year by the end of its third month of operation. Find the learning curve for the plant. What is the maximum fabrication rate of the plant?

30. With perfect efficiency, an airplane manufacturer could build a plane using only 40,000 worker-hours. However, the first plane actually built required 70,000

worker-hours, and the second required 62,000 worker-hours. Find and graph the manufacturer's learning curve.

31. A computer manufacturer is introducing a new model of computer and estimates that it will take 5600 worker-hours to build the first unit and 5400 to build the second. The manufacturer also estimates that even with perfect efficiency it would take 4000 worker-hours to build a unit. Find the manufacturer's learning curve and draw its graph. How many worker-hours will it take to make the first four units?

32. An auto manufacturer knows that a newly hired worker who has no training will take an average of 4.0 minutes to install a steering wheel on a car, but that if the worker is given 1 hour of training, the average installation time falls to 3.4 minutes. The minimum possible time in which a steering wheel can be installed is 2.2 minutes. Find a limited decline model that describes average installation time as a function of training time and draw its graph. If the assembly line runs at such a speed that steering wheels must be installed in 3 minutes or less, how much training must new employees receive?

*33. Suppose a manufacturer does not know how many worker-hours would be required to build a unit with perfect efficiency, but estimates that it will take 4000 worker-hours to build the first unit, 3500 to build the second, and 3200 to build the third. Find the learning curve. What is the minimum time required to build a unit under conditions of perfect efficiency?

Life Science

34. The number of bacteria in a culture is limited to 40,000 by the amount of nutrients present. Suppose that 2000 bacteria are placed in the culture to begin with, and that 5 hours later there are 38,000 of them. Find a limited growth model that describes the number of bacteria in the culture and sketch its graph. How long does it take for the original number of bacteria to double?

35. A pond is stocked with 60 fish, and one year later contains 150 fish. The pond can support a maximum of 1200 fish. Find a limited growth model for the size of the fish population. How many fish will the pond contain 10 years after it was stocked?

36. A farmer's crop is ready to be harvested. If it is harvested today, all of it will be saved, but if it rains and harvesting must be delayed, a smaller percentage will be saved. It is known that if harvesting is delayed one day, only 70% of the crop will be saved and that the minimum percentage salvageable under any conditions is 40%. Find a limited decline model that describes the percentage of the crop that will be saved as a function of the delay in harvesting. How much will be salvaged if harvesting is delayed for four days?

37. The average concentration of ozone in the air of a large city was 200 parts per million in 1988 and 160 ppm in 1990. The minimum possible concentration is 100 ppm. Find a limited decline model for the concentration of ozone as a function of time.

*38. A patient is injected with a drug. At the moment of injection the concentration of the drug in the patient's blood is 0 ppm, 1 minute later it is 24 ppm, and 2 minutes after injection it is 40 ppm. Find a limited growth model that describes the concentration of the drug. What is the maximum concentration possible?

Exercises 39 and 40 were suggested by material on pages 81–84 of the December 1986, *Scientific American*. They concern drug toxicity.

39. A certain drug is toxic to cancer cells and also to normal cells in such a way that a dosage of x units will kill y_c% of the cancer cells and y_n% of the normal cells. Suppose that experimentation shows that a dosage of 30 units of the drug kills 77.7% of the cancer cells and 0% of the normal cells, while a dosage of 40 units kills 86.5% of the cancer cells and 39.5% of the normal cells. Assume that as the dosage x increases, both y_c and y_n will approach 100%.
    (a) Find a limited growth model for y_c.
    (b) Find a limited growth model for y_n.
    (c) Graph these functions on the same set of axes. Is it possible for the drug to kill 90% of the cancer cells without killing any normal cells?

40. Repeat Exercise 39 if experimentation shows that a dosage of 50 units will kill 91.8% of the cancer cells and 0% of the normal cells, while a dosage of 60 units will kill 95.0% of the cancer cells and 39.5% of the normal cells.

Social Science

41. The percentage of city voters who consider themselves independents was 22% in 1970 and 29% in 1980. Find a limited growth model that describes the percentage of independents as a function of time, if the maximum possible percentage who could consider themselves independents is 100%. If this model is valid, when will a majority of the voters consider themselves independents?

Computer Science

*42. An article by James Meindl ("Chips for Advanced Computing," *Scientific American*, October 1987) offers two predictions for the maximum number of components that will be placed on computer chips in the future. Both predictions assume that the maximum number of components per chip can be modeled by a limited growth function. The first prediction assumes that the maximum number of components per chip will ultimately approach 100 billion, the second that it will ultimately approach 1 billion. Both predictions imply that in 1992 the maximum number of components per chip will be approximately 500 million. The maximum number of components per chip in 1987 was approximately 1 million.
    (a) Find the model for Meindl's first prediction.
    (b) Find the model for Meindl's second prediction.
    (c) Graph the functions of parts (a) and (b) on the same set of axes. If the first prediction holds, when will the maximum number of components per chip reach 1 billion?

Nuclear Strategy

Exercises 43 and 44 were suggested by an article of von Hippel, Levi, Postol, and Daugherty ("Civilian Casualties from Counterforce Attacks," *Scientific American*, September 1988). They concern casualties under nuclear attack. The article indicates that the approximate percentage of the U.S. population that would be killed by a nuclear attack of x hundred megatons would be between y_L and y_U, and the approximate percentage of the Soviet population that would be killed by such an attack would be between z_L and z_U, where y_L, y_U, z_L, and z_U can be described by limited growth models. The models imply that a 100-megaton attack would kill between 8.2% and 12.7% of the U.S. population and between 11.1% and 16.7% of the Soviet population, while a 500-megaton attack would kill between 34.9% and 49.3% of the U.S. population and between 44.6% and 59.9% of the Soviet population. Of course, as x increases, y_L, y_U, z_L, and z_U all approach 100.

*43. (a) Find the limited growth model for y_L.
   (b) Find the limited growth model for y_U.
   (c) Graph the functions of parts (a) and (b) on the same set of axes. Under the doctrine of "mutual assured destruction" (MAD), the capability of mounting an attack of 200 megatons will deter aggression. According to these models, what percentage of the U.S. population would such an attack kill?

*44. (a) Find the limited growth model for z_L.
   (b) Find the limited growth model for z_U.
   (c) Graph y_L, y_U, z_L, and z_U on the same set of axes. What percentage of the Soviet population would a 200-megaton attack kill?

Learning Curves Revisited

Exercise 45 through 52 were suggested by material in Chase and Aquilano, *Production and Operations Management*, fourth edition, Irwin, 1985. They are concerned with learning curves. In industry, learning curves are sometimes defined differently than we have done in this section. Production managers often refer to a "90% learning curve" or an "80% learning curve," for example. A 90% learning curve is one with the property that every time the unit number doubles, the worker-hours required to produce the unit decrease by 10%. For instance, if production follows a 90% learning curve and it takes 1000 worker-hours to produce the first unit, then it will take

$$0.9(1000) = 900 \text{ worker-hours to produce the second unit,}$$

$$0.90(900) = 810 \text{ worker-hours to produce the fourth unit,}$$

$$0.90(810) = 729 \text{ worker-hours to produce the eighth unit,}$$

and so on. Similarly, for an 80% learning curve, 1000 worker-hours for the first unit would lead to $0.80(1000) = 800$ for the second, $0.80(800) = 640$ for the fourth, and so on. When a learning curve is defined in this way, the worker-hours y required to produce the xth unit, $x \geq 1$, are given by an equation of the form

$$y = cx^r$$

for some real numbers c and r. The use of such an equation implies that as the number of units x increases, the worker-hours required to produce the xth unit approaches 0.

*45. Find the worker-hours required to produce the second, fourth, eighth, and sixteenth unit if the first unit requires 10,000 worker-hours and production follows an 85% learning curve.

*46. Graph an 85% learning curve if the first unit requires 10,000 worker-hours. Note that as x increases y approaches 0.

*47. Find the equation $y = cx^r$ that describes a 90% learning curve if the first unit produced requires 10,000 worker-hours.

*48. Suppose production follows a $100p\%$ learning curve and the first unit produced requires c_0 worker-hours. Find and graph a function that describes the number of worker-hours required to build the xth unit, $x \geq 1$.

EXPONENTIAL AND LOGARITHMIC FUNCTIONS   853

*49. Use the result of Exercise 48 to find an equation that describes an 88% learning curve if it takes 5000 worker-hours to produce the first unit. How many worker-hours will it take to produce the tenth unit?

How can a production manager tell if production follows a 100p% learning curve for some p? Note that if

$$y = cx^r,$$

then

$$\ln y = r \cdot \ln x + \ln c.$$

If we set $Y = \ln y$ and $X = \ln x$, the equation

$$\ln y = r \cdot \ln x + \ln c$$

becomes

$$Y = rX + \ln c,$$

which is a linear equation in X and Y. Therefore the production manager can proceed as follows:

1. Collect data on the number of hours y_i it takes to produce unit x_i.
2. Let $Y_i = \ln y_i$ and $X_i = \ln x_i$.
3. Determine whether the points (X_i, Y_i) lie on a line $Y = mX + b$; if they do, production follows a 100p% learning curve $y = cx^r$, where $c = e^b$, $r = m$, and $p = 2^r$.

*50. Justify statement 3 above.

*51. Suppose a production manager collects the following data concerning the worker-hours required to produce selected units:

Unit Number	Worker-Hours Required
2	4000.0
5	2977.8
7	2672.1
12	2245.8

Show that production follows a learning curve and find its equation $y = cx^r$. What percentage learning curve is it?

*52. Repeat Exercise 51 using the following data:

Unit Number	Worker-Hours Required
6	3200
10	2581
11	2479
15	2176

Logarithmic Growth

53. Use the curve-sketching procedure to sketch the graph of

$$y = \ln(at + b), \quad a > 0, \quad b > 1.$$

54. Annual aluminum production in a country was 7 million tons in 1950 and 10 million tons in 1980. Find a model that describes aluminum production in the country and predict

(a) production in 1990;
(b) the year in which production will reach 11 million tons.

Management

55. A firm's sales were $1.5 million when it spent $200,000 on advertising and $1.6 million when it spent $250,000 on advertising. Find a logarithmic growth model for sales as a function of the amount spent on advertising.

56. A firm's annual profits have been growing logarithmically; they were $2.1 million in 1988 and $2.9 million in 1989. Find and graph the function that models the firm's annual profits as a function of time. What will the firm's profit be in 1995? When will profits attain an annual level of $5 million?

57. When a company sold 1000 units of its product, its revenue was $3,135,000; when it sold 2000 units, its revenue was $3,761,000. If the company's revenues are a logarithmic function of the number of units sold, find the revenue function. What would the company's revenue be if it sold 10,000 units? How many units would it have to sell in order to have revenue of $10,000,000?

Life Science

58. If the concentration of radioactive iodine in sea water was 4.8 parts per million in 1945 and 8.1 ppm in 1955, find a logarithmic growth model that describes the concentration as a function of time. The increase in radioactive iodine was largely due to aboveground testing of nuclear weapons; if damage to some forms of sea life occurs at a concentration of 9 ppm, when would such damage have begun if aboveground atomic testing had not been halted in the early 1960s?

59. The concentration of nitrous oxide in the air of a large city is growing logarithmically. If the concentration was 2.0 ppb in 1988 and 2.2 ppb in 1989, find a function that describes the concentration as a function of time.

60. A lake that is being deacidified has its pH y increasing logarithmically. If the pH at the start of the deacidification process was 2.1 and three years later it was 2.7, find and graph a model for y as a function of time. How long will it take until the water in the lake is neutral (pH = 7)?

Social Science

61. A state's aid to higher education is growing logarithmically, and was $1.2 billion when there were 450,000 students in the system and $1.5 billion when there were 500,000 students in the system. Find a function that describes the amount of such aid as a function of the number of students and predict the amount of such aid when there are 600,000 students in the system.

62. The number of people employed in a state's bureaucracy is growing logarithmically; in 1985 it was 449,000 and in 1988 it was 472,000. Find a function that describes the number of people employed in the bureaucracy as a function of time. When will the bureaucracy number 500,000?

Nuclear Testing

*63. According to an article by Lynn Sykes and Dan Davis ("The Yields of Soviet Strategic Warheads," *Scientific American,* January 1987), if x is the yield of an under-

ground nuclear explosion, in kilotons, and y is the magnitude of the surface seismic waves generated by the explosion, then y is a logarithmically increasing function of x; furthermore, when $x = 10$, $y = 3.36$, and when $x = 25$, $y = 3.60$.
(a) Find a model that describes y as a function of x.
(b) What is the seismic magnitude of a 50-kiloton explosion?
(c) Suppose the underground test of an atomic weapon can be detected by the seismic waves it generates if the magnitude of the waves is at least 3.2. What is the size of the smallest detectable underground explosion?

SUMMARY

This summary consists of terms and symbols whose meaning you should know, facts you should know, and techniques and methods you should be able to employ.

Section 13.1

Exponential function with base b: $y = cb^{rx}$, $b > 0$, $b \neq 1$. Graphing exponential functions with base b. Domain, range, continuity, one-to-oneness, intercepts, limits at infinity of exponential functions. Rewriting an exponential function with base less than 1 as an exponential function with base greater than 1. Using exponential models. Creating exponential models: if y changes b-fold in m periods, then $y = cb^{t/m}$. The number e. Finding powers of e. Graph of $y = e^x$. Why use e as a base?. Every exponential function with base not equal to e can be rewritten as an exponential function with base e. The compound interest formula $A = P(1 + r)^{mt}$ as an exponential function of t with base $1 + r$. Continuous compounding of interest. The formula for continuous compounding of interest: $A = Pe^{rt}$. The present value formula for continuously compounded interest: $P = Ae^{-rt}$.

Section 13.2

Inverse function of a one-to-one function. Logarithmic function to the base b: $y = \log_b x$, $b > 0$, $b \neq 1$ as the inverse function of $y = b^x$. $y = \log_b x$ means $x = b^y$. Graphing $y = \log_b x$. Domain, range, continuity, one-to-oneness, intercepts, limits at infinity of logarithmic functions. The properties of logarithms. Using the properties of logarithms. The natural logarithmic function: $y = \ln x$. Finding values of $\ln x$. The graph of $y = \ln x$. Common logarithms. The change-of-base formula for exponentials. Using the change-of-base formula for exponentials to change the bases of exponential functions. The change-of-base formula for logarithms. Using the change-of-base formula for logarithms to change the bases of logarithmic functions. Solving exponential equations by taking logarithms of both sides of the equation.

Section 13.3

The existence of the derivative of $y = \ln x$ at every positive number x. The derivative of $\ln x$. The derivative of $\ln g(x)$. Finding derivatives of functions involving the natural logarithm. Sketching the graphs of functions involving the natural logarithm. The derivative of $\log_b g(x)$. Finding derivatives of functions involving logarithms to the base b, $b \neq e$. Logarithmic differentiation.

Section 13.4

The existence of the derivative of $y = e^x$ for all x. The derivative of e^x. The derivative of $e^{g(x)}$. Finding derivatives of functions that involve powers of e. Sketching the graphs of functions that involve powers of e. The derivative of $b^{g(x)}$. Finding derivatives of functions that involve b to a power, $b \neq e$.

Section 13.5

The rate of change of $y = ce^{rt}$ is directly proportional to y. A quantity whose rate of change with respect to time t is directly proportional to the amount present at time t can be modeled by an equation of the form $y = ce^{rt}$. Models of exponential growth: $y = ce^{rt}$, $r > 0$. Models of exponential decline: $y = ce^{-rt}$, $r > 0$. Graphs of exponential growth and exponential decline models. Creating and using models of exponential growth and exponential decline. Limited and unlimited growth. Limited growth model $y = a - ce^{-rt}$, $a > c > 0$, $r > 0$. Limited decline model $y = a + ce^{-rt}$, $a > 0$, $c > 0$, $r > 0$. Graphs of limited growth and limited decline models. Creating and using models of limited growth and limited decline: learning curves. Logarithmic growth model $y = \ln(at + b)$, $a > 0$, $b > 1$. Graph of logarithmic growth model. Creating and using logarithmic growth models.

REVIEW EXERCISES

In Exercises 1 through 4, graph the given function.

1. $y = 7^{2x}$
2. $y = 2\left(\dfrac{1}{3}\right)^{-2x}$
3. $y = 3e^{-x/2}$
4. $y = 2 + e^{2x}$

5. Show that $e = \displaystyle\lim_{x \to 0^+} (1 + x)^{1/x}$.

6. A firm's average cost per unit was $12 in the year the firm was founded and has been tripling every 20 years. Find an exponential equation that describes the average cost per unit as a function of time. Use base 3.

7. How much will you have in seven years if you invest $4000 at 6.5% per year compounded continuously?

8. What is the present value of $54,000 discounted for 12 years at 11% per year compounded continuously?

EXPONENTIAL AND LOGARITHMIC FUNCTIONS   857

In Exercises 9 through 20, evaluate the given expression.

9. $\log_7 7$ **10.** $\log_7 1$ **11.** $\log_7 49^3$ **12.** $\log_7 7^{200}$

13. $\log_7 \dfrac{1}{343^2}$ **14.** $\log_7 7^{-35}$ **15.** $\log_7 7^{\log_7 22}$ **16.** $\ln e^{5.2}$

17. $e^{\ln 4.2}$ **18.** $e^{-2\ln 5}$ **19.** $\log_7 \dfrac{7^4}{7^9}$ **20.** $\log_7(7^{10} \cdot 7^3)$

In Exercises 21 and 22, use only the properties of logarithms and the facts that
$$\log_{10} 2 = 0.3010, \quad \log_{10} 7 = 0.8451, \quad \text{and} \quad \log_{10} 11 = 1.0414$$
to evaluate the given expression.

21. $\log_{10} 56$ **22.** $\log_{10} \dfrac{392}{1331}$

In Exercises 23 through 26, graph the given function.

23. $y = \log_5 x$ **24.** $y = \log_5 x^3$ **25.** $y = \ln 3x$ **26.** $y = 2 \ln 1.5x$

In Exercises 27 through 30, rewrite the given function using the given base.

27. $f(x) = 7^{2x}$, using base e **28.** $y = 10^{-3x}$, using base 7

29. $y = 2 \log_{13} 3x$, using base e **30.** $g(x) = \log_8 x$, using base 5

31. Loudness is measured on a logarithmic scale known as the **decibel scale.** This is done by comparing the loudness x of a given sound with the loudness x_0 of a threshold sound (one that is barely audible); the decibel measure of x is then defined to be
$$db(x) = \log_{10} \dfrac{x}{x_0}.$$
Find the loudness in decibels of a sound that is
(a) 10   (b) 1000   (c) 1,000,000   (d) 2,500,000
times as loud as the threshold sound.
(e) A large passenger jet at full throttle produces a noise of approximately 14 db. How much louder is this than the threshold sound?

In Exercises 32 through 35, solve the given equation for x.

32. $e^{-3x} = 5$ **33.** $e^{2.3x} = 4.2$ **34.** $5^{-3x} = 40$ **35.** $(1.1)^{0.02x} = 0.65$

36. Suppose the number of fish still alive in a pond t days after it is contaminated by an oil spill is given by
$$y = 2000e^{-0.1t}.$$
(a) Graph this function.
(b) How many fish are still alive five days after the oil spill?
(c) How long will it be until half the fish have died?

37. A radioactive isotope has a half-life of eight months. Find an exponential function with base 2 that describes the amount of the isotope present as a function of time.
(a) What percentage of the isotope decays in two years?
(b) How long does it take for 75% of the isotope to decay?

38. What interest rate, compounded monthly, must be earned if $10,000 is to grow to $50,000 in 30 years?

39. How long will it take for $35,000 to triple at 10% per year compounded continuously?

40. If payments of $250 are made at the end of each month into a sinking fund that earns 8% per year compounded monthly, how long will it take until the fund contains $100,000?

In Exercises 41 through 54, find $f'(x)$.

41. $f(x) = \ln(x^3 + 3x - 2)$

42. $f(x) = \ln(x^2 + 1)^2$

43. $f(x) = \ln\dfrac{x^2}{2x - 1}$

44. $f(x) = \dfrac{\ln x^2}{\ln(2x - 1)}$

45. $f(x) = 2e^{5x}$

46. $f(x) = 4x^2 e^{-x}$

47. $f(x) = e^{x^2} \ln x^2$

48. $f(x) = \dfrac{e^x - 1}{e^x + 1}$

49. $f(x) = e^{e^{-2x}}$

50. $f(x) = x^{-x^2}$

51. $f(x) = \log_2(4x - 3)$

52. $f(x) = e^x \log_{10} e^{-2x}$

53. $f(x) = 4 \cdot 2^{-5x}$

54. $f(x) = x^2 10^{-3x}$

In Exercises 55 through 60, use the curve-sketching procedure to graph the given function.

55. $y = x + \ln x$

56. $y = x^4 \ln x$

57. $y = x^{-3} \ln x$

58. $y = x + e^{-x}$

59. $y = \dfrac{e^x + e^{-x}}{e^x - e^{-x}}$

60. $y = x^2 e^{-x^2}$

61. The money spent on social services by a state government in year t is
$$y = [\ln(3t + 2)]^{1/2}$$
million dollars. Here $t = 0$ represents 1987. Find the rate at which spending on social services was changing in 1989.

62. It is estimated that the concentration of pollutants in a lake t years from now will be
$$y = 200e^{-0.2t}$$
parts per million. Find the rate at which the concentration of pollutants in the lake will be changing five years from now.

63. Find the rate at which the size of a bacteria population is changing 2 hours after the start of an experiment if the number of bacteria present t hours after the start of the experiment is $y = 20,000 \cdot 3^{-0.45t} + 100,000$.

64. Under a runaway greenhouse effect, the earth's average temperature would increase at a rate directly proportional to the average temperature. If the average temperature of the earth is now 59.5°F and if under a runaway greenhouse effect it would be 77.4°F 100 years from now, find a model for the average temperature under a runaway greenhouse effect. Then find
  (a) the average temperature 25 years from now;
  (b) how long it would take for the average temperature to reach 120°F.

65. Insecticides have caused an exponential decline in the number of waterfowl in a wetlands region. Four years ago there were 2000 waterfowl in the wetlands, while today there are only 1340 of them. Find a exponential decline model that describes the number of waterfowl in the wetlands as a function of time.

66. According to an article by Peter Moretti and Louis Divone ("Modern Windmills," *Scientific American*, June 1986), the ideal efficiency for rotor-type windmills is approximately given by the equation
$$y = 0.593 - 0.4e^{-0.73x}.$$

Here x is the ratio of the speed of the rotor tips to the speed of the wind and y is the so-called power coefficient of the windmill: the power coefficient measures the amount of power the windmill extracts from the wind.
(a) Graph this function.
(b) Suppose the speed of the rotor tips increases while the wind speed remains constant. What can you say about the power that the windmill will extract from the wind?
(c) Same question as in part (b) if the speed of the wind increases while the speed of the rotor tips remains constant.

67. A newly hired employee can produce at the rate of 1600 units per month, whereas one with one month's experience can produce at the rate of 1700 units per month. The maximum production with perfect efficiency is 2400 units per month. Find and graph the learning curve. When will an employee be producing at the rate of 2000 units per month?

68. It takes 400 worker-hours to produce the first unit of a new computer chip and 380 worker-hours to produce the second unit. The minimum time needed to produce a chip is estimated to be 250 worker-hours. Find and graph the learning curve. How long will it take to produce the first four chips? When will production time per chip fall below 300 worker hours?

69. The pollen count during August followed a limited growth model, with the maximum daily count y being expressed as a function of the number of days since August 1. If the count on August 3 was 118 and on August 6 was 125, and if the maximum possible count is 200, find the model and find out when the count reached 150.

70. It takes a rat 350 seconds to negotiate a maze on its first attempt, 335 seconds on its second attempt, and 322 seconds on its third attempt. If the time the rat takes to get through the maze follows a limited decline model, find the model.

71. A firm's profits are growing logarithmically, and were $100,000 in 1980 and $150,000 in 1990. Predict the firm's profits in 1995 and the year in which they will reach $250,000.

ADDITIONAL TOPICS

Here are some suggestions for topics not covered in the text that you might want to investigate on your own.

1. In Section 13.2 we mentioned that 4, 5, and 6 of the properties of logarithms follow from the properties of exponents introduced in Section 1.4. Use the definition of logarithm and the properties of exponents to prove 4, 5, and 6 of the properties of logarithms.

2. Let f and g be functions. If
$$\lim_{x \to +\infty} \frac{f(x)}{g(x)} = +\infty,$$
then f **grows faster than** g. If
$$\lim_{x \to +\infty} \frac{f(x)}{g(x)} = 0,$$

then *f* **grows more slowly than** *g*. If

$$\lim_{x \to +\infty} \frac{f(x)}{g(x)} = L,$$

where $L \neq 0$ is a finite number, then *f* and *g* **grow at the same rate.**

(a) If *f* and *g* are polynomial functions, what conditions on them will cause them to grow at the same rate? What conditions will cause one of them to grow faster than the other?

(b) If *f* is an exponential function defined by $f(x) = e^{rx}$, $r > 0$, if *g* is a polynomial function, and if *h* is a logarithmic function defined by $h(x) = (\ln x)^s$, $s > 0$, show that *f* grows faster than *g* and that *g* grows faster than *h*. (*Hint:* You may wish to use L'Hôpital's Rule; see Additional Topics, chapter 12, number 6.)

3. The number defined as the limit of $(1 + 1/x)^x$ as $x \to +\infty$ is denoted by *e* in honor of the Swiss mathematician Leonhard Euler (1707–1783). Euler (pronounced "oiler") was one of the most prolific mathematicians of all time: his output of mathematical discoveries and results was immense. Find out about Euler's life and some of his contributions to mathematics.

4. Logarithms were discovered by the Scottish mathematician and inventor John Napier (1550–1617), who also labored for many years constructing the first tables of logarithms. The availability of such tables simplified many calculations and had a profound influence on the development of the sciences, astronomy in particular. Find out about the life of Napier and the early applications of logarithms by the great Danish astronomer Tycho Brahe and others. (A good place to start is H. W. Turnbull's article "The Great Mathematicians," which appears in volume 1 of *The World of Mathematics,* James R. Newman, ed., Simon & Schuster, 1956.)

SUPPLEMENT: S-CURVES

Adequate modeling of some situations involving limited growth or decline may require the use of more complicated exponential functions than the ones we introduced in Section 13.5. For instance, the growth of an individual company or industry oftens follows an **S-curve,** such as that in Figure 13–31.

Notice how the S-curve has an initial period of slow growth, then a period of rapid growth, and finally another period of slow growth. Such behavior is fairly common: a new company will typically grow quite slowly during the first few years of its existence, then "take off" and experience a period of very rapid growth. Eventually, however, the company matures and the growth slows down again. Thus the company's **life cycle** can be described by an S-curve. For this reason, S-curves are sometimes called **life-cycle curves.**

One function that has an S-curve as its graph is the **logistic function**

$$y = \frac{a}{1 + be^{-rt}}, \quad a > 0, \quad b > 1, \quad r > 0.$$

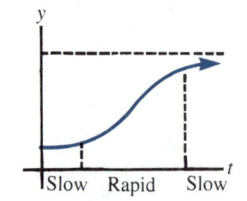

FIGURE 13–31
S-Curve

The graph of the logistic function on $[0, +\infty)$ is shown in Figure 13–32. (See Exercise 1.) The larger the value of *r*, the steeper the curve will be. (See Exercise 16.) Notice the point of inflection at $((\ln b)/r, a/2)$; to the left of this point of

Exponential and Logarithmic Functions

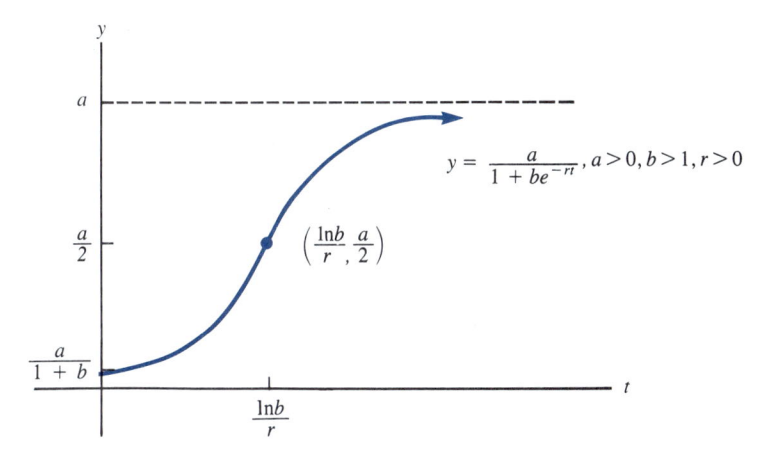

FIGURE 13–32
The Logistic Function

$$y = \frac{a}{1+be^{-rt}}, a>0, b>1, r>0$$

Key points shown: $\left(\frac{\ln b}{r}, \frac{a}{2}\right)$, asymptote at $y=a$, y-intercept at $\frac{a}{1+b}$, and $\frac{a}{2}$ on the y-axis at $t = \frac{\ln b}{r}$.

inflection, the rate of growth is increasing, while to the right of the point of inflection, the rate of growth is decreasing. (Why?) Thus the point of inflection marks the place where the *rate* of growth stops increasing and begins to decrease.

EXAMPLE 1 A new computer company was started and t years later its annual sales were S million dollars, where

$$S = \frac{20}{1+19e^{-0.6t}}.$$

The graph of this function is shown in Figure 13–33. Initially the firm's annual sales were running at a level of $1 million per year. Sales increased slowly for the first several years, then increased rapidly for several more, and finally settled down, approaching a level of $20 million per year. The rate at which sales grew increased during the first 4.9 years of the company's existence and decreased thereafter.

Inflection point: $((\ln 19)/0.6, 10) \approx (4.9, 10)$

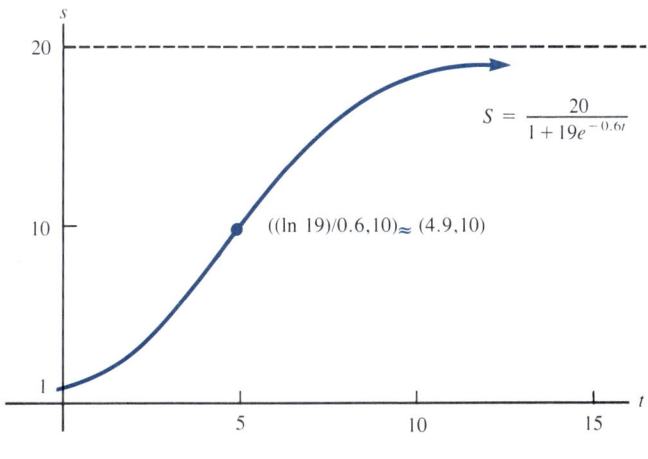

FIGURE 13–33

What were annual sales four years after the company was founded? The answer is

$$S = \frac{20}{1 + 19e^{-0.6(4)}} \approx 7.3$$

million dollars. Now let us find how many years it took until sales attained an annual level of $16 million. To do so, we must solve

$$16 = \frac{20}{1 + 19e^{-0.6t}}$$

for t. We have

$$1 + 19e^{-0.6t} = \frac{20}{16} = 1.25,$$

or

$$e^{-0.6t} = \frac{1.25 - 1}{19} = \frac{1}{76}.$$

Thus

$$\ln e^{-0.6t} = \ln(1/76) = -\ln 76,$$

and hence

$$t = \frac{\ln 76}{0.6} \approx 7.2$$

years. ☐

Another function that has an S-curve as its graph is the **Gompertz function**

$$y = ae^{-be^{-rt}}, \quad a > 0, \quad b > 1, \quad r > 0.$$

The graph of the Gompertz function on $[0, +\infty)$ is shown in Figure 13–34. (See Exercise 7.) The larger the value of r, the steeper the curve will be. (See Exercise

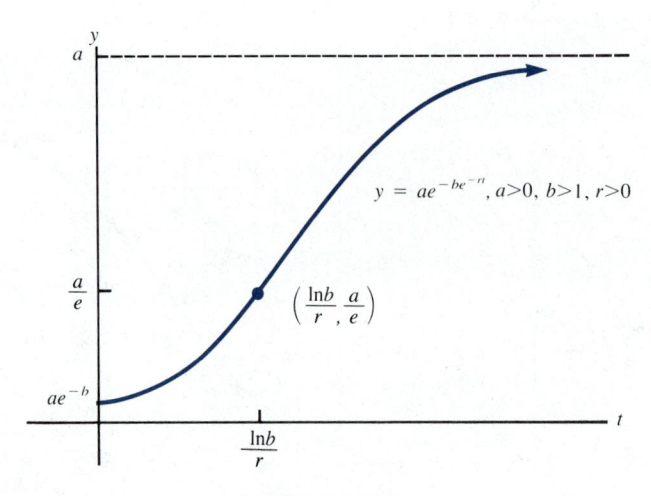

FIGURE 13–34
The Gompertz Function

EXPONENTIAL AND LOGARITHMIC FUNCTIONS

17.) Note the point of inflection at $((\ln b)/r, a/e)$: to the left of this point, the rate of growth is increasing, while to the right of it the rate of growth is decreasing.

EXAMPLE 2

A new video game is placed in arcades and t weeks later the revenue from the game is running at a level of y thousand dollars per week, where

$$y = 25e^{-5e^{-0.4t}}.$$

The graph of this function is shown in Figure 13–35. Initially revenue was running at a level of approximately $170 per week, and it remained low for the first few weeks. Weekly revenues from the game then increased rapidly, however, before finally slowing down and approaching $25,000 per week. The rate at which revenues grew increased during the first four weeks and decreased thereafter.

Suppose we wish to find the revenue five weeks after the installation of the game. We have

$$y = 25e^{-5e^{-0.4(5)}} = 25e^{-5e^{-2}} = 25e^{-5(0.1353)} \approx 12.7$$

thousand dollars.

Now suppose we wish to find out how long it would take for revenues to attain the level of $24,000 per week. Then we must solve

$$24 = 25e^{-5e^{-0.4t}}$$

for t. Thus

$$\ln(24/25) = -5e^{-0.4t},$$

so

$$\frac{-0.0408}{-5} = e^{-0.4t},$$

and hence

$$\ln(0.0408/5) = -0.4t.$$

$\left(\frac{\ln 5}{0.4}, \frac{25}{e}\right) \approx (4, 9.2)$

$y = 25e^{-5e^{-0.4t}}$

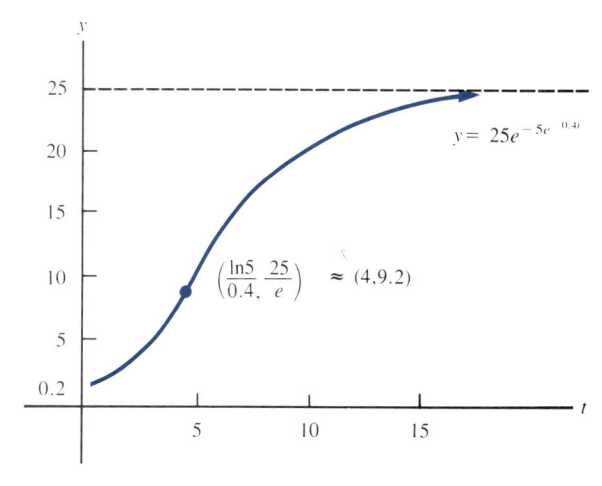

FIGURE 13–35

CHAPTER 13

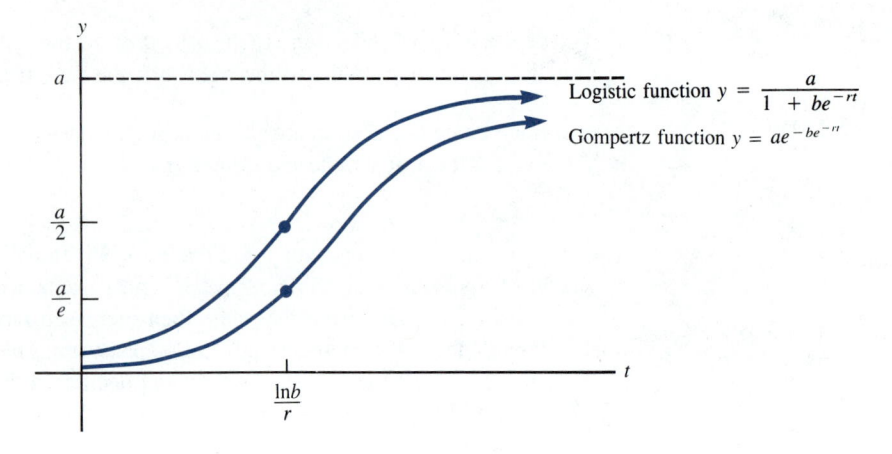

FIGURE 13-36

Logistic function $y = \dfrac{a}{1 + be^{-rt}}$

Gompertz function $y = ae^{-be^{-rt}}$

Therefore

$$t \approx 12.0 \text{ weeks.} \quad \square$$

Both the logistic function and the Gompertz function have S-curves as their graphs. The difference between them is that, for the same values of *a, b,* and *r,* the Gompertz curve increases more slowly than the logistic curve. See Figure 13–36.

Finally, we remark that sometimes decline follows a reverse S-curve, as in Figure 13–37. One function that has such a graph is the **sigmoid function**

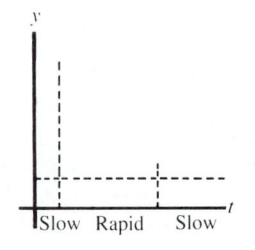

$$y = a - b(1 - e^{-rt})^n, \quad a > b > 0, \quad r > 0, \quad n \in \mathbf{N}, \quad n > 1.$$

FIGURE 13-37
Reverse S-Curve

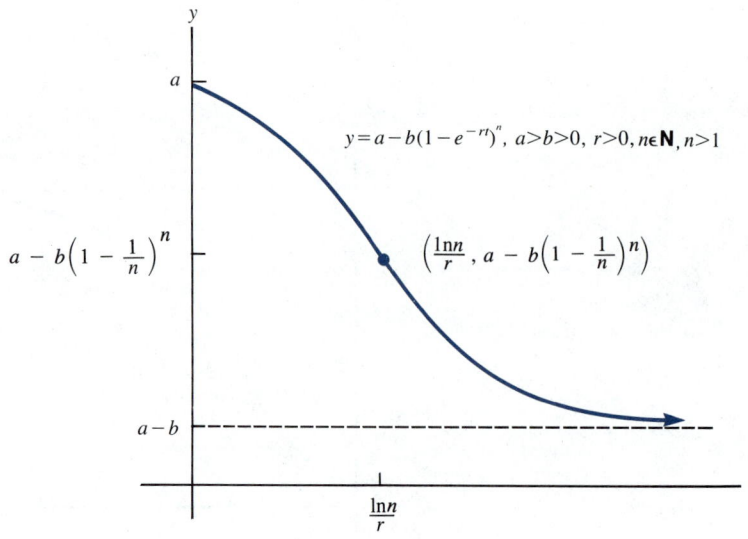

FIGURE 13-38
The Sigmoid Function

Exponential and Logarithmic Functions

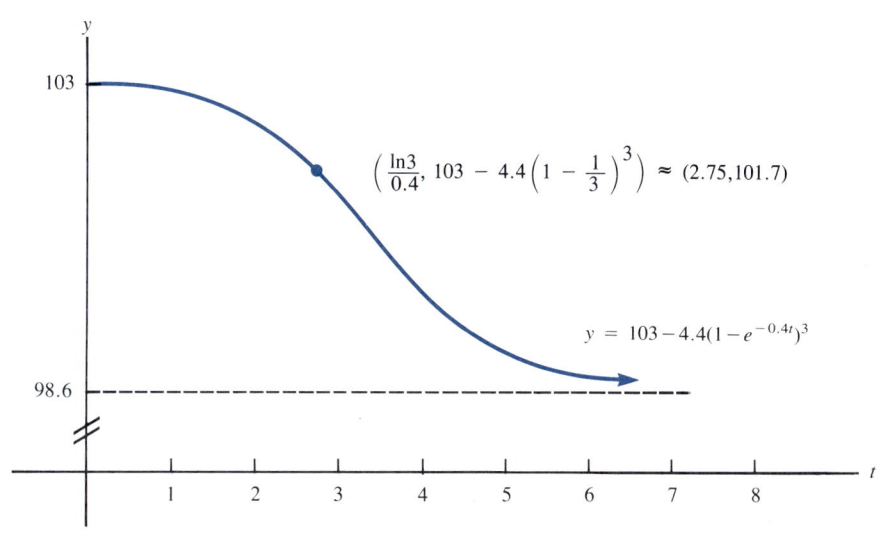

FIGURE 13-39

The graph of the sigmoid function is shown in Figure 13-38. (See Exercise 11.) The smaller n is, the steeper the curve will be. (See Exercise 18.) Notice the point of inflection at $((\ln n)/r, a - b(1 - 1/n)^n)$: the rate of decline increases to the left of the point of inflection and decreases to its right.

EXAMPLE 3 A patient is given a temperature-reducing drug and t hours later has a temperature of

$$y = 103 - 4.4(1 - e^{-0.4t})^3$$

degrees Fahrenheit. The graph of this function is shown in Figure 13-39. Notice that the drug had little effect on the patient's temperature during the first hour, but that it then brought the temperature down quite rapidly to a near-normal level before its effect began to wear off. □

■ *Computer Manual: Program ESSCURVE*

SUPPLEMENTARY EXERCISES

1. Use the curve-sketching procedure to sketch the graph of

$$y = \frac{a}{1 + be^{-rt}}, \quad a > 0, \quad b > 1, \quad r > 0$$

on $[0, +\infty)$.

2. Sales of electronic calculators are described by the logistic function

$$y = \frac{100}{1 + 49e^{-0.4t}},$$

where y is annual sales of calculators in millions and t is time in years, with $t = 0$ representing 1970.

(a) Sketch the graph of this function.
(b) What were calculator sales in 1970? in 1975? in 1985?
(c) What will sales be in 1995?
(d) When did the rate of sales growth begin to slow down? What were sales at this time?
(e) When were sales running at a level of 80 million per year?

3. The manufacturing capacity of a nation is given by
$$y = \frac{450}{1 + 300e^{-0.06t}}.$$
Here y is the volume of manufactured goods produced in the country in year t, y in billions of 1990 dollars and $t = 0$ representing A.D. 1800.
(a) Sketch the graph of this function.
(b) What was the manufacturing capacity of the nation in A.D. 1800? What will it be in the year 2000?
(c) When did the rate of growth of the manufacturing capacity begin to slow down?

4. An article on energy-efficient buildings by Arthur Rosenfeld and David Hafemeister ("Energy Efficient Buildings," *Scientific American,* April 1988) suggests that the total savings over the period 1975–2050 of making appliances, homes, and commercial buildings energy efficient would be approximately
$$y = \frac{100}{1 + 99e^{-0.21t}}$$
billion 1985 dollars. Here t is in years with $t = 0$ representing 1975.
(a) Sketch the graph of this function.
(b) When would the rate at which savings increase begin to slow down? What would total savings be at this time?
(c) When would total savings amount to \$75 billion?

5. In a town of 12,000 people the spread of an influenza epidemic is described by the function
$$y = \frac{5000}{1 + 4999e^{-0.5t}},$$
where t is time in days since the first case of flu appeared and y is the number of people who contracted the flu on or before day t.
(a) Sketch the graph of this function.
(b) When did the epidemic begin to slow down?
(c) What proportion of the town's inhabitants contracted the flu?

6. Two hundred students live in a college dormitory. One of them starts a rumor and t hours later y of them have heard it, where
$$y = \frac{200}{1 + 199e^{-1.1t}}.$$
(a) Sketch the graph of this function.
(b) How long will it take until half of the dorm residents have heard the rumor?
(c) Assuming that the resident who is the subject of the rumor will be the last one to hear it, how long will it take until everyone else has heard it?

7. Use the curve-sketching procedure to sketch the graph of
$$y = ae^{-be^{-rt}}, \quad a > 0, \quad b > 1, \quad r > 0.$$

EXPONENTIAL AND LOGARITHMIC FUNCTIONS   867

8. The growth of a cell colony inside a container is described by the Gompertz curve
$$y = 100e^{-2e^{-0.02t}}.$$
Here y is the percentage of the volume of the container that the colony fills after t days of growth.
   (a) Sketch the graph of this function.
   (b) What percentage of the container did the colony fill at the start of its growth? after 5 days of growth? after 10 days of growth?
   (c) When did the colony's rate of growth begin to slow? What percentage of the container did it fill at this time?
   (d) How long did it take for the colony to fill 90% of the container?

9. The manufacturing capacity of a nation is given by
$$y = 450e^{-300e^{-0.06t}},$$
where y is in billions of 1990 dollars and t is in years with $t = 0$ representing A.D. 1800. Repeat Exercise 3 above using this function and compare your results with those of Exercise 3.

10. Repeat Exercise 6 above if
$$y = 200e^{-199e^{-1.1t}},$$
and compare your results with those of Exercise 6.

11. Use the curve-sketching procedure to sketch the graph of
$$y = a - b(1 - e^{-rt})^n, \quad a > b > 0, \quad r > 0, \quad n \in \mathbb{N}, \quad n > 1$$
on $[0, +\infty)$.

12. Annual sales y of the buggy-whip industry are given by
$$y = 400 - 350(1 - e^{-0.1t})^3,$$
where y is in millions of 1989 dollars and t is in years since A.D. 1900.
   (a) Sketch the graph of this function.
   (b) What were annual sales of the industry in 1900? in 1910? in 1925?
   (c) When did annual sales fall to a level of $100 million?

13. The effects of a painkiller are modeled by the equation
$$y = 100[1 - (1 - e^{-0.4t})^2],$$
where t is time in minutes since the painkiller was taken and y is a discomfort index, with $y = 0$ indicating no discomfort.
   (a) Sketch the graph of this function.
   (b) What was the discomfort index at the time the painkiller was taken? What was it 5 minutes later? ten minutes later?
   (c) How long did it take for the discomfort index to fall to 10?

14. Repeat Exercise 13 if
$$y = 100[1 - (1 - e^{-0.4t})^4],$$
and compare your results with those of Exercise 13.

15. The sales of fad items can often be described by the combination of a Gompertz function and a sigmoid function. Thus, suppose that sales of a fad item are given by
$$y = \begin{cases} 200e^{-10e^{-0.5t}}, & 0 \leq t \leq 6, \\ 121.6[1 - (1 - e^{-0.5(t-6)})^2], & 6 < t \leq 12. \end{cases}$$
Graph this function and analyze the sales of the item.

16. Let a, b, r, and s be positive numbers, with $b > 1$ and $r > s$. Graph the logistic functions
$$y = \frac{a}{1 + be^{-rx}} \quad \text{and} \quad y = \frac{a}{1 + be^{-sx}}$$
on the same set of axes. This shows that the larger r is, the steeper the curve.

17. Let a, b, r, and s be positive numbers, with $b > 1$ and $r > s$. Graph the Gompertz functions
$$y = ae^{-be^{-rx}} \quad \text{and} \quad y = ae^{-be^{-sx}}$$
on the same set of axes. This shows that the larger r is, the steeper the curve.

18. Let a, b, and r be positive numbers, with $a > b$, and let $n > m > 1$. Graph the sigmoid curves
$$y = a - b(1 - e^{-rt})^n \quad \text{and} \quad y = a - b(1 - e^{-rt})^m$$
on the same set of axes. This shows that the smaller n is, the steeper the curve.

14

Integration

In Chapters 10 through 13 we were concerned with the problem of finding the rate of change of a function. We saw that the rate of change of a function is the derivative of the function, and we have exploited this fact in many interesting and important ways. Often, however, it is not the function that is known and its rate of change that must be found, but rather the rate of change that is known and the function that must be found. For instance,

- A biologist might know the rate of population growth for a species and wish to find a function that describes the size of its population.
- A physicist might know the velocity of a particle and want to find a function that describes its position.
- A manufacturer might know the marginal cost function for a product and wish to find its cost function.

In cases like these, instead of starting with a function f and differentiating it to find its rate of change f', we start with a function f that describes a rate of change and **antidifferentiate** it to find a function F such that $F' = f$. Such a function F is called an **antiderivative** of f.

We begin this chapter with two sections devoted to antiderivatives. In these sections we consider the set of all antiderivatives of a function: this set is known as the **indefinite integral** of the function. We state the rules of indefinite integrals and demonstrate some techniques for finding indefinite integrals. In the third section we introduce the **definite integral** of a function: the definite integral is a number that gives the total change of an antiderivative of the function over an interval. We then show how the definite integral may be used to find areas, and conclude by defining the definite integral as the limit of a certain type of sum.

14.1 THE INDEFINITE INTEGRAL

We have discussed the motivation for antidifferentiation. In this section we consider the mechanics of the antidifferentiation process, introduce the concept of the indefinite integral of a function, and discuss some properties of the indefinite integral.

> **The Antiderivative of a Function**
>
> If f and F are functions such that
> $$F'(x) = f(x)$$
> for all x in the domain of f, then $F(x)$ is said to be an **antiderivative** of $f(x)$.

EXAMPLE 1   Let f be defined by $f(x) = 2x + 3$ and F by $F(x) = x^2 + 3x$. Then $F(x)$ is an antiderivative of $f(x)$, because
$$F'(x) = 2x + 3 = f(x)$$
for all x. More generally, if $G(x) = x^2 + 3x + c$, where c is any number, then
$$G'(x) = 2x + 3 + 0 = f(x)$$
for all x, and therefore $G(x)$ is an antiderivative for $f(x)$. Thus
$$f(x) = 2x + 3$$
has many antiderivatives: each choice of the number c in
$$G(x) = x^2 + 3x + c$$
results in a different antiderivative of $f(x)$. ☐

As Example 1 suggests, if $f(x)$ has one antiderivative $F(x)$, then it has a whole *set* of antiderivatives, because if $F'(x) = f(x)$ for all x in the domain of f and c is any number, then
$$(F(x) + c)' = F'(x) + 0 = f(x)$$
for all x in the domain of f; hence $F(x) + c$ is an antiderivative for $f(x)$ also. On the other hand, if $F(x)$ and $G(x)$ are any two antiderivatives of $f(x)$, then there is some number c such that
$$G(x) = F(x) + c$$
for all x in the domain of f. To see this, set $H(x) = G(x) - F(x)$ for all x in the domain of f. Then
$$H'(x) = G'(x) - F'(x) = f(x) - f(x) = 0.$$

Hence the derivative of H is always zero, and therefore H has a horizontal tangent line at every point of its graph. But the only way this can happen is if the graph of H is itself a horizontal line. If this is so, then there is some number c such that $H(x) = c$ for every x in the domain of f. Thus $G(x) - F(x) = c$, or $G(x) = F(x) + c$ for all x in the domain of f.

> **Properties of the Antiderivative**
>
> If $F(x)$ is an antiderivative of $f(x)$, then
>
> 1. $F(x) + c$ is also an antiderivative of $f(x)$, for any number c.
> 2. Every antiderivative of $f(x)$ is of the form $F(x) + c$ for some number c.

Now we are ready to define the **indefinite integral** of a function f. This is just the set of all antiderivatives of $f(x)$.

> **The Indefinite Integral**
>
> The **indefinite integral** of a function f with respect to x, denoted by
>
> $$\int f(x)\, dx,$$
>
> is defined by
>
> $$\int f(x)\, dx = F(x) + c,$$
>
> where $F(x)$ is any antiderivative of $f(x)$ and c is a constant.

The symbol $\int$ is called an **integral sign,** and the expression $f(x)$ that follows the integral sign is called the **integrand.** The dx after the integrand is the differential of x discussed in Section 11.3; its presence tells us that the integral is taken with respect to the variable x. The constant c is referred to as the **constant of integration.** Note that to find the indefinite integral of f, we need only find *one* antiderivative $F(x)$ of $f(x)$.

EXAMPLE 2 Let us find

$$\int (2x + 3)\, dx.$$

Since $F(x) = x^2 + 3x$ is an antiderivative of $f(x) = 2x + 3$, we have

$$\int (2x + 3)\, dx = x^2 + 3x + c. \quad \square$$

EXAMPLE 3 Let us find

$$\int (3x^2 + 2x + 1)\, dx.$$

Since $F(x) = x^3 + x^2 + x$ is an antiderivative of $f(x) = 3x^2 + 2x + 1$ (why?), it follows that

$$\int (3x^2 + 2x + 1)\,dx = x^3 + x^2 + x + c. \quad \square$$

EXAMPLE 4 We claim that

$$\int (6t^5 - 6t^2 + 2t - 2)\,dt = t^6 - 2t^3 + t^2 - 2t + c.$$

The claim is true if the derivative of the right side with respect to t is the integrand. (Why?) Check that this is so. $\quad \square$

Since the process of integration is the reverse of the process of differentiation, every rule of differentiation gives rise to a corresponding rule of integration. The rules are traditionally formulated in terms of a variable u. We can prove each rule by showing that the derivative with respect to u of the right side is the integrand.

The Rules of Integration

1. For any number k,

$$\int k\,du = ku + c.$$

2. For any number $r \neq -1$,

$$\int u^r\,du = \frac{u^{r+1}}{r+1} + c.$$

3. $\int \frac{1}{u}\,du = \ln|u| + c$. In other words,

$$\text{if } u > 0, \quad \int \frac{1}{u}\,du = \ln u + c;$$

$$\text{if } u < 0, \quad \int \frac{1}{u}\,du = \ln(-u) + c.$$

4. $\int e^u\,du = e^u + c.$

5. For any number k and function $f(u)$,

$$\int kf(u)\,du = k\int f(u)\,du.$$

6. For any functions $f_1(u)$ and $f_2(u)$,

$$\int [f_1(u) \pm f_2(u)]\,du = \int f_1(u)\,du \pm \int f_2(u)\,du.$$

Rule 6 remains true if the integrand is the sum or difference of more than two functions.

EXAMPLE 5

$$\int 2\,du = 2u + c.$$

$$\int -3\,dx = -3x + c.$$

$$\int dt = \int 1\,dt = t + c.$$

$$\int 0\,du = c. \quad \square$$

EXAMPLE 6

$$\int u\,du = \int u^1\,du = \frac{u^{1+1}}{1+1} + c = \frac{u^2}{2} + c.$$

$$\int x^2\,dx = \frac{x^3}{3} + c.$$

$$\int \sqrt{x}\,dx = \int x^{1/2}\,dx = \frac{x^{3/2}}{3/2} + c = \frac{2}{3}x^{3/2} + c.$$

$$\int t^{-3}\,dt = \frac{t^{-3+1}}{-3+1} + c = -\frac{t^{-2}}{2} + c. \quad \square$$

EXAMPLE 7

$$\int 4x^2\,dx = 4\int x^2\,dx = 4\frac{x^3}{3} + c = \frac{4}{3}x^3 + c.$$

$$\int \frac{4}{x}\,dx = 4\int \frac{1}{x}\,dx = 4\ln|x| + c.$$

$$\int -2e^t\,dt = -2\int e^t\,dt = -2e^t + c. \quad \square$$

EXAMPLE 8

$$\int (u^2 + 3u)\,du = \int u^2\,du + \int 3u\,du$$

$$= \frac{u^3}{3} + 3\int u\,du$$

$$= \frac{u^3}{3} + \frac{3u^2}{2} + c. \quad \square$$

EXAMPLE 9

$$\int (4x^4 - 3x^{-2} + 6)\,dx = 4\int x^4\,dx - 3\int x^{-2}\,dx + 6\int dx$$

$$= 4\frac{x^5}{5} - 3\frac{x^{-1}}{-1} + 6x + c$$

$$= \frac{4}{5}x^5 + \frac{3}{x} + 6x + c. \quad \square$$

If $F'(x) = f(x)$ for all x in the domain of f, then $F(x)$ is an antiderivative of $f(x)$, so

$$\int f(x)\,dx = \int F'(x)\,dx = F(x) + c.$$

Thus if f is the derivative of a function F, we can find F by integrating $f(x)$ and then determining the value of the constant of integration. We illustrate with examples.

EXAMPLE 10 Suppose a firm's marginal cost function is given by

$$f(x) = 2x + 5.$$

Marginal cost is the derivative of cost, so we can find the firm's cost function by integrating $f(x)$ and then determining the value of the constant of integration. Thus

$$C(x) = \int f(x)\, dx = \int (2x + 5)\, dx = x^2 + 5x + c.$$

To find c, we must know the value of $C(x)$ for some value of x. For instance, suppose the firm's fixed cost is $10,000. Then

$$10{,}000 = C(0) = 0^2 + 5(0) + c = c,$$

so $c = 10{,}000$ and the cost function is given by

$$C(x) = x^2 + 5x + 10{,}000. \quad \square$$

EXAMPLE 11 Suppose that t hours after the start of an experiment a colony of bacteria is growing at the rate of $y = 300\sqrt{t}$ bacteria per hour and that there were 2000 bacteria in the colony 4 hours after the start of the experiment. Since y is the rate at which the size of the bacteria population changes with respect to time, it is the derivative of the function F that describes population size as a function of time. Therefore we may find F by integrating y. But it is easy to check that $200t^{3/2}$ is an antiderivative for y (do so), and hence

$$\int y\, dt = \int 300\sqrt{t}\, dt = 200t^{3/2} + c.$$

Thus $F(t) = 200t^{3/2} + c$ for some c. But since there were 2000 bacteria present when $t = 4$, it follows that

$$2000 = F(4) = 200(4^{3/2}) + c = 200(8) + c = 1600 + c,$$

and thus that $c = 400$. Therefore F is given by

$$F(t) = 200t^{3/2} + 400. \quad \square$$

■ *Computer Manual: Program INTPOLY*

14.1 EXERCISES

In Exercises 1 through 40, find the indefinite integral.

1. $\int 3\, dx$
2. $\int \sqrt{2}\, dx$
3. $\int 0\, dx$
4. $\int u^4\, du$
5. $\int x^{10}\, dx$
6. $\int t^{12}\, dt$
7. $\int t^{-4}\, dt$
8. $\int q^{-4/5}\, dq$
9. $\int \sqrt{x^3}\, dx$
10. $\int 2x\, dx$
11. $\int 2x^3\, dx$
12. $\int \dfrac{5}{x^4}\, dx$
13. $\int 3x^{-6}\, dx$
14. $\int -3\sqrt{y}\, dy$
15. $\int 2t^{-1/2}\, dt$

INTEGRATION

16. $\displaystyle\int 2u^{1/3}\, du$ 17. $\displaystyle\int \frac{2}{p}\, dp$ 18. $\displaystyle\int -\frac{4}{x}\, dx$ 19. $\displaystyle\int \frac{3}{t}\, dt,\ t<0$ 20. $\displaystyle\int -3e^y\, dy$

21. $\displaystyle\int 2e^x\, dx$ 22. $\displaystyle\int 12e^x\, dx$ 23. $\displaystyle\int (x+2)\, dx$ 24. $\displaystyle\int (3-2x)\, dx$

25. $\displaystyle\int (2v^2 - 3v + 1)\, dv$ 26. $\displaystyle\int (x^2 - 3x + 5)\, dx$ 27. $\displaystyle\int (2x^3 - 6x^2 + 18x - 3)\, dx$

28. $\displaystyle\int (x^3 + 3x^2 - 2x)\, dx$ 29. $\displaystyle\int (-4x^3 + 3x^2 - 2x + 7)\, dx$ 30. $\displaystyle\int (-3x^{-5} + 4x^2 + 12)\, dx$

31. $\displaystyle\int (6e^x + x)\, dx$ 32. $\displaystyle\int \left(\frac{2}{t^2} + \frac{1}{t^3} + 3\right) dt$ 33. $\displaystyle\int \left(7x^2 - 14x + \frac{2}{x} - \frac{5}{x^3}\right) dx$

34. $\displaystyle\int \left(\frac{2}{\sqrt{q}} - \frac{3}{q} + 1\right) dq,\ q>0$ 35. $\displaystyle\int (100x^{99} + 99x^{98} + 98x^{97} + \cdots + 2x + 1)\, dx$

36. $\displaystyle\int -\left(\frac{1}{t^2} + \frac{2}{t^3} + \frac{3}{t^4} + \cdots + \frac{99}{t^{100}}\right) dt$ 37. $\displaystyle\int (x+5)(2x-3)\, dx$

38. $\displaystyle\int (2x+5)^2\, dx$ 39. $\displaystyle\int (4-x)^3\, dx$ 40. $\displaystyle\int (3x+1)(x+5)^2\, dx$

41. An airplane's velocity t hours after takeoff is $v = 300t$ kilometers per hour. If the plane flies in a straight line, find the distance it travels in the 2 hours after takeoff. (*Hint:* The distance the plane has traveled at takeoff is 0 kilometers.)

42. If a rocket's acceleration t seconds after lift-off is $a(t) = 0.4t + 200$ mps², find the rocket's velocity function. (*Hint:* Its initial velocity is 0 feet per second.)

43. A rocket is launched straight up with a constant acceleration of 100 feet per second per second.
    (a) Find the rocket's velocity 10 seconds after launch.
    (b) Find its altitude at the end of 10 seconds. (*Hint:* Its initial altitude is 0 ft.)

*44. Stefan's Law says that the rate at which an object radiates energy is proportional to the fourth power of its absolute temperature: if y is the rate at which energy is radiated per unit area, then $y = kT^4$, where T is temperature in degrees Kelvin (K) and k is a constant. Find an expression for the amount of energy radiated per unit area. (*Hint:* An object will radiate no energy at $T = 0°$ K.)

Management

45. A firm's marginal cost function is given by $c(x) = 2x + 1$, its fixed cost is $5000, and its marginal revenue function is given by $r(x) = 15$. Find the firm's cost, revenue, and profit functions. (*Hint:* The revenue obtained from selling 0 units is $0.)

46. A firm's marginal cost function is given by $c(x) = 5x + 20$, the cost of making 10 units is $10,000, and its marginal revenue function is given by $r(x) = 40$. Find the firm's cost, revenue, and profit functions.

47. A firm's marginal profit function is given by $p(x) = x^2 - 100x + 1600$, and its profit from making and selling 3 units is $4341. Find the firm's profit function.

48. Suppose a firm's marginal cost function is given by
$$c(x) = 0.3x^2 + 0.5x + 2.$$
Find the firm's cost function if its fixed cost is $20,000.

49. The rate at which the quantity of a commodity demanded changes with respect to price is y units per dollar, where
$$y = -50p.$$
Find the demand function for the product if 25,000 units are demanded at a price of $100.

50. The rate at which the quantity of a product supplied to the market changes with respect to price is y units per dollar p, where
$$y = \frac{200}{p}.$$
Find the supply function for the product if 1000 units were supplied at a price of $10.

Life Science

51. A nesting pair of blue herons is introduced into a bird sanctuary where blue herons did not previously exist. It is estimated that the population of blue herons in the sanctuary will increase at the rate of y individuals per year, where $y = \sqrt{t} + 4$ and t is time in years since the species was introduced into the sanctuary. Find a function that describes the size of the blue heron population in the sanctuary in year t.

52. A population of bacteria grows at a rate of $y = 0.02e^t$ bacteria per minute. If the initial population size was one bacterium, find an equation that describes the size of the population as a function of time.

53. Ivory poachers are killing elephants in a game preserve at rate of $y = -100e^{-t}$ individuals per year t, where $t = 0$ represents the present. If there are 400 elephants in the preserve now, find a function that describes the number in the preserve as a function of time. (*Hint:* e^{-t} is an antiderivative for $-e^{-t}$.)

Social Science

54. In 1976, 68% of the voters in a state were affiliated with a political party. Since then the percentage affiliated with a party has changed at the rate of $-0.6\sqrt{t}$ percent per year, where t is in years, with $t = 0$ representing 1976. Find a function that describes the percentage of voters affiliated with a political party in year t.

55. The property tax rate in a town is expressed in terms of dollars per thousand dollars of the assessed value of the property. The tax rate has been changing at a rate of
$$y = \frac{2}{t+1}$$
dollars per thousand dollars of assessed value, and was $80 per thousand in 1989. Find a function that describes the property tax rate as a function of time. (*Hint:* $\ln|t+1|$ is an antiderivative for $(t+1)^{-1}$.)

Fick's Law

*56. Fick's Law states that if $s(t)$ is the concentration of a solute inside a cell at time t, its rate of diffusion is
$$s'(t) = \frac{kA}{V}(S - s(t)),$$

INTEGRATION

where A is the area of the cell membrane, V is the volume of the cell, S is the concentration of the solute outside the cell, and k is a constant. Find $s(t)$. (*Hint:* Write Fick's Law in the form

$$\frac{s'(t)}{S - s(t)} = \frac{kA}{V},$$

find an antiderivative for each side of this equation, and solve for $s(t)$. You may assume that $S - s(t)$ is positive for all t.)

14.2 FINDING INDEFINITE INTEGRALS

The rules of integration of Section 14.1 do not tell us how to integrate all functions of interest. For instance, none of the rules apply directly to the integrals

$$\int \sqrt{3x + 1}\, dx \quad \text{and} \quad \int x^2(x^3 + 1)^{10}\, dx.$$

In this section we discuss and illustrate two techniques for finding indefinite integrals: integration by guessing and integration by substitution.

Sometimes we can guess the general form that an integral must take, and then by differentiating this general form, find the exact expression for the integral. For instance, consider the integral

$$\int \sqrt{3x + 1}\, dx.$$

An antiderivative of $\sqrt{3x + 1}$ must have the general form $(3x + 1)^{3/2}$. However, $(3x + 1)^{3/2}$ is not quite the antiderivative we seek, because its derivative is not quite equal to $\sqrt{3x + 1}$:

$$\frac{d}{dx}(3x + 1)^{3/2} = \frac{9}{2}(3x + 1)^{1/2} = \frac{9}{2}\sqrt{3x + 1}.$$

Thus we have an unwanted constant of $9/2$. However, if we multiply $(3x + 1)^{3/2}$ by $2/9$, the unwanted constant will cancel out:

$$\frac{d}{dx}\left[\frac{2}{9}(3x + 1)^{3/2}\right] = \frac{2}{9}\frac{d}{dx}(3x + 1)^{3/2} = \frac{2}{9}\frac{9}{2}\sqrt{3x + 1} = \sqrt{3x + 1}.$$

Thus

$$\frac{2}{9}(3x + 1)^{3/2}$$

is an antiderivative of $\sqrt{3x + 1}$, so

$$\int \sqrt{3x + 1}\, dx = \frac{2}{9}(3x + 1)^{3/2} + c.$$

EXAMPLE 1 Consider the integral

$$\int \frac{1}{3x + 2}\, dx, \quad 3x + 2 > 0.$$

Since $3x + 2 > 0$, an antiderivative of $1/(3x + 2)$ must have the general form $\ln(3x + 2)$. But

$$\frac{d}{dx}(\ln(3x + 2)) = \frac{1}{3x + 2}\frac{d}{dx}(3x + 2) = \frac{3}{3x + 2}.$$

Therefore

$$\frac{d}{dx}\left(\frac{1}{3}\ln(3x + 2)\right) = \frac{1}{3}\frac{d}{dx}(\ln(3x + 2)) = \frac{1}{3}\frac{3}{3x + 2} = \frac{1}{3x + 2}.$$

Thus

$$\int \frac{1}{3x + 2}\,dx = \frac{1}{3}\ln(3x + 2) + c. \quad \square$$

EXAMPLE 2 Let us find the indefinite integral

$$\int e^{-2x}\,dx.$$

Since

$$\frac{d}{dx}(e^{-2x}) = -2e^{-2x},$$

we have

$$\int e^{-2x}\,dx = -\frac{1}{2}e^{-2x} + c. \quad \square$$

Guessing the general form of an integral does not always work, because often we cannot tell what the general form should look like. In any case, we would like to have a more systematic technique for performing integration. No technique of integration works for all integrals, but one widely used method is **integration by substitution,** which we now discuss.

Recall that in Section 11.3 we introduced the differentials dx and dy and showed that if $y = f(x)$ then $dy = f'(x)\,dx$. We shall use u rather than y as the dependent variable; thus if $u = f(x)$, then $du = f'(x)\,dx$. We can often use this relationship between du and dx to change the variable in an integral from x to u in such a way that the resulting integral in u is easy to find. The technique is called **integration by substitution,** because we substitute u for some portion of the integrand. We illustrate with an example.

EXAMPLE 3 Let us consider again the integral

$$\int \sqrt{3x + 1}\,dx.$$

Suppose we set $u = 3x + 1$. Then

$$du = (3x + 1)'\,dx = 3\,dx,$$

INTEGRATION

or

$$\frac{1}{3} du = dx.$$

If we substitute u for $3x + 1$ in the integrand and $\frac{1}{3} du$ for dx, we obtain

$$\int \sqrt{3x + 1}\, dx = \int \sqrt{u}\, \frac{1}{3} du = \frac{1}{3} \int u^{1/2}\, du$$

$$= \frac{1}{3} \frac{u^{3/2}}{3/2} + c = \frac{2}{9} u^{3/2} + c.$$

Back-substituting $3x + 1$ for u now yields

$$\int \sqrt{3x + 1}\, dx = \frac{2}{9}(3x + 1)^{3/2} + c.$$

This is the result we obtained previously by guessing. ☐

How did we know to choose $u = 3x + 1$ in Example 3? The answer is that no other choice for u simplifies the integral. For instance, if we had set $u = 3x$, then $du = 3\, dx$ and substitution would give the integral

$$\int \sqrt{u + 1}\, \frac{1}{3} du,$$

which is really no simpler than the one we started with. On the other hand, if we had let $u = \sqrt{3x + 1}$, then we would have had

$$du = \frac{3}{2\sqrt{3x + 1}}\, dx;$$

but with this expression for du, substitution does not even give an integral in the variable u. (Try it.)

EXAMPLE 4 Let us find

$$\int e^{-2x}\, dx,$$

using the technique of integration by substitution.

We set $u = -2x$. Then $du = -2\, dx$, so $-\frac{1}{2} du = dx$. Thus

$$\int e^{-2x}\, dx = \int e^u \left(-\frac{1}{2}\right) du = -\frac{1}{2} \int e^u\, du$$

$$= -\frac{1}{2} e^u + c = -\frac{1}{2} e^{-2x} + c,$$

which is the result we obtained in Example 2 by guessing. ☐

Integration by substitution is often useful in dealing with more complicated integrals, such as

$$\int x^2(x^3+1)^{10}\,dx.$$

To illustrate, let $u = x^3 + 1$. Then $du = 3x^2\,dx$, or $\frac{1}{3}du = x^2\,dx$, so

$$\int x^2(x^3+1)^{10}\,dx = \int (x^3+1)^{10} x^2\,dx = \int u^{10}\frac{1}{3}\,du = \frac{1}{3}\frac{u^{11}}{11} + c$$

$$= \frac{1}{33}(x^3+1)^{11} + c.$$

How did we know to let $u = x^3 + 1$? The answer is that we recognized part of the integrand, namely x^2, as the derivative (except for a constant factor) of another part, namely $x^3 + 1$.

> **Key to Integration by Substitution**
>
> Whenever part B of an integrand is the derivative (except possibly for a constant factor) of part A of the integrand, the substitution $u = $ part A will simplify the integral.

This is so because, under these conditions, part B of the integrand will be replaced by du when the substitution is performed.

EXAMPLE 5 Let us find $\int \frac{\ln x}{x}\,dx$.

We recognize $1/x$ as the derivative of $\ln x$. Therefore we make the substitution $u = \ln x$. Since $du = (1/x)\,dx$, we have

$$\int \frac{\ln x}{x}\,dx = \int (\ln x)\frac{1}{x}\,dx = \int u\,du = \frac{1}{2}u^2 + c = \frac{1}{2}(\ln x)^2 + c. \quad \square$$

EXAMPLE 6 Find $\int xe^{-x^2}\,dx$.

Except for a constant factor of -2, x is the derivative of $-x^2$. Therefore we set $u = -x^2$, so $du = -2x\,dx$, or $-\frac{1}{2}du = x\,dx$, and

$$\int xe^{-x^2}\,dx = \int e^{-x^2} x\,dx = \int e^u\left(-\frac{1}{2}\right)du = -\frac{1}{2}\int e^u\,du$$

$$= -\frac{1}{2}e^u + c = -\frac{1}{2}e^{-x^2} + c. \quad \square$$

INTEGRATION

We conclude this section by remarking that integration by substitution is in fact just the chain rule "in reverse." For instance, if $F(x) = (x^2 + 1)^4 + c$, then the chain rule tells us that

$$F'(x) = 4(x^2 + 1)^3 (x^2 + 1)' = 8x(x^2 + 1)^3.$$

If we now integrate $8x(x^2 + 1)^3$ by making the substitution $u = x^2 + 1$, so that $du = 2x\, dx$, we obtain

$$\int 8x(x^2 + 1)^3\, dx = \int 4(x^2 + 1)^3\, 2x\, dx$$
$$= \int 4u^3\, du$$
$$= u^4 + c$$
$$= (x^2 + 1)^4 + c.$$

Thus our integration by substitution "reversed" or "undid" the chain rule.

14.2 EXERCISES

In Exercises 1 through 42, find the integral.

1. $\int \sqrt{x+1}\, dx$
2. $\int \sqrt{2x+5}\, dx$
3. $\int \sqrt{4x-3}\, dx$
4. $\int (5t+2)^{3/2}\, dt$

5. $\int (6x-2)^{4/3}\, dx$
6. $\int \frac{1}{(2x-1)^2}\, dx$
7. $\int (7x-3)^{12}\, dx$
8. $\int (8x+1)^9\, dx$

9. $\int \frac{5}{(2x+3)^2}\, dx$
10. $\int \frac{3}{\sqrt{2-3x}}\, dx$
11. $\int 6(4-7x)^3\, dx$
12. $\int \frac{-6}{5x+2}\, dx$

13. $\int \frac{2}{1-2x}\, dx$
14. $\int \frac{3}{2x+1}\, dx$
15. $\int \frac{4}{8-3x}\, dx$
16. $\int e^{4x}\, dx$

17. $\int 2e^{3x}\, dx$
18. $\int 8e^{-2x}\, dx$
19. $\int 3e^{-7x}\, dx$
20. $\int 2x(x^2-2)^3\, dx$

21. $\int x(x^2+1)^4\, dx$
22. $\int -3x^2(x^3+1)^2\, dx$
23. $\int x^4(2x^5+10)\, dx$
24. $\int 2x\sqrt{x^2+1}\, dx$

25. $\int 5x\sqrt{2-x^2}\, dx$
26. $\int 2x^2\sqrt{x^3+1}\, dx$
27. $\int (2x+3)(x^2+3x+1)^4\, dx$

28. $\int (x-2)(2x^2-8x^5)\, dx$
29. $\int \frac{2x}{x^2+1}\, dx$
30. $\int \frac{2x}{3x^2+6}\, dx$

31. $\int \frac{-7t^2}{t^3+2}\, dt$
32. $\int \frac{2-3x^2}{x^3-2x}\, dx$
33. $\int 3xe^{x^2}\, dx$
34. $\int 4xe^{-x^2}\, dx$

35. $\int (2x+1)e^{2x^2+2x}\, dx$
36. $\int (1-2x)e^{x^2-x+1}\, dx$
37. $\int e^x e^{e^x}\, dx$
38. $\int \frac{2e^{\sqrt{x}}}{\sqrt{x}}\, dx$

39. $\int \frac{(\ln x)^2}{x}\, dx$
40. $\int \frac{1}{x \ln x}\, dx$
41. $\int \frac{1}{x(\ln x)^2}\, dx$
42. $\int \frac{\sqrt{\ln x}}{x}\, dx$

Management

43. A firm's marginal cost and marginal revenue functions are given by

$$c(x) = 3x\sqrt{x^2 + 1} \quad \text{and} \quad r(x) = \frac{7}{2}x(x^2 + 1)^{3/4},$$

respectively. Find the firm's cost function and revenue function if the fixed cost is $1000.

44. A firm's marginal revenue function is given by $r(x) = 0.2x(1.2x^2 + 4)^{1/2}$. Find the firm's revenue function.

45. A firm's marginal profit function is given by

$$p(x) = \frac{x}{\sqrt{2x^2 + 9}}.$$

Find the firm's profit function if the profit from making and selling 6 units is $52.50.

46. Suppose that the amount of information that must be transmitted by the national telephone network is increasing at a rate of

$$y = (2 \cdot 10^7)e^{0.02t}$$

bits per year t, where $t = 0$ represents the current year. If the network must transmit 10^{10} bits of information this year, find a function that describes the amount of information transmitted as a function of time.

47. With no experience, a worker can assemble 100 units per week. With t weeks of experience, the worker can assemble them at the rate of

$$f(t) = 200e^{-2t}$$

units per week. Find a function that describes the worker's weekly production.

48. A firm's marginal profit function is given by $p(x) = (0.5x + 0.75)/(x^2 + 3x + 9)$, profit in thousands of dollars. Find the firm's profit function if its fixed cost is $10,000.

Life Science

49. It is estimated that the total world population of blue whales in 1978 was 250 and that the population since then has been changing at a rate of y individuals per year, where

$$y = \frac{20t}{(2t^2 + 1)^{3/2}}.$$

Here t is in years, with $t = 0$ representing 1978. Find a function that describes the size of the blue whale population in year t.

50. An animal population is growing at the rate of y thousand individuals per year t, where $y = 8t/(6t^2 + 11)$ and $t = 0$ represents 1988. Find an expression for population size as a function of time if the population in 1988 was 60,000.

51. The concentration of a drug in a patient's bloodstream was 2.5 micrograms per milliliter (μg/mL) 3 minutes after the drug was injected, and it was decreasing at a rate of $-0.5e^{-0.25t}$ (μg/mL)/minute. Find a function that describes the concentration of the drug in the patient's bloodstream as a function of time.

INTEGRATION

52. The body produces antibodies as a reaction to vaccination. Suppose that t days after a vaccination, antibodies are being produced at a rate of

$$y = \frac{200}{5t + 4}$$

per day, where y is in thousands of antibodies. Assuming there were no antibodies present before vaccination, find a function that describes antibody production t days after vaccination.

53. Air pollution in a region is decreasing at a rate of

$$y = -\frac{20}{2t + 5}$$

parts per million per year t, where $t = 0$ represents the present. If air pollution in the region currently stands at 40 ppm, find a function that describes air pollution as a function of time.

54. The concentration of smog produced by automobiles in a large city changes with the temperature at a rate given by

$$f(x) = \frac{0.006}{(0.02x + 100)^{1/2}},$$

where x is temperature in degrees Celsius and the smog concentration is in parts per million (ppm). Find a function that describes the smog concentration if 6 ppm are produced at 0°C.

14.3 THE DEFINITE INTEGRAL

The concept of the **definite integral** of a function over an interval is one of the most important of calculus. As we shall see in the remainder of this chapter and in the next, definite integrals have many uses. In this section we introduce the definite integral of a function in its aspect as the total change in any antiderivative of the function. We also demonstrate how to apply the method of integration by substitution to definite integrals.

The Definite Integral as Total Change

The **definite integral** of a function over an interval is a number that expresses the total change in any antiderivative of the function over the interval. For instance, if

$$f(x) = 2x + 1,$$

any antiderivative of $f(x)$ is of the form

$$F(x) = x^2 + x + c.$$

The total change in F over the interval $[1, 3]$ is thus

$$F(3) - F(1) = (3^2 + 3 + c) - (1^2 + 1 + c) = 12 - 2 = 10.$$

Hence we say that the definite integral of f over the interval $[1, 3]$ is equal to 10. Note that when calculating a definite integral, we can use any antiderivative, since the constant of integration subtracts out. Therefore we omit the constant of integration when working with definite integrals.

> **The Definite Integral**
>
> If f is a function defined on the interval $[a, b]$, then the **definite integral of f over $[a, b]$,** symbolized by
>
> $$\int_a^b f(x)\, dx,$$
>
> is the total change in any antiderivative of $f(x)$ over the interval. Thus if $F(x)$ is any antiderivative of $f(x)$, then
>
> $$\int_a^b f(x)\, dx = F(b) - F(a).$$
>
> The numbers a and b are referred to as the **limits of integration.**

(The foregoing is not really the *definition* of the definite integral, but rather a *property* of the definite integral. As we shall see in Section 14.5, the definition of the definite integral identifies it as the limit of a certain type of sum, and the fact that this limit may be expressed as the total change in any antiderivative of the integrand is a theorem about the definite integral. We have chosen to introduce the definite integral through this property because we wish to develop some facility in handling definite integrals before we take up their formal definition.)

EXAMPLE 1 If f is defined by $f(x) = 4x - 5$, let us find the definite integral of f over the interval $[1, 2]$. Since $F(x) = 2x^2 - 5x$ is an antiderivative for $f(x)$, we have

$$\int_1^2 (4x - 5)\, dx = F(2) - F(1)$$
$$= (2(2)^2 - 5(2)) - (2(1)^2 - 5(1))$$
$$= 1. \quad \square$$

If $F(x)$ is an antiderivative of $f(x)$, we usually write

$$\int_a^b f(x)\, dx = F(x) \Big|_a^b = F(b) - F(a)$$

when evaluating the definite integral. For instance, we could have carried out the integration of the preceding example as follows:

$$\int_1^2 (4x - 5)\, dx = (2x^2 - 5x)\Big|_1^2 = (2(2)^2 - 5(2)) - (2(1)^2 - 5(1)) = 1.$$

INTEGRATION

Note the difference between the definite integral $\int_a^b f(x)\,dx$ and the indefinite integral $\int f(x)\,dx$: the definite integral is a *number*, while the indefinite integral is a *family of functions*.

EXAMPLE 2 Let us find $\int_{-1}^{3} (x^2 + 2x - 1)\,dx$. We have

$$\int_{-1}^{3} (x^2 + 2x - 1)\,dx = \left(\frac{x^3}{3} + x^2 - x\right)\bigg|_{-1}^{3}$$

$$= \left(\frac{3^3}{3} + 3^2 - 3\right) - \left(\frac{(-1)^3}{3} + (-1)^2 - (-1)\right)$$

$$= 15 - \frac{5}{3}$$

$$= \frac{40}{3}. \quad \Box$$

EXAMPLE 3 Let us evaluate $\int_{-4}^{4} 2e^x\,dx$. We have

$$\int_{-4}^{4} 2e^x\,dx = 2e^x \bigg|_{-4}^{4} = 2(e^4 - e^{-4}). \quad \Box$$

Note that the rules for indefinite integrals

$$\int kf(x)\,dx = k\int f(x)\,dx \qquad \text{for any number } k$$

and

$$\int [f_1(x) \pm f_2(x)]\,dx = \int f_1(x)\,dx \pm \int f_2(x)\,dx$$

imply the corresponding rules for definite integrals:

$$\int_a^b kf(x)\,dx = k\int_a^b f(x)\,dx \qquad \text{for any number } k$$

and

$$\int_a^b [f_1(x) \pm f_2(x)]\,dx = \int_a^b f_1(x)\,dx \pm \int_a^b f_2(x)\,dx.$$

Now we demonstrate how the interpretation of the definite integral as total change can be exploited.

EXAMPLE 4 Suppose a city's population is growing at the rate of

$$f(t) = 12{,}000\sqrt{t}$$

persons per year, where t is in years with $t = 0$ representing 1988. Let us find the total change in population from 1988 to 1992.

Since $f(t)$ describes the rate of change of the population, its definite integral $\int_a^b f(t)\, dt$ gives the total change in the population from time $t = a$ to time $t = b$. Therefore we must evaluate

$$\int_0^4 12{,}000\sqrt{t}\, dt.$$

But

$$\int_0^4 12{,}000\sqrt{t}\, dt = 8000 t^{3/2} \Big|_0^4 = 8000(4^{3/2} - 0^{3/2}) = 64{,}000.$$

Hence, over the period 1988–1992 the city will experience a net gain in population of 64,000 people. □

EXAMPLE 5 A firm's marginal cost function is

$$c(x) = 2x^2 - 8x + 30.$$

The firm is currently producing six units per day. What will be the additional cost if the firm decides to produce 10 units per day?

Since marginal cost is the rate of change of cost, the total change in cost due to increasing production from $x = 6$ to $x = 10$ units is

$$\int_6^{10} (2x^2 - 8x + 30)\, dx = \frac{2}{3}x^3 - 4x^2 + 30x \Big|_6^{10}$$

$$= \left(\frac{2}{3}(10)^3 - 4(10)^2 + 30(10)\right) - \left(\frac{2}{3}(6)^3 - 4(6)^2 + 30(6)\right)$$

$$= \frac{1700}{3} - \frac{540}{3}$$

$$\approx \$386.67.$$

per day. □

EXAMPLE 6 An investment in a new computer will save a company money at the rate of

$$s(t) = 800 - 0.5t^2$$

dollars per month, where t is time in months since the installation of the computer. Figure 14–1 shows the graph of $s(t)$. Note that the rate of savings declines with time, reaching zero when $t = 40$.

The definite integral $\int_a^b s(t)\, dt$ gives the savings due to the computer from time $t = a$ to time $t = b$. Therefore to find the total savings due to the computer, we must evaluate the integral from the time the savings begin until the time they end. But the savings begin when $t = 0$ and end when $t = 40$. Thus the total savings due to the computer are

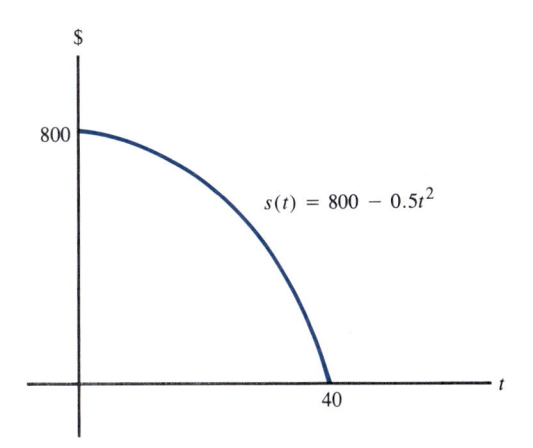

$$\int_0^{40} (800 - 0.5t^2)\, dt = 800t - \frac{t^3}{6} \Big|_0^{40}$$

$$= \left(800(40) - \frac{(40)^3}{6}\right) - \left(800(0) - \frac{0^3}{6}\right)$$

$$= \$21{,}333.33. \quad \square$$

Integration by Substitution for Definite Integrals

The definite integral $\int_a^b f(x)\, dx$ can be evaluated in two ways. In the first method we find an antiderivative for the integrand $f(x)$, either by guessing or by integration by substitution; then we use the antiderivative to evaluate the definite integral. For instance, to evaluate

$$\int_0^1 x(x^2 + 1)^4\, dx$$

by substitution, we let $u = x^2 + 1$; then $du = 2x\, dx$, and

$$\int x(x^2 + 1)^4\, dx = \frac{1}{2}\int u^4\, du = \frac{1}{10}u^5 + c = \frac{1}{10}(x^2 + 1)^5 + c.$$

Thus

$$\int_0^1 x(x^2 + 1)^4\, dx = \frac{1}{10}(x^2 + 1)^5 \Big|_0^1 = \frac{1}{10}(2^5 - 1^5) = \frac{31}{10}.$$

Alternatively, we can apply the method of integration by substitution directly to the definite integral. If we do so, however, we must be careful to change the limits of integration when we change the variable from x to u in the integral. Thus, in our example, since $u = x^2 + 1$, it follows that

$$\text{when } x = 0, \quad u = 0^2 + 1 = 1$$

and

$$\text{when } x = 1, \quad u = 1^2 + 1 = 2.$$

Applying the method of integration by substitution directly to the definite integral, we now have

$$\int_0^1 x(x^2 + 1)^4 \, dx = \frac{1}{2} \int_1^2 u^4 \, du = \frac{1}{10} u^5 \bigg|_1^2 = \frac{1}{10}(2^5 - 1^5) = \frac{31}{10}.$$

EXAMPLE 7   Let us find $\int_1^2 \frac{x + 1}{x^2 + 2x} \, dx$ by the method of substitution for definite integrals. If we let

$$u = x^2 + 2x,$$

then

$$du = 2(x + 1) \, dx.$$

Thus,

$$\text{when } x = 1, \quad u = (1)^2 + 2(1) = 3,$$
$$\text{when } x = 2, \quad u = (2)^2 + 2(2) = 8,$$

and

$$\int_1^2 \frac{x + 1}{x^2 + 2x} \, dx = \frac{1}{2} \int_3^8 \frac{du}{u} = \frac{1}{2} \ln u \bigg|_3^8$$
$$= \frac{1}{2}(\ln 8 - \ln 3) = \frac{1}{2} \ln \frac{8}{3}.$$

(Note that we did not use absolute values and write $\ln |u|$ when we integrated $1/u$ because we already knew that u was positive, since it was between 3 and 8.) ☐

■ *Computer Manual: Program DEFNTGRL*

14.3   EXERCISES

The Definite Integral as Total Change

In Exercises 1 through 22, evaluate the definite integral.

1. $\int_0^4 x \, dx$

2. $\int_1^5 x^2 \, dx$

3. $\int_{-1}^1 3x^3 \, dx$

INTEGRATION

4. $\int_{-1}^{1} x^4 \, dx$

5. $\int_{0}^{3} (x^2 - 2x + 2) \, dx$

6. $\int_{2}^{4} (3x^2 + 1) \, dx$

7. $\int_{-1}^{1} (2x^3 + 3x^2 - 2) \, dx$

8. $\int_{-2}^{0} (x^4 - 5x^2 + 3x - 1) \, dx$

9. $\int_{1}^{4} \sqrt{x} \, dx$

10. $\int_{0}^{4} (3x - \sqrt{x}) \, dx$

11. $\int_{0}^{1} (x + \sqrt[3]{x}) \, dx$

12. $\int_{1}^{64} (x^{5/3} - \sqrt{x}) \, dx$

13. $\int_{1}^{2} x^{-2} \, dx$

14. $\int_{1}^{2} -\left(\frac{1}{t^2} + \frac{1}{t^3} + \frac{1}{t^4}\right) dt$

15. $\int_{0}^{3} 3e^x \, dx$

16. $\int_{0}^{3} e^{-x} \, dx$

17. $\int_{\ln 2}^{\ln 5} 5e^t \, dt$

18. $\int_{1}^{e} \frac{1}{x} \, dx$

19. $\int_{-2}^{-1} \frac{4}{t} \, dt$

20. $\int_{e^2}^{e^3} \frac{3}{x} \, dx$

*21. $\int_{-a}^{a} x^n \, dx$, $a > 0$, n an even positive integer.

*22. $\int_{-a}^{a} x^n \, dx$, $a > 0$, n an odd positive integer.

Management

23. Addlepate, Inc., has marginal cost function given by
$$c(x) = x^2 + 5x$$
and marginal revenue function given by
$$r(x) = 3x^2 + 8x,$$
where cost and revenue are in thousands of dollars and x is in thousands of units. The corporation is currently producing and selling 5000 units. What will be the additional cost and revenue if the firm increases production and sales to 8000 units?

24. Referring to Exercise 23: find the total cost and revenue generated by the first 5000 units Addlepate produces.

25. A firm's marginal profit function is given by $p(x) = -2x + 200$. If the firm is currently producing and selling 70 units, what will its additional profit be if it produces and sells another 10 units?

26. A firm's product generates revenue at the rate of
$$r = 9600t - 100t^2$$
dollars per month, where t is the number of months it has been on the market. Sketch the graph of this function and find the total revenue generated by the product over its lifetime.

27. Apex Company has marginal revenue function given by
$$r(x) = 10x^{1/2},$$
where $x \geq 1$. If Apex currently sells 25 units per day, find the additional revenue if the firm could sell 100 units per day.

28. Maintenance costs for a building t years old increase at a rate of $y = 0.5t^{3/2}$ thousand dollars per year. Find the total maintenance cost for the building over its first four years of life.

29. An investment in new equipment will save a firm $s(t)$ dollars per year, where
$$s(t) = 20{,}000 - 2000t$$

and t is measured in years since the installation of the equipment. Find the firm's total savings due to the new equipment.

Life Science

30. The pollutants in a pond are increasing at the rate of y grams per day, where
$$y = 10t + 3$$
and t is in days with $t = 0$ representing the present. Find the total amount of pollutants that will accumulate in the pond over the next 30 days.

31. A certain strain of influenza spreads in such a way that $f(t)$ persons per day contract the disease, where
$$f(t) = -5t^2 + 300t,$$
t in days. Find the number of persons who will contract the disease in the first two weeks of an epidemic that begins on day 0.

32. A fast-growing strain of bamboo grows $f(t)$ inches per day, where
$$f(t) = 6 + \frac{2}{\sqrt{t}},$$
t being the number of days since germination, and the formula is valid for $t \geq 10$. Find the amount the bamboo will grow in the two weeks from $t = 10$ to $t = 24$.

Social Science

33. In a certain metropolitan region the number of unemployed persons is increasing at a rate of y persons per year t, where
$$y = 60t^2 + 2000$$
and $t = 0$ is the present. Forecast the increase in unemployment in the region over the next two years.

Hurricanes

*34. According to an article by Kerry Emanuel ("Toward a General Theory of Hurricanes," *American Scientist*, July–August 1988), the total pressure drop from the outer wall of a hurricane to its eye is given by
$$-\int_a^c \frac{RT}{p}\, dp,$$
where p is atmospheric pressure, in millibars, a is the pressure at the outer wall and c the pressure at the eye, T is the temperature in degrees Celsius, and R is a constant. Evaluate the integral to find an expression for the total pressure drop as a function of T.

Integration by Substitution for Definite Integrals

In Exercises 35 through 50, evaluate the definite integral by the method of integration by substitution for definite integrals.

35. $\displaystyle\int_{\ln 2}^{\ln 3} e^{3x}\, dx$ 36. $\displaystyle\int_{-2}^{3} e^{-2x}\, dx$ 37. $\displaystyle\int_{0}^{2} 2xe^{-x^2}\, dx$ 38. $\displaystyle\int_{1}^{3} xe^{x^2}\, dx$

INTEGRATION

39. $\int_0^9 \dfrac{e^{\sqrt{x}}}{\sqrt{x}}\,dx$ 40. $\int_{-1}^{e-2} \dfrac{6}{x+2}\,dx$ 41. $\int_0^4 \dfrac{1}{2x+1}\,dx$ 42. $\int_1^2 \dfrac{4}{3x-2}\,dx$

43. $\int_0^{\sqrt{2}} x(x^2-2)\,dx$ 44. $\int_0^4 x^2(x^3-1)^2\,dx$ 45. $\int_{-1}^2 2x^2\sqrt{x^3+2}\,dx$ 46. $\int_0^1 \dfrac{2x}{x^2+5}\,dx$

47. $\int_1^2 \dfrac{3x^2+x}{2x^3+x^2}\,dx$ 48. $\int_0^1 \dfrac{3x}{x^2-4}\,dx$ 49. $\int_0^9 (\sqrt{x}+1)\sqrt{\dfrac{2}{3}x^{3/2}+x}\,dx$ 50. $\int_2^4 \dfrac{x^2+2x+1}{x^3}\,dx$

Management

51. A firm intends to build a new plant. It is estimated that t years after the plant is built it will be producing at the rate of
$$f(t) = 100{,}000 e^{0.01t}$$
units per year. Find the number of units the plant will produce in its first five years. How long must the plant operate if it is to produce 2 million units?

52. Able Associates has marginal profit function given by
$$p(x) = \dfrac{2000}{0.5x+10}.$$
Find the firm's profit from making and selling its first 50 units.

53. With t weeks of experience a worker can assemble parts at a rate of
$$y = 200 - 120e^{-0.05t}$$
per week. Find the number of parts the worker can assemble over weeks 0 through 5; over weeks 5 through 10.

54. Suppose it takes y worker-hours to build the xth unit of a product, where
$$y = 4000 + 2000e^{-0.1x}, \qquad x \geq 1.$$
Find the total worker-hours needed to build units 1 through 5; to build units 1 through 100.

55. Investment in new industrial robots saves a firm money at the rate of
$$S = 100(1 - e^{-0.02t})$$
thousand dollars per year t years after purchase. Find the firm's total savings due to the robots over the period from two to five years after they were purchased.

56. Suppose that platinum is being mined at the rate of
$$y = \dfrac{2000}{0.5t+1}$$
metric tons per year t, where $t = 0$ represents the present. Find the total amount of platinum that will be mined over the next two years; over the following three years.

57. Suppose that a natural gas field is increasing production at the rate of
$$y = \dfrac{2t}{0.2t^2+1}$$
million cubic feet per year t, where $t = 0$ represents the present. Find the total amount of gas that will be produced over the next five years; over the period $t = 10$ to $t = 15$.

Life Science

58. A patient receives a solution intravenously at a rate of
$$f(t) = 10.8e^{0.06t}$$
cubic centimeters per hour, t in hours. Find the amount of solution the patient receives during the first 24 hours of treatment.

59. An animal population is increasing at a rate of
$$y = \frac{200}{t+1} + 20$$
individuals per year t. Find the total population gain from time $t = 0$ to time $t = 10$.

60. The size of a bacteria population is changing at the rate of $y = 50te^{-t^2/32}$ thousand individuals per hour t, where $t = 0$ represents the present. Find the total change in the population over the next 2 hours.

61. Cases of a new disease are increasing at a rate of
$$y = 1200t/(0.5t^2 + 800)$$
cases per week, where t is in weeks since the beginning of the year. Find the total number of new cases during the tenth week of the year ($t = 9$ to $t = 10$); during the fiftieth week of the year.

Social Science

62. A politician's campaign manager estimates that voters who support the politician can be registered at a rate of y individuals per day t, where
$$y = 2e^{0.25t},$$
with $t = 0$ being the present. How many voters can be registered over the next 30 days?

63. Research shows that people who learn a foreign language but never use it tend to forget their vocabulary at the rate of $100y\%$ per year, where
$$y = \frac{0.20}{t+1}.$$
Here t is time in years since the language was last used. How much of a language will be forgotten if it is not used for 10 years?

14.4 THE DEFINITE INTEGRAL AND AREA

Let f be a function continuous and nonnegative over the interval $[a, b]$, as in Figure 14–2, and consider the shaded region bounded by the graph of f, the x-axis, and the lines $x = a$ and $x = b$. This region has an area, which we call the **area under f from a to b**. This area and the definite integral are related as follows: the area under f from a to b is the definite integral of f from a to b:

Integration

The Definite Integral as Area

If f is continuous and nonnegative over the interval $[a, b]$, then the area under f from a to b is given by the definite integral

$$\int_a^b f(x)\,dx.$$

This remarkable fact is the subject of this section. We demonstrate why it is true and show how to use it to find the area of planar regions.

The Area Under a Curve

Let us see why the area under f from a to b is the definite integral of f from a to b. If f is continuous and nonnegative over the interval $[a, b]$, as in Figure 14–2, then associated with f is an **area function** A, which is defined as follows: for each number x in the interval $[a, b]$,

$$A(x) = \text{the area under } f \text{ from } a \text{ to } x.$$

See Figure 14–3. Note that

$$A(b) = \text{the area under } f \text{ from } a \text{ to } b$$

and that

$$A(a) = \text{the area under } f \text{ from } a \text{ to } a = 0.$$

Now we will show that $A(x)$ is an antiderivative for $f(x)$. To do this, we must show that $A'(x) = f(x)$ for every number x in $[a, b]$. But

$$A'(x) = \lim_{h \to 0} \frac{A(x+h) - A(x)}{h},$$

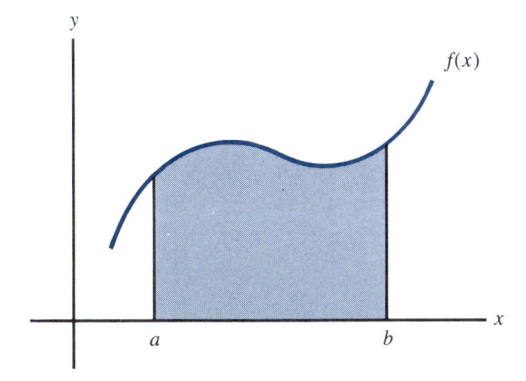

FIGURE 14–2
The Area under f from $x = a$ to $x = b$

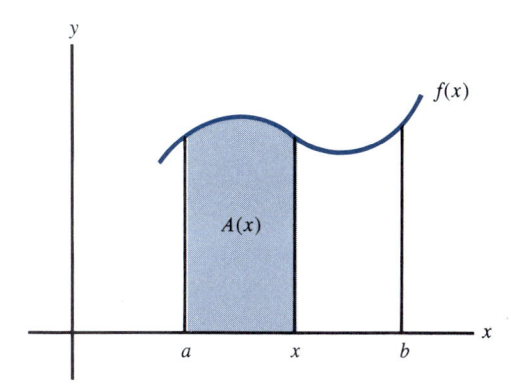

FIGURE 14–3

if the limit exists. However, if h is small enough so that $x + h$ is in the interval, then $A(x + h) - A(x)$ is just the area under f from x to $x + h$. See Figure 14–4. But as Figure 14–5 demonstrates, the area under f from x to $x + h$ can be approximated by the area of a rectangle whose height is $f(x + h)$ and whose base has length $x + h - x = h$; thus
$$A(x + h) - A(x) \approx hf(x + h).$$
Furthermore, it is clear that as $h \to 0$, the approximation becomes better and better. Therefore if we rewrite the approximation as
$$\frac{A(x + h) - A(x)}{h} \approx f(x + h)$$

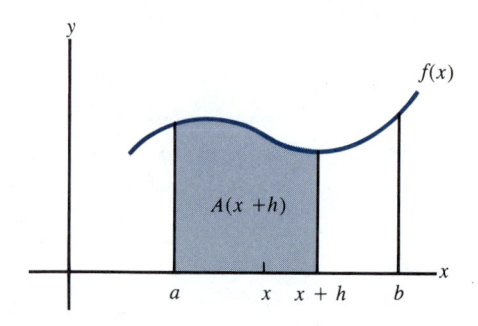

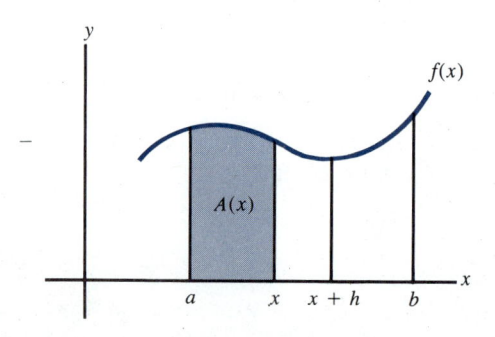

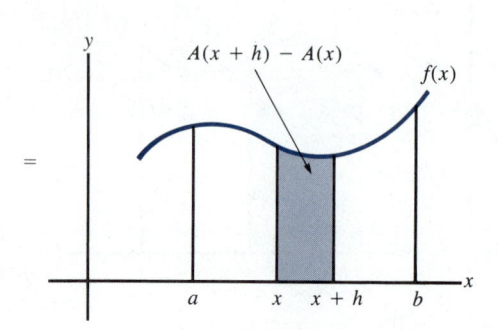

FIGURE 14–4

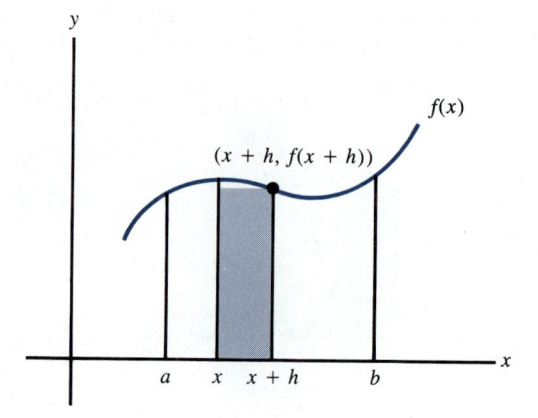

FIGURE 14–5

INTEGRATION

and take the limit as $h \to 0$, we obtain

$$\lim_{h \to 0} \frac{A(x+h) - A(x)}{h} = \lim_{h \to 0} f(x+h).$$

But since f is continuous,

$$\lim_{h \to 0} f(x+h) = f(x)$$

by the second limit definition of continuity. Therefore

$$A'(x) = \lim_{h \to 0} \frac{A(x+h) - A(x)}{h} = \lim_{h \to 0} f(x+h) = f(x),$$

which shows that $A(x)$ is an antiderivative for $f(x)$.

Now suppose that $F(x)$ is any antiderivative for $f(x)$ on $[a, b]$. Then, as we saw in Section 14-1, there is some number c such that

$$F(x) = A(x) + c$$

for all x in $[a, b]$. But then

$$F(a) = A(a) + c = 0 + c = c,$$

so

$$F(x) = A(x) + F(a)$$

for all x in $[a, b]$. This in turn implies that when $x = b$,

$$A(b) = F(b) - F(a).$$

But $A(b)$ is the area under f from a to b, and

$$\int_a^b f(x)\, dx = F(b) - F(a).$$

Therefore

$$\int_a^b f(x)\, dx = A(b),$$

so the area under f from a to b is the definite integral of f from a to b.

Finding Areas

The relationship between the definite integral and area makes it easy to find the area of many regions in the xy-plane. We illustrate with examples.

EXAMPLE 1 Let $f(x) = x$. We will find the area under f from 0 to 1. As Figure 14–6 shows, this is the area of a triangular region whose base has length 1 and whose height is also 1. Since the area of a triangle is one-half the length of its base times its height, the area in question must be $\frac{1}{2}$. We can obtain the same result by using the definite integral, since the area in question is equal to

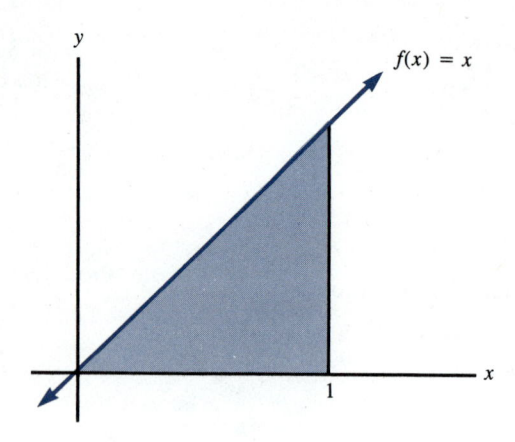

FIGURE 14–6

FIGURE 14–7

$$\int_0^1 f(x)\,dx = \int_0^1 x\,dx = \frac{1}{2}x^2\bigg|_0^1 = \frac{1}{2}(1^2 - 0^2) = \frac{1}{2}. \quad \square$$

EXAMPLE 2 Let $f(x) = x^2$. We find the area of the region shaded in Figure 14–7, which is the area under f from -1 to 2, and thus is equal to

$$\int_{-1}^{2} x^2\,dx = \frac{1}{3}x^3\bigg|_{-1}^{2} = \frac{1}{3}(2^3 - (-1)^3) = 3. \quad \square$$

If the graph of a function f lies below the x-axis from $x = a$ to $x = b$, as in Figure 14–8, then the definite integral $\int_a^b f(x)\,dx$ will be a negative number; in such a case the area of the region that lies between the graph and the x-axis from a to b (the shaded region in Figure 14–8) is equal to the absolute value of the definite integral.

EXAMPLE 3 Suppose $f(x) = x^2 - 3x$ and we want to find the area of the region bounded by the graph of f and the x-axis from $x = 0$ to $x = 3$. This is the shaded region in Figure 14–9. Since the region lies below the x-axis, the definite integral is negative:

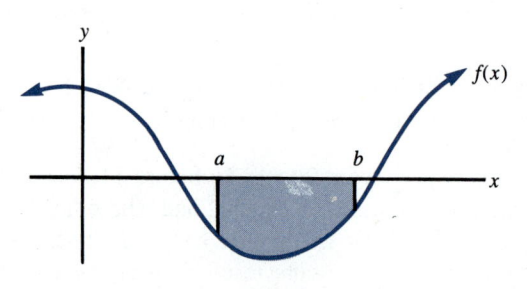

FIGURE 14–8

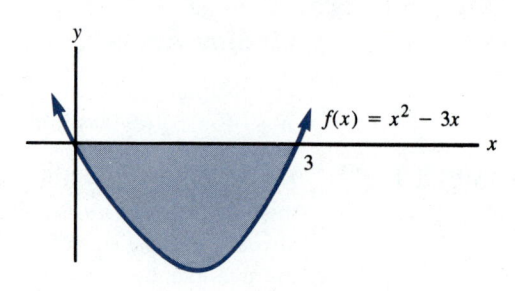

FIGURE 14–9

INTEGRATION

$$\int_0^3 f(x)\,dx = \int_0^3 (x^2 - 3x)\,dx$$

$$= \frac{1}{3}x^3 - \frac{3}{2}x^2 \Big|_0^3$$

$$= \frac{1}{3}3^3 - \frac{3}{2}3^2$$

$$= -\frac{9}{2}.$$

Therefore the region has area equal to $|-9/2| = 9/2$. ☐

If we need to find the area of a region part of which lies above the x-axis and part below, we can find the areas of the separate parts of the region and add them.

EXAMPLE 4 Let $f(x) = x^3 - 3x^2 + 2x$. Suppose we need to find the area of the shaded region in Figure 14–10. The area of the part of the region that is above the x-axis is

$$\int_0^1 (x^3 - 3x^2 + 2x)\,dx = \left(\frac{1}{4}x^4 - x^3 + x^2\right)\Big|_0^1 = \frac{1}{4}.$$

The area of the part of the region that is below the x-axis is the absolute value of

$$\int_1^2 (x^3 - 3x^2 + 2x)\,dx = \left(\frac{1}{4}x^4 - x^3 + x^2\right)\Big|_1^2 = -\frac{1}{4}.$$

Thus the area of this region is $|-1/4| = 1/4$. The area of the entire region is therefore $1/4 + 1/4 = 1/2$. ☐

The preceding examples show that when using the definite integral to find the area of a planar region, it is essential before doing the integration to determine

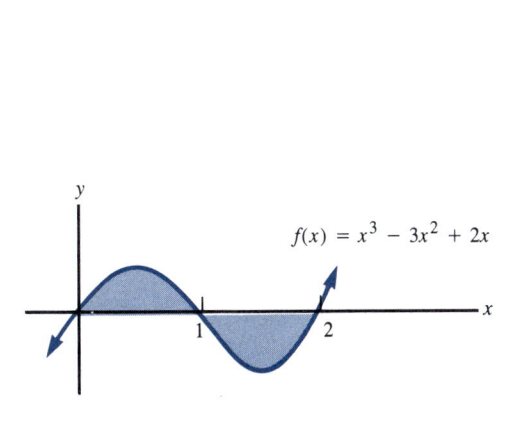

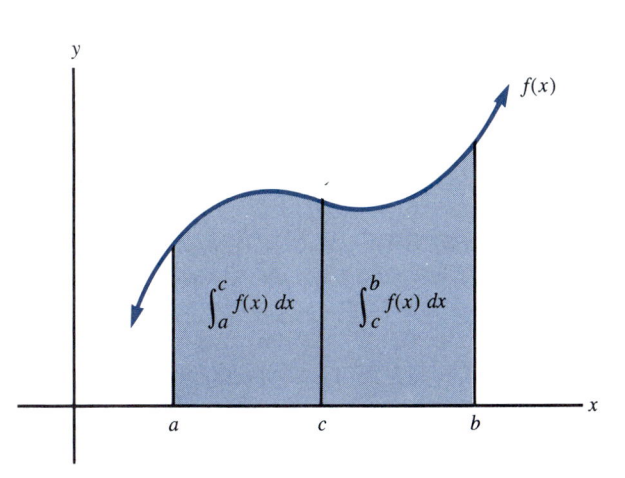

FIGURE 14–10

FIGURE 14–11

whether the region lies above the *x*-axis, below it, or partly above it and partly below it. Thus careful curve sketching is required *before* any integration is performed.

Considering the definite integral as an area leads to many interesting consequences. For instance, it is true that for any number *a*,

$$\int_a^a f(x)\,dx = 0.$$

This is so because, as noted previously, the area under *f* from *a* to *a* must be zero. Similarly, if *c* is any number such that $a < c < b$, then

$$\int_a^b f(x)\,dx = \int_a^c f(x)\,dx + \int_c^b f(x)\,dx,$$

because, as Figure 14–11 shows, the area under *f* from *a* to *b* is equal to the area under *f* from *a* to *c* plus the area under *f* from *c* to *b*.

Another consequence of regarding the definite integral as an area is that total change can be represented by area. For instance, we have seen (Example 5, Section 14–3) that since marginal cost is the rate of change of cost, the definite integral of marginal cost from *a* to *b* is the total change in cost over the interval [*a*, *b*]; but this integral is also the area under the marginal cost curve from *a* to *b*. Hence, total change in cost is represented by the area of a region under the marginal cost curve. The same reasoning applied to any rate-of-change function leads to the following conclusion:

Total Change as Area

If *f* describes the rate of change of some quantity Q over the interval [*a*, *b*], then the total change in Q over [*a*, *b*] is represented by the area under *f* from *a* to *b*.

See Figure 14–12. This is an important fact to remember, because in business, economics, and the sciences, arguments concerning total change are often presented in terms of areas under some rate-of-change function.

The Area Bounded by Two Curves

Figure 14–13 shows a region bounded by the graphs of the functions *f* and *g* and the vertical lines $x = a$ and $x = b$. The area of such a region is referred to as the **area bounded by the curves *f* and *g* from $x = a$ to $x = b$**. As Figure 14–14 indicates, we can find the area of this region by subtracting the area under the lower curve from the area under the upper one. Since the area under the lower curve in Figure 14–14 is $\int_a^b g(x)\,dx$ and that under the upper curve is $\int_a^b f(x)\,dx$, the area bounded by the curves is

$$\int_a^b f(x)\,dx - \int_a^b g(x)\,dx = \int_a^b [f(x) - g(x)]\,dx.$$

INTEGRATION

FIGURE 14–12
f Describes Rate of Change of *Q*

FIGURE 14–13
The Area Bounded by *f* and *g* from $x = a$ to $x = b$

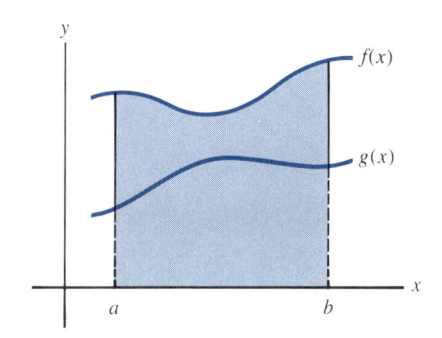

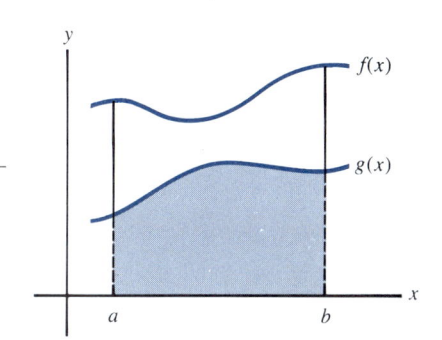

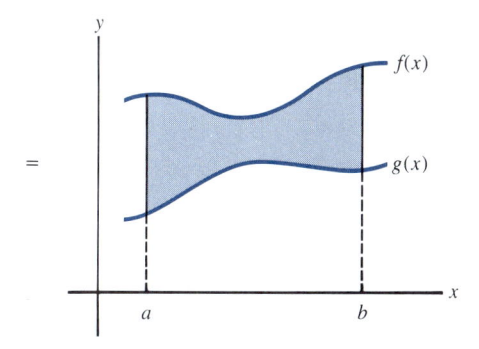

FIGURE 14–14
$$\int_a^b f(x)\, dx - \int_a^b g(x)\, dx = \int_a^b [f(x) - g(x)]\, dx$$

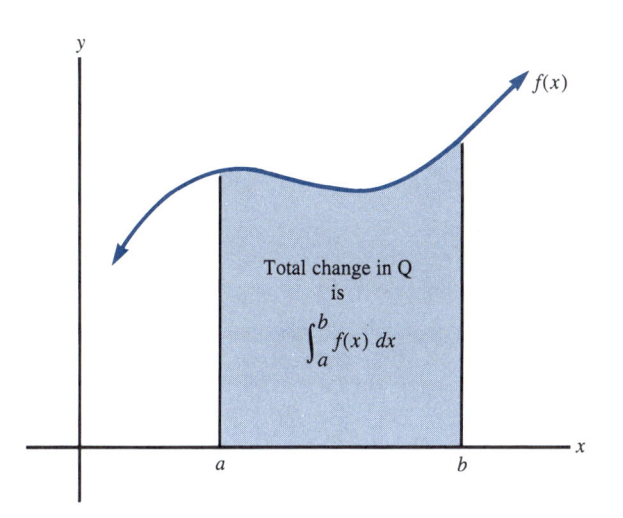

Finding the Area of a Region Bounded by Two Curves

To find the area of a region bounded by two curves from $x = a$ to $x = b$: Determine which of the curves is the upper and which the lower over the interval $[a, b]$. If f is the upper and g the lower, the area bounded by the curves is

$$\int_a^b [f(x) - g(x)] \, dx.$$

Figure 14–14 pictured the curves as nonnegative, but this is not necessary: the procedure for finding the area bounded by two curves works even if part or all of the region between them lies below the x-axis. However, it *is* necessary that in the integral $\int_a^b [f(x) - g(x)] \, dx$, f must be the upper curve and g the lower one.

EXAMPLE 5 Let us find the area of the shaded region in Figure 14–15. From the figure, we see that the upper curve over the interval $[-1, 1]$ is $y = x^2 + 1$, while the lower one is $y = -x^2$. Hence the area of the region bounded by the curves from $x = -1$ to $x = 1$ is

$$\int_{-1}^1 [(x^2 + 1) - (-x^2)] \, dx = \int_{-1}^1 [2x^2 + 1] \, dx$$

$$= \frac{2}{3}x^3 + x \Big|_{-1}^1$$

$$= \frac{10}{3}. \quad \square$$

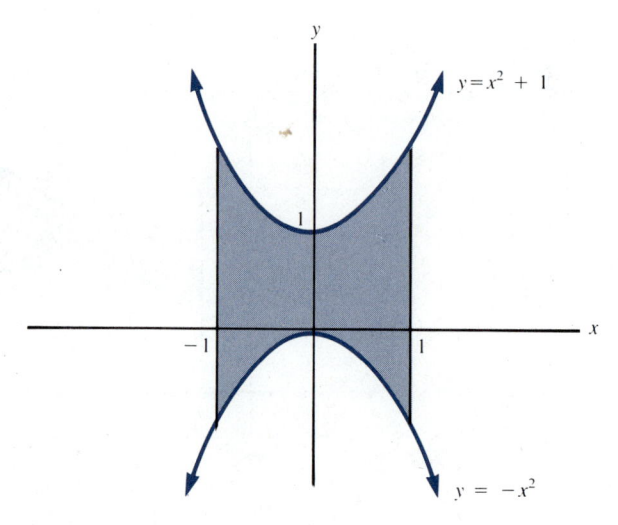

FIGURE 14–15

INTEGRATION

Suppose we wish to find the area of a region that is completely enclosed by the graphs of two functions f and g, such as the area in Figure 14–16, for instance. The area of such a region is referred to as the **area bounded by f and g,** and it is equal to

$$\int_a^b [f(x) - g(x)]\, dx.$$

The limits of integration a and b may be found by setting $f(x) = g(x)$ and solving for x.

EXAMPLE 6

Let us find the area of the region bounded by p and q, where

$$p(x) = 8x - x^2 \quad \text{and} \quad q(x) = x^2 - 4x + 10.$$

Since we must know which curve is the upper one and which the lower one, we sketch the graphs of these functions. See Figure 14–17. To find the points where the curves intersect, we set $p(x) = q(x)$ and solve for x:

$$8x - x^2 = x^2 - 4x + 10,$$

so

$$2x^2 - 12x + 10 = 0,$$

or

$$2(x - 1)(x - 5) = 0.$$

Therefore the curves intersect at $(1, p(1)) = (1, 7)$ and $(5, p(5)) = (5, 15)$. Since p is the upper curve and q the lower one over the interval $[1, 5]$, it follows that the area of the region in question is equal to

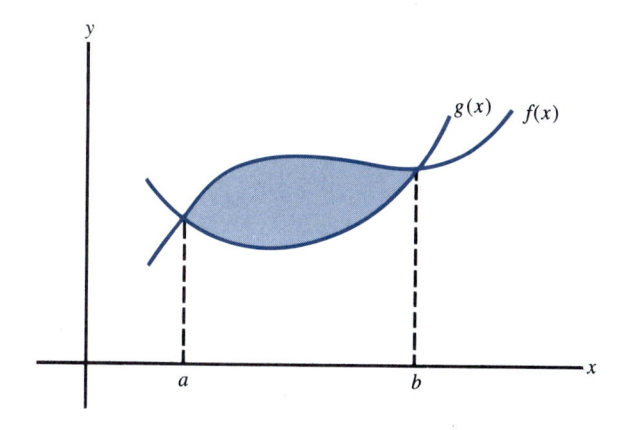

FIGURE 14–16
The Area Bounded by f and g

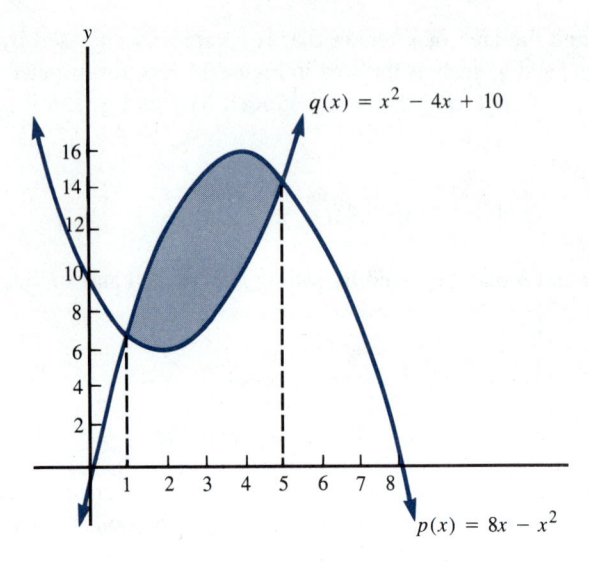

FIGURE 14–17

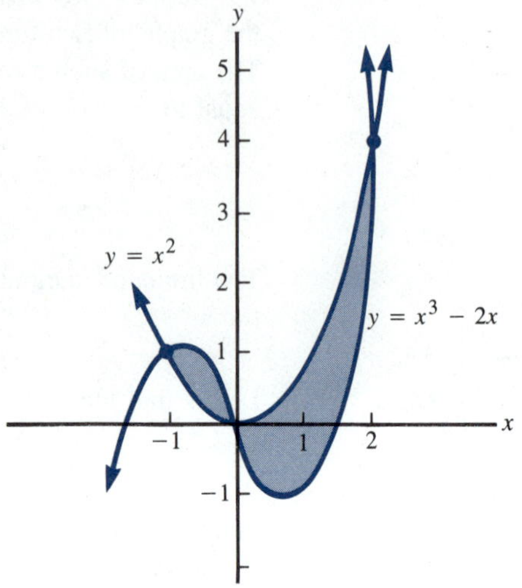

FIGURE 14–18

$$\int_1^5 [p(x) - q(x)]\, dx = \int_1^5 [(8x - x^2) - (x^2 - 4x + 10)]\, dx$$

$$= \int_1^5 (-2x^2 + 12x - 10)\, dx$$

$$= -\frac{2}{3}x^3 + 6x^2 - 10x \Big|_1^5$$

$$= \frac{64}{3}. \quad \square$$

Our procedure for finding the area bounded by two curves requires that we subtract the lower curve from the upper one. But which curve is the lower and which is the upper may change from interval to interval. Again, careful curve sketching is essential.

EXAMPLE 7 Find the area between the curves

$$y = x^2 \quad \text{and} \quad y = x^3 - 2x.$$

Setting the equations equal to each other and solving for x yields

$$x^2 = x^3 - 2x,$$

or

$$x^3 - x^2 - 2x = 0,$$

or

$$x(x + 1)(x - 2) = 0,$$

so the points of intersection are $(0, 0)$, $(-1, 1)$, and $(2, 4)$. See Figure 14–18. Note that over the interval $[-1, 0]$ the upper curve is $y = x^3 - 2x$, but over the interval $[0, 2]$ the upper curve is $y = x^2$. Therefore the area of the region bounded by the two curves is

$$\int_{-1}^{0} [(x^3 - 2x) - x^2] \, dx + \int_{0}^{2} [x^2 - (x^3 - 2x)] \, dx$$

$$= \left(\frac{1}{4}x^4 - \frac{1}{3}x^3 - x^2\right)\Big|_{-1}^{0} + \left(-\frac{1}{4}x^4 + \frac{1}{3}x^3 + x^2\right)\Big|_{0}^{2}$$

$$= 0 - \left(\frac{1}{4} + \frac{1}{3} - 1\right) + \left(-\frac{16}{4} + \frac{8}{3} + 4\right) - 0$$

$$= \frac{37}{12}. \quad \square$$

■ **Computer Manual: Program DEFNTGRL**

14.4 EXERCISES

The Area Under a Curve

In Exercises 1 through 16, find the area under f from a to b.

1. $f(x) = x^2 + 3x + 5$, $a = 1$, $b = 3$
2. $f(x) = x^2 + 3x + 5$, $a = -1$, $b = 3$
3. $f(x) = x^2 - 2x + 3$, $a = 0$, $b = 2$
4. $f(x) = x^2 - 2x + 3$, $a = 2$, $b = 4$
5. $f(x) = x^2 - 2x + 3$, $a = 0$, $b = 4$
6. $f(x) = -x^3 + 2x^2 + 15x$, $a = -3$, $b = 0$
7. $f(x) = -x^3 + 2x^2 + 15x$, $a = -3$, $b = 5$
8. $f(x) = x^3 - 3x^2 - 6x + 8$, $a = -2$, $b = 1$
9. $f(x) = x^3 - 3x^2 - 6x + 8$, $a = 1$, $b = 4$
10. $f(x) = x^3 - 3x^2 - 6x + 8$, $a = -2$, $b = 4$
11. $f(x) = \frac{1}{x}$, $a = \frac{1}{2}$, $b = e$
12. $f(x) = e^{2x}$, $a = 0$, $b = 4$
13. $f(x) = e^{-3x}$, $a = 0$, $b = 1$
14. $f(x) = \frac{2}{x + 1}$, $a = 0$, $b = 3$
15. $f(x) = (\ln x)/x$, $a = e^2$, $b = e^3$
16. $f(x) = xe^{-x^2}$, $a = 0$, $b = 3$

Management

17. The typical worker in a jewelry plant can produce $f(t)$ necklaces per hour, where
$$f(t) = -3t^2 + 36t,$$
t in hours.
(a) Graph this function.
(b) Interpret the number of necklaces the typical worker can produce in an 8-hour day as an area, and find this number.

18. A firm's marginal cost and marginal revenue functions are defined by
$$c(x) = x^2 - 80x + 3600 \quad \text{and} \quad r(x) = -x^2 + 100x,$$
respectively.
   (a) Graph these functions.
   (b) Suppose the firm is currently making and selling 20 units per day. Interpret as areas the additional cost and revenue that would result from doubling production and sales, and then find the additional cost and revenue.

19. The savings produced by new equipment accrue at a rate of
$$y = 10{,}000 - 400t$$
dollars per year, where t is the number of years since the equipment was purchased. Find the area of the region bounded by the graph of this function and the coordinate axes, and interpret your result.

20. The savings produced by a new computer accrue at a rate of
$$y = 20e^{-0.05t}$$
a thousand dollars per month, where t is in months since the purchase of the computer. Find the area of the region bounded by the graph of this function, the y-axis, and the line $t = 10$, and interpret your result.

21. With t months of experience on the job, a worker can produce at the rate of y units per month, where
$$y = 1000 - 600e^{-0.25t}.$$
Find the area under the graph of this function from $t = 0$ to $t = 12$ and interpret your result.

22. Suppose gold is being produced at a rate of
$$y = \frac{25}{t + 0.5} + 3$$
metric tons per year t, where $t = 0$ represents the present. Find the area under this function from $t = 1$ to $t = 4$ and interpret your result.

23. The savings produced by new equipment accrue at a rate of y thousand dollars per year t, where t is years since the equipment was purchased and
$$y = \begin{cases} 50(t + 4)^{-1}, & 0 \le t \le 6, \\ 12.5 - 1.25t, & t > 6. \end{cases}$$
Find the total savings due to the new equipment.

Life Science

24. A virus grows within the human body at a rate of y organisms per hour, where
$$y = 2000e^{0.02t}$$
and t is measured in hours since infection occurred.
   (a) Graph this function.
   (b) Interpret the number of viral organisms that are produced during the first 3 hours after infection as an area and find this number.
   (c) If death results when the total number of viral organisms produced since infection reaches 1 million, how long will it be from infection until death if no treatment is given?

25. The wolf population in a certain region of Alaska is increasing at a rate of
$$f(t) = 12\sqrt{0.5t + 4}$$

individuals per year t, where $t = 0$ represents the present. Find the area under f from $t = 0$ to $t = 10$ and interpret your result.

The Area Bounded by Two Curves

In Exercises 26 through 29, find the area of the region bounded by f and g from $x = a$ to $x = b$.

26. $f(x) = x^2 + 2$, $g(x) = x^2$, $a = -2$, $b = 2$
27. $f(x) = 2$, $g(x) = -x^2$, $a = 0$, $b = 1$
28. $f(x) = x^3$, $g(x) = 4 - x^2$, $a = -1$, $b = 1$
29. $f(x) = e^x$, $g(x) = x$, $a = 1$, $b = 2$

In Exercises 30 through 37, find the area bounded by the given functions.

30. $f(x) = x$, $g(x) = x^2$
31. $f(x) = x$, $g(x) = x^3$
32. $f(x) = \sqrt{x}$, $g(x) = x^2$
33. $f(x) = -x + 4$, $g(x) = -x^2 + 4x$
34. $y = x^2 - 2x$, $y = x$
35. $f(x) = x^2 - 9x + 26$, $g(x) = -x^2 + 9x - 14$
36. $y = -x^2 + 4x + 96$, $y = x^2 + 6x + 12$
37. $y = x^4 - 2x^2$, $y = 2x^2$

Exercises 38 through 47, find the area of the region bounded by the graphs of the given equations.

38. $y = e^{-x/2}$, $y = 0$, $x = 0$, $x = 4$
39. $y = 2/(x + 3)$, $y = 0$, $x = 0$, $x = 2$
40. $y = 4 + 2x$, $y = 7 - x$, $y = -3 + 4x$
41. $y = 2x + 5$, $y = 8 + x$, $y = 29 - 2x$
42. $y = x + 1$, $y = 1 - x$, $y = x - 1$, $y = 3 - x$
43. $y = x + 1$, $y = x - 5$, $y = 4 - 2x$, $y = 9 - x$
44. $y = x^2 - 4x + 14$, $y = -x^2 + 6x - 9$, $x = 1$, $x = 4$
45. $y = e^x$, $y = e^{-x}$, $x = -1$, $x = 1$, $y = 0$
46. $y = 12 - x^2$, $y = x$ for $x \geq 0$, $y = -4x$ for $x \leq 0$
47. $y = 12 - x^2$, $y = x$ for $x \leq 0$, $y = -4x$ for $x \geq 0$

Management

48. A new car brings in revenue for a car rental agency at a rate of
$$f(t) = 12{,}500$$
dollars per year, and its cost to the rental company increases at a rate of
$$g(t) = 5000 + 2000t$$
dollars per year. Find the area of the region bounded by the graphs of f and g and the y-axis, and interpret your result.

49. A firm's savings due to new equipment are given by
$$y = 6 + 2t$$
and the cost of maintaining the new equipment by
$$y = 2.4t,$$
where y is in thousands of dollars and t is in years since the equipment was purchased. Find the area of the region bounded by these curves and the y-axis, and interpret your result.

50. The savings due to a new telephone system accrue at a rate of
$$y = 10 + 1.2t$$
thousand dollars per month, where t is the number of months since the system was installed. Maintenance costs for the system t months after its installation are m thousand dollars, where
$$m = 0.8 + 0.4t.$$
Find the area of the region bounded by these curves, the y-axis, and the line $t = 5$, and interpret your result.

51. A firm's new drill press is saving s thousand dollars per month in raw materials cost, where $s = 150 - t^2$. Also, the cost of the labor needed to run the new press is increasing at the rate of $c = 0.5t^2$ thousand dollars per month. Here t is months since the press was installed. Graph these functions, interpret the firm's net savings as an area, and find the net savings.

52. Since time $t = 0$, a firm's rate of cash inflow has been $C = -t^2 + 12t + 160$ thousand dollars per month t.
    (a) Find the firm's total cash inflow.
    (b) If the firm's cash outflow since time $t = 0$ has been $E = t^2 + 70$ thousand dollars per month, find the firm's net cash flow.

Social Science

53. Emigrants are leaving a nation at the rate of $f(t)$ thousand persons per year, where $f(t) = 20e^{0.5t}$, and immigrants are entering the nation at the rate of $g(t)$ thousand persons per year, where $g(t) = 40e^{0.4t}$. Here $t = 0$ is 1985. Graph these functions, and then find the net population change due to the combination of emigration and immigration over the period
    (a) 1985–1990   (b) 1985–1995.

Lorenz Curves

Suppose a function L describes the distribution of income in a society in the following manner:

for $0 \leq x \leq 100$, $100L(x)\%$ is the percentage of the total income earned by the $x\%$ of the population that earns the least.

For instance, if $L(0.2) = 0.05$, then 5% of the total income is earned by the 20% of the population that earns the least; if $L(0.5) = 0.3$, then 30% of the total income is earned by the 50% of the population that earns the least. Such a function L is called a **Lorenz curve**. Exercises 54 through 58 are concerned with Lorenz curves.

*54. Suppose a Lorenz curve is given by $L(x) = 0.75x^2 + 0.25x$. Find $L(0.1)$, $L(0.2)$, and $L(0.9)$, and interpret your results.

*55. Show that if L is a Lorenz curve, then $L(0) = 0$ and $L(1) = 1$.

*56. The line segment $y = x$, $0 \leq x \leq 1$, is the Lorenz curve for a society in which income is equally distributed. Why?

The graph of a typical Lorenz curve is as follows:

INTEGRATION

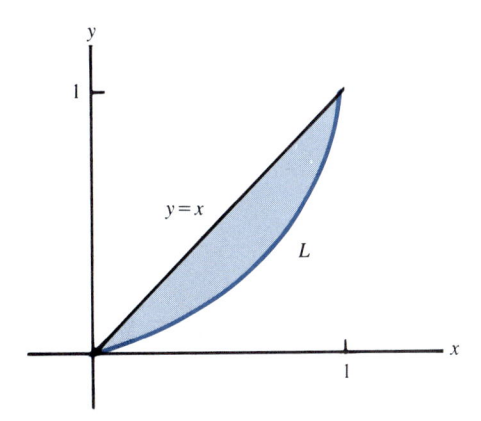

The greater the inequality of income distribution, the farther below the line $y = x$ the Lorenz curve will be, and hence the larger the shaded region will be. Economists measure inequality of income distribution by the **coefficient of inequality**

$$C = \frac{\text{Area of region bounded by } y = x \text{ and Lorenz curve } L}{\text{Area under } y = x \text{ from } x = 0 \text{ to } x = 1}.$$

Thus $0 \le C \le 1$, and the larger C is, the greater the inequality of income distribution.

*57. Show that for a given Lorenz curve L,

$$C = 2 \int_0^1 (x - L(x))\, dx.$$

*58. Graph the Lorenz curves

$$L_1(x) = \frac{1}{2}x^2 + \frac{1}{2}x$$

and

$$L_2(x) = \frac{1}{5}x^2 + \frac{4}{5}x$$

on the same set of axes and find their coefficients of inequality. Which of them describes the more unequal distribution of income?

14.5 RIEMANN SUMS

When we discussed the definite integral in Section 14.3 we remarked that we had not actually defined it but rather had introduced it by means of one of its properties. Now it is time to define the definite integral as the limit of a certain type of sum known as a **Riemann sum**. In so doing we will make use of the summation notation Σ; if you have not previously encountered the summation notation, you will find it explained in Section 8.2 of this text.

The Definition of the Definite Integral

Let n be any positive integer. Suppose we divide the interval $[a, b]$ into n subintervals of equal length. Then each of the subintervals will have length Δx, where

$$\Delta x = \frac{1}{n}(\text{length of }[a, b]) = \frac{1}{n}(b - a) = \frac{b - a}{n}$$

See Figure 14–19. Note that the endpoints of the subintervals are

$$a = a_0, a_1, a_2, \ldots, a_{n-1}, a_n = b.$$

Thus the ith subinterval is $[a_{i-1}, a_i]$, and its length is Δx.

FIGURE 14–19

Now let f be a continuous function defined on $[a, b]$. From each subinterval $[a_{i-1}, a_i]$ we choose a single number x_i, find the functional value $f(x_i)$ (see Figure 14–20), and form the sum

$$\sum_{i=1}^{n} f(x_i) \Delta x = f(x_1) \Delta x + f(x_2) \Delta x + \cdots + f(x_n) \Delta x.$$

This sum is called a **Riemann sum for f over $[a, b]$**. If f is nonnegative on $[a, b]$ and we construct a rectangle of height $f(x_i)$ on each subinterval $[a_{i-1}, a_i]$, the area of the rectangle will be $f(x_i) \Delta x$. See Figure 14–21. Thus the Riemann sum will be the sum of the areas of all the rectangles, and hence will be an approximation to the area under f from a to b. See Figure 14–22, where the Riemann sum is the sum of the areas of the rectangles.

EXAMPLE 1   Suppose we let f be defined by $f(x) = x$, let $[a, b] = [0, 1]$ and take $n = 3$. Then

$$\Delta x = \frac{1 - 0}{3} = \frac{1}{3},$$

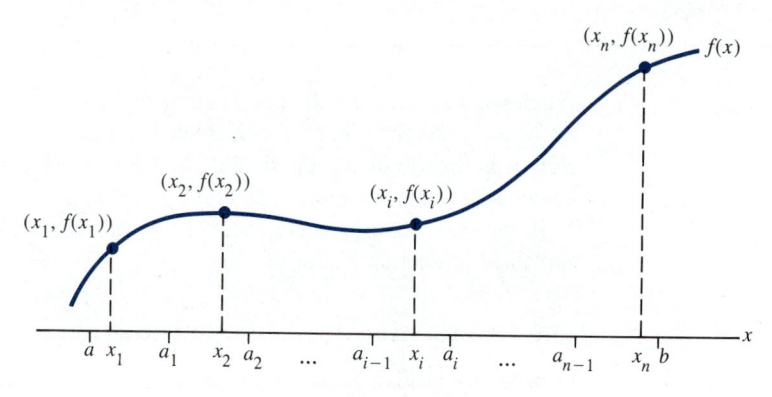

FIGURE 14–20

INTEGRATION   911

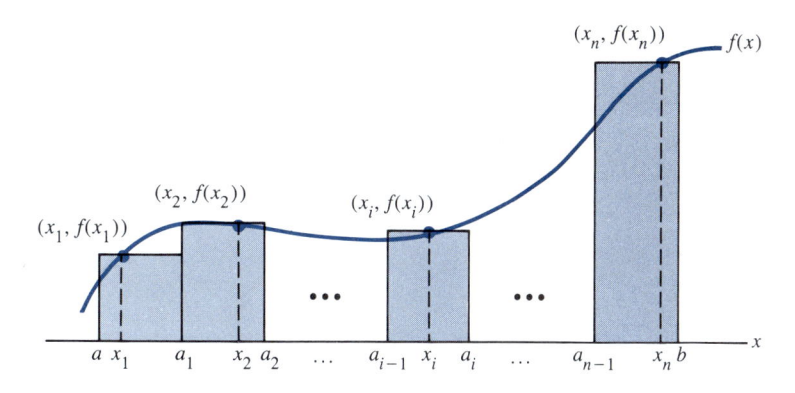

FIGURE 14–21

and

$$a_0 = 0, \quad a_1 = \frac{1}{3}, \quad a_2 = \frac{2}{3}, \quad a_3 = 1.$$

Thus a Riemann sum for f over $[0, 1]$ with $n = 3$ is a sum of the form

$$f(x_1)\frac{1}{3} + f(x_2)\frac{1}{3} + f(x_3)\frac{1}{3},$$

where x_1 may be any number in $[0, \frac{1}{3}]$, x_2 may be any number in $[\frac{1}{3}, \frac{2}{3}]$, and x_3 may be any number in $[\frac{2}{3}, 1]$. For instance, if we choose $x_1 = \frac{1}{6}$, $x_2 = \frac{2}{3}$, and $x_3 = \frac{5}{6}$, then the sum becomes

$$f\left(\frac{1}{6}\right)\frac{1}{3} + f\left(\frac{2}{3}\right)\frac{1}{3} + f\left(\frac{5}{6}\right)\frac{1}{3} = \frac{1}{6}\cdot\frac{1}{3} + \frac{2}{3}\cdot\frac{1}{3} + \frac{5}{6}\cdot\frac{1}{3} = \frac{5}{9}.$$

For this Riemann sum, we obtain the rectangles of Figure 14–23. ☐

Suppose we form a Riemann sum $\Sigma_{i=1}^{n} f(x_i) \Delta x$ for a continuous function f over an interval $[a, b]$. What will happen to this sum if we divide the interval $[a, b]$ into

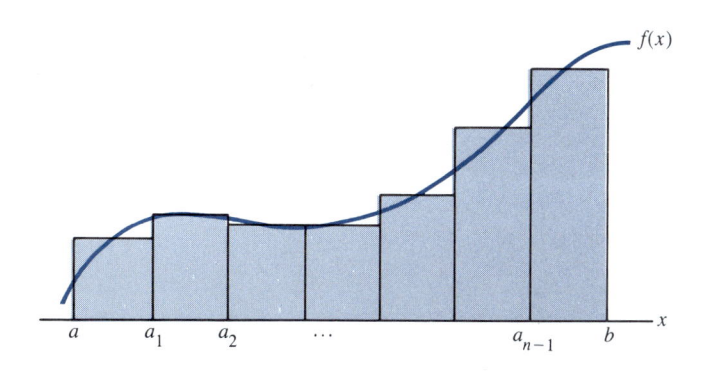

FIGURE 14–22

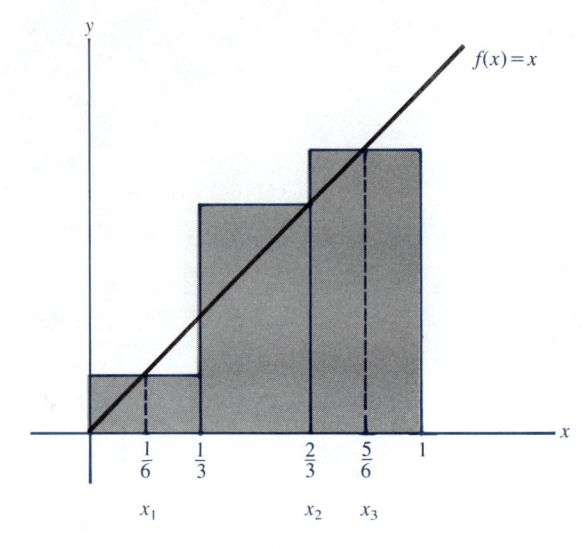

FIGURE 14–23

more and more subintervals of shorter and shorter length? That is, what will happen to the sum if we let Δx approach 0? (Note that if Δx is to approach 0, then n must approach $+\infty$.) If f is nonnegative on $[a, b]$, as $\Delta x \to 0$ the total area of the rectangles approaches the area under f from a to b (see Figure 14–24) and hence the Riemann sum, which is the total area of the rectangles, approaches a real number. It can be shown that this conclusion holds even when f is not nonnegative on $[a, b]$:

$$\lim_{\Delta x \to 0} \sum_{i=1}^{n} f(x_i)\,\Delta x$$

is a real number, regardless of the choice of $x_1, \ldots, x_n$. The definite integral of f from a to b is defined to be this limit of Riemann sums.

Definition of the Definite Integral

Let f be a function continuous on the interval $[a, b]$. Let n be a positive integer, $\Delta x = \dfrac{b-a}{n}$, and set

$$a_0 = a, \quad a_i = a + i \cdot \Delta x \quad \text{for } i = 1, \ldots, n.$$

For $i = 1, \ldots, n$, let x_i be any number in the interval $[a_{i-1}, a_i]$. Then $\int_a^b f(x)\,dx$, the **definite integral of f from a to b,** is defined by

$$\int_a^b f(x)\,dx = \lim_{\Delta x \to 0} \sum_{i=1}^{n} f(x_i)\,\Delta x.$$

INTEGRATION

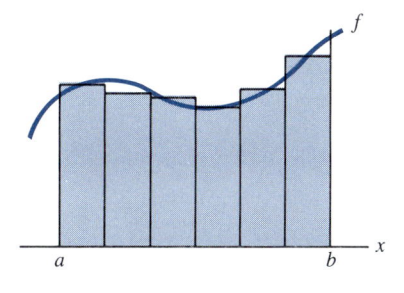

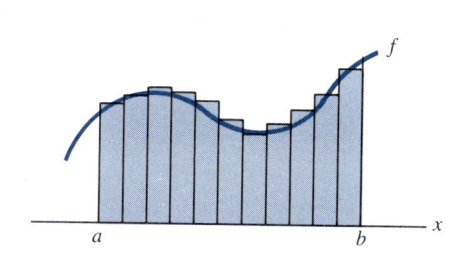

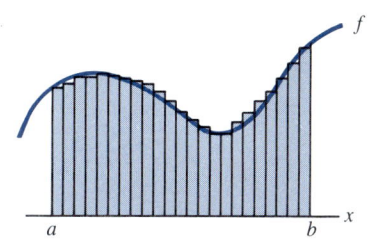

FIGURE 14–24

EXAMPLE 2 We illustrate the definition by using it to find

$$\int_0^1 f(x)\, dx$$

when $f(x) = x$. We have

$$\Delta x = \frac{1-0}{n} = \frac{1}{n}$$

for any integer n. Hence

$$a_0 = 0, \quad a_1 = \frac{1}{n}, \quad a_2 = \frac{2}{n}, \ldots, a_{n-1} = \frac{n-1}{n}, \quad a_n = 1;$$

that is, $a_i = i/n$ for $i = 1, \ldots, n$. For each i, we may choose x_i to be any number in $[a_{i-1}, a_i]$; let us choose $x_i = a_i = i/n$. Then the Riemann sum is

$$\sum_{i=1}^n f(x_i)\,\Delta x = \sum_{i=1}^n f\!\left(\frac{i}{n}\right)\frac{1}{n} = \sum_{i=1}^n \frac{i}{n}\cdot\frac{1}{n} = \frac{1}{n^2}\sum_{i=1}^n i$$

$$= \frac{1}{n^2}(1 + 2 + \cdots + n).$$

Now, there is a formula for the sum of the first n positive integers:

$$1 + 2 + \cdots + n = \frac{n(n+1)}{2}.$$

(See Exercise 33, Exercises 9.1.) Therefore

$$\sum_{i=1}^n f(x_i)\,\Delta x = \frac{1}{n^2}\frac{n(n+1)}{2} = \frac{1}{2}\frac{n+1}{n},$$

and thus

$$\int_0^1 f(x)\,dx = \int_0^1 x\,dx$$
$$= \lim_{\Delta x \to 0} \sum_{i=1}^n f(x_i)\,\Delta x$$
$$= \lim_{n \to +\infty} \frac{1}{2} \frac{n+1}{n}$$
$$= \frac{1}{2} \lim_{n \to +\infty} \frac{1 + 1/n}{1}$$
$$= \frac{1}{2} \frac{1+0}{1}$$
$$= \frac{1}{2}. \quad \square$$

Note that the definition of the definite integral as a limit of Riemann sums makes the connection between the integral and area obvious. For, if f is nonnegative on $[a, b]$, the Riemann sums for f over $[a, b]$ approximate the area under f from a to b; since this approximation becomes better and better as Δx approaches 0 (see Figure 14–24 again), it follows that the area under f from a to b is the limit of the Riemann sums as $\Delta x \to 0$, and hence by definition is equal to the definite integral of f from a to b.

From the definition of the definite integral as the limit of Riemann sums it is possible to prove that if f is continuous on $[a, b]$ and $F(x)$ is an antiderivative of $f(x)$ on $[a, b]$, then

$$\int_a^b f(x)\,dx = F(b) - F(a).$$

This fact is so important that it is known as the **Fundamental Theorem of Calculus:**

The Fundamental Theorem of Calculus

If f is continuous on $[a, b]$ and $F(x)$ is an antiderivative of $f(x)$ on $[a, b]$, then

$$\int_a^b f(x)\,dx = F(b) - F(a).$$

We introduced the definite integral in Section 14.3 by using the Fundamental Theorem of Calculus, and have until now gotten along perfectly well without knowing that the integral is in fact a limit of Riemann sums; this being the case, why do we bother to define the integral this way at all? Why not just use the Fundamental Theorem as the definition of the definite integral? There are two reasons for not doing so. The first of these is that we cannot use the Fundamental Theorem to evaluate a definite integral if we cannot find an antiderivative for its integrand. This sometimes occurs, and when it does we must fall back on the Riemann sum definition and try to evaluate the integral in the manner of Example 2. We will have more to say about this in Section 15.5. The second, and perhaps more important, reason for introducing the Riemann sum definition of the definite integral is that it allows us to use the integral as an accumulator. We illustrate the use of the integral as an accumulator by finding the average value of a function over an interval.

The Average Value of a Function

It is often useful to be able to find the average value of a function over some interval. For instance, if f is a function that describes a retailer's inventory as a function of time, we might wish to find the retailer's average inventory for a month; if g is a function that describes the temperature during a day, we might want to find the average temperature for the day.

We know that we average a collection of numbers by summing them and dividing the sum by the number of them. Now suppose f is a continuous function defined on $[a, b]$. We will show how to find the average value of f on $[a, b]$. Let M denote this average value; then M ought to be the sum of all functional values $f(x)$, for all x in $[a, b]$, divided by the number of such x's. We cannot find M directly because there are infinitely many such x's, but we can approximate M as follows: let

$$\Delta x = \frac{b - a}{n}$$

and divide $[a, b]$ into n subintervals $[a_{i-1}, a_i]$, where $a_0 = a$ and

$$a_i = a + i \cdot \Delta x$$

for $i = 1, \ldots, n$. If we choose one number x_i from each subinterval $[a_{i-1}, a_i]$ and then average the functional values of the x_i's, we will obtain an approximation for M:

$$M \approx \frac{f(x_1) + f(x_2) + \cdots + f(x_n)}{n}.$$

As we let $n \to +\infty$, the preceding approximation becomes better and better. Let us rewrite the approximation as

$$M \approx \frac{1}{b - a} \left[f(x_1) \frac{b - a}{n} + f(x_2) \frac{b - a}{n} + \cdots + f(x_n) \frac{b - a}{n} \right],$$

or

$$(b - a)M \approx \sum_{i=1}^{n} f(x_i) \Delta x.$$

The sum on the right is a Riemann sum for f over $[a, b]$, and since the approximation becomes exact as $\Delta x \to 0$ (that is, as $n \to +\infty$), we have

$$(b - a)M = \lim_{\Delta x \to 0} \sum_{i=1}^{n} f(x_i) \Delta x = \int_a^b f(x) \, dx.$$

Therefore

$$M = \frac{1}{b-a}\int_a^b f(x)\,dx.$$

The Average Value of a Function

The average value of a continuous function f over the interval $[a, b]$ is

$$M = \frac{1}{b-a}\int_a^b f(x)\,dx.$$

Note that the average value M of f over $[a, b]$ has the property that

$$(b-a)M = \int_a^b f(x)\,dx;$$

in other words, M is the value of $f(x)$ over $[a, b]$ such that a rectangle with base $[a, b]$ and height M has the same area as the region under the graph of f from $x = a$ to $x = b$. See Figure 14–25.

EXAMPLE 3 Let f be the function with defining equation $f(x) = x$. The average value of f over the interval $[0, 1]$ is

$$M = \frac{1}{1-0}\int_0^1 x\,dx = \left.\frac{x^2}{2}\right|_0^1 = \frac{1}{2}.$$

Thus the rectangle of height $\tfrac{1}{2}$ over the interval $[0, 1]$ has the same area as the region under f from $x = 0$ to $x = 1$. See Figure 14–26. □

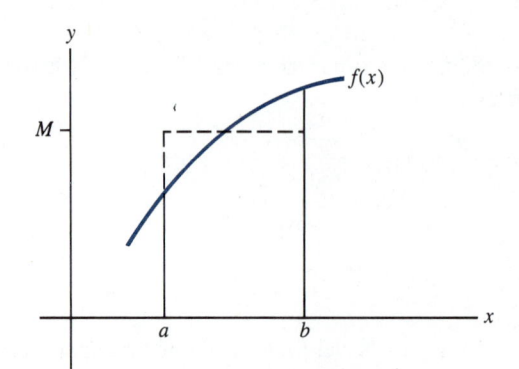

FIGURE 14–25
$(b-a)M = \int_a^b f(x)\,dx$

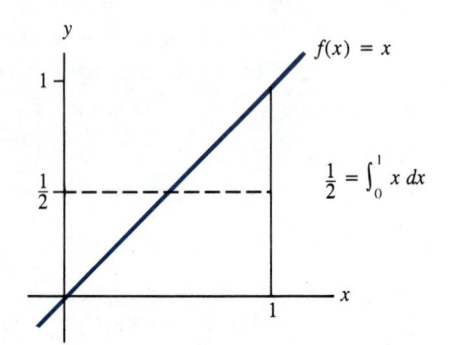

FIGURE 14–26
$\tfrac{1}{2} = \int_0^1 x\,dx$

INTEGRATION

EXAMPLE 4 Over the course of each month (30 days) a retailer's inventory declines linearly from 300 units to 60 units. What is the retailer's average monthly inventory?

If we let y be the number of units in inventory on day t, $0 \leq t \leq 30$, then y is a linear function of t such that $y = 300$ when $t = 0$ and $y = 60$ when $t = 30$. Hence

$$y = \frac{60 - 300}{30 - 0} t + 300$$

or

$$y = -8t + 300,$$

$0 \leq t \leq 30$. Therefore the average number of units in inventory each month is

$$\frac{1}{30 - 0} \int_0^{30} (-8t + 300) \, dt = \frac{1}{30} (-4t^2 + 300t) \Big|_0^{30} = \frac{1}{30} (5400) = 180. \quad \square$$

Notice the technique we used to develop the formula for the average value of a function: we approximated the quantity $(b - a)M$ by a Riemann sum and then showed that the quantity was equal to the limit of the Riemann sum as Δx approached 0, and hence that it was equal to a definite integral. This technique is what is meant by using the definite integral as an accumulator. It is a very general procedure that can be applied to many different situations; we shall consider some of them in Section 15.1

■ *Computer Manual: Program DEFNTGRL*

14.5 EXERCISES

The Definition of the Definite Integral

In Exercises 1 through 12, use the method of Example 2 of this section to find the definite integral of the given function over the given interval.

1. $f(x) = x$ over $[1, 2]$
2. $f(x) = 2x$ over $[0, 2]$
3. $f(x) = x + 1$ over $[1, 3]$
4. $f(x) = 3 - x$ over $[0, 3]$
5. $f(x) = 2x - 7$ over $[4, 8]$
6. $f(x) = |x|$ over $[-1, 1]$
7. $f(x) = x^2$ over $[0, 1]$ $\left(\text{Hint: Use the formula } 1^2 + 2^2 + \cdots + n^2 = \frac{n(n + 1)(2n + 1)}{6}.\right)$
8. $f(x) = x^2$ over $[1, 3]$
9. $f(x) = x^2 + 1$ over $[1, 2]$
10. $f(x) = 2 - x^2$ over $[0, 2]$
11. $f(x) = (x - 1)^2$ over $[0, 2]$
12. $f(x) = x^2 - x$ over $[0, 1]$

The Average Value of a Function

In Exercises 13 through 24, find the average value of the given function over the given interval.

13. $f(x) = 2x + 3$, [1, 5]
14. $f(x) = x^2$, [−2, 2]
15. $f(x) = \sqrt{x}$, [1, 4]
16. $f(x) = x^3$, [0, 1]
17. $f(x) = x^3 − 3x + 2$, [1, 2]
18. $f(x) = e^{-x}$, [0, 2]
19. $f(x) = 2e^x$, [0, 2]
20. $f(x) = x^{-2}$, [3, 10]
21. $f(x) = \dfrac{1}{x}$, $[e, e^2]$
*22. $y = mx + b$, [0, c]
*23. $y = x^n$, n an odd positive integer, [−a, a], a > 0.
*24. $y = x^n$, n an even positive integer, [−a, a], a > 0.
25. The velocity of a freely falling body t seconds after the start of its fall is $v = 32t$ feet per second. Find the average velocity of a freely falling body over the first 4 seconds of its fall.
26. If a rocket is fired straight up, its altitude t minutes after launch is
$$s = -t^2 + 40t$$
kilometers. Find the rocket's average altitude during the second 5 minutes of its flight.

Management

27. A firm's monthly sales, in dollars, in month t are given by
$$S = 120{,}000t + 52{,}000,$$
where $t = 0$ represents the beginning of the year. Find the firm's average monthly sales over the first six months of the year.
28. Dour Company's annual profits are given by the equation
$$P = 10e^{0.06t},$$
where P is in millions of dollars and t is in years, with $t = 0$ representing the present. Find Dour's average annual profit over the past five years and forecast their average annual profit over the next five years.
29. A new plant produces
$$y = 0.1t^2 + 0.5t + 400$$
units per month, where t is in months since the plant began operating. Find the plant's average monthly production during its first year of operation.
30. A retailer's inventory declines linearly from 250 units at the beginning of the month to 20 units at the end of the month. Find the retailer's average inventory for the month (30 days).
31. A retailer's inventory declines linearly from 600 units to 100 units over a 90-day period. Find the retailer's average inventory over the period.
32. A retailer's inventory declines linearly from 1000 units to 250 units over a 30-day period, then declines linearly from 250 units to 25 units over the next 60 days. Find the retailer's average inventory over the 90-day period.
33. A retailer's inventory t days after the arrival of a shipment of goods is $y = 800 − 0.5t^2$ units. Find the retailer's average inventory, assuming that a new shipment of goods arrives just as inventory reaches zero.

INTEGRATION

34. A worker can assemble compact disk players at the rate of $8\sqrt{t}$ per hour, where $0 \le t \le 8$. Find the worker's average hourly assembly rate over the 8-hour workday.

35. A manufacturer's inventory of a finished product over a 90-day period is y units, where
$$y = 100 - \frac{1}{90}t^2, \quad 0 \le t \le 90.$$
Find the manufacturer's average inventory over the period.

36. The supply function for a commodity is given by
$$q = 900\sqrt{p-1}, \quad p \ge 1.$$
Find the average quantity supplied over the price range $1 to $10.

37. The supply function for a commodity is given by
$$p = \frac{q^2}{900^2} + 1.$$
Find the average price of the commodity if the quantity supplied ranges from 0 to 300 units.

38. The demand function for a product is given by
$$q = 10{,}000e^{-0.04p}.$$
Find the average quantity demanded as the price ranges from $5 to $10 per unit.

39. Suppose that it is a publisher's policy to keep a textbook in inventory until only 100 copies of it are left and then declare it out of print. If t years after a certain book is published there are $y = 10{,}000e^{-0.8t}$ copies in the publisher's inventory and if the retail price of each copy is $39.95, find the average retail value of the publisher's inventory of the book.

40. With t weeks of experience, a worker can produce
$$y = 200 - 100e^{-0.3t}$$
units of product per week. Find the worker's average weekly productivity over his or her

(a) first month on the job;
(b) second month on the job;
(c) fifth month on the job.

Life Science

41. The amount of a drug present in a patient's liver t minutes after it has been injected into the patient's bloodstream is $y = 0.35t$ grams. Find the average amount of the drug present in the patient's liver during the first 10 minutes after injection.

42. During an experiment a colony of bacteria is treated with a bactericide and t minutes later there are $y = 100{,}000 - 22{,}000t$ bacteria present. Find the average number of bacteria present during the first 3 minutes after the application of the bactericide.

43. It is estimated that as of year t the harp seal population will be $y = 4500/(2t + 1)$ individuals. Here $t = 0$ represents 1988. Estimate the average harp seal population over the period 1988–2000.

SUMMARY

This summary consists of terms and symbols whose meaning you should know, facts you should know, and techniques and methods you should be able to employ.

Section 14.1

Antiderivative of a function f: a function F such that $F'(x) = f(x)$ for all x in the domain of f. If $F(x)$ is an antiderivative of $f(x)$, so is $F(x) + c$ for every constant c. If $F(x)$ is an antiderivative of $f(x)$, every antiderivative of $f(x)$ is of the form $F(x) + c$ for some constant c. Finding antiderivatives. The indefinite integral of $f(x)$:

$$\int f(x)\, dx = F(x) + c,$$

where $F(x)$ is any antiderivative of $f(x)$. Integral sign, integrand, constant of integration. The rules of integration. Using the rules of integration to find indefinite integrals. Determining the value of the constant of integration.

Section 14.2

Integration by guessing. Integration by substitution. Using substitution to find indefinite integrals. When a substitution will simplify an integral. Integration by substitution and the chain rule.

Section 14.3

The definite integral as the total change in any antiderivative:

$$\int_a^b f(x)\, dx = F(b) - F(a),$$

$F(x)$ any antiderivative of $f(x)$ on $[a, b]$. Limits of integration. Notation:

$$\int_a^b f(x)\, dx = F(x)\bigg|_a^b = F(b) - F(a)$$

Evaluating definite integrals. Using the definite integral to find total change. Integration by substitution for definite integrals. Evaluating definite integrals by substitution.

Section 14.4

The area under a continuous nonnegative function f from a to b. The area under a continuous nonnegative function f from a to b is the definite integral of f from $x = a$ to $x = b$. The area function A for a continuous nonnegative function f. The area function A is an antiderivative for f. Finding the area under a curve from a

INTEGRATION

to b using the definite integral. Total change as area: if f describes the rate of change of Q over $[a, b]$, the total change in Q over $[a, b]$ is represented by the area under f from a to b. The area bounded by f and g from $x = a$ to $x = b$. If f is the upper curve and g is the lower on $[a, b]$ then the area bounded by f and g from $x = a$ to $x = b$ is the definite integral of $f(x) - g(x)$ from $x = a$ to $x = b$. Using the definite integral to find the area bounded by two curves from $x = a$ to $x = b$. The area bounded by f and g. Finding the area bounded by two curves by finding their points of intersection and then integrating.

Section 14.5

Riemann sum for f over $[a, b]$. If f is continuous and nonnegative on $[a, b]$, as $\Delta x \to 0$ the Riemann sums approach the area under f from a to b. The definition of the definite integral:

$$\int_a^b f(x)\, dx = \lim_{\Delta x \to 0} \sum_{i=1}^n f(x_i)\, \Delta x$$

Finding definite integrals using Riemann sums. The Fundamental Theorem of Calculus: if f is continuous on $[a, b]$ and $F(x)$ is an antiderivative of $f(x)$ on $[a, b]$, then

$$\int_a^b f(x)\, dx = F(b) - F(a).$$

Why is it necessary to define the definite integral as a limit of Riemann sums? The average value of a function over an interval. Finding the average value of a continuous function over an interval.

REVIEW EXERCISES

In Exercises 1 through 20, find the indefinite integral.

1. $\int (-x^5 + x^3 + 9x^2 - 1)\, dx$

2. $\int (-4x^5 + 3x^2 - 8x + 14)\, dx$

3. $\int (1 + x + x^2/2 + x^3/6 + x^4/24 + x^5/120)\, dx$

4. $\int (u^{-1/2} + u^{-1/3} + u^{-1/4} + u^{-1/5})\, du$

5. $\int (-2x^{-2} + 3x^{4/5} + 4)\, dx$

6. $\int \left(-2t^{-3} + \frac{2}{t^2} + 4\sqrt{t} + 1\right) dt$

7. $\int 2e^x\, dx$

8. $\int -\frac{3}{x}\, dx$

9. $\int \left(\frac{2}{x} - 3e^x\right) dx$

10. $\int (5x^{3/5} + 6x^5 - 2x^{-1} + 3x^{-4/3} + x^{1/2} - e^{2x})\, dx$

11. $\int \sqrt{4x + 1}\, dx$

12. $\int 3x(x^2 + 4)^3\, dx$

13. $\int (x + 1)(x^2 + 2x - 8)^5\, dx$

14. $\int 5(x^2 + 2)(x^3 + 6x - 2)^{4/5}\, dx$

15. $\displaystyle\int \frac{2x}{4x^2 + 1}\, dx$

16. $\displaystyle\int \frac{6x + 2}{(3x^2 + 2x + 1)^{4/3}}\, dx$

17. $\displaystyle\int \frac{x - 2}{(x^2 - 4x + 1)^{2/3}}\, dx$

18. $\displaystyle\int 3x^2 e^{x^3}\, dx$

19. $\displaystyle\int 5xe^{-x^2}\, dx$

20. $\displaystyle\int \frac{dx}{x\sqrt{\ln x}}$

21. A firm's marginal cost and marginal revenue functions are defined by
$$c(x) = 6x + 2 \quad \text{and} \quad r(x) = 20x.$$
respectively. Find the firm's cost and revenue functions if its fixed cost is $5000.

22. During an epidemic the number of cases of influenza increases at a rate of $y = 300\sqrt{t}$ cases per day t. Find a function that describes the number of cases as a function of time if there were 2000 cases on day 4 of the epidemic.

23. An investment is made and t years later it is growing at a rate of $y = 120e^{0.12t}$ dollars per year. Find an expression that describes the worth of the investment as a function of time if the initial amount of the investment was $1000.

24. The gross national product of a country is changing at a rate of
$$y = \frac{25}{t + 1}$$
billion dollars per year t, where $t = 0$ represents 1985. Find an expression for the gross national product of the country as a function of time if the GNP in 1985 was $1000 billion.

In Exercises 25 through 36, evaluate the given definite integral.

25. $\displaystyle\int_1^3 (2 - 8x)\, dx$

26. $\displaystyle\int_{-4}^{-2} (3x^2 + 5x)\, dx$

27. $\displaystyle\int_0^2 (x^2 - 3x + 1)\, dx$

28. $\displaystyle\int_{-2}^{-1} (4x^3 - x^2 + 1)\, dx$

29. $\displaystyle\int_0^{\ln 2} e^{3x}\, dx$

30. $\displaystyle\int_0^4 \sqrt{2x + 3}\, dx$

31. $\displaystyle\int_4^9 (\sqrt{x} + x^{-2})\, dx$

32. $\displaystyle\int_{4/5}^{(e + 3)/5} \frac{10}{5x - 3}\, dx$

33. $\displaystyle\int_0^1 xe^{-x^2/2}\, dx$

34. $\displaystyle\int_0^2 \frac{x^2}{x^3 + 1}\, dx$

35. $\displaystyle\int_e^{e^2} \frac{1}{x \ln x}\, dx$

36. $\displaystyle\int_e^{e^3} \frac{1}{x(\ln x)^2}\, dx$

37. The population of a town is growing at the rate of
$$f(t) = 15\sqrt{t} + 8$$
persons per year, where t is in years with $t = 0$ representing the present. What will be the change in the town's population size over the next four years?

38. A firm's marginal cost function is given by
$$c(x) = 3x + 10 + \frac{500}{x},$$
valid for $x \geq 1$. If the firm is currently producing 10 units, find the additional cost of producing 2 more units.

39. Let $f(x) = 4x^{-3}$. Find the area under f from 1 to 4.

40. Let $f(x) = \dfrac{x}{x^2 + 1}$. Find the area under f from 0 to 2.

41. Let $f(x) = -x^2 + 13x - 36$. Find the area of the region bounded by the x-axis and the graph of f.

42. Find the area of the region bounded by the graph of
$$y = (x + 2)(x - 1)(x - 3)$$
and the x-axis.

43. Find the area bounded by the x-axis and the graph of $y = x^3 - 2x^2 - 15x$.

44. Find the area of the region bounded by p and q, where
$$p(x) = -x^2 + 6x + 2 \quad \text{and} \quad q(x) = x + 2.$$

45. Find the area of the region bounded by the graphs of $y = e^x$, $y = e^{-x}$, $x = -1$, and $x = 1$.

46. It is estimated that the number of junk bonds on the market is increasing at the rate of
$$f(t) = \frac{50}{t + 2}$$
million bonds per year. Here $t = 0$ represents 1984.
  (a) Find the area under f from 0 to b. What does this area represent?
  (b) When will there be a total of 100 million junk bonds on the market?

47. Use Riemann sums to show that $\int_0^1 (4 - 3x)\, dx = 2.5$.

48. Use Riemann sums to show that $\int_1^2 x^3\, dx = 3.75$.
$$\left(\text{Hint: } 1^3 + 2^3 + \cdots + n^3 = \frac{n^2(n + 1)^2}{4}.\right)$$

49. Find the average value of the function f defined by $f(x) = xe^{-x^2}$ over the interval $[0, 2]$.

50. A company's annual return on investment is given by
$$f(t) = 100e^{-0.05t},$$
where $f(t)$ is in thousands of dollars and $t = 0$ represents 1985. Forecast the firm's average annual return on investment over the period
  (a) 1990–1995   (b) 1999–2005

51. A merchant's inventory over a 60-day period is given by
$$y = 1000e^{-0.02t}, \quad 0 \le t \le 60.$$
Find the average inventory over the period.

52. There were y thousand riders each day on a city's transportation system, where
$$y = -0.1t^2 + 2.4t + 45.6$$
and $0 \le t \le 30$. Find the average number of riders over the 30-day period.

ADDITIONAL TOPICS

Here are some suggestions for topics not covered in the text that you might want to investigate on your own.

1. Let f, g be continuous on $[a, b]$. Show why the following properties of the definite integral are true:

$$\int_a^b f(x)\,dx \geq 0 \quad \text{if } f(x) \geq 0 \text{ on } [a, b]$$

$$\int_a^b f(x)\,dx \geq \int_a^b g(x)\,dx \quad \text{if } f(x) \geq g(x) \text{ on } [a, b]$$

$$\int_a^a f(x)\,dx = 0$$

$$\int_b^a f(x)\,dx = -\int_a^b f(x)\,dx$$

$$\int_a^b f(x)\,dx = \int_a^c f(x)\,dx + \int_c^b f(x)\,dx \quad \text{if } a \leq c \leq b$$

2. If f is continuous on $[a, b]$, let m denote the minimum value of $f(x)$ on $[a, b]$ and M the maximum value of $f(x)$ on $[a, b]$. Show that

$$m(b - a) \leq \int_a^b f(x)\,dx \leq M(b - a).$$

3. The Mean Value Theorem for integrals says that if f is continuous on $[a, b]$, there is some number c in $[a, b]$ such that

$$f(c)(b - a) = \int_a^b f(x)\,dx.$$

Find out how to prove this theorem. What is its geometric significance? What connection does it have with the average value of a function over an interval?

4. In higher mathematics functions are often defined by means of definite integrals. For instance, it is possible to define the natural logarithm function by setting

$$\ln x = \int_1^x \frac{1}{t}\,dt, \quad x > 0.$$

Using only this definition and your knowledge of the definite integral, show that

$$\ln 1 = 0, \quad \ln x_1 > \ln x_2 \text{ if } x_1 > x_2, \quad \frac{d}{dx}(\ln x) = \frac{1}{x},$$

$$\ln x \text{ is continuous on } (0, +\infty), \quad \lim_{x \to +\infty} \ln x = +\infty,$$

and

$$\lim_{x \to 0^+} \ln x = -\infty.$$

5. Riemann sums were named for the German mathematician Bernhard Riemann (1826–1866). Indeed, the definite integral we have studied in this chapter is formally known as the Riemann integral. (There are other, more general, integrals.) Besides his work with integrals, Riemann made many other important contributions to mathematics; to cite just one of these, he discovered a non-Euclidean geometry that now bears his name. Find out about Riemann's life and his contributions to mathematics.

6. The development of the definite integral was a very long process. It may properly be said to have begun in ancient Greece when the mathematicians of the time began to consider the problem of finding areas of geometric figures. Find out about the development of the definite integral. (A good place to start would be with the supplement to this chapter.)

SUPPLEMENT: THE DEVELOPMENT OF THE DEFINITE INTEGRAL

We began our study of the definite integral by introducing it as the total change in an antiderivative over an interval. We next showed how the definite integral could be used to find areas, and concluded by defining it as a limit of Riemann sums. The historical development of the definite integral followed these same steps, but in a different order: first came the problem of finding areas, then the discovery that areas could be computed as limits of what we would today call Riemann sums, and finally the realization that such limits could be evaluated by making use of the antiderivative. In this supplement we give a brief account of the historical development of the definite integral.

If integration is considered to be a type of process in which areas are obtained as the limit of sums of simpler areas (the general approach taken in Section 14.5), its beginnings go back to antiquity. The Greek mathematician Archimedes (287–212 B.C.) found the areas of many geometric figures by using a technique in which the figure under question was "filled up" with simpler figures whose areas could be easily determined. This method of integration was known as the **method of exhaustion.** We will demonstrate how the method of exhaustion may be employed to find the area P of a parabolic region such as that pictured in Figure 14–27.

We let D be the midpoint of the segment AB, and let AB serve as the base of a triangle ACB ($\triangle ACB$), where C is a point on the parabola such that CD is parallel to the axis of symmetry of the parabola. See Figure 14–28, where we have chosen C so that CD lies on the axis of symmetry. As a first approximation, we have

$$P \approx \text{area of } \triangle ACB.$$

Now we repeat the construction with E and F the midpoints of AC and CB, respectively, and G and H the points on the parabola such that GE and HF are parallel to the axis of symmetry of the parabola. See Figure 14–29. We thus obtain the triangles AGC and CHB, and

$$P \approx \text{area of } \triangle ACB + \text{area of } \triangle AGC + \text{area of } \triangle CHB.$$

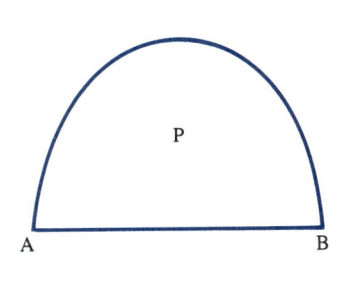

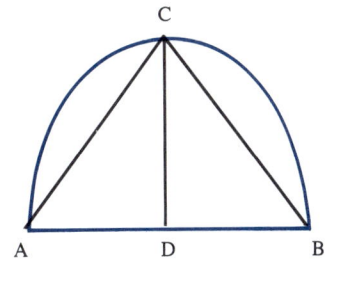

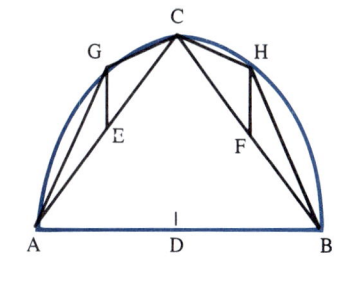

FIGURE 14–27

FIGURE 14–28

FIGURE 14–29

Clearly, this approximation is better than the previous one. It is possible to show geometrically that

$$\text{area of } \triangle AGC + \text{area of } \triangle CHB = \frac{1}{4}(\text{area of } \triangle ACB).$$

Therefore $P = \frac{5}{4}(\text{area of } \triangle ACB).$

Now we repeat the construction with segments AG, GC, CH, and HB as the bases of four new triangles, to obtain a better approximation:

$$P \approx \frac{5}{4}(\text{area of } \triangle ACB) + \frac{1}{4^2}(\text{area of } \triangle ACB)$$

$$= \frac{21}{16}(\text{area of } \triangle ACB).$$

As the process continues, the triangles "fill up" or "exhaust" the parabolic region, and hence the sum of their areas approaches the value of P. It can be shown that after n repetitions of the construction, the sum of the areas of the triangles is

$$\frac{4}{3}\left(1 - \frac{1}{4^n}\right)(\text{area of } \triangle ACB).$$

But since $\frac{1}{4^n}$ can be made arbitrarily small by taking n large enough, it follows that

$$P = \frac{4}{3}(\text{area of } \triangle ACB).$$

In modern terminology we would say that

$$P = \lim_{n \to +\infty} (\text{sum of the areas of the triangles})$$

$$= \lim_{n \to +\infty} \frac{4}{3}\left(1 - \frac{1}{4^n}\right)(\text{area of } \triangle ACB)$$

$$= \frac{4}{3}(1 - 0)(\text{area of } \triangle ACB)$$

$$= \frac{4}{3}(\text{area of } \triangle ACB).$$

We can use this result to determine the area of a parabolic segment by simply finding the length of the segments AB and CD of Figure 14–28. For instance, if AB has length 1 and CD has length $\frac{1}{4}$, then

$$\text{area of } \triangle ACB = \frac{1}{2}(1)\left(\frac{1}{4}\right) = \frac{1}{8},$$

and

$$P = \frac{4}{3} \cdot \frac{1}{8} = \frac{1}{6}.$$

INTEGRATION   927

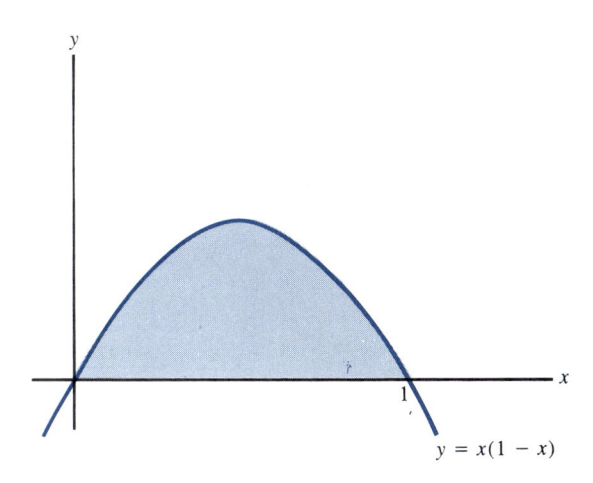

FIGURE 14-30

The method of exhaustion depends heavily on the geometric properties of the region under consideration and is difficult to apply in general situations. A better method for finding areas as the limits of sums of simpler areas became possible after the French mathematician René Descartes (1596–1650) introduced the cartesian coordinate system: This allowed mathematicians to regard curves as the graphs of equations and hence to consider regions bounded by curves as being defined by equations. This in turn made it possible to express the areas of such regions as limits of sums of areas of rectangles, much as we did in Section 14.5. Thus, for instance, suppose it is desired to find the area of the parabolic region bounded by the x-axis and the graph of $f(x) = x(1 - x)$. See Figure 14–30. We could proceed by subdividing the interval $[0, 1]$ into subintervals $[i/n, (i + 1)/n]$, $i = 0, 1, \ldots, n - 1$, and constructing on each subinterval a rectangle of height

$$f\left(\frac{i}{n}\right) = \frac{i}{n}\left(1 - \frac{i}{n}\right).$$

See Figure 14–31. Then the area P of the parabolic segment is approximated by the sum of the area of the rectangles:

$$P \approx \sum_{i=0}^{n-1} \frac{i}{n}\left(1 - \frac{i}{n}\right)\frac{1}{n}.$$

But the sum on the right-hand side simplifies to

$$\frac{1}{6}\left(1 - \frac{1}{n^2}\right).$$

Since $1/n^2$ can be made arbitrarily small by taking n large enough, it follows that $P = \frac{1}{6}$. In other words, since

$$P = \lim_{n \to +\infty} \text{(sum of the areas of the rectangles)},$$

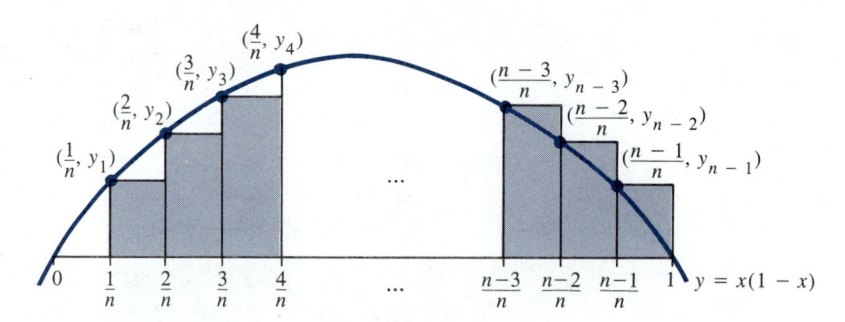

FIGURE 14–31

we have

$$P = \lim_{n \to +\infty} \frac{1}{6}\left(1 - \frac{1}{n^2}\right) = \frac{1}{6}(1 - 0) = \frac{1}{6}.$$

This technique of expressing the area of a region as a limit of sums of areas of rectangles is a very general method for finding areas, and it was known to mathematicians before the time of Newton and Leibniz. Thus if f is a function that is nonnegative over the interval $[a, b]$ and we *define* the definite integral

$$\int_a^b f(x)\,dx$$

to be the area under f from $x = a$ to $x = b$, then we can say that mathematicians possessed a technique for evaluating definite integrals (i.e., finding areas) before Newton and Leibniz. What was not appreciated was the fact that definite integrals could be evaluated by antidifferentiation. The realization that this is so is usually credited to the English mathematician Isaac Barrow (1630–1677), who was Newton's teacher; however, it was not until Newton and Leibniz put differentiation on a firm basis that this fact could be easily exploited.

Once the notion of the derivative of a function was formalized, mathematicians could easily show that definite integrals could be evaluated by finding antiderivatives for their integrands. The proof was as follows. By definition,

$$\int_a^x f(x)\,dx = \text{area under } f \text{ from } a \text{ to } x.$$

Set

$$A(x) = \int_a^x f(x)\,dx.$$

Then the Fundamental Theorem of Calculus can be proved by showing (just as we did in Section 14.4) that $A'(x) = f(x)$ and hence that if $F(x)$ is an antiderivative for $f(x)$,

$$\int_a^b f(x)\,dx = F(b) - F(a).$$

Since $A'(x) = f(x)$, the Fundamental Theorem of Calculus is often written as

$$\frac{d}{dx}\int_a^x f(x)\,dx = f(x).$$

In this form the Fundamental Theorem of Calculus deserves its name, for it makes explicit the relationship between the two great concepts of calculus—the derivative and the integral—by showing that they are inverse operations in the same sense that addition and subtraction are inverse operations: each can be used to undo the other. Thus the Fundamental Theorem is the cord that ties the two subjects of differential calculus and integral calculus together into one coherent theory.

Suggestion

If you would like to read more about the development of the integral, here are some books to consult:

Bell, E. T., *Men of Mathematics,* Simon and Schuster, New York, 1937.

Boyer, Carl B., *History of the Calculus and Its Conceptual Development,* Dover, New York, 1959.

Boyer, Carl B., *A History of Mathematics,* Wiley, New York, 1968.

Eves, Howard, *An Introduction to the History of Mathematics,* Holt, Rinehart, and Winston, New York, 1961.

Eves, Howard, *Great Moments in Mathematics before 1650,* Dolciani Mathematical Expositions, No. 5, The Mathematical Association of America, 1980.

Eves, Howard, *Great Moments in Mathematics after 1650,* Dolciani Mathematical Expositions, No. 7, The Mathematical Association of America, 1983.

15

More about Integration

In this chapter we consider some important applications and techniques of integration. We answer such questions as

- How much do consumers and producers save when the price of a product is at equilibrium?
- If income flows into a firm at a certain rate and is reinvested at another rate, how much will it be worth over a given period of time?
- How can we integrate functions defined by equations such as

$$f(x) = x^2 e^{-x} \quad \text{and} \quad g(x) = \sqrt{x^2 + 9}?$$

- How do we evaluate

$$\int_a^b f(x)\, dx$$

when we cannot find an antiderivative for $f(x)$?

In addition, we consider definite integrals whose intervals of integration are infinite, illustrate how such integrals are used in probability theory, and show how indefinite integrals are used to solve certain differential equations.

The chapter begins with a demonstration of how we can use the integral as accumulator to find producers' surplus, consumers' surplus, and the future value of an income stream. We next consider the integration technique known as integration by parts and then discuss how to use a table of integrals. This is followed by a section in which we expand the notion of the definite integral to encompass so-called improper integrals: these are integrals for which the interval of integration is infinite. We then apply our knowledge of improper integrals to the probability theory of continuous random variables. The next section is devoted to the topic of numerical integration, which considers methods for approximating the value of definite integrals. The chapter concludes with a brief discussion of differential equations.

15.1 CONSUMERS' SURPLUS, PRODUCERS' SURPLUS, AND STREAMS OF INCOME

Recall that in Section 14.5 we defined the definite integral of a continuous function f over an interval $[a, b]$ to be the limit of its Riemann sums over the interval: if

$$\Delta x = \frac{b - a}{n}, \quad a_0 = a, \quad \text{and} \quad a_i = a + i \cdot \Delta x \quad \text{for} \quad i = 1, \ldots, n,$$

then any sum of the form $\sum_{i=1}^{n} f(x_i) \Delta x$, where x_i is in $[a_{i-1}, a_i]$, is a **Riemann sum** for f over $[a, b]$, and

$$\int_a^b f(x)\, dx = \lim_{\Delta x \to 0} \sum_{i=1}^{n} f(x_i)\, \Delta x.$$

We also remarked that as a consequence of this, any quantity that is the limit of Riemann sums as Δx approaches 0 is a definite integral:

The Integral as an Accumulator

If a quantity Q can be approximated by a Riemann sum for some function f over some interval $[a, b]$, and if the approximation becomes exact as $\Delta x \to 0$, then

$$Q = \int_a^b f(x)\, dx.$$

In this section we use the integral as an accumulator to calculate producers' surplus, consumers' surplus, and the future value of a stream of income. We also briefly discuss a general procedure for setting up the integral as an accumulator.

Producers' Surplus and Consumers' Surplus

Let the equation $p = s(q)$ define a supply function for some commodity: here q is the number of units of the commodity supplied to the market, and p is the resulting unit price. Similarly, let $p = d(q)$ define the demand function for the commodity.

MORE ABOUT INTEGRATION

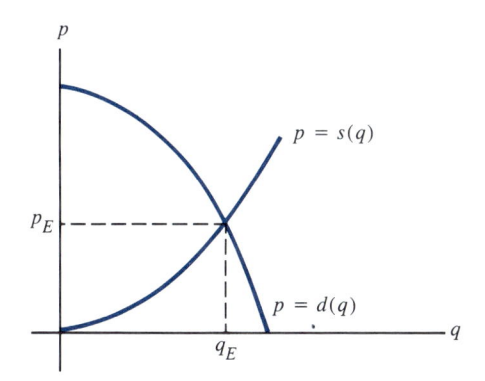

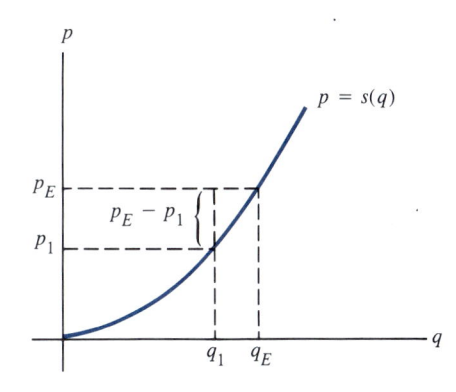

FIGURE 15-1

FIGURE 15-2

See Figure 15–1. Note that the market equilibrium quantity for the commodity is q_E, and its market equilibrium price is p_E.

For the moment, let us consider only the supply curve. Some producers would be willing to supply the commodity at prices lower than the equilibrium price p_E; these producers gain when the price is at equilibrium. For instance, Figure 15–2 indicates that some producers are willing to supply quantity q_1 of the commodity at price $p_1 < p_E$; when the price is at equilibrium, these producers gain an amount equal to $(p_E - p_1)q_1$. The total amount that producers as a group gain in this manner when the price is at equilibrium is called **producers' surplus**. We will use the integral as an accumulator to find producers' surplus.

Let us divide the interval $[0, q_E]$ on the q-axis into n subintervals, each of length Δq. For each i we select an arbitrary number q_i in the ith subinterval (see Figure 15–3); on this subinterval Δq units will be supplied at the approximate price of $s(q_i)$. Thus the total gained by producers over the ith subinterval (the area of the shaded region in Figure 15–3) is approximately $[p_E - s(q_i)] \Delta q$ (the area of the rectangle bounded by the solid lines in Figure 15–3). Hence

$$\text{Producers' surplus} \approx [p_E - s(q_1)] \Delta q + \cdots + [p_E - s(q_n)] \Delta q$$

FIGURE 15-3

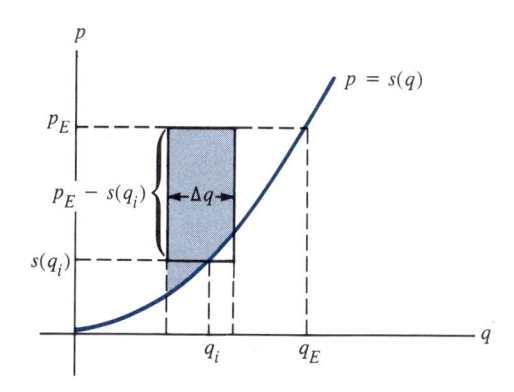

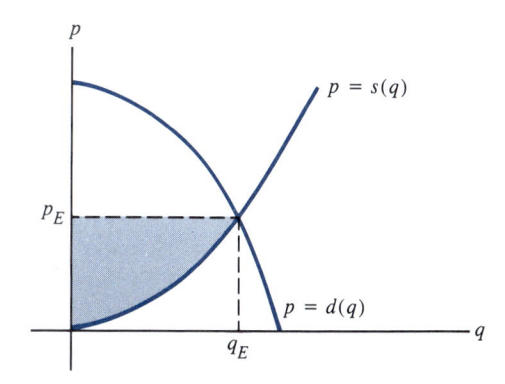

FIGURE 15-4
Producers' Surplus

on $[0, q_E]$. But the sum on the right-hand side above is a Riemann sum for $p_E - s(q)$ over $[0, q_E]$, and since the approximation becomes exact as $\Delta x \to 0$ (that is, since producers' surplus is the limit of such Riemann sums as $\Delta x \to 0$), it follows that producers' surplus is the definite integral of $p_E - s(q)$ from 0 to q_E.

Producers' Surplus

If $p = s(q)$ defines a supply function for a commodity and (q_E, p_E) is the commodity's market equilibrium point, then producers' surplus for the commodity is

$$\int_0^{q_E} [p_E - s(q)]\, dq.$$

Producers' surplus is thus the area of the region shaded in Figure 15-4.

EXAMPLE 1 Suppose the supply function for a commodity is defined by

$$s(q) = 0.009q^2$$

and its demand function by

$$d(q) = 30 - 0.003q^2.$$

Solving the equation $s(q) = d(q)$ for q yields

$$0.009q^2 = 30 - 0.003q^2,$$

or

$$q^2 = \frac{30}{0.012} = 2500.$$

Therefore $q_E = 50$ and $p_E = s(50) = 22.50$. See Figure 15-5. Thus we have

$$\text{Producers' surplus} = \int_0^{q_E} [p_E - s(q)]\, dq$$

$$= \int_0^{50} [22.50 - 0.009q^2]\, dq$$

$$= \$750. \quad \square$$

Analogous to producers' surplus is **consumer's surplus**. Figure 15-6 depicts the situation. Here the equation $p = d(q)$ defines the demand function. Consumers who would be willing to purchase the commodity at a higher price than p_E save when the price is at p_E. For example, if some consumers would have purchased q_1 units at price $p_1 > p_E$, as in the figure, then they save an amount equal to $(p_1 - p_E)q_1$ when the price is at equilibrium. **Consumers' surplus** is the total amount saved by consumers as a group when the price is at equilibrium. Just as we did for producers' surplus, let us divide the interval $[0, q_E]$ into n subintervals,

MORE ABOUT INTEGRATION

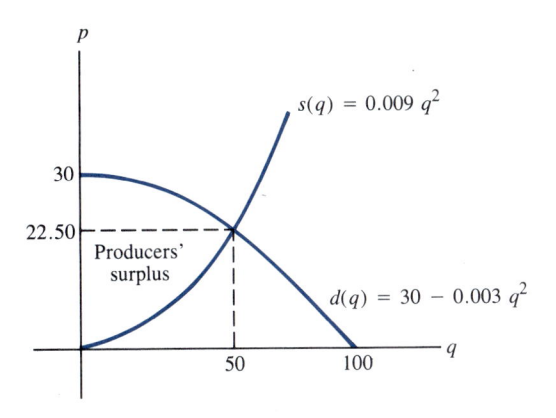

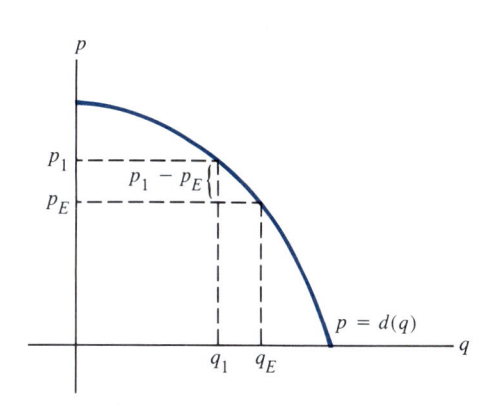

FIGURE 15-5

FIGURE 15-6

each of length Δq. On the ith subinterval we choose a number q_i and note that the total gained by consumers over the ith subinterval (the shaded region in Figure 15–7) is approximated by $[d(q_i) - p_E] \Delta q$ (the area of the rectangle bounded by the solid lines in Figure 15–7). Thus

$$\text{Consumers' surplus} \approx \sum_{i=1}^{n} [d(q_i) - p_E] \Delta q,$$

and since this approximation by Riemann sums becomes exact as $\Delta x \to 0$, it follows that consumers' surplus is the definite integral of $d(q) - p_E$ over $[0, q_E]$.

Consumers' Surplus

If $p = d(q)$ defines a demand function for a commodity and (q_E, p_E) is the commodity's market equilibrium point, then consumers' surplus for the commodity is

$$\int_0^{q_E} [d(q) - p_E] \, dq.$$

Thus consumers' surplus is the area of the region shaded in Figure 15–8.

EXAMPLE 2 Let $s(q) = 0.009q^2$ and $d(q) = 30 - 0.003q^2$, as in Example 1. Then

$$\text{Consumers' surplus} = \int_0^{q_E} [d(q) - p_E] \, dq$$

$$= \int_0^{50} [30 - 0.003q^2 - 22.50] \, dq$$

$$= \$250.$$

See Figure 15–9. ☐

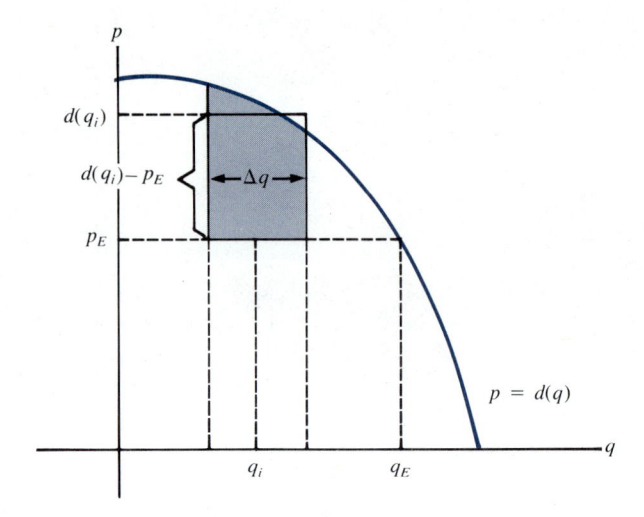

FIGURE 15–7

Income Streams

Any firm continually receives income from its business activities. This is referred to as the firm's **stream of income.** Most of the income that a firm receives is reinvested, either to keep the business going or in other ways. For purposes of analysis it is often convenient to assume that income flows into the firm continuously over the course of a year and that it is immediately reinvested at a certain interest rate that is compounded continuously. Thus we assume that at time t income is flowing into the firm at the rate of $P(t)$ dollars per year, where P is some continuous function of t, and that this income is immediately reinvested at $100r\%$ per year

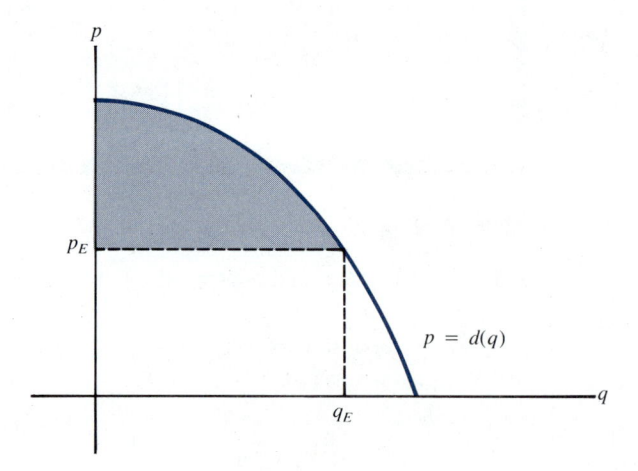

FIGURE 15–8
Consumers' Surplus

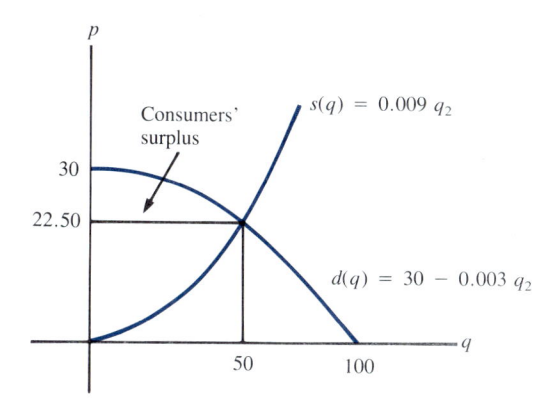

FIGURE 15-9

FIGURE 15-10

compounded continuously. We now ask the following question: if the firm continues to receive and reinvest income in this manner, what will be the value of its income stream over the next b years? To answer this question, we will again use the integral as an accumulator.

Suppose we divide the time interval $0 \leq t \leq b$ into n subintervals, each of length Δt, and select a number t_i in each subinterval. See Figure 15-10. Note that the rate at which income is flowing into the firm at time $t = t_i$ is $P(t_i)$, and that this money is reinvested at $100r$ percent per year compounded continuously for $b - t_i$ years. Hence by the formula for continuously compounded interest of Section 13.1, the money received by the firm at time $t = t_i$ will grow to

$$P(t_i)e^{r(b-t_i)}$$

dollars at the end of the b-year period. Since the total amount of income flowing into the firm over the ith subinterval (the shaded region in Figure 15-10) is approximately

$$P(t_i)\,\Delta t$$

(the area of the rectangle over the ith subinterval), and since this amount will accumulate at interest for approximately $b - t_i$ years, the future value at the end of the b-year period of the income received over the ith subinterval is approximately

$$[P(t_i)\,\Delta t]e^{r(b-t_i)}.$$

This approximation will become exact as $\Delta t \to 0$. Thus the total future value of the income stream at the end of b years can be approximated by a Riemann sum

$$P(t_1)e^{r(b-t_1)}\,\Delta t + \cdots + P(t_n)e^{r(b-t_n)}\,\Delta t,$$

and the approximation will become exact as $\Delta t \to 0$. Therefore we have the following result for the future value of a stream of income.

The Future Value of a Stream of Income

If income flows in continuously at a rate of $P(t)$ dollars per year and is reinvested at $100r\%$ per year compounded continuously, then the future value of the income stream at the end of b years is

$$\int_0^b P(t)e^{r(b-t)}\, dt.$$

EXAMPLE 3 The earnings of Zymax, Inc. are currently flowing into the firm at a constant rate of \$100,000 per year and are increasing at the rate of 6% per year. If we assume that earnings flow into the company continuously over the course of a year and are immediately reinvested at 6% per year compounded continuously, then we have $P(t) = \$100{,}000$ and $r = 0.06$. Thus the value of Zymax's earnings stream over the next four years will be

$$\int_0^b P(t)e^{r(b-t)}\, dx = \int_0^4 100{,}000 e^{0.06(4-t)}\, dt$$

$$= \int_0^4 100{,}000 e^{0.24} e^{-0.06t}\, dt$$

$$= 100{,}000 e^{0.24} \int_0^4 e^{-0.06t}\, dt$$

$$= 100{,}000 e^{0.24} \left[\frac{e^{-0.06t}}{-0.06}\right]_0^4$$

$$= 100{,}000 e^{0.24} \left[\frac{e^{-0.24} - 1}{-0.06}\right]$$

$$= -\frac{100{,}000}{0.06}(1 - e^{0.24})$$

$$= \$452{,}082. \quad \square$$

EXAMPLE 4 A new computer will save a company money at the rate of $P(t) = 2t$ thousand dollars per year, where t is in years, with $t = 0$ representing the present. Find the value of this stream of savings over the next 10 years if the company reinvests the savings at 10% per year compounded continuously. The answer is given by the integral

$$\int_0^{10} 2t e^{0.10(10-t)}\, dt = \int_0^{10} 2t e^{1-0.10t}\, dt$$

$$= 2e \int_0^{10} t e^{-0.10t}\, dt.$$

More about Integration

But, as we will show in the next section,

$$\int ue^{au}\,du = \frac{ue^{au}}{a} - \frac{e^{au}}{a^2} + c.$$

(Check this by differentiating the right side with respect to u.) Thus

$$\int_0^{10} 2te^{0.10(10-t)}\,dt = 2e\left[\frac{te^{-0.10t}}{-0.10} - \frac{e^{-0.10t}}{(-0.10)^2}\right]\Big|_0^{10}$$

$$= 2e\left[\left(-\frac{10e^{-1}}{0.10} - \frac{e^{-1}}{0.01}\right) - \left(0 - \frac{1}{0.01}\right)\right]$$

$$= 2e[-100e^{-1} - 100e^{-1} + 100]$$

$$= 200e - 400$$

$$143{,}656,$$

or \$143,656. ☐

Using the Integral as an Accumulator

In this section we have shown how to use the integral as an accumulator to find three different quantities: producers' surplus, consumers' surplus, and the value of a stream of income. If we look back at these examples, we can see that it was not actually necessary to write out the Riemann sums in each case. The essential steps in setting up a definite integral to find the value of some quantity Q over an interval [a, b] are as follows.

Procedure for Setting Up the Integral as an Accumulator

1. Divide the interval [a, b] into subintervals, each of which has length Δx.

2. Approximate the value of Q over a typical subinterval by an expression of the form $f(x)\,\Delta x$ in such a way that the approximation will become exact as $\Delta x \to 0$.

3. Then the value of Q over [a, b] is $\int_a^b f(x)\,dx$.

There are many situations other than those considered here in which this procedure can be profitably employed. Some of these are introduced in the exercises.

■ **Computer Manual: Program DEFNTGRL**

15.1 EXERCISES

Producers' Surplus and Consumers' Surplus

In Exercises 1 through 10, find producers' surplus and consumers' surplus using the given supply and demand functions.

1. supply: $p = 0.06q$
   demand: $p = -0.94q + 220$

2. supply: $p = 0.5q$
   demand: $p = 10 - 0.05q^2$

3. supply: $p = \sqrt{q}$
   demand: $p = -2q + 55$

4. supply: $p = 1.6(q + 1)^2$
   demand: $p = 200/(q + 1)$

5. supply: $p = q^2/64$
   demand: $p = 10 - 3q^2/128$

6. supply: $p = 5e^{0.8q}$
   demand: $p = 20e^{-0.2q}$

7. supply: $p = \sqrt{q + 5}$
   demand: $p = 27/(q + 5)$

8. supply: $p = 4(q - 4)/3$
   demand: $p = 400/(q + 1)$

*9. supply: $q = 0.05p + 0.25$
   demand: $q = 52.5 - 0.5p$

*10. supply: $q = \ln p$
    demand: $q = \ln(100/p)$

Income Streams

11. Find the value over its first five years of a continuous stream of income of $10,000 per year reinvested at 6% per year compounded continuously.

12. Repeat Exercise 11 if the income is reinvested at 10% per year compounded continuously.

13. A firm's earnings are flowing in continuously at a constant rate of $2.5 million per year and are reinvested at 10% per year compounded continuously. Find
    (a) the value of the firm's earnings over the next two years;
    (b) how long it will be until the value of the firm's earnings reaches $20 million.

*14. Show that if a stream of income is constant at P per year and is reinvested at $100r\%$ per year compounded continuously, then its value over the first b years is
$$\frac{P}{r}(e^{rb} - 1).$$
Use this formula to find the value over its first 20 years of a continuous stream of income of $100,000 per year reinvested at 16% per year compounded continuously.

15. A firm's earnings are flowing in continuously at $0.5t$ million dollars per year and are reinvested at 6% per year compounded continuously. Find the value of the firm's stream of earnings over its first five years. You may use the integration formula for ue^{au} that was stated in Example 4 of this section.

16. Find the value at the end of its first three years of a continuous stream of income that flows in at a rate of $P(t) = 1 + 0.2t$ million dollars per year and is reinvested at 9% per year compounded continuously.

17. Find the value at the end of its first six years of a continuous stream of income that flows in at a rate of $100 + 5t$ thousand dollars per year and is reinvested at 7% per year compounded continuously.

18. If a firm is generating profits at the rate of
$$P(t) = 20e^{0.01t}$$

million dollars per year, where t is in years and $t = 0$ is the present, find the value of the firm's stream of profits over the next three years if all profits are reinvested at 12% per year compounded continuously.

19. A firm's profits are being earned continuously at a rate of $50e^{0.01t}$ million dollars per year, where $t = 0$ represents the present. Find the value of the firm's stream of profits over the next five years if all profits are reinvested at
   (a) 8% per year compounded continuously;
   (b) 12% per year compounded continuously.

20. Suppose the earnings of a firm are flowing in continuously at a rate of $2.5 + 0.1t$ million dollars per year, where $t = 0$ represents the present.
   (a) Find a general formula for the value of the firm's earnings over the next b years.
   (b) Use the formula of part (a) to find the value of the firm's earnings over the next two years.
   (c) Use the formula of part (a) to find approximately how long it will take for earnings to reach $20 million.

21. A stream of savings is reinvested at 8% per year compounded continuously. If the savings accrue continuously at the rate of $20,000 per year, how long will it take for the value of the stream of savings to reach $250,000?

22. It is estimated that a company's investment in industrial robots will save it
$$P(t) = 200 + 12t$$
thousand dollars per year over the next 10 years, where t is time in years with $t = 0$ representing the present. Find the value of the stream of savings due to the robots if the company reinvests the savings at 10% per year compounded continuously.

23. A firm's investment in a new inventory control system will save it $P(t) = 40e^{-0.05t}$ thousand dollars per year for the next five years. If the savings are reinvested at 12% per year compounded continuously, find the value of the stream of savings.

*24. Show that if a continuous stream of income flows in at a rate of $P(t)$ dollars per time period and is *not* reinvested, then the total income produced by the stream from time $t = a$ to time $t = b$ is equal to
$$\int_a^b P(t)\, dt.$$
Use this result to find the total income over its first 10 years of a continuous stream of income that flows in at a rate of $10,000 per year.

*25. Use the result of Exercise 24 to find the total income over its first five years of a continuous stream of income that flows in at a rate of $4t$ thousand dollars per year t.

*26. Suppose a continuous stream of income flows in at a rate of $10e^{0.2t}$ thousand dollars per year t. Use the result of Exercise 24 to find the total income produced by the stream from time $t = 2$ to time $t = 6$.

Using the Integral as an Accumulator

Management

*27. **Carrying Cost.** Suppose that it costs p to carry one unit in inventory for one time period and that the number of units in inventory at time t is given by an inventory function $I(t)$. Then the total cost of carrying the inventory from time $t = 0$ to time $t = b$ is

$$\int_0^b pI(t)\, dt.$$

Use the procedure for setting up the integral as an accumulator to show that this is so.

*28. Refer to Exercise 27. If it costs a store $4 to carry one unit in inventory for one month and if the number of units in inventory at time t is given by

$$I(t) = 40{,}000 - 2000t,$$

where t is in months with $t = 0$ representing the present, find the store's total carrying cost for the next year.

*29. Suppose it costs a manufacturer $0.20 to carry a unit in inventory for one week and that the number of units in inventory as of week t is given by

$$I(t) = 12{,}000 - 150t^2,$$

where $t = 0$ represents the present. Use the result of Exercise 27 to find the manufacturer's total carrying cost over the next four weeks.

Life Science

*30. **Survival-Renewal Functions.** Suppose that $p(t)$ is the size of a population at time t, $t \geq 0$. Let $f(t)$ be the proportion of the original population that survives at time t; the function f is called a **survival function**. Let $g(t)$ denote the number of individuals added to the population at time t; the function g is called a **renewal function**. Use the integral as an accumulator to show that the size of the population at time $t = b$ is

$$p(b) = p(0)f(b) + \int_0^b g(t)f(b - t)\, dt.$$

*31. Use the result of Exercise 30 to find the size of a caribou herd six months from now if there are 2000 caribou in the herd now, the proportion of the 2000 who will survive t months from now is given by

$$f(t) = \frac{1}{1 + 0.02t},$$

and the number of caribou added to the herd is constant at 100 per month.

*32. Use the result of Exercise 30 to find the number of rabbits you will have one year from now if you start with 100 rabbits, the proportion of the original population that survives as of month t is given by

$$f(t) = e^{-0.1t},$$

and the number of rabbits being born at time t is given by $g(t) = 20t$.

*33. **Epidemiology.** According to an article by Peter J. Denning ("The Science of Computing: Computer Models of AIDS Epidemiology," *American Scientist*, July–August 1987) the relationship between infection by the human immunodeficiency virus (HIV) and the development of the disease AIDS is as follows: Let $H(t)$ denote the number of people who have been infected by the virus as of year t, let $A(t)$ denote the number who have AIDS as of year t, and let $I(x)$ denote the proportion of HIV-infected individuals whose incubation period for AIDS is x years.

Then
$$A(b) = \int_0^b I(t)H(b-t)\, dt.$$

Use the integral as an accumulator to derive the formula for $A(b)$.

*34. Refer to Exercise 33: If $I(t) = 0.3e^{-0.3t}$ and $H(t) = e^{0.01t}$, $H(t)$ in hundreds of thousands of individuals, find $A(5)$, $A(10)$, and $A(20)$, and interpret your results.

*35. Refer to Exercise 33: If $I(t) = 0.2e^{-0.2t}$ and $H(t) = e^{0.01t}$, find $A(5)$, $A(10)$, and $A(20)$, and compare your results with those of Exercise 34.

*36. Refer to Exercise 33: Some models of the AIDS epidemic assume that $I(t) = 0.125e^{-0.125t}$ and $H(t) = e^{0.01t}$. Find and graph $A(x)$, $x \geq 0$, and interpret your result.

*37. **Blood Flow.** Poiseuille's Law states that the velocity of blood in an artery depends on the distance of the blood from the center of the artery according to the equation
$$v(x) = c(b^2 - x^2).$$

Here x is the distance from the central axis of the artery, b is the radius of the artery, c is a constant, and $v(x)$ is the velocity of the blood in cubic centimeters per second. See the figure.

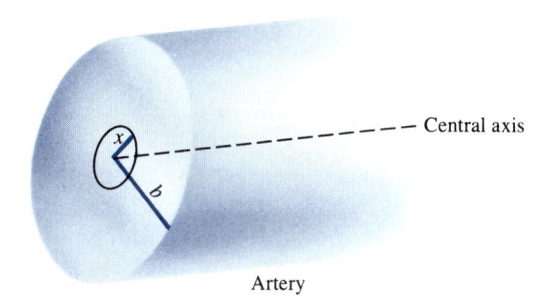

Use the integral as accumulator to show that the rate of blood flow through the artery is
$$2\pi c \int_0^b (b^2 x - x^3)\, dx = \frac{\pi c b^4}{2}$$

cubic centimeters per second. (*Hint:* Divide the artery into rings of width Δx and choose a number x_i in each ring; the approximate area of the ith ring is $2\pi x_i\, \Delta x$, and the approximate speed of blood in the ith ring is the area of the ring times $v(x_i)$.)

*38. Use the result of Exercise 37 to find the rate at which blood flows through an artery of radius 1 centimeter; of radius 2 centimeters.

Social Science

*39. A religious cult currently has 10,000 adherents. The proportion of them who will remain members of the cult t weeks from now is given by
$$f(t) = e^{-0.05t},$$

and the number of new members recruited during week t is given by
$$g(t) = 1000e^{-0.05t}.$$
Find the size of the cult one year (52 weeks) from now. (*Hint:* See Exercise 30.)

*40. A prison has space for 3500 inmates and currently has a population of 3000 inmates. The proportion of inmates who will remain in the prison t months from now is given by
$$f(t) = \frac{1}{1 + 0.005t},$$
and new prisoners are arriving at the rate of 40 per month. Will the prison be overcrowded 20 months from now? 21 months from now? (*Hint:* See Exercise 30.)

*41. **Population Density.** Suppose that the population x kilometers from the center of a region is $f(x)$ persons per square kilometer. Show that the total population within b miles of the center of the region is
$$2\pi \int_0^b xf(x)\, dx.$$
(*Hint:* Divide the region into circular rings as in Exercise 37.)

*42. Refer to Exercise 41. Find the total population within 10 kilometers of the center of a city if there are $f(x) = 1000e^{-0.1x}$ persons per square kilometer x kilometers from the center.

15.2 INTEGRATION BY PARTS AND TABLES OF INTEGRALS

To this point we have discussed only two methods for finding the indefinite integral of a function: integration by guessing an antiderivative and integration by substitution. No technique of integration is applicable to all integrals, and over the years many different techniques have been developed to handle different types of indefinite integrals. Most of these techniques of integration are beyond the scope of this book, but two of them, **integration by parts** and integration using tables of integrals, are so useful that we must present them. In this section we discuss integration by parts and demonstrate how to perform integration with the aid of a table of integrals.

Integration by Parts

If u and v are functions of x, the product rule for differentiation states that
$$\frac{d}{dx}(uv) = u\frac{dv}{dx} + v\frac{du}{dx}.$$

If we multiply both sides of this equation by the differential dx, we may write it in differential form as
$$d(uv) = u\, dv + v\, du,$$
or
$$u\, dv = d(uv) - v\, du.$$

Integrating both sides of this last equation, we obtain

$$\int u\,dv = \int d(uv) - \int v\,du.$$

But uv is an antiderivative for $d(uv)$, so we have

$$\int u\,dv = uv - \int v\,du.$$

This is known as the **integration by parts** formula.

Integration by Parts Formula

$$\int u\,dv = uv - \int v\,du.$$

EXAMPLE 1   Let us use the integration by parts formula to find

$$\int xe^{-x}\,dx.$$

We set

$$u = x \quad \text{and} \quad dv = e^{-x}\,dx.$$

Then

$$du = dx \quad \text{and} \quad v = \int e^{-x}\,dx = -e^{-x},$$

so

$$\begin{aligned}
\int xe^{-x}\,dx &= \int u\,dv \\
&= uv - \int v\,du \\
&= x(-e^{-x}) - \int(-e^{-x})\,dx \\
&= -xe^{-x} + \int e^{-x}\,dx \\
&= -xe^{-x} - e^{-x} + c.
\end{aligned}$$

(Note that when we found v we did not include the constant of integration and write

$$v = \int e^{-x}\,dx = -e^{-x} + c;$$

instead, we simply postponed adding the constant of integration until the final step.) ☐

EXAMPLE 2   How did we know in Example 1 to let $u = x$ and $dv = e^{-x}\,dx$? If we had tried the other possibility,

$$u = e^{-x} \quad \text{and} \quad dv = x\,dx,$$

then we would have obtained

$$du = -e^{-x}\,dx \quad \text{and} \quad v = \int x\,dx = \frac{x^2}{2},$$

and thus

$$\int xe^{-x}\,dx = \int u\,dv$$
$$= uv - \int v\,du$$
$$= \frac{x^2}{2}e^{-x} - \int\left(-\frac{x^2}{2}e^{-x}\right)dx.$$

But the integral on the right is more complicated than the one we started with. Therefore this choice of u and dv is not a useful one. □

Integration by parts can often be used to convert an integral that cannot be handled into one that can, as in Example 1. Examples 1 and 2 together suggest the following:

Keys to Integration by Parts

When performing integration by parts:

1. dv must include the differential dx of the original integral.

2. dv must be chosen so that it can be integrated to find v.

3. The integral $\int v\,du$ should be easy to find, or at least simpler than the integral $\int u\,dv$.

EXAMPLE 3 Let us use integration by parts to find

$$\int \ln x\,dx.$$

If we choose

$$u = \ln x \quad \text{and} \quad dv = dx,$$

then

$$du = \frac{1}{x}\,dx \quad \text{and} \quad v = \int dx = x,$$

so

$$\int v\,du = \int x\cdot\frac{1}{x}\,dx = \int 1\,dx,$$

which is certainly easy to find. Hence this is a promising choice for u and dv. We have

$$\int \ln x \, dx = \int u \, dv$$
$$= uv - \int v \, du$$
$$= x \ln x - \int 1 \, dx$$
$$= x \ln x - x + c. \quad \square$$

Sometimes it is necessary to use integration by parts more than once.

EXAMPLE 4   Let us find

$$\int x^2 e^{-x} \, dx.$$

We let

$$u = x^2 \quad \text{and} \quad dv = e^{-x} \, dx.$$

Then

$$du = 2x \, dx \quad \text{and} \quad v = \int e^{-x} \, dx = -e^{-x},$$

so

$$\int v \, du = \int (-2xe^{-x}) \, dx = -2 \int xe^{-x} \, dx.$$

But $\int xe^{-x} \, dx$ is simpler than $\int x^2 e^{-x} \, dx$; in fact, we know how to integrate xe^{-x} from Example 1. Therefore

$$\int x^2 e^{-x} \, dx = \int u \, dv$$
$$= uv - \int v \, du$$
$$= -x^2 e^{-x} - \int (-2xe^{-x}) \, dx$$
$$= -x^2 e^{-x} + 2 \int xe^{-x} \, dx.$$

But from Example 1,

$$\int xe^{-x} \, dx = -xe^{-x} - e^{-x} + c.$$

Thus

$$\int x^2 e^{-x} \, dx = -x^2 e^{-x} + 2(-xe^{-x} - e^{-x}) + c$$
$$= -e^{-x}(x^2 + 2x + 2) + c. \quad \square$$

When using integration by parts with definite integrals, the term uv must be evaluated at the limits of integration.

EXAMPLE 5   Let us evaluate

$$\int_0^2 xe^{-x}\, dx$$

using integration by parts. As in Example 1, we take

$$u = x \quad \text{and} \quad dv = e^{-x}\, dx,$$

so

$$du = dx \quad \text{and} \quad v = \int e^{-x}\, dx = -e^{-x}.$$

Thus

$$\int_0^2 xe^{-x}\, dx = -xe^{-x}\Big|_0^2 + \int_0^2 e^{-x}\, dx$$

$$= (-2e^{-2}) - (-0e^{-0}) - \left[e^{-x}\Big|_0^2\right]$$

$$= -2e^{-2} - [e^{-2} - e^{-0}]$$

$$= 1 - 3e^{-2}. \quad \square$$

Integral Tables

Many integrals cannot be found using only the techniques we have presented. For example, the integrals

$$\int \sqrt{x^2 - 4}\, dx \quad \text{and} \quad \int \frac{1}{4x^2 - 9}\, dx$$

cannot be determined by any method we have discussed so far. Fortunately there are tables of integrals that show how to integrate these and many other functions. Mathematical handbooks contain extensive tables of integrals; the integration formulas in the back endpapers of this book form an abbreviated version of such an integral table.

EXAMPLE 6   Let us use the integration formulas in the back endpapers to find

$$\int \sqrt{x^2 - 4}\, dx.$$

We search through the list of formulas until we find one in which the integrand has the form of our integrand $\sqrt{x^2 - 4}$. Formula 8, which tells us that

$$\int \sqrt{u^2 - a^2}\, du = \frac{1}{2}[u\sqrt{u^2 - a^2} - a^2 \ln|u + \sqrt{u^2 - a^2}|] + c,$$

has an integrand of the proper form if we take

$$u = x \quad \text{and} \quad a = 2.$$

More about Integration

Note that with the substitutions $u = x$ (so that $du = dx$) and $a = 2$ the integral of formula 8 exactly matches our integral. Thus,

$$\int \sqrt{x^2 - 4}\, dx = \int \sqrt{u^2 - a^2}\, du$$
$$= \frac{1}{2}[u\sqrt{u^2 - a^2} - a^2 \ln|u + \sqrt{u^2 - a^2}|] + c,$$
$$= \frac{1}{2}[x\sqrt{x^2 - 4} - 4 \ln|x + \sqrt{x^2 - 4}|] + c. \quad \square$$

When we match an integral with a formula from a table of integrals, the entire integral, including the differential, must match the formula in the table exactly. This sometimes requires making a substitution or rewriting the integral.

EXAMPLE 7 Let us find

$$\int \frac{1}{4x^2 - 9}\, dx.$$

The integrand matches that of formula 11 in the endpapers if we take

$$u = 2x \quad \text{and} \quad a = 3.$$

But if $u = 2x$, then

$$du = 2\, dx \quad \text{or} \quad \frac{1}{2}\, du = dx.$$

Thus,

$$\int \frac{1}{4x^2 - 9}\, dx = \int \frac{1}{u^2 - 3^2} \frac{1}{2}\, du$$
$$= \frac{1}{2} \int \frac{1}{u^2 - 3^2}\, du$$
$$= \frac{1}{2}\left[\frac{1}{2 \cdot 3} \ln\left|\frac{u - 3}{u + 3}\right|\right] + c$$
$$= \frac{1}{12} \ln\left|\frac{2x - 3}{2x + 3}\right| + c. \quad \square$$

EXAMPLE 8 Let us find

$$\int \frac{x}{4x^2 + 12x + 9}\, dx.$$

This integral does not appear to match any of those in the endpapers. However,
$$4x^2 + 12x + 9 = (2x + 3)^2,$$
and

$$\int \frac{x}{(2x + 3)^2}\, dx$$

matches formula 16 exactly if $u = x$ (so that $du = dx$), $a = 2$, and $b = 3$. Therefore

$$\int \frac{x}{4x^2 + 12x + 9} dx = \int \frac{x}{(2x + 3)^2} dx$$

$$= \int \frac{u}{(2u + 3)^2} du$$

$$= \frac{1}{4}\left[\ln|2u + 3| + \frac{3}{2u + 3}\right] + c$$

$$= \frac{1}{4}\left[\ln|2x + 3| + \frac{3}{2x + 3}\right] + c. \quad \square$$

■ *Computer Manual: Program DEFNTGRL*

15.2 EXERCISES

In Exercises 1 through 22, integrate by parts.

1. $\int xe^x dx$
2. $\int xe^{2x} dx$
3. $\int x^2 e^x dx$
4. $\int x^3 e^x dx$
5. $\int_0^1 x^3 e^{-x} dx$
6. $\int x^3 e^{-x^2} dx$
7. $\int x \ln x \, dx$
8. $\int_1^e x^2 \ln x \, dx$
9. $\int x^3 \ln x \, dx$
10. $\int \sqrt{x} \ln x \, dx$
11. $\int (\ln x)^2 dx$
12. $\int x(\ln x)^2 dx$
13. $\int (\ln x)^3 dx$
14. $\int_e^{e^2} \frac{\ln x}{x^2} dx$
15. $\int_1^2 x(x + 2)^5 dx$
16. $\int x^2(x + 1)^{3/2} dx$
17. $\int_1^2 x(2x + 1)^3 dx$
18. $\int_0^1 x^2(x + 3)^4 dx$
19. $\int x^2 \sqrt[3]{x + 2} \, dx$
20. $\int \frac{x}{\sqrt{x + 1}} dx$
21. $\int_{-1}^4 \frac{x^2}{\sqrt{x + 5}} dx$
22. $\int e^{\sqrt{x}} dx$ (*Hint:* Let $u^2 = x$.)

Management

23. A firm's marginal cost function is given by
$$c(x) = x\sqrt{x + 1}.$$
Find the firm's cost function if its fixed cost is $0.

24. A well produces oil at a rate of $20{,}000te^{-0.2t}$ barrels per year, where t is in years with $t = 0$ representing the present. Find the total amount of oil the well will produce over the next two years.

25. The supply function for a product is given by $p = 2 \ln(q + 1)$ and the demand function by $p = 50 - 3 \ln(q + 1)$. Find producers' surplus and consumers' surplus for the product.

Life Science

26. Suppose that the amount of CO_2 in the atmosphere is increasing at a rate of $t^2\sqrt{t+1}$ parts per million per decade. Here t is in decades, with $t = 0$ representing the present. Find out by how much the total concentration of CO_2 will increase over the next decade.

27. The concentration of a drug in a patient's bloodstream is
$$y = \frac{2\ln(t+3)}{(t+3)^2}$$
milligrams per milliliter t minutes after injection. Find the average concentration of the drug over the first 10 minutes after injection.

Integral Tables

In Exercises 28 through 49, use the integration formulas in the back endpapers to find the integral.

28. $\int \frac{1}{x^2 - 16} dx$

29. $\int \frac{1}{x\sqrt{81 - x^2}} dx$

30. $\int \frac{2}{9 - x^2} dx$

31. $\int_1^2 \frac{3}{x\sqrt{x^2 + 16}} dx$

32. $\int 2^{3x} dx$

33. $\int 5^{-3x} dx$

34. $\int x\sqrt{1 - x}\, dx$

35. $\int \frac{x}{3x + 2} dx$

36. $\int \frac{1}{x\sqrt{25 - 4x^2}} dx$

37. $\int \frac{1}{3x\sqrt{4 + 9x^2}} dx$

38. $\int \frac{1}{8 - 2x^2} dx$

39. $\int \frac{1}{x^2 + 3x} dx$

40. $\int \frac{2}{\sqrt{9x^2 - 0.25}} dx$

41. $\int \frac{3x}{36x^2 + 60x + 25} dx$

42. $\int_0^1 \frac{2x}{x^2 + 2x + 1} dx$

43. $\int \frac{3}{2x^2 + 6x} dx$

44. $\int 4\sqrt{x^2 + 5}\, dx$

45. $\int \sqrt{2x^2 - 1}\, dx$

46. $\int \sqrt{4x^2 + 9}\, dx$

47. $\int \frac{1}{3x\sqrt{2 - 4x^2}} dx$

48. $\int \frac{4}{\sqrt{2x^2 - 1}} dx$

49. $\int \log_2 2x\, dx$

Management

50. The supply function for a product is given by
$$p = \frac{5q}{50 - q}$$
and the demand function by
$$p = \frac{2000 - 45q}{50 - q}$$
Find producers' and consumers' surplus for the product.

51. A firm's marginal revenue function is given by
$$r(x) = \frac{20}{\sqrt{4x^2 - 1}}, \quad x \geq 1.$$
Find the firm's revenue function if the revenue from selling one unit is $76.93.

Life Science

52. A desert is encroaching on arable land at a rate of

$$f(t) = \frac{100}{\sqrt{t^2 + 4}}$$

square miles per year t, where $t = 0$ represents the present. Find the number of square miles of arable land the desert will claim over the next 10 years.

Social Science

53. A student who studies t hours absorbs y new concepts per hour, where

$$y = \frac{8t}{0.25t^2 + 2t + 4}.$$

Find the total number of new concepts absorbed in 4 hours of study.

15.3 IMPROPER INTEGRALS

Sometimes it is necessary to consider areas of regions that extend infinitely far to the left or right along the x-axis. For instance, the shaded region in Figure 15–11 is meant to extend infinitely far to the right along the x-axis. As we shall see, the area of such a region may or may not be finite. In this section we show how to determine if the area of such a region is finite, and how to find it if it is. In so doing, we use **improper integrals:** These are definite integrals that have at least one infinite limit of integration.

(There is another type of improper integral in which the limits of integration are finite but the integrand becomes infinite at some point of the interval of integration. We will not consider this type of improper integral.)

Consider the function f defined by $f(x) = e^{-x}$. The region under the graph of f above the x-axis and to the right of $x = 0$ (Figure 15–12) is important in probability and statistics. Let us see if this region has a finite area.

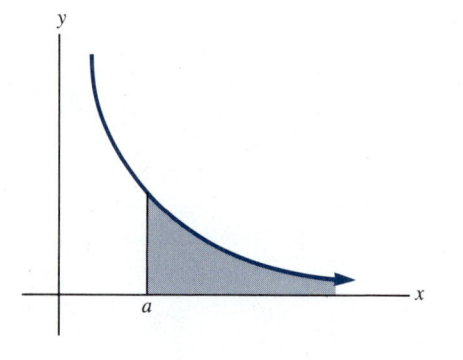

FIGURE 15–11

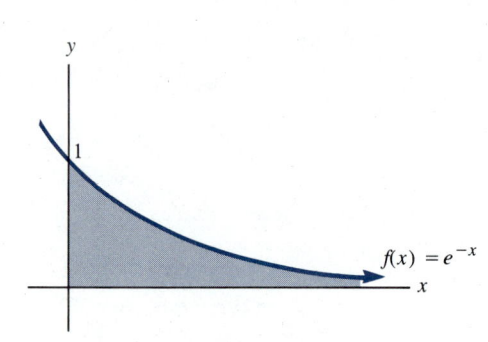

FIGURE 15–12

MORE ABOUT INTEGRATION   953

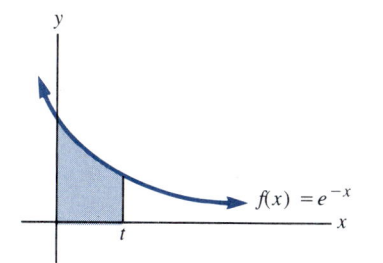

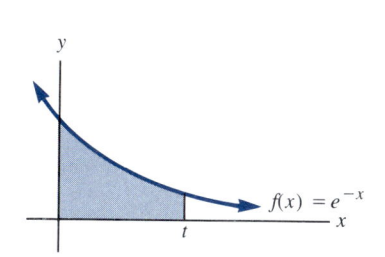

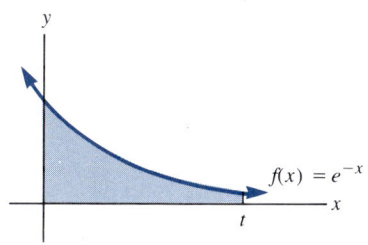

FIGURE 15–13

If t is any positive number, then the area under $f(x) = e^{-x}$ from 0 to t is certainly finite, and we know that it is equal to

$$\int_0^t e^{-x}\,dx.$$

As $t \to +\infty$, the area under $f(x)$ from 0 to t will approach the area of the region in question. See Figure 15–13. Thus, if

$$\lim_{t \to +\infty} \int_0^t e^{-x}\,dx$$

exists, then the region in question has area equal to the value of the limit; if the limit does not exist, then the area of the region is not finite. Now,

$$\lim_{t \to +\infty} \int_0^t e^{-x}\,dx = \lim_{t \to +\infty} (-e^{-x})\Big|_0^t = \lim_{t \to +\infty}(-e^{-t} + 1);$$

but, as Figure 15–12 shows,

$$\lim_{t \to +\infty} e^{-t} = 0,$$

so

$$\lim_{t \to +\infty} \int_0^t e^{-x}\,dx = 0 + 1 = 1.$$

Therefore the shaded region of Figure 15–12 has finite area equal to 1.

It is customary to write a limit such as

$$\lim_{t \to +\infty} \int_0^t e^{-x}\,dx$$

as an integral:

$$\lim_{t \to +\infty} \int_0^t e^{-x}\,dx = \int_0^{+\infty} e^{-x}\,dx.$$

The integral on the right is called an **improper integral** because one of its limits of integration is infinite.

Improper Integrals

The **improper integrals**

$$\int_a^{+\infty} f(x)\,dx \quad \text{and} \quad \int_{-\infty}^b f(x)\,dx$$

are defined by

$$\int_a^{+\infty} f(x)\,dx = \lim_{t \to +\infty} \int_a^t f(x)\,dx$$

and

$$\int_{-\infty}^b f(x)\,dx = \lim_{t \to -\infty} \int_t^b f(x)\,dx.$$

The improper integrals are said to **converge** if the limits on the right side exist; otherwise they are said to **diverge.**

EXAMPLE 1 For the improper integral

$$\int_2^{+\infty} 2x^{-3}\,dx$$

we have

$$\int_2^{+\infty} 2x^{-3}\,dx = \lim_{t \to +\infty} \int_2^t 2x^{-3}\,dx$$

$$= \lim_{t \to +\infty} (-x^{-2})\bigg|_2^t$$

$$= \lim_{t \to +\infty} \left(-\frac{1}{t^2} + \frac{1}{4}\right)$$

$$= 0 + \frac{1}{4}$$

$$= \frac{1}{4}.$$

Therefore the improper integral

$$\int_2^{+\infty} 2x^{-3}\,dx$$

converges, and the shaded region of Figure 15–14 has an area of ¼. □

EXAMPLE 2 Let us evaluate the improper integral

$$\int_{-\infty}^0 e^x\,dx.$$

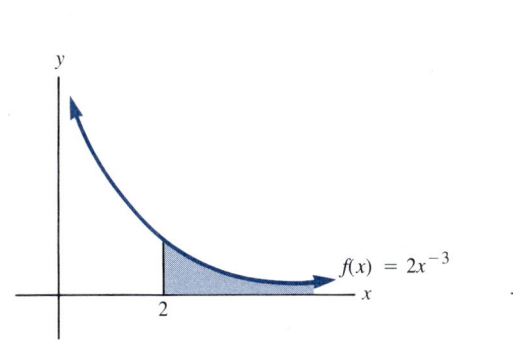

FIGURE 15–14

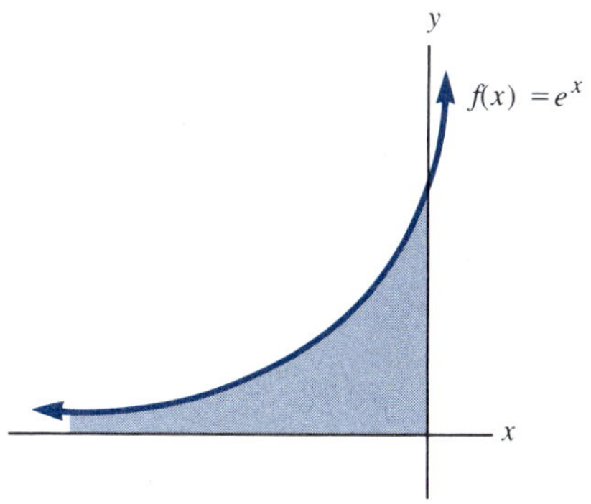

FIGURE 15–15

See Figure 15–15. We have

$$\int_{-\infty}^{0} e^x \, dx = \lim_{t \to -\infty} \int_{t}^{0} e^x \, dx = \lim_{t \to -\infty} e^x \Big|_{t}^{0} = \lim_{t \to -\infty} (1 - e^t).$$

But as Figure 15–15 shows, $\lim_{t \to -\infty} e^t = 0$, so

$$\int_{-\infty}^{0} e^x \, dx = 1 - 0 = 1.$$

Hence the integral converges, and the shaded region in Figure 15–15 has an area of 1. □

EXAMPLE 3   The improper integral

$$\int_{1}^{+\infty} \frac{1}{x} \, dx$$

diverges, because

$$\int_{1}^{+\infty} \frac{1}{x} \, dx = \lim_{t \to +\infty} \int_{1}^{t} \frac{1}{x} \, dx = \lim_{t \to +\infty} \ln x \Big|_{1}^{t} = \lim_{t \to +\infty} (\ln t),$$

and $\lim_{t \to +\infty} (\ln t) = +\infty$. Therefore the shaded area in Figure 15–16 is not finite. □

We may also define an improper integral having two infinite limits of integration by setting

$$\int_{-\infty}^{+\infty} f(x) \, dx = \int_{-\infty}^{a} f(x) \, dx + \int_{a}^{+\infty} f(x) \, dx$$

for any number a. The integral on the left converges if and only if both of the integrals on the right converge.

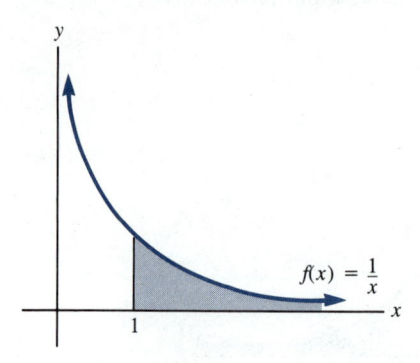

FIGURE 15–16

EXAMPLE 4   The improper integral

$$\int_{-\infty}^{+\infty} e^{-x}\, dx$$

diverges, because

$$\int_{-\infty}^{+\infty} e^{-x}\, dx = \int_{-\infty}^{0} e^{-x}\, dx + \int_{0}^{+\infty} e^{-x}\, dx,$$

and

$$\int_{-\infty}^{0} e^{-x}\, dx = \lim_{t \to -\infty} \int_{t}^{0} e^{-x}\, dx = \lim_{t \to -\infty} (-1 + e^{-t})$$

diverges. (Why?) ☐

15.3 EXERCISES

In Exercises 1 through 30, determine whether the given integral converges or diverges; if it converges, find its value and interpret your result as an area.

1. $\int_{1}^{+\infty} \dfrac{3}{x}\, dx$
2. $\int_{4}^{+\infty} \dfrac{1}{x}\, dx$
3. $\int_{1}^{+\infty} \dfrac{1}{x^2}\, dx$
4. $\int_{-\infty}^{1} \dfrac{1}{x^2}\, dx$
5. $\int_{2}^{+\infty} \dfrac{1}{x^3}\, dx$
6. $\int_{-\infty}^{-2} \dfrac{1}{x^3}\, dx$
7. $\int_{1}^{+\infty} \dfrac{1}{x^{1/3}}\, dx$
8. $\int_{1}^{+\infty} \dfrac{1}{x^{1.0001}}\, dx$
9. $\int_{1}^{+\infty} \dfrac{1}{x^{0.9999}}\, dx$
10. $\int_{-\infty}^{-1} \dfrac{1}{2-x}\, dx$
11. $\int_{1}^{+\infty} \dfrac{x}{x^2+1}\, dx$
12. $\int_{2}^{+\infty} \dfrac{x^2}{x^3+1}\, dx$
13. $\int_{0}^{+\infty} e^{x}\, dx$
14. $\int_{-\infty}^{0} e^{x}\, dx$
15. $\int_{-\infty}^{+\infty} e^{x}\, dx$
16. $\int_{-\infty}^{-1} e^{2x}\, dx$
17. $\int_{-1}^{+\infty} e^{-2x}\, dx$
18. $\int_{-\infty}^{+\infty} f(x)\, dx$, where $f(x) = \begin{cases} e^{x}, & x \le 0, \\ e^{-x}, & x \ge 0 \end{cases}$
19. $\int_{0}^{+\infty} xe^{-x^2}\, dx$
20. $\int_{-\infty}^{+\infty} xe^{-x^2}\, dx$
21. $\int_{0}^{+\infty} x^2 e^{-x^3}\, dx$
22. $\int_{1}^{+\infty} \dfrac{\ln x}{x}\, dx$
23. $\int_{2}^{+\infty} \dfrac{1}{x \ln x}\, dx$
24. $\int_{2}^{+\infty} \dfrac{1}{x(\ln x)^2}\, dx$
*25. $\int_{1}^{+\infty} \dfrac{1}{x^r}\, dx$, $0 \le r \le 1$
*26. $\int_{1}^{+\infty} \dfrac{1}{x^r}\, dx$, $r > 1$

*27. $\displaystyle\int_{a}^{+\infty} \frac{1}{x^r}\,dx, \quad 0 \le r \le 1, \quad a > 0$

*28. $\displaystyle\int_{a}^{+\infty} \frac{1}{x^r}\,dx, \quad r > 1, \quad a > 0$

*29. $\displaystyle\int_{2}^{+\infty} \frac{1}{x(\ln x)^r}\,dx, \quad r > 1$

*30. $\displaystyle\int_{2}^{+\infty} \frac{1}{x(\ln x)^r}\,dx, \quad 0 \le r \le 1$

Management

31. A firm's marginal cost function is given by
$$c(x) = \frac{20{,}000}{x^{3/2}}, \quad x \ge 1.$$
Find the firm's total cost if it makes an unlimited number of units.

32. A firm's marginal cost function is given by
$$c(x) = \frac{100{,}000}{x^{5/4}}, \quad x \ge 1.$$
Find the firm's total cost if it makes an unlimited number of units.

33. A firm's marginal profit function is given by $p(x) = 200x^{-1.1}$, profit in thousands of dollars, $x \ge 1$. Find the firm's profit if it can make and sell an unlimited number of units.

34. If a well produces oil at a rate of $10{,}000e^{-0.2t}$ barrels per year t, where $t = 0$ represents the present, find the total amount of oil the well could produce in the future.

35. Repeat Exercise 34 if the rate is $10{,}000te^{-0.2t}$ barrels per year. (*Hint:* If $r > 0$, the graph of $y = te^{-rt}$ shows that te^{-rt} approaches 0 as t approaches $+\infty$.)

36. Investment income is flowing in continuously to a blind trust at a rate of $25{,}000e^{-0.02t}$ dollars per year t. Find the value of this stream of income if it lasts forever.

37. If a stream of income flows into a firm continuously at a rate of $50e^{-0.25t}$ thousand dollars per year, find its total future value.

38. Repeat Exercise 37 if the rate is $50t^2e^{-0.25t}$ thousand dollars per year.

Life Science

39. It is estimated that the total amount of pollutants released into the atmosphere worldwide by automobiles is y million tons per decade t, where
$$y = 200(t+2)^{-2}.$$
Here $t = 0$ represents the present. Find the total amount of pollutants that will be released by automobiles in the future.

40. Repeat Exercise 39 if $y = 200t(t+2)^{-2}$.

15.4 CONTINUOUS RANDOM VARIABLES

Recall that in Section 8.4 we introduced the notion of a continuous random variable: this is a random variable that can take on all the values in some interval on the real line. For example, we considered in Section 8.4 the random variable X = IQ test scores. Then X is a continuous random variable, and, as we saw, it is described by the normal distribution, which is an example of a **continuous probability distribution.** In this section we demonstrate how calculus may be applied to the study of continuous random variables and their probability distributions.

Since a continuous random variable takes on all the values in some interval, it takes on infinitely many values, and hence the probability that it takes on a particular value will be zero. Therefore if X is a continuous random variable, we do not seek to find the probability that X equals some value, since this will be zero, but rather the probability that X falls between two given values, or that it is greater than or less than a given value. To illustrate, suppose that the operating lifetime of a certain type of transistor can be any time between zero and five years, inclusive. If we let

$$X = \text{Operating lifetime of a particular transistor,}$$

X in years, then $0 \leq X \leq 5$, and X is a continuous random variable. Thus

$$P(\text{a transistor's lifetime will be between 2 and 4 years}) = P(2 < X < 4),$$

$$P(\text{a transistor's lifetime will be greater than 3 years}) = P(X > 3),$$

and

$$P(\text{a transistor's lifetime will be less than 1 year}) = P(X < 1).$$

Note that since the probability that a continuous random variable is equal to a particular value is zero, $P(2 < X < 4)$ will be the same as $P(2 \leq X \leq 4)$; similarly $P(X > 3)$ will equal $P(X \geq 3)$ and $P(X < 1)$ will equal $P(X \leq 1)$. This is so for every continuous random variable, and therefore when working with continuous random variables we may replace $<$ by $\leq$ and $>$ by $\geq$ in probability statements whenever we wish.

If X is a continuous random variable, the probability distribution of X is usually described by a function called a **probability density function.**

Probability Density Functions

A **probability density function** is a function f such that

$$f(x) \geq 0 \quad \text{for all } x$$

and

$$\int_{-\infty}^{+\infty} f(x) \, dx = 1.$$

Note that the definition says that f must be defined and nonnegative for all x and that the area under the graph of f must be 1. See Figure 15–17.

MORE ABOUT INTEGRATION

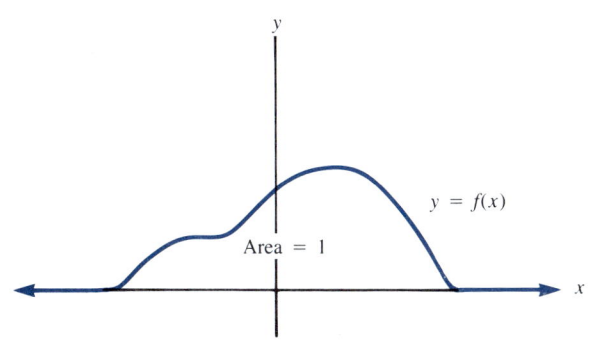

FIGURE 15–17
Probability Density Function

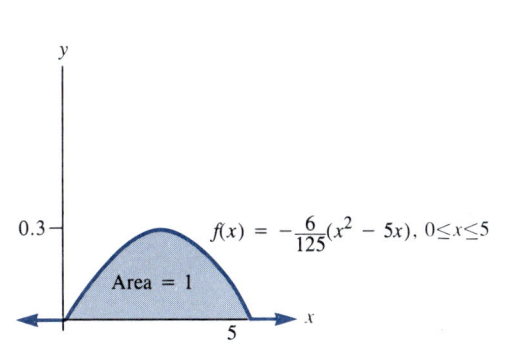

FIGURE 15–18

EXAMPLE 1 The function defined by

$$f(x) = \begin{cases} -\dfrac{6}{125}(x^2 - 5x), & 0 \le x \le 5, \\ 0, & \text{otherwise} \end{cases}$$

is a probability density function: it is defined and nonnegative for all x (Figure 15–18), and since $f(x) = 0$ when x is outside $[0, 5]$, we have

$$\int_{-\infty}^{+\infty} f(x)\, dx = \int_{0}^{5} -\frac{6}{125}(x^2 - 5x)\, dx = 1.$$

(Check this.) ☐

EXAMPLE 2 Let $f(x) = \begin{cases} 0.5e^{-x/2}, & x \ge 0, \\ 0, & x < 0. \end{cases}$

We claim that f is a probability density function. Certainly f is defined and nonnegative for all x (Figure 15–19), so it suffices to show that the improper integral

$$\int_{-\infty}^{+\infty} f(x)\, dx = \int_{0}^{+\infty} 0.5e^{-x/2}\, dx$$

is equal to 1. We have

$$\int_{0}^{+\infty} 0.5e^{-x/2}\, dx = \underset{b \to +\infty}{\text{Lim}} \int_{0}^{b} 0.5e^{-x/2}\, dx$$

$$= \underset{b \to +\infty}{\text{Lim}} \left. -e^{-x/2} \right|_{0}^{b}$$

$$= \underset{b \to +\infty}{\text{Lim}} (-e^{-b/2} + e^{0})$$

$$= \underset{b \to +\infty}{\text{Lim}} (-e^{-b/2}) + 1.$$

But $\lim_{b \to +\infty} e^{-b/2} = 0$, so

$$\int_0^{+\infty} 0.5 e^{-x/2}\, dx = \lim_{b \to +\infty} (e^{-b/2}) + 1 = 0 + 1 = 1,$$

and we are done. ☐

Given a continuous random variable X, it is usually possible to find a probability density function f such that for any numbers a and b with $a < b$,

$$P(a \le X \le b) = \text{area under } f \text{ from } a \text{ to } b = \int_a^b f(x)\, dx.$$

See Figure 15–20. When this is the case, we say that f is a **probability density function for** X. If X is a continuous random variable and f is a probability density function for X, then we can find probabilities involving X by integrating f.

EXAMPLE 3 Let X denote the operating lifetime of a transistor in years, $0 \le X \le 5$, and suppose that the probability density function for X is defined by

$$f(x) = \begin{cases} -\dfrac{6}{125}(x^2 - 5x), & 0 \le x \le 5, \\ 0, & \text{otherwise.} \end{cases}$$

Then

P(a transistor's lifetime will be between 2 and 4 years)
$= P(2 < X < 4)$
$= \displaystyle\int_2^4 -\dfrac{6}{125}(x^2 - 5x)\, dx$
$= -\dfrac{6}{125}\left(\dfrac{x^3}{3} - \dfrac{5x^2}{2}\bigg|_2^4\right)$
$= \dfrac{68}{125} = 0.544.$

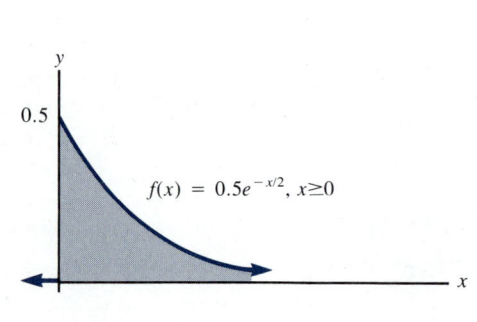

FIGURE 15–19

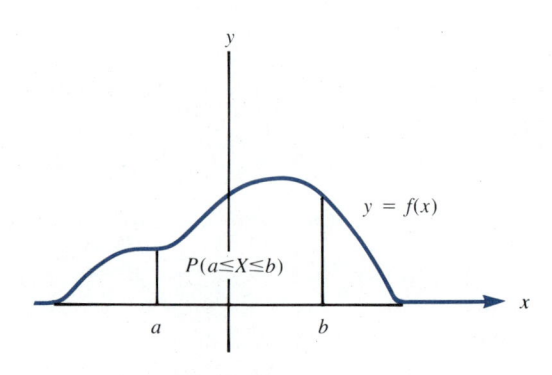

FIGURE 15–20
Probability as Area

MORE ABOUT INTEGRATION

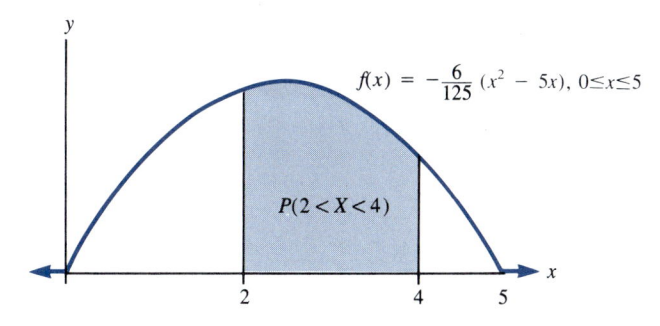

FIGURE 15–21

See Figure 15–21. Similarly,

P(a transistor's lifetime will be greater than 3 years)

$= P(X > 3)$

$= \int_3^5 -\frac{6}{125}(x^2 - 5x)\,dx$

$= \frac{44}{125} = 0.352.$

(Figure 15–22), and

P(a transistor's lifetime will be less than or equal to 1 year)

$= P(X \leq 1) = P(X < 1)$

$= \int_0^1 -\frac{6}{125}(x^2 - 5x)\,dx$

$= \frac{13}{125} = 0.104.$

(Figure 15–23). ☐

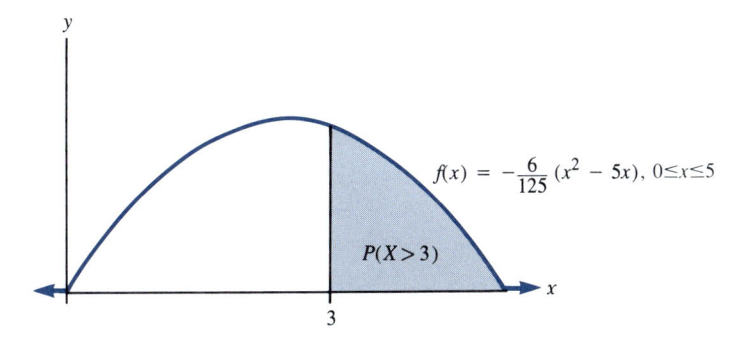

FIGURE 15–22

$$f(x) = -\frac{6}{125}(x^2 - 5x), \quad 0 \le x \le 5$$

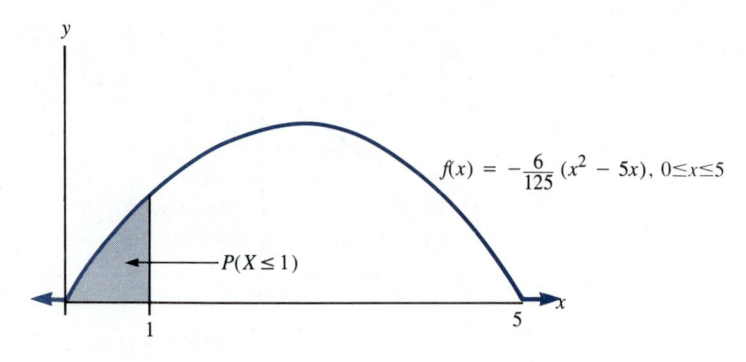

$P(X \le 1)$

FIGURE 15–23

EXAMPLE 4   Let X be a normally distributed random variable with mean μ and standard deviation σ. Then X is a continuous random variable defined on the interval $(-\infty, +\infty)$. What is its probability density function? It is the function f defined by

$$f(x) = \frac{1}{\sigma\sqrt{2\pi}} e^{-(x-\mu)^2/2\sigma^2}$$

for all x. Thus the bell curve of Figure 15–24, which describes the distribution of X and which we studied in Section 8.4, is in fact the graph of the function f. (See Exercise 42 of Exercises 13.4.) ☐

EXAMPLE 5   An automaker's quality control department tests engines by running them until they break down. Define

$$X = \text{Time it takes for an engine to break down, in months.}$$

Since the time it will take for an engine to break down is potentially unlimited, X can take on any value in the interval $[0, +\infty)$. Suppose the probability density function for X is given by

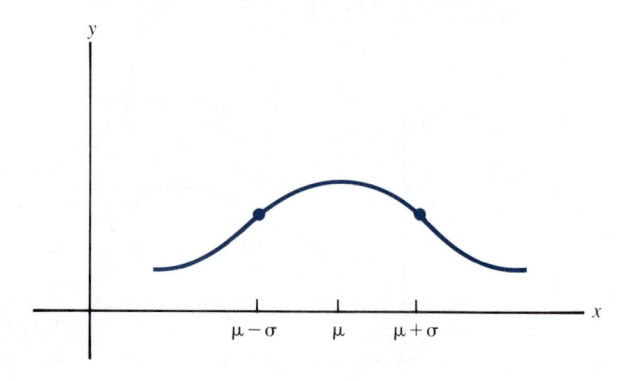

FIGURE 15–24
The Normal Curve

$$f(x) = \begin{cases} 0.5e^{-x/2}, & x \geq 0, \\ 0, & x < 0. \end{cases}$$

Then

$$P(1 < X < 3) = \int_1^3 0.5e^{-x/2}\, dx$$

$$= -e^{-x/2} \Big|_1^3 = -e^{-3/2} + e^{-1/2}$$

$$\approx 0.3834.$$

Similarly,

$$P(X \leq 2) = \int_0^2 0.5e^{-x/2}\, dx$$

$$= -e^{-x/2} \Big|_0^2 = -e^{-1} + e^0$$

$$= 1 - e^{-1} \approx 0.6321.$$

We could calculate $P(X > 2)$ by evaluating the improper integral

$$\int_2^{+\infty} 0.5e^{-x/2}\, dx;$$

however, it is easier to make use of the work we have already done and the fact that $X > 2$ is the complementary event to $X \leq 2$ and write

$$P(X > 2) = 1 - P(X \leq 2) = 1 - (1 - e^{-1}) = e^{-1} \approx 0.3679. \quad \square$$

Closely related to the probability density function of a continuous random variable is its **cumulative distribution function.**

The Cumulative Distribution Function

Let X be a continuous random variable. The **cumulative distribution function for X** is the function F defined by

$$F(x) = P(X \leq x) \text{ for all } x.$$

Note that if f is a probability density function for X, then

$$F(x) = P(X \leq x) = \int_{-\infty}^{x} f(t)\, dt$$

for all x, and thus $F(x)$ is the area under the graph of f from $-\infty$ to x. Hence F is an area function, and, just as in Section 14.4, this means that it is an antiderivative of f. Finally, if $a < b$,

$$P(a \le X \le b) = P(X \le b) - P(X \le a) = F(b) - F(a).$$

Let us summarize these properties of the cumulative distribution function:

Properties of the Cumulative Distribution Function

Let X be a continuous random variable. If f is a probability density function for X and F a cumulative distribution function for X, then

1. $F(x) = \displaystyle\int_{-\infty}^{x} f(t)\, dt \quad$ for all x.

2. $F'(x) = f(x) \quad$ for all x.

3. If $a < b$, then
$$P(a \le X \le b) = F(b) - F(a).$$

EXAMPLE 6   Suppose X is the random variable defined on $[0, +\infty)$ whose probability density function f is given by

$$f(x) = \begin{cases} 0.5 e^{-x/2}, & x \ge 0, \\ 0, & x < 0. \end{cases}$$

Let us find the cumulative distribution function F for X.

An antiderivative of f will be of the form

$$F(x) = \begin{cases} -e^{-x/2} + c_1, & x \ge 0, \\ c_2, & x < 0. \end{cases}$$

(Why?) When $x < 0$,

$$0 = P(X \le x) = F(x) = c_2,$$

so $c_2 = 0$. Also, when $x = 0$ we have

$$0 = P(X \le 0) = F(0) = -e^0 + c_1 = -1 + c_1,$$

which implies that $c_1 = 1$. Therefore the cumulative distribution function for X is defined by

$$F(x) = \begin{cases} -e^{-x/2} + 1, & x \ge 0, \\ 0, & x < 0. \end{cases}$$

Note that

$$P(1 < X < 3) = F(3) - F(1) = -e^{-3/2} + e^{-1/2}. \quad \square$$

In the previous example we began with a probability density function for the random variable and found its cumulative distribution function. Sometimes it is easier to begin with the cumulative distribution function and find the density function.

EXAMPLE 7

Let us choose a number at random from the interval [0, 10]. We let X denote the number chosen. Since each number in [0, 10] is equally likely to be selected, we conclude that there is 1 chance in 10 that the number chosen will be from the interval [0, 1], 2 chances in 10 that it will be from [0, 2], 3.5 chances in 10 that it will be from [0, 3.5], and, in general, x chances in 10 that it will be from [0, x]. Therefore we must have

$$P(0 \leq X \leq x) = \frac{x}{10}, \qquad 0 \leq x \leq 10.$$

Of course, if $x < 0$, then $P(X \leq x) = 0$, while if $x > 10$, $P(X \leq x) = 1$. Hence the cumulative distribution function F for X is defined by

$$F(x) = P(X \leq x) = \begin{cases} 0, & x < 0, \\ x/10, & 0 \leq x \leq 10, \\ 1, & x > 10, \end{cases}$$

and it follows that the probability density function for X is the function defined by

$$f(x) = F'(x) = \begin{cases} 1/10 & \text{if } 0 \leq x \leq 10, \\ 0, & \text{otherwise.} \end{cases} \quad \square$$

We conclude this section with a brief discussion of the expected value, variance, and standard deviation of a continuous random variable.

Expected Value, Variance, and Standard Deviation

If X is a continuous random variable with probability density function f, then its expected value $E(X)$ is given by

$$E(X) = \int_{-\infty}^{+\infty} x f(x) \, dx$$

and its variance $V(X)$ by

$$V(X) = \int_{-\infty}^{+\infty} x^2 f(x) \, dx - [E(X)]^2.$$

The standard deviation of X is the square root of its variance.

The interpretation of the expected value and variance are the same for continuous random variables as they are for discrete random variables: $E(X)$ is the average value of X in the long run, and $V(X)$ measures the degree to which the values of X vary from its expected value.

EXAMPLE 8

Again let $X = $ number of months a test engine runs before it fails, and suppose that the density function for X on $[0, +\infty)$ is given by

$$f(x) = \begin{cases} 0.5 e^{-x/2}, & x \geq 0, \\ 0, & x < 0. \end{cases}$$

We have

$$E(X) = \int_{-\infty}^{+\infty} xf(x)\, dx = \int_{0}^{+\infty} 0.5xe^{-x/2}\, dx$$

$$= \lim_{b \to +\infty} \int_0^b 0.5xe^{-x/2}\, dx.$$

Integration by parts gives

$$\int 0.5xe^{-x/2}\, dx = -xe^{-x/2} + \int e^{-x/2}\, dx = -xe^{-x/2} - 2e^{-x/2} + c.$$

(Check this.) Therefore

$$E(X) = \lim_{b \to +\infty} \left(-xe^{-x/2} - 2e^{-x/2} \Big|_0^b\right)$$

$$= \lim_{b \to +\infty} (-be^{-b/2} - 2e^{-b/2} + 2).$$

Since $\lim_{b \to +\infty} be^{-b/2} = 0$ and $\lim_{b \to +\infty} e^{-b/2} = 0$, we thus have $E(X) = 2$. The interpretation of this result is that on average the engines will run two months before breaking down.

We also have

$$V(X) = \int_{-\infty}^{+\infty} x^2 f(x)\, dx - [E(X)]^2 = \int_0^{+\infty} 0.5x^2 e^{-x/2}\, dx - [2]^2$$

$$= \lim_{b \to +\infty} \int_0^b x^2 (0.5) e^{-x/2}\, dx - 4$$

$$= \lim_{b \to +\infty} \left(-x^2 e^{-x/2} - 4xe^{-x/2} - 8e^{-x/2} \Big|_0^b\right) - 4$$

$$= \lim_{b \to +\infty} (-b^2 e^{-b/2} - 4be^{-b/2} - 8e^{-b/2} + 8) - 4$$

$$= 4.$$

Thus the variance of X is 4, and hence its standard deviation is $\sqrt{4} = 2$. ◻

15.4 EXERCISES

In Exercises 1 through 8, show that f is a probability density function.

1. $f(x) = \begin{cases} 1/3, & 0 \le x \le 3, \\ 0, & \text{otherwise} \end{cases}$

2. $f(x) = \begin{cases} x/2, & 0 \le x \le 2, \\ 0, & \text{otherwise} \end{cases}$

3. $f(x) = \begin{cases} \dfrac{3}{20} x(1-x)^2, & 1 \le x \le 3, \\ 0, & \text{otherwise} \end{cases}$

4. $f(x) = \begin{cases} \dfrac{1}{x^2}, & x \ge 1, \\ 0, & x < 1 \end{cases}$

More about Integration

5. $f(x) = \begin{cases} \dfrac{8}{x^3}, & x \geq 2, \\ 0, & x < 2 \end{cases}$

6. $f(x) = \begin{cases} 2e^{-2x}, & x \geq 0, \\ 0, & x < 0 \end{cases}$

7. $f(x) = \begin{cases} (k-1)x^{-k}, & k > 1, \ x \geq 1, \\ 0, & x < 1 \end{cases}$

8. $f(x) = \begin{cases} ke^{-kx}, & k > 0, \ x \geq 0, \\ 0, & x < 0 \end{cases}$

9. Let X be a continuous random variable whose probability density function is the function of Exercise 1. Find
   (a) $P(0 \leq X \leq 3)$
   (b) $P(2 \leq X \leq 3)$
   (c) $P(2.5 < X < 2.9)$
   (d) $P(X < 1.4)$
   (e) $P(X \geq 0.5)$
   (f) $P(a < X < b)$, $0 \leq a < b \leq 3$

10. Let X be a continuous random variable whose probability density function is the function of Exercise 2. Find
    (a) $P(0 < X < 0.75)$
    (b) $P(0.5 \leq X \leq 1.5)$
    (c) $P(X \leq 1.9)$
    (d) $P(X \geq 1.9)$
    (e) $P(X < b)$, $0 \leq b \leq 2$
    (f) a such that $P(X > a) = 0.10$

11. Let X be a continuous random variable whose probability density function is the function of Exercise 3. Find the cumulative distribution function for X and use it to find
    (a) $P(X \geq 2)$
    (b) $P(X < 2.5)$
    (c) $P(1.4 < X < 2.2)$
    (d) $P(1.5 \leq X \leq 2.4)$
    (e) $P(X > b)$, $1 \leq b \leq 3$
    (f) $P(X < a)$, $1 \leq a \leq 3$

12. Let X be a continuous random variable whose probability density function is the function of Exercise 4. Find
    (a) $P(1 < X < 5)$
    (b) $P(2 < X < 3)$
    (c) $P(X < 10)$
    (d) $P(X > 4)$
    (e) $P(X > b)$, $b \geq 1$
    (f) b such that $P(X < b) = 0.25$

13. Let X be a continuous random variable whose probability density function is the function of Exercise 5. Find the cumulative distribution function for X and use it to find
    (a) $P(2 < X < 4)$
    (b) $P(100 < X < 200)$
    (c) $P(X < 10)$
    (d) $P(X > 50)$
    (e) $P(a < X < b)$, $2 \leq a < b$
    (f) b such that $P(X \leq b) = 0.95$

14. Let X be a continuous random variable whose probability density function is the function of Exercise 6. Find the cumulative distribution function for X and use it to find
    (a) $P(1 < X < 2)$
    (b) $P(0.5 \leq X \leq 2.5)$
    (c) $P(X \geq 3)$
    (d) $P(X < \ln 5)$
    (e) $P(X \geq b)$, $b \geq 0$
    (f) a such that $P(X > a) = 0.50$

In Exercises 15 through 20, the given function F is a cumulative distribution function for a continuous random variable. Find the probability density function for the random variable.

15. $F(x) = \begin{cases} 0, & x < 0, \\ 3x, & 0 \leq x \leq \frac{1}{3}, \\ 1, & x > \frac{1}{3} \end{cases}$

16. $F(x) = \begin{cases} 0, & x < 0, \\ x^3, & 0 \leq x \leq 1, \\ 1, & x > 1 \end{cases}$

17. $F(x) = \begin{cases} 0, & x < 1, \\ \ln x^{1/3}, & 1 \leq x \leq e^3, \\ 1, & x > e^3 \end{cases}$

18. $F(x) = \begin{cases} 0, & x < 3, \\ \sqrt{x+1} - 2, & 3 \leq x \leq 8, \\ 1, & x > 8 \end{cases}$

19. $F(x) = \begin{cases} 0, & x < 0, \\ 1 - e^{-3x}, & x \geq 0 \end{cases}$
20. $F(x) = \begin{cases} 0, & x < 1, \\ 1 - x^{-5}, & x \geq 1 \end{cases}$

In Exercises 21 through 26, find the cumulative distribution function for the given probability distribution function f, and then use the cumulative distribution function to find the given probabilities.

21. $f(x) = \begin{cases} 1/5, & 0 \leq x \leq 5 \\ 0, & \text{otherwise} \end{cases}$, $P(1 < X < 4)$, $P(X < 2)$, $P(X > 3)$

22. $f(x) = \begin{cases} 1/x, & 1 \leq x \leq e \\ 0, & \text{otherwise} \end{cases}$, $P(X > 2)$, $P(X < 2)$, $P(1 < X < 2.5)$

23. $f(x) = \begin{cases} x^{-2}, & x \geq 1 \\ 0, & x < 1 \end{cases}$, $P(X < 2)$, $P(2 < X < 3)$, $P(X > 5)$

24. $f(x) = \begin{cases} 8x^{-3}, & x \geq 2 \\ 0, & x < 2 \end{cases}$, $P(4 < X < 6)$, $P(X > 3)$, $P(X < 4)$

25. $f(x) = \begin{cases} 3e^{-3x}, & x \geq 0 \\ 0, & x < 0 \end{cases}$, $P(X > 1)$, $P(X < 2)$, $P(3 < X < 5)$

26. $f(x) = \begin{cases} 4xe^{-2x}, & x \geq 0 \\ 0, & x < 0 \end{cases}$, $P(0 < X < 1)$, $P(X < 3)$, $P(X > 1/2)$

27. Let X denote a number chosen at random from the interval $[0, 4]$. Find the cumulative distribution function and the probability density function for X.

28. Let a point be selected at random from the circumference or interior of a circle of radius 1, and let X be the distance of the point from the center of the circle. Then X is a continuous random variable on the interval $[0, 1]$. Show that the cumulative distribution function for X is given by

$$F(x) = \begin{cases} 0, & x < 0, \\ x^2, & 0 \leq x \leq 1, \\ 1, & x > 1, \end{cases}$$

and find the probability density function for X. (*Hint:* Show that if $0 \leq x \leq 1$, then $P(0 \leq X \leq x) = x^2$.)

29. Let a point be selected at random from the perimeter or interior of the square whose vertices are at $(0, 0)$, $(0, 1)$, $(1, 1)$, and $(1, 0)$. If the point selected is (a, b), let X be the maximum of a and b. Then X is a continuous random variable on the interval $[0, 1]$. Show that the cumulative distribution function for X is given by

$$F(x) = \begin{cases} 0, & x < 0, \\ x^2, & 0 \leq x \leq 1, \\ 1, & x > 1. \end{cases}$$

In Exercises 30 through 35, find the expected value and variance of the continuous random variable whose probability density function f is given.

30. $f(x) = \begin{cases} 1/2, & 0 \leq x \leq 2, \\ 0, & \text{otherwise} \end{cases}$

31. $f(x) = \begin{cases} x/3, & 0 \leq x \leq \sqrt{6}, \\ 0, & \text{otherwise} \end{cases}$

32. $f(x) = \begin{cases} 1.2x(x + 1), & 0 \leq x \leq 1, \\ 0, & \text{otherwise} \end{cases}$

33. $f(x) = \begin{cases} x^{-2}, & x \geq 1, \\ 0, & x < 1 \end{cases}$

34. $f(x) = \begin{cases} 2e^{-2x}, & x \geq 0, \\ 0, & x < 0 \end{cases}$

35. $f(x) = \begin{cases} xe^{-x}, & x \geq 0, \\ 0, & x < 0 \end{cases}$

*36. Show that if X is a continuous random variable with probability density function f, then
$$V(X) = \int_{-\infty}^{+\infty} (x - E(X))^2 f(x)\, dx.$$

Management

Let X denote the operating lifetime of a certain type of computer disk drive, X in years, $0 \leq X \leq 4$. Suppose the probability density function for X is given by
$$f(x) = \begin{cases} \dfrac{3}{32}(4x - x^2), & 0 \leq x \leq 4, \\ 0, & \text{otherwise.} \end{cases}$$

Exercises 37 through 40 refer to this situation.

37. Find the probability that a disk drive will last less than one year.

38. Find the probability that a disk drive will last more than three years.

39. What percentage of the disk drives will have an operating lifetime of less than six months?

40. Find and interpret $E(X)$.

Let X denote the time, in weeks, between successive inventory reorders. Suppose that the probability density function for X is f, where
$$f(x) = \frac{1}{2}(x + 1)^{-3/2} \quad \text{if } x \geq 0, \qquad f(x) = 0 \quad \text{if } x < 0.$$

Exercises 41 through 44 refer to this situation.

41. Find the probability that the time between successive reorders will be at least one week.

42. If $P(X > b) = 0.05$, find b, and interpret your result.

43. If $P(X < a) = 0.1$, find a and interpret your result.

44. Find the expected time between reorders.

Let X denote the time, in weeks, between breakdowns of a production line, and suppose that X has probability distribution function f, where
$$f(x) = 2(x + 1)^{-3} \quad \text{for } x \geq 0, \qquad f(x) = 0 \quad \text{for } x < 0.$$

Exercises 45 through 48 refer to this situation.

45. Find the probability that the time between successive breakdowns will be less than one week.

46. Find the probability that the time between successive breakdowns will be more than two weeks.

47. Find the probability that the time between successive breakdowns will be more than one week but less than five weeks.

48. Find the expected time between breakdowns.

A manufacturer receives shipments of parts from a supplier at irregular intervals. Let X be the interarrival time, that is, the time between arrivals of successive shipments, with X measured in days. Suppose the probability density function f for X is given by
$$f(x) = 0.1e^{-0.1x} \quad \text{if } x \geq 0, \quad f(x) = 0 \quad \text{if } x < 0.$$

Exercises 49 through 53 refer to this situation.

49. Find the probability that the interarrival time will be between 5 and 10 days.

50. Find the probability that the interarrival time will be more than 14 days.

51. Find the probability that the interarrival time will be less than 30 days.

52. Find the time b such that 50% of the time the interarrival time will be less than or equal to b days.

53. Find the expected interarrival time.

Life Science

Let X denote the time between contraction of a certain disease and death, measured in years. Suppose that the probability distribution function for X is f, where

$$f(x) = \begin{cases} \dfrac{1}{(\ln 11)(x+1)}, & 0 \le x \le 10, \\ 0 & \text{otherwise.} \end{cases}$$

Exercises 54 through 58 refer to this situation.

54. What is the probability that a person who contracts the disease will live for more than five years?

55. What percentage of those who contract the disease die within one year?

56. What percentage of those who contract the disease survive at least two years?

57. The 5% who survive the longest after contracting the disease live how many years?

58. What is the expected survival time for those who contract the disease?

Let X denote the recovery time from surgery, measured in days, and suppose the probability density function for X is f, where

$$f(x) = 0.64xe^{-0.8x} \quad \text{for } x \ge 0, \quad f(x) = 0 \quad \text{for } x < 0.$$

Exercises 59 through 62 refer to this situation.

59. Find the probability that recovery time will be less than three days.

60. Find the percentage of patients whose recovery time will be greater than five days.

61. The 10% of the patients who recover the quickest recover in approximately how many days? (*Hint:* try 0.65 and 0.70 days.)

62. Find the expected recovery time.

According to an article by Anthony Nero, Jr. ("Controlling Indoor Pollution," *Scientific American*, May 1988), if X is the concentration of the dangerous radioactive gas radon in the air of a house, then the probability density function for X is approximately given by

$$y = \begin{cases} 0, & x < 0, \\ 13x, & 0 \le x < 0.2, \\ -12x + 5, & 0.2 \le x < 0.4, \\ 0.24e^{-0.44x}, & x \ge 0.4. \end{cases}$$

Here x is measured in hundreds of becquerels per cubic meter of air (b/m^3). Exercises 63 through 67 refer to this situation.

*63. Estimate the percentage of houses that have a radon concentration of less than 10 b/m^3.

*64. Estimate the percentage of houses that have a radon concentration of between 10 and 30 b/m³.

*65. Find the probability that a house will have a radon concentration greater than 300 b/m³.

*66. According to the article, people who live for 20 years in a house that has a radon concentration of 1000 b/m³ or more run a 2 to 3% annual risk of developing lung cancer as a result. What percentage of houses contribute to this risk?

*67. The 5% of the houses with the greatest concentration of radon have approximately what concentration?

Social Science

A sociologist studying fads among teenagers estimates that if X is the lifetime of a fad, X in months, $0 \leq X \leq 6$, then the probability density function for X is given by

$$f(x) = \begin{cases} -\frac{1}{54} x(1-x), & 0 \leq x \leq 6, \\ 0, & \text{otherwise.} \end{cases}$$

Exercises 68 through 70 refer to this situation.

68. Find the probability that a fad will last between two and four months.

69. The shortest-lived two-thirds of fads will last fewer than how many months?

70. What is the average lifetime of a fad?

15.5 NUMERICAL INTEGRATION

There are situations when it is necessary to evaluate $\int_a^b f(x)\, dx$ and it it not possible to find a simple expression for the antiderivative $F(x)$ of $f(x)$. For instance, there is no simple formula or defining equation for the antiderivative of

$$f(x) = e^{-x^2};$$

nevertheless, in statistics we must be able to evaluate integrals of the form

$$\int_a^b e^{-x^2}\, dx.$$

(See Example 4 of Section 15.4.) In a case like this we can use **numerical integration** to approximate the value of the integral. Numerical integration can be used to approximate a definite integral to any degree of accuracy. It is also useful when the integrand is defined by a collection of data points rather than by a functional expression.

Methods of numerical integration have become increasingly important with the advent of the computer. In this section we discuss and illustrate three of the most common methods: the midpoint rule, the trapezoidal rule, and Simpson's rule.

The Midpoint Rule

We saw in Section 14.5 that the definite integral $\int_a^b f(x)\,dx$ is the limit of Riemann sums:

$$\int_a^b f(x)\,dx = \lim_{\Delta x \to 0} \sum_{i=1}^{n} f(x_i)\,\Delta x,$$

where $\Delta x = \dfrac{b-a}{n}$, $a_0 = a$, $a_i = a + i\,\Delta x$ for $i = 1, \ldots, n$, and x_i may be chosen to be any number in the interval $[a_{i-1}, a_i]$. Thus

$$\int_a^b f(x)\,dx \approx \sum_{i=1}^{n} f(x_i)\,\Delta x,$$

and the approximation becomes exact as $\Delta x \to 0$, that is, as $n \to +\infty$. Since we are free to choose x_i to be any number in $[a_{i-1}, a_i]$, we may choose it to be the midpoint of this interval, which is

$$a_{i-1} + \frac{1}{2}\Delta x = a + (i-1)\,\Delta x + \frac{1}{2}\Delta x = a + \frac{2i-1}{2}\Delta x.$$

Hence

$$\int_a^b f(x)\,dx \approx \left[\sum_{i=1}^{n} f\!\left(a + \frac{2i-1}{2}\Delta x\right)\right]\Delta x;$$

that is,

$$\int_a^b f(x)\,dx \approx \left[f\!\left(a + \frac{1}{2}\Delta x\right) + f\!\left(a + \frac{3}{2}\Delta x\right) + \cdots + f\!\left(a + \frac{2n-1}{2}\Delta x\right)\right]\Delta x.$$

This is known as the **midpoint rule** for approximating definite integrals. The approximation is the area of the shaded region in Figure 15–25.

The Midpoint Rule

If n is a positive integer and $\Delta x = \dfrac{b-a}{n}$, then

$$\int_a^b f(x)\,dx \approx \left[\sum_{i=1}^{n} f\!\left(a + \frac{2i-1}{2}\Delta x\right)\right]\Delta x.$$

The approximation becomes exact as $n \to +\infty$.

EXAMPLE 1 Let us use the midpoint rule to approximate the definite integral

$$\int_1^2 x^2\,dx$$

More about Integration

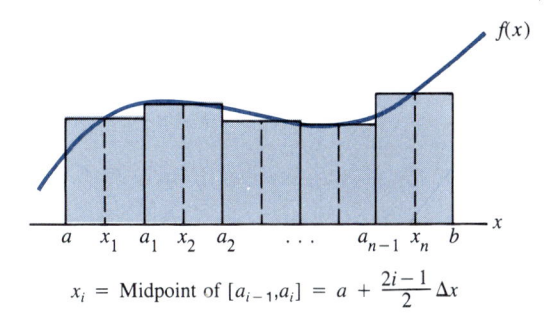

$$x_i = \text{Midpoint of } [a_{i-1}, a_i] = a + \frac{2i-1}{2}\Delta x$$

FIGURE 15-25

using $n = 5$ rectangles. Then $a = 1$, $b = 2$, and $\Delta x = (2 - 1)/5 = \frac{1}{5}$. Since $f(x) = x^2$ and

$$a + \frac{2i-1}{2}\Delta x = 1 + \frac{2i-1}{2}\frac{1}{5} = 1 + \frac{2i-1}{10}$$

for $i = 1, 2, 3, 4,$ and 5, the midpoint rule says that

$$\int_1^2 x^2\, dx \approx \left[\sum_{i=1}^{5}\left(1 + \frac{2i-1}{10}\right)^2\right]\frac{1}{5}.$$

We make a table:

i	$1 + \dfrac{2i-1}{10}$	$\left(1 + \dfrac{2i-1}{10}\right)^2$
1	$1 + 1/10 = 11/10$	$(11/10)^2 = 121/100$
2	$1 + 3/10 = 13/10$	$(13/10)^2 = 169/100$
3	$1 + 5/10 = 15/10$	$(15/10)^2 = 225/100$
4	$1 + 7/10 = 17/10$	$(17/10)^2 = 289/100$
5	$1 + 9/10 = 19/10$	$(19/10)^2 = 361/100$
		$1165/100$

Therefore

$$\int_1^2 x^2\, dx \approx \left[\frac{1165}{100}\right]\frac{1}{5} = 2.330.$$

See Figure 15-26, where the shaded area is 2.330.
    Since

$$\int_1^2 x^2\, dx = \frac{1}{3}x^3\bigg|_1^2 = \frac{7}{3} = 2.333333$$

to six decimal places, our approximation of 2.330 is accurate to two decimal places. We could obtain a better approximation to the integral by using more rectangles, that is, by using a larger value of n. ☐

The virtue of the midpoint rule is its simplicity; however, it often does not produce a good approximation for a definite integral unless the value of n is very large. Therefore other methods of numerical integration are usually preferable.

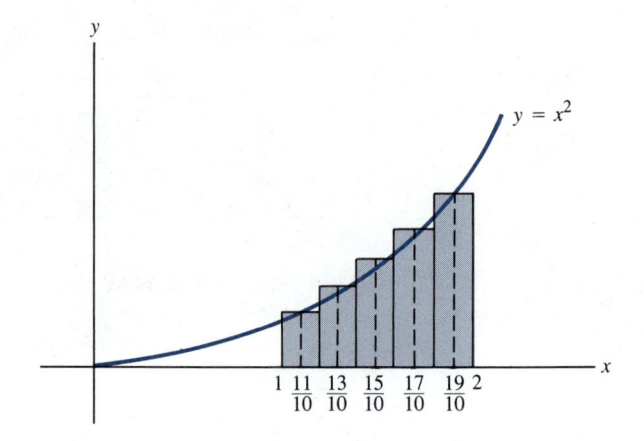

FIGURE 15–26

The Trapezoidal Rule

Since the definite integral of f from a to b may be interpreted as the area under f from a to b, any approximation to this area, whether it uses rectangles or some other means, is also an approximation to the integral. The **trapezoidal rule** approximates areas, and hence definite integrals, by means of trapezoids rather than rectangles. Figure 15–27 shows how this is done. The area under the graph of f from a to b, which is the definite integral of f from $x = a$ to $x = b$, is approximated by the sum of the areas of n trapezoids. As before, we set $\Delta x = (b - a)/n$ and partition the interval $[a, b]$ into n subintervals $[a, a_1], [a_1, a_2], \ldots, [a_{n-1}, b]$, where $a_i = a + i \Delta x$ for each i.

If a trapezoid has bases of length b_1 and b_2 and an altitude of length h, then its area is $(b_1 + b_2)h/2$. The ith trapezoid in Figure 15–27 has as its bases the line segments from $(a_{i-1}, 0)$ to $(a_{i-1}, f(a_{i-1}))$ and from $(a_i, 0)$ to $(a_i, f(a_i))$; its altitude is the interval $[a_{i-1}, a_i]$ on the x-axis. Therefore the area of the ith trapezoid is

$$\frac{1}{2}(f(a_{i-1}) + f(a_i)) \Delta x.$$

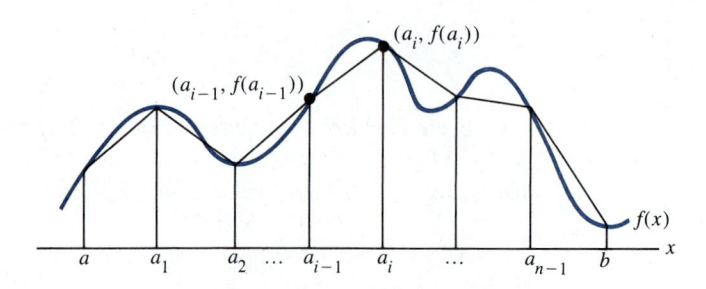

FIGURE 15–27

The sum of the areas of all n trapezoids is thus

$$\frac{1}{2}[f(a) + f(a_1)]\Delta x + \frac{1}{2}[f(a_1) + f(a_2)]\Delta x + \cdots$$

$$\cdots + \frac{1}{2}[f(a_{n-2}) + f(a_{n-1})]\Delta x + \frac{1}{2}[f(a_{n-1}) + f(b)]\Delta x,$$

which simplifies to

$$\left[\frac{1}{2}f(a) + f(a_1) + f(a_2) + \cdots + f(a_{n-1}) + \frac{1}{2}f(b)\right]\Delta x,$$

or

$$\left[\frac{1}{2}f(a) + \frac{1}{2}f(b) + \sum_{i=1}^{n-1} f(a + i\,\Delta x)\right]\Delta x.$$

(Check this.) Therefore we have

The Trapezoidal Rule

If n is a positive integer and $\Delta x = \dfrac{b-a}{n}$, then

$$\int_a^b f(x)\,dx \approx \left[\frac{1}{2}f(a) + \frac{1}{2}f(b) + \sum_{i=1}^{n-1} f(a + i\,\Delta x)\right]\Delta x.$$

This approximation becomes exact as $n \to +\infty$.

EXAMPLE 2 Use the trapezoidal rule to approximate

$$\int_1^2 \frac{1}{x}\,dx$$

with $n = 5$ trapezoids. Then $a = 1$, $b = 2$, and $\Delta x = (2-1)/5 = 1/5$, so the trapezoidal rule says that

$$\int_1^2 f(x)\,dx \approx \left[\frac{1}{2}f(1) + \frac{1}{2}f(2) + \sum_{i=1}^{4} f\left(1 + i\frac{1}{5}\right)\right]\frac{1}{5}.$$

But since $f(x) = 1/x$, we have

$$\int_1^2 \frac{1}{x}\,dx \approx \left[\frac{1}{2}\frac{1}{1} + \frac{1}{2}\frac{1}{2} + \frac{1}{1 + 1/5} + \frac{1}{1 + 2/5} + \frac{1}{1 + 3/5} + \frac{1}{1 + 4/5}\right]\frac{1}{5}$$

$$= \left[\frac{1}{2}\frac{1}{1} + \frac{1}{2}\frac{1}{2} + \frac{1}{6/5} + \frac{1}{7/5} + \frac{1}{8/5} + \frac{1}{9/5}\right]\frac{1}{5}$$

$$= \left[\frac{1}{2} + \frac{1}{4} + \frac{5}{6} + \frac{5}{7} + \frac{5}{8} + \frac{5}{9}\right]\frac{1}{5}$$

$$= 0.6956349.$$

CHAPTER 15

Since

$$\int_1^2 \frac{1}{x} dx = \ln x \Big|_1^2 = \ln 2 - \ln 1 = \ln 2 = 0.693147,$$

our approximation using the trapezoidal rule is accurate to two decimal places. We could obtain a more accurate approximation by using a larger value of n. □

In Section 14.4 we remarked that the total change in a quantity Q can be thought of as an area under the graph of the function that describes the rate of change of Q. With the aid of numerical integration we can exploit this fact to estimate total change when the function f is unknown, provided some of its functional values have been determined. We illustrate with an example.

EXAMPLE 3

Suppose a firm wishes to find the total cost of increasing its daily production from 100 units per day to 125 units per day. This cost is equal to

$$\int_{100}^{125} c(x) \, dx,$$

where $c(x)$ denotes the marginal cost of making x units. Suppose also that the firm does not know its marginal cost function, but only has information on its value at selected points, as follows:

Units Produced	Marginal Cost
100	36
105	28
110	26
115	30
120	36
125	44

See Figure 15–28. Then by the trapezoidal rule with $a = 100$, $b = 125$, and $\Delta x = 5$, we have

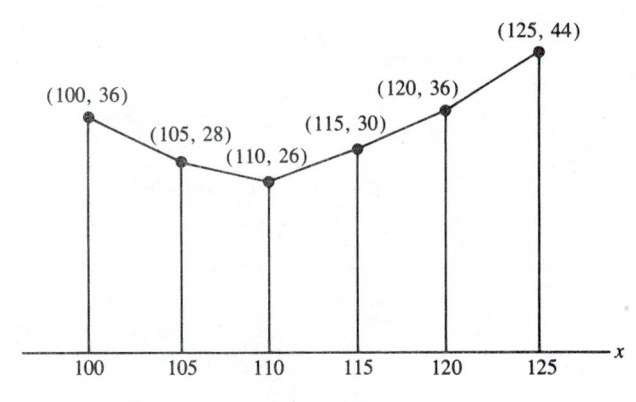

FIGURE 15–28

$$\int_{100}^{125} c(x)\,dx \approx \left[\frac{1}{2}c(100) + \frac{1}{2}c(125) + c(105) + c(110) + c(115) + c(120)\right]5$$

$$= \left[\frac{1}{2}(36) + \frac{1}{2}(44) + 28 + 26 + 30 + 36\right]5$$

$$= 800$$

dollars. ◻

Simpson's Rule

Another widely used method of numerical integration is **Simpson's rule,** which approximates the definite integral by areas of parabolic regions. Figure 15–29 shows how this is done. As usual,

$$\Delta x = (b - a)/n \quad \text{and} \quad a_i = a + i\,\Delta x$$

for each i. Here n must be an even positive integer. It can be shown that the area of the region bounded by $x = a_{i-1}$, $x = a_{i+1}$, the x-axis, and the parabolic arc that passes through the points $(a_{i-1}, f(a_{i-1}))$, $(a_i, f(a_i))$, and $(a_{i+1}, f(a_{i+1}))$ on the graph of f is

$$[f(a_{i-1}) + 4f(a_i) + f(a_{i+1})]\frac{\Delta x}{3}.$$

(See Exercise 46 of Exercises 15.5.) The sum of all such terms simplifies to

$$[f(a) + 4f(a_1) + 2f(a_2) + 4f(a_3) + \cdots + 2f(a_{n-2}) + 4f(a_{n-1}) + f(b)]\frac{\Delta x}{3}.$$

(Check this.) Therefore Simpson's rule may be stated as follows:

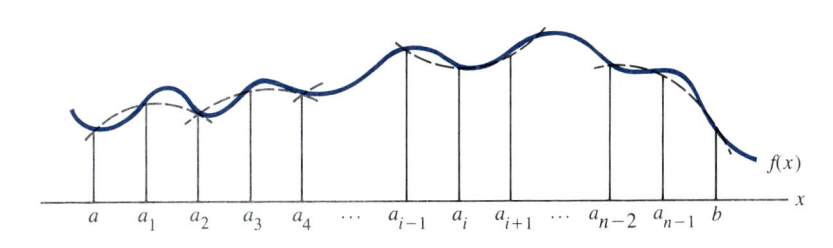

FIGURE 15–29

> **Simpson's Rule**
>
> If n is an even positive integer, $\Delta x = \dfrac{b-a}{n}$, and $a_i = a + i\,\Delta x$ for each i, then
>
> $$\int_a^b f(x)\,dx \approx [\,f(a) + 4f(a_1) + 2f(a_2) + \cdots$$
>
> $$\cdots + 2f(a_{n-2}) + 4f(a_{n-1}) + f(b)\,]\dfrac{\Delta x}{3}.$$
>
> This approximation becomes exact as $n \to +\infty$.

EXAMPLE 4 Use Simpson's rule with $n = 4$ subintervals to approximate

$$\int_1^2 \frac{1}{x}\,dx.$$

Then $a = 1$, $b = 2$, and $\Delta x = \frac{1}{4}$, so

$$a = 1, \quad a_1 = \frac{5}{4}, \quad a_2 = \frac{3}{2}, \quad a_3 = \frac{7}{4}, \quad \text{and} \quad b = 2.$$

Thus

$$\int_1^2 \frac{1}{x}\,dx \approx \left[\frac{1}{1} + 4\frac{1}{5/4} + 2\frac{1}{3/2} + 4\frac{1}{7/4} + \frac{1}{2}\right]\frac{1/4}{3} = 0.693254,$$

to six decimal places. Since

$$\int_1^2 \frac{1}{x}\,dx = \ln x \,\Big|_1^2 = \ln 2 - \ln 1 = \ln 2 = 0.693147,$$

our approximation using Simpson's rule is accurate to three decimal places. As usual, a larger value of n will produce a better approximation. ∎

■ *Computer Manual: Program NUMERINT*

15.5 EXERCISES

The Midpoint Rule

In Exercises 1 through 10, use the midpoint rule with the given value of n to approximate the given definite integral. In Exercises 1 through 6, compare your answer with the actual value of the integral.

1. $\displaystyle\int_1^5 2x\,dx, \quad n = 4$

2. $\displaystyle\int_1^5 2x\,dx, \quad n = 6$

3. $\displaystyle\int_1^2 x^2\,dx, \quad n = 6$

4. $\displaystyle\int_1^2 x^2\,dx,\quad n=8$

5. $\displaystyle\int_2^4 \frac{1}{x}\,dx,\quad n=5$

6. $\displaystyle\int_2^4 \frac{1}{x}\,dx,\quad n=6$

7. $\displaystyle\int_1^2 \frac{1}{1+\sqrt{x}}\,dx,\quad n=8$

8. $\displaystyle\int_0^1 \sqrt{1+x^2}\,dx,\quad n=8$

9. $\displaystyle\int_0^{1/2} \frac{1}{\sqrt{1-x^2}}\,dx,\quad n=10$

10. $\displaystyle\int_2^4 e^{\sqrt{x}}\,dx,\quad n=10$

11. Use the midpoint rule with $n=4$ to approximate
$$\int_0^1 \frac{1}{x^2+1}\,dx.$$
The actual value of this integral is $\pi/4$. How does your approximation compare with the actual value?

12. Use the midpoint rule with $n=4$ to approximate
$$\int_0^1 e^{-x^2}\,dx.$$

13. A building lot is bordered by a stream:

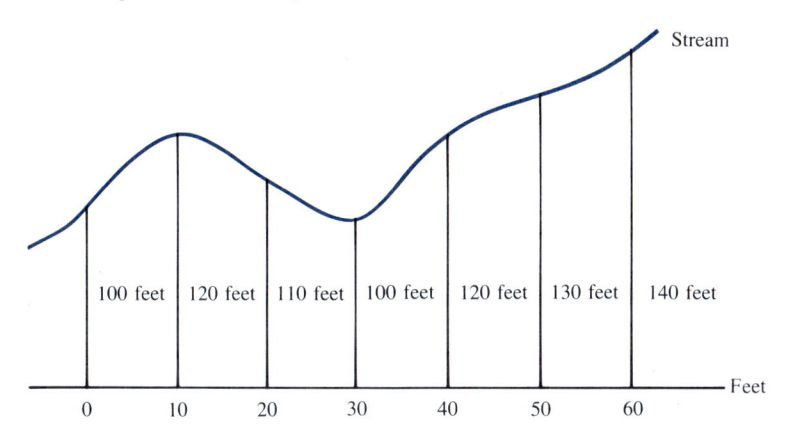

Estimate the area of the lot using the midpoint rule.

The Trapezoidal Rule

In Exercises 14 through 19, use the trapezoidal rule with the given value of n to approximate the given definite integral. In each case compare your answer with the actual value of the integral.

14. $\displaystyle\int_0^2 x\,dx,\quad n=5$

15. $\displaystyle\int_1^2 x^2\,dx,\quad n=6$

16. $\displaystyle\int_1^2 x^2\,dx,\quad n=8$

17. $\displaystyle\int_1^2 x^2\,dx,\quad n=12$

18. $\displaystyle\int_1^2 \frac{1}{x}\,dx,\quad n=5$

19. $\displaystyle\int_1^2 \frac{1}{x}\,dx,\quad n=10$

20. Use the trapezoidal rule with $n=8$ to approximate the value of
$$\int_1^2 \frac{1}{1+\sqrt{x}}\,dx.$$
Compare your answer with that of Exercise 7.

21. Use the trapezoidal rule with $n = 8$ to approximate the value of
$$\int_0^1 \sqrt{1 + x^2}\, dx.$$
Compare your answer with that of Exercise 8.

22. Use the trapezoidal rule with $n = 10$ to approximate the value of
$$\int_0^{1/2} \frac{1}{\sqrt{1 - x^2}}\, dx.$$
Compare your answer with that of Exercise 9.

23. Use the trapezoidal rule with $n = 10$ to approximate the value of
$$\int_2^4 e^{\sqrt{x}}\, dx.$$
Compare your answer with that of Exercise 10.

24. Use the trapezoidal rule with $n = 4$ to approximate the value of
$$\int_0^1 \frac{1}{x^2 + 1}\, dx.$$
Compare your answer with that of Exercise 11.

25. Use the trapezoidal rule with $n = 4$ to approximate the value of
$$\int_0^1 e^{-x^2}\, dx.$$
Compare your answer with that of Exercise 12.

*26. Show that using the trapezoidal rule to approximate
$$\int_a^b (mx + c)\, dx$$
with any n yields the exact value of the integral.

Management

27. A firm estimates its marginal cost at various production levels to be as follows:

Production (units)	10	12	14	16	18	20
Marginal Cost ($)	200	202	204	207	212	220

Use these data and the trapezoidal rule to approximate the firm's cost if it increases production from 10 to 20 units.

28. A firm's marginal profit at various production levels is as follows:

Production (units)	100	110	120	130	140	150	160
Marginal Profit ($)	48	32	12	−2	−22	−35	−45

Use the trapezoidal rule to approximate the change in profit if production is increased from 100 to 160 units.

29. An efficiency expert sampled a new worker's daily production at various times during the worker's first three weeks on the job. The result was the following graph:

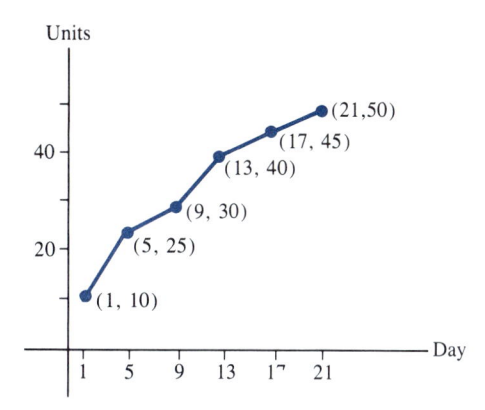

Use the trapezoidal rule to approximate the area under the graph from $x = 1$ to $x = 21$ and interpret your result.

Life Science

30. All ponds that have a surface area of more than 1 acre (43,560 square feet) qualify for stocking by the state fish and game department. Use the trapezoidal rule to determine whether or not the following pond qualifies for stocking (all figures are in feet):

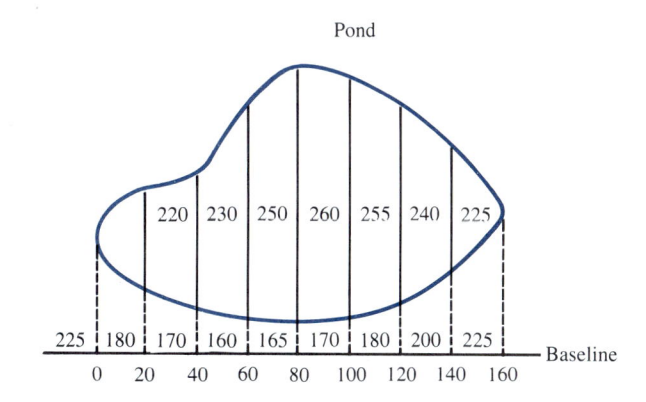

31. The concentration of a drug in a patient's bloodstream was sampled once each day, and the following graph was prepared:

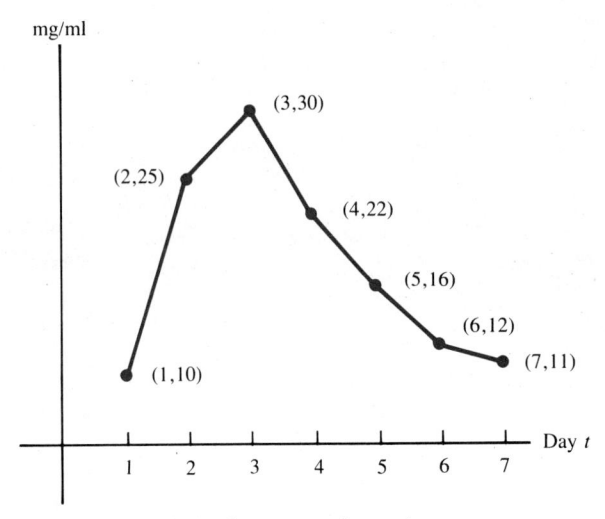

Use the trapezoidal rule to approximate the average concentration of the drug in the patient's bloodstream.

32. The rate at which an enzyme is produced by a biochemical reaction is as follows:

Time (secs)	0.0	0.1	0.2	0.3	0.4	0.5	0.6	0.7	0.8	0.9	1.0
Production (grams/sec)	0.0	0.2	0.8	1.4	2.9	4.5	3.2	1.7	0.9	0.3	0.1

Use the trapezoidal rule to estimate the total amount of the enzyme produced in 1 second.

Social Science

33. A psychologist measures the rate at which college freshmen can master new concepts in mathematics as a function of the number of weeks they have spent in college mathematics courses and finds the following:

Number of Weeks of Math Completed	5	10	15	20	25
Number of New Concepts Mastered per Week	2	6	10	13	14

Use the trapezoidal rule to approximate the total number of new concepts mastered from week 5 through week 25.

Simpson's Rule

In Exercises 34 through 39, use Simpson's rule with the given value of n to approximate the given definite integral. In each case compare your answer with the actual value of the integral.

34. $\int_{1}^{5} 2x\, dx, \quad n = 4$

35. $\int_{1}^{2} x^2\, dx, \quad n = 8$

36. $\int_{0}^{1} x^3\, dx, \quad n = 4$

37. $\int_{0}^{1} x^3\, dx, \quad n = 8$

38. $\int_{1}^{2} \frac{1}{x}\, dx, \quad n = 6$

39. $\int_{1}^{2} \frac{1}{x}\, dx, \quad n = 10$

40. Use Simpson's rule with $n = 8$ to approximate the value of
$$\int_1^2 \frac{1}{1+\sqrt{x}}\,dx.$$
Compare your answer with those of Exercises 7 and 20.

41. Use Simpson's rule with $n = 8$ to approximate the value of
$$\int_0^1 \sqrt{1+x^2}\,dx.$$
Compare your answer with those of Exercises 8 and 21.

42. Use Simpson's rule with $n = 10$ to approximate the value of
$$\int_0^{1/2} \frac{1}{\sqrt{1-x^2}}\,dx.$$
Compare your answer with those of Exercises 9 and 22.

43. Use Simpson's rule with $n = 10$ to approximate the value of
$$\int_2^4 e^{\sqrt{x}}\,dx.$$
Compare your answer with those of Exercises 10 and 23.

44. Use Simpson's rule with $n = 4$ to approximate the value of
$$\int_0^1 \frac{1}{x^2+1}\,dx.$$
Compare your answer with those of Exercises 11 and 24.

45. Use Simpson's rule with $n = 4$ to approximate the value of
$$\int_0^1 e^{-x^2}\,dx.$$
Compare your answer with those of Exercises 12 and 25.

*46. Let $f(x) = ax^2 + bx + c$ and $a_2 = a_1 + \Delta x$, $a_3 = a_1 + 2\Delta x$, $\Delta x > 0$. Show that
$$\int_{a_1}^{a_3} f(x)\,dx = [f(a_1) + 4f(a_2) + f(a_3)]\frac{\Delta x}{3}.$$

Management

47. Acme Computer Company estimates that the number of worker-hours that will be required to build the xth unit of its new computer is as follows:

Unit Number	1	4	7	10	13
Worker-hours	4000	3200	3000	2850	2750

Use Simpson's rule to approximate the total number of worker-hours that will be required to build units 1 through 13.

48. Repeat Exercise 13 using Simpson's rule and compare your result with that of Exercise 13.

49. Repeat Exercise 28 using Simpson's rule and compare your result with that of Exercise 28.

Life Science

50. The following graph shows the rate at which pollutants entered a lake at various times during one day of sampling:

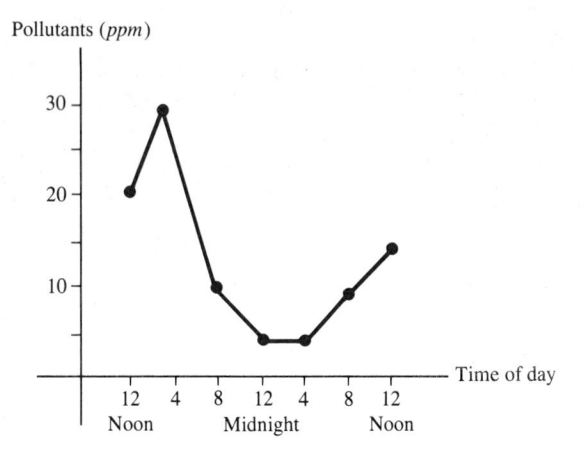

Use Simpson's rule to approximate the area under the curve and interpret your result.

51. Repeat Exercise 30 using Simpson's rule and compare your result with that of Exercise 30.

52. Repeat Exercise 32 using Simpson's rule and compare your result with that of Exercise 32.

15.6 DIFFERENTIAL EQUATIONS

A **differential equation** is an equation containing one or more derivatives: for example,

$$\frac{dy}{dx} = 8xy$$

or

$$2x\frac{d^2y}{dx^2} + 3xy^2\frac{dy}{dx} + y = 0.$$

A function is a **solution** of a differential equation if the function satisfies the equation. For instance, the function defined by

$$y = e^{4x^2}$$

is a solution of the differential equation

$$\frac{dy}{dx} = 8xy$$

because

$$\frac{dy}{dx} = \frac{d}{dx}(e^{4x^2}) = 8xe^{4x^2} = 8xy.$$

Differential equations are important, particularly in the sciences, because often we do not know the relationship between two quantities but we do know the rate of change of one of them with respect to the other. In this case we can use the fact that the derivative is the rate of change to write a differential equation involving the quantities; then we can determine the relationship between the quantities by solving this differential equation. For instance, suppose we know that an object moves in a straight line so that its velocity at time t is $v = 20t$ kilometers per hour, and we wish to find the distance $s(t)$ the object will travel in t hours. Since velocity is the derivative of distance, we may write

$$\frac{d}{dt}(s(t)) = 20t$$

and solve this differential equation to find $s(t)$. (If the distance traveled at time $t = 0$ is 0 kilometers, the solution is $s(t) = 10t^2$.)

In this section we introduce the terminology of differential equations and show how a few particularly simple differential equations can be solved and how they can be used to model some practical situations.

Suppose we wish to solve the differential equation

$$\frac{dy}{dx} = 2x + 5$$

for y. This means that we wish to find *all* functions $y = y(x)$ that satisfy the equation. If we consider the derivative to be the ratio of the differentials dy and dx, we may multiply both sides of the differential equation by dx to obtain

$$dy = (2x + 5)\, dx.$$

Then

$$\int dy = \int (2x + 5)\, dx,$$

so

$$y = x^2 + 5x + c.$$

(Strictly speaking, the integration yields the equation

$$y + c_1 = x^2 + 5x + c_2,$$

but the two constants of integration c_1 and c_2 can be combined into a single constant $c = c_2 - c_1$ by subtracting c_1 from both sides of this equation. From now on we will always combine constants of integration in this manner.) We have thus shown that every function $y = y(x)$ that satisfies the differential equation

$$\frac{dy}{dx} = 2x + 5$$

is of the form
$$y = x^2 + 5x + c,$$
and we call y the **general solution** of the differential equation. If we require that the solution of the differential equation satisfy an **initial condition,** such as $y(0) = 4$, then we can determine the value of the constant c and find the **particular solution** that satisfies the **differential equation with initial condition**
$$\frac{dy}{dx} = 2x + 5, \quad y(0) = 4.$$
Thus, if we substitute 0 for x in the general solution $y = x^2 + 5x + c$, we obtain
$$4 = y(0) = 0^2 + 5(0) + c = c.$$
Hence $c = 4$, and the particular solution of the differential equation with initial condition is therefore
$$y = x^2 + 5x + 4.$$

EXAMPLE 1 Let us find the general solution of the differential equation
$$\frac{dy}{dx} = x^3 - 3x^2 + 5x + 1.$$
We have
$$dy = (x^3 - 3x^2 + 5x + 1)\, dx,$$
so
$$\int dy = \int (x^3 - 3x^2 + 5x + 1)\, dx,$$
and thus
$$y = \frac{1}{4}x^4 - x^3 + \frac{5}{2}x^2 + x + c. \quad \square$$

EXAMPLE 2 Let us find the particular solution of the differential equation with initial condition
$$\frac{dy}{dx} = x^3 - 3x^2 + 5x + 1, \quad y(1) = 0.$$
From Example 1, the general solution is
$$y = \frac{1}{4}x^4 - x^3 + \frac{5}{2}x^2 + x + c,$$
so
$$0 = y(1) = \frac{1}{4}(1)^2 - (1)^3 + \frac{5}{2}(1)^2 + 1 + c = \frac{11}{4} + c,$$
and thus $c = -11/4$. Hence the particular solution is
$$y = \frac{1}{4}x^4 - x^3 + \frac{5}{2}x^2 + x - \frac{11}{4}. \quad \square$$

EXAMPLE 3 An object that falls freely in a vacuum under the influence of the earth's gravity accelerates at a constant rate of 32 feet per second per second. Find the velocity of the object 3 seconds after it begins to fall.

Let $v = v(t)$ denote the object's velocity t seconds after it begins to fall. Since acceleration is the derivative of velocity, we have

$$\frac{dv}{dt} = 32.$$

Hence

$$\int dv = \int 32 \, dt,$$

so

$$v = 32t + c.$$

At the instant the object begins its fall, its velocity is 0 feet per second, so $v(0) = 0$. Hence $c = 0$ and

$$v = 32t.$$

Therefore the velocity of the object at the end of 3 seconds is

$$v(3) = 32(3) = 96$$

feet per second. ☐

Situations that lead to differential equations are often expressed in the language of proportionality. Thus, if we say that the rate of change of y with respect to x is **directly proportional** to u, we mean that

$$\frac{dy}{dx} = ku,$$

where k is some number known as a **constant of proportionality.** Similarly, if the rate of change of y with respect to x is **inversely proportional** to u, then

$$\frac{dy}{dx} = \frac{k}{u},$$

where k is a constant of proportionality. We give examples to illustrate the construction of differential equations using the language of proportionality.

EXAMPLE 4 Suppose the rate of change of the concentration of pollutants in a lake is directly proportional to the number of summer cottages surrounding the lake. Let $y = y(x)$ denote the concentration of pollutants in the lake, measured in parts per million, when there are x summer cottages surrounding the lake. Then

$$\frac{dy}{dx} = kx$$

for some constant of proportionality k. Hence $dy = k \, dx$, so

$$\int dy = \int kx \, dx,$$

and thus
$$y = \frac{k}{2}x^2 + c.$$

To determine the values of k and c, we need to know the value of y for two distinct values of x. For instance, suppose that pollution was 0 ppm when there were no cottages on the lake, and 12 ppm when there were 20 cottages there; that is, suppose that $y(0) = 0$ and $y(20) = 12$. Then
$$0 = y(0) = \frac{k}{2}(0)^2 + c = c,$$
so $c = 0$ and hence
$$y = \frac{k}{2}x^2.$$

But then
$$12 = y(20) = \frac{k}{2}(20)^2 = 200k,$$
and thus $k = 12/200 = 0.06$. Therefore
$$y = 0.03x^2. \quad \square$$

EXAMPLE 5 Suppose that a firm's marginal cost is inversely proportional to the square root of the number of units produced, that its fixed cost is \$1000, and that it costs the firm \$2000 to make 4 units. If x denotes the number of units the firm produces and C is its cost function, then we have
$$\frac{dC}{dx} = \frac{k}{\sqrt{x}}, \quad C(0) = 1000, \quad C(4) = 2000, \quad k \text{ a constant.}$$
Thus
$$\int dC = \int \frac{k}{\sqrt{x}}\,dx,$$
and
$$C = 2k\sqrt{x} + c.$$
Now $C(0) = 1000$ implies that $c = 1000$, so
$$C = 2k\sqrt{x} + 1000.$$
But then $C(4) = 2000$ implies that $k = 250$ (check this), so
$$C = 500\sqrt{x} + 1000. \quad \square$$

EXAMPLE 6 Consider the differential equation
$$\frac{dy}{dx} = \frac{x}{2y}.$$

(Note that $y \neq 0$.) We can rewrite the equation as
$$2y\, dy = x\, dx.$$

Thus we have **separated the variables** of the equation: all y-terms are on the left and all x-terms on the right, and we can integrate both sides:
$$\int 2y\, dy = \int x\, dx,$$

so
$$y^2 = \frac{x^2}{2} + c.$$

We may leave the general solution in this form or solve it for y to obtain
$$y = \pm\sqrt{\frac{x^2}{2} + c}. \quad \square$$

In Section 13.5 we said that if the rate of change of y was directly proportional to y, then y was of the form $y = Ce^{rt}$, and we used this fact to construct models of exponential growth and decline. Now we can prove that if
$$\frac{dy}{dt} = ry,$$

then $y = Ce^{rt}$: if
$$\frac{dy}{dt} = ry,$$

then we may separate the variables to obtain
$$\frac{dy}{y} = r\, dt.$$

Thus
$$\int \frac{dy}{y} = \int r\, dt,$$

so
$$\ln|y| = rt + c.$$

But this implies that
$$e^{\ln|y|} = e^{rt+c},$$

or
$$|y| = e^c e^{rt},$$

or
$$y = \pm e^c e^{rt},$$

But $\pm e^c$ is a constant, so if we set $C = \pm e^c$, we have $y = Ce^{rt}$. Thus the exponential growth and decline models of Section 13.5 occur as the solutions of differential equations.

The limited growth and decline models of Section 13.5 also occur as the solutions of differential equations. We illustrate with an example.

EXAMPLE 7   Suppose there were 300 coyotes in New England in 1980 and 350 in 1985 and that the maximum coyote population that the region can support is 800. Suppose also that the rate of growth of the coyote population at time t is directly proportional to the difference between the limit of 800 and the size of the population at time t. Then if we let $y = y(t)$ denote the number of coyotes in New England in year t, where $t = 0$ represents 1980, we have

$$\frac{dy}{dt} = k(800 - y), \qquad y(0) = 300, \qquad y(5) = 350.$$

Separation of variables now yields

$$\frac{dy}{800 - y} = k\,dt,$$

so

$$\int \frac{dy}{800 - y} = \int k\,dt.$$

Since the population of coyotes is always less than 800, $800 - y$ is positive, and therefore the integration yields

$$-\ln(800 - y) = kt + c,$$

or

$$\ln(800 - y)^{-1} = kt + c,$$

But this implies that

$$e^{\ln(800-y)^{-1}} = e^{kt+c} = e^c e^{kt} = Ce^{kt},$$

where $C = e^c$. Thus

$$(800 - y)^{-1} = Ce^{kt},$$

or

$$\frac{1}{800 - y} = Ce^{kt}.$$

Solving this for y, we obtain

$$y = 800 - \frac{1}{Ce^{kt}} = 800 - \frac{1}{C}e^{-kt}.$$

Letting $b = 1/C$, we have

$$y = 800 - be^{-kt},$$

which is a limited growth model of the type introduced in Section 13.5. Since $y(0) = 300$,

$$300 = y(0) = 800 - be^{-k(0)} = 800 - b$$

and therefore $b = 500$. Thus $y = 800 - 500e^{-kt}$. Now $y(5) = 350$ implies that

$$350 = 800 - 500e^{-5k},$$

which when solved for k yields $k = 0.021$. Hence

$$y = 800 - 350e^{-0.021t}. \quad \square$$

15.6 EXERCISES

In Exercises 1 through 20, first find the general solution of the differential equation and then find the particular solution that satisfies the initial condition.

1. $\dfrac{dy}{dx} = 3x - 1, \quad y(0) = 0$
2. $\dfrac{dy}{dx} = 4 - 2x, \quad y(0) = 2$
3. $\dfrac{dy}{dx} = x^2 - 2x + 1, \quad y(1) = \dfrac{1}{3}$
4. $\dfrac{dy}{dx} = -2x^{-3}, \quad y(1) = 1$
5. $\dfrac{dy}{dx} = \dfrac{2}{x^2}, \quad y(2) = -1$
6. $\dfrac{dy}{dx} = 3x^{-3} + x, \quad y(1) = 2$
7. $\dfrac{dy}{dt} = \dfrac{4}{t}, \quad y(e) = 1$
8. $\dfrac{dy}{dx} = 2e^x, \quad y(0) = 5$
9. $\dfrac{dy}{dx} = 2e^{-x} + 1, \quad y(0) = -2$
10. $\dfrac{ds}{dt} = \dfrac{1}{t+3}, \quad s(-2) = 2$
11. $\dfrac{dy}{dx} = x\sqrt{x^2 + 1}, \quad y(0) = 1$
12. $\dfrac{dy}{dx} = \dfrac{x}{x^2+1}, \quad y(0) = 1$
13. $x\dfrac{dy}{dx} = 1, \quad y(1) = 0$
14. $\dfrac{dy}{dx} = \dfrac{2x}{y}, \quad y(0) = 2$
15. $\dfrac{dy}{dx} = \dfrac{x^2}{y}, \quad y(1) = 3$
16. $\dfrac{dy}{dx} = -\dfrac{y}{x}, \quad y(1) = 1, \quad y > 0$
17. $\dfrac{dy}{dx} = 3x^2y, \quad y(0) = 2, \quad y > 0$
18. $\dfrac{dy}{dx} = 3xy^2, \quad y(0) = 2, \quad y \neq 0$
19. $\dfrac{dy}{dx} = 2x(y - 1), \quad y(0) = 4, \quad y > 1$
20. $\dfrac{dy}{dx} = \dfrac{y^2}{x^2}, \quad y(1) = \dfrac{1}{2}, \quad y \neq 0$

21. A drag racer's car accelerates at a constant rate of 100 feet per second per second. Find the racer's velocity 2 seconds after the start of a race.

22. Repeat Exercise 21 if the car's acceleration is directly proportional to the time since the start of the race and its velocity at the end of 1 second is 100 feet per second.

23. A space capsule traveling in orbit at 18,000 mph fired its rockets in order to move to a higher orbit, thus accelerating at a constant rate of 200 mph².
    (a) Find the velocity of the capsule 4 seconds after the rockets were fired.
    (b) Find the straight-line distance traveled by the capsule if it fired its rockets for 6 seconds.

24. Repeat Exercise 23 if the capsule's acceleration was directly proportional to the time since the rockets began firing and its velocity 1 second after they began firing was 18,200 mph.

25. The acceleration of a particle is inversely proportional to the quantity $(t + 1)^2$, where t is time since the particle was set in motion. Find the particle's velocity function if its initial velocity was 10 meters per second (mps) and 5 seconds later it was 9 mps.

26. A radioactive isotope decays at a rate directly proportional to the amount of it present. Find the half-life of the isotope if there are 10 grams present to begin with and 20 days later there are 2.6 grams present.

27. Heavy-metal ions are being removed from lake water at a rate directly proportional to the amount of them present. If the concentration at the time removal began was $2.5(10^6)$ ions per liter and two years later was $6.25(10^4)$ ions per liter, find an equation that expresses the concentration as a function of time.

28. A hot steel bar is placed in a large tank of water; the rate at which the temperature of the bar changes is directly proportional to the difference between its temperature and that of the water. If the temperature of the water is 10°C and the temperatures of the bar when $t = 0$ and $t = 2$ are 600°C and 220°C, respectively, find the temperature of the bar as a function of time.

Management

29. A firm's marginal profit is directly proportional to the number of units it produces, and its profit from 0 units is \$0 while its profit from 10 units is \$50. Find the firm's profit function.

30. A computer has a useful life of five years and depreciates at a rate that is directly proportional to the difference between five years and its present age t, where $0 \leq t \leq 5$, t in years. If the computer was purchased for \$250,000 and one year later its value was \$205,000, find an equation that expresses the value of the computer as a function of the time since it was purchased, and then find its value four years after it was purchased.

31. A store's sales change at a rate directly proportional to the square root of the amount it spends on advertising. If the store spends nothing on advertising, its sales will be \$50,000; if it spends \$16,000 on advertising, its sales will be \$100,000. Find an equation that describes the store's sales as a function of the amount spent on advertising.

32. The rate of change of a company's average cost per unit is inversely proportional to the square of the number of units produced. If it costs \$50 to make the first unit and a total of \$90 to make the first two units, find a function that describes the company's average cost per unit.

33. The rate at which a company's stock price changes is directly proportional to the dividend the stock pays. Find an equation that describes stock price as a function of the dividend paid if the price was \$50 per share when the dividend was \$1 per share and \$60 per share when the dividend was \$2 per share.

34. A firm's marginal cost is directly proportional to the square root of the number of units it produces. Find its cost function C if $C(0) = \$100,000$ and $C(144) = \$215,200$.

MORE ABOUT INTEGRATION

35. The rate at which a company's average cost per unit changes is inversely proportional to the cube root of the number of units produced, and was $5 per unit when 1000 units were produced and $4 per unit when 8000 were produced. Find the company's average cost function.

36. The rate at which annual sales of a product change is directly proportional to the difference between the number of years the product has been on the market and 10 years. Suppose that when the product had been on the market for two years, sales were 1 million units per year, and that when it had been on the market for four years they were 0.75 million units per year. Find an expression that describes sales as a function of the time the product has been on the market.

37. The rate at which a firm's market share is changing is directly proportional to its market share. If its market share was 25% in 1984 and 30% in 1986, find an equation that expresses market share as a function of time. Assume that market share is always greater than 0.

38. A worker's production increases at a rate proportional to the difference between 400 units per week and the number of units the worker produces per week. If the worker can produce 200 units per week with no experience and 250 units per week with one week's experience, find an equation that describes production as a function of experience. Assume the worker's production is always less than 400 units per week.

39. A plant is to produce a new model of computer. If $y = y(x)$ is the number of worker-hours required to produce the xth unit, $x \geq 1$, it is estimated that the rate of change of y with respect to x is directly proportional to the difference between y and 4000 worker-hours. If $y(1) = 8000$ and $y(2) = 7000$, find y. Assume $y(x) > 4000$ for all x.

40. An employee's productivity changes at a rate directly proportional to the difference between the employee's current productivity and 1000 units per day. If the employee's productivity at time $t = 0$ was 1200 units per day and five months later it was 1150 units per day, find an equation that describes the employee's productivity as a function of time. Assume that the employee's productivity is always greater than 1000 units per day.

41. A plant's production rate is directly proportional to the product of its current production and the difference between current production and the maximum designed production of 100,000 units per week. Find an equation that describes the plant's production as a function of time if its production one week after it opened was 10,000 units per week and two weeks after it opened was 20,000 units per week.

Life Science

42. The concentration of the acid in lake water in a northeastern state has been increasing at a rate of 20 ppm per year and is currently 125 ppm. Find an equation that expresses the concentration of acid as a function of time.

43. The number of bacteria present in a culture changes at a rate inversely proportional to the time since an experiment began. If there were 100,000 bacteria present in the culture when the experiment began and 80,000 present 1 hour later, find an equation that describes the number of bacteria present as a function of time.

44. The concentration of pollutants in the air over a large city is directly proportional to the square root of the number of cars coming into the city each morning. If

250,000 cars cause 40 ppm of pollution and 1,000,000 cars cause 100 ppm, find the pollution that will be caused by 4,000,000 cars.

45. Beginning one day after germination, the rate of growth of a plant is inversely proportional to the time since germination. If the plant grows at a rate of 1 cm per day 1 day after germination and 0.5 cm per day eight days after germination, find an equation that describes the plant's growth as a function of time since germination.

46. Newton's Law of Cooling states that if a small object is placed in a new environment that is kept at a constant temperature, then the rate at which the temperature of the object changes is directly proportional to the difference between its temperature and that of the environment. Suppose a small cold-blooded animal whose body temperature is initially 20°C is placed in a container whose temperature is kept at 14°C. Find an equation that expresses the animal's temperature as a function of time if its temperature 1 hour after being placed in the container is 18°C. Assume that the animal's temperature is always greater than that of the container.

47. Repeat Exercise 46 if the temperature of the container is 25°C and 1 hour after being placed in the container the animal's temperature is 21°C, assuming that the animal's temperature always remains less than that of the container.

48. An influenza epidemic spreads at a rate directly proportional to the product of the number of people who have already contracted the disease and the number who have not. In a community of 10,000 people, 100 people had influenza on day 0 and 300 had it on day 1. Find an equation that describes the number of people who contracted influenza as a function of time.

Social Science

49. A subject undergoing psychological testing is allowed varying amounts of time to complete a complex task. The rate at which the subject makes errors is inversely proportional to the amount of time allowed. If the subject makes six errors when 5 minutes are allowed for completion of the task and two errors when 8 minutes are allowed, find an expression that gives the number of errors made as a function of the time allowed.

50. The percentage of eligible voters in a state who do not vote in presidential elections is changing at a rate directly proportional to the fourth root of t, where t is time in years with $t = 0$ representing 1980. If 50% of the eligible voters did not vote in 1980 and 46% did not vote in 1984, predict the percentage who will not vote in 1992.

51. The percentage of voters who are registered as Independents is changing at a rate directly proportional to the percentage who are not registered as Independents. If 25% of all voters were Independents in 1986 and 30% were Independents in 1990, find an equation that expresses the percentage of voters who are Independents as a function of time. Assume that the percentage not registered as Independents is always less than 100%.

52. The rate at which a rumor spreads is directly proportional to the product of the proportion of persons who have already heard it and the proportion who have not. Suppose that on day 0 the proportion of people who have heard a certain rumor is 0.01, and on day 4 the proportion is 0.25. Find an equation that describes the proportion who have heard the rumor as a function of time.

Transition Rates

Exercises 53 through 56 were suggested by material in Coleman, *Introduction to Mathematical Sociology*, The Free Press of Glencoe, 1964. They are concerned with transition rates as individuals migrate from one group to another. Thus, suppose individuals migrate from group A to group B at a rate directly proportional to the size of group A. If $y_A = y_A(t)$ denotes the size of group A at time t, then

$$\frac{dy_A}{dt} = -k_A y_A, \quad k_A > 0.$$

The positive constant k_A is called the **transition rate** from A to B.

*53. If members of political party A are switching to political party B at a rate directly proportional to the size of party A, and if $y_A(0) = 10$ million and $y_A(1) = 9.9$ million, find the transition rate from party A to party B.

Now suppose that $y_B = y_B(t)$ is the size of group B at time t and that there is also migration from group B to group A with transition rate k_B. Then group A loses members to group B at a rate of $k_A y_A$ and simultaneously gains members from group B at a rate of $k_B y_B$. Thus

$$\frac{dy_A}{dt} = k_B y_B - k_A y_A.$$

If we assume that the total number of individuals in the two groups is a constant N, so that $y_A + y_B = N$, then we may write

$$\frac{dy_A}{dt} = k_B(N - y_A) - k_A y_A$$

or

$$\frac{dy_A}{dt} = N k_B - (k_A + k_B) y_A.$$

*54. Show that

$$y_A = N \frac{k_A}{k_A + k_B}(1 - e^{-(k_A + k_B)t}) + c e^{-(k_A + k_B)t}$$

is a solution of the preceding differential equation.

*55. Suppose voters are migrating from political party A to political party B with a transition rate of 0.02, and from B to A with a transition rate of 0.03. Suppose also that at time $t = 0$, party A had 18 million members and party B had 12 million members, and that the total membership in the two parties will remain constant at 30 million.
   (a) Find an equation that describes the size of party A as a function of time.
   (b) Find the size of party A at time $t = 10$; what will happen to the sizes of the parties in the long run?

*56. If group A is the living and group B is the nonliving, then the transition rate k_A from A to B is the death rate and the transition rate k_B from B to A is the birth rate. Suppose that at time $t = 0$ there are y_0 living individuals. Find the solution of the differential equation

$$\frac{dy_A}{dt} = N k_B - (k_A + k_B) y_A, \quad y(0) = y_0$$

and express the solution in terms of the natural rate of increase $r = k_B/k_A$. Interpret your results.

SUMMARY

This summary consists of terms and symbols whose meaning you should know, facts you should know, and techniques and methods you should be able to employ.

Section 15.1

Riemann sum. The definite integral as a limit of Riemann sums. The definite integral as an accumulator. Producers' surplus. The formula for producers' surplus. Finding producers' surplus. Consumers' surplus. The formula for consumers' surplus. Finding consumers' surplus. Stream of income. Formula for finding the future value of a continuous stream of income reinvested at a continuously compounded interest rate. Finding the future value of a stream of income. Method for setting up the integral as an accumulator. Setting up the integral as an accumulator,

Section 15.2

The integration by parts formula: $\int u\, dv = uv - \int v\, du$. The keys to integration by parts. Using integration by parts to find integrals. Integral tables. Using integral tables to find integrals.

Section 15.3

Improper integrals:

$$\int_a^{+\infty} f(x)\, dx = \lim_{t \to +\infty} \int_a^t f(x)\, dx$$

$$\int_{-\infty}^b f(x)\, dx = \lim_{t \to -\infty} \int_t^b f(x)\, dx$$

Convergence, divergence of improper integrals of the above forms. Determining whether improper integrals converge or diverge, and evaluating them when they converge. Interpreting the value of a convergent improper integral as an area. Improper integrals of the form $\int_{-\infty}^{+\infty} f(x)\, dx$. Convergence, divergence of such integrals.

Section 15.4

Continuous random variable. Continuous probability distribution. Probability density function. Showing that a given function is or is not a probability density function. The probability density function for a continuous random variable X. If f is a probability density function for X,

$$P(a \leq X \leq b) = \int_a^b f(x)\, dx.$$

Using the probability density function to find probabilities. The cumulative distribution function F for a continuous random variable X. The properties of a cumulative distribution function. Finding the cumulative distribution function from the probability density function. Using the cumulative distribution function to find probabilities. Finding the probability distribution function from the cumulative distribution function. The expected value $E(X)$ of a continuous random variable X. The interpretation of $E(X)$. Finding the expected value of a continuous random variable. The variance $V(X)$ of a continuous random variable X. The interpretation of the variance of a continuous random variable. Finding the variance of a continuous random variable. The standard deviation of a continuous random variable.

Section 15.5

Numerical integration. The need for numerical integration. The midpoint rule as approximation by rectangles. The formula for the midpoint rule. Using the midpoint rule to approximate definite integrals. The trapezoidal rule as approximation by trapezoids. The formula for the trapezoidal rule. Using the trapezoidal rule to approximate definite integrals. Simpson's rule as approximation by parabolic segments. The formula for Simpson's rule. Using Simpson's rule to approximate definite integrals.

Section 15.6

Differential equation. A solution of a differential equation. The importance of differential equations. The general solution of a differential equation. Initial condition. Differential equation with initial condition. Particular solution of a differential equation with initial condition. Constant of proportionality. Directly proportional, inversely proportional. Setting up differential equations using the language of proportionality. Finding the general solution of a differential equation. Find the particular solution of a differential equation with initial conditions. Separation of variables. Using separation of variables to solve differential equations.

REVIEW EXERCISES

In Exercises 1 through 3, find producers' surplus and consumers' surplus for the given supply and demand functions.

1. supply: $p = 5q + 50$
   demand: $p = -2q + 120$

2. supply: $p = 200e^{0.01q}$
   demand: $p = 1000e^{-0.02q}$

3. supply: $q = 15p$
   demand: $q = 100 - p^2$

4. A firm's earnings are flowing in at the rate of $50 million per year and are compounding continuously at 15% per year. Find the value of the firm's earnings over the next 10 years.

5. Repeat Exercise 4 if the firm's earnings are flowing in at a rate of $50e^{0.01t}$ million dollars per year.

6. Repeat Exercise 4 if the firm's earnings are flowing in at a rate of $0.5t + 50$ million dollars per year.

7. Let f be a function that is positive over the interval $[a, b]$. If we revolve the graph of f about the x-axis, we obtain a solid known as a **solid of revolution**:

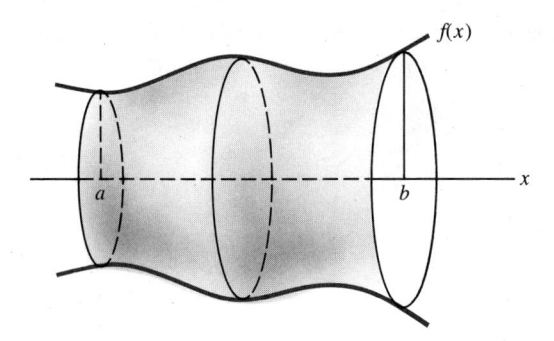

A small subinterval of $[a, b]$ of length Δx thus generates a portion of the solid of revolution whose volume can be approximated by that of a circular disk of radius $f(x)$ and width Δx:

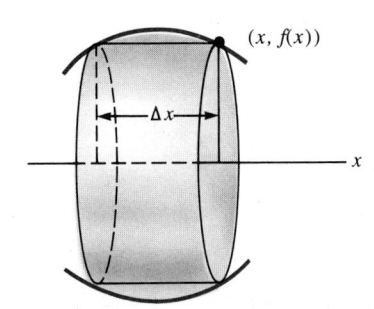

The volume of such a circular disk is $\pi[f(x)]^2 \Delta x$. Use the integral as an accumulator to show that the volume of the solid of revolution is given by

$$\pi \int_a^b [f(x)]^2 \, dx.$$

In Exercises 8 and 9, use the result of Exercise 7 to find the volume of the solid of revolution obtained by revolving the given function about the x-axis over the given interval:

8. $f(x) = x^2$, $[1, 2]$

9. $f(x) = e^{-x}$, $[0, 2]$

In Exercises 10 through 17, perform the indicated integration.

10. $\int 4xe^{-4x} \, dx$

11. $\int 3x(x + 5)^6 \, dx$

12. $\int_1^2 x^{3/2} \ln x \, dx$

13. $\int x^3(x + 1)^{4/3} \, dx$

14. $\int \dfrac{1}{2x^2 + 4x} \, dx$

15. $\int \dfrac{1}{16 - 4x^2} \, dx$

16. $\int_{1/\sqrt{2}}^{\sqrt{2}} \dfrac{1}{2x\sqrt{8 - 2x^2}} \, dx$

17. $\int \dfrac{2x}{x^2 + 4x + 4} \, dx$

In Exercises 18 through 23, determine whether the given improper integral converges or diverges, and find its value if it converges.

18. $\int_0^{+\infty} e^{-kx}\, dx, \quad k > 0$

19. $\int_3^{+\infty} \frac{2x^2}{x^3 + 1}\, dx$

20. $\int_{-\infty}^{-2} x^{-4}\, dx$

21. $\int_1^{+\infty} \frac{(\ln x)^2}{x}\, dx$

22. $\int_0^{+\infty} x^r e^{-x^{r+1}}\, dx, \quad r > -1$

23. $\int_{-\infty}^{+\infty} f(x)\, dx$, where $f(x) = \begin{cases} e^{2x} & \text{for } x \leq 0, \\ e^{-3x} & \text{for } x > 0. \end{cases}$

In Exercises 24 and 25, show that the given function is a probability density function.

24. $f(x) = \begin{cases} \frac{7}{128}x^6, & 0 \leq x \leq 2, \\ 0 & \text{otherwise} \end{cases}$

25. $f(x) = \begin{cases} \frac{\ln 2}{x(\ln x)^2}, & x \geq 2, \\ 0 & x < 2 \end{cases}$

26. What must the constant k be if
$$f(x) = \begin{cases} kx^4, & 1 \leq x \leq 3, \\ 0 & \text{otherwise} \end{cases}$$
is to be a probability density function?

Let X be a continuous random variable with probability density function f defined by
$$f(x) = \begin{cases} 0.25x^{-1/2}, & 0 \leq x \leq 4, \\ 0 & \text{otherwise.} \end{cases}$$
Exercises 27 through 32 refer to this random variable.

27. Find $P(0 \leq X \leq 2)$
28. Find $P(X < 1)$
29. Find $P(X > 3.5)$
30. Find b if $P(X > b) = 0$.
31. Find $E(X)$
32. Find $V(X)$

A company that makes air conditioners tests selected units by running them continuously until they fail. Let X denote the number of years a tested unit runs until it fails, and suppose X has probability density function f given by
$$f(x) = \begin{cases} (x + 1)^{-2}, & x \geq 0, \\ 0 & x < 0. \end{cases}$$
Exercises 33 through 37 refer to this situation.

33. Find the probability that a unit will last at least one year;
34. Find the percentage of units that will fail during the first six months of the test;
35. Find the percentage of units that will fail between two and five years.
36. The longest-lasting 10% of the units will last longer than how many years?
37. Find the expected number of years that a unit will last.
38. Use the midpoint rule with $n = 6$ to approximate
$$\int_0^1 (4 - x^4)\, dx$$
and compare your result with the actual value of the integral.
39. Repeat Exercise 38 using the trapezoidal rule.
40. Repeat Exercise 38 using Simpson's rule.

41. A criminologist sampled the police records of a city and found the following:

Day	1	5	9	13	17	21	25
Crimes per day	125	128	136	132	130	124	118

   Use the trapezoidal rule to estimate the total number of crimes reported over the 25-day period.

42. Repeat Exercise 41 using Simpson's rule and compare your answer with that of Exercise 41.

43. A beaver dam has flooded the region shown below (figures in meters):

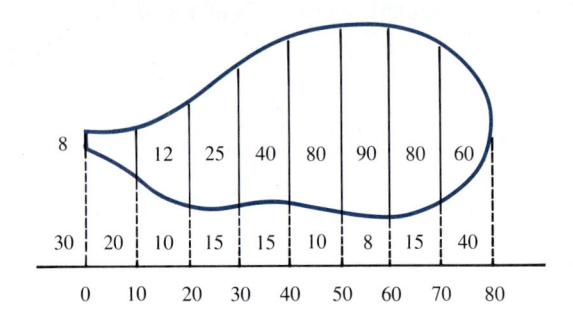

   Use the trapezoidal rule to estimate the area flooded.

44. Repeat Exercise 43 using Simpson's rule and compare your answer with that of Exercise 43.

In Exercises 45 through 50, find the solution of the given differential equation.

45. $\frac{dy}{dx} = 2x + 2, \quad y(0) = 1$

46. $\frac{dy}{dx} = \frac{2x}{1 - x^2}, \quad y(0) = 0$

47. $x\frac{dy}{dx} = \frac{1}{y}$

48. $3(y + 2)\frac{dy}{dx} = 2x - 1, \quad y(0) = 0$

49. $\frac{dy}{dx} = \frac{y^2}{x^3}, \quad y \neq 0$

50. $\frac{dy}{dx} = (y + 3)(x - 1), \quad y(0) = 2, \quad y > -3$

51. A rocket's acceleration during the first 30 seconds after launch is directly proportional to the time since launch. Find equations that give the rocket's velocity and altitude t seconds after launch, $0 \leq t \leq 30$, if its altitude 2 seconds after launch is 20 feet.

52. A firm's marginal profit is directly proportional to the cube root of the number of units it makes and sells. Find the firm's profit function if its fixed cost is $25,000 and the profit from making and selling 1000 units is $10,000.

53. A psychologist studying classroom learning estimates that students can absorb new concepts at a rate inversely proportional to $t + 10$, where t is the time in minutes since class started. If students can absorb two new concepts during the first 6 minutes of a class, and three new concepts in the first 10 minutes of the class, find an expression that relates the number of new concepts absorbed to the length of the class.

54. Suppose that the rate at which compensation is increasing in a firm is directly proportional to compensation. If the average wage in the firm now is $8 per hour and

next year it will be $8.50 per hour, find an equation that relates the average wage to time.

ADDITIONAL TOPICS

Here are some suggestions for topics not covered in the text that you might want to investigate on your own.

1. In Exercise 7 of the Review Exercises for this chapter we outlined a method of using the integral as an accumulator to find the volume of a solid of revolution. This method is known as the **disk method,** because it depends on approximating the volume of the solid by the volumes of disks. Another method of using the integral as an accumulator to find the volume of a solid of revolution is known as the **shell method.** Find out how and why the shell method works.

2. The length of a curve between any two points on it is called the **arc length** of the curve between the points. Under certain conditions arc length can be found by using the integral as an accumulator. Find out how this is done and why it works.

3. A **surface of revolution** is a surface that is obtained by revolving a curve about an axis. Under certain conditions the surface area of a surface of revolution can be found by using the integral as an accumulator. Find out how this is done and why it works.

4. In physics, work is defined as force applied times the distance through which it is applied. Find out how the integral as an accumulator can be used to calculate work.

5. A technique of integration that is very useful in integrating rational functions is the method of **partial fractions.** Find out how the method of partial fractions works.

6. We mentioned at the beginning of Section 15.3 that integrals for which the integrand becomes infinite at some point of the interval of integration are also called improper integrals. Thus integrals such as

$$\int_{-1}^{1} \frac{1}{x}\, dx \quad \text{and} \quad \int_{1}^{2} \frac{1}{x-1}\, dx$$

are improper integrals of this type: the first because $1/x$ becomes infinite at 0 and the second because $1/(x-1)$ becomes infinite at 1. Find out how this type of improper integral is integrated.

7. Some of the more common probability distributions (other than the normal) for continuous random variables are the exponential, the gamma, the beta, and the chi-squared distributions. What are the probability density functions for these distributions?

8. When using either the trapezoidal rule or Simpson's rule to approximate the value of a definite integral, it is possible to make estimates for the error of the approximation. Find out about the trapezoidal error estimate and Simpson's error estimate and how they are used.

9. A differential equation of the form

$$\frac{dy}{dx} + p(x)y = q(x)$$

is called a **linear differential equation of order one.** Find out how to solve such an equation. Also find some examples of the use of such equations.

SUPPLEMENT: THE PRESENT VALUE OF A STREAM OF INCOME

In Section 9.3 we discussed annuities and their present values. It is often useful to think of a stream of income, such as was considered in Section 15.1, as a **continuous annuity,** that is, as an annuity whose annual payment is spread continuously throughout the year. When this is done, we can apply the present value concept to the stream of income. In this supplement we show how to find the present value of a stream of income, and we illustrate some of the uses of this concept.

Suppose a stream of income flows into a firm continuously, arriving at the rate of $I(t)$ dollars per year at time t. Suppose also that money is worth $100s\%$ per year compounded continuously; that is, that $100s$ is the continuous discount rate. We will find an expression for the present value of this income stream over the interval $[0, b]$.

If we divide the time interval $[0, b]$ into small subintervals, each of width Δt, then the amount of income that flows in during any particular subinterval is approximately $I(t) \Delta t$ dollars. In Section 13.1 we saw that the present value P of an amount A due t years in the future when money is worth $100r\%$ per year compounded continuously is given by

$$P = Ae^{-rt}.$$

Therefore the present value of the amount $I(t) \Delta t$ that will arrive during the subinterval and is to be discounted at $100s\%$ per year is equal to

$$[I(t) \Delta t] e^{-st}$$

dollars. Using the integral as an accumulator, we thus find that the total present value of the stream of income over the time period $[0, b]$ is

$$\int_0^b I(t) e^{-st}\, dt.$$

> **The Present Value of a Stream of Income**
>
> If a stream of income flows in continuously, arriving at the rate of $I(t)$ dollars per year at time t, and if its future value is discounted at $100s\%$ per year compounded continuously, then the present value of the stream of income over the next b years is
>
> $$\int_0^b I(t) e^{-st}\, dt.$$

EXAMPLE 1 Suppose that each year Blink, Inc., earns $100,000 in revenue. What is the present value of Blink's revenue stream over the next four years if Blink discounts future revenues at 8% per year compounded continuously? The answer is

More about Integration

$$\int_0^4 100{,}000 e^{-0.08t}\, dt = -\frac{100{,}000}{0.08}\left(e^{-0.08t}\Big|_0^4\right)$$

$$= -\frac{100{,}000}{0.08}(e^{-0.32} - 1)$$

$$= \$342{,}313.70. \quad \square$$

The present value of a stream of income discounted over all future time is called its **capital value**.

EXAMPLE 2 Suppose Blink, Inc. of the previous example expects its revenue to continue to be \$100,000 per year forever. Then the capital value of Blink's revenue stream is

$$\int_0^{+\infty} 100{,}000 e^{-0.08t}\, dx = \lim_{b \to +\infty} \int_0^b 100{,}000 e^{-0.08t}\, dt$$

$$= -\frac{100{,}000}{0.08} \lim_{b \to +\infty} (e^{-0.08b} - 1)$$

$$= -1{,}250{,}000(0 - 1)$$

$$= \$1{,}250{,}000. \quad \square$$

Another viewpoint of the present value of an income stream takes into account the fact that the money received can be invested. Thus, suppose that income is currently arriving continuously at the rate of P dollars per year, that it compounds continuously at $100r\%$ per year, and that it is discounted at $100s\%$ per year. Then at time t in the future the amount P will have grown to

$$A(t) = Pe^{rt}$$

dollars. But the present value of the amount $A(t)$ is equal to

$$A(t)e^{-st} = Pe^{rt}e^{-st} = Pe^{(r-s)t}.$$

Therefore by the usual argument using the integral as accumulator, the present value of the income stream over the next b years is

$$\int_0^b Pe^{(r-s)t}\, dt.$$

The Present Value of an Invested Stream of Income

If a stream of income is currently flowing in at a rate of P dollars per year, and if this income is invested at $100r\%$ per year compounded continuously and its future value is discounted at $100s\%$ per year compounded continuously, then the present value of the income received over the next b years is

$$\int_0^b Pe^{(r-s)t}\, dt.$$

EXAMPLE 3 Suppose that a firm's profits are currently flowing in at a rate of $1 million per year and that it reinvests all profits. If the reinvested profits earn 8% per year compounded continuously and if the firm discounts future profits at 12% per year, then the present value of its profits over the next five years, in millions of dollars, will be

$$\int_0^5 1e^{(0.10-0.12)t}\,dt = \int_0^5 e^{-0.02t}\,dt$$

$$= -\frac{e^{-0.02t}}{0.02}\bigg|_0^5$$

$$= -\frac{e^{-0.1}-1}{0.02}$$

$$\approx 4.758,$$

or approximately $4,758,000. □

SUPPLEMENTARY EXERCISES

1. If profits are flowing into Aurora Corporation at a rate of 2 million dollars per year, and if Aurora discounts future profits at 10% per year compounded continuously, find the present value of Aurora's future profits over the next six years.

2. Refer to Exercise 1: Find the capital value of Aurora's profits.

3. A firm's revenues are flowing in at a rate of $I(t) = 2 + 5t$ million dollars per year. If the firm discounts future revenues at 7% per year compounded continuously, find the present value of the firm's stream of revenues over
   (a) the next 4 years   (b) the next 10 years

4. Refer to Exercise 3: Find the capital value of the firm's revenues.

5. Suppose Borealis Company expects to be earning profits at the rate of

$$I(t) = 1 + e^{0.05t}$$

   million dollars per year t years from now. If the company discounts future earnings at 8% per year compounded continuously, find the present value of its expected earnings
   (a) over the next 10 years   (b) over the next 50 years   (c) forever

6. An oil well brings in revenues of $10te^{-0.02t}$ thousand dollars per year t. Using a discount rate of 6% per year compounded continuously, find
   (a) the present value of the revenue stream from the well over the next five years;
   (b) the capital value of the well.

7. Camran, Ltd. is currently earning profits at the rate of $5 million per year. Camran reinvests its profits at 9% per year compounded continuously and discounts future profits at 10% per year compounded continuously. Find the present value of Camran's profits
   (a) over the next three years   (b) forever

8. Do Exercise 7 if Camran's discount rate per year is
   (a) 20%   (b) 9%

9. You are considering buying some rental property as an investment. You know that you can rent the property now for $10,000 per year and that the rent will increase at 6% per year compounded continuously. If you discount future income at 10% per year compounded continuously, find the present value of the rent you will receive if you keep the property
   (a) 10 years   (b) 40 years   (c) forever

10. Refer to Exercise 9. How long would you have to hold the property to receive rent having a present value of $100,000?

*11. Refer to Exercise 9. Suppose you intend to hold the property for 10 years and then sell it.
   (a) If you could sell it for $100,000, what is the maximum amount that you would be willing to pay for it now?
   (b) If the property will cost you $150,000, how much will you have to sell it for in order to earn more, on a present value basis, than it costs?

12. A certain stock pays a dividend of $14 per share, and this is increasing at a rate of 8% per year compounded continuously. If future income is discounted at 16% per year compounded continuously, find the present value of all future dividends you will receive if you buy the stock and hold it for
   (a) 5 years   (b) 50 years   (c) forever

*13. Refer to Exercise 12.
   (a) Would you be willing to pay $100 per share for the stock now if you knew you could sell it five years from now for $100? If you could sell it 5 years from now for $50?
   (b) Suppose the stock costs $75 a share now. How much would you have to sell it for five years from now in order to earn more, on a present value basis, than it costs?

16

Functions of Several Variables

The functions we have studied up until now have been **functions of one variable,** that is, functions with just one independent variable. However, in many applications we must consider functions with more than one independent variable. For instance, suppose a firm produces two products. Its total cost C will then depend on the number of units x of the first product *and* the number of units y of the second product that it makes, and might be given by an equation such as

$$C = 20x + 30y + 50{,}000.$$

This equation defines C as a function of the two independent variables x and y. Similarly, a farmer's crop yield Y might depend on the amount q of fertilizer and the amount w of weedkiller used on the crop, the amount r of rainfall the crop receives, and the time t that the farmer devotes to cultivating the crop, according to the equation.

$$Y = 0.5q + 2.2w + 2\sqrt{r} + 1.2t^2.$$

This equation defines crop yield Y as a function of the four independent variables q, w, r, and t. Functions such as these are called **functions of several variables.**

In this chapter we study functions of several variables. We begin by illustrating how functions of several variables are defined and interpreted, introduce the three-dimensional cartesian coordinate system, and show how functions of two variables may be graphed as surfaces in three-dimensional space. Next we discuss derivatives of functions of several variables, show how these derivatives may be interpreted, and demonstrate how to find relative maxima and minima for functions of two variables. We also present a method by means of which functions of several variables can be maximized or minimized subject to constraints. The final section of the chapter is devoted to integrating functions of two variables.

16.1 FUNCTIONS OF MORE THAN ONE VARIABLE

In this section we introduce functions of several variables and briefly discuss how functions of two variables may be graphed using the three-dimensional cartesian coordinate system.

An equation such as

$$f(x, y) = x^2 + 2xy + 5y$$

defines f as a function of two independent variables x and y. The domain of such a function f is a set of ordered pairs (a, b), where a and b are numbers. We evaluate the function at the ordered pair (a, b) by substituting a for x and b for y in the defining equation.

EXAMPLE 1 Let

$$f(x, y) = x^2 + 2xy + 5y.$$

Then

$$f(0, 0) = 0^2 + 2(0)(0) + 5(0) = 0,$$
$$f(2, 3) = 2^2 + 2(2)(3) + 5(3) = 31,$$
$$f(-1, 2) = (-1)^2 + 2(-1)(2) + 5(2) = 7,$$
$$f(4, y) = 4^2 + 2(4)y + 5y = 16 + 13y,$$
$$f(x, 1) = x^2 + 2x(1) + 5(1) = x^2 + 2x + 5,$$
$$f(x + 1, b) = (x + 1)^2 + 2(x + 1)b + 5b$$
$$= x^2 + 2x + 1 + 2xb + 2b + 5b$$
$$= x^2 + 2(1 + b)x + 7b + 1,$$

and so on. Since we may substitute any number for x and any number for y in the defining equation for f, the domain of f is the set of all ordered pairs (x, y). ◻

Similarly, a function defined by an equation such as

$$f(x, y, z) = \frac{x - y}{y - z}$$

is a function of three variables. The domain of such a function is a set of ordered triples (a, b, c), where a, b, and c are numbers. We evaluate the function at the ordered triple (a, b, c) by substituting a for x, b for y, and c for z in its defining equation.

EXAMPLE 2

Let
$$f(x, y, z) = \frac{x - y}{y - z}.$$

Then
$$f(1, 1, 0) = \frac{1 - 1}{1 - 0} = 0,$$

and
$$f(2, -1, 3) = \frac{2 - (-1)}{-1 - 3} = -\frac{3}{4}.$$

However, $f(1, 2, 2)$ is not defined. (Why not?) More generally, notice that
$$f(x, y, z) = \frac{x - y}{y - z}$$

is not defined if and only if $y = z$. Thus the domain of this function is
$$\{(x, y, z) \mid y \neq z\}. \quad \square$$

Functions of several variables can also be defined by equations that relate a dependent variable to two or more independent variables.

EXAMPLE 3

Let
$$z = xe^x + \ln y.$$

Then

when $x = 0$ and $y = 1$, $\quad z = 0e^0 + \ln 1 = 0 + 0 = 0;$

when $x = 1$ and $y = 1$, $\quad z = 1e^1 + \ln 1 = e + 0 = e;$

when $x = -2$ and $y = 2$, $\quad z = -2e^{-2} + \ln 2;$

and so on. Note that since $\ln y$ is not defined when $y \leq 0$, the function defined by this equation has domain $\{(x, y) \mid y > 0\}$. $\quad \square$

Functions of four or more variables are defined and evaluated in a manner similar to that illustrated in the preceding examples.

As we mentioned at the beginning of this chapter, functions of several variables arise naturally in many situations. We illustrate with examples.

EXAMPLE 4

Suppose that a store sells two products, X and Y, and that the sales price of one product is not affected by the sales price of the other. If each unit of product X sells for $15 and each unit of product Y for $20, then the store's revenue from selling x units of X and y units of Y is
$$R(x, y) = 15x + 20y$$

dollars. $\quad \square$

EXAMPLE 5   Tropicoid Corporation makes two models of its instant camera, the regular and the deluxe. The company's cost if it makes x units of the regular model and y units of the deluxe model is

$$C(x, y) = 40x + 60y + 0.01xy + 20{,}000$$

dollars. Note that the company's fixed cost is therefore

$$C(0, 0) = 20{,}000$$

dollars.

Now suppose Tropicoid makes a fixed number, say 1000 units, of the regular model; then its cost will be

$$C(1000, y) = 40(1000) + 60y + 0.01(1000)y + 20{,}000 = 60{,}000 + 70y$$

dollars. Therefore if production of the regular model is held fixed at 1000 units, each additional unit of the deluxe model made will cost \$70. Similarly, if production of the regular model is held fixed at 2000 units, then

$$C(2000, y) = 100{,}000 + 80y,$$

which tells us that each additional unit of the deluxe model produced will cost \$80. On the other hand, if production of the deluxe model is held fixed at 2000 units, then

$$C(x, 2000) = 140{,}000 + 60x$$

shows that each additional unit of the regular model produced will cost \$60.   □

EXAMPLE 6   Suppose a company sells both electric heaters and kerosene heaters. Let p_E be the unit price of the electric heaters and p_K the unit price of the kerosene heaters. The quantity demanded of either product may depend on both the price of that product and the price of the other; thus, if E is the quantity of electric heaters demanded, then E may depend on both p_E and p_K. For example, we might have

$$E = -100p_E + 500p_K + 10{,}000.$$

Then if electric heaters sell for \$40 each and kerosene heaters for \$60 each, the quantity of electric heaters demanded will be

$$E = -100(40) + 500(60) + 10{,}000 = 36{,}000$$

units.

Let us analyze the demand function

$$E = -100p_E + 500p_K + 10{,}000.$$

If the price p_K of kerosene heaters remains fixed at, say, $p_K = b$, then the quantity of electric heaters demanded is

$$E = -100p_E + 500b + 10{,}000.$$

Therefore each \$1 rise in the price p_E of electric heaters will result in a decrease of 100 in the quantity of electric heaters demanded by consumers. In other words,

Functions of Several Variables

if the price of kerosene heaters remains fixed, a rise in the price of electric heaters leads to a decrease in the quantity of electric heaters demanded. On the other hand, if the price of electric heaters is fixed at, say, $p_E = a$, then

$$E = -100a + 500p_K + 10{,}000,$$

which shows that each increase of \$1 in the price p_K of kerosene heaters leads to an increase of 500 in the quantity of electric heaters demanded. Thus, if the price of electric heaters remains fixed while the price of kerosene heaters goes up, the quantity of electric heaters demanded will increase. ☐

Functions of two variables can be graphed using the **three-dimensional cartesian coordinate system.** (Functions of three or more variables cannot be graphed.) The three-dimensional cartesian coordinate system is depicted in Figure 16–1: the x-, y-, and z-axes are perpendicular and intersect at the **origin;** their positive portions are shown as solid lines, their negative portions as dashed lines. Because of the difficulties with perspective that arise when the negative portions of the axes are shown, diagrams are sometimes restricted to the region where all three axes are nonnegative; this region, called the **first octant,** is shown in Figure 16–2.

Just as any ordered pair *(a, b)* of numbers can be plotted as a point on the two-dimensional cartesian coordinate system, so any ordered triple *(a, b, c)* of numbers can be plotted as a point in three-dimensional space using the three-dimensional cartesian coordinate system (Figure 16–3).

EXAMPLE 7 In Figure 16–4 we have plotted the origin (0, 0, 0) and the points

(3, 0, 0),   (0, 2, 0),   (0, 0, 4),   (3, 2, 0),   (0, 2, 3),
(1, 0, 1),   (3, 0, 4),   (3, 2, 4),   (0, 2, 4),   (1, 2, 2). ☐

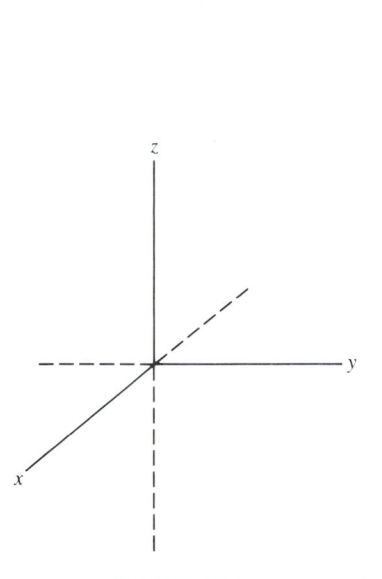

FIGURE 16–1
The Cartesian Coordinate System

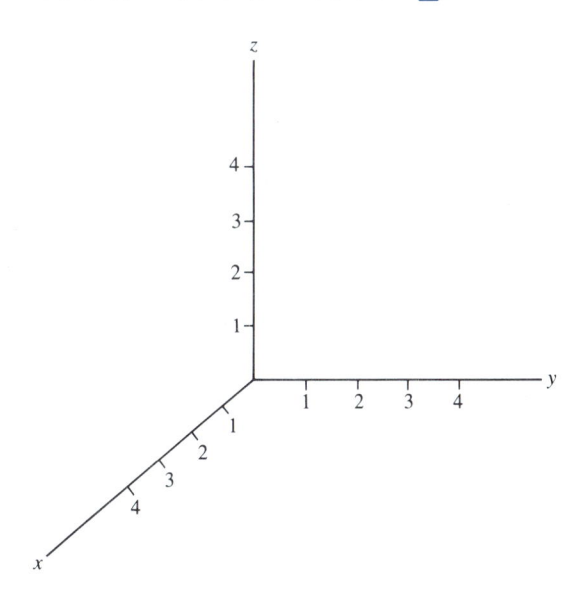

FIGURE 16–2
The First Octant

FIGURE 16-3

FIGURE 16-4

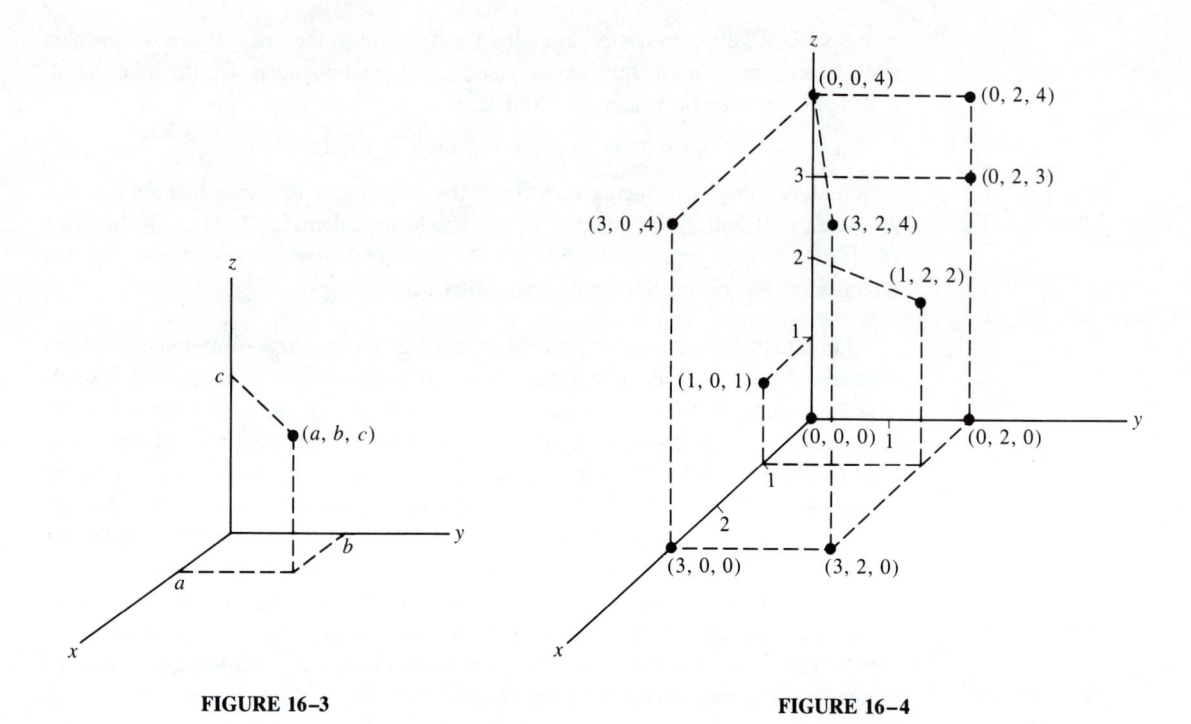

The set of all points of the form $(x, y, 0)$ is called the **xy-plane,** the set of all points of the form $(x, 0, z)$ the **xz-plane,** and the set of all points of the form $(0, y, z)$ the **yz-plane.** In Example 7, $(3, 2, 0)$ and $(0, 2, 0)$ are points in the xy-plane, $(3, 0, 4)$ and $(3, 0, 0)$ are points in the xz-plane, and $(0, 2, 4)$ and $(0, 2, 0)$ are points in the yz-plane. The xy-, xz-, and yz-planes are known collectively as the **coordinate planes.**

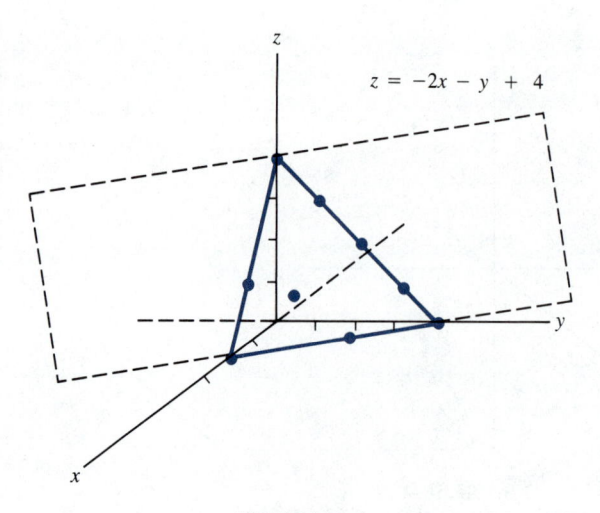

FIGURE 16-5

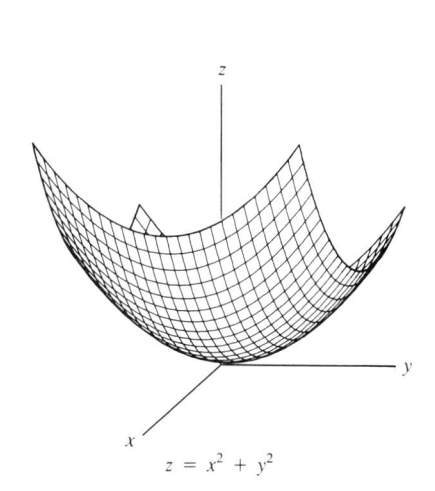

$z = x^2 + y^2$

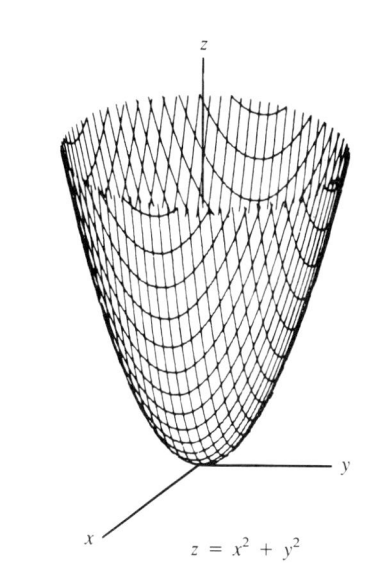

$z = x^2 + y^2$

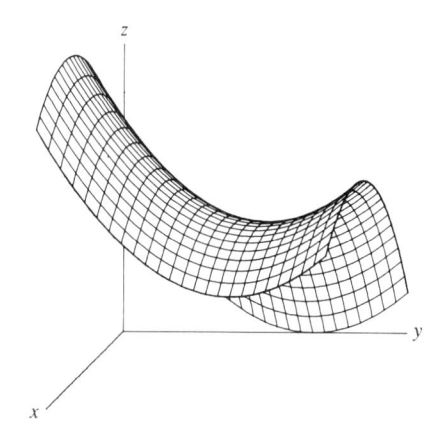

$z = 4 + (y-2)^2 - (x-2)^2$

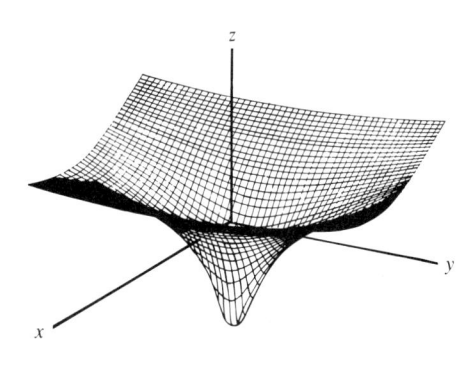

$z = \ln(x^2 + 2y^2)$

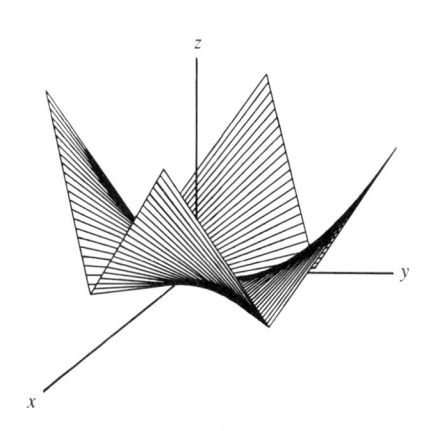

$z = 1 \, x \, 1(y)$

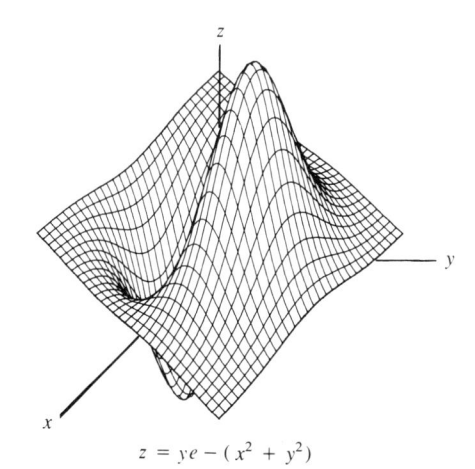

$z = ye - (x^2 + y^2)$

FIGURE 16–6

We can use the three-dimensional cartesian coordinate system to graph any function f of two variables. The technique is to set $z = f(x, y)$, give values to x and y, calculate $z = f(x, y)$, and plot the resulting point (x, y, z). The graph will be a surface.

EXAMPLE 8 Let us graph the function whose defining equation is

$$f(x, y) = -2x - y + 4.$$

We set

$$z = -2x - y + 4.$$

Then we have

x	y	$z = -2x - y + 4$	(x, y, z)
0	0	4	(0, 0, 4)
1	0	2	(1, 0, 2)
2	0	0	(2, 0, 0)
0	1	3	(0, 1, 3)
1	1	1	(1, 1, 1)
0	2	2	(0, 2, 2)
1	2	0	(1, 2, 0)
0	3	1	(0, 3, 1)
0	4	0	(0, 4, 0)

The graph of f is the plane depicted in Figure 16–5. ☐

We do not intend to place much emphasis on graphing functions of two variables: their graphs are often quite difficult to draw, and today computers can do the graphing more quickly and with better artistic results than most people can. By way of example, Figure 16–6 presents several computer-generated graphs of functions of two variables.

■ *Computer Manual: Program TWOVARS*

16.1 EXERCISES

In Exercises 1 through 10, find the domain of the function f and the indicated functional values, if possible.

1. $f(x, y) = x^2 + 2xy + y^2$, $f(0, 0), f(1, 3)$

2. $f(x, y) = \dfrac{2x}{y + 1}$, $f(-2, -3), f(4, -1)$

3. $f(p, q) = \sqrt{p} + 2\sqrt{q}$, $f(0, 4)$, $f(9, 16)$

4. $f(x, y) = \dfrac{2x^2 - 3y^2}{x + y}$, $f(1, 1), f(4, 0)$

5. $f(u, v) = \sqrt{2uv} + 1$, $f(0, 0)$, $f(3, -2)$

6. $f(x, y) = e^{-xy} + e^{x/y}$, $f(1, 0), f(2, 2)$

7. $f(x, y, z) = x^2 + y^2 + z^2$, $f(1, 2, -1), f(0, 2, 5)$

8. $f(r, s, t) = \dfrac{1}{r} + \dfrac{2}{s^2 - t^2}$, $f(1, 2, 3), f(-2, 2, 2)$

9. $f(x, y, z) = \ln x + 2 \ln y + 3 \ln z$, $f(1, 1, 1), f(e, e^2, e^3)$

10. $f(x, y, z, w) = \ln |x| + \ln |y| + \ln |z| + \ln |w|$, $f(-1, 1, 1, 1)$, $f(e, -e, -e, e)$

11. Let $z = x^2 + 2x^3y - 3y^2$. Find z when $x = 0$ and $y = 2$; when $x = -2$ and $y = -1$. What is the domain of this function?

12. Let $z = 2x^2 + \dfrac{x}{y} + 2$. Find z when $x = 0$ and $y = 1$; when $x = 2$ and $y = 2$. What is the domain of this function?

13. Let $w = u^2 - uv + (uv)^{-1}$. Find w when $u = 1$ and $v = 2$; when $u = -1$ and $v = -2$; when $u = v = a$. What is the domain of this function?

14. Let $w = e^{z-2x+3y}$. Find w when $x = -1, y = z = 2$; when $x = 0, y = a$, and $z = -a$. What is the domain of this function?

15. Let $u = z(x^2 - y^2)^{1/2}$. Find u when $x = 1, y = 2$, and $z = 3$; when $x = 3, y = 2$, and $z = 1$; when $x = a, y = 0$, and $z = b$. When is the domain of this function?

16. Let $q = (2x - 3y)^{1/2} + z^2(w - x)^{-1}$. Find q when $x = y = z = w = 1$; when $x = 1, y = -1, z = 2$, and $w = 3$; when $x = t/2, y = t/3, z = t$, and $w = -t/2$. What is the domain of this function?

17. A stockbroker charges a commission of

$$c(x, y) = 25 + 0.15x + 0.005xy$$

dollars on any trade involving x shares of stock selling at y dollars per share. Find the stockbroker's commission on a trade involving 100 shares selling at $20 per share; on 400 shares selling at $60 per share.

Management

18. A refinery produces two grades of gasoline, regular and premium. If the cost of producing one grade does not affect the cost of producing the other, if it costs $0.40 to produce each gallon of regular and $0.50 to produce each gallon of premium, and if the refinery's fixed cost is $100,000 per day, find
    (a) its cost function;
    (b) the cost of making 20,000 gallons of each grade of gasoline.

19. A firm makes two products, A and B. The profit contribution of one product is not affected by that of the other. If each unit of product A contributes $60 toward profit, each unit of product B contributes $75 toward profit, and the fixed cost is $30,000, find the profit function and the firm's profit when it makes and sells 300 units of A and 500 of B.

20. A store sells white paint and blue paint. If the prices at which the two colors are sold do not affect each other, if x is the price per gallon of the white paint and y the price per gallon of the blue paint, and if the quantity of white paint demanded is

$$W = 400 - 6x + 2y$$

gallons and that of blue paint is

$$B = 100 + 4x - 10y$$

gallons, find

(a) the function that describes the store's revenue from paint sales;
(b) the store's revenue if it sells the white paint for $5 per gallon and the blue for $7 per gallon.

Suppose the demand function for product A is given by
$$A(p_A, p_B) = 200 - 8p_A + 3p_B,$$
where p_A is the unit price of A and p_B the unit price of another product B.

Exercises 21 through 24 refer to this product.

21. Find the quantity of A demanded if its price is $10 per unit and the price of B is $12 per unit.

22. If the unit price of A is fixed at a dollars, what can you say about the quantity of A demanded?

23. If the unit price of B is fixed at b dollars, what can you say about the quantity of A demanded?

24. How can the quantity of A demanded increase? How can it decrease?

A firm makes two products, X and Y. Its cost function is given by
$$C(x, y) = 2000 + 40x + 20y$$
and its revenue function by
$$R(x, y) = 12x\sqrt{y},$$
where x is the number of units of X produced and sold and y the number of units of Y produced and sold. Exercises 25 through 33 refer to this situation.

25. What is the firm's fixed cost?

26. What is the cost if 150 units of X and 100 units of Y are produced?

27. What can you say about the cost if the number of units of X produced is held fixed? If the number of units of Y produced is held fixed?

28. Find a function that describes the firm's average cost per unit, and find the average cost per unit when 150 units of X and 100 units of Y are produced.

29. What can you say about the average cost per unit when production of X is held fixed at 150 units? When production of Y is held fixed at 100 units?

30. Find the firm's profit function and its profit if it makes and sells 150 units of X and 100 units of Y.

31. Find a function that describes the firm's average profit per unit, and find its average profit per unit if it makes and sells 150 units of X and 100 units of Y.

32. What can you say about the firm's profit when production and sales of X are held fixed at 150 units? When production and sales of Y are held fixed at 100 units?

33. What can you say about the firm's average profit per unit when production and sales of X are held fixed at 150 units? When production and sales of Y are held fixed at 100 units?

A firm makes three products, X, Y, and Z. Its cost function is given by
$$C(x, y, z) = 25x + 30y + 32z + 40{,}000,$$
where x is the number of units of X, y the number of units of Y, and z the number of units of Z made. Exercises 34 through 36 refer to this situation.

34. Find the cost of producing 50 units of X, 60 units of Y, and 40 units of Z.

35. What can you say about the firm's cost if production of X and Y are held fixed at 50 and 60 units, respectively?

36. Find a function that describes the firm's average cost per unit and find the firm's average cost per unit when $x = 50$, $y = 60$, and $z = 40$.

The supply function for commodity A is

$$S = (8p_1 + 2p_2 + 1)/(3p_3 + 1),$$

where S is in millions of units, p_1 is the unit price of A, and p_2, p_3 are unit prices of other commodities. Exercises 37 through 40 refer to this commodity.

37. Find the quantity of commodity A supplied if $p_1 = \$10$, $p_2 = \$5$, and $p_3 = \$4$.

38. What can you say about the quantity of A supplied if p_1 and p_2 remain fixed at $10 and $4, respectively?

39. How can the quantity of A supplied increase?

40. How can the quantity of A supplied decrease?

Exercises 41 through 45 were suggested by material in Niebel, *Motion and Time Study*, eighth edition, Irwin, 1988. They concern the number of machines an operator can tend. Suppose an operator tends several identical machines, moving from one to another loading and unloading them. Let

$s = $ service time (loading time plus unloading time) per machine;
$m = $ machine time (running time) per machine;
$t = $ average travel time (time spent traveling from one machine to another).

The number of machines the operator can be assigned to is then given by the **synchronous service function** f defined by

$$f(s, m, t) = \left[\frac{s + m}{s + t}\right].$$

Here $\left[\dfrac{s + m}{s + t}\right]$ denotes the largest integer less than or equal to $\dfrac{s + m}{s + t}$.

*41. What is the domain of the synchronous service function?

*42. Find the number of machines an operator can tend if loading time is 1 minute, running time is 5 minutes, unloading time is 30 seconds, and average travel time is 15 seconds.

*43. Repeat Exercise 42 if loading time is 2 minutes, running time is 10 minutes, no unloading is necessary, and average travel time is 20 seconds.

If w_1 is the operator's wage rate and w_2 is the cost of running the machine, then the total cost of one cycle (loading, running, unloading, travel to next machine) for one machine will be

$$C = \frac{(s + m)(w_1 + w_2 f(s, m, t))}{f(s, m, t)}.$$

*44. Suppose that loading time is 2 minutes, running time is 12 minutes, unloading time is 1 minute, travel time averages 30 seconds, the operator earns $12 per hour, and it costs $72 per hour to run a machine. Find the cost of one cycle for one machine.

*45. Repeat Exercise 44 if the loading time is 2 minutes 15 seconds, the running time is 3 minutes 30 seconds, the unloading time is 45 seconds, there is no travel re-

1018 CHAPTER 16

quired, the operator earns $9 per hour, and it costs $240 per hour to run the machine.

Life Science

46. Let x denote the number of dogs, y the number of mice, and z the number of cats in a neighborhood. If
$$z = 10 - 2x + 0.2y,$$
   (a) Find the number of cats in the neighborhood if there are 6 dogs and 40 mice there.
   (b) What can you say about the cat population if the dog population of the neighborhood remains constant? If the mouse population remains constant?

The approximate time it will take for a patient to recover from pneumonia is
$$f(t, d) = \frac{1.2t}{d}$$
days, where t is the patient's age in years, $t \geq 21$, and d is the patient's daily dosage of penicillin in millions of units. Exercises 47 through 50 refer to this situation.

47. Find the approximate recovery time for a patient who is 40 years old and takes 4 million units of penicillin per day.
48. What can you say about the recovery time for patients who are all the same age?
49. What can you say about the recovery time for patients all of whom take the same daily dosage of penicillin?
50. How can recovery time increase; how can it decrease?

Exercises 51 through 55 were suggested by material in Niebel, *Motion and Time Study*, eighth edition, Irwin, 1988. They concern the contrast between an object and its background. This contrast is given by the equation
$$C = \frac{L - D}{L},$$
where D is the luminance of the darker of the object and its background, L is the luminance of the lighter of the object and its background, both measured in lamberts, and C is the contrast.

51. Find the contrast if $D = 10$ and $L = 20$ lamberts.
52. Find C if $D = 20$ and $L = 10$ lamberts.
53. Find C when D is fixed at D_0. What happens to C as L increases?
54. Find C when L is fixed at L_0. What happens to C as D increases?
55. The larger the value of C, the better the contrast. How can the contrast be increased?

Social Science

Annual housing starts H in a community depend on average weekly income Y and the average annual percentage rate I for mortgage loans according to the equation
$$H = 200 + 0.2Y - 0.65I^2$$
Exercises 56 through 58 refer to this situation.

56. What will housing starts be if average income is $250 per week and the average annual percentage for mortgage loans is 12%?

57. What can you say about housing starts if I remains stable at 12% per year?

58. What can you say about housing starts if Y remains fixed at $250 per week?

16.2 PARTIAL DERIVATIVES

Just as a function of one variable may have a first derivative with respect to its independent variable, so a function of more than one variable may have several first derivatives, one with respect to each of its independent variables. These derivatives, in turn, may themselves have derivatives with respect to each independent variable, and so on. Such derivatives of functions of several variables are known as **partial derivatives,** and the process by which they are found is **partial differentiation.**

Partial derivatives are important for many reasons, not the least of which is that they can be used to find the rate of change of a function of several variables with respect to any one of its variables while the others remain fixed. Thus, for instance, if $C = C(x, y)$ is a cost function, the partial derivative of C with respect to x will tell us (approximately) how much total cost C will change as x increases by one unit while y remains the same. This section is devoted to showing how to find partial derivatives; we postpone further discussion of their applications until the following sections.

First Partial Derivatives

A function of two variables can have two first partial derivatives, one with respect to each of its independent variables.

> **First Partial Derivatives**
>
> Let f be a function of the two variables x and y.
>
> **1.** We obtain the **first partial derivative of f with respect to x,** denoted by
>
> $$f_x \quad \text{or} \quad \frac{\partial f}{\partial x},$$
>
> by differentiating f with respect to x while treating y as a constant.
>
> **2.** We obtain the **first partial derivative of f with respect to y,** denoted by
>
> $$f_y \quad \text{or} \quad \frac{\partial f}{\partial y},$$
>
> by differentiating f with respect to y while treating x as a constant.

Partial differentiation for functions of more than two variables is carried out in a similar manner.

EXAMPLE 1   Let us find the first partial derivatives f_x and f_y for the function f defined by

$$f(x, y) = x^2 + y^2 + 2x + 3y.$$

To find f_x, we must differentiate f with respect to x while treating y as a constant. To remind ourselves that y is to be treated as a constant, we replace y by c in $f(x, y)$:

$$f(x, c) = x^2 + c^2 + 2x + 3c,$$

or

$$f(x, c) = x^2 + 2x + (c^2 + 3c);$$

differentiating with respect to x, we obtain

$$f_x = 2x + 2 + 0 = 2x + 2.$$

Similarly, to find f_y we replace x by c, obtaining

$$f(c, y) = y^2 + 3y + (c^2 + 2c),$$

and differentiate with respect to y to find that

$$f_y = 2y + 3. \quad \square$$

It is not really necessary for us to replace x and y with c when finding partial derivatives. Instead, we need merely keep in mind that when we are differentiating f to find f_x, y is to be treated as a constant, and when we are differentiating f to find f_y, x is to be treated as a constant.

EXAMPLE 2   Let us find f_x and f_y if f is defined by

$$f(x, y) = 2x - 5y + x^2 y^3.$$

Differentiating with respect to x while treating y as a constant, we have

$$f_x = 2(1) - 5(0) + 2xy^3 = 2 + 2xy^3.$$

Similarly, differentiating with respect to y while treating x as a constant, we obtain

$$f_y = 2(0) - 5(1) + x^2(3y^2) = -5 + 3x^2 y^2. \quad \square$$

EXAMPLE 3   Let us find the partial derivatives $\partial g/\partial x$, $\partial g/\partial y$, and $\partial g/\partial z$ for the function g of three variables defined by

$$g(x, y, z) = 5x + 3xy - 2x^2 z + y^2.$$

To find $\partial g/\partial x$, we differentiate with respect to x while treating y and z as constants:

$$\frac{\partial g}{\partial x} = 5(1) + 3(1)y - 2(2x)z + 0 = 5 + 3y - 4xz.$$

FUNCTIONS OF SEVERAL VARIABLES   1021

Similarly, to find $\partial g/\partial y$, we differentiate with respect to y while treating x and z as constants:

$$\frac{\partial g}{\partial y} = 5(0) + 3x(1) - 2(0) + 2y = 3x + 2y.$$

To find $\partial g/\partial z$, we differentiate with respect to z while treating x and y as constants:

$$\frac{\partial g}{\partial z} = 5(0) + 3(0) - 2x^2(1) + 0 = -2x^2. \quad \square$$

EXAMPLE 4   Let us find $f_x(1, 1)$ and $f_y(2, 1)$ if

$$f(x, y) = x^2 - 7x^2y^3 + y + 1.$$

Differentiating with respect to x and treating y as a constant, we have

$$f_x = 2x - 14xy^3.$$

Therefore

$$f_x(1, 1) = 2(1) - 14(1)(1)^3 = -12.$$

Similarly,

$$f_y = -21x^2y^2 + 1,$$

so

$$f_y(2, 1) = -21(2)^2(1)^2 + 1 = -83. \quad \square$$

EXAMPLE 5   Let

$$h(x, y, z, w) = (2x + 3y^2 + 4z^3 + 5w^4)^{10}.$$

Then

$$h_x = 10(2x + 3y^2 + 4z^3 + 5w^4)^9(2)$$
$$= 20(2x + 3y^2 + 4z^3 + 5w^4)^9,$$
$$h_y = 10(2x + 3y^2 + 4z^3 + 5w^4)^9(6y)$$
$$= 60y(2x + 3y^2 + 4z^3 + 5w^4)^9,$$
$$h_z = 10(2x + 3y^2 + 4z^3 + 5w^4)^9(12z^2)$$
$$= 120z^2(2x + 3y^2 + 4z^3 + 5w^4)^9,$$

and

$$h_w = 10(2x + 3y^2 + 4z^3 + 5w^4)^9(20w^3)$$
$$= 200w^3(2x + 3y^2 + 4z^3 + 5w^4)^9.$$

(Check this.) $\square$

Notice that we have shown how to *find* first partial derivatives, but we have not *defined* them. First partial derivatives are defined by difference quotients in a manner analogous to the way we used the difference quotient to define the derivative of a function of one variable in Section 10.3. For the sake of completeness, we give these definitions here for a function of two variables.

> **Definition of Partial Derivatives**
>
> Let f be a function of two variables x and y. Then
>
> $$f_x(x, y) = \lim_{h \to 0} \frac{f(x + h, y) - f(x, y)}{h}$$
>
> and
>
> $$f_y(x, y) = \lim_{h \to 0} \frac{f(x, y + h) - f(x, y)}{h},$$
>
> provided these limits exist.

For our purposes, knowing how to find partial derivatives will be sufficient; we will not need to use their definitions.

Higher Partial Derivatives

Just as a function of one variable may have a first derivative, a second derivative, and so on, a function of several variables may have first partial derivatives, second partial derivatives, and so on. The notation for second partial derivatives of a function of two variables is as follows:

> **Second Partial Derivatives**
>
> Let f be a function of two variables x and y with first partial derivatives f_x and f_y. Then the second partial derivatives of f are f_{xx}, f_{xy}, f_{yx}, and f_{yy}, where
>
> 1. f_{xx} is the partial derivative of f_x with respect to x;
> 2. f_{xy} is the partial derivative of f_x with respect to y;
> 3. f_{yx} is the partial derivative of f_y with respect to x;
> 4. f_{yy} is the partial derivative of f_y with respect to y.

Note that if we read the subscripts that label a second partial derivative from left to right, they tell us how to find the second partial; for instance, the notation

f_{xy} tells us that we are to find the first partial of f with respect to x and then differentiate it with respect to y.

EXAMPLE 6 Let f be defined by
$$f(x, y) = 5x + y^2 + 2x^3y^4.$$
Then
$$f_x = 5 + 6x^2y^4,$$
so
$$f_{xx} = 12xy^4 \quad \text{and} \quad f_{xy} = 24x^2y^3.$$
Similarly,
$$f_y = 2y + 8x^3y^3$$
and thus
$$f_{yx} = 24x^2y^3 \quad \text{and} \quad f_{yy} = 2 + 24x^3y^2. \quad \square$$

Notice that the **mixed partials** f_{xy} and f_{yx} in Example 6 were the same. This is not the case for all functions, but it will be true for all the functions we use in this book.

The extension of the concept of second partial derivatives to functions of more than two variables is straightforward.

EXAMPLE 7 Let
$$g(x, y, z) = x^2y^3 + xyz + z^2.$$
Then
$$g_x = 2xy^3 + yz, \qquad g_y = 3x^2y^2 + xz,$$
and
$$g_z = xy + 2z.$$
Therefore there are nine second partial derivatives:
$$\begin{aligned} g_{xx} &= 2y^3, & g_{xy} &= 6xy^2 + z, & g_{xz} &= y, \\ g_{yx} &= 6xy^2 + z, & g_{yy} &= 6x^2y, & g_{yz} &= x, \\ g_{zx} &= y, & g_{zy} &= x, & g_{zz} &= 2. \end{aligned}$$
Notice that mixed partials that have the same subscripts are equal:
$$g_{xy} = g_{yx}, \quad g_{xz} = g_{zx}, \quad \text{and} \quad g_{yz} = g_{zy}. \quad \square$$

An alternative notation for second partial derivatives is as follows:
$$f_{xx} = \frac{\partial}{\partial x}\left(\frac{\partial f}{\partial x}\right) = \frac{\partial^2 f}{\partial x^2}, \qquad f_{xy} = \frac{\partial}{\partial y}\left(\frac{\partial f}{\partial x}\right) = \frac{\partial^2 f}{\partial y\, \partial x},$$
$$f_{yx} = \frac{\partial}{\partial x}\left(\frac{\partial f}{\partial y}\right) = \frac{\partial^2 f}{\partial x\, \partial y}, \qquad f_{yy} = \frac{\partial}{\partial y}\left(\frac{\partial f}{\partial y}\right) = \frac{\partial^2 f}{\partial y^2}.$$

Notice that when we use this notation, it is the order of the variables in the denominator from right to left that tells us how to find the second partial:

$$\frac{\partial^2 f}{\partial x\, \partial y} \text{ is the partial derivative of } \frac{\partial f}{\partial y} \text{ with respect to } x.$$

EXAMPLE 8   Let

$$f(x, y) = e^y + xy^2.$$

Then

$$f_y = e^y + 2xy$$

and

$$\frac{\partial^2 f}{\partial x\, \partial y} = \frac{\partial}{\partial x}(f_y) = \frac{\partial}{\partial x}(e^y + 2xy) = 2y. \quad \square$$

EXAMPLE 9   Let

$$z = x^2 + \ln xy.$$

Then

$$\frac{\partial z}{\partial x} = 2x + \frac{1}{xy}y = 2x + \frac{1}{x},$$

so

$$\frac{\partial^2 z}{\partial x^2} = 2 - \frac{1}{x^2} \quad \text{and} \quad \frac{\partial^2 z}{\partial y\, \partial x} = 0. \quad \square$$

Third and higher-order partial derivatives are found in the obvious manner.

EXAMPLE 10   Find the third partial derivative f_{xyx} for

$$f(x, y) = 3y - 2xy^2 + y \ln x.$$

The notation f_{xyx} tells us that we must differentiate first with respect to x, then with respect to y, and finally with respect to x again. Thus,

$$f_x = -2y^2 + \frac{y}{x},$$

$$f_{xy} = -4y + \frac{1}{x},$$

and

$$f_{xyx} = -\frac{1}{x^2}. \quad \square$$

16.2 EXERCISES

First Partial Derivatives

1. Find f_x and f_y if $f(x, y) = 5x - 7y + 12$.
2. Find f_x and f_y if $f(x, y) = 6x + 12y^2 - 3$.
3. Find f_x and f_y if $f(x, y) = 8xy - 2y^2$.
4. Find f_x and f_y if $f(x, y) = x^2y^3 - 2xy^2 - 1$.
5. Find $\dfrac{\partial f}{\partial x}$ and $\dfrac{\partial f}{\partial y}$ if $f(x, y) = 3x + \dfrac{2x^2}{y^2} + 6$.
6. Find $\dfrac{\partial f}{\partial x}$ and $\dfrac{\partial f}{\partial y}$ if $f(x, y) = 5x^3 + 7y^4 - 3x^2y^2 + 5xy - 2x + y$.
7. Find g_x and g_y if $g(x, y) = \sqrt{2xy}$.
8. Find $\dfrac{\partial z}{\partial x}$ and $\dfrac{\partial z}{\partial y}$ if $z = \sqrt{x^2 - y^2}$.
9. Find $\dfrac{\partial h}{\partial u}$ and $\dfrac{\partial h}{\partial v}$ if $h(u, v) = (u^2 + uv + v^3)^4$.
10. Find f_r and f_s if $f(r, s) = e^{2r} - \ln s$.
11. Find $\dfrac{\partial z}{\partial x}$ and $\dfrac{\partial z}{\partial y}$ if $z = x \ln y^2$.
12. Find $\dfrac{\partial s}{\partial p}$ and $\dfrac{\partial s}{\partial q}$ if $s = 2p^2q^4 - 5pq + 2$.
13. Find $\dfrac{\partial w}{\partial x}$ and $\dfrac{\partial w}{\partial y}$ if $w = \dfrac{2x - y}{x + y}$.
14. Find h_x and h_y if $h(x, y) = \ln \dfrac{x + y}{x - y}$.
15. Find f_x, f_y, and f_z if $f(x, y, z) = xz + x^2z^2 + yz^3 - 5y$.
16. Find h_x, h_y, and h_z if $h(x, y, z) = e^{2xz} - \ln yz + 1$.
17. Find $f_x(1, 3)$ and $f_y(-2, 1)$ if $f(x, y) = 2xy$.
18. Find $f_x(0, 0)$ and $f_y(1,1)$ if $f(x, y) = 2x + y + 4xy - x^2 + y^2$.
19. Find $f_x(1, 0)$ and $f_y(0, 1)$ if $f(x, y) = 2e^x - 3e^{-y}$.
20. Evaluate $\dfrac{\partial f}{\partial x}$ and $\dfrac{\partial f}{\partial y}$ at $(8, 3)$ if $f(x, y) = \sqrt{2x + 3y}$.
21. Evaluate $\dfrac{\partial z}{\partial x}$ and $\dfrac{\partial z}{\partial y}$ at $(0, e)$ if $z = e^{-x} \ln y$.
22. Find $\dfrac{\partial w}{\partial x}\bigg|_{(1,-1,1)}$ and $\dfrac{\partial w}{\partial y}\bigg|_{(1,-1,1)}$ if $w = \sqrt{x^2 + y^2} + 4xz$.
23. Find $f_x(1, -1, 2)$, $f_y(1, -1, 2)$, and $f_z(1, -1, 2)$ if $f(x, y, z) = x^2 + \dfrac{5y}{xz} + 10xz$.
24. Find $\dfrac{\partial p}{\partial u}\bigg|_{(0,0,2)}$, $\dfrac{\partial p}{\partial v}\bigg|_{(0,0,2)}$, and $\dfrac{\partial p}{\partial w}\bigg|_{(0,0,2)}$ if $p = (3u^2w^2 + 4v^2w - 2u + 3v + w + 1)^4$.

Higher Partial Derivatives

25. If $f(x, y) = 2x + 3y^2 + 1$, find f_{xx}, f_{xy}, f_{yx}, and f_{yy}.
26. If $f(x, y) = 4x^2 + 3xy - y^2$, find f_{xx}, f_{xy}, f_{yx}, and f_{yy}.
27. If $g(x, y) = x^2 + y^2 - 2xy^3$, find g_{xx}, g_{xy}, g_{yx}, and g_{yy}.
28. If $f(x, y) = 2y^3 - 3x^3 + 7x + 12y^2 + 1$, find f_{xx}, f_{xy}, f_{yx}, and f_{yy}.
29. If $f(x, y) = \dfrac{2x}{3y^2} + 1$, find f_{xx}, f_{xy}, f_{yx}, and f_{yy}.

30. If $z = y^2\sqrt{x} - 2x^3 + 3$, find $\dfrac{\partial^2 z}{\partial x^2}$, $\dfrac{\partial^2 z}{\partial y\,\partial x}$, $\dfrac{\partial^2 z}{\partial x\,\partial y}$, and $\dfrac{\partial^2 z}{\partial y^2}$.

31. If $z = \dfrac{4x^4}{y} + \dfrac{2y^2}{x^2}$, find $\dfrac{\partial^2 z}{\partial x^2}$, $\dfrac{\partial^2 z}{\partial y\,\partial x}$, $\dfrac{\partial^2 z}{\partial x\,\partial y}$, and $\dfrac{\partial^2 z}{\partial y^2}$.

32. If $f(x, y, z) = x^2 + 3xyz + 2yz + z^2$, find all second partial derivatives of f.

33. If $f(x, y, z) = ye^x + 2\ln x + x\ln z$, find all second partial derivatives of f.

34. If $f(x, y) = 10x^2 y^3 + x + 2y^3$, find f_{xx}, f_{xyx}, and f_{yyx}.

35. If $f(x, y) = e^{xy}$, find f_{xyxy} and f_{xyyy}.

36. Find f_{uvu} if $f(u, v) = u^4 + 2e^{u/v}$.

37. Find g_{pqq} if $g(p, q, r) = 1 + 3pqr^{-1} + q^3 p^{-1}$.

38. Find f_{xxy} if $f(x, y) = e^{x-2y}$.

39. Find f_{xxyz} if $f(x, y, z) = \ln xyz + 2x^3 y^2 z^3$.

40. If $z = \ln 2x + 3\ln y$, find $\dfrac{\partial^2 z}{\partial x^2}$, $\dfrac{\partial^3 z}{\partial x\,\partial y\,\partial x}$, and $\dfrac{\partial^3 z}{\partial x\,\partial y^2}$.

41. If $f(x, y) = 5x^2 - 3xy + y$, find $f_{xx}(1, 1)$, $f_{xy}(2, 2)$, $f_{yx}(3, -5)$, and $f_{yy}(-2, 7)$.

42. If $f(x, y) = 6x + 2ye^x$, find $f_{xx}(2, 0)$, $f_{xy}(1, -1)$, $f_{yx}(-2, 1)$, and $f_{yy}(0, 2)$.

43. If $f(x, y) = x^2 y^3 + x^4 + y^4$, find $f_{xyx}(1, 1)$ and $f_{yyx}(-1, -1)$.

44. If $f(x, y) = xy + \ln xy$, find $f_{xxyy}(2, 2)$ and $f_{xyxx}(2, 2)$.

16.3 PARTIAL DERIVATIVES AS RATES OF CHANGE

At the beginning of Section 16.2 we remarked that a first partial derivative of a function of several variables gives its rate of change with respect to one of the variables as the others remain fixed. In this section we illustrate some uses of this fact. Before considering partial derivatives as rates of change, however, we briefly explain the geometric significance of the partial derivatives of a function of two variables.

The Geometric Significance of Partial Derivatives

Just as the derivative $f'(a)$ of a function of one variable is the slope of the line tangent to the graph of f at the point $(a, f(a))$, so the partial derivatives $f_x(a, b)$ and $f_y(a, b)$ of a function of two variables are slopes of tangent lines to the graph of f at the point $(a, b, f(a, b))$. To be specific:

Slopes of Tangent Lines to Surfaces

Let f be a function of the two variables x and y, and let (a, b) be a point in the domain of f.

1. If $f_x(a, b)$ exists, it is the slope of the line parallel to the xz-plane and tangent to the surface $z = f(x, y)$ at the point $(a, b, f(a, b))$.

2. If $f_y(a, b)$ exists, it is the slope of the line parallel to the yz-plane and tangent to the surface $z = f(x, y)$ at the point $(a, b, f(a, b))$.

FUNCTIONS OF SEVERAL VARIABLES   1027

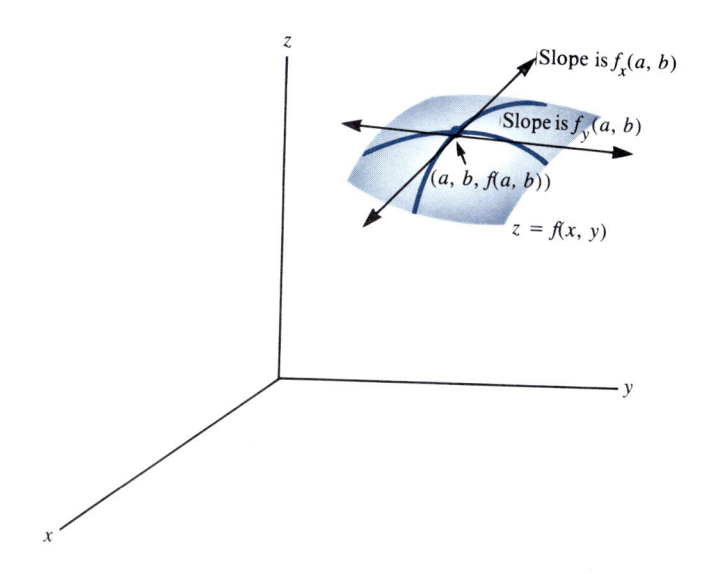

FIGURE 16–7

See Figure 16–7.

EXAMPLE 1   Let f be defined by $f(x, y) = y^2 - x^2$. When $x = 1$ and $y = 2$, $f(1, 2) = 3$, so the point $(1, 2, 3)$ is on the surface $z = y^2 - x^2$. Since

$$f_x(x, y) = -2x,$$

the slope of the line tangent to the surface at $(1, 2, 3)$ and parallel to the xz-plane is

$$f_x(1, 2) = -2(1) = -2.$$

Thus, since $f_x(1, 2)$ is negative, at the point $(1, 2, 3)$ the surface bends downward in the positive x-direction. See Figure 16–8. Similarly, since

$$f_y(x, y) = 2y,$$

the slope of the line tangent to the surface at $(1, 2, 3)$ and parallel to the yz-plane is

$$f_y(1, 2) = 2(2) = 4,$$

which is positive; hence at $(1, 2, 3)$ the surface bends upward in the positive y-direction. Again, see Figure 16–8. ☐

As Example 1 indicates, the signs of the first partial derivatives f_x and f_y evaluated at a point tell us whether at that point the surface bends upward or downward in the positive x- and y-directions, respectively. Another way to think of this is to consider what will happen if we take a "slice" through the surface parallel to the xz- or yz-plane: in either case the result will be a curve on the surface, and the signs of the first partials will tell us whether the curve is increasing or decreasing. Thus, suppose we take a slice through the surface parallel to the xz-plane: if f_x is positive

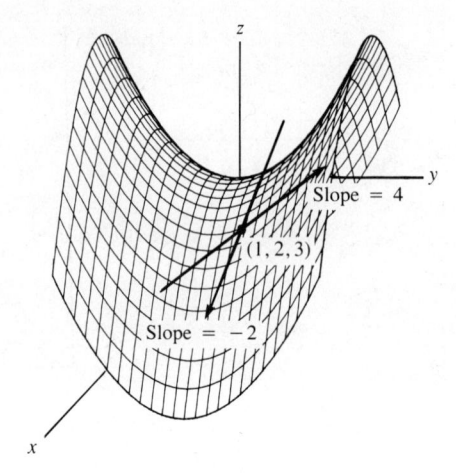

FIGURE 16-8

at a point on the curve, then at that point the curve will be increasing in the positive x-direction; if f_x is negative, the curve will be decreasing. See Figure 16–9, where $f_x(a, b) > 0$ and the curve is increasing in the positive x-direction. Similar remarks apply to f_y and slices parallel to the yz-plane.

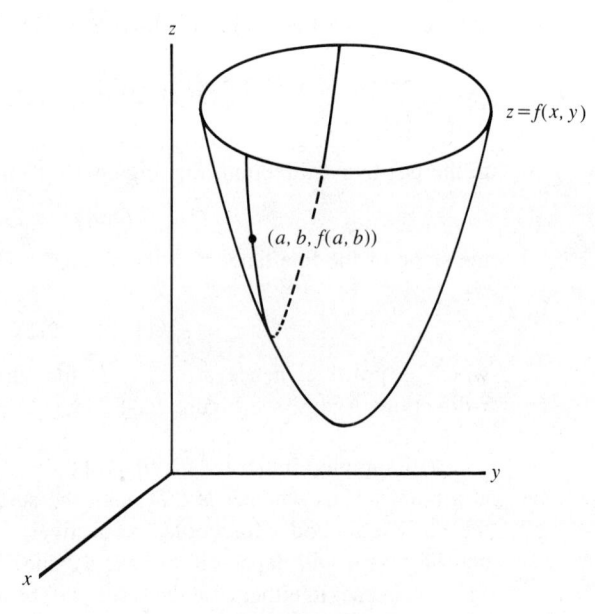

FIGURE 16-9
$f_x(a, b) > 0, f_{xx}(a, b) > 0$

The signs of the second partials f_{xx} and f_{yy} evaluated at a point tell us whether at that point the surface is concave upward or downward in the x- and y-directions, respectively. For example, if we take a slice through the surface parallel to the xz-plane and f_{xx} is positive at a point on the resulting curve, then the curve is concave upward there; if f_{xx} is negative, the curve is concave downward. Again see Figure 16–9, where $f_{xx}(a, b) > 0$ and the curve is concave upward. Similar remarks apply to f_{yy} and slices parallel to the yz-plane.

EXAMPLE 2 Let $f(x, y) = y^2 - x^2$. At any point on the surface, $f_{xx} = -2$. Hence if we take a slice through any point on the surface and parallel to the xz-plane, the resulting curve will be concave downward. Similarly, $f_{yy} = 2$ at any point on the surface, so any slice parallel to the yz-plane will result in a curve that is concave upward. See Figure 16–10. ∎

Partial Derivatives as Rates of Change

Consider Figure 16–7 again: as the figure indicates, the slope $f_x(a, b)$ may be interpreted as the instantaneous rate of change of $z = f(x, y)$ at the point

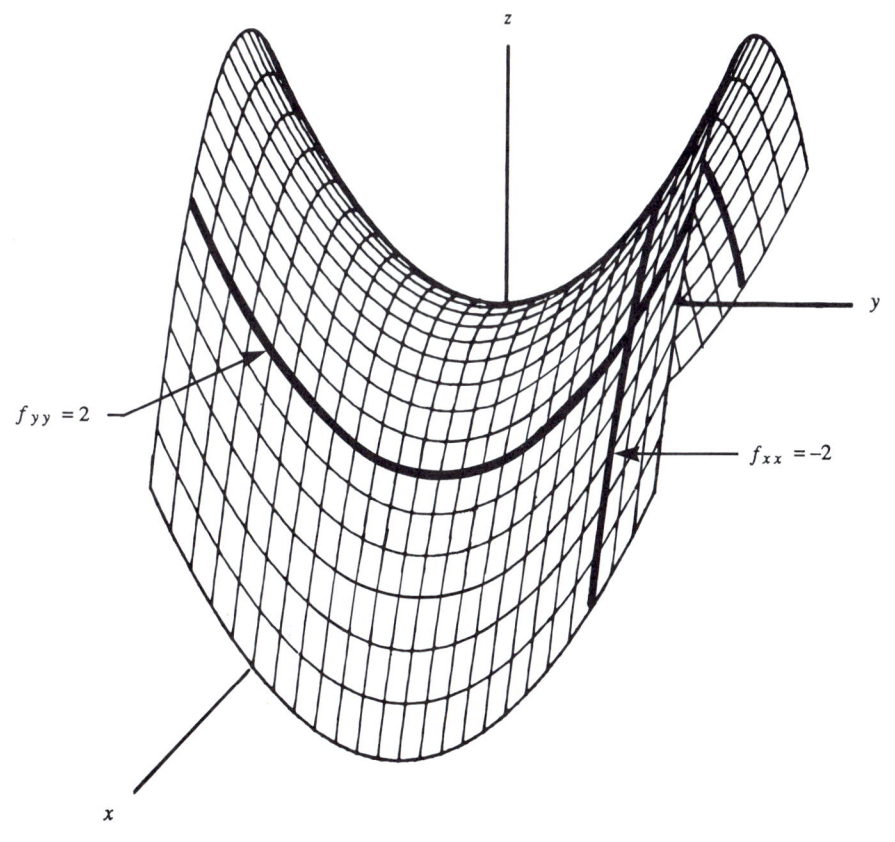

FIGURE 16–10

$(a, b, f(a, b))$ in the positive x-direction; that is, $f_x(a, b)$ is the instantaneous rate of change of f with respect to x at $(a, b, f(a, b))$ when y is held fixed at the value $y = b$. Similarly, $f_y(a, b)$ is the instantaneous rate of change of f with respect to y at $(a, b, f(a, b))$ when x is held fixed at the value $x = a$. Thus for functions of two variables, partial derivatives can be interpreted as rates of change, and this remains true for functions of more than two variables.

> **Partial Derivatives as Rates of Change**
>
> Let f be a function of several variables, and let x be any one of the independent variables of f. Then f_x gives the **instantaneous rate of change of f with respect to x** when all the other independent variables of f are fixed. Thus, when all the other independent variables of f are fixed, a small change in x will lead to a change of approximately f_x in f.

We illustrate some applications of this point of view with examples.

EXAMPLE 3 Two bactericides are tested in combination, and it is found that when x units of bactericide X and y units of bactericide Y are used together, they will destroy

$$f(x, y) = 5x + 4y$$

million bacteria. Thus if y is held fixed, the rate of change of $f(x, y)$ with respect to x is

$$f_x(x, y) = 5.$$

The interpretation is that if the number of units y of bactericide Y is held fixed, each additional unit of bactericide X used will kill approximately 5 million bacteria.

Similarly, since

$$f_y(x, y) = 4,$$

if the number of units of bactericide X is held fixed, then each additional unit of bactericide Y used will destroy approximately 4 million bacteria. ◻

EXAMPLE 4 Suppose that if a firm produces and sells x units of product A, y units of product B, and z units of product C, its profit is

$$P(x, y, z) = 2\sqrt{x} + 5y + 4\sqrt{z} + xyz$$

dollars. Then

$$P_x(x, y, z) = \frac{1}{\sqrt{x}} + yz$$

is the firm's **marginal profit with respect to x**. If y and z are held fixed, an additional unit of product A made and sold will increase the firm's profit by approximately

$$\frac{1}{\sqrt{x}} + yz$$

FUNCTIONS OF SEVERAL VARIABLES 1031

dollars. For instance, if $x = 16$, $y = 10$, and $z = 25$, then the firm's profit is
$$P(16, 10, 25) = 4068$$
dollars. If y and z are held fixed at 10 and 25, respectively, then making and selling an additional unit of product A (the seventeenth one) will increase profit by approximately
$$P_x(16, 10, 25) = \frac{1}{\sqrt{16}} + (10)(25) = \$250.25.$$
Similarly, since
$$P_y(x, y, z) = 5 + xz \quad \text{and} \quad P_z(x, y, z) = \frac{2}{\sqrt{z}} + xy,$$
holding x and z fixed at 16 and 25, respectively, and producing an additional unit of B (the eleventh one) will contribute approximately
$$P_y(16, 10, 25) = 5 + (16)(25) = \$405$$
to profit; holding x and y fixed at 16 and 10, respectively, and producing an additional unit of C (the twenty-sixth one) will contribute approximately
$$P_z(16, 10, 25) = \frac{2}{\sqrt{25}} + (16)(10) = \$160.40$$
to profit. Therefore if the firm is currently making 16 units of A, 10 of B, and 25 of C and has limited resources for expanding production, it would be wise to use its resources to increase the production of B. ∎

EXAMPLE 5 Two products are said to be **competitive** if a *decrease* in the quantity demanded of one can lead to an *increase* in the quantity demanded of the other. (This occurs if the products can be substituted for each other. For instance, beef products and pork products are competitive; if the price of beef goes up while the price of pork remains fixed, the quantity of beef demanded will decrease and the quantity of pork demanded will increase, as consumers substitute pork for beef. Similarly, if the price of pork goes up while the price of beef remains unchanged, the quantity of pork demanded will decrease while the quantity of beef demanded will increase.) On the other hand, if a *decrease* in the quantity demanded of one product can lead to a *decrease* in the quantity demanded of the other, then the products are said to be **complementary.** (Automobiles and automobile tires are examples of complementary products.)

We can tell whether products are competitive or complementary by analyzing the partial derivatives of their demand functions. To see how this is done, suppose that the price of butter is b dollars per pound and that the price of margarine is m dollars per pound. Let
$$B(b, m) = 10{,}000 - 200b + 400m \quad \text{and} \quad M(b, m) = 8000 + 300b - 500m$$
be the demand functions for butter and margarine, respectively. Then
$$B_b = -200, \quad B_m = 400, \quad M_b = 300, \quad \text{and} \quad M_m = -500.$$

Now, B_b is negative, so an increase in the price b of butter while the price m of margarine is held fixed will lead to a decrease in the quantity of butter demanded. But since M_b is positive, an increase in b when m is held fixed will also lead to an increase in the quantity of margarine demanded. Similarly, since M_m is negative and B_m is positive, an increase in m while b is held fixed leads to a decrease in the quantity of margarine demanded and an increase in the quantity of butter demanded. Thus butter and margarine are competitive products. ☐

EXAMPLE 6 If a firm uses x worker-hours per week and y dollars of capital per week, it can produce

$$f(x, y) = 4x^{3/4}y^{1/4}$$

units per week. A function such as this which relates output to input by means of an expression of the form $kx^p y^{1-p}$, $0 \leq p \leq 1$, is called a **Cobb-Douglas production function.** In this case, if the firm has 1296 worker-hours and \$10,000 in capital available this week, it can produce

$$f(1296, 10{,}000) = 4(1296)^{3/4}(10{,}000)^{1/4} = 8640$$

units. The partial derivative f_x is called the **marginal productivity of labor**; the partial derivative f_y is the **marginal productivity of capital.** Here

$$f_x(x, y) = 3x^{-1/4}y^{1/4} \quad \text{and} \quad f_y(x, y) = x^{3/4}y^{-3/4}.$$

Thus if $x = 1296$ and $y = 10{,}000$,

$$f_x(1296, 10{,}000) = 3(1296)^{-1/4}(10{,}000)^{1/4} = 5$$

and

$$f_y(1296, 10{,}000) = (1296)^{3/4}(10{,}000)^{-3/4} = 0.216.$$

Therefore if capital remains fixed at \$10,000 per week, the availability of an additional worker-hour each week would result in an increase in production of approximately five units per week. If labor remains fixed at 1296 worker-hours per week, the availability of an additional dollar of capital each week would result in an increase in production of approximately 0.216 units per week. ☐

EXAMPLE 7 Poiseuille's Law states that the volume rate of blood flow F in an artery is given by the equation

$$F = \frac{kr^4}{L},$$

where r is the radius of the artery, L is its length, and k is a positive constant that depends on the viscosity of the blood. Since

$$\frac{\partial F}{\partial r} = \frac{4kr^3}{L},$$

when length L is fixed increasing the radius slightly will cause the volume rate of blood flow F to increase (because $4k/L$ is positive) by an amount approximately

FUNCTIONS OF SEVERAL VARIABLES 1033

equal to $(4k/L)r^3$, that is, by an amount approximately proportional to the cube of the radius. On the other hand, since

$$\frac{\partial F}{\partial L} = -\frac{kr^4}{L^2},$$

when the radius r is held fixed increasing the length slightly will cause F to decrease (because $-kr^4$ is negative) by an amount approximately equal to $(kr^4)(1/L^2)$, that is, by an amount approximately inversely proportional to the square of the length. ☐

16.3 EXERCISES

The Geometric Significance of Partial Derivatives

In Exercises 1 through 8, find the slopes of the lines tangent to the given surface at the given point and parallel to the xz- and yz-planes. If slices are taken through the surface at the point and parallel to the xz- and yz-planes, state whether the resulting curves are increasing, decreasing, concave upward, or concave downward at the point.

1. $z = 20 - 2x - 5y$, (2, 1, 11)
2. $z = 2x^2 + 4y^2 + 10$, (2, 3, 54)
3. $z = 2xy + x^2y$, (1, −1, −3)
4. $z = xy^2 + 2x + 3y + 1$, (0, 2, 7)
5. $z = x^2y^3 + 2x - 3y$, (1, −1, 4)
6. $z = x^2 + y^2$, $(a, 1, a^2 + 1)$
7. $z = x^2$, $(1, b, 1)$
8. $z = e^y$, (a, b, e^b)

Partial Derivatives as Rates of Change

Management

If a firm makes and sells x units of product A and y units of product B, its cost function is given by

$$C(x, y) = 20x + 14y + 200$$

and its revenue function by

$$R(x, y) = 25x + 20y.$$

Exercises 9 through 11 refer to this firm.

9. Find the firm's marginal cost with respect to x and with respect to y. Interpret your results.
10. Find the firm's marginal revenue with respect to x and with respect to y. Interpret your results.
11. Find the firm's marginal profit with respect to x and with respect to y. Interpret your results.

If a firm makes and sells x units of product X and y units of product Y, its cost and revenue functions are given by $C(x, y) = 30x + xy + 42y + 80{,}000$ and $R(x, y) = 35x + 2xy + 50y$, respectively. Exercises 12 through 21 refer to this firm.

12. Find the firm's marginal cost functions.

13. Find the firm's marginal cost with respect to x and y when $x = 100$ and $y = 200$.
14. If y is held fixed at 200, what is the effect of making one more unit of X?
15. If x is held fixed at 100, what is the effect of making one more unit of Y?
16. In general, if y is held fixed and one more unit of X is produced, what happens to the cost? What if x is held fixed and one more unit of Y is produced?
17. Find the firm's marginal revenue functions.
18. Find the firm's marginal revenue with respect to x and y when $x = 100$ and $y = 200$.
19. If y is held fixed at 200, what is the effect of selling one more unit of X?
20. If x is held fixed at 100, what is the effect of selling one more unit of X?
21. In general, if y is fixed and one more unit of X is sold, what happens to the revenue? What if x is fixed and one more unit of Y is sold?

A company that makes and sells x units of product X and y units of product Y has cost function given by

$$C(x, y) = \sqrt{xy + 25}$$

and revenue function given by

$$R(x, y) = 2x + y.$$

Exercises 22 through 25 refer to this company.

22. Find the company's marginal cost, marginal revenue, and marginal profit with respect to x and y when $x = 20$ and $y = 10$.
23. If y is held fixed at 10, what is the effect of making and selling one more unit of X? If x is held fixed at 20, what is the effect of making and selling one more unit of Y?
24. In general, if x is held fixed and one additional unit of Y is made and sold, what happens to cost, revenue, and profit?
25. Answer the question of Exercise 24 if y is held fixed and one more unit of X is made and sold.
26. If a corporation's cost and revenue functions are given by

$$C(x, y) = 5x^2 + 3y^2 + xy + 100$$

and

$$R(x, y) = 6x^2 + 5y^2,$$

respectively, find and interpret $C_x(100, 200)$, $C_y(100, 200)$; $R_x(100, 200)$, $R_y(100, 200)$; $R_x(100, 200) - C_x(100, 200)$; and $R_y(100, 200) - C_y(100, 200)$.

27. Suppose the demand function for commodity V is given by

$$V(v, w) = 20{,}000 - 500v + 300w$$

and that for commodity W by

$$W(v, w) = 30{,}000 + 800v - 400w,$$

where v is the unit price of V and w is the unit price of W. Find V_v, V_w, W_v, and W_w, and interpret your results. Are the commodities competitive or complementary?

28. Repeat Exercise 27 if
$$V(v, w) = 20{,}000 - 200v - 100w,$$
and
$$W(v, w) = 30{,}000 - 400v - 300w.$$

29. Suppose the demand functions for products X and Y are given by
$$f(x, y) = 200{,}000 - 200x^2 + 50xy - 10y^2$$
and
$$g(x, y) = 300{,}000 - 400y^2 + 100xy - 30x^2,$$
respectively, with x the unit price of X and y the unit price of Y. Find $f_x(10, 10)$, $f_y(10, 10)$, $g_x(10, 10)$, and $g_y(10, 10)$ and interpret your results. Are the commodities competitive at these prices?

30. Repeat Exercise 29 for $f_x(10, 20)$, $f_y(10, 20)$, $g_x(10, 20)$, and $g_y(10, 20)$.

31. Suppose the supply function for commodity X is given by
$$f(x, y) = 200x - 300y - 10{,}000$$
while that for commodity Y is given by $g(x, y) = -400x + 200y - 12{,}000$. Here x is the unit price of X and y the unit price of Y. Find f_x, f_y, g_x, and g_y and interpret your results.

32. Suppose the supply functions for products X and Y are given by
$$A(x, y) = 2x^2 + 10xy - 3y^2$$
and $B(x, y) = -5x^2 + 20xy + 8y^2$, respectively, where x is the unit price of X and y that of Y. Find A_x, A_y, B_x, B_y when $x = 5$ and $y = 10$ and interpret your results.

33. Repeat Exercise 32 using $x = 7$ and $y = 3$.

34. The value of a company's daily output is z dollars, where $z = 10x^{2/5}y^{3/5}$. Here x is labor input in worker-hours per day and y is capital input in dollars per day. Find z, z_x, and z_y when $x = 32{,}768$ and $y = 100{,}000$, and interpret your results.

35. Refer to Exercise 34: In general, what happens to production if labor input is fixed and capital input increases by $1? if capital input is fixed and labor input increases by 1 worker-hour?

36. A firm's production function is given by
$$f(x, y) = 200\sqrt{xy},$$
where x is labor input in worker-hours per week, y is capital input in dollars per week, and output is in units per week.
   (a) Find the firm's weekly output if it uses 1600 worker-hours and $10,000 of capital per week.
   (b) Find $f_x(1600, 10{,}000)$ and $f_y(1600, 10{,}000)$, and interpret your results.

37. Using the notation of Exercise 36, suppose the production function is defined by
$$f(x, y) = 12x^{1/4}y^{3/4}.$$
Find $f(81, 625)$, $f_x(81, 625)$, and $f_y(81, 625)$, and interpret your results.

38. Suppose a production function is given by
$$f(x, y) = xe^{0.01y},$$
where x is labor input in thousands of worker-hours per week, y is capital input in thousands of dollars per week, and output is in thousands of units per week.

Evaluate the first partial derivatives of f with respect to x and y at
(a) $x = y = 50$ (b) $x = y = 100$ (c) $x = y = 200$
and interpret your results.

39. Refer to Exercise 38: What happens to production if labor input is fixed and capital input increases by $1000 per week? if capital input is fixed and labor input increases by 1000 worker-hours per week?

Exercises 40 and 41 were suggested by material in McGann and Russell, *Advertising Media*, second edition, Irwin, 1988. They are concerned with the problem of measuring the net audience reached by an advertisement when there is audience duplication because the advertisement appears in several different outlets. A model for net audience reached due to J. M. Agostini is

$$C = \frac{A^2}{A + 1.25D},$$

where

A = total audience of all outlets used,
D = sum of the duplicated audiences for all pairs of outlets used,
C = net (unduplicated) audience.

*40. Find $\partial C/\partial A$ and interpret your result.

*41. Find $\partial C/\partial D$ and interpret your result.

Life Science

42. If a farmer irrigates a field x times, weeds it y times, and fertilizes it z times during the growing season, the crop yield will be

$$w = 4x + 5y + 2z + 400$$

bushels. Find the first partial derivatives of w with respect to x, y, and z, and interpret your results.

43. The number of bacteria present in a culture t hours after an experiment has begun depends on t, on the amount of nutrient provided, and on the temperature at which the experiment is run. Suppose that

$$p = 100e^{0.2x} + 2xT + 10T^2 + 10{,}000t,$$

where p is the number of bacteria in the culture, x is nutrient provided in grams, and T is temperature in degrees Celsius. Evaluate the first partial derivatives of p with respect to x, T, and t when $x = 2$, $T = 20$, and $t = 3$, and interpret your results.

44. Suppose that if there are x rabbits for coyotes to prey on and y hunters to kill the coyotes, then the size of the coyote population is

$$f(x, y) = 4\frac{\sqrt{x}}{y}$$

individuals. Find $f_x(400, 2)$ and $f_y(400, 2)$ and interpret your results.

45. Refer to Exercise 44: In general, what happens to the coyote population if the number of rabbits is fixed and the number of hunters increases by one? if the number of hunters is fixed and the number of rabbits increases by one?

46. If microorganisms are supplied with p grams of nutrient and sprayed with q milliliters of a weak bactericide, there will be $w = 1000e^{p/2q}$ of them present. Find $w_p(5, 8)$ and $w_q(5, 8)$ and interpret your results.

Social Science

47. Ridership on a city's public transportation system averages

$$f(x, y) = \frac{0.1x}{1 + \ln y}$$

thousand people per day, where x is the population of the city in thousands and y is the price of a ticket. The population of the city is currently 100,000 people and the price of a ticket is \$1. Find and interpret $f_x(100, 1)$ and $f_y(100, 1)$.

48. It is estimated that the proportion p of the population who apply for welfare benefits in a community during a year depends on the unemployment rate x and the average years of education y in the community according to the equation

$$p(x, y) = e^{0.08x} - e^{0.01y}.$$

Here $5 \leq x \leq 20$ and $10 \leq y \leq 15$. Find $p_x(8, 12)$ and $p_y(8, 12)$ and interpret your results.

49. Refer to Exercise 48: What happens to the proportion who apply for benefits if the average years of education remains fixed and the unemployment rate increases by 1%?

50. If a politician's campaign spends x thousand dollars on TV ads and the opposition campaign spends y thousand dollars on TV ads, the politician will get $z = 50{,}000 \ln[(x + 1)(y + 2)^{-1}]$ votes. Find $z_x(20, 80)$ and $z_y(20, 80)$ and interpret your results.

16.4 OPTIMIZATION

In this section we discuss **relative maxima, relative minima,** and **saddle points** of functions of two variables and show how to find such points using an analog of the second derivative test for functions of one variable that we introduced in Section 12.2. (Most of the results and techniques presented in this section can be extended to functions of more than two variables, but since the two-variable case illustrates the essential ideas with the fewest possible complications, we will confine ourselves to it.)

We begin by indicating what it means for a function of two variables to have a relative maximum or minimum at a point. If f is such a function, we say that f has a **relative maximum** at the point $(a, b, f(a, b))$ if the surface $z = f(x, y)$ has a peak at $(a, b, f(a, b))$, as shown in Figure 16–11; it has a **relative minimum** at $(a, b, f(a, b))$ if the surface has a pit at $(a, b, f(a, b))$, as shown in Figure 16–12. Also, Figure 16–13 (p. 1039) depicts a function that has a **saddle point** at $(a, b, f(a, b))$. Note that a saddle point is a relative minimum in one direction and a relative maximum in another direction, and hence is itself neither a relative maximum nor a relative minimum.

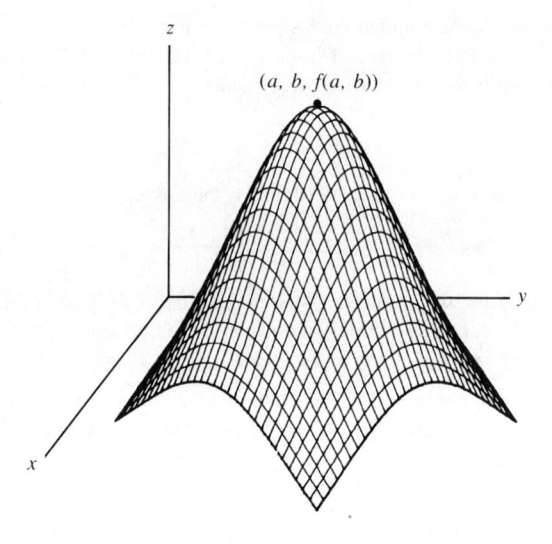

FIGURE 16–11

Suppose that a function of two variables f has a relative maximum at $(a, b, f(a, b))$ and that the line parallel to the xz-plane and tangent to the surface $z = f(x, y)$ at $(a, b, f(a, b))$ exists, as in Figure 16–14. Then this line must be horizontal, and its slope, which is $f_x(a, b)$, must be zero. Similarly, the line tangent to the surface at $(a, b, f(a, b))$ and parallel to the yz-plane must also have a slope of zero, so $f_y(a, b)$ must be zero. Therefore if f has a relative maximum at $(a, b, f(a, b))$ and $f_x(a, b)$ and $f_y(a, b)$ exist, then

$$f_x(a, b) = 0 \quad \text{and} \quad f_y(a, b) = 0.$$

Similar reasoning holds for relative minima and saddle points.

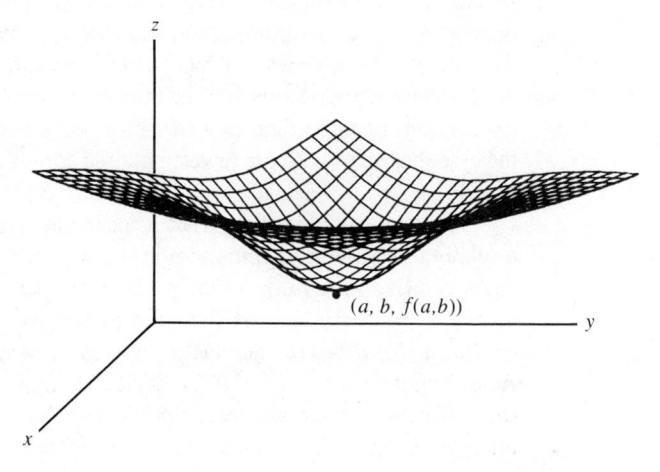

FIGURE 16–12

FIGURE 16–13

FIGURE 16–14

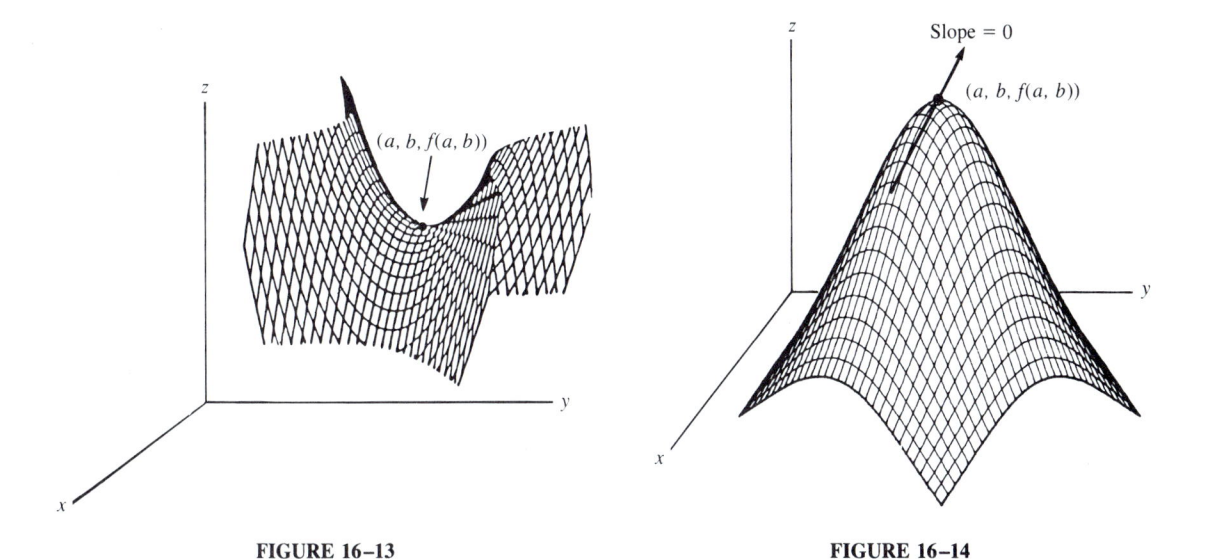

Relative Maxima, Relative Minima, and Saddle Points

Let f be a function of x and y. If

$$f_x(a, b) = 0 \quad \text{and} \quad f_y(a, b) = 0.$$

then f has either a relative maximum, a relative minimum, or a saddle point at $(a, b, f(a, b))$.

Thus we can find at least some of the points at which such a function has a relative maximum, a relative minimum, or a saddle point by setting $f_x(x, y) = 0$ and $f_y(x, y) = 0$ and then solving these equations simultaneously for x and y. (We say *some* of the points because there may be relative maxima, relative minima, or saddle points at places where f_x or f_y fail to exist.) Unfortunately, knowing that $x = a$, $y = b$ is a solution of these equations is not enough to tell us whether $(a, b, f(a, b))$ yields a relative maximum, a relative minimum, or a saddle point.

EXAMPLE 1 Consider the function f defined by

$$f(x, y) = -x^2 + 2x - y^2 + 4y.$$

Since

$$f_x = -2x + 2 \quad \text{and} \quad f_y = -2y + 4,$$

setting $f_x = 0$ and $f_y = 0$ yields

$$-2x + 2 = 0 \quad \text{and} \quad -2y + 4 = 0,$$

and thus

$$x = 1 \quad \text{and} \quad y = 2.$$

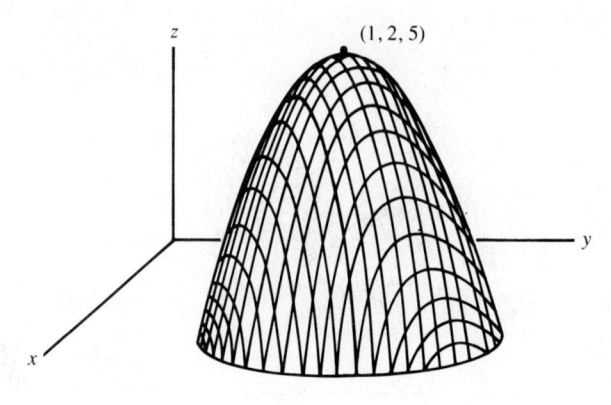

FIGURE 16–15

Therefore $(1, 2, f(1, 2)) = (1, 2, 5)$ is either a relative maximum, a relative minimum, or a saddle point, but we cannot tell which of these it is unless we know something else about the function. In this case, the graph of the function is shown in Figure 16–15, so it is clear from the graph that the point $(1, 2, 5)$ is a relative maximum. ☐

Fortunately, there is a method of testing a point (a, b) for which $f_x(a, b) = 0$ and $f_y(a, b) = 0$ to find out whether it yields a relative maximum, a relative minimum, or a saddle point. The method, which we will not prove, is called the **second derivative test.**

The Second Derivative Test

To find relative maxima, relative minima, and saddle points for a function f of x and y, proceed as follows:

1. Find the partial derivatives f_x, f_y, f_{xx}, f_{xy}, and f_{yy}.

2. Set $f_x = 0$ and $f_y = 0$ and solve these equations simultaneously for x and y to obtain all pairs of numbers (a, b) such that $f_x(a, b) = 0$ and $f_y(a, b) = 0$.

3. Let $D(x, y) = f_{xy}^2 - f_{xx}f_{yy}$. Then
   a. If $D(a, b) < 0$ and $f_{xx}(a, b) < 0$, the function has a relative maximum at $(a, b, f(a, b))$.
   b. If $D(a, b) < 0$ and $f_{xx}(a, b) > 0$, the function has a relative minimum at $(a, b, f(a, b))$.
   c. If $D(a, b) > 0$, the function has a saddle point at $(a, b, f(a, b))$.
   d. If $D(a, b) = 0$, the test fails to yield information about the behavior of the function at $(a, b, f(a, b))$.

EXAMPLE 2 Let us apply the second derivative test to the function f defined by
$$f(x, y) = -x^2 + 2xy + 2y^2 - 4x - 14y + 1.$$

1. We have
$$f_x = -2x + 2y - 4, \quad f_y = 2x + 4y - 14,$$
$$f_{xx} = -2, \quad f_{xy} = 2, \quad \text{and} \quad f_{yy} = 4.$$

2. We solve the equations
$$-2x + 2y - 4 = 0 \quad \text{and} \quad 2x + 4y - 14 = 0$$
simultaneously for x and y: from the first equation,
$$2y = 2x + 4,$$
or
$$y = x + 2.$$
Substituting this expression for y into the second equation, we obtain
$$2x + 4(x + 2) - 14 = 0,$$
or
$$6x = 6.$$
Therefore $x = 1$ and $y = 1 + 2 = 3$, and we must test the single pair $(1, 3)$.

3. To test the pair $(1, 3)$, we must evaluate
$$D(x, y) = f_{xy}^2 - f_{xx}f_{yy} = 2^2 - (-2)(4) = 12$$
at $(1, 3)$. But if $D(x, y) = 12$, then
$$D(1, 3) = 12 > 0,$$
so the function has a saddle point at $(1, 3, f(1, 3)) = (1, 3, -22)$. □

EXAMPLE 3 Let us apply the second derivative test to the function whose defining equation is
$$f(x, y) = x^4 - 2x^2 + y^2 - 2y + 1.$$

1. We have
$$f_x = 4x^3 - 4x, \quad f_y = 2y - 2,$$
$$f_{xx} = 12x^2 - 4, \quad f_{xy} = 0, \quad \text{and} \quad f_{yy} = 2.$$

2. We solve
$$4x^3 - 4x = 0$$
$$2y - 2 = 0$$

or
$$4x(x-1)(x+1) = 0$$
$$2(y-1) = 0$$
simultaneously for x and y, obtaining
$$x = 0, \quad x = 1, \quad x = -1, \quad \text{and} \quad y = 1.$$
Therefore we must test the pairs $(0, 1)$, $(1, 1)$, and $(-1, 1)$.

3. We test the pairs $(0, 1)$, $(1, 1)$, and $(-1, 1)$ using
$$D(x, y) = 0^2 - (12x^2 - 4)(2) = -8(3x^2 - 1).$$
Since
$$D(0, 1) = -8(3(0)^2 - 1) = 8 > 0,$$
the function has a saddle point at $(0, 1, f(0, 1)) = (0, 1, 0)$.
Since
$$D(1, 1) = -8(3(1)^2 - 1) = -16 < 0,$$
and
$$f_{xx}(1, 1) = 12(1)^2 - 4 = 8 > 0,$$
the function has a relative minimum at $(1, 1, f(1, 1)) = (1, 1, -1)$.
Since
$$D(-1, 1) = -8(3(-1)^2 - 1) = -16 < 0$$
and
$$f_{xx}(-1, 1) = 12(-1)^2 - 4 = 8 > 0,$$
the function has another relative minimum at $(-1, 1, -1)$. ∎

EXAMPLE 4 Let us apply the second derivative test to the function with defining equation
$$f(x, y) = x^4 + (x - y)^4.$$

1. We have
$$f_x = 4x^3 + 4(x - y)^3, \quad f_y = -4(x - y)^3,$$
$$f_{xx} = 12x^2 + 12(x - y)^2, \quad f_{xy} = -12(x - y)^2,$$
and
$$f_{yy} = 12(x - y)^2.$$

2. Solving
$$4x^3 + 4(x - y)^3 = 0$$
$$-4(x - y)^3 = 0$$

FUNCTIONS OF SEVERAL VARIABLES 1043

simultaneously for x and y, we obtain
$$x = y = 0$$
as the only solution. (Check this.)

3. Since
$$D(x, y) = (-12(x - y)^2)^2 - (12x^2 + 12(x - y)^2)(12(x - y)^2)$$
$$= 144(x - y)^4 - (144x^2(x - y)^2 + 144(x - y)^4)$$
$$= -144x^2(x - y)^2,$$

we have $D(0, 0) = 0$. Therefore the second derivative test fails; it cannot tell us whether $(0, 0, f(0, 0)) = (0, 0, 0)$ is a relative maximum, a relative minimum, or a saddle point. In a case such as this we examine the behavior of the function near the point in question. Since
$$f(0, 0) = 0$$
while
$$f(x, y) = x^4 + (x - y)^4 > 0 \quad \text{if } (x, y) \neq (0, 0),$$
it follows that the point $(0, 0, 0)$ is a relative minimum. ∎

Now we can use the second derivative test to solve optimization problems involving functions of two variables.

EXAMPLE 5 Suppose we wish to construct a rectangular box with a volume of 1000 cubic inches. Let us minimize the cost of the box if the material for it costs 10 cents per square inch.

The box will look like this:

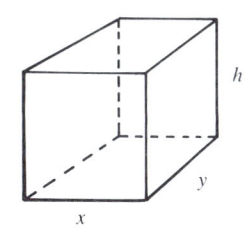

Its volume is $xyh = 1000$ cubic inches, and its cost is
$$C = 2(0.10xy) + 2(0.10xh) + 2(0.10yh)$$
dollars. Since
$$h = \frac{1000}{xy},$$
the cost function may be written as a function of two variables:
$$C(x, y) = 0.2xy + 0.2x\frac{1000}{xy} + 0.2y\frac{1000}{xy},$$

or

$$C(x, y) = 0.2xy + \frac{200}{y} + \frac{200}{x}.$$

The second derivative test shows that $C(x, y)$ has a minimum at $(10, 10, 60)$. (Check this.) Therefore the dimensions of the box of minimum cost are

$$x = 10 \text{ inches}, \quad y = 10 \text{ inches}, \quad h = 10 \text{ inches},$$

and the minimum cost is $60. ◻

EXAMPLE 6 A producer who sells the same product to different classes of customers at different prices is said to practice **price discrimination.** To see how this can occur, suppose that the demand functions for a certain product are different in two regions. For instance, suppose that in region X the demand function for a product is given by

$$p_X = 100 - \frac{x}{5},$$

but in region Y it is given by

$$p_Y = 200 - \frac{y}{15}.$$

Here x is quantity demanded in region X, y is quantity demanded in region Y, and p_X and p_Y are the unit prices in the two regions.

Let us find the producer's total revenue from sales in the two regions. Since x units are sold in region X at price p_X and y are sold in region Y at price p_Y, the revenue function is given by

$$R(x, y) = x\left(100 - \frac{x}{5}\right) + y\left(200 - \frac{y}{15}\right).$$

The second derivative test shows that the revenue function has a maximum at

$$(250, 1500, R(250, 1500)) = (250, 1500, 162{,}500).$$

(Check this.) Therefore to obtain maximum revenue of $162,500, the producer will sell 250 units in region X at a price of $p_X = \$50$ each and 1500 units in region Y at a price of $p_Y = \$100$ each, and thus will practice price discrimination. ◻

16.4 EXERCISES

In Exercises 1 through 15, use the second derivative test to find relative maxima, relative minima, and saddle points for the function.

1. $f(x, y) = 2xy$
2. $f(x, y) = x^2 + y^2 - 4$
3. $f(x, y) = 9 - 2x^2 - 3y^2$
4. $f(x, y) = x - (x + y)^2$
5. $f(x, y) = 3x^2 + 2y^2 - 18x + 8y - 2$
6. $f(x, y) = 4x^2 + y^2 - 16x - 6y + 3$

7. $f(x, y) = 4x^2 + 5y^2 + 24x - 100y + 50$
8. $f(x, y) = x^3 + y^3 - 3xy$
9. $f(x, y) = x^3 - y^3 - 3xy$
10. $f(x, y) = 4x^3 + 6x^2 - 72x + 3y^4 - 44y^3 + 180y^2$
11. $f(x, y) = x^3 + 2y^2 - 8xy + 5x + 2$
12. $f(x, y) = 2y^3 + 100y + 10xy - x^2$
13. $f(x, y) = x^4 - 8x^2 + 6y^2 - 12y + 3$
14. $f(x, y) = x^4 - (x + y)^4$
15. $f(x, y) = xy(x + y - 1) - x^2$

16. Find the dimensions of the rectangular box, without a top, of minimum cost if the material for the box costs $2 per square foot and the box must have a volume of 144 cubic feet.

17. Find the dimensions of the rectangular box, without a top, of maximum volume if the surface area of the box must be 12 square meters.

18. Find the dimensions of the rectangular box of maximum volume if the sum of the lengths of its edges must be 96 inches.

19. A carrier for pets is in the shape of a rectangular box that has an opening in its front face. The opening has an area that is two-thirds the area of the front face. Find the dimensions that minimize the surface area of the box if its volume must be 9 cubic feet.

20. The material for a rectangular box with a top and three partitions (see the figure) costs $1 per square foot, and its volume must be 1 cubic foot. Find the dimensions that minimize the cost of building the box.

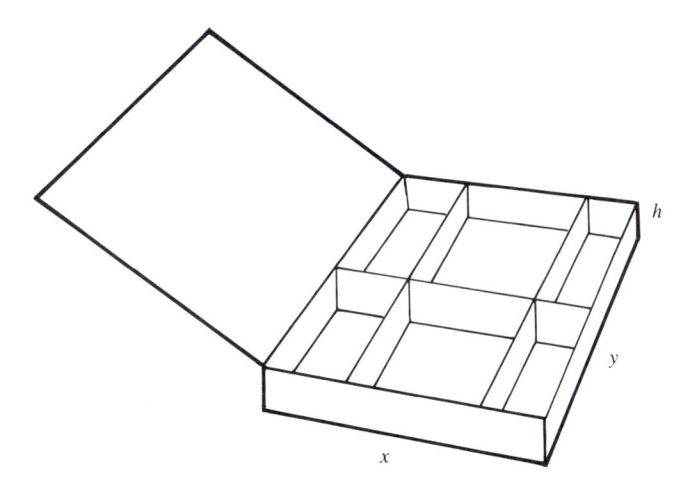

Management

21. If a firm makes and sells x units of one product and y units of another, its profit function is given by
$$P(x, y) = 20x + 30y - 2x^2 - 3y^2.$$
Find the firm's maximum profit.

22. If a company makes x units of one product and y of another, its average cost per unit is given by
$$A(x, y) = 50 + \frac{80}{x} + \frac{216}{y} + 5x + 6y.$$

Find the company's minimum average cost per unit.

23. If a firm makes and sells x thousand units of product X and y thousand units of product Y, its profit is
$$P(x, y) = -x^2 + 40x + xy + 10y - y^2 - 10$$
thousand dollars. Find the firm's maximum profit and the number of units of each product that must be made and sold in order to attain it.

24. The demand functions for a commodity in regions X and Y are given by
$$p_X = 2000 - \frac{x}{20} \quad \text{and} \quad p_Y = 1500 - \frac{y}{15},$$
respectively, with x and y the quantities demanded in the regions, and p_X, p_Y the unit prices in the regions. Find the prices that will maximize revenue. Will the supplier practice price discrimination?

25. Repeat Exercise 24 if
$$p_X = 500 - \frac{x}{4} \quad \text{and} \quad p_Y = 500 - \frac{y}{3}.$$

26. Refer to Exercise 24: if it costs $20 to make each unit sold in region X and $30 to make each unit sold in region Y, find the prices that will maximize profit.

27. The demand functions for two products A and B are as follows:
$$\text{product A:} \quad q_A = 100 - 5p_A - p_B$$
$$\text{product B:} \quad q_B = 96 - p_A - 4p_B.$$
Here p_A, p_B are the unit prices of A and B, respectively. Maximize total revenue from sales of the two products.

28. A retailer sells x units of a product at a unit price of $p_x = 9600 - 0.02x^2$ at one store, and y units at a unit price of $p_y = 6075 - 0.01y^2$ at another store. Find the maximum revenue that the retailer can obtain from the product and the unit prices required in order to attain it. Does the retailer practice price discrimination?

29. A firm can produce
$$f(x, y) = -x^2 - 1.5y^2 + 12x + 9y$$
thousand units of output using x units of labor and y units of capital. Find the firm's maximum output.

*30. A retailer wishes to build a new warehouse at a location that will minimize the sum of the distances from the warehouse to three stores. If the stores are located at the points (10, 20), (−20, 15), and (4, −8), where should the warehouse be located? (*Hint:* The distance from (x, y) to (a, b) is $[(x - a)^2 + (y - b)^2]^{1/2}$.)

31. A firm's sales depend on the amount x it spends on television commercials and the amount y it spends on newspaper advertisements according to the equation
$$S = 100{,}000x + 120{,}000y - x^2 - 4y^2 - 1.6xy.$$
Find the amounts that must be spent on each type of advertising in order to maximize sales.

Life Science

32. If a farmer applies x tons of fertilizer and y tons of pesticide to a crop, the yield will be

$$f(x, y) = 20x + 50y - 10x^2 - 30y^2$$

bushels. How much fertilizer and how much pesticide should be applied if the crop yield is to be maximized?

33. A patient undergoing chemotherapy who receives x units of drug A and y units of drug B per day suffers z units of discomfort, where

$$z = 4x^2 + 5y^2 - 8x - 6y - 5xy + 15$$

and discomfort is measured on a scale ranging from 0 to 15. What drug dosage will minimize discomfort?

16.5 LAGRANGE MULTIPLIERS

In this section we consider the problem of finding maxima and minima for functions of several variables subject to side conditions known as **constraints**. For instance, suppose we wish to find the minimum value of

$$f(x, y) = x^2 + y^2$$

subject to the condition that

$$x + y - 2 = 0.$$

In this case the **constraint equation** $x + y - 2 = 0$ is easily solved for y in terms of x; if we do so, we find that $y = -x + 2$, and this expression may be substituted for y in $f(x, y)$ to obtain

$$f(x, y) = x^2 + (-x + 2)^2 = 2x^2 - 4x + 4.$$

But now f is a function of the single variable x, so it may be written as

$$f(x) = 2x^2 - 4x + 4,$$

and we can then use either the first or second derivative test of Chapter 12 to conclude that the function has a relative minimum of 2 when $x = 1$.

Unfortunately, when we attempt to maximize or minimize a function of several variables subject to a constraint, it is not always possible to solve the constraint equation for one of the variables in terms of the others, as we did above. Therefore we need a method for finding the maximum or minimum values of $z = f(x, y)$ subject to a constraint equation $g(x, y) = 0$ without having to solve $g(x, y) = 0$ for one of the variables. In this section we present such a method: it is known as the **Lagrange multiplier method**.

Suppose we wish to find the maximum or minimum value of $f(x, y)$ subject to the constraint $g(x, y) = 0$. We begin by setting

$$h(x, y) = f(x, y) + \lambda g(x, y).$$

The symbol λ is the Greek letter lambda, which in this context is called the **Lagrange multiplier**. If we consider λ to be a specific number, then we know from our work in Section 16.4 that we can find relative maxima and minima for the function h by solving

$$h_x = 0 \quad \text{and} \quad h_y = 0$$

simultaneously for x and y. If we do this and at the same time ensure that the solutions satisfy $g(x, y) = 0$, then since

$$h(x, y) = f(x, y) \quad \text{when} \quad g(x, y) = 0,$$

it will follow that the solutions that maximize or minimize h and satisfy the constraint equation will also maximize or minimize f and satisfy the constraint equation. But now let us consider λ to be another independent variable, so that h is a function of x, y, and λ. Then

$$h(x, y, \lambda) = f(x, y) + \lambda g(x, y),$$

so h_x and h_y are the same as before. Taking the first partial derivative of h with respect to λ shows that

$$h_\lambda = g(x, y).$$

Therefore we can ensure that the solutions of $h_x = 0$ and $h_y = 0$ satisfy the constraint equation if we take λ to be an independent variable and require that $h_\lambda = 0$.

The Lagrange Multiplier Method

Let f be a function of x and y. To maximize or minimize f subject to the constraint $g(x, y) = 0$:

1. Let

$$h(x, y, \lambda) = f(x, y) + \lambda g(x, y)$$

and find h_x, h_y, and $h_\lambda = g(x, y)$.

2. Set $h_x = 0$, $h_y = 0$, and $h_\lambda = 0$, and solve these equations simultaneously for x and y.

3. The result of step 2 will be pairs of numbers $x = a$, $y = b$ that maximize or minimize f and satisfy the constraint $g(x, y) = 0$. To decide whether a particular pair (a, b) yields a maximum or a minimum for f, compare the value $f(a, b)$ with values of $f(x, y)$ for points (x, y) that are near (a, b) and satisfy the constraint equation $g(x, y) = 0$.

(The simplest way to solve the equations of step 2 simultaneously is usually to first solve $h_x = 0$ and $h_y = 0$ for λ in terms of x and y, equate the two expressions for λ, solve for either x or y in terms of the other, and substitute the result into the equation $h_\lambda = 0$.)

EXAMPLE 1   Let us find maxima and minima for

$$f(x, y) = x^2 + y^2$$

subject to the constraint

$$x + y - 2 = 0.$$

FUNCTIONS OF SEVERAL VARIABLES   **1049**

1. We form
$$h(x, y, \lambda) = x^2 + y^2 + \lambda(x + y - 2),$$
Then
$$h_x = 2x + \lambda, \qquad h_y = 2y + \lambda,$$
and
$$h_\lambda = x + y - 2.$$

2. We must solve the equations
$$2x + \lambda = 0, \quad 2y + \lambda = 0, \quad \text{and} \quad x + y - 2 = 0$$
simultaneously for x and y. Solving the first two of these equations for λ, we obtain
$$\lambda = -2x \quad \text{and} \quad \lambda = -2y.$$
Thus
$$-2x = -2y,$$
or
$$x = y.$$
Substituting x for y in the equation
$$x + y - 2 = 0$$
now yields
$$2x - 2 = 0,$$
so
$$x = 1 \quad \text{and} \quad y = 1.$$

3. Therefore the function has either a constrained maximum or a constrained minimum when $x = 1$, $y = 1$. To find out whether $f(1, 1) = 2$ is a constrained maximum or a constrained minimum, we compare it with $f(x, y)$ for points (x, y) near $(1, 1)$ that also satisfy $x + y - 2 = 0$. For instance, we have
$$f(1, 1) = 2,$$
$$f(1.05, 0.95) = 2.005,$$
$$f(1.1, 0.9) = 2.02,$$
$$f(0.8, 1.2) = 2.08,$$
and so on. Hence it is clear that $f(1, 1) = 2$ is a constrained minimum for f. □

EXAMPLE 2   Let us optimize
$$f(x, y) = 12 - x^2 - y^2$$
subject to the constraint
$$xy = 4.$$

1. We have
$$h(x, y, \lambda) = 12 - x^2 - y^2 + \lambda(xy - 4).$$

Note that the constraint equation was not originally in the form $g(x, y) = 0$, and thus we had to rewrite it in this form before we constructed the function h. Now,
$$h_x = -2x + \lambda y, \quad h_y = -2y + \lambda x, \quad \text{and} \quad h_\lambda = xy - 4.$$

2. We must solve the equations
$$-2x + \lambda y = 0, \quad -2y + \lambda x = 0, \quad \text{and} \quad xy - 4 = 0$$

simultaneously for x and y. From the first two equations,
$$\lambda = \frac{2x}{y} = \frac{2y}{x},$$

or
$$x^2 = y^2.$$

Therefore
$$y = \pm x.$$

If we substitute x for y in
$$xy - 4 = 0,$$

we find that
$$x^2 = 4,$$

so $x = \pm 2$, and we obtain the points $(2, 2)$, $(-2, -2)$. If we substitute $-x$ for y in
$$xy - 4 = 0.$$

we find that
$$x^2 = -4,$$

which has no solutions.

3. Thus $(2, 2)$ and $(-2, -2)$ are the only points that yield constrained maxima or minima for f. Since

$$f(\pm 2, \pm 2) = 4, \quad f\left(\pm 2.1, \frac{4}{\pm 2.1}\right) \approx 3.96, \quad f\left(\pm 2.2, \frac{4}{\pm 2.2}\right) \approx 3.85,$$

and so on, it is clear that $(2, 2)$ and $(-2, -2)$ both yield constrained maxima for f. □

Many practical situations require constrained optimization of functions of several variables. We illustrate with an example.

EXAMPLE 3   Consider the production function defined by

$$f(x, y) = 300x^{1/3}y^{2/3}.$$

Here x denotes the number of units of labor and y the number of units of capital required to produce $f(x, y)$ units of product. Suppose that each unit of labor costs $50 and each unit of capital costs $100, and assume that the production budget is $15,000. Then we would like to maximize production $f(x, y)$ subject to the budgetary constraint

$$50x + 100y = 15,000.$$

Using the Lagrange multiplier method, we have

$$h(x, y, \lambda) = 300x^{1/3}y^{2/3} + \lambda(50x + 100y - 15,000).$$

Thus

$$h_x = 100x^{-2/3}y^{2/3} + 50\lambda = 0,$$

$$h_y = 200x^{1/3}y^{-1/3} + 100\lambda = 0,$$

and

$$h_\lambda = 50x + 100y - 15,000 = 0,$$

from which we find that

$$\lambda = -2x^{-2/3}y^{2/3} = -2x^{1/3}y^{-1/3}.$$

Therefore

$$x^{-2/3}y^{2/3} = x^{1/3}y^{-1/3}.$$

Multiplying this last equation by $x^{2/3}y^{1/3}$ yields

$$y = x.$$

Substituting x for y in the equation

$$50x + 100y = 15,000,$$

we find that

$$x = y = 100.$$

Therefore the maximum production subject to the budgetary constraint is

$$f(100, 100) = 30,000$$

units.

Note that $x = y = 100$ implies that

$$\lambda = -2(100)^{1/3}(100)^{-1/3} = -2.$$

The absolute value of λ is called the **marginal productivity of money;** it tells us approximately how many additional units could be produced with one additional dollar. In this example, $|\lambda| = |-2| = 2$, so approximately two additional units could be produced if one additional dollar were available for production. ☐

EXAMPLE 4

The Lagrange multiplier method can be used with functions of more than two variables: we simply set all first partial derivatives of $h = f + \lambda g$ equal to zero and solve simultaneously.

Let us find three positive numbers such that the first plus twice the second plus three times the third is 36 and such that their product is a maximum. Thus we wish to maximize

$$f(x, y, z) = xyz$$

subject to the constraint

$$x + 2y + 3z = 36.$$

We have

$$h(x, y, z, \lambda) = xyz + \lambda(x + 2y + 3z - 36),$$

so

$$h_x = yz + \lambda, \quad h_y = xz + 2\lambda, \quad h_z = xy + 3\lambda,$$

and

$$h_\lambda = x + 2y + 3z - 36.$$

Solving

$$h_x = 0, \quad h_y = 0, \quad \text{and} \quad h_z = 0,$$

we find that

$$y = \frac{x}{2} \quad \text{and} \quad z = \frac{x}{3}.$$

Substituting these expressions for y and z in the equation $h_\lambda = 0$, we have $3x = 36$, from which it follows that $x = 12$, $y = 6$, and $z = 4$. Therefore the maximum product is $f(12, 6, 4) = 288$. ☐

16.5 EXERCISES

In Exercises 1 through 12, use the Lagrange multiplier method to find either a maximum or a minimum for the function subject to the given constraint. In each case state whether the value you find is a maximum or a minimum.

1. $f(x, y) = x^2 + y^2 + 2$, subject to $x + y - 3 = 0$.
2. $f(x, y) = 2x^2 + 3y^2$, subject to $x + y - 5 = 0$.
3. $f(x, y) = 9 - x^2 - y^2$, subject to $x + y = 6$.
4. $f(x, y) = 100 - 2x^2 - 4y^2$, subject to $6x + 3y = 12$.
5. $f(x, y) = xy$, subject to $x + 2y = 6$.
6. $f(x, y) = xy$, subject to $x^2 + y^2 = 8$.

Functions of Several Variables

7. $f(x, y) = x^3 - y^3$, subject to $x - y = 2$.
8. $f(x, y) = x^3 - y^3$, subject to $xy = 1$.
9. $f(x, y, z) = 2x - 4y + 2z - x^2 - y^2$, subject to $x + y + z = 0$.
10. $f(x, y, z) = x^2 + y^2 + z^2 - x - 2y - 3z$, subject to $10 - x - y - z = 0$.
11. $f(x, y, z) = 2x + 3y - 4z + x^2 - y^2$, subject to $x + 2y + 3z = 0$.
12. $f(x, y, z) = x^2 + y^2 + 2z^2 - x - y + 3z$, subject to $3x + y + 2z = \frac{1}{2}$.
13. Maximize $f(x, y) = x^3 y^2$ subject to $12x + 6y = 24$.
14. Minimize $f(x, y) = x^4 y^4$ subject to $2x - y = 4$.
15. Minimize $f(x, y) = x^2 + y^2 - 3$ subject to $xy = 16$.
16. Minimize $f(x, y) = x^3 + y^3$ subject to $xy = 1$.
17. Minimize $f(x, y) = x^3 + y^3$ subject to $x + y = 1$.
18. Find two positive numbers whose product is 25 and whose sum is a minimum.
19. Find two numbers whose sum is 20 and whose product is a maximum.
20. Find three positive numbers whose sum is 20 and whose product is a maximum.
21. Find two numbers whose product is 64 and such that the square of one plus twice the other is a minimum.
22. Find two numbers such that 3 times one plus $\frac{1}{3}$ of the other equals 24 and the product of their squares is a maximum.
23. Find three numbers whose sum is 33 and such that the square of the first plus twice the square of the second plus 3 times the square of the third is a minimum.
24. Find the minimum distance from the line $x + y = 4$ to the origin $(0, 0)$. (*Hint:* The distance from (x, y) to $(0, 0)$ is $\sqrt{x^2 + y^2}$.)
25. Find the minimum distance from the plane $z = 20 - x - 2y$ to the origin. (*Hint:* The distance from (x, y, z) to $(0, 0, 0)$ is $(x^2 + y^2 + z^2)^{1/2}$.)
26. Find the point on the plane $x + y + z = 1$ that is at a minimum distance from the origin.
27. Repeat Exercise 26 for the plane $2x + 5y + z = 10$.
*28. Find the rectangle of largest area that can be inscribed in a circle of radius r.
*29. Find the rectangular box of largest volume that can be inscribed in a sphere of radius r. (The equation of such a sphere is $x^2 + y^2 + z^2 = r^2$.)

Management

30. A firm can produce
$$f(x, y) = 4000 x^{2/5} y^{3/5}$$
units by employing x units of labor and y units of capital. If each unit of labor costs $200, each unit of capital costs $300, and the production budget is $16,000, find the firm's maximum production. Also find and interpret the marginal productivity of money.

31. Repeat Exercise 30 if
$$f(x, y) = x^{1/4} y^{3/4}$$

and each unit of labor costs $20, each unit of capital costs $40, and the production budget is $24,000.

32. A firm can product $800x^{1/4}y^{3/4}$ units of output using x units of labor and y units of capital. If each unit of labor costs $50 and each unit of capital costs $80 and the production budget for the firm is $200,000, find the maximum production and the number of units of labor and capital needed to attain it.

33. Suppose the cost of producing x units of product X and y units of product Y is

$$C(x, y) = x^2 + 2y^2 + xy + 200$$

dollars. If it takes 5 worker-hours to make each unit of X, 7 worker-hours to make each unit of Y, and there are 3456 worker-hours available, how many units of each product must be made in order to minimize cost?

34. If a store purchases x radio commercials and y newspaper ads each week, its advertising reaches

$$z = 20x\sqrt{10y + 20}$$

thousand people. If it costs $2000 to run each radio commercial and $1000 to run each newspaper ad, and if the firm budgets $7000 per week for advertising expenses, find the maximum number of people reached by its ads.

35. If a utility spends x million dollars on smokestack scrubbers and y million dollars on equipment that will burn low-sulfur coal, it can remove $z = 600xy - 20x^2 - 30y^2$ particulates per cubic meter from its smoke. If the utility's budget for reducing particulates from its smoke is $65 million, find the values of x and y that will minimize the particulates.

36. Suppose a company's profit function is given by

$$P(x, y, z) = xy + xz + yz + x + y + z - 10,000.$$

If each unit of X requires 2 kilograms (kg) of steel, each unit of Y requires 2 kg, and each unit of Z requires 3 kg, and there are 7999 kg of steel available, find the maximum profit.

Life Science

37. If a farmer applies x tons of fertilizer and y tons of weedkiller to a crop, the yield is

$$z = 10x + 30y - 2x^2 - 5y^2$$

bushels per acre. If each ton of fertilizer costs $500 and each ton of weedkiller costs $800, and if the farmer has exactly $3650 to spend on these products, find the maximum crop yield.

16.6 DOUBLE INTEGRALS

We have seen how functions of several variables may be differentiated by the process of partial differentiation; often they may also be integrated by a process known as **multiple integration.** In this section we illustrate multiple integration for functions of two variables. The integral of a function of two variables is called a **double integral.** Double integrals have many uses, most of which are beyond the scope of

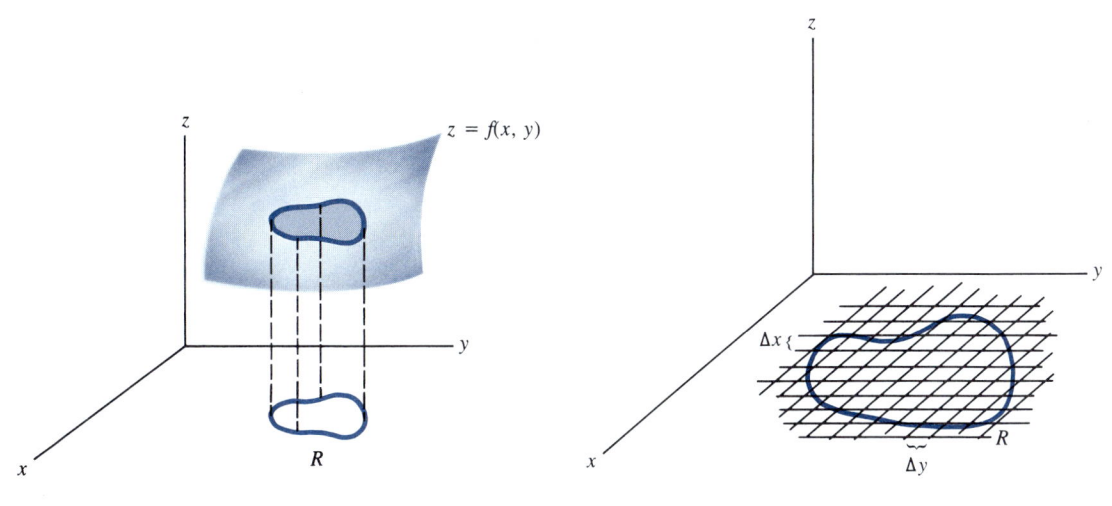

FIGURE 16–16

FIGURE 16–17

this book. However, we will demonstrate how they may be used to find volumes, areas, and the average value of a function of two variables.

Let f be a function of the two variables x and y whose domain includes a region R in the xy-plane, and such that $f(x, y) \geq 0$ for all (x, y) in R, as in Figure 16–16. Suppose we place a grid of rectangles on R, with each rectangle having width Δx and length Δy. See Figure 16–17. In each rectangle that overlaps R, we choose a point (x_i, y_i) that is also in R, and construct a rectangular box whose base is the rectangle and whose height is $f(x_i, y_i)$. Figure 16–18 depicts the situation for one rectangle. The rectangular box will touch the surface $z = f(x, y)$ at the point $(x_i, y_i, f(x_i, y_i))$, and its volume will be $f(x_i, y_i) \Delta x \Delta y$. If there are n rectangles that overlap R, then there will be n such rectangular boxes, and the sum of their volumes may be written as the **Riemann sum**

$$\sum_{i=1}^{n} f(x_i, y_i) \Delta x \Delta y = [f(x_1, y_1) + f(x_2, y_2) + \cdots + f(x_n, y_n)] \Delta x \Delta y.$$

This Riemann sum is therefore an approximation to the volume of the solid whose base is R and whose top is the portion of the surface $z = f(x, y)$ that lies over R (Figure 16–19). If we make the grid on R finer and finer by letting Δx and Δy approach 0, the Riemann sum will approach the volume of this solid; in other words,

$$\lim_{\substack{\Delta x \to 0 \\ \Delta y \to 0}} \sum_{i=1}^{n} f(x_i, y_i) \Delta x \Delta y$$

will exist and will be equal to the volume of the solid. We call this limit the **double integral of f over R**. [We have assumed that $f(x, y) \geq 0$ over the region R, but a similar argument shows that if $f(x, y) \leq 0$ over R then the limit of the Riemann sums as Δx and Δy approach 0 is the negative of the volume of the solid bounded by R and the surface $z = f(x, y)$.]

FIGURE 16–18

FIGURE 16–19

The Double Integral

If f is a function of x and y whose domain includes the region R in the xy-plane, the **double integral of f over R,** denoted by

$$\iint_R f(x, y)\, dA,$$

is defined to be the limit

$$\lim_{\substack{\Delta x \to 0 \\ \Delta y \to 0}} \sum_{i=1}^{n} f(x_i, y_i)\, \Delta x\, \Delta y,$$

if it exists. If $f(x, y) \geq 0$ for all (x, y) in R, then

$$\iint_R f(x, y)\, dA$$

is the volume of the cylindrical solid bounded by R and the surface $z = f(x, y)$.

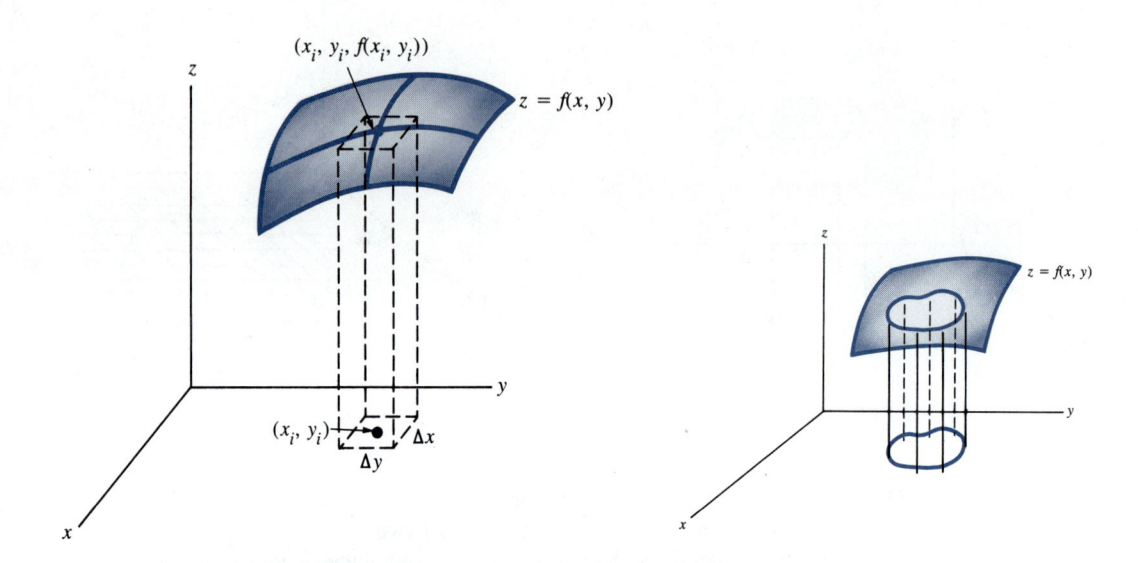

FUNCTIONS OF SEVERAL VARIABLES

Fortunately, we will not need to evaluate double integrals by actually taking limits of Riemann sums; instead, we will evaluate them as **iterated integrals** of the form

$$\int_c^d \int_a^b f(x, y)\, dx\, dy \quad \text{or} \quad \int_a^b \int_c^d f(x, y)\, dy\, dx.$$

Such iterated integrals are integrated from the inside out. Thus, to evaluate

$$\int_c^d \int_a^b f(x, y)\, dx\, dy,$$

we first find the inside integral

$$\int_a^b f(x, y)\, dx$$

by integrating $f(x, y)$ with respect to x from $x = a$ to $x = b$ while treating y as a constant. Since each occurrence of x in the antiderivative will be replaced by a and b, the result of this integration will be a function of y, which we can then integrate from $y = c$ to $y = d$, as the outside integral requires. We illustrate with an example.

EXAMPLE 1 Let us evaluate the iterated integral

$$\int_0^2 \int_0^1 (2x + y)\, dx\, dy.$$

We first evaluate the inside integral by integrating $f(x, y) = 2x + y$ with respect to x while treating y as a constant. Thus

$$\int_0^1 (2x + y)\, dx = x^2 + yx \Big|_{x=0}^{x=1} = [1^2 + y(1)] - [0^2 + y(0)] = 1 + y.$$

Now

$$\int_0^2 \int_0^1 (2x + y)\, dx\, dy = \int_0^2 \left(\int_0^1 (2x + y)\, dx \right) dy$$

$$= \int_0^2 (1 + y)\, dy$$

$$= y + \frac{y^2}{2} \Big|_0^2$$

$$= 4. \quad \square$$

We can interchange the order of integration in an iterated integral without affecting the value of the integral.

EXAMPLE 2 Let us evaluate the double integral

$$\int_0^1 \int_0^2 (2x + y)\, dy\, dx.$$

This is the integral of Example 1 with the order of integration reversed. We have

$$\int_0^1 \int_0^2 (2x + y) \, dy \, dx = \int_0^1 \left(\int_0^2 (2x + y) \, dy \right) dx$$

$$= \int_0^1 \left(2xy + \frac{y^2}{2} \Big|_{y=0}^{y=2} \right) dx$$

$$= \int_0^1 \left(\left[2x \cdot 2 + \frac{2^2}{2} \right] - \left[2x \cdot 0 + \frac{0^2}{2} \right] \right) dx$$

$$= \int_0^1 (4x + 2) \, dx$$

$$= 2x^2 + 2x \Big|_0^1$$

$$= 4. \quad \square$$

Now we return to our consideration of double integrals. It turns out that if the region R over which we are integrating is bounded by lines and curves, then we can evaluate the double integral as an iterated integral.

Evaluating Double Integrals

1. If R is the region bounded by the lines $x = a$ and $x = b$ and the graphs of the equations $y = g_1(x)$ and $y = g_2(x)$, as shown in Figure 16–20, then

$$\iint_R f(x, y) \, dA = \int_a^b \int_{g_1(x)}^{g_2(x)} f(x, y) \, dy \, dx.$$

2. If R is the region bounded by the lines $y = c$ and $y = d$ and the graphs of the equations $x = h_1(y)$ and $x = h_2(y)$, as shown in Figure 16–21, then

$$\iint_R f(x, y) \, dA = \int_c^d \int_{h_1(y)}^{h_2(y)} f(x, y) \, dx \, dy.$$

Note that in the integral

$$\int_a^b \int_{g_1(x)}^{g_2(x)} f(x, y) \, dy \, dx$$

we integrate first with respect to y, and y ranges from the lower boundary $g_1(x)$ to the upper boundary $g_2(x)$ of R. The result of the inner integration will thus be a

FUNCTIONS OF SEVERAL VARIABLES

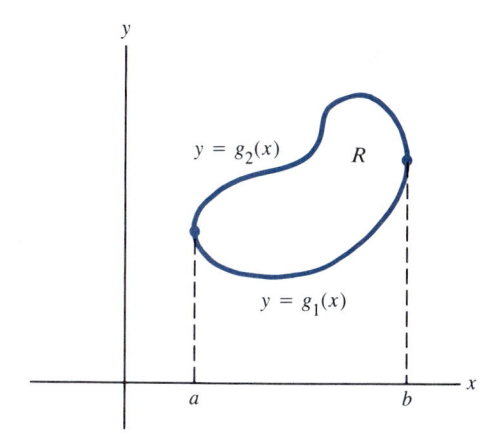

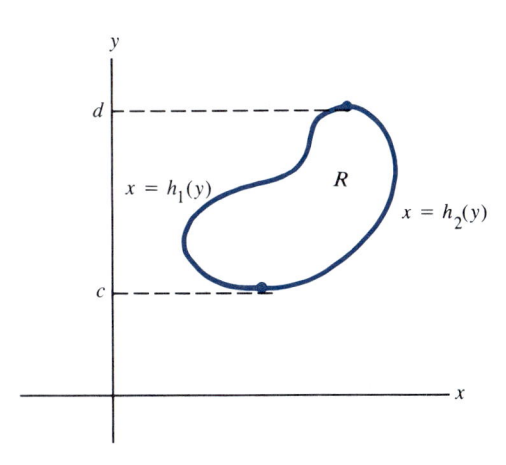

FIGURE 16–20

FIGURE 16–21

function of the single variable x, which we then integrate from $x = a$ to $x = b$, as required by the outer integral. Similar remarks apply to the integral

$$\int_c^d \int_{h_1(y)}^{h_2(y)} f(x, y) \, dx \, dy.$$

EXAMPLE 3 Let $f(x, y) = 12 - x - y$, and let R be the rectangle shown in Figure 16–22. Since R is bounded by the lines $x = 1$, $x = 2$, $y = 2$, and $y = 4$, we have

$$\iint_R (12 - x - y) \, dA = \int_1^2 \int_2^4 (12 - x - y) \, dy \, dx$$

$$= \int_1^2 \left(12y - xy - \frac{y^2}{2} \Big|_{y=2}^{y=4} \right) dx$$

$$= \int_1^2 [(48 - 4x - 8) - (24 - 2x - 2)] \, dx$$

$$= \int_1^2 (18 - 2x) \, dx$$

$$= 18x - x^2 \Big|_1^2$$

$$= 15.$$

Thus the volume of the solid pictured in Figure 16–23 is 15. Note that we can also evaluate the double integral of $f(x, y) = 12 - x - y$ over the rectangle R as

$$\iint_R (12 - x - y) \, dA = \int_2^4 \int_1^2 (12 - x - y) \, dx \, dy.$$

This gives the same result. (Check that this is so.) ☐

FIGURE 16-22

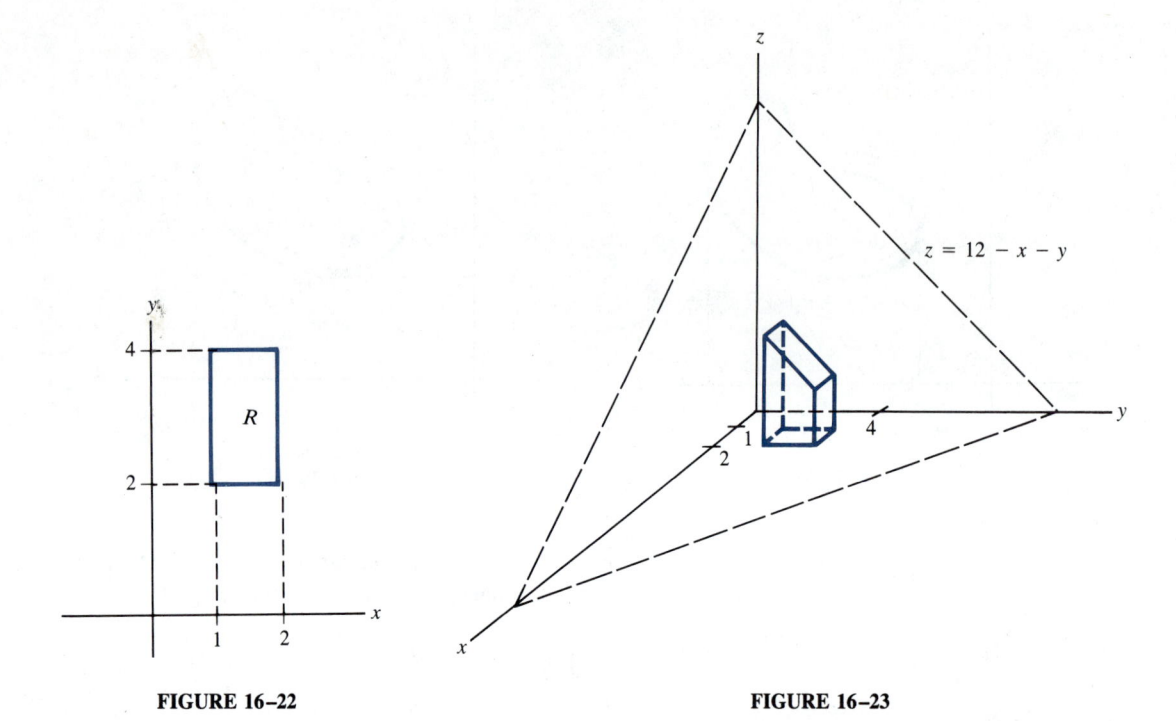

FIGURE 16-23

EXAMPLE 4 Let us find the volume of the solid whose base is the triangle bounded by the x-axis, the y-axis, and the line $x + y = 1$, and whose top is the surface $z = x^2 + y^2$. See Figure 16–24.

The volume of the solid is the double integral

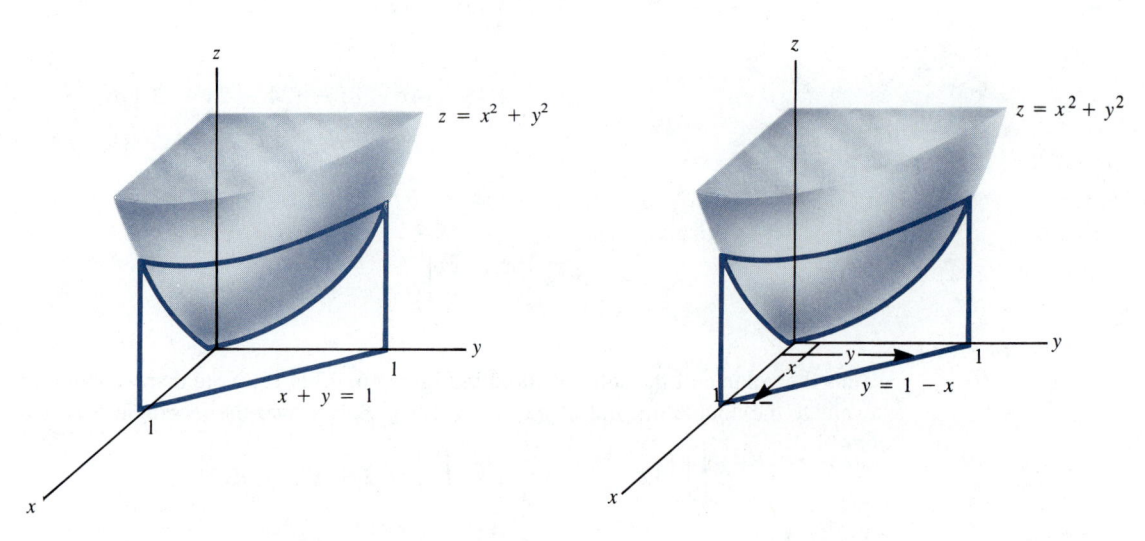

FIGURE 16-24

FIGURE 16-25

$$\iint_R (x^2 + y^2)\, dA,$$

where R is the given triangle. We seek to evaluate this double integral as an iterated integral of the form

$$\int_a^b \int_{g_1(x)}^{g_2(x)} (x^2 + y^2)\, dy\, dx.$$

Here we are integrating first with respect to y, and hence y must range from the x-axis ($y = 0$) to the line $y = 1 - x$, and x must range from 0 to 1 (Figure 16–25). Thus the volume of the solid is

$$\int_0^1 \int_0^{1-x} (x^2 + y^2)\, dy\, dx = \int_0^1 \left(x^2 y + \frac{1}{3} y^3 \right) \bigg|_{y=0}^{y=1-x} dx$$

$$= \int_0^1 \left[x^2(1-x) + \frac{1}{3}(1-x)^3 \right] dx$$

$$= \int_0^1 \frac{1}{3}(-4x^3 + 6x^2 - 3x + 1)\, dx$$

$$= \frac{1}{6}.$$

(Check the integration.) □

If we had wished, we could have set up the iterated integral in the previous example so that we integrated first with respect to x and then with respect to y; in this case the integral would have been

$$\int_0^1 \int_0^{1-y} (x^2 + y^2)\, dx\, dy,$$

since now x must range from the y-axis to the line $x = 1 - y$, and y must range from 0 to 1. See Figure 16–26. Sometimes one way of setting up the iterated integral leads to an easier integration than the other.

EXAMPLE 5 Consider the iterated integral

$$\int_0^1 \int_x^1 e^{y^2}\, dy\, dx.$$

Here we are integrating first with respect to y, so y ranges from $y = x$ to $y = 1$, and then x ranges from $x = 0$ to $x = 1$. Therefore the region R over which we are integrating is the triangle shown in Figure 16–27. If we attempt to evaluate this integral in the order given, we must first find

$$\int_x^1 e^{y^2}\, dy;$$

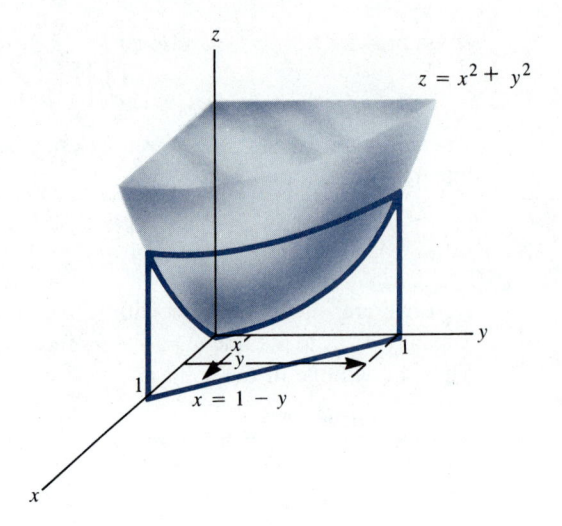

FIGURE 16–26

but since we do not know an antiderivative for e^{y^2}, we cannot carry out the integration. However, if we reverse the order of integration, we have

$$\int_0^1 \int_0^y e^{y^2} \, dx \, dy = \int_0^1 \left(xe^{y^2} \Big|_{x=0}^{x=y} \right) dy$$

$$= \int_0^1 ye^{y^2} \, dy.$$

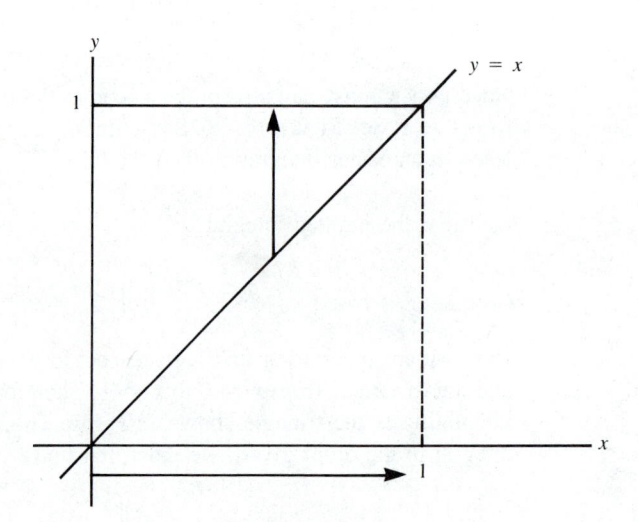

FIGURE 16–27

FUNCTIONS OF SEVERAL VARIABLES

If we now let $u = y^2$, so that $du = 2y\, dy$, then

$$\int_0^1 ye^{y^2}\, dy = \frac{1}{2}\int_0^1 e^u\, du = \frac{1}{2}(e - 1). \quad \square$$

Suppose we integrate the function f defined by $f(x, y) = 1$ over a region R in the xy-plane. We obtain the volume of the cylinder whose base is the region R and whose top is the plane $z = 1$. See Figure 16–28. But the volume of this cylinder is equal to the area of its base times its height, that is, to the area of R times 1, or the area of R. Thus we can find the area of a region R in the xy-plane by integrating $f(x, y) = 1$ over R:

The Double Integral and Area

If R is a region in the xy-plane, the area of R is given by

$$\iint_R dA.$$

EXAMPLE 6   Find the area of the region R pictured in Figure 16–29. The area in question is the value of the double integral of 1 over R, and thus is equal to

$$\iint_R dA = \int_0^3 \int_x^{-x^2+4x} dy\, dx$$

$$= \int_0^3 \left(y \Big|_{y=x}^{y=-x^2+4x} \right) dx$$

$$= \int_0^3 (-x^2 + 3x)\, dx$$

$$= -\frac{x^3}{3} + \frac{3}{2}x^2 \Big|_0^3$$

$$= \frac{9}{2}.$$

Note that we could have found this area by the method of Section 14.4, in which case we would have obtained the same result by evaluating the single integral

$$\int_0^3 (-x^2 + 4x - x)\, dx. \quad \square$$

We conclude this section by illustrating how the double integral can be used to find the average value of a function over a region. By means of an argument analogous to that employed in Section 14.5, it is possible to show the following:

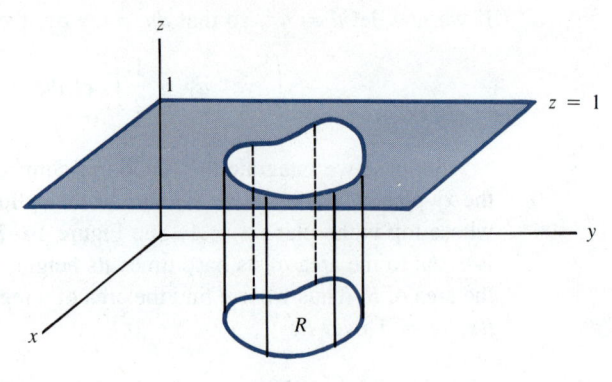

FIGURE 16-28

The Average Value of a Function over a Region

Let f be a function of the two variables x and y, and let R be a region in the domain of f. Then

$$\text{Average value of } f \text{ over } R = \frac{1}{\text{area of } R} \iint_R f(x, y)\, dA.$$

EXAMPLE 7 Let us find the average value of the function f defined by $f(x, y) = 2x + 3y$ over the triangle bounded by the coordinate axes and the line $x + y = 1$. See Figure 16-30. Since the area of this triangle is $\frac{1}{2}$, we have

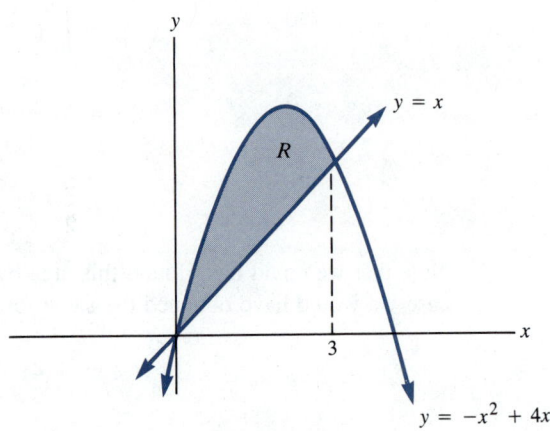

FIGURE 16-29

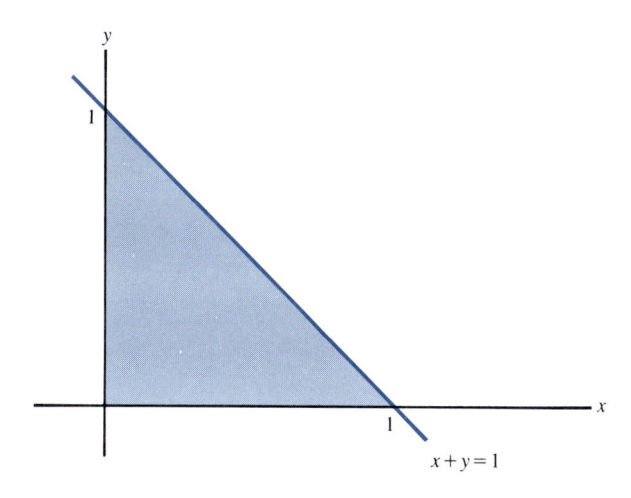

FIGURE 16-30

$$\text{Average value of } f = \frac{1}{1/2} \int_0^1 \int_0^{1-x} (2x + 3y) \, dy \, dx$$

$$= 2 \int_0^1 \left(2xy + \frac{3}{2} y^2 \bigg|_0^{1-x} \right) dx$$

$$= 2 \int_0^1 \left[2x(1 - x) + \frac{3}{2}(1 - x)^2 \right] dx$$

$$= 2 \int_0^1 \left(-\frac{x^2}{2} - x + \frac{3}{2} \right) dx$$

$$= 2 \left(-\frac{x^3}{6} - \frac{x^2}{2} + \frac{3x}{2} \bigg|_0^1 \right)$$

$$= \frac{5}{3}. \quad \square$$

16.6 EXERCISES

In Exercises 1 through 6, evaluate the iterated integral.

1. $\int_0^1 \int_0^1 x \, dy \, dx$

2. $\int_0^1 \int_0^1 2y \, dx \, dy$

3. $\int_0^2 \int_{-1}^1 (2x + 3y^2) \, dx \, dy$

4. $\int_{-2}^3 \int_{-2}^1 xy \, dy \, dx$

5. $\int_0^1 \int_0^2 xe^{-y} \, dy \, dx$

6. $\int_0^2 \int_0^1 ye^{-x} \, dx \, dy$

In Exercises 7 through 12, sketch the region of integration and evaluate the integral. (In some cases you may find it helpful to reverse the given order of integration.)

7. $\int_0^2 \int_0^1 x^2 \, dy \, dx$

8. $\int_1^2 \int_{-1}^2 (2x - y) \, dx \, dy$

9. $\int_1^2 \int_{-2}^x (x + y) \, dy \, dx$

10. $\int_1^2 \int_{y^2}^y (y^2 + 1) \, dx \, dy$

11. $\int_0^1 \int_y^{\sqrt{y}} y^{1/2} \, dx \, dy$

12. $\int_0^1 \int_{e^{-x}}^{e^x} e^x \, dy \, dx$

In Exercises 13 through 21, evaluate the double integral over the indicated region R in the xy-plane.

13. $\iint_R (x + y) \, dA$, R the rectangle bounded by the x-axis, the y-axis, and the lines $x = 1$, $y = 2$.

14. $\iint_R (2x - 4y) \, dA$, R the square bounded by the lines $x = 1$, $x = 3$, $y = 1$, $y = 3$.

15. $\iint_R (2x - y) \, dA$, R the square with vertices $(1, 1)$, $(3, 1)$, $(3, 3)$, and $(1, 3)$.

16. $\iint_R (x - 2y) \, dA$, R the triangle bounded by the y-axis and the lines $x - y = 0$, $y = 2$.

17. $\iint_R (x - 2y) \, dA$, R the triangle bounded by the x-axis and the lines $x - y = 0$, $x = 2$.

18. $\iint_R (x + y^2) \, dA$, R the triangle bounded by the lines $y = x$, $x = 3$, and $y = 1$.

19. $\iint_R x^2 y \, dA$, R the region bounded by $y = x^2$ and $y = 1 - x^2$.

20. $\iint_R \frac{1}{xy} \, dA$, R the region bounded by $y = e^{-x}$ and the lines $y = 1$, $x = 1$, $x = 2$.

21. $\iint_R x e^y \, dA$, R the region bounded by $y = \ln x$, $y = 0$, and $x = 2$.

In Exercises 22 through 31, find the volume of the solid bounded by the given surface and the region R.

22. The plane $z = 2$; the rectangle R bounded by the x-axis, the y-axis, and the lines $x = 2$ and $y = 4$.

23. The plane $z = 8 - x - y$; the rectangle R bounded by the lines $x = 1$, $y = 1$, $x = 2$, and $y = 3$.

24. The plane $z = y$; the square R bounded by $x = 0$, $y = 0$, $x = 2$, and $y = 2$.

25. The plane $z = 20 - 2x - 5y$, R the triangle bounded by the lines $x + y = 1$, $y = 0$, and $x = 0$.

26. The plane $z = 6 - 2x - 3y$; the triangle R bounded by $x = 0$, $y = 0$, and $-2x - 3y = 6$.

27. The plane $z = 4x$; the region R bounded by the nonnegative x-axis, the nonnegative y-axis, and the circle $x^2 + y^2 = 1$.

28. The plane $z = x$, R the region bounded by the nonnegative x-axis, the nonnegative y-axis, and the circle $x^2 + y^2 = 1$.

29. The surface $z = x^2 + y^2$, R the square bounded by the lines $x = 0$, $x = 1$, $y = 0$, and $y = 1$.

30. The surface $z = x^2 + y^2$, R the triangle bounded by the lines $y = x$, $y = 1$, and $x = 0$.

31. The surface $z = e^{-y}$; the region R bounded by the x-axis, $x = e$, and $y = \ln x$.

In Exercises 32 through 35, use a double integral to find the area of the region R.

32. R is bounded by the x-axis and $y = 2x - x^2$.

33. R is bounded by the y-axis, $x = \sqrt{y}$, and $y = 1$.

34. R is bounded by $y = x^2$ and $y = x^3$.

35. R is bounded by $y = x^2 - 4x$ and $y = -x^2 + 4x$.

In Exercises 36 through 39, find the average value of the given function on the given region.

36. $f(x, y) = x + y$, on the rectangle bounded by the lines $x = 1$, $x = 3$, $y = 2$, $y = 5$.

37. $f(x, y) = 2x + y$, on the triangle bounded by the coordinate axes and the line $x + y = 2$.

38. $f(x, y) = x^2 - y^2$, on the region bounded by the x-axis and the parabola $y = 1 - x^2$.

39. $f(x, y) = xe^y$, on the region bounded by the x-axis, the y-axis, the line $y = 1$, and the curve $y = \ln x$.

Management

40. A firm can produce $f(x, y) = 4000x^{0.4}y^{0.6}$ units by employing x thousand worker-hours of labor and y million dollars of capital. If $0 \leq x \leq 10$ and $0 \leq y \leq 5$, find the firm's average production.

41. A firm can produce $f(x, y) = 1000x^{1/4}y^{3/4}$ units by employing x units of labor and y units of capital. If the firm's total budget for each of capital and labor must be no more than 500 units, find the firm's average production.

42. If a store runs x newspaper ads and y television commercials during a month, its sales will be $S = 200x + 220y$ thousand dollars. Suppose newspaper ads cost $1000 each and it costs $5000 to run each TV commercial. If the total monthly budget for both types of advertising combined varies from $0 to $100,000, find the store's average monthly sales.

43. Repeat Exercise 42 if in addition to the budget restriction given there, the store always runs at least 20 newspaper ads and at least 5 TV commercials every month.

44. Repeat Exercise 42 if the store's monthly budget for the two types of advertising combined varies from $50,000 to $100,000.

Life Science

45. The time it takes a patient of age t years who receives d units of penicillin per day to recover from pneumonia is

$$f(t, d) = \frac{250t}{d}$$

days, where $21 \leq t \leq 60$ and $500 \leq d \leq 1000$. Find the average amount of time it takes for a patient to recover from pneumonia.

46. The pollution produced by a coal-fired power plant is given by

$$z = \frac{20x}{y + 1}$$

where z is pollution in parts per million, x is the rated capacity of the plant in kilowatt-hours, and y is the amount spent during construction of the plant on pollution-abating equipment. Find the average pollution produced by coal-fired power plants if $100 \leq x \leq 1000$ and $0 \leq y \leq 25$.

47. An animal habitat is rectangular in shape:

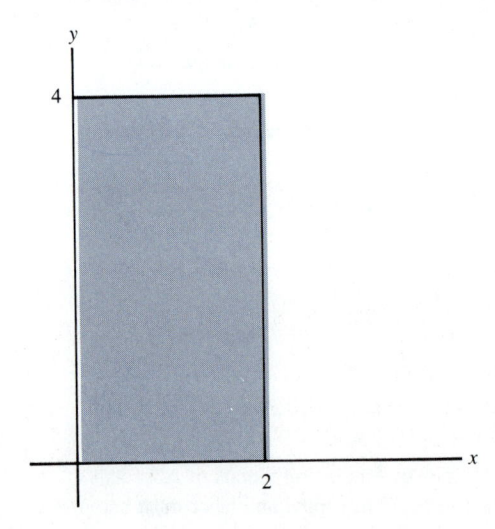

If the density of the animal population at point (x, y) is

$$f(x, y) = 10 - (x - 1)(y - 2)$$

individuals per square kilometer, find the average population density within the habitat.

*48. An animal habitat is circular in shape:

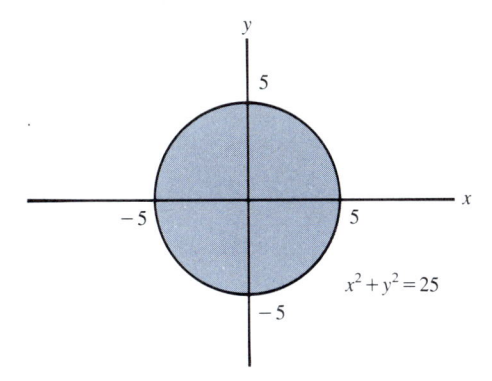

If the density of the animal population at point (x, y) is $f(x, y) = 25 - x - y$ individuals per square kilometer, find the average population density within the habitat.

Social Science

49. Ridership on city mass transit systems is given by

$$r = \frac{100x}{2 + y}.$$

Here r is daily ridership, in thousands, x is the population of the city, in thousands, and y is the price of a ticket, in dollars. Find the average daily ridership for cities whose populations vary from 100,000 to 120,000 and whose ticket prices range from $0.75 to $1.00.

50. The proportion p of the population in a region who will apply for welfare benefits at some time during a month is given by $p = e^{0.08x} - e^{0.01y}$, where x is the monthly unemployment rate and y is the average years of education of the adults in the region. Find the average proportion who will apply for welfare benefits if the unemployment rate ranges between 5 and 20% and the average years of education range between 10 and 12 years.

SUMMARY

This summary consists of terms and symbols whose meaning you should know, facts you should know, and techniques and methods you should be able to employ.

Section 16.1

Functions of two variables. The domain of a function of two variables as a set of ordered pairs. Functions of more than two variables. The domain of a function of n variables as a set of ordered n-tuples. Evaluating a function of several variables at a point. The three-dimensional cartesian coordinate system. The origin, the first octant. Plotting points in the three-dimensional cartesian coordinate system. Graphing functions of two variables: the graph as a surface.

Section 16.2

First partial derivatives: the partial derivative with respect to x, the partial derivative with respect to y. Notation: f_x, f_y. Finding first partial derivatives. Evaluating a first partial derivative at a point. Notation: $\partial f/\partial x$, $\partial f/\partial y$. The definition of first partial derivatives. Second partial derivatives: f_{xx}, f_{xy}, f_{yx}, f_{yy}. Finding second partial derivatives. Mixed partials f_{xy}, f_{yx}. Equality of mixed partials. Notation: $\partial^2 f/\partial x^2$, $\partial^2 f/\partial y^2$, $\partial^2 f/\partial x \partial y$, $\partial^2 f/\partial y \partial x$. Third and higher partial derivatives. Finding third and higher partial derivatives.

Section 16.3

The first partial derivatives f_x, f_y as the slopes of lines tangent to the surface $z = f(x, y)$. The significance of the sign of f_x, f_y. The significance of the sign of f_{xx}, f_{yy}. Using partial derivatives to find slopes of tangent lines to surfaces. Using partial derivatives to determine whether the curves generated by slices parallel to the xz-plane or yz-plane are increasing, decreasing, concave upward, or concave downward. The first partial derivative with respect to x as the rate of change of the function when all variables except x are held fixed. Using partial derivatives to find rates of change. Marginality. Competitive and complementary products. Cobb-Douglas production function: the marginal productivity of labor and of capital.

Section 16.4

Relative maximum, relative minimum, saddle point for a function of two variables. Finding relative maxima, relative minima, saddle points by setting the first partial derivatives equal to zero. The second derivative test. Using the second derivative test to find relative maxima, relative minima, and saddle points. The failure of the second derivative test. Price discrimination.

Section 16.5

Optimization subject to constraints. Constraint equation. The Lagrange multiplier method of optimizing a function of two variables subject to a constraint. Using the Lagrange multiplier method to optimize functions of two variables subject to constraints. The marginal productivity of money. Using the Lagrange multiplier method to optimize a function of three variables subject to a constraint.

Section 16.6

Region R in the xy-plane. Grid of rectangles on a region. Riemann sum for a function of two variables over a region R. The double integral as a limit of Riemann sums:

$$\iint_R f(x, y)\, dA = \lim_{\substack{\Delta x \to 0 \\ \Delta y \to 0}} \sum_{i=1}^{n} f(x_i, y_i)\, \Delta x\, \Delta y.$$

The double integral is the volume of the cylindrical solid bounded by the region R and the surface $z = f(x, y)$. Iterated integrals of the forms

$$\int_a^b \int_c^d f(x, y)\, dy\, dx, \quad \int_c^d \int_a^b f(x, y)\, dx\, dy.$$

Evaluating iterated integrals of the above forms. The double integral as an iterated integral of the form

$$\int_a^b \int_{g_1(x)}^{g_2(x)} f(x, y)\, dy\, dx \quad \text{or} \quad \int_c^d \int_{h_1(y)}^{h_2(y)} f(x, y)\, dx\, dy.$$

Setting up double integrals as iterated integrals. Evaluating double integrals as iterated integrals. Changing the order of integration. Using double integrals to find volumes. Using double integrals to find the area of a region. Using double integrals to find the average value of a function over a region.

REVIEW EXERCISES

1. Find the domain of the function f defined by

$$f(x, y) = x + \frac{2xy}{y - x},$$

and find $f(0, 0)$, $f(1, 0)$, $f(0, -1)$, and $f(3, 4)$, if possible.

2. Find the domain of the function defined by

$$w = e^{-xyz} + \ln 2xy,$$

and, if possible, find w when

   (a) $x = y = z = 0$ (b) $x = y = z = 2$
   (c) $x = -1, y = 3, z = 2$ (d) $x = 2, y = 1, z = -3$

3. Reddin Toothenclaw's pet store sells both quails and snails. If quails are priced at p_1 dollars each, Reddin can sell

$$q_1 = 200 - 5p_1$$

of them; if snails are priced at p_2 dollars each, Reddin can sell

$$q_2 = 750 - 20p_2$$

of them. Assuming that the prices of quails and snails do not affect one another, express Reddin's revenue as a function of p_1 and p_2 and find the revenue if quails sell for $10 each and snails for $2 each.

4. If a firm makes x units of X and y units of Y its profit is

$$P(x, y) = -8000 + 40x + 30y - 0.05xy$$

dollars. Find the firm's profit if it makes and sells 400 units of X and 300 units of Y. If x is fixed at 400, what can you say about the contribution to profit made by each unit of Y?

5. Find f_x, f_y, f_{xx}, f_{xy}, f_{yx}, and f_{yy} if

$$f(x, y) = 8x^2y - 2xy^3 + 12x + 2y - 5.$$

6. Find $f_x(2, 1), f_y(-1, 0), f_{xx}(2, 2), f_{yy}(-3, 4)$ for the function f of Exercise 5.

7. Let $g(p, q, r) = e^{pq} + e^{-qr} + 2p + r^4$. Find $g_{pq}, g_{qqpp},$ and g_{qppp}.

8. Find $\dfrac{\partial z}{\partial x}, \dfrac{\partial z}{\partial y}, \dfrac{\partial z^2}{\partial x \partial y}, \dfrac{\partial z^2}{\partial y \partial x}, \dfrac{\partial z^2}{\partial x^2}$, and $\dfrac{\partial z^2}{\partial y^2}$, if
$$z = (x^2 + 2xy - y^3)^6.$$

9. Let $f(x, y) = 2x + 3y + 5x^2y^2$. Find the slope of the line tangent to the graph of f at the point $(2, 2, 90)$ and parallel to the
   (a) xz-plane   (b) yz-plane

10. If the number of wolves in a region is given by
$$f(x, y) = x^{1/4} y^{-2},$$
where x is the number of animals the wolves can prey on and y is the number of animals that can prey on the wolves, find $f_x(16, 1)$ and $f_y(16, 1)$ and interpret your results.

11. Suppose a firm's revenue from selling x units of X and y units of Y is
$$R(x, y) = 20x + 25y + 0.1\sqrt{xy}$$
dollars. Find $R_x(25, 100)$ and $R_y(25, 100)$ and interpret your results.

12. Find the relative maxima, relative minima, and saddle points of f if
$$f(x, y) = x^2 + 2y^2 - 1.$$

13. Repeat Exercise 12 if
$$f(x, y) = -2x^4 + 16x^2 + y^2 - 4y + 8.$$

14. Find the dimensions of a rectangular box that is to have a volume of 10.4 cubic meters if the bottom is to be hardwood, which costs $10 per square meter, the sides are to be softwood, which costs $5 per square meter, the top is to be plastic, which costs $3 per square meter, and the cost of the box is to be minimized.

15. The demand functions for a commodity in regions X and Y are given by
$$p_X = 800 - \dfrac{x}{10} \quad \text{and} \quad p_Y = 500 - \dfrac{y}{5},$$
respectively. Find the prices that maximize revenue.

16. Find the maximum or minimum value of $f(x, y) = x^3 y^3$ subject to the constraint $6x + 2y = 12$, and state whether your answer is a maximum or a minimum.

17. Repeat Exercise 16 for the function defined by
$$f(x, y, z) = x^2y + yz + z^2$$
subject to the constraint $xyz = 64$.

18. Find two positive numbers such that the product of twice the first and three times the second is 384 and their sum is a minimum.

19. A company can produce $f(x, y) = 1000x^{1/5}y^{4/5}$ units of its product, where x is units of labor and y is units of capital. If each unit of labor costs $20 and each unit of capital costs $100, and if the company's production budget is $1000, find the maximum production. Also find and interpret the marginal productivity of money.

20. Evaluate the integral
$$\iint_R x^2 \, dA,$$

where R is the region bounded by the graphs of $y = x$, $y = 5 - x$, $y = 1$, and the x-axis.

21. Find the volume of the cylindrical solid whose base is the triangle in the xy-plane bounded by the lines $y = 2 - x$, $y = x$, and $x = 0$, and whose top is the plane $z = 4 - 2x$.

22. Use double integrals to find the area of the region bounded by $y = e^x$, $y = e^{-x}$, and $x = 1$.

23. Refer to the company of Exercise 19: Suppose the amount of labor available varies from 15 to 30 units and the amount of capital available varies from 4 to 10 units; find the average output.

ADDITIONAL TOPICS

Here are some suggestions for topics not covered in the text that you might want to investigate on your own.

1. Just as there is a chain rule for functions of one variable, so there are chain rules for functions of several variables. For instance, if $z = f(x, y)$, $x = g(u, v)$ and $y = h(u, v)$, then one such chain rule states that

$$\frac{\partial z}{\partial u} = \frac{\partial z}{\partial x}\frac{\partial x}{\partial u} + \frac{\partial z}{\partial y}\frac{\partial y}{\partial u}.$$

Find out about chain rules for partial derivatives and how to use them.

2. In Section 11.3 we developed the formula

$$f(x + \Delta x) \approx f(x) + f'(x)\Delta x$$

for approximation by differentials. There is a corresponding formula for functions of two variables, namely

$$f(x + \Delta x, y + \Delta y) \approx f(x, y) + f_x(x, y)\Delta x + f_y(x, y)\Delta y.$$

The expression $f_x(x, y)\Delta x + f_y(x, y)\Delta y$ is called the **total differential** of f. Find out about total differentials and how they are used to approximate functions and interpret the Lagrange multiplier λ.

3. Calculus in three dimensions is often best studied in the framework of **vectors**. Find out about vectors and how they are used in calculus. In particular, find out about the **directional derivative** and **gradient** of a function, how they are defined and computed, and their meanings.

4. In the Additional Topics at the end of Chapter 2 we suggested that you find out about the **polar coordinate system.** If you have not done so, do so now and then find out how to set up and evaluate double integrals in polar coordinates.

5. **Triple integrals** are analogous to double integrals: the triple integral of a function f of the variables x, y, and z is an integral over the interior of a three-dimensional solid V that lies within the domain of f, and has the form

$$\iiint_V f(x, y, z)\, dV.$$

Triple integrals are evaluated as threefold iterated integrals of the form

$$\int_a^b \int_c^d \int_p^q f(x, y, z)\, dz\, dy\, dx.$$

Find out how to set up and evaluate triple integrals and find some of their uses.

6. In addition to the cartesian or rectangular coordinate system in three dimensions, in which points are located by their distances from the x-, y-, and z-axes, there are two other useful three-dimensional coordinate systems: **cylindrical coordinates** and **spherical coordinates**. Find out how these coordinate systems work and how to set up triple integrals in each of them.

7. The technique of Lagrange multipliers of Section 16.5 is due to the French mathematician Joseph-Louis Lagrange (1736–1813). Lagrange made contributions to many areas of mathematics and physics, and wrote a famous and influential book entitled *Analytical Mechanics*. He was also caught up in the French revolution, was instrumental in the development of the metric system, and was made a count by Napoleon. Find out about Lagrange's life and his contributions to mathematics and physics.

SUPPLEMENT: LEAST-SQUARES REGRESSION

In Section 8.5 we introduced the **least-squares regression** line $Y = mX + b$ for a set of data points $\{(x_i, y_i) | 1 \leq i \leq n\}$. Recall that this is the line that best fits the data in the sense that it minimizes the sum

$$f(m, b) = \sum_{i=1}^{n} (mx_i + b - y_i)^2.$$

We stated without proof in Section 8.5 that the coefficients m and b can be found by means of the formulas

$$m = \frac{n \sum_{i=1}^{n} x_i y_i - \left(\sum_{i=1}^{n} x_i\right)\left(\sum_{i=1}^{n} y_i\right)}{n \sum_{i=1}^{n} x_i^2 - \left(\sum_{i=1}^{n} x_i\right)^2}$$

and

$$b = \frac{\sum_{i=1}^{n} y_i - m \sum_{i=1}^{n} x_i}{n}.$$

FUNCTIONS OF SEVERAL VARIABLES

In this supplement we show how the least-squares regression line can be obtained using the methods of calculus. This will allow us to derive the formulas for m and b given above; more importantly, the method we use will also make it possible for us to fit curves other than lines to data sets.

We begin with a numerical example.

EXAMPLE 1 Consider the following data that relate students' grades on a final examination to their number of absences from class during the semester:

x = Number of Absences: 1 2 6 7
y = Grade on Final Exam: 91 86 74 73

A scatter diagram (Figure 16–31) shows that the relationship between x and y appears to be linear. Therefore we seek to fit a least-squares regression line $Y = mX + b$ to the data. But by definition, this is the line such that m and b minimize

$$f(m, b) = \sum_{i=1}^{4} (mx_i + b - y_i)^2.$$

(Note that m and b are the independent variables here: the points (x_i, y_i) are known.) Since $f(m, b)$ is a sum of squares, it has no maxima or saddle points; hence if we set its partial derivatives

$$f_m(m, b) = \sum_{i=1}^{4} 2(mx_i + b - y_i)x_i$$

and

$$f_b(m, b) = \sum_{i=1}^{4} 2(mx_i + b - y_i)$$

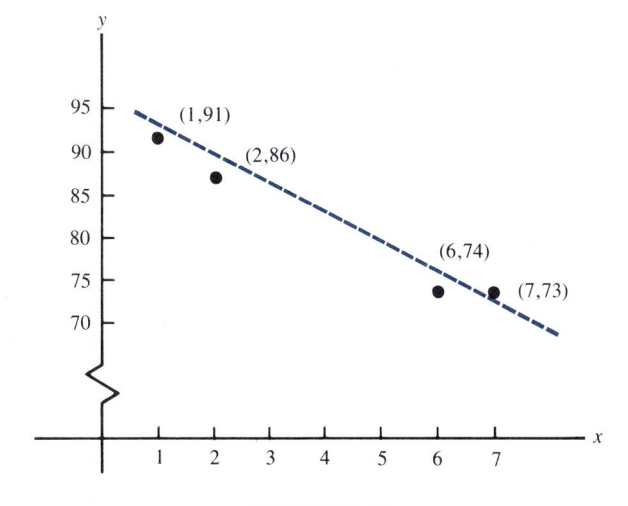

FIGURE 16–31

equal to zero and solve, we will find its minimum. Thus we must solve the system of equations

$$\sum_{i=1}^{4} x_i(mx_i + b - y_i) = 0,$$

$$\sum_{i=1}^{4} (mx_i + b - y_i) = 0$$

for m and b. Substituting the points (x_i, y_i) into these equations, we obtain

$(m + b - 91) + 2(2m + b - 86) + 6(6m + b - 74) + 7(7m + b - 73) = 0$

and

$(m + b - 91) + (2m + b - 86) + (6m + b - 74) + (7m + b - 73) = 0,$

which reduces to the system of equations

$$90m + 16b = 1218$$
$$16m + 4b = 324.$$

The solution of this system is $m = -3$, $b = 93$. Therefore the equation of the least-squares regression line is

$$Y = -3X + 93.$$

As noted in Section 8.5, we use the least-squares regression line to predict the *average* value of y for a given value of x. For instance, using the regression line, we can predict that the average grade on the final examination for those students who had four absences during the semester will be

$$Y = -3(4) + 93 = 81. \quad \square$$

If we take the data of Example 1 and apply the formulas for m and b which were stated at the beginning of this supplement, we will again find that $m = -3$ and $b = 93$. Thus the method for finding m and b illustrated in Example 1 gives the same answer as do the formulas. This is not surprising, because the formulas can be derived using the method of the example, as we will now demonstrate.

Suppose $\{(x_i, y_i) | 1 \leq i \leq n\}$ is a set of data points and we wish to find the coefficients m and b of the least-squares regression line $Y = mX + b$. Then we must minimize the sum

$$f(m, b) = \sum_{i=1}^{n} (mx_i + b - y_i)^2.$$

Proceeding as in Example 1, we set the partial derivatives of f equal to 0 and solve for m and b. Thus we have

$$f_m = \sum_{i=1}^{n} 2x_i (mx_i + b - y_i) = 0$$

and

$$f_b = \sum_{i=1}^{n} 2(mx_i + b - y_i) = 0.$$

FUNCTIONS OF SEVERAL VARIABLES

From the second of these equations,

$$m \sum_{i=1}^{n} x_i + \sum_{i=1}^{n} b - \sum_{i=1}^{n} y_i = 0,$$

or

$$m \sum_{i=1}^{n} x_i + nb - \sum_{i=1}^{n} y_i = 0.$$

Therefore

$$b = \frac{\sum_{i=1}^{n} y_i - m \sum_{i=1}^{n} x_i}{n}.$$

If we now rewrite b in the form

$$b = \frac{\sum_{i=1}^{n} y_i}{n} - m \frac{\sum_{i=1}^{n} x_i}{n} = \bar{y} - m\bar{x}$$

and then substitute $\bar{y} - m\bar{x}$ for b in the equation $f_m = 0$, we obtain

$$\sum_{i=1}^{n} x_i (mx_i + \bar{y} - m\bar{x} - y_i) = 0,$$

or

$$m \sum_{i=1}^{n} x_i^2 + \bar{y} \sum_{i=1}^{n} x_i - m\bar{x} \sum_{i=1}^{n} x_i - \sum_{i=1}^{n} x_i y_i = 0,$$

or

$$m \left(\sum_{i=1}^{n} x_i^2 - \bar{x} \sum_{i=1}^{n} x_i \right) = \sum_{i=1}^{n} x_i y_i - \bar{y} \sum_{i=1}^{n} x_i.$$

Therefore

$$m = \frac{\sum_{i=1}^{n} x_i y_i - \bar{y} \sum_{i=1}^{n} x_i}{\sum_{i=1}^{n} x_i^2 - \bar{x} \sum_{i=1}^{n} x_i} = \frac{\sum_{i=1}^{n} x_i y_i - \left(\frac{1}{n}\sum_{i=1}^{n} y_i\right)\left(\sum_{i=1}^{n} x_i\right)}{\sum_{i=1}^{n} x_i^2 - \left(\frac{1}{n}\sum_{i=1}^{n} x_i\right)\left(\sum_{i=1}^{n} x_i\right)}$$

$$= \frac{n \sum_{i=1}^{n} x_i y_i - \left(\sum_{i=1}^{n} x_i\right)\left(\sum_{i=1}^{n} y_i\right)}{n \sum_{i=1}^{n} x_i^2 - \left(\sum_{i=1}^{n} x_i\right)^2}.$$

The technique we have used here to find the least-squares line can also be used to fit other least-squares curves to data sets. We illustrate with an example.

EXAMPLE 2 Consider the following data relating average cost per unit y to the number of units produced x:

x (in thousands of units):	1	2	3	4	5
y (in dollars):	5.90	3.10	2.00	2.90	6.10

A scatter diagram (Figure 16–32) suggests that a parabola will fit the data better than a line. Thus we seek to find the equation of the least-squares parabola

$$Y = aX^2 + bX + c.$$

This will be the parabola that minimizes

$$f(a, b, c) = \sum_{i=1}^{5} (ax_i^2 + bx_i + c - y_i)^2.$$

We minimize f by setting its partial derivatives equal to zero and solving for a, b, and c. We have

$$f_a = \sum_{i=1}^{5} 2x_i^2(ax_i^2 + bx_i + c - y_i) = 0,$$

$$f_b = \sum_{i=1}^{5} 2x_i(ax_i^2 + bx_i + c - y_i) = 0,$$

and

$$f_c = \sum_{i=1}^{5} 2(ax_i^2 + bx_i + c - y_i) = 0.$$

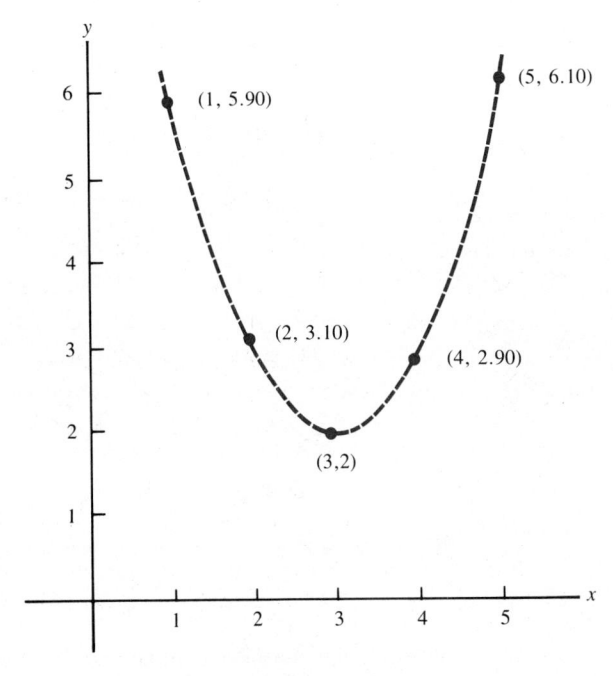

FIGURE 16–32

Functions of Several Variables

Substituting in the values for x_i and y_i and simplifying, we obtain the system of equations

$$979a + 225b + 55c = 235.2$$
$$225a + 55b + 15c = 60.2$$
$$55a + 15b + 5c = 20.0.$$

This system may be solved by the Gauss-Jordan method of Section 4.1 to obtain $a = 1.00$, $b = -5.98$, and $c = 10.94$. Therefore the least-squares quadratic is

$$Y = X^2 - 5.98X + 10.94. \quad \square$$

We can similarly use the method of this supplement to fit cubic, quartic (fourth-power) and other least-squares curves to data sets. We can also use it to fit least-squares equations to three-dimensional data sets. Again, we illustrate with an example.

EXAMPLE 3 Suppose we have the following data that relate savings to income and family size:

x = Annual Income (in 1000s of $):	20	30	40	50
y = Family Size:	2	2	4	3
z = Annual Savings (in 1000s of $):	2.10	2.90	3.20	4.25

Note that the data are three-dimensional: the data set is of the form

$$\{(x_i, y_i, z_i) | 1 \leq i \leq 4\}.$$

We wish to express savings z as a function of income x and family size y. Suppose we decide to use a model of the form

$$Z = aX + bY + c.$$

Then the equation of this form that best fits the data will be the one that minimizes

$$f(a, b, c) = \sum_{i=1}^{4} (ax_i + by_i + c - z_i)^2.$$

Hence we must have

$$f_a = \sum_{i=1}^{4} 2x_i(ax_i + by_i + c - z_i) = 0.$$

$$f_b = \sum_{i=1}^{4} 2y_i(ax_i + by_i + c - z_i) = 0,$$

and

$$f_c = \sum_{i=1}^{4} 2(ax_i + by_i + c - z_i) = 0.$$

Substituting the values of x_i, y_i, and z_i into the equations and simplifying, we obtain the system

$$5400a + 410b + 140c = 469.50$$
$$410a + 33b + 11c = 35.55$$
$$140a + 11b + 4c = 12.45.$$

The solution of this system is $a = 0.08$, $b = -0.25$, and $c = 1$. Therefore the least-squares regression equation is

$$Z = 0.08X - 0.25Y + 1.$$

This equation can be used in the usual manner to predict the average value of z for given values of x and y. For instance, we can predict that the average savings for families of four whose income is $25,000 per year will be

$$Z = 0.08(25) - 0.25(4) + 1 = 2$$

thousand dollars per year. □

■ *Computer Manual: Program LESTSQRS*

SUPPLEMENTARY EXERCISES

1. According to an article by J. Larry Brown ("Hunger in the U.S." *Scientific American*, February 1987), participation in the food stamp program as a percentage y of people who are living at or below the poverty level has been as follows:

Year:	1980	1981	1982	1983	1984	1985
% y:	68	65	60	61	61	59

   Using the method of Example 1, find the least-squares regression line for y as a function of time. Then check your answer by using the formulas for m and b. What are the significances of the slope and y-intercept of the least-squares line?

2. Find the quadratic least-squares regression equation for the following data:

x:	1	3	4	6	10
y:	8.8	5.1	6.0	13.9	53.5

3. A firm has the following data on its average cost per unit A:

Units Made x:	10	20	45	50	65	70
Avg Cost/Unit A ($):	120	55	10	20	85	115

   Find a quadratic least-squares model for A as a function of x. At what level of production is average cost per unit the lowest?

4. According to the article by Brown cited in Exercise 1, distribution of food to the poor by U.S. foodbanks during the period 1979–1985 was as follows:

Year:	'79	'80	'81	'82	'83	'84	'85
Food Distributed: (millions of tons)	4	10	18	30	44	70	100

   Find a quadratic least-squares model for the amount of food distributed as a function of time, and use it to "predict" the amount distributed in 1986.

*5. Prove that if $\{(x_i, y_i) | 1 \le i \le n\}$ is a data set and $Y = aX^2 + bX + c$ is the least-squares quadratic for the data, then a, b, and c can be found by solving the system of equations

Functions of Several Variables

$$a \sum_{i=1}^{n} x_i^4 + b \sum_{i=1}^{n} x_i^3 + c \sum_{i=1}^{n} x_i^2 = \sum_{i=1}^{n} x_i^2 y_i$$

$$a \sum_{i=1}^{n} x_i^3 + b \sum_{i=1}^{n} x_i^2 + c \sum_{i=1}^{n} x_i = \sum_{i=1}^{n} x_i y_i$$

$$a \sum_{i=1}^{n} x_i^2 + b \sum_{i=1}^{n} x_i + nc = \sum_{i=1}^{n} y_i.$$

Then use these equations to check your answers for Exercises 2 and 3.

6. Find the least-squares regression equation of the form
$$Y = aX_1 + bX_2 + c$$
for the data

x_1:	8	10	14	15	21
x_2:	42	36	32	24	15
y:	200	170	140	100	30

7. A firm has the following cost data for the total cost C of making its two products X_1 and X_2:

Units of X_1:	50	60	80	100	120	125	140
Units of X_2:	20	20	40	50	40	50	70
Cost C ($):	2900	3600	4500	5200	6300	6700	7700

Find a least-squares model of the form $C = aX_1 + bX_2 + c$ for this set of data. What are the significances of the values obtained for a, b, and c?

*8. As in Exercise 5, find a system of equations that when solved will yield the coefficients a, b, and c in the least-squares regression equation $Y = aX_1 + bX_2 + c$. Then use these equations to check your answers for Exercises 6 and 7.

*9. According to an article by Simons, Sherrod, Collopy, and Jenkins ("Restoring the Bald Eagle," *American Scientist*, May–June 1988), the probability p that a colony of eagles will become extinct if it was founded by x individuals is as follows:

x:	10	20	30	40
p:	0.47	0.20	0.09	0.04

Find an exponential least-squares equation of the form $p = ae^{bx}$ for this set of data and use it to predict the probability of extinction if a colony is founded by 25 individuals. (*Hint:* $\ln p = \ln a + bx$.)

*10. The following table appeared on page 342 of the July–August 1988, issue of *American Scientist*: it gives the cost of completing the Shoreham, NY, nuclear power plant, as this cost was estimated in the given year.

Year	Estimated Cost C
1966	$ 75 million
1971	$200 million
1976	$600 million
1981	$1.5 billion
1986	$4.5 billion
1988	$5.3 billion

Find an exponential least-squares model of the form $C = ae^{bx}$ and use it to "predict" the cost of completing the plant in 1990.

*11. This Exercise outlines a proof of the formula for the correlation coefficient r that was presented in Section 8.5.

(a) Let M denote the minimum value of

$$f(m, b) = \sum_{i=1}^{n} (mx_i + b - y_i)^2.$$

By substituting $\bar{y} - m\bar{x}$ for b, and using the facts that the variances s_x^2, s_y^2 for x and y are given by

$$s_x^2 = \frac{\sum_{i=1}^{n} (x_i - \bar{x})^2}{n-1} \quad \text{and} \quad s_y^2 = \frac{\sum_{i=1}^{n} (y_i - \bar{y})^2}{n-1},$$

show that

$$M = m^2(n-1)s_x^2 - 2m \sum_{i=1}^{n} (x_i - \bar{x})(y_i - \bar{y}) + (n-1)s_y^2.$$

(b) Show that

$$\sum_{i=1}^{n} (x_i - \bar{x})(y_i - \bar{y}) = \sum_{i=1}^{n} x_i y_i - n\bar{x}\bar{y}.$$

(c) Show that

$$m = \frac{\sum_{i=1}^{n} x_i y_i - n\bar{x}\bar{y}}{(n-1)s_x^2}.$$

Then use this fact and the results of (a) and (b) to show that

$$M = (n-1)(s_y^2 - m^2 s_x^2).$$

(d) Let $r = m/s_x s_y$ and show that $M = (n-1)(1 - r^2)$.

(e) Use the result of (d) to show that:
if $r = \pm 1$, the least-squares regression line fits the data perfectly;
if r is close to ± 1, the least-squares regression line fits the data very closely;
the closer r is to 0, the worse the fit of the least-squares regression line to the data.

(f) Show that

$$r = \frac{n \sum_{i=1}^{n} x_i y_i - \left(\sum_{i=1}^{n} x_i\right)\left(\sum_{i=1}^{n} y_i\right)}{\sqrt{\left[n \sum_{i=1}^{n} x_i^2 - \left(\sum_{i=1}^{n} x_i\right)^2\right]\left[n \sum_{i=1}^{n} y_i^2 - \left(\sum_{i=1}^{n} y_i\right)^2\right]}}$$

Tables

Table 1: Compound Interest: Values of $(1 + r)^k$

Table 2: Present Value: Values of $(1 + r)^{-k}$

Table 3: Annuities: Values of $s_{\overline{k}|r} = \dfrac{(1 + r)^k - 1}{r}$

Table 4: Present Values of Annuities: Values of $a_{\overline{k}|r} = \dfrac{1 - (1 + r)^{-k}}{r}$

Table 5: Binomial Probabilities

Table 6: Areas under the Standard Normal Curve

Table 7: Powers of e

Table 8: Natural Logarithms

TABLE 1 Compound Interest: Values of $(1 + r)^k$

k	0.005	0.0075	0.01	0.015	0.02	0.025	0.03	0.04	0.06	0.08	0.10
1	1.005000	1.007500	1.010000	1.015000	1.020000	1.025000	1.030000	1.040000	1.060000	1.080000	1.100000
2	1.010025	1.015056	1.020100	1.030225	1.040400	1.050625	1.060900	1.081600	1.123600	1.166400	1.210000
3	1.015075	1.022669	1.030301	1.045678	1.061208	1.076891	1.092727	1.124864	1.191016	1.259712	1.331000
4	1.020151	1.030339	1.040604	1.061364	1.082432	1.103813	1.125509	1.169859	1.262477	1.360489	1.464100
5	1.025251	1.038067	1.051010	1.077284	1.104081	1.131408	1.159274	1.216653	1.338226	1.469328	1.610510
6	1.030378	1.045852	1.061520	1.093443	1.126162	1.159693	1.194052	1.265319	1.418519	1.586874	1.771561
7	1.035529	1.053696	1.072135	1.109845	1.148686	1.188686	1.229874	1.315932	1.503630	1.713824	1.948717
8	1.040707	1.061599	1.082857	1.126493	1.171659	1.218403	1.266770	1.368569	1.593848	1.850930	2.143589
9	1.045911	1.069561	1.093685	1.143390	1.195093	1.248863	1.304773	1.423312	1.689479	1.999005	2.357918
10	1.051140	1.077583	1.104622	1.160541	1.218994	1.280085	1.343916	1.480244	1.790848	2.158925	2.593743
11	1.056396	1.085664	1.115668	1.177949	1.243374	1.312087	1.384234	1.539454			
12	1.061678	1.093807	1.126825	1.195618	1.268242	1.344889	1.425761	1.601032			
13	1.066986	1.102010	1.138093	1.213552	1.293607	1.378511	1.468534	1.665074			
14	1.072321	1.110276	1.149474	1.231756	1.319479	1.412974	1.512590	1.731676			
15	1.077683	1.118603	1.160969	1.250232	1.345868	1.448298	1.557967	1.800944			
16	1.083071	1.126992	1.172579	1.268986	1.372786	1.484506	1.604706	1.872981			
17	1.088487	1.135445	1.184304	1.288020	1.400241	1.521618					
18	1.093929	1.143960	1.196148	1.307341	1.428246	1.559659					
19	1.099399	1.152540	1.208109	1.326951	1.456811	1.598650					
20	1.104896	1.161184	1.220190	1.346855	1.485947	1.638616					
21	1.110420	1.169893	1.232392	1.367058	1.515666	1.679582					

22	1.115972	1.178667	1.244716	1.545980	1.721571
23	1.121552	1.187587	1.257163	1.576899	1.764611
24	1.127160	1.196414	1.269735	1.608437	1.808726
25	1.132796	1.205387	1.282432		
26	1.138460	1.214427	1.295256		
27	1.144152	1.223535	1.308209		1.996495
28	1.149873	1.232712	1.321291	1.741024	
29	1.155622	1.241957	1.334504		
30	1.161400	1.251272	1.347849		
31	1.167207	1.260656	1.361327		2.203757
32	1.173043	1.270111	1.374941	1.884541	
33	1.178908	1.279637	1.388690		
34	1.184803	1.289234	1.402577		
35	1.190727	1.298904	1.416603		2.432535
36	1.196681	1.308645	1.430769	2.039887	
37	1.202664	1.318460	1.445077		
38	1.208677	1.328349	1.459527		
39	1.214721	1.338311	1.474123		2.685064
40	1.220794	1.348349	1.488864	2.208040	
41	1.226898	1.358461	1.503752		
42	1.233033	1.368650	1.518790		

TABLE 2 Present Value: Values of $(1 + r)^{-k}$

k	0.005	0.0075	0.01	0.015	0.02	0.025	0.03	0.04	0.06	0.08	0.10
1	0.995025	0.992556	0.990099	0.985222	0.980392	0.975610	0.970874	0.961538	0.943396	0.925926	0.909091
2	0.990075	0.985167	0.980296	0.970662	0.961169	0.951814	0.942596	0.924556	0.889996	0.857339	0.826446
3	0.985149	0.977833	0.970590	0.956317	0.942322	0.928599	0.915142	0.888996	0.839619	0.793832	0.751315
4	0.980248	0.970554	0.960980	0.942184	0.923845	0.905951	0.888487	0.854804	0.792094	0.735030	0.683013
5	0.975371	0.963329	0.951466	0.928260	0.905731	0.883854	0.862609	0.821927	0.747258	0.680583	0.620921
6	0.970518	0.956158	0.942045	0.914542	0.887971	0.862297	0.837484	0.790315	0.704961	0.630170	0.564474
7	0.965690	0.949040	0.932718	0.901027	0.870560	0.841265	0.813092	0.759918	0.665057	0.583490	0.513158
8	0.960885	0.941975	0.923483	0.887711	0.853490	0.820747	0.789409	0.730690	0.627412	0.540269	0.466507
9	0.956105	0.934963	0.914340	0.874592	0.836755	0.800728	0.766417	0.702587	0.591898	0.500249	0.424098
10	0.951348	0.928003	0.905287	0.861667	0.820348	0.781198	0.744094	0.675564	0.558395	0.463193	0.385543
11	0.946615	0.921095	0.896324	0.848933	0.804263	0.762145	0.722421	0.649581			
12	0.941905	0.914238	0.887449	0.836387	0.788493	0.743556	0.701380	0.624597			
13	0.937219	0.907432	0.878663	0.824027	0.773033	0.725420	0.680951	0.600574			
14	0.932556	0.900677	0.869963	0.811849	0.757875	0.707727	0.661118	0.577475			
15	0.927917	0.893973	0.861349	0.799852	0.743015	0.690466	0.641862	0.555265			
16	0.923300	0.887318	0.852821	0.788031	0.728446	0.673625	0.623167	0.533908			
17	0.918707	0.880712	0.844377	0.776385	0.714163	0.657195	0.605016	0.513373			
18	0.914136	0.874156	0.836017	0.764912	0.700159	0.641166	0.587395	0.493628			
19	0.909588	0.867649	0.827740	0.753607	0.686431	0.625528	0.570286	0.474642			
20	0.905063	0.861190	0.819544	0.742470	0.672971	0.610271	0.553676	0.456387			
21	0.900560	0.854779	0.811430	0.731498	0.659776	0.595386					

22	0.896080	0.848416	0.803396	0.720688	0.646839	0.580865
23	0.891622	0.842100	0.795442	0.710037	0.634156	0.566697
24	0.887186	0.835831	0.787566	0.699544	0.621721	0.552875
25	0.882772	0.829609	0.779768	0.689206		
26	0.878380	0.823434	0.772048	0.679021		
27	0.874010	0.817304	0.764404	0.668986		0.500878
28	0.869662	0.811220	0.756836	0.659099	0.574375	
29	0.865335	0.805181	0.749342	0.649359		
30	0.861030	0.799187	0.741923	0.639762		
31	0.856746	0.793238	0.734577	0.630308		0.453771
32	0.852484	0.787333	0.727304	0.620993	0.530633	
33	0.848242	0.781472	0.720103			
34	0.844022	0.775654	0.712973			
35	0.839823	0.769880	0.705914			0.411094
36	0.835645	0.764149	0.698925	0.585090	0.490223	
37	0.831487	0.758461	0.692005			
38	0.827351	0.752814	0.685153			
39	0.823235	0.747210	0.678370			0.372431
40	0.819139	0.741648	0.671653	0.551262	0.452890	
41	0.815064	0.736127	0.665003			
42	0.811009	0.730647	0.658419			

TABLE 3   Annuities: Values of $s_{\overline{k}|r} = \dfrac{(1+r)^k - 1}{r}$

k	r=0.005	r=0.01
1	1.000000	1.000000
2	2.005000	2.010000
3	3.015025	3.030100
4	4.030100	4.060401
5	5.050251	5.101005
6	6.075502	6.152015
7	7.105879	7.213535
8	8.141409	8.285670
9	9.182116	9.368527
10	10.228026	10.462213
11	11.279166	11.566835
12	12.335562	12.682503
15	15.536547	16.096895
18	18.785788	19.614748
21	22.084011	23.239194
24	25.431955	26.973465
27	28.830370	30.820888
30	32.280016	34.784892
33	35.781667	38.869008
36	39.336105	43.076878
39	42.944126	47.412251
42	46.606540	51.878989
45	50.324164	56.481075
48	54.097832	61.222608
54	61.816691	71.141047
60	69.770030	81.669670
66	77.964972	92.846015
72	86.408855	104.70993
78	95.109243	117.30372

k	r=0.015	r=0.02
1	1.000000	1.000000
2	2.015000	2.020000
3	3.045225	3.060400
4	4.090903	4.121608
5	5.152267	5.204040
6	6.229551	6.308121
7	7.322994	7.434283
8	8.432839	8.582969
9	9.559332	9.754628
10	10.702722	10.949721
11	11.863262	12.168715
12	13.041211	13.412090
14	15.450382	15.973938
16	17.932370	18.639285
18	20.489376	21.412312
20	23.123667	24.297370
24	28.633521	30.421862
28	34.481479	37.051210
32	40.688288	44.227030
36	47.275969	51.994367
40	54.267894	60.401983
44	61.688868	69.502657
48	69.565219	79.353519
52	77.924891	90.016409
56	86.797543	101.55826
60	96.214652	114.05154
64	106.20963	127.57466
68	116.81793	142.21252
72	128.07720	158.05702

k	r=0.03	r=0.04	r=0.06	r=0.08	r=0.10
1	1.000000	1.000000	1.000000	1.000000	1.000000
2	2.030000	2.040000	2.060000	2.080000	2.100000
3	3.090900	3.121600	3.183600	3.246400	3.310000
4	4.183627	4.246464	4.374616	4.506112	4.641000
5	5.309136	5.416323	5.637093	5.866601	6.105100
6	6.468410	6.632975	6.975319	7.335929	7.715610
7	7.662462	7.898295	8.393838	8.922803	9.487171
8	8.892336	9.214226	9.897468	10.636628	11.435888
9	10.159106	10.582795	11.491316	12.487558	13.579477
10	11.463879	12.006107	13.180795	14.486562	15.937425
12	14.192030	15.025805	16.869941	18.977126	21.384284
14	17.086324	18.291911	21.015066	24.214920	27.974983
15	18.598914	20.023588	23.275970	27.152114	31.772482
16	20.156881	21.824531	25.672528	30.324283	35.949730
18	23.414435	25.645413	30.905653	37.450244	45.599173
20	26.870374	29.778079	36.785591	45.761964	57.274999

TABLE 4 Present Values of Annuities: Values of $a_{\overline{k}|r} = \dfrac{1-(1+r)^{-k}}{r}$

k	r=0.005	r=0.01	k	r=0.015	r=0.02	k	r=0.03	r=0.04	r=0.06	r=0.08	r=0.10
1	0.995025	0.990099	1	0.985222	0.980392	1	0.970874	0.961538	0.943396	0.925926	0.909091
2	1.985099	1.970395	2	1.955883	1.941561	2	1.913470	1.886095	1.833393	1.783265	1.735537
3	2.970248	2.940985	3	2.912200	2.883883	3	2.828611	2.775091	2.673012	2.577097	2.486852
4	3.950496	3.901965	4	3.854385	3.807729	4	3.717098	3.629895	3.465106	3.312127	3.169865
5	4.925866	4.853431	5	4.782645	4.713460	5	4.579707	4.451822	4.212364	3.992710	3.790787
6	5.896384	5.795476	6	5.697187	5.601431	6	5.417191	5.242137	4.917324	4.622880	4.355261
7	6.862074	6.728194	7	6.598214	6.471991	7	6.230283	6.002055	5.582381	5.206370	4.868419
8	7.822959	7.651678	8	7.485925	7.325481	8	7.019692	6.732745	6.209794	5.746639	5.334926
9	8.779064	8.566017	9	8.360517	8.162237	9	7.786109	7.435332	6.801692	6.246888	5.759024
10	9.730412	9.471305	10	9.222185	8.982585	10	8.530203	8.110896	7.360087	6.710081	6.144567
11	10.677027	10.367628	11	10.071118	9.786848	12	9.954004	9.385074	8.383844	7.536078	6.813692
12	11.618932	11.255077	12	10.907505	10.575341	14	11.296073	10.563123	9.294984	8.244237	7.366688
15	14.416624	13.865052	14	12.543381	12.106249	15	11.937935	11.118387	9.712249	8.559479	7.606080
18	17.172768	16.398269	16	14.131264	13.577709	16	12.561102	11.652296	10.105895	8.851369	7.823709
21	19.887979	18.856983	18	15.672561	14.992031	18	13.753513	12.659297	10.827603	9.371887	8.201412
24	22.562866	21.243387	20	17.168639	16.351433	20	14.877475	13.590326	11.469921	9.818147	8.513564
27	25.198028	23.559608	24	20.030405	18.913926						
30	27.794054	25.807708	28	22.726717	21.281272						
33	30.351526	27.989693	32	25.267139	23.468335						
36	32.871016	30.107505	36	27.660684	25.488842						
39	35.353089	32.163033	40	29.915845	27.355479						
42	37.798300	34.158108	44	32.040622	29.079963						
45	40.207196	36.094508	48	34.042554	30.673120						
48	42.580318	37.973959	52	35.928742	32.144950						
54	47.221352	41.568664	56	37.705879	33.504694						
60	51.725561	44.955038	60	39.380269	34.760887						
66	56.096976	48.145156	64	40.957853	35.921415						
72	60.339514	51.150391	68	42.444228	36.993563						
78	64.456973	53.981459	72	43.844667	37.984063						

TABLE 5 Binomial Probabilities
(Entries That Are 0 to 4 or More Decimal Places Have Been Omitted)

$n = 5$:

X	.05	.10	.20	.25	.30	.40	.50
0	.7738	.5905	.3277	.2373	.1681	.0778	.0313
1	.2036	.3281	.4096	.3955	.3602	.2592	.1563
2	.0214	.0729	.2048	.2637	.3087	.3456	.3125
3	.0011	.0081	.0512	.0879	.1323	.2304	.3125
4		.0005	.0064	.0146	.0284	.0768	.1563
5			.0003	.0010	.0024	.0102	.0313

$n = 10$:

X	.05	.10	.20	.25	.30	.40	.50
0	.5987	.3487	.1074	.0563	.0282	.0060	.0010
1	.3151	.3874	.2684	.1877	.1211	.0403	.0098
2	.0746	.1937	.3020	.2816	.2335	.1209	.0439
3	.0105	.0574	.2013	.2503	.2668	.2150	.1172
4	.0010	.0112	.0881	.1460	.2001	.2508	.2051
5	.0001	.0015	.0264	.0584	.1029	.2007	.2461
6		.0001	.0055	.0162	.0368	.1115	.2051
7			.0008	.0031	.0090	.0425	.1172
8			.0001	.0004	.0014	.0106	.0439
9					.0001	.0016	.0098
10						.0001	.0010

$n = 15$:

X	.05	.10	.20	.25	.30	.40	.50
0	.4633	.2059	.0352	.0134	.0047	.0005	
1	.3658	.3432	.1319	.0668	.0305	.0047	.0005
2	.1348	.2669	.2309	.1559	.0916	.0219	.0032
3	.0307	.1285	.2501	.2252	.1700	.0634	.0139
4	.0049	.0428	.1876	.2252	.2186	.1268	.0417
5	.0006	.0105	.1032	.1651	.2061	.1859	.0916
6		.0019	.0430	.0917	.1472	.2066	.1527
7		.0003	.1038	.0393	.0811	.1771	.1964
8			.0035	.0131	.0348	.1181	.1964
9			.0007	.0034	.0116	.0612	.1527
10			.0001	.0007	.0030	.0245	.0916
11				.0001	.0006	.0074	.0417
12					.0001	.0016	.0139
13						.0003	.0032
14							.0005
15							

FUNCTIONS OF SEVERAL VARIABLES

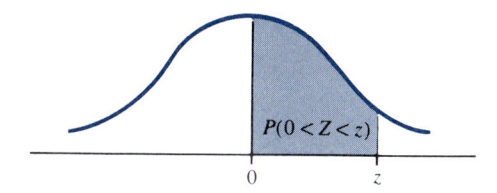

$P(0 < Z < z)$

TABLE 6 Areas under the Standard Normal Curve

Normal Deviate z	.00	.01	.02	.03	.04	.05	.06	.07	.08	.09
0.0	.0000	.0040	.0080	.0120	.0160	.0199	.0239	.0279	.0319	.0359
0.1	.0398	.0438	.0478	.0517	.0557	.0596	.0636	.0675	.0714	.0753
0.2	.0793	.0832	.0871	.0910	.0948	.0987	.1026	.1064	.1103	.1141
0.3	.1179	.1217	.1255	.1293	.1331	.1368	.1406	.1443	.1480	.1517
0.4	.1554	.1591	.1628	.1664	.1700	.1736	.1772	.1808	.1844	.1879
0.5	.1915	.1950	.1985	.2019	.2054	.2088	.2123	.2157	.2190	.2224
0.6	.2257	.2291	.2324	.2357	.2389	.2422	.2454	.2486	.2517	.2549
0.7	.2580	.2611	.2642	.2673	.2704	.2734	.2764	.2794	.2823	.2852
0.8	.2881	.2910	.2939	.2967	.2995	.3023	.3051	.3078	.3106	.3133
0.9	.3159	.3186	.3212	.3238	.3264	.3289	.3315	.3340	.3365	.3389
1.0	.3413	.3438	.3461	.3485	.3508	.3531	.3554	.3577	.3599	.3621
1.1	.3643	.3665	.3686	.3708	.3729	.3749	.3770	.3790	.3810	.3830
1.2	.3849	.3869	.3888	.3907	.3925	.3944	.3962	.3980	.3997	.4015
1.3	.4032	.4049	.4066	.4082	.4099	.4115	.4131	.4147	.4162	.4177
1.4	.4192	.4207	.4222	.4236	.4251	.4265	.4279	.4292	.4306	.4319
1.5	.4332	.4345	.4357	.4370	.4382	.4394	.4406	.4418	.4429	.4441
1.6	.4452	.4463	.4474	.4484	.4495	.4505	.4515	.4525	.4535	.4545
1.7	.4554	.4564	.4573	.4582	.4591	.4599	.4608	.4616	.4625	.4633
1.8	.4641	.4649	.4656	.4664	.4671	.4678	.4686	.4693	.4699	.4706
1.9	.4713	.4719	.4726	.4732	.4738	.4744	.4750	.4756	.4761	.4767
2.0	.4772	.4778	.4783	.4788	.4793	.4798	.4803	.4808	.4812	.4817
2.1	.4821	.4826	.4830	.4834	.4838	.4842	.4846	.4850	.4854	.4857
2.2	.4861	.4864	.4868	.4871	.4875	.4878	.4881	.4884	.4887	.4890
2.3	.4893	.4896	.4898	.4901	.4904	.4906	.4909	.4911	.4913	.4916
2.4	.4918	.4920	.4922	.4925	.4927	.4929	.4931	.4932	.4934	.4936
2.5	.4938	.4940	.4941	.4943	.4945	.4946	.4948	.4949	.4951	.4952
2.6	.4953	.4955	.4956	.4957	.4959	.4960	.4961	.4962	.4963	.4964
2.7	.4965	.4966	.4967	.4968	.4969	.4970	.4971	.4972	.4973	.4974
2.8	.4974	.4975	.4976	.4977	.4977	.4978	.4979	.4979	.4980	.4981
2.9	.4981	.4982	.4982	.4983	.4984	.4984	.4985	.4985	.4986	.4986
3.0	.4987	.4987	.4987	.4988	.4988	.4989	.4989	.4989	.4990	.4990

Adapted by permission from Ernest Kurnow, Gerald J. Glasser, and Frederick R. Ottman. *Statistics for Business Decisions* (Homewood, Ill.: Richard D. Irwin, 1959), p. 501. © 1959 by Richard D. Irwin, Inc.

TABLE 7 Powers of e

x	e^x	e^{-x}	x	e^x	e^{-x}	x	e^x	e^{-x}	x	e^x	e^{-x}
0.00	1.00000	1.00000	0.42	1.52196	0.65705	0.84	2.31637	0.43171	1.26	3.52542	0.28365
0.01	1.01005	0.99005	0.43	1.53726	0.65051	0.85	2.33965	0.42741	1.27	3.56085	0.28083
0.02	1.02020	0.98020	0.44	1.55271	0.64404	0.86	2.36316	0.42316	1.28	3.59664	0.27804
0.03	1.03045	0.97045	0.45	1.56831	0.63763	0.87	2.38691	0.41895	1.29	3.63279	0.27527
0.04	1.04081	0.96079	0.46	1.58407	0.63128	0.88	2.41090	0.41478	1.30	3.66930	0.27253
0.05	1.05127	0.95123	0.47	1.59999	0.62500	0.89	2.43513	0.41066	1.31	3.70617	0.26982
0.06	1.06184	0.94176	0.48	1.61607	0.61878	0.90	2.45960	0.40657	1.32	3.74342	0.26714
0.07	1.07251	0.93239	0.49	1.63232	0.61263	0.91	2.48432	0.40252	1.33	3.78104	0.26448
0.08	1.08329	0.92312	0.50	1.64872	0.60653	0.92	2.50929	0.39852	1.34	3.81904	0.26185
0.09	1.09417	0.91393	0.51	1.66529	0.60050	0.93	2.53451	0.39455	1.35	3.85743	0.25924
0.10	1.10517	0.90484	0.52	1.68203	0.59452	0.94	2.55998	0.39063	1.36	3.89619	0.25666
0.11	1.11628	0.89583	0.53	1.69893	0.58861	0.95	2.58571	0.38674	1.37	3.93535	0.25411
0.12	1.12750	0.88692	0.54	1.71601	0.58275	0.96	2.61170	0.38289	1.38	3.97490	0.25158
0.13	1.13883	0.87810	0.55	1.73325	0.57695	0.97	2.63794	0.37908	1.39	4.01485	0.24908
0.14	1.15027	0.86936	0.56	1.75067	0.57121	0.98	2.66446	0.37531	1.40	4.05520	0.24660
0.15	1.16183	0.86071	0.57	1.76827	0.56553	0.99	2.69123	0.37158	1.41	4.09596	0.24414
0.16	1.17351	0.85214	0.58	1.78604	0.55990	1.00	2.71828	0.36788	1.42	4.13712	0.24171
0.17	1.18531	0.84366	0.59	1.80399	0.55433	1.01	2.74560	0.36422	1.43	4.17870	0.23931
0.18	1.19722	0.83527	0.60	1.82212	0.54881	1.02	2.77320	0.36059	1.44	4.22070	0.23693
0.19	1.20925	0.82696	0.61	1.84043	0.54335	1.03	2.80107	0.35701	1.45	4.26312	0.23457
0.20	1.22140	0.81873	0.62	1.85893	0.53794	1.04	2.82922	0.35345	1.46	4.30596	0.23224
0.21	1.23368	0.81058	0.63	1.87761	0.53259	1.05	2.85765	0.34994	1.47	4.34924	0.22993
0.22	1.24608	0.80252	0.64	1.89648	0.52729	1.06	2.88637	0.34646	1.48	4.39295	0.22764
0.23	1.25860	0.79453	0.65	1.91554	0.52205	1.07	2.91538	0.34301	1.49	4.43710	0.22537
0.24	1.27125	0.78663	0.66	1.93479	0.51685	1.08	2.94468	0.33960	1.50	4.4817	0.22313
0.25	1.28403	0.77880	0.67	1.95424	0.51171	1.09	2.97427	0.33622	1.55	4.7115	0.21225
0.26	1.29693	0.77105	0.68	1.97388	0.50662	1.10	3.00417	0.33287	1.60	4.9530	0.20190
0.27	1.30996	0.76338	0.69	1.99372	0.50158	1.11	3.03436	0.32956	1.65	5.2070	0.19205
0.28	1.32313	0.75578	0.70	2.01375	0.49659	1.12	3.06485	0.32628	1.70	5.4739	0.18268
0.29	1.33643	0.74826	0.71	2.03399	0.49164	1.13	3.09566	0.32303	1.75	5.7546	0.17377
0.30	1.34986	0.74082	0.72	2.05443	0.48675	1.14	3.12677	0.31982	1.80	6.0496	0.16530
0.31	1.36343	0.73345	0.73	2.07508	0.48191	1.15	3.15819	0.31664	1.85	6.3598	0.15724
0.32	1.37713	0.72615	0.74	2.09594	0.47711	1.16	3.18993	0.31349	1.90	6.6859	0.14957
0.33	1.39097	0.71892	0.75	2.11700	0.47237	1.17	3.22199	0.31037	1.95	7.0287	0.14227
0.34	1.40495	0.71177	0.76	2.13828	0.46767	1.18	3.25437	0.30728	2.00	7.3891	0.13534
0.35	1.41907	0.70469	0.77	2.15977	0.46301	1.19	3.28708	0.30422	2.05	7.7679	0.12873
0.36	1.43333	0.69768	0.78	2.18147	0.45841	1.20	3.32012	0.30119	2.10	8.1662	0.12246
0.37	1.44773	0.69073	0.79	2.20340	0.45384	1.21	3.35348	0.29820	2.15	8.5849	0.11648
0.38	1.46228	0.68386	0.80	2.22554	0.44933	1.22	3.38719	0.29523	2.20	9.0250	0.11080
0.39	1.47698	0.67706	0.81	2.24791	0.44486	1.23	3.42123	0.29229	2.25	9.4877	0.10540
0.40	1.49182	0.67032	0.82	2.27050	0.44043	1.24	3.45561	0.28938	2.30	9.9742	0.10026
0.41	1.50682	0.66365	0.83	2.29332	0.43605	1.25	3.49034	0.28650	2.35	10.4856	0.09537

x	e^x	e^{-x}	x	e^x	e^{-x}
2.40	11.0232	0.09072	5.80	330.300	0.00303
2.45	11.5883	0.08629	5.90	365.038	0.00274
2.50	12.1825	0.08209	6.00	403.429	0.00248
2.55	12.8017	0.07808	6.10	445.858	0.00224
2.60	13.4637	0.07427	6.20	492.749	0.00203
2.65	14.1540	0.07065	6.30	544.572	0.00184
2.70	14.8797	0.06721	6.40	601.845	0.00166
2.75	15.6426	0.06393	6.50	665.142	0.00150
2.80	16.4447	0.06081	6.60	735.09	0.00136
2.85	17.2878	0.05784	6.70	812.41	0.00123
2.90	18.1741	0.05502	6.80	897.85	0.00111
2.95	19.1060	0.05234	6.90	992.27	0.00101
3.00	20.0855	0.04979	7.00	1096.63	0.00091
3.05	21.1153	0.04736	7.10	1211.97	0.00083
3.10	22.1980	0.04505	7.20	1339.43	0.00075
3.15	23.3361	0.04285	7.30	1480.30	0.00068
3.20	24.533	0.04076	7.40	1635.98	0.00061
3.25	25.790	0.03877	7.50	1808.04	0.00055
3.30	27.113	0.03688	7.50	1808.04	0.00055
3.35	28.503	0.03508	7.60	1998.20	0.00050
3.40	29.964	0.03337	7.70	2208.35	0.00045
3.50	33.115	0.03020	7.80	2440.60	0.00041
3.60	36.598	0.02732	7.90	2697.28	0.00037
3.70	40.447	0.02472	8.00	2980.96	0.00034
3.80	44.701	0.02237	8.10	3294.47	0.00030
3.90	49.402	0.02024	8.20	3640.95	0.00027
4.00	54.598	0.01832	8.30	4023.87	0.00025
4.10	60.340	0.01657	8.40	4447.07	0.00022
4.20	66.686	0.01500	8.50	4914.77	0.00020
4.30	73.700	0.01357	8.60	5431.66	0.00018
4.40	81.451	0.01228	8.70	6002.91	0.00017
4.50	90.017	0.01111	8.80	6634.25	0.00015
4.60	99.484	0.01005	8.90	7331.97	0.00014
4.70	109.947	0.00910	9.00	8103.08	0.00012
4.80	121.510	0.00823	9.10	8955.29	0.00011
4.90	134.290	0.00745	9.20	9897.13	0.00010
5.00	148.413	0.00674	9.30	10938.0	0.00009
5.10	164.022	0.00610	9.40	12088.4	0.00008
5.20	181.272	0.00552	9.50	13359.7	0.00007
5.30	200.337	0.00499	9.60	14764.8	0.00007
5.40	221.406	0.00452	9.70	16317.6	0.00006
5.50	244.692	0.00409	9.80	18033.7	0.00006
5.60	270.427	0.00370	9.90	19930.4	0.00005
5.70	298.867	0.00335			

TABLE 8 Natural Logarithms

x	ln x	x	ln x	x	ln x	x	ln x	x	ln x	x	ln x	x	ln x
0.01	−4.6052	0.51	−0.6733	1.01	0.0100	1.51	0.4121	2.01	0.6981	2.55	0.9361	5.05	1.6194	7.55	2.0215	10.05	2.3076
0.02	−3.9120	0.52	−0.6539	1.02	0.0198	1.52	0.4187	2.02	0.7031	2.60	0.9555	5.10	1.6292	7.60	2.0281	10.10	2.3125
0.03	−3.5066	0.53	−0.6349	1.03	0.0296	1.53	0.4253	2.03	0.7080	2.65	0.9746	5.15	1.6390	7.65	2.0347	10.15	2.3175
0.04	−3.2189	0.54	−0.6162	1.04	0.0392	1.54	0.4318	2.04	0.7130	2.70	0.9933	5.20	1.6487	7.70	2.0412	10.20	2.3224
0.05	−2.9957	0.55	−0.5978	1.05	0.0488	1.55	0.4383	2.05	0.7178	2.75	1.0116	5.25	1.6582	7.75	2.0477	10.25	2.3273
0.06	−2.8134	0.56	−0.5798	1.06	0.0583	1.56	0.4447	2.06	0.7227	2.80	1.0296	5.30	1.6677	7.80	2.0541	10.30	2.3321
0.07	−2.6593	0.57	−0.5621	1.07	0.0677	1.57	0.4511	2.07	0.7275	2.85	1.0473	5.35	1.6771	7.85	2.0605	10.35	2.3370
0.08	−2.5257	0.58	−0.5447	1.08	0.0770	1.58,	0.4574	2.08	0.7324	2.90	1.0647	5.40	1.6864	7.90	2.0669	10.40	2.3418
0.09	−2.4079	0.59	−0.5276	1.09	0.0862	1.59	0.4637	2.09	0.7372	2.95	1.0818	5.45	1.6956	7.95	2.0732	10.45	2.3466
0.10	−2.3026	0.60	−0.5108	1.10	0.0953	1.60	0.4700	2.10	0.7419	3.00	1.0986	5.50	1.7047	8.00	2.0794	10.50	2.3514
0.11	−2.2073	0.61	−0.4943	1.11	0.1044	1.61	0.4762	2.11	0.7467	3.05	1.1151	5.55	1.7138	8.05	2.0857	10.55	2.3561
0.12	−2.1203	0.62	−0.4780	1.12	0.1133	1.62	0.4824	2.12	0.7514	3.10	1.1314	5.60	1.7228	8.10	2.0919	10.60	2.3609
0.13	−2.0402	0.63	−0.4620	1.13	0.1222	1.63	0.4886	2.13	0.7561	3.15	1.1474	5.65	1.7317	8.15	2.0980	10.65	2.3656
0.14	−1.9661	0.64	−0.4463	1.14	0.1310	1.64	0.4947	2.14	0.7608	3.20	1.1632	5.70	1.7405	8.20	2.1041	10.70	2.3702
0.15	−1.8971	0.65	−0.4308	1.15	0.1398	1.65	0.5008	2.15	0.7655	3.25	1.1787	5.75	1.7492	8/25	2.1102	10.75	2.3749
0.16	−1.8326	0.66	−0.4155	1.16	0.1484	1.66	0.5068	2.16	0.7701	3.30	1.1939	5.80	1.7579	8.30	2.1163	10.80	2.3795
0.17	−1.7720	0.67	−0.4005	1.17	0.1570	1.67	0.5128	2.17	0.7747	3.35	1.2090	5.85	1.7664	8.35	2.1223	10.85	2.3842
0.18	−1.7148	0.68	−0.3857	1.18	0.1655	1.68	0.5188	2.18	0.7793	3.40	1.2238	5.90	1.7750	8.40	2.1282	10.90	2.3888
0.19	−1.6607	0.69	−0.3711	1.19	0.1740	1.69	0.5247	2.19	0.7839	3.45	1.2384	5.95	1.7834	8.45	2.1342	10.95	2.3933
0.20	−1.6094	0.70	−0.3567	1.20	0.1823	1.70	0.5306	2.20	0.7885	3.50	1.2528	6.00	1.7918	8.50	2.1401	11.00	2.3979
0.21	−1.5606	0.71	−0.3425	1.21	0.1906	1.71	0.5365	2.21	0.7930	3.55	1.2669	6.05	1.8001	8.55	2.1459	11.05	2.4024
0.22	−1.5141	0.72	−0.3285	1.22	0.1989	1.72	0.5423	2.22	0.7975	3.60	1.2809	6.10	1.8083	8.60	2.1518	11.10	2.4069
0.23	−1.4697	0.73	−0.3147	1.23	0.2070	1.73	0.5481	2.23	0.8020	3.65	1.2947	6.15	1.8165	8.65	2.1576	11.15	2.4114
0.24	−1.4271	0.74	−0.3011	1.24	0.2151	1.74	0.5539	2.24	0.8065	3.70	1.3083	6.20	1.8245	8.70	2.1633	11.20	2.4159
0.25	−1.3863	0.75	−0.2877	1.25	0.2231	1.75	0.5596	2.25	0.8109	3.75	1.3218	6.25	1.8326	8.75	2.1691	11.25	2.4204

TABLE 8 *(continued)*

x	$\ln x$	x	$\ln x$	x	$\ln x$	x	$\ln x$	x	$\ln x$	x	$\ln x$	x	$\ln x$
0.26	−1.3471	0.76	−0.2744	1.26	0.2311	1.76	0.5653	2.26	0.8154	3.80	1.3350	6.30	1.8405	8.80	2.1748	11.30	2.4248
0.27	−1.3039	0.77	−0.2614	1.27	0.2390	1.77	0.5710	2.27	0.8198	3.85	1.3481	6.35	1.8485	8.85	2.1804	11.35	2.4292
0.28	−1.2730	0.78	−0.2485	1.28	0.2469	1.78	0.5766	2.28	0.8242	3.90	1.3610	6.40	1.8563	8.90	2.1861	11.40	2.4336
0.29	−1.2379	0.79	−0.2357	1.29	0.2546	1.79	0.5822	2.29	0.8286	3.95	1.3737	6.45	1.8641	8.95	2.1917	11.45	2.4380
0.30	−1.2040	0.80	−0.2231	1.30	0.2624	1.80	0.5878	2.30	0.8329	4.00	1.3863	6.50	1.8718	9.00	2.1972	11.50	2.4423
0.31	−1.1712	0.81	−0.2107	1.31	0.2700	1.81	0.5933	2.31	0.8372	4.05	1.3987	6.55	1.8795	9.05	2.2028	13.0	2.5649
0.32	−1.1394	0.82	−0.1985	1.32	0.2776	1.82	0.5988	2.32	0.8416	4.10	1.4110	6.60	1.8871	9.10	2.2083	17.0	2.8332
0.33	−1.1087	0.83	−0.1863	1.33	0.2852	1.83	0.6043	2.33	0.8459	4.15	1.4231	6.65	1.8946	9.15	2.2138	19.0	2.9444
0.34	−1.0788	0.84	−0.1744	1.34	0.2927	1.84	0.6098	2.34	0.8502	4.20	1.4351	6.70	1.9021	9.20	2.2192	23.0	3.1355
0.35	−1.0498	0.85	−0.1625	1.35	0.3001	1.85	0.6152	2.35	0.8544	4.25	1.4469	6.75	1.9095	9.25	2.2246	29.0	3.3673
0.36	−1.0217	0.86	−0.1508	1.36	0.3075	1.86	0.6206	2.36	0.8587	4.30	1.4586	6.80	1.9169	9.30	2.2300	31.0	3.4340
0.37	−0.9943	0.87	−0.1393	1.37	0.3148	1.87	0.6259	2.37	0.8629	4.35	1.4702	6.85	1.9242	9.35	2.2354	37.0	3.6109
0.38	−0.9676	0.88	−0.1278	1.38	0.3221	1.88	0.6313	2.38	0.8671	4.40	1.4816	6.90	1.9315	9.40	2.2407	41.0	3.7136
0.39	−0.9416	0.89	−0.1165	1.39	0.3293	1.89	0.6366	2.39	0.8713	4.45	1.4929	6.95	1.9387	9.45	2.2460	43.0	3.7612
0.40	−0.9163	0.90	−0.1054	1.40	0.3365	1.90	0.6419	2.40	0.8755	4.50	1.5041	7.00	1.9459	9.50	2.2513	47.0	3.8501
0.41	−0.8916	0.91	−0.0943	1.41	0.3436	1.91	0.6471	2.41	0.8796	4.55	1.5151	7.05	1.9530	9.55	2.2565	53.0	3.9703
0.42	−0.8675	0.92	−0.0834	1.42	0.3507	1.92	0.6523	2.42	0.8838	4.60	1.5261	7.10	1.9601	9.60	2.2618	59.0	4.0775
0.43	−0.8440	0.93	−0.0726	1.43	0.3577	1.93	0.6575	2.43	0.8879	4.65	1.5369	7.15	1.9671	9.65	2.2670	61.0	4.1109
0.44	−0.8210	0.94	−0.0619	1.44	0.3646	1.94	0.6627	2.44	0.8920	4.70	1.5476	7.20	1.9741	9.70	2.2721	67.0	4.2047
0.45	−0.7985	0.95	−0.0513	1.45	0.3716	1.95	0.6678	2.45	0.8961	4.75	1.5581	7.25	1.9810	9.75	2.2773	71.0	4.2627
0.46	−0.7765	0.96	−0.0408	1.46	0.3784	1.96	0.6729	2.46	0.9002	4.80	1.5686	7.30	1.9879	9.80	2.2824	73.0	4.2905
0.47	−0.7550	0.97	−0.0305	1.47	0.3853	1.97	0.6780	2.47	0.9042	4.85	1.5790	7.35	1.9947	9.85	2.2875	79.0	4.3694
0.48	−0.7340	0.98	−0.0202	1.48	0.3920	1.98	0.6831	2.48	0.9083	4.90	1.5892	7.40	2.0015	9.90	2.2925	83.0	4.4188
0.49	−0.7134	0.99	−0.0101	1.49	0.3988	1.99	0.6881	2.49	0.9123	4.95	1.5994	7.45	2.0082	9.95	2.2976	89.0	4.4886
0.50	−0.6931	1.00	0.0000	1.50	0.4055	2.00	0.6931	2.50	0.9163	5.00	1.6094	7.50	2.0149	10.00	2.3026	97.0	4.5747

Solutions to Exercises

1.1 EXERCISES

1. integers, rational, real **3.** rational, real **5.** irrational, real **7.** rational, real **9.** rational, real

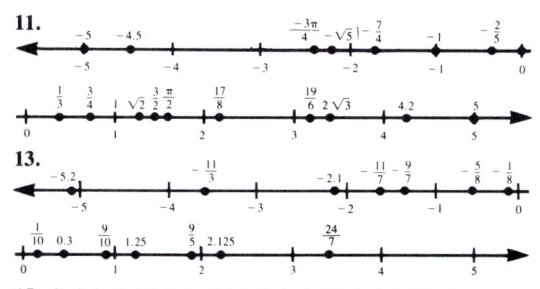

11. [number line]

13. [number line]

[number line]

15. 0, 0.1, 0.15, 0.3, 0.35, 0.5, 0.55, 0.6, 0.75, 0.9, 0.95, 1 **17.** True **19.** False **21.** True **23.** True **25.** True **27.** True **29.** True **31.** False **33.** $S \geq \$1{,}500{,}000$ **35.** $S > \$1{,}500{,}000$ **37.** $E \geq 100$ tons **39.** $0.10 < R < 0.125$ **41.** $\$0.20 < E < \$0.22$ **43.** $r < 0.5M$ **45.** (4, 8) **47.** $(-\infty, 2/3)$ **49.** $[-3, 3/4]$ **51.** $[4, \infty]$ **53.** $-6 < x \leq 3$

[number line]

55. $-1/2 \leq x \leq 3$ **57.** $x > 2$

[number lines]

59. $0, 292/293, 0.025, 1/\sqrt{2}, \pi/4, 0.9999\ldots, 1$

61. $C = \begin{cases} \$0.40 & \text{if } x < 1000 \\ \$0.35 & \text{if } x \geq 1000 \end{cases}$

63. $P = \begin{cases} \$4.00 & \text{if } x < 20 \\ \$4.50 & \text{if } 20 \leq x \leq 40 \\ \$5.00 & \text{if } x > 40 \end{cases}$

65. $d = \begin{cases} 200 & \text{if } x \leq 14 \\ 120 & \text{if } 15 \leq x \leq 30 \\ 100 & \text{if } x > 30 \end{cases}$ **67.** 3 **69.** 4/3 **71.** 5

73. 0 **75.** -8 **77.** 14 **79.** -14 **81.** Change if |actual speed $-$ desired speed| ≥ 1 **83.** $|220 - w| \leq 0.25$

1.2 EXERCISES

1. 2 **3.** -8 **5.** 1 **7.** 2.5 **9.** 1 **11.** 3 **13.** 2 **15.** $-92/25$ **17.** 8 **19.** $-1/8$ **21.** $-25/17$ **23.** $\$1320$ **25.** $\$847.45$ **27.** 2.5 years **29.** 12.5% **31.** 100°C **33.** $55/3$°C **35.** 212°F **37.** 50°F **39.** $l = 10, w = 5$ **41.** $l = 33, w = 18$ **43.** Bette $\$6.00$; Al $\$6.25$ **45.** $\$23{,}500$ in stocks; $\$26{,}500$ in CDs **47.** 10 P.M. **49.** 5.4375 hours after first train leaves **51.** 134.4 mph **53.** 8 gallons of 15%; 12 gallons of 10% **55.** 3 liters **57.** 2.9 lb of hazelnuts **59.** $\$348{,}000$ of profit spent on product 1; $\$452{,}000$ of profit spent on product 2

1.3 EXERCISES

1. $x < 3; (-\infty, 3)$ **3.** $x \leq -2; (-\infty, -2]$ **5.** $7 < x; (7, +\infty)$ **7.** $x \leq -2; (-\infty, -2]$ **9.** $15/6 < x; (15/6, +\infty)$ **11.** $6 < x; (6, +\infty)$ **13.** $-5 \geq x; (-\infty, -5)$ **15.** $0 \leq 1; (-\infty, +\infty)$ **17.** $-2/11 \geq x; (-\infty, -2/11]$ **19.** $x > -3/2; (-3/2, +\infty)$ **21.** $-84 \leq x; [-84, +\infty)$ **23.** $x > 200; (200, +\infty)$ **25.** $x > \dfrac{202{,}500}{7.25}$ **27.** 284,374 **29.** $[0, 5/2)$ **31.** $[0, 2.84)$ **33.** $x \geq 800$ **35.** $[0, 11]$ **37.** $x \geq 39.2$ **39.** $x \geq 10.625$ **41.** $x = 16$ **43.** Plan A is cheaper for 42 minutes or less; Plan B is cheaper for 43 minutes or more

1.4 EXERCISES

1. 32 **3.** 1/3 **5.** 1 **7.** 1 **9.** 81/16 **11.** 0.729 **13.** 256 **15.** 1/343 **17.** 9/16 **19.** 144 **21.** 125 **23.** 390,625 **25.** 1/36 **27.** 1/225 **29.** 16 **31.** $125/8$ **33.** $35x^5$ **35.** $\dfrac{36x^3}{y \cdot z^6}$ **37.** $\dfrac{3y^2z}{2z^3}$ **39.** $4 \cdot x^8$ **41.** $\dfrac{1}{2x^2y^3z}$ **43.** 2 **45.** 1/8 **47.** $\sqrt{5}$ **49.** 1/729 **51.** 1024 **53.** 1/3 **55.** 27 **57.** 531,441 **59.** 20 **61.** $\dfrac{10}{x}$ **63.** $\dfrac{y}{2x^{1/3}}$ **65.** $\dfrac{x^2 \cdot y^{5/2}}{z}$ **67.** 5000 **69.** 41.2 **71.** $40\sqrt{5}$ **73.** 1.84 times more active **75.** 1.15 times more active **77.** $0.4690 < P < 0.5310$ **79.** No

1.5 EXERCISES

1. $5x - 2$ **3.** $3x^2 + 9x + 4$ **5.** $7t^2 + 5t - 5$
7. $-2x^3 + 8x^2 + 2x + 7$ **9.** $-3x^2 + 4x$ **11.** $t^2 - 4t - 21$ **13.** $6x^2 + 11x - 7$ **15.** $5p^3 - 2p^2 + 5p - 2$
17. $12y^3 + y^2 - y$ **19.** $2x^3 - x^2 - 2x + 1$ **21.** $x^4 + x^3 - 7x^2 + 7x - 2$ **23.** $2(x + 2)$ **25.** $18(y + 3)$
27. $2x + 1$ **29.** $y^2(y + 1)$ **31.** $15x^3(14x^2 + 1)$ **33.** $x(x - 5)$ **35.** $(x - 4)(x + 4)$ **37.** $(x + 2)(x - 7)$ **39.** $(p + 3)^2$ **41.** $(x + 3)(x - 9)$ **43.** $(-x + 2)(x - 9)$
45. $x^2 + 1$ **47.** $3(x + 5)(x - 4)$ **49.** $2(x + 8)(x - 6)$
51. $(2x + 1)(x - 1)$ **53.** $(x - 2)(2x + 3)$ **55.** $(2x - 3)(2x + 1)$ **57.** $(x - 1)(x^2 + x + 1)$ **59.** $(x + 2)(x^2 - 2x + 4)$ **61.** $x(x + 5)^2$ **63.** $x(x - 11)(x + 2)$
65. $5x(10 - x)(10 + x)$ **67.** $5x^2(x - 6)(x + 2)$ **69.** $x^2(x - 4)^2$ **71.** $10x^2(x - 1)(x + 1)$ **73.** $x(3x + 5)(9x^2 - 15x + 25)$ **75.** $x^2(4x + 3)(3x^2 + 4)$

1.6 EXERCISES

1. $x = -2, x = +2$ **3.** no solution **5.** $x = -3$ **7.** $x = 2, x = -5$ **9.** $p = 0, p = -65$ **11.** $x = 2, x = -\frac{1}{2}$ **13.** $x = 0, x = 1$ **15.** $x = 1$ **17.** $x = 0, x = 1$
19. $x = 0$ **21.** $x = 6, x = -4$ **23.** $x = 4, x = -4$
25. $x = -4, x = 2$ **27.** $x = -5, x = -\frac{3}{2}$ **29.** $x = -\frac{2}{3}, x = \frac{1}{3}$ **31.** $x = \frac{1}{2}$ **33.** no solution **35.** no solution. **37.** $x = 3 \pm 2\sqrt{2}$ **39.** $x = 1$ **41.** $x = -2 \pm \sqrt{7}$ **43.** 5.625 seconds **45.** 2.8125 seconds **47.** 85 units **49.** $l = 80, w = 40$ **51.** $x = 5$ inches **53.** $p = \$112$ or $p = \$88$

1.7 EXERCISES

1. $\frac{x}{6}$ **3.** $4x$ **5.** $\frac{x+2}{x}$ **7.** $\frac{q+1}{q^2-2}$ **9.** $\frac{x^2-1}{2x}$
11. $\frac{x+3}{x-1}$ **13.** $-\frac{1}{x}$ **15.** $\frac{3x+4}{x-1}$ **17.** $\frac{x^2}{15}$ **19.** $\frac{x-2}{x(x+3)}$
21. $\frac{x^2+1}{4(x-2)}$ **23.** $\frac{5}{3}$ **25.** $\frac{49}{8x}$ **27.** $\frac{x+2}{x-3}$ **29.** $\frac{5x}{6}$
31. $\frac{12-x^2}{6x}$ **33.** $\frac{2x+7}{x^2}$ **35.** $\frac{x(3x-1)}{(x-1)(x+1)}$
37. $\frac{(x+1)(x^2+3x-1)}{x(x-1)(x+2)}$ **39.** $x = 0$ **41.** $x = \frac{3}{2}$ **43.** $x = \pm 1$ **45.** $x = 3, x = 1$ **47.** $x = -7, x = 5$ **49.** $x = -\frac{1}{3}$ **51.** $t = 45$ **53a.** $x = 35{,}000$ **b.** $x = 21{,}000{,}000$ **55a.** $t = 2$ **b.** $t = 8$

REVIEW EXERCISES

1. True **2.** False **3.** True **4.** False **5.** True **6.** False
7. False **8.** True **9.** True **10.** False **11.** False
12. True **13.** True **14.** True **15.** True **16.** True
17. $(-2, 17]$
18. $(-\infty, \frac{4}{9}]$
19. $(-2, 8)$
20. $(\sqrt{2}, \infty)$
21. $5 < x \le 6$
22. $x > -2$
23. $-\infty < x \le 3$
24. $-4 < x < 4$
25. $x \le 25$ **26.** $T \ge 20°$ **27.** $p < \$2$ **28.** $\$2 \le p < \$2.25$ **29.** $x = 5$ **30.** $x = -\frac{1}{2}$ **31.** $x = \frac{4}{5}$ **32.** $x = 3$ **33.** $x = -3$ **34.** $x = -\frac{43}{6}$ **35.** $x = 460$
36. $\$1250$ on card charging 1.3%; $\$750$ on card charging 1.4% **37.** $x = 300$ **38.** 72 seconds after second racer crosses line; 3.7 miles **39.** $x < 1; (-\infty, 1)$ **40.** $x \le -2; (-\infty, -2]$ **41.** $2 \le x; [2, +\infty]$ **42.** $x < -4; (-\infty, -4)$
43. $\frac{21}{43} < x; (\frac{21}{43}, +\infty)$ **44.** $x \le -14; (-\infty, -14]$
45. $5 < 26{,}000; [0, 26{,}000)$ **46.** $x \ge 58{,}400$ units
47. $50 < x$ **48.** 1.25 seconds, 12.5 meters **49.** 0.04
50. $\frac{1}{9}$ **51.** 9 **52.** $\frac{234}{4}$ **53.** $\frac{3}{64}$ **54.** $\frac{2}{125}$ **55.** 1331
56. $\frac{1}{3600}$ **57.** $\frac{21}{x^2}$ **58.** $\frac{48z^9}{xy^6}$ **59.** $\frac{4x}{3y^6}$ **60.** $\frac{6xy^3}{5z^2}$ **61.** $\frac{9x^4z^2}{y^6}$
62. $\frac{x^6w^3}{8y^{12}z^3}$ **63.** $\frac{x^2}{zy^{1/2}}$ **64.** $\frac{3xz^{3/5}}{2y^{5/3}}$ **65.** $13x^2 - x + 10$
66. $9x^3 - 5x^2 - 11x - 1$ **67.** $12x^2 + 33x - 9$
68. $10x^3 - 17x^2 + 5x - 3$ **69.** $-3x(x - 2)$
70. $2x^2(6x^2 + 1)$ **71.** $(x + 3)(x^2 - 3x + 9)$ **72.** $(3x - 5)(9x^2 + 15x + 25)$ **73.** $(x - 7)(x - 7)$ **74.** $(3x + 5)(3x - 5)$ **75.** $(x + 6)(x - 4)$ **76.** $(x + 9)(x - 2)$
77. $(2x + 5)(x - 3)$ **78.** $x(3x - 2)(2x + 5)$ **79.** no solution **80.** $x = 8, x = -8$ **81.** $x = 0, x = 8$ **82.** $x = 11, x = -9$ **83.** $x = -13, x = -4$ **84.** $x = \frac{5}{2}, x = -\frac{3}{2}$ **85.** $x = -9$ **86.** $x = 3 \pm \sqrt{11}$ **87.** no solution **88.** $x = 1 \pm \sqrt{6}$ **89.** $l \approx 110.5$ meters; $w \approx 90.5$ meters **90.** 3 inches **91.** $p = \$15, p = \$5$
92. $\frac{3x+1}{2}$ **93.** $\frac{x-3}{2}$ **94.** $\frac{7x+2}{3x-1}$ **95.** $\frac{x+2}{x^2}$
96. $\frac{x+3}{x+4}$ **97.** $2x - 3$ **98.** $\frac{-7x}{15}$ **99.** $\frac{7}{x}$ **100.** $\frac{x^3-6}{3x}$
101. $\frac{5x^4+3}{x^2}$ **102.** $\frac{(2x+1)(x+1)}{x(x-1)}$
103. $\frac{x^4-5x^2+6x-10}{x^2(x^2+2)}$ **104.** $\frac{x^7}{15}$ **105.** $\frac{6}{x^3}$ **106.** $\frac{x+5}{x-1}$

107. $\dfrac{(x+4)(3x+1)}{(x-1)(2x+1)}$ **108.** $7/2$ **109.** $\dfrac{9x^3}{4}$
110. $\dfrac{x^2+5x+4}{2}$ **111.** $\dfrac{(x+1)^2}{x+4}$ **112.** $x = -7$ **113.** $x = 3$ **114.** $x = -9$ **115.** $x = -9, x = -2$ **116.** $x = -3$ **117.** no solution **118.** $20/3$ years from now

2.1 EXERCISES

1. True **3.** True **5.** False **7.** True **9a.** 2, 106 **b.** 2, -1, 0, 106 **11.** $\{1, 3, 5, 7, \ldots\}$ **13.** $D = \{$Jan, Feb, Mar$\}$ **15.** $\{1, 2, 3, 4, 5, 6, 7\}$ **17.** $\{$The first six months of the year$\}$ **19.** $\{$The natural numbers less than 5$\}$ **21.** $\{$The natural numbers less than 9$\}$ **23.** $A = \{x | x$ is a female college student in the U.S.A.$\}$ **25.** $\{x | x \in \mathbb{Z}, -2 \le x \le 3\}$ **25.** $\{x | x \in \mathbb{Z}, -2 \le x \le 3\}$ **31.** $\{x | x$ is a college that experienced a decline in fundraising in 1986–1987$\}$ **33.** $\{x | x$ is an insolvent insurer$\}$ **35.** False **37.** True **39.** True **41.** False **43.** False **45.** True **47.** True **49.** False **51.** True **53.** False **55.** True **57.** False **59.** False **61.** False **63.** True **65.** True **67.** False **69.** $\emptyset$ **71.** $\emptyset, \{a\}, \{b\}, \{a, b\}$ **73.** $\emptyset, \{a\}, \{b\}, \{c\}, \{d\}, \{a, b\} \{a, c\} \{a, d\}, \{b, c\}, \{b, d\} \{c, d\}. \{a, b, c\} \{a, b, d\} \{a, c, d\}, \{b, c, d\} \{a, b, c, d\}$ **75a.** 128 **b.** $\{$I, T, N, V$\}$, $\{$I, T, S, N, V$\}$, $\{$I, T, S, F, N, V$\}$, $\{$I, T, F, N, V$\}$, $\{$I, S, N, V$\}$, $\{$I, S, F, N, V$\}$, $\{$I, F, N, V$\}$, $\{$R, T, N, V$\}$, $\{$R, T, S, N, V$\}$, $\{$R, T, S, F, N, V$\}$, $\{$R, T, F, N, V$\}$, $\{$R, S, N, V$\}$, $\{$R, S, F, N, V$\}$, $\{$R, F, N, V$\}$

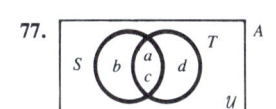

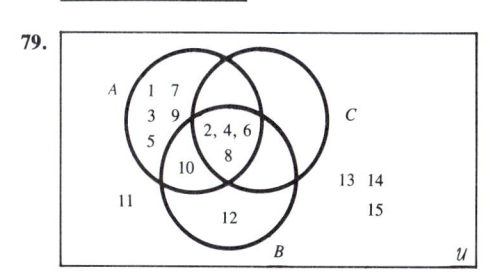

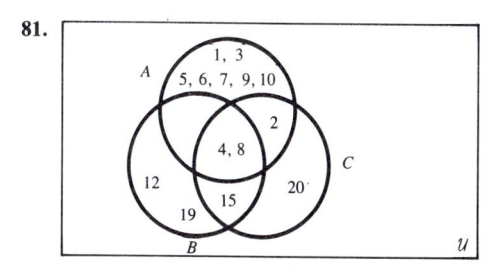

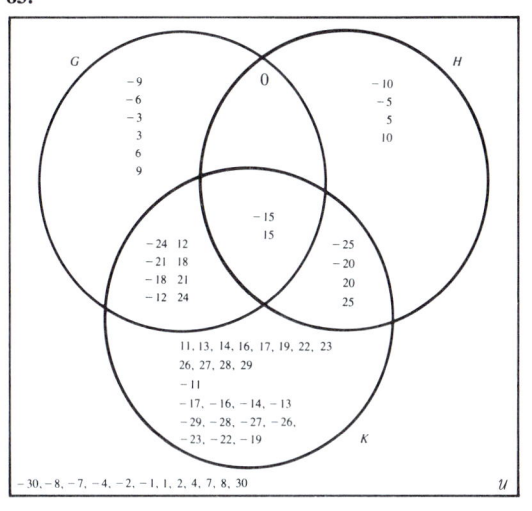

77. [Venn diagram: set S contains b; set T contains d; intersection contains a, c; universe U, outside set A]

79. [Venn diagram with sets A, B, C in universe U: A only: 1, 7, 3, 9, 5; $A \cap C$: 2, 4, 6; $A \cap B \cap C$: 8; $A \cap B$: 10; B only: 11, 12; C only: 13, 14; $B \cap C$: 15]

81. [Venn diagram with sets A, B, C in universe U: A only: 1, 3; $A \cap C$: 5, 6, 7, 9, 10; C only: 2; $A \cap B \cap C$: 4, 8; B only: 12; $B \cap C$: 15; C only: 20; outside: 19]

83. [Venn diagram with sets G, H, K in universe U: G only: $-9, -6, -3, 3, 6, 9$; $G \cap H$: 0; H only: $-10, -5, 5, 10$; $G \cap K$: $-24, 12, -21, 18, -18, 21, -12, 24$; $G \cap H \cap K$: $-15, 15$; $H \cap K$: $-25, -20, 20, 25$; K only: 11, 13, 14, 16, 17, 19, 22, 23, 26, 27, 28, 29, -11, $-17, -16, -14, -13$, $-29, -28, -27, -26$, $-23, -22, -19$; outside: $-30, -8, -7, -4, -2, -1, 1, 2, 4, 7, 8, 30$]

2.2 EXERCISES

1. $\{b, e, f, g, h, i\}$ **3.** $\{a, b, c, d, e, f\}$ **5.** $\{e, h, i\}$ **7.** $\{a, b, f\}$ **9.** $\{a, b, c, d, e, f, g\}$ **11.** $\{a, c, d, h, i\}$ **13.** $\{1, 3, 5, 7, 9, 11, 12, 13, 14, 15\}$ **15.** $\{4, 6, 10\}$ **17.** $\{4, 6\}$ **19.** $\{1, 2, 3, 4, 5, 6, 7, 8, 10, 11, 12, 13\}$ **21.** $\{2, 3, 4, 5, 6, 8, 10\}$ **23.** $\{3, 5, 10, 11, 12, 13\}$ **25.** $\{10\}$ **27.** $[8, \infty)$ **29.** $\emptyset$ **31.** $(10, 20]$ **33.** $(-\infty, 10]$ **35.** $(-\infty, 5)$ **37.** $\{x | x$ is a physician who is not an internist$\}$ **39.** $\{x | x$ is a physician who is an internist or is retired$\}$ **41.** $\{x | x$ is a physician who is not an internist or is retired$\}$ **43.** $\{x | x$ is a physician who is neither an internist nor retired$\}$ **45.** $\{x | x$ is a college student not taking an art course$\}$ **47.** $\{x | x$ is a college student taking both a biology course and a chemistry course$\}$ **49.** $\{x | x$ is a college student taking an art, biology, or chemistry course$\}$ **51.** $\{x | x$ is a college student taking an art course and is not taking a chemistry course$\}$ **57.** $A' = \{$Employees not paid an hourly rate$\}$; $C' = \{$Employees who have two or more years experience with the company$\}$ **59.** $(A \cap C)' = \{$Employees who are not paid an hourly wage and have two or more years service with the company$\}$; $(B \cup D)' = \{$Employees not paid a monthly salary who do not have two through five years service with the company$\}$ **61.** $(A \cap C)'$ **63.** $B \cap C'$ **65.** $\{$Persons not creative in the visual arts$\}$; $\{$Persons who are not immigrants$\}$ **67.** $\{$Persons not creative in the musical arts or are not native-born with both parents native-born$\}$; $\{$Persons not creative in the visual arts or persons who are not native-born but have at least one parent who is an immigrant$\}$ **69.** $V \cap M'$ **71.** $V \cap M \cap I$ **73.** $V' \cap M' \cap I'$

75a. 750 **b.** 500 **c.** 350 **d.** 50 **e.** 300. **f.** No. Only $1/7$ of Dudster residents who do not own a car favor the services.

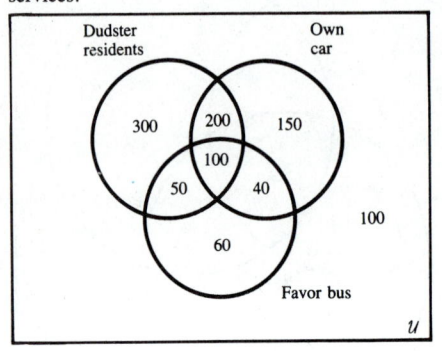

77a. 5 **b.** 10 **c.** 5 **d.** 15

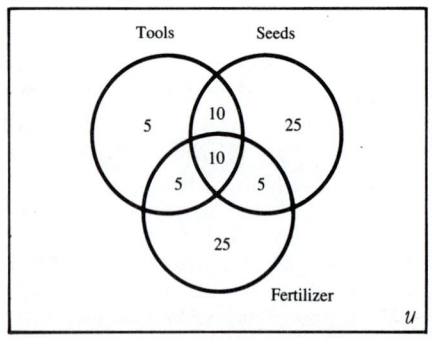

79a. 6 **b.** 11 **c.** 20 **d.** 6 **e.** 16

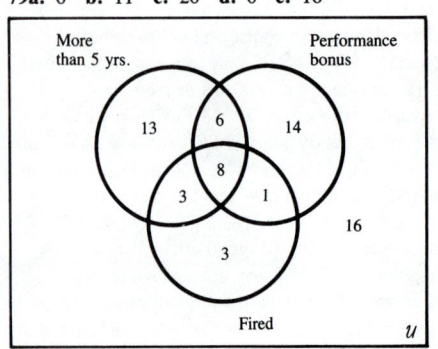

81a. 965 **b.** 360 **c.** 605

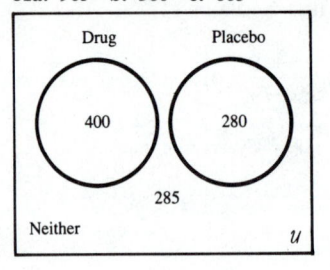

83a. 400 **b.** 420 **c.** 560

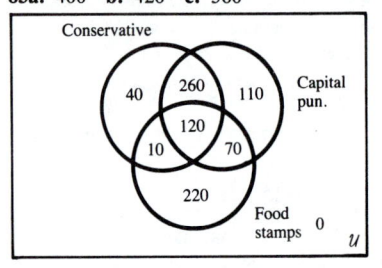

2.3 EXERCISES

1. No **3.** Yes, $R: \{2, 4\}$ **5.** Yes, $R: \{2, 4, 8\}$ **7.** No
9. Yes, $R: \{1\}$ **11.** -2 **13.** -14 **15.** $14/3$ **17.** $4b - 2$
19. $4t^2 - 6$ **21.** $g(1) = 1$, $g(4) = 1/4$, $g(2) = 1/2$, $g(100)$
$= 1/100$ **23.** $g(c) = \dfrac{1}{c}$, $g(c)^3 = \dfrac{1}{c^3}$, $g\left(\dfrac{1}{c}\right) = c$, $c \neq 0$
25. $h(0) = 2$, $h(2) = 0$, $h(1) = 0$, $h(-2) = 12$ **27.** $h(a) = a^2 - 3a + 2$, $h(a + 1) = a^2 - a$, $h(a^2) = a^4 - 3a + 2$ **29.** $y = 24$ **31.** $y = 6\%_{16}$ **33.** $y = q^2 - 2q$
35. $y(-3) = 0.5$, $y(5) = 2.5$, $y(4) = 4$, $y(0) = 0$
37. Case 1: $x = b$; $y = \dfrac{b}{b - 3}$, $b \neq 3$; Case 2: $x = \dfrac{1}{b}$, $b \neq 0$; $y = \dfrac{1}{1 - 3b}$, $b \neq 1/3$ **39.** $k(0) = 2$, $k(3) = 4$, $k(8) = 6$, $k(-1) = 0$ **41.** $k(c - 1) = 2\sqrt{c}$, $c \neq 0$; $k(c^2 - 1) = 2|c|$ **43.** $q(1) = 2$, $q(4) = 1$, $q(9) = 2/3$, $q(1/4) = 4$, $q(1/100) = 20$ **45.** $p = \dfrac{2}{\sqrt{b}}$; $p = \dfrac{2}{b}$, $b > 0$ **47.** $30,000
49. $32,000 **51.** $-$200,000 **53.** $25,000 **55.** 192,000
57. 80,000 **59.** $432.00 **61.** 32.04 **63.** 205,000
65. 480 **67.** $4000/9$ **69.** $f(\text{very poor}) = 1$; $f(\text{poor}) = 2$; $f(\text{average}) = 3$; $f(\text{above average}) = 4$; $f(\text{excellent}) = 5$
71. 2.66 **73.** $2 - \sqrt{2}$ **75.** 0.083 **77.** 0.111

2.4 EXERCISES

1a.

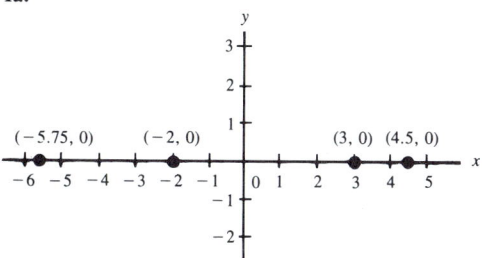

b.

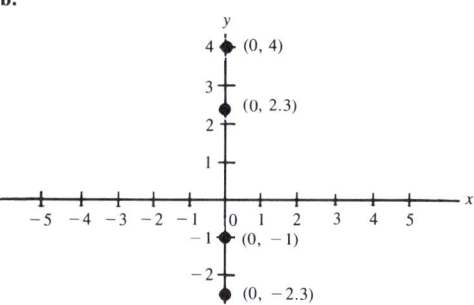

c.

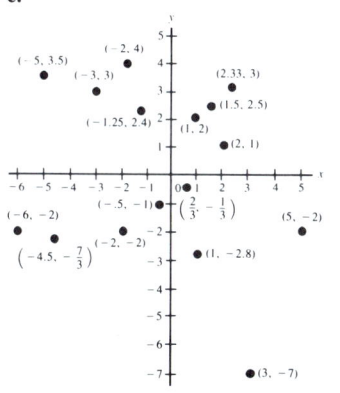

d.

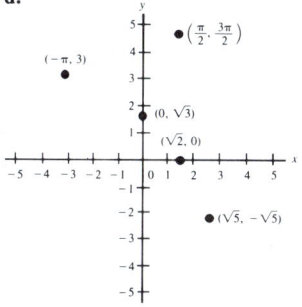

3. y-intercept $(0, -3)$; x-intercept $(1.5, 0)$

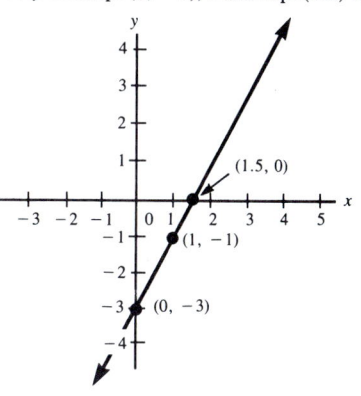

5. y-intercept $(0, 2)$

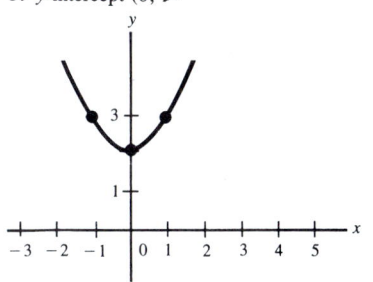

7. intercepts at $(0, 0)$

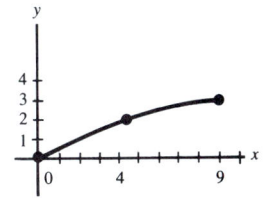

9. intercepts at (0, 0)

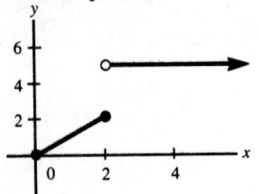

11. x-intercepts (0, 0), (6, 0); y-intercept (0, 0)

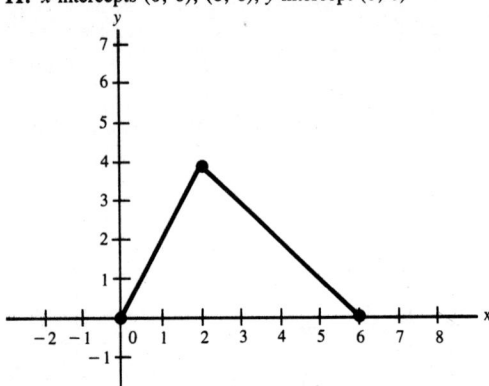

13. Yes **15.** No **17.** Yes

19. y is a function of x

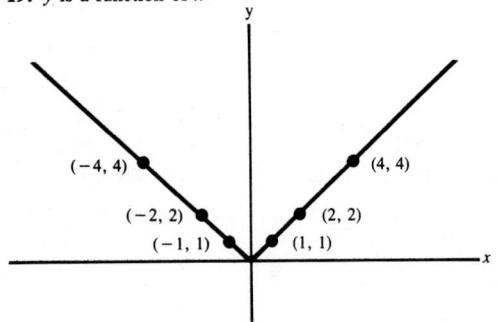

21. y is not a function of x

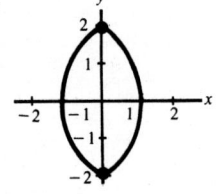

23. y is a function of x

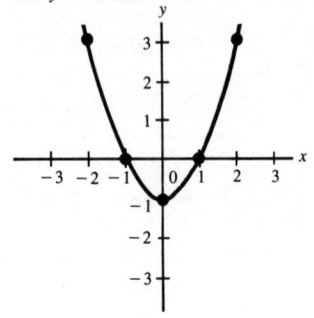

25. symmetric about x-axis **27.** no symmetries exist **29.** symmetric about x-axis, y-axis, and origin **31.** symmetric about the origin **33.** symmetric about the y-axis **35.** symmetric about origin **37.** symmetric about the origin, x-axis, and y-axis **39.** No

41. Yes

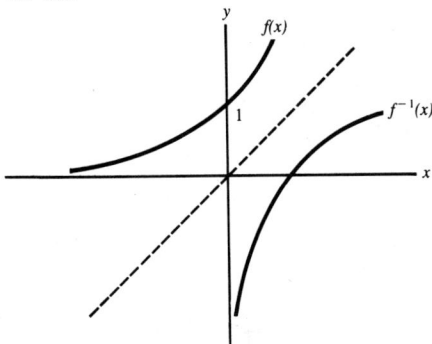

43. Yes

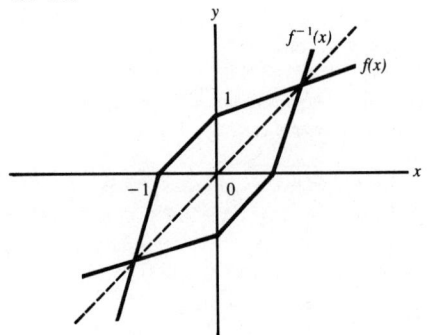

45. f is one-to-one; $y = \dfrac{x-1}{3}$ **47.** f is one-to-one; $y = -x$ **49.** Yes; $f^{-1}(x) = \sqrt[3]{x}$ **51.** Yes; $f^{-1}(x) = x^2$ **53a.** $100,000 **b.** $125,000 **c.** 200 units **d.** Alfa's total cost increases **55a.** loss of $1 million **b.** $5,000,000

c. $1,000,000   d. 10,000 units; 30,000 units
e. $7,000,000 at 100,000 units   f. increases from $-$$1
million at 0 units to $5 million at 20,000 units, decreases to
$1 million at 50,000 units, increases to $7 million at 100,000
units   **57a.**

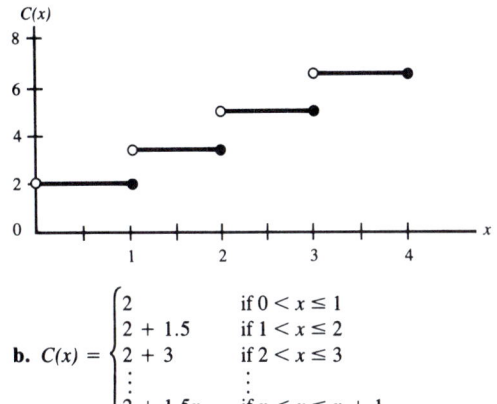

C(x) graph with step function values at 2, 3, 5, 6 approximately

b. $C(x) = \begin{cases} 2 & \text{if } 0 < x \leq 1 \\ 2 + 1.5 & \text{if } 1 < x \leq 2 \\ 2 + 3 & \text{if } 2 < x \leq 3 \\ \vdots & \vdots \\ 2 + 1.5n & \text{if } n < x \leq n+1 \end{cases}$

59a. 70%   **b.** 30%   **c.** 4 minutes   **d.** As the time increased, the number of bacteria decreased.   **61a.** 98.6°F
b. 97.5°F   **c.** after 12 or 13 hours   **d.** 97.5°F; 6 hours after
e. 102°F; 15 hours after   **f.** The temperature fell from 98.6° at the onset to a minimum of 97.5° after 6 hours, rose to 102° at 15 hours, and fell to 98.6° after 24 hours   **63a.** 10%
b. 90%   **c.** 4 weeks   **d.** After 10 weeks almost all persons appear to have accepted the fad

REVIEW EXERCISES

1. False   **2.** True   **3.** False   **4.** True   **5.** False   **6.** True
7. False   **8.** False   **9.** True   **10.** False   **11.** True
12. False   **13.** True   **14.** False   **15.** True   **16.** $A = \{a, b, c, d, e\}$ **17.** $C = \{10, 11, 12, \ldots\}$ **18.** $B = \{$even nonnegative integers$\}$ **19.** $C = \{$integers less than -10 or greater than 10$\}$ **20.** $A = \{x | x$ is a state in the United States$\}$ **21.** $C = \{x | x \in \mathbf{Z}, -5 \leq x \leq 2\}$

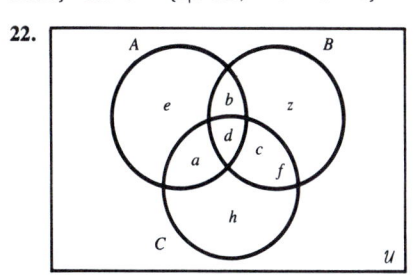

22. Venn diagram with sets A, B, C and elements e, b, z, a, d, c, f, h in universe u

23. $A' = \{c, f, g, h, z\}$ **24.** $B \cap C = \{c, f\}$ **25.** $A \cup B = \{a, b, c, d, e, f, z\}$ **26.** $A \cap B \cap C = \{d\}$ **27.** $A \cup B \cup C = \{a, b, c, d, e, f, z, h\}$ **28.** $A' \cap C = \{c, f, h\}$

29. $A' \cup B = \{b, c, d, f, g, h, z\}$ **30.** $A \cap (B \cup C)' = \{e\}$ **31.** $A' \cup (B \cap C') = \{c, b, f, g, h, z\}$ **32.** $S' = \{x | x$ is a registered voter in Ohio who is not a registered Democrat$\}$ **33.** $\{x | x$ is a registered voter in Ohio and is a Democrat under 30 years old$\}$ **34.** $S \cup V = \{x | x$ is a registered voter in Ohio who is a registered Democrat or a female$\}$ **35.** $\{x | x$ is a registered voter in Ohio who is a registered Democrat or is under 30 years old or is female$\}$
36. $S \cap T \cap V = \{x | x$ is a registered voter in Ohio who is a female, registered Democrat under 30 years old$\}$ **37.** $\{x | x$ is a registered voter in Ohio who is not a registered Democrat or is under 30 years old$\}$ **38.** $S \cap V' = \{x | x$ is a registered voter in Ohio who is a male registered Democrat$\}$
39. $\{x | x$ is a registered voter in Ohio who is not a female registered Democrat under 30 years old$\}$ **40.** $S' \cap (T \cup V)' = \{x | x$ is a registered voter in Ohio who is not a registered Democrat and is male or under 30 years old$\}$ **41.** 64
42. 20   **43.** 254   **44.** 70   **45.** No   **46.** Yes; domain = $\{x, y, z, w\}$, range = $\{a, b, c\}$ **47.** 5   **48.** -1 **49.** 1
50. $5 + 3a$ **51.** -14 **52.** -14 **53.** 0   **54.** $2 - b$
55. $2 - \dfrac{1}{b}$ **56.** 4   **57.** $-\frac{1}{2}$ **58.** does not exist   **59.** -1

60. $\dfrac{t+2}{t-4}$ **61.** $y(0) = 1$, $y(5) = \frac{3}{2}$, $y(11) = \dfrac{\sqrt{15}}{2}$, $y(21) = 2.5$ **62.** $y(-4) = 0$, $y(-3) = 0.5$, $y(-2) = \frac{1}{2}\sqrt{2}$, $y(\frac{9}{16}) = \frac{1}{8}\sqrt{73}$, $y(4) = \sqrt{2}$ **63.** $y(-3) = 7$, $y(-1) = 1$, $y(-\frac{2}{3}) = 0$, $y(0) = 2$, $y(\frac{2}{3}) = 4$, $y(1) = 5$, $y(3) = 11$
64. $y(0) = 0$, $y(1) = 1$, $y(-1) = -1$, $y(8) = 2$, $y(-27) = -3$, $y(9) \approx 2.08$, $y(-16) = \sqrt[3]{-16} = -2.52$
65. $y(-2) = 0$, $y(-1) = 0$, $y(-\frac{1}{2}) = \frac{1}{2}$, $y(0) = 1$, $y(\frac{9}{5}) = 2.8$, $y(2) = 3$, $y(\frac{13}{6}) = 3$, $y(3) = 3$ **66.** domain = $\{x | x \in \mathbf{R}\}$

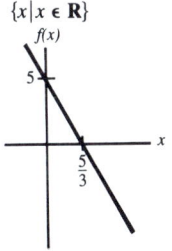

Graph of $f(x)$: line through $(0, 5)$ and $(\frac{5}{3}, 0)$

67. domain = $\{x | x \in \mathbf{R}\}$

Graph of downward parabola with vertex near $(0, 2)$, x-intercepts at -1 and 1

68. domain = $\{x | x \in \mathbf{R}, x \geq -4\}$

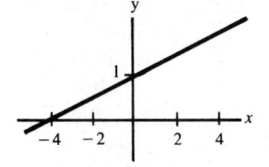

69. domain = $\{x | x \in \mathbf{R}\}$

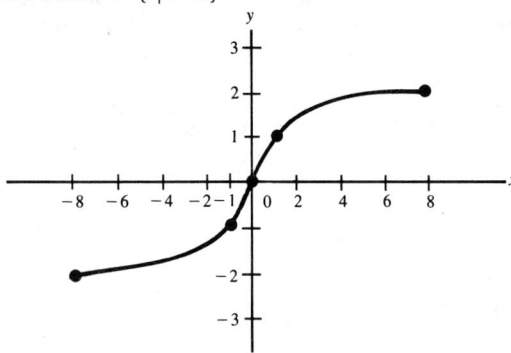

70. domain = $\{x | x \in \mathbf{R}\}$

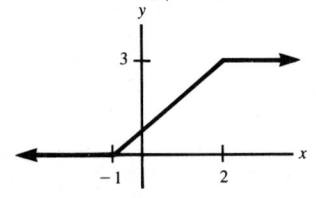

71. symmetric about y-axis **72.** symmetric about origin
73. symmetric about y-axis **74.** symmetric about origin, x-axis, and y-axis **75.** not one-to-one **76.** Yes, $f^{-1}(x) = x^3$
77. not one-to-one **76.** Yes, $f^{-1}(y) = \dfrac{x-b}{a}$

SUPPLEMENTARY EXERCISES

1. The nonmembers of the fuzzy set f are p, q, and u. The partial members are r, s, and t. **3.** $f(x)$ is not a fuzzy set because its range is not contained in the interval [0, 1]

5. full members = $\emptyset$; nonmembers = $\{1\}$; partial members = $\{x | x \in \mathbf{R}, x > 1\}$

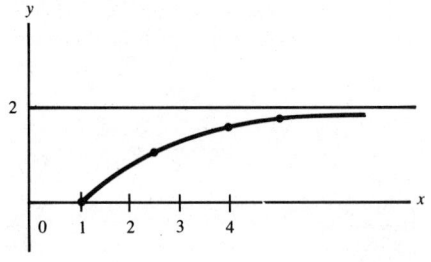

7a. old cars: $\{t | t \geq 5\}$; partially old cars: $\{t | 0 < t < 5\}$; new cars: $\{t | t = 0\}$ **c.** old cars: $\{\emptyset\}$; partially old cars: $\{t | t > 1\}$; new cars: $\{t | t \leq 1\}$ **d.** old cars: $\{t | t \geq 6\}$; partially old cars: $\{t | 2 < t < 6\}$; new cars: $\{t | t \leq 2\}$ **13.** $g'(p) = 0.8$, $g'(q) = 0$, $g'(r) = 0.6$, $g'(s) = 0.3$, $g'(t) = 1$, $g'(u) = 0$
15. $(f \cap g)(p) = 0.2$, $(f \cap g)(q) = 0.1$, $(f \cap g)(r) = 0.4$, $(f \cap g)(s) = 0.5$, $(f \cap g)(t) = 0$, $(f \cap g)(u) = 0$ **17.** $(f' \cap g')(p) = 0$, $(f' \cap g')(q) = 0$, $(f' \cap g')(r) = 0.2$, $(f' \cap g')(s) = 0.3$, $(f' \cap g')(t) = 0.1$, $(f' \cap g')(u) = 0$

19.

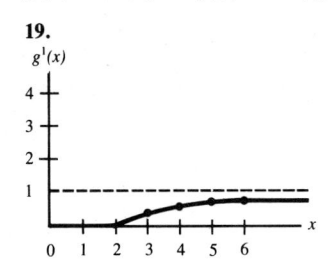

21.

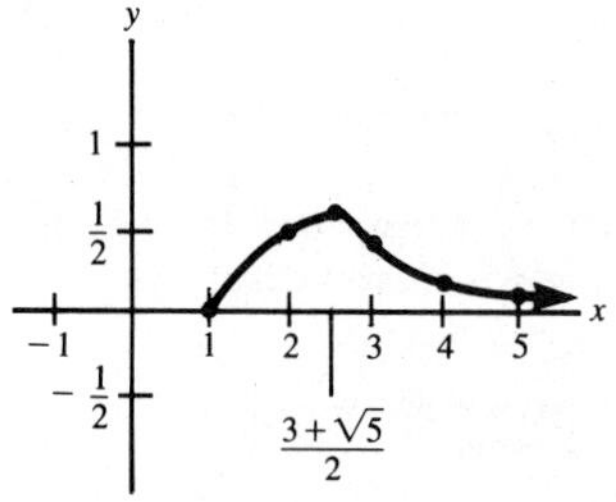

23.

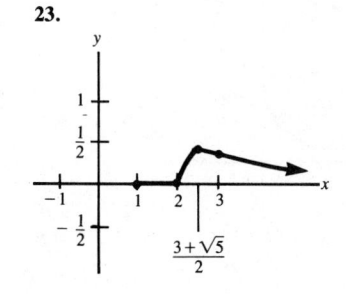

3.1 EXERCISES

1. $m = 2$, $y = 2x - 5$, $\left(\frac{5}{2}, 0\right)$, $(0, -5)$

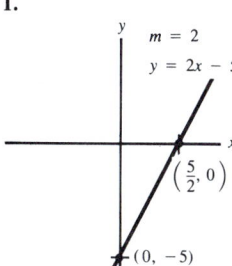

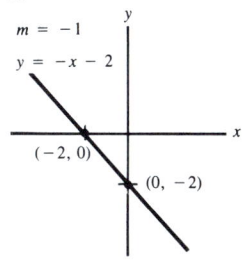

3. $m = -1$, $y = -x - 2$, $(-2, 0)$, $(0, -2)$

5. $m = 2$, $y = 2x$, $(1, 2)$, $(0, 0)$

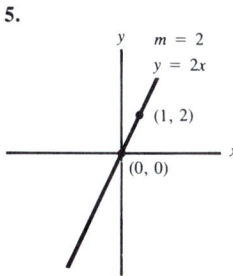

7. $m = 0$, no x-intercept, $(0, 5)$, $(2, 5)$, $y = 5$

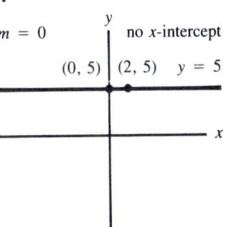

9. $x = -1$, no slope, no y-intercept, $(-1, 3)$, $(-1, 0)$

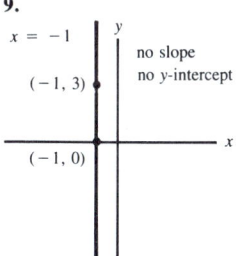

11. $m = -1$, $(0, 8)$, $(8, 0)$, $x + y = 8$

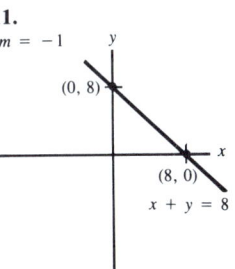

13. $m = \frac{3}{8}$, $-3x + 8y = 24$, $(0, 3)$, $(-8, 0)$

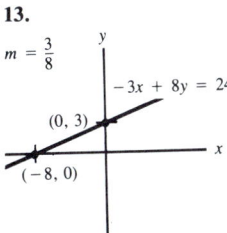

15. $m = -1$, $(0, 2)$, $(2, 0)$, $\frac{x}{2} + \frac{y}{2} = 1$

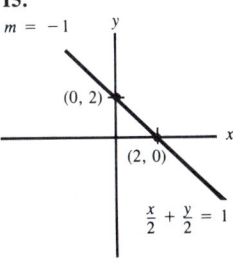

17. $m = -\frac{1}{2}$, $(0, 0)$, $(2, -1)$, $2x + 4y = 0$

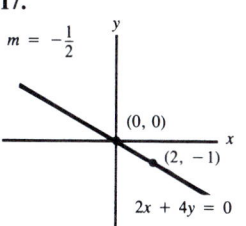

19. $(3, 10.4)$, $(0, 2)$, $y = 2.8x + 2$

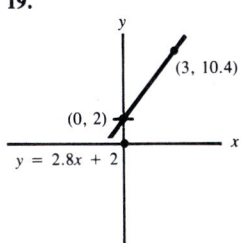

21. $\frac{2}{3}$ of deluxe, yes, yes, no, 3000, 2000; $(0, 2000)$, $(3000, 0)$, $20x + 30y = 6000$

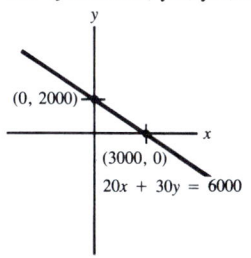

23. $ROE = ROA$ (since $L = 0$) $= (1 - t)E/OE$

25. ROE is constant; $ROE = 2$, $(0, 2)$

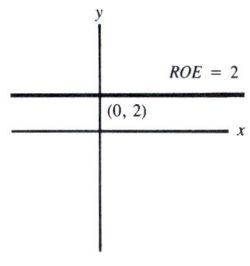

27a. increase L **b.** ROE is constant **c.** decrease L

29. 0.25, 0.005, 50 mg/L; 0.25, .05, 10, 50, $y = 0.25 - .005x$

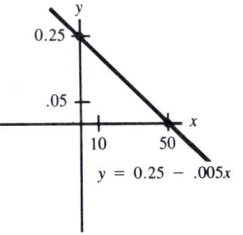

31. 915 milliseconds, 1500 milliseconds, minimal decision time, rate of increase in decision time as distance increases; $y = 850 + 26x$, $(25, 1500)$, $(2.5, 915)$, 300, 5

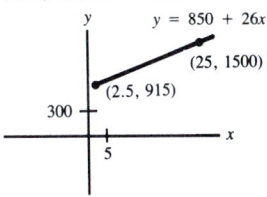

33. 630, increase in search time per distractor symbol added

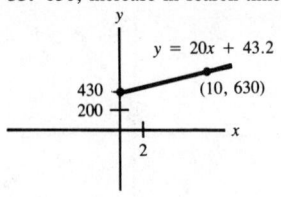

$y = 20x + 43.2$; points (10, 630), 430, 200, 2

35. 0.35, 0.45, 0.47, increases at rate of 0.01 per year

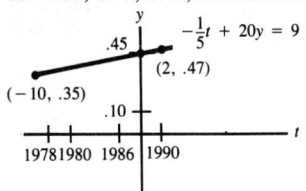

$-\frac{1}{5}t + 20y = 9$; points (-10, .35), (2, .47), .45, .10; 1978 1980 1986 1990

37. $y = 4x + 1$ **39.** $y = 6$ **41.** $y = \frac{1}{2}x + 3$ **43.** $y = \frac{3}{2}x - 1$ **45.** $y - 3 = -5(x - 1)$, $y = -5x + 8$ **47.** $y + 2 = 0(x - 8)$, $y = -2$ **49.** $y + 1 = -7(x + 4)$ or $y + 8 = -7(x + 3)$, $y = -7x - 29$ **51.** $y - 0 = \frac{3}{7}(x - 0)$ or $y - 3 = \frac{3}{7}(x - 7)$, $y = \frac{3}{7}x$ **53.** $x = 2$ **55.** $3x - y = 7$, $y = 3x - 7$ **57.** $2x - 3y = -9$, $y = \frac{2}{3}x + 3$ **59.** $4x - 3y = 0$, $y = \frac{4}{3}x$ **61.** $x = 7$ **63.** $y = -0.015x + 0.457$ **a.** 38.2%   **b.** 1996   **65.** $y = 0.40x$ **a.** $56,000   **b.** $250,000   **67.** $y = 28,000 - 4000x$ **a.** $15,000   **b.** 2.875 years   **69.** $y = 204,800 - 300x$ **b.** rate of decrease in sales per unit change in price; maximum possible sales   **c.** 192,800; $16   **71.** $y = 12.5x$; increase in total sales per additional store, 24   **73.** $80x + 45y = 18,000$; no, yes, 225,400
75. $C = P(mt + b)$

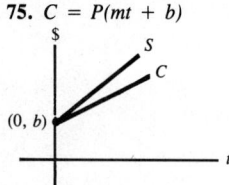

77. $7.02 million   **79.** $y = 2.1x + 53$ **a.** 62.45 cm   **b.** 5.24 months   **81.** $y = 0.6x + 59.5$ **a.** 25 years   **b.** 75 years   **83.** 0.02 g/L per year, 0.05 g/L   **85.** $80x + 125y = 2000$; 25, 12.5   **87.** $y = 160x + 2240$ **a.** 3200   **b.** 4.75 years from now   **89.** Yes, $y = 2000x + 15,000$ intercept = minimum sales slope = increase in sales per point aptitude   **a.** $205,000   **b.** score less than 67.5

3.2 EXERCISES

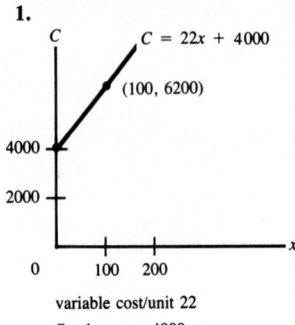

1. $C = 22x + 4000$; (100, 6200); variable cost/unit 22, fixed cost = 4000

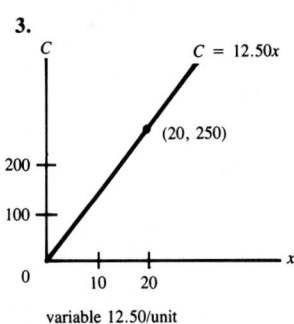

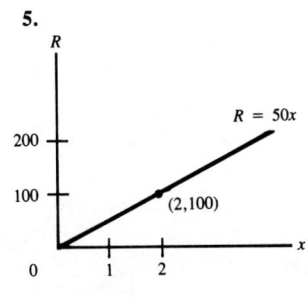

3. $C = 12.50x$; (20, 250); variable 12.50/unit, fixed 0

5. $R = 50x$; (2, 100); price 50/unit, no revenue if none sold

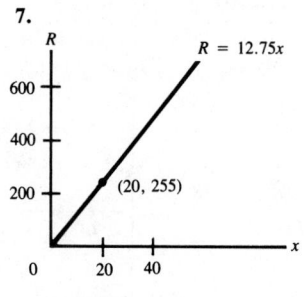

7. $R = 12.75x$; (20, 255); price 12.75 per unit, no revenue if none sold

SOLUTIONS TO EXERCISES

9.

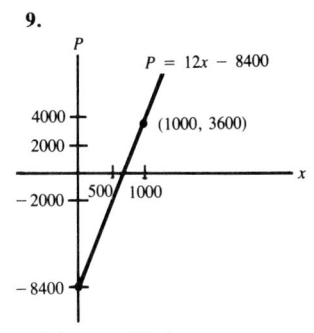

$P = 12x - 8400$

(1000, 3600)

profit increases 12/unit
fixed cost of $8400

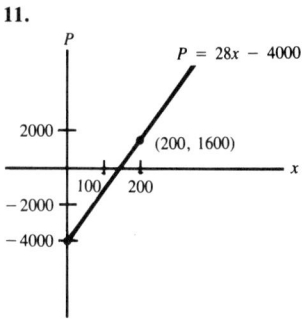

11.

$P = 28x - 4000$

(200, 1600)

profit increased 28/unit
fixed cost of $4000

13.

$P = .25x$

(200, 50)

profit increases .25 per unit
no fixed cost

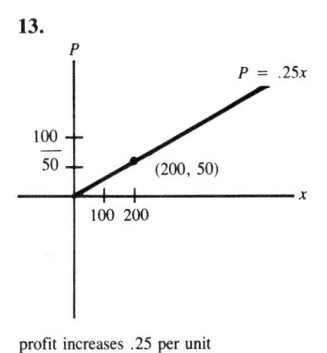

15a. $C = 140x + 60,000$; $R = 155x$; $P = 15x - 60,000$
b. $C = 480,000, 620,000, 760,000$; $R = 465,000, 620,000, 775,000$; $P = -15,000, 0, 15,000$

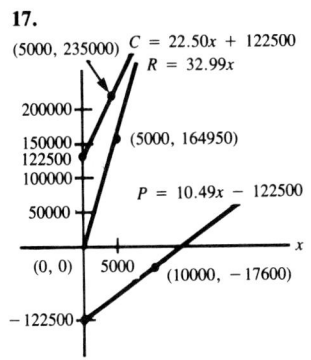

17.

(5000, 235000) $C = 22.50x + 122500$
$R = 32.99x$

(5000, 164950)

$P = 10.49x - 122500$

(0, 0)   5000   (10000, −17600)

−122500

x-intercept of P will be where C and R intersect

19. $C = 10.30x + 25,250$; $R = 15.25x$; $P = 4.95x - 25,250$ **a.** 76,750, 76,250, −500 **b.** 81,900, 83,875, 1975 each contributes $4.95 **21.** 4000 **23.** 684 **25.** 45,000 units, $350,000 profit **27.** 36,111 **29.** yes **31.** $C = 20x + 50,000$, variable $20/unit, fixed $50,000, 2500 units **33.** $P = 16.5x - 162,500$, fixed cost of $162,500, 15,910 units

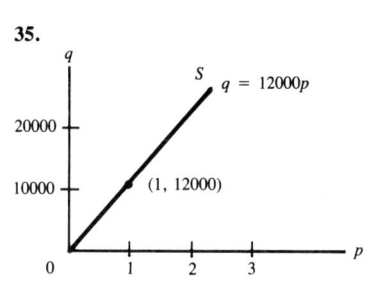

35.

S $q = 12000p$

(1, 12000)

production goes up 12000 per unit increase in price
minimum price is 0

37.

S $q = 23000p - 4140$

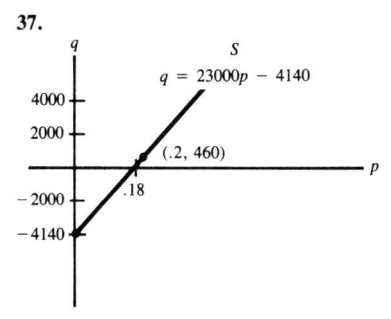

(.2, 460)

production goes up 23000 per unit increase in price
minimum price of .18

39.

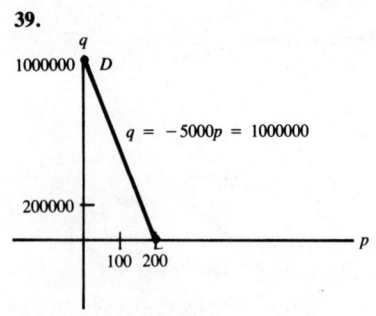

demand goes down by 5000 per unit increase in price
maximum demand of 1000000

Graph: $q = -5000p + 1000000$, D-intercept at (0, 1000000), passes through (200000 at some p), x-intercept at p = 200.

41.

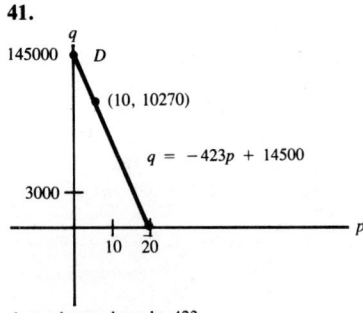

demand goes down by 423 per unit increase in price
maximum demand of 14500

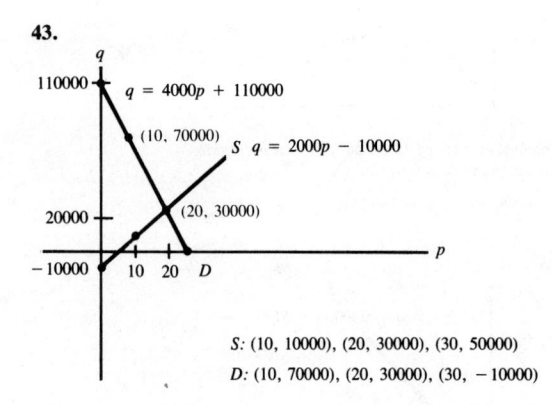

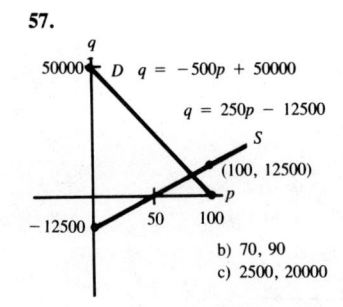

Graph: $q = -423p + 14500$, point (10, 10270), intercept 145000 (q-axis shows 145000, should be 14500), q = 3000 marked.

43.

$q = 4000p + 110000$ (D)
point (10, 70000)
$S\ q = 2000p - 10000$
point (20, 30000)
q-axis: 110000, 20000, -10000

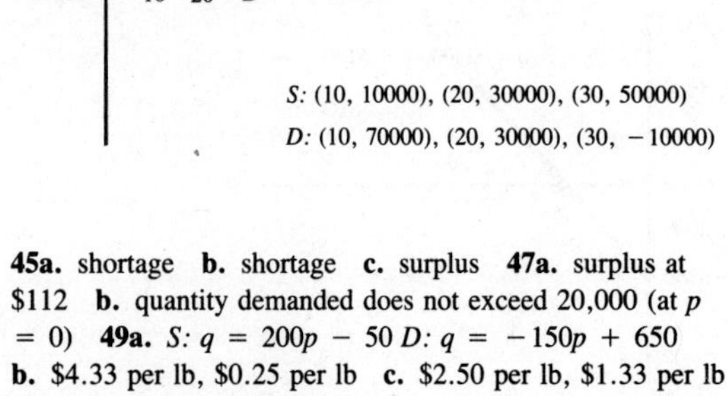

S: (10, 10000), (20, 30000), (30, 50000)
D: (10, 70000), (20, 30000), (30, -10000)

45a. shortage **b.** shortage **c.** surplus **47a.** surplus at $112 **b.** quantity demanded does not exceed 20,000 (at $p = 0$) **49a.** $S: q = 200p - 50$ $D: q = -150p + 650$ **b.** $4.33 per lb, $0.25 per lb **c.** $2.50 per lb, $1.33 per lb **51.** $S: q = 2000p - 200$; $0.10 zero S; $D: q = -4000p + 16{,}000$; $4.00 zero D; neither (at $p = \$2.70$) surplus nor shortage

53.

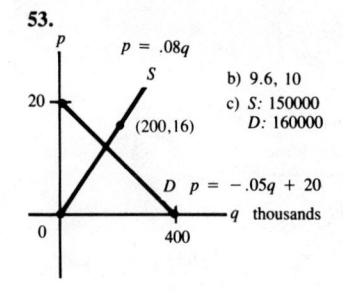

$p = .08q$ (S)
(200, 16)
$D\ \ p = -.05q + 20$
q thousands, 400
p-axis: 20

b) 9.6, 10
c) S: 150000 D: 160000

55.

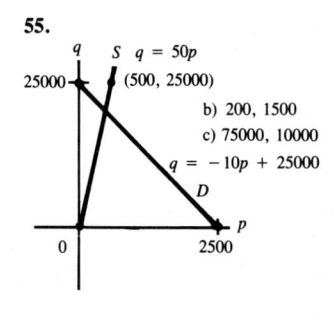

$S\ \ q = 50p$
(500, 25000)
$q = -10p + 25000$ (D)
q-axis: 25000, p-axis: 2500

b) 200, 1500
c) 75000, 10000

57.

$D\ \ q = -500p + 50000$
$q = 250p - 12500$
(100, 12500)
S
q-axis: 50000, -12500; p-axis: 50, 100

b) 70, 90
c) 2500, 20000

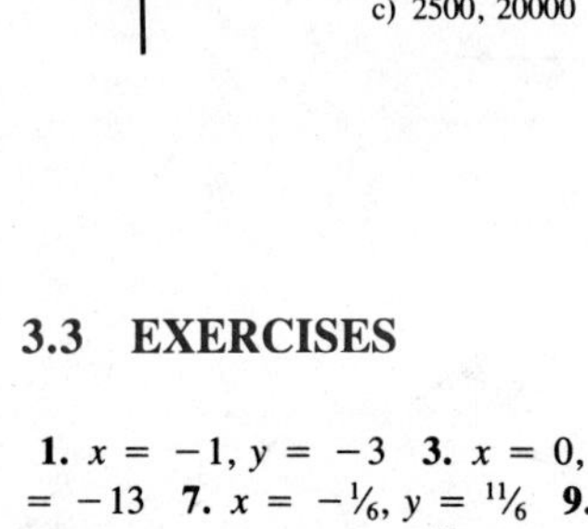

3.3 EXERCISES

1. $x = -1, y = -3$ **3.** $x = 0, y = 9$ **5.** $x = -4, y = -13$ **7.** $x = -\frac{1}{6}, y = \frac{11}{6}$ **9.** $x = 2, y = 1$ **11.** $x = 36, y = 20$ **13.** $x = \frac{10}{17}, y = -\frac{38}{17}$ **15.** none **17.** $x = \frac{1}{2}, y = \frac{11}{4}$ **19.** $x = 2, y = 0, z = -4$ **21.** $x = 6, y = 2, z = 4$ **23.** no solution **25.** small = 30, large = 22

27.

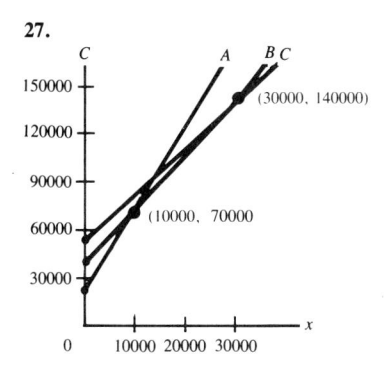

a. $C_B = 3.50x + 35{,}000$, $C_C = 3.00x + 50{,}000$ **c.** A from 0 to 10,000, B from 10,000 to 30,000, C from 30,000 to 100,000 **29.** by 2017 **31.** 20.2 years **33.** 1997 **35.** $C = 6x + 100{,}000$, $R = 10x$, $(25{,}000, 250{,}000)$

37. fixed cost = $280,000

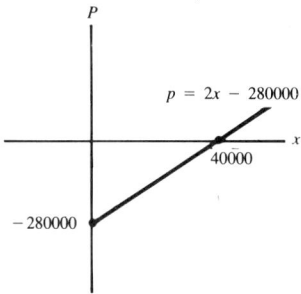

39. (28,750, 92,000)

41.

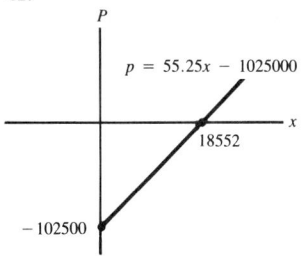

43. $C = 80x + 400{,}000$, $R = 85x$, $P = 5x - 400{,}000$
45. $P = 3.75x - 150{,}000$

47.

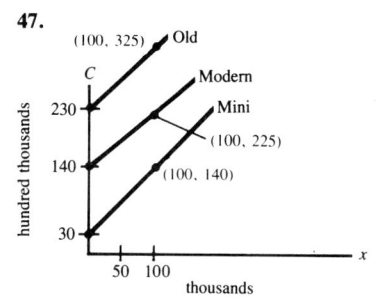

Break-even points: old $x = 148{,}400$, insufficient capacity; modern $x = 84{,}850$, sufficient; mini $x = 21{,}430$, sufficient; Old $C = 230(100{,}000) + 95x$; Modern $C = 140(100{,}000) + 85x$; Mini $C = 30(100{,}000) + 110x$ **49.** old: $p \geq \$325$; modern: $p \geq \$225$; mini: $p \geq \$140$ **51.** old: $p \geq \$400$; modern: $p \geq \$305$; mini: $p \geq \$310$

53.

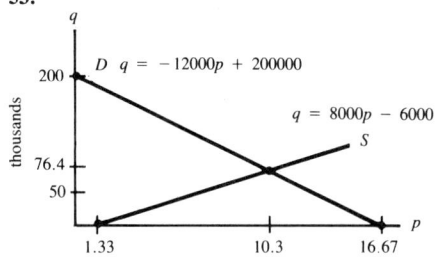

55. $p = 8$, $q = 62{,}000$ **57.** S: $q = 4000p - 18{,}000$, D: $q = -3000p + 48{,}500$

59.

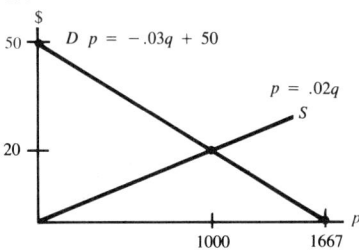

61.

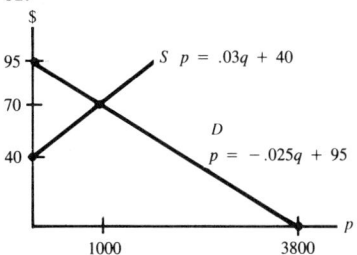

3.4 EXERCISES

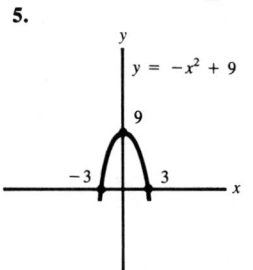

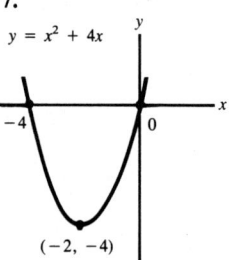

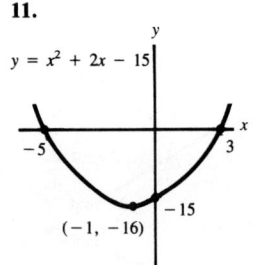

1. Graph of $y = 2x^2$ with points $(-1, 2)$ and $(1, 2)$.

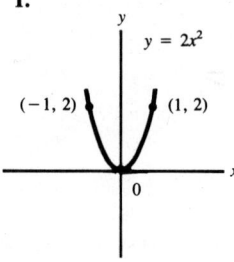

3. Graph of $y = \frac{x^2}{2}$ with points $(-1, \frac{1}{2})$ and $(1, \frac{1}{2})$.

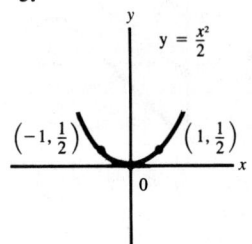

5. Graph of $y = -x^2 + 9$ with vertex at 9, x-intercepts -3 and 3.

7. Graph of $y = x^2 + 4x$ with x-intercepts -4 and 0, vertex $(-2, -4)$.

9. Graph of $y = 3x^2 - 6x$ with x-intercepts 0 and 2, vertex $(1, -3)$.

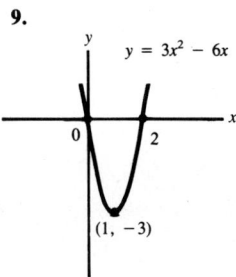

11. Graph of $y = x^2 + 2x - 15$ with x-intercepts -5 and 3, y-intercept -15, vertex $(-1, -16)$.

13. Graph of $y = 2x^2 - x - 6$ with x-intercepts $-\frac{3}{2}$ and 2, y-intercept -6, vertex $(\frac{1}{4}, -\frac{49}{8})$.

15. Graph of $y = x^2 + x + 1$ with points $(-1, 1)$, $(-\frac{1}{2}, \frac{3}{4})$, and 1.

17. Graph with x-intercepts $2 - \frac{1}{2}\sqrt{34}$ and $2 + \frac{1}{2}\sqrt{34}$, y-intercept 9, vertex $(2, 17)$.

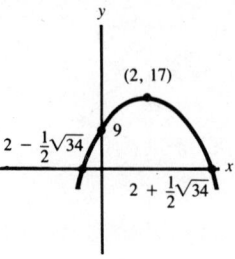

19. Graph with x-intercepts $-\sqrt{20}$ and $\sqrt{20}$, y-intercept 60.

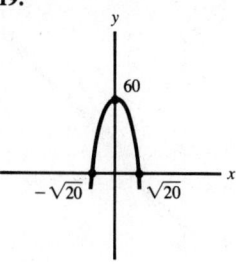

21. Graph with y-intercept 19, vertex $(5.5, 3.875)$, point $(11, 19)$.

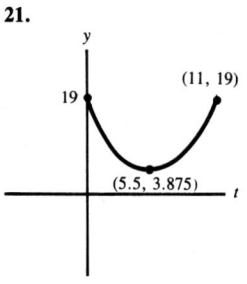

23. Graph with t-intercepts 25 and 105, y-intercept 50, vertex $(52.5, 220.5)$.

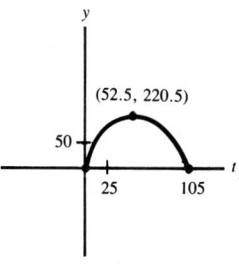

25a. $R = 150x$

b. Graph showing C line and R curve with points $(50, 7500)$, $(75, 625)$, intersections at $50, 75, 100$; P curve with minimum near -5000.

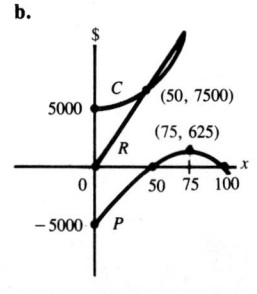

c. $x = 50, 100$

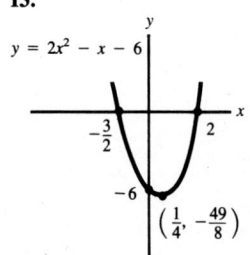

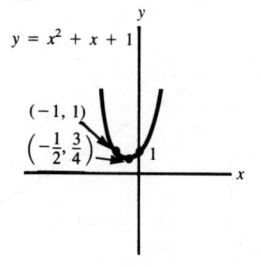

d. $P = -x^2 + 150x - 5000$ **e.** 625 at 75 units sold

27a. Graph showing C line with point $(16, 128)$, R curve, and P curve with vertex $(3, -84.5)$, x-intercepts -10 and 16, y-intercept -80.

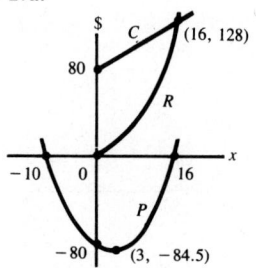

b. 16 **c.** $P = 0.5x^2 - 3x - 80$ **d.** no maximum

29a. 30 to 45 units
b. 1.125 at 37.5 units sold

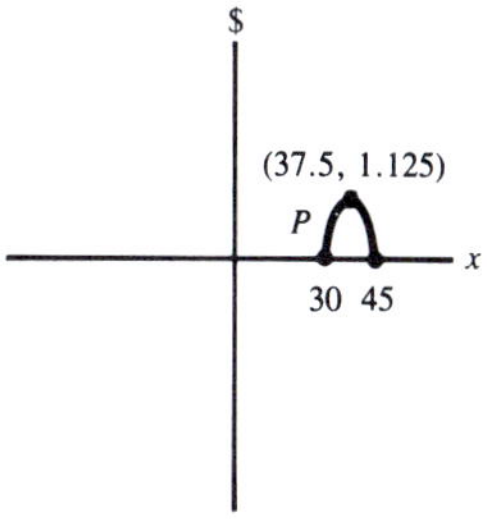

31.

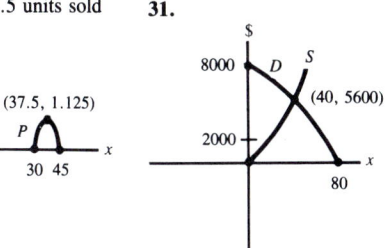

33. $p = 1.25$, 5 million units **35.** $R = \frac{1}{5}x^2 + 2x$
37a. 9.6° at 30 minutes **b.** 70 minutes **39.** 32.4 at 36 hours, 72 hours **41.** $y = -t^2 - 12t + 100$; 5.66 days
43. 4 mailings, $480,000 **45.** 79.2% after 8 months $y = -0.3x^2 + 4.8x + 60$, x in months

3 REVIEW EXERCISES

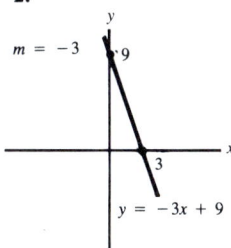

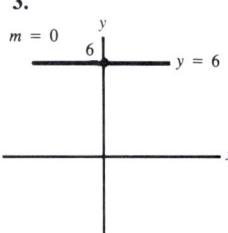

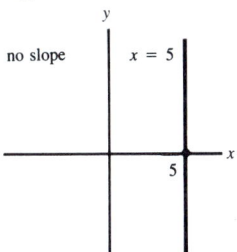

1. $m = 4$, $y = 4x - 3$

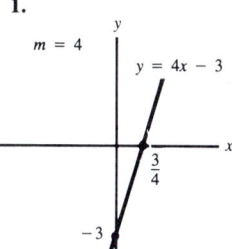

2. $m = -3$, $y = -3x + 9$

3. $m = 0$, $y = 6$

4. no slope, $x = 5$

5. $m = \frac{2}{3}$, $4x + 6y = -12$

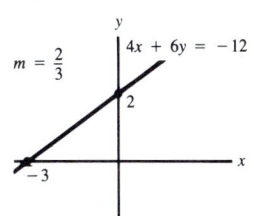

6. $m = -\frac{2}{5}$, $2x + 5y = 20$

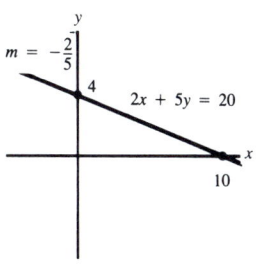

7. $m = -2$, $y = -2x$

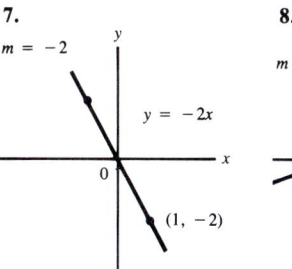

8. $m = \frac{3}{11}$, $3x - 11y = -4$

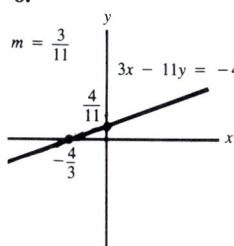

9. $m = \frac{5}{3}$

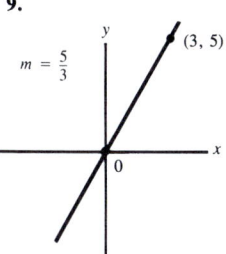

10. $m = -\frac{5}{3}$, $-.05x - .03y = 3$

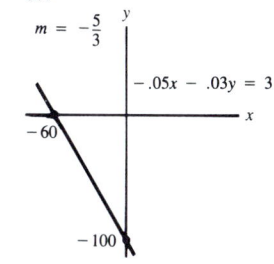

11. $y = \frac{1}{3}x + 6$ **12.** $y = 6$ **13.** $y = -3x - 1$ **14.** $y = -5x + 7$ **15.** $y = -2x + 30$ **16.** $y = 5x - 3$
17. $y = \frac{1}{3}x + \frac{2}{3}$ **18.** $y = 9$ **19.** $y = -\frac{63}{47}x - \frac{19}{47}$
20. $y = 10$ **21.** $y = \frac{1}{5}x + \frac{13}{5}$ **22.** $y = -\frac{2}{3}x - \frac{5}{3}$
23. $y = -6000t + 120,000$, 20 hours **24.** $y = 12,400t + 36,800$ **a.** 36,800 **b.** 1994 **25.** fixed 36,000, variable 12 per unit, sells at 17 per unit,

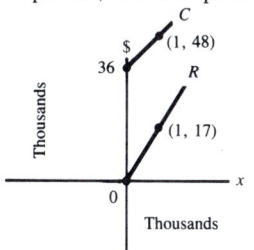

26a. $C = 120x + 84,000$, $R = 136x$, $P = 16x - 84,000$,

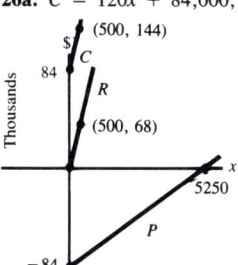

b. C, P intercept − fixed costs, slopes C, R prices, P profit per unit sold **c.** $C = 804,000$, $R = 816,000$, P 12,000
27a. S: $q = 2000p - 10,000$ D: $q = -6000p + 400,000$

b. slopes: change per unit change in price intercepts: D − maximal demand   **c.** S: 90,000 D: 100,000   **d.** 55, 33.33   **28.** $x = 9, y = 33$ **29.** $x = {}^{63}/_{26}, y = {}^{6}/_{26}$ **30.** no solution   **31.** all points on line $27x - 2y = 21$ **32.** 2.5, 1.4   **33.** (5750, 356,500)   **34.** 13.5, 68,750

35.

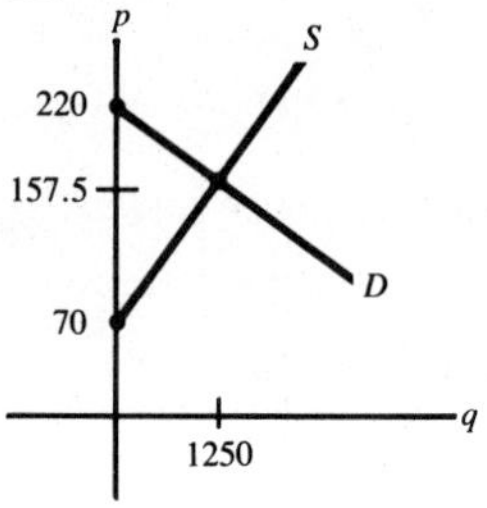

36.

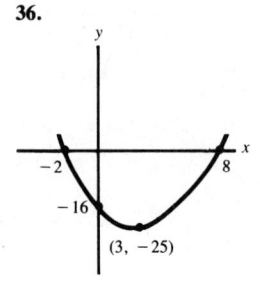

37.

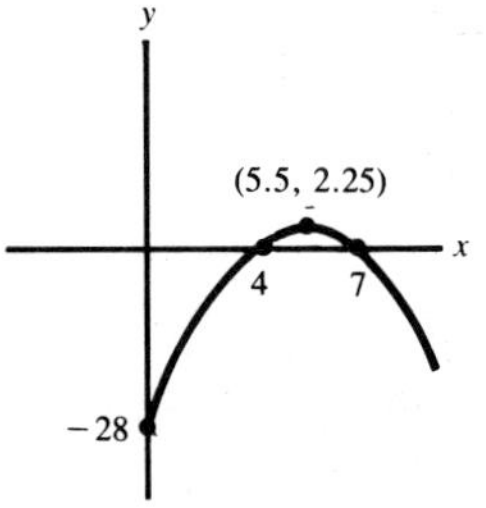

38.

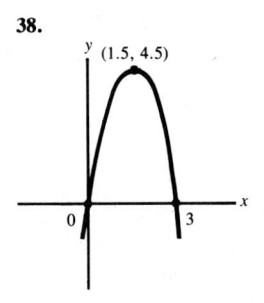

39.

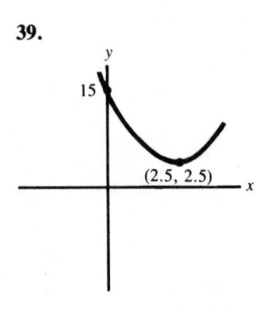

40.

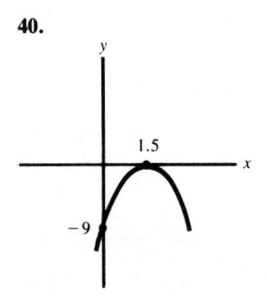

41.

42a. 30, 70   **b.** $p = -\frac{1}{2}x^2 + 50x - 1050$,

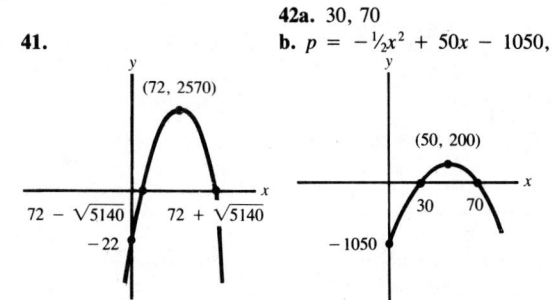

c. 200 at 50 units sold   **43a.** 2000 after 15 hours   **b.** 50 hours
44. $q = 800p^2 - 4000p - 2000, \frac{1}{2}(5 + \sqrt{35})$ or 5.46

4.1 EXERCISES

1. $x_1 = -13, x_2 = 4$ **3.** $x_1 = 3 + \frac{1}{2}x_2$ **5.** $x_1 = \frac{1}{2} - \frac{1}{4}x_2$ **7.** $x_1 = 1, x_2 = -1, x_3 = 0$ **9.** $x_1 = \frac{1}{2}, x_2 = -\frac{1}{2}, x_3 = 2$ **11.** $x_1 = 2 - \frac{1}{3}x_3, x_2 = -2 + \frac{2}{3}x_3, x_3 = x_3$ **13.** $x_1 = 2, x_2 = 0, x_3 = -2, x_4 = 1$ **15.** $x_1 = 1, x_2 = 1, x_3 = 1, x_4 = -1$ **17.** $x_1 = 2 + \frac{1}{2}x_3, x_2 = 3 - \frac{3}{2}x_3$ **19.** $x_1 = 2, x_2 = -5$ **21.** no solution   **23.** $x_1 = 2x_4 - 4x_5 + 8; x_2 = -x_4 - x_5 - 3; x_3 = 4x_4 - 4x_5 + 14$ **25.** no solution   **27.** $42,000, $36,000, $22,000   **29.** many solutions for 30 large, any combination of small and medium totaling 30   **31.** S: $104p^2 - 70p$ D: $-2p^2 + 10,000$, eq. $10.05   **33.** 42, 84, 63, 56   **35.** $S = -40 + 2L, M = 100 - 3L$, any number of large from 20 to 33   **37.** $6500   **39.** $200p^2 + 4000p$ **41.** 4, 2, 2   **43.** less than five Z's, $X = 4 - \frac{1}{5}Z, Y = {}^{19}/_{6} - \frac{2}{3}Z$ **45.** 800, 700, 1200   **47.** no solution

4.2 EXERCISES

1. $1 \times 4, 2 \times 2, 3 \times 2, 2 \times 1, 1 \times 1, 1 \times 2, 3 \times 1, 3 \times 3, 2 \times 3, 4 \times 4, 1 \times 3, 4 \times 1, 5 \times 4, 3 \times 5, 2 \times 2$ **3.** F, G, J   **5.** row: A, E, F, L; column: D, E, G, M   **7.** 5, 21, 13, 40, 45   **9.** no, Mom's making dinner   **11.** set low

13. Males $\begin{array}{c} \text{Wks} \\ 2 \\ 4 \\ 6 \end{array} \begin{bmatrix} \text{Birth} & \text{Discharge} \\ 41.2 & 43.4 \\ 35.4 & 42.6 \\ 32.0 & 41.8 \end{bmatrix}$ Females $\begin{array}{c} \text{Wks} \\ 2 \\ 4 \\ 6 \end{array} \begin{bmatrix} \text{Birth} & \text{Discharge} \\ 38.2 & 42.2 \\ 34.3 & 42.8 \\ 32.1 & 42.2 \end{bmatrix}$

15. highest C, F; lowest D, E; E has highest formal status
17. yes   **19.** yes   **21.** no   **23.** no   **25.** $\begin{bmatrix} 2 & 3 \\ 5 & 5 \end{bmatrix}$

27. $\begin{bmatrix} 3 \\ 4 \\ 2 \end{bmatrix}$ **29.** $\begin{bmatrix} 16 & 7 \\ 1 & -1 \\ 2 & 10 \end{bmatrix}$ **31.** not possible   **33.** not possible   **35.** not possible   **37.** $\begin{bmatrix} x+r & y+s \\ z+t & w+u \end{bmatrix}$

39. $\begin{bmatrix} x+3y & 0 \\ 4y-5z & 8z+5x+2y \end{bmatrix}$ **41.** $A + B = B + A = \begin{bmatrix} a+e & b+f \\ c+g & d+h \end{bmatrix} = \begin{bmatrix} e+a & f+b \\ g+c & h+d \end{bmatrix}$; $A + (B + C) = (A + B) + C = \begin{bmatrix} a+e+i & b+f+j \\ c+g+k & d+h+l \end{bmatrix}$

43. $\begin{bmatrix} 12 & -6 \\ 6 & 18 \end{bmatrix}$ **45.** $\begin{bmatrix} -4 & 2 \\ -2 & -6 \end{bmatrix}$ **47.** $\begin{bmatrix} 2 & -1 \\ 1 & 3 \end{bmatrix}$

49. $\begin{bmatrix} -2/3 & 2/3 \\ -4/3 & 0 \\ -2 & -10/3 \end{bmatrix}$ **51.** $\begin{bmatrix} 14 & -3 & 10 & 25 \\ 10 & 6 & 18 & 2 \end{bmatrix}$

53. $\begin{bmatrix} 23 & 39/2 & -8 & 17/2 \\ 30 & -1 & 16 & 6 \end{bmatrix}$ **55.** $\begin{bmatrix} 12 & 0 & 6 & 18 \\ 10 & 4 & 14 & 2 \end{bmatrix}$

SOLUTIONS TO EXERCISES 1111

57. $\begin{bmatrix} 0 & 0 & 0 & 0 \\ 0 & 0 & 0 & 0 \end{bmatrix}$ **59.** $(c+d)A = \begin{bmatrix} cx+dx & cy+dy \\ cz+dz & cw+dw \end{bmatrix}$

$= \begin{bmatrix} cx & cy \\ cz & cw \end{bmatrix} + \begin{bmatrix} dx & dy \\ dz & dw \end{bmatrix} = cA + dA;$

$c(A+B) = c\begin{bmatrix} x+r & y+s \\ z+t & w+u \end{bmatrix} = \begin{bmatrix} cx+cr & cy+cs \\ cz+ct & cw+cu \end{bmatrix}$

$= \begin{bmatrix} cx & cy \\ cz & cw \end{bmatrix} + \begin{bmatrix} cr & cs \\ ct & cu \end{bmatrix} = cA + cB;$

$A + (-1)B = \begin{bmatrix} x & y \\ t & w \end{bmatrix} + \begin{bmatrix} -r & -s \\ -t & -u \end{bmatrix} =$

$\begin{bmatrix} x-r & y-s \\ z-t & w-u \end{bmatrix} = A - B$ **61.** $\begin{bmatrix} 14 & 1 & 6 & 5 \\ 1 & 8 & 1 & 4 \\ 3 & 4 & 6 & 0 \end{bmatrix}$

63. $\begin{bmatrix} 3 & 2 & 1 & 1 \\ 4 & 0 & 2 & 8 \\ 6 & 3 & 5 & 3 \end{bmatrix}$

65. $\begin{bmatrix} 4.9 & 24.5 & 4.9 \\ 9.8 & 3.92 & 1.96 \\ 13.72 & 0.98 & 0 \\ 7.84 & 2.94 & 1.96 \end{bmatrix}, \begin{bmatrix} 48.02 & 24.01 & 4.80 \\ 9.60 & 3.84 & 1.92 \\ 13.45 & 0.96 & 0 \\ 7.68 & 2.88 & 1.92 \end{bmatrix}$

67. change during monitoring $\begin{bmatrix} 7 & 100 & -20 \\ 3 & 250 & 10 \\ -30 & -20 & -20 \end{bmatrix}$

69. B **71.** B **73.** stocks **75.** money market

4.3 EXERCISES

1. 20 **3.** $\begin{bmatrix} 2 & 10 \\ -1 & -3 \end{bmatrix}$ **5.** none **7.** $\begin{bmatrix} 14 \\ -6 \end{bmatrix}$ **9.** $\begin{bmatrix} 10 \\ 8 \\ 4 \end{bmatrix}$

11. $[3 \ 4]$ **13.** none **15.** $\begin{bmatrix} 2 & 5 & 7 \\ 6 & -7 & -12 \\ 0 & 4 & 6 \end{bmatrix}$

17. $\begin{bmatrix} 12 & 24 \\ 4 & 8 \end{bmatrix}$ **19.** $\begin{bmatrix} 3 & 2 \\ -2 & -1 \end{bmatrix}$ **21.** none

23. $\begin{bmatrix} 11 & 26 \\ 19 & 38 \end{bmatrix}$ **25.** $\begin{bmatrix} 9 & 1 \\ -4 & 1 \end{bmatrix}$ **27.** 11 **29.** $\begin{bmatrix} 2 & 3 \\ -5 & 17 \end{bmatrix}$

31. $\begin{bmatrix} a & b & c \\ d & e & f \\ g & h & i \end{bmatrix}$ **33.** $\begin{bmatrix} ra & rb & rc \\ rd & re & rf \\ rg & rh & ri \end{bmatrix}$

35. $\begin{bmatrix} ea+fc & eb+fd \\ 0 & 0 \end{bmatrix}$ **37.** $\begin{bmatrix} 1 & -3 \\ 9 & -2 \end{bmatrix}$

39. $\begin{bmatrix} -28 & 3 \\ -9 & -23 \end{bmatrix}$ **41.** $\begin{bmatrix} 1 & 0 & 0 \\ 0 & 1 & 0 \\ 0 & 0 & 1 \end{bmatrix}$ **43.** $\begin{bmatrix} -2 & -6 \\ 182 & 96 \end{bmatrix}$

45. $\begin{bmatrix} 130 & -90 \\ 62 & -36 \end{bmatrix}$ **49.** $[484 \ 128 \ 234]$

51. $[39.98 \ 32.80 \ 41.09 \ 37.68]$ **53.** 7850 **55.** 925

57. $\begin{bmatrix} 1 & 0 & 0 \\ 0 & .80 & 0 \\ 0 & 0 & 1 \end{bmatrix}$ **59.** projected number of sets sold by

brand, projected revenue $[3580 \ 4200 \ 4840]$

61a. $\begin{bmatrix} 0 & 1 & 1 & 0 & 0 \\ 0 & 0 & 0 & 1 & 0 \\ 0 & 1 & 0 & 1 & 1 \\ 0 & 0 & 0 & 0 & 1 \\ 0 & 0 & 0 & 0 & 0 \end{bmatrix}$ **b.** $\begin{bmatrix} 0 & 1 & 0 & 2 & 1 \\ 0 & 0 & 0 & 0 & 1 \\ 0 & 0 & 0 & 1 & 1 \\ 0 & 0 & 0 & 0 & 0 \\ 0 & 0 & 0 & 0 & 0 \end{bmatrix},$

2-step

$\begin{bmatrix} 0 & 0 & 0 & 1 & 2 \\ 0 & 0 & 0 & 0 & 0 \\ 0 & 0 & 0 & 0 & 1 \\ 0 & 0 & 0 & 0 & 0 \\ 0 & 0 & 0 & 0 & 0 \end{bmatrix}, \begin{bmatrix} 0 & 0 & 0 & 0 & 1 \\ 0 & 0 & 0 & 0 & 0 \\ 0 & 0 & 0 & 0 & 0 \\ 0 & 0 & 0 & 0 & 0 \\ 0 & 0 & 0 & 0 & 0 \end{bmatrix}$

3-step 4-step

c. $\begin{bmatrix} 0 & 2 & 1 & 3 & 4 \\ 0 & 0 & 0 & 1 & 1 \\ 0 & 1 & 0 & 2 & 3 \\ 0 & 0 & 0 & 0 & 1 \\ 0 & 0 & 0 & 0 & 0 \end{bmatrix}$ 4 paths 1 to 5
3 paths 3 to 5

63a. $G^2 = \begin{bmatrix} 0 & 0 & 0 & 0 & 0 & 1 \\ 0 & 0 & 0 & 0 & 0 & 2 \\ 0 & 0 & 0 & 0 & 0 & 1 \\ 0 & 0 & 0 & 0 & 0 & 0 \\ 0 & 0 & 0 & 0 & 0 & 0 \\ 0 & 0 & 0 & 0 & 0 & 0 \end{bmatrix}$

b. Ⓐ → Ⓔ → Ⓕ
Ⓑ → Ⓓ
↑
Ⓒ

65. Ranking 4, 1, 2, 5 and 6 tied, 3

4.4 EXERCISES

9. $\begin{bmatrix} -7 & 4 \\ 2 & -1 \end{bmatrix}$ **11.** none **13.** $\frac{1}{7}\begin{bmatrix} 5 & -4 \\ -2 & 3 \end{bmatrix}$

15. $\frac{1}{4}\begin{bmatrix} 1 & 3 & 5 \\ 1 & -1 & 1 \\ 1 & -1 & -3 \end{bmatrix}$ **17.** none

19. $\frac{1}{9}\begin{bmatrix} 4 & -1 & 3 \\ 1 & 2 & -6 \\ -3 & 3 & 0 \end{bmatrix}$

21. $\frac{1}{22}\begin{bmatrix} -8 & -16 & 8 & 10 \\ 15 & 8 & -4 & -5 \\ -9 & 4 & -2 & 3 \\ 12 & 2 & 10 & -4 \end{bmatrix}$

23. $\frac{1}{2}\begin{bmatrix} -5 & 13 & 13 & -3 \\ -6 & 14 & 12 & -2 \\ 1 & 1 & 1 & -1 \\ 1 & -3 & -3 & 1 \end{bmatrix}$ **25.** none

31. $\begin{bmatrix} 1 & -4 \\ 0 & 1 \end{bmatrix}$ **33.** $\begin{bmatrix} 1 & 2 & 10 \\ 0 & 1 & 5 \\ 0 & 0 & 1 \end{bmatrix}$

35. $\begin{bmatrix} 1 & -1 & -3 & -2 \\ 0 & 1 & 1 & -3 \\ 0 & 0 & 1 & 1 \\ 0 & 0 & 0 & 1 \end{bmatrix}$ **37.** $x_1 = -6, x_2 = 12$

39. $x_1 = -\frac{18}{19}, x_2 = \frac{50}{19}$ **41.** $x_1 = -4, x_2 = -23, x_3 = 9$ **43.** $x_1 = \frac{11}{3}, x_2 = \frac{1}{5}, x_3 = -\frac{13}{15}$ **45.** 3, -2, 1, 1
47. $\frac{1}{2}$, 3, $-\frac{1}{2}$, 5 **49a.** 16, 6 **b.** 24, 24 **c.** $139\frac{1}{5}, 11\frac{1}{5}$
51a. 3, 4, 1 **b.** 2, 0, 4 **c.** $0.009a - 0.01s - 0.0003c$, $-0.012a + 0.04s + 0.0002c$, $0.004a - 0.02s + 0c$
53a. 2, 1, 2 **b.** 0, $\frac{5}{4}, \frac{5}{2}$ **c.** $0a + \frac{1}{3}b - c$, $-\frac{1}{4}a + \frac{5}{12}b - \frac{1}{2}c$, $\frac{1}{2}a - \frac{5}{6}b + 2c$

4.5 EXERCISES

1. 24, 15, 57, 10, 27, 80, 49, 20, 88, 7, 5, 20, 8, 14, 41, 44, 25, 92, 42, 27, 102 **3.** 45, 1, 35, 51, 9, 60, 42, 20, 75, 13, 15, 53, 26, 27, 82, 15, 14, 56, 26, 5, 39, 28, 19, 64, 3, 21, 60 **5.** RUN FOR YOUR LIFE **7.** $\begin{bmatrix} 236.4 \\ 327.3 \end{bmatrix}$, GNP = 563.7 **9.** Internal demand: 15, 5 Total input/output: 50, 30, GNP = 80 **11.** $\begin{bmatrix} 1.257 & 0.134 \\ 0.053 & 1.070 \end{bmatrix}$

13. Internal demand: 8, 14 Total input/output: 50, 60 GNP = 110 **15.** $\begin{bmatrix} 1.029 & 0.147 \\ 0.221 & 1.103 \end{bmatrix}$ **17.** Internal demand: 75, 60, 105 Total input/output: 175, 260, 405, GNP = 840

	M	A	S	Internal	External	Total Input
	19.4	5.5	44.5	71	105	176
19.	10.6	11.1	37.1	57	220	277
	29.9	16.6	55.7	101	270	371
					GNP	824

21. $\begin{bmatrix} 16{,}000 \\ 2000 \\ 4000 \\ 2000 \\ 1000 \end{bmatrix}$ **23.** $\begin{bmatrix} 0 & 0 & 4 & 6 & 0 & 0 \\ 0 & 0 & 2 & 5 & 0 & 0 \\ 0 & 0 & 0 & 1 & 8 & 2 \\ 0 & 0 & 0 & 0 & 12 & 3 \\ 0 & 0 & 0 & 0 & 0 & 4 \\ 0 & 0 & 0 & 0 & 0 & 0 \end{bmatrix}$

4 REVIEW EXERCISES

1. 8, 6, -3 **2.** no solution **3.** $x_1 = \frac{1}{2} - \frac{9}{2}x_4$; $x_2 = 1 + \frac{1}{2}x_4$; $x_3 = -\frac{1}{2} - \frac{5}{2}x_4$ **4.** 2, -3, 4 **5.** $x_1 = 17 + 5x_4$; $x_2 = 22 + 8x_4$; $x_3 = -27 - 11x_4$ **6.** $x_1 = 170 - 5x_3$, $x_2 = -305 + 11x_3$, $x_3 = x_3$ with $28 \le x_3 \le 34$
7. 20, 25, 30 **8.** no, only 100 gallons of paint will be used up **9.** 2×3, 2×3, 2×2, 2×2, 3×2, 4×1, 1×3, 3×2; F, G; C, D **10a.** yes **b.** no **11.** $\begin{bmatrix} 7 & 8 \\ -1 & 1 \end{bmatrix}$

12. $\begin{bmatrix} -1 & 4 \\ -3 & 1 \end{bmatrix}$ **13.** not defined **14.** $\begin{bmatrix} 4 & 12 & -2 \\ 6 & 0 & 8 \end{bmatrix}$

15. $\begin{bmatrix} -3 & -15 & -3 \\ 0 & 6 & -6 \end{bmatrix}$ **16.** $\begin{bmatrix} 10 & 34 & -2 \\ 12 & -4 & 20 \end{bmatrix}$

17. $\begin{bmatrix} -1 & 7 & 8 \\ -9 & -10 & -2 \end{bmatrix}$

18. $\begin{bmatrix} 2 & 2 \\ 0 & 8 \end{bmatrix}$ **19.** $\begin{bmatrix} 10 & 2 & 4 \\ 6 & -1 & 31 \end{bmatrix}$ **20.** none **21.** none

22. $\begin{bmatrix} -14 \\ 38 \end{bmatrix}$ **23.** 18 **24.** $\begin{bmatrix} -4 & -2 & -6 \\ 2 & 1 & 3 \\ 14 & 7 & 21 \end{bmatrix}$

25. $\begin{bmatrix} 3 & 7 \\ -7 & 24 \end{bmatrix}$ **26.** $\begin{bmatrix} -1 & 38 \\ -38 & 113 \end{bmatrix}$ **27.** $\begin{bmatrix} -40 & 189 \\ -189 & 527 \end{bmatrix}$

28. $5400 **29.** $\begin{bmatrix} 9 & -7 \\ -5 & 4 \end{bmatrix}$ **30.** none

31. $\frac{1}{41}\begin{bmatrix} 7 & 2 \\ -3 & 5 \end{bmatrix}$ **32.** none **33.** $\frac{1}{36}\begin{bmatrix} -2 & 18 & -8 \\ 3 & -9 & 12 \\ -5 & -9 & 16 \end{bmatrix}$

34. $\frac{1}{12}\begin{bmatrix} 8 & 6 & -2 & -4 \\ -8 & -3 & 5 & 4 \\ 8 & 3 & 7 & -4 \\ -6 & -6 & -6 & 6 \end{bmatrix}$

35. -24, 14 **36.** $\frac{7}{12}, -\frac{5}{12}$ **37.** $\frac{26}{12}, -\frac{15}{12}, -\frac{7}{12}$
38a. 2, 2, 1 **b.** 4, 5, 3 **c.** 2, 6, 4 **39.** 57, 57, 113, 39, 50, 40, 47, 18, 74, 47, 27, -9, 103, 79, 60, 11, 90, 78, 133, 36, 73, 78, 96, 42, 14, 38, 91, 40, 32, 40, 123, 28, 67, 54, 56, 22, 35, 49, 107, 35, 89, 84, 100, 34, 8, 6, 7, 0
40. NO MORE CRYPTOGRAPHY PROBLEMS
41. $\begin{bmatrix} 4523.8 \\ 5238.1 \end{bmatrix}$, GNP = 9761.9

	Input	Output	Internal Demand	External Demand	Total Input
42.	943	2571	3514	1200	4714
	2829	514	3343	1800	5143
				GNP	9857

4 SUPPLEMENTARY EXERCISES

1. $0 \le h_1 \le 4000$, $h_2 = 250$, $h_3 = 250$, $h_4 = 500$ **3.** $0 \le h_1 \le 1000$, $h_2 = 0$, $h_3 = 100$, $h_4 = 400$, $h_5 = 500$

5. $k = 3$, profile = $\begin{bmatrix} 6666.67 \\ 3333.33 \\ 0 \\ 0 \end{bmatrix}$, $H = \begin{bmatrix} -6666.67 \\ 0 \\ 666.67 \\ 0 \end{bmatrix}$,

maximum profit = \$33,333.33

5.1 EXERCISES

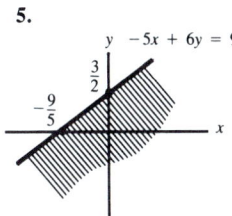

1. [graph: $x + y = 1$]

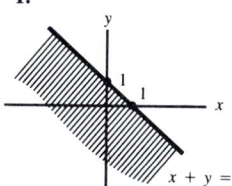

3. [graph: $3x - 2y = 6$]

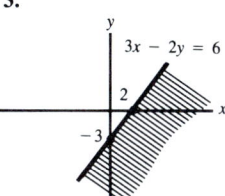

5. [graph: $-5x + 6y = 9$]

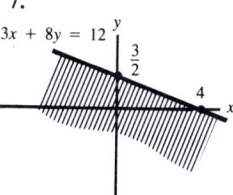

7. [graph: $3x + 8y = 12$]

9. [graph: $x - 4y = 0$, point $(4, 1)$]

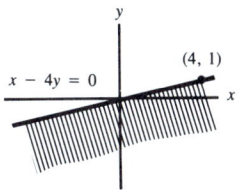

11. [graph: $4x - 2y = 10$]

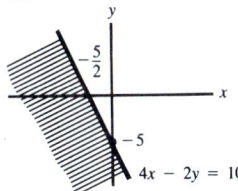

13. [graph: $y = \frac{3}{2}$]

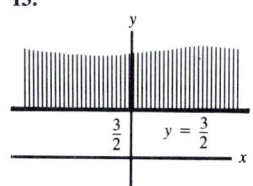

15. [graph: $x = -3$]

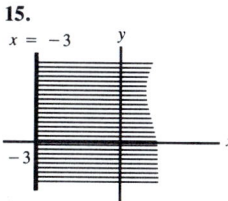

17. [graph: $y = 3x$, point $(1, 3)$]

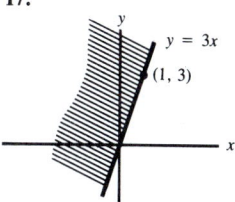

19. [graph]

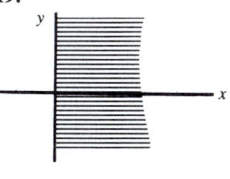

21. [graph: $x + y = 0$, point $(1, -1)$]

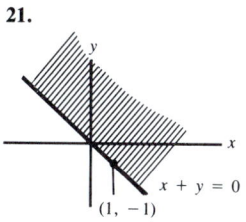

23. [graph: $y = 2$, $2x + 5y = 20$, point $(5, 2)$]

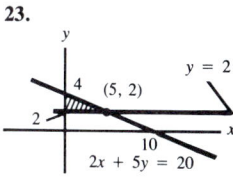

25. [graph: $3x + y = 15$, $2x + 4y = 40$, point $(2, 9)$]

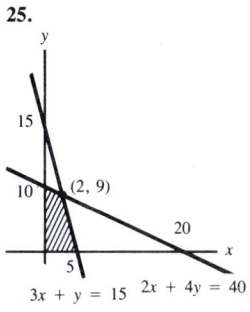

27. [graph: $x - y = 0$, $3x + 2y = 24$, $3x + 8y = 48$, points $\left(\frac{48}{11}, \frac{48}{11}\right)$, $\left(\frac{16}{3}, 4\right)$]

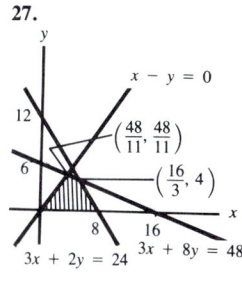

29. [graph: $3x + 3y = 36$, $2x + 6y = 36$, $x + 6y = 24$, points $(9, 3)$, $(12, 2)$]

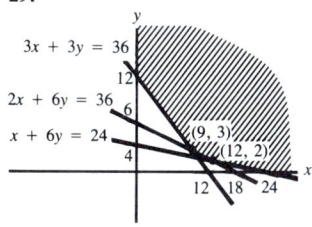

31. [graph: $4x + 5y = 80$, $x + y = 8$, $2x + 3y = 36$]

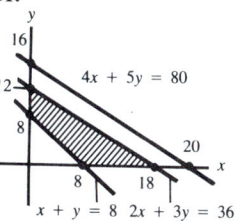

33. no feasibility region

35. [graph: $x - y = 4$, $x + y = 12$, $x + y = 4$, point $(8, 4)$]

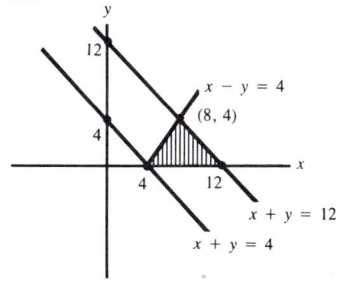

37.

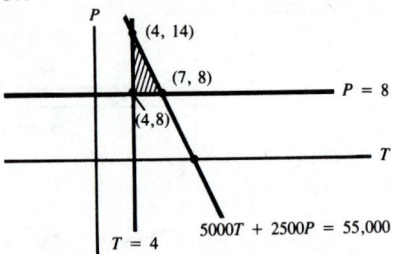

(figure: P vs T graph with points (4, 14), (7, 8), (4, 8); lines P = 8, T = 4, 5000T + 2500P = 55,000)

39.

(figure: S vs W graph with points (0, 10/3), (10, 0); lines 6S + 2W = 20, 6S + 12W = 120)

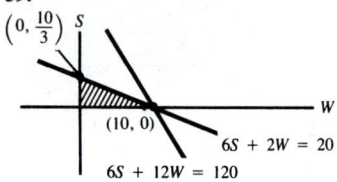

41. no feasibility region

43.

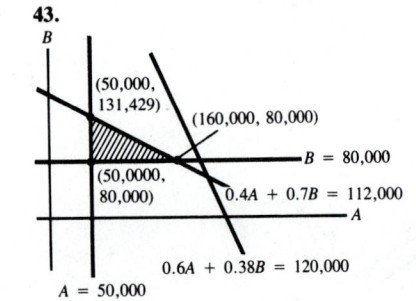

(figure: B vs A graph with points (50,000, 131,429), (160,000, 80,000), (50,000, 80,000); lines B = 80,000, 0.4A + 0.7B = 112,000, 0.6A + 0.38B = 120,000, A = 50,000)

45.

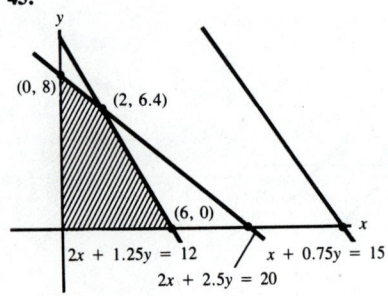

(figure: y vs x graph with points (0, 8), (2, 6.4), (6, 0); lines 2x + 1.25y = 12, 2x + 2.5y = 20, x + 0.75y = 15)

47.

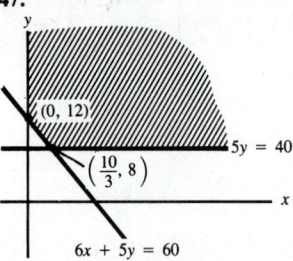

(figure: y vs x graph with points (0, 12), (10/3, 8); lines 5y = 40, 6x + 5y = 60)

49.

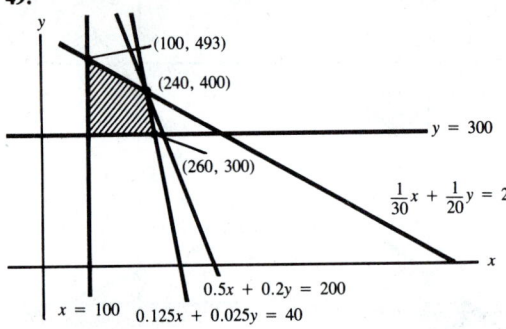

(figure: y vs x graph with points (100, 493), (240, 400), (260, 300); lines y = 300, $\frac{1}{30}x + \frac{1}{20}y = 28$, 0.5x + 0.2y = 200, x = 100, 0.125x + 0.025y = 40)

51.

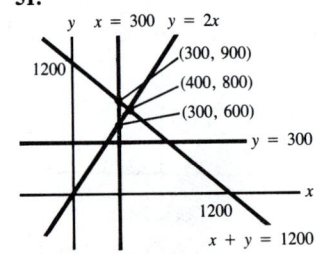

(figure: y vs x graph with points (300, 900), (400, 800), (300, 600); lines x = 300, y = 2x, y = 300, x + y = 1200)

5.2 EXERCISES

1. max $P = 0.095x + 0.082y$; $x + y \geq 10{,}000$; $x + y \leq 25{,}000$; $x \leq 16{,}000$; $x, y \geq 0$ **3.** min $C = 0.69x + 0.99y + 1.19z$; $x + y + z \geq 24$; $x, y, z \geq 0$; $-0.02x + 0.04y + 0.06z = 0$; $y \leq 4$ **5.** max $P = 0.12x + 0.08y + 0.10z$; $x + y + z = 30{,}000$; $x \geq 5000$, $y \geq 5000$, $z \geq 5000$ **7.** min $C = 1.50x + 1.00y + 1.25z$; $x + y + z \geq 200$; $x \leq 80$; $0.75x - 0.25y - 0.25z \geq 0$; $x, y, z \geq 0$; $y \leq 100$; $-0.40x + 0.60y - 0.40z \leq 0$; $z \leq 60$; $-0.20x - 0.20y + 0.80z \geq 0$ **9.** max $P = 12.25x + 8.75y$; $8x + 4y \leq 560$; $x, y \geq 0$; $6x + 6y \leq 630$ **11.** min $C = 8x + 10y$; $8x + 4y \leq 560$; $6x + 6y \leq 630$; $x \geq 40$ **13.** max $P = 1.00x + 1.00y$; $2x + y \leq 160$; $x, y \geq 0$; $3x + 4y \leq 340$ **15.** min $C = 10{,}000x + 15{,}000y$; $400x + 1200y \geq 24{,}000$; $x, y \geq 0$; $400x + 400y \geq 12{,}000$ **17.** min $C = 4x$

$+ 2y + 6z$; $x + y + z = 400$; $x, y, z \geq 0$; $-0.04x - 0.02y + 0.04z = 0$ **19.** max $E = 2.4x + 3.2y + 1.6z$; $200x + 400y + 300z = 12{,}000$; $200x \leq 4800$, $400y \leq 4800$; $300z \leq 4800$; $x, y, z \geq 0$ **21.** min $C = 4x_1 + 8y_1 + 10.5x_2 + 17.5y_2 + 6x_3 + 9y_3$; $0.2x_1 + 0.4y_1 \leq 40$; $x_1 + x_2 + x_3 = 300$; $x_1, x_2, x_3, y_1, y_2, y_3 \geq 0$; $0.3x_2 + 0.5y_2 \leq 40$; $y_1 + y_2 + y_3 = 100$; $0.2x_3 + 0.3y_3 \leq 40$ **23.** min $C = 0.24x_1 + 0.20y_1 + 0.28z_1 + 0.32x_2 + 0.24y_2 + 0.32z_2 + 0.36x_3 + 0.20y_3 + 0.32z_3$; $x_1 + y_1 + z_1 \leq 40$; $30x_1 + 25x_2 + 20x_3 = 1200$; $x_1, x_2, x_3, y_1, y_2, y_3, z_1, z_2, z_3 \geq 0$; $x_2 + y_2 + z_2 \leq 40$; $20y_1 + 20y_2 + 25y_3 = 800$; $x_3 + y_3 + z_3 \leq 40$; $25z_1 + 30z_2 + 20z_3 = 900$ **25.** min $C = 0.18x + 0.22y + 0.20z$; $250 \leq 100x + 50y + 100z \leq 400$; $y + z \geq 1$; $100 \leq 50x + 100y + 50z \leq 200$; $x, y, z \geq 0$; $350 \leq 100x + 100y + 50z \leq 500$

5.3 EXERCISES

1. $F = 12$ at $(6, 0)$ **3.** $F = 48$ on line joining $(^{16}/_3, 4)$ and $(8, 0)$ **5.** 26 at $(4, 2)$ **7.** 42 at $(9, 3)$ **9.** 116 at $(4, 8)$ **13.** $16,000 money market, $9000 bonds, return $2258 **15.** 35 of X, 70 of Y **17.** 4 TV, 14 newspaper **19.** 16 scientific, no business **21.** 49, 12 business, 10 scientific **23.** 35, 2 sockets, 9 wickets **25.** 15 days each **27.** $108,000, 1200 bias-ply, 2800 radial **29.** 140,000 A, 80,000 B **31.** 132 regular, 168 deluxe **33.** 4 TV, 8 radio **35.** 3000 from X, 1500 from Y **37.** 2000 medium, 5600 large **39.** 500 of each **41.** 12 of Y, no X **43.** 4 of X, 6 of Y **45.** 100 bales of hay, $733\frac{1}{3}$ bushels of corn **47.** 300 men, 600 women

5.4 EXERCISES

1.
	x_1	x_2	s_1	s_2	sol
s_1	4	5	1	0	20
s_2	3	1	0	1	10
F	-2	-5	0	0	0

3.
	x_1	x_2	x_3	s_1	sol
s_1	5	1	1	1	100
F	-3	-1	-4	0	0

5.
	x_1	x_2	x_3	x_4	s_1	s_2	sol
s_1	0	3	1	2	1	0	8
s_2	5	1	0	0	0	1	10
F	-2	-1	-3	1	0	0	0

7. max $F = 2x_1 + 5x_2$; $2x_1 + x_2 \leq 10$; $3x_1 \leq 5$; $x_1, x_2 \geq 0$
7. max $F = 2x_1 + 5x_2$; $2x_1 + x_2 \leq 10$; $3x_1 \leq 5$; $x_1, x_2 \geq 0$
$3x_3 \leq 30$; $5x_2 \leq 40$; $x_1, x_2, x_3 \geq 0$ **11.** ent. x_2, dep. s_1
13. ent. x_3, dep. s_1 **15.** ent. x_3, dep. s_1 **17.** ent. x_2, dep. s_1 **19.** ent. x_3, dep. s_2 **21.** 20 at $x_1 = 10$, $x_2 = 0$ **23.** $x_1 = 4$, $x_2 = 6$, $F = 36$ **25.** $x_1 = x_3 = 0$, $x_2 = {}^{15}\!/_2$, $F = 30$ **27.** $x_1 = \frac{1}{7}$, $x_2 = {}^{50}\!/_7$, $x_3 = \frac{2}{7}$, $F = {}^{270}\!/_7$ **29.** 30 small, 20 large **31.** send 10 to territory B **33.** 0 15 seconds, 14 30 seconds, 16 45 seconds **35.** no regular, 212,000 gallons premium, 20,000 gallons super premium

37. no tulips, 15 jonquils, 30 hyacinths (dozens) **39.** 5 of X, 45 of Y, 35 of Z **41.** 3 extra, 5 normal, 0 noncompetitive

5.5 EXERCISES

1. $x_1 = 20$, $x_2 = 0$, $F = 60$ **3.** $x_1 = {}^{11}\!/_7$, $x_2 = {}^{15}\!/_7$, $F = {}^{41}\!/_7$ **5.** $x_1 = 20$, $x_2 = 30$, $x_3 = 40$, $F = 280$ **7.** $x_1 = x_3 = 8$, $x_2 = 0$, $F = 16$ **9.** $10,000 money market, $30,000 bond **11.** $2250 Amnity, $1500 B & T, $750 Community **13.** 4600 shares of B, 100 each of A, C, D **15.** 1 each to A, B, 48 to C **17.** 100 tons from each of Paydirt and Richstrike, 0 from Goldstar, revenue of $105,000 **19.** 24 radio, 12 TV, 8 news **21.** 10 of first, 70 of second **23.** 120 juniors, 150 freshmen **25.** 30 at $x_1 = 10$, $x_2 = 0$ **27.** $x_1 = 0$, $x_2 = x_3 = 10$, $F = 20$ **29.** $x_1 = 10$, $x_2 = 50$, $x_3 = 0$, $F = 100$ **31.** $x_1 = x_2 = 1$, $x_3 = x_4 = 0$, $F = 3$ **33.** $\frac{100{,}000}{3}$ high, $\frac{200{,}000}{3}$ low **35.** 10 of A, 20 B, 30 C **37.** 10 lb Brazil, 25 lb peanuts, 15 lb cashews **39.** 4000 low, 2000 medium, 600 high **41.** x_1: 8 and 8; x_2: 8 and 10; x_3: 8 and 12; x_4: 10 and 10; min $C = 4x_1 + 2x_2$; $x_1 + x_2 + x_3 + x_4 \leq 350$; $2x_1 + x_2 + x_3 = 275$; $x_2 + 2x_4 = 100$; $x_3 = 250$; plus condition x's are integers ≥ 0. **43.** take 8 B's **45.** 10 northern males, 20 northern females, 20 southern males, 0 southern females, cost of $7000

5.6 EXERCISES

1. $\begin{bmatrix} 1 & -2 \\ 3 & 9 \end{bmatrix}$ **3.** $\begin{bmatrix} 4 & -1 & 0 \\ 2 & 3 & 1 \\ 3 & 2 & 0 \end{bmatrix}$ **5.** max $20x_1 + 60x_2$; $2x_1 + 3x_2 \leq 1$; $x_1 + 4x_2 \leq 2$ **7.** max $100x_1 + 80x_2 + 220x_3$; $2x_1 + x_2 \leq 4$; $x_1 + 2x_3 \leq 2$; $x_1 + x_2 + 3x_3 \leq 1$; $2x_2 + 5x_3 \leq 4$ **9.** max $25x_1 - 125x_2 + 40x_3$; $4x_1 - 3x_2 \leq 3$; $2x_1 + 6x_3 \leq 1$; $-5x_2 + 2x_3 \leq 2$ **11.** min $50x_1 + 90x_2 + 10x_3$; $3x_1 + 4x_2 + x_3 \geq 5$; $2x_1 + 3x_2 \geq 2$ **13.** 10 at $x_1 = 0$, $x_2 = 10$ **15.** $x_1 = 0$, $x_2 = x_3 = 100$, $F = 200$ **17.** F arbitrarily negative by taking large x_3 **19.** $x_1 = 30$, $x_2 = 10$, $F = 50$ **21.** 75 A, 100 B **23.** 50 of F, 350 of A or B **25.** 0 of first, 48 lb of second, 72 lb of third **27.** 40 of A, 70 of B, 90 of C

5 REVIEW EXERCISES

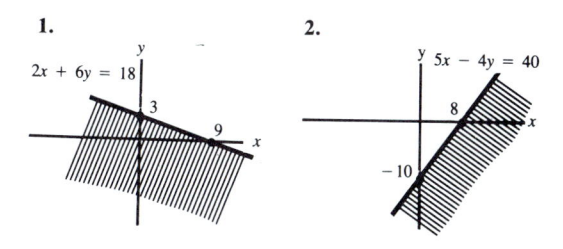

1.

2.

3.

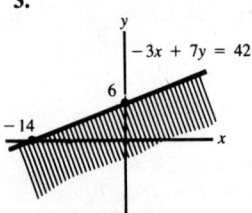

[Graph showing $-3x + 7y = 42$, with intercepts at -14 and 6]

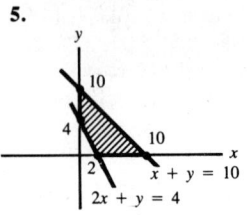

4.

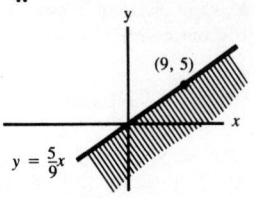

[Graph showing $y = \frac{5}{9}x$ through $(9, 5)$]

5.

[Graph showing $x + y = 10$ and $2x + y = 4$, with intercepts at 4, 10, 2, 10]

6.

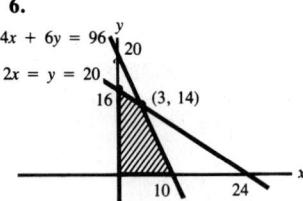

[Graph showing $4x + 6y = 96$, $2x - y = 20$, with point $(3, 14)$, values 20, 16, 10, 24]

7.

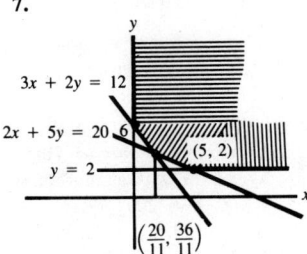

[Graph showing $3x + 2y = 12$, $2x + 5y = 20$, $y = 2$, with point $(5, 2)$ and $\left(\frac{20}{11}, \frac{36}{11}\right)$]

8.

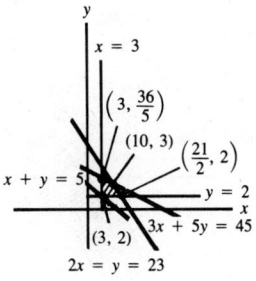

[Graph showing $x = 3$, $x + y = 5$, $y = 2$, $3x + 5y = 45$, $2x - y = 23$, with points $\left(3, \frac{36}{5}\right)$, $(10, 3)$, $\left(\frac{21}{2}, 2\right)$, $(3, 2)$]

9.

[Graph showing $x = 5$, $100x + 125y = 2000$, $4x + 3y = 72$, with points $(5, 12)$, $(15, 4)$, values 24, 16, 5, 18, 20]

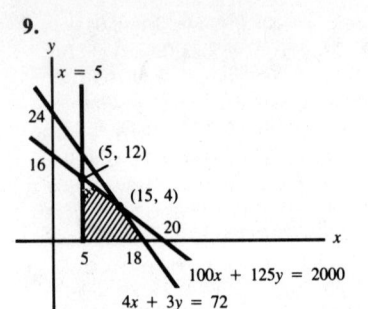

10. maximum of 40 at (0, 10) **11.** maximum of 57 at (3, 14) **12.** minimum of $^{244}/_{11}$ at $(^{20}/_{11}, ^{36}/_{11})$ **13.** minimum of 38 at (3, 2); maximum of 90 at any point on the line $3x + 5y = 45$ between $(3, ^{36}/_5)$ and (10, 3), inclusive
14. maximum of 77,400 if buy 15 grinding machines and 4 polishing machines **15.** minimum of $195,000 with 15 ads in *Bugle*, 20 in *Clarion* **16.** maximum of 25.5 when $x_1 = 3$, $x_2 = 6$, $x_3 = 12$ **17.** maximum profit of $4620 when make 0 small, 90 medium, 110 large **18.** maximum of $^{40}/_3$ at $x_1 = x_2 = ^{8}/_3$ **19.** minimum of 4 when $x_1 = 0$, $x_2 = 10$, $x_3 = 6$ **20.** minimum cost of $130 when stock 10 of Brand A, 20 of Brand B, 0 of Brand C **21.** $x_1 = 0.3$, $x_2 = 0.8$, $F = 24$ **22.** $x_1 = ^{1}/_2$, $x_2 = ^{2}/_3$, $x_3 = 0$, $F = 3.50$
23. $F = 95$, any values such that $y_3 - y_2 = 2$, $y_4 - y_5 = 9$

5 SUPPLEMENTARY EXERCISES

	Shadow	Range		Shadow	Range
1.	1	100, 150.	3.	3.5	$^{70}/_9$, 10
	2	40, 60		0.005	7000, 9000

	Shadow	Range		Shadow	Range
				1.5	0, 3
5.	1.52	112,000, $261^{1}/_3 \times 1000$	7.	0.5	3, 8
	0.92	$^{360}/_7 \times 1000$, 120,000		0.5	2, 7
				0	4, ∞

	Shadow	Range
9.	0.1	6000, ∞
	0.02	0, 10

	Shadow	Range
11.	0.2	152, ∞ (one ≤ constraint)

	Shadow	Range
13.	0.25	−50, 50 (one ≤ constraint)

15. no slack variables

6.1 EXERCISES

1. {2}; $^{1}/_6$ **3.** {3, 5}; $^{1}/_3$ **5.** {1, 2, 3, 4, 5, 6}; 1 **7.** {(1, 1)}; $^{1}/_{36}$ **9.** {(1, 3), (2, 2), (3, 1)}; $^{1}/_{12}$ **11.** {(1, 1), (1, 2), (1, 3), (1, 4), (1, 5), (2, 1), (2, 2), (2, 3), (2, 4), (3, 1), (3, 2), (3, 3), (4, 1), (4, 2), (5, 1)}; $^{15}/_{36}$ **13.** {*hh*}; $^{1}/_4$ **15.** {*tt*}; $^{1}/_4$ **17.** {*ht, th, tt*}; $^{3}/_4$ **19.** {*ht, th, tt*}; $^{3}/_4$ **21.** {KS, KH, KD, KC}; $^{1}/_{13}$ **23.** {AH}; $^{1}/_{52}$ **25.** {KS, KH, KD, KC, QS, QH, QD, QC}; $^{2}/_{13}$ **27.** {KH}; $^{1}/_{52}$ **29.** ∅; 0 **31.** 2 to 1

33. 1 to 1 **35.** 9 to 1 **37.** 11 to 9 **39.** $2/5$ **41.** $1/8$
43. $5/7$ **45.** $8/19$ **47.** 3 to 2 **49a.** 9 to 1 **b.** $1/10$
51. motor: 4999 to 1; air: 4,999,993 to 7; fall: 624,971 to 29; drown: 9,999,783 to 217; fire: 9,999,803 to 197 **53.** B, A, C, D **55.** $P(A) = 2/3$, $P(B) = 1/3$ **57.** $0 \le P \le 0.4$ **59.** $0 \le P \le 2/9$ **61.** 0.75 **63.** 0.005 **65.** 0.055 **67.** 0.600 **69.** $7/12$ **71.** $11/19$ **73.** $15/38$ **75.** P(the product will be successful) = 1 **77.** P(the product will be successful) = 0.40 **79.** P(the product will be successful) x, $0.40 < x < 4/9$ **81.** $2/3$ **83.** 0.605 **85.** 0.315 **87.** 0.445 **89.** 0.247 **91.** 0.485

6.2 EXERCISES

1. $525/1065$ **3.** $170/1065$ **5.** $735/1065$ **7.** $200/1065$ **9.** $100/1065$ **11.** $500/1065$ **13.** $14/72$; restaurant is in the North **15.** $57/72$; restaurant is not in the South **17.** $3/72$; restaurant is a Pizza House in the East **19.** $7/24$ **21.** $43/72$; restaurant is either in the East or the West **23.** $65/72$ **25.** $2/3$ **27.** 0.377 **29.** 0; A' is impossible **31.** $12/13$ **33.** $9/13$ **35.** 0.30 **37.** $6/7$ **39.** no, for then $P(A \cup B) = 1.1 > 1$, which is impossible **41.** $7/11$ **43.** 0.56 **45.** no, $P(A \cup B) \ne P(A) + P(B)$ **47.** $8/52$ **49.** $26/52$ **51.** $25/52$ **53.** $1/4$ **55.** $1/2$ **57.** 0.04 **59.** 0.75 **61.** 0.12, an account for $100 through $500 is paid late **63.** 0.64, an account has less than $100 or more than $500 **65.** $8/235$; allowed 8 hours sleep and recalled none of list **67.** $150/235$; allowed 4 hours sleep or recalled part of list **69.** $141/235$; allowed 0 hours sleep or recalled entire list or none of it **71.** 1

6.3 EXERCISES

1. $200/540$ **3.** $175/335$ **5.** $10/27$ **7.** $8/21$ **9.** $12/23$; affiliated with Pizza House if it is in the West **13.** $9/52$; affiliated with Quakey's if it is not in the East **15.** $35/46$ **17.** $P(A|B) = 0.75$, $P(B|A) = 0.50$ **19.** no **21.** 0.6 **23.** 0.4 **25.** 0.4 **27.** $3/5$ **29.** $2/5$ **31.** $4/15$ **33.** $6/37$ **35.** $36/37$ **37.** $1/4$, the account will be paid late if it is between $100 and $500 **39.** $3/7$, the account is less than or equal to $500 given that it is never paid **41.** $42/117$; allowed 8 hours sleep given that recalled part of list **43.** $33/118$; allowed 4 hours sleep given that recalled entire list or none of it **45.** $67/189$, a person has no sleep given that did not recall any of the list **47.** $7/18$, a person has either 8 hours or no sleep given that does not recall any of the list **49.** $8/25$ **51.** 0.06 **53.** $1/9$ **55.** $3/28$ **57.** $3/8$ **59.** $9/64$ **61.** $3/8$ **63.** $1/27$ **65.** $5/12$ **67.** $1/55$ **69.** $5/12$ **71.** 0.9975 **73.** 0.9510 **75.** 0.27 **77.** $1/2$ **79.** $1/2$ **81.** $16/81$ **83.** $8/27$ **85.** 0 **87.** $1/2$ **89.** $3/4$ **91.** $1/4$ **93.** $1/4$ **95.** $3/4$

6.4 EXERCISES

1. 5040 **3.** 362,880 **5.** 47,520 **7.** 360 **9.** 1 **11.** 60 **13.** 720 **15.** 120 **17.** 720 **19.** 10,000 **21.** 10,000,000 **23.** 9,900,000 **25.** 17,576,000 **27.** 15,625,000 **29.** 252 **31.** 20 **33.** 10 **35.** 220 **37.** 1 **39.** 1 **41.** 35

43. 4950×10^{11} **45.** 1 **47.** 210 **49.** 56 **51.** 495 **53.** ≈ 6.35 **55.** $1/90$ **57.** $1/15$ **59.** $3/10$ **61.** $8/15$ **63.** $7/15$ **65.** $22/425$ **67.** $\dfrac{28}{1105}$ **69.** $\dfrac{9}{1,000,000}$ **71.** $\dfrac{891}{1,000,000}$ **73.** $\dfrac{899,100}{1,000,000}$ **75.** $\dfrac{180}{1,947,792}$ **77.** $\dfrac{1,941,086}{1,947,792}$ **79.** 0.507 **81.** $1/18$ **83.** $5/36$ **85.** $160/429$ **87.** $202/429$ **89.** $3/55$ **91.** $5/26$ **93.** $7/13$ **95.** $1/9$

6.5 EXERCISES

1. Bernoulli; $n = 10$, $p = 1/2$ **3.** not Bernoulli; more than 2 outcomes possible on each trial **5.** Bernoulli; $n = 5$, $p = 1/13$ **7.** not Bernoulli; P(success) not the same on every trial **9.** $1/16$ **11.** $3/8$ **13.** 0.0010 **15.** 0.9990 **17.** 0.2252 **19.** 0.9992 **21.** $26/27$ **23.** $1/27$ **25.** 0.2066 **27.** 0.3902 **29.** 0.8291 **31.** 0.7164 **33.** 0.9656 **35.** 0.5016 **37.** 0.8145 **39.** 0.000475 **41.** 0.0563 **43.** 0.0035 **45.** $8/27$ **47.** $26/27$ **49.** 0.4095 **51.** 0.9186 **53.** 0.005 **55.** 0.0950 **57.** 0.4003

6.6 EXERCISES

1.

X	0	1	2	3
P(X)	$1/8$	$3/8$	$3/8$	$1/8$

3.

X	0	1
P(X)	$3/4$	$1/4$

5.

X	0	1	2
P(X)	$57/102$	$39/102$	$6/102$

7.

X	0	1	2	3
P(X)	$1/30$	$3/10$	$1/2$	$1/6$

9. 2 **11.** $3/2$; heads will occur 1.5 times more often in the long run. **13.** $1/4$; a heart will be chosen 1 in 4 times in the long run. **15.** $1/2$; a heart will be drawn once every two trials in the long run. **17.** 1.8; a woman will hold office once every 5 elections in the long run. **19.** $-\$0.50$ **21.** $21/11$ **23.** $3/4$ **25.** $12/13$ **27.** $-\$1.25$, no **29.** $\$1.50$ **31.** $-2/38 \approx -\$0.053$; $10/9 \approx \$1.11$ **33.** $-1/37 \approx -\$0.027$ **35.** $-\$0.59$ **37.** $82/97 \approx 0.85$ **39.** either job **41.** 5 dozen **43.** $3350 **45.** 5.48 ppm **47.** 2 **49.** 1 **51.** 2 **53.** 2 **55.** 0.683 **57.** 0.007 **59.** 0.89; the probability that a family will have fewer than 3 males who marry **61.** 1.151; the average number of males in a family who will marry **63.** 48.7% **65.** $7.1 million, $-4.1 million, $1.1 million, $-0.7 million; build small new plant **67.** $6.2 million; the decision maker should be willing to pay $6.2 to get perfect information about the state of nature.

6 REVIEW EXERCISES

1. $\mathcal{U} = \{hhh, hht, hth, htt, thh, tht, tth, ttt\}$ **2.** $\{hhh\}$ **3.** $\{hhh, hht, hth, thh\}$ **4.** $\{htt, tht, tth, ttt\}$ **5.** $\{ttt\}$

6. {htt, tht, tth, thh, hth, hht} 7. $\frac{1}{8}$ 8. $\frac{1}{2}$ 9. $\frac{1}{2}$ 10. $\frac{1}{8}$
11. $\frac{3}{4}$ 12. $\frac{330}{400}$ 13. $\frac{60}{400}$ 14. $\frac{70}{400}$ 15. $\frac{250}{400}$
16. $P(A) = \frac{3}{13}$, $P(B) = \frac{6}{13}$, $P(C) = \frac{3}{13}$, $P(D) = \frac{1}{13}$
17. $\frac{4}{21}$ 18. $\frac{185}{315}$ 19. $\frac{165}{315}$ 20. $\frac{210}{315}$ 21. $\frac{95}{315}$
22. $\frac{205}{315}$; not a compact car 23. $\frac{20}{315}$; a full-sized 2-door sedan 24. $\frac{295}{315}$; not a full-sized 2-door sedan 25. $\frac{185}{315}$; a 2-door sedan or a station wagon 26. $\frac{130}{315}$; a 4-door sedan
27. $\frac{125}{315}$; a station wagon or compact car 28. $\frac{20}{35}$
29. $\frac{20}{110}$ 30. $\frac{30}{55}$; a 4-door sedan if it is a full-sized car
31. $\frac{90}{185}$; a subcompact car if it is not a 4-door sedan
32. $\frac{6}{10}$ 33. $\frac{7}{10}$ 34. $\frac{2}{10}$ 35. 0.1 36. 0.9 37. $\frac{8}{10}$
38. $\frac{1}{169}$ 39. $\frac{24}{169}$ 40. $\frac{1}{13}$ 41. $\frac{25}{169}$ 42. $\frac{1}{221}$ 43. $\frac{32}{221}$
44. $\frac{1}{13}$ 45. $\frac{33}{221}$ 46. 100,000; 0.0001 47a. 32,760
b. $\frac{4}{15}$ 48a. 1365 b. $\frac{4}{15}$ 49. $\frac{1}{4}$ 50. $\frac{41}{55}$ 51. 0.1681
52. 0.8319 53. 0.9976 54. 0.0308 55. 0.8012
56. 1.5001 57. $\frac{1}{81}$ 58. $\frac{80}{81}$ 59. $\frac{8}{9}$ 60. $\frac{65}{81}$ 61. $\frac{16}{81}$
62. $\frac{20}{9}$

63.

X	0	1	2
P(X)	$\frac{105}{221}$	$\frac{96}{221}$	$\frac{20}{221}$

, $E(X) = \frac{8}{13}$

64. $-\$0.20$; no; $3 for winning 65. 220.4 66. $88,160
67. yes, expected profit from RVs is $107,200

6 SUPPLEMENTARY EXERCISES

1. choose a sample of size 99; if 4 or less are defective accept, otherwise reject.

3. $OC(x) = \sum_{k=0}^{4} \binom{99}{k} (0.01x)^k (1 - 0.01x)^{99-k}$

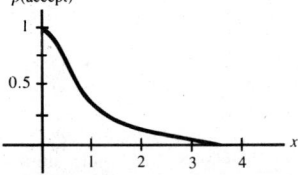

p(accept)

5a. $x = \#$ defective out of 8, $P(\text{accept lot}) = P(0$ of 3 are defective) $P(\text{accept lot}) =$ binomial probability $P\left(0 \text{ of } 3, p = \frac{x}{8}\right) = \left(1 - \frac{x}{8}\right)^3$

x	P(accept lot)
0	1.000
1	0.700
2	0.422
3	0.244
4	0.125
5	0.053
6	0.016
7	0.002
8	0.000

b. $x = \#$ defective out of 8; $P(\text{accept lot}) = P(0$ of 3 are defective) $= P(1\text{st not}) P(2\text{nd not} | 1\text{st not}), P(3\text{rd not} | 1\text{st \& 2nd not})$

x	P(accept lot)
0	1.000
1	0.625
2	0.357
3	0.179
4	0.071
5	0.018
6	0
7	0
8	0

7. Sampling plan is Test 71, accept lot if 1 or 0 are defective; $P(\text{accept lot}) = \sum_{k=0}^{1} \binom{71}{h} \left(\frac{x}{100}\right)^h \left(1 - \frac{x}{100}\right)^{71-k}$,

$x = \%$ of lot that is defective

x	P(accept lot)	AOQ
0	1.000	0.000
1	0.8412	0.0084
2	0.5835	0.0117
3	0.3676	0.0110
4	0.2182	0.0087
5	0.1241	0.0062

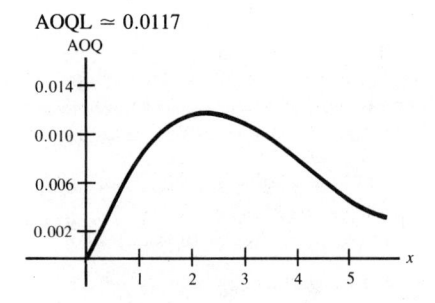

AOQL ≃ 0.0117

7.1 EXERCISES

1. Invest in savings account for expected return of $613.68
3a. $p < 0.409$ b. $p > 0.409$ c. $p = 0.409$

5. market A

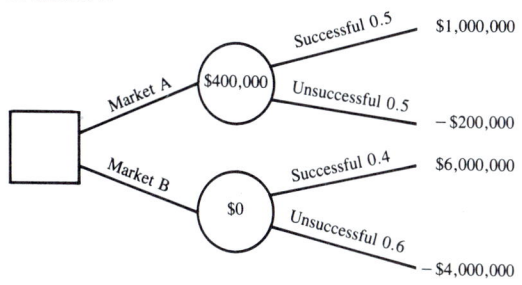

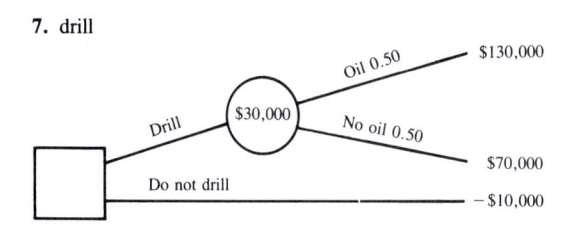

7. drill

9. Produce 60,000 for expected return of $590,000
11. Refine 80% for expected profit of 59,460,000 **13.** Buy 3000 bikinis for expected profit of $36,000 **15.** Do not make vaccine, expected cost of $14 billion **17.** Wait until next summer to decide about contract for expected cost of $33,000 **19a.** $p > 5/6$ **b.** $p < 5/6$ **c.** $p = 5/6$
21. Make either bet expected utility of 0

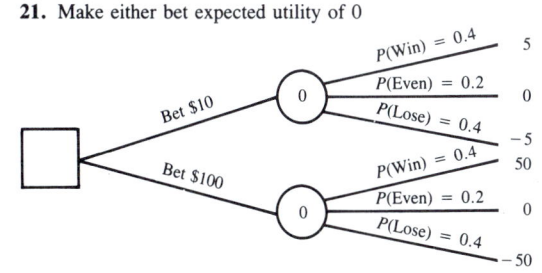

23. Choose either treatment for expected utility of -0.25

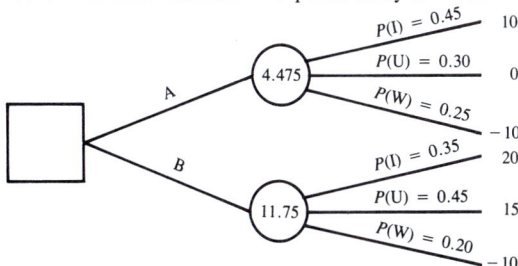

25. Choose treatment B for expected utility of 11.75

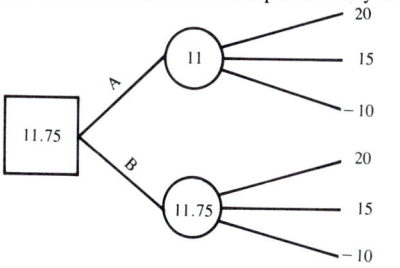

7.2 EXERCISES

1. $V_2 = [16, 14]$; $V_3 = [14.8, 15.2]$; $V_4 = [15.04, 14.96]$
3. $V_2 = [80, 240, 80]$; $V_3 = [80, 256, 64]$; $V_4 = [83.2, 259.2, 57.6]$ **5.** $V_2 = [3000, 1000, 2000]$; $V_3 = [2000, 3000, 1000]$; $V_4 = [1000, 2000, 3000]$ **7.** [15, 15]

9. $\left[\dfrac{600}{7}, \dfrac{1800}{7}, \dfrac{400}{7}\right]$ **11.** [2000, 2000, 2000]

13. $\begin{array}{c} \\ K \\ L \end{array} \begin{array}{cc} K & L \\ \begin{bmatrix} 0.8 & 0.2 \\ 0.3 & 0.7 \end{bmatrix} \end{array}$ **15.** [210, 140] **17.** [800, 450, 250], [810, 435, 255], [813.9, 429.15, 256.95]

19. Switch from $\begin{array}{c} \\ A \\ S \\ T \end{array}$ switched to $\begin{array}{ccc} A & S & I \\ \begin{bmatrix} 0.9 & 0.04 & 0.06 \\ 0.03 & 0.94 & 0.03 \\ 0.01 & 0.02 & 0.97 \end{bmatrix} \end{array}$

21. $[A, S, T] = [14.29, 28.57, 57.14]$ **23.** Coke: 30.85%; Pepsi: 36.65%; other: 32.51% **25.** Coke: 39.77%; Pepsi: 51.14%; other: 9.09% **27.** Superior: 29.98%, average: 42.94%, poor: 36.08% **29.** Good: 36,800, fair: 22,125, poor: 31,075 **31.** For example, the 40% of the patients in IC, are still in IC the next day. No patients who go out on one day are in IC the next day. **33.** $\begin{array}{c} \\ C \\ S \end{array} \begin{array}{cc} C & S \\ \begin{bmatrix} 0.94 & 0.06 \\ 0.02 & 0.98 \end{bmatrix} \end{array}$

35. [80,000, 240,000] **37.** Democrats: $42\tfrac{6}{7}\%$, Republican: $57\tfrac{1}{7}\%$ **39.** Overachievers: 43.1%, achievers: 41.4%, underachievers: 15.5% **41.** $p = 0.4$; $p = 0.54$

43. $\begin{array}{c} \\ NT \\ T \\ N \end{array} \begin{array}{ccc} NT & T & N \\ \begin{bmatrix} 0.5 & 0.4 & 0.1 \\ 0 & 0.9 & 0.1 \\ 0.9 & 0.1 & 0 \end{bmatrix} \end{array}$

7.3 EXERCISES

1. strictly determined; saddle point is (A_1, B_2); outcome is (A_1, B_2), $+2$ for A, -2 for B **3.** not strictly determined **5.** not strictly determined **7.** not strictly determined

9. not strictly determined **11.** strictly determined; saddle points are (A_4, B_3), (A_4, B_4); outcomes are (A_4, B_3), (A_4, B_4), $+2$ for A, -2 for B **13.** Gamma raises its price, Delta lowers its price; $+\$10$ million for Gamma, $-\$10$ million for Delta **15.** strictly determined; saddle point (1, 2); Gamma will raise price; Delta lowers; Gamma gains 20, Delta loses 20 **17.** A will choose A_4 and gain 2%, B will choose B_3 and lose 2% **19.** A: discuss issues or attack; B: discuss issues; B wins with 58% of the vote **21.** The convoy will choose the northern route and the allied commander will send heavy reconnaissance on the northern route. **23.** A: play A_1 $\frac{1}{2}$ of the time, A_2 $\frac{1}{2}$ of the time; B: play B_1 $\frac{7}{10}$ of the time, B_2 $\frac{3}{10}$ of the time; expected returns: $+\frac{1}{2}$ for A, $-\frac{1}{2}$ for B **25.** A: play A_2 $\frac{4}{5}$ of the time, A_3 $\frac{1}{5}$ of the time; B: play B_1 $\frac{1}{5}$ of the time, B_2 $\frac{4}{5}$ of the time; expected returns: $+\frac{9}{5}$ for A, $-\frac{9}{5}$ for B **27.** A: play A_2 $\frac{5}{8}$ of the time and A_3 $\frac{3}{8}$ of the time for expected return of $\frac{11}{8}$; B: play B_1 $\frac{1}{8}$ of time and B_3 $\frac{7}{8}$ of time for expected return of $-\frac{11}{8}$ **29.** Alpha: promote through magazines $\frac{4}{11}$ of the time, through retail outlets $\frac{7}{11}$ of the time; Beta: promote through retail outlets $\frac{7}{11}$ of the time, through combination $\frac{4}{11}$ of the time; expected returns: $-\frac{5}{11}$% for Alpha, $+\frac{5}{11}$% for Beta **31.** A: choose A_3 $\frac{3}{5}$ of the time and A_4 $\frac{2}{5}$ of the time for expected return of $\frac{22}{5}$; B: choose B_2 $\frac{4}{5}$ of the time and B_3 $\frac{1}{5}$ of the time for expected loss of $\frac{22}{5}$ **33.** A will attack B $\frac{1}{2}$ of the time and discuss issues $\frac{1}{2}$ of the time for expected gain in polls of $\frac{1}{2}$% per week. **35.** If they both confess, they each go to prison for 2 years. If one confesses and the other does not, the confessor goes free and the nonconfessor spends 5 years in jail. If neither confesses, they both go to jail for 1 year. The rational decision is to confess. **37.** Expected payoff for either to increase spending is 45, assuming equal chances of increasing and decreasing for each ration, with the same assumption, expected payoff of decreasing spending is -20.

7.4 EXERCISES

1. 0.3125 **3.** 0.5 **5.** 0.7766, 0.2234 **7.** 0.5 **9.** 0.917 **11.** 48/179 **13.** 161/179 **15.** $\frac{6}{13}$ **17.** $\frac{4}{7}$ **19.** $\frac{4}{9}$ **21.** $\frac{8}{21}$ **23.** $\frac{5}{7}$ **25.** $\frac{7}{12}$ **27.** $\frac{4}{13}$ **29.** Power supply: 0.68; disk drive: 0.18; CPU 0.14 **31.** 0.75 **33.** 0.3103; newspaper advertising **35.** $p = 0.444$ **37.** $p = 0.062$ **39a.** 0.124 **b.** 0.00025 **41.** 22.7% **43.** 0.156 **45.** $p = 0.271$

7.5 EXERCISES

1. Invester should buy the forecast and buy stocks if it says the market will be up. Otherwise, invest in savings.
3. do the survey; if survey says demand will be high, market the product; otherwise do not market it

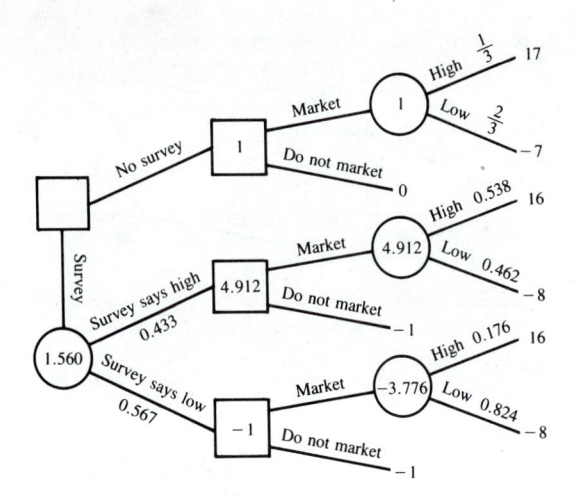

5. do not test, do not drill

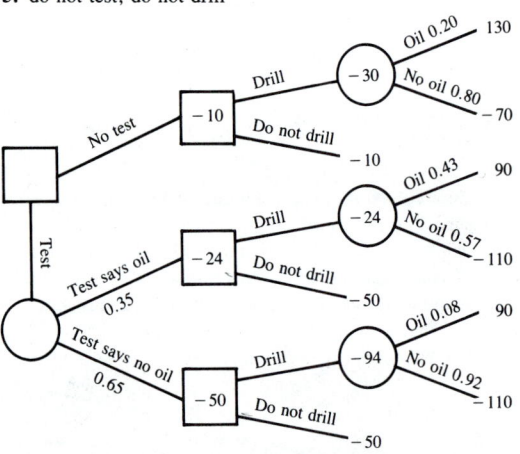

7. Do not hire the survey, produce 60,000 **9.** Oil company should buy the forecast and refine 100,000 if it says mild, 120,000 for normal, 150,000 for harsh **11.** The experts should not be hired and the flu vaccine should be produced **13.** call the specialist and follow his advice **15.** hire the consultant; if the consultant predicts success for the spots, use them, but if not, conduct the registration drive instead **17.** The town should make a more accurate estimate and contract now to spray the mosquitoes if it says the population will be large. Otherwise wait.

7 REVIEW EXERCISES

1. $V_2 = [3250, 4350, 5400]$; $V_3 = [4067.5, 3297.5, 5635]$; $V_4 = [3792.625, 3677.625, 5529.75]$ **2.** $V_2 = [400, 100]$; $V_3 = [100, 400]$; $V_4 = [400, 100]$ **3.** $V_2 = [1000, 2000,$

1000, 2000]; V_3 = [1000, 1500, 2000, 1500]; V_4 = [750, 2000, 1500, 1750] **4.** [3866.171, 3576.206, 5557.621]
5. No steady state vector. **6.** $\left[\frac{6000}{7}, \frac{12000}{7}, \frac{12000}{7}, \frac{12000}{7}\right]$
7.
$$\begin{array}{c c} & \text{cable TV} \quad \text{no cable TV} \\ \text{cable TV} & \begin{bmatrix} 0.95 & 0.05 \\ 0.10 & 0.90 \end{bmatrix} = T \\ \text{no cable TV} & \end{array}$$

8. 32.95% **9.** 67.67% **10.** strictly determined; outcome is (A_2, B_3), $+1$ for A, -1 for B **11.** A: play A_2 2/9 of the time, A_3 7/9 of the time; B: play B_1 1/9 of the time, B_3 8/9 of the time expected returns: $-11/9$ for A, $+11/9$ for B
12. Nation A will use diplomacy 7/12 and economic 5/12 and lose 1/12; Nation B will use diplomacy 7/12 and economic 1/12
13. ship 125,000

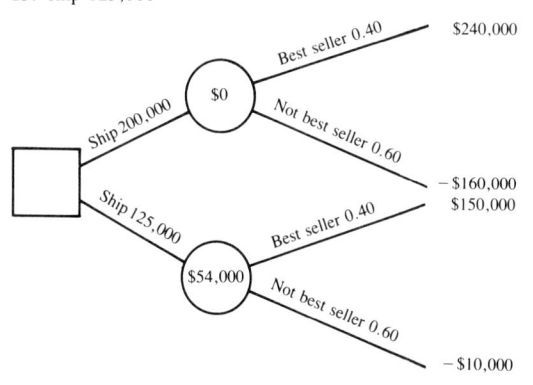

15. choose moderation

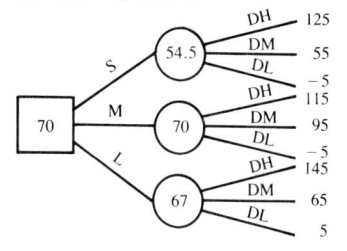

16. choose lowballing

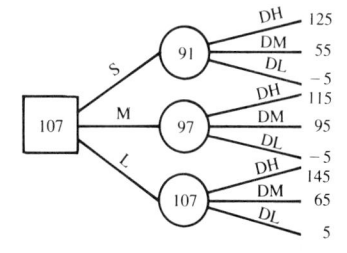

17. 0.4856 **18.** 0.7193 **19.** $P(A_3|B) = 0.048$
20. $P(\text{greenhouse}|\text{hot summers}) = 2/3$ **21a.** $P = 0.226$
b. $P = 0.200$ **c.** $P = 0.574$

14a. $p < 5/8$ **b.** $p > 5/8$

22. do not survey, ship 125,000

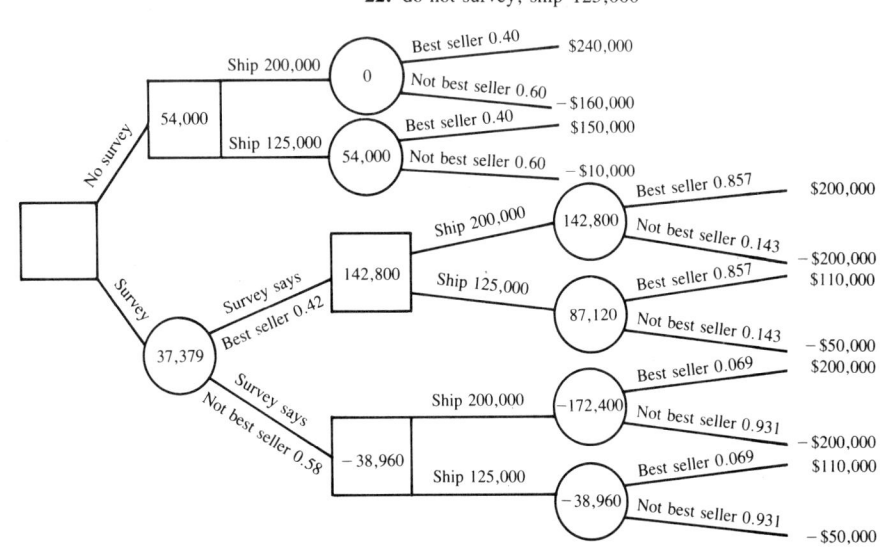

23. manufacturer should go with the survey. If it says demand low or high then lowball; if it says demand medium, then moderation.

7 SUPPLEMENTARY EXERCISES

NOTE: p_i = probability that row player plays row i; q_j = probability that column player plays column j

1. $p_1 = p_2 = \frac{1}{2}$; $q_1 = q_2 = \frac{1}{2}$ **3.** $p_1 = p_2 = p_3 = \frac{1}{3}$; $q_1 = \frac{1}{6}$, $q_2 = \frac{1}{3}$, $q_3 = \frac{1}{2}$ **5.** $p_1 = \frac{5}{21}$, $p_2 = \frac{1}{3}$, $p_3 = \frac{3}{7}$, $p_4 = 0$; $q_1 = q_5 = 0$, $q_2 = \frac{3}{4}$, $q_3 = \frac{10}{21}$, $q_4 = \frac{2}{21}$ **7.** $p_1 = p_3 = 0$, $p_2 = \frac{5}{9}$, $p_4 = \frac{4}{9}$; $q_1 = \frac{1}{9}$, $q_2 = \frac{8}{9}$, $q_3 = q_4 = 0$.

3.

Class	Relative Frequency	Cumulative Frequency
250–279.9	162/506	162
280–309.9	128/506	290
310–349.9	96/506	386
350–389.9	66/506	452
390–469.9	43/506	495
470–549.9	11/506	506

8.1 EXERCISES

1.

Class	Relative Frequency	Cumulative Frequency
0–4	8/62	8
5–9	12/62	20
10–14	22/62	42
15–19	13/62	55
20–24	7/62	62

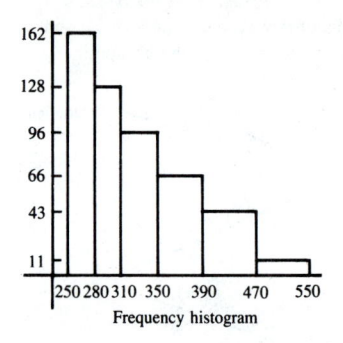

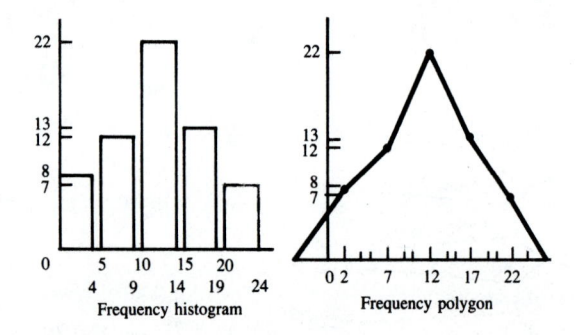

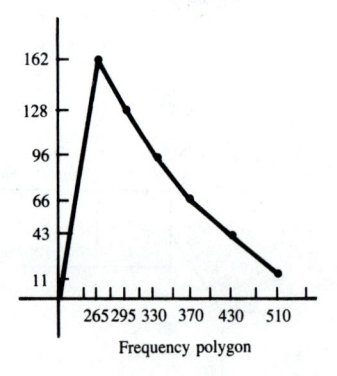

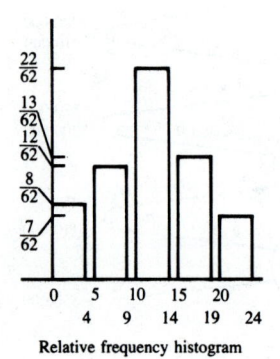

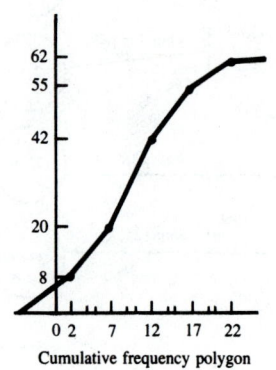

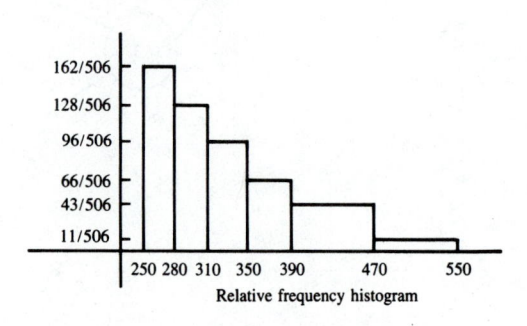

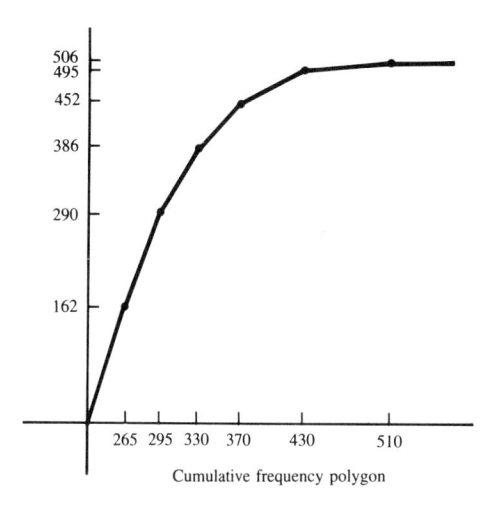

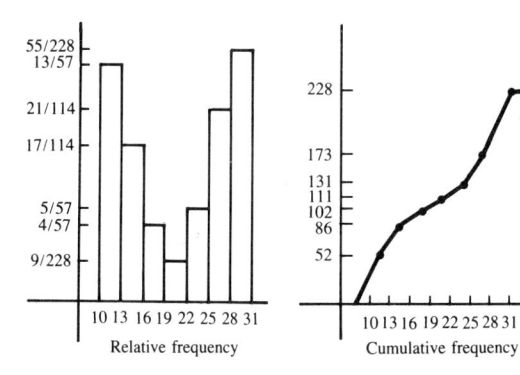

7.

Class	Frequency	Relative Frequency	Cumulative Frequency
6–11	2	$1/15$	2
12–17	4	$2/15$	6
18–23	9	$3/10$	15
24–29	11	$11/30$	26
30–35	4	$2/15$	30

9.

Class	Frequency
3.00–3.99	3
4.00–4.99	5
5.00–5.99	6
6.00–6.99	11
7.00–7.99	9
8.00–8.99	8
9.00–9.99	5
10.00–10.99	3

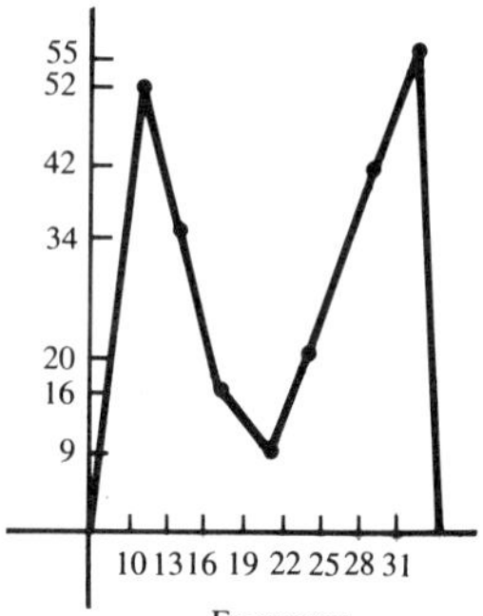

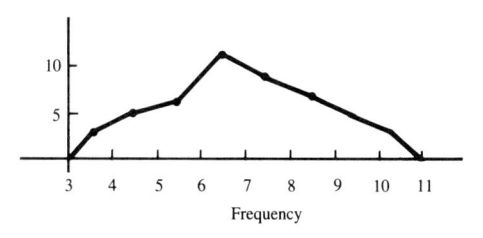

5.

Class	Frequency	Relative Frequency	Cumulative Frequency
10.00–12.99	52	$13/57$	52
13.00–15.99	34	$17/114$	86
16.00–18.99	16	$4/57$	102
19.00–21.99	9	$9/228$	111
22.00–24.99	20	$5/57$	131
25.00–27.99	42	$21/114$	173
28.00–30.99	55	$55/228$	228

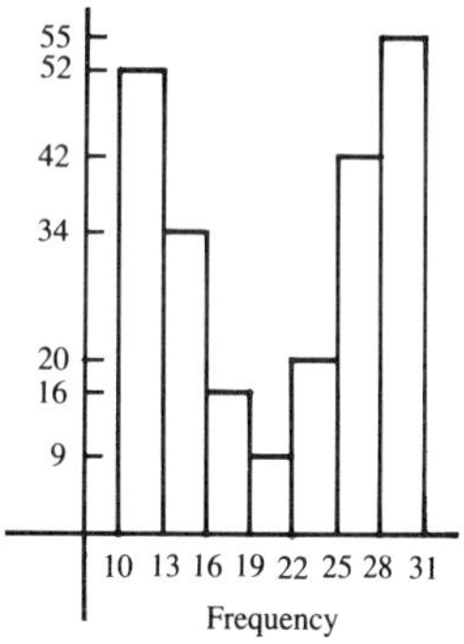

11.

Class	Frequency	Relative Frequency
6.0–9.9	4	$1/8$
10.0–13.9	3	$3/32$
14.0–17.9	5	$5/32$
18.0–21.9	9	$9/32$
22.0–25.9	8	$1/4$
26.0–29.9	3	$3/32$

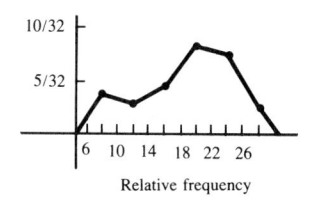

13.

Class	Frequency
1.0–1.4	3
1.5–1.9	6
2.0–2.4	11
2.5–2.9	18
3.0–3.4	6
3.5–3.9	3
4.0–4.4	2
4.5–4.9	1

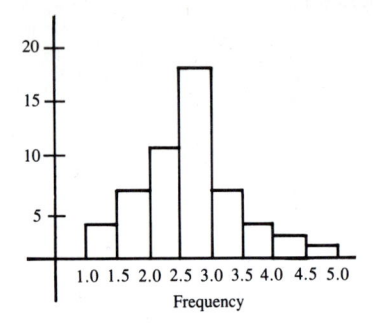

15.

Class	Frequency
0–601	2
601–700	3
701–800	10
801–900	13
901–1000	7
1001–1100	4
1101–1200	4
1201–1300	2
Over 1300	5

44.17 **39.** Stream A (based on coefficient of variation) **41.** $V_1 = 18.9$, $V_2 = 30.39$; the second one **43.** $V_{1970} = 220$, $V_{1977} = 199$, $V_{1980} = 207$; In 1977, the net worth of U.S. families was less variable than in 1970, and it has increased in 1980. **45.** 0.89 **47.** 0.80 **49.** 0.11 **51.** 0.095 **53.** 67.3% **55.** 84% **57.** 19.8% **59.** 13.2% **61.** [16.06, 33.94] **63.** [−15, 65] **65.** [14.43, 81.07] **67.** 19 weeks **69.** $\bar{x} = \$550$, $\sigma = \$4455$ **71.** the investments of exercise 70

8.4 EXERCISES

1. 0.50 **3.** 0.9735 **5.** 16% **7.** 2.5% **9.** 99.85% **11.** 50% **13.** 81.5% **15a.** 2.65% **b.** 2.58 cm **17.** 13.5% **19.** 0.15% **21a.** 0.5000 **b.** 0.5000 **23a.** 0.1700 **b.** 0.3888 **c.** 0.4974 **25a.** 0.0173 **b.** 0.3002 **c.** 0.0396 **27a.** 0.7166 **b.** 0.9909 **c.** 0.4226 **29a.** 0.6985 **b.** 0.8438 **c.** 0.9932 **31a.** 0.9980 **b.** 0.9564 **c.** 0.7389 **33.** 1.645 **35.** −0.675 **37.** −0.823 **39.** 1.285 **41.** 0.6730 **43.** 0.0401 **45.** 9.35 **47.** 66.75 **51a.** 0.0808 **b.** 0.2420 **c.** 0.6554 **d.** 0.9641 **53a.** 47.72% **b.** 43.32% **c.** 6.06% **d.** 62.47% **55a.** ≈ 2.0128 cm **b.** ≈ 1.9948 cm **57a.** ≈ 0 **b.** 0.9082 **c.** 0.0918 **59a.** 0.62% **b.** 9.94% **c.** 78.88% **d.** 9.94% **e.** 0.62% **f.** 99.38% **g.** 89.44% **61.** 0.0426 **63.** 0.7287 **65.** 0.0125 **67.** 1000, 156, 1156 **69.** 1800, 478, 2278

8.2 EXERCISES

1. 28 **3.** 38 **5.** 15 **7.** 168 **9.** 34 **11.** −⅙ **13.** 50 **15.** $\bar{x} = 4.46$ **17.** $\mu = 61.75$ **19.** $\mu = 2.132$ **21.** $\bar{x} \approx 70.0$ **23.** $\mu \approx 14.64$ **25.** $\bar{x} = \$29,420$ **27.** $\bar{x} = \$121,000$ **29.** $\bar{x} = 12.4$ **31.** $\bar{x} = 3.51$ years **33.** $\bar{x} = 2.55$ **35.** $\mu = 110.4$, so scores are too high **37.** 1970: $\bar{x} = \$43,225$; 1977: $\bar{x} = \$149,650$; 1983: $\bar{x} = \$254,300$ **39.** 316.5 **41.** 18 **43.** 5708.5 **45.** \$27,200 **47.** 5.05 million **49.** 2 **51.** no mode **53.** 50, 62, 73 **55.** 12 **57.** median: 39,600 **59.** two modes: \$16,000, \$42,000 **61.** mode: Brand X **63a.** mean: 56.4; median: 57.5; mean **b.** mean: 9.58; median: 10.2; median **65a.** first diet: mean = 18.95, median = 16.55; second diet: mean = 18.50, median = 22.0 **b.** second diet **67.** median: ≈ \$4.52 million; mode: ≈ \$3.745 million **69.** median: ≈ 108.5; mode: ≈ 104.5 **71.** $\bar{x} \approx 3.014$, $\bar{x}_T \approx 3.336$, $\bar{x}_B \approx 2.72$; no

8.3 EXERCISES

1. 79 **3.** 1.1 **5.** $\sigma^2 = 8.1$, $\sigma = 2.8$ **7.** $s^2 = 0$, $s = 0$ **9.** $\sigma^2 = 2.68$, $\sigma = 1.64$ **11.** $s^2 \approx 34.6$, $s = 5.9$ **13.** $\sigma^2 \approx 62.9$, $\sigma = 7.9$ **15.** $V_A = 31.93$, $V_B = 43.27$; B **17.** $V_A = 9.67$, $V_B = 378.02$; B **19.** $V = 49.17$ **21.** $s^2 = \$0.597$ million, $s = \$0.773$ million **23.** $\sigma^2 = \$31.02$ thousand, $\sigma = \$5.57$ thousand **25.** $s^2 = 0.79\%$, $s = 0.88\%$ **27.** $V_S = 23.3$, $V_P = 307.85$; profits **29.** $\sigma^2 \approx 35.2$, $\sigma \approx 5.93$ **31.** $V_A = 3.14$, $V_B = 2.77$; A **33.** $\sigma^2 = 3.19$, $\sigma = 1.786$ **35.** $\sigma^2 = 1251.6$, $\sigma = 35.38$ **37.** $V =$

8.5 EXERCISES

1. $\bar{y} = 4.2x + 5.5$; $r = 0.9899$ The line provides a good fit since $|r|$ is close to 1.

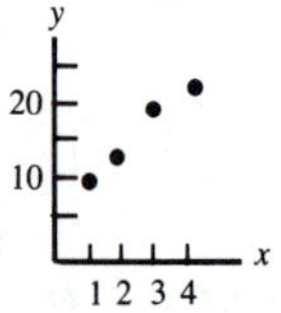

3. $\bar{y} = 0.10x + 8.4$; $r = 0.096$ The line does not provide a good fit since $|r|$ is close to 0.

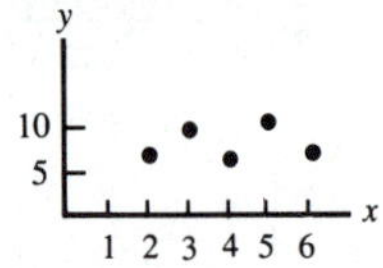

5. $\bar{y} = -1.973x + 82.87$; $r = -0.895$ The line provides a good fit.

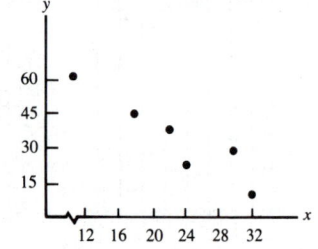

7. $\bar{y} = -0.06x + 36.07$; $r = -0.94$ The line provides a good fit.

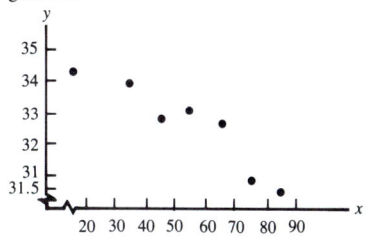

9a. $\bar{y} = -3x + 93.8$ **b.** $r = -0.966$; very good fit **c.** for each day missed, a student can expect the grade to drop 3 points **d.** 78.8 **11.** $\bar{y} = 0.000481x + 0.1887$ where y is in millions of babies and x is in pairs of brooding storks; $r = 0.989$; this seems to suggest that newborn babies are dependent on brooding storks. **13a.** $\bar{y} = 26.61 - 9.53x$ **b.** sample variation in price per gallon explains 96% of sample variation in weekly sales **c.** each increase of \$1 in price per gallon leads to decrease of 953 gallons in average weekly sales **d.** 1278 gallons **15.** $\bar{y} = -0.375x + 9.87$, where x is price in dollars per barrel and y is profit in billions of dollars; for each dollar change in price, the profit changes by $-\$0.375$ billion **b.** $r = -0.89$; 79.2% of change in profit is related to change in price **c.** \$3.95 billion **17.** $\bar{y} = 5.86x + 56.95$, where x is in years with $x = 1$ representing 1971; $r = 0.88$; on average, for each passing year there is an increase of 5.86 months construction time

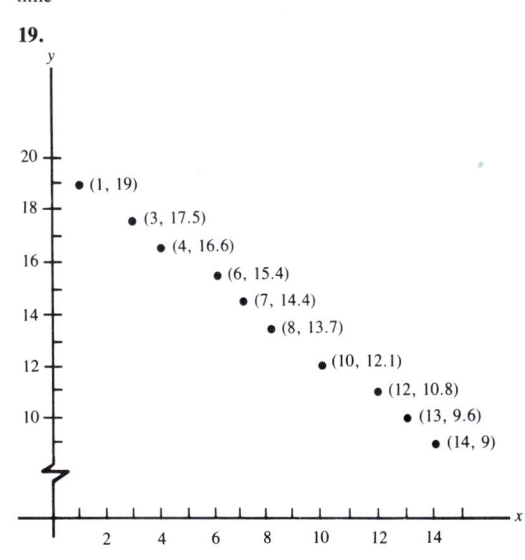

19.

$-0.77X + 19.81$; time until tumor disappears is decreased by approximately 0.77 days **21.** $\bar{y} = -1.98x + 65.06$, where $x = 1$ in 1960, $x = 2$ in 1964, ..., $x = 7$ in 1984 **23.** $\bar{y} = 0.35x + 4.757$, where $x = 1$ in 1960, $x = 2$ in 1964, ..., $x = 7$ in 1984. **25.** $r = -0.93$; there is a significant inverse relationship between voter participation and hours of TV watching **27.** $\beta = 1.04$; S_2 is less volatile than S_1.

8 REVIEW EXERCISES

1.

Score	Frequency	Relative Frequency	Cumulative Frequency
50–59	2	$1/15$	2
60–69	12	$2/5$	14
70–79	4	$2/15$	18
80–89	10	$1/3$	28
90–99	2	$1/15$	30

2. $\bar{x} = 10.25$, median $= 10$, modes $= 8, 12$ **3.** $\bar{x} = 8.67$, median $= 8.5$, mode $= 8.0$ **4.** $\bar{x} = 270.19$ **5.** $\bar{x}_1 = 262.18$, $\bar{x}_2 = 270.23$, $\bar{x}_3 = 269.05$, $\bar{x}_4 = 279.3$ **6.** \$1,240,498 **7.** $\bar{x} = 3.86$ minutes **8.** mean $=$ median $= 5.5$, mode $= 3.0$; mean or median **9.** mean $= 2.25$, median $= 2.0$, modes $= 2.0, 3.0$; modes **10.** mean $= 7.0\%$, median $= 9.5\%$, mode $= 10\%$; median **11.** range $= 10$, $\sigma^2 = 10.69$, $\sigma = 3.270$ **12.** range $= 10$, $s^2 = 12.03$, $s = 3.468$ **13.** $s^2 = 5.697$, $s = 2.387$ **14.** retailers **15.** $s^2 = 1.09$, $s = 1.04$ **16a.** 93.75% **b.** [23, 51] **17.** $\approx [0, 1,311,000]$ **18.** 0.4974 **19.** 0.6469 **20.** 0.2216 **21.** 0.8855 **22.** 0.9987 **23.** 0.0179 **24.** 0.0643 **25.** 0.9332 **26.** 28.25 **27.** 13.665 **28.** 21.275 **29.** 16.625 **30.** 17.09 **31.** 13.3 **32.** 20.625 **33.** 16.75 **34a.** 0.13% **b.** 0.8790% **c.** 0.8413% **d.** 9.05% **e.** 0.0078 **f.** 32.08% **35a.** ≈ 3.736 **b.** ≈ 2.296 **36.** $\bar{y} = 6.04x + 14.06$; $r = 0.9996$; the line provides a very good fit.

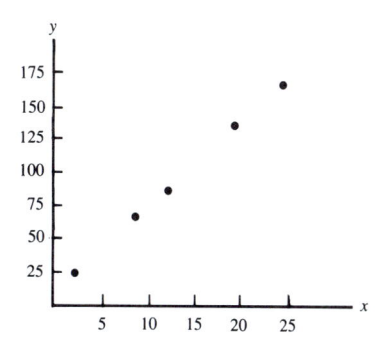

37. $\bar{y} = -0.374x + 56.01$; $r = -0.124$; the line does not provide a good fit.

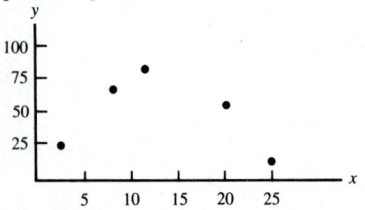

38a. $\bar{y} = 114.4x + 210.4$ **b.** $r = 0.994$ **c.** on average, for each increase of 1 ton of fertilizer used per acre, corn yields will increase by 114.4 bushels per acre **d.** 359.1

8 SUPPLEMENTARY EXERCISES

1. [40.36, 43.65]; [40.04, 43.96]; [39.43, 44.58] **3.** 393
5. [11.914, 12.046]; [11.9016, 12.0584]; [11.8768, 12.0832] **7.** 89 **9.** 99.74% **11.** 543 **13.** 1777

9.1 EXERCISES

1. 12, 19, 26, 33; arithmetic; 7 **3.** 3, 6, 11, 18; neither
5. 3, 12, 48, 192; geometric; 4 **7.** 2, $\frac{3}{2}$, $\frac{4}{3}$, $\frac{5}{4}$; neither
9. 2, 4, 8, 16; geometric; 2 **11.** neither **13.** arithmetic; $\frac{2}{3} + \frac{7}{3}(n-1)$ **15.** neither **17.** geometric; 3^n **19.** neither
21. arithmetic; $68 - 52(n-1)$ **23.** $S_{12} = 618$ **25.** $S_7 = \frac{127}{64}$ **27.** $\frac{110}{27}$ **29.** $S_{100} = 5050$ **31.** $S_{1000} = 2,100,500$ **33.** $S_n = \frac{(n)(n+1)}{2}$ **35.** $A = \$6800$
37. $45,455 **39.** 20 years **41.** 50% **43.** $43,000; $331,000 **45.** $365; $66,795 **47.** 420 days **49a.** $a_n = 20,000 (0.75)^{n-1}$ **b.** $a_7 = \$3559.58$ **c.** $69,321.29

9.2 EXERCISES

1. $16,200 **3.** $16,236.48 **5.** $22,472 **7.** $22,536.50
9. $9011.14 **11.** $73,466.40 **13.** $74,297.37
15. $9414.22 **17.** $31,476.91 **19.** $33,082.55 **21.** 12%
23. 12.55% **25.** 6.14% **27.** 10.2% **29.** Bank A **31.** $i > 8.3\%$ **33.** $6755.64 **35.** $11,467.94 **37.** $9505.12
39. $9456.37 **41.** $23,709.44 **43.** $20,266.93
45. $34,946.25 **47.** $31,952.75 **49.** yes **51.** yes
53. purchase Z **55.** invest in A and B **57.** invest in A and C **59.** $92,723.11 **61.** they would pay the same amount on either loan **63.** $9891.47 **65.** $16,746.41

9.3 EXERCISES

1. $2467.11 **3.** $7867.22 **5.** $4291.48 **7.** $22,113.51
9. $1348.67 **11.** $3305.39 **13.** $55,906.37
15. $1,073,278 **17.** $64,272.60 **19.** $2162.29
21. $1860.79 **23.** $508.44 **25.** $2.19 **27.** $21.87
29. $2479.45 **31.** $1362.16 **33.** $26,104.59
35. $5,727,500 **37.** $5652.44 **39.** $786,844.42
41. $8175.72 **43.** $8584.32 **45.** $42,580.32
47. $43,257.31 **49.** $44,632.42 **51.** $1903.86
53. $2,454,536.90 **55.** $8339.23 **57.** $48,448.19
59. $7598.21 **61.** $1,228,913.40 **63.** yes **65.** Z
67. none **69.** $i \leq 17.7\%$ per year compounded monthly; $i \geq 23.35\%$ per year compounded monthly **71.** $4386.12
73. $8622.57 **75.** 9.68%

9 REVIEW EXERCISES

1. $10,774.84; $12,525.45 **2.** $132,664.89; $142,231.48
3. 12.68% **4.** 12.96% compounded monthly **5.** $i > 14.752\%$ **6.** $7991.87 **7.** $2268.85 **8.** purchase Beta
9. Bank X **10.** $9131.10 **11.** $22,880.98 **12.** $2387.55
13. $60,332,833 **14.** $872,981.23 **15.** $3412.08 **16.** N
17. N **18.** none **19.** $24,837.58 **20.** $5250.11

9 SUPPLEMENTARY EXERCISES

1.

Payment	Unpaid Balance before Payment	Monthly Payment	To Interest	To Principal	Unpaid Balance after Payment
1	$5,000.00	$953.81	$200.00	$753.81	$4,246.19
2	$4,246.19	$953.81	$169.85	$783.96	$3,462.23
3	$3,462.23	$953.81	$138.49	$815.32	$2,646.91
4	$2,646.91	$953.81	$105.88	$847.93	$1,798.97
5	$1,798.97	$953.81	$ 71.96	$881.85	$ 917.12
6	$ 917.12	$953.81	$ 36.68	$917.12	$ 0.00

Solutions to Exercises

3.

Payment	Unpaid Balance before Payment	Monthly Payment	To Interest	To Principal	Unpaid Balance after Payment
1	$10,000.00	$1,574.99	$250.00	$1,324.99	$8,675.01
2	$ 8,675.01	$1,574.99	$216.88	$1,358.11	$7,316.90
3	$ 7,316.90	$1,574.99	$182.92	$1,392.07	$5,924.83
4	$ 5,924.83	$1,574.99	$148.12	$1,426.87	$4,497.96
5	$ 4,497.96	$1,574.99	$112.45	$1,462.54	$3,035.42
6	$ 3,035.42	$1,574.99	$ 75.89	$1,499.10	$1,536.31
7	$ 1,536.31	$1,574.72	$ 38.41	$1,536.31	$ 0.00

5.

Month	Unpaid Balance before Payment	Monthly Payment	To Interest	To Principal	Unpaid Balance after Payment
1.	$100,000.00	$1,101.09	$1,000.00	$ 101.09	$99,898.91
2.	99,898.91	1,101.09	998.99	102.10	99,796.81
3.	99,796.81	1,101.09	997.97	103.12	99,693.69
4.	99,693.69	1,101.09	996.94	104.15	99,589.54
.	.	.	.	.	.
.	.	.	.	.	.
.	.	.	.	.	.
100.	83,036.74	1,101.09	830.37	270.72	82,766.02
101.	82,766.02	1,101.09	827.66	273.43	82,492.59
102.	82,492.59	1,101.09	824.93	276.16	82,216.43
.	.	.	.	.	.
.	.	.	.	.	.
.	.	.	.	.	.
238.	3,234.46	1,101.09	32.34	1,068.75	2,165.71
239.	2,165.71	1,101.09	21.66	1,079.43	1,086.28
240.	1,086.28	1,086.28	0.00	1,086.28	0.00

7.

Month	Unpaid Balance before Payment	Finance Charge	Total Owed	Payment	Unpaid Balance after Payment
1.	$500.00	$7.50	$507.50	$126.88	$380.62
2.	380.62	5.71	386.33	96.58	289.75
3.	289.75	4.35	294.10	73.53	220.57
4.	220.57	3.31	223.88	55.97	167.91
5.	167.91	2.52	170.43	50.00	120.43
6.	120.43	1.81	122.24	50.00	72.24
7.	72.24	1.08	73.32	73.32	0.00

10.1 EXERCISES

1.

x	2.8	2.9	2.99	2.999	2.9999
$f(x)$	−1.16	−0.59	−0.0599	−0.005999	−0.0005999

x	3.2	3.1	3.01	3.001	3.0001
$f(x)$	1.24	0.61	0.0601	0.006001	0.0006000

$\lim_{x \to 3} (x^2 - 9) = 0$

3.

x	−0.1	−0.01	−0.001	−0.0001	−0.00001
$f(x)$	−10	−100	−1000	−10,000	−100,000

x	0.1	0.01	0.001	0.0001	0.00001
$f(x)$	10	100	1000	10,000	100,000

$\lim_{x \to 0} \frac{1}{x}$ does not exist.

5. does not exist **7.** 2 **9.** 2 **11.** does not exist **13.** 2
15. −1 **17.** does not exist **19.** −2 **21.** 2 **23.** 5
25. $\sqrt{3}$ **27.** 8 **29.** ⅓ **31.** 8 **33.** 243 **35.** does not exist **37.** 14 **39.** 0 **41.** 18 **43.** −17 **45.** 25
47. ³²¹⁄₁₆ **49.** −8 **51.** −36 **53.** does not exist **55.** ¹¹⁄₃
57. −8 **59.** ½ **61.** does not exist **63.** ¹⁶⁄₁₅

65. a. does not exist **b.** 0 **c.** 1

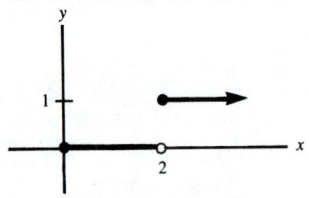

67. a. 3 **b.** $a, 6 - a$
c. does not exist 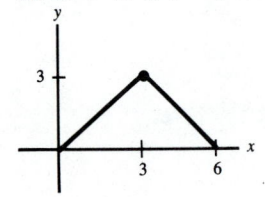

69. a. 4 **b.** 4 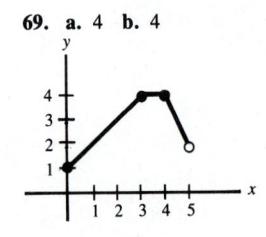
c. does not exist
d. $1 + a, 4, 12 - 2a, 0$

71.

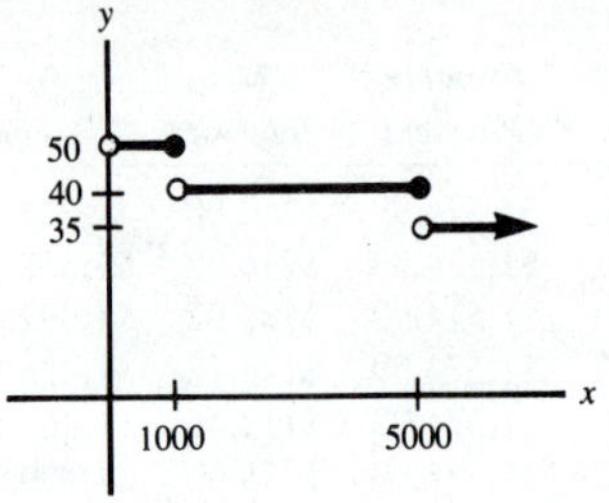

a. does not exist, does not exist **b.** 50, 40, 35

73.

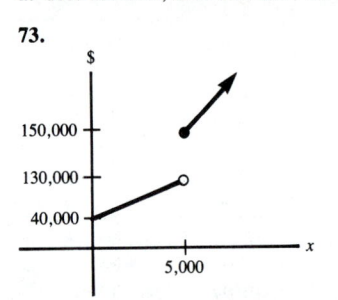

$40,000 + 18a$, does not exist, $50,000 + 20a$

75.

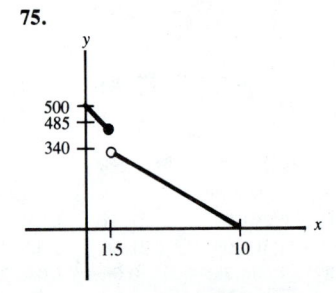

$500 - 10a$, does not exist, $400 - 40a$

10.2 EXERCISES

1. 1, −1, does not exist **3.** 2, 2, 2 **5.** 1, 2, does not exist, −1, 1, does not exist **7.** 2 **9.** 19 **11.** 0 **13.** 0
15. 0 **17.** 0, −1, does not exist **19.** 0, 0, 0 **21.** 1, −1, does not exist **23.** 20,000, 20,000, 31,000, 35,000
25. 10.50, 10.50, 46.50, 46.50 **27.** $+\infty, +\infty, x = 1$
29. $-\infty, -\infty, x = 3$ **31.** $+\infty, -\infty$ **33.** $-\infty, +\infty$
35. ½, ½ **37.** $-\infty, +\infty$ **39.** $+\infty, -\infty$ **41.** $+\infty, -\infty$
43. $300 billion, all can be removed at a cost of $300 billion
45. $+\infty$, cannot remove as much as 50% **47.** $+\infty$, all cannot be obtained **49.** $\lim_{x \to +\infty} f(x) = -1$, $\lim_{x \to -\infty} f(x) = 2$
51. $\lim_{x \to +\infty} f(x) = 3$, $\lim_{x \to -\infty} f(x) = 3$ **53.** $\lim_{x \to +\infty} f(x) = +\infty$, $\lim_{x \to -\infty} f(x) = 0$ **55.** $+\infty$ **57.** 0 **59.** 0 **61.** 1 **63.** $-\frac{2}{3}$

65. $+\infty$ **67.** 0 **69.** $5/11$ **71.** $+\infty$ **73.** vertical: $x = 2/3$; horizontal: $y = 2$; oblique: none **75.** vertical: $x = 3$; horizontal: $y = 5/2$; oblique: none **77.** vertical: $x = 0$; horizontal: $y = 3$; oblique: none **79.** vertical: $x = \pm 2$; horizontal: $y = 0$; oblique: none **81.** vertical: $x = 0$; horizontal: none; oblique: $y = 2x$ **83.** 5; as production increases avg. cost/unit approaches \$5 **85.** 300; as production increases worker-hrs required to build a unit approaches 300 **87.** 2; as time passes, numbers recalled approaches 2

10.3 EXERCISES

1. 1 **3.** 3 **5.** $-2x$ **7.** $3x^2$ **9.** $-3/x^2$ **11.** $1 - x^{-2}$ **13.** $2(x + 1)^{-2}$ **15.** $0.5(x + 1)^{-1/2}$ **17.** $y = x$ **19.** $y = 3x + 2$ **21.** $y = -6x + 10$ **23.** $y = 3x + 2$ **25.** $y = -3x/4 + 3$ **27.** $y = 2$ **29.** does not exist **31.** does not exist **33.** $x^{-2/3}/3$ **35.** all are differentiable everywhere **37a.** $(-\infty, -1), (-1, +\infty)$ **b.** $x = -1$ **c.** $x = -1, x = 2$ **39a.** $(-\infty, +\infty)$ **b.** none **c.** $x = \pm 2$ **41a.** $(-\infty, +\infty)$ **b.** none **c.** none **43a.** $(-\infty, 1), (1, +\infty)$ **b.** $x = 1$ **c.** $x = \pm 1$ **45a.** $(-\infty, 2), (2, +\infty)$ **b.** $x = 2$ **c.** $x = \pm 2, x = 0$ **47.** continuous for all x **49.** continuous on $(-\infty, 2), (2, +\infty)$, continuous where defined **51.** continuous on $(-\infty, -2), (-2, +\infty)$, continuous where defined **53.** continuous on $[0, 1), (1, 2), (2, 3]$, discontinuous at $x = 1, x = 2$ **55.** continuous on $[0, 3), (3, +\infty)$, discontinuous at $x = 3$ **57.** continuous on $(-\infty, 0), (0, +\infty)$, discontinuous at $x = 0$ **59.** $f(x) = \begin{cases} 0, & x = 0 \\ 12, & 0 < x \le 1 \\ 24, & 1 < x \le 2 \\ 36, & 2 < x \le 3 \\ 48, & 3 < x \le 4 \end{cases}$

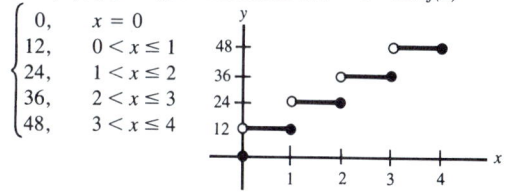

continuous on $(0, 1), (1, 2), (2, 3), (3, 4]$; discontinuous at $x = 0, 1, 2, 3$; nondifferentiable at $x = 0, 1, 2, 3, 4$ **61.** continuous on $(0, 1000), (1000, 5000), (5000, +\infty)$; discontinuous at $x = 1000, x = 5000$; nondifferentiable at $x = 1000, x = 5000$ **63.** continuous on $[0, +\infty)$; discontinuous nowhere; nondifferentiable at $p = \$1.50$

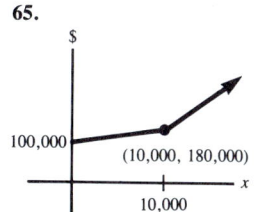

65.

continuous on $[0, +\infty)$; discontinuous nowhere; nondifferentiable at $x = 10,000$

67. $C(x) = \begin{cases} 9.00x, & 0 < x < 10 \\ 8.60x, & 10 \le x \le 24 \\ 8.20x, & x > 24 \end{cases}$

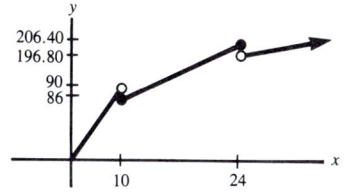

continuous on $(0, 10), (10, 24), (24, +\infty)$; discontinuous at $x = 10, x = 24$; nondifferentiable at $x = 10, x = 24$

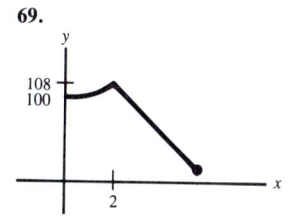

69.

continuous on $[0, 6]$; discontinuous nowhere; nondifferentiable at $x = 2$

10.4 EXERCISES

1. 0 **3.** 0 **5.** $7x^6$ **7.** $-2x^{-3}$ **9.** $-5x^{-6}$ **11.** $7x^{2/5}/5$ **13.** $-2.7x^{-3.7}$ **15.** $x^{-4/5}/5$ **17.** 3 **19.** $-6x^2$ **21.** $3x^{1/2}$ **23.** $-2x^{-1/2}$ **25.** $6x^{-3}$ **27.** $-x^{-4/3}$ **29.** -8 **31.** $2x - 5$ **33.** $16x - 12$ **35.** $3x^2 - 14x + 9$ **37.** $4x^3 - 9x^2 - 6x + 5$ **39.** $-200x^{99} + 297x^{98}$ **41.** $1.5x^{-1/2} - 9x^{-5/2}$ **43.** $-4x^{-2} - 2x^{7/5}/5$ **45.** $2x^{-2/3} + 5x^{-1/2}/2 + 2x^{-3/2}$ **47.** $12x - 3$ **49.** $-x^{-2}$ **51.** $y = 5x - 3$ **53.** $y = -7x + 10$ **55.** $y = 12x + 15$ **57.** $y = -13x/3 + 6$ **59.** $y = x/2 - 1$

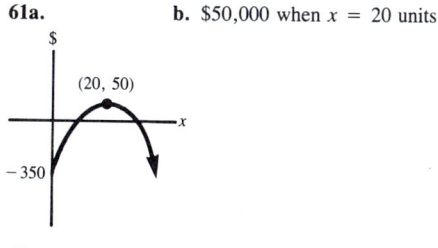

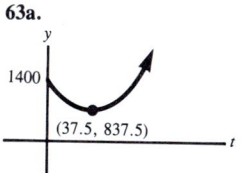

61a.   **b.** \$50,000 when $x = 20$ units

63a.

b. 837.5 units/day when $t = 37.5$ months **65.** $x = 10$, avg. cost = $200/unit **67.** $t = 4$, $t = 8$ months (April, August)

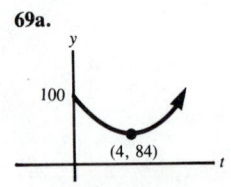

69a.

(graph showing y-axis with 100 marked, curve through point (4, 84), t-axis)

b. 84 cubs in 1984 **71.** $(2, 269.0\overline{6})$, $(5, 56.\overline{6})$ **73.** $(9, 6.7991)$, $(40, 3.82)$ **75.** ≈ -9.42, ≈ -11.11, ≈ -26.19; the approximate percentage changes brought about by a 1% rise in interest rates

10.5 EXERCISES

1. $v(t) = 40$ mph, $a(t) = 0$ mph^2 **3a.** 0.2, 6.4, 48.6 m **b.** 0.5, 4, 13.5 mps **c.** 0.75, 1.5, 2.25 mps^2 **5a.** 134, 206 fps **b.** 24 fps^2 for both **7.** 50 ft, 0 fps, neither (at peak), -4 fps^2 **9.** 0 ft, -20 fps, is hitting the ground, -4 fps^2 **11.** 233.75 ft, 3 fps, up, -2 fps^2 **13.** 233.75 ft, -3 fps, down, -2 fps^2 **15.** 236 ft when $t = 4$ sec **17.** $s(t) = 16t^2$ **19.** 64 ft, 64 fps, 32 fps^2 **21.** 1.6 mps^2 **23.** 0.968 m, 1.76 mps, 1.6 mps^2 **25.** 15.488 m, 7.04 mps, 1.6 mps^2 **27.** 0.5, 5, 50 $/unit **29a.** $104,000, $-$1000/month, decreasing **b.** $64,000, $-$5000/month, decreasing **31.** 12,000 units/$, 2,002,000 units/$ **33a.** 11 million units/$ **b.** 11.5 million units/$ **c.** $11\frac{2}{3}$ million units/$ **d.** 11.75 million units/$ **e.** 11.8 million units/$ **35a.** $216/unit **b.** $72/unit **c.** $0/unit **d.** $-$8/unit **e.** $-$8/unit **f.** $0/unit **g.** $27/unit **h.** $72/unit **37.** $t = 1$: -63 units/wk^2, decreasing; $t = 4$: 0 units/wk^2, neither; $t = 6$: 12 units/wk^2, increasing; $t = 8$: 0 units/wk^2, neither; $t = 9$: -15 units/wk^2, decreasing **39.** $t = 0.5$: 0.24 gps; $t = 1$: 0.12 gps; $t = 1.5$: 0.08 gps; $t = 2$: 0.04 gps **41a.** 0.6 ppm/million cars **b.** 2.4 ppm/million cars **c.** 5.4 ppm/million cars **d.** 6.3375 ppm/million cars **e.** 7.35 ppm/million cars **43.** $t = 1$: 72,000/day, increasing; $t = 2$: 0/day, neither; $t = 3$: $-80,000$/day, decreasing; $t = 4$: $-108,000$/day, decreasing; $t = 5$: 0/day, neither; $t = 6$: 352,000/day, increasing **45.** $t = 5$: forgets at rate of 0.2/min; $t = 10$: forgets at rate of 0.4/min; $t = 15$: forgets at rate of 0.6/min; $t = 25$: forgets at rate of 1/min **47.** $t = 5$: 0.219/month, increasing; $t = 9$: 0/month, neither; $t = 30$: -0.126/month, decreasing; $t = 40$: 0, neither **49.** $t = 10$: -0.001/day; $t = 5$: -0.004/day; $t = 1$: -0.1/day

10.6 EXERCISES

1. $145,000, $180,000, $35,000 **3.** $600 **5.** $80,000, $x = 600$ **7.** $532,500, $4400 **9.** $40,000, $x = 60$ **11.** $100 **13.** $-$300 **15.** $120,000, $120,000, $0 **17.** $400 **19.** $22,500, $x = 350$, $p = 525 **21.** $2,850,000, $45,500 **23.** $3,795,500, $x = 110$, $p = $39,600$ **25.** $300,000, $500 **27.** $-$212,500, $x = 50$, $p = 6000 **29.** $3,750,000, $x = 1250$, $p = 9875 **31.** $180,000, $x = 100$, $p = 2500 **33.** marg. avg. cost: $-$2/unit; marg. avg. rev.: $0/unit; marg. avg. profit: $2/unit **35.** marg. avg. cost: $-$195/unit; marg. avg. rev.: $-$230/unit; marg. avg. profit: $-$35/unit **37.** -1.0; $p = 20$: 60,000, $1,200,000; $p = 19: 63,000, $1,197,000; $p = 21: 57,000, $1,197,000 **39.** -1.5; $p = 150: 10,000, $1,500,000; $p = 153: 9700, $1,484,100; $p = 147: 10,300, $1,514,100 **41.** $-\frac{2}{3}$; $p = 100: 15,000, $1,500,000; $p = 102: 14,800, $1,509,600; $p = 98: 15,200, $1,489,600 **43.** -1; $p = 20: 40,000, $800,000; $p = 20.20: 39,598, $799,880; $p = 19.80: 40,398, $799,880 **45.** -0.2, -5.0 **47.** $p = 300; $p = 315: 1425, $448,875; $p = 285: 1575, $448,875 **49.** elastic: $p > 14.50; inelastic: $0 < p < 14.50; unit elasticity: $p = 14.50 **51.** $E(p) = -p(900 - p)^{-1}$ **53.** elastic, so increase in price will decrease revenue, decrease in price will increase revenue **55.** $E(p) = -2p^2(7500 - p^2)^{-1}$ **57.** inelastic, so increase in price will increase revenue, decrease in price will decrease revenue **59.** elastic: $p > 2; inelastic: $0 < p < 2; unit elasticity: $p = 2; revenue will increase **61a.** accounting **b.** neither **c.** marketing **63.** $E(q) = (q - 160,000)/q$ **65.** $-\frac{1}{3}$ **67.** -7, $-\frac{1}{3}$ **69.** -0.3

10.7 EXERCISES

1. $2x - 5$ **3.** $3x^2 - 10x + 6$ **5.** $2.5x^{-1/2} - 4x^{-1/3}/3$ **7.** $21t^6 - 2t^{-3}$ **9.** $-2x^{-2} + 12x^{-4}$ **11.** 9 **13.** $\frac{19}{2}$ **15.** $\frac{265}{6144}$ **17.** $3a^2 + 4a$ **19.** $-\frac{3}{2}$, $-\frac{2}{3}$ **21.** $-0.25t^{-2}$, $-0.25y^{-2}$ **23.** $6x + 2$, 6 **25.** $60x^3 - 72x$ **27.** 720 **29.** $12x^2 - 24x$, $24x - 24$ **31.** $12x^{-4} - 2x^{-3}$, $240x^{-6} - 24x^{-5}$ **33.** $12x^{-5}$, $-60x^{-6}$ **35.** $-2520x^{-8}$ **37.** 0 **39.** 0 **41.** 8, -18 **43.** $1680a^4 + 24$

10 REVIEW EXERCISES

1. does not exist **2.** $+\infty$ **3.** $-\infty$ **4.** 2 **5.** 2 **6.** 2 **7.** -1 **8.** 1 **9.** does not exist **10.** 0 **11.** 0 **12.** 0 **13.** -1 **14.** -1 **15.** -1 **16.** 0 **17.** 0 **18.** 0 **19.** $+\infty$ **20.** 1 **21.** 0 **22.** 4 **23.** 24 **24.** 13 **25.** $\frac{106}{9}$ **26.** -1 **27.** 1 **28.** does not exist **29.** $\frac{5}{4}$ **30.** $+\infty$ **31.** $+\infty$ **32.** does not exist **33.** $\frac{5}{2}$ **34.** $+\infty$ **35.** $\frac{1}{5}$ **36.** 0 **37.** $\frac{7}{2}$ **38.** $+\infty$ **39.** $\frac{5}{7}$ **40.** $+\infty$ **41.** $\frac{1}{4}$ **42.** 0 **43.** $x = 2$, $y = -\frac{3}{2}$ **44.** $x = 4$ **45.** $x = 0$, $y = x - 3$ **46.** $15x^2$, $y = 135x + 270$ **47.** $(3x)^{-2/3}$, $x = 0$ **48a.** $x = -1, 0, 3, 5$ **b.** $x = -1, 3$ **49.** continuous on $(-\infty, +\infty)$, no discontinuities, differentiable everywhere

50. continuous on $(-\infty, -5)$, $(-5, -4)$, $(-4, +\infty)$, no discontinuities, differentiable everywhere **51.** continuous on $(-\infty, 0)$, $(0, 2)$, $(2, +\infty)$, discontinuous at $x = 0$ and $x = 2$, nondifferentiable at $x = -1, 0, 1, 2$ **52.** 0 **53.** $\sqrt{3}$ **54.** $6x + 6$ **55.** $-6x^5 + 20x^4 - 36x^2 + 32x - 8$ **56.** $-6x^{-4} + x^{-2}$ **57.** $-8x^{-3} - 6x^{-4}$ **58.** $2x^{-1/2}$ **59.** $4x^{-1/5} + x^{-3/2} - 2x^{-5/3}$ **60.** $y = 2\sqrt{3}$ **61.** $y = 18x - 34$ **62.** $y = 3x + 12$ **63.** $y = -x/2 + 7/2$ **64.** $y = 24x/5 - 128/5$ **65.** $y = -x/729 + 34/27$ **66.** $t = 3$: vel $= 120$ mps, acc $= -10$ mps^2; $t = 6$: vel $= 90$ mps, acc $= -10$ mps^2; max altitude $= 1125$ m at 30 sec **67.** $-\$96,000$/yr, $-\$48,000$/yr, $-\$48/\sqrt{2}$ thousand/yr $\approx -\$33,941.13$/yr **68.** $\$72$, $x = 14$, $p = \$42$ **69a.** $-\frac{1}{3}$, -1, -3 **b.** $\$1,800,000$, $\$2,400,000$, $\$1,800,000$ **c.** $\$1,914,000$, $\$2,376,000$, $\$1,386,000$ **70.** elastic: $p > \$20$; inelastic: $0 < p < \$20$; unit elasticity: $p = \$20$; decrease, increase **71.** $8x^3 - 10x$, $24x^2 - 10$, 96 **72.** $x^{-1/2} - x^{-3/2}$, $-0.5x^{-3/2} + 1.5x^{-5/2}$ **73.** $9x^2 - 16x^{-1/5}/5$, $18x + 16x^{-6/5}/25$, $2154/125$ **74.** $4u + 3b + u^{-1/2}$, $4 - u^{-3/2}/2$ **75.** $8/11$, $11/8$

11.1 EXERCISES

1. $12x + 7$ **3.** $15x^2 + 2x - 10$ **5.** $4x^3 + 4x$ **7.** $-4/(x + 2)^2$ **9.** $-3/2\sqrt{x}(\sqrt{x} - 2)^2$ **11.** $-1/(2x - 1)^2$ **13.** $-22/(5x - 1)^2$ **15.** $(x^2 + 4x + 6)/(x + 2)^2$ **17.** $(7 + 6x - x^2)/3(3 - x)^2$ **19.** $4x(x^2 + 6x + 8)$ **21.** $-24/x^4$ **23.** $2(5x^{-2} + 8x^{-3})$ **25.** $-(3x^2 + 4x + 6)/(x - 2)^2$ **27.** $\dfrac{2x^2 + 6x + 2}{x^4 - 2x^2 + 1}$ **29.** $6x^5 + 5x^4 + 4x^3 + 6x^2 + 2x + 1$ **31.** $-(10x^2 + 24x + 45)/3(2x - 1)^2(x + 3)^2$ **33.** $(-30x^3 - 79x^2 + 88x + 159)/(5x + 11)^2$ **35.** $(-3x^6 + 8x^5 - x^4 + 4x^3 + 4x^2 + 4x + 4)/(x^2 + 1)^2(x^3 + 2)^2$ **37.** $6x^5 + 5x^4 + 4x^3 + 6x^2 + 1$ **39.** $(-311x^2 - 162x - 31)/(7x + 2)^2(x - 1)^2$ **41.** $t = 1$: $\approx \$3.3$ million/wk; $t = 2$: $\approx \$2.77$ million/wk; $t = 5$: $\approx \$1.77$ million/wk; total approaches $\$40$ million **43.** $x = 1$: $\approx \$0.494$/unit; $x = 10$: $\approx \$0.004$/unit; $x = 100$: $\approx \$0.00004$/unit; avg. cost per unit approaches $\$4$ **45a.** $-\$11.944$/unit, $-\$4.625$/unit, $\approx -\$0.227$/unit **b.** cost per unit approaches $+\infty$ **47.** $q = 5$: $-\$15.80$/unit; $q = 25$: $\approx -\$22.39$/unit; $q = 100$: $\approx -\$24.08$/unit; price approaches $\$0.25$ **49.** $t = 0$: ≈ 77.1 symbols/month; $t = 6$: ≈ 10.5 symbols/month; $t = 12$: ≈ 3.9 symbols/month; total number of symbols approaches 270 **51.** $x = 0$: 70%/repetition; $x = 2$: $\approx 7.78\%$/repetition; $x = 5$: $\approx 1.94\%$/repetition

11.2 EXERCISES

1. $20x - 5$; $20x + 13$ **3.** $9x^2 + 12x + 3$; $3x^2 + 6x + 1$ **5.** $8x^6 + 8x^3 + 2$; $16x^6 + 1$ **7.** $\sqrt{5x + 14}$; $\sqrt{5x + 4}$ **9.** $\dfrac{2x^4}{9}$; $\dfrac{3x^4}{4}$ **11.** $y = 5t^2 + 23$ **13.** $y = 2 - 9x + 9x^2 - 3x^3$ **15.** $u = 17w^{3/2} + 2w^{3/8} + 3$ **17.** $f(g(x))$, $g(x) = 3x + 5$, $f(x) = x^{10}$ **19.** $f(g(x))$, $g(x) = 5 - 7x$, $f(x) = x^{-4}$ **21.** $f(g(x))$, $g(x) = x^2 - 2x$, $f(x) = x^5$ **23.** $f(g(x))$, $g(x) = 4x - 1$, $f(x) = \sqrt{x}$ **25.** $f(g(x))$, $g(x) = x^2 + 7$, $f(x) = x^{3/2}$ **27.** $f(g(x))$, $g(x) = x^2 + 3$, $f(x) = x^{1/5}$ **29.** C $= 0.1b^{1/3}$; 1 **31.** $q = \dfrac{10,000}{0.02t + 5}$, ≈ 1953 units **33.** $R = \dfrac{10,000t + 2,500,000}{t + 250}$ **35.** $f(t) = \dfrac{100}{1.92t + 4.52}$ **37.** $30(3x + 5)^9$ **39.** $28(5 - 7x)^{-5}$ **41.** $10(x - 1)(x^2 - 2x)^4$ **43.** $100(10x - 3)(5x^2 - 3x + 1)^{99}$ **45.** $-3/2\sqrt{1 - 3x}$ **47.** $2x/3(x^2 + 1)^{2/3}$ **49.** $-3/x^2(1 - 2x^{-1})^{5/2}$ **51.** $x^2(5x^2 - 1)(35x^2 - 3)$ **53.** $x(2x^2 + 1)^{10}(x^3 - 1)^8(98x^3 + 27x - 44)$ **55.** $x^2(4x + 3)^{-2/3}(40x/3 + 9)$ **57.** $x(2x^3 - x - 1)/(x^2 - 1)^{1/2}(x^3 - 1)^{2/3}$ **59.** $-3(4\sqrt{x} - 3x^{-2})^{-4}(4x^{4/3} + 10)^{-4}(88x^{5/6}/3 + 8x^{-5/3} + 20x^{-1/2} + 60x^{-3})$ **61.** $2x(x^2 - 1)/\sqrt{(x^2 + 1)(x^2 - 3)}$ **63.** $-16(x + 2)^3/(x - 2)^5$ **65.** $(9 - x)/(x^2 + 3)^{3/2}$ **67.** $9x^{1/2}(1 + x^{3/2})^2/(1 - x^{3/2})^4$ **69.** $100t + 5$ **71a.** $\dfrac{4t^3}{(t^4 + 1)^{1/2}}$ **b.** $4t^3 + 4t$ **73a.** $-3w(w^2 + 1)^{-3/2}$ **b.** $\dfrac{-3}{w^4(1 - 5w^{-3})^2}$ **75.** s **77.** $-12V(3V^2 + 29)^{-2}$ **79.** MC $= 0.0477$, MR $= 0.9535$, MP $= 0.906$; the 11th unit would cost approximately 4.77¢, generate 95.35¢ revenue and result in a profit of 90.6¢ **81.** 150 **83.** -7.63 **85.** $\dfrac{34}{13^{3/2}18^{1/2}} \approx 171$ and $\dfrac{34}{50^{1/2} + 21^{3/2}} \approx 50$ individuals per yr; $\lim_{t \to +\infty} y = 2$, so population size will approach 2000 **87.** $t = 1$: $300,000/\sqrt{10,300} \approx 2956$ bacteria/hr; $t = 2$: $600,000/\sqrt{11,200} \approx 5669$ bacteria/hr; $t = 10$: $3,000,000/\sqrt{40,000} = 15,000$ bacteria/hr **89.** $\approx -0.486\%$ per year **91.** $t = 8$: ≈ -0.017/election; $t = 9$: ≈ -0.015/election **93a.** $\$771.84$/employee **b.** $-\$304.64$/employee

11.3 EXERCISES

1. $df = (6x - 2)dx$ **3.** $dy = 8(x^3 - 5x + 2)^7(3x^2 - 5)dx$ **5.** $dy = (5x^4 + 8x^3 + 8x + 8)dx$ **7.** $dy = [(-x^4 + 5x^2 + 2)/(x^3 + 2x)^2]dx$ **9.** $\Delta f \approx 3x^2\Delta x$, $\Delta f \approx 0.6$ **11.** $\Delta f \approx -x^{-2}\Delta x$, $\Delta f \approx -0.005$ **13.** $\Delta f \approx \Delta x/2\sqrt{x}$, $\Delta f \approx -0.025$ **15.** $\Delta f \approx x^{-2/3}\Delta x/3$, $\Delta f \approx -1.6$ **19.** $\Delta f/f \approx \Delta x/x$, ≈ 0.05, $\approx 5\%$ **21.** $\Delta f/f \approx 3\Delta x/x$, ≈ -0.071, $\approx -7.1\%$ **23.** $\Delta f/f \approx \Delta x/2x$, ≈ 0.0125, $\approx 1.25\%$ **25a.** $\Delta x/x$ ± 0.01, $\pm 1\%$ **27a.** $2\Delta r/r$ **b.** ± 0.02, $\pm 2\%$ **29a.** ± 0.000333, $\pm 0.0333\%$ **b.** ± 0.0005, $\pm 0.05\%$ **31.** 0.03, 3% **33.** -0.03, -3% **35a.** within $\pm 1\%$ of the true value **b.** within $\pm 2\%$ of the true value **37a.** within $\pm 0.05\%$ of the true value **b.** within $\pm 0.033\%$ of the true value **39a.** $\Delta p \approx -2500 \, q^{-3/2}\Delta q$ **b.** $\Delta p \approx \$0.231$ **41.** $\approx \$750$ **43a.** $\Delta P/P \approx (-2x + 80)\Delta x/(-x^2 + 80x - 200)$ **b.** 0.04, 4% **45.** ≈ -0.078, -7.8% **47a.** $\Delta y \approx -1200(400 - 120t)^{-1/2}$ **b.** ≈ -120 **49.** ≈ -10 mg/cm^3 **51.** $\Delta y \approx 100\Delta t/\sqrt{t}$ **b.** ≈ 33 **57.** ≈ 0.235 or 23.5% **59.** by increasing debt; yes, if try to grow too fast can take on too much debt **61.** ≈0.075 or 7.5%

11.4 EXERCISES

1. $dy/dx = 1/2y$ **3.** $dy/dx = 2x/3y^2$ **5.** $dy/dx = -6x^2/(y-1)$ **7.** $dy/dx = 2x/3y^2$ **9.** $dy/dx = y/2x$ **11.** $dy/dx = y(x + y^3)/x(x + 2y^3)$ **13.** $dy/dx = -x$ **15.** $dy/dx = \sqrt{y}(2y\sqrt{x} + x^2\sqrt{y} - 4x^{5/2})/x^{3/2}(2\sqrt{y} - x^{3/2} - 6x\sqrt{y})$ **17.** $y = 0.4x - 2$ **19.** $y = x$ **21.** $y = 2x - 4$ **23.** $x = 1$ **25.** $dy/dx = -25x/48 + 97/48$ **27.** $dy/dt = -(x/y)(dx/dt)$ **29.** $dy/dt = (2x/y)(dx/dt)$ **31.** $dy/dt = [(2x - y^2 + 2)/(2xy + 3)](dx/dt)$ **33.** $d^2y/dx^2 = -y^{-3}$ **35.** $d^2y/dx^2 = (2y^6 - 12y^3 + 18)/9x^2y^5$ **37.** $d^2y/dx^2 = (24x^3y^2 - 8x^3y - 16x^6 + y^4 - 2y^3)/4x^2y^3$ **39a.** -20 **b.** -20 **41a.** -0.2 units/$ **b.** -0.105 units/$

11.5 EXERCISES

1. 128π cm³/sec **3.** 2400 cm³/sec **5.** $2/14.7 \approx 0.136$ in³/min **7.** 1.25 fps **9.** $-40\sqrt{2} + 30 \approx -26.6$ mph **11.** $20,000/\sqrt{38,400} \approx 102.1$ mph **13.** $-1/4\pi$ fps **15.** $\pm 5\sqrt{63}/2 \approx \pm 19.84$ mph ($-$ if ship is approaching observer, $+$ if it is receding from observer) **17.** $-\$103.50$/day **19.** -79.2 month **21.** $29,963.40$/year **23.** $\$25/7 \approx \$3.57$ thousand/ad **25.** $3,840,000\pi$ ft²/hr **27.** $dy/dt = 500/\sqrt{x}$ **29.** $0.4k$ mm/min

11 REVIEW EXERCISES

1. $30x^4 - 8x^3 + 48x - 8$ **2.** $f'(x) = -\dfrac{44}{(5x - 3)^2}$ **3.** $f'(x) = \dfrac{4x^2 + 12x - 3}{2(2x + 3)^2}$ **4.** $f'(x) = 9x^2 + 7x^{5/2} - x^{-1/2}$ **5.** $f'(x) = \dfrac{5(2x^4 - 3x^2 - 4x + 3)}{(2x^2 + 1)^2}$ **6.** $(-5x^4 + 65x^2 + 20x + 20)/(5x + 2)^2 (x^2 + 1)^2$ **7.** $(6x^6 + 2x^5 + 10x^4 + 4x^3 + x^2 - 4x - 1)/(x^2 + 1)^2$ **8.** $(4x^7 - 8x^5 + 2x^4 + 4x^3 - 6x^2 + 2x)/(x + 1)^2(x - 1)^2$ **9.** $70(7x - 9)^9$ **10.** $12(6x + 5)(3x^2 + 5x + 1)^{11}$ **11.** $(9x^2 - 1)/\sqrt{6x^3 - 2x + 3}$ **12.** $-(5x^2 + 4x + 10)/2(5x + 2)^{1/2}(x^2 - 3)^{3/2}$ **13.** $4(2x + 3)^5(x^2 - 5x)^7(11x^2 - 23x - 30)$ **14.** $10(6x^2 - 2x + 4)(2x - 3)^{19}(x^2 + 1)^4$ **15.** $f'(x) = \dfrac{7(3x - 1)^6(-3x^2 + 2x + 6)}{(x^2 + 2)^8}$ **16.** $f'(x) = -\dfrac{5(3x + 4)^4(9x^2 + 32x + 3)}{(5x^2 - 1)^5}$ **17.** $384t^3 - 880t$ **18.** $\dfrac{dy}{dt} = 128t - 88$ **19.** $\dfrac{du}{dw} = 11\left(7 - \dfrac{3}{2\sqrt{w}}\right)(4v^3 - 2)(v^4 - 2v + 11)^{10}$ **20.** $\dfrac{118125(120)^{-1/4}}{2} \approx 17,845$ units/dollars **21a.** $\Delta p \approx -400,000(10q)^{-3/2}\Delta q$ **b.** $-\$2$ **22a.** $\Delta p/p \approx -5(10q)^{-1}\Delta q$ **b.** 0.004, 0.4% **23.** ± 0.005, $\pm 0.5\%$ **24.** within $\pm 0.5\%$ of the true value **25.** $dy/dx = 1/8y$ **26.** $dy/dx = (2xy - y^3)/(3xy^2 - x^2)$ **27.** $dy/dx = -(2xy + y)/(2y^3 - x^2)$ **28.** $dy/dx = -(2x^3 + y)/(2xy - 1)$ **29.** $y = -x/2 + 3$ **30.** $y = 17x/2 + 57/2$ **31.** $d^2y/dx^2 = 2y/x^2$ **32.** $d^2y/dx^2 = (2y - 10xy^{3/2} + 6x^2y^2)x^{-2}$ **33.** $\dfrac{dy}{du} = \dfrac{y(2 - 3x^2y)}{2x(3x^2y + 1)} \dfrac{dx}{du}$ **34.** -50 units/month **35.** $2815/\sqrt{1489}$ mph

11 SUPPLEMENTARY EXERCISES

1. in both cases, 0.6; $0.60 of each additional dollar of income is spent **3a.** 0.5; when $Y = \$400$ billion, $0.50 of each additional dollar of income is spent **b.** $\frac{1}{3}$; when $Y = \$900$ billion, $\frac{1}{3}$ of each additional dollar of income is spent **5a.** 1; when $Y = \$125$ million, each additional dollar of income is spent **b.** 0.25; when $Y = \$1000$ million, $0.25 of each additional dollar of income is spent **c.** 0.0625; when $Y = \$8000$ million, $0.0625 of each additional dollar of income is spent **7.** ≈1.47; each additional dollar of savings generates ≈$1.47 additional income **9a.** $\frac{4}{3}$; each additional dollar of savings generates ≈$1.33 additional income **b.** $\frac{10}{9}$; each additional dollar of savings generates ≈$1.11 additional income **11.** 0.4; of each additional dollar of income, $0.40 is saved **13.** $Y = \$400$ billion: 0.5; of each additional dollar of income, $0.50 is saved; $Y = \$900$ billion: $\frac{2}{3}$; of each additional dollar of income, ≈$0.67 is saved **15.** $Y = \$125$ million: 0; of each additional dollar of income, none is saved; $Y = \$1000$ million: 0.75; of each additional dollar of income, $0.75 is saved; $Y = \$8000$ million: 0.9375; of each additional dollar of income $0.9375 is saved

12.1 EXERCISES

1. crit values: $x = -1, 1, 2, 3$; rel max: $(-1, 2), (2, 1)$; rel min: $(1, -1), (3, -2)$ **3.** crit values: $x = -2, 0, 2, 3, 4$; rel max: $(-2, 2), (2, 2), (4, 2)$; rel min: $(0, 0), (3, 1)$ **5.** crit values: $x = 2$; rel max: $(2, 2)$ no rel min **7.** decreasing: $(-\infty, 5)$; increasing: $(5, +\infty)$ **9.** increasing: $(-\infty, 0), (0, +\infty)$ **11.** decreasing: $(-\infty, -5), (2, +\infty)$; increasing: $(-5, 2)$ **13.** increasing: $(-\infty, \frac{1}{2}), (\frac{5}{2}, +\infty)$; decreasing: $(\frac{1}{2}, \frac{5}{2})$ **15.** decreasing: $(-\infty, -2), (0, 2)$; increasing: $(-2, 0), (2, +\infty)$ **17.** increasing: $(0, +\infty)$ **19.** decreasing: $(-\infty, 0), (0, +\infty)$ **21.** decreasing: $(-\infty, 2), (2, +\infty)$ **23.** rel min: $(2, -1)$ **25.** rel min: $(-6, -18)$ **27.** rel max: $(0, 0)$; rel min: $(\frac{4}{3}, -\frac{32}{27})$ **29.** rel min: $(-2, -16)$; rel max: $(2, 16)$ **31.** rel max: $(-3, 27)$; rel min $(1, -5)$ **33.** no rel max or min **35.** rel max: $(-6, 560)$; rel min: $(3, -169)$ **37.** rel min: $(-1, -1), (1, -1)$; rel max: $(0, 0)$ **39.** rel min: $(-2, -32), (1, -5)$; rel max: $(0, 0)$ **41.** rel max: $(-2, 74)$; rel min: $(2, -54)$

43. rel min: $(-3, -729)$, $(3, -729)$; rel max: $(0, 0)$
45. no rel max or min   **47.** rel min: $(-3, -1/6)$; rel max: $(-1, -1/2)$ **49.** rel max: $(-2, -4)$; rel min: $(0, 0)$
51. rel max: $(-1, -2)$; rel min: $(1, 2)$ **53.** rel max: $(-2, -32/3)$; rel min: $(2, 32/3)$ **55.** increasing; $[0, 18)$; decreasing: $(18, +\infty)$; max of 166.2 units when $t = 18$ yrs
57. increasing: $[0, 20)$; decreasing: $(20, 40]$; max of $152,000 when spend $20,000   **59.** increasing: $[1, 6)$; decreasing: $(6, 12]$; max of $23/60 \approx 38.3\%$ when $t = 6$
61a. increasing: $[0, 5)$; decreasing: $(5, 10]$ **b.** $25,000 when $p = \$5$ **c.** inelastic: $[0, 5)$; elastic: $(5, 10]$; unit elasticity: 5   **65.** increasing: $[0, 2)$; decreasing: $(2, +\infty)$; max population of 2.5 million 2 months from now
67. increasing: $[0, 3)$, $(15, 20]$; decreasing: $(3, 15)$; max of 6.6281 million in 1993, min of 1.9625 million in 2005

12.2 EXERCISES

1. concave upward: $(-\infty, +\infty)$ **3.** concave upward: $(0, +\infty)$; concave downward: $(-\infty, 0)$ **5.** concave upward: $(2, +\infty)$; concave downward: $(-\infty, 2)$ **7.** concave upward: $(-\infty, +\infty)$ **9.** concave upward: $(-2, 3)$; concave downward: $(-\infty, -2)$, $(3, +\infty)$ **11.** concave upward: $(-\infty, 2)$; concave downward: $(2, +\infty)$ **13.** rel max: $(3, 14)$
15. rel max: $(-6, 108)$; rel min: $(0, 0)$ **17.** rel max: $(-4, 521)$; rel min: $(9, -1676)$ **19.** rel min: $(-3, -108)$; rel max: $(7, 392)$ **21.** rel min: $(\sqrt{2}, 6)$, $(-\sqrt{2}, 6)$; rel max: $(0, 10)$ **23.** rel max: $(1, 9)$, $(3, 9)$; rel min: $(2, 8)$ **25.** no rel max or min   **27.** rel min: $(0, -50)$, $(2, -42)$; rel max: $(1, -31)$, $(3, -23)$ **29.** rel min: $(0, 0)$ **31.** no rel max or min   **33.** rel min: $(-3, -1/6)$; rel max: $(1, 1/2)$ **35.** \$72/unit when $x = 30$ **37.** \$280,000 at $x = 60$ **41.** 8000 when $t = 3$ **43.** 75/thousand in 1995

12.3 EXERCISES

1a. abs max: 3 at $x = 0$; abs min: -2 at $x = 3$ **b.** abs max: 4 at $x = 6$; abs min: -2 at $x = 3$ **c.** abs max: 3 at $x = 0$; abs min: none   **d.** abs max: none; abs min: -2 at $x = 3$ **3.** on $[0, 5]$: abs max: -5 at $x = 0$; abs min: -32 at $x = 3$; on $[6, 8]$: abs max: 43 at $x = 8$; abs min: -5 at $x = 6$; on $[0, +\infty)$: abs max: none; abs min: -32 at $x = 3$; on $(-\infty, +\infty)$: abs max: none; abs min: -32 at $x = 3$
5. on $[0, 6]$: abs max: 144 at $x = 6$; abs min: -16 at $x = 2$; on $[-1, 1]$: abs max: 11 at $x = -1$; abs min: -11 at $x = 1$; on $[-6, 6]$: abs max: 144 at $x = 6$; abs min: -144 at $x = -6$; on $(-\infty, +\infty)$: abs max: none; abs min: none
7. on $[0, 12]$: abs max: 200 at $x = 0$; abs min: -8 at $x = 4$; on $[0, 5]$: abs max: 200 at $x = 0$; abs min: -8 at $x = 4$; on $[5, 12]$: abs max: 100 at $x = 10$; abs min: 0 at $x = 5$; on $(-\infty, +\infty)$: abs max: none; abs min: none   **9.** on $(-\infty, 0]$: abs max: none; abs min: 0 at $x = 0$; on $[-3, 3]$: abs max: 177 at $x = -3$; abs min: -48 at $x = 2$; on $[0, 4]$: abs max: 128 at $x = 4$; abs min: -48 at $x = 2$; on $[0, +\infty)$: abs max: none; abs min: -48 at $x = 2$ **11.** on $(-\infty, 1)$: abs max: none; abs min: none; on $(1, +\infty)$: abs max: none; abs

min: none; on $[0, 2]$: abs max: none; abs min: none; on $[2, 4]$: abs max: $1/3$ at $x = 4$; abs min: -1 at $x = 2$ **13.** on $(-3, 3)$: abs max: 0 at $x = 0$; abs min: none; on $(3, +\infty)$: abs max: none; abs min: none; on $[-2, 2]$: abs max: 0 at $x = 0$; abs min: $-4/5$ at $x = 2$, -2; on $(-\infty, +\infty)$: abs max: none; abs min: none   **15.** 30 mph   **21.** 160 ft by 80 ft
23. 120 m   **25.** no cut, all into square   **27.** $5 - 3^{-1/2} \approx 4.42$ miles from A **29.** base 2 ft by 4 ft, height $40/9$ ft
31. radius = $3/\sqrt[3]{\pi}$ in, height = $6/\sqrt[3]{\pi}$ in   **33a.** \$66,400
b. \$54,000   **35.** \$50,000,000; \$500/unit; \$48,000,000; \$600/unit   **37.** 200 boxes, 1 order every other week, \$52
41. \$605,000 when sell 11,000 tickets at \$55 each
43. make 60 regular and 70 deluxe for max profit of \$1310
45. max revenue of \$1920 when price is \$16/bouquet
47. at 10-yr mark   **49.** 34,225 eggs, 185 chickens
51. max of ≈ 1333 in 1985, min of ≈ 771 in 1970
53. velocity is greatest on the central axis

12.4 EXERCISES

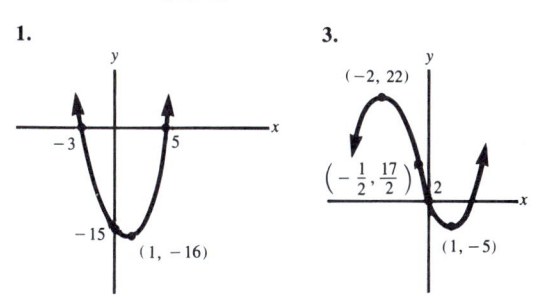

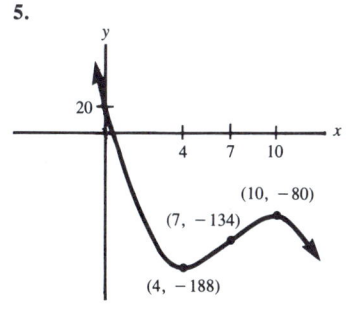

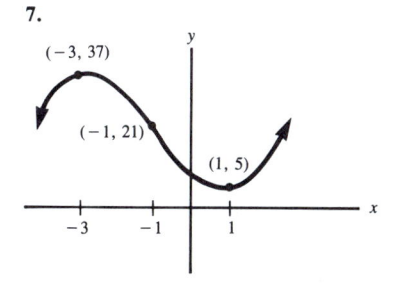

1.

[Graph showing parabola-like curve with points -3, 5 on x-axis, -15 on y-axis, and minimum at $(1, -16)$]

3.

[Graph showing cubic-like curve with points $(-2, 22)$, $(-\frac{1}{2}, \frac{17}{2})$, 2 on x-axis, and $(1, -5)$]

5.

[Graph showing curve with 20 on y-axis, points 4, 7, 10 on x-axis, $(10, -80)$, $(7, -134)$, and $(4, -188)$]

7.

[Graph showing curve with $(-3, 37)$, $(-1, 21)$, $(1, 5)$, and points -3, -1, 1 on x-axis]

9.

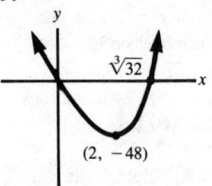

Graph with labeled points $\sqrt[3]{32}$ and $(2, -48)$.

11.

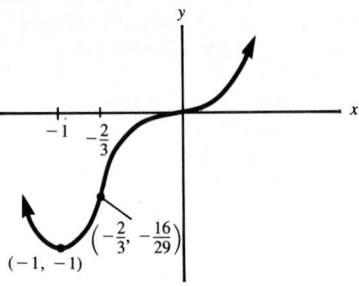

Graph with labeled points $(-1, -1)$, $\left(-\frac{2}{3}, -\frac{16}{29}\right)$, and marks at -1, $-\frac{2}{3}$.

13.

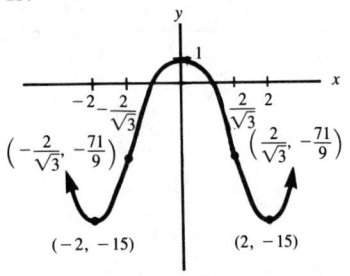

Graph with labeled points $(-2, -15)$, $(2, -15)$, $\left(-\frac{2}{\sqrt{3}}, -\frac{71}{9}\right)$, $\left(\frac{2}{\sqrt{3}}, -\frac{71}{9}\right)$, and marks at -2, $-\frac{2}{\sqrt{3}}$, $\frac{2}{\sqrt{3}}$, 2.

15.

Graph with labeled points $\left(-\sqrt{7}, \frac{28\sqrt{7}}{15}\right)$, $\left(1, \frac{68}{15}\right)$, $\left(-2, \frac{14}{15}\right)$, $\left(-1, -\frac{68}{15}\right)$, $\left(2, -\frac{14}{15}\right)$, $\left(\sqrt{7}, -\frac{28\sqrt{7}}{15}\right)$, and marks at $-\sqrt{7}$, -2, -1, 1, 2, $\sqrt{7}$.

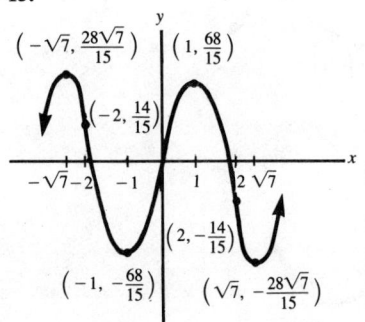

17.

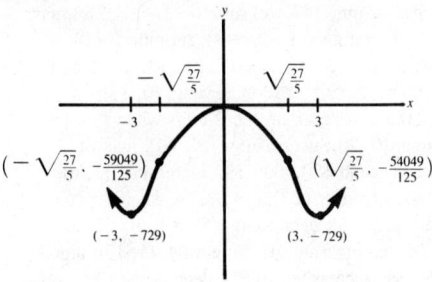

Graph with labeled points $-\sqrt{\frac{27}{5}}$, $\sqrt{\frac{27}{5}}$, $\left(-\sqrt{\frac{27}{5}}, -\frac{59049}{125}\right)$, $\left(\sqrt{\frac{27}{5}}, -\frac{54049}{125}\right)$, $(-3, -729)$, $(3, -729)$, marks at -3, 3.

19.

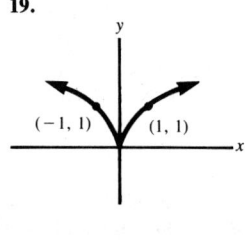

Graph with labeled points $(-1, 1)$ and $(1, 1)$.

21.

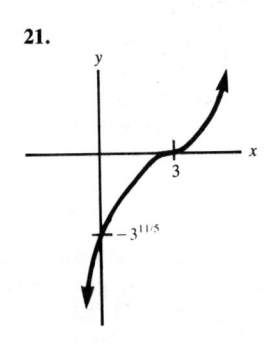

Graph with labeled points 3 on x-axis and $-3^{11/5}$ on y-axis.

23.

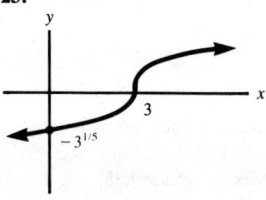

Graph with labeled point 3 on x-axis and $-3^{1/5}$ on y-axis.

25.

Graph with labeled points $(1, 3)$ and $(-1, -3)$.

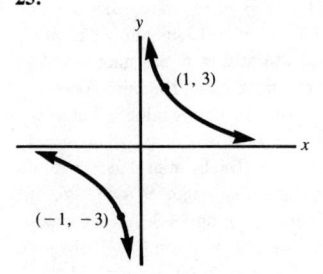

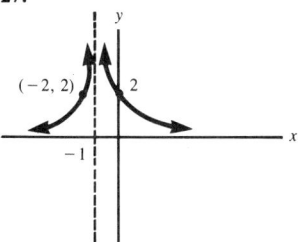

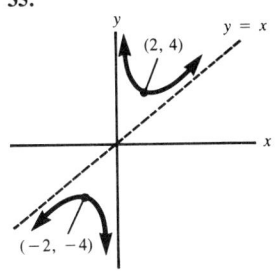

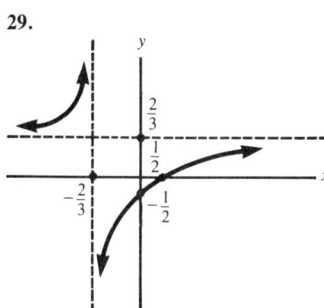

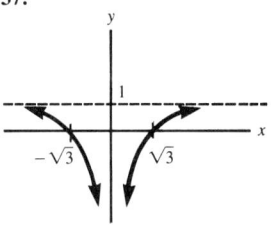

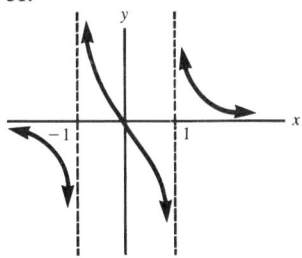

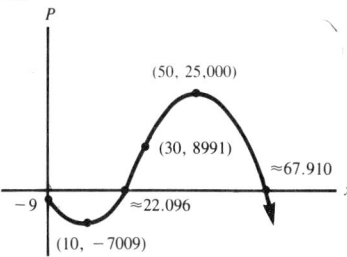

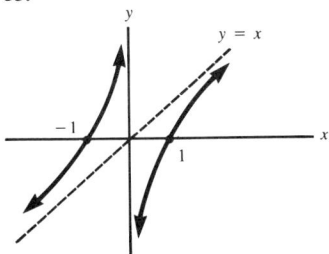

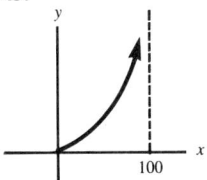

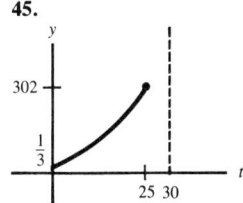

27.

29.

31.

33.

35.

37.

41.

43.

45.

47.

49.

51.

53.

55.

57.

59a.

b.

c.

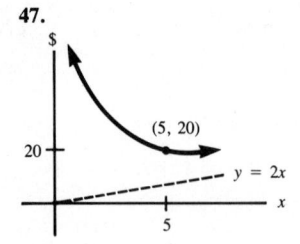

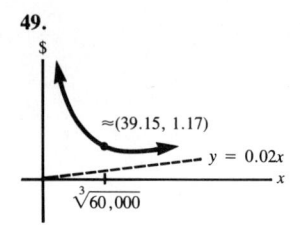

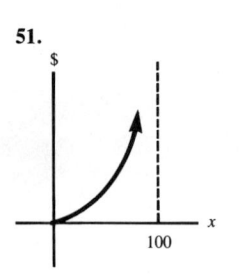

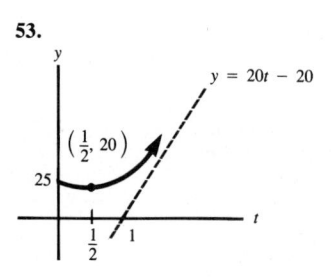

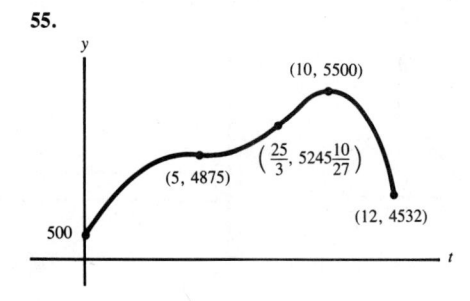

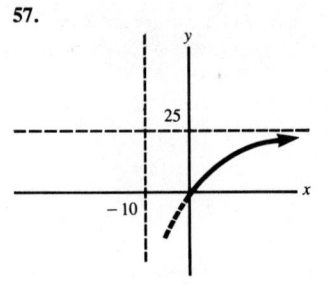

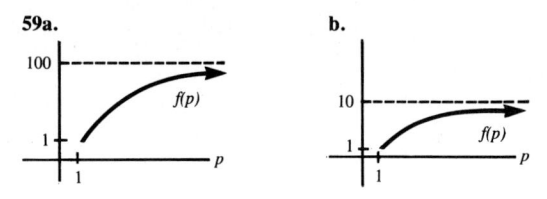

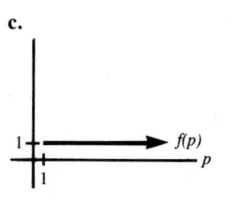

12 REVIEW EXERCISES

1. rel max: (1, 5), (5, 3), (7, 5); rel min: (0, 4), (4, 2)
2. $x = -1, 0, 1, 4, 5, 6, 7, 8$ **3.** increasing: (0, 1), (4, 5), (6, 7); decreasing: $(-\infty, -1), (-1, 0), (1, 3), (3, 4), (5, 6), (7, 8), (8, +\infty)$ **4.** concave upward: $(-\infty, -1), (3, 4.5), (5.5, 6), (6, 7), (7, 8)$; concave downward: $(-1, 0), (0, 3), (4.5, 5.5), (8, +\infty)$ **5.** pts of inflection: $(-1, 5), (4.5, 2.5), (5.5, 2.5), (8, 4)$ **6.** rel max: $(7/2, 33/4)$ **7.** rel max: $(-10, 200)$; rel min: $(6, -1848)$ **8.** rel min: $(-3, -27)$ **9.** rel max: $(-\sqrt{8}, 200 + 448\sqrt{8})$; rel min: $(\sqrt{8}, 200 - 448\sqrt{8})$ **10.** no rel max or min **11.** rel min: $((-1 - \sqrt{3})/2, -2\sqrt{33}/(33 - \sqrt{33}))$; rel max: $((-1 + \sqrt{3})/2, 2\sqrt{33}/(33 + \sqrt{33}))$ **12.** \$78.50/unit **13.** $7/12$ of the trees **14.** rel max: $(9/4, 57/8)$ **15.** rel max $(-5, 100)$; rel min: $(1, -8)$ **16.** rel max: $(0, 1)$; rel min: $(3/2, -65/16)$, $(-3/2, -65/16)$ **17.** rel max: $(0, 1)$; rel min: $(2, -43/5)$
18. rel min: $(0, 0)$ **19.** rel min: $(0, 0)$
20. \$8,000,000,000 **21.** abs max: 4 at $x = 5$; abs min: 1 at $x = 1$ **22.** abs max: 3 at $x = 3$; abs min: 1 at $x = 1$ **23.** abs max: 4 at $x = 5$; abs min: -2 at $x = -2$ **24.** abs max: 3 at $x = 3$; abs min: -2 at $x = -2$ **25.** abs max: 4 at $x = 5$; abs min: none **26.** abs max: 4 at $x = 5$; abs min: none **27.** abs max: 0 at $x = -3$; abs min: none **28.** abs max: 2 at $x = 0$; abs min: none **29a.** abs max: none; abs min: -256 at $x = \pm 4$ **b.** abs max: 0 at $x = 0$; abs min:

−31 at $x = \pm 1$ **c.** abs max: 144 at $x = 6$; abs min: −256 at $x = \pm 4$ **d.** abs max: 144 at $x = 6$; abs min: −256 at $x = 4$ **30.** $10,000, $64,000 **31.** 20 cars, $4,840,000 **32.** radius $= 1/\sqrt{\pi}$ ft, height $= 2/\sqrt{\pi}$ ft

33.

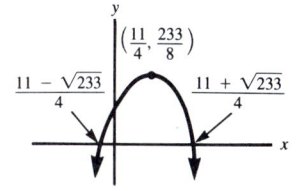

34.

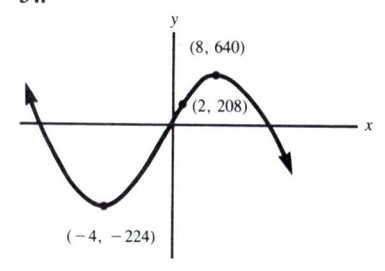

35.

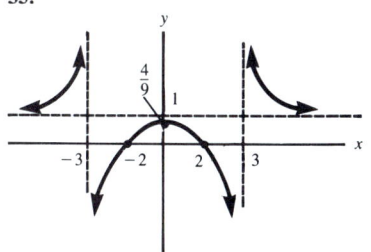

36.

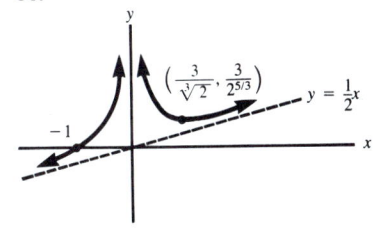

37.

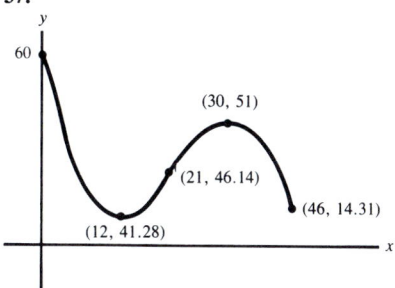

38.

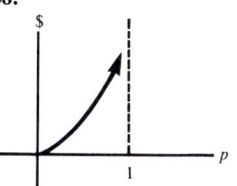

39.

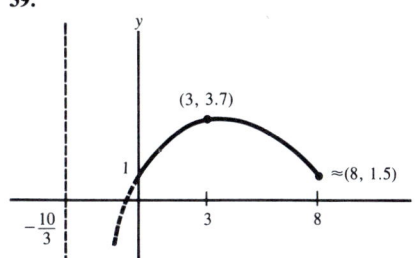

12 SUPPLEMENTARY EXERCISES

1.

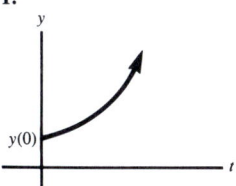

3.

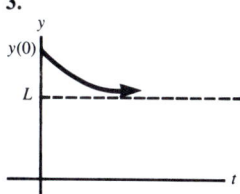

5.

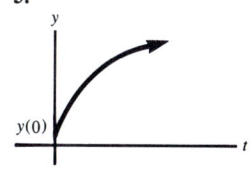

7.

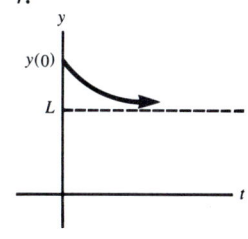

13.1 EXERCISES

1. Graph of $y = 3^x$ through $(-1, \frac{1}{3})$ and $(1, 3)$.

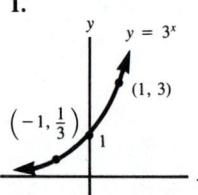

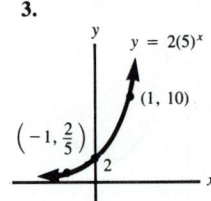

3. Graph of $y = 2(5)^x$ through $(-1, \frac{2}{5})$ and $(1, 10)$.

5. Graph of $y = 3^{x/2}$ through $(-2, \frac{1}{3})$ and $(2, 3)$.

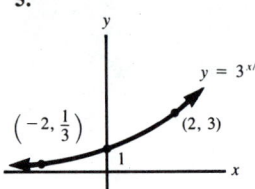

7. Graph of $y = 4^{2x/3}$ through $(-\frac{3}{2}, \frac{1}{4})$ and $(\frac{3}{2}, 4)$.

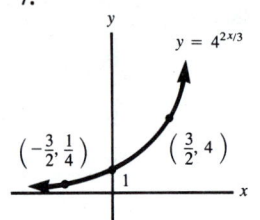

9. Graph of $y = 2(3)^{-x/3}$ through $(-3, 6)$ and $(3, \frac{2}{3})$.

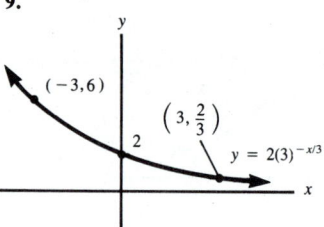

11. Graph of $y = -5\left(\frac{2}{5}\right)^{x/2}$ through $\left(-2, -\frac{25}{2}\right)$ and $(2, -2)$.

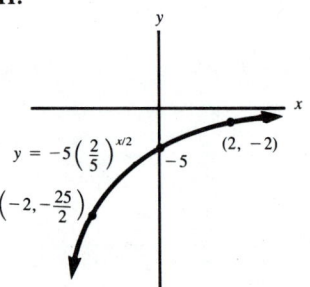

13. Graph of $y = 2^{0.2x}$ through $\left(-5, \frac{1}{2}\right)$ and $(5, 2)$.

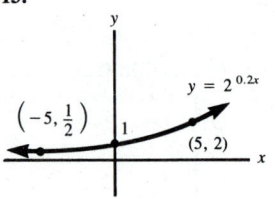

15. Graph of $y = 2(3)^{-0.3x}$ through $\left(-\frac{10}{3}, 6\right)$ and $\left(\frac{10}{3}, \frac{2}{3}\right)$.

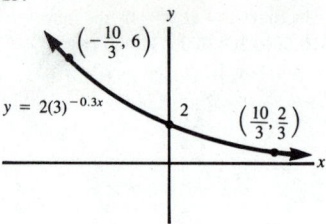

17. 1.061837 **19.** 2.117 **21.** 0.367879 **23.** 0.030197

25. Graph of $y = e^{-x}$ through $(-1, e)$ and $(1, e^{-1})$.

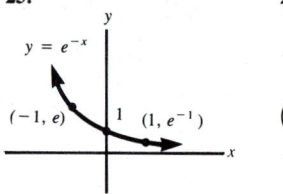

27. Graph of $y = e^{-2x}$ through $\left(-\frac{1}{2}, e\right)$ and $\left(\frac{1}{2}, e^{-1}\right)$.

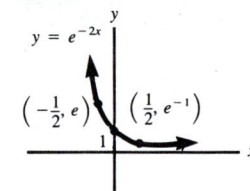

29. Graph of $y = e^{x/2}$ through $(-2, e^{-1})$ and $(2, e)$.

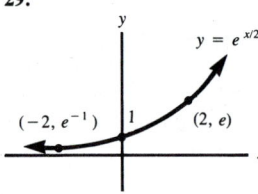

31. Graph of $y = 1.4e^{0.5x}$ through $(-2, 1.4e^{-1})$ and $(2, 1.4e)$.

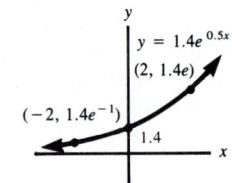

33. Graph of $y = 3e^{-0.05x}$ through $(-20, 3e)$ and $(20, 3e^{-1})$.

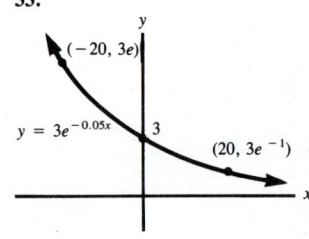

35. $y = 3e^{0.2772t}$ **37.** $y = e^{-2.0793x}$ **39.** $y = 2e^{-8.7888x}$

41a. Graph of $y = 200e^{-0.05t}$ through $(20, 200e^{-1})$.

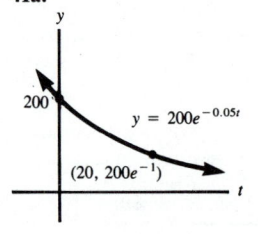

b. ≈ 121.306 yr **c.** ≈ 13.862 yr **d.** 0; in the long run y approaches 0 **43.** $100e^{-0.05776t}$ **45.** $S = 5e^{0.0692t}$
47. $22.50e^{1.0986t}$, where $t = 0$ in 1985

49. a. $300,000

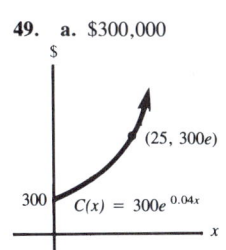

$C(x) = 300e^{0.04x}$, point $(25, 300e)$, y-intercept 300

b. $447,547.41 **51.** $D = \$10,000(0.6)^{t-1}$, $3600, $1296
53. $D = \$80,000(14/15)^{t-1}$, $80,000, $42,945.30, $15,274.70

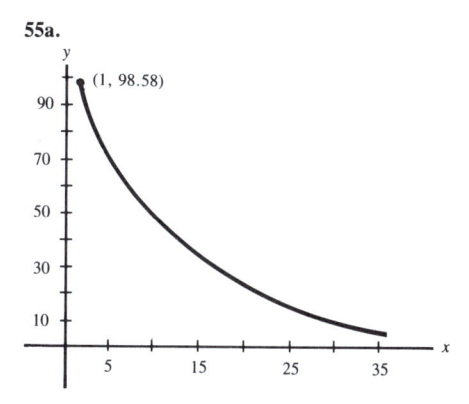

55a.

Graph with point $(1, 98.58)$, decreasing curve.

b. 90.8%; 20.8% **57.** $10,832.87 **59.** $20,247.88
61.

Compoundings/yr	Future Value of $10,000 at 6% for 1 yr
1	$10,600.00
2	$10,609.00
4	$10,613.64
12	$10,616.79
continuous	$10,618.37

63. $7408.18 **65.** $67,032.01
67.

Compoundings/yr	Present Value of $10,000 at 9% for 1 yr
1	$9174.31
2	$9157.30
4	$9148.43
12	$9142.38
continuous	$9139.31

69. $963,194.42 **71.** present value − cost ≈ $39,500 for X, $55,800 for Y, so buy Y **73.** present value − cost ≈ −$683 for X, −$12,000 for Y, so buy neither

13.2 EXERCISES

1. 4, 7, −2, 2, −3 **3.** 0, 1, 0, 1, −1 **5.** 16 **7.** 6
9. 13 **11.** −3 **13.** 5.1 **15.** 4.25 **17.** 125 **19.** e^3

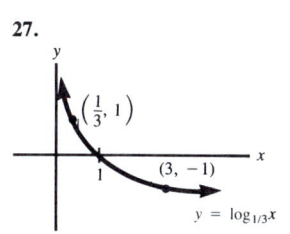

21. 4/9 **23.** −3

25. Graph of $y = \log_3 x$ through $(3, 1)$ and $(\frac{1}{3}, -1)$.

27. Graph of $y = \log_{1/3} x$ through $(\frac{1}{3}, 1)$ and $(3, -1)$.

29. $y = e^{2.30259x}$ **31.** $y = 2e^{7.78364x}$ **33.** $y = 1.4427(\ln x)$
35. $y = -6.21335(\ln x)$ **37a.** 2 **b.** 4 **c.** 7
39. between 10^8 and 10^9 times

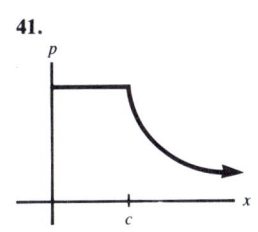

41.

Graph of p vs x with value 1 up to c, then declining toward 0.

when income is $\leq c$, 100% do not purchase the good within 1 yr; as income grows past c, the percentage who do not purchase it within 1 yr declines toward 0 **43.** $15,000: 100%; $30,000: $100e^{-0.5}$% ≈60.7%; $50,000: $100e^{-1.5}$% ≈22.3% **45.** For a given income greater than c, the proportion who do not purchase the good within 1 yr becomes smaller as a increases **47.** $x = -0.7993$ **49.** $x = 1.73287$ **51.** $x = 1.58496$ **53.** $x = 4.09564$ **55.** $x = 5.89097$ **57.** $y = 2^{-6.64386x}$ **59.** $y = 1.09861(\log_3 x)$
61. $11.91917(\log_{11} x)$ **63.** 34.6574 days, 115.1293 days
65. 7.7016 yrs **67.** 18.4472 yrs **69.** 18.3258%
71. 7.7494% **73.** 12 years **75.** 5½ years **77.** 10.26%
79. $t = 32.189$ (A.D. 2017) **81.** 13.288 hrs **83.** 198.97 yrs after 1988 (A.D. 2187); 352.48 yrs after 1988 (2340 A.D.)
85. IRR $= (A/P)^{1/t} - 1$ **87.** $i \leq 10.24948$
89. 0.0570086% **91.** 0.0302778%

13.3 EXERCISES

1. $3/x$ **3.** $7/(7x + 2)$ **5.** $(2x + 5)/(x^2 + 5x)$ **7.** $16x(x^2 + 1)/(x^4 + 2x^2 + 200)$ **9.** $2(\ln 3x + 1)$ **11.** $2x^2(3 \ln x + 1)$ **13.** $2/(2x + 1) - 3/(3x + 2)$ **15.** $\ln(6/5)/x(\ln 6x)^2$ **17.** $3(\ln x)^2/x$ **19.** $(3x^2 + 6x + 4)[\ln(x^3 + 3x^2 + 4x + 3)]^{-1/2}/2(x^3 + 3x^2 + 4x + 3)$ **21.** $1/x(\ln x)(\ln(\ln x))$

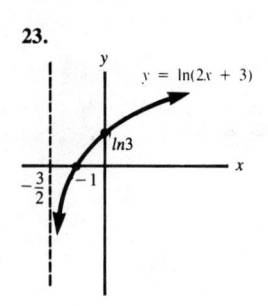

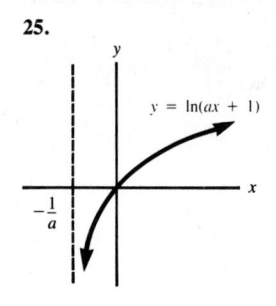

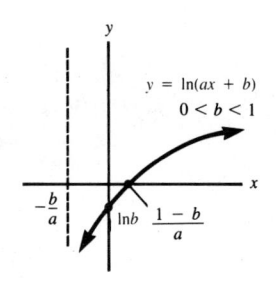

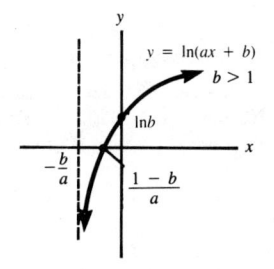

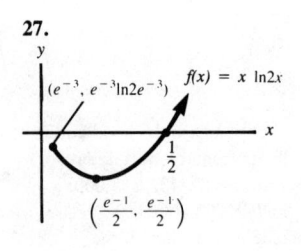

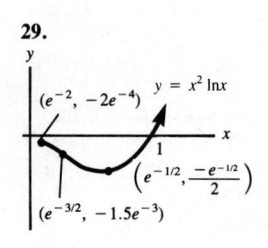

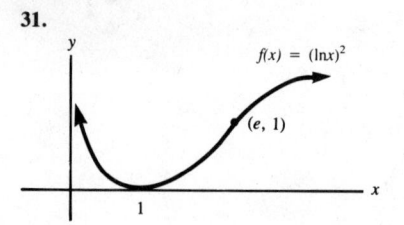

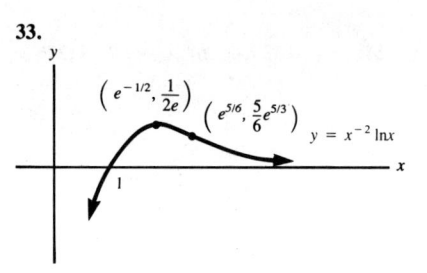

23. [graph: $y = \ln(2x + 3)$, $\ln 3$, $-3/2$, -1]

25. [graph: $y = \ln(ax + 1)$, $-1/a$]

27. [graph: $f(x) = x \ln 2x$, $(e^{-3}, e^{-3}\ln 2e^{-3})$, $1/2$, $\left(\frac{e^{-1}}{2}, \frac{e^{-1}}{2}\right)$]

[graph: $y = \ln(ax + b)$, $0 < b < 1$, $-b/a$, $\ln b$, $(1-b)/a$]

[graph: $y = \ln(ax + b)$, $b > 1$, $-b/a$, $\ln b$, $(1-b)/a$]

29. [graph: $y = x^2 \ln x$, $(e^{-2}, -2e^{-4})$, $(e^{-1/2}, -e^{-1/2}/2)$, $(e^{-3/2}, -1.5e^{-3})$]

31. [graph: $f(x) = (\ln x)^2$, $(e, 1)$, 1]

33. [graph: $y = x^{-2}\ln x$, $(e^{-1/2}, \frac{1}{2e})$, $(e^{5/6}, \frac{5}{6}e^{5/3})$, 1]

35. $1/(\ln 2)x$ **37.** $3/(\ln 7)x$ **39.** $8(x + 2)/(\ln 3)x(x + 4)$ **41.** $2x[\log_{10}(2x - 1) + x/(\ln 10)(2x - 1)]$ **43.** $5/24 \approx 0.20833$ million \$/yr; $5/114 \approx 0.04386$ million \$/yr **45.** $48/\sqrt{15}(\sqrt{15} + 3) \approx 1.80323$ million tons/yr, $48/\sqrt{25}(\sqrt{25} + 3) \approx 1.2$ million tons/yr, $48/\sqrt{35}(\sqrt{35} + 3) \approx 0.90998$ million tons/yr

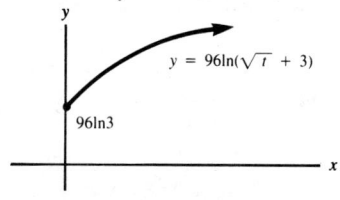

[graph: $y = 96\ln(\sqrt{t} + 3)$, $96\ln 3$]

47. $100e^{-1} + 20 \approx \$56.78794$ million, when $t = 10(e - 0.1) \approx 26.182$ (A.D. 1976) **49.** $2/7$ pH/yr, $1/6$ pH/yr

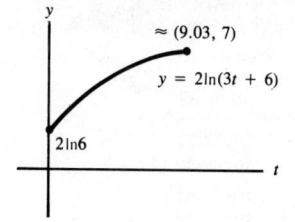

[graph: $y = 2\ln(3t + 6)$, $\approx (9.03, 7)$, $2\ln 6$]

51. $36(10^{12/5})/(20 + 60(10^{2/5})) \approx 53$ persons/yr

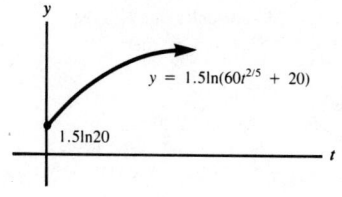

[graph: $y = 1.5\ln(60t^{2/5} + 20)$, $1.5\ln 20$]

Solutions to Exercises

53. $dr/dx = 1/(\ln 10)x$ **55.** $f(x)[1/x + 9x^2/(x^3 + 1) + 25x^4/(x^5 + 1)]$ **57.** $f(x)[6x^2/(x^3 + 1) + 6x/(x^2 + 1) - 8(x + 2)/(x^2 + 4x + 1)]$ **59.** $2x^{2x}(\ln x + 1)$
61. $-x^{-x}(\ln x + 1)$ **63.** $2x^{2(x+1)}[(x + 1)/x + \ln x]$
65. $(\ln x)^x[\ln(\ln x) + 1/\ln x]$

13.4 EXERCISES

1. $2e^{2x}$ **3.** $-2e^{-x}$ **5.** $(e^x + e^{-x})/2$ **7.** $5(2x + 3)e^{x^2+3x+4}$
9. $6x^2 e^{x^3}$ **11.** $3x^2$ **13.** $2e^{-4x}(1 - 4x)$ **15.** $e^x(x - 1)/x^2$
17. $e^x(\ln x + 1/x)$ **19.** $e^{2x}(2 \ln 3x + 1/x)$ **21.** $2xe^{x^2}e^{e^{x^2}}$
23. $e x^{e-1}$

25.

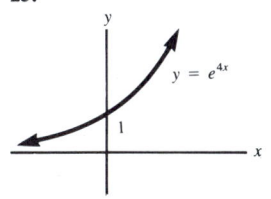

27.

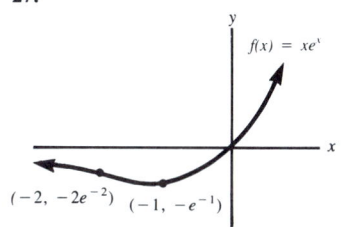

29.

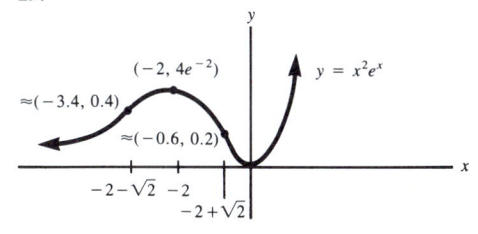

31.

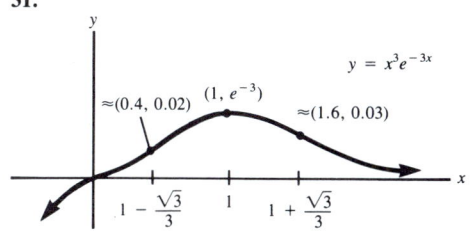

33.

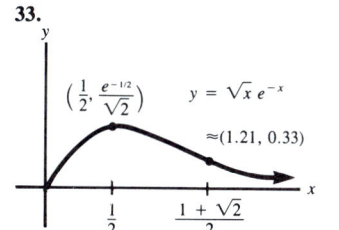

35.

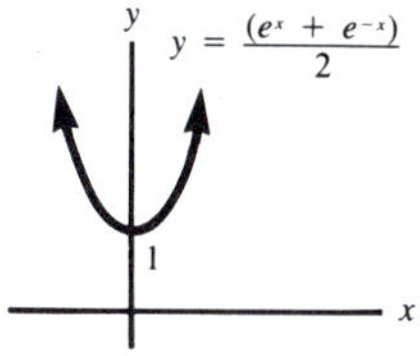

37.

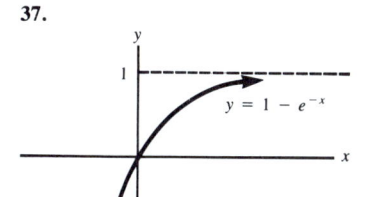

39.

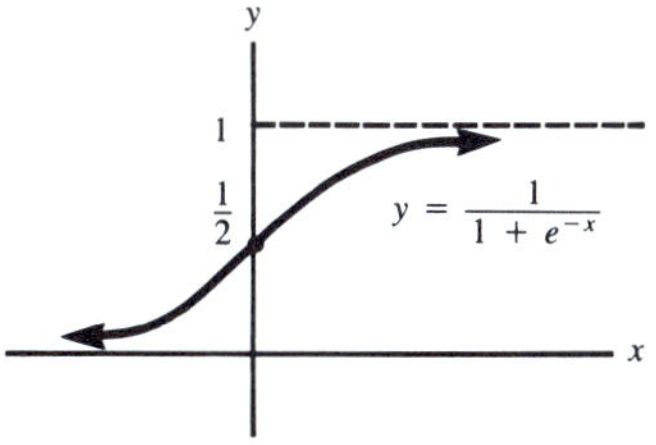

41.

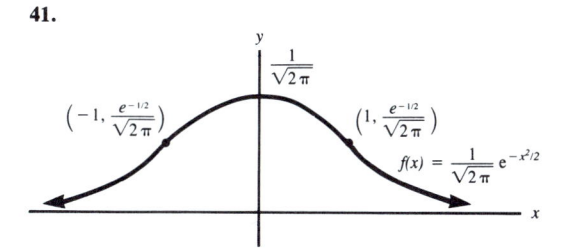

43. $(\ln 2)2^x$ **45.** $-6(\ln 5)5^{-2x}$ **47.** $x^{5x}(2 + x \ln 5)$
49. $3^{2x}[(2 \ln 3)/x - 1/x^2]$ **51.** \$1193.46/yr, \$1780.43/yr
53. rPe^{rt}, $r = i/100$ **55.** -20 g/min, -7.358 g/min

57. −$7357.59/yr, −$2706.71/yr **59.** −$0.110/unit, $0.027/unit, $0.137/unit

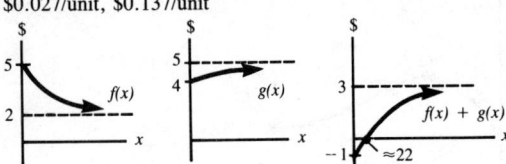

61. ≈21.75 days **63.** 10/9 ≈ 1.111 µg/mL, −(4/9)ln 3 ≈ −0.488 µg/mL **65.** 1000(ln 5) ≈ 1609, 1000(ln 5)$5^{-1.2}$ ≈ 233, 1000(ln 5)$5^{-2.4}$ ≈ 34

67.

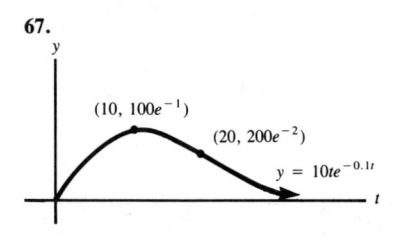

max is $100e^{-1}$ ≈ 36.8 mg/mL when $t = 10$ **69.** ≈1.099 students/hr, ≈213.761 students/hr, ≈68.781 students/hr **71.** yes; 9 yrs **73.** the value of t for which $dI/dt = rI$

13.5 EXERCISES

1. $y = 10{,}000e^{0.08t}$ **3.** $y = P_0 e^{rt}$, $r = \dfrac{\ln(P_n/P_0)}{n}$ **5a.** $y = ce^{-0.043t}$ **b.** ≈65.1% **c.** ≈53.5 days **7.** $y = ce^{-rt}$, $r = 0.6931/T$ **9a.** $y = ce^{-0.0001209t}$ **b.** ≈2379.5 yrs old **c.** ≈13,312 yrs old **11a.** $y = 100{,}000e^{0.4t}$ **b.** $1.644 million **c.** mid-1992 ($t ≈ 12.5$) **13a.** $y = 2.5e^{-0.193t}$ **b.** $0.363 million **c.** end of 1992 ($t ≈ 11.9$) **15.** 2008 ($t ≈ 28$) **17.** $q = 20{,}000e^{0.04p}$, $p ≈ 17.88 **19.** ≈211,458 **21.** $V = 0.0267e^{-0.1438t}$, ≈0.634%

23.

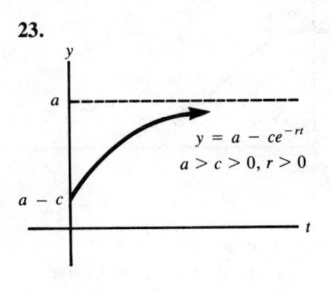

25. $y = 200 − 140e^{-0.113t}$;

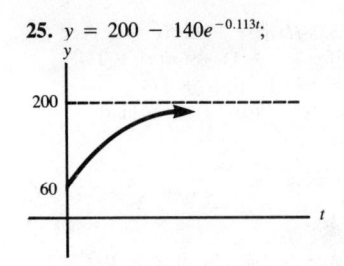

≈7.5 wks **27.** $y = 100{,}000 − 80{,}000e^{-0.144t}$ **29.** $y = 120 − 62.5e^{-0.223t}$, y in thousands; 120,000 units/yr **31.** $y = 4000 + 1828.57e^{-0.1335x}$;

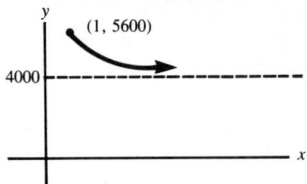

≈21,296.5 hrs **33.** $y = 2750 + 2082e^{-0.51x}$, 2750 hrs **35.** $y = 1200 − 1140e^{-0.082t}$, ≈698 **37.** $y = 100(1 + e^{-0.255t})$, $t = 0$ represents 1988 **39a.** $y_c = 100(1 − e^{-0.05x})$ **b.** $y_n = 100(1 − e^{-0.05(x-30)})$
c.

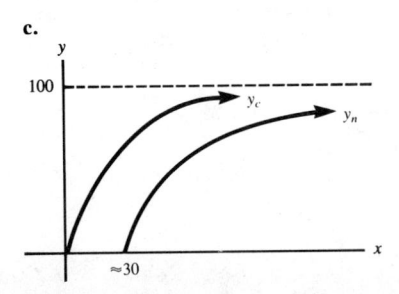

No **41.** $y = 100 − 78e^{-0.0094t}$, $t = 0$ represents 1970; $t ≈ 47.3$ yrs (A.D. mid-2017) **43a.** $y_L = 100(1 − e^{-0.086x})$ **b.** $y_U = 100(1 − e^{-0.136x})$
c.

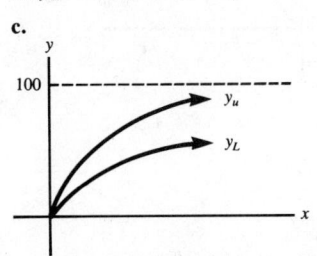

between 15.8% and 23.8%

45. 8500, 7225, 6141.25, 5270.0625 hrs **47.** $y = 10{,}000x^{-0.152}$ **49.** $y = 5000x^{-0.184}$, ≈3273 worker-hrs **51.** $y = 5000x^{-0.3221}$, 80%
53.

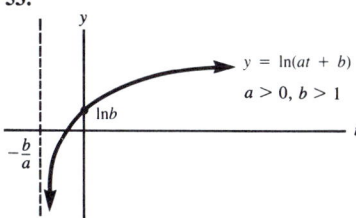

[Graph: $y = \ln(at + b)$, $a > 0, b > 1$; intercepts $\ln b$, $-b/a$]

55. $y = \ln(9.43x + 2.6)$, x, y in millions of dollars **57.** $y = \ln(20x + 3)$, x in thousands of units, y in millions of dollars; ≈$5,313,000; ≈1,101,000 units **59.** $y = \ln(1.64t + 7.39)$, $t = 0$ represents 1988 **61.** $y = \ln(23.231x - 7.134)$, x in millions of students, y in billions of dollars; ≈$1.92 billion **63a.** $y = \ln(0.52x + 23.58)$ **b.** ≈3.90 **c.** ≈1.8 kilotons

13 REVIEW EXERCISES

1. [Graph: $y = 7^{2x}$, points $(-\tfrac{1}{2}, \tfrac{1}{7})$, $(\tfrac{1}{2}, 7)$]

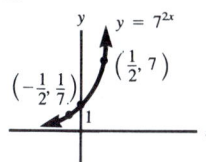

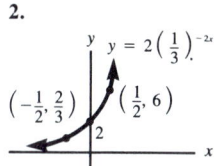

2. [Graph: $y = 2(\tfrac{1}{3})^{-2x}$, points $(-\tfrac{1}{2}, \tfrac{2}{3})$, $(\tfrac{1}{2}, 6)$]

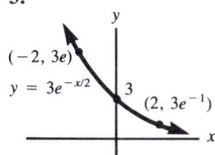

3. [Graph: $y = 3e^{-x/2}$, points $(-2, 3e)$, $(2, 3e^{-1})$]

4. [Graph: $y = 2 + e^{2x}$, points $(-1, 2+e^{-2})$, $(1, 2+e^2)$]

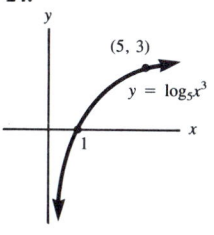

25. [Graph: $y + \ln 3x$, points $(\tfrac{e}{3}, 1)$, $(\tfrac{e^{-1}}{3}, -1)$]

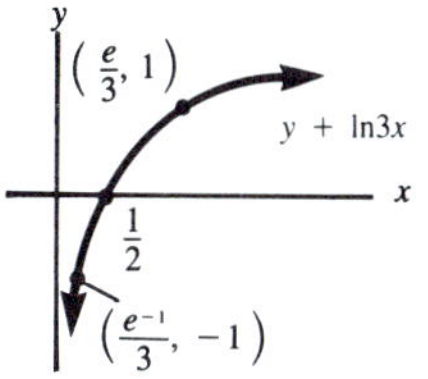

26. [Graph: $y = \ln 1.5x$, points $(\tfrac{2e}{3}, 2)$, $(\tfrac{2e^{-1}}{3}, -2)$]

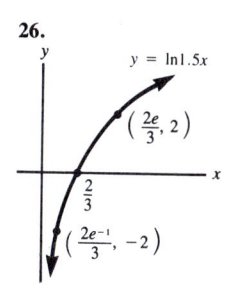

27. $f(x) = e^{3.8918x}$ **28.** $y = 7^{-3.55x}$ **29.** $y = 0.7797(\ln 3x)$
30. $g(x) = 0.774\,(\log_5 x)$ **31a.** 1 **b.** 3 **c.** 6 **d.** ≈6.4
e. 10^{14} **32.** $x = -0.536479$ **33.** $x = 0.623950$
34. $x - 0.764010$ **35.** $x = -225.98998$
36a. [Graph: $y = 2000e^{-0.5t}$, point 2000]

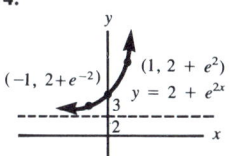

6. $y = 12(3)^{t/20}$ **7.** $6304.69 **8.** $14,425.31 **9.** 1
10. 0 **11.** 6 **12.** 200 **13.** -6 **14.** -35 **15.** $\log_7 22$
16. 5.2 **17.** 4.2 **18.** $\tfrac{1}{25}$ **19.** -5 **20.** 13 **21.** 1.7481
22. -0.531

23. [Graph: $y = \log_5 x$, points $(5, 1)$, $(\tfrac{1}{5}, -1)$]

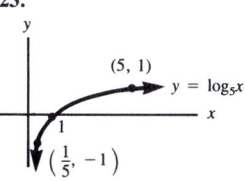

24. [Graph: $y = \log_5 x^3$, point $(5, 3)$]

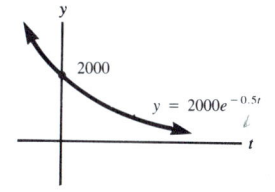

b. ≈1213 **c.** ≈6.9 days **37a.** 87.5% **b.** 16 months
38. 5.3768% **39.** 10.9861 yrs **40.** 195 months **41.** $3(x^2 + 1)/(x^3 + 3x - 2)$ **42.** $4x/(x^2 + 1)$ **43.** $2/x - 2/(2x - 1)$ **44.** $[(2/x)\ln(2x - 1) - (4/(2x - 1))\ln x]/(\ln(2x - 1))^2$ **45.** $10e^{5x}$ **46.** $4xe^{-x}(2 - x)$ **47.** $2e^{x^2}(2x \ln x + 1/x)$
48. $2e^x/(e^x + 1)^2$ **49.** $-2e^{-2x}e^{e^{-2x}}$ **50.** $-x^{1-x^2}(2 \ln x + 1)$ **51.** $4/(\ln 2)(4x - 3)$ **52.** $e^x \left(\dfrac{-2}{\ln 10}\right) \log_{10} e^{-2x}$
53. $-20(\ln 2)2^{-5x}$ **54.** $x(10^{-3x})(2 - 3x \ln 10)$
55. [Graph: $y = x + \ln x$, $y = x$, $y = \ln x$, point $(1, 1)$, ≈0.57]

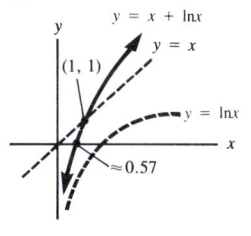

56. [Graph: $y = x^4 \ln x$, points $(e^{-1/4}, -e^{-1/4})$, $(e^{-7/12}, -\tfrac{7e^{-7/3}}{12})$]

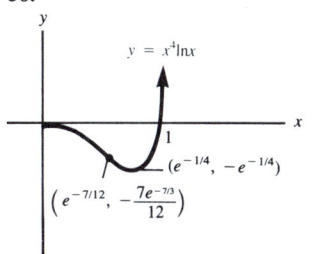

57.

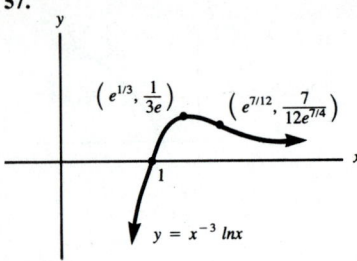

[Graph: $y = x^{-3} \ln x$ with points $(e^{1/3}, \frac{1}{3e})$ and $(e^{7/12}, \frac{7}{12e^{7/4}})$]

58.

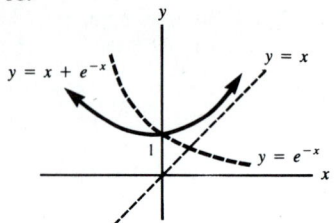

[Graph showing $y = x + e^{-x}$, $y = x$, and $y = e^{-x}$]

59.

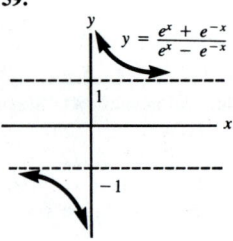

[Graph of $y = \dfrac{e^x + e^{-x}}{e^x - e^{-x}}$ with asymptotes at $y = 1$ and $y = -1$]

60.

[Graph of $y = x^2 e^{-x^2}$ with points $(-1, e^{-1})$, $(1, e^{-1})$, $\approx(\pm 1.51, 0.23)$, $\approx(\pm 0.47, 0.18)$; x-intercepts $-\left(\frac{5+\sqrt{17}}{4}\right)^{-1/2}$, $-\left(\frac{5-\sqrt{17}}{4}\right)^{-1/2}$, $\left(\frac{5-\sqrt{17}}{4}\right)^{-1/2}$, $\left(\frac{5+\sqrt{17}}{4}\right)^{-1/2}$]

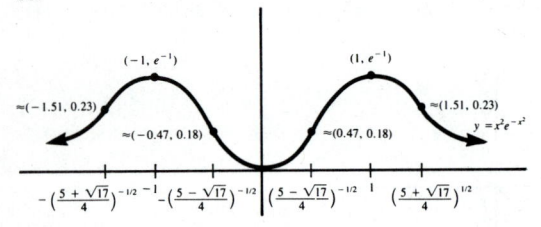

61. $\dfrac{3}{16}(\ln 8)^{-1/2} \approx \$0.13$ million/yr **62.** $-40e^{-1} \approx -14.7$ ppm/yr **63.** ≈ -3678.6/hr **64.** $y = 59.5 e^{0.00263t}$
a. $\approx 63.5°$F **b.** ≈ 266.7 yrs **65.** $y = 2000 e^{-0.1t}$

66a.

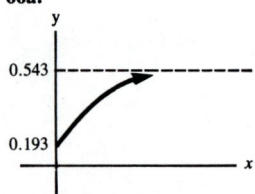

[Graph with horizontal asymptotes at 0.543 and 0.193]

b. will increase toward 0.593 **c.** will decrease toward 0.193 **67.** $y = 2400 - 800 e^{-0.1335t}$

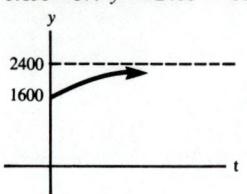

[Graph approaching 2400, through 1600]

in ≈ 5.2 months
68. $y = 250 + 173 e^{-0.1431x}$

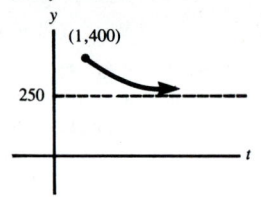

[Graph through $(1, 400)$ decreasing toward 250]

≈ 1490 hrs, $\approx 26/3$ units **69.** $y = 200 - 87 e^{-0.0297t}$, midday Aug. 19 ($t \approx 18.6$) **70.** $y = 237.5 + 129.8 e^{-0.1431t}$ **71.** $y = \ln(0.1763t + e)$, y in hundreds of thousands of dollars, $t = 0$ representing 1980, A.D. mid-2033 ($t \approx 53.7$)

13 SUPPLEMENTARY EXERCISES

3a.

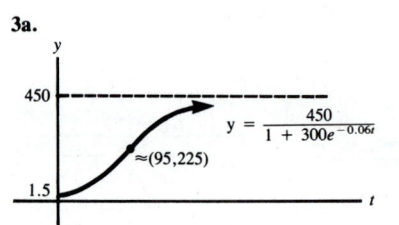

[Graph of $y = \dfrac{450}{1 + 300 e^{-0.06t}}$ through $\approx(95, 225)$, starting at 1.5, asymptote 450]

b. $\approx \$1.5$ billion, $\approx \$449$ billion **c.** ≈ 1895 A.D.

5a.

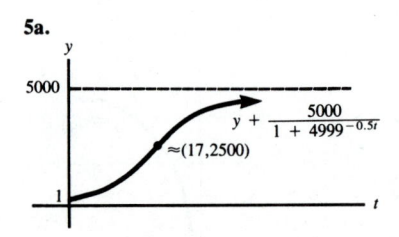

[Graph of $y = \dfrac{5000}{1 + 4999 e^{-0.5t}}$ through $\approx(17, 2500)$, asymptote 5000]

b. ≈ 17 days after it started **c.** $5/12 \approx 41.7\%$

9a.

(graph: y = 450e^{-300e^{-0.06t}}, point ≈(95, 165.5), asymptote y = 450)

b. ≈0; ≈$449 billion   **c.** ≈1895 A.D.

13a.

(graph: y = 100[1 − (1 − e^{−0.4t})^2], point ≈(1.73, 75))

b. 100; ≈25.2, ≈3.6   **c.** t ≈ 7.42 mins

15.

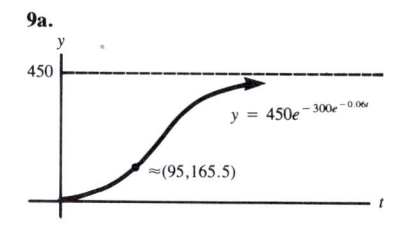

(graph with points ≈(4.6, 73.6), ≈(6, 121.6), ≈(7.4, 91.2))

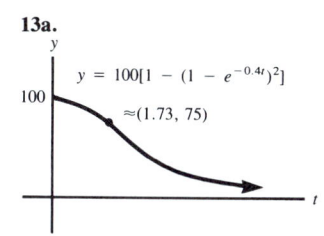

17.

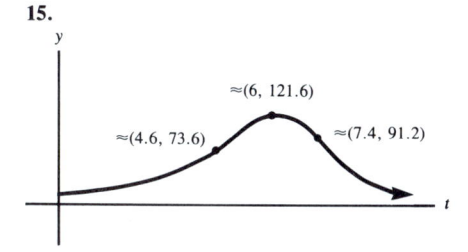

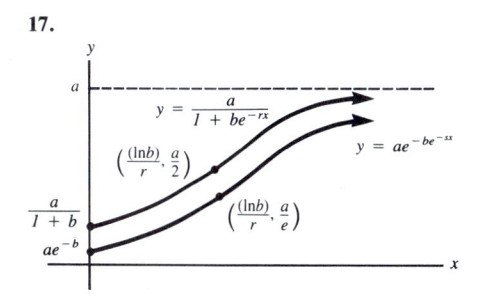

(graph showing $y = \frac{a}{1+be^{-rx}}$ and $y = ae^{-be^{-sx}}$ with points $\left(\frac{\ln b}{r}, \frac{a}{2}\right)$ and $\left(\frac{\ln b}{r}, \frac{a}{e}\right)$, asymptotes y = a, y = a/(1+b), y = ae^{−b})

14.1 EXERCISES

1. $3x + c$ **3.** c **5.** $x^{11}/11 + c$ **7.** $-t^{-3}/3 + c$
9. $2x^{5/2}/5 + c$ **11.** $x^4/2 + c$ **13.** $-3x^{-5}/5 + c$ **15.** $4t^{1/2} + c$ **17.** $2\ln|p| + c$ **19.** $3\ln(-t) + c$ **21.** $2e^x + c$
23. $x^2/2 + 2x + c$ **25.** $2v^3/3 - 3v^2/2 + v + c$ **27.** $x^4/2 - 2x^3 + 9x^2 - 3x + c$ **29.** $-x^4 + x^3 - x^2 + 7x + c$
31. $6e^x + x^2/2 + c$ **33.** $7x^3/3 - 7x^2 + 2\ln|x| + 5x^{-2}/2 + c$ **35.** $x^{100} + x^{99} + x^{98} + \cdots + x^2 + x + c$
37. $2x^3/3 + 7x^2/2 - 15x + c$ **39.** $-(4-x)^4/4 + c$

41. 600 km   **43a.** 1000 fps   **b.** 5000 ft   **45.** $C(x) = x^2 + x + 5000$, $R(x) = 15x$, $P(x) = -x^2 + 14x - 5000$
47. $P(x) = x^3/3 - 50x^2 + 1600x - 18$ **49.** $q = -25p^2 + 275,000$ **51.** $F(t) = 2t^{3/2}/3 + 4t + 2$ **53.** $F(t) = 100e^{-t} + 300$ **55.** $F(t) = 2\ln|t+1| + 80$

14.2 EXERCISES

1. $2(x+1)^{3/2}/3 + c$ **3.** $(4x-3)^{3/2}/6 + c$ **5.** $(6x-2)^{7/3}/14 + c$ **7.** $(7x-3)^{13}/91 + c$ **9.** $-5(2x+3)^{-1/2} + c$ **11.** $-3(4-7x)^4/14 + c$ **13.** $-\ln|1-2x| + c$
15. $-4\ln|8-3x|/3 + c$ **17.** $2e^{3x}/3 + c$ **19.** $-3e^{-7x}/7 + c$ **21.** $(x^2+1)^5/10 + c$ **23.** $(2x^5+10)^2/20 + c$
25. $-5(2-x^2)^{3/2}/3 + c$ **27.** $(x^2+3x+1)^5/5 + c$
29. $\ln(x^2+1) + c$ **31.** $-7\ln|t^3+2|/3 + c$ **33.** $3e^{x^2}/2 + c$ **35.** $e^{2x^2+2x}/2 + c$ **37.** $e^{e^x} + c$ **39.** $(\ln x)^3/3 + c$
41. $-(\ln x)^{-1} + c$ **43.** $C(x) = (x^2+1)^{3/2} + 999$, $R(x) = (x^2+1)^{7/4} - 1$ **45.** $P(x) = (2x^2+9)^{1/2}/2 + 48$
47. $F(t) = -100e^{-2t} + 200$ **49.** $F(t) = -10(2t^2+1)^{-1/2} + 260$ **51.** $F(t) = 2e^{-0.25t} + 1.56$ **53.** $F(t) = -10\ln(2t+5) + 56.09$

14.3 EXERCISES

1. 8   **3.** 0   **5.** 6   **7.** −2   **9.** 14/3   **11.** 5/4   **13.** 1/2
15. $3(e^3 - 1)$ **17.** 15   **19.** $-4 \cdot \ln 2$ **21.** $2a^{n+1}/(n+1)$
23. cost: $226,500; rev: $543,000   **25.** $500
27. $5833.33   **29.** $100,000   **31.** 24,827 people
33. 4160 people   **35.** 19/3   **37.** $1 - e^{-4}$ **39.** $2(e^3 - 1)$
41. $\ln 3$ **43.** −1   **45.** $4(10^{3/2} - 1)/9$ **47.** $(0.5)\ln(20/3)$
49. $54\sqrt{3}$ **51.** ≈512,711 units; ≈18.2 yrs   **53.** 469.1 units, 586.6 units   **55.** $20,239.89   **57.** 5 ln 6 ≈ 8.96 million ft³; 5 ln(46/21) ≈ 3.92 million ft³   **59.** 200(1 + ln 11) ≈ 680 animals   **61.** (1200) ln(850/840.5) ≈ 13; (1200) ln(2050/2000.5) ≈ 29   **63.** (0.2) ln 11 ≈ 47.96%

14.4 EXERCISES

1. 92/3   **3.** 14/3   **5.** 52/3   **7.** 863/6   **9.** 81/4   **11.** $1 + \ln 2$ **13.** $(1 - e^{-3})/3$ **15.** 5/2

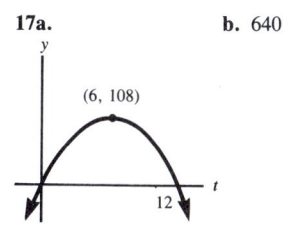

17a.   **b.** 640

(graph with vertex (6, 108), crossing t-axis at 12)

19. $125,000; total savings   **21.** 9719.489 units; total production for year   **23.** 50 ln(5/2) + 10 ≈ $55,814.54
25. 304; total population gain during period   **27.** 7/3

29. $e^2 - e - 3/2$ **31.** $1/2$ **33.** $9/2$ **35.** $1/3$ **37.** $128/15$ **39.** $2 \cdot \ln(5/3)$ **41.** 6 **43.** 21 **45.** $2(1 - e^{-1})$ **47.** $320/3$ **49.** $\$45{,}000$; total net savings **51.** $\$1000$ **53a.** $\approx 191{,}606$ people **b.** $\approx -536{,}711$ people

14.5 EXERCISES

1. 28 **3.** 19 **5.** 15 **7.** 168 **9.** 34 **11.** 12 **13.** $3/2$ **15.** 6 **17.** 20 **19.** $1/3$ **21.** $10/3$ **23.** $2/3$ **25.** 9 **27.** $14/9$ **29.** $5/4$ **31.** $e^2 - 1$ **33.** $1/e(e-1)$ **35.** 0 **37.** 64 fps **39.** $\$412{,}000$ **41.** 407.8 units **43.** 350 units **45.** $1600/3$ units **47.** 70 units **49.** $\approx \$1.04$ per unit **51.** $\approx \$85{,}882.82$ **53.** 1.75 g **55.** $(187.5)\ln 25 \approx 604$ animals

14 REVIEW EXERCISES

1. $-x^6/6 + x^4/4 + 3x^3 - x + c$ **2.** $-2x^6/3 + x^3 - 4x^2 + 14x + c$ **3.** $x + x^2/2 + x^3/6 + x^4/24 + x^5/120 + x^6/720 + c$ **4.** $2u^{1/2} + 3u^{2/3}/2 + 4u^{3/4}/3 + 5u^{4/5}/4 + c$ **5.** $2x^{-1} + 15x^{9/5}/9 + 4x + c$ **6.** $t^{-2} - 2t^{-1} + 8t^{3/2}/3 + t + c$ **7.** $2e^x + c$ **8.** $-3\ln|x| + c$ **9.** $2\ln|x| - 3e^x + c$ **10.** $25/8x^{8/5} + x^6 - 2\ln|x| - 9x^{-1/3} + 2x^{3/2}/3 - e^{2x}/2 + c$ **11.** $(4x+1)^{3/2}/6 + c$ **12.** $3(x^2+4)^4/8 + c$ **13.** $(x^2 + 2x - 8)^6/12 + c$ **14.** $25(x^3 + 6x - 2)^{9/5}/27 + c$ **15.** $0.25 \ln(4x^2 + 1) + c$ **16.** $-3(3x^2 + 2x + 1)^{-1/3} + c$ **17.** $3(x^2 - 4x + 1)^{1/3}/2 + c$ **18.** $e^{x^3} + c$ **19.** $-5 e^{-x^2/2} + c$ **20.** $2(\ln x)^{1/2} + c$ **21.** $C(x) = 3x^2 + 2x + 5000, R(x) = 10x^2$ **22.** $F(t) = 200t^{3/2} + 400$ **23.** $F(t) = 1000e^{0.12t}$ **24.** $F(t) = 25 \cdot \ln(t+1) + 1000$ **25.** -28 **26.** 26 **27.** $-4/3$ **28.** $-49/3$ **29.** $7/3$ **30.** $(11\sqrt{11} - 3\sqrt{3})/3$ **31.** $\dfrac{461}{36}$ **32.** 2 **33.** $1 - e^{-1/2}$ **34.** $\dfrac{1}{3}\ln 9$ **35.** $\ln 2$ **36.** $2/3$ **37.** 112 **38.** $\approx \$177.16$ **39.** $\dfrac{15}{8}$ **40.** $\dfrac{1}{2}\ln 5$ **41.** $\dfrac{125}{6}$ **42.** $253/12$ **43.** $\dfrac{863}{6}$ **44.** $\dfrac{125}{6}$ **45.** $-4 + 2e + 2e^{-1}$ **46a.** $50 \ln \dfrac{b+2}{2}$; total increase in junk bonds over period from 1984 to $1984 + b$ **b.** last half of 1996 ($t \approx 12.8$) **49.** $(1 - e^{-4})/4$ **50a.** $\$68{,}908$ **b.** $\$47{,}730$ **51.** 582.3 units **52.** $51{,}600$

15.1 EXERCISES

1. PS $= \$1452$, CS $= \$22{,}748$ **3.** PS $= \$41.67$, CS $= \$625$ **5.** PS $= \$42.67$, CS $= \$64$ **7.** PS $= \$1.45$, CS $= \$3.87$ **9.** PS $= \$250$, CS $= \$25$ **11.** $\$58{,}309.80$ **13a.** $\approx \$5.535$ million **b.** ≈ 5.88 yrs **15.** $\approx \$6.925$ million **17.** $\approx \$849{,}702$ **19a.** $\approx \$314{,}681{,}000$ **b.** $\$350{,}385{,}000$ **21.** ≈ 8.66 yrs **23.** $\approx \$245{,}488$ **25.** $\$50{,}000$ **29.** $\$8960$ **31.** ≈ 2353 caribou **35.** $A(5) \approx 65{,}085$, $A(10) \approx 92{,}365$, $A(20) \approx 114{,}580$ individuals **39.** 4605 individuals

15.2 EXERCISES

1. $e^x(x-1) + c$ **3.** $x^2e^x - 2xe^x + 2e^x + c$ **5.** $6 - 16e^{-1}$ **7.** $0.5x^2(\ln x - 0.5) + c$ **9.** $0.25x^4(\ln x - 0.25) + c$ **11.** $x(\ln x)^2 - 2x(\ln x) + 2x + c$ **13.** $x(\ln x)^3 - 3x(\ln x)^2 + 6x(\ln x) - 6x + c$ **15.** $19{,}022/21$ **17.** 110.1 **19.** $3x^2(x+2)^{4/3}/4 - 9x(x+2)^{7/3}/14 + 63(x+2)^{10/3}/140 + c$ **21.** $116/15$ **23.** $C(x) = 2(x+1)^{3/2}(3x-2)/15 + 4/15$ **25.** PS $= 2e^{10} - 22 \approx \$44{,}031$, CS $= 3e^{10} - 33 \approx \$66{,}046$ **27.** $(1 + \ln 3)/15 - (1 + \ln 13)/65 \approx 0.085$ mg/mL **29.** $-\dfrac{1}{9}\ln\left|\dfrac{9 + \sqrt{81-x^2}}{x}\right| + c$ **31.** $\dfrac{3}{4}[\ln(4 + \sqrt{17}) - \ln(2 + \sqrt{5})] \approx 0.4883$ **33.** $-\dfrac{5^{-3x}}{3(\ln 5)} + c$ **35.** $\dfrac{x}{3} - \dfrac{2\ln|3x+2|}{9} + c$ **37.** $-\dfrac{1}{6}\ln\left|\dfrac{2+\sqrt{4+9x^2}}{x}\right| + c$ **39.** $\dfrac{1}{3}\ln\left|\dfrac{x}{x+3}\right| + c$ **41.** $\dfrac{1}{12}[\ln|6x+5| + 5(6x+5)^{-1}] + c$ **43.** $\dfrac{1}{2}\ln\left|\dfrac{x}{x+3}\right| + c$ **45.** $\dfrac{1}{2\sqrt{2}}[\sqrt{2}\,x\sqrt{2x^2-1} - \ln|\sqrt{2}\,x + \sqrt{2x^2-1}|] + c$ **47.** $-\dfrac{1}{3\sqrt{2}}\ln\left|\dfrac{\sqrt{2} + \sqrt{2-4x^2}}{2x}\right| + c$ **49.** $x(\log_2 2x - 1/\ln 2) + c$ **51.** $R(x) = 10\ln(2x + \sqrt{4x^2-4}) + 70$ **53.** $32(\ln 2 - 0.5) \approx 6.2$ concepts

15.3 EXERCISES

1. divergent **3.** convergent; 1 **5.** convergent; $1/8$ **7.** divergent **9.** divergent **11.** divergent **13.** divergent **15.** divergent **17.** convergent; $e^2/2$ **19.** convergent; $1/2$ **21.** convergent; $1/3$ **23.** divergent **25.** divergent **27.** divergent **29.** convergent; $(\ln 2)^{1-r}/(r-1)$ **31.** $\$40{,}000$ **33.** $\$2$ million **35.** $250{,}000$ bbls **37.** $\$200{,}000$ **39.** 100 million tons

15.4 EXERCISES

9a. 1 **b.** $\dfrac{1}{3}$ **c.** $\dfrac{2}{15}$ **d.** $\dfrac{7}{15}$ **e.** $\dfrac{5}{6}$ **f.** $(b-a)/3$ **11.** $F(x) = \dfrac{3}{20}\left(\dfrac{1}{2}x^2 - \dfrac{2}{3}x^3 + \dfrac{1}{4}x^4\right) - \dfrac{1}{80}$ **a.** $\dfrac{73}{80}$ **b.** $\dfrac{459}{1280}$ **c.** 0.16 **d.** 0.27 **e.** $1 - \dfrac{3}{20}\left(\dfrac{1}{2}b^2 - \dfrac{2}{3}b^3 + \dfrac{1}{4}b^4\right)$ **f.** $\dfrac{3}{20}\left(\dfrac{1}{2}a^2 - \dfrac{2}{3}a^3 + \dfrac{1}{4}a^4\right)$ **13.** $F(x) = 1 - \dfrac{4}{x^2}$ **a.** $\dfrac{3}{4}$ **b.** $3/10{,}000$ **c.** $\dfrac{24}{25}$ **d.** $\dfrac{1}{625}$ **e.** $4\left(\dfrac{1}{a^2} - \dfrac{1}{b^2}\right)$ **f.** $4\sqrt{5}$ **15.** $f(x) = \begin{cases} 3, & 0 \le x \le 1/3 \\ 0 & \text{otherwise} \end{cases}$

17. $f(x) = \begin{cases} \frac{1}{3}x, & 1 \le x \le e^3 \\ 0 & \text{otherwise} \end{cases}$

19. $f(x) = \begin{cases} 3e^{-3x}, & x \ge 0 \\ 0, & x < 0 \end{cases}$

21. $F(x) = \begin{cases} 0, & x < 0 \\ x/5, & 0 \le x \le 5 \\ 1, & x > 5 \end{cases}$, $\frac{3}{5}, \frac{2}{5}, \frac{2}{5}$

23. $F(x) = \begin{cases} 0, & x < 1 \\ 1 - 1/x, & x \ge 1 \end{cases}$, $\frac{1}{2}, \frac{1}{6}, \frac{1}{5}$

25. $F(x) = \begin{cases} 0, & x < 0 \\ 1 - e^{-3x}, & x \ge 0 \end{cases}$, $e^{-3}, 1 - e^{-6}, e^{-9} - e^{-15}$

27. $F(x) = \begin{cases} 0, & x < 0 \\ x/4, & 0 \le x \le 4 \\ 1, & x > 4 \end{cases}$,

$f(x) = \begin{cases} 0, & x < 0 \\ \frac{1}{4}, & 0 \le x \le 4 \\ 0, & x > 4 \end{cases}$ **31.** $E(x) = 6\sqrt{6}/9$, $V(x) = \frac{1}{3}$

33. $E(x) = V(x) = +\infty$ **35.** $E(x) = 2$, $V(x) = 2$ **37.** $\frac{5}{32}$
39. $\frac{11}{256}$ $100\% \approx 4.3\%$ **41.** $1/\sqrt{2} \approx 0.7071$ **43.** 235
45. $\frac{3}{4}$ **47.** $\frac{2}{9}$ **49.** $e^{-0.5} - e^{-1} \approx 0.2387$ **51.** $1 - e^{-3} \approx 0.9502$ **53.** 10 days **55.** $(\ln 2/\ln 11) 100\% \approx 28.9\%$
57. $11^{0.95} - 1 \approx 8.76$ yrs **59.** 0.6916 **61.** ≈ 0.66 day
63. 6.5% **65.** $\approx 14.6\%$ **67.** ≈ 543 b/m³ **69.** $\approx 5\frac{1}{3}$ months

15.5 EXERCISES

1. 24 **3.** 2.33102 **5.** 0.69191 **7.** 0.45193 **9.** 0.52352
11. 0.78670 **13.** ≈ 7000 ft² (using $n = 3$) **15.** 2.33796
17. 2.33449 **19.** 0.69377 **21.** 1.14871 **23.** 11.37189
25. 0.74298 **27.** $\approx \$2070$ **29.** ≈ 680; worker's total production **31.** 19.25 mg/mL **33.** ≈ 185 concepts
35. 2.33333 **37.** 0.25 **39.** 0.69315 **41.** 1.14779
43. 11.37058 **45.** 0.74686 **47.** $\approx 36,950$ worker-hrs
49. $\approx -\$123.33$ **51.** 35,067 ft²

15.6 EXERCISES

1. $y = 3x^2/2 - x + c$, $y = 3x^2/2 - x$ **3.** $y = x^3/3 - x^2 + x + c$, $y = x^3/3 - x^2 + x$ **5.** $y = -2/x + c$, $y = -2/x$ **7.** $y = 4 \ln|t| + c$, $y = 4 \ln|t| - 3$ **9.** $y = -2e^{-x} + x + c$, $y = -2e^{-x} + x$ **11.** $y = \frac{1}{3}(x^2 + 1)^{3/2} + c$, $y = \frac{1}{3}(x^2 + 1)^{3/2} + \frac{2}{3}$ **13.** $y = \ln|x| + c$, $y = \ln|x|$
15. $y^2 = \frac{2}{3}x^3 + c$, $y^2 = \frac{2}{3}x^3 + \frac{25}{3}$ **17.** $y = e^{x^3} \cdot e^c$, $y = 2e^{x^3}$ **19.** $y = 1 + ke^{x^2}$, $y = 1 + 3e^{x^2}$ **21.** 200 ft/sec

23a. 18,800 mph **b.** 111,600 miles **25.** $v = \frac{6}{5(t+1)} + \frac{44}{5}$ **27.** $y(t) = (2.5 \times 10^6) e^{-1.84t}$ **29.** $p = x^2/2$ **31.** $S = 0.7813 A^{3/2} + 50$ **33.** $p = \frac{10}{3} d^2 + \frac{140}{3}$ **35.** $A = -\frac{1}{300} x^{2/3} + \frac{16}{3}$ **37.** $M = 25e^{0.09t}$ **39.** $y = 5333 e^{-0.29t} + 4000$ **41.** $y = \frac{100{,}000}{1 + 20.25e^{-0.811t}}$ **43.** $I = Ne^{-0.034t}$
45. $y = 1 - 0.24 \ln t$, ≥ 1, y in cm. **47.** $T = 25 - 5e^{-0.22t}$ **49.** $y = -8.51 \ln t + 19.70$ **51.** $I = 100 - 75e^{-0.07t}$ **53.** $k_A = 0.010$ **55.** **a.** $Y_A = 18{,}000{,}000$ **b.** $Y_A = 18{,}000{,}000$

15 REVIEW EXERCISES

1. PS = \$250, CS = \$100 **2.** PS = \$32,546.85, CS = \$14,552.91 **3.** PS = \$187.50, CS = \$208.33
4. $\approx \$1{,}160{,}563{,}000$ **5.** $\approx \$361{,}770{,}000$
6. $\approx \$1{,}249{,}045{,}000$ **8.** $31 \pi/5$ **9.** $\pi(1 - e^{-4})/2$
10. $-e^{-4x}(x + 0.25) + c$ **11.** $3(7x - 5)(x + 5)^{7}/56 + c$
12. $2^{7/2}(\ln 2 - 0.4)/5 + \frac{4}{25}$ **13.** $3x^3(x + 1)^{7/3}/7 - 27x^2(x + 1)^{10/3}/70 + 81x(x + 1)^{13/3}/455 - 243(x + 1)^{16/3}/7280$
14. $\frac{1}{2} \ln \left| \frac{x}{x + 2} \right| + c$ **15.** $-\frac{1}{16} \ln \left| \frac{x - 2}{x + 2} \right| + c$
16. $-\frac{1}{4\sqrt{2}} \ln \left(\frac{1 + \sqrt{2}}{2\sqrt{2} + \sqrt{7}} \right)$ **17.** $\frac{1}{6} [\ln |x + 2| + 2(x + 2)^{-1}] + c$ **18.** convergent; $1/k$ **19.** divergent
20. convergent; $\frac{1}{24}$ **21.** divergent **22.** convergent; $1/(r + 1)$ **23.** convergent; $\frac{5}{6}$ **24.** 1, $f(x) \ge 0$ **25.** 1, $f(x) \ge 0$
26. $k = \frac{5}{242}$ **27.** $\frac{\sqrt{2}}{2}$ **28.** $\frac{1}{2}$ **29.** $1 - \frac{1}{2}\sqrt{3.5}$

30. 1 **31.** $\frac{4}{3}$ **32.** $\frac{16}{5}$ **33.** $\frac{1}{2}$ **34.** $\frac{1}{3}$ **35.** $\frac{1}{6}$ **36.** 9 years **37.** diverges **38.** 3.80461 **39.** 3.79077
40. 3.79990 **41.** ≈ 3554 crimes **42.** ≈ 3081 crimes
43. ≈ 3910 ft² **44.** ≈ 3953.33 ft²

15 SUPPLEMENTARY EXERCISES

1. $\approx \$9.024$ million **3a.** $\approx \$40.240$ million **b.** $\approx \$46.180$ million **5a.** $\approx \$15.523$ million **b.** $\approx \$38.167$ million **c.** $\approx \$45.833$ million **7.** **a.** $\approx \$14.777$ million **b.** \$500 million **9a.** $\approx \$82{,}420$ **b.** $\approx \$199{,}526$ **c.** \$250,000
11a. $\approx \$119{,}208$ **b.** $\ge \$183{,}701$ **13a.** for \$100: yes, for \$50: no **b.** $\ge \$38.52$ per share

16.1 EXERCISES

1. $\{(x,y) \mid x\in\mathbf{R}, y\in\mathbf{R}\}$, 0, 16 **3.** $\{(p,q) \mid p\in\mathbf{R}, q\in\mathbf{R}, p \ge 0, q \ge 0\}$, 4, 11 **5.** $\{(u,v) \mid u\in\mathbf{R}, v\in\mathbf{R}, uv \ge 0\}$, 1, undefined **7.** $\{(x,y,z) \mid x\in\mathbf{R}, y\in\mathbf{R}, z\in\mathbf{R}\}$, 6, 29 **9.** $\{(x,y,z) \mid x\in\mathbf{R}, y\in\mathbf{R}, z\in\mathbf{R}, x > 0, y > 0, z > 0\}$, 0, 14 **11.** $-12; 17; \{(x,y) \mid x\in\mathbf{R}, y\in\mathbf{R}\}$ **13.** $-\frac{1}{2}; -\frac{1}{2}; 1/a^2; \{(x,y) \mid x\in\mathbf{R}, y\in\mathbf{R}, x \ne 0,$

$y \neq 0\}$ **15.** undefined; $\sqrt{5}$; $b |a|$; $\{(x,y,z) | x\in\mathbf{R}, y\in\mathbf{R}, z\in\mathbf{R}, |x| \geq |y|\}$ **17.** $50; $205 **19.** $P = 60A + 75B - 30,000$; $P = $25,500$ **21.** 156 units **23.** it is $200 - 8p_A + 3b$, so will decrease by 8 units for every $1 increase in p_A **25.** $2000 **27.** each additional unit of Y produced costs $20; each additional unit of X produced costs $40 **29.** avg cost per unit if make 150 units of X and y units of Y is $(8000 + 20y)/(150 + y)$ dollars; avg cost per unit if make 100 units of Y and x units of X is $(4000 + 40x)/(100 + x)$ dollars **31.** $A(x,y) = (12xy^{1/2} - 40x - 20y - 2000)/(x + y)$; $32 **33.** $A(150,y) = (1800y^{1/2} - 20y - 8000)/(150 + y)$; $A(x, 100) = (80x - 4000)/(x + 100)$ **35.** $C = 32z + 43,050$ **37.** 7 million units **39.** if p_1, p_2 increase, p_3 decreases **41.** $\{(s,m,t) | s\in\mathbf{R}, m\in\mathbf{R}, t\in\mathbf{R}, s \geq 0, m \geq 0, t \geq 0, s + t \neq 0\}$ **43.** 5 **45.** $26.4875 **47.** 12 days **49.** $f(t,b) = 1.2 t/b$; recovery time depends directly on age **51.** ½ **53.** $C = 1 - D_0/L$; increases toward maximum value 1. **55.** decrease D, increase L **57.** $H = 106.4 + 0.2Y$; increases by 0.2 for every $1 increase in avg income/wk

16.2 EXERCISES

1. $f_x = 5, f_y = -7$
3. $f_x = 8y, f_y = 8x - 4y$
5. $\dfrac{\partial f}{\partial x} = 3 + \dfrac{4x}{y^2}, \dfrac{\partial f}{\partial y} = -\dfrac{4x^2}{y^3}$
7. $g_x = \sqrt{\dfrac{y}{2x}}, g_x = \sqrt{\dfrac{x}{2y}}$
9. $\dfrac{\partial h}{\partial u} = 4(2u + v)(u^2 + uv + v^3)^3$,
$\dfrac{\partial h}{\partial v} = 4(u + 3v^2)(u^2 + uv + v^3)^3$
11. $\dfrac{\partial z}{\partial x} = 2 \ln y, \dfrac{\partial z}{\partial y} = \dfrac{2x}{y}$
13. $\dfrac{\partial w}{\partial x} = \dfrac{3y}{(x + y)^2}, \dfrac{\partial w}{\partial y} = -\dfrac{3x}{(x + y)^2}$
15. $f_x = z + 2xz^2, f_y = z^3 - 5, f_z = x + 2x^2 z + 3yz^2$
17. $f_x(1,3) = 6, f_y(-2,1) = -4$ **19.** $f_x(1,0) = 2e, f_y(0,1) = 3e^{-1}$ **21.** $\dfrac{\partial z}{\partial x}\bigg|_{(0,e)} = -1, \dfrac{\partial z}{\partial y}\bigg|_{(0,e)} = e^{-1}$ **23.** $f_x(1,-1,2) = 24.5, f_y(1,-1,2) = 2.5, f_z(1,-1,2) = 11.25$ **25.** $f_{xx} = 0, f_{xy} = f_{yx} = 0, f_{yy} = 6$ **27.** $g_{xx} = 2, g_{xy} = g_{yx} = -6y^2, g_{yy} = 2 - 12xy$
29. $f_{xx} = 0, f_{xy} = f_{yx} = -\dfrac{4}{3y^3}, f_{yy} = \dfrac{4x}{y^4}$
31. $\dfrac{\partial^2 z}{\partial x^2} = \dfrac{48x^2}{y} + \dfrac{12y^2}{x^4}, \dfrac{\partial^2 z}{\partial y \partial x} = \dfrac{\partial^2 z}{\partial x \partial y} =$
$-\dfrac{16x^3}{y^2} - \dfrac{8y}{x^3}, \dfrac{\partial^2 z}{\partial y^2} = \dfrac{8x^4}{y^3} + \dfrac{4}{x^2}$
33. $f_{xx} = ye^x - \dfrac{2}{x^2}, f_{xy} = f_{yx} = e^x, f_{xz} = f_{zx} = \dfrac{1}{z}$,
$f_{yy} = 0, f_{yz} = f_{zy} = 0, f_{zz} = -\dfrac{x}{z^2}$
35. $f_{xyxy} = e^{xy}(2 + 4xy + x^2y^2), f_{xyyy} = x^2 e^{xy}(3 + xy)$
37. $g_{pqq} = -6qp^{-2}$ **39.** $f_{xxyz} = 72xyz^2$
41. $f_{xx}(1,1) = 10, f_{xy}(2,2) = -3, f_{yx}(3, -5) = -3, f_{yy}(-2,7) = 0$
43. $f_{xyx}(1,1) = 6, f_{yyx}(-1, -1) = 12$

16.3 EXERCISES

1. slope of line parallel to xz-plane and tangent to surface at $(2,1,11)$ is -2; curve on surface parallel to xz-plane is decreasing and is neither concave upward nor concave downward at $(2,1,11)$ (it is a straight line); slope of line parallel to yz-plane and tangent to surface at $(2,1,11)$ is -5; curve on surface parallel to yz-plane is decreasing and is neither concave upward nor concave downward at $(2,1,11)$ (it is a straight line) **3.** slope of line parallel to xz-plane and tangent to surface at $(1,-1,3)$ is -4; curve on surface parallel to xz-plane is decreasing and concave downward at $(1,-1,3)$; slope of line parallel to yz-plane and tangent to surface at $(1,-1,3)$ is 3; curve on surface parallel to yz-plane is increasing and neither concave upward nor concave downward at $(1,-1,3)$ **5.** slope of line parallel to xz-plane and tangent to surface at $(1,-1,4)$ is 0; curve on surface parallel to xz-plane is neither increasing nor decreasing and is concave downward at $(1,-1,4)$; slope of line parallel to yz-plane and tangent to surface at $(1,-1,4)$ is 0; curve on surface parallel to yz-plane is neither increasing nor decreasing and is concave downward at $(1,-1,4)$ **7.** slope of line parallel to xz-plane and tangent to surface at $(1,b,1)$ is 2; curve on surface parallel to xz-plane is increasing and concave upward at $(1,b,1)$; slope of line parallel to yz-plane and tangent to surface at $(1,b,1)$ is 0; curve on surface parallel to yz-plane is neither increasing nor decreasing and is neither concave upward nor concave downward at $(1,b,1)$ **9.** $C_x = 20, C_y = 14$: if y is held fixed, each additional unit of A made will cost $20; if x is held fixed, each additional unit of B made will cost $14 **11.** $P_x = 5, P_y = 6$: if y is held fixed, each additional unit of A made and sold will contribute $5 to profit; if x is held fixed, each additional unit of B made and sold will contribute $6 to profit **13.** $C_x = $230, C_y = 142 **15.** increases cost by $\approx 142 **17.** $R_x = 35 + 2y, R_y = 50 + 2x$ **19.** increases revenue by $\approx 435 **21.** increases by $35 + 2x$; increases by $50 + 2y$ **23.** cost increases by $\approx $5/(10x + 25)^{1/2}$, revenue by $2, profit by $\approx \$2 - \$5/(10x + 25)^{1/2}$; cost increases by $\approx \$10/(20y + 25)^{1/2}$, revenue by $\$1$, and profit by $\approx \$1 - \$10/(20y + 25)^{1/2}$ **25.** cost increases by $\approx \$y/2(xy + 25)^{1/2}$, revenue by $\$2$, profit by $\approx \$2 - \$y/2(xy + 25)^{1/2}$ **27.** $V_v = -500$, $V_w = 300$, $W_v = 800$, $W_w = -400$; increase of $\$1$ in price of V while price of W is held fixed results in quantity of V demanded decreasing by 500 units and quantity of W demanded increasing by 800 units; increase of $\$1$ in price of W while price of V is held fixed results in quantity of V demanded increasing by 300 units and quantity of W demanded decreasing by 400 units; products are competitive **29.** $f_x(10,10) = -3500$, $f_y(10,10) = 300$, $g_x(10,10) = 400$, $g_y(10,10) = -7000$; when y is fixed at $\$10$, an increase of $\$1$ in the price of X will result in quantity of X demanded decreasing by ≈ 3500 units and quantity demanded of Y increasing by ≈ 400 units; when x is fixed at $\$10$, an increase of $\$1$ in the price of Y will result in quantity of X demanded increasing by ≈ 300 units and quantity demanded of Y decreasing by ≈ 7000 units; the products are competitive at this combination of prices **31.** $f_x = 200$, $f_y = -300$, $g_x = -400$, $g_y = 200$; if y is held fixed, an increase of $\$1$ in the price of X will result in quantity of X supplied increasing by 200 units and quantity of Y supplied decreasing by 400 units; if x is held fixed, an increase of $\$1$ in the price of Y will result in quantity of X supplied decreasing by 300 units and quantity of Y supplied increasing by 200 units **33.** $A_x(7,3) = 58$, $A_y(7,3) = 52$, $B_x(7,3) = -10$, $B_y(7,3) = 188$; if y is held fixed at $\$3$, an increase of $\$1$ in the price of X will result in quantity of X supplied increasing by ≈ 58 units and quantity of Y supplied decreasing by ≈ 10 units; if x is held fixed at $\$7$, an increase of $\$1$ in the price of Y will result in quantity of X supplied increasing by ≈ 52 units and quantity of Y supplied increasing by ≈ 188 units **35.** increases by $\approx 6x^{2/5}y^{-2/5}$ dollars; increases by $\approx 4x^{-3/5}y^{3/5}$ dollars **37.** $f(81,625) = 4500$, $f_x(81,625) = {}^{125}\!/_9$, $f_y(81,625) = {}^{27}\!/_5$; when 81 worker-hrs of labor and $\$625$ of capital are available each week, production is 4500 units per week; if capital is held fixed at $\$625$ and an additional 1 worker-hr per week is made available, production will increase by $\approx {}^{125}\!/_9$ units per week; if labor is held fixed at 81 worker-hrs per week and an additional $\$1$ of capital is made available each week, production will increase by $\approx {}^{27}\!/_5$ units per week **39.** increases by $\approx 0.01xe^{0.01y}$ thousand units/wk; increases by $\approx e^{0.01y}$ thousand units/wk **41.** $\partial C/\partial D = -1.25A^2(A + 1.25D)^{-2}$; if total audience A is fixed, increasing the sum of the duplicated audiences by 1 will result in a change in the net (unduplicated) audience of $\approx -1.25A^2(A + 1.25D)^{-2}$ **43.** $P_x(2,20,3) = 20e^{0.4} + 40 \approx 69.8$, $P_T(2,20,3) = 404$, $P_t(2,20,3) = 10,000$; if T is fixed at 20° and an additional 1 gram of nutrient present at time $t = 3$ will result in an increase of ≈ 70 bacteria; if x is held fixed at 2 grams, an increase of 1° in the temperature at time $t = 3$ will result in an increase of ≈ 404 bacteria; if x is held fixed at 2 grams and T at 20°, running the experiment 1 additional hour will result in an increase of 10,000 bacteria **45.** decreases by $\approx -4\sqrt{x}/y^2$; increases by $\approx 2/y\sqrt{x}$ **47.** $f_x(100,1) = 0.1$, $f_y(100,1) = -10$; if the price of a ticket is held fixed at $\$1$, an increase of 1000 in the city's population will result in an increase in ridership of ≈ 100 persons per day; if the population of the city remains fixed at 100,000, an increase of $\$1$ in the price of a ticket will result in a decrease in ridership of $\approx 10,000$ per day **49.** increases by $\approx 0.08e^{0.08x}$

16.4 EXERCISES

1. saddle pt: (0,0,0) **3.** rel max: (0,0,9)

5. rel min: $(3, -2, -37)$

7. rel min: $(-3, 10, -486)$

9. saddle pt: (0,0,0), rel max: $(-1, 1, 1)$

11. rel min: $(5, 10, -48)$, saddle pt: $\left(\frac{1}{3}, \frac{2}{3}, \frac{76}{27}\right)$

13. saddle pt: $(0,1,-6)$, rel min: $(\pm 2, 1, -19)$

15. saddle pt: $(0,1,0)$, $(-1,1,0)$, $(0,0,0)$, rel max: $\left(-\frac{1}{3}, \frac{2}{3}, \frac{1}{27}\right)$

17. base 2 m by 2 m, height 1 m

19. width $= \dfrac{3}{\sqrt[3]{2}}$ ft, length $= \sqrt[3]{4}$ ft, height $= \dfrac{3}{\sqrt[3]{2}}$ ft

21. $\$125$ **23.** $\$690,000$, $x = 30,000$ and $y = 20,000$ units **25.** $p_A = p_B = \$250$; no **27.** $p_A = \$8$, $p_B = \$10$, max revenue $= \$880$ **29.** 49,500 units **31.** $x = \$95,238.10$, $y = \$5952.38$ **33.** 2 units of A, 1.6 units of B

16.5 EXERCISES

1. min: $f\left(\dfrac{3}{2}, \dfrac{3}{2}\right) = \dfrac{13}{2}$ **3.** max: $f(3,3) = -9$

5. max: $f\left(3, \dfrac{3}{2}\right) = \dfrac{9}{2}$

7. min: $f(1, -1) = 2$ **9.** max: $f(0, -3, 3) = 9$ **11.** min: 5.25 when $x = -\frac{5}{3}$, $y = \frac{17}{6}$, $z = -\frac{4}{3}$ **13.** $f\left(\dfrac{6}{5}, \dfrac{8}{5}\right) = \dfrac{13,824}{3125}$ **15.** $f(\pm 4, \pm 4) = 29$ **17.** $f\left(\dfrac{1}{2}, \dfrac{1}{2}\right) = \dfrac{1}{4}$ **19.** 10,10 **21.** 4,16 **23.** 18, 9, 6 **25.** $10\sqrt{6}/3$ **27.** $(\frac{2}{3}, \frac{5}{3}, \frac{1}{3})$ **29.** cube of side $2r/\sqrt{3}$ **31.** max production is $f(300,450) = 300^{1/4} 450^{3/4} \approx 406.6$ units; marginal productivity of money $= \dfrac{3}{160}\left(\dfrac{300}{450}\right)^{1/4} \approx 0.0169$

units; 1 additional dollar could be used to produce approximately 0.0169 additional units **33.** 351 of X, 243 of Y **35.** $x = \$33$ million, $y = \$32$ million **37.** 57.5 bushels/acre

16.6 EXERCISES

1. $\dfrac{1}{2}$ **3.** 16 **5.** $\dfrac{1}{2}(1 - e^{-2})$

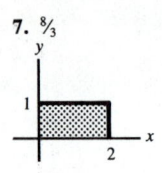

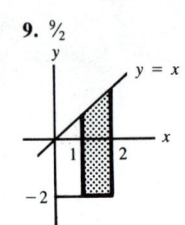

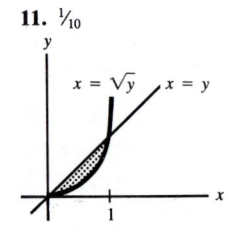

7. $\dfrac{8}{3}$

9. $\dfrac{9}{2}$

11. $\dfrac{1}{10}$

13. 3
15. 8 **17.** 0 **19.** $\sqrt{2}/30$ **21.** $\dfrac{5}{6}$ **23.** 9 **25.** $\dfrac{53}{6}$
27. $\dfrac{4}{3}$ **29.** $\dfrac{2}{3}$ **31.** $e - 2$ **33.** $\dfrac{2}{3}$ **35.** $\dfrac{64}{3}$ **37.** 2
39. $(e - 3)/(e - 1)$ **41.** $\approx 228{,}571$ units
43. $\approx\$9{,}573{,}000$ **45.** $502{,}875 (\ln 2)/19{,}500 \approx 14.04$ days
47. 10 animals/km² **49.** $\approx 3{,}828{,}500$ riders

16 REVIEW EXERCISES

1. $\{(x,y) \mid x \ne y\}$; undefined, 1, 0, 27 **2.** $\{(x,y,z) \mid xy > 0\}$
a. undefined **b.** $e^{-8} + \ln 8$ **c.** undefined **d.** $e^6 + \ln 4$
3. $R(p_1,p_2) = 200p_1 - 5p_1^2 + 750p_2 - 20p_2^2$; $\$2920$
4. $\$11{,}000$; contributes $\$10$ to profit **5.** $f_x = 16xy - 2y^3 + 12$, $f_y = 8x^2 - 6xy^2 + 2$, $f_{xx} = 16y$, $f_{xy} = f_{yx} = 16x - 6y^2$, $f_{yy} = -12xy$ **6.** $f_x(2,1) = 42$, $f_y(-1,0) = 10$, $f_{xx}(2,2) = 32$, $f_{yy}(-3,4) = 144$ **7.** $g_{pq} = e^{pq}(pq + 1)$, $g_{qqpp} = e^{pq}(p^2q^2 + 4pq + 2)$, $g_{qppp} = q^2 e^{pq}(pq + 3)$
8. $\dfrac{\partial z}{\partial x} = 12(x + y)(x^2 + 2xy - y^3)^5$, $\dfrac{\partial z}{\partial y} = 6(2x - 3y^2)(x^2 + 2xy - y^3)^5$, $\dfrac{\partial^2 z}{\partial x\,\partial y} = 12(x^2 + 2xy - y^3)^4(11x^2 + 12xy - 15xy^2 - 16y^3)$, $\dfrac{\partial^2 z}{\partial y\,\partial x} = \dfrac{\partial^2 z}{\partial x\,\partial y}$, $\dfrac{\partial^2 z}{\partial x^2} = 12(x^2 + 2xy - y^3)^4(11x^2 + 22xy + 10y^2 - y^3)$, $\dfrac{\partial^2 z}{\partial y^2} = 6(x^2 + 2xy - y^3)^4(-6x^2y - 72xy^2 + 51y^4 + 20x^2)$ **9a.** 82
b. 83 **10.** $f_x(16,1) = \dfrac{1}{32}$, $f_y(16,1) = -4$; if predators remain fixed at 1, an additional prey animal (the 17th one) would cause the number of wolves to increase by approximately $\dfrac{1}{32}$; if prey remains fixed at 16, an additional predator (the 2nd one) would cause the number of wolves to decrease by approximately 4

11. $R_x(25, 100) = \$20.10$, $R_y(25, 100) = \$25.025$; if y is held fixed at 100 units, an additional unit of x (the 26th) could be sold for approximately $\$20.10$; if x is held fixed at 25 units, an additional unit of y (the 101st) could be sold for approximately $\$25.025$

12. rel min: $(0, 0, -1)$

13. rel max: $(0, 2, 4)$, saddle pt: $(\pm 2, 2, 36)$

14. base is 2 m by 2 m, height is 2.6 m

15. $p_X = \$400$, $p_Y = \$250$

16. max: $f(1, 3) = 27$

17. min: $f(2, 8, 4) = 80$

18. 8, 8

19. max productivity is $f(10, 8) = 1000(10^{1/5}8^{4/5}) \approx 8365$; marginal productivity of money $= 2(10^{4/5}8^{-4/5}) \approx 8.3625$; an additional $\$1$ could be used to increase production by approximately 8.365 units

20. $\dfrac{92}{3}$ **21.** $\dfrac{10}{3}$

22. $e - e^{-1} - 2$ **23.** $1250(10^{9/5} - 4^{9/5})(30^{6/5} - 15^{6/5})/243 \approx 8770$ units

16 SUPPLEMENTARY EXERCISES

1. $Y = 1.6X + 66.3$ **3.** $Y = 0.1228X^2 - 9.853X + 204.94$; ≈ 39.5 units **7.** $Y = 47.92X_1 + 5.2168X_2 + 434.54$; fixed cost is $\approx \$435$, variable cost per unit of X_1 when X_2 is held fixed is $\approx \$48$, variable cost per unit of X_2 when X_1 is held fixed is $\approx \$5$ **9.** $P = 1.051e^{-0.0819x}$; ≈ 0.136

Applications Index

Management

Accounts receivable, 361–62, 372–73, 455, 510
Acid rain, 53–54
Advertising, 11, 137, 267–68, 277, 286–87, 298–300, 315, 326, 328, 336, 338–39, 343, 455, 726, 749, 756, 768, 835–36, 847, 849, 854, 992, 1036, 1046, 1054, 1067–68
Airline ridership, 202
Annuities
  future value, 582, 587
  net present value, 579–80, 587–88
  present value, 553
Approximation by differentials, 708, 711–14, 729
Assignment problems
  jobs, 283–84, 316
  machines, 288, 299
  salespeople, 315, 325, 343
Audience share, 750
Auditing, 550
Auto emissions, 622
Average
  cost, 30, 47, 49, 88, 102–3, 624, 645, 665, 690, 749–50, 756, 767, 781, 785, 801, 807, 835, 848, 856, 992–93, 1016–17, 1045–46, 1078–79, 1080
  demand, 719
  inventory, 917–19, 923
  production, 1067, 1073
  profit, 665, 1016
  revenue, 665
  sales, 1067–68

Blending problems, 278, 281–82, 286, 299–300, 316, 325–27, 336–37, 343
Bond issues, 582, 587
Borrowing, 570, 572, 587, 809, 820
Break-even analysis, 156–58, 162–63, 171, 174–75, 179–80, 328

Budgeting, 21, 23, 137, 148–49, 202, 267–68, 277, 286–87, 298–300, 315, 325–26, 343, 403
Business
  failures, 354
  strategy, 211, 217, 405, 444–45, 447, 449, 476

Cable TV, 467
Car rentals, 434–35
Carrying cost, 941–42
Cash flow, 908
Charge accounts, 455, 510
Collective bargaining, 386–87
Comparative costs, 152–53, 160–61
Comparative pricing, 480–81
Competitive, complementary products, 1031–32, 1034–35
Compound interest, 570–71, 587, 809, 820
Consumer choices, 384, 386
Consumer purchases and income, 818
Consumers' surplus, 934–35, 940, 950–51, 997
Consumption function, 731–33
Control charts, 550
Cost, 7, 11, 28, 46–47, 51, 53–54, 84, 88, 102, 104, 152–53, 160–61, 283–84, 286–88, 336, 622, 638, 663–65, 708, 712, 725–26, 768, 781, 786, 888, 891, 906, 922, 957, 976, 980, 1007, 1010, 1015–18, 1054, 1081
  advertising, 299, 328, 336, 338–39
  average, 30, 47, 49, 88, 102–3, 624, 645, 665, 690, 749–50, 756, 767, 781, 785, 801, 807, 835, 848, 856, 992–93, 1016–17, 1045–46, 1078–79, 1080
  functions, 130, 141, 147–49, 162, 171, 174–75, 179–80, 202, 536, 620, 622, 638, 653, 656, 676, 807, 871, 876–77, 884, 922, 950, 988, 992, 1015, 1081

Cost—Cont.
  of goods sold, 138
  inventory, 339, 763–64, 767–68, 786
  marginal, 657–58, 663–65, 690, 702, 708, 750, 756, 1033–34
  production, 281–82, 286–87, 294–95, 298, 300, 326–27, 336–37, 638
  replacement, 768–69
  shipping, 300, 620
  variable, 609, 637
Counting problems, 72–73, 77–78, 381, 384, 386–87
Credit bureau ratings, 436
Cross-classification problems, 75, 361–62, 372–73
Curve-sketching problems, 778–79, 781, 786, 827, 835, 867
Customer purchases, 77–78

Defective production, 389–90, 392, 455–56, 527
Demand, 23, 52, 88, 137, 420–21, 425–26, 463–64, 609, 637–38, 667, 835, 848, 878, 1010–11, 1016, 1031–32, 1034–35
  average, 719
  elasticity of, 660–63, 665–68, 676–77
  functions, 145–46, 149, 150–51, 163, 171–72, 175, 179–80, 202, 609, 622, 637–38, 653–54, 659, 665–68, 676, 690, 693, 699, 701, 711, 718–19, 721–22, 726, 729
Depreciation, 42, 136–37, 562, 676, 992
  double-declining balance, 808
Direct mail, 495, 500
Discounted loans, 572, 587, 809, 820
Dividends, 11

Earnings, 587, 940
  per share, 10

Effective interest rate, 587
Elasticity of demand, 660–63, 665–68, 676–77, 750
Employee
  classifications, 75, 78
  errors, 452–53, 455
  firings, 354
  productivity, 645, 650, 654, 749, 842–44, 849–50, 859, 981, 993
  promotions, 381, 387
  ratings, 435–36
  sick days, 486, 542
  training, 103–4, 392, 850
  wages and salaries, 11, 486, 496, 511, 1000–1001
  years of service, 542–43

Factoring accounts receivable, 572–73, 587, 809, 820
Feasibility problems, 274–75, 277–78, 338

Gross national product, 499, 510
Gross profit, 138, 229

Health insurance, 46–47

Income streams
  future value, 936–41, 957, 997–98
  present value, 1002–5
Information transmittal, 884
Input-output analysis, 246–53, 258
Insolvent insurers, 64
Internal rate of return, 820–21
Inventory, 205–6, 210–11, 214–15
  average, 917–19, 923
  carrying cost for, 941–42
  control, 136, 339, 528, 969
  and interarrival time for orders, 969–70
  optimal reorder quantity, 763–64, 767–68, 786
  valuation, 215–16, 226–30, 299
Investment, 19, 553, 570–72, 820

Learning curves, 820, 835–36, 842–44, 849–50, 852–53, 859, 884, 893, 906, 919, 993

Machine
  reliability, 374, 393
  tending time, 1017
Marginal
  average cost, 665
  cost, 657–58, 663–65, 750, 756, 1033–34
  productivity of capital/labor, 1032
  productivity of money, 1051, 1053, 1072
  profit, 657–58, 663–65, 702–3, 1030–31, 1033–34
  propensity to consume, 731–33
  propensity to save, 733
  revenue, 657–58, 663–65, 689–90, 702–3, 750, 827, 1033–36, 1072
  revenue product, 704
Market equilibrium, 158–59, 163–64, 171–72, 175, 179, 202, 835, 848
Marketing
  of new product, 78, 424–25, 463
  strategy, 441, 468–69
Market share, 136, 430–31, 433, 435, 441, 445, 447, 467, 476, 993
Market survey, 72–73
Maximization problems
  advertising effectiveness, 286–87, 315, 326
  audience, 267–68, 298–300, 326, 750, 768, 1054
  employee productivity, 650, 749
  inventory value, 299
  plant productivity, 645, 850
  production, 338, 1045, 1051, 1053–54, 1072
  profit, 174–75, 179–80, 264–65, 279–80, 282–83, 285–87, 289–90, 292–94, 298–300, 303–11, 314–16, 325, 337, 339, 644, 658–59, 663–65, 676, 702–3, 749, 756, 760, 767, 786, 827, 835–36, 1045–46, 1054
  revenue, 299, 316, 325–26, 750, 756, 767–68, 785, 1044–45, 1072
  sales, 315, 325, 749, 1046
Minimization problems
  average cost, 645, 749–50, 756, 767, 785, 1045–46
  cost, 281–84, 286–88, 294–95, 298–300, 326–28, 336–39, 768, 786, 1054
  distance, 1046
  inventory, 763–64, 767–68, 786

Minimization problems—*Cont.*
  plant productivity, 644
  pollution by utility, 1054
  production time, 850
  replacement cost, 768–69
  trim loss, 288, 328
Money market funds, 498
Multiplier effect, 732–33

Net audience, 1036
Net income, 29–30
  per share, 19
Nuclear power plant, 537–38, 622, 781, 1081

Oil
  drilling, 374, 393, 425, 463, 609, 622
  production, 808, 950, 957
Operating lifetime of product, 958, 960–63, 965–66, 969, 999

Package fill, 545–49
Parts
  assembly, 231
  production of, 253–54
Plant
  construction, 10, 161, 217, 405, 537–38
  productivity, 644, 849–50
Platinum use, 137
Present value, 568–69, 571–72, 583–84, 587–88, 809–10, 820–21
Price discrimination, 1044–45
Pricing, 40–41, 43, 53, 211, 444–45, 447, 449, 609, 637, 768
Producers' surplus, 932–34, 940, 950–51, 997
Product
  success of, 355
  testing, 550, 962–63, 965–66, 969, 999
  usage, 6–7
Production, 10, 21, 23, 49, 463, 465, 524, 526, 884, 893, 905–6, 918–19, 993, 1045, 1051, 1053–54, 1072
  average, 1067, 1073
  breakdowns, 969
  functions, 711, 726, 1032, 1035–36, 1051, 1053–54, 1067, 1072
  process, 12

Production—Cont.
  scheduling, 239–40, 255, 278, 343
  time, 835, 844, 849–50, 859, 893, 983, 993
Product mix, 131–33, 137–38, 160, 199–202, 243, 274–75, 277–80, 282–83, 285–87, 289–90, 292–95, 298–300, 303–11, 314–16, 325–28, 336–37, 339, 343
Profit, 23, 42, 52, 80, 88, 93, 103, 179–80, 265–66, 279–80, 282–83, 285–86, 287, 289–90, 292–94, 298–300, 303–11, 314–16, 325, 337, 339, 410, 425–26, 463–64, 468, 499, 510, 537, 658–59, 663–65, 676, 750, 756, 760, 827, 835–36, 848, 854, 859, 891, 893, 907, 918, 940–41, 957, 980–81, 983, 1004, 1015–16, 1046, 1054, 1071
  average, 665, 1016
  functions, 142–43, 147–49, 162, 171, 174–75, 611, 620, 644, 653, 656, 663–65, 700, 702–3, 712, 726, 749, 760, 767, 778–80, 786, 877, 884, 992, 1015–16
  marginal, 657–58, 663–64, 702–3, 1030–31, 1033–34
  per share, 51
  stream of, 584
Proportional/percentage change, 712, 729
Publishing, 467–68, 469
Purchasing
  cars, 215
  equipment, 23, 571–72, 582–84, 587, 809–10
  firms, 587–88
  ink, 620
  machinery, 243–44, 257, 338, 568–69, 579–80
  software, 287–88
  swimsuits, 426

Quality control, 411–416

Related rates problems, 721–22, 725–26
Return
  on assets, 133
  on equity, 133
  on investment, 10, 402, 511, 923

Revenue, 40–41, 43, 53, 299, 325–26, 403, 542–43, 665–68, 676–77, 689, 704, 750, 756, 767–68, 785, 848, 854, 863–64, 891, 1009, 1015–16, 1071–72
  average, 665
  functions, 141, 147–48, 162, 171, 174–75, 179–80, 654, 656, 663–65, 701, 712, 725, 827, 877, 884, 922, 951, 1002–4, 1015–16
  marginal, 657–58, 663–64, 689–90, 702–3, 827, 1033–34, 1072
  stream of, 582–83
Right-hand-side ranging, 342–43

Salaries, 495–96, 498
Sales, 10, 49, 80, 103–4, 126–27, 132, 137–38, 178–79, 215, 256, 349–50, 354–55, 397–98, 402, 410, 495, 498–99, 509–11, 514, 536, 543, 645, 653–54, 714, 726, 749, 781, 795, 807, 820, 827, 833, 847–48, 854, 861–62, 865–67, 918, 992–93, 1046
  average, 1067–68
  commissions, 136, 403
  per employee, 496, 510
Salesforce, 136
Savings, 888–89, 891–92, 906–8, 941
Securities brokers, 64, 104
Shadow prices, 341–43
Shipping, 195–96, 202, 243, 277, 300, 620
Shortage, 23, 52, 150–51
Sinking funds, 582, 587
Software, 609, 637
Steel mills, 162–63
Stock prices, 10, 390–91, 992
Stream of income, 936–41, 957, 997–98, 1002–5
Stream of profits, 584
Stream of revenues, 582–83
Supply, 23, 52, 835, 848, 878, 919, 1017, 1035
  functions, 144–45, 149–51, 163, 171–72, 175, 179–80, 202–3, 536, 622, 653–54, 708, 711, 725, 729
Surplus, 23, 150–51
Sustainable growth rate, 713–14
Synchronous service function, 1017

Taxes, 455
Tool rental, 429–30, 432

Value of industry's production, 827
Wages, 11, 486, 511, 1000–1001
Worker-hours, 624

Yield on government securities, 781

Life Science
AIDS, 456, 496, 500, 808–9, 942–43
Anesthesia, 848–49
Animal activity, 30
Antibody production, 885
Antivenin shot, 624
Approximation by differentials, 712
Artery, volume of, 654, 712

Bactericide, effects of, 134, 178, 645, 655, 919, 1030, 1037
Birch trees, 785
Births, 645
Blending problems, 316, 343
Blood
  coagulation, 551
  pressure, 727
  velocity of, 727, 943, 1032–33

Cell growth, 139, 867
Chemical reactions, 161, 654
Chemotherapy, 1047
Competing species, 792
Concentration of solute in a cell, 878–79
Condor eggs, 514
Cost, 288–89, 295–96, 300–301, 328, 613, 622, 782
Counting problems, 79, 387
Crop yield, 278, 301, 457–62, 487, 496, 498, 544, 769, 850, 1007, 1036, 1046–47, 1054
Curve-sketching problems, 781–82, 788–92, 827, 836, 865–67

Deacidification of lake, 827, 854, 993
Desert, encroachment by, 952
Diet
  experiment, 486, 499
  planning, 278, 301
  supplements, 179, 278, 300, 328

Disease, cases of, 848, 894, 942–43, 994
Doctor's recommendations, 355
Drug(s)
  blending, 316, 343
  concentration, 176, 512, 688, 712, 820, 836, 850, 884, 919, 951, 981
  dosage, 8, 11, 24, 106, 139, 154–55, 316, 326, 328, 343, 538, 691
  effectiveness, 79, 356, 393, 542, 865
  effects of, 654, 867
  overdose, 65
  testing, 355, 387
  toxicity, 851

Egg production, 769
Endemic disease, 792
Enzyme production, 982, 984
Epidemic, 808–9, 836, 866, 892, 922, 942–43, 994
Extinction of eagles, 808, 1081

Feasibility problems, 278
Fick's law, 878–79
Fish eggs, 134
Fishing, 512
Flu vaccine, 426, 447–48, 464

Genetics, 79, 375–76, 393, 403–4
Greenhouse effect, 139, 469, 858
Growing time, 526–27

Heart disease, 456
HIV, 456, 942–43
Hospital
  accreditation, 64
  patient movement within, 436

Immunization, 782
Infant growth, 138
Infant mortality, 11, 24, 355–56, 393
Intravenous solution, 894

Litter size, 403
Lung cancer, 456, 538

Maximization problems
  air pollution, 176
  crop yield, 769, 1046–47, 1054
  egg production, 769
  drug concentration, 176, 850
  drug effectiveness, 316, 326
  inoculations, 751
  nutritional value, 301
  population size, 301, 645, 750, 769
  proportion of trees affected by disease, 785
  temperature, 172–73, 176, 756
  velocity of blood, 769
Mental imagery, 134
Minimization problems
  cost, 288–89, 295–96, 300–301, 328
  discomfort, 1047
  drug dosage, 328
  population size, 176, 645, 756, 769
  white blood count, 180

Newton's law of cooling, 994

Ozone depletion, 138, 538–39, 700–701, 703

Patient
  movement within hospital, 436
  treatment, 428, 464–65
Ph.D.s in life sciences, 161
Plant growth, 526–27, 892, 994
Poiseuille's law, 727, 769, 943, 1032–33
Pollen count, 859
Pollution
  air, 176, 216, 512, 613, 622, 654, 701, 703, 726–27, 758, 836, 850, 854, 885, 951, 957, 970–71, 993–94, 1068
  water, 105–6, 134, 139, 403, 407, 712, 726, 800–801, 857–58, 892, 984, 987, 993
Population
  density, 1068–69
  gain, 894, 906
Population size, 89, 161, 788–91, 871
  bacteria, 90–91, 104–5, 134, 178, 562, 638, 645, 654, 690, 703, 712, 756, 808, 820, 836, 848, 850, 858, 876, 878, 884, 919, 993, 1036
  birds, 49, 176, 244, 301, 750, 782, 808, 858, 878, 1081

Population size—*Cont.*
  fish, 512, 712, 769, 848, 850, 858
  furbish lousewort, 176
  gypsy moths, 496, 498, 750
  and the logistic equation, 180–81, 184–85
  mammals, 85–86, 139, 203, 511–12, 645, 703, 781–82, 800–801, 838, 841–42, 878, 884, 919, 942, 990–91
Predator-prey relationships, 700, 790–92, 1018, 1036, 1072
Premature babies, 211–12
Proportional/percentage changes, 712
Public health, 426, 447–48, 464, 727, 751, 782
  bilharzia, 703
  influenza, 836, 866, 892, 922

Radioactive iodine, 827, 854
Radiation therapy, 24
Radon concentration, 970–71
Recovery time, 970, 1018, 1068
Related rates problems, 726–27

Scar, visibility of, 655
Sickle-cell anemia, 375–76, 403–4
Signing by chimpanzee, 691
Snake bite, 624
Spatial perception test, 487
Species competition, 161
Stimulus, smallest, 134
Survival-renewal functions, 942
Survival time, 970
  AIDS, 496, 500
  liver cancer, 355
Sustainable yields, 259–65

Temperature
  animal, 10, 30, 994
  human, 172–73, 176, 756, 865
Timber harvest, 259–65, 792

Vision, 134–35
Visual
  contrast, 1018
  processing, 134–35
Vitamins, 161, 203, 244, 288–89, 295–96, 300–301, 328

Weather, 457–62
Weber's law, 134
Weed growth, 655
Weight loss, 486, 499
White blood count, 180

Social Science

Approximation by differentials, 712–13
Aptitude test, 140

Ballot order, 379
Birth/death rates, 995
Budgeting, 317
Bureaucracy, size of, 799–800, 828, 836, 854
Busing, 301

Campaign
  spending, 1037
  strategy, 404, 426–27, 445–46, 448, 465, 476
Charity contributions, 176, 437
Competition between nations, 467
Cost, 301, 328, 782
Counting problems, 79–80, 109, 379–80, 384, 387
Course evaluation, 89
Creativity, 75–76
Crime rate, 757, 1000
Cross-classification problems, 75–76, 109, 357, 362, 373
Curve-sketching problems, 782, 786, 828, 836, 867

Debt service, 691
Desegregation, 301

Education, 139, 356, 487, 497–98, 836, 854
Elderly, 456
Election outcome, 380
Emigration, 908
Employee relationships, 118
Exam grading, 527

Fads, 106–7, 867, 971
Feasibility problems, 279
Foodbanks, 1080
Food stamp program, 1080
Fund-raising, 64

Group dynamics, 316–17

High school
  construction, 422–24, 469
  dropout rate, 139, 487
  enrollment, 735
Housing starts, 782, 1018–19

Income distribution, 908–9
Immigration, 908
IQ tests, 487, 497, 500, 517–18, 522–23, 551

Language
  retention, 713, 895
  comprehension, 655
Learning, 203, 512, 952, 982, 1000
Likert scales, 89, 500–501
Literacy campaign, 49, 139–40
Lobbyists, 64
Lorenz curves, 909–9

Marriage, 404
Mathematics test, 89
Maximization problems
  approval rating, 177
  charity contributions, 176
  cohort of 18–25 year olds, 757
  number of participants, 326
  problem-solving ability, 316–17
  school enrollment, 735, 751
  social service benefits, 317
  voter interest, 177
  votes, 751
Memory, 362, 373, 624–25, 655, 713
Military spending, 449
Minimization problems
  cohort of 18–25 year olds, 757
  cost, 301, 328
  crime rate, 757
  pupils bused, 301
  school enrollment, 735, 751
  voter approval, 170–71, 646
Municipal budget, 691

Net worth of families, 497–98, 513

Organization chart, 231

Party loyalty, 456
Political
  convention, 328
  polls, 30, 80, 279, 301, 393
  issues, 500–501
Population
  changes, 436–37, 887–88, 908, 922
  density, 944
  size, 757
Poverty, 456
Prison overcrowding, 944
Proportional, percentage changes, 713
Psychological testing, 994
Psychology experiments, 326, 362, 373, 512, 691, 849, 859

Reading test, 512–13
Related rates problem, 727
Religious cult, 943–44
Rumors, 712–13, 836, 866–67, 994

SAT test, 161–62
School enrollment, 735, 751, 782
Social services, 317, 828, 858
Social workers, 727
Socioeconomic class, 384, 387
Status, 212
Student classification, 437
Surveys, 79–80, 394

Tax burden, 487, 878
Television viewing, 539
Tenure, 437
Tests, 89, 140, 161–62, 487, 497, 500, 512–13, 517–18, 522–23, 527, 551
Town contract, 427, 465
Transition rates, 995
Transit system ridership, 923, 1037, 1069

Unemployment, 514, 828, 892
  rate, 646, 655, 782–83

Voter(s)
  affiliation, 135, 162, 203, 356–57, 393, 437, 851, 878, 994–95
  ages, 357, 456–57, 491
  approval ratings, 170–71, 177, 497–500, 645–46, 786

Voter(s)—*Cont.*
  ideology, 456–57
  interest, 177
  participation, 139, 539, 703, 994
  preferences, 30
  support for candidates, 656, 751, 895
Votes missed by representatives, 524–25

Welfare, 356, 655, 1037, 1069

Other
Acceleration problems, 650–53, 676
Accidental death, 353
Age, 18
Altitude, 40, 42, 174, 648–50, 652–53, 676, 1000
Amortization, 588–93
Annuities
  future value, 553, 573–77, 580–81, 587–88
  present value, 577–80, 582–85, 587–88
Architectural competition, 399
Area, 897–905, 907, 922–23, 979, 983, 1000, 1063, 1067, 1073
Automatic pilot, 9
Average daily yield, 821

Birth rate, 535–36
Blending problems, 16, 18–19, 51, 284–85
Bonds, 579, 584–85, 588, 646, 821
Books, 285, 383
Borrowing, 553, 561–62, 566, 569–70, 586, 589–90, 592, 809
Boyle's law, 724

Cards, 352–53, 358–59, 361, 364, 382–83, 385, 400–401, 409–10
Cars, 117, 408, 468, 581
Chemical reaction, 51
Class absences, 535, 1075–76
Classification problems, 8, 69–71, 408
Coin-tossing, 346–48, 352, 388–89, 391–92, 394–98, 400, 407, 527
Cold fusion, 786
College
  education, 576
  students, 68–71, 360, 363–64, 371

Committees, 382, 384–85, 400–401, 409
Comparative cost
  of apartments, 24
  of appliances, 22, 24, 52
  of telephone services, 24–25
Compound interest, 553, 563–73, 586–87, 804–5, 809, 816–17, 819, 835, 856–57
Computer
  downtime, 455
  programming, 369–70
  science, 89–90, 625, 783, 851
Consumer
  preferences, 377, 392, 499
  purchases, 451–52
Cost, 284–85, 327, 634, 637
  comparative, 22, 24–25, 52
Counting problems, 74–75, 377–78, 381–86, 409–10
Course(s)
  enrollment, 74–75
  evaluation, 499
Credit cards, 51, 592–93
Crime, 448–49
Cross-classification problems, 360, 363–64, 371–72
Cruise control, 12
Cryptography, 245–46, 251, 257–58
Curve-sketching problems, 780, 786, 792, 828, 866–67

Decibel scale, 857
Dice, 346–48, 351–52, 391, 394–95, 400–401
Dimension problems, 17–18, 40, 42, 53
Distance-rate-time problems, 13, 18, 51–52
Dividends, 324–25
Drawing cards from deck, 352–53, 358–59, 361, 364, 391–92, 400–401, 409
Drawing marbles from vase, 361, 367–69, 372–74, 401, 408, 454

Effective interest rate, 566, 570, 587
Employee relationships, 118
Energy
  efficiency, 866
  radiation of, 877
Exams, 454–55, 535, 1075–76

Family income and savings, 529, 531–34
Fashion models, 117–118
Fuzzy sets, 113–14, 117–18

Gambling, 353, 385–86, 398–99, 401–2, 410, 427–28, 477
Gold, 535
Grade point averages, 117
Grades and absences, 1075–76
Gross national product, 922

Hours worked, 18, 21–22, 24
Hurricanes, 892

Income, 493, 529, 531–34
Inflation, 780, 819, 847
Insurance, 581
Interest
  charges, 51
  compound, 563–73, 586–87
  rebate, 588
  simple, 17, 556–58, 561–62
  true annual rate, 588
Investments, 18, 132, 201, 216–17, 243, 280–81, 284–85, 297, 324–25, 343, 405, 421–22, 424, 462–63, 543, 553, 561–65, 567, 569–71, 579–81, 586, 804–5, 809, 816–17, 819, 821, 835, 846, 857, 922
  optimum holding period for, 836–37
  risk of, 132, 514–15
Investor survey, 75

Job-hunting, 353–54, 361, 402
Junk bonds, 923

Keno, 410

Lawyers, 71
Library, 285
License plates, 377–78, 384
Light bulbs, 354, 402, 526
Lost luggage, 354
Lotteries, 385–86, 402, 583

Magazines, 117
Manufacturing capacity, 845–46, 853–54, 866–67

Applications Index

Marbles drawn from vases, 361, 367–69, 372–74, 401, 408, 454
Maximization problems
   altitude, 174, 649, 652, 676
   area, 765–66, 1053
   customer satisfaction, 285
   dividends, 324–25
   return on investment, 280–81, 284–85, 297, 324, 837
   volume, 761–73, 786, 1045, 1053
Measurement errors
   area, 709–11
   volume, 710–11, 729
Military, 446
Minimization problems
   cost, 284–85, 327, 765–66, 1043, 1045, 1072
   distance, 1053
   perimeter, 760–61, 766
   sunspots, 174
   surface area, 767, 1045
Mortgages, 590–92
Movies, 114
Music, 377
Mutual funds, 201

Nuclear
   strategy, 851–52
   testing, 828, 854–55
Numbers, 401

Officers of club, 377, 381, 385, 392, 400–401, 409

Package fill, 524
Paper-scissors-rock, 476–77
Patent applications, 792

Position problems, 871
Postal function, 634
Present value, 566–69, 571, 582–85, 587, 806, 809, 819, 856
Pressure, 12, 51
Product
   lifetime, 354, 524, 526
   ranking, 494
Proportional/percentage changes, 709–11

Radioactivity, 801–3, 807, 815–16, 819, 835, 839–40, 847, 857, 992
Radiocarbon dating, 847
Raffles, 383, 401, 409
Related rates problems, 720, 722–25, 729–30
Rental income, 1005
Restaurants, 360, 371–72
Retirement nest egg, 576–77
Return on investment, 18, 280–81, 284–85, 297, 324, 462–63, 543
Richter scale, 817–18, 828
Right-hand-side ranging, 343
Roulette, 353, 398–99, 401

SAT test, 491–92
Salaries, 499, 562
Savings, 529, 531–34, 562, 866, 1079–80
Selection problems, 377, 381–82, 384–85, 392, 400–401, 409
Service times, 542–44
Shadow prices, 343
Simple interest, 17, 556–58, 561–62
Sinking funds, 576–77, 581, 819, 858
Sports, 189–90, 231–32, 349, 392, 409–10, 527, 724

Stefan's law, 653, 877
Stockbroker's commissions, 1015
Stocks, 18, 297, 324, 343, 539–40, 571, 1005
Storks, 535–36
Strategy
   for investment, 216–17, 405
   military, 440
Sunspots, 174
Superconductivity, 749
Surveys, 75

Telephone
   calls, 637
   numbers, 383
Temperature, 17, 51, 749, 992
Tests, 130–31, 390, 392, 490–92, 494
   and grading, 80, 499, 541
Trade surplus, 19
Transportation, 75
Trust fund, 562

Utilities, 427–28

VCR rental, 637
Velocity, 646–53, 676, 877, 991–92, 1000
Volatility, 539–40
Volume, 1059–60, 1066–67, 1073
   of solid of revolution, 998

Wage rates, 13–14, 18, 24
Weather, 113–14, 353–54, 370, 450
Weight distribution, 515
Windmills, 858–59

Subject Index

Absolute
　maximum, minimum, 757–64
　value, 8–9
Acceleration, 650
Acceptance quality level, 412
Addition
　of matrices, 207–8
　of polynomials, 31
　of rational expressions, 44–45
Algorithm
　for Gauss-Jordan elimination, 194
　Karmarkar's, 302, 340
　Khachian's, 340
　linear programming, 292
Amortization, 588–92
　schedule, 589
Annuity, 553
　continuous, 1002
　due, 577
　formula, 574
　net present value of, 579–80
　ordinary, 573–77
　present value of, 577–80
　sinking fund, 576–77
Antiderivative, 871–73
Antidifferentiation, 871–72
Approximation
　of definite integrals, 971–78, 1001
　by differentials, 706–9
Archimedes, 925
Arc length, 1001
Area
　bounded by two curves, 900–905
　under a curve, 894–99
　and double integrals, 1063
　function, 895
Arithmetic sequence, 555–58
Artificial variable, 318
Assignment problem, 340
Asymptote
　horizontal, 614
　oblique, 617
　vertical, 612–13
Augmented matrix, 190–200
　row operations on, 192–93

Average
　daily yield, 821
　outgoing quality, 416
Average value of a function
　of one variable, 915–17
　of two variables, 1063–65
Axes, 91

Barrow, Isaac, 928
Base, 25
Bayes' law, 420, 449–53
　and decision trees, 420, 457–62
　generalized, 452
Bell curve, 516
Bernoulli process, 387–88
Binomial
　coefficient, 380
　distribution, 388–91
　formula, 389
　normal approximation to, 527
Bond yields, 584–85
Break-even
　analysis, 156–58, 171
　point, 156
　quantity, 156, 656

Calculus
　development of, 678–81
　fundamental theorem of, 914
Capital
　marginal productivity of, 1032
　value, 1003
Cardano, Girolamo, 54
Cartesian
　coordinate system, 91, 1011
　product, 111
Cauchy's mean value theorem, 787
Central tendency, 489–94, 500
　mean, 489–92
　median, 492
　mode, 493
Chain rule, 694–99, 1073
　first formulation, 694
　second formulation, 698

Change-of-base formulas, 813
Chaos, 180–85
Chebyshev's theorem, 513–14
Chi-square distribution, 544
Class mark, 481
Cobb-Douglas production function, 1032
Coefficient
　of inequality, 909
　of a polynomial, 31
　of variation, 508
Combinations, 379–80
　and probability, 381–82
Common factor, 32
Common logarithm, 813
Competitive products, 1031
Complementary event, 356, 358
Complementary products, 1031
Complement of a fuzzy set, 115
Complement of a set, 67–68
Complex numbers, 54
Composite function, 691–93
Compound events, 356–59, 366–70
Compound interest, 553, 563–66
　continuous, 804–6
　future value formulas, 564, 805
　and present value, 566–69, 806
Concavity, 164, 752
Conditional events, 362–66
Conditional probabilities, 365
Confidence intervals, 545–49
Conic sections, 111
Constant
　function, 639
　of integration, 873
　multiple rule, 641
　of proportionality, 987
　rule, 639
Constraint equation, 1047
Constraints, 276, 280, 297
Consumer's risk, 412
Consumers' surplus, 934–35
Consumption, 731–32
　function, 731

Continuity, 631–35
  and differentiability, 634–35
  limit definitions of, 633
Continuous
  annuity, 1002
  compounding, 805
  distribution, 515
  function, 631
  probability distribution, 958
  random variable, 957–58
Control chart, 550
Convergent improper integral, 954
Conversion formula, 521–24
Coordinate planes, 1012
Coordinates
  cartesian, 91, 1011
  cylindrical, 1074
  polar, 111, 1073
  spherical, 1074
Correlation, 532–34
  coefficient, 533
Cost
  function, 140–41, 143, 656
  of goods sold, 138
  marginal, 657
Counting problems
  and probabilities, 377–82
  and sets, 72–73
Cramer's rule, 258
Critical value, 738
Cryptography, 245–46
Cubic formula, 54
Cumulative
  distribution function, 963–65
  frequency, 481
Curve sketching, 769–779
  procedure, 771

Dantzig, George, 302
Decibel scale, 857
Decimals, 54
Decision
  matrix, 216–17, 404–5
  theory, 470
  trees, 420–24, 457–62
Decline, exponential, 837–40
Decreasing function, 739
Definite integral, 871
  as area, 895–905
  development of, 924–29
  as limit of Riemann sums, 912–15
  as total change in antiderivative, 885–89
Degree of a polynomial, 31

Demand, 145, 158–59, 660–63
  elastic, 661
  external, 247–49
  inelastic, 661
  internal, 247–49
Departing variable, 307–8
Dependent event, 365
Derivative, 595, 625–30
  definition of, 627
  higher, 670–71
  notation for, 669–70
  partial, 1019–24
  as rate of change, 646–51
  rules, 639–42, 684–88, 694–99, 822–24, 829, 833
  second, 670
  sign of, 739
  and three-step rule, 629
Descartes, Rene, 111, 927
Descriptive method, 59
Descriptive statistics, 479–80
Determinant, 258
Difference
  quotient, 628
  of two cubes, 33
  of two squares, 33
Differentiability, 630–31
  and continuity, 634–35
Differentiable function, 630
Differential equation, 788, 984–91, 1001
  general solution of, 986
  with initial condition, 986
  particular solution of, 986
  and proportionality, 987–91
  qualitative solution of, 788–91
  solution of, 788, 984
Differential notation, 669–70
Differentials, 679, 704–9
  approximation by, 706–9
Differentiation
  of exponential functions, 829–34
  implicit, 714–17
  logarithmic, 825–26
  of logarithmic functions, 821–25
  rules of, 638–43, 684–88, 694–99, 822–24, 829, 833
Direct proportion, 987
Directed graph, 230–31
Directional derivative, 1073
Discontinuous function, 631
Discounted loan, 572
Discounting, 567–68
Discrete distribution, 515
Disjoint events, 359

Disjoint sets, 71–72
Disk method, 1001
Dispersion, 501–8
  range, 502
  standard deviation, 503–4
  variance, 503–4
Distance function, 647
Divergent improper integral, 954
Domain, 81, 84–85
Dominance matrix, 190, 231–32
Dominating plays, 438
Double
  inequality, 6–7
  integrals, 1054–65
Duality, 329–34
  theorem, 332
Dual problem, 329

e, 802
Echelon matrix, 258
Effective interest rate, 566
Elasticity of demand, 660–63
  formula, 661
  and revenue, 662
Element, 58
Empty set, 61
Entering variable, 306–7
Equality
  of matrices, 206
  of sets, 60
Equation(s)
  constraint, 1047
  exponential, 815
  of a line, 127–31
  linear in one variable, 12–16
  Lotka-Volterra, 790
  matrix, 238
  parametric, 730
  polynomial, 37–41
  quadratic, 38–41
  rational, 46–47
  simultaneous linear, 152–56
Equivalence relation, 111
Euler, Leonhard, 860
Event(s), 346–50, 356–59, 362–70
  certain, 348
  complementary, 356, 358
  compound, 356–59, 362–70
  conditional, 362–66
  dependent, 365
  disjoint, 359
  impossible, 348
  independent, 365
  intersection of, 357, 366–70

Subject Index

Event(s)—Cont.
  mutually exclusive, 359
  probability of, 347–50, 356–59, 362–70
  union of, 357–59
Exhaustion, method of, 925–26
Expected value, 394, 396–99, 405, 965
Experiment, 346
Explicit function, 714
Exponential
  equations, 815–17
  expression, 25
  function, 795
  growth, decline models, 837–44, 860–65
Exponential functions, 796–804
  with base b, 797
  with base e, 802
  differentiation of, 829–34
  properties of, 799
Exponentials, 796–804
  change-of-base formula, 814
Exponents, 25–28
  integral, 25
  properties of, 26
  rational, 27
Extended mean value theorem, 787
Extrapolation, 532

Factorial, 376
Factoring
  accounts receivable, 572
  polynomials, 32–35
Fair game, 398
Feasibility region, 273–76
  unbounded, 297
Feasible combination, 273
Fermat, Pierre, 411
First derivative test, 742
First partial derivatives, 1019–22
Fixed cost, 140
Fluents, 678
Fluxions, 678
Fontana, Nicolo, 54
Frequency
  of a class, 480
  cumulative, 481
  distribution, 480–85
  relative, 481
Function, 57, 80–86, 90–99
  antiderivative of, 871–72
  area, 895
  average value of, 916

Function—Cont.
  Cobb-Douglas production, 1032
  composite, 692
  concavity of, 752
  constant, 639
  consumption, 731
  continuous, 631
  cost, 140, 143, 656
  cumulative distribution, 963
  decreasing, 739
  demand, 145
  derivative of, 627
  differentiable, 630
  discontinuous, 631
  distance, 647
  domain of, 81, 84–85
  explicit, 714
  exponential, 795, 797
  Gompertz, 862
  graph of, 92–94
  implicit, 714
  increasing, 739
  inverse, 98–99, 810
  linear, 121, 122–31
  linear approximation to, 787
  logarithmic, 795, 810
  logistic, 860
  marginal, 657, 665
  natural logarithm, 813
  nondifferentiable, 630
  one-to-one, 98
  power, 639
  probability density, 958
  profit, 142–43, 656
  quadratic, 121, 164–73
  range of, 81
  and relation, 82
  renewal, 942
  revenue, 141, 143, 656
  of several variables, 1007–14
  sigmoid, 864
  supply, 144
  survival, 942
  synchronous service, 1017
  vertical line test for, 95
Functional notation, 82
Functional value, 83
Fundamental property of fractions, 43
Fundamental theorem of calculus, 914, 928–29
Fundamental theorem of linear programming, 320
Future stream of payments, 577
Future value, 557, 564

Fuzzy sets, 111–16
  membership in, 112
  operations on, 115

Game(s)
  blackjack, 411
  chuck-a-luck, 411
  craps, 410
  fair, 398
  nonzero-sum, 448–49
  roulette, 398–99
  strictly determined, 438–41
  two-person zero-sum, 438–43
  zero-sum, 438–43, 470–76
Game theory, 420, 438–49
  and linear programming, 470–76
Gauss, Carl Friedrich, 258
Gauss-Jordan elimination, 192–200
  algorithm for, 194
  and inverse matrices, 238–40
General form of a line, 131
General power rule, 697
General term of a sequence, 554
Geometric sequence, 558–560
Gompertz function, 862–64
Gradient, 1073
Graph
  directed, 230–31
  of a function, 92
  of a relation, 94
  symmetries of, 96–97, 99
Graphing, 90–99, 769–79
  linear function, 122–27
  quadratic function, 164–70
  relation, 94–95
Gross profit, 138
Growth
  exponential, 837–40
  logarithmic, 844–46
Growth rate of functions, 859–60

Half-life, 801
Histogram, 481

Identity matrix, 224–25
Implicit differentiation, 715–17
Implicit function, 714
Income, stream of, 936–39
Increasing function, 739
Independent events, 365
Inequalities, 6–7, 19–22
Inferential statistics, 479–80

Infinite limits, 612–13
Infinity symbols, 7, 612
Inflection point, 770
Initial condition, 986
Input-output
 analysis, 246–50
 matrix, 247
 model, 249
 table, 246
Insanity, 852
Integers, 4
Integral
 as accumulator, 915, 932, 939
 and consumers' surplus, 934–35
 definite, 871, 885–90
 double, 1054–65
 improper, 952–56
 indefinite, 871, 873–76, 879–83
 iterated, 1057
 and producers' surplus, 932–34
 sign, 873
 and stream of income, 936–37, 1002–4
Integrand, 873
Integration
 constant of, 873
 by guessing, 879–80
 limits of, 886
 numerical, 971–78
 by parts, 944–48
 and Riemann sums, 909–14
 rules of, 874
 by substitution, 880–83, 889–90
 using tables of integrals, 948–50
Intercepts, 92, 123, 168
Interest
 compound, 553, 563–69, 804–6
 simple, 556–57
Interest rate
 effective, 566
 nominal, 563
 true annual, 588
Interest rebate, 588
Internal rate of return, 820
Intersection
 of events, 357, 366–70
 of fuzzy sets, 115
 of lines, 152–54
 of sets, 70–71
Intervals, 7–8
Inventory
 control, 527–28
 matrix, 205–6, 226–27
Inverse function, 98–99, 810
 rule, 730

Inverse matrices, 233–38
 and matrix equations, 238–40
Inverse proportion, 987
Investment risk, 514–15
Irrational numbers, 4
Iterated integral, 1057

Karmarkar, Narendra, 302, 340
Keys to
 counting outcomes, 381
 integration by parts, 946
 integration by substitution, 882
 related rates problems, 721
 solving applied problems, 15
 symmetries, 96
Khachian, L. G., 340

Labor, marginal productivity of, 1032
Lagrange, Joseph-Louis, 1074
Lagrange multiplier, 1047
 method, 1047–52
Lead time, 528
Learning curves, 842–44
Least common divisor, 44–45
Least-squares line, 529, 531, 1074–77
Leibniz, Gottfried Wilhelm, 677–80, 928
Length of a line segment, 180
Leontief matrix, 249
Leverage, 133
L'Hopital's rule, 787–88
Life cycle curve, 860
Likert scale, 89, 500–501
Limit(s)
 and continuity, 633
 of a function, 595–600, 677
 infinite, 612–13
 at infinity, 614–18
 of integration, 886
 one-sided, 610–12
 of polynomial functions, 602
 properties of, 600–604, 677
 of rational functions, 602–4
 of Riemann sums, 912
Limited growth, decline, 840–44
Line(s)
 finding equation of, 127–32
 finding intersection of, 152–54
 graphing, 122–27
 parallel, 128
 perpendicular, 180
 secant, 625
 tangent, 625
 vertical, 124

Linear approximation, 787
Linear correlation, 532–34
Linear differential equation, 1001
Linear equation(s), 12–16
 simultaneous, 152–59, 190–200
Linear functions
 cost, revenue, profit models, 140–43
 graphing, 122–27
 slope of, 125, 129
 supply, demand models, 144–45
Linear inequalities
 in one variable, 19–22
 in two variables, 268–76
Linear programming
 algorithm, 292
 fundamental theorem of, 290
 and game theory, 470–76
 model, 280
 and the simplex method, 302–33
 two-dimensional, 289–96
Linear programming problem, 279
 constraints of, 279–84
 dual of, 330
 inconsistent, 297
 matrix of, 330
 primal, 330
 setting up, 279–84
Linear regression, 529–32
Line diagram, 7
Logarithmic differentiation, 825–26
Logarithmic function, 795, 810–11, 821–25
Logarithmic growth, 844–46
Logarithms, 810–17
 change-of-base formula, 814
 common, 813
 and exponential equations, 815–17
 natural, 813
 properties of, 812
Logistic equation, 180
Logistic function, 860–62, 864
Lorenz curve, 908–9
Lot tolerance percentage defective, 412

Marginal
 average cost, 665
 cost, 657
 productivity of capital, labor, 1032
 productivity of money, 1051
 profit, 657, 1030
 propensity to consume, 731
 propensity to save, 733
 revenue, 657
 revenue product, 704

Subject Index

Market equilibrium, 158–59
Markov chain, 420, 428–33
Matrix, 189, 190, 204–6
　addition, 207–8
　augmented, 190
　decision, 216–17, 404–5
　determinant of, 258
　of a directed graph, 230
　direct-requirements, 247
　dominance, 231–32
　echelon, 258
　entries of, 190
　equality, 206
　growth, 260
　identity, 224–25
　input-output, 247
　inventory, 205–6, 226–27
　inverse, 223–34
　Leontief, 249
　$m \times n$, 204
　multiplication, 217–27
　nonsingular, 234
　order of, 204
　parts, 253
　payoff, 211, 420, 438
　product, 219
　scalar multiplication, 208–9
　singular, 234
　square, 205
　subtraction, 207–8
　technological, 247
　transition, 429
　transpose of, 329
　zero, 204
Maximum
　absolute, 757
　relative, 736, 1037
Mean, 489–492
　of grouped data, 491
Mean value theorem
　for derivatives, 787
　for integrals, 924
Median, 492
　of grouped data, 500
Midpoint rule, 972–73
Minimum
　absolute, 757
　relative, 736, 1037
Mixed partials, 1023
Mixed strategies, 441–43, 470–76
Mode, 493
　of grouped data, 500
Model
　cost, revenue, profit, 140–43
　input-output, 249

Model—*Cont.*
　quadratic, 170–73
　supply and demand, 143–46
Multiplication
　of matrices, 217–27
　of polynomials, 32
　principle, 377–78
　of rational expressions, 44
　scalar, 208–9
Multiplier, 732
　effect, 730
　Lagrange, 1047
Mutually exclusive
　events, 359
　sets, 71

Napier, John, 860
Natural logarithm, 813
Natural numbers, 4
Net present value, 568–69, 579–80
Newton, Isaac, 677–80, 928
Newton's method, 677
Nominal interest rate, 563
Nondifferentiable function, 630
Nonsingular matrix, 234
Nonzero-sum game, 448–49, 470
Normal approximation to binomial, 527
Normal curve, 516–19
Normal distribution, 516–24, 527–28
n-person zero-sum game, 469
nth
　root, 27
　term of a sequence, 554
Null set, 61
Number systems, 4
Numerical integration, 971–78
　errors of, 1001

Objective function, 280
Octant, 1011
Odds, 348–49
One-sided limit, 610
One-to-one function, 98
Operating characteristic curve, 413
Operations
　for solving equations, 13
　for solving inequalities, 20
　on sets, 66–73
Optimal reorder quantity, 763
Orbit, 181
Order of a matrix, 204
Origin, 4, 91, 1011
Outcome, 346

Parabola, 164–67
Parallel lines, 128, 180
Partial derivatives
　definition of, 1022
　first, 1019–22
　geometric significance of, 1026–29
　higher, 1022–24
　mixed, 1023
　as rates of change, 1029–33
　second, 1022
Partial fractions, 1001
Partial pivoting, 258
Parts
　diagram, 253–54
　integration by, 944–48
　matrix, 253
Pascal, Blaise, 411
Payoff matrix, 211, 420, 438
Pearson, Karl, 545
Percentage-of-sales, 138
Perfect
　cube, 33
　elasticity, 668
　inelasticity, 668
　square, 33
Permutations, 378–79
　and probabilities, 380–81
Perpendicular lines, 180
Pivot, 194
Plotting points, 91
Point of inflection, 770
Point-slope form, 129–30
Poisson process, 411
Polar coordinates, 111
Polygon, 481
Polynomial(s)
　arithmetic of, 31–32
　division, 54
　equations, 37–41
　factoring of, 32–35
　functions, 602, 613, 616
　quadratic, 34
　special types, 33
Population, 479
Power
　function, 639
　rule, 640
Present value, 553, 566–69
　of annuity, 577–80
　continuous compounding, 806
　formulas, 567, 806
　net, 568–69, 579–80
　of stream of income, 1002–4
Pretableau, 319

Price
  discrimination, 1044
  vector, 226
Primal problem, 329
Prime notation, 669
Principal, 556
Prisoner's dilemma, 448
Probability
  binomial, 389
  of compound events, 356–59, 365–67
  density function, 958–63
  of an event, 345–48
  formulas, 370
  and odds, 348–49
  relative frequency, 349–50
  rules of, 350
  subjective, 350
Probability distribution, 394–96
  binomial, 389, 396
  chi-square, 544
  continuous, 515, 958
  discrete, 515
  exponential, 544
  normal, 515–24
  Poisson, 411
  rectangular, 544
  uniform, 544
Producer's risk, 412
Producers' surplus, 932–34
Product rule, 684–85
Profile, 259
Profit
  function, 142–43, 656
  marginal, 657
  maximization, 658–60
  and percentage-of-sales, 138
Properties
  of antiderivatives, 873
  of cumulative distribution function, 964
  of the definite integral, 923–24
  of exponential functions, 799
  of exponents, 26
  of limits, 600
  of logarithms, 812
  of the normal curve, 516
Proportional change, 708–9
Proportionality, 987

Quadratic
  equations, 38–41
  formula, 35

Quadratic—*Cont.*
  function, 121, 164–73
  models, 170–73
Quality control, 411–16
Quartic formula, 55
Quotient rule, 686–88

Radiocarbon dating, 847
Random variable, 394
  continuous, 515, 957–66
  cumulative distribution function for, 963
  discrete, 515
  expected value of, 396–99, 965–66
  probability density function for, 960
  probability distribution of, 394–96
  standard deviation of, 965–66
  variance of, 965–66
Range
  of data, 482, 502
  of a function, 81
Rates of change, 646–51
  and partial derivatives, 1029–33
  and related rates, 719–24
Rational
  equations, 46–47
  exponents, 27–28
  expressions, 43–47
  functions, 602–4, 613, 616–17
  numbers, 4
Real line, 4–5
Real numbers, 4–5
Reducing to lowest terms, 43–44
Related rates, 719–24
Regression
  linear, 529–32, 1074–77
  multiple, 545, 1079–80
  quadratic, 1078–79
Regular Markov chain, 433
Relation, 82, 94, 111
Relative frequency, 481
  probability, 349–50
Relative maxima, minima, 736–47, 751–55, 1037–44
Reorder point, 528
Revenue
  and elasticity of demand, 662
  function, 141, 143, 656
  marginal, 657
Riemann, Bernhard, 924
Riemann sum, 909–10, 1055
Right-hand side ranging, 342–43
Rolle's theorem, 787
Roster method, 59

Row operations, 192
Rule(s)
  of differentiation, 639–41, 685–86, 694, 697–98, 730, 822–24, 829, 833
  for finding sustainable yields, 262
  L'Hopital's, 787–88
  of matrix arithmetic, 208–9, 226
  midpoint, 972
  of probability, 350
  of rational arithmetic, 44
  of seventy eight, 588
  Simpson's, 978
  three-step, 629
  trapezoidal, 975

Saddle point, 1037
  of a game, 440
Safety stock, 528
Sample, 479
Sample space, 346
Sampling
  distribution of the mean, 546–47
  plan, 411
Scalar multiplication, 208–9
Scatter diagram, 529
S-curves, 860–65
Secant line, 625
Second derivative, 670
  test, 753, 1040
Second partials, 1022–24
Seed, 181
Sensitivity to initial data, 184
Separation of variables, 988–91
Sequence, 554–60
  arithmetic, 555
  geometric, 558
Set(s), 57–63
  cartesian product of, 111
  containment, 60–61
  and counting problems, 72–73
  disjoint, 71–72
  element of, 58
  empty, 61–62
  equality, 60
  fuzzy, 111–16
  inclusion, 60–61
  mutually exclusive, 71–72
  notation, 59–60
  operations on, 66–73
  universal, 62
  Venn diagrams of, 62–63
  well-defined, 111
Set-builder method, 59–60

Shadow prices, 340–43
Shell method, 1001
Sigmoid function, 864–65
Simple interest, 556–57
Simplex method, 268
   and duality, 329–33
   flow chart for, 311
   and maximization, 302–13, 317–22
   and minimization, 322–23, 332–33
   and right-hand side ranging, 342–43
   and shadow prices, 340–42
Simplex tableau
   initial, 302–5
   pretableau, 319
   transformation of, 306–13
Simpson's rule, 977–78
Simultaneous equations, 152–59, 190–200, 238–40
Singular matrix, 234
Sinking fund, 576–77
Slope of a line, 124–27, 129
Slope-intercept form, 127
Solid of revolution, 998
Square matrix, 205
Standard deviation, 502–8
   of a continuous random variable, 965
   of grouped data, 507
Standard normal distribution, 519–524
   conversion formula, 521
States of nature, 420
Statistics, 479–80
Steady state, 432
Stock volatility, 539–40
Stockout, 528
Strange attractor, 185
Stream of income
   capital value of, 1003
   future value of, 936–39
   present value of, 1002–4
Strictly determined game, 438–41
Subjective probability, 350
Subset, 61
Substitution, integration by, 880–83, 889–90
Subtraction
   of matrices, 207–8
   of polynomials, 31–32
   of rational expressions, 44–45
Summation notation, 488–89
Sum rule, 641–42

Supply, 144–45, 158–59
Surface of revolution, 1001
Surplus variable, 318
Sustainable growth rate, 713–14
Sustainable yields, 259–63
Symmetry, 96, 99, 164–65
Synchronous service function, 1017
Systems
   of equations, 152–59, 190–200, 238–40
   of inequalities, 272–76

Tangent line, 625, 1026
Techniques of integration
   by guessing, 879–80
   by parts, 944–48
   by substitution, 880–83
   using tables, 948–50
Technological matrix, 247
Terms of a polynomial, 31
Theory of the firm, 656–63
Three-step rule, 629
Total differential, 1073
Transition
   matrix, 429
   rate, 995
Transportation problem, 340
Transpose, 329
Trapezoidal rule, 974–77
Tree diagram, 367, 420–24, 457–62
Triple integral, 1073–74
Two-person game, 438

Union
   of events, 357, 358–59
   of fuzzy sets, 115
   of sets, 68–70
Universal set, 62
Utility, 427

Valuation of inventory, 226–27
Value added, 247
Variable
   artificial, 318
   basic, 304
   departing, 307
   dependent, 85

Variable—Cont.
   entering, 306
   independent, 83
   nonbasic, 304
   slack, 303
   surplus, 318
Variable cost per unit, 140
Variance, 502–8
   of a continuous random variable, 965–66
   of grouped data, 507
Variation, coefficient of, 508
Vector, 1073
   column, 205
   harvest, 261
   price, 226
   profile, 259
   replacement, 261
   row, 205
Velocity, 647–48
Venn diagram, 62–63
Vertex
   of a feasibility region, 273
   of a parabola, 164–67
Vertical line, 124, 126
   test, 95
Volume, 1055–56, 1059–61
Von Neumann, John, 470

Work, 1001

x-intercept, 92
xy-plane, xz-plane, 1012

Yield
   curve, 781
   to maturity, 584
y-intercept, 92
yz-plane, 1012

Zero
   matrix, 204
   product law, 37
Zero-sum game
   mixed strategies for, 441–43, 470–76
   strictly determined, 438–41

Rules of Differentiation

Constant Rule

$$\frac{dc}{dx} = 0$$

Power Rule

$$\frac{d}{dx}(x^r) = rx^{r-1}, \qquad r \neq 0$$

Constant-Multiple Rule

$$\frac{d}{dx}(cf) = c\frac{df}{dx}$$

Sum Rule

$$\frac{d}{dx}(f \pm g) = \frac{df}{dx} \pm \frac{dg}{dx}$$

Product Rule

$$\frac{d}{dx}(fg) = f\frac{dg}{dx} + g\frac{df}{dx}$$

Quotient Rule

$$\frac{d}{dx}\left(\frac{f}{g}\right) = \frac{g(df/dx) - f(dg/dx)}{g^2}$$

General Power Rule

$$\frac{d}{dx}(u^r) = ru^{r-1}\frac{du}{dx}, \qquad r \neq 0$$

Chain Rule

$$\frac{df}{dx} = \frac{df}{du}\frac{du}{dx}$$

Derivatives of Exponential Functions

$$\frac{d}{dx}(e^u) = e^u\frac{du}{dx}$$

$$\frac{d}{dx}(b^u) = (\ln b)b^u\frac{du}{dx}$$

Derivatives of Logarithmic Functions

$$\frac{d}{dx}(\ln u) = \frac{1}{u}\frac{du}{dx}$$

$$\frac{d}{dx}(\log_b u) = \frac{1}{\ln b}\frac{1}{u}\frac{du}{dx}$$